Quantum Information and Quantum Signal Processing

François Chapeau-Blondeau

Quantum Information and Quantum Signal Processing

Springer

François Chapeau-Blondeau
University of Angers
Angers, France

ISBN 978-3-032-27498-4 ISBN 978-3-032-27499-1 (eBook)
https://doi.org/10.1007/978-3-032-27499-1

Responsible Editor: Mary James

This Springer imprint is published by the registered company Springer Nature Switzerland AG
The registered company address is: Gewerbestrasse 11, 6330 Cham, Switzerland

"Seek, and you will find."

"Nothing is given. Everything is constructed."

Preface

This book proposes a self-contained presentation of the domain of quantum information and quantum computation with a signal processing or signals-and-systems orientation. The focus on quantum signal processing is a distinctive feature of the present textbook. In the book, essential notions forming a ground to quantum information and quantum computation, or quantum information processing at large, are consistently and pedagogically reviewed. At the same time, the book applies, at the quantum level, standard concepts from classical signal processing or signals-and-systems theory, and develops them in the quantum realm, so as to serve for a coherent exposition of quantum information processing, and to substantiate the more specific field of quantum signal processing. Accordingly, the book, in addition to the general presentation it provides of quantum information processing, may strongly appeal to readers with a background in classical signal processing or signals-and-systems theory (although this is by no means a necessity), since the book offers an opportunity to transfer and extend their knowledge to the quantum world.

The book in particular features:

- A self-contained presentation of the fundamental principles of quantum theory, on state, process, and measurement, with a signals-and-systems theoretic orientation.
- Basics of quantum gates and circuits, for computation and processing.
- Landmark quantum algorithms for computation, with the Deutsch–Jozsa algorithm for parallelism, Grover search algorithm, Shor factoring algorithm, and HHL algorithm for linear systems.
- Quantum correlation, entanglement, and entropy, related to EPR experiment and Bell inequalities, teleportation, cryptography, and quantum repeater.
- Quantum noise modeling, with noise models for qubit or multiple-qubit states or signals, and their circuit implementations.
- Quantum detection theory and state discrimination, with optimal detectors and the square-root detector.
- Quantum estimation theory, with the Fisher information, classical and quantum, and worked-out examples, especially for phase estimation, up to quantum tomography on state and process.

- Quantum Fourier transform with applications.
- Quantum information theory, with quantum source compression and channel coding, classical and quantum information capacities of quantum channels, and error correction.
- A summary terminates each chapter, recapping the essential ideas and notions of the chapter, and specially useful to assist assimilation by the readers.

In general, this textbook is adapted for a readership starting at the advanced undergraduate or graduate levels, up to the research level for more advanced readers who would like to enter the quantum field. Some basic notions in linear algebra and probabilities will be helpful, but beyond, no specialized background is required. When they stand as helpful references to guide the quantum exposition, some core notions of classical signal processing are briefly reviewed at the onset of each corresponding chapter. The textbook is well suited for curricula in electrical engineering, or in computing and information sciences and technologies, and more broadly in applied physics or in applied mathematics. In its construction, the book is designed primarily to serve as a didactic text for students and scientists new to the domain, but also to serve as a reference text consistently assembling in one place many useful results of the domain. Finally, the book can bring the reader, equipped with the fundamental concepts and their illustrations presented throughout, to initiate an autonomous research in the quantum field, and in particular toward quantum signal processing. In this direction, beyond the situations treated in the book, all the issues that are commonly addressed by classical signal processing, can probably be formulated and examined at the quantum level, when the information carriers are quantum objects, and many explorations remain to be undertaken in this way.

Angers, France
February 2026

François Chapeau-Blondeau

Acknowledgements I would like to thank all the colleagues from the LARIS laboratory at the University of Angers, France, who, in some way or another, participated in or expressed interest and encouragement to our quantum activities, and among them specially David Rousseau, Étienne Belin, Nicolas Gillard, and the late and regretted Nicolas Delanoue. I also want to mention the successive doctoral schools devoted at the University of Angers to the domain of information sciences and technologies, where I could teach for several years a course on quantum information and quantum computation, which played a part in helping me prepare this book. I am also grateful to the colleagues from the research group |QuantInG⟩ (Quantum Information Group) at GdR CNRS IASIS, especially Angélique Drémeau, Thierry Chonavel, Cédric Herzet, and Jérôme Idier, for fruitful interactions and for giving me the opportunity to lecture on quantum signal processing at |QuantInG⟩ from June 2024 to February 2025.

François Chapeau-Blondeau

Competing Interests The author has no competing interests to declare that are relevant to the content of this manuscript.

Contents

About the Author

François Chapeau-Blondeau received the master-level Engineer Diploma in electronics from the Engineering School ESEO in Angers, France, in 1982, and the Ph.D. degree in electrical engineering from the University Pierre and Marie Curie in Paris, France, in 1987, followed by a year of postdoctoral research in the USA. He has then developed an academic career, with teaching and research in these scientific directions. He is now Professor Emeritus at the University of Angers, France, from the Department of Physics at the Faculty of Sciences and from the LARIS research laboratory. Throughout his career, he has carried out teaching and research activities at the interface between physics and information sciences and technologies, especially signal processing, classical and quantum as well. In these areas, he has authored or coauthored more than 160 research articles in international journals and several contributed book chapters. From 2015, he has lectured for several years on quantum information and quantum computation in the doctoral school at the University of Angers. From June 2024 to February 2025, he lectured on quantum signal processing in the research group $|\text{QuantInG}\rangle$ (Quantum Information Group) at GdR CNRS IASIS, in Paris, France.

Chapter 1
Introduction

Quantum information and quantum computation constitute a rather new field of science and technology, aiming at exploiting quantum objects, phenomena, and properties, in order to contribute to information processing and computation. In the realm of information sciences and technologies, one is naturally driven to the quantum level by the progress of information processing devices toward their fundamental physical limits. Reduction in size and increased integration, with elemental physical objects acting as information carriers, lead to a point where quantum behaviors have to be explicitly taken into account. In addition, at this fundamental quantum level, novel and specific properties arise, not present at the classical (non-quantum) level, and offering distinctive potentialities for enhanced information processing performance—this being the case for instance with quantum superposition and with quantum entanglement. Animated by this double motivation, the field of quantum information and quantum computation is today experiencing rapid progress. Its ongoing development concerns both science and technology, particularly physics, engineering, and computing, with potential benefits to many application areas [1, 2].

This book aims to provide a self-contained presentation for the domain of **quantum information and quantum computation**, or quantum information processing at large. Accordingly, essential notions forming a ground to the domain are consistently reviewed. At the same time, a special orientation for the presentation is provided by the field of **signal processing**, or signals-and-systems theory, as it exists classically. We show how standard concepts from classical signal processing or signals-and-systems theory can equally be traced and developed at the quantum level. These concepts can be considered as delineating a more specialized topic inside quantum information and computation, that is further explored; but also they can serve as a common thread for a coherent exposition of quantum information processing. This more specific signal-processing orientation we adopt does not change the essential notions of quantum information and computation that are to be presented, but in addition it provides a possibility to transfer prior classical signal-processing conceptions to help tackle the quantum field. Accordingly, the book can serve first as a text

F. Chapeau-Blondeau, *Quantum Information and Quantum Signal Processing*,
https://doi.org/10.1007/978-3-032-27499-1_1

presenting the basic notions of quantum information and quantum computation, and no special background in signals-and-systems theory is needed for this. In addition, for readers possessing such background, the book can be taken as an opportunity to recognize and to extend their knowledge, in a suitable form, in the quantum realm.

After the present Introduction chapter, the bulk of the book begins with three chapters (Chaps. 2, 3, 4) presenting the three general postulates or fundamental principles forming a foundation for quantum theory, and to be useful throughout the book. **Chapter** 2 presents the first general postulate or fundamental principle, dealing with the concept of quantum state. The common notions attached to the quantum state (as a state vector or a density operator) are reviewed in Chap. 2. In addition, for our signal-processing perspective, the quantum state here is assimilated to the notion of quantum signal, as it can be appreciated that the quantum state exhibits suitable properties to assume the role of an information-carrying signal and act as such for its implication in quantum information and computation. **Chapter** 3 presents the second general postulate or fundamental principle, dealing with the concept of quantum process, to describe the evolution of quantum states (or signals). The common notions attached to quantum processes (unitary for closed evolutions, or non-unitary for open evolutions) are reviewed in Chap. 3. Within our signals-and-systems theoretic perspective, the quantum process transforming quantum states can be viewed as the operative system or functional block acting to transform or process some quantum signal of interest. **Chapter** 4 presents the third general postulate or fundamental principle, dealing with the concept of quantum measurement, with its essential and non-intuitive specificities at the quantum level. The common notions attached to quantum measurement are reviewed in Chap. 4, with its marked distinction relative to its classical counterpart, and to encompass both the standard projective von Neumann measurement and generalized quantum measurements.

Once the three fundamental quantum postulates are in place, on state, process, and measurement, the chapters that follow, starting with Chap. 5, present essential notions or phenomena or situations standing as core references of the field of quantum information and quantum computation. These chapters, dealing with quantum topics, are also organized to recognize, as soon as their title, that the reference notions they address equally relate to the field of signal processing. This substantiates our vision that signal processing stands as a possible thread for a coherent exposition of core concepts from quantum information and quantum computation, as we undertake here; or in the other direction, that core concepts from signal processing are equally meaningful at the quantum level. **Chapters from** 5 **to** 8, dealing with quantum circuits and computation, with quantum correlation, with quantum entropy, and with quantum information compression, mainly focus on important quantum phenomena not predominantly affected by quantum noise or decoherence degrading quantum states or signals. The important notion of quantum noise itself is presented in **Chap.** 9, standing essentially as the effect of decoherence, when quantum states get degraded, often as a result of their interaction with an uncontrolled environment. The following chapters are then organized to address significant signal-processing tasks mainly concerning (quantum) signals in the presence of (quantum) noise. **Chapter** 10 with quantum detection and **Chap.** 11 with quantum estimation connect to the two reference tasks of signal processing with statistical detection and estimation theory. **Chapter** 12 concerns, in the quantum realm, the central signal-processing tool of the Fourier transform, with a prominent application for quantum phase estimation, assessable with standard notions of statistical signal processing as the mean-squared estimation error and the Fisher information (introduced in Chap. 11). **Chapter** 13 on quantum communication essentially deals with the transmission of (quantum) signals over noisy quantum channels, with meaningful tools for performance evaluation supplied by statistical information theory (with entropy-based quantities such as the mutual information or the information capacity) extended to the quantum domain. Finally, **Chap.** 14 introduces the domain of quantum coding and error correction, to combat the noise or decoherence, defaults or failures, affecting quantum communication and computation. Each chapter terminates with a summary recapping the essential ideas and notions of the chapter.

In its style of presentation, the book is rather intended as a book of physics and engineering. The book tries to convey both concrete physical insight and formal mathematical reasoning, aiming at physical understanding and intuition backed up by relevant mathematical formalism. Mathematical proofs are given for important results and as representative examples of mathematical argumentation useful to the quantum field. But no exhaustive mathematical proofs, if ever possible, are targeted, and especially when proofs become too technical (some stand as substantial pieces of algebra combined with probability theory) the important results are merely stated and referred for proof to the literature. No specialized background is required to embark on reading the book. But some familiarity with basic concepts of linear algebra and probability theory will be helpful, with in addition essential notions thereof briefly recalled in the Appendix. Some core notions of classical signal processing are briefly reviewed at the onset of each chapter, when they stand as helpful references to guide the quantum exposition.

The book can serve as an introductory text to quantum information and quantum computation, or quantum information processing at large, as it presents and illustrates the fundamental concepts and essential notions of the domain. At the same time, the book delineates contents to substantiate the more specific field of quantum signal processing, understood as the standard approach of signal processing extended to the quantum realm. Finally, the book with its fundamental concepts, methodologies, and their illustrations, provides a basis for the reader to undertake an autonomous research, to contribute to the ongoing quantum endeavor.

Chapter 2
Quantum Signal

This chapter essentially presents the first general postulate or fundamental principle for quantum theory, with the notion of quantum state. For our signal-processing orientation, the quantum state here is assimilated to the notion of quantum signal, to be represented by a state vector or by a density operator. Accordingly, this chapter will allow us to verify that the conceptualization generally attached to the notion of signal, information-carrying signal, can be transferred to the quantum state, and reciprocally, that a quantum state can be conferred the nature of a signal. This will be a perspective for the book: a quantum state can be viewed as a signal, and the role of signal can be assumed by the quantum state. This will not change the standard conceptions commonly ascribed to the quantum state, but they will be presented within this additional interpretative framework we adopt here for a quantum signal-processing perspective.

2.1 Quantum State

As a fundamental postulate or principle, to represent a quantum object or system and describe its implication in physical phenomena, quantum theory resorts to the notion of **quantum state**. A quantum state is defined as a vector, commonly denoted $|\psi\rangle$, belonging to a complex Hilbert space $\mathcal{H}$. The notation $|\cdot\rangle$ involving a closing angular bracket is due to the physicist Paul Dirac [3]; beyond a mere notation, we will see that it is helpful to handle and interpret algebraic expressions and derivations involving quantum states; and a state vector $|\psi\rangle$ can be called a "ket" vector.

For our purpose here, inspired by digital signal processing, the Hilbert space $\mathcal{H}$ can be assigned the finite dimension N and is denoted $\mathcal{H} \equiv \mathcal{H}_N$. The N-dimensional vector space $\mathcal{H}_N$ can be referred to an orthonormal basis, established as the standard or computational basis, and denoted $\{|1\rangle, |2\rangle, \dots, |N\rangle\}$. A state vector $|\psi\rangle$ can then be expressed as the linear combination:

F. Chapeau-Blondeau, *Quantum Information and Quantum Signal Processing*,
https://doi.org/10.1007/978-3-032-27499-1_2

$$|\psi\rangle = \sum_{n=1}^{N} \alpha_n |n\rangle \in \mathcal{H}_N , \tag{2.1}$$

with the coordinates $\alpha_n \in \mathbb{C}$ which are complex numbers. One can also introduce another (any other) orthonormal basis for $\mathcal{H}_N$, through a unitary change of basis from the computational basis. A linear combination as in Eq. (2.1), affording versatile superpositions of quantum states, already contains or suggests the possibility of quantum parallelism from computing with quantum states, a faculty we shall examine in detail later in this book.

The state vector $|\psi\rangle$ has the squared norm $\langle\psi|\psi\rangle$ defined by Eq. (A.4) of the Appendix. Quantum theory, for reasons related to quantum measurement that will be specified shortly, also imposes that any valid quantum state $|\psi\rangle$ must be a normalized or unit-norm vector, i.e., satisfying the normalization constraint:

$$\langle\psi|\psi\rangle = \sum_{n=1}^{N} |\alpha_n|^2 = 1 . \tag{2.2}$$

The state of a quantum object or system is thus represented by a unit-norm vector $|\psi\rangle$ belonging to a complex Hilbert space. And for our signal-processing perspective, the state vector $|\psi\rangle$ provides a convenient notion of quantum signal.

2.2 Measurement of a Quantum State

Before we can proceed with more elaboration on the notion of quantum state, and to motivate the unit-norm constraint on the state vector, we need now to introduce, in its essence, the fundamental postulate or principle ruling quantum measurement, which is quite specific and central to quantum theory and its formulation.

In essence, for a quantum measurement on a quantum system, the state $|\psi\rangle$ of the system to be measured has to be referred to (expressed in) an orthonormal basis of representation, as for instance the basis $\{|n\rangle\}$ of Eq. (2.1). In practice, it is the measurement setup, the measuring instrument, which selects, defines, the reference basis. Then, the quantum measurement takes place as an orthogonal projection of the (pre-measurement) state vector $|\psi\rangle$ onto one of the states $|n\rangle$ of the reference basis. Moreover, this projection occurs at random, with a probability of projecting $|\psi\rangle$ onto $|n\rangle$ which is given by the squared modulus $|\alpha_n|^2$ of the complex coordinate α_n of the pre-measurement state $|\psi\rangle$ on the basis vector $|n\rangle$ of the reference basis. This probabilistic interpretation for the state coordinates in relation to quantum measurement is known as the **Born rule**. Through the measurement probabilities $|\alpha_n|^2$, the Born rule explains in quantum theory the normalization constraint of Eq. (2.2) on the squared moduli of the state coordinates, which represent a probability distribution summing to 1. The complex coordinates α_n of Eq. (2.1) are sometimes called probability amplitudes. There is in this way an intrinsic probabilistic character to quantum measurement. "God plays dice" here, as far as we know today, to phrase it in a picturesque terminology in use during the historical[1] elaboration of quantum theory. In addition, the quantum measurement also implies, in general, an irreversible destructive or disturbing effect on the measured system, in the sense that its pre-measurement state $|\psi\rangle$ gets replaced by (projected onto) one basis state $|n\rangle$ occurring at random, while $|\psi\rangle$ is irremediably lost in the process. Finally, a quantum measurement does not give access directly to the complete state $|\psi\rangle$ characterizing the measured system, but only to some observation outcome (partially and probabilistically) related to $|\psi\rangle$. To summarize, quantum measurement at this fundamental level exhibits four essential specificities: the measurement outcome depends on the measuring setup by the choice of the measurement basis $\{|n\rangle\}$, and it comes with probabilistic, destructive, and partial aspects attached to it. This (non-intuitive) character of quantum measurement essentially results as a datum from physical reality and stands as a fundamental feature to be taken into account in the workings of quantum theory.

The quantitative connection of the theoretical description (the state vector) of the quantum system with the experimental observation (the measurement result) then goes through repeating the experiment where the system is prepared in the same state $|\psi\rangle$ and subsequently measured in the same basis $\{|n\rangle\}$, so as to observe the measurement outcome $|n\rangle$ with an empirical frequency to match the probability $\Pr\{|n\rangle\} = |\alpha_n|^2$ predicted by the theory.

We note that $\alpha_n = \langle n|\psi\rangle$ and $\alpha_n^* = \langle n|\psi\rangle^* = \langle\psi|n\rangle$ so that the probability $\Pr\{|n\rangle\} = |\alpha_n|^2 = \alpha_n\alpha_n^*$ is also

$$\Pr\{|n\rangle\} = |\alpha_n|^2 = \left|\langle n|\psi\rangle\right|^2 = \langle n|\psi\rangle\,\langle\psi|n\rangle = \langle\psi|n\rangle\,\langle n|\psi\rangle\ . \tag{2.3}$$

We will continue later, in Chap. 4, with more development on this central notion of quantum measurement. But for now we return to more elaboration on the notion of quantum state or quantum signal.

2.3 The Quantum Bit

The quantum bit or **qubit** (in the terminology introduced in [4]) is a system whose quantum state $|\psi\rangle$ belongs to a two-dimensional Hilbert space $\mathcal{H}_2$. When the vector space $\mathcal{H}_2$ is referred to the (orthonormal) computational basis $\{|0\rangle, |1\rangle\}$, the qubit state $|\psi\rangle$ can be expressed, according to Eq. (2.1), as the linear combination:

$$|\psi\rangle = \alpha_0\,|0\rangle + \alpha_1\,|1\rangle \in \mathcal{H}_2\ , \tag{2.4}$$

with the normalization constraint $|\alpha_0|^2 + |\alpha_1|^2 = 1$. By extending the terminology, the quantum system with an N-dimensional state in Eq. (2.1) with $N > 2$, is conveniently referred to as a **qudit**.

[1] A previous wisdom on the subject, "Nihil est sine ratione seu nullus effectus sine causa."

The qubit represents a fundamental object or system for quantum information and quantum signal processing. It stands as an analogue to the classical bit forming the foundation of digital electronics or digital signal processing. The qubit, with its two basis states, is an adequate way of representing, for instance, an electron with its two states of spin, or a photon with its two states of polarization. The qubit, as such, is an elementary or elemental object of quantum information and well suited for founding a digital version of quantum signal processing. Despite such elementary character, the qubit, however, is not necessarily immediately given as a natural object directly accessible in the physical world. The qubit rather comes as the result of a purposeful elaboration related to some physical reality. In a comparable way, a classical bit or binary signal of digital electronics is not as such directly accessible in Nature. It requires, for instance, an underlying analog voltage intentionally limited to two meaningful values, and the effective control that any relevant physical evolution and its exploitation are compatible with this two-state limitation and interpretation. Accordingly, a photon with its two basis states of polarization will be adequately represented by a qubit, if it is ensured that the polarization and its evolution are appropriately decoupled from other physical properties of the photon, such as its energy, momentum, or spatial position, which are controlled to play no meaningful role in the evolution and exploitation of the polarization qubit. In a similar way, an atom or an ion that can evolve between two quantum states—one ground state and one excited state—will offer an adequate materialization of a qubit, if it is ensured that no other states (say additional excited states at higher energy levels) can ever be reached by the atom (the atomic qubit) in its evolution. In this respect, the qubit, rather than a preexisting "natural" entity (as long as any exists in science), can be seen as an "engineered" entity, or an engineering entity, which exists through some conceptualization involving a purposeful control and interpretative framework applied to physical reality.

For the qubit state $|\psi\rangle$ of Eq. (2.4), a graphical representation is displayed in Fig. 2.1 in the vector plane $\mathcal{H}_2$.

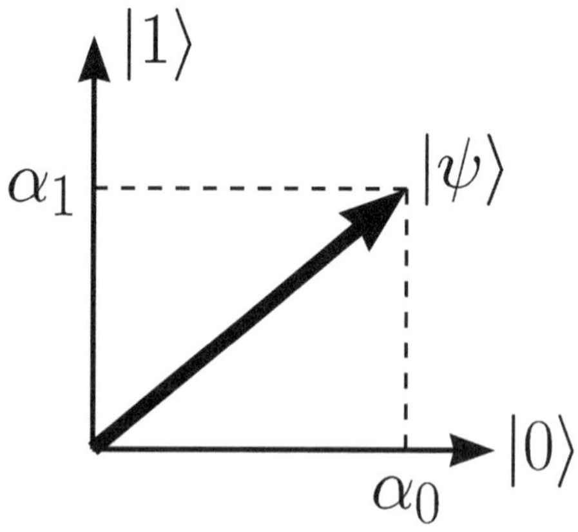

Fig. 2.1 A graphical representation of the qubit state $|\psi\rangle$ of Eq. (2.4) as a unit-norm vector in the complex vector plane $\mathcal{H}_2$

The representation of Fig. 2.1, however, does not always adequately convey the variability and degrees of freedom accessible to the qubit state $|\psi\rangle$, because, in general, the two coordinates (α_0, α_1) are complex numbers. A useful alternative representation exists for the qubit state. It is based on the possibility in Eq. (2.4), for the squared moduli of the two complex coordinates satisfying $|\alpha_0|^2 + |\alpha_1|^2 = 1$, of the parameterization $|\alpha_0|^2 = \cos^2(\theta')$ and $|\alpha_1|^2 = \sin^2(\theta')$ with the necessary and sufficient condition of an angle $\theta' \in [0, \pi/2]$, or equivalently $|\alpha_0|^2 = \cos^2(\theta/2)$ and $|\alpha_1|^2 = \sin^2(\theta/2)$ with an angle $\theta \in [0, \pi]$, to cover all the feasible configurations. This leads for the two complex coordinates themselves to $\alpha_0 = \cos(\theta/2)e^{i\varphi_0}$ and $\alpha_1 = \sin(\theta/2)e^{i\varphi_1}$ with two arbitrary angles φ_0 and φ_1 in $[0, 2\pi[$. Then Eq. (2.4) becomes

$$|\psi\rangle = e^{i\varphi_0}\Big[\cos(\theta/2)\,|0\rangle + e^{i(\varphi_1-\varphi_0)}\sin(\theta/2)\,|1\rangle\Big] . \tag{2.5}$$

In Eq. (2.5), the prefactor $e^{i\varphi_0}$ is a global phase factor which is usually not physically accessible or meaningful. This follows from the postulate for quantum measurement introduced in Sect. 2.2. A global phase factor does not affect the modulus of the state coordinates that determine (probabilistically) the measurement results. Accordingly, two quantum states differing by a global phase factor, like $|\psi\rangle$ and $e^{i\varphi_0}|\psi\rangle$, are indistinguishable by measurement, in any reference basis, and in this respect can be considered as two equivalent representations of the same physical quantum state. In the terminology of signals-and-systems theory, such a global phase factor $e^{i\varphi_0}$ in Eq. (2.5) is a non-identifiable parameter (inaccessible by measurement or observation). In this way, we can select $\varphi_0 = 0$, which amounts to the faculty of fixing a physically non-critical origin for counting the phases of the complex coordinates of a quantum state. This fixes a real coordinate α_0 and a relative phase difference $\varphi_1 - \varphi_0 = \varphi$ between the two coordinates of $|\psi\rangle$, which can then be parameterized as

$$\boxed{|\psi\rangle = \cos(\theta/2)\,|0\rangle + e^{i\varphi}\sin(\theta/2)\,|1\rangle\ ,} \tag{2.6}$$

with the angles $\theta \in [0, \pi]$ and $\varphi = [0, 2\pi[$ to represent all the configurations physically meaningful for a qubit state $|\psi\rangle$. With the parameterization of Eq. (2.6) we have two independent real coordinates (θ, φ) to represent the qubit state $|\psi\rangle$, while in Eq. (2.4) we have two complex coordinates (α_0, α_1) connected by the normalization constraint.

With the parameterization of Eq. (2.6) by the two angles (θ, φ), the (normalized) qubit state $|\psi\rangle$ can be associated, in a one-to-one fashion, with a point resting on the surface of the sphere of unit radius in the three-dimensional real space $\mathbb{R}^3$. This is the Bloch sphere, and the **Bloch representation** of the qubit state $|\psi\rangle$, as illustrated in Fig. 2.2.

As a rule, based on the parameterization of Eq. (2.6), any two qubit states that are orthogonal in the Hilbert plane $\mathcal{H}_2$ are represented by two antipodal points on the Bloch sphere of Fig. 2.2.

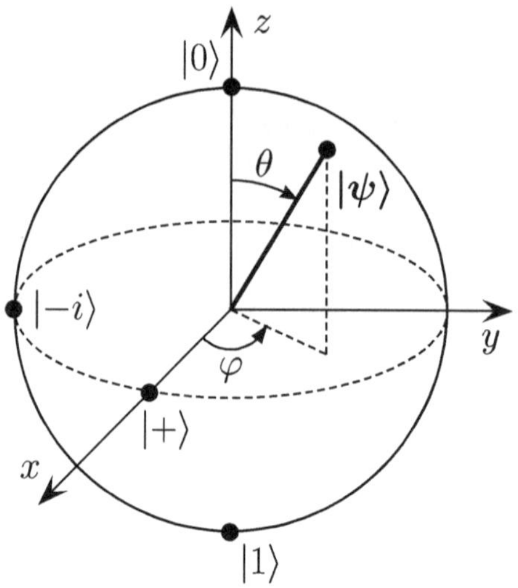

Fig. 2.2 The Bloch sphere of unit radius in $\mathbb{R}^3$ and a representation for the qubit state $|\psi\rangle$ of Eq. (2.6) parameterized by the two angles $(\theta, \varphi) \in [0, \pi] \times [0, 2\pi[$ as a point on the surface of the Bloch sphere

For instance, the basis state $|0\rangle$ is located at the north pole of the Bloch sphere (with $\theta = 0$) while the basis state $|1\rangle$ is located at the south pole of the Bloch sphere (with $\theta = \pi$).

There exists another orthonormal basis often convenient for the qubit, defined by the two states

$$\begin{cases} |+\rangle = \dfrac{1}{\sqrt{2}}\big(|0\rangle + |1\rangle\big) , & (2.7) \\ |-\rangle = \dfrac{1}{\sqrt{2}}\big(|0\rangle - |1\rangle\big) , & (2.8) \end{cases}$$

as depicted in Fig. 2.3 in the vector plane $\mathcal{H}_2$.

The basis $\{|+\rangle , |-\rangle\}$ from Eqs. (2.7) and (2.8) for the qubit is known as the **Hadamard basis**. It is involved in the reciprocal relations:

$$\begin{cases} |0\rangle = \dfrac{1}{\sqrt{2}}\big(|+\rangle + |-\rangle\big) , & (2.9) \\ |1\rangle = \dfrac{1}{\sqrt{2}}\big(|+\rangle - |-\rangle\big) . & (2.10) \end{cases}$$

In the angular parameterization of Eq. (2.6), the state $|+\rangle$ is characterized by $(\theta = \pi/2,\ \varphi = 0)$ and on the Bloch sphere of Fig. 2.2 it rests where the positive part of the x axis intersects the Bloch sphere; meanwhile, the state $|-\rangle$ is characterized by $(\theta = \pi/2,\ \varphi = \pi)$ and rests in the antipodal location where the negative part of the x axis intersects the Bloch sphere.

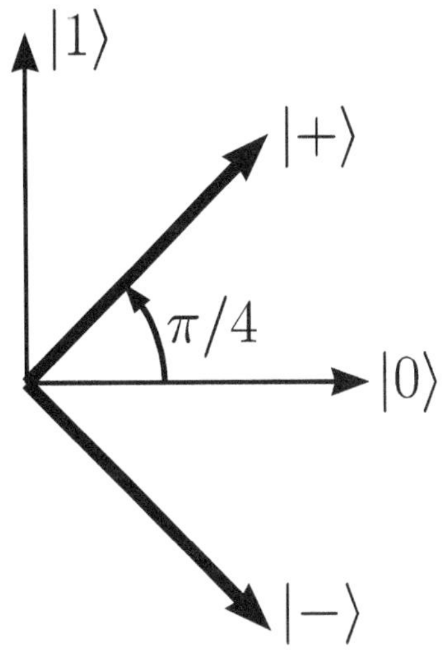

Fig. 2.3 In relation to the computational basis $\{|0\rangle, |1\rangle\}$, the Hadamard basis $\{|+\rangle, |-\rangle\}$ for the qubit, as defined by Eqs. (2.7) and (2.8)

In the Hadamard basis $\{|+\rangle, |-\rangle\}$, the generic qubit state of Eq. (2.6) becomes

$$|\psi\rangle = \frac{1}{\sqrt{2}}\big[\cos(\theta/2) + e^{i\varphi}\sin(\theta/2)\big]\,|+\rangle + \frac{1}{\sqrt{2}}\big[\cos(\theta/2) - e^{i\varphi}\sin(\theta/2)\big]\,|-\rangle \tag{2.11}$$

$$= \alpha_+\,|+\rangle + \alpha_-\,|-\rangle \ . \tag{2.12}$$

When the qubit state $|\psi\rangle$ is measured in the computational basis $\{|0\rangle, |1\rangle\}$, we have according to Eq. (2.3) the probabilities $\Pr\{|0\rangle\} = \cos^2(\theta/2)$ and $\Pr\{|1\rangle\} = \sin^2(\theta/2)$ insensitive to the relative phase φ. Meanwhile, when $|\psi\rangle$ is measured in the Hadamard basis $\{|+\rangle, |-\rangle\}$, we have by Eqs. (2.11) and (2.12) the probabilities $\Pr\{|\pm\rangle\} = |\alpha_\pm|^2 = \big[1 \pm \cos(\varphi)\sin(\theta)\big]/2$ which are now sensitive to the relative phase φ. Different bases for measurement reveal different aspects of the quantum state. More generally, when the qubit state $|\psi\rangle$ is measured in the orthonormal basis $\{|m\rangle, |m^\perp\rangle\}$ with $|m\rangle = m_0\,|0\rangle + m_1\,|1\rangle$ and $|m^\perp\rangle = m_1^*\,|0\rangle - m_0^*\,|1\rangle$, we have the probability $\Pr\{|m\rangle\} = |\langle m|\psi\rangle|^2 = |m_0^*\cos(\theta/2) + m_1^* e^{i\varphi}\sin(\theta/2)|^2$. Different selections of the measurement coordinates (m_0, m_1) lead to a probability $\Pr\{|m\rangle\}$ that probes differently the two coordinates (θ, φ) of the qubit state $|\psi\rangle$.

In this direction, despite the probabilistic and destructive character of one single measurement on a quantum state $|\psi\rangle$, as introduced in Sect. 2.2, it is nevertheless possible to envisage a complete characterization of the full quantum state $|\psi\rangle$, if one has access to many identical copies of $|\psi\rangle$. In this circumstance, appropriately selected quantum measurements of several kinds (like several projective bases) repeated on identical preparations of $|\psi\rangle$, will enable one to reconstruct the full quantum state $|\psi\rangle$ within a statistical estimation framework, common to standard signal processing, and to be examined in more detail in Chap. 11 and Sect. 11.6.1.

Another useful orthonormal basis for the qubit is defined by the two states

$$\begin{cases} |+i\rangle = \dfrac{1}{\sqrt{2}}\big(|0\rangle + i\,|1\rangle\big)\,, & (2.13) \\ |-i\rangle = \dfrac{1}{\sqrt{2}}\big(|0\rangle - i\,|1\rangle\big)\,, & (2.14) \end{cases}$$

of the vector plane $\mathcal{H}_2$. In the angular parameterization of Eq. (2.6), the state $|+i\rangle$ is characterized by $(\theta = \pi/2,\ \varphi = \pi/2)$ and on the Bloch sphere of Fig. 2.2 it rests where the positive part of the y axis intersects the Bloch sphere; meanwhile, the state $|-i\rangle$ is characterized by $(\theta = \pi/2,\ \varphi = 3\pi/2)$ and rests in the antipodal location where the negative part of the y axis intersects the Bloch sphere.

2.4 Composite Quantum States and Entanglement

We consider a quantum system A with Hilbert space $\mathcal{H}^A$ and another quantum system B with Hilbert space $\mathcal{H}^B$. The two systems A and B may interact, and we want to have a joint description of the two systems as forming a whole system in itself. The two systems A and B form a composite (or bipartite) quantum system AB whose description resorts to a joint quantum state belonging to the **tensor-product** Hilbert space $\mathcal{H}^A \otimes \mathcal{H}^B$, also considered in Sect. A.4 of the Appendix.

For $\mathcal{H}^A$ of dimension N_A with the computational basis $\{|a\rangle\,;\, a = 1, 2, \ldots, N_A\}$ and $\mathcal{H}^B$ of dimension N_B with the computational basis $\{|b\rangle\,;\, b = 1, 2, \ldots, N_B\}$, the product space $\mathcal{H}^A \otimes \mathcal{H}^B$ is the vector space with dimension $N_A N_B$ spanned by the orthonormal basis states $\{|a\rangle \otimes |b\rangle\,;\, a = 1, 2, \ldots, N_A;\, b = 1, 2, \ldots, N_B\}$. Convenient alternative notations are $|a\rangle \otimes |b\rangle \equiv |a\rangle\,|b\rangle \equiv |ab\rangle$.

From a state vector

$$|A\rangle = \sum_{a=1}^{N_A} \alpha_a^A\,|a\rangle \in \mathcal{H}^A\,, \tag{2.15}$$

and a state vector

$$|B\rangle = \sum_{b=1}^{N_B} \alpha_b^B\,|b\rangle \in \mathcal{H}^B\,, \tag{2.16}$$

the tensor-product state follows as

$$|A\rangle \otimes |B\rangle = \sum_{a=1}^{N_A}\sum_{b=1}^{N_B} \alpha_a^A \alpha_b^B\,|a\rangle \otimes |b\rangle \in \mathcal{H}^A \otimes \mathcal{H}^B\,. \tag{2.17}$$

Any normalized vector $|AB\rangle$ of the product space $\mathcal{H}^A \otimes \mathcal{H}^B$ represents, in principle, a valid quantum state for the composite system AB. Such a valid quantum state $|AB\rangle$ for AB is obtained as any normalized linear combination of the $N_A N_B$ basis vectors $|ab\rangle = |a\rangle \otimes |b\rangle$ of $\mathcal{H}^A \otimes \mathcal{H}^B$ taking the form:

$$|AB\rangle = \sum_{a=1}^{N_A} \sum_{b=1}^{N_B} \alpha_{ab} \, |a\rangle \otimes |b\rangle \; \in \mathcal{H}^A \otimes \mathcal{H}^B \; . \tag{2.18}$$

In this way, a joint or composite state $|AB\rangle$ as in Eq. (2.18) is parameterized by a set of $N_A N_B$ coordinates α_{ab}. Among such states $|AB\rangle$ is the subset of states that can be factored as $|A\rangle \otimes |B\rangle$ as in Eq. (2.17), and a joint state from this subset is parameterized by $N_A + N_B$ coordinates (N_A for $|A\rangle$ and N_B for $|B\rangle$). It results that, in general, there are (many) joint states $|AB\rangle \in \mathcal{H}^A \otimes \mathcal{H}^B$ as in Eq. (2.18) that cannot be factored as $|A\rangle \otimes |B\rangle$ as in Eq. (2.17) with $|A\rangle \in \mathcal{H}^A$ and $|B\rangle \in \mathcal{H}^B$. Such non-factorizable or non-separable states are **entangled states**, to follow the terminology introduced in [5].

Any quantum state of the product space $\mathcal{H}^A \otimes \mathcal{H}^B$ that cannot be factored as the product of a state of $\mathcal{H}^A$ and a state of $\mathcal{H}^B$ defines an entangled state. Entanglement appears as a specifically quantum property, with no classical analogue. From a physical standpoint, when the composite system AB is in an entangled quantum state, then no definite quantum state can be separately assigned to each of the two parts A and B individually. Only the composite system AB can be assigned a definite quantum state. It results that the physical system AB behaves as a whole, a global and delocalized whole. Nevertheless, each part A and B, although entangled, in general remains physically individually materialized and accessible, such as with two photons in an entangled pair. Some local action individually on one part, say A, can well be effected. This will affect the whole global system AB and may lead to what appears as an instantaneous change to the other part B, no matter how distant from A it may be. Quantum entanglement plays an important role in quantum phenomenology, and it will be encountered on many occasions in the sequel, associated with useful behaviors meaningful to quantum information and signal processing.

This description with tensor-product Hilbert spaces extends to any number of quantum systems assembled to form a composite system amenable to a joint description. For quantum information and quantum signal processing, a very common and useful practice is to assemble a number L of qubits to form a composite or multipartite system of L qubits, very similar to a word of L bits in classical digital processing. Individual qubits have their quantum state in the two-dimensional Hilbert space $\mathcal{H}_2$; accordingly a system of L qubits is characterized by a joint quantum state belonging to the tensor-product Hilbert space $\mathcal{H}_2^{\otimes L}$ with dimension 2^L. Any N-dimensional Hilbert space $\mathcal{H}_N$ of N-dimensional quantum states as in Eq. (2.1) can be obtained by assembling a sufficient number L of qubits satisfying $2^L \geq N$, when $\mathcal{H}_N$ is realized as an N-dimensional subspace of $\mathcal{H}_2^{\otimes L}$. This faculty confers to the qubit the status of an elementary or elemental building block for quantum technology. In an engineering perspective, for quantum technological design, this represents the methodological advantage of the possibility of a modular approach from standardized elemental units (qubits), assembled to gradually raise the dimensionality while controlling the elaboration and its growth in complexity.

A qubit pair, with $L = 2$, is the simplest quantum system giving access to the property of entanglement. In the computational basis $\{|00\rangle, |01\rangle, |10\rangle, |11\rangle\}$ of $\mathcal{H}_2^{\otimes 2}$, a valid state for a qubit pair AB is, for example,

$$|AB\rangle = |+\rangle \otimes |+\rangle = \frac{1}{\sqrt{2}}\big(|0\rangle + |1\rangle\big) \otimes \frac{1}{\sqrt{2}}\big(|0\rangle + |1\rangle\big) = \frac{1}{2}\big(|00\rangle + |01\rangle + |10\rangle + |11\rangle\big) . \tag{2.19}$$

This is a separable state or product state of the pair AB, by construction, made by assembling two qubits A and B each individually prepared in the one-qubit state $|+\rangle$ of Eq. (2.7). Another valid state for the qubit pair AB is

$$|AB\rangle = \frac{1}{\sqrt{2}}\big(|00\rangle + |11\rangle\big) . \tag{2.20}$$

This is an entangled state for the qubit pair AB, not factorizable as the product of two one-qubit states. As a result, no definite state exists separately for each individual qubit A and B; only the pair AB as a whole has a definite quantum state in Eq. (2.20).

For qubit pairs in $\mathcal{H}_2^{\otimes 2}$, aside the computational basis mentioned previously, there exists another convenient orthonormal basis, known as the **Bell basis**, and defined by the relations:

$$\begin{cases} |\beta_{00}\rangle = \dfrac{1}{\sqrt{2}}\big(|00\rangle + |11\rangle\big) , & (2.21) \\ |\beta_{01}\rangle = \dfrac{1}{\sqrt{2}}\big(|01\rangle + |10\rangle\big) , & (2.22) \\ |\beta_{10}\rangle = \dfrac{1}{\sqrt{2}}\big(|00\rangle - |11\rangle\big) , & (2.23) \\ |\beta_{11}\rangle = \dfrac{1}{\sqrt{2}}\big(|01\rangle - |10\rangle\big) , & (2.24) \end{cases}$$

invertible as

$$\begin{cases} |00\rangle = \dfrac{1}{\sqrt{2}}\big(|\beta_{00}\rangle + |\beta_{10}\rangle\big) , & (2.25) \\ |01\rangle = \dfrac{1}{\sqrt{2}}\big(|\beta_{01}\rangle + |\beta_{11}\rangle\big) , & (2.26) \\ |10\rangle = \dfrac{1}{\sqrt{2}}\big(|\beta_{01}\rangle - |\beta_{11}\rangle\big) , & (2.27) \\ |11\rangle = \dfrac{1}{\sqrt{2}}\big(|\beta_{00}\rangle - |\beta_{10}\rangle\big) . & (2.28) \end{cases}$$

The Bell state $|\beta_{11}\rangle$ of Eq. (2.24) is sometimes referred to as the singlet state (the sole antisymmetric state accessible to two paired fermions, such as a pair of electrons of an atomic orbital). The Bell basis $\{|\beta_{00}\rangle, |\beta_{01}\rangle, |\beta_{10}\rangle, |\beta_{11}\rangle\}$ from Eqs. (2.21)–(2.24) is made of four *entangled* (Bell) states of $\mathcal{H}_2^{\otimes 2}$, while the computational basis $\{|00\rangle, |01\rangle, |10\rangle, |11\rangle\}$ is made of four *separable* states of $\mathcal{H}_2^{\otimes 2}$.

When referred to the Hadamard basis $\{|+\rangle, |-\rangle\}$ of Eqs. (2.7) and (2.8), and by exploiting Eqs. (2.9) and (2.10), the Bell basis is also

$$
\begin{cases}
|\beta_{00}\rangle = \dfrac{1}{\sqrt{2}}\big(|{+}{+}\rangle + |{-}{-}\rangle\big)\,, & (2.29)\\[2ex]
|\beta_{01}\rangle = \dfrac{1}{\sqrt{2}}\big(|{+}{+}\rangle - |{-}{-}\rangle\big)\,, & (2.30)\\[2ex]
|\beta_{10}\rangle = \dfrac{1}{\sqrt{2}}\big(|{+}{-}\rangle + |{-}{+}\rangle\big)\,, & (2.31)\\[2ex]
|\beta_{11}\rangle = \dfrac{1}{\sqrt{2}}\big(|{-}{+}\rangle - |{+}{-}\rangle\big)\,. & (2.32)
\end{cases}
$$

2.5 Mixed Quantum State

2.5.1 *Definition of the Density Operator*

Up to now we have seen that quantum theory describes a quantum system by assigning to it a normalized state vector $|\psi\rangle$ belonging to a complex Hilbert space. Most importantly, this quantum state determines the connection of the theory with the physical (observable) reality through the process of quantum measurement introduced in Sect. 2.2. When a measurement of a quantum system in the state $|\psi\rangle$ of Eq. (2.1) is performed in the orthonormal basis $\{|n\rangle\}$, then the system is found (projected) in the basis state $|n\rangle$ with the probability $\Pr\{|n\rangle\} = |\langle n|\psi\rangle|^2$ of Eq. (2.3). When repeated measurements are performed on identical copies of $|\psi\rangle$, the resulting empirical frequencies experimentally observed for projecting onto each $|n\rangle$ are predictable from the theoretical probabilities $\Pr\{|n\rangle\}$. This is the situation when a definite, precisely determined, state vector $|\psi\rangle$ can be assigned to the quantum system, which is then said to be described by a **pure state** $|\psi\rangle$.

More generally, however, situations exist where such a precisely determined state vector $|\psi\rangle$ cannot be assigned to a quantum system of interest Q. This may occur, for instance, when the initial preparation of Q—which, in general, still remains a delicate task—comes with some imprecision on the prepared state, or when some statistical fluctuations may arise in the parameters determining the quantum state of Q. Imprecision may also occur when Q couples to an uncontrolled or unobserved environment. By such coupling interaction, the quantum system of interest Q may become entangled with the uncontrolled environment, so that Q can no longer be ascribed a well-defined pure state of its own. This imprecision or uncertainty resulting on the quantum state represents a loss of coherence or a decoherence of the quantum state [6, 7], which can equivalently be seen as the effect of some quantum noise. Such situations with uncertainty on the quantum state can be uniformly handled with the notion of density operator. To introduce this notion, a statistical ensemble $\big\{(p_j, |\psi_j\rangle)\big\}$ is considered, for specifying that the quantum system of interest Q can be in the pure state $|\psi_j\rangle$ with some probability p_j, for $j = 1$ to J.

In a quantum measurement on Q in the orthonormal basis $\big\{|n\rangle\big\}$, to characterize the probability $\Pr\big\{|n\rangle\big\}$ of observing (projecting onto) the basis state $|n\rangle$, we first consider the conditional probability $\Pr\big\{|n\rangle \,\big|\, |\psi_j\rangle\big\}$ of projecting onto $|n\rangle$ when the quantum system Q is in state $|\psi_j\rangle$, which by Eq. (2.3) is $\Pr\big\{|n\rangle \,\big|\, |\psi_j\rangle\big\} = \langle n|\psi_j\rangle\,\langle\psi_j|n\rangle$. The total probability for projecting onto $|n\rangle$ then follows as the weighted sum $\Pr\big\{|n\rangle\big\} = \sum_{j=1}^{J} p_j \Pr\big\{\,|n\rangle \,\big|\, |\psi_j\rangle\,\big\}$, which is

$$\Pr\big\{|n\rangle\big\} = \sum_{j=1}^{J} p_j\,\langle n|\psi_j\rangle\,\langle\psi_j|n\rangle \tag{2.33}$$

$$= \langle n|\bigg(\sum_{j=1}^{J} p_j\,|\psi_j\rangle\,\langle\psi_j|\bigg)|n\rangle\;, \tag{2.34}$$

where Eq. (2.34) is deduced by exploiting properties of linear algebra. Leaning on the notion of outer product from Eq. (A.8) of the Appendix, it is convenient to introduce the operator ρ defined as

$$\boxed{\rho = \sum_{j=1}^{J} p_j\,|\psi_j\rangle\,\langle\psi_j|\;,} \tag{2.35}$$

so that the probability of Eq. (2.34) becomes

$$\Pr\big\{|n\rangle\big\} = \langle n|\rho|n\rangle = \operatorname{tr}\big(\rho\,|n\rangle\,\langle n|\big)\;, \tag{2.36}$$

with the last form in Eq. (2.36) resulting from the identity of Eq. (A.13). The operator ρ of Eq. (2.35) defines the **density operator** associated with the statistical ensemble $\{(p_j, |\psi_j\rangle)\}_{j=1}^{J}$ of J pure quantum states. It provides a complete representation of the state of Q described by the statistical ensemble of states $\{(p_j, |\psi_j\rangle)\}_{j=1}^{J}$, in the sense that the density operator ρ completely determines the result of measuring Q : by Eq. (2.36), in the measurement, Q ends up projected onto state $|n\rangle$ with the probability $\Pr\big\{|n\rangle\big\}$. Upon repeating the measurement on Q successively prepared in an initial state drawn from the statistical ensemble $\{(p_j, |\psi_j\rangle)\}_{j=1}^{J}$, the resulting empirical frequencies experimentally observed for projecting onto each $|n\rangle$ are predictable from the theoretical probabilities $\Pr\big\{|n\rangle\big\}$ of Eq. (2.36). In this respect, the density operator ρ of Eq. (2.35) stands as a complete representation for the state of a quantum system described by a statistical ensemble like $\{(p_j, |\psi_j\rangle)\}_{j=1}^{J}$. The quantum system is then said to be in a **mixed state** represented by the density operator ρ.

Additionally, if a system is prepared in a mixed state of density operator ρ_k with probability p_k, for $k = 1$ to K, then the state of this system is represented by the overall density operator $\rho = \sum_{k=1}^{K} p_k \rho_k$. This naturally follows through standard probabilistic reasoning by considering for each mixed state ρ_k a statistical ensemble of pure states, with each such ensemble receiving the corresponding probabilistic weight p_k to constitute the overall density operator ρ.

The density operator is an important notion in quantum theory. It provides a universal representation for quantum states—pure or mixed—since the notion applies equally to a pure state $|\psi\rangle$ associated with the (rank-one) density operator $\rho = |\psi\rangle\langle\psi|$ when Eq. (2.36) coincides with Eq. (2.3). And from a pure state $|\psi\rangle$, a global phase change yielding $e^{i\phi}|\psi\rangle$ does not alter the density operator $\rho = |\psi\rangle\langle\psi|$, which conveniently incorporates the physical equivalence indicated after Eq. (2.5). Another benefit for state representation when moving from a vector $|\psi\rangle$ (a simpler mathematical object) to an operator ρ (a more complicated mathematical object) is that the measurement probability, which is a *nonlinear* (quadratic) function of the vector state $|\psi\rangle$ in Eq. (2.3), becomes a *linear* function of the operator state ρ in Eq. (2.36). This linearity will be useful on many occasions in the sequel. The density operator, generalizing the state vector to provide a universal representation for quantum states, will often stand, for our quantum signal-processing perspective, as our preferential notion to assume the role of **quantum signal**, information-carrying signal. The density operator will be particularly relevant in our signal-processing perspective when addressing quantum noise and its effect on quantum states, which is essentially to transform pure states into mixed states, uniformly described by density operators representing the quantum signal, with its possible noisy versions of central concern in signal processing.

2.5.2 *Properties of the Density Operator*

For a quantum system with N-dimensional Hilbert space $\mathcal{H}_N$, the density operator ρ defined by Eq. (2.35) is a linear operator from $\mathcal{H}_N$ onto $\mathcal{H}_N$ and belonging to the vector space $\mathcal{L}(\mathcal{H}_N)$. It generally owns the properties that follow.

- ρ is Hermitian. Since $\big(|\psi_j\rangle\langle\psi_j|\big)^\dagger = |\psi_j\rangle\langle\psi_j|$ and the p_j are real, then $\rho^\dagger = \rho$.
- ρ has unit trace. This follows from linearity of the trace, and from Eq. (A.14) yielding $\mathrm{tr}(|\psi_j\rangle\langle\psi_j|) = \langle\psi_j|\psi_j\rangle = 1$, and since the p_j are probabilities which sum to 1.
- ρ is a positive operator. For any state $|\psi\rangle$ of $\mathcal{H}_N$, the scalar $\langle\psi|\rho|\psi\rangle = \sum_j p_j \langle\psi|\psi_j\rangle\langle\psi_j|\psi\rangle = \sum_j p_j |\langle\psi|\psi_j\rangle|^2 \geq 0$ which is always real positive establishes ρ as a positive operator.

Since ρ is a Hermitian operator of $\mathcal{L}(\mathcal{H}_N)$, it can always be endowed with N real eigenvalues λ_n (possibly with multiplicity higher than 1) which can be associated with N (normalized) eigenstates $|\lambda_n\rangle$ forming an orthonormal basis of the space $\mathcal{H}_N$. Since ρ is positive with unit trace, its N eigenvalues λ_n are positive and sum to 1. A diagonal form, or eigendecomposition, results for ρ as

$$\rho = \sum_{n=1}^{N} \lambda_n \, |\lambda_n\rangle \, \langle\lambda_n| \ . \tag{2.37}$$

Conversely, any positive unit-trace operator ρ is always Hermitian and can always be decomposed according to Eq. (2.37) with positive coefficients λ_n summing to 1 and orthonormal states $|\lambda_n\rangle$; therefore, any such ρ constitutes a valid density operator, generated by the statistical ensemble $\{(\lambda_n, |\lambda_n\rangle)\}$ formed by the N orthogonal eigenstates $|\lambda_n\rangle$ of ρ each probabilistically weighted by the corresponding positive eigenvalue λ_n.

Also, any density operator ρ defined according to Eq. (2.35) from the statistical ensemble $\{(p_j, |\psi_j\rangle)\}$ comprising any number J of pure states $|\psi_j\rangle$ of $\mathcal{H}_N$ can equivalently be interpreted as generated by the alternative statistical ensemble $\{(\lambda_n, |\lambda_n\rangle)\}$ of its N eigenstates.

This indicates that distinct statistical ensembles of pure states can generate the same density operator. Moreover, it can be shown that, in general, any density operator $\rho \in \mathcal{L}(\mathcal{H}_N)$ can be generated by an infinite set of distinct statistical ensembles, connected by a property of **isometric freedom** in the following way [8, 9]. For two distinct statistical ensembles $\{(p_j, |\psi_j\rangle)\}_{j=1}^{J}$ and $\{(q_k, |\phi_k\rangle)\}_{k=1}^{K}$, we introduce the unnormalized vectors of $\mathcal{H}_N$ defined as $|\widetilde{\psi}_j\rangle = \sqrt{p_j}\,|\psi_j\rangle$ and $|\widetilde{\phi}_k\rangle = \sqrt{q_k}\,|\phi_k\rangle$. The two statistical ensembles $\{(p_j, |\psi_j\rangle)\}_{j=1}^{J}$ and $\{(q_k, |\phi_k\rangle)\}_{k=1}^{K}$ with $J \leq K$ generate the same density operator $\rho = \sum_{j=1}^{J} |\widetilde{\psi}_j\rangle\,\langle\widetilde{\psi}_j| = \sum_{k=1}^{K} |\widetilde{\phi}_k\rangle\,\langle\widetilde{\phi}_k|$ if and only if the two sets of unnormalized vectors $\{|\widetilde{\psi}_j\rangle\}_{j=1}^{J}$ and $\{|\widetilde{\phi}_k\rangle\}_{k=1}^{K}$ can be connected by an isometric relation:

$$|\widetilde{\phi}_k\rangle = \sum_{j=1}^{J} v_{kj}\, |\widetilde{\psi}_j\rangle \ , \tag{2.38}$$

where the complex numbers $[v_{kj}] = V$ form a $K \times J$ matrix V defining an isometry, satisfying $V^{\dagger}V = I_J$, with I_J the $J \times J$ identity matrix, giving $\sum_{k=1}^{K} v_{kj}^{*} v_{kj'} = \delta_{jj'}$ the Kronecker delta. In other words, the J columns of V define J unit vectors with dimension $K \geq J$ and mutually orthogonal. With $\langle\widetilde{\phi}_k| = \sum_{j'=1}^{J} v_{kj'}^{*}\,\langle\widetilde{\psi}_{j'}|$, we indeed obtain

$$\sum_{k=1}^{K} |\widetilde{\phi}_k\rangle\,\langle\widetilde{\phi}_k| = \sum_{j=1}^{J}\sum_{j'=1}^{J}\sum_{k=1}^{K} v_{kj} v_{kj'}^{*}\, |\widetilde{\psi}_j\rangle\,\langle\widetilde{\psi}_{j'}| \tag{2.39}$$

$$= \sum_{j=1}^{J}\sum_{j'=1}^{J} \delta_{jj'}\, |\widetilde{\psi}_j\rangle\,\langle\widetilde{\psi}_{j'}| = \sum_{j=1}^{J} |\widetilde{\psi}_j\rangle\,\langle\widetilde{\psi}_j| \ , \tag{2.40}$$

as two reconstructions of the same density operator ρ.

The smallest size J of a statistical ensemble for the density operator ρ is given by the rank of ρ. From there, all equivalent statistical ensembles with arbitrary size $K \geq J$ can be isometrically deduced via Eq. (2.38). Upon measurement governed by Eq. (2.36), all these distinct statistical ensembles isometrically related and generating the same density operator ρ lead to the same measurement statistics and are therefore indistinguishable and physically equivalent in this respect.

The density operators ρ of $\mathcal{L}(\mathcal{H}_N)$ form a **convex set**, in the sense that when ρ and ρ' are two density operators of $\mathcal{L}(\mathcal{H}_N)$, then the convex sum $\rho'' = p\rho + (1-p)\rho'$ is also a density operator of $\mathcal{L}(\mathcal{H}_N)$, for any real $p \in [0, 1]$, since ρ'' remains positive with unit trace.

The operator ρ^2 has the eigenvalues λ_n^2. The positive scalar

$$\mathrm{tr}\big(\rho^2\big) = \sum_{n=1}^{N} \lambda_n^2 \tag{2.41}$$

is known as the **purity** of the quantum state ρ. It provides a convenient numerical criterion to decide whether a given density operator ρ represents a pure state or a mixed state, since $\mathrm{tr}\big(\rho^2\big) = 1$ for a pure state and $\mathrm{tr}\big(\rho^2\big) < 1$ for a mixed state.

A pure state has a rank-one density operator $\rho = |\psi\rangle\langle\psi| \in \mathcal{L}(\mathcal{H}_N)$ which is idempotent (that is, $\rho^2 = \rho$) and is always endowed with an eigenvalue $+1$ associated with the eigenstate $|\psi\rangle$. The $N-1$ other eigenvalues of ρ are 0 (thus having multiplicity $N-1$), with associated eigenstates formed by any $N-1$ mutually orthogonal states of $\mathcal{H}_N$ spanning the $(N-1)$-dimensional subspace orthogonal to $|\psi\rangle$. So a pure state has the maximum purity $\mathrm{tr}\big(\rho^2\big) = \mathrm{tr}(\rho) = 1$. By contrast, a mixed state ρ will necessarily have (some) eigenvalues λ_n differing from 0 and 1 satisfying $\lambda_n^2 < \lambda_n$, whence $\mathrm{tr}\big(\rho^2\big) < 1$ for a mixed state.

The constrained minimization of the purity $\mathrm{tr}\big(\rho^2\big) = \sum_n \lambda_n^2$ can be addressed by the Lagrange multiplier method with the Lagrangian $\mathcal{L} = \sum_n \lambda_n^2 - \mu\big(\sum_n \lambda_n - 1\big)$ and a Lagrange multiplier μ. The resulting canonical Lagrange equations are then solved by $\lambda_n = 1/N$ for $n = 1$ to N. This establishes that the minimal purity is achieved as $\mathrm{tr}\big(\rho^2\big) = 1/N$ by the density operator $\rho = \mathrm{I}_N/N$, representing the maximally mixed state, also called fully depolarized state, with I_N the identity operator on $\mathcal{H}_N$.

2.5.3 Purification of a Mixed State

For a mixed quantum state represented by the density operator ρ of $\mathcal{L}(\mathcal{H}_N)$, a purification is an operation that introduces a pure state of a larger Hilbert space, whose reduction by partial tracing delivers the density operator ρ. The reason for such a partial tracing operation is that, for a bipartite quantum system AB in a pure state $|AB\rangle \in \mathcal{H}^A \otimes \mathcal{H}^B$ as in Sect. 2.4, when only the subsystem A is accessible for measurement, the results upon measuring A can be completely described by assigning to A the mixed state represented by the (reduced) density operator $\rho_A = \text{tr}_B\big(|AB\rangle\langle AB|\big)$ obtained by partial tracing over the subsystem B. This stands as an additional motivation for the density operator, supplementing the notion of statistical ensemble of pure states considered in Sect. 2.5.1: a density operator naturally arises in this way by the partial tracing operation for describing the (local) state of a quantum system A which is part of a larger quantum system AB in a pure state $|AB\rangle$ yet with no access or control on the other part B. This quantum state reduction by partial tracing $\rho_A = \text{tr}_B\big(|AB\rangle\langle AB|\big)$ has a classical analogue, with two joint random variables (A, B) with joint probability distribution conveniently written $P(A, B)$, when B is a hidden or latent variable, and A alone is described by the marginalized probability distribution $P(A) = \sum_B P(A, B)$. We shall come back later on this important operation of quantum state reduction by partial tracing, in Sect. 3.2.1 with Eq. (3.35) when describing the evolution of an open quantum system, and also with the notion of partial measurement, in Sect. 4.1.3 by a von Neumann projective measurement, and in Sect. 4.4 by a generalized measurement. For the moment, quantum state reduction by partial tracing is applied for the purification of a mixed quantum state.

We consider a principal quantum system Q, with the N-dimensional Hilbert space $\mathcal{H}^Q \equiv \mathcal{H}_N$, placed in a mixed quantum state defined by the statistical ensemble $\{(p_j, |\psi_j\rangle)\}_{j=1}^J$, and therefore represented by the density operator

$$\rho_Q = \sum_{j=1}^{J} p_j\,|\psi_j\rangle\langle\psi_j| \tag{2.42}$$

of $\mathcal{L}(\mathcal{H}^Q)$, as in the generic situation of Eq. (2.35). For the purification of ρ_Q, we introduce an auxiliary quantum system, or reference system, R, with a Hilbert space $\mathcal{H}^R \equiv \mathcal{H}_J$ whose dimension is the cardinal J of the statistical ensemble for Q. We introduce an orthonormal basis $\{|e_j\rangle\}_{j=1}^J$ of the auxiliary Hilbert space $\mathcal{H}^R$. The bipartite pure state

$$|QR\rangle = \sum_{j=1}^{J} \sqrt{p_j}\,|\psi_j\rangle \otimes |e_j\rangle \tag{2.43}$$

of $\mathcal{H}^Q \otimes \mathcal{H}^R$ offers a purification for ρ_Q, since we have the reduced state $\rho_Q = \text{tr}_R\big(|QR\rangle\langle QR|\big)$. This follows from the bipartite density operator which is

$$|QR\rangle\langle QR| = \sum_{j=1}^{J}\sum_{k=1}^{J} \sqrt{p_j p_k}\,|\psi_j\rangle\langle\psi_k| \otimes |e_j\rangle\langle e_k| \;, \tag{2.44}$$

so that

$$\mathrm{tr}_R\big(|QR\rangle\langle QR|\big) = \sum_{j=1}^{J}\sum_{k=1}^{J} \sqrt{p_j p_k}\, |\psi_j\rangle\langle\psi_k|\, \mathrm{tr}\big(|e_j\rangle\langle e_k|\big) \tag{2.45}$$

$$= \sum_{j=1}^{J}\sum_{k=1}^{J} \sqrt{p_j p_k}\, |\psi_j\rangle\langle\psi_k|\, \langle e_k|e_j\rangle \tag{2.46}$$

$$= \sum_{j=1}^{J} p_j\, |\psi_j\rangle\langle\psi_j| = \rho_Q\,, \tag{2.47}$$

as targeted, since $\langle e_k|e_j\rangle = \delta_{jk}$ the Kronecker delta.

In addition, the reference quantum system R alone is described by the reduced density operator

$$\rho_R = \mathrm{tr}_Q\big(|QR\rangle\langle QR|\big) = \sum_{j=1}^{J}\sum_{k=1}^{J} \sqrt{p_j p_k}\, \langle\psi_k|\psi_j\rangle\, |e_j\rangle\langle e_k| \tag{2.48}$$

of $\mathcal{L}(\mathcal{H}_J)$. If on the reference system R in state ρ_R a quantum measurement were performed in the orthonormal basis $\{|e_j\rangle\}_{j=1}^{J}$ of $\mathcal{H}_J$, the random projection of R would occur onto $|e_j\rangle$ with the probability $p_j = \langle e_j|\rho_R|e_j\rangle = \mathrm{tr}\big(\rho_R\,|e_j\rangle\langle e_j|\big)$, according to Eq. (2.36), also forcing Q in the state $|\psi_j\rangle$ as expected from the pre-measurement joint state of Eq. (2.43).

As we have seen in Sect. 2.5.2 with the property of isometric freedom, two statistical ensembles of pure states generate the same density operator if and only if they can be related by an isometry as in Eq. (2.38). A comparable **isometric freedom** exists for the different pure bipartite states purifying a given density operator ρ_Q. This can be seen in the following way. We consider an alternative statistical ensemble $\{(q_k, |\phi_k\rangle)\}_{k=1}^{K}$ of size $K \geq J$ generating the same density operator $\rho_Q = \sum_{j=1}^{J} p_j\,|\psi_j\rangle\langle\psi_j| = \sum_{k=1}^{K} q_k\,|\phi_k\rangle\langle\phi_k|$ for Q, with according to Eq. (2.38) the isometric relation:

$$\sqrt{q_k}\,|\phi_k\rangle = \sum_{j=1}^{J} v_{kj}\sqrt{p_j}\,|\psi_j\rangle\,, \tag{2.49}$$

where the isometry $[v_{kj}]$ acts on the states for Q. Adhering to Eq. (2.43), a purifying pure state for ρ_Q follows as

$$|QR'\rangle = \sum_{k=1}^{K} \sqrt{q_k}\,|\phi_k\rangle \otimes |e'_k\rangle\,, \tag{2.50}$$

where we have introduced an orthonormal basis $\{|e'_k\rangle\}_{k=1}^{K}$ for the K-dimensional Hilbert space of a reference system R'. By Eq. (2.49) the purifying state of Eq. (2.50) is also

$$|QR'\rangle = \sum_{k=1}^{K}\sum_{j=1}^{J} v_{kj}\sqrt{p_j}\,|\psi_j\rangle \otimes |e'_k\rangle \tag{2.51}$$

$$= \sum_{j=1}^{J} \sqrt{p_j}\,|\psi_j\rangle \otimes \sum_{k=1}^{K} v_{kj}\,|e'_k\rangle \tag{2.52}$$

$$= \sum_{j=1}^{J} \sqrt{p_j}\,|\psi_j\rangle \otimes |e''_j\rangle\ , \tag{2.53}$$

where we have introduced the J vectors

$$|e''_j\rangle = \sum_{k=1}^{K} v_{kj}\,|e'_k\rangle \tag{2.54}$$

resulting from the action of the isometry $[v_{kj}]$ on the K-dimensional reference system R'. We deduce $\langle e''_{j'}| = \sum_{k'=1}^{K} v^*_{k'j'}\,\langle e'_{k'}|$ so that we have the inner product

$$\langle e''_{j'}|e''_j\rangle = \sum_{k'=1}^{K}\sum_{k=1}^{K} v^*_{k'j'}v_{kj}\,\langle e'_{k'}|e'_k\rangle \tag{2.55}$$

$$= \sum_{k'=1}^{K}\sum_{k=1}^{K} v^*_{k'j'}v_{kj}\delta_{kk'} \tag{2.56}$$

$$= \sum_{k=1}^{K} v^*_{kj'}v_{kj} = \delta_{jj'}\ , \tag{2.57}$$

with Eq. (2.57) resulting from $[v_{kj}]$ being an isometry. This establishes that $\{|e''_j\rangle\}_{j=1}^{J}$ forms a set of J mutually orthogonal states, defining an orthonormal basis for an alternative J-dimensional reference (sub)space, and Eq. (2.53) restores for ρ_Q a purifying state under the form of Eq. (2.43) on a J-dimensional reference.

All bipartite pure states purifying a given density operator ρ_Q are isometrically related in this way [8, 10–12], with the purifying reference systems connected by an isometry under the form of Eq. (2.54). This is similar to the isometric connection of Eq. (2.49) between the statistical ensembles of pure states representing the density operator ρ_Q. A same isometry acts indifferently on the statistical ensemble for Q in Eq. (2.49), or on the auxiliary reference system for purification in Eq. (2.54). The minimal size J fixing the minimal dimension of the purifying reference system R is the minimal cardinal J of a statistical ensemble for ρ_Q, and it is given by the rank of the density operator ρ_Q.

In the purification operation, the purifying pure state as in Eqs. (2.43) or (2.50) is, in general, an entangled state of the bipartite Hilbert space, to entail by partial tracing a mixed reduced state ρ_Q for Q. The reduced state for Q would result as a pure state $\rho_Q = |Q\rangle\,\langle Q|$ if and only if the purifying bipartite pure state is a separable state $|Q\rangle \otimes |R\rangle$ between the space for Q and the space for the auxiliary reference R. As we have seen, a mixed state ρ_Q represents some imprecise knowledge concerning the state of a quantum system Q. The purification operation provides an insightful figuration of an important mechanism by which such an imprecise mixed quantum state may arise. This is the situation we evoked above in the second paragraph of Sect. 2.5.1, when a quantum system of interest Q is entangled with another (reference or environment) system R upon which no control or capability of observation is accessible. As a result, the system Q is to be described by a mixed quantum state ρ_Q. This is often the mechanism at work when a quantum system of interest Q is described by a mixed quantum state ρ_Q, materializing some imprecise knowledge, for the reason that the system Q is entangled with an uncontrolled environment. The ensuing degradation of the quantum state into a mixed state is interpreted as a quantum noise or a decoherence phenomenon, or conversely, a quantum noise or decoherence phenomenon degrading a quantum state often results from its entanglement with an uncontrolled environment. Noise degrading an information-carrying signal is a central concern of signal processing, classical and quantum alike. Accordingly, these important phenomena of quantum noise and decoherence associated with mixed quantum states will be examined in more detail in the sequel, in accordance with our quantum signal-processing perspective.

A given density operator ρ for a quantum system Q can arise from a statistical ensemble for preparing Q, as in Eq. (2.35), or in the canonical form of Eq. (2.37) with the statistical ensemble formed from the eigendecomposition of ρ. By the purification mechanism, the same density operator ρ can be obtained by partial tracing in Eqs. (2.45)–(2.47) after entanglement with an auxiliary system. The resulting (reduced) density operator ρ, in turn, can receive a statistical ensemble from its eigendecomposition. Statistical ensemble and reduction by partial tracing are therefore two alternative descriptions always available for a given density operator ρ. The underlying constitutive scenarios—existence of a statistical ensemble or entanglement with another system—when they produce the same density operator ρ remain indistinguishable by quantum measurement alone operating on ρ.

2.5.4 Density Operator of the Qubit

When a qubit is in a pure state $|\psi\rangle$ of the space $\mathcal{H}_2$, the parameterizations of respectively Eqs. (2.4) or (2.6) lead to two expressions for the representation matrix of the density operator $\rho = |\psi\rangle\langle\psi| \in \mathcal{L}(\mathcal{H}_2)$ as

$$\rho = |\psi\rangle\langle\psi| = \begin{bmatrix} |\alpha_0|^2 & \alpha_0\alpha_1^* \\ \alpha_0^*\alpha_1 & |\alpha_1|^2 \end{bmatrix} = \frac{1}{2}\begin{bmatrix} 1+\cos(\theta) & e^{-i\varphi}\sin(\theta) \\ e^{i\varphi}\sin(\theta) & 1-\cos(\theta) \end{bmatrix}. \tag{2.58}$$

In the general situation of a mixed state, the density operator ρ, when referred to the computational basis $\{|0\rangle, |1\rangle\}$ of $\mathcal{H}_2$, can be represented by the matrix:

$$\rho = \begin{bmatrix} \rho_{00} & \rho_{01} \\ \rho_{10} & \rho_{11} \end{bmatrix} . \tag{2.59}$$

This is equivalent, in the standard basis $\{|0\rangle\langle 0|, |0\rangle\langle 1|, |1\rangle\langle 0|, |1\rangle\langle 1|\}$ of $\mathcal{L}(\mathcal{H}_2)$, to the linear combination

$$\rho = \rho_{00}|0\rangle\langle 0| + \rho_{01}|0\rangle\langle 1| + \rho_{10}|1\rangle\langle 0| + \rho_{11}|1\rangle\langle 1| . \tag{2.60}$$

Some constraints follow on the four (complex) matrix elements in Eq. (2.59) to ensure that the density operator ρ is Hermitian positive with unit trace. The real positive diagonal elements ρ_{00} and ρ_{11}, satisfying $\text{tr}(\rho) = \rho_{00} + \rho_{11} = 1$, are called the "populations", and represent the probabilities $\langle 0|\rho|0\rangle = \rho_{00}$ and $\langle 1|\rho|1\rangle = \rho_{11}$ of finding ρ respectively in state $|0\rangle$ or $|1\rangle$ when measured in the computational basis $\{|0\rangle, |1\rangle\}$. The off-diagonal elements $\rho_{01} = \langle 0|\rho|1\rangle$ and $\rho_{10} = \langle 1|\rho|0\rangle = \rho_{01}^*$, called the "coherences", satisfy $|\rho_{01}| = |\rho_{10}| \le \sqrt{\rho_{00}\rho_{11}} \le 1/2$ for positivity of ρ, with the upper bound $\sqrt{\rho_{00}\rho_{11}}$ reached for pure states. When measuring in the computational basis $\{|0\rangle, |1\rangle\}$, the probabilities $\Pr\{|0\rangle\} = \rho_{00}$ and $\Pr\{|1\rangle\} = \rho_{11} = 1 - \rho_{00}$ are governed by the population ρ_{00}; meanwhile, when measuring in the Hadamard basis $\{|+\rangle, |-\rangle\}$ of Eqs. (2.7) and (2.8), the probabilities $\Pr\{|\pm\rangle\} = \langle\pm|\rho|\pm\rangle = (1/2) \pm \text{Re}(\rho_{01})$ are governed by the (real part of the) coherence ρ_{01}. In other words, the populations can be probed by measuring the qubit in the computational basis $\{|0\rangle, |1\rangle\}$, while the coherences can be probed by measuring the qubit in the Hadamard basis $\{|+\rangle, |-\rangle\}$.

There exists another quite useful representation of the qubit density operator $\rho \in \mathcal{L}(\mathcal{H}_2)$, which relies on the Pauli basis $\{\mathrm{I}_2, \sigma_x, \sigma_y, \sigma_z\}$ of $\mathcal{L}(\mathcal{H}_2)$ as introduced in Sect. A.5 of the Appendix, to express ρ as the linear combination, known as the **Bloch representation**:

$$\boxed{\rho = \frac{1}{2}\left(\mathrm{I}_2 + r_x\sigma_x + r_y\sigma_y + r_z\sigma_z\right) = \frac{1}{2}\left(\mathrm{I}_2 + \vec{r}\cdot\vec{\sigma}\,\right) ,} \tag{2.61}$$

especially using the concise notation introduced in Eq. (A.44) with the formal vector $\vec{\sigma} = [\sigma_x, \sigma_y, \sigma_z]$. The constraints on the four coordinates of ρ, in the standard basis of $\mathcal{L}(\mathcal{H}_2)$ in Eq. (2.60) or in the Pauli basis in Eq. (2.61), as imposed by the unit-trace positivity of ρ, express differently, and in a very convenient way with Eq. (2.61). The form of Eq. (2.61) is justified because the three Pauli operators have the trace zero and $\text{tr}(\mathrm{I}_2) = 2$, so that $\text{tr}(\rho) = 1$ requires always a coefficient $1/2$ on I_2 in the linear combination of Eq. (2.61). In addition, since all four basis operators $\{\mathrm{I}_2, \sigma_x, \sigma_y, \sigma_z\}$ are themselves Hermitian, the Hermiticity property $\rho = \rho^\dagger$ requires three *real* coefficients (r_x, r_y, r_z) in Eq. (2.61). These three coefficients are assembled to form the real vector $\vec{r} = [r_x, r_y, r_z]^\top \in \mathbb{R}^3$, which is known as the **Bloch vector**. Via the Bloch representation of Eq. (2.61), the Bloch vector $\vec{r} \in \mathbb{R}^3$ completely characterizes the qubit density operator ρ of $\mathcal{L}(\mathcal{H}_2)$. One has in particular $\vec{r} = \text{tr}(\rho\,\vec{\sigma}\,) = [\text{tr}(\rho\sigma_x),\ \text{tr}(\rho\sigma_y),\ \text{tr}(\rho\sigma_z)]^\top$.

From Eq. (2.61) and the representation matrices of the Pauli operators in Eqs. (A.36)–(A.38), the representation matrix of ρ referred to the computational basis $\{|0\rangle, |1\rangle\}$ of $\mathcal{H}_2$ follows as

$$\rho = \frac{1}{2}\begin{bmatrix} 1 + r_z & r_x - ir_y \\ r_x + ir_y & 1 - r_z \end{bmatrix}, \tag{2.62}$$

to be identified with Eq. (2.59), with in addition the reciprocal relations between the coordinates of ρ:

$$\begin{cases} r_x = \rho_{01} + \rho_{10} = 2\,\mathrm{Re}(\rho_{10}), & (2.63) \\ r_y = i(\rho_{01} - \rho_{10}) = 2\,\mathrm{Im}(\rho_{10}), & (2.64) \\ r_z = \rho_{00} - \rho_{11}. & (2.65) \end{cases}$$

From Eq. (2.62), the eigendecomposition of the density operator ρ can be explicitly worked out by solving the characteristic equation $\det(\rho - \lambda \mathrm{I}_2) = 0$. Two eigenvalues for ρ follow as

$$\lambda_\pm = \frac{1}{2}\left(1 \pm \|\vec{r}\,\|\right). \tag{2.66}$$

Necessarily $\lambda_\pm \geq 0$ since ρ is a positive operator. It results that any valid density operator ρ is parameterized in Eq. (2.61) by a vector $\vec{r}$ of Euclidean norm $\|\vec{r}\,\| \leq 1$.

The special case of a pure state $\rho = |\psi\rangle\langle\psi|$ comes with the eigenvalue $\lambda_+ = 1$ associated with the eigenstate $|\psi\rangle$ and the eigenvalue $\lambda_- = 0$ associated with the eigenstate $|\psi^\perp\rangle$ orthogonal to $|\psi\rangle$ in the vector plane $\mathcal{H}_2$, so that $\langle\psi|\psi^\perp\rangle = 0$. Consistently, from Eq. (2.66), any pure state $\rho = |\psi\rangle\langle\psi|$ of a qubit is always parameterized in Eq. (2.61) by a vector $\vec{r}$ of unit norm $\|\vec{r}\,\| = 1$. In addition, by identifying Eq. (2.62) with Eq. (2.58) the three coordinates of $\vec{r}$ in $\mathbb{R}^3$ follow as

$$\vec{r}\;\left|\;\begin{array}{ll} r_x = \sin(\theta)\cos(\varphi), & (2.67) \\ r_y = \sin(\theta)\sin(\varphi), & (2.68) \\ r_z = \cos(\theta). & (2.69) \end{array}\right.$$

It results that such an $\vec{r}$ representing a pure qubit state $|\psi\rangle$ is a unit vector of $\mathbb{R}^3$ which points to the surface of the Bloch sphere of unit radius, at the two polar angles (θ, φ), just as in Fig. 2.2.

If we return to the general case of a mixed state ρ, the two eigenvalues $\lambda_\pm$ in Eq. (2.66) are distinct from 0 and 1, yet are positive and sum to 1, so that they are necessarily associated with $\|\vec{r}\,\| < 1$. Such a mixed qubit state ρ is thus represented in Eq. (2.61) by a vector $\vec{r}$ pointing in the interior of the Bloch sphere of unit radius or inside the Bloch ball.

In this way, any density operator ρ for a qubit can be characterized via Eq. (2.61) by a Bloch vector $\vec{r} \in \mathbb{R}^3$, which is on the surface of the Bloch ball (i.e., with $\|\vec{r}\,\| = 1$)

for a pure state, and in the interior of the Bloch ball (i.e., with $\|\vec{r}\,\| < 1$) for a mixed state.

Finally, it is possible to characterize the two (orthogonal) eigenstates $|\lambda_\pm\rangle$ of ρ associated with the two eigenvalues $\lambda_\pm$ from Eq. (2.66). This can be done by algebraic calculation of the eigenvectors of the 2×2 matrix for ρ. This can also be done by a more geometrical reasoning in the Bloch sphere picture. From the two orthogonal pure states $|\lambda_\pm\rangle$ of $\mathcal{H}_2$, the two projectors $|\lambda_\pm\rangle\langle\lambda_\pm|$ are valid density operators that can be characterized in Eq. (2.61) by two unit-norm antipodal Bloch vectors $\vec{r}_+$ and $\vec{r}_- = -\vec{r}_+$. For ρ, the diagonal form $\rho = \lambda_+|\lambda_+\rangle\langle\lambda_+| + \lambda_-|\lambda_-\rangle\langle\lambda_-|$ implies $\vec{r} = \lambda_+\vec{r}_+ + \lambda_-\vec{r}_- = (\lambda_+ - \lambda_-)\vec{r}_+$. And since Eq. (2.66) gives $\lambda_+ - \lambda_- = \|\vec{r}\,\|$, it comes out that the eigenstates $|\lambda_\pm\rangle$ are the two orthogonal pure states of $\mathcal{H}_2$ characterized by the two unit Bloch vectors $\vec{r}_\pm = \pm\vec{r}/\|\vec{r}\,\|$ to represent the two projectors

$$|\lambda_\pm\rangle\langle\lambda_\pm| = \frac{1}{2}\left(\mathrm{I}_2 \pm \frac{\vec{r}}{\|\vec{r}\,\|}\cdot\vec{\sigma}\right) \tag{2.70}$$

onto the two eigenstates $|\lambda_\pm\rangle$.

From Eq. (2.61), one also has the determinant $\det(\rho) = \big(1 - \|\vec{r}\,\|^2\big)/4 \geq 0$ and $\rho^2 = \big(1 + \|\vec{r}\,\|^2\big)\mathrm{I}_2/4$, so that the purity $\mathrm{tr}\big(\rho^2\big) = \big(1 + \|\vec{r}\,\|^2\big)/2$ verifies the criterion that $\mathrm{tr}\big(\rho^2\big) = 1$ for a pure state and $\mathrm{tr}\big(\rho^2\big) < 1$ for a mixed state.

2.5.5 *An Example of Qubit Density Operator*

As an example of density operator we consider for a qubit the statistical ensemble comprising the three states

$$\left\{\left(p_1 = \frac{1}{7},\ |\phi_1\rangle = |0\rangle\right);\ \left(p_2 = \frac{2}{7},\ |\phi_2\rangle = |1\rangle\right);\ \left(p_3 = \frac{4}{7},\ |\phi_3\rangle = |+\rangle\right)\right\}. \tag{2.71}$$

According to Eq. (2.35), the corresponding density operator follows as

$$\rho = \frac{1}{7}|0\rangle\langle 0| + \frac{2}{7}|1\rangle\langle 1| + \frac{4}{7}|+\rangle\langle +| = \frac{1}{7}\begin{bmatrix} 3 & 2 \\ 2 & 4 \end{bmatrix}. \tag{2.72}$$

The two eigenvalues of ρ are $\lambda_+ = \big(1 + \sqrt{17}/7\big)/2 \approx 0.795$ associated with the eigenstate $|\lambda_+\rangle = \cos(\theta/2)|0\rangle + \sin(\theta/2)|1\rangle \approx 0.615|0\rangle + 0.788|1\rangle$, and $\lambda_- = \big(1 - \sqrt{17}/7\big)/2 \approx 0.205$ associated with the eigenstate $|\lambda_-\rangle = \sin(\theta/2)|0\rangle - \cos(\theta/2)|1\rangle \approx 0.788|0\rangle - 0.615|1\rangle$, with the angle $\theta = 2\arccos\big[\big(8/(17 + \sqrt{17})\big)^{1/2}\big] \approx 1.816\,\mathrm{rad} \approx 0.578\pi$ rad.

As a result, the same density operator ρ can be obtained from the two-state statistical ensemble

$$\Big\{\big(\lambda_+, |\lambda_+\rangle\big);\ \big(\lambda_-, |\lambda_-\rangle\big)\Big\}, \tag{2.73}$$

as

$$\rho = \lambda_+ |\lambda_+\rangle\langle\lambda_+| + \lambda_- |\lambda_-\rangle\langle\lambda_-| = \frac{1}{7}\begin{bmatrix} 3 & 2 \\ 2 & 4 \end{bmatrix} . \tag{2.74}$$

The three-state ensemble of Eq. (2.71) and the two-state ensemble of Eq. (2.73), generating the same density operator ρ, are equivalent in this respect and indistinguishable by quantum measurement.

The three qubit states $|\phi_j\rangle$ of the initial statistical ensemble of Eq. (2.71) can be expressed in the eigenbasis $\{|\lambda_+\rangle, |\lambda_-\rangle\}$ of the qubit Hilbert space $\mathcal{H}_2$ as

$$\begin{cases} |\phi_1\rangle \approx 0.615\,|\lambda_+\rangle + 0.788\,|\lambda_-\rangle \,, & (2.75) \\ |\phi_2\rangle \approx 0.788\,|\lambda_+\rangle - 0.615\,|\lambda_-\rangle \,, & (2.76) \\ |\phi_3\rangle \approx 0.993\,|\lambda_+\rangle + 0.122\,|\lambda_-\rangle \,. & (2.77) \end{cases}$$

As for Eq. (2.38), the three unnormalized vectors

$$\begin{cases} |\widetilde{\phi}_1\rangle = \sqrt{p_1}\,|\phi_1\rangle = \sqrt{1/7}\,|0\rangle \approx 0.378\,|0\rangle \,, & (2.78) \\ |\widetilde{\phi}_2\rangle = \sqrt{p_2}\,|\phi_2\rangle = \sqrt{2/7}\,|1\rangle \approx 0.535\,|1\rangle \,, & (2.79) \\ |\widetilde{\phi}_3\rangle = \sqrt{p_3}\,|\phi_3\rangle = \sqrt{2/7}\Big(|0\rangle + |1\rangle\Big) = \sqrt{2}\,|\widetilde{\phi}_1\rangle + |\widetilde{\phi}_2\rangle & (2.80) \end{cases}$$

can be expressed as a function of the two unnormalized vectors $|\widetilde{\lambda}_+\rangle = \sqrt{\lambda_+}\,|\lambda_+\rangle$ and $|\widetilde{\lambda}_-\rangle = \sqrt{\lambda_-}\,|\lambda_-\rangle$ as

$$\begin{cases} |\widetilde{\phi}_1\rangle \approx 0.261\,|\widetilde{\lambda}_+\rangle + 0.657\,|\widetilde{\lambda}_-\rangle \,, & (2.81) \\ |\widetilde{\phi}_2\rangle \approx 0.473\,|\widetilde{\lambda}_+\rangle - 0.726\,|\widetilde{\lambda}_-\rangle \,, & (2.82) \\ |\widetilde{\phi}_3\rangle \approx 0.842\,|\widetilde{\lambda}_+\rangle + 0.204\,|\widetilde{\lambda}_-\rangle \,. & (2.83) \end{cases}$$

Equations (2.81)–(2.83) explicitly exhibit the 3×2 isometry matrix $[v_{kj}] = V$ defined by Eq. (2.38) and accomplishing the isometric connection between the two statistical ensembles generating the same qubit density operator ρ here.

2.5.6 *Noisy Preparation of a Qubit*

We consider another example showing that the notion of density operator is closely related to the notion of quantum noise. A qubit is intended to be prepared in the pure state $|\psi\rangle = |0\rangle$. There is, however, some imprecision in the preparation. This could occur, for instance, in the preparation of a photon in the polarization state $|0\rangle$, yet in the presence of some random mechanical vibrations in the optical setup supposed to select the intended polarization and causing some misalignment of the selected polarization. As a consequence, the qubit instead is prepared in the "noisy" state

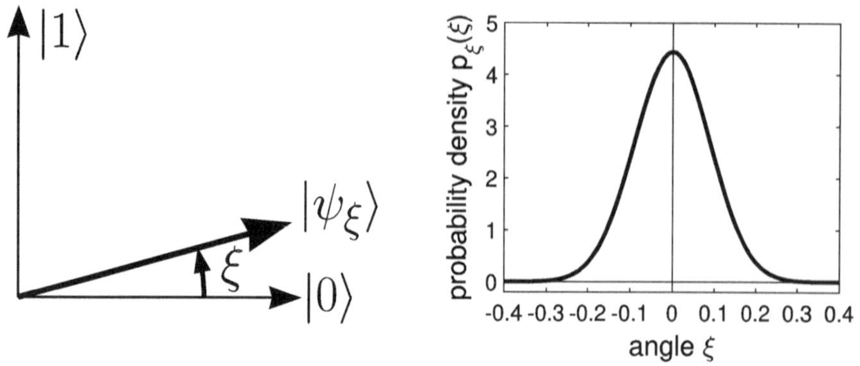

Fig. 2.4 The noisy preparation of a qubit in state $|\psi_\xi\rangle$ of Eq. (2.84) with the random angle ξ distributed according to the probability density $p_\xi(\xi)$

$$|\psi_\xi\rangle = \cos(\xi)\,|0\rangle + \sin(\xi)\,|1\rangle \ , \tag{2.84}$$

where ξ is a small angle which fluctuates randomly around 0 at each preparation experiment, as depicted in Fig. 2.4. The angle ξ is modeled by a continuously distributed random variable with the probability density function $p_\xi(\xi)$, which for convenience is assumed to be even, as also depicted in Fig. 2.4.

This imprecision on the prepared state can be represented by the density operator

$$\rho = \int_\xi p_\xi(\xi)\,|\psi_\xi\rangle\,\langle\psi_\xi|\,d\xi \ , \tag{2.85}$$

which is the natural extension of the discrete sum of Eq. (2.35) to a statistical ensemble of continuously distributed states $|\psi_\xi\rangle$.

From Eq. (2.84), the outer product $|\psi_\xi\rangle\,\langle\psi_\xi|$ evaluates as

$$|\psi_\xi\rangle\,\langle\psi_\xi| = \begin{bmatrix} \cos^2(\xi) & \cos(\xi)\sin(\xi) \\ \cos(\xi)\sin(\xi) & \sin^2(\xi) \end{bmatrix} , \tag{2.86}$$

so that averaging according to Eq. (2.85) leads to the density operator

$$\rho = \begin{bmatrix} \big\langle\cos^2(\xi)\big\rangle & 0 \\ 0 & \big\langle\sin^2(\xi)\big\rangle \end{bmatrix} = \big\langle\cos^2(\xi)\big\rangle\,|0\rangle\,\langle 0| + \big\langle\sin^2(\xi)\big\rangle\,|1\rangle\,\langle 1| \ , \tag{2.87}$$

where $\langle\cdot\rangle$ in Eq. (2.87) stands for the statistical average according to the probability density $p_\xi(\xi)$. The even density $p_\xi(\xi)$ affords the convenience of a diagonal density operator ρ in Eq. (2.87), although this is not critical.

The noisy preparation over an infinite statistical ensemble $\big\{\big(p_\xi(\xi), |\psi_\xi\rangle\big)\big\}$ of continuously distributed noisy states $|\psi_\xi\rangle$ is thus equivalent to the preparation over the discrete two-state statistical ensemble $\{(\langle\cos^2(\xi)\rangle, |0\rangle), (\langle\sin^2(\xi)\rangle, |1\rangle)\}$ consisting in a qubit prepared in state $|0\rangle$ with probability $\langle\cos^2(\xi)\rangle$ or in state $|1\rangle$ with probability $\langle\sin^2(\xi)\rangle$. These two scenarios for the preparation—although one may look more physically plausible than the other—are indistinguishable by quantum measurement. They may be quite distinct at the level of intimate physical mechanisms determining the quantum state of the qubit, yet this intimate constitution remains hidden to the observer.

This example also illustrates the versatility of the noise scenarios that can cause imprecision and loss of coherence in a quantum state and which can be captured by the uniform notion of density operator. We have here an example of a pure quantum state $|\psi_\xi\rangle$ endowed with random coordinates parameterized by a random variable ξ that, when associated with a probability distribution $p_\xi(\xi)$ for ξ, gives rise to the statistical ensemble $\{(p_\xi(\xi), |\psi_\xi\rangle)\}$ described by the density operator ρ to represent the quantum state with random parameters. Every measurement repeated on successive preparations of $|\psi_\xi\rangle$ occurs at a given realization of the angle ξ sampled from the distribution $p_\xi(\xi)$, so that the resulting measurement statistics experimentally observed are predictable from the density operator ρ. Such scenarios of **randomization in the state coordinates**, with a pure state prepared with random coordinates or parameters, give rise to a statistical ensemble of states and are uniformly representable by a density operator.

2.6 The Wave Function

It is interesting to connect the notion of quantum state presented in this chapter in the context of quantum information, to another notion also relating to quantum state in quantum physics or quantum mechanics, which is the wave function. This chapter set up a discrete finite-dimensional notion of quantum state, well suited to a presentation of quantum signal processing relating to digital signal processing. The wave function arises when the notion of quantum state installed in this chapter is considered in the continuous infinite-dimensional limit.

In (classical) mechanics, a fundamental entity appearing at the forefront (and at the foundation) is the material particle—point particle or point mass—with its position in space. This is the same in quantum mechanics, which, to a particle that can take its position continuously in one spatial dimension[2], assigns a quantum state expressible as

$$|\psi\rangle = \int_{-\infty}^{\infty} \psi(x)\,|x\rangle\,dx \in \mathcal{H}\,. \tag{2.88}$$

This is the same notion of quantum state as in Eq. (2.1), yet taken to a continuous limit. In general, the position of a material particle, when it is measured, can be observed in a continuously infinite set of possible configurations. Accordingly, the quantum

[2] The approach readily extends to three spatial dimensions.

state $|\psi\rangle$ of Eq. (2.88) is defined in a continuous infinite-dimensional Hilbert space $\mathcal{H}$, referred to a continuous infinite-dimensional standard basis $\{|x\rangle\}$ with x varying over the real line $\mathbb{R}$ to represent the positions at which the particle can be found when measured. The orthonormality of the computational basis is expressed by

$$\langle x|x'\rangle = \delta(x - x') , \tag{2.89}$$

a continuous analogue to Eq. (A.7) of the Appendix, with $\delta(x)$ the Dirac delta function, with also a completeness relation

$$\int_{-\infty}^{\infty} |x\rangle\,\langle x|\,dx = \mathsf{Id} , \tag{2.90}$$

analogous to Eq. (A.10), when Id is the identity operator on the Hilbert space $\mathcal{H}$.

In the standard basis $\{|x\rangle\}$, the coordinate of the quantum state $|\psi\rangle$ is provided by the function $\psi(x)$ in Eq. (2.88), which is the **wave function**. It is the continuous analogue of the discrete set of coordinates $\{\alpha_n\}_{n=1}^{N}$ from Eq. (2.1), and similarly $\psi(x)$ completely determines the quantum state $|\psi\rangle$. The wave function $\psi(x)$ satisfies a normalization constraint

$$\langle\psi|\psi\rangle = \int_{-\infty}^{\infty} |\psi(x)|^2 dx = 1 , \tag{2.91}$$

analogous to Eq. (2.2). A similar probabilistic interpretation of the squared modulus of the coordinate applies as in Sect. 2.2, yet extended to the continuous setting. Upon measuring the particle, the probability density is $|\psi(x)|^2$ for observing the particle at the continuous position $x \in \mathbb{R}$.

Such a physical perspective somehow recognizes to the material particle, with its position in space, the status of a natural elemental object imposed or suggested by physical reality and serving as a primitive notion for the elaboration of classical as well as quantum mechanics. Adhering to this physical perspective, quantum mechanics, when it starts its elaboration from such material particle as a primitive notion, has to face the methodological complication of having to cope right from the start with a quantum state of continuously infinite dimension, for which the wave function offers a common representation. Comparatively, quantum information, with an engineering perspective, to put it this way, proposes to start at the other extreme of the dimensionality scale, from a less natural but more "engineered" elemental object—as argued at the beginning of Sect. 2.3—the qubit, as the simplest nontrivial object with the minimal finite dimension of two. This especially presents, as we already noted in Sect. 2.4, the methodological advantage of the possibility of a modular approach in the elaboration and technological design, by gradually increasing the dimensionality. Accordingly, for quantum information here, the finite-dimensional quantum state, as $|\psi\rangle$ in Eq. (2.1), will be the central prevalent notion, rather than the infinite-dimensional continuous wave function $\psi(x)$.

2.7 Summary on Quantum Signal

This chapter focused on the **first postulate** (among three) or fundamental principle for quantum theory, stating essentially that a quantum object or system is described by a quantum state formed by a vector of a complex Hilbert space, with various important aspects of this first postulate that were further examined.

A quantum system in a **pure state** (known with certainty) is described by a state vector $|\psi\rangle$ with unit norm from a complex Hilbert space, as in Eq. (2.1). A **mixed state** describes the broader situation where imprecise knowledge may exist about the quantum state, which is handled with the more general notion of a density operator ρ (positive with unit trace) as in Eq. (2.35) to represent a statistical ensemble $\{(p_j, |\psi_j\rangle)\}_{j=1}^{J}$ of feasible pure states. Alternatively, by the purification mechanism of Sect. 2.5.3, a density operator ρ arises from partial tracing to represent a quantum system of interest entangled with an inaccessible environment. When describing a quantum state, both the state vector $|\psi\rangle$ and the density operator ρ can be assigned the role of **quantum signal**, and chiefly the density operator ρ which is a universal representation able to capture the notion of noisy quantum state or signal, central to our signal-processing perspective.

The quantum state or signal represented by $|\psi\rangle$ or ρ connects to the measurement results via the probabilities of Eqs. (2.3) or (2.36). This in particular manifests that the quantum state in itself, although standing as a fundamental descriptive concept, is not in general completely and directly accessible in a quantum measurement, but only partially via a probabilistic influence.

Composite quantum systems formed with several parts or subsystems are described in a **tensor-product** Hilbert space. Two subsystems are **entangled** when their joint bipartite state cannot be factored as a product of two separate states for each subsystems, and as a consequence no definite pure quantum state can be assigned individually to each subsystem, only the composite system as a whole can receive a well-defined pure state of its own.

A fundamental system for quantum information is the two-dimensional **qubit**, with a quantum state, pure or mixed, always representable by a density operator $\rho = (\mathrm{I}_2 + \vec{r} \cdot \vec{\sigma})/2$ in the Bloch representation of Eq. (2.61), parameterized by a Bloch vector $\vec{r} \in \mathbb{R}^3$ with norm $\|\vec{r}\| = 1$ for a pure state and $\|\vec{r}\| < 1$ for a mixed state. By tensoring L qubits, any N-dimensional quantum states as in Eq. (2.1) can receive a materialization for all sizes $N < 2^L$, and accordingly qubits form fundamental building blocks for quantum technologies.

Chapter 3
Quantum Process

This chapter essentially deals with the second general postulate or fundamental principle for quantum theory, concerning the evolution of quantum states (or quantum signals). The operation describing the evolution of a quantum state—pure or mixed—is represented as a quantum process. In analogy with signals-and-systems theory, when the quantum state—represented by a state vector or by a density operator—is considered as a signal, the quantum process can be conferred the role of the operative system or functional block, acting to transform or process the quantum signal. This chapter presents this notion of quantum process, for the evolution of quantum states, with unitary processes for closed evolutions, and with non-unitary processes for open evolutions, especially applicable to quantum decoherence or quantum noise of great importance for the sequel of our signal-processing perspective.

3.1 Unitary Quantum Process

3.1.1 Unitary Evolution of a Quantum State

As we have seen in Chap. 2, a valid state for a quantum system can be represented by a state vector $|\psi\rangle$ with unit norm (possibly from a statistical ensemble with a probabilistic weight if some uncertainty exists) belonging to a complex Hilbert space $\mathcal{H}$. When involved in some physical phenomenon, the quantum system can undergo an evolution of its state. When in the evolution the system does not get coupled to some other system, it forms an isolated or **closed quantum system**. Its state remains represented by a state vector of the Hilbert space $\mathcal{H}$ always keeping a unit norm as a valid state, for consistency with the first fundamental postulate of Sect. 2.1. Preserving unit norm for the state vector is not the only property ensured in a quantum evolution. In addition, the evolution of the state vector in $\mathcal{H}$ is always found to be linear, taking the form of a linear map from $\mathcal{H}$ onto $\mathcal{H}$ belonging to the

F. Chapeau-Blondeau, *Quantum Information and Quantum Signal Processing*,
https://doi.org/10.1007/978-3-032-27499-1_3

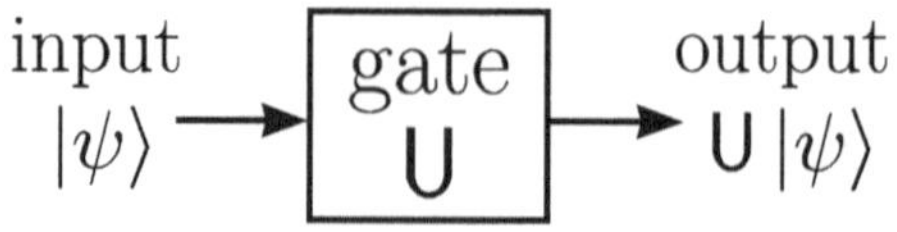

Fig. 3.1 The evolution by the unitary operator U of Eq. (3.1), defining the (unitary linear) quantum process U represented as a quantum gate or circuit for U, receiving as input signal the quantum state $|\psi\rangle$ and delivering as output signal the quantum state $\mathsf{U}\,|\psi\rangle$, via the input–output transformation $|\psi\rangle \mapsto \mathsf{U}\,|\psi\rangle$ of Eq. (3.1)

vector space $\mathcal{L}(\mathcal{H})$. This stands as a fundamental principle or postulate[1] of quantum theory: the evolution of the state of a closed quantum system with Hilbert space $\mathcal{H}$ always occurs as a linear map on $\mathcal{H}$ preserving unit norm. A norm-preserving linear map from $\mathcal{H}$ onto $\mathcal{H}$ is a unitary operator, so that the evolution of a closed quantum system with state $|\psi\rangle \in \mathcal{H}$ is represented by a unitary linear operator U from $\mathcal{H}$ onto $\mathcal{H}$ performing the state transformation:

$$\boxed{|\psi\rangle \longmapsto \mathsf{U}\,|\psi\rangle\ ,} \tag{3.1}$$

with $\mathsf{U}^\dagger = \mathsf{U}^{-1}$ for unitarity.

The transformation of Eq. (3.1), with a unitary operator $\mathsf{U} \in \mathcal{L}(\mathcal{H})$ to transform or evolve the state of a closed quantum system here, represents a **quantum process** transforming quantum states. Such notion of quantum process fits well within the perspective of signals-and-systems theory, where the process performing the unitary evolution by U in Eq. (3.1) can be represented by a quantum gate or circuit or functional block, receiving as input signal the quantum state $|\psi\rangle$ before the evolution takes place, and delivering as output signal the quantum state $\mathsf{U}\,|\psi\rangle$ after the evolution has taken place, as depicted in Fig. 3.1.

Since a unitary operator U is invertible as $\mathsf{U}^{-1} = \mathsf{U}^\dagger$, it results that a quantum gate implementing a unitary process U as in Fig. 3.1 is always an invertible gate.

The picture of Fig. 3.1 and evolution of Eq. (3.1) include the possibility where the input signal $|\psi\rangle$ is a composite state belonging to a tensor-product Hilbert space, with a unitary operator U acting on the product space, and connecting an input space and an output space with same dimension, both isomorphic and identified as $\mathcal{H}$. This will be the case, for instance, with a quantum gate or circuit as in Fig. 3.1, receiving L input qubits to support the input state, and necessarily delivering a same number L of qubits on its output, for invertibility. Multiple-qubit gates will be examined in more detail in Sect. 3.1.5.

[1] This is an empirical principle or postulate, motivated by experimental evidence, as is the rule at the root of a physical theory. We could as well have envisaged some nonlinear evolution preserving unit norm for quantum states, for instance, with a non-norm-preserving linear map F supplemented by a renormalization for the evolution $|\psi\rangle \mapsto \mathsf{F}\,|\psi\rangle \,/\, \langle\psi|\mathsf{F}^\dagger\mathsf{F}|\psi\rangle^{1/2}$ that preserves unit norm but is nonlinear, and therefore is not permitted here. Empirical evidence shows that a valid evolution of a closed quantum system is always linear. At the quantum level, God plays dice and remains linear.

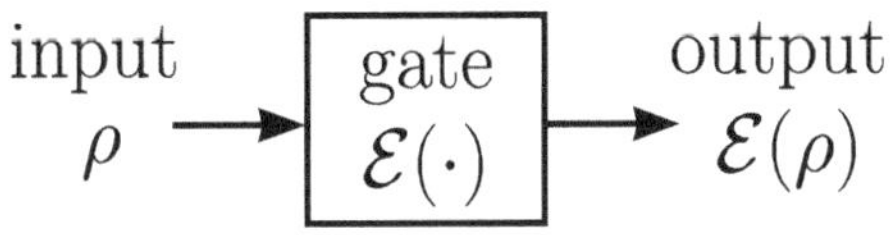

Fig. 3.2 The quantum process $\mathcal{E}(\cdot)$ receiving as input signal the density operator ρ and delivering as output signal the density operator $\mathcal{E}(\rho)$, via the input–output transformation $\rho \mapsto \mathcal{E}(\rho)$ defined by the unitary operator U acting in Eq. (3.2)

When the state of a quantum system is imprecisely known and described as in Sect. 2.5 by a statistical ensemble $\{(p_j, |\psi_j\rangle)\}$ represented by the density operator $\rho = \sum_j p_j |\psi_j\rangle \langle\psi_j|$ of Eq. (2.35), then by a unitary transformation U each member evolves according to Eq. (3.1) as $|\psi_j\rangle \mapsto \mathsf{U}|\psi_j\rangle$. Consequently, the quantum system after the unitary transformation by U is described by the statistical ensemble $\{(p_j, \mathsf{U}|\psi_j\rangle)\}$, represented by the density operator $\sum_j p_j \mathsf{U}|\psi_j\rangle\langle\psi_j|\mathsf{U}^\dagger = \mathsf{U}\sum_j p_j |\psi_j\rangle\langle\psi_j|\mathsf{U}^\dagger = \mathsf{U}\rho\mathsf{U}^\dagger$. It results that in the process the density operator ρ evolves according to

$$\boxed{\rho \longmapsto \mathsf{U}\rho\mathsf{U}^\dagger = \mathcal{E}(\rho)\ .} \tag{3.2}$$

The map $\rho \mapsto \mathcal{E}(\rho)$ defined by Eq. (3.2) is a linear operator from $\mathcal{L}(\mathcal{H})$ onto $\mathcal{L}(\mathcal{H})$, mapping an input density operator $\rho \in \mathcal{L}(\mathcal{H})$ into an output density operator $\mathcal{E}(\rho) \in \mathcal{L}(\mathcal{H})$, so that $\mathcal{E}(\cdot)$ can be referred to as a **superoperator**. As it should, the superoperator $\mathcal{E}(\cdot)$ preserves the unit trace of valid density operators, since $\mathrm{tr}(\mathcal{E}(\rho)) = \mathrm{tr}(\mathsf{U}\rho\mathsf{U}^\dagger) = \mathrm{tr}(\rho\mathsf{U}^\dagger\mathsf{U}) = \mathrm{tr}(\rho) = 1$ by circularity of the trace. The transformation of Eq. (3.2) is another materialization of the notion of quantum process, for the evolution of quantum states, when the states are represented by density operators.

As we mentioned in Sect. 2.5, the density operator provides a universal representation for a quantum state—pure or mixed—so that Eq. (3.2) can be seen as a universal representation for the evolution of a closed quantum system by means of the (unitary linear) quantum process $\mathcal{E}(\cdot)$, picturable by a quantum gate as shown in Fig. 3.2.

3.1.2 *The Schrödinger Equation*

As we have done in Sect. 2.6 for the wave function, it is interesting to connect the description of a quantum evolution via a unitary operator in Eq. (3.1), with another approach to the description of a quantum evolution, quite common in quantum physics, which is via the Schrödinger equation. This alternative approach describes the evolution of a quantum state in continuous time by means of a differential equation, and it can be deduced from our fundamental postulate of unitary evolution of Sect. 3.1.1 as follows.

Equation (3.1) dictates a unitary evolution as $|\psi(t)\rangle = \mathsf{U}(t, t_0)\,|\psi(t_0)\rangle$ for any pure quantum state $|\psi\rangle \in \mathcal{H}$ between two times t_0 and $t \geq t_0$. If we envisage an evolution in continuous time, we can introduce an infinitesimal time interval dt, and from t to $t+dt$ the state evolves unitarily as $|\psi(t+dt)\rangle = \mathsf{U}(t+dt, t)\,|\psi(t)\rangle$. With sufficient smoothness enabling differentiability in the continuous-time evolution, we assume that the infinitesimal variation of the state can be expressed as $|\psi(t+dt)\rangle = |\psi(t)\rangle + \frac{d}{dt}|\psi(t)\rangle\,dt$. In addition, for the unitary evolution operator U, we assume the infinitesimal variation $\mathsf{U}(t+dt, t) = \mathsf{Id} + \mathsf{G}dt$, with Id the identity operator on the Hilbert space $\mathcal{H}$, since when dt completely vanishes as $dt = 0$ we need $\mathsf{U}(t+dt, t) = \mathsf{U}(t, t) = \mathsf{Id}$ for a trivial evolution. With $\mathsf{U}(t+dt, t)$ a linear operator, G is also a linear operator on $\mathcal{H}$, with the physical dimension of $[\text{time}]^{-1}$. We deduce the vector $\frac{d}{dt}|\psi(t)\rangle = \mathsf{G}\,|\psi(t)\rangle$. It follows that $\frac{d}{dt}\langle\psi(t)| = \langle\psi(t)|\,\mathsf{G}^\dagger$. The evolution must preserve the norm of the state vector so that $\frac{d}{dt}\langle\psi(t)|\psi(t)\rangle = 0$. Since $\frac{d}{dt}\langle\psi(t)|\psi(t)\rangle = \big(\frac{d}{dt}\langle\psi(t)|\big)\,|\psi(t)\rangle + \langle\psi(t)|\,\big(\frac{d}{dt}|\psi(t)\rangle\big)$, we deduce that $\frac{d}{dt}\langle\psi(t)|\psi(t)\rangle = \langle\psi(t)|(\mathsf{G}^\dagger + \mathsf{G})|\psi(t)\rangle = 0$ for any state $|\psi(t)\rangle$ and time t. This imposes an anti-Hermitian $\mathsf{G} = -\mathsf{G}^\dagger$. It is then convenient to introduce the operator $\mathsf{H} = i\hbar\mathsf{G}$. The scalar $\hbar$ is the reduced Planck constant, with the physical dimension of [energy]×[time]. With G anti-Hermitian, the operator H results as a Hermitian operator, with the physical dimension of [energy], known as the Hamiltonian or energy operator attached to the quantum system. The time evolution $\frac{d}{dt}|\psi(t)\rangle = \mathsf{G}\,|\psi(t)\rangle$ then becomes

$$\frac{d}{dt}|\psi(t)\rangle = -\frac{i}{\hbar}\mathsf{H}\,|\psi(t)\rangle\ , \tag{3.3}$$

which is the **Schrödinger equation**, governing the evolution in continuous time of the state of a closed quantum system. With this approach, starting from the unitary evolution of Eq. (3.1), the differential Schrödinger equation of Eq. (3.3) follows as a consequence, with the (Hermitian) Hamiltonian operator H deduced from the unitary evolution operator U.

On an N-dimensional Hilbert space $\mathcal{H} \equiv \mathcal{H}_N$, the Hermitian Hamiltonian operator H can be endowed with N (real) eigenvalues E_n with the physical dimension of [energy], associated with N eigenstates $|E_n\rangle$ forming an orthonormal eigenbasis for the space $\mathcal{H}_N$, and affording the diagonal form $\mathsf{H} = \sum_{n=1}^N E_n\,|E_n\rangle\langle E_n|$. For a time-invariant Hamiltonian H with time-independent eigenstates $|E_n\rangle$ and eigenvalues E_n, if at time $t_0 = 0$ the quantum system is initialized in one energy eigenstate $|\psi(0)\rangle = |E_n\rangle$, then, at any time $t \geq 0$, the evolved state $|\psi(t)\rangle = \exp(-iE_nt/\hbar)\,|\psi(0)\rangle = \exp(-iE_nt/\hbar)\,|E_n\rangle$ is solution to the Schrödinger equation (3.3). For a time-invariant Hamiltonian, by Eq. (3.3), an energy eigenstate thus evolves as $|E_n\rangle \mapsto \exp(-i\omega_nt)\,|E_n\rangle$, with an angular frequency $\omega_n = E_n/\hbar$ adhering to the Planck relation. The time evolution of the energy eigenstate only builds up a global phase factor $\exp(-i\omega_nt)$ as time passes, to represent a physically invariant quantum state characterized by a constant energy $E_n = \hbar\omega_n$. In general, an arbitrary initialization $|\psi(0)\rangle$ can always be expressed in the energy or Hamiltonian eigenbasis as $|\psi(0)\rangle = \sum_{n=1}^N \alpha_n\,|E_n\rangle$, so that by Eq. (3.3) with a time-invariant Hamiltonian the state evolves as $|\psi(t)\rangle = \sum_{n=1}^{N} \alpha_n \exp(-i\omega_n t)\, |E_n\rangle$, showing a generic and explicit form of the evolution by referring the quantum state to the Hamiltonian eigenbasis.

With the above approach, starting from the postulation of the unitary evolution of Eq. (3.1) for the quantum state, we have shown how to deduce the continuous-time evolution of the Schrödinger equation in Eq. (3.3). The converse is also feasible: if the continuous-time evolution by the Schrödinger equation in Eq. (3.3) is first postulated, the unitary evolution of Eq. (3.1) can be deduced, in the following way.

Starting from the differential equation of Eq. (3.3), a formal solution can be written to express the evolution of the quantum state between an initial time t_0 and a final time $t \geq t_0$, taking the form[2]

$$|\psi(t)\rangle = \exp\left(-\frac{i}{\hbar}\int_{t_0}^{t} \mathsf{H}\, dt'\right) |\psi(t_0)\rangle = \mathsf{U}(t, t_0)\, |\psi(t_0)\rangle \ . \tag{3.4}$$

In Eq. (3.4), the operator

$$\mathsf{U}(t, t_0) = \exp\left(-\frac{i}{\hbar}\int_{t_0}^{t} \mathsf{H} dt'\right) \tag{3.5}$$

is the evolution operator of the quantum state between times t_0 and $t \geq t_0$. It is unitary when H is Hermitian, which is always the case for physical systems, since a Hermitian H is required for Eq. (3.3) to preserve the norm of the state vector. For a given physical system, the quantum Hamiltonian or energy operator H, commonly, can be constructed from a classical scalar Hamiltonian or energy function for the system, via appropriate correspondence rules. The unitary operator $\mathsf{U}(t, t_0)$ obtained in Eq. (3.5) is the unitary operator U introduced in Eq. (3.1). As we have seen in Sect. 3.1.1, such an operator U in Eq. (3.1) provides the possibility of a "functional-block" description—quite common in signals-and-systems theory—for the evolution of the quantum state, between an initial input state when some processing begins and a final output state when it ends, as depicted in Fig. 3.1 or 3.2. By contrast, the Hamiltonian H in Eq. (3.3) serves to a continuous-time description of the evolution, from the internal physical working of the system.

For the situation considered above of a time-invariant Hamiltonian H, the unitary evolution operator of Eq. (3.5) becomes

$$\mathsf{U}(t, t_0) = \exp\left(-i(t - t_0)\frac{\mathsf{H}}{\hbar}\right) . \tag{3.6}$$

A time-invariant Hamiltonian H with time-independent eigenstates $|E_n\rangle$ and eigenvalues $E_n = \hbar\omega_n$ gives $\mathsf{H}/\hbar = \sum_{n=1}^{N} \omega_n\, |E_n\rangle \langle E_n|$. When an initial state $|\psi(0)\rangle =$

[2] Some care should be taken in the case of a time-dependent Hamiltonian $\mathsf{H}(t)$ that does not commute at two distinct times, when Eqs. (3.4) and (3.5) should then be understood via the notion of time-ordered exponential and Dyson series.

$\sum_{n=1}^{N} \alpha_n |E_n\rangle$ from time $t_0 = 0$ is evolved at $t \geq 0$ by the Schrödinger equation of Eq. (3.3) as $|\psi(t)\rangle = \sum_{n=1}^{N} \alpha_n \exp(-i\omega_n t) |E_n\rangle$, the corresponding unitary evolution operator from Eq. (3.6) is $\mathsf{U}(t) = \sum_{n=1}^{N} \exp(-i\omega_n t) |E_n\rangle \langle E_n|$, to yield $|\psi(t)\rangle = \mathsf{U}(t) |\psi(0)\rangle$.

In general, when the evolution of Eq. (3.1) is written $|\psi(t)\rangle = \mathsf{U}(t, t_0) |\psi(t_0)\rangle$, we have the time derivative $\frac{d}{dt} |\psi(t)\rangle = \frac{d}{dt}\mathsf{U}(t, t_0) |\psi(t_0)\rangle$. And since Eq. (3.3) is also $\frac{d}{dt} |\psi(t)\rangle = -(i/\hbar)\mathsf{H}\mathsf{U}(t, t_0) |\psi(t_0)\rangle$, by identification we obtain a Schrödinger equation for the change in time of the unitary evolution operator:

$$\frac{d}{dt}\mathsf{U}(t, t_0) = -\frac{i}{\hbar}\mathsf{H}\,\mathsf{U}(t, t_0)\,. \tag{3.7}$$

For the density operator $\rho = |\psi\rangle \langle\psi|$ we have $\frac{d}{dt}\rho = \left(\frac{d}{dt} |\psi\rangle\right) \langle\psi| + |\psi\rangle \frac{d}{dt} \langle\psi|$, and by using Eq. (3.3) and its complex conjugate we obtain $\frac{d}{dt}\rho = -(i/\hbar)\big(\mathsf{H} |\psi\rangle \langle\psi| - |\psi\rangle \langle\psi| \mathsf{H}\big) = -(i/\hbar)[\mathsf{H}, \rho]$ involving the commutator $[\mathsf{H}, \rho]$. More generally, a property of convexity-linearity applies, since any density operator ρ for a mixed state, as seen in Eq. (2.35), is expressible as a convex sum of pure state density operators similar to $|\psi\rangle \langle\psi|$, so that, by linearity, the change in time of a density operator $\rho(t)$ can be expressed as

$$\frac{d}{dt}\rho(t) = -\frac{i}{\hbar}\big[\mathsf{H}, \rho(t)\big]\,, \tag{3.8}$$

which is known as the **Liouville–von Neumann equation**.

The Schrödinger equation (3.3) describes the instantaneous change in continuous time of the quantum state vector $|\psi\rangle$, by means of a temporal differential equation, which is quite common in physics. Meanwhile, the quantum process of Eq. (3.1) describes the global change of the state vector $|\psi\rangle$ between an initial configuration at time t_0 and a final configuration at time $t \geq t_0$, by means of an evolution operator $\mathsf{U}(t, t_0)$ which can be understood as the input–output transformation by a functional block as in Fig. 3.1, quite common in signals-and-systems theory. The same applies equally for the evolution of an arbitrary density operator ρ, with a differential equation in Eq. (3.8) or a functional-block transformation in Eq. (3.2) and Fig. 3.2. When the Hamiltonian happens to be a trivial operator $\mathsf{H} = H_0\mathsf{Id}$ proportional to the identity operator Id, with H_0 a scalar constant, the evolution operator in Eq. (3.6) reduces to $\mathsf{U}(t, t_0) = \exp\big(-i(t - t_0)H_0/\hbar\big)\mathsf{Id}$, leading for the state vector to the temporal evolution $|\psi(t)\rangle = \exp\big(-i(t - t_0)H_0/\hbar\big) |\psi(t_0)\rangle$. As time passes, the state $|\psi(t)\rangle$ only builds up a global phase factor $\exp\big(-i(t - t_0)H_0/\hbar\big)$ associated with an invariant density operator $|\psi(t)\rangle \langle\psi(t)| = |\psi(t_0)\rangle \langle\psi(t_0)|$ maintaining a physically equivalent quantum state. This could represent for a quantum process the situation of the output state, which would remain in such an invariant state, when after an input–output transformation has taken place it experiences no further physical influence other than a trivial Hamiltonian $\mathsf{H} = H_0\mathsf{Id}$. A quantum state stored in a quantum memory would be (ideally) in such a situation.

If, for the sake of schematically structuring the matters, a boundary were to be drawn somewhere between quantum physics and quantum information, one could pass here. On one side would be quantum physics, aiming at describing the internal mechanisms at work in natural phenomena evolving in continuous time. This would involve working out the Hamiltonian H characterizing the physical situation, and by engaging it in the Schrödinger differential equation of Eq. (3.3) the evolution would ensue. On the other side of the boundary would be quantum information, which would start from functional blocks implementing global transformations by means of evolution operators or processes like U, so as to exploit and combine such blocks to perform operations interpretable as information-processing tasks. The constituent details underpinning the evolution processes (the interior of the block) would be the realm of quantum physics, while their exploitation as functional blocks for information processing would be the realm of quantum information.

We now return to the unitary evolutions of quantum states by means of quantum processes as in Eqs. (3.1) and (3.2), and in Figs. 3.1 and 3.2, that we preferentially employ here for our quantum signal-processing perspective.

3.1.3 No-Cloning Theorem

Among interesting unitary processes, one could envisage the possibility of a quantum process or circuit intended to copy or duplicate any arbitrary unknown quantum state $|\psi\rangle$ belonging to the Hilbert space $\mathcal{H}$. The circuit should use one single copy of the quantum state $|\psi\rangle \in \mathcal{H}$ at the input and produce two copies under the form $|\psi\rangle \otimes |\psi\rangle = |\psi\rangle\,|\psi\rangle \in \mathcal{H} \otimes \mathcal{H}$ at the output, for any arbitrary $|\psi\rangle \in \mathcal{H}$. To match the output dimensionality, the circuit should arrange for an extra input register initialized in some "blank" quantum state $|\emptyset\rangle \in \mathcal{H}$ independent of the unknown state $|\psi\rangle$ to be copied, and intended to receive the extra copy of $|\psi\rangle$ at the output. One would then seek a unitary operator U from $\mathcal{H} \otimes \mathcal{H}$ onto $\mathcal{H} \otimes \mathcal{H}$ to realize the input–output transformation

$$|\psi\rangle\,|\emptyset\rangle \longmapsto \mathsf{U}\big(|\psi\rangle\,|\emptyset\rangle\big) = |\psi\rangle\,|\psi\rangle\ , \tag{3.9}$$

for any $|\psi\rangle \in \mathcal{H}$, as depicted in Fig. 3.3.

The transformation of Eq. (3.9) preserves unit norm for quantum states, since both $|\psi\rangle\,|\emptyset\rangle$ and $|\psi\rangle\,|\psi\rangle$ are vectors of $\mathcal{H} \otimes \mathcal{H}$ with unit norm for any $|\psi\rangle \in \mathcal{H}$.

Fig. 3.3 A putative quantum process or circuit envisaged to copy an arbitrary unknown quantum state $|\psi\rangle$ by performing the input–output transformation U of Eq. (3.9)

The precluding feature is that the transformation of Eq. (3.9) is in general a nonlinear mapping, displaying some quadratic character in the state $|\psi\rangle$, so that no linear map U can generally perform the transformation of Eq. (3.9). This can be seen as follows.

Consider two arbitrary quantum states $|\psi_1\rangle$ and $|\psi_2\rangle$ in $\mathcal{H}$. By Eq. (3.9) one has $\mathsf{U}(|\psi_1\rangle|\emptyset\rangle) = |\psi_1\rangle|\psi_1\rangle$ and $\mathsf{U}(|\psi_2\rangle|\emptyset\rangle) = |\psi_2\rangle|\psi_2\rangle$. For the input superposition $|\psi\rangle = \alpha_1|\psi_1\rangle + \alpha_2|\psi_2\rangle$, a linear U would give

$$\mathsf{U}(|\psi\rangle|\emptyset\rangle) = \mathsf{U}\Big(\alpha_1|\psi_1\rangle|\emptyset\rangle + \alpha_2|\psi_2\rangle|\emptyset\rangle\Big) \tag{3.10}$$

$$= \alpha_1\mathsf{U}(|\psi_1\rangle|\emptyset\rangle) + \alpha_2\mathsf{U}(|\psi_2\rangle|\emptyset\rangle) \tag{3.11}$$

$$= \alpha_1|\psi_1\rangle|\psi_1\rangle + \alpha_2|\psi_2\rangle|\psi_2\rangle\ . \tag{3.12}$$

However, the targeted output from the copying circuit should be

$$\mathsf{U}(|\psi\rangle|\emptyset\rangle) = |\psi\rangle|\psi\rangle \tag{3.13}$$

$$= \alpha_1^2|\psi_1\rangle|\psi_1\rangle + \alpha_1\alpha_2|\psi_1\rangle|\psi_2\rangle + \alpha_1\alpha_2|\psi_2\rangle|\psi_1\rangle + \alpha_2^2|\psi_2\rangle|\psi_2\rangle\ . \tag{3.14}$$

In general, Eq. (3.14) differs from Eq. (3.12), so that no linear U exists that could generally perform the transformation of Eq. (3.9). In general, the transformation U of Eq. (3.9) does not either conserve the scalar inner product, since at the input the scalar product is $\langle\psi_1|\psi_2\rangle\langle\emptyset|\emptyset\rangle = \langle\psi_1|\psi_2\rangle$, when at the output it is $\langle\psi_1|\psi_2\rangle\langle\psi_1|\psi_2\rangle = \langle\psi_1|\psi_2\rangle^2$ generally differing from $\langle\psi_1|\psi_2\rangle$, so that no unitary U exists that could generally perform the transformation of Eq. (3.9).

This is the **no-cloning theorem** [13, 14], which states that no (linear unitary) quantum process exists that is able to copy or duplicate any arbitrary quantum state; or in other words, that an arbitrary unknown quantum state in general cannot be copied or cloned. This property, resulting from the linear unitary character of quantum evolution, imposes a severe limitation to quantum processing and computing compared with their classical counterparts. One can note that the possibility of exact cloning of quantum states would not be consistent with the random projective destructing character of quantum measurement as introduced in Sect. 2.2. If cloning were available, upon measuring a quantum state $|\psi\rangle$, the measurement apparatus could first make an arbitrary large number of copies of $|\psi\rangle$, whose cumulated measurements would enable arbitrarily precise statistical reconstruction of the complete quantum state $|\psi\rangle$ from the single initial copy of $|\psi\rangle$. In this way, quantum cloning would be associated with the possibility of precise measurement of the complete quantum state. This is not so.

The picture resulting from the no-cloning theorem gets nevertheless mitigated in different ways. It is quite possible to copy known quantum states. For instance, known quantum states that are purposely prepared to serve as signaling states for quantum communication can well be copied (prepared in multiple copies) to get involved in some repetition code intended to combat noise in the transmission. Also, if exact cloning is not possible, various forms of approximate cloning are accessible and may be sufficient to serve various useful purposes [14, 15]. Finally, the no-cloning theorem underlies the security of quantum cryptographic protocols by ensuring that no information can be stolen (copied) in a way that could go unnoticed. These aspects will be explored further in later chapters.

3.1.4 Single-Qubit Unitary Process

The closed evolution of a qubit state is generally governed by a unitary operator $\mathsf{U} \in \mathcal{L}(\mathcal{H}_2)$. Such a unitary operator can be represented by a 2×2 unitary matrix referred to an orthonormal basis of $\mathcal{H}_2$. A specially useful parameterization is as follows. As it results from Sect. A.2 of the Appendix, a unitary operator U of $\mathcal{L}(\mathcal{H}_2)$ can always be expressed as $\mathsf{U} = \exp(i\mathsf{H})$ where H is a Hermitian operator of $\mathcal{L}(\mathcal{H}_2)$. This operator H can be expressed in the Pauli basis $\{\mathrm{I}_2, \sigma_x, \sigma_y, \sigma_z\}$ of $\mathcal{L}(\mathcal{H}_2)$ introduced in Sect. A.5 and previously employed in Eq. (2.61), through the linear combination conveniently written as

$$\mathsf{H} = c\mathrm{I}_2 - \frac{\xi}{2}\vec{n} \cdot \vec{\sigma} = c\mathrm{I}_2 - \frac{\xi}{2}\big(n_x\sigma_x + n_y\sigma_y + n_z\sigma_z\big) . \tag{3.15}$$

Equation (3.15) uses the same notation $\vec{\sigma} = [\sigma_x, \sigma_y, \sigma_z]$ as in Eq. (A.44) and in Eq. (2.61), and a vector $\vec{n} = [n_x, n_y, n_z]^\top$. Since the Pauli operators are themselves Hermitian, having H Hermitian imposes four *real* coefficients $\big(c, \xi\vec{n}\,\big)$ in the linear combination of Eq. (3.15). Moreover, it is always possible to take $\vec{n}$ as a *unit* vector of $\mathbb{R}^3$, with c and ξ two arbitrary real parameters. The specific form for the linear combination in Eq. (3.15) is motivated by the fact that the parameters it contains receive an insightful interpretation, as we shall see shortly.

From the general parameterization of a Hermitian operator H of $\mathcal{L}(\mathcal{H}_2)$ in Eq. (3.15) follows a general parameterization for a unitary operator $\mathsf{U} = \exp(i\mathsf{H})$ of $\mathcal{L}(\mathcal{H}_2)$ as

$$\mathsf{U} = e^{ic}\exp\bigg(-i\,\frac{\xi}{2}\vec{n} \cdot \vec{\sigma}\bigg) . \tag{3.16}$$

Equation (3.16) is obtained because for two operators A and B that commute[3] one has $\exp(\mathsf{A} + \mathsf{B}) = \exp(\mathsf{A})\exp(\mathsf{B})$. In addition, $\exp(ic\mathrm{I}_2)\mathsf{B}$ is the same as $\exp(ic)\mathsf{B}$. Moreover, the factor $\exp(ic)$ acting as a global phase will often be irrelevant. When acting on a density operator ρ according to Eq. (3.2), the unitary operator of Eq. (3.16) written $\mathsf{U} = e^{ic}\mathsf{U}_\xi$ produces the transformed state $\mathsf{U}\rho\mathsf{U}^\dagger = \mathsf{U}_\xi\rho\mathsf{U}_\xi^\dagger$ where c plays no role.

A relevant parameterization of a qubit unitary operator can therefore be taken under the general form:

[3] And only for them. When A and B do not commute, to expand $\exp(\mathsf{A} + \mathsf{B})$ one has to resort instead to the Zassenhaus formula or the Baker–Campbell–Hausdorff formula involving the nonzero commutator $[\mathsf{A}, \mathsf{B}] = \mathsf{AB} - \mathsf{BA}$.

$$\mathsf{U}_\xi = \exp\left(-i\,\frac{\xi}{2}\vec{n}\cdot\vec{\sigma}\right), \tag{3.17}$$

with ξ an arbitrary real parameter and $\vec{n}$ a unit vector of $\mathbb{R}^3$. In addition, there exists a useful equivalent form as

$$\boxed{\mathsf{U}_\xi = \exp\left(-i\,\frac{\xi}{2}\vec{n}\cdot\vec{\sigma}\right) = \cos\left(\frac{\xi}{2}\right)\mathrm{I}_2 - i\sin\left(\frac{\xi}{2}\right)\vec{n}\cdot\vec{\sigma}\,.} \tag{3.18}$$

For any unit vector $\vec{n}$ of $\mathbb{R}^3$, one has $(\vec{n}\cdot\vec{\sigma})^2 = \mathrm{I}_2$, as it can be verified by performing the corresponding products of 2×2 matrices, or by using the relationships between the Pauli operators as mentioned in Sect. A.5. Next, by using the power-series expansion of the exponential function in Eq. (3.17), one recognizes the expansions of the cosine and sine and arrives at Eq. (3.18). We now work out a useful interpretation for the two parameters ξ and $\vec{n}$ specifying the general parameterization of the qubit unitary U_ξ in Eq. (3.18).

Starting from a qubit state ρ having the Bloch representation of Eq. (2.61), the transformed qubit state $\rho' = \mathsf{U}_\xi\rho\mathsf{U}_\xi^\dagger$ follows as

$$\rho' = \frac{1}{2}\Big[\mathrm{I}_2 + \mathsf{U}_\xi(\vec{r}\cdot\vec{\sigma})\mathsf{U}_\xi^\dagger\Big]. \tag{3.19}$$

To characterize the product $\mathsf{U}_\xi(\vec{r}\cdot\vec{\sigma})\mathsf{U}_\xi^\dagger$, we can exploit the second form of Eq. (3.18) for U_ξ, and the identities of Eqs. (A.45) and (A.46) for the products like $(\vec{n}\cdot\vec{\sigma})(\vec{r}\cdot\vec{\sigma})$, which in general do not commute. We finally obtain $\mathsf{U}_\xi(\vec{r}\cdot\vec{\sigma})\mathsf{U}_\xi^\dagger = \vec{r}\,'\cdot\vec{\sigma}$, which yields the Bloch vector $\vec{r}\,'$ for the transformed state of Eq. (3.19) as

$$\vec{r}\,' = \cos(\xi)\vec{r} + \sin(\xi)\,\vec{n}\times\vec{r} + \big[1-\cos(\xi)\big](\vec{r}\,\vec{n}\,)\vec{n}\,. \tag{3.20}$$

Equation (3.20) receives a geometrical interpretation in $\mathbb{R}^3$ offering a characterization of the action of U_ξ on a qubit state with Bloch vector $\vec{r}$. For $\vec{r} \;/\!/\; \vec{n}$, then $\vec{n}\times\vec{r} = \vec{0}$ and $(\vec{r}\,\vec{n}\,)\vec{n} = \vec{r}$, so that Eq. (3.20) leads to $\vec{r}\,' = \vec{r}$, i.e., a quantum state ρ with a Bloch vector $\vec{r}$ parallel to $\vec{n}$ remains invariant by U_ξ. For $\vec{r}\perp\vec{n}$, then $\vec{r}\,\vec{n} = 0$ so that Eq. (3.20) leads to $\vec{r}\,' = \cos(\xi)\vec{r} + \sin(\xi)\vec{n}\times\vec{r}$. In the orthogonal basis $\{\vec{n},\,\vec{r},\,\vec{n}\times\vec{r}\}$ of $\mathbb{R}^3$, a vector $\vec{r}$ perpendicular to $\vec{n}$ gets rotated by the angle ξ in the plane $(\vec{r},\,\vec{n}\times\vec{r})$ orthogonal to $\vec{n}$. As a consequence, the action of the unitary U_ξ of Eq. (3.18) on a quantum state ρ is to rotate its Bloch vector $\vec{r}$ in $\mathbb{R}^3$ by the angle ξ around the axis $\vec{n}$. This is the insightful geometrical interpretation that motivates the general parameterization by ξ and $\vec{n}$ of the qubit unitary U_ξ in Eq. (3.18).

The transformation of the Bloch vector in Eq. (3.20) can be expanded into a matrix form $\vec{r}\,' = U_\xi\vec{r}$ with the 3×3 real matrix

$$U_\xi = \begin{bmatrix} \cos(\xi) + [1-\cos(\xi)]n_x^2 & [1-\cos(\xi)]n_x n_y - \sin(\xi)n_z & [1-\cos(\xi)]n_x n_z + \sin(\xi)n_y \\ [1-\cos(\xi)]n_x n_y + \sin(\xi)n_z & \cos(\xi) + [1-\cos(\xi)]n_y^2 & [1-\cos(\xi)]n_y n_z - \sin(\xi)n_x \\ [1-\cos(\xi)]n_x n_z - \sin(\xi)n_y & [1-\cos(\xi)]n_y n_z + \sin(\xi)n_x & \cos(\xi) + [1-\cos(\xi)]n_z^2 \end{bmatrix}. \tag{3.21}$$

This matrix U_ξ represents in $\mathbb{R}^3$ the rotation by the angle ξ around the axis $\vec{n} = [n_x, n_y, n_z]^\top$. It is a real orthogonal matrix, i.e., satisfying $U_\xi^{-1} = U_\xi^\top$, with determinant $\det(U_\xi) = 1$.

For any $\vec{n} = [n_x, n_y, n_z]^\top$, the operator

$$\vec{n} \cdot \vec{\sigma} = \begin{bmatrix} n_z & n_x - in_y \\ n_x + in_y & -n_z \end{bmatrix} \tag{3.22}$$

has the determinant $\det(\vec{n} \cdot \vec{\sigma}\,) = -\vec{n}^{\,2}$ and two eigenvalues $\pm\sqrt{\vec{n}^{\,2}}$. A unit norm $\vec{n} \in \mathbb{R}^3$ leads to $\det(\vec{n} \cdot \vec{\sigma}\,) = -1$, and two eigenvalues ± 1 associated with the two orthogonal eigenstates $|\pm\vec{n}\,\rangle$ which are two pure states of $\mathcal{H}_2$ determined by the two unit Bloch vectors $\pm\vec{n} \in \mathbb{R}^3$. As a consequence, the unitary operator U_ξ of Eq. (3.18) has the two eigenvalues $\lambda_\pm = \exp(\mp i\xi/2)$, associated with the same two orthogonal eigenstates $|\lambda_\pm\rangle = |\pm\vec{n}\,\rangle$ in $\mathcal{H}_2$, and U_ξ has the expanded form:

$$\mathsf{U}_\xi = \begin{bmatrix} \cos(\xi/2) - i\sin(\xi/2)n_z & -\sin(\xi/2)(n_y + in_x) \\ \sin(\xi/2)(n_y - in_x) & \cos(\xi/2) + i\sin(\xi/2)n_z \end{bmatrix}. \tag{3.23}$$

For example, we have in $\mathcal{L}(\mathcal{H}_2)$ the qubit operator

$$\mathsf{R_x}(\xi) = \exp\left(-i\frac{\xi}{2}\sigma_x\right) = \cos\left(\frac{\xi}{2}\right)\mathrm{I}_2 - i\sin\left(\frac{\xi}{2}\right)\sigma_x = \begin{bmatrix} \cos(\xi/2) & -i\sin(\xi/2) \\ -i\sin(\xi/2) & \cos(\xi/2) \end{bmatrix}, \tag{3.24}$$

of two eigenstates $|\lambda_\pm\rangle = (|0\rangle \pm |1\rangle)/\sqrt{2}$ in $\mathcal{H}_2$, and associated in $\mathbb{R}^3$ with the rotation of angle ξ around the Ox axis of unit vector $\vec{n} = [n_x = 1,\ n_y = 0,\ n_z = 0]^\top$, represented in Eq. (3.21) by the matrix:

$$R_x(\xi) = \begin{bmatrix} 1 & 0 & 0 \\ 0 & \cos(\xi) & -\sin(\xi) \\ 0 & \sin(\xi) & \cos(\xi) \end{bmatrix}. \tag{3.25}$$

This gives in particular $\mathsf{R_x}(\xi = \pi) = -i\sigma_x$, connecting in this way the Pauli operator σ_x of $\mathcal{L}(\mathcal{H}_2)$ with a rotation by π around the Ox axis in $\mathbb{R}^3$.

Similarly, we have in $\mathcal{L}(\mathcal{H}_2)$ the qubit operator

$$\mathsf{R_y}(\xi) = \exp\left(-i\frac{\xi}{2}\sigma_y\right) = \cos\left(\frac{\xi}{2}\right)\mathrm{I}_2 - i\sin\left(\frac{\xi}{2}\right)\sigma_y = \begin{bmatrix} \cos(\xi/2) & -\sin(\xi/2) \\ \sin(\xi/2) & \cos(\xi/2) \end{bmatrix}, \tag{3.26}$$

of two eigenstates $|\lambda_{\pm}\rangle = \big(|0\rangle \pm i\,|1\rangle\big)/\sqrt{2}$ in $\mathcal{H}_2$, and associated in $\mathbb{R}^3$ with the rotation of angle ξ around the Oy axis, represented in Eq. (3.21) by the matrix

$$R_y(\xi) = \begin{bmatrix} \cos(\xi) & 0 & -\sin(\xi) \\ 0 & 1 & 0 \\ \sin(\xi) & 0 & \cos(\xi) \end{bmatrix}. \tag{3.27}$$

The matrix $R_y(\xi)$ in Eq. (3.27), as a special instance of U_ξ in Eq. (3.21), is always a rotation in the real three-dimensional space $\mathbb{R}^3$, acting on the qubit Bloch vector. Here in addition, the operator $\mathsf{R_y}(\xi) \in \mathcal{L}(\mathcal{H}_2)$ of Eq. (3.26) is also a genuine rotation in the complex vector plane $\mathcal{H}_2$, by the angle $\xi/2$, acting on the qubit state vector. We also have $\mathsf{R_y}(\xi = \pi) = -i\sigma_y$, connecting the Pauli operator σ_y of $\mathcal{L}(\mathcal{H}_2)$ with a rotation by π around the Oy axis on the qubit Bloch vector in $\mathbb{R}^3$, or a rotation by $\pi/2$ on the qubit state vector in $\mathcal{H}_2$.

Finally, in the same vein we have in $\mathcal{L}(\mathcal{H}_2)$ the qubit operator

$$\mathsf{R_z}(\xi) = \exp\left(-i\,\frac{\xi}{2}\sigma_z\right) = \cos\left(\frac{\xi}{2}\right)\mathrm{I}_2 - i\,\sin\left(\frac{\xi}{2}\right)\sigma_z = \begin{bmatrix} e^{-i\xi/2} & 0 \\ 0 & e^{i\xi/2} \end{bmatrix}, \tag{3.28}$$

of two eigenstates $|\lambda_{+}\rangle = |0\rangle$ and $|\lambda_{-}\rangle = |1\rangle$ in $\mathcal{H}_2$, and associated in $\mathbb{R}^3$ with the rotation of angle ξ around the Oz axis, represented in Eq. (3.21) by the matrix

$$R_z(\xi) = \begin{bmatrix} \cos(\xi) & -\sin(\xi) & 0 \\ \sin(\xi) & \cos(\xi) & 0 \\ 0 & 0 & 1 \end{bmatrix}, \tag{3.29}$$

and giving in particular $\mathsf{R_z}(\xi = \pi) = -i\sigma_z$, connecting the Pauli operator σ_z of $\mathcal{L}(\mathcal{H}_2)$ with a rotation by π around the Oz axis in $\mathbb{R}^3$.

The unitary U_ξ of Eq. (3.18), by the product of its two eigenvalues $\exp(\mp i\xi/2)$, has the determinant $\det(\mathsf{U}_\xi) = 1$. With unit determinant, the set of unitary operators U_ξ of Eq. (3.18) on the complex vector plane $\mathcal{H}_2$ has the structure of an algebraic group known as the SU(2) group (special unitary). Meanwhile, the set of rotation matrices U_ξ of Eq. (3.21) in the real vector space $\mathbb{R}^3$ has the structure of an algebraic group known as the SO(3) group (special orthogonal). The properties and correspondences between U_ξ and U_ξ we have seen above express that these two groups SU(2) and SO(3) are homomorphic.

In its orthonormal eigenbasis $\big\{|+\vec{n}\,\rangle, |-\vec{n}\,\rangle\big\}$ of $\mathcal{H}_2$, the unitary operator U_ξ of Eq. (3.18) is represented by the 2×2 diagonal matrix $\mathsf{U}_\xi = \mathrm{diag}\big[\exp(-i\xi/2), \exp(i\xi/2)\big]$, also factorizable as

$$\mathsf{U}_\xi = \exp\left(-i\,\frac{\xi}{2}\right)\begin{bmatrix} 1 & 0 \\ 0 & e^{i\xi} \end{bmatrix}. \tag{3.30}$$

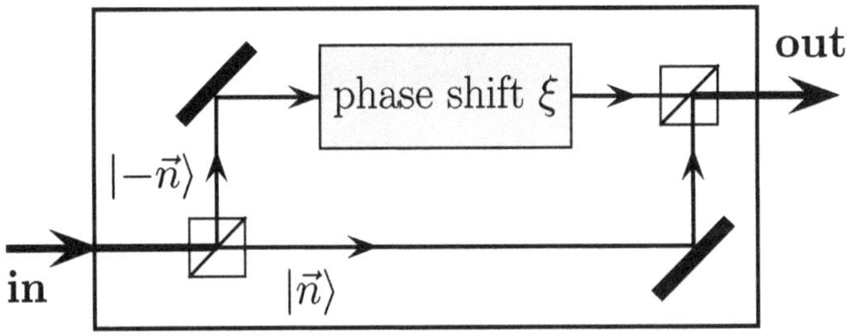

Fig. 3.4 A Mach–Zehnder interferometer featuring two polarizing beam splitters and two mirrors, to realize a photonic implementation of the qubit unitary process U_ξ from Eq. (3.18)

Again, the prefactor $\exp(-i\xi/2)$ in Eq. (3.30) acting as a global phase on quantum states has no physically detectable effect on their evolution. Accordingly, the evolution by U_ξ of Eq. (3.18), when referred to its eigenbasis $\{|+\vec{n}\rangle, |-\vec{n}\rangle\}$ of $\mathcal{H}_2$, is equivalently described by the matrix

$$\mathsf{U}_\xi = \begin{bmatrix} 1 & 0 \\ 0 & e^{i\xi} \end{bmatrix}. \tag{3.31}$$

This matrix of Eq. (3.31) representing the unitary operator U_ξ of Eq. (3.18) in its eigenbasis $\{|+\vec{n}\rangle, |-\vec{n}\rangle\}$ conveys the essence of a unitary evolution of a qubit state. The unit vector $\vec{n} \in \mathbb{R}^3$ fixes an orthonormal basis $\{|+\vec{n}\rangle, |-\vec{n}\rangle\}$ of $\mathcal{H}_2$, and the action of U_ξ is essentially to introduce a relative phase shift of ξ between the two components of the qubit state in this basis, via $|+\vec{n}\rangle \mapsto |+\vec{n}\rangle$ and $|-\vec{n}\rangle \mapsto e^{i\xi}\,|-\vec{n}\rangle$. The unitary transformation of Eq. (3.31) acting in the qubit Hilbert space $\mathcal{H}_2$ is equivalent on the Bloch vector in $\mathbb{R}^3$ to a rotation by the angle ξ around the Oz axis carried by the direction $\vec{n}$.

When it comes to a physical implementation, this unitary process U_ξ is, for instance, the basic operation that is carried out by an optical interferometer [16, 17] acting on individual photons according to their polarization and materializing polarization qubits. This is illustrated in Fig. 3.4.

The Mach–Zehnder interferometer of Fig. 3.4 uses a polarizing beam splitter at the input in order to separate an incident photon state according to the two orthogonal polarizations of the basis $\{|\vec{n}\rangle \equiv |+\vec{n}\rangle, |-\vec{n}\rangle\}$. The component along $|-\vec{n}\rangle$ of the incident state experiences a phase shift of ξ relative to the component along $|\vec{n}\rangle$. The two components are then recombined by the conjugated beam splitter (or polarization beam combiner) at the output. The eigenbasis $\{|\vec{n}\rangle, |-\vec{n}\rangle\}$ can be selected by rotating the beam splitters.

For an evolution of the qubit state in continuous time as in Sect. 3.1.2, an arbitrary qubit Hamiltonian H can always be expressed in the Pauli basis of $\mathcal{L}(\mathcal{H}_2)$ as $\mathsf{H} = a_0\mathrm{I}_2 + \vec{a}\cdot\vec{\sigma}$, with real parameters $(a_0, \vec{a}) \in \mathbb{R} \times \mathbb{R}^3$ for Hermiticity of H. From a qubit density operator ρ having the Bloch representation of Eq. (2.61), by using the identity of Eq. (A.45), we obtain the commutator $[\mathsf{H}, \rho] = i(\vec{a} \times \vec{r})\cdot\vec{\sigma}$. With

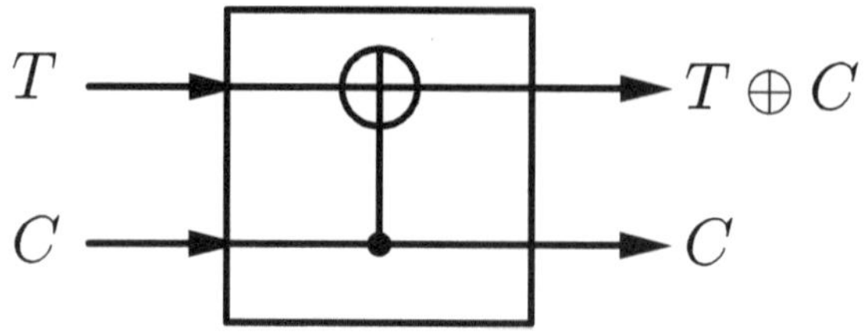

Input $\lvert CT\rangle$	Output $\lvert C, T \oplus C\rangle$
$\lvert 00\rangle$	$\lvert 00\rangle$
$\lvert 01\rangle$	$\lvert 01\rangle$
$\lvert 10\rangle$	$\lvert 11\rangle$
$\lvert 11\rangle$	$\lvert 10\rangle$

Fig. 3.5 The controlled-Not or Cnot gate acting on two-qubit quantum states, and a truth table defining how the Cnot gate unitarily transforms the computational basis of $\mathcal{H}_2^{\otimes 2}$

$d\rho/dt = (1/2)\big(d\vec{r}/dt\big) \cdot \vec{\sigma}$ the Liouville–von Neumann equation (3.8) then takes the form:

$$\frac{d\vec{r}}{dt} = \vec{\omega} \times \vec{r}\,, \tag{3.32}$$

which describes in $\mathbb{R}^3$ the instantaneous rotation of the qubit Bloch vector $\vec{r}(t)$ with the angular velocity $\vec{\omega} = 2\vec{a}/\hbar$. For the evolution of the qubit state, the rotation of the Bloch vector $\vec{r}$ is described in continuous time by Eq. (3.32), while the matrix U_ξ of Eq. (3.21) describes the global effect of the rotation. A time-invariant Hamiltonian H applied from the initial time $t_0 = 0$ realizes in Eq. (3.6) the unitary evolution operator $\mathsf{U}(t) = \exp(-i\mathsf{H}t/\hbar)$, which is equivalent to the generic form of Eq. (3.18) with a unit axis $\vec{n} = \vec{a}/\|\vec{a}\|$ and a rotation angle $\xi = \|\vec{\omega}\|t$ linearly increasing over time.

3.1.5 Multiple-Qubit Unitary Process

A quantum system constituted by L qubits, with a state $|\psi\rangle$ in the 2^L-dimensional Hilbert space $\mathcal{H}_2^{\otimes L}$, has a closed evolution generally governed by a unitary operator U from $\mathcal{H}_2^{\otimes L}$ onto $\mathcal{H}_2^{\otimes L}$. Such a unitary process can be materialized by a quantum gate or circuit receiving the L qubits as input. Such a quantum gate is always invertible as it implements a bijective unitary input–output relationship, and it therefore generally exhibits also L qubits as output.

An important example is when $L = 2$ for qubit pairs. A very useful two-qubit gate is the controlled-Not gate or **Cnot gate** depicted in Fig. 3.5.

As for any quantum circuit representing a unitary operator U, the Cnot gate of Fig. 3.5 can be completely characterized by specifying how it transforms the states of the computational basis. For the two-qubit Cnot gate, the four states of the computational basis of $\mathcal{H}_2^{\otimes 2}$ can be numbered with two bits, a control bit C and a target bit T. For the transformation of the computational basis, if the control bit is $C = 0$ then the target bit T is unchanged, meanwhile if the control bit is $C = 1$ then the target bit T is reversed. To summarize, the target bit T undergoes the transformation $T \mapsto T \oplus C$ as the exclusive-Or or Xor operation with the (unchanged) control bit

C. This transformation of the computational basis by the Cnot gate is specified by a truth table in Fig. 3.5.

Equivalently, the Cnot gate implements the unitary operator $\mathsf{U}_{\mathsf{Cnot}}$ represented in the computational basis of $\mathcal{H}_2^{\otimes 2}$ by the matrix:

$$\mathsf{U}_{\mathsf{Cnot}} = \begin{bmatrix} 1 & 0 & 0 & 0 \\ 0 & 1 & 0 & 0 \\ 0 & 0 & 0 & 1 \\ 0 & 0 & 1 & 0 \end{bmatrix}. \tag{3.33}$$

It can be verified from Eq. (3.33) that the unitary operator $\mathsf{U}_{\mathsf{Cnot}}$ is also Hermitian, so that $\mathsf{U}_{\mathsf{Cnot}} = \mathsf{U}_{\mathsf{Cnot}}^{\dagger} = \mathsf{U}_{\mathsf{Cnot}}^{-1}$. Alternatively, we also have $\mathsf{U}_{\mathsf{Cnot}} = |0\rangle\langle 0| \otimes \mathrm{I}_2 + |1\rangle\langle 1| \otimes \sigma_x$.

Once the transformation of the computational basis is specified, the operation of the quantum gate is completely defined. In addition, for superpositions of basis states, a quantum gate—as it implements a unitary linear operator—obeys the standard principle of superposition of linear algebra. For instance, the input to the Cnot gate can be placed in the superposition $|CT\rangle = |+\rangle|0\rangle = (|00\rangle + |10\rangle)/\sqrt{2}$. This is a separable or product state at the input, with the control qubit in the superposed state $|+\rangle$ of Eq. (2.7). From the transformation of basis states expressed by the truth table of Fig. 3.5 and the principle of superposition, the output state results as $(|00\rangle + |11\rangle)/\sqrt{2}$. This is an entangled state at the output. This illustrates a significant property of the Cnot gate that it is able to create entanglement between two qubits. In this regard, the Cnot gate is important to quantum computation as it gives access to (or creates) the specifically quantum resource of entanglement.

Any N-dimensional quantum states can be realized by assembling a sufficient number L of qubits satisfying $2^L \geq N$. Then, for processing N-dimensional quantum states supported by L qubits, any unitary operator U_N acting on an N-dimensional Hilbert space will act in an N-dimensional subspace $\mathcal{H}_N$ of the 2^L-dimensional space $\mathcal{H}_2^{\otimes L}$ of the L qubits. Such unitary U_N maps the N-dimensional subspace $\mathcal{H}_N$ of $\mathcal{H}_2^{\otimes L}$ into a subspace of $\mathcal{H}_2^{\otimes L}$ with same dimension N. The complementary space $\mathcal{H}_N^{\perp}$ orthogonal in $\mathcal{H}_2^{\otimes L}$ to $\mathcal{H}_N$ gives the direct sum $\mathcal{H}_2^{\otimes L} = \mathcal{H}_N \oplus \mathcal{H}_N^{\perp}$. The unitary U_N specified on $\mathcal{H}_N$ for some definite processing on N-dimensional states represented by L-qubit states of $\mathcal{H}_2^{\otimes L}$ can be further specified over the whole space $\mathcal{H}_2^{\otimes L}$ of the L qubits, simply by a trivial unitary extension acting on the orthogonal complementary subspace $\mathcal{H}_N^{\perp}$ to be unitarily mapped into the subspace orthogonal complement in $\mathcal{H}_2^{\otimes L}$ to the transformed subspace $\mathsf{U}_N\mathcal{H}_N$. In this way, any N-dimensional quantum states can be realized by assembling a sufficient number L of qubits, and any unitary transformation on such N-dimensional quantum states can be realized by a unitary transformation on L qubits.

In addition, for L-qubit states, it can be shown that any unitary transformation on L qubits, represented by a $2^L \times 2^L$ unitary matrix, can always be realized [18–20] by combining one-qubit gates, represented by 2×2 unitary matrices, with the Cnot gate, represented by the 4×4 unitary matrix of Eq. (3.33). An essential step is that any unitary matrix can be decomposed into a product of two-level unitary matrices, which are unitary matrices that act non-trivially only on two or fewer vector components; then, any two-level unitary operation on L qubits can be realized by combining one-qubit gates and the two-qubit Cnot gate [9, 18–20]. It results that any quantum gate or circuit on L qubits can always be realized by combining one-qubit gates and the two-qubit Cnot gate [20]. In this respect, the one-qubit gates presented in Sect. 3.1.4 supplemented by the two-qubit Cnot gate of Fig. 3.5 form **a universal set of quantum gates** for computation on quantum states. Important consequences follow, both of principle and practical. *In principle*, we indicated previously that any valid evolution of a closed quantum system is governed by a unitary operator; now we know that the reciprocal also holds: any unitary operator (at least on a finite-dimensional Hilbert space) can be realized as a valid quantum evolution by a physical system materialized via the universal set of gates. *In practice*, it is enough to have physical implementations for one-qubit gates and for the two-qubit Cnot gate, to be able to implement any conceivable unitary evolution of an arbitrary number of qubits or of any finite-dimensional quantum states. In addition, we learn that pairwise interactions of qubits, as that taking place in the two-qubit Cnot gate, are enough to implement any entangled state or any entanglement configuration among an arbitrary number of qubits. Moreover, coming back to the frontier evoked in the penultimate paragraph of Sect. 3.1.2 between quantum physics (focusing on H) and quantum information (focusing on U), for implementing any unitary evolution U, the existence of a universal set of gates dispenses from further (deeper) reference to a specific underlying Hamiltonian H.

A quantum circuit on L qubits can, in principle, always be fed at its input by a superposition containing all feasible basis states, for instance, as $\big((|0\rangle + |1\rangle)/\sqrt{2}\big)^{\otimes L} = |+\rangle^{\otimes L}$. As a result, by linearity, the quantum circuit will compute simultaneously in a single pass, all the responses corresponding to all the superposed inputs, and will deliver them as a superposition in the output state. This affords the possibility of parallelism in the computation by the quantum circuit. This faculty holds the potential for enhanced performance by parallel computation with quantum circuits. We will examine in more detail in Chap. 5 such useful exploitation of quantum circuits for computation, especially affording enhanced performance unattainable by classical means.

3.2 Non-Unitary Quantum Process

When a quantum system is not closed, but interacts with its environment, it represents an **open quantum system**. The evolution of an open quantum system can be described as a non-unitary (yet still linear) transformation of its density operator. Such situation of an open quantum system and its non-unitary evolution is central to the description of the phenomenon of quantum decoherence. Quantum decoherence arises as the alteration of quantum states when they interact with an uncontrolled environment [6, 21, 22]. Quantum decoherence can also be understood as quantum noise and is an essential factor impacting the functioning and development of quantum technologies. Noise is a central element in signal processing, and as such quantum noise will be examined in a specialized chapter to come later, and then it will be encountered on many other occasions in following chapters of the book. Before, in the present section, we will introduce basic elements for describing the non-unitary evolution of open quantum systems.

3.2.1 *Stinespring Dilated Unitary Representation*

We consider a quantum system of interest Q of Hilbert space $\mathcal{H}^Q$ which is coupled to an environment E with Hilbert space $\mathcal{H}^E$. The environment is delineated so that the composite bipartite system QE can be regarded as a closed quantum system, having as Hilbert space the tensor-product space $\mathcal{H}^Q \otimes \mathcal{H}^E$. The joint evolution of the principal system Q in interaction with its environment E then takes place as a unitary evolution of the (closed) composite bipartite system QE, governed by a unitary operator acting in $\mathcal{H}^Q \otimes \mathcal{H}^E$.

More detailedly, we consider the principal system Q initially prepared in the pure state $|Q\rangle$ of $\mathcal{H}^Q$. The environment E can always be assumed to start in some pure state $|e_0\rangle$ of $\mathcal{H}^E$ (possibly via a purification step as in Sect. 2.5.3 enlarging the dimensionality of $\mathcal{H}^E$). The composite system QE starts in the initial separable state $|\psi_{QE}\rangle = |Q\rangle \otimes |e_0\rangle$ of $\mathcal{H}^Q \otimes \mathcal{H}^E$. Then, the composite state $|\psi_{QE}\rangle$ undergoes the evolution by the bipartite unitary operator $\mathsf{U}_{QE} \in \mathcal{L}(\mathcal{H}^Q \otimes \mathcal{H}^E)$ under the form:

$$|\psi_{QE}\rangle = |Q\rangle \otimes |e_0\rangle \xmapsto{\mathsf{U}_{QE}} \mathsf{U}_{QE}\,|\psi_{QE}\rangle = \mathsf{U}_{QE}\,|Q\rangle \otimes |e_0\rangle = |\psi'_{QE}\rangle \in \mathcal{H}^Q \otimes \mathcal{H}^E\,. \tag{3.34}$$

In general, the unitary evolution of the composite system QE results in a principal system Q becoming entangled with the environment E. This is represented by the composite system QE terminating in a transformed state $|\psi'_{QE}\rangle$ which usually is an entangled state of $\mathcal{H}^Q \otimes \mathcal{H}^E$. Because of this entanglement with the environment E, the principal system Q can no longer be assigned a well-defined pure state of $\mathcal{H}^Q$, but a density operator for Q is feasible. Despite its entanglement with the environment E, the principal system Q, from a material standpoint, remains a physical entity individually accessible. For instance, Q can be a photon that has traveled across an optical fiber, or an electron with its spin embedded in the earth magnetic field. A separate measurement can therefore well be envisaged on the principal system Q alone, while no access or control is assumed on the environment E. We are going to see that it is generally possible to obtain a density operator $\rho'_Q \in \mathcal{L}(\mathcal{H}^Q)$ characterizing the quantum state of the principal system Q alone, after its joint evolution with the environment E, with ρ'_Q enabling to describe all the measurement results accessible upon measuring Q after it has evolved.

For a principal system Q initially in state $|Q\rangle\langle Q|$, with no access or control on the environment E, the final state ρ'_Q for Q is obtained, from the joint state $|\psi'_{QE}\rangle$ of the composite system QE, by tracing out the degrees of freedom of the uncontrolled environment E. This is realized by **partial tracing** of the joint density operator $|\psi'_{QE}\rangle\langle\psi'_{QE}|$ over the environment E, so as to obtain for Q the (reduced) density operator:

$$\rho'_Q = \mathcal{E}\big(|Q\rangle\langle Q|\big) = \mathrm{tr}_E\Big(|\psi'_{QE}\rangle\langle\psi'_{QE}|\Big) = \mathrm{tr}_E\Big(\mathsf{U}_{QE}\big(|Q\rangle\langle Q| \otimes |e_0\rangle\langle e_0|\big)\mathsf{U}^\dagger_{QE}\Big)\,. \tag{3.35}$$

This operation of partial tracing was previously encountered in Sect. 2.5.3 for the purification of a mixed state. We will further motivate this operation of partial tracing in Sect. 4.1.3 with Eq. (4.8) and in Sect. 4.4, when more elaboration becomes available concerning the notion of quantum measurement. Since both U_{QE} and the partial tracing are linear operations, Eq. (3.35) defines a linear map $\mathcal{E}(\cdot)$, or superoperator, comparable to the quantum process of Fig. 3.2, which maps an initial density operator $|Q\rangle\langle Q| \in \mathcal{L}(\mathcal{H}^Q)$ into a final density operator $\rho'_Q = \mathcal{E}\big(|Q\rangle\langle Q|\big) \in \mathcal{L}(\mathcal{H}^Q)$, so as to describe the evolution of the quantum state of the principal system Q alone. Together the joint unitary evolution of Eq. (3.34), terminated by the partial tracing operation of Eq. (3.35), realize what is known as a **dilated or Stinespring unitary representation** of the evolution of the open quantum system Q resulting from its interaction with an uncontrolled environment E.

The superoperator $\mathcal{E}(\cdot)$ of Eq. (3.35) is **trace-preserving**, in the sense that for any unit-trace input $|Q\rangle\langle Q|$, the output $\rho'_Q = \mathcal{E}\big(|Q\rangle\langle Q|\big)$ is also with unit trace, since $\mathrm{tr}_Q(\rho'_Q) = \mathrm{tr}_{QE}\big(|\psi'_{QE}\rangle\langle\psi'_{QE}|\big) = \langle\psi'_{QE}|\psi'_{QE}\rangle = 1$. The superoperator of Eq. (3.35) is also **positive**: to any positive operator $|Q\rangle\langle Q|$ it assigns $\rho'_Q = \mathcal{E}\big(|Q\rangle\langle Q|\big)$ which is also a positive operator. This can be seen by introducing, for the environment space $\mathcal{H}^E$ of dimension K, an orthonormal basis $\{|e_k\rangle\}_{k=1}^K$ enabling in Eq. (3.35) to express the trace over $\mathcal{H}^E$ as

$$\rho'_Q = \mathcal{E}\big(|Q\rangle\langle Q|\big) = \sum_{k=1}^{K} \langle e_k|\psi'_{QE}\rangle\langle\psi'_{QE}|e_k\rangle\,. \tag{3.36}$$

It follows that, for any state vector $|Q'\rangle$ of $\mathcal{H}^Q$, one has $\langle Q'|\rho'_Q|Q'\rangle = \sum_{k=1}^K |\langle\psi_k|\psi'_{QE}\rangle|^2$, with the vectors $|\psi_k\rangle = |Q'\rangle \otimes |e_k\rangle \in \mathcal{H}^Q \otimes \mathcal{H}^E$, which obviously is a positive scalar, establishing ρ'_Q as a positive operator.

We invoke again the property of convexity-linearity as done above with Eq. (3.8): When the principal system Q is initially in a mixed state with density operator ρ_Q, this one, as seen for instance with Eq. (2.37), is expressible as a convex sum of pure state density operators similar to $|Q\rangle\langle Q|$. Then, by linearity of the superoperator of Eq. (3.35), from the final density operator $\rho'_{QE} = \mathsf{U}_{QE}\big(\rho_Q \otimes |e_0\rangle\langle e_0|\big)\mathsf{U}^\dagger_{QE}$ of the compound QE, the final state of the principal system Q is obtained as

$$\boxed{\rho'_Q = \mathcal{E}(\rho_Q) = \mathrm{tr}_E\big(\rho'_{QE}\big) = \mathrm{tr}_E\Big(\mathsf{U}_{QE}\big(\rho_Q \otimes |e_0\rangle\langle e_0|\big)\mathsf{U}^\dagger_{QE}\Big)\,,} \tag{3.37}$$

and the linear map of Eq. (3.37) results as well as a positive trace-preserving superoperator (a convex sum of positive trace-preserving maps is a positive trace-preserving map). The map or superoperator $\mathcal{E}(\cdot)$ of Eq. (3.37) transforms any initial density operator $\rho_Q \in \mathcal{L}(\mathcal{H}^Q)$ into a final density operator $\rho'_Q = \mathcal{E}(\rho_Q) \in \mathcal{L}(\mathcal{H}^Q)$, and in this way it defines a quantum process comparable to that of Fig. 3.2.

The superoperator $\mathcal{E}(\cdot)$ defined by Eq. (3.37) is also a little more than positive, it is **completely positive**. This is meaningful when the principal system Q, when it starts to interact with the environment E, happens to be entangled with another quantum system R of Hilbert space $\mathcal{H}^R$ of arbitrary dimension, so that the compound RQ is initially in the joint state ρ_{RQ} on $\mathcal{H}^R \otimes \mathcal{H}^Q$. Next, when the principal system Q interacts with the environment E through a joint unitary evolution as before, while nothing is done on the system R, and after partial tracing over the environment E, then through such a non-unitary evolution the compound RQ terminates in a valid quantum state of $\mathcal{H}^R \otimes \mathcal{H}^Q$. This is obtained because the evolution of Eq. (3.37) satisfies the property that $\big(\mathcal{I}_R \otimes \mathcal{E}\big)(\rho_{RQ})$ is a valid (positive unit-trace) density operator on $\mathcal{H}^R \otimes \mathcal{H}^Q$, for any density operator ρ_{RQ} on $\mathcal{H}^R \otimes \mathcal{H}^Q$, where $\mathcal{I}_R$ is the identity superoperator for operators on $\mathcal{H}^R$. This establishes $\mathcal{E}(\cdot)$ of Eq. (3.37) as a completely positive superoperator. At the root, this follows from the evolution $\big(\mathcal{I}_R \otimes \mathcal{U}_{QE}\big)\big(\rho_{RQ} \otimes |e_0\rangle\langle e_0|\big) = \rho'_{RQE}$ with the superoperator $\mathcal{U}_{QE}(\rho_{QE}) = \mathsf{U}_{QE}\rho_{QE}\mathsf{U}^\dagger_{QE}$, and the partial tracing $\rho'_{RQ} = \mathrm{tr}_E\big(\rho'_{RQE}\big)$, on which similar arguments as before apply establishing $\rho'_{RQ} = \big(\mathcal{I}_R \otimes \mathcal{E}\big)(\rho_{RQ})$ as a positive unit-trace operator on $\mathcal{H}^R \otimes \mathcal{H}^Q$ for any positive unit-trace operator ρ_{RQ} on $\mathcal{H}^R \otimes \mathcal{H}^Q$.

As a supplement, to anticipate on what we shall see in Sect. 3.2.10, if we consider an arbitrary superoperator (comparable to $\mathcal{E}(\cdot)$ but not necessarily obtained via partial tracing), a convenient criterion for complete positivity is the positivity of its associated Choi–Jamiołkowski state.

The superoperator $\mathcal{E}(\cdot)$, defined in Eq. (3.37) through a joint unitary evolution on a composite system QE followed by partial tracing over E, therefore arrives as a **completely positive trace-preserving** linear map for Q. This map or superoperator defines a quantum process $\mathcal{E}(\cdot)$ which provides a general characterization of the way the initial (pure or mixed) quantum state ρ_Q of the open system Q gets transformed by its interaction with the environment [6, 9, 12].

Reciprocally, a result known as the **Stinespring dilation theorem** stipulates that any completely positive trace-preserving linear map from $\mathcal{L}(\mathcal{H}^Q)$ onto $\mathcal{L}(\mathcal{H}^Q)$ can always be put under the form of Eq. (3.37), with tensoring to a larger (dilated) space followed by unitary evolution and finally partial tracing [23–25], and in this respect such a map represents a valid quantum evolution.

The characterization of the quantum process $\mathcal{E}(\cdot)$ via the dilated unitary representation of Eq. (3.37) however requires some (at least partial) knowledge of the environment E, especially concerning its initial state $|e_0\rangle$ and the joint evolution

U_{QE}. Yet, such details may be difficult to specify for the environment, and they may finally end up uncritical or superfluous, because wiped out in the partial tracing operation over E in Eq. (3.37). To alleviate this difficulty of specifying details that may finally prove irrelevant (or inaccessible to begin with), there exists an alternative way of characterizing the quantum process $\mathcal{E}(\cdot)$ describing the non-unitary evolution of the open system Q, addressed via the notion of operator-sum or Kraus representation that we now consider.

3.2.2 *Operator-Sum or Kraus Representation*

We proceed with a more explicit evaluation of the partial trace over the environment E involved in the Stinespring dilated unitary representation. When the environment space $\mathcal{H}^E$ of dimension K is referred to the orthonormal basis $\{|e_k\rangle\}_{k=1}^K$, we introduce the K operators Λ_k from $\mathcal{H}^Q$ onto $\mathcal{H}^Q$, defined for $k = 1$ to K, through the notion of partial inner product [7, 25, 26], by

$$\Lambda_k = \langle e_k| \mathsf{U}_{QE} |e_0\rangle \ . \tag{3.38}$$

This is equivalent to

$$\Lambda_k |Q\rangle = \langle e_k| \mathsf{U}_{QE} |Q\rangle \otimes |e_0\rangle \ , \tag{3.39}$$

defining each operator Λ_k by its action on any state vector $|Q\rangle$ of $\mathcal{H}^Q$, or also equivalent to

$$\langle Q'|\Lambda_k|Q\rangle = \langle e_k| \otimes \langle Q'| \mathsf{U}_{QE} |Q\rangle \otimes |e_0\rangle \ , \tag{3.40}$$

defining each Λ_k by its (scalar) matrix element between any two state vectors $|Q\rangle$ and $|Q'\rangle$ of $\mathcal{H}^Q$.

By using the orthonormal basis $\{|e_k\rangle\}_{k=1}^K$ of $\mathcal{H}^E$ for expressing the partial trace, Eq. (3.35) becomes

$$\mathcal{E}(|Q\rangle\langle Q|) = \sum_{k=1}^{K} \langle e_k| \mathsf{U}_{QE}\big(|Q\rangle\langle Q| \otimes |e_0\rangle\langle e_0|\big)\mathsf{U}_{QE}^\dagger |e_k\rangle \tag{3.41}$$

$$= \sum_{k=1}^{K} \langle e_k| \mathsf{U}_{QE} |Q\rangle \otimes |e_0\rangle \langle Q| \otimes \langle e_0| \mathsf{U}_{QE}^\dagger |e_k\rangle \tag{3.42}$$

$$= \sum_{k=1}^{K} \Lambda_k |Q\rangle\langle Q| \Lambda_k^\dagger \ . \tag{3.43}$$

So when the principal system Q is initially in a pure state $|Q\rangle\langle Q|$, its final state is expressible as

$$\rho'_Q = \mathcal{E}\big(|Q\rangle\langle Q|\big) = \sum_{k=1}^{K} \Lambda_k |Q\rangle\langle Q| \Lambda_k^\dagger . \tag{3.44}$$

Again, by convexity-linearity, when the principal system Q is initially in a mixed state ρ_Q, this one is expressible as a convex sum of pure states similar to $|Q\rangle\langle Q|$, so that by linearity of Eq. (3.44) the resulting final state of Q is

$$\boxed{\rho'_Q = \mathcal{E}(\rho_Q) = \sum_{k=1}^{K} \Lambda_k \rho_Q \Lambda_k^\dagger .} \tag{3.45}$$

Equation (3.45) stands as an equivalent form to the dilated unitary representation of Eq. (3.37) for the completely positive trace-preserving superoperator $\mathcal{E}(\cdot)$. The form of Eq. (3.45) constitutes an **operator-sum or Kraus representation** for the completely positive trace-preserving superoperator $\mathcal{E}(\cdot)$ describing the evolution of the open quantum system Q.

The superoperator $\mathcal{E}(\cdot)$ of Eq. (3.37) preserving the unit trace implies for Eq. (3.45) to satisfy, by circularity of the trace,

$$1 = \mathrm{tr}(\rho'_Q) = \mathrm{tr}\left(\sum_{k=1}^{K} \Lambda_k^\dagger \Lambda_k \rho_Q\right) \tag{3.46}$$

for all $\rho_Q \in \mathcal{L}(\mathcal{H}^Q)$. This holds especially for all pure states $\rho_Q = |Q\rangle\langle Q|$ yielding the trace $1 = \langle Q| \sum_k \Lambda_k^\dagger \Lambda_k |Q\rangle$ for all normalized vector $|Q\rangle$. This constrains to 1 each diagonal matrix element of the Hermitian operator $\sum_k \Lambda_k^\dagger \Lambda_k$, especially when it is diagonal in an orthonormal eigenbasis, so that it coincides with the identity operator I^Q on $\mathcal{H}^Q$. This implies that the K operators Λ_k defined by Eq. (3.38) satisfy a **completeness relation** under the form:

$$\sum_{k=1}^{K} \Lambda_k^\dagger \Lambda_k = \mathrm{I}^Q , \tag{3.47}$$

and this is granted as a consequence of the definition of Eq. (3.38). In an operator-sum representation like Eq. (3.45), the operators $\Lambda_k \in \mathcal{L}(\mathcal{H}^Q)$ are known as **Kraus operators**. In general, they need not be unitary or Hermitian.

It can be noted that the unitary evolution of Eq. (3.2) for the closed quantum system is a special case of the non-unitary evolution of Eq. (3.45) for the open system, with in Eq. (3.2) a single Kraus operator $\Lambda_1 \equiv \mathsf{U}$. With more than one Kraus operator in Eq. (3.45), we are in the generic situation of a non-unitary evolution of an open quantum state, which nevertheless remains linear.

Such a completely positive trace-preserving linear superoperator $\mathcal{E}(\cdot)$, specified either in the dilated unitary form of Eq. (3.37) or in the operator-sum or Kraus form

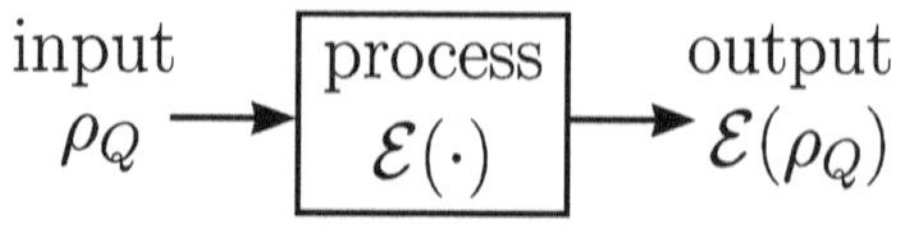

Fig. 3.6 The evolution of a quantum system Q, closed or open, evolving unitarily or non-unitarily, is described by a completely positive trace-preserving linear superoperator or quantum operation $\mathcal{E}(\cdot)$, expressible either in the dilated unitary form of Eq. (3.37) or in the operator-sum or Kraus form of Eq. (3.45). This defines a quantum process or quantum channel $\mathcal{E}(\cdot)$, receiving as input signal the density operator ρ_Q and delivering as output signal the density operator $\mathcal{E}(\rho_Q)$, via the input–output transformation $\rho_Q \mapsto \mathcal{E}(\rho_Q)$

of Eq. (3.45), represents what is also known as a **quantum operation**. The evolution of a quantum system Q, closed or open, evolving unitarily or non-unitarily, can be described by a quantum operation like $\mathcal{E}(\cdot)$. Such a quantum operation $\mathcal{E}(\cdot)$ also represents our notion of **quantum process**, as illustrated in Fig. 3.6, offering a general description for the evolution of a quantum state described by the density operator or quantum signal ρ_Q. In addition, in a more concrete terminology, the process $\mathcal{E}(\cdot)$ can be referred to as a **quantum channel**, taking as input the density operator ρ_Q and delivering as output the density operator $\mathcal{E}(\rho_Q)$.

From the Stinespring dilation theorem mentioned at the end of Sect. 3.2.1, it follows that a linear superoperator $\mathcal{E}(\cdot)$ is completely positive if and only if [9, 12, 25, 27] it can be expressed under an operator-sum representation as in Eq. (3.45), with the operators Λ_k satisfying the completeness relation of Eq. (3.47) to confer the property of trace preservation.

It can be noted that, although the superoperator $\mathcal{E}(\cdot)$ is defined in Eq. (3.45) for density operators ρ_Q on $\mathcal{H}^Q$, by linearity of $\mathcal{E}(\cdot)$ the same definition applies to any operators A on $\mathcal{H}^Q$. This can be shown explicitly by introducing the orthogonal basis $\{|n\rangle\langle m|\}$ for operators on $\mathcal{H}^Q$. Each basis operator $|n\rangle\langle m|$ can be expressed as a linear combination of density operators as

$$|n\rangle\langle m| = |o_+\rangle\langle o_+| + i\,|o_i\rangle\langle o_i| - \frac{1+i}{2}\Big(|n\rangle\langle n| + |m\rangle\langle m|\Big), \tag{3.48}$$

with here the two states $|o_+\rangle = \big(|n\rangle + |m\rangle\big)/\sqrt{2}$ and $|o_i\rangle = \big(|n\rangle + i\,|m\rangle\big)/\sqrt{2}$ of $\mathcal{H}^Q$, and by linearity of $\mathcal{E}(\cdot)$ in Eq. (3.45) it gets transformed as

$$\mathcal{E}\big(|n\rangle\langle m|\big) = \sum_{k=1}^{K} \Lambda_k\,|n\rangle\langle m|\,\Lambda_k^\dagger\,. \tag{3.49}$$

As a result, any operator A on $\mathcal{H}^Q$ transforms as

$$\mathcal{E}(\mathsf{A}) = \sum_{k=1}^{K} \Lambda_k \mathsf{A} \Lambda_k^\dagger\,, \tag{3.50}$$

and from Eq. (3.47) it inherits the trace preservation $\mathrm{tr}\big(\mathcal{E}(\mathsf{A})\big) = \mathrm{tr}(\mathsf{A})$.

3.2.3 Non-Unitary Quantum Process and Irreversibility

A non-unitary quantum process $\mathcal{E}(\cdot)$ as in Fig. 3.6 in general represents a contractive dynamics, implementing an irreversible evolution of quantum states [9, 12, 28, 29], as we now explain. This requires us first to define a suitable distance between two density operators.

The **trace norm** or ℓ_1-norm of an operator $\mathsf{A} \in \mathcal{L}(\mathcal{H}_N)$ is defined as $\|\mathsf{A}\|_1 = \mathrm{tr}\big(|\mathsf{A}|\big)$, with the absolute-value positive operator $|\mathsf{A}| = \sqrt{\mathsf{A}^\dagger \mathsf{A}}$. Any density operator $\rho \in \mathcal{L}(\mathcal{H}_N)$ has unit-trace norm $\|\rho\|_1 = 1$. The **trace distance** of two density operators ρ_1 and ρ_2 of $\mathcal{L}(\mathcal{H}_N)$ is defined as

$$D_1(\rho_1, \rho_2) = \frac{1}{2}\|\rho_1 - \rho_2\|_1 = \frac{1}{2}\,\mathrm{tr}\big(|\rho_1 - \rho_2|\big)\,. \tag{3.51}$$

By the triangle inequality on the trace norm, we have $0 \le D_1(\rho_1, \rho_2) \le \big(\|\rho_1\|_1 + \|\rho_2\|_1\big)/2 = 1$, and $D_1(\rho_1, \rho_2) = 0$ if and only if $\rho_1 = \rho_2$, while $D_1(\rho_1, \rho_2) = 1$ if and only if ρ_1 and ρ_2 have orthogonal supports in the N-dimensional space $\mathcal{H}_N$ or equivalently $\rho_1\rho_2 = \mathsf{0}$ the null operator.

Another useful characterization of the trace distance $D_1(\rho_1, \rho_2)$ of two density operators is via the maximization:

$$D_1(\rho_1, \rho_2) = \max_{\mathsf{0} \le \mathsf{F} \le \mathsf{I}_N} \mathrm{tr}\big((\rho_1 - \rho_2)\mathsf{F}\big)\,. \tag{3.52}$$

This can be proved by means of the orthogonal decomposition of the Hermitian operator $\mathsf{A} = \rho_1 - \rho_2$ as $\mathsf{A} = \mathsf{A}_+ - \mathsf{A}_-$, where A_+ and A_- are two positive operators of $\mathcal{L}(\mathcal{H}_N)$ with orthogonal supports in $\mathcal{H}_N$ and defined as follows. The Hermitian operator $\mathsf{A} \in \mathcal{L}(\mathcal{H}_N)$ has N real eigenvalues a_n associated with N eigenstates $|a_n\rangle$ forming an orthonormal basis of $\mathcal{H}_N$. From the eigen- or diagonal decomposition of A as

$$\mathsf{A} = \sum_{n=1}^{N} a_n\,|a_n\rangle\,\langle a_n|\,, \tag{3.53}$$

we define the positive operator

$$\mathsf{A}_+ = \sum_{n:\,a_n \ge 0} a_n\,|a_n\rangle\,\langle a_n| \tag{3.54}$$

collecting the positive eigenvalues a_n, and the positive operator

$$\mathsf{A}_- = \sum_{n:\,a_n<0} -a_n \,|a_n\rangle\,\langle a_n| \tag{3.55}$$

collecting the strictly negative eigenvalues a_n.

From $\mathsf{A} = \rho_1 - \rho_2$ we deduce $\mathrm{tr}(\mathsf{A}) = \mathrm{tr}(\rho_1) - \mathrm{tr}(\rho_2) = 0$, and from $\mathsf{A} = \mathsf{A}_+ - \mathsf{A}_-$ we deduce $0 = \mathrm{tr}(\mathsf{A}) = \mathrm{tr}(\mathsf{A}_+) - \mathrm{tr}(\mathsf{A}_-)$ so that $\mathrm{tr}(\mathsf{A}_+) = \mathrm{tr}(\mathsf{A}_-)$. For the absolute-value operator $|\mathsf{A}| = \sum_n |a_n|\,|a_n\rangle\,\langle a_n|$ we have $|\mathsf{A}| = \mathsf{A}_+ + \mathsf{A}_-$, giving $\|\mathsf{A}\|_1 = \mathrm{tr}(|\mathsf{A}|) = \mathrm{tr}(\mathsf{A}_+) + \mathrm{tr}(\mathsf{A}_-) = 2\,\mathrm{tr}(\mathsf{A}_+)$. The trace distance of Eq. (3.51) can then be expressed as $D_1(\rho_1, \rho_2) = \|\mathsf{A}\|_1/2 = \mathrm{tr}(\mathsf{A}_+)$.

We also introduce the (positive) projector

$$\Pi_+ = \sum_{n:\,a_n\geq 0} |a_n\rangle\,\langle a_n| \;, \tag{3.56}$$

so that $\mathsf{A}_+ = \Pi_+\mathsf{A} = \mathsf{A}\Pi_+$. This gives the trace distance

$$D_1(\rho_1, \rho_2) = \mathrm{tr}(\mathsf{A}_+) = \mathrm{tr}(\mathsf{A}\Pi_+) = \mathrm{tr}\big((\rho_1 - \rho_2)\Pi_+\big) \;. \tag{3.57}$$

For the maximization in Eq. (3.52), we have

$$\mathsf{AF} = \sum_{n=1}^{N} a_n\,|a_n\rangle\,\langle a_n|\,\mathsf{F} \;, \tag{3.58}$$

and, by linearity of the trace and the trace identity of Eq. (A.13),

$$\mathrm{tr}(\mathsf{AF}) = \sum_{n=1}^{N} a_n\,\langle a_n|\,\mathsf{F}\,|a_n\rangle = \sum_{n:\,a_n\geq 0} a_n\,\langle a_n|\,\mathsf{F}\,|a_n\rangle - \sum_{n:\,a_n<0} |a_n|\,\langle a_n|\,\mathsf{F}\,|a_n\rangle \;. \tag{3.59}$$

In Eq. (3.52) for operators F satisfying $0 \leq \mathsf{F} \leq \mathrm{I}_N$ we have for every n the inequalities

$$0 \leq \langle a_n|\,\mathsf{F}\,|a_n\rangle \leq \langle a_n|\,\mathrm{I}_N\,|a_n\rangle \;, \tag{3.60}$$

$$0 \leq \langle a_n|\,\mathsf{F}\,|a_n\rangle \leq 1 \;, \tag{3.61}$$

so that an upper bound for $\mathrm{tr}(\mathsf{AF})$ in Eq. (3.59) is

$$\mathrm{tr}(\mathsf{AF}) \leq \sum_{n:\,a_n\geq 0} a_n = \mathrm{tr}(\mathsf{A}_+) = D_1(\rho_1, \rho_2) \;. \tag{3.62}$$

The upper bound in Eq. (3.62) along with the maximizer $\mathsf{F} = \Pi_+$ obtained in Eq. (3.57) establishes the characterization of the trace distance $D_1(\rho_1, \rho_2)$ as the maximum in Eq. (3.52).

The important result for us here is that any quantum process $\mathcal{E}(\cdot)$ is **contractive** in the sense that, for any two density operators ρ_1 and ρ_2 of $\mathcal{L}(\mathcal{H}_N)$, it realizes

$$\boxed{D_1\big(\mathcal{E}(\rho_1), \mathcal{E}(\rho_2)\big) \le D_1(\rho_1, \rho_2)\,,} \tag{3.63}$$

as we are going to prove now.

With the Hermitian operator $\mathsf{A} = \rho_1 - \rho_2 = \mathsf{A}_+ - \mathsf{A}_-$ and $D_1(\rho_1, \rho_2) = \mathrm{tr}\big(\mathsf{A}_+\big)$, the two operators A_+ and A_- are positive. The (linear) superoperator $\mathcal{E}(\cdot)$ is completely positive and trace-preserving, so that $\mathcal{E}\big(\mathsf{A}_+\big)$ and $\mathcal{E}\big(\mathsf{A}_-\big)$ also are two positive operators, with especially the invariant trace $\mathrm{tr}\big(\mathcal{E}(\mathsf{A}_+)\big) = \mathrm{tr}\big(\mathsf{A}_+\big)$. We therefore have equally the distance $D_1(\rho_1, \rho_2) = \mathrm{tr}\big(\mathsf{A}_+\big) = \mathrm{tr}\big(\mathcal{E}(\mathsf{A}_+)\big)$.

Now we consider the distance $D_1\big(\mathcal{E}(\rho_1), \mathcal{E}(\rho_2)\big)$, and we introduce $\mathsf{F}_{\mathrm{opt}}$ the maximizer achieving in Eq. (3.52) the distance $D_1\big(\mathcal{E}(\rho_1), \mathcal{E}(\rho_2)\big) = \mathrm{tr}\big(\big[\mathcal{E}(\rho_1) - \mathcal{E}(\rho_2)\big]\mathsf{F}_{\mathrm{opt}}\big)$. We know from Eqs. (3.56) and (3.57) that $\mathsf{F}_{\mathrm{opt}}$ is the projector onto the eigensubspace associated with the positive eigenvalues of the Hermitian operator $\mathcal{E}(\rho_1) - \mathcal{E}(\rho_2)$. Therefore $\mathsf{F}_{\mathrm{opt}}$ is a positive operator of $\mathcal{L}(\mathcal{H}_N)$ satisfying $0 \le \mathsf{F}_{\mathrm{opt}} \le \mathrm{I}_N$. We then deduce the inequality

$$D_1(\rho_1, \rho_2) = \mathrm{tr}\big(\mathcal{E}(\mathsf{A}_+)\big) \ge \mathrm{tr}\big(\mathcal{E}(\mathsf{A}_+)\mathsf{F}_{\mathrm{opt}}\big)\,, \tag{3.64}$$

which follows because from the positive density operator $\mathcal{E}(\mathsf{A}_+) = \sum_{n=1}^{N} s_n \,|s_n\rangle\,\langle s_n|$ in diagonal form, we obtain $\mathrm{tr}\big(\mathcal{E}(\mathsf{A}_+)\mathsf{F}_{\mathrm{opt}}\big) = \sum_{n=1}^{N} s_n\,\langle s_n|\mathsf{F}_{\mathrm{opt}}|s_n\rangle$ where each weight $\langle s_n|\mathsf{F}_{\mathrm{opt}}|s_n\rangle$ is a positive scalar in $[0, 1]$, so that the weighted sum $\mathrm{tr}\big(\mathcal{E}(\mathsf{A}_+)\mathsf{F}_{\mathrm{opt}}\big)$ cannot exceed $\sum_{n=1}^{N} s_n = \mathrm{tr}\big(\mathcal{E}(\mathsf{A}_+)\big)$.

In a comparable way, since $\mathcal{E}(\mathsf{A}_-)$ is a positive operator, $\mathrm{tr}\big(\mathcal{E}(\mathsf{A}_-)\mathsf{F}_{\mathrm{opt}}\big)$ is a positive scalar. The inequality of Eq. (3.64) can then be supplemented as

$$\begin{aligned}
D_1(\rho_1, \rho_2) &\ge \mathrm{tr}\big(\mathcal{E}(\mathsf{A}_+)\mathsf{F}_{\mathrm{opt}}\big) &(3.65)\\
&\ge \mathrm{tr}\big(\mathcal{E}(\mathsf{A}_+)\mathsf{F}_{\mathrm{opt}}\big) - \mathrm{tr}\big(\mathcal{E}(\mathsf{A}_-)\mathsf{F}_{\mathrm{opt}}\big) = \mathrm{tr}\Big(\mathcal{E}\big(\mathsf{A}_+ - \mathsf{A}_-\big)\mathsf{F}_{\mathrm{opt}}\Big) &(3.66)\\
&= \mathrm{tr}\big(\mathcal{E}(\rho_1 - \rho_2)\mathsf{F}_{\mathrm{opt}}\big) = \mathrm{tr}\Big(\big[\mathcal{E}(\rho_1) - \mathcal{E}(\rho_2)\big]\mathsf{F}_{\mathrm{opt}}\Big) &(3.67)\\
&= D_1\big(\mathcal{E}(\rho_1), \mathcal{E}(\rho_2)\big)\,, &(3.68)
\end{aligned}$$

which concludes the proof of the contractive inequality of Eq. (3.63).

For two pure states $|\psi_1\rangle$ and $|\psi_2\rangle$, the trace distance reduces to

$$D_1\big(|\psi_1\rangle\,\langle\psi_1|\,,\,|\psi_2\rangle\,\langle\psi_2|\big) = \sqrt{1 - |\langle\psi_1|\psi_2\rangle|^2}\,, \tag{3.69}$$

which can be established by performing the eigendecomposition of the Hermitian operator $|\psi_1\rangle\,\langle\psi_1| - |\psi_2\rangle\,\langle\psi_2|$ in (and outside) the plane $\big(|\psi_1\rangle\,,\,|\psi_2\rangle\big)$ of the space $\mathcal{H}_N$, as realized in a comparable way in Sect. 10.3.

In general for a unitary process $\rho \mapsto \mathcal{E}(\rho) = \mathsf{U}\rho\mathsf{U}^\dagger$ the trace distance is conserved with equality in Eq. (3.63) as

$$D_1\big(\mathsf{U}\rho_1\mathsf{U}^\dagger, \mathsf{U}\rho_2\mathsf{U}^\dagger\big) = D_1(\rho_1, \rho_2) , \tag{3.70}$$

since for any Hermitian operator A both A and $\mathsf{U}\mathsf{A}\mathsf{U}^\dagger$ share the same eigenvalues and therefore the same trace norm $\|\mathsf{A}\|_1 = \|\mathsf{U}\mathsf{A}\mathsf{U}^\dagger\|_1$. Otherwise, for a non-unitary process $\mathcal{E}(\cdot)$ the trace distance in Eq. (3.63) is strictly contractive. This means that a non-unitary quantum process $\mathcal{E}(\cdot)$ in general implements an **irreversible evolution** of quantum states. There is no quantum process that could invert $\mathcal{E}(\cdot)$, as it then should expand the trace distance, which is not feasible.

A non-unitary process $\mathcal{E}(\cdot)$ as a strictly contractive map exhibits a unique fixed point or stationary state $\rho_{\rm st}$ satisfying $\mathcal{E}\big(\rho_{\rm st}\big) = \rho_{\rm st}$. On an N-dimensional Hilbert space $\mathcal{H}_N$, a quantum process satisfying $\mathcal{E}(\mathrm{I}_N) = \mathrm{I}_N$ has the maximally mixed state I_N/N as fixed point and is then called **unital**. Unital quantum processes form a special class with interesting properties [30–33]. When a quantum process satisfies $\mathcal{E}(\mathrm{I}_N) \neq \mathrm{I}_N$ it is **non-unital**.

For two qubit density operators ρ_1 and ρ_2 according to Eq. (2.61), the trace distance

$$D_1(\rho_1, \rho_2) = \frac{1}{2}\big\|\vec{r}_1 - \vec{r}_2\big\| \tag{3.71}$$

is governed via the Euclidean norm by their respective Bloch vectors $\vec{r}_1$ and $\vec{r}_2$ in $\mathbb{R}^3$.

3.2.4 *Isometric Freedom in the Kraus Representation*

For a given quantum process $\mathcal{E}(\cdot)$, the set of Kraus operators acting in Eq. (3.45) is however not unique, but it comes with a property of **isometric freedom**, much like the various statistical ensembles generating a same density operator as examined via Eq. (2.38). Any two sets of Kraus operators $\{\Lambda_k\}_{k=1}^K$ and $\{\Lambda'_\ell\}_{\ell=1}^L$ of $\mathcal{L}\big(\mathcal{H}^\mathcal{Q}\big)$, with $K \le L$, represent the same quantum process $\mathcal{E}(\cdot)$ if and only if they can be connected [9, 25] by an isometric relation:

$$\Lambda'_\ell = \sum_{k=1}^K v_{\ell k}\Lambda_k , \tag{3.72}$$

where the complex numbers $\big[v_{\ell k}\big] = V$ form an $L \times K$ matrix V defining an isometry, satisfying $V^\dagger V = I_K$, with I_K the $K \times K$ identity matrix. In particular, we have $\sum_{\ell=1}^L v_{\ell k}^* v_{\ell k'} = \delta_{kk'}$. The K columns of V define K unit vectors with dimension $L \ge K$ that are mutually orthogonal. Also, Eq. (3.72) inverts as $\Lambda_k = \sum_{\ell=1}^L v_{\ell k}^* \Lambda'_\ell$. In such conditions, the operator-sum operation of Eq. (3.45) with the isometrically transformed operators of Eq. (3.72) gives

$$\sum_{\ell=1}^{L} \Lambda'_\ell \rho_Q \Lambda'^\dagger_\ell = \sum_{\ell=1}^{L} \left(\sum_{k=1}^{K} v_{\ell k} \Lambda_k \right) \rho_Q \left(\sum_{k'=1}^{K} v^*_{\ell k'} \Lambda^\dagger_{k'} \right) \tag{3.73}$$

$$= \sum_{k=1}^{K} \sum_{k'=1}^{K} \Lambda_k \rho_Q \Lambda^\dagger_{k'} \sum_{\ell=1}^{L} v_{\ell k} v^*_{\ell k'} \tag{3.74}$$

$$= \sum_{k=1}^{K} \sum_{k'=1}^{K} \Lambda_k \rho_Q \Lambda^\dagger_{k'} \delta_{kk'} \tag{3.75}$$

$$= \sum_{k=1}^{K} \Lambda_k \rho_Q \Lambda^\dagger_k \,, \tag{3.76}$$

establishing the equivalence of the two operator-sum representations. In addition, for the completeness relation according to Eq. (3.47), we have

$$\sum_{\ell=1}^{L} \Lambda'^\dagger_\ell \Lambda'_\ell = \sum_{\ell=1}^{L} \left(\sum_{k=1}^{K} v^*_{\ell k} \Lambda^\dagger_k \right) \left(\sum_{k'=1}^{K} v_{\ell k'} \Lambda_{k'} \right) \tag{3.77}$$

$$= \sum_{k=1}^{K} \sum_{k'=1}^{K} \Lambda^\dagger_k \Lambda_{k'} \delta_{kk'} \tag{3.78}$$

$$= \sum_{k=1}^{K} \Lambda^\dagger_k \Lambda_k = \mathrm{I}^Q \,. \tag{3.79}$$

So the isometrically transformed set $\{\Lambda'_\ell\}_{\ell=1}^L$ as well as the initial set $\{\Lambda_k\}_{k=1}^K$ simultaneously constitutes two valid sets of Kraus operators representing the same quantum process $\mathcal{E}(\rho_Q) = \sum_{k=1}^K \Lambda_k \rho_Q \Lambda_k^\dagger = \sum_{\ell=1}^L \Lambda'_\ell \rho_Q \Lambda'^\dagger_\ell$.

As we indicated, the transformation of Eq. (3.72) mapping a Kraus set with size K into a larger Kraus set with size $L \geq K$ involves an $L \times K$ isometric matrix V whose K columns are K mutually orthogonal unit vectors with dimension $L \geq K$. An available option is to supplement the K columns of V by $L - K$ additional L-dimensional unit vectors all chosen mutually orthogonal, via a Gram–Schmidt orthonormalization procedure, for instance, so as to transform V into an $L \times L$ unitary matrix. We can then enlarge the smaller Kraus set $\{\Lambda_k\}_{k=1}^K$ by adjoining to it $L - K$ null operators. We then obtain two sets of Kraus operators $\{\Lambda_k\}_{k=1}^L$ and $\{\Lambda'_\ell\}_{\ell=1}^L$ with same size L that can be unitarily related by

$$\Lambda'_\ell = \sum_{k=1}^{L} v_{\ell k} \Lambda_k \Longleftrightarrow \Lambda_k = \sum_{\ell=1}^{L} v^*_{\ell k} \Lambda'_\ell \,, \tag{3.80}$$

with $V = \left[v_{\ell k} \right]$ now an $L \times L$ unitary matrix, and one of the two sets, here $\{\Lambda_k\}_{k=1}^L$, containing $L - K$ null operators, that altogether can be omitted when expressing the action of the quantum process $\mathcal{E}(\cdot)$ as in Eq. (3.45).

If the environment space $\mathcal{H}^E$, instead of the basis $\{|e_k\rangle\}_{k=1}^K$, is referred to an alternative orthonormal basis $\{|e'_{k'}\rangle\}_{k'=1}^K$ defined by

$$|e'_{k'}\rangle = \sum_{k=1}^{K} u^*_{k'k} |e_k\rangle \Longleftrightarrow |e_k\rangle = \sum_{k'=1}^{K} u_{k'k} |e'_{k'}\rangle , \tag{3.81}$$

with the complex numbers $\left[u_{k'k}\right]$ forming a $K \times K$ unitary matrix, then the corresponding Kraus operators following from Eq. (3.38) are

$$\Lambda'_{k'} = \langle e'_{k'}| \mathsf{U}_{QE} |e_0\rangle = \sum_{k=1}^{K} u_{k'k} \langle e_k| \mathsf{U}_{QE} |e_0\rangle = \sum_{k=1}^{K} u_{k'k}\Lambda_k . \tag{3.82}$$

The two sets of Kraus operators $\{\Lambda_k\}_{k=1}^K$ and $\{\Lambda'_{k'}\}_{k'=1}^K$ are therefore unitarily related via

$$\Lambda'_{k'} = \sum_{k=1}^{K} u_{k'k}\Lambda_k \Longleftrightarrow \Lambda_k = \sum_{k'=1}^{K} u^*_{k'k}\Lambda'_{k'} , \tag{3.83}$$

which is the form of the isometric relation of Eq. (3.80) for two sets of equal size K. The two sets of Kraus operators $\{\Lambda_k\}_{k=1}^K$ and $\{\Lambda'_{k'}\}_{k'=1}^K$ therefore represent the same quantum process $\mathcal{E}(\cdot)$.

The smallest size K for a set of Kraus operators for $\mathcal{E}(\cdot)$ is given by the Choi rank of $\mathcal{E}(\cdot)$, as further discussed in Sect. 3.2.10. From there, all alternative sets of Kraus operators representing $\mathcal{E}(\cdot)$ and having a size $L \geq K$ can be isometrically deduced via Eq. (3.72).

Furthermore, any quantum process $\mathcal{E}(\cdot)$ acting as in Eq. (3.45) on quantum states ρ_Q with Hilbert space $\mathcal{H}^Q$ of dimension $\dim(\mathcal{H}^Q)$ can always be represented (in a non-unique way) with at most $[\dim(\mathcal{H}^Q)]^2$ Kraus operators [9, 12, 25], since $[\dim(\mathcal{H}^Q)]^2$ represents the maximal value possible for the Choi rank of $\mathcal{E}(\cdot)$. This in particular accords with the cardinality of a set of linearly independent operators acting on $\mathcal{H}^Q$, which cannot exceed the dimension $[\dim(\mathcal{H}^Q)]^2$ of the vector space $\mathcal{L}(\mathcal{H}^Q)$ of operators on $\mathcal{H}^Q$. The density operators ρ_Q and Kraus operators Λ_k all belong to the space $\mathcal{L}(\mathcal{H}^Q)$ with dimension $[\dim(\mathcal{H}^Q)]^2$. The partial trace and the dilated unitary U_{QE} of the Stinespring representation of Eq. (3.37), and the Kraus operator-sum representation of Eq. (3.45), are all linear operations, to define a linear superoperator $\mathcal{E}(\cdot)$. Such a (linear) superoperator $\mathcal{E}(\cdot)$ from $\mathcal{L}(\mathcal{H}^Q)$ onto $\mathcal{L}(\mathcal{H}^Q)$ belongs to a vector space $\mathcal{L}(\mathcal{L}(\mathcal{H}^Q))$ with dimension $[\dim(\mathcal{H}^Q)]^4$. Quite naturally, in terms of degrees of freedom, to characterize $\mathcal{E}(\cdot)$ with its $[\dim(\mathcal{H}^Q)]^4$ degrees of freedom, it is enough to resort to a number $[\dim(\mathcal{H}^Q)]^2$ of operators Λ_k each with $[\dim(\mathcal{H}^Q)]^2$ degrees of freedom. This important property can be proved more formally as follows.

We consider a principal quantum system Q with N-dimensional Hilbert space $\mathcal{H}^Q \equiv \mathcal{H}_N$, and we shall show that $\boldsymbol{N^2}$ **Kraus operators suffice** to represent any quantum process $\mathcal{E}(\cdot)$ on Q as in Eq. (3.45). The principal quantum system Q is coupled to an environment E with Hilbert space $\mathcal{H}^E \equiv \mathcal{H}_K$ taken with a (possibly very large) dimension $K \geq N^2$ and implementing on Q the quantum process $\mathcal{E}(\cdot)$ defined by the set $\{\Lambda_k\}_{k=1}^K$ of the K Kraus operators from Eq. (3.38). From the Kraus set $\{\Lambda_k\}_{k=1}^K$ determined by the environment E, we shall show (construct) for $\mathcal{E}(\cdot)$ an equivalent Kraus set containing no more than N^2 Kraus operators. This will be accomplished by exhibiting a $K \times K$ unitary matrix $U = \left[u_{\ell k}\right]$ acting as V in Eq. (3.80) to connect the two Kraus sets.

We consider the $K \times K$ matrix $W = \left[w_{kk'}\right]$ defined by the matrix elements

$$w_{kk'} = \mathrm{tr}\left(\Lambda_k^\dagger \Lambda_{k'}\right) = \sum_{n=1}^{N} \langle n|\Lambda_k^\dagger \Lambda_{k'}|n\rangle \ , \tag{3.84}$$

since, to explicitly compute the trace, the K Kraus operators Λ_k are operators of $\mathcal{L}(\mathcal{H}_N)$ from $\mathcal{H}_N$ onto $\mathcal{H}_N$ referred here to the orthonormal basis $\{|n\rangle\}_{n=1}^N$ of $\mathcal{H}_N$. We now prove that the rank of W is no larger than N^2. Each operator Λ_k is a vector from the vector space $\mathcal{L}(\mathcal{H}_N)$ with dimension N^2. Therefore, there cannot be more than N^2 linearly independent such vectors Λ_k. Among the $K \geq N^2$ operators Λ_k, we select a subset of N^2 operators collectively spanning $\mathcal{L}(\mathcal{H}_N)$. Such a subset necessarily exists, otherwise the operators Λ_k could not collectively satisfy the completeness relation of Eq. (3.47). In this spanning subset, the N^2 Kraus operators Λ_k it contains are assumed to be numbered from $k = 1$ to N^2, this being always feasible with no loss of generality. The remaining $K - N^2$ Kraus operators can be expressed as the linear combinations $\Lambda_\ell = \sum_{k=1}^{N^2} c_{\ell k}\Lambda_k$, for $\ell = N^2 + 1$ to K. This induces the relations $w_{j\ell} = \mathrm{tr}\left(\Lambda_j^\dagger \Lambda_\ell\right) = \sum_{k=1}^{N^2} c_{\ell k}\, \mathrm{tr}\left(\Lambda_j^\dagger \Lambda_k\right) = \sum_{k=1}^{N^2} c_{\ell k} w_{jk}$, for every row index $j = 1$ to K, expressing that the (last) $K - N^2$ columns of W indexed by $\ell = N^2 + 1$ to K are linear combinations of the (first) N^2 columns of W indexed by $k = 1$ to N^2. To be explicit, we can write for the column ℓ from $N^2 + 1$ to K, the linear combination $\mathrm{column}(\ell) = \sum_{k=1}^{N^2} c_{\ell k}\, \mathrm{column}(k)$. This proves that the rank of the $K \times K$ matrix W is no larger than N^2.

The matrix W from Eq. (3.84) is Hermitian, since $w_{kk'}^* = \mathrm{tr}\left((\Lambda_k^\dagger \Lambda_{k'})^\dagger\right) = \mathrm{tr}\left(\Lambda_{k'}^\dagger \Lambda_k\right) = w_{k'k}$. Consequently, there exist a $K \times K$ diagonal matrix D and a $K \times K$ unitary matrix $U = \left[u_{k\ell}\right]$ such that $W = UDU^\dagger \Leftrightarrow D = U^\dagger W U$. The diagonal matrix $D = \mathrm{diag}\left[\lambda_1, \lambda_2, \ldots, \lambda_K\right]$ is formed with the K eigenvalues λ_k of W, with at least $K - N^2$ such eigenvalues that are zero since W has a rank no larger than N^2. The unitary matrix U has its K columns formed by an orthonormal eigenbasis of W. The matrix relation $D = U^\dagger W U$ can be expressed component-wise as

$$\lambda_m \delta_{m\ell} = \sum_{k=1}^{K}\sum_{k'=1}^{K} u_{km}^* w_{kk'} u_{k'\ell} \tag{3.85}$$

$$= \sum_{k=1}^{K}\sum_{k'=1}^{K} u_{km}^* \sum_{n=1}^{N} \langle n|\Lambda_k^\dagger \Lambda_{k'}|n\rangle\ u_{k'\ell} \ , \tag{3.86}$$

with the Kronecker delta $\delta_{m\ell}$. We introduce the new set of K Kraus operators of $\mathcal{L}(\mathcal{H}_N)$ defined by

$$\Lambda'_m = \sum_{k=1}^{K} u_{km} \Lambda_k \ , \tag{3.87}$$

for $m = 1$ to K. The unitary connection of Eq. (3.87) is similar to Eq. (3.80), since for a unitary matrix U, the adjoint $U^\dagger$, but also the transpose $U^\top$ and the conjugate U^* are all unitary matrices. The unitary connection of Eq. (3.87) already ensures that the new Kraus set $\{\Lambda'_m\}_{m=1}^K$ is an equivalent Kraus set for the quantum process $\mathcal{E}(\cdot)$, as well as $\{\Lambda_k\}_{k=1}^K$. It remains to show that the new Kraus set $\{\Lambda'_m\}_{m=1}^K$ contains no more than N^2 non-null Kraus operators.

By using Eq. (3.87) in Eq. (3.86) we deduce

$$\lambda_m \delta_{m\ell} = \sum_{n=1}^{N} \langle n|\Lambda_m'^\dagger \Lambda'_\ell|n\rangle \ . \tag{3.88}$$

And by inserting $\sum_{n'=1}^N |n'\rangle\langle n'| = \mathrm{I}_N$ in Eq. (3.88), we obtain

$$\lambda_m \delta_{m\ell} = \sum_{n=1}^{N} \sum_{n'=1}^{N} \langle n|\Lambda_m'^\dagger|n'\rangle \langle n'|\Lambda'_\ell|n\rangle \ . \tag{3.89}$$

For the diagonal terms when $m = \ell$, this gives

$$\lambda_m = \sum_{n=1}^{N} \sum_{n'=1}^{N} \langle n|\Lambda_m'^\dagger|n'\rangle \langle n'|\Lambda'_m|n\rangle = \sum_{n=1}^{N} \sum_{n'=1}^{N} \left|\langle n'|\Lambda'_m|n\rangle\right|^2 . \tag{3.90}$$

As we have seen, there are at least $K - N^2$ eigenvalues λ_m that are zero. When $\lambda_m = 0$, Eq. (3.90) implies that all the matrix elements $\langle n'|\Lambda'_m|n\rangle$ of the Kraus operator Λ'_m are zero, revealing that such a Λ'_m is the null operator. In the alternative Kraus set $\{\Lambda'_m\}_{m=1}^K$ defined by Eq. (3.87), there are therefore no more than N^2 Kraus operators that are non-null, as we wanted to prove.

Moreover, for two distinct Kraus operators Λ'_m and Λ'_ℓ with $m \neq \ell$, Eq. (3.88) indicates that the Hilbert–Schmidt inner product $\mathrm{tr}\big(\Lambda_m'^\dagger \Lambda'_\ell\big) = 0$. Such vanishing inner products indicate that the non-null Kraus operators Λ'_m resulting from Eq. (3.87) are mutually orthogonal operators of $\mathcal{L}(\mathcal{H}_N)$. The rank of W from Eq. (3.84), which is no larger than N^2, constitutes an upper bound, just like the Choi rank, to the minimal number of Kraus operators for a given quantum process $\mathcal{E}(\cdot)$.

The operator-sum or Kraus representation of Eq. (3.45) for representing the quantum process $\mathcal{E}(\cdot)$ offers a significant advantage compared with the dilated unitary representation of Eq. (3.37). The unitary representation of Eq. (3.37) refers to an environment E which can be of very large dimension, with a huge number of degrees of freedom, all far from being completely characterized. Yet, the environment influences a quantum system of interest Q whose state ρ_Q is limited by the dimensionality $[\dim(\mathcal{H}^Q)]^2$. The advantage is that, for representing the transformation $\rho_Q \mapsto \mathcal{E}(\rho_Q)$, the operator-sum representation of Eq. (3.45) involves a number of degrees of freedom (contained in the Kraus operators) that can be matched to the dimensionality of ρ_Q and is not directly influenced by the dimensionality of the environment itself. Many degrees of freedom of the environment get somehow wiped out by the partial tracing operation over E in Eq. (3.37). For instance, for a qubit with two-dimensional Hilbert space, only four Kraus operators at most, of size 2×2, are enough to describe its interaction with any environment, be it a thermal bath with a quasi-infinite number of degrees of freedom.

3.2.5 Probabilistic Interpretation of the Kraus Representation

There also exists an interesting probabilistic interpretation [6, 9, 24] to the Kraus or operator-sum representation of Eq. (3.45). The action of the quantum process $\rho_Q \mapsto \mathcal{E}(\rho_Q)$ from Eq. (3.45) is equivalent to a random application of the operators Λ_k on the input quantum state in the following way.

On a pure input state $|Q\rangle$, the random application of a given operator Λ_k acts to produce the output state

$$|Q_k\rangle = \frac{\Lambda_k |Q\rangle}{\sqrt{\langle Q|\Lambda_k^\dagger \Lambda_k|Q\rangle}} = \frac{\Lambda_k |Q\rangle}{\sqrt{p(k)}}, \tag{3.91}$$

which is a valid (normalized) pure state from $\mathcal{H}^Q$ of density operator $\rho_k = |Q_k\rangle\langle Q_k|$, with Λ_k acting with the probability

$$p(k) = \langle Q|\Lambda_k^\dagger \Lambda_k|Q\rangle = \mathrm{tr}\big(\Lambda_k |Q\rangle\langle Q| \Lambda_k^\dagger\big) . \tag{3.92}$$

A mixed input state can be described with the input statistical ensemble $\{(p_j, |Q^j\rangle)\}$ to represent the density operator $\rho_Q = \sum_j p_j |Q^j\rangle\langle Q^j|$ according to Eq. (2.35). Each pure input state $|Q^j\rangle$ occurs with the probability p_j, and on each state $|Q^j\rangle$ the random application of a given Λ_k produces the output state

$$|Q_k^j\rangle = \frac{\Lambda_k |Q^j\rangle}{\sqrt{\langle Q^j|\Lambda_k^\dagger \Lambda_k|Q^j\rangle}} = \frac{\Lambda_k |Q^j\rangle}{\sqrt{p(k|j)}}, \tag{3.93}$$

with the conditional probability

$$p(k|j) = \langle Q^j|\Lambda_k^\dagger \Lambda_k|Q^j\rangle = \mathrm{tr}\big(\Lambda_k |Q^j\rangle\langle Q^j| \Lambda_k^\dagger\big) . \tag{3.94}$$

With an input over the statistical ensemble $\{(p_j, |Q^j\rangle)\}$, the total probability for a given Λ_k to act is

$$p(k) = \sum_j p(k|j)p_j \tag{3.95}$$

$$= \sum_j p_j \operatorname{tr}\big(\Lambda_k |Q^j\rangle \langle Q^j| \Lambda_k^\dagger\big) \tag{3.96}$$

$$= \operatorname{tr}\big(\Lambda_k \rho_Q \Lambda_k^\dagger\big) . \tag{3.97}$$

When a given Λ_k has acted at random, the output terminates in a statistical ensemble of states $\{(p(j|k), |Q_k^j\rangle)\}$, with $p(j|k)$ the conditional probability that it was the input $|Q^j\rangle$ that experienced the action of this Λ_k to yield the output $|Q_k^j\rangle$ of Eq. (3.93). This output statistical ensemble is represented by the density operator:

$$\rho_k = \sum_j p(j|k) |Q_k^j\rangle \langle Q_k^j| = \sum_j \frac{p(j|k)}{p(k|j)} \Lambda_k |Q^j\rangle \langle Q^j| \Lambda_k^\dagger . \tag{3.98}$$

By the Bayes rule we have $p(k|j)p_j = p(j|k)p(k)$ or $p(j|k)/p(k|j) = p_j/p(k)$, so that Eq. (3.98) is also

$$\rho_k = \frac{1}{p(k)} \sum_j p_j \Lambda_k |Q^j\rangle \langle Q^j| \Lambda_k^\dagger \tag{3.99}$$

$$= \frac{1}{p(k)} \Lambda_k \rho_Q \Lambda_k^\dagger . \tag{3.100}$$

On a pure or mixed input state, the random application of one or the other of the operators Λ_k remains hidden, so that the output state is described by the statistical ensemble $\{(p(k), \rho_k)\}$ represented by the density operator:

$$\sum_{k=1}^{K} p(k)\rho_k = \sum_{k=1}^{K} \Lambda_k \rho_Q \Lambda_k^\dagger , \tag{3.101}$$

which is precisely the operator-sum representation of the output state $\mathcal{E}(\rho_Q)$ of Eq. (3.45).

To summarize, the Kraus or operator-sum representation of the quantum process $\rho_Q \mapsto \mathcal{E}(\rho_Q)$ from Eq. (3.45) can be interpreted as equivalent to a random application of the Kraus operators Λ_k, to replace the pure input state $|Q\rangle$ by the output state $|Q_k\rangle$ of Eq. (3.91) with the probability $p(k)$ of Eq. (3.92), or to replace the mixed input state ρ_Q by the output state ρ_k of Eq. (3.100) with the probability $p(k)$ of Eq. (3.97). In addition, a concrete approach to implement such a probabilistic procedure will be found at the end of Sect. 4.2.3, through the notion of unread measurement.

Following this probabilistic interpretation, another interesting property of the operator-sum representation of Eq. (3.45) is that the Kraus operators Λ_k can possibly be guessed or postulated empirically, phenomenologically, heuristically, according to the type of environment and its effect envisaged on the quantum system of interest Q.

For example, one can consider the situation where the quantum system of interest Q is a qubit representing an electronic spin. Such spin may suffer an interaction with the earth magnetic field that could cause some flip of the qubit state along the direction of the magnetic field. Such an interaction altering the spin state ρ_Q could be modeled, as an instance of Eq. (3.45), by

$$\rho'_Q = \mathcal{E}(\rho_Q) = (1-p)\rho_Q + p\sigma_x\rho_Q\sigma_x^\dagger \,, \tag{3.102}$$

involving the two Kraus operators $\{\Lambda_1 = \sqrt{1-p}\,\mathrm{I}_2,\ \Lambda_2 = \sqrt{p}\sigma_x\}$ especially satisfying the completeness relation of Eq. (3.47) as they should. The parameter $p \in [0, 1]$ is interpretable as a probability that empirically quantifies the strength of the coupling of the spin with the magnetic field. With some probability p some flip of the qubit state occurs by the action of $\Lambda_2 = \sqrt{p}\sigma_x$, while with probability $1-p$ the qubit state is left unchanged as expressed by the identity operator in $\Lambda_1 = \sqrt{1-p}\,\mathrm{I}_2$.

A second example is the noisy preparation of the qubit described in Sect. 2.5.6, which can also be viewed as the random action of a bit-flip Pauli operator σ_x. Instead of the noise-free preparation intended in the pure (input) state $|0\rangle$, the noisy preparation results in the mixed (output) state of Eq. (2.87) involving a state $|0\rangle$ that can be randomly flipped into $|1\rangle = \sigma_x |0\rangle$ with the flip probability $p = \langle \sin^2(\xi) \rangle$. This random alteration can be described as the transformation $|0\rangle\langle 0| \mapsto (1-p)\,|0\rangle\langle 0| + p\sigma_x\,|0\rangle\langle 0|\,\sigma_x^\dagger$, and it is accomplished in Eq. (3.45) with the two Kraus operators $\{\Lambda_1 = \sqrt{1-p}\,\mathrm{I}_2,\ \Lambda_2 = \sqrt{p}\sigma_x\}$ defining an operator-sum representation for modeling the noisy preparation of the qubit.

A third example could be the situation of a qubit representing a photon with its two basis states of polarization, which could suffer some isotropic scattering while traveling in some propagating medium. Such an interaction altering the photon state ρ_Q could be modeled, as an instance of Eq. (3.45), by

$$\rho'_Q = \mathcal{E}(\rho_Q) = (1-p)\rho_Q + \frac{p}{3}\left(\sigma_x\rho_Q\sigma_x^\dagger + \sigma_y\rho_Q\sigma_y^\dagger + \sigma_z\rho_Q\sigma_z^\dagger\right) . \tag{3.103}$$

Leaning on the orthonormal Pauli basis $\{\mathrm{I}_2, \sigma_x, \sigma_y, \sigma_z\}$ for the qubit operators of $\mathcal{L}(\mathcal{H}_2)$, we have here four Kraus operators $\{\Lambda_1 = \sqrt{1-p}\,\mathrm{I}_2,\ \Lambda_2 = \sqrt{p/3}\sigma_x,\ \Lambda_3 = \sqrt{p/3}\sigma_y,\ \Lambda_4 = \sqrt{p/3}\sigma_z\}$ satisfying the completeness relation of Eq. (3.47). The parameter $p \in [0, 1]$ is interpretable as a probability that empirically quantifies the strength of the scattering. With the probability $1-p$ the photon remains unaltered, while with the uniform probability $p/3$ the photon state is altered by random application of one of the three Pauli operators related to the three directions in $\mathbb{R}^3$.

3.2.6 Stinespring Isometric Extension Operator

Also, by means of the Kraus operators Λ_k, it is possible to avoid to refer to U_{QE} and $|e_0\rangle$ for the evolution $|\psi'_{QE}\rangle = \mathsf{U}_{QE}\,|Q\rangle \otimes |e_0\rangle$ of the composite system QE in Eq. (3.34). Both U_{QE} and $|e_0\rangle$ are somehow included in the Kraus operators Λ_k defined in Eq. (3.38), and this is enough to our purpose. We define Λ, an isometric (i.e., norm-preserving) operator from $\mathcal{H}^Q$ onto $\mathcal{H}^Q \otimes \mathcal{H}^E$, as

$$\Lambda = \mathsf{U}_{QE}\,|e_0\rangle \ , \tag{3.104}$$

satisfying $\Lambda^\dagger\Lambda = \mathrm{I}^Q$. The **Stinespring isometric extension operator** Λ of Eq. (3.104) takes any state vector $|Q\rangle$ of $\mathcal{H}^Q$ and sends it in $\mathcal{H}^Q \otimes \mathcal{H}^E$ into the state vector $\Lambda\,|Q\rangle = \mathsf{U}_{QE}\,|Q\rangle \otimes |e_0\rangle$ which is precisely the evolved composite state $|\psi'_{QE}\rangle$ produced by the dilated unitary evolution of Eq. (3.34). Since U_{QE} is unitary, Λ is an isometry transforming any orthonormal basis of $\mathcal{H}^Q$ into an orthonormal set of states in $\mathcal{H}^Q \otimes \mathcal{H}^E$. There exists an alternative definition for the Stinespring isometric extension operator Λ not referring to U_{QE} and $|e_0\rangle$ but instead to the Kraus operators Λ_k, as also addressed, for instance, in [23, 25]. We define the operator

$$\mathsf{A} = \sum_{k=1}^{K} \Lambda_k \otimes |e_k\rangle \ , \tag{3.105}$$

which is just as Λ an operator from $\mathcal{H}^Q$ onto $\mathcal{H}^Q \otimes \mathcal{H}^E$, taking any state vector $|Q\rangle$ of $\mathcal{H}^Q$ into the vector $\mathsf{A}\,|Q\rangle = \sum_{k=1}^{K} \Lambda_k\,|Q\rangle \otimes |e_k\rangle$ of $\mathcal{H}^Q \otimes \mathcal{H}^E$. This operator A can be completely characterized by its scalar matrix elements from an arbitrary state $|Q\rangle$ of $\mathcal{H}^Q$ to an arbitrary state $|Q'\rangle \otimes |e_\ell\rangle$ of $\mathcal{H}^Q \otimes \mathcal{H}^E$, reading

$$\langle e_\ell|\,\langle Q'|\,\mathsf{A}\,|Q\rangle = \langle e_\ell|\,\langle Q'|\sum_{k=1}^{K} \Lambda_k\,|Q\rangle\,|e_k\rangle \ . \tag{3.106}$$

From Eq. (3.38) to express Λ_k, we have

$$\langle e_\ell|\,\langle Q'|\,\mathsf{A}\,|Q\rangle = \sum_{k=1}^{K} \langle e_\ell|\Big(\langle e_k|\,\langle Q'|\,\mathsf{U}_{QE}\,|Q\rangle\,|e_0\rangle\Big)\,|e_k\rangle \ . \tag{3.107}$$

The pair of parentheses in Eq. (3.107) encloses a scalar so that we also have

$$\langle e_\ell|\,\langle Q'|\,\mathsf{A}\,|Q\rangle = \sum_{k=1}^{K}\Big(\langle e_k|\,\langle Q'|\,\mathsf{U}_{QE}\,|Q\rangle\,|e_0\rangle\Big)\,\langle e_\ell|e_k\rangle \tag{3.108}$$

$$= \langle e_\ell|\,\langle Q'|\,\mathsf{U}_{QE}\,|Q\rangle\,|e_0\rangle \ , \tag{3.109}$$

since $\langle e_\ell | e_k \rangle = \delta_{\ell k}$. And by using Eq. (3.104) to recognize Λ we finally obtain

$$\langle e_\ell | \langle Q' | \mathsf{A} | Q \rangle = \langle e_\ell | \langle Q' | \Lambda | Q \rangle \ , \tag{3.110}$$

indicating that A and Λ share all the same matrix elements and are therefore the same operator, therefore leading for Λ to the alternative definition:

$$\Lambda = \sum_{k=1}^{K} \Lambda_k \otimes |e_k\rangle \ . \tag{3.111}$$

By means of Eq. (3.111), we avoid the reference to U_{QE} and $|e_0\rangle$ for the evolution of the composite state of Eq. (3.34), which is now expressed as

$$|Q\rangle \longmapsto |\psi'_{QE}\rangle = \Lambda\, |Q\rangle = \sum_{k=1}^{K} \Lambda_k\, |Q\rangle \otimes |e_k\rangle \ . \tag{3.112}$$

The corresponding density operators $|Q\rangle\langle Q|$ and $|\psi'_{QE}\rangle\langle\psi'_{QE}|$ evolve according to

$$|\psi'_{QE}\rangle\langle\psi'_{QE}| = \Lambda\, |Q\rangle\langle Q|\, \Lambda^\dagger = \sum_{k=1}^{K}\sum_{k'=1}^{K} \Lambda_k\, |Q\rangle\langle Q|\, \Lambda_{k'}^\dagger \otimes |e_k\rangle\langle e_{k'}| \ . \tag{3.113}$$

Again, by convexity-linearity, when the principal system Q is initially in a mixed state ρ_Q, this one is expressible as a convex sum of pure states similar to $|Q\rangle\langle Q|$, so that by linearity of Eq. (3.113), the composite system QE evolves according to

$$\boxed{\rho'_{QE} = \Lambda \rho_Q \Lambda^\dagger = \sum_{k=1}^{K}\sum_{k'=1}^{K} \Lambda_k \rho_Q \Lambda_{k'}^\dagger \otimes |e_k\rangle\langle e_{k'}| \ .} \tag{3.114}$$

This Eq. (3.114) is a powerful equation, expressing the evolution of the whole composite system QE, as a function of the properties of the environment encapsulated in the Kraus operators Λ_k from Eq. (3.38), and as a function of the initial state ρ_Q of the principal system Q.

From Eq. (3.114), by partial tracing over the environment E, we obtain the evolution of the principal system Q alone as

$$\rho'_Q = \mathcal{E}(\rho_Q) = \mathrm{tr}_E\big(\Lambda \rho_Q \Lambda^\dagger\big) \tag{3.115}$$

$$= \sum_{k=1}^{K} \sum_{k'=1}^{K} \mathrm{tr}_E\Big(\Lambda_k \rho_Q \Lambda_{k'}^\dagger \otimes |e_k\rangle\langle e_{k'}|\Big) \tag{3.116}$$

$$= \sum_{k=1}^{K} \sum_{k'=1}^{K} \Lambda_k \rho_Q \Lambda_{k'}^\dagger \,\mathrm{tr}\big(|e_k\rangle\langle e_{k'}|\big) \tag{3.117}$$

$$= \sum_{k=1}^{K} \Lambda_k \rho_Q \Lambda_k^\dagger \,, \tag{3.118}$$

recovering with Eq. (3.118) the operator-sum representation of Eq. (3.45).

When expressed as a function solely of the Stinespring isometric extension operator Λ, the evolutions of Eq. (3.114) for the compound QE and of Eq. (3.115) for the principal system Q can be referred to as a Stinespring isometric representation of the evolutions [23, 25].

3.2.7 *Evolution of the Environment*

As we have just seen, not only the evolution of the principal system Q via Eq. (3.45), but also the evolution of the full composite system QE via Eq. (3.114) can be completely characterized by specifying the Kraus operators $\{\Lambda_k\}_{k=1}^{K}$. This makes possible a characterization of the final state ρ'_E of the environment E, especially as a function of the initial state ρ_Q of the principal system Q. This is obtained from the final composite state ρ'_{QE} by partial tracing over the principal system Q. This defines a dual quantum channel $\rho'_E = \widetilde{\mathcal{E}}(\rho_Q) = \mathrm{tr}_Q\big(\rho'_{QE}\big)$ complementary or conjugate to the primal channel $\rho'_Q = \mathcal{E}(\rho_Q) = \mathrm{tr}_E\big(\rho'_{QE}\big)$, and which follows [12, 25] from Eq. (3.114) as

$$\rho'_E = \widetilde{\mathcal{E}}(\rho_Q) = \mathrm{tr}_Q\big(\rho'_{QE}\big) = \sum_{k=1}^{K} \sum_{k'=1}^{K} \mathrm{tr}\big(\Lambda_k \rho_Q \Lambda_{k'}^\dagger\big) |e_k\rangle\langle e_{k'}| \,, \tag{3.119}$$

generally valid for a pure or a mixed state ρ_Q. Equation (3.119) can be seen for the density operator ρ'_E on $\mathcal{H}^E$ as a decomposition in the standard basis $\big\{|e_k\rangle\langle e_{k'}|\big\}$ for operators on the space $\mathcal{H}^E$. Therefore, the final state of the environment ρ'_E is a density operator on $\mathcal{H}^E$ represented by the $K \times K$ matrix W with elements

$$w_{kk'} = \mathrm{tr}\big(\Lambda_k \rho_Q \Lambda_{k'}^\dagger\big) \,, \tag{3.120}$$

determined by both the initial state ρ_Q and the Kraus operators Λ_k of the interaction [7, 25]. We note, in particular, by observing Eq. (3.120), that the final state ρ'_E of the environment has no dependence on an initial state $|e_0\rangle$ for the environment. The Kraus operators Λ_k themselves, via Eq. (3.38), may depend on such initial state

$|e_0\rangle$, but once the Kraus operators Λ_k are introduced to define a quantum process, this wipes out the details of the environment such as $|e_0\rangle$.

As we shall see later in Chap. 13 on quantum communication, Eq. (3.119) is in particular useful to investigate the possible leakage of information occurring from the principal quantum system Q into the environment E during their interaction.

3.2.8 *Simulating the Environment*

When a set of Kraus operators Λ_k satisfying the completeness relation of Eq. (3.47) is given, it defines, via the operator-sum representation of Eq. (3.45), a completely positive trace-preserving map or quantum process $\mathcal{E}(\cdot)$. The Stinespring dilation theorem mentioned at the end of Sect. 3.2.1 ensures that such a quantum process $\mathcal{E}(\cdot)$ can be realized via a dilated unitary representation as in Eq. (3.37). This has an important physical significance, as it indicates that any non-unitary evolution of a quantum state ρ_Q described as $\mathcal{E}(\rho_Q)$ in the operator-sum representation of Eq. (3.45), can always be obtained from a unitary evolution of a larger (dilated) quantum system resulting from coupling with another auxiliary quantum system playing the role of the environment E, and that we now seek to characterize further.

We assume that K Kraus operators Λ_k are given that satisfy the completeness relation of Eq. (3.47) to define the quantum process $\mathcal{E}(\rho_Q)$ of Eq. (3.45). We want to characterize a model for an environment E offering a dilated unitary representation as in Eq. (3.37) for $\mathcal{E}(\rho_Q)$. This includes specifying the Hilbert space $\mathcal{H}^E$ and its dimension, the initial state $|e_0\rangle$ for E and the joint unitary evolution U_{QE}, so as to satisfy Eq. (3.37). There might exist many possibilities. We describe a "canonical" one now [6, 9]. An appropriate environment E can be realized by an auxiliary quantum system with Hilbert space $\mathcal{H}^E$ of dimension taken to be K, the number of Kraus operators. The composite system in $\mathcal{H}^Q \otimes \mathcal{H}^E$ starts in the initial pure separable state $|\psi_{QE}\rangle = |Q\rangle \otimes |e_0\rangle$ and undergoes the transformation U_{QE} defined by

$$\mathsf{U}_{QE}\,|\psi_{QE}\rangle = \mathsf{U}_{QE}\,|Q\rangle \otimes |e_0\rangle = \sum_{k=1}^{K} \Lambda_k\,|Q\rangle \otimes |e_k\rangle = |\psi'_{QE}\rangle\ , \tag{3.121}$$

where $\{|e_k\rangle\}_{k=1}^{K}$ is an orthonormal basis of the K-dimensional space $\mathcal{H}^E$ representing the environment E. Equation (3.121) has similarities with Eq. (3.112) for the Stinespring isometric extension operator Λ from $\mathcal{H}^Q$ onto $\mathcal{H}^Q \otimes \mathcal{H}^E$. What we need to verify now is that Eq. (3.121) indeed defines a valid unitary operator U_{QE} from $\mathcal{H}^Q \otimes \mathcal{H}^E$ onto $\mathcal{H}^Q \otimes \mathcal{H}^E$. Any two states of the form $|Q_1\rangle \otimes |e_0\rangle$ and $|Q_2\rangle \otimes |e_0\rangle$ of the product space $\mathcal{H}^Q \otimes \mathcal{H}^E$ have inner product $\big(|Q_1\rangle \otimes |e_0\rangle\,, |Q_2\rangle \otimes |e_0\rangle\big) = \langle Q_1|Q_2\rangle\,\langle e_0|e_0\rangle = \langle Q_1|Q_2\rangle$. We have for the states transformed by U_{QE} of Eq. (3.121) the inner product

$$\Big(\mathsf{U}_{QE}\,|Q_1\rangle\otimes|e_0\rangle\,,\mathsf{U}_{QE}\,|Q_2\rangle\otimes|e_0\rangle\Big)=\sum_{k=1}^{K}\sum_{k'=1}^{K}\langle Q_1|\Lambda_k^\dagger\Lambda_{k'}|Q_2\rangle\,\langle e_k|e_{k'}\rangle \quad (3.122)$$

$$=\langle Q_1|\sum_{k=1}^{K}\Lambda_k^\dagger\Lambda_k\,|Q_2\rangle \quad (3.123)$$

$$=\langle Q_1|Q_2\rangle\;, \quad (3.124)$$

since $\langle e_k|e_{k'}\rangle=\delta_{kk'}$ the Kronecker delta, and $\sum_{k=1}^{K}\Lambda_k^\dagger\Lambda_k=\mathrm{I}^Q$. Equations (3.122)–(3.124) express conservation of the inner product, which guarantees that U_{QE} of Eq. (3.121) indeed defines a unitary transformation on $\mathcal{H}^Q\otimes|e_0\rangle$ and can be extended so as to completely define a unitary operator U_{QE} on $\mathcal{H}^Q\otimes\mathcal{H}^E$. The input subspace $\mathcal{H}^Q\otimes\mathrm{span}\big(|e_0\rangle\big)$ is unitarily mapped by U_{QE} into an output subspace of $\mathcal{H}^Q\otimes\mathcal{H}^E$ with same dimension. The orthogonal complements to these two subspaces with same dimension, at the input and at the output, can be unitarily mapped one into the other for extending U_{QE}. The extension of U_{QE} can even be achieved in an infinite number of ways, yet non-critical and irrelevant to our purpose here.

The density operator associated with the composite state $|\psi'_{QE}\rangle$ of Eq. (3.121) is

$$\rho'_{QE}=|\psi'_{QE}\rangle\,\langle\psi'_{QE}|=\sum_{k=1}^{K}\sum_{k'=1}^{K}\Lambda_k\,|Q\rangle\,\langle Q|\,\Lambda_{k'}^\dagger\otimes|e_k\rangle\,\langle e_{k'}|\;, \quad (3.125)$$

which upon partial tracing over the environment E provides the reduced density operator for the principal quantum system Q as

$$\rho'_Q=\mathrm{tr}_E\big(\rho'_{QE}\big)=\sum_{k=1}^{K}\Lambda_k\,|Q\rangle\,\langle Q|\,\Lambda_k^\dagger\,. \quad (3.126)$$

By convexity-linearity, when the quantum system Q is initially in a mixed state ρ_Q, this one is expressible as a convex sum of pure states similar to $|Q\rangle\,\langle Q|$, so that by linearity in Eq. (3.126) the resulting transformed state is $\rho'_Q=\sum_{k=1}^{K}\Lambda_k\rho_Q\Lambda_k^\dagger$ which is the standard operator-sum representation of Eq. (3.45) targeted for $\mathcal{E}(\rho_Q)$.

We have here a canonical procedure, for simulating or synthesizing an arbitrary quantum process $\mathcal{E}(\cdot)$, via a dilated unitary representation. From a given set of K Kraus operators defining a quantum process $\mathcal{E}(\cdot)$, the procedure prescribes a model for an environment E with dimension K offering via the joint unitary evolution of Eq. (3.121) a dilated unitary representation for the quantum process $\mathcal{E}(\cdot)$. As we have seen previously, any finite-dimensional quantum states can be materialized by assembling a sufficient number of qubits. Also, any unitary evolutions of an arbitrary number of qubits can always be implemented by means of the universal set of one-qubit gates supplemented by the two-qubit Cnot gate, as we indicated in Sect. 3.1.5. Therefore, the dilated unitary evolution of Eq. (3.121) can always be implemented by means of unitary qubit gates, to simulate or synthesize the arbitrary quantum process $\mathcal{E}(\cdot)$. This in particular reinforces the status of the qubit as a fundamental building block for quantum information processing. Moreover, it shows that the set of qubit gates universal for quantum synthesis is not limited to the synthesis of unitary quantum processes, but can equally manage the synthesis of any non-unitary quantum processes $\mathcal{E}(\cdot)$, via the canonical procedure we described in this section. This procedure for simulating an arbitrary quantum process $\mathcal{E}(\cdot)$ will be used repeatedly in Chap. 9 to come, in order to construct environment models, under the form of quantum circuits, to simulate quantum noises of interest affecting quantum states or signals.

We have indicated in Sect. 3.2.4 that, for the same quantum process $\mathcal{E}(\cdot)$, the set of Kraus operators is not unique. Changing the set of Kraus operators according to the isometry of Eq. (3.72) provides, via the same canonical procedure, another model of environment E offering an alternative (equivalent) dilated unitary representation for the same quantum process $\mathcal{E}(\cdot)$.

In addition, we have shown in Sect. 3.2.4 that at most $[\dim(\mathcal{H}^Q)]^2$ Kraus operators are sufficient to represent any quantum process $\mathcal{E}(\cdot)$ for the principal system Q via the operator sum of Eq. (3.45). Therefore, an environment with dimension $[\dim(\mathcal{H}^Q)]^2$ is also sufficient to represent any quantum process $\mathcal{E}(\cdot)$ via a dilated unitary representation as in Eq. (3.37). This amounts to an environment with Hilbert space $\mathcal{H}^E$ isomorphic to $\mathcal{H}^Q \otimes \mathcal{H}^Q$, or loosely speaking an environment at most twice as big as the principal system Q. This provides the advantage of a model for the environment that is matched to the size of the principal system Q itself, and that is enough to simulate, in an indistinguishable manner, any larger environment E that may interact with Q. For instance, when the principal system Q is a single qubit, the environment model does not need to encompass more than two qubits. This is enough to properly simulate the effect on the qubit Q of any conceivable environment, be it again a thermal bath with a quasi-infinite number of degrees of freedom.

3.2.9 Processes with Different Input and Output Spaces

We have defined a quantum process or quantum channel by a map $\rho_Q \mapsto \mathcal{E}(\rho_Q)$ where the input density operator $\rho_Q \in \mathcal{L}(\mathcal{H}^Q)$ and output density operator $\mathcal{E}(\rho_Q) \in \mathcal{L}(\mathcal{H}^Q)$ belong to the same Hilbert space. This situation is often quite adequate, but it can nevertheless be extended to the case where the input and output spaces are distinct. This could be motivated, for instance, with the example of the quantum erasure channel [12, 25]. A quantum erasure channel can represent a noisy optical fiber transmitting photons. At the input, the (polarization) state of a photon belongs to a two-dimensional Hilbert space $\mathcal{H}_2$ with the two basis states $\{|0\rangle, |1\rangle\}$. The photon can suffer alteration of its state while traveling across the noisy fiber. But it may also occur that the photon gets absorbed in the fiber, so that at the output an additional state may be required to accommodate this eventuality of an absorbed photon. This can be achieved with a three-dimensional output space $\mathcal{H}_3 = \mathcal{H}_2 \oplus \{|\times\rangle\}$ spanned by the three basis states $\{|0\rangle, |1\rangle, |\times\rangle\}$, with $|\times\rangle$ a quantum state orthogonal to $|0\rangle$ and $|1\rangle$ devised to represent an absorbed photon.

A second example is provided by partial tracing of a density operator. Consider a bipartite quantum system AB with joint density operator $\rho_{AB} \in \mathcal{L}(\mathcal{H}^A \otimes \mathcal{H}^B)$. The partial tracing operation $\rho_A = \mathrm{tr}_B(\rho_{AB}) = \mathcal{E}(\rho_{AB}) \in \mathcal{L}(\mathcal{H}^A)$ defines a completely positive trace-preserving linear map $\mathcal{E}(\cdot)$ from $\mathcal{L}(\mathcal{H}^A \otimes \mathcal{H}^B)$ onto $\mathcal{L}(\mathcal{H}^A)$, i.e., a quantum process. This quantum process can be represented by the Kraus operators Λ_b which are operators from $\mathcal{H}^A \otimes \mathcal{H}^B$ onto $\mathcal{H}^A$, acting as

$$\rho_A = \mathcal{E}(\rho_{AB}) = \mathrm{tr}_B(\rho_{AB}) = \sum_{b=1}^{\dim(\mathcal{H}^B)} \Lambda_b \rho_{AB} \Lambda_b^\dagger , \tag{3.127}$$

and which can be characterized as follows. Let $\{|a\rangle\}$ with $a = 1$ to $\dim(\mathcal{H}^A)$ an orthonormal basis of $\mathcal{H}^A$, and $\{|b\rangle\}$ with $b = 1$ to $\dim(\mathcal{H}^B)$ an orthonormal basis of $\mathcal{H}^B$. We have $\Lambda_b = \mathrm{I}^A \otimes \langle b|$, with I^A the identity operator on $\mathcal{H}^A$, or equivalently via its action on the basis states of $\mathcal{H}^A \otimes \mathcal{H}^B$, we have $\Lambda_b(|a\rangle \otimes |b'\rangle) = \delta_{bb'} |a\rangle$. It is also easily verified that $\sum_b \Lambda_b^\dagger \Lambda_b = \mathrm{I}^A \otimes \mathrm{I}^B$ as required by the completeness relation of Eq. (3.47). So Eq. (3.127) provides an example of a quantum process or channel, in the Kraus representation, with an input space of dimension $\dim(\mathcal{H}^A)\dim(\mathcal{H}^B)$ and an output space of dimension $\dim(\mathcal{H}^A)$.

A third example is the measurement channel to come in Eq. (4.40), with an N-dimensional input space $\mathcal{H}_N$ and an M-dimensional output space $\mathcal{H}_M$.

The considerations of this Sect. 3.2 extend naturally to this situation of different input and output spaces. For an input space $\mathcal{H}^Q$ and an output space $\mathcal{H}^{Q'}$, a quantum process as in Eq. (3.45) will now define a map $\rho_Q \mapsto \mathcal{E}(\rho_Q)$ from $\mathcal{L}(\mathcal{H}^Q)$ onto $\mathcal{L}(\mathcal{H}^{Q'})$. Similar results carry over that such a map from $\mathcal{L}(\mathcal{H}^Q)$ onto $\mathcal{L}(\mathcal{H}^{Q'})$ is a linear completely positive trace-preserving map if and only if it can be put under an operator-sum representation as in Eq. (3.45). The Kraus operators Λ_k are now operators from $\mathcal{H}^Q$ onto $\mathcal{H}^{Q'}$, with a $\dim(\mathcal{H}^{Q'}) \times \dim(\mathcal{H}^Q)$ representation matrix, and satisfying the same completeness relation as in Eq. (3.47) with I^Q again the identity operator on the input space $\mathcal{H}^Q$. And the number of Kraus operators need not be larger than $\dim(\mathcal{H}^Q)\dim(\mathcal{H}^{Q'})$ to represent all feasible maps [6, 12].

The set of quantum processes $\rho_Q \mapsto \mathcal{E}(\rho_Q)$ from $\mathcal{L}(\mathcal{H}^Q)$ onto $\mathcal{L}(\mathcal{H}^{Q'})$ forms a **convex set**, in the sense that when $\mathcal{E}_1(\cdot)$ and $\mathcal{E}_2(\cdot)$ are two quantum processes from $\mathcal{L}(\mathcal{H}^Q)$ onto $\mathcal{L}(\mathcal{H}^{Q'})$, then the convex combination realized as $\mathcal{E}(\rho_Q) = p\mathcal{E}_1(\rho_Q) + (1-p)\mathcal{E}_2(\rho_Q)$, with $p \in [0, 1]$, represents a valid quantum process from $\mathcal{L}(\mathcal{H}^Q)$ onto $\mathcal{L}(\mathcal{H}^{Q'})$.

3.2.10 Choi–Jamiołkowski state

There exists a possibility of representing a quantum process $\mathcal{E}(\cdot)$ as introduced earlier in Sect. 3.2, by means of a quantum state. This is achieved by the Choi–Jamiołkowski representation, establishing what is identified also as a duality or isomorphism between quantum processes and quantum states [12, 27, 34, 35]. To introduce the Choi–Jamiołkowski representation, we can consider a quantum system A endowed with an input density operator $\rho_A \in \mathcal{L}(\mathcal{H}^A)$ experiencing the quantum process $\rho_A \mapsto \mathcal{E}_{A\to B}(\rho_A) = \rho_B$ delivering an output density operator $\rho_B \in \mathcal{L}(\mathcal{H}^B)$, possibly with two different input space $\mathcal{H}^A$ with dimension N_A and output space $\mathcal{H}^B$ with dimension N_B as in Sect. 3.2.9. An auxiliary reference quantum system R with a Hilbert space $\mathcal{H}^R$ isomorphic to the space $\mathcal{H}^A$ is introduced, with dimension $\dim(\mathcal{H}^R) = \dim(\mathcal{H}^A) = N_A$. A maximally entangled pure state belonging to the bipartite space $\mathcal{H}^R \otimes \mathcal{H}^A$ is considered as

$$|RA\rangle = \frac{1}{\sqrt{N_A}} \sum_{j=1}^{N_A} |j_R\rangle \otimes |j_A\rangle \in \mathcal{H}^R \otimes \mathcal{H}^A , \tag{3.128}$$

with $\{|j_R\rangle\}$ and $\{|j_A\rangle\}$, for $j = 1$ to N_A, forming two orthonormal bases for $\mathcal{H}^R$ and $\mathcal{H}^A$, respectively. The corresponding density operator is

$$\rho_{RA} = |RA\rangle\langle RA| = \frac{1}{N_A} \sum_{j=1}^{N_A} \sum_{k=1}^{N_A} |j_R\rangle\langle k_R| \otimes |j_A\rangle\langle k_A| \in \mathcal{L}(\mathcal{H}^R \otimes \mathcal{H}^A) . \tag{3.129}$$

In the bipartite system RA, nothing is done on the reference system R, while the quantum process $\mathcal{E}_{A\to B}(\cdot)$ is applied on system A, achieving the transformation:

$$\boxed{\mathcal{I}_R \otimes \mathcal{E}_{A\to B}(\rho_{RA}) = \frac{1}{N_A} \sum_{j=1}^{N_A} \sum_{k=1}^{N_A} |j_R\rangle\langle k_R| \otimes \mathcal{E}_{A\to B}\big(|j_A\rangle\langle k_A|\big) = \rho_{RB} \in \mathcal{L}(\mathcal{H}^R \otimes \mathcal{H}^B) ,} \tag{3.130}$$

with $\mathcal{I}_R$ the identity superoperator on $\mathcal{L}(\mathcal{H}^R)$.

The resulting quantum state $\rho_{RB} \in \mathcal{L}(\mathcal{H}^R \otimes \mathcal{H}^B)$ of Eq. (3.130) defines the **Choi–Jamiołkowski state** associated with the quantum process $\mathcal{E}_{A\to B}(\cdot)$. The Choi–Jamiołkowski state ρ_{RB} is a density operator that completely characterizes the quantum process $\mathcal{E}_{A\to B}(\cdot)$. From the constitution of ρ_{RB} in Eq. (3.130), we observe that ρ_{RB} contains all the N_A^2 images $\mathcal{E}_{A\to B}\big(|j_A\rangle\langle k_A|\big)$ of the N_A^2 basis states $\{|j_A\rangle\langle k_A|\}$ of $\mathcal{L}(\mathcal{H}^A)$, which together completely characterizes $\mathcal{E}_{A\to B}(\cdot)$. Each of these N_A^2 images $\mathcal{E}_{A\to B}\big(|j_A\rangle\langle k_A|\big)$ remains separately identified in ρ_{RB}, since each is separately addressed by means of a set of N_A^2 orthogonal addresses $|j_R\rangle\langle k_R|$ over $\mathcal{L}(\mathcal{H}^R)$ isomorphic to $\mathcal{L}(\mathcal{H}^A)$.

The Choi–Jamiołkowski state is a useful tool for which interesting properties can be proved. A linear map from $\mathcal{L}(\mathcal{H}^A)$ onto $\mathcal{L}(\mathcal{H}^B)$ is **completely positive** if and only if its Choi–Jamiołkowski state is a positive operator [12, 35], which is always the case for any valid quantum process $\mathcal{E}_{A\to B}(\cdot)$. In this way, establishing complete positivity does not require to consider a reference system R of arbitrary dimension and arbitrary entanglement with the principal system A (as might be suggested by its definition in the paragraph just after Eq. (3.37)), but it can be established with a reference system R having the dimension of A and maximum entanglement with A as in Eq. (3.128). Also, a quantum process $\mathcal{E}_{A\to B}(\cdot)$ is said to be **entanglement breaking** if and only if its Choi–Jamiołkowski state ρ_{RB} is separable [36] between B and R. This guarantees that any prior entanglement of A with any reference system R of arbitrary dimension gets separated at the output of an entanglement-breaking process or channel acting on A.

To recover the quantum process $\mathcal{E}_{A\to B}(\cdot)$ from its Choi–Jamiołkowski state ρ_{RB}, for any pure state $|A\rangle = \sum_{j=1}^{N_A} a_j\,|j_A\rangle$ of $\mathcal{H}^A$ we associate the pure conjugate state $|R_A\rangle = \sum_{j=1}^{N_A} a_j^*\,|j_R\rangle$ in $\mathcal{H}^R$, so as to obtain

$$\langle R_A|\rho_{RB}|R_A\rangle = \frac{1}{N_A}\sum_{j',k'}^{N_A}\sum_{j,k}^{N_A} a_{j'}a_{k'}^*\,\langle j_R'|j_R\rangle\,\langle k_R|k_R'\rangle \otimes \mathcal{E}_{A\to B}\big(|j_A\rangle\,\langle k_A|\big) \tag{3.131}$$

$$= \frac{1}{N_A}\sum_{j,k}^{N_A} a_j a_k^* \mathcal{E}_{A\to B}\big(|j_A\rangle\,\langle k_A|\big) \tag{3.132}$$

$$= \frac{1}{N_A}\mathcal{E}_{A\to B}\big(|A\rangle\,\langle A|\big)\ . \tag{3.133}$$

Furthermore, for any two basis states $|\ell_R\rangle$ and $|m_R\rangle$ of $\mathcal{H}^R$, we have $\langle \ell_R|\rho_{RB}|m_R\rangle = \mathcal{E}_{A\to B}\big(|\ell_A\rangle\,\langle m_A|\big)/N_A$ and $\langle \ell_R|\rho_{RB}|m_R\rangle$ is also $\langle \ell_R|\rho_{RB}|m_R\rangle = \mathrm{tr}_R\big[\big(|m_R\rangle\,\langle \ell_R| \otimes \mathrm{I}^B\big)\rho_{RB}\big]$, with I^B the identity operator on $\mathcal{H}^B$. This leads to

$$\mathcal{E}_{A\to B}\big(\rho_A\big) = N_A\,\mathrm{tr}_R\big(\big(\rho_A^\top \otimes \mathrm{I}^B\big)\rho_{RB}\big)\ , \tag{3.134}$$

for any pure or mixed state $\rho_A \in \mathcal{L}\big(\mathcal{H}^A\big)$, providing the targeted recovery formula of $\mathcal{E}_{A\to B}(\cdot)$ from ρ_{RB}, involving the transposed density operator $\rho_A^\top$.

The Choi–Jamiołkowski density operator ρ_{RB} is a positive operator whose rank K, between 1 and $\dim(\mathcal{H}^A)\dim(\mathcal{H}^B) = N_A N_B$, defines the **Choi rank** of the quantum process $\mathcal{E}_{A\to B}(\cdot)$ and fixes the minimal number of Kraus operators required for an operator-sum representation of $\mathcal{E}_{A\to B}(\cdot)$. We introduce its eigendecomposition $\rho_{RB} = \sum_{k=1}^{K} \lambda_k\,|\lambda_k\rangle\,\langle\lambda_k|$, so as to define the map acting on any pure state $|A\rangle$ of $\mathcal{H}^A$ via the partial inner product:

$$\Lambda_k\,|A\rangle = \sqrt{N_A\lambda_k}\,\langle R_A|\lambda_k\rangle\ , \tag{3.135}$$

defining a linear map Λ_k form $\mathcal{H}^A$ onto $\mathcal{H}^B$, for any integer $k = 1$ to K. Next, we have

$$\sum_{k=1}^{K} \Lambda_k |A\rangle \langle A| \Lambda_k^\dagger = N_A \sum_{k=1}^{K} \lambda_k \langle R_A|\lambda_k\rangle \langle \lambda_k|R_A\rangle \tag{3.136}$$

$$= N_A \langle R_A|\rho_{RB}|R_A\rangle \tag{3.137}$$

$$= \mathcal{E}_{A\to B}\big(|A\rangle \langle A|\big) , \tag{3.138}$$

verified for any pure state $|A\rangle \langle A| \in \mathcal{L}(\mathcal{H}^A)$. By the property of convexity-linearity already invoked several times in this chapter, for any pure or mixed state $\rho_A \in \mathcal{L}(\mathcal{H}^A)$, we therefore have

$$\sum_{k=1}^{K} \Lambda_k \rho_A \Lambda_k^\dagger = \mathcal{E}_{A\to B}(\rho_A) . \tag{3.139}$$

Since Eq. (3.139) is true for any state $\rho_A \in \mathcal{L}(\mathcal{H}^A)$ and $\mathcal{E}_{A\to B}(\cdot)$ preserves the unit trace, the K operators Λ_k satisfy the completeness relation as in Eqs. (3.46) and (3.47). In this way, the set of K operators $\{\Lambda_k\}_{k=1}^K$ defined by Eq. (3.135) via the eigendecomposition of the Choi–Jamiołkowski operator ρ_{RB} of Eq. (3.130) constitutes a set of Kraus operators for the quantum process $\mathcal{E}_{A\to B}(\cdot)$. It comes with the minimal size K defined by the rank of the Choi–Jamiołkowski operator ρ_{RB}. Moreover, since the eigenstates $|\lambda_k\rangle$ are mutually orthogonal in $\mathcal{H}^R \otimes \mathcal{H}^B$, from Eq. (3.135) acting on a basis state $|j_A\rangle$ of $\mathcal{H}^A$ in place of $|A\rangle$, the Hilbert–Schmidt inner product $\mathrm{tr}(\Lambda_k^\dagger \Lambda_\ell)$ is found to satisfy $\mathrm{tr}(\Lambda_k^\dagger \Lambda_\ell) = N_A \lambda_k \delta_{k\ell}$, so that the minimal set $\{\Lambda_k\}_{k=1}^K$ forms a set of K mutually orthogonal Kraus operators from $\mathcal{H}^A$ onto $\mathcal{H}^B$. The cardinal K of this minimal Kraus set cannot exceed the dimension $N_A N_B$ of the space of operators from $\mathcal{H}^A$ onto $\mathcal{H}^B$. This minimal Kraus set $\{\Lambda_k\}_{k=1}^K$ can be unitarily transformed into another Kraus set of the same size K as realized with Eq. (3.80), or it can be isometrically enlarged to a larger size as realized with Eq. (3.72).

Any alternative decomposition of ρ_{RB} under a form $\rho_{RB} = \sum_k \lambda_k |\lambda_k\rangle \langle \lambda_k|$, yet not involving the eigenstates of ρ_{RB} but a necessarily larger set of states $\{|\lambda_k\rangle\}$, would induce via Eq. (3.135) another (larger) set of (nonorthogonal) Kraus operators $\{\Lambda_k\}$ for the quantum process $\mathcal{E}_{A\to B}(\cdot)$. Finally, another orthonormal basis $\{|j_A\rangle\}$ for $\mathcal{H}^A$ would induce in Eq. (3.130) a distinct Choi–Jamiołkowski state ρ_{RB}, yet having the same rank K and same essential properties. Its eigendecomposition in Eq. (3.135) would induce yet another equivalent orthogonal Kraus set $\{\Lambda_k\}_{k=1}^K$ for $\mathcal{E}_{A\to B}(\cdot)$, having the same minimal size K and unitarily related to the former minimal Kraus set.

3.2.11 Non-Unitary Qubit Process

For the qubit, the general form of the quantum process of Eq. (3.45) can be used to obtain a characterization of the non-unitary evolution in Bloch representation. For the qubit, each Kraus operator Λ_k in Eq. (3.45) is an operator of $\mathcal{L}(\mathcal{H}_2)$ which can be expressed in the Pauli basis of $\mathcal{L}(\mathcal{H}_2)$ as

$$\Lambda_k = b_k \mathrm{I}_2 + \vec{a}_k \cdot \vec{\sigma} \ , \tag{3.140}$$

with $(b_k, \vec{a}_k)$ a set of four complex scalar parameters. We then have $\Lambda_k^\dagger = b_k^* \mathrm{I}_2 + \vec{a}_k^* \cdot \vec{\sigma}$, and with the identity of Eq. (A.45) in the Appendix,

$$\Lambda_k^\dagger \Lambda_k = \left(b_k^* b_k + \vec{a}_k^* \vec{a}_k\right)\mathrm{I}_2 + \left(b_k^* \vec{a}_k + b_k \vec{a}_k^* + i\vec{a}_k^* \times \vec{a}_k\right) \cdot \vec{\sigma} \ . \tag{3.141}$$

The Kraus operators Λ_k satisfying the completeness relation of Eq. (3.47) therefore induce the relations:

$$1 = \sum_{k=1}^{K} \left(b_k^* b_k + \vec{a}_k^* \vec{a}_k\right) , \tag{3.142}$$

$$\vec{0} = \sum_{k=1}^{K} \left(b_k^* \vec{a}_k + b_k \vec{a}_k^* + i\vec{a}_k^* \times \vec{a}_k\right) . \tag{3.143}$$

From a qubit with density operator $\rho = \left(\mathrm{I}_2 + \vec{r} \cdot \vec{\sigma}\,\right)/2$ in Bloch representation as in Eq. (2.61), again with the identity of Eq. (A.45), we obtain

$$\begin{aligned} \Lambda_k \rho \Lambda_k^\dagger = &\left[\left(b_k^* b_k + \vec{a}_k^* \vec{a}_k\right) + \left(b_k^* \vec{a}_k + b_k \vec{a}_k^* + i\vec{a}_k^* \times \vec{a}_k\right)\vec{r}\,\right] \frac{\mathrm{I}_2}{2} \\ &+ \Big[b_k^* \vec{a}_k + b_k \vec{a}_k^* + i\vec{a}_k \times \vec{a}_k^* + \left(b_k b_k^* - \vec{a}_k \vec{a}_k^*\right)\vec{r} \\ &+ \left(\vec{a}_k \vec{r}\,\right)\vec{a}_k^* + \vec{a}_k\left(\vec{r}\,\vec{a}_k^*\right) + i\left(b_k^* \vec{a}_k - b_k \vec{a}_k^*\right) \times \vec{r}\,\Big] \cdot \frac{\vec{\sigma}}{2} \ . \end{aligned} \tag{3.144}$$

Then, by performing the sum over k prescribed in the operator-sum representation of Eq. (3.45), and exploiting the relations of Eqs. (3.142) and (3.143), we arrive at the transformed state in Bloch representation:

$$\rho' = \mathcal{E}(\rho) = \sum_{k=1}^{K} \Lambda_k \rho \Lambda_k^\dagger = \frac{1}{2}\left(\mathrm{I}_2 + \vec{r}\,' \cdot \vec{\sigma}\right) . \tag{3.145}$$

The transformed Bloch vector $\vec{r}\,'$ is obtained by the affine map

$$\boxed{\vec{r}\,' = A\vec{r} + \vec{c}\ ,} \tag{3.146}$$

deduced from Eq. (3.144) involved in the sum over k. In particular, this gives the constant vector

$$\vec{c} = i\,2 \sum_{k=1}^{K} \vec{a}_k \times \vec{a}_k^{\,*}\,, \tag{3.147}$$

which always satisfies $\vec{c}^{\,*} = \vec{c}$ so that $\vec{c}$ is a real vector of $\mathbb{R}^3$. In Eq. (3.146) also A is the 3×3 matrix defined by

$$A\vec{r} = \sum_{k=1}^{K} \Big[\big(b_k b_k^* - \vec{a}_k\vec{a}_k^{\,*}\big)\vec{r} + \big(\vec{a}_k\vec{r}\,\big)\vec{a}_k^{\,*} + \vec{a}_k\big(\vec{r}\,\vec{a}_k^{\,*}\big) + i\big(b_k^*\vec{a}_k - b_k\vec{a}_k^{\,*}\big) \times \vec{r}\, \Big]\,. \tag{3.148}$$

This is indeed a linear function of $\vec{r}$, which moreover satisfies $A\vec{r} = \big(A\vec{r}\,\big)^* = A^*\vec{r}$ for any $\vec{r} \in \mathbb{R}^3$, establishing A as a real 3×3 matrix.

The affine map of Eq. (3.146) represents a mapping of the Bloch ball into itself. It maps an input Bloch vector $\vec{r}$ of the Bloch ball, into an output Bloch vector $\vec{r}\,'$ of the Bloch ball, specifying the Bloch representation of the qubit state resulting from the quantum process $\mathcal{E}(\cdot)$ of Eq. (3.145).

When the vector $\vec{c} = \vec{0}$ in Eq. (3.146), then the Bloch vector $\vec{r} = \vec{0}$ (characterizing the fully depolarized qubit density operator $\rho = \mathrm{I}_2/2$) is mapped into $\vec{r}\,' = \vec{0}$, so that the quantum process in Eq. (3.145) satisfies $\mathcal{E}(\mathrm{I}_2) = \mathrm{I}_2$ having the density operator $\mathrm{I}_2/2$ as a fixed point and is therefore unital. In particular, when all K vectors $\vec{a}_k$ are real, we obtain $\vec{c} = \vec{0}$ in Eq. (3.147).

When a set of Kraus operators Λ_k is given in Eq. (3.145) for the qubit, this determines a valid affine map in Eq. (3.146). Conversely, when an affine map as in Eq. (3.146) mapping the Bloch ball into itself is given, it is usually possible to find an associated set of Kraus operators Λ_k for Eq. (3.145). These Kraus operators can be parameterized as in Eq. (3.140). Matching the 9 entries of A fixes 9 scalar constraints on these parameters; there are 3 more scalar constraints to match the given $\vec{c}$, and Eqs. (3.142) and (3.143) add 4 scalar constraints. This total of 16 scalar constraints can usually be matched with at least four 2×2 Kraus operators of $\mathcal{L}(\mathcal{H}_2)$. But a larger number of Kraus operators could as well be used for implementing the same quantum process $\mathcal{E}(\cdot)$ and matching the 16 constraints, which could then be, in principle, reduced to four qubit operators, following the considerations explained with Eq. (3.72) and in Sect. 3.2.10.

A geometric interpretation exists for the transformation of the Bloch ball in $\mathbb{R}^3$ by the affine map of Eq. (3.146). By the (left) polar decomposition, the 3×3 real matrix A in Eq. (3.146) can be written as $A = UG$, where U is a 3×3 real unitary or orthogonal matrix ($U^{-1} = U^\top$), and G a 3×3 real positive symmetric matrix ($G = G^\top$). The matrix G always has, associated with three real positive eigenvalues $\{g_1, g_2, g_3\}$, three eigenvectors $\{\vec{g}_1, \vec{g}_2, \vec{g}_3\}$ forming an orthonormal basis of $\mathbb{R}^3$. The matrix G is uniquely determined as the positive square root $G = \sqrt{A^\top A}$, and when A is invertible, so is G, and U is uniquely determined as $U = AG^{-1}$. The transformation by Eq. (3.146) of the Bloch ball in $\mathbb{R}^3$ is thus a deformation (necessarily a contraction) by G along the three orthogonal axes $\{\vec{g}_1, \vec{g}_2, \vec{g}_3\}$ followed by a real unitary transformation U and a translation by $\vec{c}$. The Bloch ball is thus, in general, transformed by Eq. (3.146) into a tilted ellipsoid, centered at $\vec{c}$, and fitted inside the (initial) Bloch ball. An equivalent alternative would follow from the right polar decomposition under the form $A = GU$.

By the quantum process of Eq. (3.146), a pure state with unit-norm Bloch vector $\|\vec{r}\| = 1$ usually experiences a contraction to the reduced norm $\|\vec{r}\,'\| \leq 1$ to become a mixed state, which is a characteristic trait of quantum decoherence to turn a pure state into a mixed state.

3.2.12 Beyond the Quantum-Operation Formalism

An important point in this Sect. 3.2 is that for the modeling of an open quantum system Q as a part of a larger closed system, we assumed, as soon as the joint evolution of Eq. (3.34), that the environment E starts in a pure state $|e_0\rangle$ independent and separable from the initial state of the principal system Q. This is an essential assumption to obtain a reduced evolution for Q via a quantum operation $\rho_Q \mapsto \mathcal{E}(\rho_Q)$ acting on the state of Q alone and representing our notion of quantum process $\mathcal{E}(\cdot)$ of Fig. 3.6. If initial correlation or entanglement exists between Q and its environment E, their subsequent joint evolution may not allow a reduction to a quantum operation for Q alone, but Q instead could remain influenced by details of the environment and their dynamics. This would be so unless the delimitation of Q and of its surrounding could be extended, possibly going back before the interaction started, so as to recover an initial separation of Q and its environment E and restore the possibility of a quantum operation $\rho_Q \mapsto \mathcal{E}(\rho_Q)$ for Q alone. The assumption of initial separation of the quantum states of Q and E is, however, quite reasonable in many practical situations of interest. The separation is commonly enforced by applications in quantum information and computation that usually start with a deliberate and controlled preparation of Q for some useful purpose. Also, large environments with many degrees of freedom often tend to rapidly dissipate putative prior correlations and restore independent states for Q and E. Other assumptions may be implied by the quantum-operation formalism, such as some stationarity in the environment E to enable a fixed quantum operation $\mathcal{E}(\cdot)$ attached to some definite conditions. More elements for the modeling of open quantum dynamics beyond the quantum-operation formalism can be found, for instance, in [37–41] and in the monographs [42, 43]. In any case, the quantum-operation formalism offers a systematic and powerful approach, quite often relevant and of great value to gain insight on the effect of quantum noise and decoherence in open quantum systems. Accordingly, this formalism will be central to us in the sequel to develop our perspective of quantum signal processing in the presence of noise.

Another point is that in the modeling of the evolution of an open quantum system by means of a quantum operation $\rho_Q \mapsto \mathcal{E}(\rho_Q)$, the temporal dynamics remains unresolved. The evolution of the principal system Q is described by the quantum process of Fig. 3.6, implementing the quantum operation $\rho_Q \mapsto \mathcal{E}(\rho_Q)$, seen as a functional block transforming the input state ρ_Q into the output state $\mathcal{E}(\rho_Q)$, without explicitly specifying the time course of the input–output evolution. The quantum operation $\mathcal{E}(\cdot)$ encapsulates the end-to-end transformation over time, from the time when the input state ρ_Q is applied, to the time when the output state $\mathcal{E}(\rho_Q)$ becomes accessible for subsequent use. Such an input–output transformation $\mathcal{E}(\cdot)$ as a functional block, displaying a signals-and-systems theoretic orientation, is well suited for many applications in quantum information processing. A time-resolved modeling of the open evolution of a quantum state ρ_Q, in continuous time, can nevertheless be envisaged, and this will be examined in more detail in Chap. 9 dealing with the modeling of quantum noise, and especially its Sect. 9.14. A significant approach thereof is by means of a temporal differential equation involving $\rho_Q(t)$ at each time t, to replace the end-to-end quantum operation $\rho_Q \mapsto \mathcal{E}(\rho_Q)$. This differential equation governing the state $\rho_Q(t)$ of an open quantum system at each time t is known as the **Lindblad equation** [44, 45]. It somehow extends to an open quantum system the Schrödinger equation (3.3) applying to a closed quantum system. The Lindblad differential equation will be presented in more detail in Sect. 9.14.2. The quantum-operation formalism $\rho_Q \mapsto \mathcal{E}(\rho_Q)$ will however be preferred throughout the book for our signals-and-systems theoretic perspective.

3.3 Summary on Quantum Process

This chapter focused on the **second postulate** (among three) or fundamental principle for quantum theory, essentially stating that the state of a closed or isolated quantum system evolves through a unitary transformation. Several important consequences of this second postulate were examined, especially for open quantum systems, treated as a part from a larger closed system involving coupling with an uncontrolled environment.

The evolution of a quantum system of interest Q can be closed or open, whether Q is free or not from interaction with its environment E. In both situations, the evolution of Q can be described by a transformation of its density operator $\rho_Q \mapsto \mathcal{E}(\rho_Q)$, taking the form a completely positive trace-preserving linear superoperator or quantum operation $\mathcal{E}(\cdot)$, and representing our notion of **quantum process** to transform quantum signals, as depicted in Fig. 3.6.

For an isolated or **closed quantum system** Q the superoperator $\mathcal{E}(\cdot)$ reduces to $\mathcal{E}(\rho_Q) = \mathsf{U}\rho_Q\mathsf{U}^\dagger$ as in Eq. (3.2), with U a unitary operator, implementing a reversible evolution of Q. For any finite-dimensional system Q, any unitary evolution by U can always be physically materialized by means of a **universal set of quantum gates** (such as the one-qubit gates supplemented by the two-qubit Cnot gate) as in Sect. 3.1.5, acting on a sufficient number of qubits, to form a ground for quantum computation.

For an **open quantum system** Q coupled to an uncontrolled environment E, the superoperator $\mathcal{E}(\cdot)$ can be obtained through a (Stinespring) dilated unitary representation of the evolution of the system–environment compound QE, terminated by partial tracing over the environment E, according to Eq. (3.37). Alternatively, the superoperator $\mathcal{E}(\cdot)$ can be obtained through an operator-sum or Kraus representation, according to Eq. (3.45).

The operator-sum representation $\rho_Q \mapsto \mathcal{E}(\rho_Q) = \sum_k \Lambda_k \rho_Q \Lambda_k^\dagger$ of Eq. (3.45) is specially appealing as it gives the possibility to describe the evolution of the density operator $\mathcal{L}(\mathcal{H}^Q) \ni \rho_Q \mapsto \mathcal{E}(\rho_Q) \in \mathcal{L}(\mathcal{H}^{Q'})$ with **Kraus operators** Λ_k from $\mathcal{H}^Q$ onto $\mathcal{H}^{Q'}$ in number K no larger than the Choi rank of $\mathcal{E}(\cdot)$, which is itself no larger than $\dim(\mathcal{H}^Q)\dim(\mathcal{H}^{Q'})$. In addition, from a given set of K Kraus operators for a quantum process $\mathcal{E}(\cdot)$, an environment model of the same dimension K can be designed, following the procedure of Sect. 3.2.8, to obtain a dilated unitary evolution simulating $\mathcal{E}(\cdot)$. Via this dilated unitary evolution of Eq. (3.121), any quantum process $\mathcal{E}(\cdot)$ can always be implemented or simulated by means of unitary qubit gates.

From a perspective of signals- and-systems theory, when the density operator ρ_Q assumes the role of a quantum signal, the quantum process $\rho_Q \mapsto \mathcal{E}(\rho_Q)$ naturally assumes the role of the operative system or functional block acting to transform the signal, this enabling to transfer insight from signals-and-systems theory to the domain of quantum information.

In the important situation where Q is a **qubit**, a quantum process $\mathcal{E}(\cdot)$ generally takes the form of an affine transformation $\vec{r} \mapsto A\vec{r} + \vec{c}$ of the qubit Bloch vector $\vec{r}$ in $\mathbb{R}^3$, as given in Eq. (3.146), generically contracting the Bloch ball into an internal ellipsoid. Alternatively, an environment model with no more than two qubits is enough to obtain a dilated unitary evolution implementing or simulating any quantum process $\mathcal{E}(\cdot)$ affecting the qubit Q. In the special circumstance of a closed evolution, the qubit Q undergoes a unitary transformation, which can be generically parameterized as U_ξ from Eq. (3.18), and by which the qubit Bloch vector $\vec{r}$ experiences a rotation in $\mathbb{R}^3$ with no contraction.

Chapter 4
Quantum Measurement

This chapter essentially deals with the third general postulate or fundamental principle for quantum theory, concerning quantum measurement. The chapter presents the notion of quantum measurement, in the first place with the standard von Neumann projective measurement, and then with also the generalized measurement, and finally with the slightly restrictive but yet very useful positive operator-valued (POVM) measure.

4.1 Standard Von Neumann Projective Measurement

4.1.1 Principle

As anticipated in Sect. 2.2, as a fundamental postulate or principle of quantum theory, a measurement for a quantum system with N-dimensional Hilbert space $\mathcal{H}_N$ refers to an arbitrary orthonormal basis $\{|n\rangle\}$ of $\mathcal{H}_N$ and takes place as a random projection of the pre-measurement state $|\psi\rangle$ onto one particular basis state $|n\rangle$. This projection occurs on state $|n\rangle$ with the probability $\Pr\{|n\rangle\} = P(n)$ given by Eq. (2.3) when the pre-measurement state is the pure state $|\psi\rangle \in \mathcal{H}_N$. More generally the probability $\Pr\{|n\rangle\} = P(n)$ is given by Eq. (2.36) when the pre-measurement state is the mixed state of density operator $\rho \in \mathcal{L}(\mathcal{H}_N)$, with Eq. (2.36) also applying for a pure state when $\rho = |\psi\rangle\langle\psi|$. In addition, the probability $\Pr\{|n\rangle\} = P(n)$ of Eq. (2.36) can also be written as

$$P(n) = \langle n|\rho|n\rangle = \mathrm{tr}(\rho\Pi_n) , \tag{4.1}$$

where $\Pi_n = |n\rangle\langle n|$ is the orthogonal projector onto the state $|n\rangle$ in $\mathcal{H}_N$.

By the completeness relation of Eq. (A.10) from the Appendix, the N projectors Π_n satisfy

F. Chapeau-Blondeau, *Quantum Information and Quantum Signal Processing*,
https://doi.org/10.1007/978-3-032-27499-1_4

$$\sum_{n=1}^{N} \Pi_n = \mathrm{I}_N \tag{4.2}$$

and are therefore N positive operators realizing a resolution or a decomposition of the identity operator I_N on $\mathcal{H}_N$, with Eq. (4.2) also ensuring that the probabilities $P(n)$ for $n = 1$ to N sum to 1 for any pre-measurement state ρ.

This represents, for a quantum system with Hilbert space $\mathcal{H}_N$, what is known as **a standard projective measurement, or von Neumann measurement**. It is defined, via an orthonormal basis $\{|n\rangle\}$ of $\mathcal{H}_N$, by a set of N orthogonal projectors $\Pi_n = |n\rangle \langle n|$ which together satisfy the completeness relation of Eq. (4.2), so as to implement a measurement with N possible outcomes or results ruled by the probabilities $P(n)$ of Eq. (4.1).

Such standard von Neumann measurement can also be presented in a slightly different (but equivalent) manner in order to prepare for the possibility of a generalized quantum measurement, and as follows.

For a quantum system with Hilbert space $\mathcal{H}_N$, a von Neumann measurement in the orthonormal basis $\{|n\rangle\}$ of $\mathcal{H}_N$ takes place as the random application of a definite projector $\Pi_n = |n\rangle \langle n|$ acting on the pre-measurement state $|\psi\rangle \in \mathcal{H}_N$. The measurement result n reveals that the projector Π_n (randomly) acted in the measurement to produce the transformed vector $\Pi_n |\psi\rangle$ which is an unnormalized vector of $\mathcal{H}_N$ with squared norm $\big\| \Pi_n |\psi\rangle \big\|^2 = \langle\psi| \Pi_n^\dagger \Pi_n |\psi\rangle$. This squared norm is nothing else than $|\alpha_n|^2$, the squared modulus of the complex coordinates α_n of the pre-measurement state $|\psi\rangle$ on the normalized basis vector $|n\rangle$, so that according to Eq. (2.3) it is also the probability $P(n) = \Pr\{|n\rangle\} = \big\| \Pi_n |\psi\rangle \big\|^2$ which is

$$P(n) = \langle\psi| \Pi_n^\dagger \Pi_n |\psi\rangle = \mathrm{tr}\big(|\psi\rangle \langle\psi| \Pi_n^\dagger \Pi_n\big) , \tag{4.3}$$

quantifying the probability that it is the projector Π_n that (randomly) acts in the measurement.

The requirement $\sum_{n=1}^{N} P(n) = 1$ that the probabilities of Eq. (4.3) sum to 1 for any state $|\psi\rangle$ entails that

$$1 = \langle\psi| \sum_{n=1}^{N} \Pi_n^\dagger \Pi_n |\psi\rangle = \mathrm{tr}\left(|\psi\rangle \langle\psi| \sum_{n=1}^{N} \Pi_n^\dagger \Pi_n \right) \tag{4.4}$$

should be verified for any $|\psi\rangle$, which in turn imposes the completeness or closure relation:

$$\sum_{n=1}^{N} \Pi_n^\dagger \Pi_n = \mathrm{I}_N . \tag{4.5}$$

The condition of Eq. (4.5) is granted for the N projectors $\Pi_n = |n\rangle \langle n|$ which satisfy $\Pi_n^\dagger = \Pi_n$ and $\Pi_n^\dagger \Pi_n = \Pi_n \Pi_n = \Pi_n$ as idempotent operators. In this situation, Eq. (4.5) is ensured by (and equivalent to) the completeness relation of Eq. (4.2).

In addition, the post-measurement state $|\phi_n^{\text{post}}\rangle$ is the normalized basis vector $|n\rangle = |\phi_n^{\text{post}}\rangle$, but as an equivalent alternative it is also the unnormalized transformed vector $\Pi_n |\psi\rangle$ after it has been normalized into the valid quantum state $|\phi_n^{\text{post}}\rangle = \Pi_n |\psi\rangle / \| \Pi_n |\psi\rangle \| = \Pi_n |\psi\rangle / \sqrt{P(n)} = |n\rangle$, or equivalently

$$|\phi_n^{\text{post}}\rangle = \frac{\Pi_n |\psi\rangle}{\sqrt{\langle\psi| \Pi_n^\dagger \Pi_n |\psi\rangle}} . \tag{4.6}$$

This (equivalent) manner of presenting the standard von Neumann measurement in $\mathcal{H}_N$ is interesting in that it specially focuses on N operators Π_n satisfying the completeness condition of Eq. (4.5) to define the measurement. The measurement result n identifies the operator Π_n that randomly acted. From a pre-measurement state $|\psi\rangle \in \mathcal{H}_N$, the unnormalized transformed vector $\Pi_n |\psi\rangle$ determines both the measurement probability $P(n)$ as its squared norm in Eq. (4.3) and the post-measurement state after its normalization in Eq. (4.6). It is also a form that better matches with the notion of generalized quantum measurement, to come shortly, in Sect. 4.2.

4.1.2 *Physical Implementation for a Qubit*

Two examples in Fig. 4.1 illustrate the von Neumann projective measurement of a qubit. For a qubit materialized by a photon with its two basis states of polarization, a standard von Neumann projective measurement can be implemented by a polarizing beam splitter, as shown in Fig. 4.1(a). The polarizing beam splitter deflects the incoming photon according to its polarization. Depending on its orientation, the polarizing beam splitter separates in two directions at right angle the two polarizations $|0\rangle$ and $|1\rangle$. This implements the random projection of the incoming photon either onto the basis state $|0\rangle$ or $|1\rangle$. The measurement result is revealed when the photon is detected by one of the two detectors after the beam splitter. Rotating the orientation of the polarizer around the incoming photon beam allows one to rotate the measurement basis $\{|0\rangle, |1\rangle\}$.

For a qubit materialized by a particle with two basis states of spin (electron, atom), a standard von Neumann projective measurement can be implemented by a Stern–Gerlach apparatus, as shown in Fig. 4.1(b). The Stern–Gerlach apparatus features a magnet to deflect the incoming particle according to its spin. The magnetic field separates upward or downward the two spin orientations $|\uparrow\rangle$ and $|\downarrow\rangle$. This implements the random projection of the incoming spin either onto the basis state $|\uparrow\rangle$ or $|\downarrow\rangle$. The measurement result is revealed by the spot where the deflected particle hits the screen. Rotating the orientation of the magnet around the incoming particle beam allows one to rotate the measurement basis $\{|\uparrow\rangle, |\downarrow\rangle\}$.

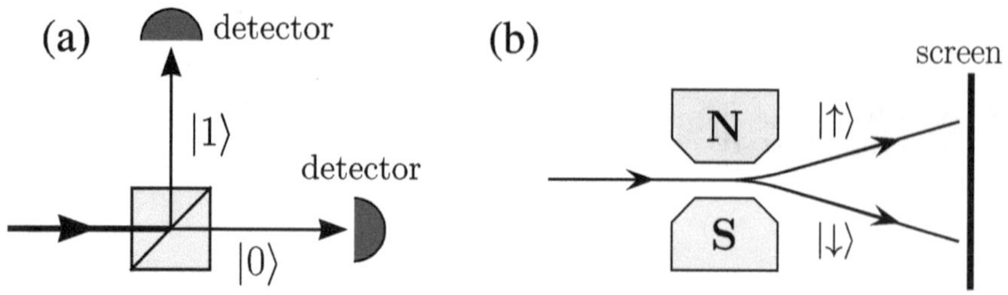

Fig. 4.1 Physical implementation of a two-state von Neumann projective measurement, by a polarizing beam splitter for a photonic qubit in (**a**), or by a Stern–Gerlach magnet apparatus for a spin qubit in (**b**)

4.1.3 Partial Measurement on a Bipartite System

We consider a quantum system A with the N_A-dimensional Hilbert space $\mathcal{H}^A$ and another quantum system B with the N_B-dimensional Hilbert space $\mathcal{H}^B$, both forming the composite bipartite quantum system AB with Hilbert space $\mathcal{H}^A \otimes \mathcal{H}^B$ of dimension $N_A N_B$.

A von Neumann quantum measurement of the bipartite system AB can be performed according to the fundamental postulate of Sect. 4.1.1. The bipartite space $\mathcal{H}^A \otimes \mathcal{H}^B$ is referred to an orthonormal basis $\{|m\rangle\}$ for $m = 1$ to $N_A N_B$, and the von Neumann measurement operates via the $N_A N_B$ rank-one projectors $\Pi_m = |m\rangle\langle m|$. Such a measurement will have a number $N_A N_B$ of possible outcomes. When performed on the pre-measurement state $|AB\rangle \in \mathcal{H}^A \otimes \mathcal{H}^B$ of the system AB, the outcome m is obtained with the probability $P(m) = \langle AB|\Pi_m|AB\rangle = \mathrm{tr}\big(\Pi_m |AB\rangle\langle AB|\big)$, while AB terminates in the post-measurement state $\Pi_m |AB\rangle / \sqrt{P(m)} = |m\rangle \in \mathcal{H}^A \otimes \mathcal{H}^B$. If the preparation of AB is in the mixed state ρ_{AB}, the probability of projecting onto $|m\rangle$ is $P(m) = \mathrm{tr}\big(\Pi_m \rho_{AB}\big)$.

Another possibility arises from the fact that in the bipartite system AB, the two subsystems A and B are in general physically materialized separately, like one or the other photon belonging to an entangled photon pair. One can then choose to measure the system A with a standard von Neumann measurement while leaving the system B untouched. This may be called a **partial measurement**, or an incomplete measurement, of the bipartite system AB. The von Neumann measurement of A is performed by referring $\mathcal{H}^A$ to an orthonormal basis $\{|a\rangle\}$ for $a = 1$ to N_A. This measurement can be described with the N_A projectors $\Pi_a = |a\rangle\langle a| \otimes \mathrm{I}^B$ acting on the bipartite system AB in state $|AB\rangle \in \mathcal{H}^A \otimes \mathcal{H}^B$, where I^B is the identity operator on the Hilbert space $\mathcal{H}^B$, for a system B untouched in the measurement. Such a measurement will have a number N_A of possible outcomes. When performed on the pre-measurement state $|AB\rangle$ of the system AB, the outcome where A gets projected onto $|a\rangle$ is obtained with the probability $P(a) = \langle AB|\Pi_a|AB\rangle$, while AB terminates in the post-measurement state $\Pi_a |AB\rangle / \sqrt{P(a)} \in \mathcal{H}^A \otimes \mathcal{H}^B$. The probability $P(a)$ is also

$$P(a) = \langle AB|\Pi_a|AB\rangle = \mathrm{tr}\big(\Pi_a\,|AB\rangle\,\langle AB|\big) = \mathrm{tr}\Big(\big(|a\rangle\,\langle a| \otimes \mathrm{I}^B\big)\rho_{AB}\Big)\;, \tag{4.7}$$

with $\rho_{AB} = |AB\rangle\,\langle AB|$. The same probability $P(a) = \mathrm{tr}\big(\Pi_a\rho_{AB}\big) = \mathrm{tr}\big((|a\rangle\,\langle a| \otimes \mathrm{I}^B)\rho_{AB}\big)$ applies when the preparation of AB is in the mixed state ρ_{AB}. For $\mathrm{tr}\big((|a\rangle\,\langle a| \otimes \mathrm{I}^B)\rho_{AB}\big)$, which is also $\mathrm{tr}_{AB}\big((|a\rangle\,\langle a| \otimes \mathrm{I}^B)\rho_{AB}\big) = \mathrm{tr}_A\,\mathrm{tr}_B\big((|a\rangle\,\langle a| \otimes \mathrm{I}^B)\rho_{AB}\big)$, we can use the general property of the partial trace given in Eq. (A.24) of the Appendix to obtain $\mathrm{tr}_B\big((|a\rangle\,\langle a| \otimes \mathrm{I}^B)\rho_{AB}\big) = |a\rangle\,\langle a|\,\mathrm{tr}_B(\rho_{AB})$. We are thus led to introduce the operator of $\mathcal{L}(\mathcal{H}^A)$ defined as

$$\rho_A = \mathrm{tr}_B(\rho_{AB})\;, \tag{4.8}$$

which is a positive operator of $\mathcal{L}(\mathcal{H}^A)$ since $\langle\psi_A|\rho_A|\psi_A\rangle = \mathrm{tr}_B\big(\langle\psi_A|\rho_{AB}|\psi_A\rangle\big) = \sum_{b=1}^{N_B}\langle b|\,\langle\psi_A|\rho_{AB}|\psi_A\rangle\,|b\rangle \geq 0$ for any $|\psi_A\rangle \in \mathcal{H}^A$, and ρ_A is with unit trace. The probability of Eq. (4.7) can now be obtained as

$$P(a) = \mathrm{tr}\big(|a\rangle\,\langle a|\,\rho_A\big) = \langle a|\rho_A|a\rangle\;. \tag{4.9}$$

In this way, ρ_A of Eq. (4.8) plays the role of a density operator of $\mathcal{L}(\mathcal{H}^A)$, which can be used alone in Eq. (4.9), to obtain the probabilities of the different outcomes upon measuring the system A in the basis $\big\{|a\rangle\big\}$ of $\mathcal{H}^A$ while B is left untouched.

The operator ρ_A of Eq. (4.8) is the **reduced density operator** for system A, obtained by tracing the bipartite density operator ρ_{AB} over the discarded system B, and ρ_A enables one to obtain the probabilities of the different outcomes upon measuring A alone. The same argument applies and the same reduced density operator ρ_A of Eq. (4.8) works equally when the bipartite system AB is prepared in a pure or mixed state ρ_{AB}. The reduced density operator is an important notion, already encountered in Sect. 3.2.1 with Eq. (3.35), it will be further motivated in Sect. 4.4, and it will be exploited on many other occasions in the sequel.

In addition here, for a more complete description of the von Neumann measurement on the bipartite system AB, we can refer the bipartite space $\mathcal{H}^A \otimes \mathcal{H}^B$ to the orthonormal basis $\{|a\rangle\}_{a=1}^{N_A} \otimes \{|b\rangle\}_{b=1}^{N_B}$, where $\{|a\rangle\}_{a=1}^{N_A}$ is the basis selected for measuring A while $\{|b\rangle\}_{b=1}^{N_B}$ is any orthonormal basis of $\mathcal{H}^B$. We have the pre-measurement state

$$|AB\rangle = \sum_{a=1}^{N_A}\sum_{b=1}^{N_B}\alpha_{ab}\,|a\rangle \otimes |b\rangle\;. \tag{4.10}$$

The measurement of A projects on a definite basis state $|a\rangle$ of $\mathcal{H}^A$, to produce for the bipartite system AB the post-measurement (unnormalized) state

$$\Pi_a\,|AB\rangle = \big(|a\rangle\,\langle a| \otimes \mathrm{I}^B\big)\,|AB\rangle = |a\rangle \otimes \sum_{b=1}^{N_B}\alpha_{ab}\,|b\rangle\;, \tag{4.11}$$

with the probability

$$P(a) = \langle AB|\Pi_a|AB\rangle = \sum_{b=1}^{N_B} |\alpha_{ab}|^2 . \tag{4.12}$$

Alternatively, from Eq. (4.10) we deduce for A the reduced density operator

$$\rho_A = \mathrm{tr}_B\big(|AB\rangle\langle AB|\big) = \sum_{a,a'=1}^{N_A} \sum_{b=1}^{N_B} \alpha_{ab}\alpha^*_{a'b}\,|a\rangle\langle a'| \ , \tag{4.13}$$

and then

$$\langle a|\rho_A|a\rangle = \sum_{b=1}^{N_B} |\alpha_{ab}|^2 = P(a) \ , \tag{4.14}$$

recovering the probability $P(a)$ of Eq. (4.12) as anticipated from Eq. (4.9), also providing further motivation to the notion of reduced density operator of Eq. (4.8).

The post-measurement state of Eq. (4.11) for AB is always a separable state $|a\rangle \otimes |B'\rangle$ of $\mathcal{H}^A \otimes \mathcal{H}^B$, with A terminating in the state $|a\rangle$ of $\mathcal{H}^A$ and B terminating in the (normalized) state $|B'\rangle = \sum_{b=1}^{N_B} \alpha_{ab}\,|b\rangle\,/\sqrt{P(a)}$ of $\mathcal{H}^B$.

We note that instead of doing nothing on system B with the identity operator I^B, we could as well implement a regular von Neumann measurement on B with the set of N_B orthogonal projectors $\{|b\rangle\langle b|\}_{b=1}^{N_B}$ decomposing the identity I^B of $\mathcal{H}^B$. This would correspond on the compound AB to a regular von Neumann measurement with the set of $N_A N_B$ projectors $\{|a\rangle\langle a|\}_{a=1}^{N_A} \otimes \{|b\rangle\langle b|\}_{b=1}^{N_B}$, and as a result each definite configuration $|a\rangle \otimes |b\rangle$ would be obtained with the probability $|\alpha_{ab}|^2$. The probability $P(a)$ of Eq. (4.12) then would correspond to the situation where the measurement result for B remains unknown, so that the probability of getting $|a\rangle$ for A with an indefinite (unknown) state for B would be the sum of Eq. (4.12) over all the configurations $|b\rangle$ of B associated with a definite $|a\rangle$ for A. So what we describe as a partial measurement of the compound AB with the N_A projectors $\big\{|a\rangle\langle a| \otimes \mathrm{I}^B\big\}$ could as well be described as a complete measurement of AB with the $N_A N_B$ (rank-one) projectors $\big\{|a\rangle\langle a| \otimes |b\rangle\langle b|\big\}$ yet with the measurement result for B remaining unknown or unobserved, a notion sometimes referred to as an **unread measurement** of B.

A pre-measurement state $|AB\rangle \in \mathcal{H}^A \otimes \mathcal{H}^B$ of the bipartite system separable as $|AB\rangle = |A\rangle \otimes |B\rangle$, with a state $|A\rangle \in \mathcal{H}^A$ and a state $|B\rangle \in \mathcal{H}^B$, indicates that each of the subsystems A and B individually starts in a well-defined state. In the measurement via the N_A projectors $\Pi_a = |a\rangle\langle a| \otimes \mathrm{I}^B$, the probability $P(a)$ reduces to $P(a) = \langle AB|\Pi_a|AB\rangle = \langle A|a\rangle\langle a|A\rangle = |\langle a|A\rangle|^2$, while AB terminates in the separable post-measurement state $\Pi_a\,|AB\rangle\,/\sqrt{P(a)} = |a\rangle \otimes |B\rangle$, i.e., A gets projected onto $|a\rangle$ while B remains in state $|B\rangle$.

Such a separable description of the fate of A and B is no longer feasible when the bipartite system AB is initially prepared in an *entangled* state $|AB\rangle \in \mathcal{H}^A \otimes \mathcal{H}^B$, with neither of the subsystems A and B having separately a well-defined state. After the local measurement on A alone, the system B will terminate in a final state

$|B'\rangle \in \mathcal{H}^B$ as introduced above, that in general will depend on the measurement outcome $|a\rangle$ obtained (at random, via the Born rule) on system A. Although the system B is untouched in the measurement, B terminates in a state $|B'\rangle$ determined by the measurement result on A. And this occurs no matter the spatial distance that may exist between the two parts A and B initially entangled. In this way, quantum measurement on an entangled bipartite system AB leads to what may appear as nonlocal instantaneous interaction between the subsystems A and B. These are remarkable behaviors manifesting the unconventional character of quantum entanglement and the unexpected correlations it may induce. These important consequences of quantum entanglement can be further illustrated with the following example, upon measuring an entangled qubit pair.

Measuring an Entangled Qubit Pair

We consider a qubit A with Hilbert space $\mathcal{H}^A \equiv \mathcal{H}_2$ and a qubit B with Hilbert space $\mathcal{H}^B \equiv \mathcal{H}_2$, with the qubit pair AB prepared in the entangled bipartite state:

$$|AB\rangle = \frac{1}{\sqrt{2}}\Big(|0\rangle \otimes |0\rangle + |1\rangle \otimes |1\rangle\Big) \in \mathcal{H}_2^{\otimes 2}\,, \tag{4.15}$$

which is the Bell state $|\beta_{00}\rangle$ of Eq. (2.21).

The qubit B is left untouched while the qubit A is measured locally in the basis $\big\{|0\rangle\,, |1\rangle\big\}$ by means of the two projectors $\Pi_0 = |0\rangle\langle 0| \otimes \mathrm{I}_2$ and $\Pi_1 = |1\rangle\langle 1| \otimes \mathrm{I}_2$ applied on the pair AB in state $|AB\rangle$. It is then found that the outcome where A is projected onto $|0\rangle$ is obtained with the probability $P(0) = \langle AB|\Pi_0|AB\rangle = 1/2$ when the pair AB terminates in the post-measurement state $\Pi_0\,|AB\rangle/\sqrt{P(0)} = |0\rangle \otimes |0\rangle$. It is a separable product state of the pair, where each qubit A and B is locally in state $|0\rangle$. Meanwhile, the other measurement outcome of A projected onto $|1\rangle$ is obtained with the probability $P(1) = \langle AB|\Pi_1|AB\rangle = 1/2$ with the pair AB terminating in the separable post-measurement state $\Pi_1\,|AB\rangle/\sqrt{P(1)} = |1\rangle \otimes |1\rangle$. The situation is symmetric upon measuring the qubit B alone locally in the basis $\big\{|0\rangle\,, |1\rangle\big\}$ while leaving the qubit A untouched.

So before any measurement, because of the entangled preparation of Eq. (4.15), none of the qubits A and B has a definite state; only the pair AB has a definite quantum state. It follows that when measured locally in the basis $\big\{|0\rangle\,, |1\rangle\big\}$, each qubit A or B can be found (projected) either in state $|0\rangle$ or in state $|1\rangle$, equiprobably. However, as we have seen, as soon as the qubit A is locally measured, for instance, in state $|0\rangle$, the qubit B which is untouched gets instantly projected in the same state $|0\rangle$, and this occurs no matter how spatially distant the two qubits may be. Before measurement neither of the two qubits has a definite state, and once a local measurement fixes the state of one qubit of the entangled pair, this instantly determines the state of the other qubit of the pair, no matter how distant it is.

Still reinforcing this impression of instantaneous action at a distance, one can choose alternatively to measure the qubit A locally in the Hadamard basis $\big\{|+\rangle\,, |-\rangle\big\}$

instead of the computational basis $\{|0\rangle, |1\rangle\}$. By Eq. (2.29), the entangled preparation of Eq. (4.15) is equally

$$|AB\rangle = \frac{1}{\sqrt{2}}\Big(|0\rangle \otimes |0\rangle + |1\rangle \otimes |1\rangle\Big) = \frac{1}{\sqrt{2}}\Big(|+\rangle \otimes |+\rangle + |-\rangle \otimes |-\rangle\Big), \quad (4.16)$$

so that when A is locally measured in the state $|+\rangle$ or $|-\rangle$, respectively, the qubit B is instantly placed in the same state $|+\rangle$ or $|-\rangle$. In this way, when the qubit A is locally measured in the basis $\{|+\rangle, |-\rangle\}$, it appears as if the qubit B was instantly informed that the measurement of A took place in $\{|+\rangle, |-\rangle\}$, and not in $\{|0\rangle, |1\rangle\}$, and that B should instantly adjust its state as $|+\rangle$ or $|-\rangle$ according to the measurement result on A.

These behaviors are consequences of standard quantum measurement, when performed on an entangled pair of systems, it gives rise to what appears as an instantaneous interaction between two distant parts. Such behavior seems to conflict with the theory of relativity which forbids superluminal propagation. Such behavior has long remained a puzzle in the elaboration of quantum theory. As an attempt to circumvent this difficulty of an instantaneous interaction at a distance, it has been suggested that quantum theory might at least be incomplete. Some putative hidden variables, yet to be uncovered, might exist that would be fixed as soon as the preparation stage of Eq. (4.15) when the pair AB is created, and that would completely determine, in advance, the states of each of the two parts A and B and their subsequent behavior upon any conceivable measurement. This predetermination by hidden variables would be the cause of the coordinated behavior of A and B upon measurement, with no need to exchange or propagate (superluminally) any influence between them. Yet, a sequence of arguments have progressively been elaborated to rule out the possibility of such hidden variables. This will be addressed in more detail later in Sect. 6.1 examining quantum correlations.

It remains that such coordinated behavior upon quantum measurement in the presence of entanglement cannot actually be used to communicate (a fortiori superluminally) controlled information between the two parts of the pair. The reason is that the state determination (projection) from a measurement performed at one end occurs at random and cannot be steered to transmit chosen or controlled information to the other end. In addition, there is no material or physically tangible element that can be identified to propagate between the two parts of the entangled pair and accompanying the measurement. The current understanding of the situation is that the entangled pair forms and behaves as a whole, a nonlocal delocalized whole, and it reacts as a whole globally when one part is acted upon, with nothing actually propagating from one part to the other. Although entanglement cannot be employed for direct communication of information between the two distant parts, we will see in the sequel that entanglement, as a specifically quantum property, can actually be exploited for useful information processing and to achieve performances that remain unattainable by classical means.

4.2 Generalized Quantum Measurement

4.2.1 Definition

A quantum system with Hilbert space $\mathcal{H}_N$ can be measured by means of a generalized measurement, which can be regarded as an extension of the von Neumann projective measurement presented in Sect. 4.1.1. A generalized measurement is defined by a set of M measurement operators $\mathsf{M}_m \in \mathcal{L}(\mathcal{H}_N)$, which do not need to be projectors or Hermitian, for $m = 1$ to M. The generalized measurement acting on a quantum system in a state $|\psi\rangle \in \mathcal{H}_N$ takes place as a random application of one definite measurement operator M_m of the set. This outcome produces the transformed vector $\mathsf{M}_m |\psi\rangle$ which is an unnormalized vector of $\mathcal{H}_N$ with squared norm $\big\|\mathsf{M}_m |\psi\rangle\big\|^2 = \langle\psi| \mathsf{M}_m^\dagger \mathsf{M}_m |\psi\rangle$. This squared norm represents the probability

$$P(m) = \langle\psi| \mathsf{M}_m^\dagger \mathsf{M}_m |\psi\rangle = \mathrm{tr}\big(|\psi\rangle\langle\psi| \mathsf{M}_m^\dagger \mathsf{M}_m\big) \tag{4.17}$$

that it is the measurement operator M_m that acts in the measurement, as an extension to Eq. (4.3). Such a generalized measurement has thus M possible outcomes or results, with M possibly differing (by being larger or smaller) from the dimension N of the Hilbert space $\mathcal{H}_N$.

The requirement $\sum_{m=1}^{M} P(m) = 1$ that the probabilities of Eq. (4.17) sum to 1 for any state $|\psi\rangle$ entails that

$$1 = \langle\psi| \sum_{m=1}^{M} \mathsf{M}_m^\dagger \mathsf{M}_m |\psi\rangle = \mathrm{tr}\left(|\psi\rangle\langle\psi| \sum_{m=1}^{M} \mathsf{M}_m^\dagger \mathsf{M}_m\right) \tag{4.18}$$

should be satisfied for any $|\psi\rangle$. This in turn imposes the completeness or closure relation

$$\sum_{m=1}^{M} \mathsf{M}_m^\dagger \mathsf{M}_m = \mathrm{I}_N \tag{4.19}$$

that should be verified collectively by the measurement operators M_m, as an extension to Eq. (4.5).

In addition, the post-measurement state $|\phi_m^{\mathrm{post}}\rangle$ is given by the unnormalized transformed vector $\mathsf{M}_m |\psi\rangle$ after it has been normalized into the valid quantum state $|\phi_m^{\mathrm{post}}\rangle = \mathsf{M}_m |\psi\rangle / \|\mathsf{M}_m |\psi\rangle\| = \mathsf{M}_m |\psi\rangle / \sqrt{P(m)}$, or equivalently:

$$|\phi_m^{\mathrm{post}}\rangle = \frac{\mathsf{M}_m |\psi\rangle}{\sqrt{\langle\psi|\mathsf{M}_m^\dagger \mathsf{M}_m|\psi\rangle}}, \tag{4.20}$$

as an extension to Eq. (4.6).

As a special case, the standard von Neumann measurement in $\mathcal{H}_N$ we described in Sect. 4.1.1 can be seen as a particular generalized measurement, with $M = N$ measurement operators as the N rank-one projectors $\mathsf{M}_n = \Pi_n = |n\rangle\langle n|$, for any orthonormal basis $\{|n\rangle\}$ of $\mathcal{H}_N$.

Generally in this way, a generalized measurement is defined by any set of M measurement operators $\mathsf{M}_m \in \mathcal{L}(\mathcal{H}_N)$ that satisfy the completeness relation of Eq. (4.19). What we will show now is that the possibility of such a generalized measurement follows from the postulate of quantum measurement in its standard von Neumann projective form presented in Sect. 4.1.1. The generalized measurement in $\mathcal{H}_N$ is the manifestation in $\mathcal{H}_N$ of a standard von Neumann projective measurement taking place in an auxiliary Hilbert space with which the initial quantum system has been purposely entangled. In this respect, no extra postulate or principle need be added beyond that already contained in Sect. 4.1.1 to justify the quantum measurement, as we now explain.

4.2.2 *Justification and Implementation of the Generalized Measurement*

On an N-dimensional quantum system Q in an arbitrary state $|\psi\rangle \in \mathcal{H}_N$, suppose we want to perform (and justify) a generalized measurement defined by the set of M measurement operators $\mathsf{M}_m \in \mathcal{L}(\mathcal{H}_N)$ satisfying the completeness relation of Eq. (4.19). We introduce an auxiliary quantum system A with dimension M associated with the M-dimensional Hilbert space $\mathcal{H}_M$ referred to the orthonormal basis $\{|m\rangle\}$ for $m = 1$ to M. The joint evolution of systems Q and A we are going to describe can be seen as a process taking place in the physical apparatus or instrument performing the generalized measurement.

Letting $|e_0\rangle$ be any fixed state of the auxiliary space $\mathcal{H}_M$, we consider the (putatively unitary) operator U from $\mathcal{H}_N \otimes \mathcal{H}_M$ onto $\mathcal{H}_N \otimes \mathcal{H}_M$ acting on states of the form $|\psi\rangle \otimes |e_0\rangle \in \mathcal{H}_N \otimes \mathcal{H}_M$ according to

$$|\psi\rangle \otimes |e_0\rangle \xmapsto{\mathsf{U}} \mathsf{U}|\psi\rangle \otimes |e_0\rangle = \sum_{m=1}^{M} \mathsf{M}_m |\psi\rangle \otimes |m\rangle \ . \tag{4.21}$$

Following an argument similar to that of Eqs. (3.122)–(3.124), two arbitrary states $|\psi_1\rangle \otimes |e_0\rangle$ and $|\psi_2\rangle \otimes |e_0\rangle$ of $\mathcal{H}_N \otimes \mathcal{H}_M$ have inner product $\big(|\psi_1\rangle \otimes |e_0\rangle, |\psi_2\rangle \otimes |e_0\rangle\big) = \langle\psi_1|\psi_2\rangle\langle e_0|e_0\rangle = \langle\psi_1|\psi_2\rangle$. Their images by U have inner product

$$\Big(\mathsf{U}\,|\psi_1\rangle\otimes|e_0\rangle\,,\,\mathsf{U}\,|\psi_2\rangle\otimes|e_0\rangle\Big)=\sum_{m=1}^{M}\sum_{m'=1}^{M}\langle\psi_1|\mathsf{M}_m^\dagger\mathsf{M}_{m'}|\psi_2\rangle\,\langle m|m'\rangle \tag{4.22}$$

$$=\langle\psi_1|\sum_{m=1}^{M}\mathsf{M}_m^\dagger\mathsf{M}_m\,|\psi_2\rangle \tag{4.23}$$

$$=\langle\psi_1|\psi_2\rangle\ , \tag{4.24}$$

where Eq. (4.23) follows from $\langle m|m'\rangle=\delta_{mm'}$ by orthonormality of the basis $\big\{|m\rangle\big\}$, and Eq. (4.24) follows from the completeness relation of Eq. (4.19) satisfied by the M measurement operators M_m. This conservation of the inner product by U indicates that the transformation of Eq. (4.21), characterized in Eq. (4.21) by its action over the N-dimensional subspace $\mathcal{H}_N\otimes\operatorname{span}\big(|e_0\rangle\big)$ of $\mathcal{H}_N\otimes\mathcal{H}_M$, is indeed compatible with a unitary U and can be extended so as to completely define a unitary U acting on the whole space $\mathcal{H}_N\otimes\mathcal{H}_M$. The N-dimensional input subspace $\mathcal{H}_N\otimes\operatorname{span}\big(|e_0\rangle\big)$ is unitarily mapped by U into an output subspace of $\mathcal{H}_N\otimes\mathcal{H}_M$ with same dimension N. The orthogonal complements to these two subspaces, at the input and at the output, can be unitarily mapped one into the other for extending a unitary U. The extension of U can even be achieved in an infinite number of ways, yet non-critical and irrelevant to our purpose here (but possibly affording some useful flexibility at the time of a physical implementation for U).

The effect of the unitary transformation U in Eq. (4.21) is in general to entangle the system Q with the auxiliary system A. After the action by U in Eq. (4.21), a standard von Neumann projective measurement is performed on the auxiliary system A with the M projectors $|m\rangle\langle m|$, while the system Q remains untouched. This is equivalent to performing a measurement of the bipartite system QA with the M projectors $\mathrm{I}_N\otimes|m\rangle\langle m|$, where I_N is the identity operator on the Hilbert space $\mathcal{H}_N$ of system Q. This also conforms with the notion of partial measurement on QA from Sect. 4.1.3. Such a measurement acting on the pre-measurement state $\mathsf{U}\,|\psi\rangle\otimes|e_0\rangle$ of Eq. (4.21) randomly projects the bipartite system QA onto one unnormalized vector $\mathsf{M}_m\,|\psi\rangle\otimes|m\rangle$ of $\mathcal{H}_N\otimes\mathcal{H}_M$ having the squared norm $\big\|\mathsf{M}_m\,|\psi\rangle\otimes|m\rangle\big\|^2=\langle\psi|\mathsf{M}_m^\dagger\mathsf{M}_m|\psi\rangle$. This squared norm represents the probability

$$P(m)=\langle\psi|\,\mathsf{M}_m^\dagger\mathsf{M}_m\,|\psi\rangle=\operatorname{tr}\big(|\psi\rangle\langle\psi|\,\mathsf{M}_m^\dagger\mathsf{M}_m\big) \tag{4.25}$$

of obtaining the outcome m among M possible, as announced in Eq. (4.17).

Normalization of the projected vector $\mathsf{M}_m\,|\psi\rangle\otimes|m\rangle$ leads for the bipartite system QA to a final post-measurement state

$$\frac{\mathsf{M}_m\,|\psi\rangle}{\sqrt{\langle\psi|\mathsf{M}_m^\dagger\mathsf{M}_m|\psi\rangle}}\otimes|m\rangle \tag{4.26}$$

of $\mathcal{H}_N\otimes\mathcal{H}_M$. This is a separable product state for the bipartite system QA, with the auxiliary system A in state $|m\rangle$, and the system Q separately terminating in the post-measurement state

$$|\phi_m^{\text{post}}\rangle = \frac{\mathsf{M}_m\,|\psi\rangle}{\sqrt{\langle\psi|\mathsf{M}_m^\dagger\mathsf{M}_m|\psi\rangle}} \tag{4.27}$$

of $\mathcal{H}_N$, as announced in Eq. (4.20).

In this way, the generalized measurement on Q, through coupling with an auxiliary system A, is obtained as a standard von Neumann projective measurement, delegated to the auxiliary system A, and taking place in accordance with the principle of Sect. 4.1.1. At the same time, the generalized measurement of the system Q by means of the M measurement operators $\mathsf{M}_m \in \mathcal{L}(\mathcal{H}_N)$, as it results, can be described solely from the point of view of Q. The system Q starts in an arbitrary state $|\psi\rangle \in \mathcal{H}_N$, and after its entanglement with the measuring apparatus involving A and taking place in the enlarged space $\mathcal{H}_N \otimes \mathcal{H}_M$ via U, the generalized measurement delivers the outcome m among M with the probability $P(m)$ of Eq. (4.25), while Q terminates in the separable state $|\phi_m^{\text{post}}\rangle \in \mathcal{H}_N$ of Eq. (4.27). The system Q is acted upon while measuring A because of the prior entanglement of QA established by U. The generalized measurement is nothing else than a standard von Neumann projective measurement, yet delegated to an auxiliary system and preceded by a unitary entanglement into an enlarged Hilbert space as part of the measuring protocol. This result is sometimes referred to as the Naimark (or Neumark) theorem. The argument of this section establishes two essential things concerning the generalized measurement: its feasibility based on the standard measurement principle of Sect. 4.1.1, and an approach for its practical implementation via the entangling interaction of Eq. (4.21).

We note that the auxiliary system with M-dimensional Hilbert space $\mathcal{H}_M$ can be realized by assembling K qubits, so as to offer an auxiliary Hilbert space with dimension $2^K \geq M$, referable to an orthonormal basis $\{|m\rangle\}$ with $m = 1$ to 2^K. The extra dimensions indexed by $m = M + 1$ to 2^K can be associated in Eq. (4.21) with (fictitious) null operators M_m that will have in Eq. (4.25) a probability zero of being measured, so as to be with no effect in the protocol of the generalized measurement.

As we have explained at the end of Sect. 3.1.5, any unitary operator, such as U in Eq. (4.21), is in principle always physically realizable, ensuring the practical feasibility of the generalized measurement envisaged here. For illustration, Chap. 14 with Fig. 14.3, on the occasion of a quantum error-detection circuit, will show an explicit example of coupling to an auxiliary system A to implement a generalized measurement. In addition, a simpler example of the inner working of a generalized measurement can be found in Chap. 9 with the quantum circuit of Fig. 9.8 implementing $\mathsf{U} \equiv \mathsf{U}_{QE}$ from Eq. (9.75), where the von Neumann projective measurement in the basis $\{|0\rangle, |1\rangle\}$ of the environment qubit $E \equiv A$ at the output would perform a generalized measurement of the qubit Q by means of the two projectors $\{\mathsf{M}_0 = |0\rangle\langle 0|, \mathsf{M}_1 = |1\rangle\langle 1|\}$, which is equivalent to a direct von Neumann measurement of Q in the basis $\{|0\rangle, |1\rangle\}$. A significant difference though is that the qubit Q would avoid being destroyed or absorbed in the generalized measurement, as it would have been with a direct measurement of Q as in Fig. 4.1; instead, it is the environment qubit $E \equiv A$ that would incur destruction to reveal the result $|0\rangle$ or $|1\rangle$ of measuring Q.

In general, on a quantum system of interest Q with N-dimensional Hilbert space $\mathcal{H}_N$, a generalized measurement with M measuring operators $\mathsf{M}_m \in \mathcal{L}(\mathcal{H}_N)$ affords the interesting possibility of a number M of measurement outcomes larger or smaller than the dimension N of the measured system Q. As a rule, a size $M > N$ may afford more variability and probing capability in the measurement results, for better efficiency in various information-processing tasks; on the other hand, $M < N$ would represent more sparing or innocuous measurements with lesser disturbance to the measured system. Examples of both conditions will be found in the chapters to come. A quantum signal- or information-processing task typically will end with a quantum measurement, which in general can be sought and optimized as a generalized measurement, with the aim of maximizing some criterion of efficiency relevant to the task.

An additional feature is that, in the generalized measurement, the pre-measurement state $|\psi\rangle$ undergoes the transformation $|\psi\rangle \mapsto |\phi_m^{\text{post}}\rangle$ determined in Eq. (4.27). This is in general a **nonlinear transformation** of $|\psi\rangle$, realized in Eq. (4.27) by a generally non-unitary (non-norm-preserving) linear operator M_m supplemented by a renormalization to unit norm, to obtain a valid (normalized) post-measurement state $|\phi_m^{\text{post}}\rangle$. This is the type of nonlinear evolution that was ruled out, in the footnote of Sect. 3.1.1, for the state of a closed quantum system. It can be obtained here because the principal system Q is an open, not a closed, quantum system. By coupling (entangling) the principal system Q to an auxiliary quantum system A through Eq. (4.21), and by measuring the auxiliary system A, the quantum state $|\psi\rangle$ of the principal system Q undergoes the nonlinear evolution $|\psi\rangle \mapsto |\phi_m^{\text{post}}\rangle$ of Eq. (4.27), yet in a probabilistic manner governed by the probability $P(m)$ of Eq. (4.25). The measurement outcome m reveals which nonlinear transformation $|\psi\rangle \mapsto |\phi_m^{\text{post}}\rangle$ took place. In this perspective, the set of operators $\{\mathsf{M}_m\}$ can sometimes be designed, not so much for measurement, but to implement useful nonlinear (probabilistic) evolutions to quantum states. This enlarges the type of evolutions that can be envisaged for quantum states, from deterministic linear evolutions for closed systems, to probabilistic nonlinear evolutions for open systems. Interesting applications of such nonlinear quantum dynamics, including quantum chaos, have been reported, for instance, in [46–48].

4.2.3 Generalized Measurement of a Mixed Quantum State

Instead of a pure state $|\psi\rangle$, we now consider that the pre-measurement state of the quantum system to be measured is a mixed state associated with the statistical ensemble $\{(p_j, |\psi_j\rangle)\}$ represented by the density operator $\rho = \sum_j p_j |\psi_j\rangle\langle\psi_j|$ according to Eq. (2.35). Under the condition of a system in state $|\psi_j\rangle$, a generalized measurement with the set of M operators $\{\mathsf{M}_m\}$ would yield the outcome m with, according to Eq. (4.17), the (conditional) probability

$$P(m|j) = \langle\psi_j|\mathsf{M}_m^\dagger\mathsf{M}_m|\psi_j\rangle = \mathrm{tr}\big(|\psi_j\rangle\langle\psi_j|\,\mathsf{M}_m^\dagger\mathsf{M}_m\big)\,. \tag{4.28}$$

Consequently, with a preparation over the statistical ensemble, the total probability that the generalized measurement yields the outcome m is

$$P(m) = \sum_j P(m|j)p_j \tag{4.29}$$

$$= \sum_j p_j\,\mathrm{tr}\big(|\psi_j\rangle\langle\psi_j|\,\mathsf{M}_m^\dagger\mathsf{M}_m\big)\,, \tag{4.30}$$

and finally

$$\boxed{P(m) = \mathrm{tr}\big(\rho\mathsf{M}_m^\dagger\mathsf{M}_m\big)} \tag{4.31}$$

by linearity of the trace.

Next, the post-measurement state. A pure pre-measurement state $|\psi_j\rangle$ yielding the measurement outcome m terminates, according to Eq. (4.20), with the measured system in the pure post-measurement state:

$$|\phi_m^j\rangle = \frac{\mathsf{M}_m\,|\psi_j\rangle}{\sqrt{\langle\psi_j|\mathsf{M}_m^\dagger\mathsf{M}_m|\psi_j\rangle}} = \frac{\mathsf{M}_m\,|\psi_j\rangle}{\sqrt{P(m|j)}}\,. \tag{4.32}$$

According to the rules governing the generalized measurement as we explained, a measurement outcome m is obtained when it is the measurement operator M_m that acts on the pre-measurement state. So when the measurement outcome m is obtained, the system terminates in the post-measurement state $|\phi_m^j\rangle$ of Eq. (4.32) whenever the pre-measurement state is $|\psi_j\rangle$, an eventuality that can be assigned the probability $P(j|m)$. Therefore, when the measurement outcome m is obtained (that is, when M_m has acted), the system terminates in a statistical ensemble of states $\big\{\big(P(j|m), |\phi_m^j\rangle\big)\big\}$, with $P(j|m)$ the conditional probability that it was the preparation $|\psi_j\rangle$ that produced this outcome m associated with the post-measurement state $|\phi_m^j\rangle$. This statistical ensemble is represented by the density operator

$$\rho_m = \sum_j P(j|m)\,|\phi_m^j\rangle\langle\phi_m^j|\,, \tag{4.33}$$

which by Eq. (4.32) is also

$$\rho_m = \sum_j \frac{P(j|m)}{P(m|j)}\,\mathsf{M}_m\,|\psi_j\rangle\langle\psi_j|\,\mathsf{M}_m^\dagger\,. \tag{4.34}$$

By the Bayes rule $P(m|j)p_j = P(j|m)P(m)$ or $P(j|m)/P(m|j) = p_j/P(m)$, so that Eq. (4.34) is also

$$\rho_m = \frac{1}{P(m)} \sum_j p_j \mathsf{M}_m |\psi_j\rangle \langle\psi_j| \mathsf{M}_m^\dagger = \frac{1}{P(m)} \mathsf{M}_m \rho \mathsf{M}_m^\dagger \tag{4.35}$$

yielding

$$\boxed{\rho_m = \frac{\mathsf{M}_m \rho \mathsf{M}_m^\dagger}{\mathrm{tr}\left(\mathsf{M}_m \rho \mathsf{M}_m^\dagger\right)}} \tag{4.36}$$

as the density operator of the post-measurement state, when the system starts in the pre-measurement state ρ and the outcome m is delivered by the generalized measurement governed by the probability $P(m)$ of Eq. (4.31).

Together Eqs. (4.31) **and** (4.36) **can be taken as the most general forms that govern any quantum measurement**, as they encompass both the standard von Neumann and the generalized measurement, acting on a pure or on a mixed state. They also consolidate the status of the density operator ρ as the generally adequate representation of the pre-measurement quantum state in order to describe or predict the outcome of a quantum measurement, both in terms of probabilities and post-measurement states.

In addition, it is sometimes meaningful to consider a generalized measurement with the set of M measurement operators $\{\mathsf{M}_m\}$ that effectively takes place, but with the outcome it delivers which remains unknown, this situation being also described as an **unread measurement**, as already encountered in Sect. 4.1.3 when addressing partial measurement on a bipartite system. Here, with an unknown outcome of the generalized measurement, the post-measurement state is described by the statistical ensemble $\{(P(m), \rho_m)\}_{m=1}^{M}$ represented by the density operator:

$$\rho^{\text{post}} = \sum_{m=1}^{M} P(m)\rho_m = \sum_{m=1}^{M} \mathsf{M}_m \rho \mathsf{M}_m^\dagger . \tag{4.37}$$

This shows that such an unread measurement can be viewed as a quantum process with the M Kraus operators M_m. Conversely, any quantum process with K Kraus operators Λ_k of Eqs. (3.45) and (3.47) can be viewed as an unread generalized measurement with K measurement operators $\mathsf{M}_k = \Lambda_k$ in Eqs. (4.37) and (4.19), as also discussed for instance in [6, 9]. The procedure described in Sect. 3.2.5 to provide a probabilistic interpretation for the Kraus or operator-sum representation of a quantum process $\mathcal{E}(\cdot)$ as in Eq. (3.45) can be concretely realized as an unread generalized measurement with the measurement operators $\mathsf{M}_k = \Lambda_k$. With the protocol of Sect. 4.2.2 to implement a generalized measurement by coupling to an auxiliary system, the probabilistic procedure of Sect. 3.2.5 for a quantum process, seen as an unread measurement, coincides with the canonical procedure of Sect. 3.2.8 aiming at simulating a quantum process via a model for its environment, with Eq. (4.21) to implement the generalized measurement which is similar to Eq. (3.121) to simulate the quantum process via an environment model. As long as the measurement result is unread, the generalized measurement of Sect. 4.2.2 where the auxiliary system is actually measured is equivalent to the process simulation of Sect. 3.2.8 where the environment is not measured. And we can find an analogy: a quantum state is a superposition $|\psi\rangle = \sum_n \alpha_n |n\rangle$ and a (von Neumann) measurement is its random collapse into one $\alpha_n |n\rangle$ properly renormalized, while a quantum process is a superposition $\mathcal{E}(\rho) = \sum_k \Lambda_k \rho \Lambda_k^\dagger$ and a (generalized) measurement is its random collapse into one $\Lambda_k \rho \Lambda_k^\dagger$ properly renormalized.

4.2.4 Positive Operator-Valued Measure

The presentation we have realized of the measurement of a quantum system, culminating in the most general forms of Eqs. (4.31) and (4.36), describes both the probability of each measurement outcome via Eq. (4.31), and the post-measurement state via Eq. (4.36). In a quantum measurement, it is often the case that the measured system gets destroyed or absorbed in the measurement apparatus and is no longer needed or accessible for subsequent processing, so that the post-measurement state of Eq. (4.36) does not really matter (and is often nothing more than a random projection on a stereotyped basis state). This is the case, for instance, in the two examples of quantum measurement of Fig. 4.1. The random projection inherent to the measurement is performed in Fig. 4.1 by the beam splitter for the photonic qubit or by the magnet for the spin qubit, but to read out the projection result the projected or deflected qubit is finally absorbed in a detector or by the screen. Furthermore, quite often for quantum signal or information processing, the performance is assessed by means of a statistical quantitative criterion, which is determined by the probabilities of the different measurement outcomes from Eq. (4.31), but for which the post-measurement states of Eq. (4.36) are irrelevant. This we shall witness on many occasions in chapters to come, for instance when the performance criterion is the probability of error in quantum detection, or the mean-squared error in quantum parameter estimation, or the channel capacity in quantum communication.

Accordingly, the probability distribution $P(m)$ of Eq. (4.31) for the measurement outcomes is often the sole quantity of interest to characterize the performance of the quantum measurement. In such circumstance, the characterization of the measurement via a set of M measurement operators $\{\mathsf{M}_m\}$ to act as in Eq. (4.36) is unnecessarily detailed. Their products $\mathsf{M}_m^\dagger \mathsf{M}_m = \mathsf{E}_m$, for $m = 1$ to M, are sufficient to determine the measurement probabilities $P(m)$ of Eq. (4.31).

For any operator $\mathsf{M}_m \in \mathcal{L}(\mathcal{H}_N)$, the product $\mathsf{E}_m = \mathsf{M}_m^\dagger \mathsf{M}_m$ is a positive operator of $\mathcal{L}(\mathcal{H}_N)$, since $\langle\psi|\mathsf{E}_m|\psi\rangle = \langle\psi|\mathsf{M}_m^\dagger \mathsf{M}_m|\psi\rangle = \left\|\mathsf{M}_m |\psi\rangle\right\|^2 \geq 0$ for any $|\psi\rangle \in \mathcal{H}_N$. In this way is defined a **positive operator-valued measure (POVM)** by means of a set of M positive operators E_m of $\mathcal{L}(\mathcal{H}_N)$, known as POVM elements. They need, according to Eq. (4.19), to satisfy the completeness relation

$$\sum_{m=1}^{M} \mathsf{E}_m = \mathrm{I}_N \,. \tag{4.38}$$

Measurement by such a POVM of a quantum system in state ρ delivers the outcome m with the probability

$$\boxed{P(m) = \mathrm{tr}\big(\rho \mathsf{E}_m\big)\,.} \tag{4.39}$$

In this way, to represent the effect of the quantum measurement on the statistical performance criteria evoked above, requires and involves only the POVM elements $\{\mathsf{E}_m\}_{m=1}^{M}$, via Eq. (4.39) ruling the probabilities of the different measurement outcomes. Conversely, to devise a quantum measurement in order to optimize such statistical criteria will set requirements only on the POVM elements $\{\mathsf{E}_m\}_{m=1}^{M}$.

The statistical performance attached to a quantum measurement is thus determined by the POVM associated with this measurement. Beyond such statistical performance, a further issue that one may consider for a given POVM $\{\mathsf{E}_m\}_{m=1}^{M}$ arises if one needs to specify the physical implementation of the quantum measurement and the characterization of the post-measurement states. This demands, for a quantum measurement characterized by its M POVM elements $\mathsf{E}_m \in \mathcal{L}(\mathcal{H}_N)$ (the sole to influence its statistical performance), to determine a set of M measurement operators $\mathsf{M}_m \in \mathcal{L}(\mathcal{H}_N)$ satisfying $\mathsf{M}_m^\dagger \mathsf{M}_m = \mathsf{E}_m$ for $m = 1$ to M. This can be accomplished in an infinite number of ways, as we shall see. The choice of a feasible set $\{\mathsf{M}_m\}_{m=1}^{M}$ will be needed to physically implement the quantum measurement, via Eq. (4.21), but this choice will not affect the statistical performance for a same given POVM $\{\mathsf{E}_m\}_{m=1}^{M}$.

Since each POVM operator $\mathsf{E}_m \in \mathcal{L}(\mathcal{H}_N)$ is positive, it is also Hermitian, and therefore diagonalizable in the orthonormal basis of its N eigenstates $\{|\lambda_n\rangle\}$ in $\mathcal{H}_N$. There is, in general, at least one such eigenbasis $\{|\lambda_n^{(m)}\rangle\}$ for each E_m but for simplicity we omit to label it by m. For E_m, the matrix representing the operator E_m in the computational basis of $\mathcal{H}_N$, one has $E_m = UDU^\dagger$ with the diagonal matrix $D = \mathrm{diag}[\lambda_1, \lambda_2, \ldots, \lambda_N]$ of the (positive) eigenvalues of E_m, and U an $N \times N$ unitary matrix. The square root $M_m = \sqrt{E_m} = U\sqrt{D}\,U^\dagger$, with $\sqrt{D} = \mathrm{diag}\big[\sqrt{\lambda_1}, \sqrt{\lambda_2}, \ldots, \sqrt{\lambda_N}\big]$, defines an operator M_m (which is positive) verifying $\mathsf{M}_m^\dagger \mathsf{M}_m = \mathsf{M}_m^2 = \mathsf{E}_m$. This provides one set of M measurement operators $\{\mathsf{M}_m = \sqrt{\mathsf{E}_m}\}$ associated with the M POVM elements $\{\mathsf{E}_m\}$. With $\{\mathsf{V}_m\}$ an arbitrary set of M unitary operators of $\mathcal{L}(\mathcal{H}_N)$, the alternative M measurement operators $\{\mathsf{V}_m\mathsf{M}_m\}$ provide another feasible set associated with the same POVM $\{\mathsf{E}_m\}$.

In addition, when the measurement operators M_m are projectors acting in $\mathcal{H}_N$, they are both Hermitian and idempotent satisfying $\mathsf{M}_m^\dagger \mathsf{M}_m = \mathsf{M}_m\mathsf{M}_m = \mathsf{M}_m$, so that $\mathsf{E}_m = \mathsf{M}_m$ and the POVM elements E_m coincide with the measurement operators M_m. Conversely, when the POVM elements E_m are given as projectors $\Pi_m = \mathsf{E}_m$, then a feasible set of measurement operators is given by $\mathsf{M}_m = \Pi_m = \mathsf{E}_m$. But, in general, the two sets of measurement operators $\{\mathsf{M}_m\}$ and associated POVM elements $\{\mathsf{E}_m = \mathsf{M}_m^\dagger \mathsf{M}_m\}$ are distinct. The POVM $\{\mathsf{E}_m\}_{m=1}^{M}$ is enough to characterize the statistical performance of the measurement, while a set of measurement operators $\{\mathsf{M}_m\}_{m=1}^{M}$ is usually required for its physical implementation via Eq. (4.21) and the characterization of the post-measurement states via Eq. (4.36).

The POVMs on $\mathcal{L}(\mathcal{H}_N)$ form a **convex set** [49], in the sense that, from two POVMs $\{\mathsf{E}_m\}$ and $\{\mathsf{E}'_m\}$ of $\mathcal{L}(\mathcal{H}_N)$, when the convex combinations $\mathsf{E}''_m = p\mathsf{E}_m + (1-p)\mathsf{E}'_m$ are obtained by a convexly weighted sum of the POVM operators (with possible adjunctions of the null operator to equalize the two POVM sizes), then the resulting set of operators $\{\mathsf{E}''_m\}$ is also a POVM on $\mathcal{L}(\mathcal{H}_N)$.

It is possible, and sometimes useful, to represent an M-element POVM $\{\mathsf{E}_m\}_{m=1}^M$ as a quantum process, or quantum channel, $\mathcal{M}(\cdot)$, defined for any state $\rho \in \mathcal{L}(\mathcal{H}_N)$ by

$$\mathcal{M}(\rho) = \sum_{m=1}^{M} \operatorname{tr}\big(\rho\mathsf{E}_m\big) |m\rangle \langle m| \ , \tag{4.40}$$

with $\{|m\rangle\}_{m=1}^M$ an orthonormal basis of some auxiliary Hilbert space $\mathcal{H}_M$ with dimension M. Each auxiliary state $|m\rangle$ keeps a record of the classical measurement outcome m that occurs with the probability $\operatorname{tr}(\rho\mathsf{E}_m)$ of Eq. (4.39). Equation (4.40) defines a **measurement channel** which is an instance of a quantum-to-classical channel, mapping the pre-measurement quantum state ρ into the post-measurement state $|m\rangle$ representing the classical outcome m. The probability in Eq. (4.39) for obtaining m is equally $P(m) = \langle m|\mathcal{M}(\rho)|m\rangle = \operatorname{tr}(\rho\mathsf{E}_m)$.

To verify that Eq. (4.40) indeed defines a valid quantum process or channel, as those defined by the Kraus or operator-sum representation of Eq. (3.45) and depicted in Fig. 3.6, we start by evaluating the trace as

$$\operatorname{tr}\big(\rho\mathsf{E}_m\big) = \operatorname{tr}\Big(\sqrt{\mathsf{E}_m}\,\rho\sqrt{\mathsf{E}_m}\Big) = \sum_{n=1}^{N} \langle n|\sqrt{\mathsf{E}_m}\,\rho\sqrt{\mathsf{E}_m}\,|n\rangle \ , \tag{4.41}$$

so as to obtain

$$\mathcal{M}(\rho) = \sum_{m=1}^{M}\sum_{n=1}^{N} \langle n|\sqrt{\mathsf{E}_m}\,\rho\sqrt{\mathsf{E}_m}\,|n\rangle\,|m\rangle\langle m| \tag{4.42}$$

$$= \sum_{m=1}^{M}\sum_{n=1}^{N} |m\rangle\langle n|\,\sqrt{\mathsf{E}_m}\,\rho\sqrt{\mathsf{E}_m}\,|n\rangle\langle m| \tag{4.43}$$

$$= \sum_{m=1}^{M}\sum_{n=1}^{N} \Lambda_{mn}\rho\Lambda_{mn}^{\dagger} \ . \tag{4.44}$$

The operators $\Lambda_{mn} = |m\rangle\langle n|\sqrt{\mathsf{E}_m}$ satisfy the completeness relation

$$\sum_{m=1}^{M}\sum_{n=1}^{N} \Lambda_{mn}^{\dagger}\Lambda_{mn} = \sum_{m=1}^{M}\sum_{n=1}^{N} \sqrt{\mathsf{E}_m}\, |n\rangle \langle m|m\rangle \langle n|\, \sqrt{\mathsf{E}_m} \tag{4.45}$$

$$= \sum_{m=1}^{M} \sqrt{\mathsf{E}_m} \left(\sum_{n=1}^{N} |n\rangle \langle n| \right) \sqrt{\mathsf{E}_m} \tag{4.46}$$

$$= \sum_{m=1}^{M} \sqrt{\mathsf{E}_m}\sqrt{\mathsf{E}_m} = \sum_{m=1}^{M} \mathsf{E}_m = \mathrm{I}_N\,, \tag{4.47}$$

establishing that $\mathcal{M}(\cdot)$ of Eq. (4.40) is a valid quantum process with Kraus operators Λ_{mn}, adhering to the general operator sum of Eq. (3.45).

4.2.5 Generalized Measurements for the Qubit

A possible generalized measurement for the qubit is formed by the set

$$\mathsf{M}_m = \sqrt{\frac{2}{M}}\, |e_m\rangle \langle e_m|\,, \quad \text{for } m = 0, 1, \ldots, M-1, \tag{4.48}$$

of $M > 2$ measurement operators $\mathsf{M}_m \in \mathcal{L}(\mathcal{H}_2)$, with the M (normalized) state vectors of $\mathcal{H}_2$,

$$|e_m\rangle = \cos\left(2\pi \frac{m}{M}\right) |0\rangle + \sin\left(2\pi \frac{m}{M}\right) |1\rangle\;. \tag{4.49}$$

Each measurement operator M_m in Eq. (4.48) is a rank-one operator proportional to the rank-one projector $|e_m\rangle \langle e_m|$. It can be verified that $\sum_{m=0}^{M-1} \mathsf{M}_m^{\dagger}\mathsf{M}_m = \mathrm{I}_2$ as demanded by the completeness relation of Eq. (4.19). For illustration, Fig. 4.2 represents three sets of state vectors $|e_m\rangle$ from Eq. (4.49) defining via Eq. (4.48) three distinct generalized measurements for the qubit, for three sizes M.

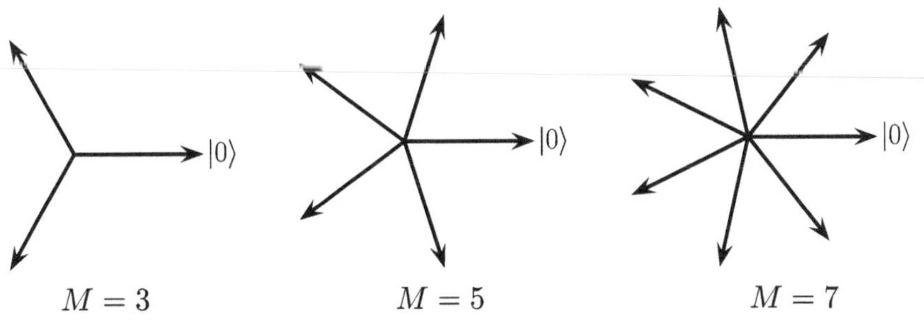

Fig. 4.2 Three sets of state vectors $|e_m\rangle$ from Eq. (4.49) in the vector plane $\mathcal{H}_2$, defining via Eq. (4.48) three distinct "star" generalized quantum measurements for the qubit

This family can be referred to as the "star" generalized measurements of the qubit. The generalized measurement with $M = 3$ is sometimes referred to as the "trine" measurement, and with $M = 5$ the "Chrysler" measurement [12]. By symmetry in the vector sets of Eq. (4.49), for even M the generalized measurements degenerate into measurements with a smaller number $M/2$ of distinct operators M_m, since $|e_m\rangle$ and $-|e_m\rangle$ are equally present in the set to give a same (degenerate) M_m occurring twice.

When the measured qubit is initially in a state with density operator ρ, the generalized measurement delivers the outcome m, corresponding to the action of the operator M_m, with, according to Eq. (4.31), the probability

$$P(m) = \mathrm{tr}\big(\rho \mathsf{M}_m^\dagger \mathsf{M}_m\big) = \frac{2}{M}\,\mathrm{tr}\big(\rho\,|e_m\rangle\langle e_m|\big) = \frac{2}{M}\,\langle e_m|\,\rho\,|e_m\rangle\ ; \tag{4.50}$$

and according to Eq. (4.36) the post-measurement state follows as

$$\rho_m = \frac{1}{P(m)}\mathsf{M}_m \rho \mathsf{M}_m^\dagger = |e_m\rangle\langle e_m|\ , \tag{4.51}$$

representing the pure state $|e_m\rangle$ since M_m of Eq. (4.48) is proportional to the orthogonal projector onto $|e_m\rangle$ in $\mathcal{H}_2$.

According to Sect. 4.2.4, the associated POVM is defined by the M positive operators

$$\mathsf{E}_m = \mathsf{M}_m^\dagger \mathsf{M}_m = \frac{2}{M}\,|e_m\rangle\langle e_m| \tag{4.52}$$

of $\mathcal{L}(\mathcal{H}_2)$. For a qubit POVM, this is a generic form, with each POVM element E_m proportional to a projector, when the POVM elements are imposed to be of rank one. The POVM elements E_m completely determine via Eq. (4.39) the measurement probabilities $P(m) = \mathrm{tr}(\rho \mathsf{E}_m) = (2/M)\langle e_m|\rho|e_m\rangle$ of Eq. (4.50).

Larger flexibility can be obtained with the following general POVM for the qubit. In the Pauli basis of $\mathcal{L}(\mathcal{H}_2)$, the POVM elements can be generally parameterized as

$$\mathsf{E}_m = b_m \mathrm{I}_2 + \vec{a}_m \cdot \vec{\sigma}\ , \quad \text{for } m = 0, 1, \dots M-1. \tag{4.53}$$

For a valid M-element POVM, each E_m in Eq. (4.53) is a positive operator of $\mathcal{L}(\mathcal{H}_2)$ with $b_m \in \mathbb{R}$ and $\vec{a}_m \in \mathbb{R}^3$ satisfying $\sum_{m=0}^{M-1} b_m = 1$ and $\sum_{m=0}^{M-1} \vec{a}_m = \vec{0}$, so as to realize $\sum_{m=0}^{M-1} \mathsf{E}_m = \mathrm{I}_2$ for the completeness relation of Eq. (4.38). In addition, for all m, one should choose $\|\vec{a}_m\| \le b_m \le 1 - \|\vec{a}_m\|$ which is required to ensure $0 \le \mathsf{E}_m \le \mathrm{I}_2$, this implying $\|\vec{a}_m\| \le 1/2$. Furthermore, since as it results from Sect. A.5 we have for E_m the determinant $\det(\mathsf{E}_m) = b_m^2 - \|\vec{a}_m\|^2$ and the two eigenvalues $\lambda_m^\pm = b_m \pm \|\vec{a}_m\|$, for E_m to be of (full) rank 2 requires $\lambda_m^\pm \ne 0$ or equivalently $b_m^2 \ne \|\vec{a}_m\|^2$ for $\det(\mathsf{E}_m) \ne 0$.

When the measured qubit is initially in a state of density operator ρ characterized in Eq. (2.61) by the Bloch vector $\vec{r}$, the POVM delivers the outcome m with the probability:

$$P(m) = \mathrm{tr}\big(\rho \mathsf{E}_m\big) = b_m + \vec{a}_m \vec{r} \ . \tag{4.54}$$

The post-measurement state cannot be specified until we have defined M measurement operators M_m associated with the M POVM elements E_m, as explained in Sect. 4.2.4, for instance, by $\mathsf{M}_m = \sqrt{\mathsf{E}_m}$. However, as we indicated, this step can usually be performed in a infinite number of ways and is often unnecessary, at least for a theoretical characterization of the performance.

The POVM element $\mathsf{E}_m = (2/M)\,|e_m\rangle\langle e_m|$ associated with the states $|e_m\rangle$ from Eq. (4.49) is obtained with $b_m = 1/M$ and $\vec{a}_m$ the vector in the plane (Ox, Oz) of $\mathbb{R}^3$ with norm $1/M$ and coelevation angle $\theta_m = 4\pi m/M$ with Oz, or equivalently with coordinates $\vec{a}_m = [\sin(\theta_m), 0, \cos(\theta_m)]^\top/M$, for $m = 0$ to $M-1$. As expected, each operator E_m has rank one, with $b_m^2 = \|\vec{a}_m\|^2 = 1/M^2$ and $\det(\mathsf{E}_m) = 0$.

4.3 Measurement of a Quantum Observable

4.3.1 Definition for a Quantum Observable

As we have seen in Sect. 4.1.1, the definition of a standard von Neumann measurement rests on the introduction of an orthonormal (projective) basis $\{|n\rangle\}_{n=1}^N$ of $\mathcal{H}_N$. An alternative way for specifying an orthonormal basis of $\mathcal{H}_N$ is by introducing a Hermitian operator on $\mathcal{H}_N$. As indicated in Sect. A.2 of the Appendix, a Hermitian operator on $\mathcal{H}_N$ always comes with N eigenstates forming an orthonormal basis of $\mathcal{H}_N$. In this way, any Hermitian operator on $\mathcal{H}_N$ via N mutually orthogonal eigenstates specifies an orthonormal projective basis of $\mathcal{H}_N$. Conversely, any orthonormal projective basis $\{|n\rangle\}_{n=1}^N$ of $\mathcal{H}_N$ can always be associated with a Hermitian operator, via its diagonal form of Eq. (A.9) and a free choice of N real eigenvalues.

Quantum theory stipulates that any physical quantity that is classically defined and classically measurable, can be associated for its quantum representation with a Hermitian operator on the Hilbert space attached to the quantum system. Such Hermitian operator is known as a quantum observable. Measuring such physical quantity, at the quantum level, is equivalent to performing a projective measurement defined by the mutually orthogonal projectors projecting onto the mutually orthogonal eigensubspaces of the observable. The result of the physical measurement reveals the projector that acted (at random) during the measurement, and equivalently it reveals the associated (real) eigenvalue, which is taken as the measured value for the physical quantity represented by the observable. In the non-degenerate situation where all the eigenvalues are of multiplicity 1, each eigensubspace is a one-dimensional subspace and the measurement operators are rank-one projectors defining a standard von Neumann projective measurement projecting onto the eigenstates of the observable, and the measurement result then also reveals the post-measurement state which is a definite eigenstate of the observable. Otherwise, upon degeneracy of the measured eigenvalue, some uncertainty usually remains on the post-measurement state, which is confined to an eigensubspace of dimension >1 given by the multiplicity of the associated degenerate eigenvalue returned when measuring the observable.

A close connection therefore exists between a von Neumann projective measurement defined by an orthonormal basis and an observable (a Hermitian operator). Depending on the context and prospect, a von Neumann quantum measurement can therefore be defined by specifying an orthonormal projective basis, or by referring to an observable. Beyond, to draw gross partitioning to organize the matters, we could say that observables are not necessarily of primary interest to quantum information. Observables are important to quantum physics aiming at extending at the quantum level the description of classical physics with its familiar quantities. All is however conditioned by the quantum state. Accordingly, quantum information primarily focuses on the quantum state, as an information carrier, and its manipulation by quantum processes, and its measurement. The main purpose is to perform useful information processing, and quantum observables are brought into play when they serve this purpose.

4.3.2 *Average of an Observable*

A quantum observable $\Omega \in \mathcal{L}(\mathcal{H}_N)$ with Π_ℓ the (orthogonal) projector onto the eigensubspace of Ω with (real) eigenvalue ω_ℓ receives the diagonal form:

$$\Omega = \sum_\ell \omega_\ell \Pi_\ell \,. \tag{4.55}$$

When Ω is measured on a quantum system in a state described by the density operator $\rho \in \mathcal{L}(\mathcal{H}_N)$, then the projector Π_ℓ will act with the probability $P(\ell) = \mathrm{tr}\big(\rho\Pi_\ell\big)$ from Eq. (4.31) since $\Pi_\ell = \Pi_\ell^\dagger = \Pi_\ell^\dagger \Pi_\ell$, to deliver the eigenvalue ω_ℓ as the measured value of the physical quantity represented by the observable Ω.

When a series of measurements of the observable Ω are repeated on the same preparation ρ of the system, the average of the measured values is given as the real scalar

$$\langle \Omega \rangle = \sum_\ell \omega_\ell P(\ell) \tag{4.56}$$

$$= \sum_\ell \omega_\ell \,\mathrm{tr}\big(\rho\Pi_\ell\big) \tag{4.57}$$

$$= \mathrm{tr}\Big(\rho \sum_\ell \omega_\ell \Pi_\ell\Big) \,, \tag{4.58}$$

by linearity of the trace. Then, by using Eq. (4.55) in Eq. (4.58), the average value of the observable Ω when measured on state ρ follows as

$$\langle \Omega \rangle = \mathrm{tr}\big(\rho\Omega\big) \,. \tag{4.59}$$

When the pre-measurement state happens to be the pure state $\rho = |\psi\rangle\langle\psi|$, then Eq. (4.59) equivalently gives the average value:

$$\langle \Omega \rangle = \langle \psi|\Omega|\psi\rangle \,. \tag{4.60}$$

In the non-degenerate situation of N distinct eigenvalues ω_n, the associated N orthogonal eigenstates $|\omega_n\rangle$ define N rank-one projectors $\Pi_n = |\omega_n\rangle\langle\omega_n|$, leading from Eq. (4.55) to the diagonal form:

$$\Omega = \sum_{n=1}^{N} \omega_n \, |\omega_n\rangle\langle\omega_n| \,, \tag{4.61}$$

and the post-measurement state $|\omega_n\rangle$ when the eigenvalue ω_n is measured, with the probability $P(n) = \mathrm{tr}\big(\rho\Pi_n\big) = \langle\omega_n|\rho|\omega_n\rangle$ from Eq. (4.1). Then a pure state $\rho = |\psi\rangle\langle\psi|$ consistently gives the probability $P(n) = \mathrm{tr}\big(\rho\Pi_n\big) = |\langle\omega_n|\psi\rangle|^2$ and the average $\langle\Omega\rangle = \sum_{n=1}^{N} \omega_n|\langle\omega_n|\psi\rangle|^2$.

On a quantum system in the generic state $\rho \in \mathcal{L}(\mathcal{H}_N)$ we can envisage a generalized measurement by an M-element POVM $\{\mathsf{E}_m\}_{m=1}^{M}$ as in Sect. 4.2.4. Each POVM element E_m is a positive, hence Hermitian operator of $\mathcal{L}(\mathcal{H}_N)$, and can therefore be measured separately as an observable. Its average value follows as $\langle\mathsf{E}_m\rangle = \mathrm{tr}(\rho\mathsf{E}_m)$, which is also $\langle\mathsf{E}_m\rangle = P(m)$ by Eq. (4.39). In other words, when measuring the state ρ with the POVM $\{\mathsf{E}_m\}_{m=1}^{M}$, the probability $P(m)$ of obtaining each outcome m matches the average value $\langle\mathsf{E}_m\rangle$ that would be obtained upon measuring the POVM element E_m separately as an observable.

4.3.3 Heisenberg Uncertainty Relation

There exists a fundamental limitation when measuring two distinct observables on a given quantum system in some definite state, identified as the Heisenberg uncertainty relation [50–52], and that we now examine.

For two operators A and B of $\mathcal{L}(\mathcal{H}_N)$, from their commutator $[\mathsf{A},\mathsf{B}] = \mathsf{AB} - \mathsf{BA}$ defined by Eq. (A.15) in the Appendix, and their anticommutator $\{\mathsf{A},\mathsf{B}\} = \mathsf{AB} + \mathsf{BA}$ from Eq. (A.16), we have

$$\mathsf{AB} = \frac{1}{2}[\mathsf{A},\mathsf{B}] + \frac{1}{2}\{\mathsf{A},\mathsf{B}\} \,. \tag{4.62}$$

When A and B are two Hermitian operators of $\mathcal{L}(\mathcal{H}_N)$, then their commutator $[\mathsf{A},\mathsf{B}]$ is anti-Hermitian, i.e., $[\mathsf{A},\mathsf{B}]^\dagger = -[\mathsf{A},\mathsf{B}]$; meanwhile, their anticommutator $\{\mathsf{A},\mathsf{B}\}$ is Hermitian, i.e., $\{\mathsf{A},\mathsf{B}\}^\dagger = \{\mathsf{A},\mathsf{B}\}$. These two properties are, respectively, equivalent to $i\,\mathbb{R} \ni \langle\psi|\,[\mathsf{A},\mathsf{B}]\,|\psi\rangle = -\,\langle\psi|\,[\mathsf{A},\mathsf{B}]\,|\psi\rangle^*$ pure imaginary, and $\mathbb{R} \ni \langle\psi|\,\{\mathsf{A},\mathsf{B}\}\,|\psi\rangle = \langle\psi|\,\{\mathsf{A},\mathsf{B}\}\,|\psi\rangle^*$ real, for any $|\psi\rangle \in \mathcal{H}_N$. This allows us to identify, from Eq. (4.62), for the scalar $\langle\psi|\mathsf{AB}|\psi\rangle$, the real and imaginary parts from

$$\langle\psi|\mathsf{AB}|\psi\rangle = \frac{1}{2}\underbrace{\langle\psi|\,[\mathsf{A},\mathsf{B}]\,|\psi\rangle}_{\text{imaginary (part)}} + \frac{1}{2}\underbrace{\langle\psi|\,\{\mathsf{A},\mathsf{B}\}\,|\psi\rangle}_{\text{real (part)}} . \tag{4.63}$$

Then, by the Pythagorean theorem, it follows that

$$\big|\langle\psi|\mathsf{AB}|\psi\rangle\big|^2 \geq \frac{1}{4}\big|\langle\psi|\,[\mathsf{A},\mathsf{B}]\,|\psi\rangle\big|^2 . \tag{4.64}$$

In addition, for the two vectors $\mathsf{A}\,|\psi\rangle$ and $\mathsf{B}\,|\psi\rangle$ of $\mathcal{H}_N$, the Cauchy–Schwarz inequality of Eq. (A.6) gives

$$\big|\langle\psi|\mathsf{AB}|\psi\rangle\big|^2 \leq \langle\psi|\mathsf{A}^2|\psi\rangle\,\langle\psi|\mathsf{B}^2|\psi\rangle , \tag{4.65}$$

so that, by combining with Eq. (4.64), we obtain the inequality:

$$\langle\psi|\mathsf{A}^2|\psi\rangle\,\langle\psi|\mathsf{B}^2|\psi\rangle \geq \frac{1}{4}\big|\langle\psi|\,[\mathsf{A},\mathsf{B}]\,|\psi\rangle\big|^2 . \tag{4.66}$$

The inequality of Eq. (4.66) is satisfied by any pair of Hermitian operators A and B of $\mathcal{L}(\mathcal{H}_N)$. A physically significant consequence is when A and B are two observables measured on the same quantum system in state $|\psi\rangle$ of $\mathcal{H}_N$, as we further explain below.

The average (real scalar) $\langle\mathsf{A}\rangle = \langle\psi|\mathsf{A}|\psi\rangle$, according to Eq. (4.60), represents the average of the values obtained when measuring the observable A in state $|\psi\rangle$. It enables us to define the centered or dispersion operator:

$$\widetilde{\mathsf{A}} = \mathsf{A} - \langle\mathsf{A}\rangle\,\mathrm{I}_N , \tag{4.67}$$

which is also Hermitian as A itself, and with vanishing average $\langle\widetilde{\mathsf{A}}\rangle = 0$. Its squaring leads to $\widetilde{\mathsf{A}}^2 = \mathsf{A}^2 - 2\,\langle\mathsf{A}\rangle\,\mathsf{A} + \langle\mathsf{A}\rangle^2\,\mathrm{I}_N$, which is also Hermitian as A^2 itself. Then the average $\langle\widetilde{\mathsf{A}}^2\rangle = \langle\mathsf{A}^2\rangle - \langle\mathsf{A}\rangle^2$ is the scalar variance which physically represents the squared dispersion of the values obtained when measuring the observable A on state $|\psi\rangle$.

With a similar treatment on another observable B of $\mathcal{L}(\mathcal{H}_N)$, the commutators satisfy $\big[\widetilde{\mathsf{A}},\widetilde{\mathsf{B}}\big] = [\mathsf{A},\mathsf{B}]$, so that the inequality of Eq. (4.66) yields

$$\langle\widetilde{\mathsf{A}}^2\rangle\,\langle\widetilde{\mathsf{B}}^2\rangle \geq \frac{1}{4}\big|\langle[\mathsf{A},\mathsf{B}]\rangle\big|^2 . \tag{4.68}$$

With the scalar dispersion $\Delta A = \left(\langle \widetilde{\mathsf{A}}^2 \rangle\right)^{1/2}$ and $\Delta B = \left(\langle \widetilde{\mathsf{B}}^2 \rangle\right)^{1/2}$, we obtain equivalently

$$\boxed{\Delta A\, \Delta B \geq \frac{1}{2} \left| \langle [\mathsf{A}, \mathsf{B}] \rangle \right| = \frac{1}{2} \left| \langle \psi | \, [\mathsf{A}, \mathsf{B}] \, | \psi \rangle \right| .} \tag{4.69}$$

Equation (4.69) is known as the Heisenberg inequality or **Heisenberg uncertainty relation**. It expresses a fundamental limitation in the joint dispersions of the measurement results obtained for two observables A and B that can be measured on a quantum system prepared in a given quantum state $|\psi\rangle$. It reflects an intrinsic indeterminacy in the quantum quantities, not a limitation of the measuring instrument.

We already know, from the probabilistic character of quantum measurement as seen as early as Sect. 2.2 that, when repeatedly preparing the quantum system in the same state $|\psi\rangle$ and repeatedly measuring the same observable A, the measurement results generally occur at random and therefore exhibit some intrinsic statistical dispersion ΔA. So the novelty in Eq. (4.69) is not so much about the uncertainty of quantum measurement in itself, which was previously known. We also already know that the quantum measurement of an observable A usually exhibits a destructive or disturbing character, in the sense that the pre-measurement state $|\psi\rangle$ gets replaced by (projected onto) an eigenstate of A occurring at random. Subsequently, as long as the system is no longer in the state $|\psi\rangle$, it will usually not be possible to follow with measuring another observable B on the same quantum system in the same pre-measurement state $|\psi\rangle$ as in the conditions of Eq. (4.69). So the novelty in Eq. (4.69) is not so much either about the destructive or disturbing character of quantum measurement in itself, which was previously known. More appropriately, Eq. (4.69) applies when the quantum system is repeatedly prepared in the same state $|\psi\rangle$ and the observable A is repeatedly measured to evaluate its dispersion ΔA, and when in addition the quantum system is repeatedly prepared in the same state $|\psi\rangle$ and the observable B is repeatedly measured to evaluate its dispersion ΔB ; then the dispersions ΔA and ΔB that ensue are related by Eq. (4.69).

For two observables A and B that commute, the commutator [A, B] vanishes, so that Eq. (4.69) is $\Delta A\, \Delta B \geq 0$ which adds nothing for two dispersions ΔA and ΔB defined as two non-negative scalars. Two Hermitian operators A and B of $\mathcal{L}(\mathcal{H}_N)$ commute if and only if they share a common orthonormal eigenbasis in $\mathcal{H}_N$. In such circumstance of a common eigenbasis, two commuting observables A and B of $\mathcal{L}(\mathcal{H}_N)$ can be measured simultaneously. A same measurement will randomly project $|\psi\rangle$ onto one eigenstate common to A and B and will simultaneously reveal both the corresponding eigenvalue of A as the measurement result for A and the corresponding eigenvalue of B as the measurement result for B. Therefrom, the two dispersions $\Delta A \geq 0$ and $\Delta B \geq 0$ are simultaneously accessible, none of them should a priori be expected to vanish in general, and they trivially satisfy $\Delta A\, \Delta B \geq 0$. When the quantum system happens to be prepared in a (common) eigenstate $|\psi\rangle = |\lambda\rangle$, then the repeated measurements of the commuting observables A and B always project onto $|\lambda\rangle$ and always return the same value λ for the measurement result, so that $\Delta A = 0$ and $\Delta B = 0$; otherwise, one will have $\Delta A > 0$ and $\Delta B > 0$.

A vanishing commutator $[\mathsf{A}, \mathsf{B}]$ is a sufficient condition for a vanishing lower bound in the right-hand side of Eq. (4.69), but it is not a necessary condition. With a non-vanishing commutator $[\mathsf{A}, \mathsf{B}]$, the lower bound in the right-hand side of Eq. (4.69) may also vanish, depending on the state $|\psi\rangle$ of the measured system, provided that the commutator averaged on the state $|\psi\rangle$ vanishes. This happens, for instance, with two non-commuting observables A and B of $\mathcal{L}(\mathcal{H}_N)$ each time the state $|\psi\rangle$ of the measured system is an eigenstate of either A or B. For instance, if the quantum system is in the eigenstate $|\psi\rangle = |\lambda_\mathsf{A}\rangle$ of A, then $\mathsf{A}\,|\psi\rangle = \mathsf{A}\,|\lambda_\mathsf{A}\rangle = \lambda_\mathsf{A}\,|\lambda_\mathsf{A}\rangle$ and $\langle\psi|\,\mathsf{A} = \langle\lambda_\mathsf{A}|\,\mathsf{A} = \lambda_\mathsf{A}\,\langle\lambda_\mathsf{A}|$, so that the average of the commutator is

$$\langle[\mathsf{A}, \mathsf{B}]\rangle = \big\langle\psi\big|\,[\mathsf{A}, \mathsf{B}]\,\big|\psi\big\rangle \tag{4.70}$$
$$= \langle\psi|\mathsf{AB}|\psi\rangle - \langle\psi|\mathsf{BA}|\psi\rangle \tag{4.71}$$
$$= \lambda_\mathsf{A}\,\langle\lambda_\mathsf{A}|\mathsf{B}|\lambda_\mathsf{A}\rangle - \lambda_\mathsf{A}\,\langle\lambda_\mathsf{A}|\mathsf{B}|\lambda_\mathsf{A}\rangle \tag{4.72}$$
$$= 0\,. \tag{4.73}$$

So each time the measured system is in an eigenstate of either A or B, the Heisenberg inequality of Eq. (4.69) reduces to $\Delta A\,\Delta B \geq 0$ even with non-commuting observables A or B. This is consistent with the fact that one of the observables, say A in the situation of Eqs. (4.70)–(4.73), has vanishing dispersion $\Delta A = 0$; meanwhile, Eq. (4.69) does not tell anything significant concerning the other dispersion ΔB.

The Heisenberg inequality of Eq. (4.69) is specially significant for two non-commuting observables A and B, and a measured system not in one of their eigenstates. Both observables will usually display non-vanishing dispersions ΔA and ΔB jointly lower bounded according to Eq. (4.69), expressing in this way some form of joint uncertainty or some limitation in the joint precision in the measurement of the quantities A and B on the same quantum system.

In quantum physics, there exist important quantities, coming in pair, known as canonical conjugate quantities or complementary quantities [51, 52]. This is for instance the case with position and momentum, or angle and angular momentum, or energy (energy is like frequency $\times\ \hbar$) and time (life time). Such canonical conjugate quantities are represented in quantum theory by two observables A and B connected by the (canonical) commutation relation:

$$[\mathsf{A}, \mathsf{B}] = i\hbar\,\mathrm{Id}\,. \tag{4.74}$$

Since the average of the identity operator is $\langle\mathrm{Id}\rangle = 1$ for any state $|\psi\rangle$, the Heisenberg inequality of Eq. (4.69) indicates that the dispersions of two canonical conjugate quantities are always lower bounded as

$$\Delta A\,\Delta B \geq \frac{\hbar}{2}\,, \tag{4.75}$$

involving a common fixed lower bound $\hbar/2$ in the right-hand side for any pair of canonical conjugate quantities.

This is the case, for instance, for a material particle with one-dimensional position operator $\mathsf{A} \equiv \mathsf{X}$ and associated momentum operator $\mathsf{B} \equiv \mathsf{P}$. If the spatial dispersion ΔX of the particle gets narrower, necessarily the dispersion ΔP of its momentum increases, or conversely, so as to satisfy Eq. (4.75). At the limit $\Delta X \to 0$, necessarily $\Delta P \to +\infty$, or conversely, to ensure a finite limit of their product in Eq. (4.75). The dispersion ΔX of the position of the particle is controlled by a wave function $\psi(x)$ in position representation, as introduced in Sect. 2.6. Meanwhile, the dispersion ΔP of its momentum is controlled by a wave function $\Psi(p)$ in momentum representation. Quantum theory stipulates that two such wave functions, for two canonical conjugate quantities, are connected by a Fourier transform.[1] In such circumstance, the Heisenberg inequality of Eq. (4.75) becomes similar to the Gabor–Heisenberg inequality [54, 55] known in signal processing to lower-bound the joint dispersions of a signal and of its Fourier transform.

Extension to mixed states: The Heisenberg inequality of Eq. (4.69), for observables measured on a pure state $|\psi\rangle$, involves averages computed according to $\langle \mathsf{A}\rangle = \langle\psi|\mathsf{A}|\psi\rangle$ from Eq. (4.60). The Heisenberg inequality can be extended to observables measured on a mixed state ρ, when the averages are computed according to $\langle \mathsf{A}\rangle = \mathrm{tr}(\rho\mathsf{A})$ from Eq. (4.59). This extension cannot, however, be directly derived by linearity when a mixed state ρ is expressed as a convex sum of pure states like $|\psi\rangle\langle\psi|$. The reason is that, contrary to other important relations that happen to be linear in the quantum state represented by a density operator, such as those involved to obtain Eq. (3.8) or Eq. (3.37) or Eq. (3.45) or Eq. (3.114), the terms involved in the Heisenberg inequality are nonlinear in the state, and therefore do not lend themselves to a linear superposition principle. Instead, the extension of the Heisenberg inequality to mixed states proceeds as follows. It relies on the Hilbert–Schmidt inner product $\mathrm{tr}(\mathsf{A}^\dagger\mathsf{B})$ of two operators A and B of $\mathcal{L}(\mathcal{H}_N)$, as defined in the Appendix at the end of Sect. A.2. For two Hermitian operators A and B of $\mathcal{L}(\mathcal{H}_N)$ and ρ a density operator of $\mathcal{L}(\mathcal{H}_N)$, the two Hermitian operators $\sqrt{\rho}\mathsf{B}$ and $\sqrt{\rho}\mathsf{A}$ of $\mathcal{L}(\mathcal{H}_N)$ have inner product $\mathrm{tr}\big((\sqrt{\rho}\mathsf{B})^\dagger\sqrt{\rho}\mathsf{A}\big) = \mathrm{tr}(\mathsf{B}\rho\mathsf{A}) = \mathrm{tr}(\rho\mathsf{A}\mathsf{B})$, and this inner product engaged in the Cauchy–Schwarz inequality of Eq. (A.17) gives

$$\big|\mathrm{tr}(\rho\mathsf{A}\mathsf{B})\big|^2 \le \mathrm{tr}(\rho\mathsf{A}^2)\ \mathrm{tr}(\rho\mathsf{B}^2)\ , \tag{4.76}$$

by circular invariance of the trace. In similarity with Eq. (4.63), we now have $\mathrm{tr}(\rho\mathsf{A}\mathsf{B}) = \mathrm{tr}(\rho[\mathsf{A},\mathsf{B}])/2 + \mathrm{tr}(\rho\{\mathsf{A},\mathsf{B}\})/2$ as a sum of an imaginary component and a real component, so that again by the Pythagorean theorem, we have

$$\big|\mathrm{tr}(\rho\mathsf{A}\mathsf{B})\big|^2 \ge \frac{1}{4}\big|\mathrm{tr}(\rho[\mathsf{A},\mathsf{B}])\big|^2\ , \tag{4.77}$$

[1] This is rooted in the wave mechanics initiated by Louis de Broglie [53], which (by postulation) to a particle with a well-defined value p for the momentum associates a wave function for the position as the plane wave $\psi(x) \propto \exp(ipx/\hbar)$, which then plays the role of a Green function or an impulse response introducing the Fourier transform. A wave packet with weight $\Psi(p)$ for the distribution of the momentum p gives the weighted sum $\psi(x) \propto \int \Psi(p)\exp(ipx/\hbar)dp$.

which combined with the Cauchy–Schwarz inequality of Eq. (4.76) gives

$$\mathrm{tr}\big(\rho \mathsf{A}^2\big)\ \mathrm{tr}\big(\rho \mathsf{B}^2\big) \geq \frac{1}{4}\big|\mathrm{tr}\big(\rho[\mathsf{A},\mathsf{B}]\big)\big|^2 , \tag{4.78}$$

as an extension to Eq. (4.66).

The average value of the observable A measured on the mixed state ρ is $\langle \mathsf{A}\rangle = \mathrm{tr}(\rho \mathsf{A})$. The centered or dispersion operator $\widetilde{\mathsf{A}} = \mathsf{A} - \langle \mathsf{A}\rangle\, \mathrm{I}_N$ from Eq. (4.67) gives the average $\langle \widetilde{\mathsf{A}}\rangle = \mathrm{tr}\big(\rho\widetilde{\mathsf{A}}\big) = 0$ and the scalar variance or squared dispersion $\langle \widetilde{\mathsf{A}}^2\rangle = \mathrm{tr}\big(\rho\widetilde{\mathsf{A}}^2\big) = \langle \mathsf{A}^2\rangle - \langle \mathsf{A}\rangle^2$. For two observables A and B of $\mathcal{L}(\mathcal{H}_N)$ and associated dispersion operators $\widetilde{\mathsf{A}}$ and $\widetilde{\mathsf{B}}$, the commutators still satisfy $\big[\widetilde{\mathsf{A}},\widetilde{\mathsf{B}}\big] = [\mathsf{A},\mathsf{B}]$. The inequality of Eq. (4.78) applied to $\widetilde{\mathsf{A}}$ and $\widetilde{\mathsf{B}}$, with the scalar dispersion $\Delta A = \big(\langle \widetilde{\mathsf{A}}^2\rangle\big)^{1/2}$ and $\Delta B = \big(\langle \widetilde{\mathsf{B}}^2\rangle\big)^{1/2}$, leads to

$$\Delta A\, \Delta B \geq \frac{1}{2}\big|\langle [\mathsf{A},\mathsf{B}]\rangle\big| = \frac{1}{2}\big|\mathrm{tr}\big(\rho[\mathsf{A},\mathsf{B}]\big)\big| , \tag{4.79}$$

as an extension to mixed states of the Heisenberg inequality of Eq. (4.69).

Still other extensions have been proposed to the Heisenberg uncertainty relation, especially relying on entropic quantities to assess the joint uncertainty or dispersion in quantum measurement, as addressed, for instance, in [56–58].

4.3.4 Observables of the Qubit

An observable for the qubit is a Hermitian operator Ω of $\mathcal{L}(\mathcal{H}_2)$; it can be expressed in Bloch representation as $\Omega = \omega_0 \mathrm{I}_2 + \vec{\omega}\cdot\vec{\sigma}$, with four *real* scalar coordinates $(\omega_0, \vec{\omega})$ so as to satisfy $\Omega^\dagger = \Omega$. The observable Ω has two real eigenvalues $\lambda_\pm = \omega_0 \pm \|\vec{\omega}\|$, associated with two orthogonal eigenstates $|\lambda_\pm\rangle \in \mathcal{H}_2$ with unit Bloch vectors $\pm\vec{\omega}/\|\vec{\omega}\|$. The trace is $\mathrm{tr}(\Omega) = \lambda_+ + \lambda_- = 2\omega_0$ and the determinant is $\det(\Omega) = \lambda_+\lambda_- = \omega_0^2 - \|\vec{\omega}\|^2$. Measuring the observable Ω is equivalent to performing a von Neumann measurement in its orthonormal projective eigenbasis $\big\{|\lambda_+\rangle, |\lambda_-\rangle\big\}$ of $\mathcal{H}_2$ with the two projectors $|\lambda_\pm\rangle\langle\lambda_\pm| = \big(\mathrm{I}_2 \pm \vec{\omega}\cdot\vec{\sigma}/\|\vec{\omega}\|\big)/2$.

On a qubit in state $\rho = \big(\mathrm{I}_2 + \vec{r}\cdot\vec{\sigma}\big)/2$, measuring such an observable Ω yields two measurement outcomes with probabilities $\Pr\{\lambda_\pm\} = \big(1 \pm \vec{r}\,\vec{\omega}/\|\vec{\omega}\|\big)/2$. It results that ω_0 has no effect on the measurement probabilities, but only on the scalar value $\lambda_\pm = \omega_0 \pm \|\vec{\omega}\|$ which can be taken as the quantitative result returned by the measurement.

An important observable measurable on the qubit has the form $\Omega = \vec{\omega}\cdot\vec{\sigma}$ with $\|\vec{\omega}\| = 1$ and is known as a **spin observable**, performing a spin measurement in the direction $\vec{\omega}$ of $\mathbb{R}^3$. It yields as possible outcomes the two eigenvalues $\lambda_\pm = \pm\|\vec{\omega}\| = \pm 1$, with the probabilities

$$\Pr\{\pm 1\} = \frac{1}{2}\big(1 \pm \vec{r}\,\vec{\omega}\big) . \tag{4.80}$$

Such a spin measurement of Ω is equivalent to a von Neumann projective measurement with the two projectors $|\lambda_\pm\rangle\langle\lambda_\pm| = (\mathrm{I}_2 \pm \Omega)/2$. Any von Neumann projective measurement of a qubit can always be described as a spin measurement for some unit measurement vector $\vec{\omega} \in \mathbb{R}^3$, taken as the unit Bloch vector defining the two orthogonal basis states of the projective basis of the von Neumann measurement. For example, measuring the Pauli observable σ_z is equivalent to a von Neumann projective measurement in the basis $\{|0\rangle, |1\rangle\}$, the Pauli observable σ_x is equivalent to a von Neumann projective measurement in the basis $\{|+\rangle, |-\rangle\}$, and similarly for σ_y in the projective basis $\{|+i\rangle, |-i\rangle\}$ from Eqs. (2.13)–(2.14).

4.4 Reduced State by Partial Tracing

We have indicated in Sect. 3.2.1 that, for a quantum system Q entangled with an uncontrolled environment E, it is possible to obtain a (reduced) density operator ρ_Q for Q by partial tracing over the environment E according to Eqs. (3.35) or (3.37). The so-called reduced density operator ρ_Q obtained in this way in Eqs. (3.35) or (3.37) in itself provides a complete description of the measurement results that can be obtained upon performing any quantum measurement on the system Q alone, while the environment E is left untouched—a procedure referred to as a partial measurement in Sect. 4.1.3. For a *von Neumann projective measurement* on system Q, this reduction by partial tracing received justification in Sect. 4.1.3, through Eq. (4.8) and an explicit calculation of the partial trace in some representation basis with Eq. (4.13).

The justification can be extended for a *generalized measurement* on system Q. For this purpose, we consider that the system Q of Hilbert space $\mathcal{H}^Q$ is measured by means of a generalized measurement as in Sect. 4.2, defined by the M measurement operators $\mathsf{M}_m \in \mathcal{L}(\mathcal{H}^Q)$, associated with the M POVM elements $\mathsf{E}_m = \mathsf{M}_m^\dagger \mathsf{M}_m \in \mathcal{L}(\mathcal{H}^Q)$, satisfying the completeness relation of Eq. (4.19) or Eq. (4.38). Meanwhile, nothing is done on the environment E, so that the generalized measurement of Q can be viewed as a generalized measurement on the bipartite system QE defined by the M measurement operators $\mathsf{M}_m \otimes \mathrm{I}^E$, with I^E the identity operator on the environment space $\mathcal{H}^E$, associated with the M POVM elements $\mathsf{E}_m \otimes \mathrm{I}^E$, that indeed form a valid generalized measurement for QE with a completeness relation analogous to Eq. (4.19) or (4.38) still satisfied.

The compound QE is a closed quantum system to which can be assigned a quantum (pre-measurement) state represented by the density operator ρ_{QE} of $\mathcal{L}(\mathcal{H}^Q \otimes \mathcal{H}^E)$. The measurement by the POVM $\{\mathsf{E}_m \otimes \mathrm{I}^E\}$ operating on QE produces the outcome m with the probability $P(m)$ given by Eq. (4.39) as

$$P(m) = \mathrm{tr}_{QE}\big((\mathsf{E}_m \otimes \mathrm{I}^E)\rho_{QE}\big) = \mathrm{tr}_Q\Big(\mathrm{tr}_E\big((\mathsf{E}_m \otimes \mathrm{I}^E)\rho_{QE}\big)\Big) . \tag{4.81}$$

As a general property of the partial trace, by Eq. (A.24) of the Appendix, we have

$$\mathrm{tr}_E\big((\mathsf{A}\otimes \mathrm{I}^E)\rho_{QE}\big) = \mathsf{A}\,\mathrm{tr}_E(\rho_{QE})\,, \tag{4.82}$$

verified by any operator A of $\mathcal{L}(\mathcal{H}^Q)$. We apply Eq. (4.82) to the positive operator $\mathsf{E}_m \equiv \mathsf{A}$, and define the operator

$$\rho_Q = \mathrm{tr}_E\big(\rho_{QE}\big)\,, \tag{4.83}$$

which is a positive operator of $\mathcal{L}(\mathcal{H}^Q)$ with unit trace. The probability of Eq. (4.81) can now be obtained as

$$P(m) = \mathrm{tr}\big(\mathsf{E}_m \rho_Q\big)\,. \tag{4.84}$$

Therefore, partial tracing of the joint state ρ_{QE} over the discarded environment E generates in Eq. (4.83) a (reduced) density operator $\rho_Q \in \mathcal{L}(\mathcal{H}^Q)$ which completely determines, via Eq. (4.84), the probabilities of all possible measurement outcomes upon measuring Q with the generalized measurement $\{\mathsf{M}_m\}_{m=1}^M$ associated with the POVM $\{\mathsf{E}_m = \mathsf{M}_m^\dagger \mathsf{M}_m\}_{m=1}^M$, while the environment E is left untouched.

From a mathematical standpoint, with the reduced density operator $\rho_Q = \mathrm{tr}_E\big(\rho_{QE}\big)$ of Eq. (4.83), the coincidence of Eqs. (4.81) and (4.84) for expressing the measurement probabilities

$$P(m) = \mathrm{tr}\big(\mathsf{E}_m \rho_Q\big) = \mathrm{tr}\big((\mathsf{E}_m \otimes \mathrm{I}^E)\rho_{QE}\big) \tag{4.85}$$

follows from the general (mathematical) property of Eq. (4.82) for the partial trace. From a physical standpoint, this coincidence in Eq. (4.85) expresses a **strong consistency** in the measurement probabilities evaluated via two equivalent alternative scenarios: Either the measuring POVM $\{\mathsf{E}_m\}$ acts locally on Q alone in state ρ_Q, or the joint POVM $\big\{\mathsf{E}_m \otimes \mathrm{I}^E\big\}$ acts globally on the bipartite system QE in state ρ_{QE}. This physical (strong) consistency on the probabilities is granted by the mathematical property of Eq. (4.82) for the partial trace.

In addition, when the measurement of the compound QE by the joint POVM $\big\{\mathsf{E}_m \otimes \mathrm{I}^E\big\}$ produces the outcome m, the post-measurement state of QE is given, according to Eqs. (4.35)–(4.36), by

$$\rho_{QE}^{\mathrm{post}}(m) = \frac{1}{P(m)}\big(\mathsf{M}_m \otimes \mathrm{I}^E\big)\rho_{QE}\big(\mathsf{M}_m \otimes \mathrm{I}^E\big)^\dagger\,. \tag{4.86}$$

By a similar operation of partial tracing of the joint state $\rho_{QE}^{\mathrm{post}}(m)$ over the environment E, as in Eq. (4.83), we obtain a reduced density operator $\rho_Q^{\mathrm{post}}(m) = \mathrm{tr}_E\big(\rho_{QE}^{\mathrm{post}}(m)\big)$ of $\mathcal{L}(\mathcal{H}^Q)$ characterizing the post-measurement state of system Q after outcome m is obtained. It follows as

$$\rho_Q^{\mathrm{post}}(m) = \mathrm{tr}_E\big(\rho_{QE}^{\mathrm{post}}(m)\big) = \frac{1}{P(m)}\,\mathrm{tr}_E\Big(\big(\mathsf{M}_m \otimes \mathrm{I}^E\big)\rho_{QE}\big(\mathsf{M}_m^\dagger \otimes \mathrm{I}^E\big)\Big)\,. \tag{4.87}$$

By the same general property of the partial trace in Eq. (A.24), we have $\mathrm{tr}_E\big((\mathsf{M}_m \otimes \mathrm{I}^E)\rho_{QE}(\mathsf{M}_m^\dagger \otimes \mathrm{I}^E)\big) = \mathsf{M}_m\big[\mathrm{tr}_E(\rho_{QE})\big]\mathsf{M}_m^\dagger = \mathsf{M}_m\rho_Q\mathsf{M}_m^\dagger$. From Eq. (4.87), we therefore deduce

$$\rho_Q^{\mathrm{post}}(m) = \frac{1}{P(m)}\mathsf{M}_m\rho_Q\mathsf{M}_m^\dagger \,. \tag{4.88}$$

We obtain in Eq. (4.88) the post-measurement state $\rho_Q^{\mathrm{post}}(m)$ for system Q, simply by involving the pre-measurement state ρ_Q of Q in the regular relations of Eqs. (4.35)–(4.36).

Therefore, the reduced state ρ_Q defined by partial tracing in Eq. (4.83) provides a complete description of the system Q involved in the partial (generalized) measurement, as it gives access by regular relations to both the outcome probabilities via Eq. (4.84) and post-measurement state via Eq. (4.88).

In addition, measuring any observable Ω of the system Q, while nothing is done on the environment E, amounts to measuring the observable $\Omega \otimes \mathrm{I}^E$ on the bipartite system QE. Repeating the measurement allows one to collect measured values for the observable Ω of Q, while nothing is returned about the environment E. Following Eq. (4.59), the average value $\langle\Omega\rangle$ of the observable Ω is accessible as

$$\langle\Omega\rangle = \mathrm{tr}_Q\big(\Omega\rho_Q\big) = \mathrm{tr}_{QE}\big((\Omega \otimes \mathrm{I}^E)\rho_{QE}\big) \,. \tag{4.89}$$

(Some superfluous subscripts in the traces may sometimes be written as here for more clarity.) From a mathematical standpoint, Eq. (4.89) is independently deduced from the general property of the partial trace in Eq. (4.82) applied to the Hermitian operator $\Omega \equiv \mathsf{A}$ to give $\mathrm{tr}_E\big((\Omega \otimes \mathrm{I}^E)\rho_{QE}\big) = \Omega\,\mathrm{tr}_E(\rho_{QE}) = \Omega\rho_Q$. From a physical standpoint, Eq. (4.89) expresses some **weak consistency** (to call it this way) in the result of two equivalent alternative scenarios for evaluating a similar average $\langle\Omega\rangle$, by averaging the values returned when measuring either the local observable Ω separately on Q alone in state ρ_Q, or the joint observable $\Omega \otimes \mathrm{I}^E$ on the bipartite system QE in the joint in state ρ_{QE}. Again, this physical (weak) consistency on the average is granted by the mathematical property of Eq. (4.82) for the partial trace.

If we adapt the derivation in Eqs. (4.56)–(4.59), we find that when (strong) consistency according to Eq. (4.85) holds on any POVM $\{\mathsf{E}_m\}_{m=1}^M$ for the measurement probabilities $P(m)$, then (weak) consistency follows for the average $\langle\Omega\rangle$ on any observable Ω as in Eq. (4.89). But conversely, when consistency according to Eq. (4.89) holds for any observable Ω, it also holds for any POVM element E_m as in Eq. (4.85), because a POVM element E_m is a positive operator on $\mathcal{H}^Q$, therefore a Hermitian operator, therefore an observable of Q, with average $\langle\mathsf{E}_m\rangle = P(m)$. In this respect, both consistency relations in Eq. (4.85) and Eq. (4.89) are equivalent. Strong consistency on the probabilities as in Eq. (4.85) and weak consistency on the average as in Eq. (4.89) are equivalent. This may seem natural, if not necessary, as they both follow from the mathematical property of Eq. (4.82) for the partial trace. Both consistency relations point to the same reduced density operator $\rho_Q = \mathrm{tr}_E\big(\rho_{QE}\big)$ for a local description of the quantum system Q coupled to an environment E.

The partial trace $\mathrm{tr}_E(\cdot)$ of Eq. (4.83) satisfies both consistency requirements of Eq. (4.85) and of Eq. (4.89) (which are equivalent) as a consequence of the mathematical property of Eq. (4.82). Moreover, it can be shown [9, 59] that the partial trace $\rho_Q = \mathrm{tr}_E(\rho_{QE})$ is the unique option to satisfy the consistency requirement of Eq. (4.89) (equivalent to Eq. (4.85)). As a proof, we can assume an alternative map $f_E(\cdot)$ on density operators from $\mathcal{L}(\mathcal{H}^Q \otimes \mathcal{H}^E)$ onto $\mathcal{L}(\mathcal{H}^Q)$, other than the partial trace $\mathrm{tr}_E(\cdot)$, and we impose that this map satisfy the (consistency) relation similar to Eq. (4.89) as

$$\mathrm{tr}\big(\Omega f_E(\rho_{QE})\big) = \mathrm{tr}\big((\Omega \otimes \mathrm{I}^E)\rho_{QE}\big) , \tag{4.90}$$

for any observable $\Omega \in \mathcal{L}(\mathcal{H}^Q)$. This is equivalent to imposing Eq. (4.90) for any positive operator $\mathsf{E}_m \in \mathcal{L}(\mathcal{H}^Q)$, because if Eq. (4.90) holds for any observable Ω it also holds for any positive operator E_m since such E_m is a Hermitian operator (an observable). Conversely, if Eq. (4.90) holds for any positive operator E_m it holds for any observable Ω. This follows because Eq. (4.90) is a linear constraint in Ω, and when it holds for any positive operator E_m it also holds for any negative operator $-\mathsf{E}_m$, and therefore it also holds for any Hermitian operator Ω which is always the sum of a positive and of a negative operator (this can be shown from the diagonal form of Ω by retaining the positive eigenvalues of Ω for the positive operator and the negative eigenvalues of Ω for the negative operator, as done more explicitly with Eqs. (3.53)–(3.55) above). We then introduce a Hermitian orthonormal basis $\{\mathsf{B}_j\}$ of $\mathcal{L}(\mathcal{H}^Q)$, i.e., formed by Hermitian operators of $\mathcal{L}(\mathcal{H}^Q)$, and such a basis always exists as we will argue in Sect. 11.6.1. In this basis, by referring to the Hilbert–Schmidt inner product $\mathrm{tr}(\mathsf{A}^\dagger \mathsf{B})$ for the coordinates, any operator A of $\mathcal{L}(\mathcal{H}^Q)$ has the unique decomposition $\mathsf{A} = \sum_j \mathrm{tr}(\mathsf{B}_j^\dagger \mathsf{A})\mathsf{B}_j$. For ρ_Q from the partial trace this gives the decomposition $\rho_Q = \mathrm{tr}_E(\rho_{QE}) = \sum_j \mathrm{tr}(\mathsf{B}_j^\dagger \rho_Q)\mathsf{B}_j = \sum_j \mathrm{tr}\big((\mathsf{B}_j^\dagger \otimes \mathrm{I}^E)\rho_{QE}\big)\mathsf{B}_j$ since all the operators B_j are Hermitian and therefore satisfy the consistency relation of Eq. (4.89). The alternative map $f_E(\cdot)$ gives $f_E(\rho_{QE}) = \sum_j \mathrm{tr}\big(\mathsf{B}_j^\dagger f_E(\rho_{QE})\big)\mathsf{B}_j = \sum_j \mathrm{tr}\big((\mathsf{B}_j^\dagger \otimes \mathrm{I}^E)\rho_{QE}\big)\mathsf{B}_j$ since $f_E(\cdot)$ is assumed to satisfy the consistency relation of Eq. (4.90). We arrive at the same decomposition $\sum_j \mathrm{tr}\big((\mathsf{B}_j^\dagger \otimes \mathrm{I}^E)\rho_{QE}\big)\mathsf{B}_j$ in the basis $\{\mathsf{B}_j\}$ for both $\mathrm{tr}_E(\rho_{QE})$ and $f_E(\rho_{QE})$, proving that a putative alternative map $f_E(\cdot)$ that would satisfy the consistency relation of Eq. (4.90) necessarily coincides with the partial trace $\mathrm{tr}_E(\cdot)$ satisfying Eq. (4.89).

The reduced density operator $\rho_Q = \mathrm{tr}_E(\rho_{QE})$ is a central notion, providing a local description of a quantum system Q coupled to an uncontrolled or unaccessed environment E. And the environment E can equivalently represent any quantum system to which Q is coupled and that remains unobserved or unaccessed. Also, in the present derivation of the reduced density operator $\rho_Q = \mathrm{tr}_E(\rho_{QE})$ for the system Q coupled to an uncontrolled environment E, the narrative was that while Q is measured locally, nothing is done on E or E is left untouched. But it turns out that while Q is measured locally, any quantum evolution represented by the quantum process $\mathcal{E}_E(\cdot)$ can take place on E. Provided that $\mathcal{E}_E(\cdot)$ acts locally, separately, on E alone, this does not change the reduced density operator $\rho_Q = \mathrm{tr}_E(\rho_{QE})$ obtained locally for Q. This follows from the no-signaling theorem, which is going to be examined in Sect. 6.3, and ultimately it rests on general properties of the partial trace upon a local operation $\mathcal{E}_E(\cdot)$ on E. The reduced density operator plays an essential role in the study of the decoherence of quantum states as a result of their coupling to an uncontrolled environment, which represents a fundamental source of quantum noise altering quantum signals. Accordingly, reduced density operators will be encountered on many occasions to come in the book.

4.5 Irreversibility and Nonlinearity from Measuring the Environment

To describe the evolution of a quantum system Q coupled to an uncontrolled environment E, we have considered in Sect. 3.2.1 that the compound QE initially in the joint state $\rho_{QE} = \rho_Q \otimes |e_0\rangle\langle e_0|$ is unitarily evolved into $\rho'_{QE} = \mathsf{U}_{QE}\rho_{QE}\mathsf{U}^\dagger_{QE} = \mathsf{U}_{QE}\big(\rho_Q \otimes |e_0\rangle\langle e_0|\big)\mathsf{U}^\dagger_{QE}$. To obtain the final state ρ'_Q for Q, instead of the partial trace operation $\rho'_Q = \mathrm{tr}_E\big(\rho'_{QE}\big) = \mathcal{E}(\rho_Q)$ as in Eq. (3.37), it is possible to consider the equivalent scenario that follows [6, 25].

On the bipartite system QE in the evolved joint state ρ'_{QE} under the Stinespring isometric form of Eq. (3.114), we consider performing a measurement with the projectors $\Pi_k = \mathrm{I}^Q \otimes |e_k\rangle\langle e_k|$, with $\{|e_k\rangle\}_{k=1}^K$ the orthonormal basis of the environment space $\mathcal{H}^E$ appearing in Eq. (3.114) and used in Eq. (3.38) to define the Kraus operators $\{\Lambda_k\}_{k=1}^K$. Such a set of operators $\{\Pi_k\}_{k=1}^K$ forms a licit generalized measurement (and also a POVM) for the bipartite system QE. This amounts to performing a von Neumann measurement on the environment E in the projective basis $\{|e_k\rangle\}_{k=1}^K$ while doing nothing on the principal system Q. When the outcome k is obtained, the projector Π_k has acted on the bipartite system QE and projected its state onto the joint state associated with the unnormalized density operator $\Pi_k\rho'_{QE}\Pi_k^\dagger$, which follows from Eq. (3.114) as

$$\Pi_k\rho'_{QE}\Pi_k^\dagger = \Pi_k\Lambda\rho_Q\Lambda^\dagger\Pi_k^\dagger \tag{4.91}$$

$$= \big(\Lambda_k \otimes |e_k\rangle\big)\rho_Q\big(\Lambda_k \otimes |e_k\rangle\big)^\dagger \tag{4.92}$$

$$= \Lambda_k\rho_Q\Lambda_k^\dagger \otimes |e_k\rangle\langle e_k| \; . \tag{4.93}$$

And according to Eq. (4.31) for a generalized measurement with Π_k acting, the measurement outcome k is obtained with the probability

$$p_k = \mathrm{tr}\big(\Pi_k\rho'_{QE}\Pi_k^\dagger\big) = \mathrm{tr}\big(\Lambda_k\rho_Q\Lambda_k^\dagger\big) \; . \tag{4.94}$$

In this way, after the measurement outcome k is obtained, the bipartite system QE terminates in the joint quantum state (properly normalized to unit trace):

$$\frac{1}{p_k}\Pi_k \rho'_{QE} \Pi_k^\dagger = \frac{\Lambda_k \rho_Q \Lambda_k^\dagger}{\mathrm{tr}\big(\Lambda_k \rho_Q \Lambda_k^\dagger\big)} \otimes |e_k\rangle \langle e_k| \ . \tag{4.95}$$

This is a separable product state, with the environment E projected onto the pure state $|e_k\rangle$ and the principal system Q placed into the generally mixed state:

$$\rho_k = \frac{\Lambda_k \rho_Q \Lambda_k^\dagger}{\mathrm{tr}\big(\Lambda_k \rho_Q \Lambda_k^\dagger\big)} \ . \tag{4.96}$$

This is the same state ρ_k of Eq. (3.100) occurring with the probability p_k similar to Eq. (3.97) as already encountered in Sect. 3.2.5 when devising for the operator-sum operation on Q a probabilistic interpretation that now receives a more explicit materialization via a von Neumann measurement performed on the environment E coupled to Q. The random replacement by ρ_k of Eq. (3.100) as evoked rather abstractly in Sect. 3.2.5 is here produced by the von Neumann measurement on the environment E.

The distribution of the measurement probabilities p_k in Eq. (4.94) is dependent on the initial state ρ_Q of the principal system Q. Thus the measurement taking place in the environment E coupled to Q can be seen as information about the system Q leaking out into the environment, this being accompanied by an alteration of the initial state ρ_Q of Q forced into a random final state ρ_k from Eq. (4.96).

If the measurement outcome k is unknown, this is what we have called an unread measurement. The principal system Q is then in a statistical ensemble of states $\{(p_k, \rho_k)\}_{k=1}^K$ represented by the density operator $\sum_{k=1}^K p_k \rho_k = \sum_{k=1}^K \Lambda_k \rho_Q \Lambda_k^\dagger$, which is precisely the operator-sum or Kraus representation of the final state $\rho'_Q = \mathcal{E}(\rho_Q)$ for Q as in Eq. (3.45). This establishes that the action of an uncontrolled environment in determining the evolution $\rho_Q \mapsto \mathcal{E}(\rho_Q) = \rho'_Q$ of Q can equivalently be interpreted as the effect of an unread measurement taking place on the environment E. A non-unitary irreversible evolution $\mathcal{E}(\cdot)$ ensues, altering the state of Q and leaking information about Q into the environment E. This occurs unless $\mathcal{E}(\cdot)$ is a unitary reversible (closed) evolution of Q, with only one non-null (unitary) Kraus operator Λ_1, no entanglement occurring between Q and E in their bipartite evolution by Eq. (3.114), and the environment always measured in the same state $|e_1\rangle$ with probability $p_1 = 1$ in Eq. (4.94), independently of the initial state ρ_Q of Q, leaking no information about Q into the environment E. In a closed (reversible unitary) evolution of Q, the system Q remains informationally isolated from its environment, while in an open (irreversible) evolution of Q, the system Q leaks information into its environment [6, 7].

If the von Neumann measurement of the environment E takes place in another orthonormal basis $\{|e'_k\rangle\}_{k=1}^K$, this brings into play in Eqs. (4.91)–(4.93) an alternative set of Kraus operators $\{\Lambda'_k\}_{k=1}^K$ which is unitarily related to the previous Kraus set $\{\Lambda_k\}_{k=1}^K$ associated with the former basis $\{|e_k\rangle\}_{k=1}^K$. These two unitarily related sets of Kraus operators generate the same quantum evolution $\mathcal{E}(\cdot)$ of the principal system Q, as we have seen previously with Eqs. (3.81)–(3.83). Any (unread) von Neumann measurement of the environment E coupled to Q generates the same evolution $\mathcal{E}(\cdot)$ for Q.

Interaction of an open quantum system Q with an uncontrolled environment leads to a non-unitary irreversible alteration of Q leaking information about Q into the environment. This points at the quantum level to connections between physical irreversibility and information loss, reminiscent of thermodynamics and Maxwell's demon [60–64].

Returning to the final post-measurement state ρ_k of Eq. (4.96), when the principal system Q in the Hilbert space $\mathcal{H}^Q$ is initially prepared in the separable pure state $\rho_Q = |Q\rangle\langle Q|$, the final state of Eq. (4.96) can be expressed as $\rho_k = |Q_k\rangle\langle Q_k|$ with

$$|Q_k\rangle = \frac{\Lambda_k |Q\rangle}{\sqrt{\langle Q|\Lambda_k^\dagger \Lambda_k|Q\rangle}}\,, \tag{4.97}$$

which is a valid separable pure state of $\mathcal{H}^Q$ as in Eq. (3.91). After measuring the environment E in the basis state $|e_k\rangle$, the principal system Q gets placed in the pure state $|Q_k\rangle$ of Eq. (4.97), and this outcome occurs with the probability $p_k = \langle Q|\Lambda_k^\dagger \Lambda_k|Q\rangle$ from Eq. (4.94). The evolution $|Q\rangle \mapsto |Q_k\rangle$ defined by Eq. (4.97) comes as a *nonlinear* evolution of the quantum state vector $|Q\rangle$ of the principal system Q, essentially determined here by the environment E via the Kraus operator Λ_k that randomly acts. This is the same type of nonlinear evolution $|\psi\rangle \mapsto |\phi_m^{\text{post}}\rangle$ that was observed with Eq. (4.27) in Sect. 4.2.2 when motivating a generalized measurement on Q via coupling to an auxiliary system A, yet with an evolution of Q determined in Eq. (4.27) by the measurement operator M_m rather than by the Kraus operator Λ_k in Eq. (4.97). These two points of view are closely related, as we now know, from our discussion at the end of Sect. 4.2.3 showing that a non-unitary evolution of Q can be viewed as an (unread) generalized measurement on Q, and from Sect. 4.2.2 motivating a generalized measurement on Q by showing it is equivalent to a standard von Neumann measurement on an auxiliary system coupled to Q.

4.6 Summary on Quantum Measurement

This chapter focused on the **third postulate** (among three) of fundamental principle for quantum theory, stating that measurement of a quantum system essentially occurs as a random projection in an orthonormal basis of the Hilbert space, with several important aspects of this third postulate that were further examined, especially for generalized measurements (as a von Neumann projective measurement in an auxiliary Hilbert space).

A standard or **von Neumann projective measurement**, on a system in a state (pure or mixed) $\rho \in \mathcal{L}(\mathcal{H}_N)$, is defined from N mutually orthogonal projectors $\Pi_n = |n\rangle\langle n|$ constructed from an orthonormal basis $\{|n\rangle\}_{n=1}^N$ of $\mathcal{H}_N$, and satisfying the completeness relation $\sum_{n=1}^N \Pi_n^\dagger \Pi_n = \mathrm{I}_N$ of Eq. (4.5). The measurement takes place as a random projection by Π_n with the probability $P(n) = \langle n|\rho|n\rangle = \mathrm{tr}\big(\rho\,|n\rangle\,\langle n|\big) = \mathrm{tr}\big(\rho\Pi_n^\dagger\Pi_n\big)$ from Eq. (4.1), to yield the pure post-measurement state $|n\rangle$ with density operator $|n\rangle\,\langle n| = \Pi_n\rho\Pi_n^\dagger/P(n)$.

A **generalized quantum measurement** is defined by M measurement operators $\mathsf{M}_m \in \mathcal{L}(\mathcal{H}_N)$, which satisfy the completeness relation $\sum_{m=1}^{M}\mathsf{M}_m^\dagger\mathsf{M}_m = \mathrm{I}_N$ of Eq. (4.19). When measuring an N-dimensional quantum system Q in state $\rho \in \mathcal{L}(\mathcal{H}_N)$, this affords the possibility of $M \neq N$ distinct random outcomes, with for outcome m the probability $P(m) = \mathrm{tr}\big(\rho\mathsf{M}_m^\dagger\mathsf{M}_m\big)$ from Eq. (4.31), associated with the post-measurement state $\rho_m = \mathsf{M}_m\rho\mathsf{M}_m^\dagger/P(m)$ from Eq. (4.36). A generalized measurement on a system Q with M measurement operators can always be realized as a standard von Neumann measurement delegated to an M-dimensional auxiliary system appropriately entangled with Q. The generalized quantum measurement of Q encompasses the standard von Neumann measurement on Q when $\mathsf{M}_n = \Pi_n = |n\rangle\,\langle n|$, for $n = 1$ to $N = M$.

When only the M measurement probabilities $P(m)$ are of interest, and not the post-measurement states ρ_m, the M positive operators $\mathsf{E}_m = \mathsf{M}_m^\dagger\mathsf{M}_m$ are enough to characterize the quantum measurement via the probabilities $P(m) = \mathrm{tr}\big(\rho\mathsf{E}_m\big)$, and they define a **POVM**, as presented in Sect. 4.2.4. When a POVM $\{\mathsf{E}_m\}_{m=1}^{M}$ is fixed, the set of M positive measurement operators $\mathsf{M}_m = \sqrt{\mathsf{E}_m}$ verifying $\mathsf{M}_m^\dagger\mathsf{M}_m = \mathsf{E}_m$ offers one possibility for a physical implementation.

In a bipartite quantum system QE with joint density operator ρ_{QE}, if Q alone is measured while E remains unaccessed, the measurement results observed on Q are completely characterized by the **reduced density operator** $\rho_Q = \mathrm{tr}_E\big(\rho_{QE}\big)$ from Eq. (4.83).

For N-dimensional quantum systems, a **quantum observable** Ω is a Hermitian operator of $\mathcal{L}(\mathcal{H}_N)$. On a quantum system in state $\rho \in \mathcal{L}(\mathcal{H}_N)$, measuring the observable Ω is equivalent to performing a projective measurement by means of the projectors onto the mutually orthogonal eigensubspaces of Ω to reveal the corresponding eigenvalue, as explained in Sect. 4.3.

As discussed at the end of Sect. 4.2.3, any **unread** generalized measurement of Q with M operators $\{\mathsf{M}_m\}_{m=1}^{M}$ is a quantum process with the M Kraus operators M_m. Conversely, any quantum process with K Kraus operators $\{\Lambda_k\}_{k=1}^{K}$ can be seen as an unread generalized measurement of Q with K measurement operators $\mathsf{M}_k = \Lambda_k$.

When an open quantum system Q is entangled with an auxiliary system A or environment E, a von Neumann measurement of A or E can provide the possibility of a **nonlinear evolution** of the state of Q, yet in a probabilistic manner, as discussed at the end of Sect. 4.2.2 and in Sect. 4.5.

Chapter 5
Quantum Circuits and Computation

In this chapter, we present essential notions on quantum circuits materializing the physical hardware exploitable for quantum processing and computation. As illustrations, we also present landmark quantum algorithms with broad applicability and manifesting specific capabilities to achieve performance unattainable by classical means.

5.1 Universal Quantum Gates

Quantum circuits are useful to the physical implementation of quantum processing and computation [65]. For their materialization, several technologies are currently being developed, offering different trade-offs between desirable properties, and based, for example, on photonic devices [66, 67], superconducting circuits [68, 69], spin qubits [70, 71], trapped ions [72, 73], neutral atoms [74, 75], or other technological modalities [76–78]. We will not go into the details and specificities of these concurrent technologies of physical implementation, which are today in rapid evolution and constant progress.

Independently of their technological materialization, quantum circuits exploit important generic properties in common. As we have indicated in Sect. 2.4, any finite-dimensional quantum states can be realized by assembling qubits in sufficient number, and accordingly qubit circuits play a central role for quantum processing and computation. In addition, as indicated in Sect. 3.1.5, any unitary[1] evolution of an arbitrary number of qubits can be realized [20] by quantum circuits assembling one-qubit unitary gates and two-qubit Cnot gates. In this way, one-qubit gates supplemented by the two-qubit Cnot gate form **a universal set of quantum gates** for synthesizing circuits for quantum processing and computation. It is even possible to

[1] And non-unitary as well, via the dilated unitary evolution of Eq. (3.121) and the considerations of Sect. 3.2.8.

F. Chapeau-Blondeau, *Quantum Information and Quantum Signal Processing*,
https://doi.org/10.1007/978-3-032-27499-1_5

reduce the one-qubit gates to a small subset of a few well-chosen one-qubit gates, to maintain, with the Cnot, a universal set of gates.

For example, for a one-qubit unitary operator $\mathsf{U} \in \mathcal{L}(\mathcal{H}_2)$ we can consider the most general mathematical parameterization under the form:

$$\mathsf{U} = e^{ic} \exp\left(-i\,\frac{\xi}{2}\vec{n}\cdot\vec{\sigma}\right) = e^{ic}\begin{bmatrix} \cos(\xi/2) - i\sin(\xi/2)n_z & -\sin(\xi/2)\,(n_y + in_x) \\ \sin(\xi/2)\,(n_y - in_x) & \cos(\xi/2) + i\sin(\xi/2)n_z \end{bmatrix}, \tag{5.1}$$

which is similar to Eq. (3.16). Such a qubit unitary process U can be synthesized [20] by means of the product of three rotation operators, as

$$\mathsf{U} = e^{i\alpha}\mathsf{R_z}(\beta)\mathsf{R_y}(\gamma)\mathsf{R_z}(\delta)\,, \tag{5.2}$$

giving

$$\mathsf{U} = e^{i\alpha}\begin{bmatrix} e^{-i(\beta+\delta)/2}\cos(\gamma/2) & -e^{-i(\beta-\delta)/2}\sin(\gamma/2) \\ e^{i(\beta-\delta)/2}\sin(\gamma/2) & e^{i(\beta+\delta)/2}\cos(\gamma/2) \end{bmatrix}, \tag{5.3}$$

with especially three appropriate rotation angles β, γ, and δ selected so that Eq. (5.3) matches Eq. (5.1), and reminiscent of the Euler angles of solid mechanics. By this "Euler-angle" decomposition of Eq. (5.2), any qubit unitary operator U can be implemented by cascading two types of qubit rotation gates, the rotation gate $\mathsf{R_z}(\cdot)$ from Eq. (3.28) rotating in $\mathbb{R}^3$ the qubit Bloch vector around the Oz axis, and the rotation gate $\mathsf{R_y}(\cdot)$ from Eq. (3.26) rotating the qubit Bloch vector around the Oy axis (and also possibly a gate $e^{i\alpha}\mathrm{I}_2$ if the global phase matters).

With a given implementation technology, it is therefore enough to manage the physical realization of a universal set of quantum gates, to be able, in principle, to materialize a quantum circuit performing any conceivable unitary process on an arbitrary number of qubits. For instance, with photonic qubits, for a universal set of gates, a generic one-qubit unitary operation under the form of Eq. (3.18) can be realized with the interferometric device of Fig. 3.4. A photonic two-qubit Cnot gate requires a coupling between two photons, by means of a nonlinear interaction, such as the Kerr effect, for instance, in optics, with examples discussed in [79–81].

A useful example is the quantum circuit from Fig. 5.1 which transforms the two-qubit separable computational basis $\{|ab\rangle = |a\rangle \otimes |b\rangle\}$ into the entangled Bell basis $\{|\beta_{ab}\rangle\}$, for $a, b \in \{0, 1\}$, according to Eqs. (2.21)–(2.24).

The quantum circuit of Fig. 5.1 uses the two-qubit Cnot gate from Fig. 3.5, and the one-qubit Hadamard gate H from Eq. (A.43) also obtainable in Eq. (5.1) with the parameters $c = \pi/2$, $\xi = \pi$ and $\vec{n} = [1, 0, 1]^\top/\sqrt{2}$. Both the Hadamard gate H and the Cnot gate are self-inverse gates, verifying $\mathsf{H}^{-1} = \mathsf{H}$ and $\mathsf{Cnot}^{-1} = \mathsf{Cnot}$. The operation of the Bellization circuit of Fig. 5.1 is therefore simple to invert. The reverse cascade with the Cnot gate followed by the Hadamard gate H inverses the transformation of Fig. 5.1 and transforms the Bell basis $\{|\beta_{ab}\rangle\}$ back into the computational basis $\{|ab\rangle = |a\rangle \otimes |b\rangle\}$, for $a, b \in \{0, 1\}$, according to Eqs. (2.25)–(2.28). Such a procedure of basis transformation may be generally useful as a preprocessing for an equivalent implementation of a Von Neumann projective measurement in a

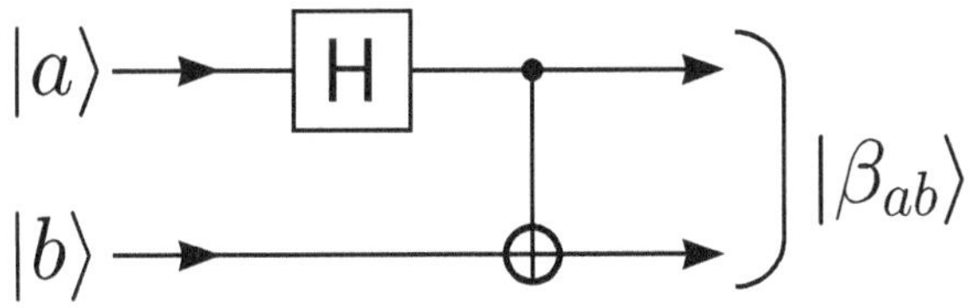

Fig. 5.1 Bellization gate or circuit using a one-qubit Hadamard gate H and a two-qubit Cnot gate, to transform the two-qubit computational basis $\{|ab\rangle = |a\rangle \otimes |b\rangle\}$ into the Bell basis $\{|\beta_{ab}\rangle\}$, for $a, b \in \{0, 1\}$, according to Eqs. (2.21)–(2.24)

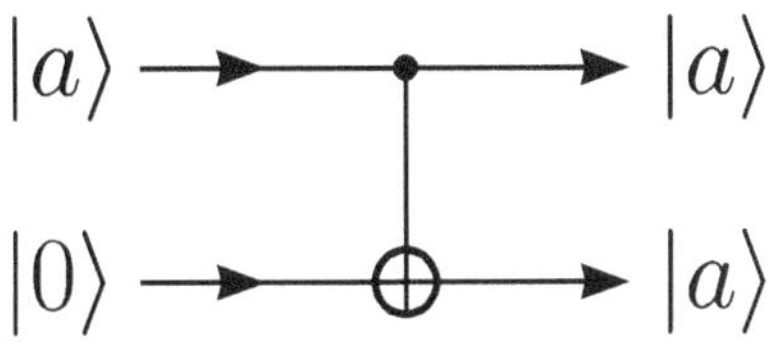

Fig. 5.2 The fan-out quantum gate, with $a = 0$ or 1, duplicates the two basis states $|0\rangle$ and $|1\rangle$ of the qubit

more favorable basis. A Bell basis is an entangled basis, and direct measurement of a qubit pair in the Bell basis $\{|\beta_{ab}\rangle\}$ would require a collective joint processing of the two qubits of the pair. If the inverse Bellization circuit is applied to the qubit pair prior to its measurement, the measurement in the Bell basis $\{|\beta_{ab}\rangle\}$ is turned into an equivalent measurement of the pair in the two-qubit computational basis $\{|ab\rangle = |a\rangle \otimes |b\rangle\}$, which is a separable basis, so that now—more favorably—each qubit of the pair can be measured separately in the one-qubit computational basis $\{|0\rangle, |1\rangle\}$. More generally in this direction, any entangled orthonormal measurement basis on L qubits can always be unitarily transformed into the separable computational basis of L qubits, where each of the L qubits can be measured separately for an equivalent measurement. Accordingly, quantum processors enabling implementation of arbitrary unitary transformations on L qubits, by means of a universal set of qubit gates, can be confined to separable measurements of individual qubits in the one-qubit computational basis $\{|0\rangle, |1\rangle\}$, this without loss of generality.

The quantum circuit of Fig. 5.1 also illustrates an important general feature of multiple-qubit circuits, already encountered with the two-qubit Cnot gate of Fig. 3.5: Although at the input may be applied a multiple-qubit separable state where each qubit has a well-defined state of its own, entanglement may be created by the circuit, as in Fig. 5.1 or Fig. 3.5, to produce at the output a multiple-qubit entangled state where each qubit is no longer endowed with an individual state of its own.

Finally, the Bellization circuit of Fig. 5.1 concisely features two elementary gates giving access to the two crucial and distinctive resources of quantum computation: the Hadamard gate creates superposition of quantum states, while the Cnot gate creates entanglement.

Another useful quantum circuit is the fan-out gate of Fig. 5.2, which duplicates the two basis states $|0\rangle$ and $|1\rangle$ of the qubit.

It is important to note that the fan-out gate of Fig. 5.2 duplicates *only* the two basis states $|0\rangle$ and $|1\rangle$. It does not conflict with the no-cloning theorem of Sect. 3.1.3

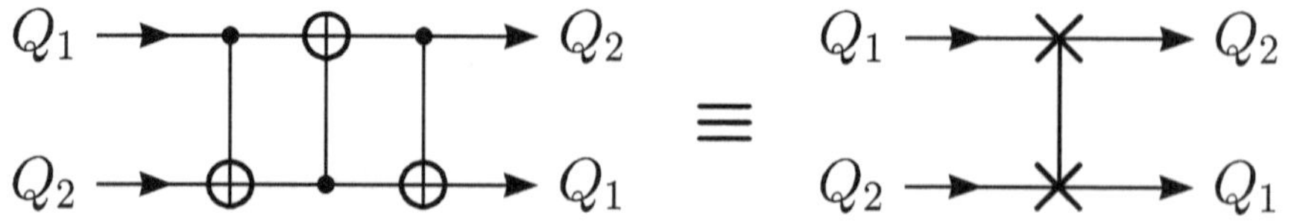

Fig. 5.3 The swap quantum gate which swaps the states of two qubits Q_1 and Q_2, with its equivalent schematic symbol

because it does not duplicate any arbitrary qubit state. If the input marked $|a\rangle$ in the fan-out gate of Fig. 5.2 is fed with the general qubit state $|\psi_a\rangle = \alpha_0 |0\rangle + \alpha_1 |1\rangle$, by linearity, the output state is the two-qubit entangled state $\alpha_0 |00\rangle + \alpha_1 |11\rangle$, which in general differs from $|\psi_a\rangle \otimes |\psi_a\rangle$ (except precisely when $|\psi_a\rangle = |0\rangle$ or $|1\rangle$). The fan-out quantum gate of Fig. 5.2 duplicates the Boolean variable a encoded in a quantum form by the qubit basis state $|a\rangle$, with $a = 0$ or 1. Such quantum encoding of Boolean variables, and their being duplicable, will be relevant to us shortly, in the next section. By cascading several fan-out gates, the two basis states $|0\rangle$ and $|1\rangle$ of the qubit can be duplicated in an arbitrary number of copies.

Another example of useful quantum circuit is the swap gate of Fig. 5.3. The swap quantum gate of Fig. 5.3 implements a pipeline cascading three Cnot gates. It can be easily verified from the action of the Cnot gate that the swap gate of Fig. 5.3, for any two states of two qubits Q_1 and Q_2 at the input, swaps or interchanges these two states on the output.

When the input state to the swap gate of Fig. 5.3 is the general two-qubit state

$$|\psi_{\text{in}}\rangle = \alpha_{00} |00\rangle + \alpha_{01} |01\rangle + \alpha_{10} |10\rangle + \alpha_{11} |11\rangle \ , \tag{5.4}$$

the output state is the two-qubit state

$$|\psi_{\text{out}}\rangle = \alpha_{00} |00\rangle + \alpha_{01} |10\rangle + \alpha_{10} |01\rangle + \alpha_{11} |11\rangle \ . \tag{5.5}$$

Quantum circuits, as they implement unitary processes, are reversible circuits, performing a bijective input–output transformation. This reversibility of quantum circuits holds the promise, in principle, of computation with arbitrarily small energy dissipation [20, 82]. In addition, any (non-reversible) computation performed classically with Boolean logic circuits can be implemented reversibly with quantum circuits. This can be obtained with the Toffoli quantum gate as we now explain.

5.2 Toffoli Gate

The Toffoli gate, also known as the controlled-controlled-Not gate, is a three-qubit quantum gate, with two control qubits and one target qubit. It is represented in Fig. 5.4, where the three binary variables a, b, and c, when 0 or 1, are used to express the transformation of the computational basis by the Toffoli gate.

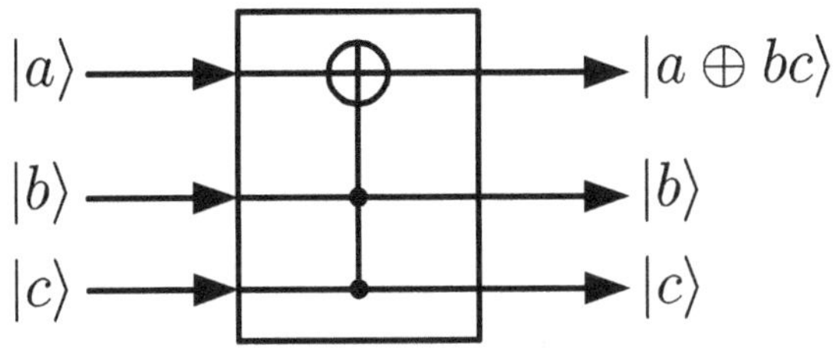

Fig. 5.4 Toffoli or controlled-controlled-Not quantum gate, where the target qubit prepared in the computational-basis state $|a\rangle$ is reversed by application of a Not gate, if and only if the two control qubits are in states $|b\rangle = |1\rangle$ and $|c\rangle = |1\rangle$

In the Toffoli gate, as expressed in Fig. 5.4 by means of the bitwise Xor operation $\oplus$, the state of the target qubit prepared in the computational-basis state $|a\rangle$ is reversed by application of a Not gate, if and only if the states of the two control qubits are $|b\rangle = |1\rangle$ and $|c\rangle = |1\rangle$, otherwise the target qubit is left unchanged. By describing the transformation of the states of the three-qubit computational basis, Fig. 5.4 provides a complete characterization of the Toffoli gate. In addition, as we already suggested at the end of Sect. 3.1.5, a powerful and specifically quantum feature lies in the possibility of operating multiple-qubit circuits like the Toffoli gate of Fig. 5.4, with a linear superposition of basis states at the input. In this circumstance, the quantum state at the output will result from the standard superposition principle of linear algebra, but it may generally come as a multiple-qubit entangled state, usually not separable as a tensor product of one-qubit states. Accordingly, the notations with three separable quantum states displayed in Fig. 5.4, and in Fig. 5.5, and the like, are intended to illustrate the transformation of the computational-basis states only.

Leaning on the universal set of quantum gates indicated in Sect. 5.1, a circuit for synthesizing the Toffoli gate is given in [83] (its Fig. 9) with 6 Cnot gates and 8 one-qubit gates; a similar circuit is also given in [9] (its Fig. 4.9). In passing, the decomposition or synthesis of a given quantum circuit by means of a universal set of quantum gates is not today a necessarily straightforward task benefiting from an off-the-shelf systematic and optimized resolution procedure, especially because it may depend on the criterion chosen for optimality, on the algorithmic cost of the procedure, and on practical resources and hardware constraints [20, 84–87]. Quantum compilers are devised for automatic handling of this task [86, 88–90].

The Toffoli gate is specially interesting as it gives the capability of synthesizing, by means of a reversible quantum circuit, the non-reversible classical logic gates forming the foundation of Boolean circuits of digital electronics. Figure 5.5 shows a quantum synthesis of a classical Nand operation $a \uparrow b$ between two Boolean variables a and b.

In the quantum Nand circuit of Fig. 5.5, the binary variables a and b, when 0 or 1, are again used to label the two states $|0\rangle$ and $|1\rangle$ of the qubit computational basis. By the operation of the Toffoli gate, the result of the Boolean Nand $a \uparrow b$ is obtained at the output as the computational-basis state $|a \uparrow b\rangle$. To ensure reversibility,

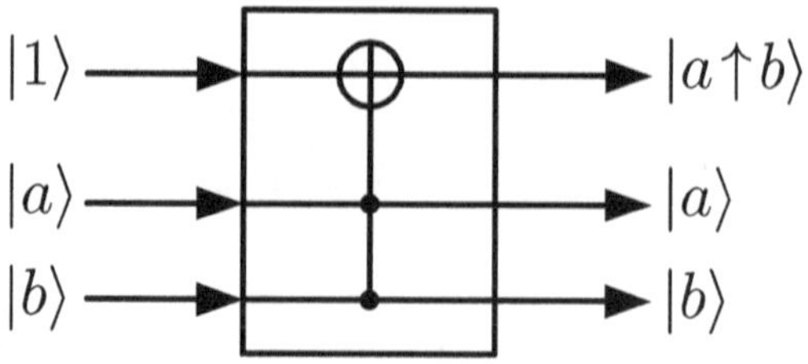

Fig. 5.5 A Toffoli quantum gate realizing a reversible synthesis of a classical Nand operation $a \uparrow b$ between the two Boolean variables a and b

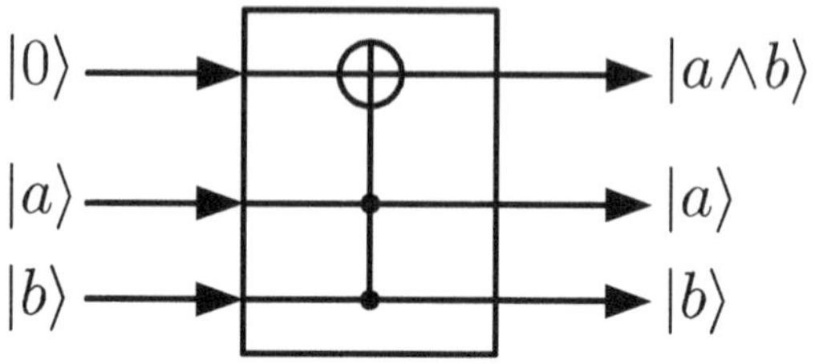

Fig. 5.6 A Toffoli quantum gate realizing a reversible synthesis of a classical And operation $a \wedge b$ between the two Boolean variables a and b

the quantum circuit in Fig. 5.5 arranges the same number of qubits (three here) at the input and at the output, with their three-qubit states connected by a bijective input–output relationship. The circuit transfers the two inputs $|a\rangle$ and $|b\rangle$ to the output and exploits an auxiliary qubit initialized in state $|1\rangle$ to support the result $|a \uparrow b\rangle$ at the output. From the full specification of Fig. 5.5 for the quantum Nand circuit by specifying how it transforms the states of the three-qubit computational basis, the interesting and specifically quantum supplement, as we indicated before, is that the circuit is also operable on any superposition of basis states.

Since the two-variable Nand operation represents alone a universal gate for classical Boolean functions, the Nand implementation of Fig. 5.5 provides the elemental brick enabling quantum synthesis of any conceivable Boolean function on an arbitrary number of Boolean variables. For such universal Boolean synthesis, the quantum fan-out gate of Fig. 5.2 duplicating basis states may also be useful, but is itself realizable from a Toffoli gate, since when in Fig. 5.4 the input $|c\rangle$ is forced at $|1\rangle$ the Toffoli gate realizes the Cnot gate on its two other inputs, which is enough to realize the fan-out gate of Fig. 5.2. In the quantum implementation of the Boolean Nand in Fig. 5.5, since the two Boolean input variables are forwarded to the output, it can be envisaged to realize any combinations of Nand operations so as to compute any Boolean function of the Boolean variables. From such quantum Boolean circuits synthesized in this way, the extra feature of interest lies in the possibility of evaluating the Boolean functions simultaneously in parallel on superposition of basis states. We will examine in more detail, in the next section, this important possibility of parallel evaluation of a function, central to quantum parallel computation.

For the moment, as an alternative, Fig. 5.6 shows the quantum synthesis, with a Toffoli gate, of a classical And operation $a \wedge b$ between two Boolean variables a and b.

The two-variable And operation supplemented by the logical Not together forms another universal set of gates for classical Boolean functions. The quantum And implementation of Fig. 5.6 supplemented by the quantum or Pauli σ_x Not gate, and possibly by the quantum fan-out circuit of Fig. 5.2, together also provides the possibility of a quantum synthesis for any conceivable Boolean function on an arbitrary number of variables.

5.3 Parallel Evaluation of a Function

We consider an arbitrary (classical, so to speak) Boolean function $f(\cdot)$ from L input bits $\vec{a}$ into one output bit $f(\vec{a})$, as

$$\vec{a} \in \{0, 1\}^L \longmapsto f(\vec{a}) \in \{0, 1\} . \tag{5.6}$$

The 2^L values of the Boolean variable $\vec{a}$ receive a quantum encoding by the 2^L basis states $|\vec{a}\,\rangle$ of the orthonormal computational basis of a system of L qubits. The evaluation of the Boolean function $f(\cdot)$, to obtain $f(\vec{a})$ for any input argument $\vec{a}$, can be executed by means of a unitary operator U_f implemented by the (reversible) quantum circuit depicted in Fig. 5.7.

To ensure reversibility, the quantum circuit of Fig. 5.7 arranges a same number of qubits ($L + 1$ here) at the input and at the output, with their states connected by a bijective input–output relationship. The variable $\vec{a} \in \{0, 1\}^L$ in Fig. 5.7 is an L-bit word used to label the 2^L quantum states $|\vec{a}\,\rangle$ of the computational basis of L input qubits to the circuit. The circuit transfers the L input qubits labeled by $\vec{a}$ to the output. The circuit in Fig. 5.7 also exploits an auxiliary input qubit, with its two basis states $|b\rangle$ labeled by $b \in \{0, 1\}$, and intended to receive at the output the binary result $f(\vec{a})$. To maintain the overall reversibility, it is convenient to produce at the output the binary variable $b \oplus f(\vec{a})$, which delivers (reversibly) the binary output $f(\vec{a})$ from the input $b = 0$, and the binary complement $\overline{f(\vec{a})}$ from the input $b = 1$. The possibility of such a quantum circuit as in Fig. 5.7, for computing any Boolean

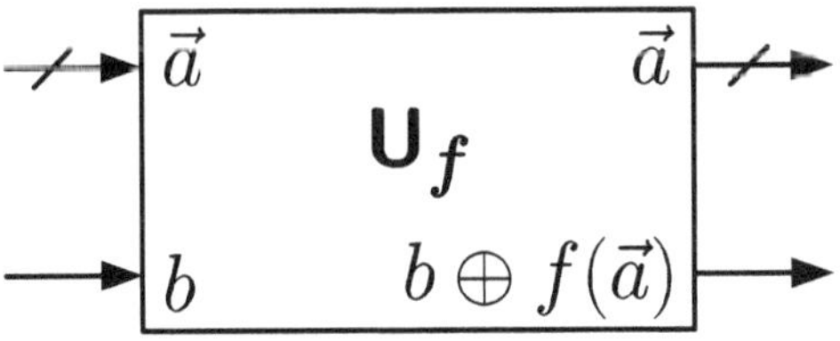

Fig. 5.7 A quantum circuit to realize a reversible evaluation of an arbitrary Boolean function $\vec{a} \mapsto f(\vec{a})$ from L input bits $\vec{a}$ into one output bit $f(\vec{a})$. The circuit performs the unitary transformation U_f from $L + 1$ input qubits into $L + 1$ output qubits

function $f(\vec{a})$ while forwarding the input argument $\vec{a}$ to the output, is granted by the universal Boolean synthesis afforded by the Toffoli gate as we explained in Sect. 5.2.

Furthermore, any multicomponent Boolean function from $L_{\rm in}$ input bits into $L_{\rm out}$ output bits can always be expressed as $L_{\rm out}$ Boolean functions from $L_{\rm in}$ input bits into one output bit, similar to $f(\cdot)$ handled in Fig. 5.7. By combining quantum circuits as in Fig. 5.7, any multicomponent Boolean function from $L_{\rm in}$ input bits into $L_{\rm out}$ output bits can therefore receive an implementation by a reversible quantum circuit. The quantum fan-out gate of Fig. 5.2 may be useful to duplicate the basis states serving as inputs to the computation of each of the $L_{\rm out}$ Boolean components. Alternatively, as long as the input argument remains accessible or forwarded on the output in the circuit of Fig. 5.7, the successive components of a multicomponent Boolean function can be computed sequentially one after the other for the same input argument.

The quantum circuit of Fig. 5.7, for any L-qubit state $|\vec{a}\,\rangle$ from the computational basis with $\vec{a} \in \{0, 1\}^L$, realizes the input–output transformations

$$|\vec{a}\,\rangle\,|b = 0\rangle \xmapsto{\;\mathsf{U}_f\;} |\vec{a}\,\rangle\,|f(\vec{a})\rangle \;, \tag{5.7}$$

$$|\vec{a}\,\rangle\,|b = 1\rangle \longmapsto |\vec{a}\,\rangle\,\Big|\overline{f(\vec{a})}\Big\rangle \;. \tag{5.8}$$

When the auxiliary qubit identified by b in Fig. 5.7 is initialized at the input in the superposed state $|+\rangle = \big(|0\rangle + |1\rangle\big)/\sqrt{2}$, by linearity the circuit realizes the input–output transformation

$$|\vec{a}\,\rangle\,|+\rangle \xmapsto{\;\mathsf{U}_f\;} |\vec{a}\,\rangle\,\frac{1}{\sqrt{2}}\Big(|f(\vec{a})\rangle + \Big|\overline{f(\vec{a})}\Big\rangle\Big) = |\vec{a}\,\rangle\,|+\rangle \;, \tag{5.9}$$

and in this circumstance the information about $f(\vec{a})$ is lost at the output. More interesting is to initialize the auxiliary qubit at the input in the superposition $|-\rangle = \big(|0\rangle - |1\rangle\big)/\sqrt{2}$, to obtain the input–output transformation

$$|\vec{a}\,\rangle\,|-\rangle \xmapsto{\;\mathsf{U}_f\;} |\vec{a}\,\rangle\,\frac{1}{\sqrt{2}}\Big(|f(\vec{a})\rangle - \Big|\overline{f(\vec{a})}\Big\rangle\Big) = |\vec{a}\,\rangle\,(-1)^{f(\vec{a})}\,|-\rangle \;. \tag{5.10}$$

This is a separable output state for the auxiliary qubit terminating in state $|-\rangle$, so that the action of the quantum circuit of Fig. 5.7 can be separated as performing on $|\vec{a}\,\rangle$ the input–output transformation

$$|\vec{a}\,\rangle \longmapsto (-1)^{f(\vec{a})}\,|\vec{a}\,\rangle \;, \tag{5.11}$$

where the basis state $|\vec{a}\,\rangle$ experiences a change of sign (or not) according to the binary value $f(\vec{a})$. This change of sign amounts to a *global* phase factor on the quantum state $|\vec{a}\,\rangle$ which alone is not directly exploitable. But it will become a useful *relative* phase factor with superposed input arguments to $f(\cdot)$.

We consider the superposed input argument

$$|+\rangle^{\otimes L} = \left(\frac{1}{\sqrt{2}}\right)^L \sum_{\vec{a}\in\{0,1\}^L} |\vec{a}\,\rangle \ , \tag{5.12}$$

which represents an equiweighted superposition of the 2^L states of the computational basis for the L data qubits at the input. This separable product state is easily obtained from L one-qubit Hadamard gates as $(\mathsf{H}\,|0\rangle)^{\otimes L} = \mathsf{H}^{\otimes L}\,|0\rangle^{\otimes L} = |+\rangle^{\otimes L}$. With the auxiliary qubit initialized as $|0\rangle$, the quantum circuit of Fig. 5.7, from Eq. (5.7) by the standard superposition principle of linear algebra, realizes the input–output transformation

$$|+\rangle^{\otimes L}\,|0\rangle \stackrel{\mathsf{U}_f}{\longmapsto} \left(\frac{1}{\sqrt{2}}\right)^L \sum_{\vec{a}\in\{0,1\}^L} |\vec{a}\,\rangle\,|f(\vec{a})\rangle \ . \tag{5.13}$$

The output state of Eq. (5.13) is in general an entangled state of the $L+1$ qubits, formed as the coherent superposition of all the 2^L values $f(\vec{a})$ of the Boolean function $f(\cdot)$, for all the 2^L arguments $\vec{a} \in \{0, 1\}^L$, with each value $f(\vec{a})$ perfectly identified by a set of 2^L orthogonal addresses $|\vec{a}\,\rangle$. This output state in Eq. (5.13) is obtained by one single pass of the superposed input of Eq. (5.12) across the quantum circuit of Fig. 5.7, or one single evaluation of the unitary operator U_f. This is the **parallelism** for computation afforded by the quantum circuit.

One then needs to figure out what to do with all the values of $f(\vec{a})$ simultaneously obtained and present in superposition in the output quantum state of Eq. (5.13). If one were so impatient as to rush to a quantum measurement of this output quantum state in Eq. (5.13), by a projective measurement in the computational basis one would obtain nothing more than one single value of $f(\vec{a})$, moreover for an argument $\vec{a}$ selected at random by the Born rule, which would be a rather disappointing outcome of the promise of the parallel evaluation.

Instead, some skillful quantum processing should better be envisaged on the superposed state of Eq. (5.13), which will afford the possibility of parallel computation simultaneously on all the values of $f(\vec{a})$ for all arguments $\vec{a}$. Such parallel processing will have to operate through quantum unitary operations, to be specifically designed to accomplish some useful task of interest, implementable with a universal set of quantum gates as indicated in Sect. 5.1. At minimum, any classical computations expressible by Boolean logic can be accomplished quantumly in this way, with the added possibility of running them in parallel. Additional auxiliary qubits may be introduced if useful to the computation. At the end of the (parallel) processing, a quantum measurement will usually take place, to collect the result of the computation. The computation result can even be collected deterministically, when each feasible outcome is encoded as a basis state of the projective measurement basis, so as to come out with probability 1 to deterministically reveal the computation result. This is an important property available. Quantum measurement is intrinsically probabilistic, in general, in the sense indicated early in Sect. 2.2. However, quantum computation or processing at large, which need to terminate by a quantum measurement to collect some useful results, in principle, can be made perfectly deterministic, when each meaningful result is encoded as a basis state of the projective measurement basis. Meanwhile, probabilistic aspects may also subsist in useful quantum computations. This important possibility of deterministic quantum computation will be illustrated as soon as the next section, presenting the Deutsch–Jozsa quantum algorithm, to deterministically solve some meaningful information-processing task through parallel processing.

The possibility of parallel quantum computation may be specially suited when large amounts of data $\big($the 2^L values of $\vec{a}$ and $f(\vec{a})\big)$ need to be processed to determine or to extract a small amount of meaningful information, codable on a few bits as the computation result. For signal processing, one may think, for instance, of an input digital image with millions of pixels, to be processed to decide whether the image belongs to one of two, or a few, predefined classes. Such possibility of quantum parallel computation, with potential applicability to many sectors, for signal and information processing and beyond, today remains largely to be explored and substantiated. On each occasion the task is to design a quantum algorithm, operating on data qubits and possibly auxiliary qubits useful to the computation, proceeding through unitary transformations implementable from (universal) quantum gates, and associated with a final quantum measurement[2] to collect the result. This is the essence of quantum computation. Many quantum algorithms have already been proposed [93, 94]. However, no systematic or generic methodology exists today to handle such design, or even to predetermine which information-processing tasks may or may not benefit from a quantum enhancement, and to what extent, and a fortiori via which approach, and accordingly much remains to be done in this direction.

For parallel evaluation of the Boolean function $f(\cdot)$, the quantum circuit of Fig. 5.7 can alternatively be operated with the auxiliary qubit initialized at the input in the state $|-\rangle$ instead of $|0\rangle$ in Eq. (5.13), so as to realize on the $L+1$ qubits the input–output transformation

$$|+\rangle^{\otimes L}\,|-\rangle \xmapsto{\;\mathsf{U}_f\;} \left(\frac{1}{\sqrt{2}}\right)^L \sum_{\vec{a}\in\{0,1\}^L} (-1)^{f(\vec{a})}\,|\vec{a}\,\rangle\,|-\rangle\;. \tag{5.14}$$

As in Eq. (5.10), the state $|-\rangle$ of the auxiliary qubit can be separated in Eq. (5.14), so as to obtain for the L data qubits the input–output transformation:

$$|+\rangle^{\otimes L} \longmapsto \left(\frac{1}{\sqrt{2}}\right)^L \sum_{\vec{a}\in\{0,1\}^L} (-1)^{f(\vec{a})}\,|\vec{a}\,\rangle\;. \tag{5.15}$$

[2] Intermediate measurements can also be useful, as for measurement-based computation, which however is shown equivalent to unitary evolutions terminated by a single final measurement, but may offer a valuable option for practical implementation [91, 92].

In a similar way as in Eq. (5.13), the output state of Eq. (5.14) or of Eq. (5.15) contains in a coherent superposition all the 2^L values $f(\vec{a})$ for all the 2^L arguments $\vec{a} \in \{0, 1\}^L$, however the information of $f(\vec{a})$ is rather encoded in the relative phases $(-1)^{f(\vec{a})}$ of the 2^L orthogonal components addressed by $|\vec{a}\,\rangle$. For each argument $\vec{a}$ the binary value $f(\vec{a})$ is encoded in Eq. (5.13) as a qubit state $|f(\vec{a})\rangle$ involved in a tensor product $|\vec{a}\,\rangle\,|f(\vec{a})\rangle$ with the corresponding argument $|\vec{a}\,\rangle$. By contrast in Eq. (5.14) or (5.15), for each argument $\vec{a}$ the binary value $f(\vec{a})$ is encoded into a scalar coefficient $(-1)^{f(\vec{a})}$, or sign, or phase, applied to the corresponding argument $|\vec{a}\,\rangle$, that is, a scalar encoding $(-1)^{f(\vec{a})}$ instead of encoding in a quantum state $|f(\vec{a})\rangle$. These are two alternative encodings of the information contained in the function $f(\cdot)$ that may have different utilities for computation. We encounter here a generic issue for quantum processing and computation, which lies in the way to encode or represent some function, or some signal or image or data, into quantum states or quantum objects, so as to gain access to some useful processing capabilities. Representative examples for quantum image processing can be found, for instance, in [95–97]. The answer to this encoding issue is not unique, but depends on the context and prospect.

To illustrate quantum parallel computation, we will now describe two quantum algorithms, the Deutsch–Jozsa algorithm and the Grover search algorithm, which are two pioneer algorithms demonstrating useful exploitations of a quantum parallel evaluation of a function, to achieve a processing performance unattainable by classical means.

5.4 Deutsch–Jozsa Quantum Algorithm

The **Deutsch–Jozsa algorithm** [98, 99] is a quantum algorithm to decide, in parallel, whether a Boolean function $f(\cdot)$ as in Eq. (5.6) belongs to one or the other of two predefined classes. The Deutsch–Jozsa algorithm was first introduced in 1985 by [98] for Boolean functions on two-bit inputs, and extended in 1992 by [99] to Boolean functions on L-bit inputs with arbitrary integer size $L \geq 2$, to appreciate in passing the pace of advancement of these early quantum algorithms. The Deutsch–Jozsa algorithm is described in [98, 99], for its principle, on generic Boolean functions. Here, within our signal-processing perspective, we will cast the Deutsch–Jozsa algorithm under the form of a task of digital image processing, as proposed in [100]. This will enable us to convey the original principle of the Deutsch–Jozsa algorithm and, at the same time, to illustrate the potential of parallel quantum computation for digital signal and image processing.

The Boolean function $f(\cdot)$ from Eq. (5.6) is taken here to represent a binary digital image. The input L-bit argument $\vec{a} \in \{0, 1\}^L$ is taken as the binary representation of the coordinates or address of a pixel, over the two-dimensional array supporting a digital image with a total of 2^L pixels. The value $f(\vec{a})$ at address $\vec{a}$ represents the binary intensity (black or white) of the image at pixel $\vec{a}$. For the Deutsch–Jozsa algorithm, the image $f(\cdot)$ is assumed to belong to one or the other of two predefined image classes. Namely, the image can be a constant image, with all its 2^L pixels

Fig. 5.8 The two constant binary images with size $256 \times 256 = 2^{16}$ pixels, to be classified by the Deutsch–Jozsa quantum algorithm against the balanced images of Fig. 5.9

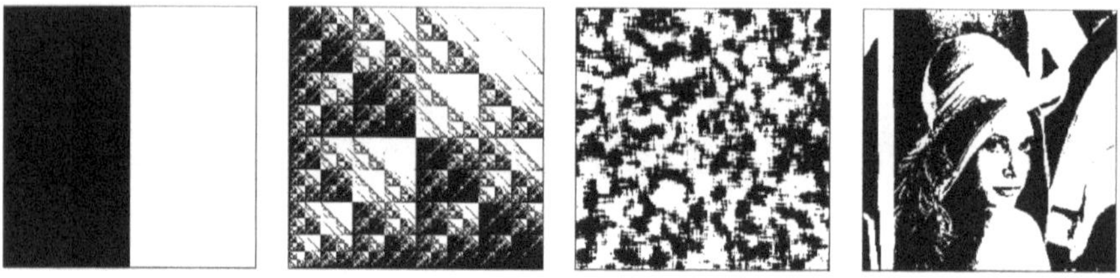

Fig. 5.9 Four balanced binary images with size $256 \times 256 = 2^{16}$ pixels and equal numbers of black and white pixels, to be classified by the Deutsch–Jozsa quantum algorithm against the constant images of Fig. 5.8. Each binary image, in its quantum representation supported by the superposed state $|\Phi\rangle$ of Eq. (5.17), can be stored in a quantum register of 16 qubits exploitable for parallel processing

at 0 or all at 1, so that there are only two possible constant images, represented in Fig. 5.8. Alternatively, the image can be a balanced image, with half of its 2^L pixels at 0 and half at 1, so that there are many possible balanced images, with some examples represented[3] in Fig. 5.9. Even more 2^L-pixel binary images are neither constant nor balanced, but it is assumed that the image $f(\cdot)$ that is input to the algorithm is either constant or balanced. The task then is, by testing the image value in various pixels, or evaluating the value $f(\vec{a})$ at various addresses $\vec{a}$, to determine whether the candidate image is constant or balanced. Classically, depending on the candidate image, a total number between 2 and $2^L/2 + 1$ of pixels or evaluations of $f(\cdot)$ must be tested consecutively in order to decide between a constant or a balanced image. By contrast, with a quantum approach, all the 2^L pixels can be tested simultaneously, via a single evaluation of $f(\cdot)$ in parallel, to be in position to decide, by means of an appropriate quantum processing, between a constant or a balanced image, as we are going to see.

The quantum circuit for the image-processing Deutsch–Jozsa algorithm is represented in Fig. 5.10.

[3] The visual appearance of the image representing a Boolean function $f(\vec{a})$ is tied to the specific spatial arrangement that is chosen to convert an L-bit argument $\vec{a}$ into a pixel address over the two-dimensional array supporting the image, but once fixed this choice is not critical for the operation of the quantum algorithm.

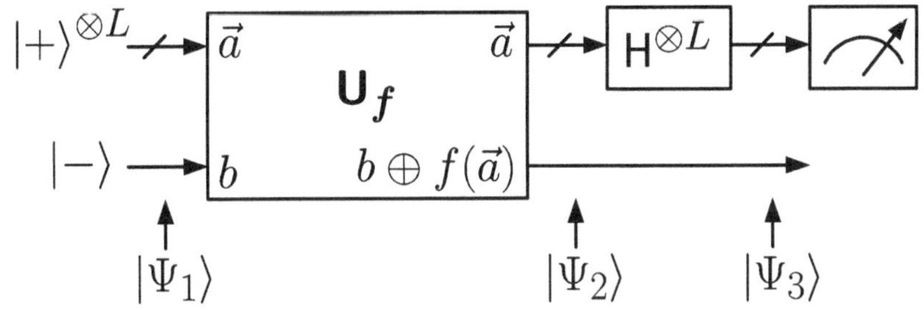

Fig. 5.10 The quantum circuit implementing the image-processing Deutsch–Jozsa algorithm to decide whether a binary image $f(\cdot)$ is constant as in Fig. 5.8 or balanced as in Fig. 5.9, through processing simultaneously in parallel all the pixels of the image

The first layer of the Deutsch–Jozsa circuit in Fig. 5.10 is an instance of the quantum circuit of Fig. 5.7 materializing the description of the binary image $f(\cdot)$. This quantum circuit for $f(\cdot)$ here can be regarded as a processor capable of computing the image intensity $f(\vec{a})$ at any pixel address $\vec{a}$ applied as input, or alternatively it can be regarded as a memory retrieving the intensity $f(\vec{a})$ at any pixel address $\vec{a}$. The distinctive and specifically quantum property is that this device, processor or memory, is able to operate in parallel: with several addresses like $\vec{a}$ applied in superposition at its input, it delivers a coherent superposition of the corresponding intensities $f(\vec{a})$ at its output. This is precisely what is done in the computation by the circuit of Fig. 5.10, all the 2^L pixel addresses are input in superposition by means of the quantum state $|+\rangle^{\otimes L}$, so as to form the $(L+1)$-qubit input state $|\Psi_1\rangle = |+\rangle^{\otimes L}\,|-\rangle$ in Fig. 5.10. By Eq. (5.14), the first circuit layer in Fig. 5.10 delivers the $(L+1)$-qubit state

$$|\Psi_2\rangle = \left(\frac{1}{\sqrt{2}}\right)^L \sum_{\vec{a}\in\{0,1\}^L} (-1)^{f(\vec{a})}\,|\vec{a}\,\rangle\,|-\rangle\ . \tag{5.16}$$

This state $|\Psi_2\rangle$ is separable as $|\Psi_2\rangle = |\Phi\rangle\,|-\rangle$, with the auxiliary qubit remaining in state $|-\rangle$, and the L data qubits placed in the L-qubit state $|\Phi\rangle$ given by

$$|\Phi\rangle = \left(\frac{1}{\sqrt{2}}\right)^L \sum_{\vec{a}\in\{0,1\}^L} (-1)^{f(\vec{a})}\,|\vec{a}\,\rangle\ . \tag{5.17}$$

At this stage of the circuit in Fig. 5.10, the quantum state $|\Phi\rangle$ contains in a coherent superposition all the values $f(\vec{a})$ of the binary image at all the 2^L pixels $\vec{a}$, within the coherent superposition $|\Phi\rangle$ each image value $f(\vec{a})$ coded as a phase $(-1)^{f(\vec{a})}$ on each address $|\vec{a}\,\rangle$. This encoding of the complete binary image with its 2^L pixels is supported by the quantum state $|\Phi\rangle$ of the L data qubits. Instead of these L qubits for the quantum encoding, by contrast, a classical encoding of the same binary image would require 2^L classical bits. This is an illustration of the advantage offered by quantum superposition for compact coding of images, and many variants thereof are conceivable, depending on the targeted properties [95, 96, 101, 102]. For the binary images of Figs. 5.8–5.9 with their $256 \times 256 = 2^{16}$ pixels, a quantum register of $L = 16$ qubits placed in the superposed state $|\Phi\rangle$ of Eq. (5.17) conveys a whole image of $2^{16} = 65536$ pixels, and image processing on the whole image can be performed in parallel by processing the 16-qubit register. The Deutsch–Jozsa algorithm, as we describe it, is an example of such parallel processing, for image classification.

We note that in this quantum encoding of a 2^L-pixel binary image by an L-qubit register placed in the superposed state $|\Phi\rangle$ of Eq. (5.17), an image and its binary inverse are encoded by two equivalent quantum states differing only by a global change of sign. In this respect, such two images are indistinguishable from their quantum encoding or, in other words, the quantum encoding is invariant in the binary inversion of an image. If needed, this invariance could be easily suppressed with an extra qubit, so as to use a register with $L + 1$ qubits to encode the 2^L-pixel binary images with no invariance, for instance, with an $(L + 1)$-qubit state as in the right-hand side of Eq. (5.13). But the invariance here is profitable to our classification task, because an image and its binary inverse both belong to the same class, constant or balanced, and need not be distinguished.

For image classification, in the circuit of Fig. 5.10, from the state $|\Psi_2\rangle = |\Phi\rangle\,|-\rangle$, on the L data qubits resting in state $|\Phi\rangle$ of Eq. (5.17), a Hadamard transformation $\mathsf{H}^{\otimes L}$ is applied, leading to the $(L + 1)$-qubit output state:

$$|\Psi_3\rangle = \Big(\mathsf{H}^{\otimes L}\,|\Phi\rangle\Big)\,|-\rangle = \left(\frac{1}{\sqrt{2}}\right)^L \sum_{\vec{a}\in\{0,1\}^L} (-1)^{f(\vec{a})}\Big(\mathsf{H}^{\otimes L}\,|\vec{a}\,\rangle\Big)\,|-\rangle\ . \tag{5.18}$$

As it follows from Eq. (A.43) in the Appendix, the one-qubit Hadamard transformation is defined by $\mathsf{H}\,|0\rangle = |+\rangle$ and $\mathsf{H}\,|1\rangle = |-\rangle$, which can be written concisely as

$$\mathsf{H}\,|a\rangle = \frac{1}{\sqrt{2}}\Big(|0\rangle + (-1)^a\,|1\rangle\Big) = \frac{1}{\sqrt{2}}\sum_{z\in\{0,1\}} (-1)^{az}\,|z\rangle\ , \tag{5.19}$$

for $a \in \{0, 1\}$ used as a binary label over the qubit computational basis. For any state $|\vec{a}\,\rangle$ of the L-qubit computational basis, labeled by the L-bit binary word $\vec{a} = a_1a_2\cdots a_L \in \{0, 1\}^L$, we therefore obtain the L-qubit Hadamard transformation defined as

$$\mathsf{H}^{\otimes L}\,|\vec{a}\,\rangle = \mathsf{H}\,|a_1\rangle \otimes \mathsf{H}\,|a_2\rangle \otimes\cdots\otimes \mathsf{H}\,|a_L\rangle = \left(\frac{1}{\sqrt{2}}\right)^L \sum_{\vec{z}\in\{0,1\}^L} (-1)^{\vec{a}\,\vec{z}}\,|\vec{z}\,\rangle\ , \tag{5.20}$$

with the scalar inner product $\vec{a}\,\vec{z} = a_1z_1 \oplus\cdots\oplus a_Lz_L = a_1z_1 + \cdots + a_Lz_L$ modulo 2.

We can then insert Eq. (5.20) into Eq. (5.18) to obtain the $(L + 1)$-qubit output state

$$|\Psi_3\rangle = \left(\frac{1}{2}\right)^L \sum_{\vec{a}\in\{0,1\}^L} \sum_{\vec{z}\in\{0,1\}^L} (-1)^{f(\vec{a})}(-1)^{\vec{a}\,\vec{z}}\, |\vec{z}\,\rangle\, |-\rangle\ . \tag{5.21}$$

This state $|\Psi_3\rangle$ is separable as $|\Psi_3\rangle = |\Psi\rangle\,|-\rangle$, with the auxiliary qubit terminating (remaining) in state $|-\rangle$, and the L data qubits terminating in a state $|\Psi\rangle = \mathsf{H}^{\otimes L}\,|\Phi\rangle$ that can be written as

$$|\Psi\rangle = \sum_{\vec{z}\in\{0,1\}^L} w(\vec{z}\,)\,|\vec{z}\,\rangle\ , \tag{5.22}$$

with the scalar weight

$$w(\vec{z}\,) = \frac{1}{2^L} \sum_{\vec{a}\in\{0,1\}^L} (-1)^{f(\vec{a})\oplus\vec{a}\,\vec{z}}\ . \tag{5.23}$$

The weight $w(\vec{z}\,)$ successively applies in Eq. (5.22) to all the 2^L basis states $|\vec{z}\,\rangle$ of the L data qubits, and each weight $w(\vec{z}\,)$ is determined in Eq. (5.23) by the configuration of the image $f(\vec{a})$ at its 2^L pixels $\vec{a}$. The key property is that a constant image or a balanced image, respectively, sets the 2^L weights $w(\vec{z}\,)$ from Eq. (5.23) to place the output state $|\Psi\rangle$ of Eq. (5.22) in two orthogonal configurations that can be perfectly distinguished by measuring $|\Psi\rangle$ in the L-qubit computational basis. This can be seen as follows.

We can concentrate on the weight $w(\vec{z}=\vec{0}\,)$ at $\vec{z}=\vec{0}=0\cdots 0$, which is

$$w\big(\vec{z}=\vec{0}\,\big) = \frac{1}{2^L} \sum_{\vec{a}\in\{0,1\}^L} (-1)^{f(\vec{a})}\ . \tag{5.24}$$

It results that when the image $f(\vec{a})$ is constant over its 2^L pixels $\vec{a}$, we obtain $w(\vec{z}=\vec{0}\,)=\pm 1$, and since the output state $|\Psi\rangle$ in Eq. (5.22) is with unit norm, the value $w(\vec{z}=\vec{0}\,)=\pm 1$ implies that $w(\vec{z}\neq\vec{0}\,)=0$ for all $\vec{z}\neq\vec{0}$. In the output superposition $|\Psi\rangle$ of Eq. (5.22), only the basis state $|\vec{z}=\vec{0}\,\rangle$ is therefore present, with a nonzero weight $w(\vec{z}=\vec{0}\,)=\pm 1$. In this circumstance, when the L-qubit output state $|\Psi\rangle$ is measured by projection in the L-qubit computational basis, the projection takes place onto the state $|\vec{0}\,\rangle$ with the probability $|w(\vec{0}\,)|^2=1$. The projective basis is an L qubit separable basis, so that each of the L data qubits at the output can be measured separately in its one-qubit projective basis $\big\{|0\rangle\,,\,|1\rangle\big\}$, and the result is that the L data qubits are all measured in state $|0\rangle$, deterministically, when the image $f(\vec{a})$ is constant.

By contrast, when the image $f(\vec{a})$ is balanced, half of its 2^L pixels give $(-1)^{f(\vec{a})}=+1$ and the other half give $(-1)^{f(\vec{a})}=-1$, so that from Eq. (5.24) we deduce $w(\vec{z}=\vec{0}\,)=0$, indicating that the basis state $|\vec{z}=\vec{0}\,\rangle$ is absent or with zero weight in the output superposition $|\Psi\rangle$ of Eq. (5.22). Upon measurement of $|\Psi\rangle$ in the L-qubit computational basis, the L data qubits are never found all simultaneously in the qubit state $|0\rangle$. When the image $f(\vec{a})$ is balanced, at least one data qubit at the output is measured in state $|1\rangle$.

This is the quantum parallel processing to classify the images from Figs. 5.8 and 5.9. A single evaluation of all the pixel values is realized through a single pass across the circuit of Fig. 5.10, or a single query of the image memory, and all the 2^L pixels are processed in parallel, to obtain deterministically by a final measurement of the L data qubits at the output, the binary result of the classification. This quantum algorithm is emblematic of the potential of quantum parallelism to process a large amount of data and extract (deterministically) a few essential pieces of information as the processing result.

What is accomplished by this Deutsch–Jozsa quantum algorithm expressed as an image-processing task can be understood from standard concepts of signal processing. The weights $w(\vec{z})$ defined by Eq. (5.23) for all $\vec{z} \in \{0, 1\}^L$ can be viewed as representing a Hadamard transform for the binary image $(-1)^{f(\vec{a})}$ defined with ± 1 patterns for all $\vec{a} \in \{0, 1\}^L$, this in the sense of the Hadamard or Walsh–Hadamard transform [103] used in signal and image processing. The Walsh–Hadamard transform decomposes an image over an orthonormal basis of rectangular waveforms oscillating at various frequencies, much like the Fourier transform decomposes an image over an orthonormal basis of harmonic waveforms oscillating at various frequencies. The Hadamard coefficient $w(\vec{z})$ quantifies the weight of the Hadamard rectangular wave at frequency $\vec{z}$ in the decomposition of the binary image $(-1)^{f(\vec{a})}$, much like a Fourier coefficient does in the harmonic frequency plane of an image. The Hadamard coefficient $w(\vec{z} = \vec{0})$ quantifies the component at DC or at zero frequency of the binary image $(-1)^{f(\vec{a})}$. The constant images have only this DC coefficient $w(\vec{z} = \vec{0})$ which is nonzero, while the balanced images have no component at DC with $w(\vec{z} = \vec{0}) = 0$. Equivalently, the balanced images have a zero average, since the coefficient $w(\vec{z} = \vec{0})$ is the average intensity over the $(-1)^{f(\vec{a})}$ image. The Deutsch–Jozsa algorithm decides its classification of an image $f(\vec{a})$ by testing the DC Hadamard coefficient of the binary image $(-1)^{f(\vec{a})}$. The Hadamard coefficients $w(\vec{z})$ are directly accessible in the quantum measurement since their squared moduli $|w(\vec{z})|^2$ are the probabilities of the various measurement outcomes of projecting the output state $|\Psi\rangle$ onto the basis state $|\vec{z}\rangle$. Other classes of images could easily be introduced in the classification, defined from disjoint subsets dividing the set of Hadamard coefficients $w(\vec{z})$, subsets that would remain perfectly distinguishable by the final measurement. For instance, a third class could be introduced as the class of half-constant binary images that would alternate a black half and a white half, regularly, top-bottom, or left-right as exemplified by the left-most image in Fig. 5.9, and possibly their phase-shifted and oblique versions. Such half-constant images would come with a well-identified signature in their Hadamard coefficients $w(\vec{z})$ corresponding to the first (lowest) nonzero frequencies of the Hadamard rectangular waves. The half-constant images could be the images with nonzero Hadamard coefficients only at these first (lowest) nonzero Hadamard frequencies. By comparison, constant images are the images with nonzero Hadamard coefficients only at DC. The half-constant images are originally a subclass of the original class of balanced images. Three classes could be redefined, as the class of the constant images, the class of half-constant images, and the class of balanced images that are not half-constant. A similar quantum algorithm could in parallel classify a candidate binary image over these three classes. Furthermore, instead of the Hadamard (or Fourier) transform, any other unitary transformation could be applied, decomposing over an orthonormal basis (such as wavelets, or otherwise) in which the images to be classified have their coefficients distributed across disjoint subsets distinguishable by the final quantum measurement.

We can also note that instead of a Von Neumann projective measurement via a complete set of 2^L projectors onto each of the 2^L basis states $|\vec{z}\,\rangle$ of the computational basis of the L data qubits, it can be realized that only two projectors are enough to extract the targeted binary response to the classification of the input image as constant or balanced. This would correspond to a generalized measurement, as presented in Sect. 4.2, defined by two projectors: the rank-one projector $\Pi_0 = |\vec{z} = \vec{0}\,\rangle\langle\vec{z} = \vec{0}\,|$ onto the state $|\vec{z} = \vec{0}\,\rangle$, and the complementary projector $\Pi_1 = \mathsf{Id} - \Pi_0$ of rank $2^L - 1$ with Id the identity operator on the 2^L-dimensional space of the L data qubits. The L-qubit output state $|\Psi\rangle$ of Eq. (5.22) is expected to belong either to the one-dimensional subspace $\text{span}\big(|\vec{z} = \vec{0}\,\rangle\big)$ or to its complementary orthogonal subspace $\text{span}\big(|\vec{z} = \vec{0}\,\rangle\big)^{\perp}$ with dimension $2^L - 1$, whether, respectively, the input image is constant or balanced. The subspace of $|\Psi\rangle$ is revealed by the binary result delivered by the generalized measurement, as it uses two projectors that project onto these two complementary subspaces. Practically, such a two-outcome generalized measurement of $|\Psi\rangle$ can be accomplished following the procedure of Sect. 4.2.2. It would involve an additional auxiliary qubit prepared as $|e_0\rangle$, and a unitary operator U on $L+1$ qubits realizing the entangling transformation $|\Psi\rangle \otimes |e_0\rangle \mapsto \mathsf{U}\,|\Psi\rangle \otimes |e_0\rangle = \Pi_0\,|\Psi\rangle \otimes |0\rangle + \Pi_1\,|\Psi\rangle \otimes |1\rangle$ as in Eq. (4.21), and such a unitary U can always be implemented with the universal set of qubit gates presented in Sect. 5.1. It then suffices to measure the auxiliary qubit alone at the output, to obtain the binary result of the classification. An important difference with this generalized measurement is that the output image carried by $|\Psi\rangle$ is not destroyed, but is left untouched in its original subspace. Only the one projector acting among two is revealed as the response to the binary classification, but the measured image is left invariant and not destroyed by the projector actually acting. By contrast, a complete Von Neumann measurement with 2^L projectors will destroy the output state $|\Psi\rangle$, by projecting it onto one (selected at random by the Born rule) of the 2^L basis states $|\vec{z}\,\rangle$ of the L-qubit space. Such a complete Von Neumann measurement could extract more information, if needed, for instance, in the classification problem with three classes we evoked above, but it is also more harmful to the measured quantum state. This is a general trade-off with quantum measurements, generally when more information is extracted, at the same time more alteration to the measured state is incurred. This issue connects to what is known as strong or weak quantum measurements [104, 105]. This trade-off is handled with maximum flexibility and efficiency by exploiting generalized measurements supplementing standard projective Von Neumann measurements.

Fig. 5.11 Five examples of binary images with size $256 \times 256 = 2^{16}$ pixels generated by the image model of Eq. (5.25), for five configurations of the model parameters $\{\vec{c}, b\}$ with $\vec{c}$ here a 16-bit binary word. For such images with $L = 16$, the model of Eq. (5.25) generates a set of $2^{16} \times 2 \approx 10^5$ distinct binary images of size 256×256 pixels. By processing an image, the task is to identify the parameter configuration $\{\vec{c}, b\}$ that served to its generation. Each binary image, in its quantum representation based on the superposed state of Eq. (5.17), can be supported by a quantum register of $L = 16$ qubits exploitable for parallel processing

Another related algorithm of quantum parallel processing is the **Bernstein–Vazirani algorithm** [106, 107]. It also considers Boolean functions $f(\cdot)$ as in Eq. (5.6), from an L-bit argument $\vec{a}$ into a one-bit output $f(\vec{a})$. But when the Deutsch–Jozsa algorithm considers two classes, constant or balanced, for possible functions $f(\cdot)$, the Bernstein–Vazirani algorithm considers functions $f(\cdot)$ defined by the model

$$f(\vec{a}) = \vec{c}\,\vec{a} \oplus b\,, \tag{5.25}$$

where $\vec{c}$ is an L-bit binary word (like $\vec{a}$) and $\vec{c}\,\vec{a}$ their modulo 2 scalar inner product as $\vec{a}\,\vec{z}$ in Eq. (5.20), while b is a one-bit parameter. The task then is, by testing or evaluating the function $f(\cdot)$ for various arguments $\vec{a}$, to determine the value of the parameter set $\{\vec{c}, b\}$ of the model. Again through parallel computation, the Bernstein–Vazirani quantum algorithm solves the problem with a single evaluation of $f(\cdot)$, on an appropriate superposition of input arguments (more precisely, one evaluation for determining the L-bit parameter $\vec{c}$ and one for b).

To describe the quantum algorithm, it is interesting as before to interpret the generic Boolean functions $f(\cdot)$ from Eq. (5.25), as defining a model of binary images with intensities $f(\vec{a})$ over a two-dimensional array of 2^L pixels addressed by $\vec{a}$, as done in [100]. Each binary image is generated from a configuration of the parameters $\{\vec{c}, b\}$. From 2^{L+1} distinct parameter configurations, the image model of Eq. (5.25) generates an image (sub)set containing 2^{L+1} distinct binary images, out of the set of 2^{2^L} binary images with 2^L pixels. The one-bit parameter b ensures that for every binary image in the image set (obtained with $b = 0$), its binary complement (with $b = 1$) is also a binary image of the set. Several examples of binary images generated by the image model of Eq. (5.25) are shown in Fig. 5.11. We face the task of digital image processing consisting, for binary images generated by the image model of Eq. (5.25), in identifying the model parameters $\{\vec{c}, b\}$, by testing the image intensities $f(\vec{a})$ at various pixels $\vec{a}$. This is done as a parallel image processing by the quantum algorithm that follows.

The quantum algorithm uses the same circuit of Fig. 5.10 processing $L + 1$ qubits. After the Hadamard transformation $\mathsf{H}^{\otimes L}$, the L data qubits at the output are measured in the computational basis, but the measurement result is now processed differently. The L data qubits at the output terminate in the same quantum state $|\Psi\rangle$ of Eq. (5.22). When the function $f(\cdot)$ is modeled according to Eq. (5.25), the scalar weight $w(\vec{z}\,)$ of Eq. (5.23) becomes

$$w(\vec{z}\,) = \frac{1}{2^L} \sum_{\vec{a}\in\{0,1\}^L} (-1)^{\vec{c}\,\vec{a}\,\oplus b\oplus\vec{a}\,\vec{z}} \,, \tag{5.26}$$

still interpretable as the coefficient, at the Hadamard frequency $\vec{z} \in \{0, 1\}^L$, of a Walsh–Hadamard transform for the binary image $(-1)^{f(\vec{a})} = (-1)^{\vec{c}\,\vec{a}\,\oplus b}$. We now concentrate on the Hadamard coefficient at frequency $\vec{z} = \vec{c}$, which by $\vec{c}\,\vec{a} \,\oplus\, \vec{a}\,\vec{c} = 0$ follows as

$$w(\vec{z} = \vec{c}\,) = \frac{1}{2^L} \sum_{\vec{a}\in\{0,1\}^L} (-1)^{0\oplus b} = \pm 1 \,. \tag{5.27}$$

Again, since the output state $|\Psi\rangle$ in Eq. (5.22) is with unit norm, the value $w(\vec{z} = \vec{c}\,) = \pm 1$ implies that $w(\vec{z} \neq \vec{c}\,) = 0$ for all $\vec{z} \neq \vec{c}$. The binary image $(-1)^{f(\vec{a})}$ resulting from the model of Eq. (5.25) has only one nonzero Hadamard component, at the Hadamard frequency $\vec{z} = \vec{c}$. Accordingly, upon measuring the L-qubit output state $|\Psi\rangle$ of Eq. (5.22) in the computational basis, the L-bit word $\vec{c}$ is obtained with probability $|w(\vec{c}\,)|^2 = 1$. This deterministically identifies the value of the parameter $\vec{c}$ acting in the model of Eq. (5.25), by processing with a single pass across the quantum circuit of Fig. 5.10 evaluating $f(\cdot)$, or a single query of the image memory. By contrast, a classical approach to obtain $\vec{c}$ would require at least L successive evaluations of the function $f(\cdot)$, at L adequately chosen arguments $\vec{a}$. In addition, one more evaluation of $f(\cdot)$, at $\vec{a} = \vec{0}$, yields the parameter b, since always $f(\vec{a} = \vec{0}\,) = b$ for any $f(\cdot)$ in the model class of Eq. (5.25). The Bernstein–Vazirani quantum algorithm to solve the identification problem in parallel essentially locates the nonzero Hadamard component of the binary image, as singled out by the quantum measurement at the output.

5.5 Grover Quantum Search Algorithm

Another quantum algorithm illustrating the power of parallel quantum computation is offered by the **Grover algorithm**. The Grover algorithm, introduced in 1996 by [108, 109], can be described as a search in an unsorted database, but its principle incorporates enough flexibility to be applicable to many other comparable tasks. A database contains N items, numbered by the integer index $n = 1$ to N. Each item n can be tested on some criterion represented by a binary-valued function $f(\cdot)$, which returns $f(n) = 1$ if item n satisfies the criterion, and $f(n) = 0$ if not. A device capable of returning the value $f(n)$ of the criterion for any input item or argument n is sometimes called an "oracle." The database among its N items is assumed to contain one item, numbered $n = n_0$, satisfying the criterion $f(\cdot)$. Only one successful item $n = n_0$ is considered to exist in the database for the moment, but extension to several ones will later be feasible. The task is to find out the index n_0 of the successful item. With the Kronecker delta, we can also write $f(n) = \delta_{nn_0}$. Since the database is unsorted or unstructured, the successful item n_0 may be anywhere in the database, and a classical approach has no other choice than to successively evaluate the criterion $f(n)$ for each candidate n between 1 and N, as long as $f(n) = 0$ is obtained, and until some $n = n_0$ giving $f(n_0) = 1$ is found. Whatever the scanning schedule for $n \in \{1, 2, \ldots, N\}$, such a classical search will require a number of iterations or algorithmic complexity of order $O(N)$, linear in the size N of the database. By contrast, a quantum algorithm that can query the database, or evaluate the criterion $f(\cdot)$, in parallel, on several input items in superposition, can resolve the search in $O(\sqrt{N})$ iterations, as we are going to see.

By the universality of quantum circuits for Boolean logic reported in Sect. 5.2, it is always possible to introduce a quantum circuit (the oracle) to realize an evaluation of the binary-valued criterion function $f(\cdot)$ for any input index n. The input index $n \in \{1, 2, \ldots, N\}$ is encoded as a basis state $|n\rangle$ of the orthonormal computational basis of an N-dimensional quantum system with Hilbert space $\mathcal{H}_N$, and this is always feasible by employing qubits in a sufficiently large number L satisfying $2^L \geq N$. The binary function $f(\cdot)$ from $n \in \{1, 2, \ldots, N\}$ into $f(n) \in \{0, 1\}$ can be implemented by a quantum circuit as in Fig. 5.7, where the port marked $\vec{a}$ now carries the quantum state $|n\rangle$, supplemented by an auxiliary qubit b, so as to realize the unitary transformation U_f acting on the basis states of $\mathcal{H}_N \otimes \mathcal{H}_2$ by

$$|n\rangle\,|b\rangle \xmapsto{\ \mathsf{U}_f\ } |n\rangle\,|b \oplus f(n)\rangle \ . \tag{5.28}$$

When the auxiliary qubit marked b is placed in the superposed state $|-\rangle$, as for Eq. (5.10), we obtain the input–output transformation

$$|n\rangle\,|-\rangle \xmapsto{\ \mathsf{U}_f\ } (-1)^{f(n)}\,|n\rangle\,|-\rangle \ . \tag{5.29}$$

The auxiliary qubit remains in a separable state $|-\rangle$, so that the quantum circuit gives us access, as for Eq. (5.11), to an input–output relationship on the index state $|n\rangle$ as

$$|n\rangle \xmapsto{\ \overline{\mathsf{U}}_f\ } (-1)^{f(n)}\,|n\rangle \ , \tag{5.30}$$

defining a unitary operator $\overline{\mathsf{U}}_f$ on $\mathcal{H}_N$ by means of its action on the N orthonormal basis states $|n\rangle$ of $\mathcal{H}_N$. Since the criterion function realizes $f(n) = \delta_{nn_0}$, we have $\overline{\mathsf{U}}_f\,|n\rangle = |n\rangle$ for any $n \neq n_0$, and $\overline{\mathsf{U}}_f\,|n_0\rangle = -\,|n_0\rangle$, so that we have the alternative expression for the unitary $\overline{\mathsf{U}}_f$ as

$$\overline{\mathsf{U}}_f = \mathrm{I}_N - 2\,|n_0\rangle\,\langle n_0| \ , \tag{5.31}$$

with I_N the identity operator on $\mathcal{H}_N$. So the quantum circuit that is able to evaluate the binary criterion $f(\cdot)$ via Eq. (5.30) equivalently computes the unitary transformation $\overline{\mathsf{U}}_f$ of Eq. (5.31). Focusing narrowly on the expression of $\overline{\mathsf{U}}_f$ in Eq. (5.31) involving $|n_0\rangle$ might suggest that for computing $\overline{\mathsf{U}}_f$ one needs to know the solution $|n_0\rangle$ in the first place. But this is not the case, since the previous expression of Eq. (5.30) for the action of $\overline{\mathsf{U}}_f$ indicates that it is enough to be able to evaluate the criterion $f(n)$ for any input argument n, to be able to evaluate $\overline{\mathsf{U}}_f$, and this is exactly what the oracle quantum circuit does.

For parallel computation, it is interesting to introduce the equiweighted superposition of all the N possible index states, or the average state, as

$$|\overline{n}\,\rangle = \frac{1}{\sqrt{N}} \sum_{n=1}^{N} |n\rangle \ , \tag{5.32}$$

in particular, giving the inner product $\langle \overline{n}\,|n\rangle = 1/\sqrt{N}$ for any basis state $|n\rangle$ of $\mathcal{H}_N$. Inputting this average state $|\overline{n}\,\rangle$ to the quantum circuit computing $\overline{\mathsf{U}}_f$ of Eqs. (5.30)–(5.31), amounts to simultaneously test all the N candidate items to the search problem. The quantum circuit outputting $\overline{\mathsf{U}}_f\,|\overline{n}\,\rangle$ indicates the solution n_0 by a change of sign or phase to the basis state $|n_0\rangle$ in the output superposition. Yet this cue is not detectable by a measurement of $\overline{\mathsf{U}}_f\,|\overline{n}\,\rangle$ at the output (in the computational basis of $\mathcal{H}_N$), which would deliver nothing more than one of the N basis states $|n\rangle$ selected at random by the Born rule with an equal probability $1/N$. Nevertheless, as we are going to see, it is possible to exploit $\overline{\mathsf{U}}_f$ so as to design some unitary processing of $|\overline{n}\,\rangle$ to obtain a final quantum state that, when measured, will reveal the solution $|n_0\rangle$ with quasi-unit probability.

For this purpose we consider the quantum state

$$|n_0^{\perp}\rangle = \frac{1}{\sqrt{N-1}} \sum_{n\neq n_0}^{N} |n\rangle \ , \tag{5.33}$$

which is a (normalized) state of $\mathcal{H}_N$ orthogonal to the solution state $|n_0\rangle$. We note that this state $|n_0^{\perp}\rangle$ is not a state that we actually need to materialize for the operation of a quantum circuit or algorithm, since precisely the solution n_0 is unknown when we embark in the resolution of the search problem, when we only have access to a quantum circuit that is able to evaluate the binary criterion $f(n)$ for any n. The state $|n_0^{\perp}\rangle$ of Eq. (5.33) is merely a theoretical reference useful to define and assess some effective quantum processing that itself will be actually realizable with n_0 unknown, much like $\overline{\mathsf{U}}_f$ in Eq. (5.31). The average state $|\overline{n}\,\rangle$ from Eq. (5.32) is expressible as

$$|\overline{n}\,\rangle = \frac{\sqrt{N-1}}{\sqrt{N}}\,|n_0^{\perp}\rangle + \frac{1}{\sqrt{N}}\,|n_0\rangle \ , \tag{5.34}$$

indicating that $|\overline{n}\rangle$ lies in the vector plane $\big(|n_0\rangle\,,\,|n_0^\perp\rangle\big)$ of the space $\mathcal{H}_N$. And, in spite of its expression of Eq. (5.34) involving $|n_0\rangle$, this state $|\overline{n}\rangle$ is perfectly accessible and materializable as in Eq. (5.32) when we embark in the resolution of the search problem with an unknown n_0. The criterion operator $\overline{\mathsf{U}}_f$ of Eqs. (5.30)–(5.31) realizes $\overline{\mathsf{U}}_f\,|n_0^\perp\rangle = |n_0^\perp\rangle$ and $\overline{\mathsf{U}}_f\,|n_0\rangle = -\,|n_0\rangle$, so that in the plane $\big(|n_0\rangle\,,\,|n_0^\perp\rangle\big)$ the action of the operator $\overline{\mathsf{U}}_f$ is to perform a reflection about the state $|n_0^\perp\rangle$.

We also consider, with the same status as $|n_0^\perp\rangle$ of a theoretical reference, the quantum state

$$|\overline{n}^\perp\rangle = \frac{1}{\sqrt{N}}\,|n_0^\perp\rangle - \frac{\sqrt{N-1}}{\sqrt{N}}\,|n_0\rangle\;, \tag{5.35}$$

which is a (normalized) state of the plane $\big(|n_0\rangle\,,\,|n_0^\perp\rangle\big)$ orthogonal to the state $|\overline{n}\rangle$ (also in this same plane). We define the operator

$$\mathsf{U}_\perp = 2\,|\overline{n}\rangle\langle\overline{n}| - \mathrm{I}_N \tag{5.36}$$

on the space $\mathcal{H}_N$, which for any basis state $|n\rangle$ gives $\mathsf{U}_\perp\,|n\rangle = 2\,|\overline{n}\rangle/\sqrt{N} - |n\rangle$. It follows that for any two basis states $|n\rangle$ and $|n'\rangle$ we have the (conserved) inner product $\langle n'|\mathsf{U}_\perp^\dagger\mathsf{U}_\perp|n\rangle = \langle n'|n\rangle$ proving that $\mathsf{U}_\perp$ is a unitary operator on $\mathcal{H}_N$. Next we have $\mathsf{U}_\perp\,|\overline{n}\rangle = |\overline{n}\rangle$ and $\mathsf{U}_\perp\,|\overline{n}^\perp\rangle = -\,|\overline{n}^\perp\rangle$, indicating that in the plane $\big(|\overline{n}\rangle\,,\,|\overline{n}^\perp\rangle\big)$, which is the same plane as $\big(|n_0\rangle\,,\,|n_0^\perp\rangle\big)$, the action of the operator $\mathsf{U}_\perp$ is to perform a reflection about the average state $|\overline{n}\rangle$. And by its definition of Eq. (5.36), the unitary operator $\mathsf{U}_\perp$ is always accessible, independently of the search problem and its criterion $f(\cdot)$, and is in principle implementable by a suitable quantum circuit.

In the plane $\big(|n_0\rangle\,,\,|n_0^\perp\rangle\big)$, the composition of the two reflections $\overline{\mathsf{U}}_f$ and $\mathsf{U}_\perp$ defines a rotation $\mathsf{G} = \mathsf{U}_\perp\overline{\mathsf{U}}_f$, also known as the **Grover amplification operator**, since as we are going to see, this rotation operator G can be exploited to amplify the probability amplitude of hitting the solution state $|n_0\rangle$ from an initial uniform value $1/\sqrt{N}$ to quasi 1. The Grover rotation or amplification operator G verifies, in particular:

$$\mathsf{G}\,|n_0\rangle = \mathsf{U}_\perp\overline{\mathsf{U}}_f\,|n_0\rangle = -\mathsf{U}_\perp\,|n_0\rangle = |n_0\rangle - \frac{2}{\sqrt{N}}\,|\overline{n}\rangle\;. \tag{5.37}$$

Just like $|\overline{n}\rangle$, the state $\mathsf{G}\,|n_0\rangle$ lies in the vector plane $\big(|n_0\rangle\,,\,|n_0^\perp\rangle\big)$. In this plane $\big(|n_0\rangle\,,\,|n_0^\perp\rangle\big)$, the rotation G rotates by an angle ξ, which is the angle between $|n_0\rangle$ and $\mathsf{G}\,|n_0\rangle$ and is characterized by the inner product $\langle n_0|\mathsf{G}|n_0\rangle$ between $|n_0\rangle$ and $\mathsf{G}\,|n_0\rangle$, yielding $\cos(\xi) = \langle n_0|\mathsf{G}|n_0\rangle = 1 - 2/N$. In the regime of interest when searching a large database of size $N \gg 1$, we deduce that $\cos(\xi)$ is close to 1 and can be approximated as

$$\cos(\xi) = \langle n_0|\mathsf{G}|n_0\rangle = 1 - \frac{2}{N} \approx 1 - \frac{\xi^2}{2}\;, \tag{5.38}$$

so that we deduce the rotation angle of G as

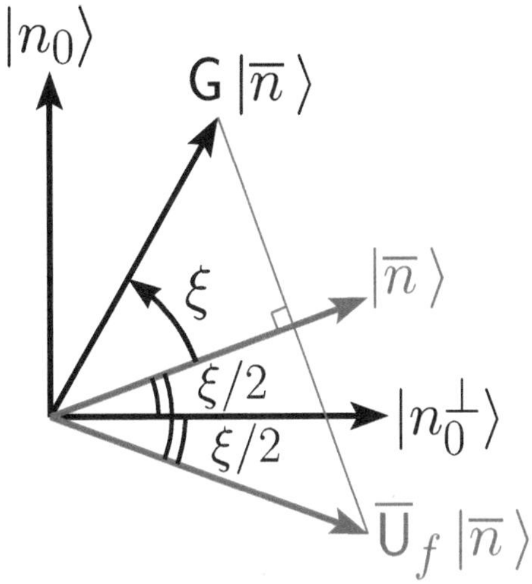

Fig. 5.12 The Grover amplification operator $\mathsf{G} = \mathsf{U}_\perp \overline{\mathsf{U}}_f$ composing the two unitaries $\overline{\mathsf{U}}_f$ from Eqs. (5.30)–(5.31) and $\mathsf{U}_\perp$ from Eq. (5.36) in the vector plane $\big(|\overline{n}\,\rangle\,,|n_0\rangle\big)$ performs a rotation of the average quantum state $|\overline{n}\,\rangle$ of Eq. (5.32) by the angle ξ of Eq. (5.39), to bring the initial state $|\overline{n}\,\rangle$ closer to the solution state $|n_0\rangle$ of the search problem

$$\xi \approx \frac{2}{\sqrt{N}}\;. \tag{5.39}$$

We however emphasize that the regime of large $N \gg 1$, although convenient for the analysis and quite realistic in practice, is not necessary to deduce ξ from $\cos(\xi) = \langle n_0|\mathsf{G}|n_0\rangle = 1 - 2/N$ in Eq. (5.38) and to execute the Grover search algorithm.

An important point is to note that the Grover operator $\mathsf{G} = \mathsf{U}_\perp \overline{\mathsf{U}}_f$ is in principle implementable by a quantum circuit as soon as we have access to the quantum circuit evaluating the binary criterion $f(\cdot)$, in order to compute $\overline{\mathsf{U}}_f$ as in Eq. (5.30). This evaluation of $f(\cdot)$ computing $\overline{\mathsf{U}}_f$ does not (consistently) require to know the solution n_0 nor even the number of possible solutions to the search problem. Meanwhile, for G, the implementation of $\mathsf{U}_\perp$ according to Eq. (5.36) is generic and independent of the search problem and its criterion $f(\cdot)$.

The first step of the quantum search is to apply the Grover operator G to the average quantum state $|\overline{n}\,\rangle$ of Eq. (5.32), to obtain

$$\mathsf{G}\,|\overline{n}\,\rangle = \mathsf{U}_\perp \overline{\mathsf{U}}_f\,|\overline{n}\,\rangle = \mathsf{U}_\perp\bigg(|\overline{n}\,\rangle - \frac{2}{\sqrt{N}}\,|n_0\rangle\bigg) = \bigg(1-\frac{4}{N}\bigg)\,|\overline{n}\,\rangle + \frac{2}{\sqrt{N}}\,|n_0\rangle\;. \tag{5.40}$$

We have the two vector planes $\big(|n_0\rangle\,,|n_0^\perp\rangle\big)$ and $\big(|\overline{n}\,\rangle\,,|\overline{n}^\perp\rangle\big)$ which coincide in the space $\mathcal{H}_N$, and also coincide with the plane $\big(|\overline{n}\,\rangle\,,|n_0\rangle\big)$. The initial state $|\overline{n}\,\rangle$ and transformed state $\mathsf{G}\,|\overline{n}\,\rangle$ are both in this plane $\big(|\overline{n}\,\rangle\,,|n_0\rangle\big)$. In this plane, the action of G is a planar rotation by the angle ξ evaluated in Eq. (5.39). From $|\overline{n}\,\rangle$, the action of G in Eq. (5.40) reduces the coordinate on $|\overline{n}\,\rangle$ and increases the coordinate on $|n_0\rangle$. We therefore deduce that by the planar rotation G, the initial state $|\overline{n}\,\rangle$ is rotated closer to the solution state $|n_0\rangle$ by the angle ξ. This action of the rotation G in the plane $\big(|\overline{n}\,\rangle\,,|n_0\rangle\big)$ is depicted in Fig. 5.12.

The application of G can be repeated on the state $\mathsf{G}\,|\overline{n}\,\rangle$, to produce $\mathsf{G}^2\,|\overline{n}\,\rangle$ which still remains in the same plane $\big(|\overline{n}\,\rangle\,,|n_0\rangle\big)$ and represents the initial state $|\overline{n}\,\rangle$ rotated by the angle 2ξ toward the solution $|n_0\rangle$. The total angle Ξ between the initial state $|\overline{n}\,\rangle$ and the targeted solution state $|n_0\rangle$ is such that $\cos(\Xi) = \langle n_0|\overline{n}\,\rangle = 1/\sqrt{N}$, the same for any $|n_0\rangle$, whence $\Xi = \arccos\big(1/\sqrt{N}\big)$. Consequently, the Grover rotation operator G which rotates by an angle ξ has to be applied a number of times $K = \Xi/\xi \approx \arccos\big(1/\sqrt{N}\big)\sqrt{N}/2$, so as to cover the total angle Ξ and bring the initial state $|\overline{n}\,\rangle$ onto the solution state $|n_0\rangle$. In the interesting regime $N \gg 1$, when $\Xi \approx \pi/2 - 1/\sqrt{N}$, we obtain the number of Grover rotations by G as

$$K \approx \frac{\pi}{4}\sqrt{N}\,. \tag{5.41}$$

These K rotations by G brought the quantum state from $|\overline{n}\,\rangle$ onto $|n_0\rangle$, so that a final measurement in the computational basis of $\mathcal{H}_N$ projects on the solution $|n_0\rangle$ almost surely. One rotation by G costs one evaluation of the criterion $f(\cdot)$ by the quantum circuit implementing $\overline{\mathsf{U}}_f$ of Eq. (5.30), so that the solution n_0 is obtained after $K \sim \sqrt{N}$ evaluations of the criterion $f(\cdot)$. A verification can be obtained by an extra evaluation of $f(\cdot)$ on the solution state returned for $|n_0\rangle$ by the quantum algorithm. This is the **algorithmic complexity $O\big(\sqrt{N}\,\big)$** afforded by the Grover quantum algorithm for the search in an unsorted database with size N, compared with the higher complexity $O(N)$ for classical approaches. Moreover, this complexity $O\big(\sqrt{N}\,\big)$ can be proven optimal in the sense that it cannot be further reduced [110].

For large databases with size $N \gg 1$, the number of iterations K from Eq. (5.41) for the Grover operator G is uniform and independent of the search criterion $f(\cdot)$. Most importantly, it is the definition of G, via $\overline{\mathsf{U}}_f$ from Eqs. (5.30)–(5.31), that is specific to the search criterion $f(\cdot)$ and that crucially determines the vector plane $\big(|\overline{n}\,\rangle\,,|n_0\rangle\big)$ where a uniform number K of rotations by G have to take place.

For the accuracy of the algorithm, in practice, the number of iterations of G has to be an integer, so that K of Eq. (5.41) is in practice rounded to the closest integer denoted $\lfloor K \rceil$. This will entail an inaccuracy (by excess or default) by an angle no larger than $\xi/2$, by which the solution state $|n_0\rangle$ will be (slightly) missed at the end of the $\lfloor K \rceil$ iterated rotations of G in the plane $\big(|n_0\rangle\,,|n_0^\perp\rangle\big)$. The final rotated state at the maximum angle inaccuracy $\xi/2$ would be a state of the plane $\big(|n_0\rangle\,,|n_0^\perp\rangle\big)$ departing from $|n_0\rangle$ by an angle $\pm\xi/2$ and receiving the form:

$$\mathsf{G}^{\lfloor K \rceil}\,|\overline{n}\,\rangle = \cos(\xi/2)\,|n_0\rangle \pm \sin(\xi/2)\,|n_0^\perp\rangle \approx \left(1 - \frac{1}{2N}\right)|n_0\rangle \pm \frac{1}{\sqrt{N}}\,|n_0^\perp\rangle\,, \tag{5.42}$$

with $\xi/2 \approx 1/\sqrt{N}$ taken from Eq. (5.39) in the regime of large size $N \gg 1$. The final measurement of $\mathsf{G}^{\lfloor K \rceil}\,|\overline{n}\,\rangle$ in the computational basis of $\mathcal{H}_N$ will then project on the targeted solution $|n_0\rangle$ with the probability $\Pr\{|n_0\rangle\} \approx 1 - 1/N$ at least, and on another basis state with the complementary probability $\Pr\{|n\rangle \neq |n_0\rangle\} \approx 1/N$ at most. So with quasi-unit probability, the correct solution $|n_0\rangle$ is returned by the algorithm in $O\big(\sqrt{N}\,\big)$ computation. A very improbable incorrect measurement outcome

$|n\rangle \neq |n_0\rangle$ would be detected by a final verification by $f(\cdot)$ of the solution returned by the algorithm, with the faculty to run the algorithm anew.

With $M \geq 1$ solutions to the search problem, the Grover quantum algorithm is modified as follows [111, 112]. We assume for the moment that the number $M \geq 1$ of solutions to the search problem is known. For an item n that is among the M solutions, we use the notation $n \in \{M\}$, and the criterion is $f(n) = 1$ for all $n \in \{M\}$, and $f(n) = 0$ for all $n \notin \{M\}$. The unitary operator $\overline{\mathsf{U}}_f$ on $\mathcal{H}_N$ implementing the criterion $f(\cdot)$ keeps the same definition as in Eq. (5.30), and it now receives the alternative expression:

$$\overline{\mathsf{U}}_f = \mathrm{I}_N - 2 \sum_{n \in \{M\}} |n\rangle \langle n| \,, \tag{5.43}$$

as an extension of Eq. (5.31). We have $\overline{\mathsf{U}}_f\,|n\rangle = -\,|n\rangle$ for all $n \in \{M\}$, and $\overline{\mathsf{U}}_f\,|n\rangle = |n\rangle$ for all $n \notin \{M\}$. We consider the average solution state

$$|\overline{n}_0\rangle = \frac{1}{\sqrt{M}} \sum_{n \in \{M\}} |n\rangle \,, \tag{5.44}$$

and the state

$$|\overline{n}_0^{\perp}\rangle = \frac{1}{\sqrt{N-M}} \sum_{n \notin \{M\}} |n\rangle \,, \tag{5.45}$$

orthogonal to $|\overline{n}_0\rangle$. It follows that $\overline{\mathsf{U}}_f\,|\overline{n}_0\rangle = -\,|\overline{n}_0\rangle$ and $\overline{\mathsf{U}}_f\,|\overline{n}_0^{\perp}\rangle = |\overline{n}_0^{\perp}\rangle$, so that in the vector plane $\big(|\overline{n}_0\rangle\,,\,|\overline{n}_0^{\perp}\rangle\big)$ of $\mathcal{H}_N$ the action of the operator $\overline{\mathsf{U}}_f$ of Eq. (5.43) is to perform a reflection about the state $|\overline{n}_0^{\perp}\rangle$.

The average quantum state $|\overline{n}\,\rangle$ of Eq. (5.32) is expressible as

$$|\overline{n}\,\rangle = \frac{\sqrt{N-M}}{\sqrt{N}}\,|\overline{n}_0^{\perp}\rangle + \frac{\sqrt{M}}{\sqrt{N}}\,|\overline{n}_0\rangle \,, \tag{5.46}$$

so that it also belongs to the plane $\big(|\overline{n}_0\rangle\,,\,|\overline{n}_0^{\perp}\rangle\big)$ of $\mathcal{H}_N$. In particular, we have the inner product $\langle \overline{n}\,|\overline{n}_0\rangle = \sqrt{M}/\sqrt{N}$. We also consider the state

$$|\overline{n}^{\perp}\rangle = \frac{\sqrt{M}}{\sqrt{N}}\,|\overline{n}_0^{\perp}\rangle - \frac{\sqrt{N-M}}{\sqrt{N}}\,|\overline{n}_0\rangle \,, \tag{5.47}$$

which is a state orthogonal to $|\overline{n}\,\rangle$ in the plane $\big(|\overline{n}_0\rangle\,,\,|\overline{n}_0^{\perp}\rangle\big)$, and replacing Eq. (5.35) with which it coincides at $M = 1$, but only at $M = 1$ when $|\overline{n}_0\rangle$ coincides with $|n_0\rangle$. We also have

$$|\overline{n}_0\rangle = \frac{\sqrt{M}}{\sqrt{N}}\,|\overline{n}\,\rangle - \frac{\sqrt{N-M}}{\sqrt{N}}\,|\overline{n}^{\perp}\rangle \,. \tag{5.48}$$

The operator $\mathsf{U}_\perp$ from Eq. (5.36) is unchanged, and it gives as before $\mathsf{U}_\perp |\overline{n}^\perp\rangle = -|\overline{n}^\perp\rangle$, so that in the plane $(|\overline{n}\rangle, |\overline{n}^\perp\rangle)$, which is the same as the plane as $(|\overline{n}_0\rangle, |\overline{n}_0^\perp\rangle)$, the action of $\mathsf{U}_\perp$ is (again) to perform a reflection about the average state $|\overline{n}\rangle$.

The same Grover operator $\mathsf{G} = \mathsf{U}_\perp \overline{\mathsf{U}}_f$ in the plane $(|\overline{n}_0\rangle, |\overline{n}_0^\perp\rangle) \equiv (|\overline{n}\rangle, |\overline{n}^\perp\rangle)$ is the composition of two reflections, so that it acts as a rotation, with a rotation angle ξ such that $\cos(\xi) = \langle \overline{n}_0 | \mathsf{G} | \overline{n}_0 \rangle$. From

$$\mathsf{G}\,|\overline{n}_0\rangle = \mathsf{U}_\perp \overline{\mathsf{U}}_f\,|\overline{n}_0\rangle = -\mathsf{U}_\perp\,|\overline{n}_0\rangle = |\overline{n}_0\rangle - 2\frac{\sqrt{M}}{\sqrt{N}}\,|\overline{n}\rangle\;, \tag{5.49}$$

we deduce

$$\cos(\xi) = \langle \overline{n}_0 | \mathsf{G} | \overline{n}_0 \rangle = 1 - 2\frac{M}{N}\;. \tag{5.50}$$

In the interesting regime of a large database with $N \gg M \geq 1$, in Eq. (5.50) we have $\cos(\xi)$ close to 1 which can be approximated as $\cos(\xi) \approx 1 - \xi^2/2$, so that we deduce for the rotation by G the angle

$$\xi \approx 2\frac{\sqrt{M}}{\sqrt{N}}\;. \tag{5.51}$$

The rotation by G in the plane $(|\overline{n}\rangle, |\overline{n}_0\rangle)$, can be iterated to gradually rotate the initial state $|\overline{n}\rangle$ onto the average solution state $|\overline{n}_0\rangle$. In this respect, we have

$$\mathsf{G}\,|\overline{n}\rangle = \mathsf{U}_\perp \overline{\mathsf{U}}_f\,|\overline{n}\rangle = \mathsf{U}_\perp\Bigg(|\overline{n}\rangle - 2\sum_{n\in\{M\}} \langle n|\overline{n}\rangle\,|n\rangle\Bigg) = \left(1 - 4\frac{M}{N}\right)|\overline{n}\rangle + 2\frac{\sqrt{M}}{\sqrt{N}}\,|\overline{n}_0\rangle\;, \tag{5.52}$$

and from Eq. (5.48) also

$$\mathsf{G}\,|\overline{n}\rangle = \left(1 - 2\frac{M}{N}\right)|\overline{n}\rangle - 2\frac{\sqrt{(N-M)M}}{N}\,|\overline{n}^\perp\rangle\;, \tag{5.53}$$

showing how the rotation by G rotates $|\overline{n}\rangle$ in direction of the average solution $|\overline{n}_0\rangle$. The total angle Ξ to be covered is such that $\cos(\Xi) = \langle \overline{n}_0 | \overline{n}\rangle = \sqrt{M/N}$, giving

$$\Xi = \arccos\left(\sqrt{M/N}\right)\;, \tag{5.54}$$

with the convenient (but not critical) approximation $\Xi \approx \pi/2 - \sqrt{M/N}$ in the regime of interest where $N \gg M \geq 1$. The number K of rotations by G to bring $|\overline{n}\rangle$ onto $|\overline{n}_0\rangle$ is therefore

$$K = \frac{\Xi}{\xi} \approx \frac{\pi}{4}\sqrt{\frac{N}{M}}\;, \tag{5.55}$$

with the approximation in the regime of interest $N \gg M \geq 1$. After these K rotations the final state $\mathsf{G}^K |\overline{n}\rangle$ terminates in the state $|\overline{n}_0\rangle$, which is the equiweighted superposition of the M solutions to the search problem, as given in Eq. (5.44). Measurement of this final superposed state in the computational basis of $\mathcal{H}_N$ delivers one of the M solutions, randomly with equal probability $1/M$. Essentially, with M solutions to the search problem, the quantum algorithm, with a complexity $O(\sqrt{N})$ in the size N of the database, requires $\sqrt{M}$ less Grover iterations to return one of the M solutions.

For the accuracy of the algorithm, rounding the number K of Grover rotations G to the closest integer $\lfloor K \rceil$, as before, produces an angular inaccuracy on the total rotation of at most $\xi/2 \approx \sqrt{M/N}$, leading to the final quantum state:

$$\mathsf{G}^{\lfloor K \rceil} |\overline{n}\,\rangle = \cos(\xi/2)\,|\overline{n}_0\rangle \pm \sin(\xi/2)\,|\overline{n}_0^\perp\rangle \approx \left(1 - \frac{M}{2N}\right)|\overline{n}_0\rangle \pm \frac{\sqrt{M}}{\sqrt{N}}\,|\overline{n}_0^\perp\rangle \ , \tag{5.56}$$

which coincides with Eq. (5.42) when $M = 1$. Accordingly, the final measurement of $\mathsf{G}^{\lfloor K \rceil} |\overline{n}\,\rangle$ in the computational basis of $\mathcal{H}_N$ will return a solution with a probability $\approx 1 - M/N$ at least, and it may fail with the complementary probability $\approx M/N$ at most. Such improbable failure would be detected by a final verification by $f(\cdot)$ so as to possibly repeat the algorithm.

Finally, **when the number M of solutions is unknown** for the search problem [112–114], the same quantum algorithm can be applied provided that, for the Grover rotation $\mathsf{G} = \mathsf{U}_\perp \overline{\mathsf{U}}_f$ which remains accessible in the same way via Eq. (5.30) for $\overline{\mathsf{U}}_f$, the angle ξ of Eqs. (5.50)–(5.51) can be independently evaluated beforehand. The evaluation of the angle ξ can be accomplished by exploiting the quantum Fourier transform through the methodology of quantum phase estimation that will be presented in detail in Sect. 12.3 of Chap. 12 to come. To anticipate, this methodology is able to estimate, via a single quantum measurement, the unknown phase ξ characterizing an eigenvalue $e^{i\xi}$ of a unitary operator U_ξ, provided we are able to prepare an excitation signal to U_ξ in the eigenstate of U_ξ associated with the eigenvalue $e^{i\xi}$.

Here for the Grover algorithm, the unitary process U_ξ is the Grover operator G, which acts as a rotation of angle ξ in the plane $\big(|\overline{n}_0\rangle\,,\,|\overline{n}_0^\perp\rangle\big) \equiv \big(|\overline{n}\,\rangle\,,\,|\overline{n}_0\rangle\big)$ of the complex vector space $\mathcal{H}_N$. As a consequence, G has two orthogonal eigenstates $|g\rangle$ and $|g^\perp\rangle$ in the plane $\big(|\overline{n}_0\rangle\,,\,|\overline{n}_0^\perp\rangle\big)$, respectively, associated with the eigenvalues $e^{i\xi}$ and $e^{-i\xi} = e^{i(2\pi-\xi)}$. To act as an excitation signal to G we use the average quantum state $|\overline{n}\,\rangle$ of Eq. (5.32), which is known by Eq. (5.46) to belong to the plane $\big(|\overline{n}_0\rangle\,,\,|\overline{n}_0^\perp\rangle\big)$, and therefore also to the plane $\big(|g\rangle\,,\,|g^\perp\rangle\big)$, so that we can write

$$|\overline{n}\,\rangle = \alpha_g\,|g\rangle + \alpha_g^\perp\,|g^\perp\rangle \ . \tag{5.57}$$

Compared with the standard quantum phase estimation procedure of Sect. 12.3, here the excitation signal $|\overline{n}\,\rangle$ is not prepared in an eigenstate of $\mathsf{U}_\xi \equiv \mathsf{G}$, but in a superposition of two eigenstates of G, as written in Eq. (5.57). We are unable to decompose this superposition, to excite G with $|g\rangle$ alone, since we do not know (yet) the angle ξ and the vector $|\overline{n}_0\rangle$. For phase estimation, with an auxiliary qubit prepared in the state $|+\rangle = \big(|0\rangle + |1\rangle\big)/\sqrt{2}$, the controlled transformation occurring in Eq. (12.42) is now replaced by

$$\frac{1}{\sqrt{2}}\big(|0\rangle + |1\rangle\big)\,|\overline{n}\,\rangle \longmapsto \frac{1}{\sqrt{2}}\Big[\big(|0\rangle + e^{i\xi}\,|1\rangle\big)\alpha_g\,|g\rangle + \big(|0\rangle + e^{-i\xi}\,|1\rangle\big)\alpha_g^{\perp}\,|g^{\perp}\rangle\Big]\,. \tag{5.58}$$

The same processing as in Sect. 12.3.1 is now applied with the superposed state from Eq. (5.58), including the final inverse quantum Fourier transform of Eq. (12.48). This is end to end a linear processing, ruled by the superposition principle of linear algebra, for its action on the superposed state from Eq. (5.58). As a result, on the multiple-qubit signal serving to probe the phase ξ, the final quantum measurement at the output after the inverse quantum Fourier transform delivers either the phase angle ξ or $-\xi \equiv 2\pi - \xi$, respectively, with two non-critical probabilities controlled by α_g. Since for the Grover algorithm we expect[4] in Eqs. (5.50)–(5.51) a positive rotation angle ξ no larger than $\pi/2$, either measurement result $\pm\xi$ is enough to evaluate ξ. Via Eq. (5.50) or (5.51), when the size N of the database is known, an estimate for the angle ξ also estimates the number M of solutions. The result is even capable of identifying a search problem with no solution, when a vanishing estimate for the angle ξ is returned. The phase estimation thus also solves the problem of counting the number of solutions to the search. From the angle $\xi \neq 0$ for the rotation G, the total angle Ξ follows via Eq. (5.54), and the number $K = \Xi/\xi$ of iterations is deduced. And K iterations of G are performed as before to resolve the search.

From Eq. (5.51) we deduce that the phase estimation must be performed with an accuracy or angular resolution of order $O\big(1/\sqrt{N}\,\big)$. By the phase estimation methodology of Sect. 12.3 such an accuracy can be obtained [112–114] with $O\big(\sqrt{N}\,\big)$ evaluations of the Grover rotation G. The overall algorithmic complexity of $O\big(\sqrt{N}\,\big)$ for solving the search problem is therefore preserved in these more stringent conditions where the number M of solutions is unknown. As an alternative to the Fourier-transform phase estimation, the work in [111] reports an iterative quantum heuristic solving the Grover search with an unknown number of solutions also in $O\big(\sqrt{N}\,\big)$ complexity.

As we indicated, the Grover quantum algorithm, although presented as a search in an unsorted database, can in fact be applied or extended to address many types of problems (see, for instance, [93, 115–118]). The input index $n \in \{1, 2, \ldots, N\}$ can, in general, point to any type of data or object. For example, the Grover algorithm can be exploited to factor integers into prime numbers, in relation to the RSA cryptosystem which uses as a cryptographic key k_{RSA} a product of two (large) prime numbers. To break the key by factoring k_{RSA} can be performed by searching for an odd integer (indexed by n) between 3 and $\sqrt{k_{\mathrm{RSA}}}$ dividing k_{RSA}, with a search criterion $f(\cdot)$

[4] This when $M \leq N/2$. Otherwise, if we need to handle the unlikely circumstance of large $M > N/2$ that would make, as indicated by Eq. (5.53), the rotation angle ξ exceed $\pi/2$, it is possible to enlarge the size of the database from N to $2N$ by appending N extra items that are not solutions, so as to enforce a rotation angle ξ by G no larger than $\pi/2$.

essentially given by the Euclidean division. Classically this would entail $O(\sqrt{k_{RSA}})$ iterations, while the Grover quantum algorithm would allow a quadratically smaller $O(\sqrt{\sqrt{k_{RSA}}})$ number of iterations, by applying in parallel a Euclidean division algorithm to several candidate integers. Many comparable search or inversion problems over a finite (large) set of inputs can benefit from such quantum parallel computation.

As the Grover rotation G is iterated, the Grover algorithm gradually amplifies the probability amplitude of a solution state in the quantum superposition that is finally measured. Such quantum amplitude amplification is extended in [112, 113] to show, in particular, that if for a search problem a classical heuristic is known to find a solution in $O(K)$ complexity, then with a quantum approach a solution can be found in $O(\sqrt{K})$ complexity. Also, amplitude amplification in [112, 113] is used for counting the number of solutions to many types of search problems, with the usual quadratic improvement over classical approaches. In the same vein, leaning on the Grover quantum search algorithm, the works in [119, 120] present an algorithm to find the minimum (or maximum) of an unstructured set of N numbers with $O(\sqrt{N})$ calls to an oracle function which marks each number of the set inferior to a prescribed threshold that is iteratively reduced to reach the minimum.

5.6 Controlled Quantum Operations

Controlled quantum operations are useful to provide access to conditional processing in quantum computation. Such faculty can be obtained from the possibility of controlled qubit operations. For quantum computation, as we indicated in Sect. 5.1, the one-qubit unitary gates supplemented by the two-qubit Cnot gate form a universal set of quantum gates, enabling to implement any unitary process on any number of qubits. For quantum computation, it is possible (and useful) to implement an arbitrary qubit unitary operator U in a controlled manner, in the following specific sense, illustrated by Fig. 5.13. The control procedure of Fig. 5.13 extends to an arbitrary qubit unitary gate U, the control of the qubit inversion gate σ_x or Cnot gate previously described in Fig. 3.5.

In the quantum circuit of Fig. 5.13, an auxiliary qubit C is employed to act as a control on the unitary qubit gate U. The procedure is such that when the control qubit C is in the basis state $|0\rangle$ the target qubit T is left unchanged, and when the control qubit C is in the basis state $|1\rangle$ the target qubit T experiences the unitary

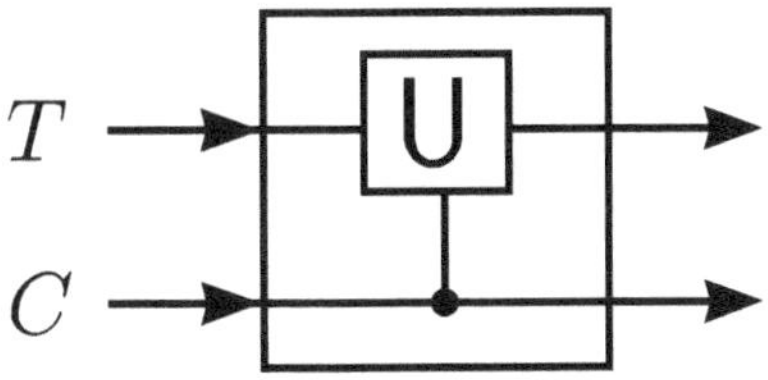

Fig. 5.13 Quantum circuit realizing a controlled-U unitary operation on the target qubit T driven by the control qubit C, as defined by Eqs. (5.59)–(5.60)

transformation U. The circuit realizes the input–output unitary transformation defined by

$$\begin{cases} |T\rangle \otimes |C=0\rangle \longmapsto |T\rangle \otimes |C=0\rangle \ , & (5.59) \\ |T\rangle \otimes |C=1\rangle \longmapsto \mathsf{U}\,|T\rangle \otimes |C=1\rangle \ . & (5.60) \end{cases}$$

Since the circuit of Fig. 5.13, through Eqs. (5.59)–(5.60), defines a two-qubit unitary gate, we are confident that it can be implemented via a universal set of elementary qubit gates. We shall accomplish this explicitly, by showing how to synthesize such a two-qubit controlled-U gate by means of one-qubit unitary gates and two-qubit Cnot gates (the universal set). For this synthesis, we consider for the one-qubit unitary process $\mathsf{U} \in \mathcal{L}(\mathcal{H}_2)$ the general form given by the Euler-angle decomposition of Eq. (5.2) as

$$\mathsf{U} = e^{i\alpha}\mathsf{R_z}(\beta)\mathsf{R_y}(\gamma)\mathsf{R_z}(\delta) \ . \tag{5.61}$$

In addition, the product of the three rotations above can be decomposed [20] as

$$\mathsf{R_z}(\beta)\mathsf{R_y}(\gamma)\mathsf{R_z}(\delta) = \mathsf{A}\sigma_x\mathsf{B}\sigma_x\mathsf{C} \ , \tag{5.62}$$

with the three one-qubit unitary operators

$$\begin{cases} \mathsf{A} \equiv \mathsf{R_z}(\beta)\,\mathsf{R_y}\left(\dfrac{\gamma}{2}\right) , & (5.63) \\ \mathsf{B} \equiv \mathsf{R_y}\left(-\dfrac{\gamma}{2}\right)\mathsf{R_z}\left(-\dfrac{\delta+\beta}{2}\right) , & (5.64) \\ \mathsf{C} \equiv \mathsf{R_z}\left(\dfrac{\delta-\beta}{2}\right) . & (5.65) \end{cases}$$

To prove the decomposition of Eq. (5.62), we first establish by direct matrix multiplication involving Eqs. (3.26) and (3.28) that $\sigma_x\mathsf{R_y}(\xi)\sigma_x = \mathsf{R_y}(-\xi)$ and $\sigma_x\mathsf{R_z}(\xi)\sigma_x = \mathsf{R_z}(-\xi)$, for any angle ξ. Since $\sigma_x\sigma_x = \mathrm{I}_2$, we have

$$\sigma_x\mathsf{B}\sigma_x = \sigma_x\mathsf{R_y}\left(-\frac{\gamma}{2}\right)\sigma_x\sigma_x\mathsf{R_z}\left(-\frac{\delta+\beta}{2}\right)\sigma_x \tag{5.66}$$

$$= \mathsf{R_y}\left(\frac{\gamma}{2}\right)\mathsf{R_z}\left(\frac{\delta+\beta}{2}\right) , \tag{5.67}$$

inserted in the right-hand side of Eq. (5.62) to give

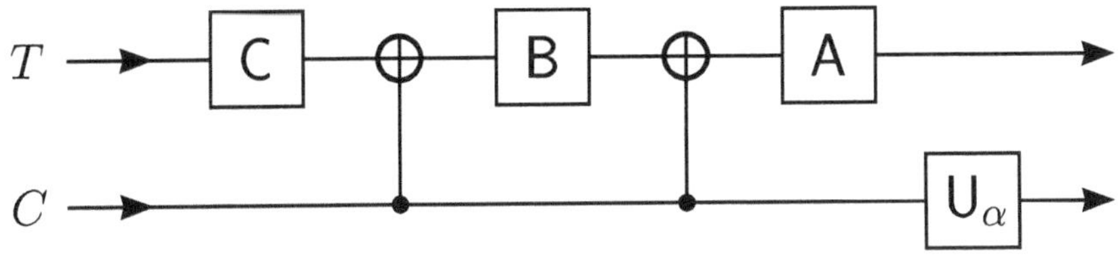

Fig. 5.14 Quantum circuit using one-qubit unitary gates and two two-qubit Cnot gates, so as to realize a two-qubit controlled gate implementing on the target qubit T a controlled version of the qubit unitary operator U from Eq. (5.61), under the control of the auxiliary qubit C

$$\mathsf{A}\sigma_x\mathsf{B}\sigma_x\mathsf{C} = \overbrace{\mathsf{R_z}(\beta)\,\mathsf{R_y}\left(\frac{\gamma}{2}\right)}^{\mathsf{A}}\overbrace{\mathsf{R_y}\left(\frac{\gamma}{2}\right)\mathsf{R_z}\left(\frac{\delta+\beta}{2}\right)}^{\sigma_x\mathsf{B}\sigma_x}\overbrace{\mathsf{R_z}\left(\frac{\delta-\beta}{2}\right)}^{\mathsf{C}} . \tag{5.68}$$

(with underbraces: $\mathsf{R_y}\left(\frac{\gamma}{2}\right)\mathsf{R_y}\left(\frac{\gamma}{2}\right) = \mathsf{R_y}(\gamma)$ and $\mathsf{R_z}\left(\frac{\delta+\beta}{2}\right)\mathsf{R_z}\left(\frac{\delta-\beta}{2}\right) = \mathsf{R_z}(\delta)$)

At any two angles γ and δ, in $\mathcal{L}(\mathcal{H}_2)$ the rotation operators $\mathsf{R_y}(\gamma)$ and $\mathsf{R_y}(\delta)$ commute, since they have the same eigenstates which are the eigenstates of σ_y, and they factorize as $\mathsf{R_y}(\gamma)\mathsf{R_y}(\delta) = \mathsf{R_y}(\gamma+\delta)$. Similarly, $\mathsf{R_z}(\gamma)\mathsf{R_z}(\delta) = \mathsf{R_z}(\gamma+\delta)$. These two factorizations lead for Eq. (5.68) to

$$\mathsf{A}\sigma_x\mathsf{B}\sigma_x\mathsf{C} = \mathsf{R_z}(\beta)\mathsf{R_y}(\gamma)\mathsf{R_z}(\delta) , \tag{5.69}$$

proving the decomposition of Eq. (5.62).

By similar factorization rules for commuting rotation operators of $\mathcal{L}(\mathcal{H}_2)$, the three operators of Eqs. (5.63)–(5.65) also satisfy $\mathsf{ABC} = \mathrm{I}_2$. This reduction to the identity I_2 is a key property, along with the decomposition of Eq. (5.62), for our task of implementing a controlled version of the qubit unitary U from Eq. (5.61). In the decomposition of Eq. (5.62), a useful step is to replace each inversion Pauli gate σ_x by a controlled inversion or Cnot gate, driven by a control qubit C. When the control qubit C is in state $|0\rangle$, the product in Eq. (5.62) reduces to $\mathsf{ABC} = \mathrm{I}_2$, and when C is in state $|1\rangle$, the product in Eq. (5.62) is $\mathsf{A}\sigma_x\mathsf{B}\sigma_x\mathsf{C} = \mathsf{R_z}(\beta)\mathsf{R_y}(\gamma)\mathsf{R_z}(\delta)$. To complete the synthesis of the controlled unitary U from Eq. (5.61), we also need the one-qubit unitary gate

$$\mathsf{U}_\alpha = \begin{bmatrix} 1 & 0 \\ 0 & e^{i\alpha} \end{bmatrix} \tag{5.70}$$

as in Eq. (3.31). The final two-qubit controlled gate implementing a controlled version of the qubit unitary U from Eq. (5.61) is obtained by the quantum circuit represented in Fig. 5.14.

In the quantum circuit of Fig. 5.14, when the control qubit C is in state $|0\rangle$, we obtain the input–output transformation:

$$|T\rangle \otimes |C=0\rangle \longmapsto \mathsf{ABC}\,|T\rangle \otimes |C=0\rangle = |T\rangle \otimes |C=0\rangle , \tag{5.71}$$

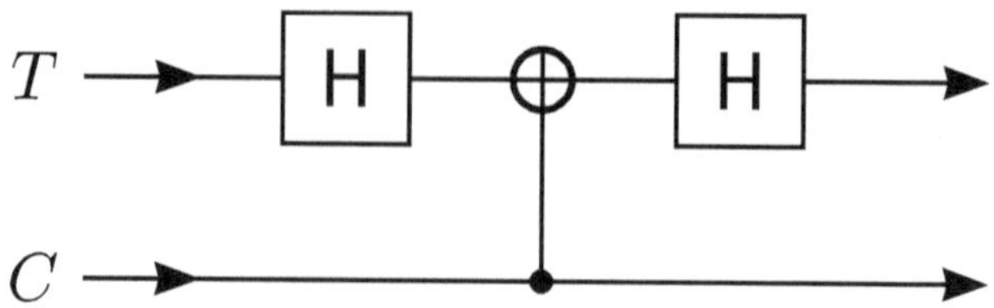

Fig. 5.15 Quantum circuit using two one-qubit Hadamard gates H and one two-qubit Cnot gate, to implement a two-qubit controlled Pauli gate σ_z on the target qubit T under the control of the auxiliary qubit C

and when the control qubit C is in state $|1\rangle$, we have the input–output transformation:

$$|T\rangle \otimes |C=1\rangle \longmapsto \mathsf{A}\sigma_x\mathsf{B}\sigma_x\mathsf{C}\,|T\rangle \otimes e^{i\alpha}\,|C=1\rangle = e^{i\alpha}\mathsf{R}_\mathsf{z}(\beta)\mathsf{R}_\mathsf{y}(\gamma)\mathsf{R}_\mathsf{z}(\delta)\,|T\rangle \otimes |C=1\rangle$$
$$= \mathsf{U}\,|T\rangle \otimes |C=1\rangle\ , \qquad (5.72)$$

as targeted for a controlled version of the qubit unitary operator U from Eq. (5.61).

For example, we have the Pauli gate $\sigma_z = e^{i\pi/2}\mathsf{R}_\mathsf{z}(\beta=\pi/2)\mathsf{R}_\mathsf{y}(\gamma=0)\mathsf{R}_\mathsf{z}(\delta=\pi/2)$, and this decomposition, as an instance of Eq. (5.61), can be used for implementing a controlled-σ_z gate with the structure of Fig. 5.14. For specific gates as here, there may exist a simpler alternative, resting on a similar mechanism, but obtained from a reduced decomposition. Namely here we have $\sigma_z = \mathsf{H}\sigma_x\mathsf{H}$, with H the Hadamard qubit unitary operator of Eq. (A.43), especially satisfying $\mathsf{HH} = \mathrm{I}_2$. This simpler decomposition for σ_z exhibits, when σ_x is replaced by a Cnot, the same behavior as the general decomposition from Eq. (5.62), to yield the controlled-σ_z gate of Fig. 5.15.

For another comparable example, we have the Pauli gate $\sigma_y = e^{i\pi/2}\mathsf{R}_\mathsf{z}(\beta=0)\mathsf{R}_\mathsf{y}(\gamma=\pi)\mathsf{R}_\mathsf{z}(\delta=0)$, with this decomposition according to Eq. (5.61) which can be used for implementing a controlled-σ_y gate as in Fig. 5.14. A simpler alternative approach is to exploit the decomposition $\sigma_y = \mathsf{S}\sigma_x\mathsf{S}^\dagger$, with the qubit unitary operator

$$\mathsf{S} = \begin{bmatrix} 1 & 0 \\ 0 & i \end{bmatrix} \qquad (5.73)$$

satisfying $\mathsf{SS}^\dagger = \mathrm{I}_2$, and leading to the controlled-σ_y gate of Fig. 5.16.

Controlled unitary gates form a special class of unitary gates, so the possibility of their synthesis as in Fig. 5.14, or in Figs. 5.15 and 5.16, is no surprise and follows from the universal character of the set of gates consisting of one-qubit gates supplemented by the two-qubit Cnot gate, as indicated in Sect. 5.1. The handling of controlled unitary gates here also illustrates how clever methodologies may be implied for synthesizing quantum circuits. For supplementing the controlled version of arbitrary one-qubit gates obtained here, a controlled Cnot gate is accessible as the Toffoli gate of Sect. 5.2, so that the universal set of quantum gates of Sect. 5.1 is also accessible in a controlled version. From such universal set of controlled gates, in principle, any

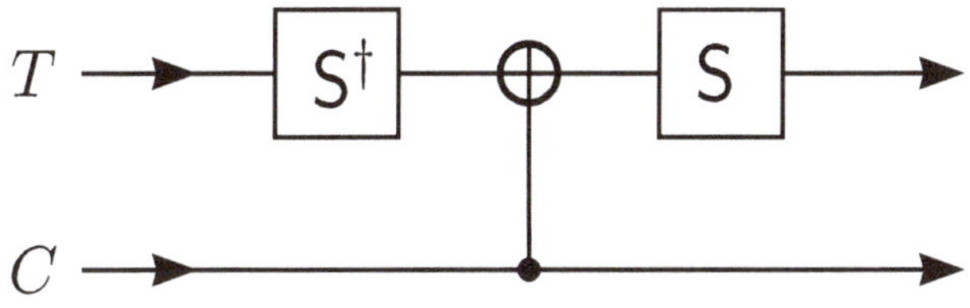

Fig. 5.16 Quantum circuit using the one-qubit phase gate S from Eq. (5.73) and one two-qubit Cnot gate, to implement a two-qubit controlled Pauli gate σ_y on the target qubit T under the control of the auxiliary qubit C

Fig. 5.17 Controlled-swap quantum gate or Fredkin gate which swaps the state of two target qubits Q_1 and Q_2 conditioned on the control qubit C

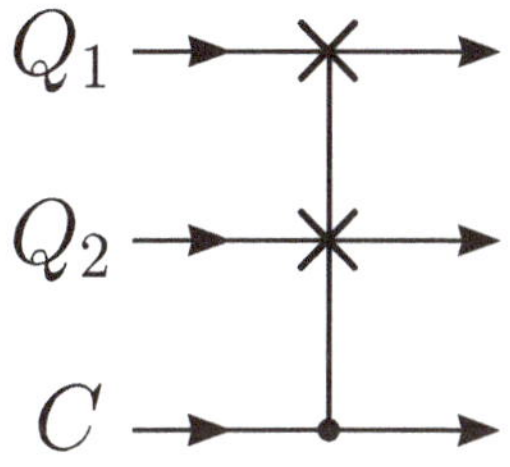

unitary gates on any number of qubits can be obtained in a controlled form. The control can be driven by a single control qubit, as in Fig. 5.13, or by any number of control qubits, with their switching condition (on or off) being decoded by means of the logical synthesis capabilities of the Toffoli gate as seen in Sect. 5.2. And since qubits can be used to materialize any finite-dimensional quantum states, any controlled unitary operation can be obtained also for them.

As another example, the two-qubit swap gate of Fig. 5.3 can be obtained in a controlled version, also known as the **Fredkin gate** [121], representing the three-qubit gate depicted in Fig. 5.17. Physical implementations of the Fredkin gate from two-qubit gates are discussed, for instance, in [122, 123].

The controlled-swap or Fredkin gate of Fig. 5.17 swaps the two target qubits Q_1 and Q_2 when the control qubit C is in the basis state $|1\rangle$, otherwise when the control qubit C is in the basis state $|0\rangle$ the two target qubits Q_1 and Q_2 are left untouched. The Fredkin gate is self-inverse. A pipeline cascading two consecutive Fredkin gates acts trivially on the three qubits.

The Fredkin gate shares with the Toffoli gate of Sect. 5.2 the same property of universality for synthesizing any Boolean functions on any number of Boolean variables. Especially, Nand, And, Not, and fan-out functions can be obtained from the Fredkin gate.

Controlled gates are important as they embody the possibility of conditional operations in algorithms harnessing unitary processes for quantum computation. In addition, controlled gates will be useful to us in the sequel, on several important occasions, for instance, in Chap. 9 to design circuit models for quantum noise by simulating non-unitary decohering evolutions, or in Chap. 12 to address in its Sect. 12.3 the methodology exploiting the quantum Fourier transform for quantum phase estimation as evoked for the Grover algorithm at the end of Sect. 5.5.

5.7 Summary on Quantum Circuits and Computation

Quantum circuits for computation can execute algorithms under the form of unitary evolutions of quantum states. Any finite-dimensional quantum states can be materialized as multiple-qubit states, and their unitary (and non-unitary as well via a dilated unitary) evolutions, on an arbitrary number of qubits, can always be implemented by combining one-qubit gates and the two-qubit Cnot gate, which together form a **universal set** of quantum gates. Quantum circuits generally implement unitary reversible transformations, yet by means of the three-qubit **Toffoli gate** of Sect. 5.2, any **multicomponent Boolean function** from L_{in} input bits into L_{out} output bits can receive an implementation by a reversible quantum circuit. Accordingly, any classical computations expressible by Boolean logic can be executed quantumly.

As seen in Sect. 5.3, any quantum circuit can be operated on a linear superposition of input states so as to simultaneously compute all the corresponding responses delivered in superposition at the output, to form a ground for **parallel computation** with potential for enhanced performance unattainable by classical means. Such quantum parallelism is specially appealing in areas where large amounts of data have to be processed in order to extract a few pieces of relevant information. As landmark examples, the **Deutsch–Jozsa quantum algorithm** of Sect. 5.4 performs parallel classification of Boolean functions (or digital images) from one single evaluation, while the **Grover quantum algorithm** of Sect. 5.5 solves the search in an unsorted database containing N items in $O(\sqrt{N})$ complexity instead of $O(N)$ classically.

As we have seen, for digital image processing for example, a binary image with 2^L pixels can be supported by a quantum register of $L + 1$ qubits, placed in a superposed state as for instance in the right-hand side of Eq. (5.13), whereas 2^L classical bits would be required instead (for a list of 2^L binary intensities with implicit addressing). In addition to such compact encoding, the quantum representation also provides access to parallel processing, as illustrated by the Deutsch–Jozsa and the Bernstein–Vazirani quantum algorithms of Sect. 5.4. Many possibilities remain today open for exploration in these areas of quantum digital image processing and quantum parallel computation.

The **controlled gates** of Sect. 5.6 provide the possibility of conditional operations in quantum computation. Controlled quantum gates will be useful to us in the sequel, for circuit models to simulate quantum noise in Chap. 9, and useful also in Chap. 12 in relation to the quantum Fourier transform. Chapter 12 will especially further illustrate the power of quantum computation, by applying the quantum Fourier transform to solve various problems of interest, such as the factorization of integers into prime numbers by the Shor quantum algorithm, with enhanced performance unattainable classically.

Quantum circuits for computation, quantum processors and computers, and associated quantum algorithms today remain quite active areas in progress for research and technological developments [118, 124, 125].

Chapter 6
Quantum Correlation

Quantum correlation, broadly speaking, can be understood as any type of statistical dependence that may be supported by quantum states or signals, or also more specifically, that may arise between the measurement outcomes that can be obtained from them. In this respect, this matches the notion of statistical correlation associated with signals in classical signal processing. However, in the quantum domain, specific and distinctive properties, related to entanglement, induce behaviors in the correlation that have no classical counterpart. This offers novel capabilities relevant to quantum signal and information processing. This chapter presents important reference phenomena and properties associated with such quantum correlation, with relevance to quantum signal and information processing.

6.1 EPR Correlations and Bell Inequalities

This section describes a fundamental reference scenario manifesting a specific and unconventional character of quantum correlation. It consists in a so-called Einstein–Podolsky–Rosen (EPR) experiment, initially discussed in [126] as a thought experiment challenging the nature and interpretation of (the then relatively nascent) quantum theory, and later related to Bell inequalities [127–129] constraining "classical reality", inequalities which were finally experimentally proved violated by "quantum reality" [130–132]. We propose here, for the EPR experiment and associated arguments, a formulation devised to be in tune with our signal and information processing perspective.

We begin by setting a classical reference of statistical correlation, by considering four binary random variables A_1, A_2, B_1, and B_2, assuming the values ± 1, randomly. We form the four-variable quantity

F. Chapeau-Blondeau, *Quantum Information and Quantum Signal Processing*,
https://doi.org/10.1007/978-3-032-27499-1_6

$$\Gamma = (A_1 - A_2)B_1 - (A_1 + A_2)B_2 = A_1B_1 - A_2B_1 - A_1B_2 - A_2B_2 = \pm 2 , \tag{6.1}$$

where we anticipated that Γ has access only to the two values ± 2, randomly. The reason is that, since A_1 and A_2 are ± 1, either $(A_1 - A_2)B_1 = 0$ or $(A_1 + A_2)B_2 = 0$, and in each case the remaining term for Γ is ± 2. Importantly, this limitation $\Gamma = \pm 2$ holds whatever the statistical dependence or correlation that may exist among the four binary random variables (A_1, A_2, B_1, B_2), which may even involve some deterministic dependence since this is a special case of statistical dependence. We now envisage computing a statistical average $\langle\Gamma\rangle$ for Γ, over the joint probability distribution of the four random variables (A_1, A_2, B_1, B_2), whatever it may be (expressing any conceivable dependence, even deterministic connection, statistical correlation, or independence). By linearity of the statistical average, from Eq. (6.1) we obtain

$$\langle\Gamma\rangle = \big\langle A_1B_1 - A_2B_1 - A_1B_2 - A_2B_2\big\rangle = \langle A_1B_1\rangle - \langle A_2B_1\rangle - \langle A_1B_2\rangle - \langle A_2B_2\rangle \ . \tag{6.2}$$

Since $\Gamma = \pm 2$, its average is limited as $-2 \le \langle\Gamma\rangle \le 2$, giving

$$-2 \le \langle\Gamma\rangle = \langle A_1B_1\rangle - \langle A_2B_1\rangle - \langle A_1B_2\rangle - \langle A_2B_2\rangle \le 2 \ . \tag{6.3}$$

Equation (6.3) is an example of a **Bell inequality** [127], or under this form it can be referred to also as a CHSH inequality [128, 133]. The Bell inequality of Eq. (6.3) is necessarily satisfied by a set of four binary random variables, whatever their structure of correlation or dependence, based on what we should call classical reasoning.

Next, we consider obtaining the four random variables A_1, A_2, B_1, and B_2, as the outcomes observed when performing measurements on a quantum object, namely a qubit pair. It is customary here to describe the protocol [7, 9] by involving two protagonists, Alice and Bob. They share a qubit pair denoted AB, Alice having access to qubit A while Bob has access to qubit B. Alice and Bob on their respective qubit of the pair choose to measure observables of the form

$$\Omega(\xi) = \sin(\xi)\sigma_x + \cos(\xi)\sigma_z \ , \tag{6.4}$$

parameterized by the angle ξ. These are spin observables of the qubit, as defined in Sect. 4.3.4. Each observable $\Omega(\xi)$ has the two eigenvalues $\lambda_\pm = \pm 1$ associated with the two orthogonal eigenstates $|\lambda_+(\xi)\rangle = [\cos(\xi/2), \sin(\xi/2)]^\top$ and $|\lambda_-(\xi)\rangle = [-\sin(\xi/2), \cos(\xi/2)]^\top$. Measuring on a qubit the observable $\Omega(\xi)$ is equivalent to performing a von Neumann projective measurement in the orthonormal eigenbasis $\big\{|\lambda_+(\xi)\rangle\, , |\lambda_-(\xi)\rangle\big\}$. The measurement will deliver as an outcome, one of the two eigenvalues ± 1, at random. In practice, such an experiment can be performed on photons [131, 132], and rotating a polarizing beam splitter around the axis of the incident photon beam allows one to select the angle ξ, as illustrated in Fig. 4.1(a).

On her qubit A, Alice can choose to measure the observable $\mathsf{A} \equiv \Omega(\xi = \alpha)$ at an angle $\xi = \alpha$, and in this way she obtains realizations of a binary random variable $A = \pm 1$. In a comparable way, on his qubit B, Bob can choose to measure the observable $\mathsf{B} \equiv \Omega(\xi = \beta)$ at an angle $\xi = \beta$, and he obtains realizations of a binary random variable $B = \pm 1$. Alice and Bob operate on a pair of qubits prepared in the joint bipartite quantum state $|AB\rangle$. The four joint probabilities for their measurement outcomes can be evaluated as

$$P(A = \pm 1, B = \pm 1) = \left|\langle \lambda_\pm(\alpha) \otimes \lambda_\pm(\beta) \mid AB \rangle\right|^2 , \tag{6.5}$$

conforming to Eq. (2.3).

We consider the situation where the qubit pair is prepared in the maximally entangled state

$$|AB\rangle = \frac{1}{\sqrt{2}}\big(|01\rangle - |10\rangle\big) , \tag{6.6}$$

which is the Bell state $|\beta_{11}\rangle$ or singlet state of Eq. (2.24). This entangled state for the preparation induces correlation between the two qubits of the pair AB. Some aspects were already evoked in Sect. 4.1.3 when discussing measurement on an entangled qubit pair. If each qubit were measured locally in the computational basis $\big\{|0\rangle , |1\rangle\big\}$, there would be an equal probability $1/2$ of finding it in each basis state $|0\rangle$ or $|1\rangle$. There would however exist strong correlation (complete anticorrelation) between the measurement results: if qubit A is measured in state $|0\rangle$ then qubit B is necessarily measured in state $|1\rangle$, and conversely. As an alternative, as we mentioned, Alice and Bob measure their respective qubit, not in the computational basis $\big\{|0\rangle , |1\rangle\big\}$, but in bases formed as $\big\{|\lambda_+(\xi)\rangle , |\lambda_-(\xi)\rangle\big\}$ from their local choice of a spin observable specified by an angle ξ. As we shall see, their measurement results performed on an entangled qubit pair will exhibit correlation with distinctive non-classical properties, especially in relation to the Bell inequality of Eq. (6.3). With $|AB\rangle$ prepared in the entangled Bell state of Eq. (6.6), the probabilities of Eq. (6.5) amount to

$$P(A = +1, B = +1) = P(A = -1, B = -1) = \frac{1}{4}\big[1 - \cos(\alpha - \beta)\big] , \tag{6.7}$$

and

$$P(A = +1, B = -1) = P(A = -1, B = +1) = \frac{1}{4}\big[1 + \cos(\alpha - \beta)\big] . \tag{6.8}$$

In particular, from the joint probabilities of Eqs. (6.7)–(6.8), by summation, the marginal probabilities follow as

$$P(A = +1) = P(A = -1) = P(B = +1) = P(B = -1) = \frac{1}{2} . \tag{6.9}$$

This is the same (marginal) equiprobability, observed by Alice and Bob for their two measurement outcomes, when locally measuring their respective qubit in the basis $\big\{|0\rangle , |1\rangle\big\}$ or in any rotated basis $\big\{|\lambda_+(\xi)\rangle , |\lambda_-(\xi)\rangle\big\}$.

We also have access to the pair correlation between Alice's and Bob's measurements, as they occur in the Bell inequality of Eq. (6.3), through the average $\langle AB\rangle = (+1)(+1)P(A=+1, B=+1) + (-1)(-1)P(A=-1, B=-1) + (+1)(-1)P(A=+1, B=-1) + (-1)(+1)P(A=-1, B=+1)$, giving

$$\langle AB\rangle = -\cos(\alpha-\beta)\ . \tag{6.10}$$

Alternatively, Eq. (4.60) would provide a more direct access to the same cross-correlation $\langle AB\rangle$ as the observable average $\langle AB\rangle = \langle\Omega(\alpha)\otimes\Omega(\beta)\rangle = \langle AB\,|\,\Omega(\alpha)\otimes\Omega(\beta)\,|\,AB\rangle$, without explicitly exhibiting the joint probability distribution $P(A=\pm1, B=\pm1)$ of Eqs. (6.7)–(6.8). In Eq. (6.10), the configuration $\alpha=\beta=0$ amounts for Alice and Bob to measuring their respective qubit of the pair in the computational basis $\{|0\rangle\,,|1\rangle\}$, yielding $\langle AB\rangle=-1$ for completely anticorrelated measurement results as anticipated above. Complete anticorrelation with $\langle AB\rangle=-1$ in Eq. (6.10) is similarly obtained when Alice and Bob measure their respective qubit at an arbitrary but identical angle $\alpha=\beta$.

The next step is to obtain four random variables as in the Bell inequality of Eq. (6.3). On her qubit A, Alice can randomly switch between measuring the observable $\mathsf{A}_1\equiv\Omega(\xi=\alpha_1)$ at an angle $\xi=\alpha_1$ to obtain realizations of the binary random variable $A_1=\pm1$, or measuring the observable $\mathsf{A}_2\equiv\Omega(\xi=\alpha_2)$ at an angle $\xi=\alpha_2$ to obtain realizations of the binary random variable $A_2=\pm1$. In a similar way, on his qubit B, Bob independently can randomly switch between measuring the observable $\mathsf{B}_1\equiv\Omega(\xi=\beta_1)$ at an angle $\xi=\beta_1$ to obtain realizations of the binary random variable $B_1=\pm1$, or measuring $\mathsf{B}_2\equiv\Omega(\xi=\beta_2)$ at an angle $\xi=\beta_2$ for realizations of the binary random variable $B_2=\pm1$. The switching schedules for Alice and Bob are considered independent, but are otherwise non-critical, provided the statistics of the four joint configurations A_1B_1, A_1B_2, A_2B_1, and A_2B_2 can be sampled experimentally. For each of the four measurement configurations, Eq. (6.10) gives access to the theoretical averages $\langle A_1B_1\rangle$, $\langle A_2B_1\rangle$, $\langle A_1B_2\rangle$, and $\langle A_2B_2\rangle$ as they appear in Eq. (6.2), so as to yield the theoretical average

$$\langle\Gamma\rangle = -\cos(\alpha_1-\beta_1)+\cos(\alpha_2-\beta_1)+\cos(\alpha_1-\beta_2)+\cos(\alpha_2-\beta_2)\ . \tag{6.11}$$

It then turns out that when Alice selects the angles $\alpha_1=0$ and $\alpha_2=\pi/2$ while Bob selects $\beta_1=3\pi/4$ and $\beta_2=\pi/4$, the average of Eq. (6.11) follows as $\langle\Gamma\rangle = -\cos(3\pi/4)+\cos(\pi/4)+\cos(\pi/4)+\cos(\pi/4)=2\sqrt{2}>2$. This is a **violation the Bell inequality** of Eq. (6.3). This is at least the theoretical prediction of quantum theory, to entail violation of the Bell inequality, and this quantitative prediction has been known since [127]. At this point we are facing a contradiction, between the classical prediction of the Bell inequality of Eq. (6.3), and the quantum prediction from Eq. (6.11) leading to a violation of the Bell inequality.

The theoretical prediction of Eq. (6.11) for the average $\langle\Gamma\rangle$ can be tested against an experimental evaluation. In a long sequence of repeated experiments where the qubit pair is prepared in the same entangled Bell state $|AB\rangle$ of Eq. (6.6), Alice on her qubit randomly chooses between measuring A_1 or A_2 to collect the realizations $A_1=\pm1$ or

$A_2 = \pm 1$, while Bob on his qubit randomly chooses between measuring B_1 or B_2 to collect the realizations $B_1 = \pm 1$ or $B_2 = \pm 1$. By gathering their measurement results attached to every repeated preparation of the qubit pair, Alice and Bob can obtain a large number of joint realizations ± 1 for each of the four products $A_1 B_1$, $A_1 B_2$, $A_2 B_1$ and $A_2 B_2$. For example, in the subset of experiments where Alice chooses to measure A_1 while Bob chooses to measure B_1, they both collect realizations $A_1 = \pm 1$ and $B_1 = \pm 1$, and they gather these realizations to obtain realizations for the product $A_1 B_1 = \pm 1$ and compute an estimate of the average product $\langle A_1 B_1 \rangle$. In a similar way for the four feasible joint configurations, Alice and Bob can compute empirical averages corresponding to the four quantities $\langle A_1 B_1 \rangle$, $\langle A_2 B_1 \rangle$, $\langle A_1 B_2 \rangle$ and $\langle A_2 B_2 \rangle$ as they appear in Eq. (6.2). This gives access to an empirical estimation of the average quantity $\langle \Gamma \rangle$ involved in the Bell inequality of Eq. (6.3). It was not until [131, 132] that a firm experimental test was actually carried out, with a physical (a photonic) experiment of the EPR setting enabling an empirical evaluation of the statistical average $\langle \Gamma \rangle$ of Eq. (6.11). It led to an experimental confirmation of the violation of the Bell inequality, as predicted by quantum theory.

Other measurement configurations by Alice and Bob can lead to violation of the Bell inequality of Eq. (6.3). Instead of four distinct observables $\Omega(\xi)$ from Eq. (6.4), Alice and Bob can choose among three distinct observables $\Omega(\xi)$. With the angles $\alpha_1 = \pi/4$, $\alpha_2 = 0$, and $\beta_1 = -\pi/4$, $\beta_2 = 0$, they choose among the three observables $\Omega(\pm\pi/4)$ and $\Omega(0) \equiv \sigma_z$. In this circumstance, the variables A_2 and B_2 are obtained by measuring the same observable $\Omega(0) \equiv \sigma_z$ by Alice and Bob on their respective qubit of the pair; their measurement results are always completely anticorrelated to give $\langle A_2 B_2 \rangle = -\cos(\alpha_2 - \beta_2) = -1$. Now Eq. (6.11) leads to the average $\langle \Gamma \rangle = -\cos(\pi/2) + \cos(\pi/4) + \cos(\pi/4) + 1 = \sqrt{2} + 1 > 2$, also violating the Bell inequality of Eq. (6.3), admittedly by a smaller amount when exploiting less variability in the observables.

The pending question now is why is it so? Why is it that the four binary random variables (A_1, A_2, B_1, B_2) collected by Alive and Bob in their EPR experiment do not satisfy the Bell inequality of Eq. (6.3)? What is going odd with the "classical" reasoning of Eqs. (6.1)–(6.3) arriving at the Bell inequality of Eq. (6.3), and the "quantum" reasoning of Eqs. (6.5)–(6.11) arriving at a violation of the Bell inequality experimentally verified? In the framework of the problem as formulated here in this chapter, the reader can take a moment to figure out what it may be. The question is in fact related to deep and subtle debates that have taken place for years, over the development of quantum theory, on its interpretation, and the impact it casts on the nature of physical reality [5, 126, 127, 134]; the question was not settled until the experimental verdict of [131, 132], at least for the physical side of it, let aside maybe a more philosophical one. Without accounting for all the many facets of the question, we can propose an explanation as follows.

We note that the "classical" reasoning accompanying Eqs. (6.1)–(6.3), to arrive at the Bell inequality of Eq. (6.3), envisages that the four random variables (A_1, A_2, B_1, B_2) jointly exist simultaneously, and accordingly that a four-variable joint probability distribution $P(a_1, a_2, b_1, b_2)$ can be associated with them, to describe their feasible joint configurations, whatever they may be. Then, the average quantity $\langle\Gamma\rangle$ of Eq. (6.2) follows along with the constraint upon it set by the Bell inequality of Eq. (6.3). By contrast, to arrive at the same average quantity $\langle\Gamma\rangle$, the "quantum" reasoning of Eqs. (6.5)–(6.11) does not refer to simultaneous joint configurations of the four random variables (A_1, A_2, B_1, B_2) and the four-variable joint probability distribution $P(a_1, a_2, b_1, b_2)$ they imply. Such a four-variable joint distribution $P(a_1, a_2, b_1, b_2)$ is somehow overabundant to gain access to the average $\langle\Gamma\rangle$ of Eq. (6.2). Instead, it is enough to invoke the two-variable probability distributions $P(a_1, b_1)$, $P(a_2, b_1)$, $P(a_1, b_2)$, and $P(a_2, b_2)$, for the corresponding pairs of random variables, as they result from Eqs. (6.7)–(6.8). The two-variable distributions are enough to compute the pair averages or pair correlations like $\langle AB\rangle$, which inserted in Eq. (6.2) give access to the average $\langle\Gamma\rangle$ of Eq. (6.2) or Eq. (6.11). If a four-variable joint distribution $P(a_1, a_2, b_1, b_2)$ exists in the first place, then from it by marginalization, for instance as $P(a_1, b_1) = \sum_{a_2, b_2} P(a_1, a_2, b_1, b_2)$, the four two-variable distributions $P(a_1, b_1)$, $P(a_2, b_1)$, $P(a_1, b_2)$, and $P(a_2, b_2)$ follow. Yet, since the quantum reasoning based on the two-variable distributions leads to violation of the Bell inequality of Eq. (6.3), we have to conclude that the two-variable probability distributions established by quantum theory via Eqs. (6.7)–(6.8) cannot be associated with, or derived from, an underlying four-variable joint probability distribution $P(a_1, a_2, b_1, b_2)$, otherwise the Bell inequality would hold. So within quantum theory, there cannot exist a four-variable joint probability distribution $P(a_1, a_2, b_1, b_2)$ ruling the four random variables A_1, A_2, B_1, and B_2 obtained by Alice and Bob as measurement results in their EPR experiment; this is mathematical logic. But why should it be so, from a physical standpoint?

Classical reasoning would argue that each of the four random variables A_1, A_2, B_1, and B_2 can be granted the status of a physical reality on its own, in the following sense. Each quantity A_1, A_2, B_1, or B_2 is a quantity that can be conceived, conceptualized, defined separately, for instance the quantity A_1 is the result of measuring the observable $\mathsf{A}_1 \equiv \Omega(\alpha_1)$ on qubit A; each quantity can be actually submitted to a physical (experimental) measurement so as to collect a measurement result or a value. We can even give up the restriction that the measurement result should be predictable in advance, but it could realize as a random variable, especially for fundamental reasons as the probabilistic character of measurement at the quantum level. This is the situation of each quantity A_1, A_2, B_1, and B_2: it can be conceptualized (defined) and it can measured, this is enough for each of them to exist as a physical reality, affording the possibility to consider jointly the values these four quantities may assume, so as to describe their collective situation with a joint probability distribution $P(a_1, a_2, b_1, b_2)$, as implied by the classical reasoning leading to the Bell inequality. Quantum reasoning leads us to recognize that no, this may not be enough, and instead suggests that if the quantity is not *actually* measured, it should not be granted the status of a physical reality, otherwise $P(a_1, a_2, b_1, b_2)$ should ensue and the Bell inequality would hold.

We have seen that in quantum theory the measurement generally has a destructive effect. When the observable A_1 is measured on qubit A, the measurement usually destroys the state of the qubit, and another observable $\mathsf{A}_2 \neq \mathsf{A}_1$ in general cannot be measured on the same qubit A. This we understand, since the state has been destroyed when measuring A_1. But the observations in the EPR experiment lead us one step further in the interpretation of quantum theory. If A_1 is measured on qubit A, then A_2 cannot be measured on qubit A since the state has been destroyed; but in addition, if A_1 is measured at the value A_1 on qubit A, the result A_2 that could have been obtained by measuring A_2 on qubit A cannot even be conceived as having a physical reality: if it is A_1 that is measured at A_1, a putative value A_2 for A_2 simply cannot exist as a physical quantity. In this respect, the two observables A_1 and A_2 that cannot be simultaneously measured are incompatible; the quantities A_1 and A_2 that could result from measurement of such two incompatible observables cannot exist simultaneously. When A_1 exists by being measured, A_2 that cannot be measured simply does not exist as a physical reality, and as a consequence it cannot be involved, along with A_1, in some reasoning as done with Eqs. (6.1)–(6.3), otherwise the reasoning faces to be contradicted by physical observation. This is what happened with the violation of Bell inequalities, as observed experimentally in [131, 132], that in this respect invalidates the classical reasoning and leads us to revise or adjust some of our conceptions about physical reality, as we undertake here. The Bell inequality of Eq. (6.3) does not apply to the quantities delivered by quantum measurement in the EPR experiment; the quantities A_1 and A_2 that cannot be measured simultaneously or in conjunction cannot exist simultaneously or in conjunction; the argument invoked just after Eq. (6.1), that either $(A_1 - A_2)B_2 = 0$ or $(A_1 + A_2)B_1 = 0$, does not hold physically. Ultimately, the variable Γ of Eq. (6.1) does not even exist as a physical reality, as it involves incompatible quantities in its definition. Only the statistical average $\langle\Gamma\rangle$ involving pairwise averaging of compatible (since they are measured on two distinct qubits A and B) quantities evaluated separately as in Eq. (6.11) can be assigned a physical reality. And $\langle\Gamma\rangle$ is not constrained by the Bell inequality of Eq. (6.3) derived from a classical reasoning that does not apply to it. Accordingly, in the framework of such quantum reasoning, we are led to recognize that it is the observation or measurement that creates the physical reality, or determines what pertains to the physical reality and what does not.

Another element in the picture is related to an objection, historically addressed to quantum theory, that it could be incomplete [126]. This was already evoked in Sect. 4.1.3 when discussing measurement on an entangled qubit pair. As we have seen in Chap. 4, quantum theory stipulates that the result of a measurement generally occurs at random, and when one part of an entangled bipartite system is measured this may affect instantly the other part. To avoid such weird properties of indeterminism and instant action at a distance, it has been suggested that hidden variables might exist, yet to be discovered, that would be settled as soon as the creation of the quantum state, and would then completely determine the subsequent behavior of the system and that of its parts upon measurement. For the qubit pair AB in the EPR experiment, a set of hidden variables denoted $\aleph$, would be fixed at the creation of the qubit pair in the initial state $|AB\rangle$, and would prescribe in advance the result of any conceivable measurement that could be made on the pair AB, especially when Alice chooses to measure A_1 or A_2 and when Bob chooses to measure B_1 or B_2 on their respective qubit. With such hidden variables, there would be no need to resort to instant action at a distance and indeterminism in the cause—the cause now lying in (or receding to) the hidden variables. If such hidden variables $\aleph$ preexist, capable of predicting in advance the measurement results A_1, A_2, B_1, and B_2, then this situation can be described by some conditional probability distribution that we write for convenience $P(a_1, a_2, b_1, b_2 \mid \aleph)$. This is a rather versatile model, accommodating the possibility of a deterministic or a probabilistic influence of $\aleph$ to determine (A_1, A_2, B_1, B_2). In addition, if the conditions for preparing the quantum system AB are perfectly known and controlled, one may assume the same hidden variables $\aleph = \aleph_{\text{fixed}}$ at each repetition of the preparation, representable by a degenerate peaked prior probability distribution $p(\aleph) = \delta(\aleph - \aleph_{\text{fixed}})$. Alternatively, since admittedly the hidden variables are not fully controlled, they may differ between repetitions of the preparation, and be representable by a non-degenerate prior probability distribution $p(\aleph)$. In any case, with the existence of hidden variables $\aleph$ allowing to predict all four quantities (A_1, A_2, B_1, B_2), one would have access to a four-variable probability distribution $P(a_1, a_2, b_1, b_2) = \int_\aleph P(a_1, a_2, b_1, b_2 \mid \aleph) p(\aleph) d\aleph$ for the measurement results (A_1, A_2, B_1, B_2) of Alice and Bob on the pair AB. Violation of the Bell inequality, which precludes the existence of a four-variable probability distribution $P(a_1, a_2, b_1, b_2)$, also rules out the possibility of such hidden variables.

An important message here to single out, that concerns random variables obtained as the results of quantum measurements, can be summarized as follows. On a given quantum system, two observables A and B can be measured simultaneously if and only if they commute or equivalently share a common eigenbasis, as we have already seen in Sect. 4.3.3. If they do not commute, the two observables are incompatible: when A is measured then B cannot be measured on the same system (instead, what is feasible is to repeat anew the same preparation in succession, so as to measure A or B successively on renewed identical preparations, and the Heisenberg inequality of Sect. 4.3.3 deals with this situation). But it is not only because the state of the system has been destroyed when measuring A that no measurement result or value can be ascribed simultaneously to B. We have to recognize, more fundamentally, that it is also because, when A is measured and B is not, the putative result of measuring B on the same system or a value that could be ascribed to B, has no physical existence or reality; and the mere fact of postulating one to involve it in some physical reasoning may lead to a conclusion invalidated by physical observation (as a Bell inequality that is subsequently violated by experiment). Two incompatible quantities cannot simultaneously exist as physical reality.

Physical observation at the quantum level leads us to revise or refine our classical conceptions on determinism, on locality and on realism. At the quantum level, physical reality is intrinsically non-deterministic (this we know from the probabilistic character of quantum measurement reported as early as Sect. 2.2); it is nonlocal, as revealed by the behavior from entangled states presented earlier in Sects. 2.4 and 4.1.3. Concerning realism, from what we have learned from the EPR experiment in this section, to put it abruptly, reality has no independent intrinsic existence, it is observation that creates physical reality.[1] More circumspectly, quantities that cannot be simultaneously measured have no simultaneous physical reality.

To obtain distinctive quantum correlations capable of violating the Bell inequality, in addition to variables obtained from measurement on incompatible quantum quantities, another essential ingredient is the presence of quantum entanglement. In the EPR experiment above, it can be noted that if the qubit pair AB is prepared in a separable quantum state $|AB\rangle = |A\rangle \otimes |B\rangle$, then quantum theory stipulates that the results of the measurements performed by Alice and Bob are statistically independent, with joint probabilities for (A, B) in Eq. (6.5) factorizing as $P(a, b) = P(a)P(b)$. The four variables representing the measurement results assume averages $\langle A_1 \rangle$, $\langle A_2 \rangle$, $\langle B_1 \rangle$, and $\langle B_2 \rangle$ always lying in $[-1, 1]$. From this statistical independence, the average for Γ of Eq. (6.1) follows as $\langle \Gamma \rangle = \big(\langle A_1 \rangle - \langle A_2 \rangle\big) \langle B_1 \rangle - \big(\langle A_1 \rangle + \langle A_2 \rangle\big) \langle B_2 \rangle$. With each of the four marginal averages like $\langle A_1 \rangle$ lying in $[-1, 1]$, it mathematically follows that the stipulation of quantum theory is an average $\langle \Gamma \rangle$ always lying in $[-2, 2]$ and not outside. For an experimental test, when the observables A_1 and A_2 are incompatible, one should consider measuring the variables A_1 and A_2 necessarily in distinct experiments, whose repetitions give separate access to the averages $\langle A_1 \rangle$ and $\langle A_2 \rangle$, and similarly for $\langle B_1 \rangle$ and $\langle B_2 \rangle$, all four lying in $[-1, 1]$ and yielding $\langle \Gamma \rangle$ in $[-2, 2]$, producing no violation of the Bell inequality of Eq. (6.3), in accordance with quantum theory. In this way, separable product states $|AB\rangle = |A\rangle \otimes |B\rangle$ of the pair AB never produce violation of the Bell inequality of Eq. (6.3). Violation of the Bell inequality of Eq. (6.3), revealing the failure of classical reasoning applied to the variables (A_1, A_2, B_1, B_2), necessitates an entangled quantum state $|AB\rangle$ of the qubit pair AB. Following the same approach as with Eq. (6.6), it can be verified that any entanglement between the qubits A and B, however small but non-vanishing, can produce a violation of the Bell inequality of Eq. (6.3) for appropriate measurement configurations [135]. The special correlations between the random variables A_1, A_2, B_1, and B_2 delivered by quantum measurement from an entangled pair and exhibiting the specific "quantum" behavior manifested by a violation of the Bell inequality are critically tied to the presence of entanglement. In this respect, quantum entanglement is a property that can produce non-classical correlations between the results of measurements performed by two distant parties; and these correlations, when issued from measurements on incompatible quantum quantities, may escape classical constraints that otherwise would apply to classical variables. In the next section we will examine another prototypical experiment that also sheds light on the special correlations associated with quantum entanglement and meaningful to quantum signal and information processing.

The EPR experiment and its interpretation essential to quantum theory and to our apprehension of physical reality have been consolidated and expanded in many directions. Experimental confirmations of the violation of Bell inequalities have been

[1] "Do you really think that the moon exists only when someone looks at it?" could it be argued. For quantum theory, an observation of a quantum system is any interaction with its environment that leaks out information about the system. As soon as a photon is emitted by the moon and absorbed somewhere outside to produce a change in its environment, this is an observation of the moon.

obtained in different settings [136–138]. For detecting entanglement-induced non-classical correlations, information-theoretic Bell-type inequalities have been derived [139–141] to assess statistical dependence beyond the linear cross-correlation of Eqs. (6.10)–(6.11), and exploiting the Shannon entropy [139, 140], or generalized entropies with enhanced sensitivity to entanglement [141]. Extensions to applications and to multipartite entanglement have also been considered, as for instance in [142–145].

6.2 Entanglement, Local Measurements, and Classical Communication

To investigate other significant properties of correlation associated with quantum entanglement, we consider a qubit pair AB initially prepared in a bipartite pure quantum state $|AB\rangle$ chosen as either one of the four entangled Bell states of Eqs. (2.21)–(2.24) denoted here

$$|\phi_\pm\rangle = \frac{1}{\sqrt{2}}\big(|00\rangle \pm |11\rangle\big) , \tag{6.12}$$

$$|\psi_\pm\rangle = \frac{1}{\sqrt{2}}\big(|01\rangle \pm |10\rangle\big) , \tag{6.13}$$

and forming an orthonormal basis of the two-qubit Hilbert space $\mathcal{H}_2 \otimes \mathcal{H}_2$. For convenience, the two states $|\phi_\pm\rangle$ in Eq. (6.12) will be said with even parity, and the two states $|\psi_\pm\rangle$ in Eq. (6.13) with odd parity; in addition, in Eqs. (6.12)–(6.13) the two states with a $+$ sign have positive phase, and the two states with a $-$ sign have negative phase. By selecting one among the four states of Eqs. (6.12)–(6.13), the initial preparation $|AB\rangle$ can store two bits of classical information in the qubit pair AB. The qubit A is then sent to Alice and the qubit B to Bob, and we want to examine how Alice and Bob can recover the information stored in the qubit pair AB, from local measurements on their respective qubit.

As it follows from Sect. 4.4, locally Alice's qubit A is described by the reduced state $\rho_A = \mathrm{tr}_B\big(|AB\rangle\langle AB|\big) = \mathrm{I}_2/2$, which is the maximally mixed one-qubit state, invariably for anyone of the four preparations from Eqs. (6.12)–(6.13) for $|AB\rangle$. As a result, when Alice performs any two-outcome von Neumann measurement on her qubit A, she invariably obtains anyone of the two possible outcomes at random with probability $1/2$. Consequently, Alice through a local measurement of her qubit A cannot get any information concerning the initial preparation $|AB\rangle$. The same is true locally for Bob, upon measuring his qubit B. There is no information Alice and Bob can extract locally from their qubit concerning the initial preparation $|AB\rangle$. There is however information in the correlation between their measurement results.

Alice and Bob can exchange classical information to combine their measurement results. Alice and Bob agree to measure their respective qubit in the orthonormal basis $\{|0\rangle, |1\rangle\}$ of $\mathcal{H}_2$. If their two measurement outcomes coincide, then they know that the initial preparation $|AB\rangle$ was one of the two states $|\phi_\pm\rangle$ of Eq. (6.12); if their two measurement outcomes differ, then they know that the initial preparation $|AB\rangle$ was one of the two states $|\psi_\pm\rangle$ of Eq. (6.13). In this way, Alice and Bob can know the parity of the initial preparation $|AB\rangle$, among two possible parities, which amounts to extracting one bit of classical information concerning the initial preparation $|AB\rangle$.

As an alternative, Alice and Bob can agree to measure their respective qubit in the Hadamard basis $\{|+\rangle, |-\rangle\}$ of Eqs. (2.7)–(2.8). By referring to Eqs. (2.29)–(2.32) expressing the four Bell states of Eqs. (6.12)–(6.13) in the Hadamard basis, Alice and Bob know that if their two measurement outcomes coincide, then the initial preparation $|AB\rangle$ was one of the two states with a $+$ sign in Eqs. (6.12)–(6.13); if their two measurement outcomes differ, they know that the initial preparation $|AB\rangle$ was one of the two states with a $-$ sign in Eqs. (6.12)–(6.13). In this way, Alice and Bob can know the phase of the initial preparation $|AB\rangle$, among two possible phases, which amounts to extracting one bit of classical information concerning the initial preparation $|AB\rangle$.

In addition, when Alice measures in the basis $\{|0\rangle, |1\rangle\}$ and Bob in $\{|+\rangle, |-\rangle\}$, or conversely, all four preparations give the same factorizable probability distribution for the outcomes, so that Alice and Bob cannot get any information concerning the initial preparation, and whether or not they share their measurement results makes no difference.

Alice and Bob cannot measure their qubit simultaneously in the bases $\{|0\rangle, |1\rangle\}$ and $\{|+\rangle, |-\rangle\}$; these are two incompatible measurements, equivalent to measuring the two non-commuting spin observables σ_z and σ_x respectively. Accordingly, Alice and Bob cannot know simultaneously the parity and the phase of the initial preparation $|AB\rangle$. Alice and Bob through local measurements connected by classical communication cannot completely recover the initial preparation $|AB\rangle$ and the two bits of information it stores. The information stored in an *entangled* qubit pair $|AB\rangle$ has a special character, in a sense not completely amenable to classical processing. The information is stored in an entangled pair AB behaving as a nonlocal whole, having a definite state only as a whole delocalized pair, and no definite states exist locally or separately for each part A and B. As a result, the two bits of classical information stored in the entangled qubit pair AB can only be recovered by a collective non-separable measurement acting globally on the pair AB, and not by local measurements separately on A and B supplemented by classical communication.

On the contrary, if the initial preparation $|AB\rangle$ selects among the four non-entangled states $\{|00\rangle, |01\rangle, |10\rangle, |11\rangle\}$ also forming an orthonormal basis of $\mathcal{H}_2 \otimes \mathcal{H}_2$, then Alice via a local measurement of her qubit A in the basis $\{|0\rangle, |1\rangle\}$, can extract one bit of information concerning the initial preparation $|AB\rangle$. In the same way, Bob, via local measurement of his qubit B in the basis $\{|0\rangle, |1\rangle\}$, can extract one bit of information concerning the initial preparation $|AB\rangle$. By sharing their measurement results via classical communication, Alice and Bob can then determine completely the initial preparation $|AB\rangle$ and the two bits of classical information stored in the pair AB. This is feasible here with a non-entangled, separable, preparation $|AB\rangle$.

The properties reported above indicate that the type of correlation existing in an entangled qubit pair displays distinctive nonlocal properties, not reducible to classical processing. Here, if isolated measurements are performed locally on the two parts A and B, no information whatsoever can be extracted from the entangled pair AB. If the local measurements of the two parts A and B are combined via classical communication, only one bit of information can be extracted from the two bits stored in the entangled pair AB. Solely a nonlocal non-separable collective measurement of the entangled pair AB as a whole can recover the two bits of stored information. Such quantum correlations arising from quantum entanglement appear as distinctive resources, which may find useful applications for quantum signal and information processing. Several important reference scenarios of this kind, where quantum entanglement plays a central role relevant to signal and information processing, will be presented below in this chapter, while some others may certainly remain to be discovered. Before we come to this presentation, we exhibit in the next section, in a formal manner, an important limitation however existing, precluding the use of quantum entanglement for direct transmission of information or direct signaling between two distant parties.

6.3 No-Signaling Theorem

As it already appeared in Sect. 4.1.3 when discussing measurement on an entangled qubit pair, quantum entanglement, and the correlations it can induce between two distant parties, cannot be used to directly transmit information between them two, since the measurement results generally occur at random out of control. The no-signaling theorem or principle—a theorem that we are going to prove here—expresses formally that (or how) quantum entanglement cannot be used to directly transmit information between two quantum systems A and B. As with Alice and Bob above, we consider two quantum systems A and B placed in a joint bipartite state represented by the density operator ρ_{AB} on the product Hilbert space $\mathcal{H}^A \otimes \mathcal{H}^B$. When system B alone is accessible for measurement, the results when measuring B are determined solely by the reduced density operator $\rho_B = \mathrm{tr}_A(\rho_{AB})$. While nothing is done on system B, some separate processing can be accomplished on system A, that we generally model by the quantum process or operation $\mathcal{E}_A$ acting locally on A and represented by the set of Kraus operators $\{\mathsf{A}_j\}$ on $\mathcal{H}^A$ satisfying the completeness relation $\sum_j \mathsf{A}_j^\dagger \mathsf{A}_j = \mathrm{I}^A$, with I^A the identity operator on $\mathcal{H}^A$. This quantum process $\mathcal{E}_A$ can stand for a unitary or a non-unitary transformation performed locally on system A to represent some local imprint of information on A. The joint state ρ_{AB} of the compound AB as a result gets transformed as $\rho_{AB} \mapsto \mathcal{E}_A \otimes \mathcal{I}_B(\rho_{AB}) = \rho'_{AB}$, with $\mathcal{I}_B$ the identity superoperator for B. As nothing is done on B, a set of Kraus operators for the compound AB is $\{\mathsf{A}_j \otimes \mathrm{I}^B\}$ with I^B the identity operator on $\mathcal{H}^B$, and the joint transformed state of the compound AB is expressible as the operator sum

$$\rho'_{AB} = \mathcal{E}_A \otimes \mathcal{I}_B(\rho_{AB}) = \sum_j (\mathsf{A}_j \otimes \mathrm{I}^B)\rho_{AB}(\mathsf{A}_j^\dagger \otimes \mathrm{I}^B) \; . \tag{6.14}$$

After this transformation, the reduced state of B becomes

$$\rho'_B = \mathrm{tr}_A(\rho'_{AB}) \tag{6.15}$$

$$= \mathrm{tr}_A\Big(\sum_j (\mathsf{A}_j \otimes \mathrm{I}^B)\rho_{AB}(\mathsf{A}_j^\dagger \otimes \mathrm{I}^B)\Big) \tag{6.16}$$

$$= \sum_j \mathrm{tr}_A\Big((\mathsf{A}_j \otimes \mathrm{I}^B)\rho_{AB}(\mathsf{A}_j^\dagger \otimes \mathrm{I}^B)\Big) \; , \tag{6.17}$$

by linearity of the partial trace. By circular invariance of the partial trace over the traced space, according to Eq. (A.26) of the Appendix, we obtain

$$\rho'_B = \sum_j \mathrm{tr}_A\Big(\rho_{AB}(\mathsf{A}_j^\dagger \otimes \mathrm{I}^B)(\mathsf{A}_j \otimes \mathrm{I}^B)\Big) \tag{6.18}$$

$$= \sum_j \mathrm{tr}_A\Big(\rho_{AB}(\mathsf{A}_j^\dagger \mathsf{A}_j \otimes \mathrm{I}^B)\Big) \tag{6.19}$$

$$= \mathrm{tr}_A\Big(\rho_{AB}\Big(\sum_j \mathsf{A}_j^\dagger \mathsf{A}_j \otimes \mathrm{I}^B\Big)\Big) \tag{6.20}$$

$$= \mathrm{tr}_A(\rho_{AB}) = \rho_B \; , \tag{6.21}$$

with the completeness relation on the Kraus operators A_j to go from Eq. (6.20) to Eq. (6.21). This expresses that the reduced state ρ_B of B is unaffected by the process $\mathcal{E}_A$ local to A.

If instead of applying to A the quantum process $\mathcal{E}_A$ a generalized measurement with operators $\{\mathsf{M}_m\}$ is performed locally on A while B remains untouched, then the measurement result m is obtained with the probability $P(m) = \mathrm{tr}\big((\mathsf{M}_m \otimes \mathrm{I}^B)\rho_{AB}(\mathsf{M}_m^\dagger \otimes \mathrm{I}^B)\big)$, and the compound AB terminates in the joint post-measurement state $\rho_{AB}^m = (\mathsf{M}_m \otimes \mathrm{I}^B)\rho_{AB}(\mathsf{M}_m^\dagger \otimes \mathrm{I}^B)/P(m)$, as it follows from Sect. 4.2.3 on generalized measurement. When the measurement result m is unread or unknown, the compound AB is represented by the average density operator $\rho'_{AB} = \sum_m P(m)\rho_{AB}^m = \sum_m (\mathsf{M}_m \otimes \mathrm{I}^B)\rho_{AB}(\mathsf{M}_m^\dagger \otimes \mathrm{I}^B)$. This implements on AB the same joint transformation as in Eq. (6.14) when the Kraus operators $\{\mathsf{A}_j\}$ local to A are replaced by the measurement operators $\{\mathsf{M}_m\}$ acting on A. This conforms with the observation made at the end of Sect. 4.2.3 that an unread measurement is equivalent to a quantum process. When B is unaware of the result m by the unread measurement, the state of B is represented by the reduced density operator $\mathrm{tr}_A(\rho'_{AB}) = \mathrm{tr}_A\big(\rho_{AB}(\sum_m \mathsf{M}_m^\dagger \mathsf{M}_m \otimes \mathrm{I}^B)\big) = \mathrm{tr}_A(\rho_{AB}) = \rho_B$, expressing that the reduced state ρ_B of B is unaffected by the local measurement performed on A. Any local measurement on B is governed by ρ_B and is therefore incapable of detecting the measurement performed on A. This situation has to be contrasted with the situation we described in Sect. 4.1.3, reporting that measuring qubit A of an entangled qubit pair AB can instantly affect qubit B. This change on B however cannot be detected by locally measuring B, since the local state ρ_B of B remains the same. Only when the result m of the local measurement on A is communicated to B does the situation change. With m known to B, the state of B is represented by the reduced density operator $\mathrm{tr}_A(\rho_{AB}^m) = \mathrm{tr}_A(\rho_{AB}(\mathsf{M}_m^\dagger \mathsf{M}_m \otimes \mathrm{I}^B))/P(m) = \rho_B^m$ indexed by m, that in general differs from the initial state ρ_B local to B. It is the ensemble average $\sum_m \rho_B^m P(m)$ that matches ρ_B. In this informed condition the state of B has been changed from ρ_B to ρ_B^m by measuring A, and locally measuring B can then detect that something occurred to A.

Therefore, whatever is done on system A separately through a quantum process or a quantum measurement local to A, the reduced state $\rho'_B = \rho_B$ local to system B remains unchanged. This is the **no-signaling theorem**: when A and B are in a joint (entangled) quantum state, no processing locally acting on A separately can be detected by measuring B alone.

There exists an even broader version of the no-signaling theorem, stipulating that for an entangled bipartite system AB, local processings separately on system A and on system B, do not affect the local reduced states of system A and of system B. To make it more explicit, we consider that for the same pair of two quantum systems A and B in the joint state ρ_{AB} as above, some separate processing is also performed on system B, modeled by the quantum process $\mathcal{E}_B$ acting locally on B and represented by the set of Kraus operators $\{\mathsf{B}_k\}$ on $\mathcal{H}^B$ satisfying $\sum_k \mathsf{B}_k^\dagger \mathsf{B}_k = \mathrm{I}^B$. With $\mathcal{E}_A$ acting locally on A and $\mathcal{E}_B$ acting locally on B, the joint state ρ_{AB} of the compound AB gets transformed according to $\rho_{AB} \mapsto \mathcal{E}_A \otimes \mathcal{E}_B(\rho_{AB}) = \rho'_{AB}$. After this joint transformation, locally the reduced state of system A is $\rho'_A = \mathrm{tr}_B(\rho'_{AB})$, and follows as

$$\rho'_A = \mathrm{tr}_B\big(\mathcal{E}_A \otimes \mathcal{E}_B(\rho_{AB})\big) \tag{6.22}$$

$$= \mathrm{tr}_B\Big(\sum_{j,k} (\mathsf{A}_j \otimes \mathsf{B}_k)\rho_{AB}(\mathsf{A}_j \otimes \mathsf{B}_k)^\dagger\Big) \tag{6.23}$$

$$= \mathrm{tr}_B\Big(\sum_{j,k} (\mathsf{A}_j \otimes \mathsf{B}_k)\rho_{AB}(\mathsf{A}_j^\dagger \otimes \mathsf{B}_k^\dagger)\Big) \tag{6.24}$$

$$= \mathrm{tr}_B\Big(\sum_{j,k} (\mathrm{I}^A \otimes \mathsf{B}_k)(\mathsf{A}_j \otimes \mathrm{I}^B)\rho_{AB}(\mathsf{A}_j^\dagger \otimes \mathrm{I}^B)(\mathrm{I}^A \otimes \mathsf{B}_k^\dagger)\Big) \tag{6.25}$$

$$= \sum_{j,k} \mathrm{tr}_B\Big((\mathrm{I}^A \otimes \mathsf{B}_k)(\mathsf{A}_j \otimes \mathrm{I}^B)\rho_{AB}(\mathsf{A}_j^\dagger \otimes \mathrm{I}^B)(\mathrm{I}^A \otimes \mathsf{B}_k^\dagger)\Big), \tag{6.26}$$

by linearity of the partial trace. By circular invariance of the partial trace over the traced space, according to Eq. (A.27) of the Appendix, we obtain

$$\rho'_A = \sum_{j,k} \mathrm{tr}_B\Big(\big(\mathsf{A}_j \otimes \mathrm{I}^B\big)\rho_{AB}\big(\mathsf{A}_j^\dagger \otimes \mathrm{I}^B\big)\big(\mathrm{I}^A \otimes \mathsf{B}_k^\dagger\big)\big(\mathrm{I}^A \otimes \mathsf{B}_k\big)\Big) \tag{6.27}$$

$$= \sum_{j,k} \mathrm{tr}_B\Big(\big(\mathsf{A}_j \otimes \mathrm{I}^B\big)\rho_{AB}\big(\mathsf{A}_j^\dagger \otimes \mathrm{I}^B\big)\big(\mathrm{I}^A \otimes \mathsf{B}_k^\dagger \mathsf{B}_k\big)\Big) \tag{6.28}$$

$$= \sum_{j} \mathrm{tr}_B\Big(\big(\mathsf{A}_j \otimes \mathrm{I}^B\big)\rho_{AB}\big(\mathsf{A}_j^\dagger \otimes \mathrm{I}^B\big)\Big(\mathrm{I}^A \otimes \sum_k \mathsf{B}_k^\dagger \mathsf{B}_k\Big)\Big) \tag{6.29}$$

$$= \sum_{j} \mathrm{tr}_B\Big(\big(\mathsf{A}_j \otimes \mathrm{I}^B\big)\rho_{AB}\big(\mathsf{A}_j^\dagger \otimes \mathrm{I}^B\big)\Big) . \tag{6.30}$$

By the separability property of the partial trace according to Eq. (A.24), we obtain

$$\rho'_A = \sum_j \mathsf{A}_j \, \mathrm{tr}_B(\rho_{AB}) \mathsf{A}_j^\dagger \tag{6.31}$$

$$= \sum_j \mathsf{A}_j \rho_A \mathsf{A}_j^\dagger = \mathcal{E}_A(\rho_A) . \tag{6.32}$$

So Eq. (6.32) expresses that, whatever occurs to system B of the compound AB, provided A and B evolve through separable local processes, the reduced state of A is affected solely by the local process $\mathcal{E}_A$ on A and is not affected by the remote evolution $\mathcal{E}_B$ local to B. By symmetry, the reduced state $\rho_B = \mathrm{tr}_A(\rho_{AB})$ of B is evolved locally in a comparable way. As before, the local quantum processes $\mathcal{E}_A$ and $\mathcal{E}_B$ can include generalized measurements local to A and to B, that, as long as the local measurement results are not exchanged and remain unknown to the other distant party, behave as quantum processes, to lead to similar no-signaling.

Such no-signaling theorem precludes in this way direct transmission of information between two parties sharing quantum entanglement. However, there exist alternative ways to exploit the specific correlations established by quantum entanglement between two or more parties, so as to obtain effective benefits with significance to information processing. We now present several important reference scenarios of this kind.

6.4 GHZ Quantum Correlation of Three Parties

The EPR experiment with violation of the Bell inequality in Sect. 6.1, shows us that classical constraints restraining classical variables, may not apply when these variables are obtained from measurements of incompatible quantum quantities on entangled quantum systems. Such non-classical behavior is obtained, in the EPR experiment of Sect. 6.1, as a statistical effect (violation of the Bell inequality on the average $\langle \Gamma \rangle$) inaccessible classically. We shall now describe another manifestation of such non-classical behavior attached to correlation from quantum entanglement, this

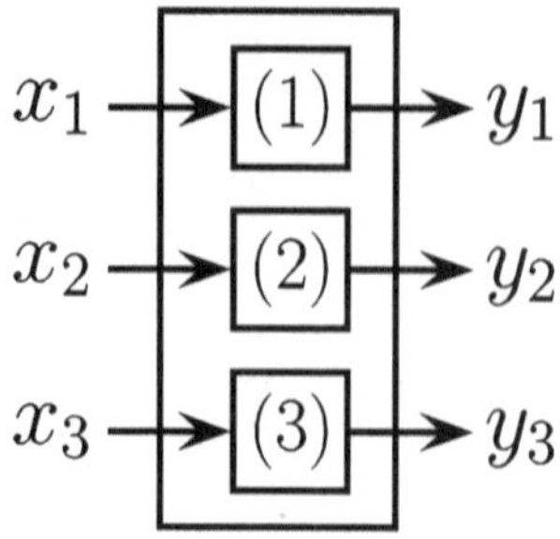

Fig. 6.1 Each player (1), (2), and (3) receives a binary input x_j and has to produce a binary output $y_j = y_j(x_j)$, so that the joint output $y_1y_2y_3$ satisfies the prescribed requirement (i)–(ii)

time as a deterministic effect among correlated variables, also inaccessible classically [146–148].

The scenario can be described as an experiment, or a game, involving three players (1), (2) and (3), each of them receiving a binary input, respectively x_1, x_2, and x_3, where each x_j is a classical binary variable with value 0 or 1. The joint input $x_1x_2x_3$ that can be given or applied to the players is randomly drawn from the set $\{000, 011, 101, 110\}$, where we recognize in this set of licit inputs the four (among eight) 3-bit binary words $x_1x_2x_3$ with even parity. After receiving its input x_j, each player has to produce an output y_j, where each y_j is also a classical binary variable assuming value 0 or 1, for $j = 1, 2, 3$. The output response y_j of player (j) can only be a function of its local input x_j, but not of the inputs of the two other players, which remain unknown to (j). The (collaborative) game is won, collectively, by the three players, when they manage to produce a joint output $y_1y_2y_3$ that satisfies the following two-fold requirement (with $\oplus$ the binary Xor or addition modulo 2):
(i) when the joint input is $x_1x_2x_3 = 000$, the joint output $y_1y_2y_3$ must satisfy $y_1 \oplus y_2 \oplus y_3 = 0$, that is, the joint output must conserve the even parity of the input;
(ii) when the joint input $x_1x_2x_3 \in \{011, 101, 110\}$, the joint output $y_1y_2y_3$ must satisfy $y_1 \oplus y_2 \oplus y_3 = 1$, that is, the joint output must reverse the parity of the input.
Before they receive their input, the players can agree on a collective strategy for producing their output response; but after they have received their input, they can no longer communicate to produce their output. The structure of the game is depicted in Fig. 6.1.

The collective strategy would aim to induce sufficient correlation between the players and their behavior so that their joint response $y_1y_2y_3$ matches the requirement (i)-(ii). One can try to figure out what could be a good strategy for the players, so as to win as frequently as possible in a sequence of plays where the joint input $x_1x_2x_3$ is drawn at random from the set of four licit configurations. There is an argument showing that no strategy exists that could win on all four licit input configurations, at least when exploiting classical resources or correlations, should we say. A (classical) strategy winning on the four licit input configurations would consist of three input–output functions $y_j(x_j)$, one for each player, that should satisfy the four constraints written on lines (a)–(d) in Fig. 6.2.

$$
\begin{array}{rcl}
y_1(0) \oplus y_2(0) \oplus y_3(0) = 0 & & \text{(a)} \\
y_1(0) \oplus y_2(1) \oplus y_3(1) = 1 & & \text{(b)} \\
y_1(1) \oplus y_2(0) \oplus y_3(1) = 1 & & \text{(c)} \\
y_1(1) \oplus y_2(1) \oplus y_3(0) = 1 & & \text{(d)} \\
\hline
0 \;\oplus\; 0 \;\oplus\; 0 \;= 1 & & \text{(e)} \\
\Longrightarrow \qquad\qquad 0 \;= 1 & & \text{(f)}
\end{array}
$$

Fig. 6.2 A winning strategy would have to satisfy the four constraints of lines (**a**)–(**d**), whose sum modulo 2 leads to line (**e**) equivalent to line (**f**). The final contradiction $0 = 1$ in line (f) indicates that the four constraints (**a**)–(**d**) are incompatible and cannot be simultaneously satisfied

Next in Fig. 6.2, the four constraints on lines (a)–(d) are summed modulo 2 respectively in their left and right sides. Since $y_j(0) \oplus y_j(0) \oplus y_j(1) \oplus y_j(1) = 0$ for any j, we deduce line (e), equivalent to line (f), with a final contradiction in line (f) indicating that the four constraints (a)–(d) are incompatible and the four of them cannot be satisfied simultaneously. This demonstrates that no (classical) strategy exists that could win on all four input configurations, so that any (classical) strategy is bound to fail on some input configuration(s).

We will now show the possibility of a strategy, exploiting quantum resource under the form of quantum correlation contained in an entangled quantum state, so as to win on all four input configurations. The quantum strategy proceeds as follows. Before they receive their input x_j, the players prepare a triplet of qubits $Q_1Q_2Q_3$ in the tripartite entangled state

$$
|Q_1Q_2Q_3\rangle = \frac{1}{2}\Big(|000\rangle - |011\rangle - |101\rangle - |110\rangle\Big) , \tag{6.33}
$$

and each player (j) gets possession of the qubit Q_j of the entangled triplet, for $j = 1, 2, 3$. The quantum state of Eq. (6.33) is an example of a so-called GHZ state (after Greenberger, Horne, Zeilinger), establishing entanglement between three parties [146, 147, 149]. The three players then agree to behave according to the following specification, settled *before* the game starts and the inputs x_j start to be distributed:

If player (j) receives the input $x_j = 0$, then it produces its output y_j as the result of measuring its qubit Q_j in the computational basis $\big\{|0\rangle , |1\rangle\big\}$.

Otherwise, if player (j) receives the input $x_j = 1$, then it applies a Hadamard transformation H to its qubit Q_j, and then it produces its output y_j as the result of measuring its transformed qubit in the basis $\big\{|0\rangle , |1\rangle\big\}$.

To prove that this quantum strategy wins on the four input configurations, we recall from Eq. (A.43) of the Appendix that $\mathsf{H}|0\rangle = |+\rangle$ and $\mathsf{H}|1\rangle = |-\rangle$. We also have the two useful identities, which are similar to Eqs. (2.30) and (2.31),

$$|01\rangle + |10\rangle = |{+}{+}\rangle - |{-}{-}\rangle \ , \tag{6.34}$$

$$|00\rangle - |11\rangle = |{+}{-}\rangle + |{-}{+}\rangle \ , \tag{6.35}$$

with the standard notations where $|01\rangle \equiv |0\rangle\,|1\rangle \equiv |0\rangle \otimes |1\rangle$, and $|{+}{-}\rangle \equiv |+\rangle\,|-\rangle \equiv |+\rangle \otimes |-\rangle$, and the like. We now examine the behavior of the quantum strategy on the four licit configurations of the input $x_1x_2x_3$.

• When $x_1x_2x_3 = 000$, each player directly measures its qubit in the basis $\{|0\rangle\,,|1\rangle\}$. Since the three qubits are jointly prepared in the entangled state $|Q_1Q_2Q_3\rangle$ of Eq. (6.33), the measurement result is to project the triplet onto either $|000\rangle$ or $|011\rangle$ or $|101\rangle$ or $|110\rangle$ equiprobably. The measurement result interpreted as $|y_1y_2y_3\rangle \equiv |y_1\rangle\,|y_2\rangle\,|y_3\rangle$ allows each player to locally obtain its binary output y_j. In every case, the joint output $y_1y_2y_3$ satisfies the expected requirement $y_1 \oplus y_2 \oplus y_3 = 0$, so that the quantum strategy wins.

• When $x_1x_2x_3 = 011$, players (2) and (3) apply the Hadamard operator H to their respective qubit. As a consequence, the state of the qubit triplet becomes

$$\mathrm{I}_2 \otimes \mathsf{H} \otimes \mathsf{H}\,|Q_1Q_2Q_3\rangle = \frac{1}{2}\Big(|0++\rangle - |0--\rangle - |1+-\rangle - |1-+\rangle\Big) \tag{6.36}$$

$$= \frac{1}{2}\Big[|0\rangle\Big(|{+}{+}\rangle - |{-}{-}\rangle\Big) - |1\rangle\Big(|{+}{-}\rangle + |{-}{+}\rangle\Big)\Big] \tag{6.37}$$

$$= \frac{1}{2}\Big[|0\rangle\Big(|01\rangle + |10\rangle\Big) - |1\rangle\Big(|00\rangle - |11\rangle\Big)\Big] \tag{6.38}$$

$$= \frac{1}{2}\Big(|001\rangle + |010\rangle - |100\rangle + |111\rangle\Big)\ , \tag{6.39}$$

where Eq. (6.38) is obtained from the identities of Eqs. (6.34)–(6.35). Then, when each of the three players measures its qubit in the basis $\{|0\rangle\,,|1\rangle\}$, the measurement result is to project the triplet either onto $|001\rangle$ or $|010\rangle$ or $|100\rangle$ or $|111\rangle$ equiprobably. In every case, the joint output $y_1y_2y_3$ satisfies the expected requirement $y_1 \oplus y_2 \oplus y_3 = 1$, so that the quantum strategy wins.

• When $x_1x_2x_3 = 101$, players (1) and (3) apply the Hadamard operator H to their respective qubit. As a consequence, the state of the triplet becomes

$$\mathsf{H} \otimes \mathrm{I}_2 \otimes \mathsf{H}\,|Q_1Q_2Q_3\rangle = \frac{1}{2}\Big(|{+}0{+}\rangle - |{+}1{-}\rangle - |{-}0{-}\rangle - |{-}1{+}\rangle\Big)\ . \tag{6.40}$$

With comparable factorization as before, this transformed state is also

$$\mathsf{H} \otimes \mathrm{I}_2 \otimes \mathsf{H}\,|Q_1Q_2Q_3\rangle = \frac{1}{2}\Big(|001\rangle + |100\rangle - |010\rangle + |111\rangle\Big)\ . \tag{6.41}$$

As before, the measurement result will match the expected requirement $y_1 \oplus y_2 \oplus y_3 = 1$, so that the quantum strategy wins.

• Finally, when $x_1x_2x_3 = 110$, players (1) and (2) apply the Hadamard operator H to their respective qubit. As a consequence, the state of the triplet becomes

$$\mathsf{H} \otimes \mathsf{H} \otimes \mathrm{I}_2 \, |Q_1 Q_2 Q_3\rangle = \frac{1}{2}\Big(|{+}{+}0\rangle - |{+}{-}1\rangle - |{-}{+}1\rangle - |{-}{-}0\rangle\Big) . \quad (6.42)$$

With comparable factorization as before, this transformed state is also

$$\mathsf{H} \otimes \mathsf{H} \otimes \mathrm{I}_2 \, |Q_1 Q_2 Q_3\rangle = \frac{1}{2}\Big(|010\rangle + |100\rangle - |001\rangle + |111\rangle\Big) . \quad (6.43)$$

As before, the measurement result will match the expected requirement $y_1 \oplus y_2 \oplus y_3 = 1$, so that the quantum strategy wins.

So the quantum strategy wins on all four input configurations, while this is impossible to a classical strategy, as we have seen with the argument of Fig. 6.2. Yet, one could wonder why the argument of Fig. 6.2, with its generality, does not apply to the quantum strategy. The reason is in some sense the same as for the violation of the Bell inequality in Sect. 6.1. This can be explained by noting that, in the quantum strategy, on the qubit Q_j, applying the operator H and then measuring in the basis $\{|0\rangle, |1\rangle\}$, is equivalent to directly measuring the qubit Q_j in the basis $\{|+\rangle, |-\rangle\}$. Thus, equivalently in the quantum strategy, when the input is $x_j = 0$ the output $y_j = y_j(x_j = 0)$ is obtained by measuring Q_j in the basis $\{|0\rangle, |1\rangle\}$, and when the input is $x_j = 1$ the output $y_j = y_j(x_j = 1)$ is obtained by measuring Q_j in the basis $\{|+\rangle, |-\rangle\}$. Measuring in the basis $\{|0\rangle, |1\rangle\}$ or $\{|+\rangle, |-\rangle\}$ are two incompatible quantum measurements, equivalent to measuring on the qubit respectively the non-commuting spin observables σ_z or σ_x, and that cannot be performed simultaneously on the same qubit Q_j. So the corresponding outputs $y_j(0)$ and $y_j(1)$ cannot have physical reality simultaneously. The argument used in Fig. 6.2 that $y_j(0) \oplus y_j(0) \oplus y_j(1) \oplus y_j(1) = 0$ does not apply to the quantum strategy. The quantum strategy, and the variables produced by quantum measurements it exploits, are not constrained by the limitation of the classical strategy as expressed by Fig. 6.2.

Thus we find that exploitation of a quantum resource under the form of the tripartite entangled state $|Q_1 Q_2 Q_3\rangle$ of Eq. (6.33) provides the possibility of an always-winning strategy, what is not feasible classically. We note that the entangled state $|Q_1 Q_2 Q_3\rangle$ of Eq. (6.33) is a fixed stereotyped resource that remains the same at each play of the game. In particular, $|Q_1 Q_2 Q_3\rangle$ does not convey any information about the input $x_1 x_2 x_3$ applied to the players. Nevertheless, $|Q_1 Q_2 Q_3\rangle$ establishes sufficient correlation between the three players, so that for each input $x_1 x_2 x_3$, they are able to coordinate their response so as to collectively produce an output $y_1 y_2 y_3$ matched to the specific input $x_1 x_2 x_3$ they received to satisfy the requirement (i)-(ii). This is a significant achievement enabled by quantum entanglement, and meaningful to information processing, at least in principle. Even though direct information transmission does not take place, an effective benefit in the input–output processing is obtained from correlation by quantum entanglement. Correlations between variables obtained from measurements of incompatible quantum quantities on entangled systems may escape classical constraints to provide novel capabilities for information processing. Which incompatible quantities, which entanglement, and for which scenarios, are key issues to benefit from such capabilities, with possibly countless configurations to envisage.

With classical resources, for an always-winning strategy, it would be enough to have a one-way communication channel from player (2) to player (1), so that player (2) could transmit the value of its input x_2 to player (1). From $x_1 x_2$ player (1) would know exactly which of the four licit inputs $x_1 x_2 x_3$ was applied. In this classical strategy, players (2) and (3) would be instructed to never change their bit so that always $y_2 = x_2$ and $y_3 = x_3$, and player (1) would know exactly which output y_1 to produce in order to satisfy the requirement (i)-(ii) and win the game. By a similar argument, a one-way communication channel between any two players would be enough to obtain an always-winning classical strategy. Thus, as the win of the game is concerned, the quantum strategy exploiting shared prior entanglement with no communication between the players, performs as well as the classical strategy with one-way communication between two players. Quantum entanglement here performs as a one-way classical communication. An essential difference though, is that when the classical strategy wins the game, one of the players has obtained complete knowledge of the input $x_1 x_2 x_3$; while this never occurs when the quantum strategy wins the game.

This type of non-classical correlations stemming from quantum entanglement again stand as a specific resource exhibiting potential for information processing with performance unattainable classically. Applications can be found for instance in quantum security [150]. Another application of useful correlation from quantum entanglement is superdense coding, that we now describe.

6.5 Superdense Coding

Two distant protagonists Alice and Bob are connected by a quantum communication channel for transmitting qubits, for instance an optical fiber enabling to transmit individual photons. In addition, Alice and Bob share an entangled qubit pair AB, with Alice in possession of qubit A and Bob of qubit B. The qubit pair is prepared in the maximally entangled state $|AB\rangle = \big(|00\rangle + |11\rangle\big)/\sqrt{2} = |\beta_{00}\rangle$, which is the Bell state of Eq. (2.21). Alice in her qubit A chooses to encode some information, by applying to her qubit a unitary operator U_A selected among the four qubit operators $\{\mathrm{I}_2, \sigma_x, i\sigma_y, \sigma_z\}$, this set forming, like the Pauli basis, an orthogonal basis for $\mathcal{L}(\mathcal{H}_2)$. The choice by Alice among four possible operators can be indexed by, and in this respect is equivalent to, two bits of classical information encoded by Alice in her choice. By contrast, at this stage Bob does nothing on his qubit B. After this encoding operation by Alice, the entangled qubit pair AB gets placed in the transformed state $\mathsf{U}_A \otimes \mathrm{I}_2 |AB\rangle = \mathsf{U}_A \otimes \mathrm{I}_2 |\beta_{00}\rangle$. According to the choice for U_A by Alice, we obtain

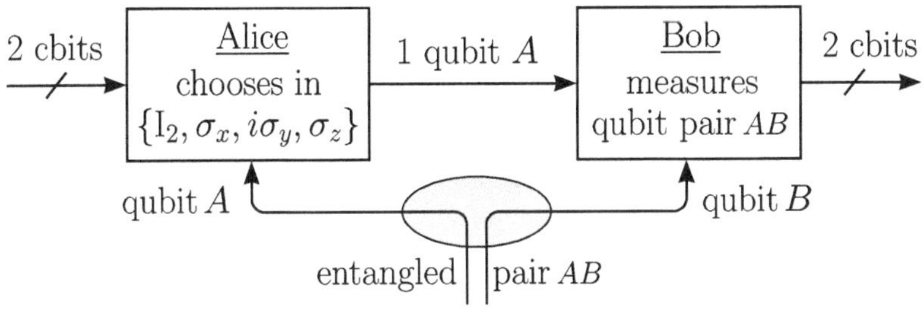

Fig. 6.3 In the protocol for superdense coding, an entangled qubit pair AB in the Bell state $|\beta_{00}\rangle$ is shared between Alice and Bob. Alice from two classical bits selects one operator among four, to apply to her qubit A that she subsequently sends to Bob. Bob, by measuring the pair AB, recovers the two classical bits of Alice

$$I_2 \otimes I_2 |AB\rangle = \frac{1}{\sqrt{2}}\Big(|00\rangle + |11\rangle\Big) = |\beta_{00}\rangle \ , \tag{6.44}$$

$$\sigma_x \otimes I_2 |AB\rangle = \frac{1}{\sqrt{2}}\Big(|10\rangle + |01\rangle\Big) = |\beta_{01}\rangle \ , \tag{6.45}$$

$$i\sigma_y \otimes I_2 |AB\rangle = \frac{1}{\sqrt{2}}\Big(-|10\rangle + |01\rangle\Big) = |\beta_{11}\rangle \ , \tag{6.46}$$

$$\sigma_z \otimes I_2 |AB\rangle = \frac{1}{\sqrt{2}}\Big(|00\rangle - |11\rangle\Big) = |\beta_{10}\rangle \ , \tag{6.47}$$

representing the four orthonormal Bell states of Eqs. (2.21)–(2.24).

After she has applied her transformation U_A, Alice sends her qubit A to Bob across the quantum communication channel. When Bob receives qubit A, he is in possession of the two qubits of the pair AB, resting in a state $U_A \otimes I_2 |AB\rangle$ from Eqs. (6.44)–(6.47). Bob then measures the qubit pair AB in the Bell basis $\big\{|\beta_{00}\rangle, |\beta_{01}\rangle, |\beta_{10}\rangle, |\beta_{11}\rangle\big\}$ of Eqs. (2.21)–(2.24), and from the measurement result Bob identifies exactly the transformation U_A selected by Alice, or equivalently Bob recovers the two classical bits encoded by Alice. The communication protocol is depicted in Fig. 6.3.

The information carrier, encoded by Alice and sent to Bob, is a single qubit. A single qubit, as an information carrier, when measured alone in a projective orthonormal basis, can lead to two measurement outcomes, able to convey no more than one classical bit of information.[2] Nevertheless, by the present entanglement-assisted communication protocol, each time one qubit it sent across the channel, two classical bits of information are transmitted from Alice to Bob. This is **superdense coding** [151, 152]. Each transmission across the channel consumes one pair of entangled qubits, shared between the sender and the receiver. This pair however is always prepared in the same stereotyped Bell state $|AB\rangle = |\beta_{00}\rangle$ conveying no information about the two-bit encoding of Alice. A side channel from a quantum server can

[2] And this would be no different if instead of the von Neumann projective measurement, a generalized measurement was envisaged on the qubit, as we shall see in Sect. 7.2.3.

distribute the stereotyped entangled qubit pair to Alice and Bob. Such a server of entangled qubit pairs is a typical resource that can be expected from a quantum internet [153, 154]. With access to such general-purpose quantum resource, the dedicated information channel between Alice and Bob will be able to communicate two classical bits of information per transmitted qubit. In this way, quantum correlation via shared entanglement between Alice and Bob provides a means to enhance the performance over that of classical information communication. The superdense coding described here is an instance of the more general technique of entanglement-assisted information communication over a quantum channel, which will be further examined in Sect. 13.4.3.

6.6 Quantum Teleportation

Quantum teleportation is another significant phenomenon enabled by quantum entanglement and with relevance to quantum signal and information processing [155, 156]. It can be seen as some form of reciprocal of superdense coding, with a complete quantum state communicated from Alice to Bob simply via transmitting a few bits of classical information.

Two distant protagonists Alice and Bob are connected by a classical communication channel. Alice is in possession of a qubit Q, which can be in an arbitrary pure state $|Q\rangle = \alpha_0 |0\rangle + \alpha_1 |1\rangle$. To communicate her full quantum state $|Q\rangle$ to Bob, Alice can apply the following protocol. Alice and Bob share an entangled qubit pair AB, prepared in the maximally entangled state $|AB\rangle = \big(|00\rangle + |11\rangle\big)/\sqrt{2} = |\beta_{00}\rangle$, which again is the Bell state $|\beta_{00}\rangle$ of Eq. (2.21); Alice is in possession of qubit A and Bob of qubit B. Alice measures her qubit pair QA in the Bell basis $\big\{|\beta_{00}\rangle, |\beta_{01}\rangle, |\beta_{10}\rangle, |\beta_{11}\rangle\big\}$ of Eqs. (2.21)–(2.24). The measurement result of Alice can be indexed by two classical binary digits ab, when the measurement by Alice projected onto the Bell state $|\beta_{ab}\rangle$. The two binary digits ab are sent by Alice to Bob across the classical communication channel between them. And when Bob receives ab, he can exactly reconstruct the arbitrary quantum state $|Q\rangle$, as we shall see.

The triplet of the three qubits Q, A, and B is initially in the pure state $|QAB\rangle = |Q\rangle \otimes |\beta_{00}\rangle$ which can be expanded as

$$|QAB\rangle = \frac{1}{\sqrt{2}}\Big(\alpha_0 |000\rangle + \alpha_0 |011\rangle + \alpha_1 |100\rangle + \alpha_1 |111\rangle\Big), \tag{6.48}$$

and then factorized conveniently as

$$
\begin{aligned}
|QAB\rangle = \frac{1}{2}\Bigg[\\
&\frac{1}{\sqrt{2}}\Big(|00\rangle + |11\rangle\Big)\Big(\alpha_0\,|0\rangle + \alpha_1\,|1\rangle\Big) + \frac{1}{\sqrt{2}}\Big(|01\rangle + |10\rangle\Big)\Big(\alpha_0\,|1\rangle + \alpha_1\,|0\rangle\Big) + \\
&\frac{1}{\sqrt{2}}\Big(|00\rangle - |11\rangle\Big)\Big(\alpha_0\,|0\rangle - \alpha_1\,|1\rangle\Big) + \frac{1}{\sqrt{2}}\Big(|01\rangle - |10\rangle\Big)\Big(\alpha_0\,|1\rangle - \alpha_1\,|0\rangle\Big)\Bigg].
\end{aligned}
\tag{6.49}
$$

The motivation for this factorization is that we equivalently have, in terms of the four Bell states $|\beta_{ab}\rangle$,

$$
\begin{aligned}
|QAB\rangle = \frac{1}{2}\Big[&|\beta_{00}\rangle\Big(\alpha_0\,|0\rangle + \alpha_1\,|1\rangle\Big) + |\beta_{01}\rangle\Big(\alpha_0\,|1\rangle + \alpha_1\,|0\rangle\Big) + \\
&|\beta_{10}\rangle\Big(\alpha_0\,|0\rangle - \alpha_1\,|1\rangle\Big) + |\beta_{11}\rangle\Big(\alpha_0\,|1\rangle - \alpha_1\,|0\rangle\Big)\Big].
\end{aligned}
\tag{6.50}
$$

The form of Eq. (6.50) makes explicitly visible the effect of the projective measurement of the qubit pair QA realized by Alice in the Bell basis $\{|\beta_{ab}\rangle\}$. For instance, when Alice's measurement projects the pair QA onto the Bell state $|\beta_{00}\rangle$ (this occurs with the probability $1/4$), at the same time Bob's qubit B gets projected onto the separable state $|B\rangle = \alpha_0\,|0\rangle + \alpha_1\,|1\rangle$, which is exactly the initial state of the qubit Q previously prepared on Alice's side. This result is associated with the two classical bits $ab = 00$ that Alice sends to Bob. When Bob receives the two classical bits $ab = 00$, he knows from Eq. (6.50) that he is in possession of a qubit B whose quantum state is $|B\rangle = \alpha_0\,|0\rangle + \alpha_1\,|1\rangle$, exactly reproducing the initial arbitrary state $|Q\rangle$ that was previously on Alice's side, and which in this way has been teleported to Bob's side.

To encompass the four possible measurement results, Bob can implement the postselection rule or postprocessing as follows. As a rule Bob, when he receives the two classical bits ab from Alice, can apply the sequence of Pauli operators $\sigma_z^a\sigma_x^b$ to his qubit B, and this will place his qubit B exactly in the arbitrary state $|Q\rangle = \alpha_0\,|0\rangle + \alpha_1\,|1\rangle$ that was initially on Alice's side.

When Alice sends $ab = 00$, then $\sigma_z^0\sigma_x^0 = \mathrm{I}_2\mathrm{I}_2 = \mathrm{I}_2$, so that Bob does nothing on his qubit B which is directly placed in the state $|Q\rangle = \alpha_0\,|0\rangle + \alpha_1\,|1\rangle$ by Alice's measurement. Alternatively, when Alice's measurement projects the pair QA onto the Bell state $|\beta_{01}\rangle$ (this occurs with the probability $1/4$), Bob's qubit B gets projected onto the separable state $|B\rangle = \alpha_0\,|1\rangle + \alpha_1\,|0\rangle$, as indicated by Eq. (6.50). Bob receives the two classical bits $ab = 01$ from Alice, and he applies $\sigma_z^0\sigma_x^1 = \mathrm{I}_2\sigma_x = \sigma_x$ on his qubit B. This places Bob's qubit B in the transformed state $\sigma_x\,|B\rangle = \alpha_0\,|0\rangle + \alpha_1\,|1\rangle$ which is again the initial arbitrary state $|Q\rangle$ teleported to Bob's side. Similarly, for the four equiprobable measurement results accessible to Alice, the transformation $\sigma_z^a\sigma_x^b$ by Bob acts as displayed in Fig. 6.4.

$$QA \text{ measured in } |\beta_{00}\rangle \text{ then } B \text{ is in } |B\rangle = \alpha_0\,|0\rangle + \alpha_1\,|1\rangle \xmapsto{\mathrm{I}_2} \cdot \xmapsto{\mathrm{I}_2} |Q\rangle$$
$$QA \text{ measured in } |\beta_{01}\rangle \text{ then } B \text{ is in } |B\rangle = \alpha_0\,|1\rangle + \alpha_1\,|0\rangle \xmapsto{\sigma_x} \cdot \xmapsto{\mathrm{I}_2} |Q\rangle$$
$$QA \text{ measured in } |\beta_{10}\rangle \text{ then } B \text{ is in } |B\rangle = \alpha_0\,|0\rangle - \alpha_1\,|1\rangle \xmapsto{\mathrm{I}_2} \cdot \xmapsto{\sigma_z} |Q\rangle$$
$$QA \text{ measured in } |\beta_{11}\rangle \text{ then } B \text{ is in } |B\rangle = \alpha_0\,|1\rangle - \alpha_1\,|0\rangle \xmapsto{\sigma_x} \cdot \xmapsto{\sigma_z} |Q\rangle$$

Fig. 6.4 The qubit triplet QAB is initially in the pure state $|QAB\rangle$ of Eq. (6.50). Alice measures the qubit pair QA in the Bell basis $\{|\beta_{ab}\rangle\}$. According to the measurement result, from four equiprobable configurations, Alice sends two classical bits ab to Bob. Then, on his qubit B projected onto the separable state $|B\rangle$ by Alice's measurement, Bob applies the transformation $\sigma_z^a\sigma_x^b\,|B\rangle$ which always places qubit B in the arbitrary quantum state $|Q\rangle = \alpha_0\,|0\rangle + \alpha_1\,|1\rangle$ teleported from Alice to Bob

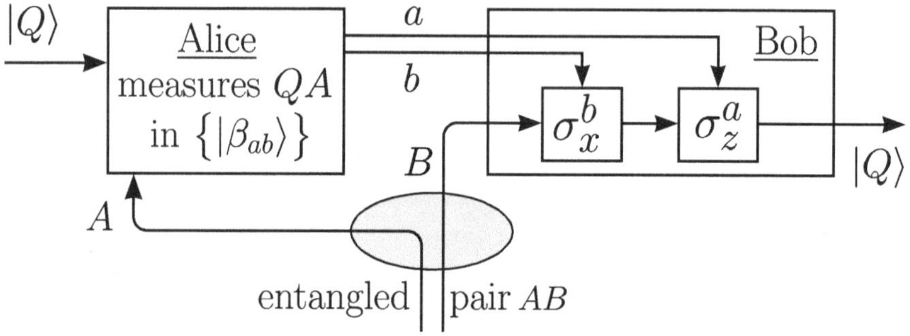

Fig. 6.5 In the teleportation protocol, an entangled qubit pair AB in the Bell state $|\beta_{00}\rangle$ is shared between Alice and Bob. A qubit Q in an arbitrary quantum state $|Q\rangle$ is initially on Alice's side. Alice measures the qubit pair QA in the Bell basis $\{|\beta_{ab}\rangle\}$, and according to the measurement result Alice sends two classical bits ab to Bob. Then, on his qubit B Bob applies the transformation $\sigma_z^a\sigma_x^b$ which places the qubit B in the arbitrary state $|Q\rangle$ teleported from Alice to Bob

As observed in Fig. 6.4, in the four feasible configurations, after the transformation $\sigma_z^a\sigma_x^b$ by Bob, the arbitrary quantum state $|Q\rangle = \alpha_0\,|0\rangle + \alpha_1\,|1\rangle$ is always obtained on Bob's side. The complete teleportation protocol is sketched in Fig. 6.5.

In the teleportation protocol of Fig. 6.5, the initial qubit state $|Q\rangle$ is destroyed by Alice in her measurement. The two classical bits ab that result are enough for Bob to reconstruct the complete quantum state $|Q\rangle$, provided he can exploit an entangled qubit pair AB he shares with Alice. This realizes the teleportation of an arbitrary quantum state $|Q\rangle = \alpha_0\,|0\rangle + \alpha_1\,|1\rangle$. Such a quantum state, with continuously varying coordinates α_0 and α_1, has access to a continuously infinite number of configurations, and in this respect it carries an infinite amount of information. The entangled qubit pair AB shared between Alice and Bob, and standing as an essential ingredient of the teleportation protocol, is always prepared in the same stereotyped Bell state $|AB\rangle = \big(|00\rangle + |11\rangle\big)/\sqrt{2} = |\beta_{00}\rangle$ conveying no information about the arbitrary quantum state $|Q\rangle = \alpha_0\,|0\rangle + \alpha_1\,|1\rangle$ to be teleported. Nevertheless, the correlation introduced between Alice and Bob by the entangled pair AB is such that it needs to be supplemented by only two classical bits of information ab and this is enough for Bob to reconstruct on his side the arbitrary (continuously-varying) state $|Q\rangle = \alpha_0 |0\rangle + \alpha_1 |1\rangle$ destroyed by Alice on her side. After the measurement by Alice of the qubit pair QA involved in the entangled state $|QAB\rangle$ of Eq. (6.50), as long as Bob does not know the measurement result ab from Alice, the quantum state Bob can assign to his qubit B is the mixed state corresponding to the statistical ensemble of four pure states $\{|Q\rangle, \sigma_x |Q\rangle, \sigma_z |Q\rangle, i\sigma_y |Q\rangle\}$ each with probability $1/4$, which is described by the qubit density operator $\mathrm{I}_2/2$ independent of the initial state $|Q\rangle$ of qubit Q. It is only when Bob receives the measurement result ab so as to apply to his qubit B the transformation $\sigma_z^a \sigma_x^b$ that the teleportation of $|Q\rangle$ gets completed, and at a transportation rate that is controlled by the classical communication of ab. In addition, after the teleportation protocol is completed, Bob knows that he owns the full arbitrary pure state $|Q\rangle$ that was previously on Alice's side and that has now disappeared from Alice's side (no cloning is involved), but beyond this, Bob does not know from the protocol which state precisely he has in hand.

The teleportation protocol again illustrates the power of quantum entanglement to establish non-classical correlation between two distant parties. Even though such correlation by entanglement cannot serve for direct (and moreover, superluminal) information communication, as we indicated, it can nevertheless lead via more elaborated schemes to significant and distinctive advantage for information communication or processing, as exemplified throughout this chapter.

Instead of a pure state, a mixed qubit state owned by Alice can equally be teleported to Bob by the same protocol. A mixed qubit state on Alice's side can be represented by a statistical ensemble $\{(p_j, |Q_j\rangle)\}$ of pure states $|Q_j\rangle$. With the probability p_j, the pure state $|Q_j\rangle$ is produced on Alice's side, and by the teleportation protocol, Bob obtains the same state $|Q_j\rangle$ on his side, so that Bob ends up with the same statistical ensemble $\{(p_j, |Q_j\rangle)\}$ representing the same mixed state, that has been teleported.

In a comparable way, a prior entanglement of Alice's qubit Q with some reference system R will equally be teleported to Bob by the same protocol. The bipartite system RQ can be taken in the initial entangled state $|RQ\rangle = |R_0\rangle \otimes \alpha_0 |0\rangle + |R_1\rangle \otimes \alpha_1 |1\rangle$, where $|R_0\rangle$ and $|R_1\rangle$ are two arbitrary states for the reference system R. The quadripartite system $RQAB$ is then initialized in the state $|RQAB\rangle = |RQ\rangle \otimes |\beta_{00}\rangle$, which through a factorization comparable to that of Eq. (6.50) is also

$$\begin{aligned}
|RQAB\rangle = \frac{1}{2}\Big[\, & |R_0\rangle\, |\beta_{00}\rangle\, \alpha_0 |0\rangle + |R_1\rangle\, |\beta_{00}\rangle\, \alpha_1 |1\rangle + |R_0\rangle\, |\beta_{01}\rangle\, \alpha_0 |1\rangle + \\
& |R_1\rangle\, |\beta_{01}\rangle\, \alpha_1 |0\rangle + |R_0\rangle\, |\beta_{10}\rangle\, \alpha_0 |0\rangle - |R_1\rangle\, |\beta_{10}\rangle\, \alpha_1 |1\rangle + \\
& |R_0\rangle\, |\beta_{11}\rangle\, \alpha_0 |1\rangle - |R_1\rangle\, |\beta_{11}\rangle\, \alpha_1 |0\rangle \Big] . \qquad (6.51)
\end{aligned}$$

In the teleportation protocol, from the pre-measurement state from Eq. (6.51), when Alice measures the qubit pair QA in the Bell basis as before, QA gets projected onto $|\beta_{00}\rangle$ with the probability $1/4$, while the system RB terminates in the state $|R_0\rangle\, \alpha_0 |0\rangle + |R_1\rangle\, \alpha_1 |1\rangle$; alternatively, when QA gets projected onto $|\beta_{01}\rangle$ with the probability $1/4$, the system RB terminates in the state $|R_0\rangle\,\alpha_0\,|1\rangle + |R_1\rangle\,\alpha_1\,|0\rangle$, and similarly for the two other measurement outcomes. Bob then performs on his qubit B the same postprocessing as in Fig. 6.4, and this places his qubit B in the final state $|RB\rangle = |R_0\rangle\,\alpha_0\,|0\rangle + |R_1\rangle\,\alpha_1\,|1\rangle$, sharing the same (teleported) entanglement that Q had with the reference system R. At the end of the teleportation protocol, the initial entanglement $|RQ\rangle$ between R and Q, is now converted as the same entanglement $|RB\rangle$ between R and B. If at the end of the protocol the reference system R remains with Alice, then between Alice and Bob the initial shared entanglement contained in the stereotyped Bell pair state $|\beta_{00}\rangle$ has been converted into the arbitrary entanglement contained in the initial state $|RQ\rangle$, and this with only local operation (the measurement by Alice) and classical communication. This is an example of what is called entanglement dilution, when between two distant parties, a stereotyped maximally entangled Bell state is converted into an arbitrary entanglement, by means of only local operations and classical communication, to offer a useful protocol for handling the important resource of quantum entanglement.

Furthermore, a quantum state of arbitrary dimension, for a qudit Q, can be teleported, by sharing a maximally entangled qudit pair between Alice and Bob, and for Alice by measuring her qudit pair in a maximally entangled orthonormal basis [155, 157].

6.7 Quantum Cryptography

Cryptography, or the establishment of secure information communication between two distant parties, is a field that can also receive specific distinctive benefits from quantum resources, as we now illustrate.

6.7.1 A Reference Problem in Cryptography

A standard problem in cryptography, standing as an important reference for digital communication, can be described as follows. A message X under the form of a sequence of classical bits or binary digits 0 or 1 is to be communicated securely between two distant protagonists Alice and Bob. A cryptographic key K is introduced, as a completely random string of bits 0 or 1 (with the same length as the message X) independent and identically distributed with equal probability $1/2$. The cryptogram or encrypted message is the bit string C obtained by the bitwise Xor operation as $C(X, K) = X \oplus K$. This is the so-called Vernam cipher or one-time pad [158, 159]. This cryptographic protocol has provably perfect security, in the sense that the entropy-based[3] mutual information vanishes as $H(C : X) = H(X) - H(X|C) = 0$, that is, the cryptogram C statistically reveals

[3] More on entropy and mutual information is to come, starting in Sect. 7.1 of Chap. 7.

strictly no information about the message X. In addition, the cryptogram C exhibits an entropy and an entropy rate at a maximum, indicating that the cryptogram has the statistical appearance of a string of independent equiprobable random bits, whatever the statistical structure and correlations in the message X. In principle, the cryptographic key K is shared between Alice and Bob as a secret between them two. Then, for secure communication of the message X, Alice can send the cryptogram C to Bob, over an insecure classical communication channel. When Bob receives the cryptogram C from Alice, he uses the key K to reconstruct the message as $C \oplus K = X$. Any eavesdropper that could have spied the transmission and intercepted a copy of the cryptogram C, would gain no information about the message X, as long as this eavesdropper does not know the key K. This secure communication protocol critically rests on the ability to establish a private secret key K between Alice and Bob. Quantum signals exchanged between Alice and Bob can help establishing such a secret key between them, again with distinctive properties not accessible by classical means, as we now explain.

6.7.2 BB84 Protocol for Quantum Key Distribution

An important quantum cryptographic protocol [160, 161], known as BB84 (as it was proposed [160] by Bennett and Brassard in 1984), operates in the following way so as to perform a quantum key distribution. To establish securely a cryptographic key K with a length of L random bits between Alice and Bob, Alice starts with a string of $4L$ independent and equiprobable random bits. For transmission to Bob, Alice chooses to encode each classical bit with a qubit state. The two states 0 or 1 for each classical bit are encoded by the two states of an orthonormal basis for the qubit. For each classical bit from $\{0, 1\}$, Alice chooses equiprobably at random to encode either with the basis states of $\{|0\rangle, |1\rangle\}$ or with the basis states of $\{|+\rangle, |-\rangle\}$. Alice sends her string of $4L$ encoded qubits to Bob across a possibly insecure qubit channel. For each qubit received from Alice, Bob chooses equiprobably at random to measure the qubit either in the basis $\{|0\rangle, |1\rangle\}$ or in the basis $\{|+\rangle, |-\rangle\}$, so as to decode a string of $4L$ classical bits. When the whole string of $4L$ bits has been transmitted, Alice and Bob can publicly disclose the sequence of their basis choices, so as to identify where they coincide. Their basis choices will coincide in approximately $2L$ places. At each position where their basis choices coincide, Alice and Bob are sure to have the same random bit. So Alice and Bob keep only the positions where their basis choices coincide, and they obtain in this way a shared key of random bits of length approximately $2L$.

A putative eavesdropper, conventionally named Eve, may have spied the transmission channel. If Eve wants to extract information about the transmission, she has to measure Alice's qubits. Eve can measure each intercepted qubit in the basis $\{|0\rangle, |1\rangle\}$ or $\{|+\rangle, |-\rangle\}$. As we have seen, quantum measurement has a destructive or disturbing effect on the state. If Eve wants to have a chance that her measurement go unnoticed, ideally she would like to forward or retransmit to Bob a qubit in the exact state she intercepted from Alice. But prior to her measurement, the no-cloning theorem of Sect. 3.1.3 forbids Eve to make an exact copy of the unknown state from Alice. And after the measurement by Eve, Alice's state is destroyed. So Eve can only forward to Bob a qubit placed in her (known) measured (post-measurement) state. And in this way, approximately half of the times, due to a mismatch between Alice's and Eve's basis choices, Eve will forward an incorrect (differing from Alice's) state to Bob. From this, approximately half of the times Bob will decode an incorrect bit value. To protect their transmission against such eavesdropping, Alice, from their $2L$ coinciding bits, classically sends to Bob the position and value of L bits selected at random. In case of eavesdropping, around $L/4$ of these L test bits will differ between Alice and Bob. On the contrary, when the L test bits coincide, this in principle ensures that no eavesdropping has taken place across the transmission. With their L remaining bits, Alice and Bob possess a shared secret key, which is guaranteed to be known to both of them alone, and which has been established between them across insecure communication channels. The guarantee essentially results from the no-cloning theorem and the destructive or disturbing character of quantum measurement, ensuring that no eavesdropper observing the transmission can go undetected. Conversely, when no such detection occurs, the quantum protocol brings the guarantee that no eavesdropping has taken place on the transmission, and that the shared key is indeed secret.

It is a remarkable property, obtained by exploiting quantum resources, that a secret key can be guaranteed between Alice and Bob by means of information transmission across insecure communication channels. This BB84 protocol however requires a quantum communication channel between Alice and Bob for transmitting qubits (for example an optical fiber capable of reliably transmitting individual photons). There exists an alternative cryptographic protocol, that also exploits quantum resources to achieve the same goal of guaranteeing a secret key, but with only (insecure) classical communication channels between Alice and Bob. This alternative protocol for quantum key distribution exploits entanglement and is known as the EPR protocol [161, 162], and it operates in the following way.

6.7.3 *Entanglement for Quantum Key Distribution*

In the EPR protocol for quantum key distribution, Alice and Bob need to have access to a quantum server dispatching pairs of entangled qubits AB prepared in the stereotyped Bell state $|AB\rangle = \big(|00\rangle + |11\rangle\big)/\sqrt{2} = |\beta_{00}\rangle$ of Eq. (2.21), with Alice receiving qubit A and Bob qubit B. When Alice and Bob measure their respective qubit of the pair AB in the basis $\big\{|0\rangle\,,\,|1\rangle\big\}$, they obtain or decode a bit 0 or a bit 1. The two possible measurement outcomes occur at random equiprobably with probability $1/2$, but Alice and Bob always obtain the same result, 0 or 1. The random number generation to construct the key, which was handled as a separate step in the BB84 protocol, is here automatically accomplished by the quantum measurement itself via its intrinsic random behavior. After measuring in this way a sequence of L successive entangled pairs received from the quantum server, Alice and Bob share a perfectly fit cryptographic key K of L random bits. The protocol however needs to be complicated a little in order to protect the key against an eavesdropper that could intervene on the distribution of the entangled pair AB from the quantum server. By referring to Eq. (2.29), we know that the stereotyped Bell state $|AB\rangle = (|00\rangle + |11\rangle)/\sqrt{2} = |\beta_{00}\rangle$ dispatched by the server, is equally $|AB\rangle = (|{+}{+}\rangle + |{-}{-}\rangle)/\sqrt{2} = |\beta_{00}\rangle$. So if Alice and Bob measure their respective qubit of the pair AB in the basis $\{|+\rangle, |-\rangle\}$, they also always obtain the same result. Alice and Bob choose to alternate randomly between measuring their respective qubit of each pair AB, either in the basis $\{|0\rangle, |1\rangle\}$ or in the basis $\{|+\rangle, |-\rangle\}$, so as to obtain their respective random bit 0 or 1. After a sequence of $4L$ entangled pairs AB from the server have been measured, Alice and Bob can publicly disclose the sequence of their basis choices. At each position where their basis choices coincide, Alice and Bob are sure to have decoded the same random bit. So Alice and Bob keep only the positions where their basis choices coincide, and they obtain in this way a shared key of length approximately $2L$. To detect a putative eavesdropping, as in the BB84 protocol, Alice sends to Bob for verification the position and value of L test bits randomly selected from their $2L$ coinciding bits. In case of eavesdropping by interception and resend on the transmission from the quantum server, mismatch will ensue on approximately $L/4$ test bits. No mismatch in the L test bits in principle ensures that no eavesdropping has taken place, and Alice and Bob with their L remaining bits possess a shared secret key. In this EPR protocol for quantum key distribution, Alice and Bob exchange information only across an insecure classical channel. In addition, they need access to a quantum server dispatching entangled qubit pairs, yet prepared in the same stereotyped Bell state that in itself carries no information.

The BB84 and EPR protocols illustrate how specific quantum properties, such as no-cloning, the destructive or disturbing character of quantum measurement, entanglement, can be exploited to establish a cryptographic key between two distant parties, with a specific guarantee that the key is indeed secret and cannot have been intercepted by a third party. These reference protocols for quantum key distribution have received numerous variations and extensions and are still expanding [161, 163, 164]. They form a foundation for quantum cryptography, which today stands as a relatively mature field of application of quantum technologies, with already commercial products offering various forms of quantum cryptography, and still in progress.

6.8 Quantum Repeater

The quantum repeater [165, 166] is a quantum device quite useful for quantum communication over long distances. The quantum repeater enables to propagate the important resource of quantum entanglement over long distances so as to entangle two very distant parties that have no direct interaction.

Two distant parties, Alice and Bob, are connected through a quantum repeater in the following way. Alice owns a qubit A entangled to a qubit R owned by the repeater. The qubit pair AR is prepared in the entangled singlet state of Eq. (2.24) that we note $|\beta_{11}\rangle_{AR}$ here. Bob owns a qubit B entangled to another qubit R' also owned by the repeater. The qubit pair $R'B$ is also prepared in the same entangled singlet state of Eq. (2.24) that we note $|\beta_{11}\rangle_{R'B}$ here. The notations here are specially adapted (or adopted) so as to keep track, among the four qubits involved, of which qubit is in which quantum state, with A, R, R', and B as labels to indicate the position of each qubit in the multipartite states. The quadruplet $ARR'B$ of the four qubits is in a joint quantum state that can be denoted as

$$|\Psi_4\rangle = |\beta_{11}\rangle_{AR} \otimes |\beta_{11}\rangle_{R'B} \tag{6.52}$$

$$= \frac{1}{\sqrt{2}}\Big(|01\rangle - |10\rangle\Big)_{AR} \otimes \frac{1}{\sqrt{2}}\Big(|01\rangle - |10\rangle\Big)_{R'B} \tag{6.53}$$

$$= \frac{1}{2}\Big(|0101\rangle - |0110\rangle - |1001\rangle + |1010\rangle\Big)_{ARR'B} . \tag{6.54}$$

By referring to the two-qubit Bell states of Eqs. (2.21)–(2.24), and by keeping track of the qubit positions via the labels in the states, the four-qubit state $|\Psi_4\rangle$ can be factorized as

$$|\Psi_4\rangle = \frac{1}{2}\Big(|\beta_{01}\rangle_{AB}\,|\beta_{01}\rangle_{RR'} + |\beta_{10}\rangle_{AB}\,|\beta_{10}\rangle_{RR'} - |\beta_{00}\rangle_{AB}\,|\beta_{00}\rangle_{RR'} - |\beta_{11}\rangle_{AB}\,|\beta_{11}\rangle_{RR'}\Big) . \tag{6.55}$$

The repeater then realizes a measurement of its qubit pair RR' in the Bell basis $\big\{|\beta_{00}\rangle, |\beta_{01}\rangle, |\beta_{10}\rangle, |\beta_{11}\rangle\big\}$ of Eqs. (2.21)–(2.24). The measurement randomly projects the qubit pair RR' onto one of the four Bell states, with equal probability $1/4$. The factorization in Eq. (6.55) of the quadruplet state $|\Psi_4\rangle$ renders easily visible the fate of the qubit pair AB after the measurement by the repeater. Each possible outcome $|\beta_{ab}\rangle$ for the qubit pair RR' is associated in Eq. (6.55) with the same Bell state $|\beta_{ab}\rangle$ for the qubit pair AB. If Alice and Bob have no knowledge of the measurement result obtained by the repeater, they are in possession of a qubit pair AB that is in a mixed quantum state corresponding to the statistical ensemble of the four Bell states each with equal probability $1/4$. Such a mixed state is described by the density operator $\rho_{AB} = (\mathrm{I}_2/2) \otimes (\mathrm{I}_2/2)$ representing a maximally mixed separable product state for the two qubits AB, which in the operation so far remain completely uncorrelated.

The useful step further is for the repeater to communicate to Alice and Bob the result of its measurement, and this can be done through a classical communication channel that needs no quantum resource. When Alice and Bob are informed that the repeater measured the Bell state $|\beta_{ab}\rangle$, at the same time, from the structure of Eq. (6.55), they know that they share a qubit pair AB placed in the same entangled quantum state $|\beta_{ab}\rangle$, as a result of the measurement by the repeater. The operation of the quantum repeater is sketched in Fig. 6.6.

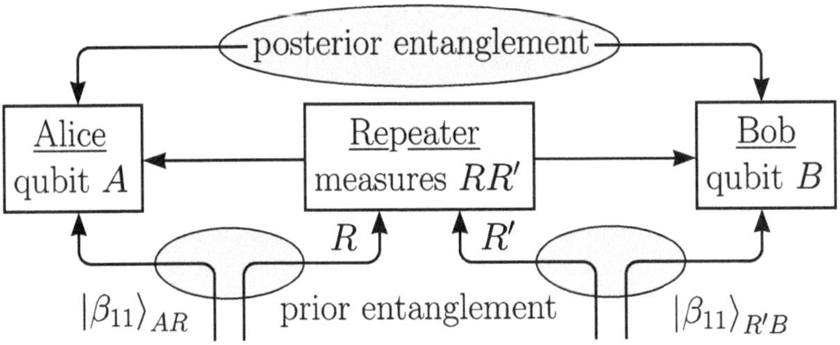

Fig. 6.6 In the quantum repeater, prior entanglement is shared between qubit A of Alice and qubit R of the repeater, and between qubit R' of the repeater and qubit B of Bob. The repeater measures the qubit pair RR' and (classically) communicates its result to Alice and Bob. As a consequence, the qubit pair AB becomes tied by posterior entanglement

The operation of the repeater has placed the two distant qubits AB in a maximally entangled pure state, without them two interacting directly. We note that any Bell state $|\beta_{ab}\rangle$ of the pair AB, can be transformed into any other of the four Bell states, simply by local unitary operations, essentially bit-flip σ_x or phase-flip σ_z, applied solely to one qubit of the pair (as illustrated by Eqs. (6.44)–(6.47) of the superdense coding in Sect. 6.5). By contrast, if a qubit pair AB were initially in a separable product state, no local operations on A and B would be capable of placing the pair in an entangled state. The quantum repeater manages to entangle two distant qubits A and B that do not directly interact. For this, the quantum repeater relies on a connection shared with Alice, and another connection shared with Bob, to the general-purpose quantum resource of a server dispatching stereotyped entangled qubit pairs, supplemented by a classical communication channel with Alice, and one with Bob, as depicted in Fig. 6.6.

By reproducing the protocol of the repeater, entanglement can in principle be propagated to arbitrary long distances, to establish between two distant parties quantum correlation useful to various purposes over quantum communication networks. Numerous extensions and elaborations can be brought to the basic protocol of the quantum repeater presented above, as well as practical realizations [166–169]. Quantum repeaters can also be seen as performing some quantum version of signal amplification, especially useful to combat attenuation or loss in the propagation of quantum signals.

6.9 Summary on Quantum Correlation

Non-classical correlations can arise from the specifically quantum property of entanglement [170]. Non-conventional properties may result offering novel capabilities inaccessible classically and bearing relevance to quantum signal and information processing.

For instance, correlations between variables obtained from measurements of **incompatible** (non-jointly measurable) quantum quantities on entangled systems may escape classical constraints, to give access to novel capabilities. This is essentially what we can see at work with the violation of the (classical) **Bell inequalities** in Sect. 6.1, or in the realization of an always-winning quantum strategy with the GHZ entangled triplet of Sect. 6.4.

Quantum entanglement induces correlation with a marked nonlocal character between distant entities, which cannot be reduced to classical properties, as exemplified by the distributed information storing of Sect. 6.2. Quantum correlation via entanglement cannot however be directly exploited for instant information communication between two distant parties. This is expressed by the **no-signaling theorem** of Sect. 6.3, stipulating that in an entangled bipartite system, no processing performed locally on one subsystem can change the reduced state local to the other subsystem. Nevertheless, specific quantum benefits from entanglement are accessible for information communication and processing, with several reference scenarios thereof presented in this chapter: Shared entanglement between two distant parties A and B can be used to double the transmission rate of a classical information channel between A and B, as with **superdense coding** in Sect. 6.5; or it can be used to transport a complete quantum state from A to B simply by transmitting a few bits of classical information between them, as with the **teleportation** of Sect. 6.6.

Quantum signals can also be used for **cryptography**, to establish secure communication between two distant parties, by exploiting the no-cloning theorem and the destructive or disturbing character of quantum measurement, to guarantee that the transmission has not been intercepted by an eavesdropper spying on the channel, as seen in Sect. 6.7. **Quantum repeaters** described in Sect. 6.8 can be used to propagate the specific resource of quantum entanglement between two distant parties that do not directly interact and can sit an arbitrary long distance apart.

These reference phenomena of **non-classical correlation related to quantum entanglement** we presented in this chapter have received numerous extensions, and they still point today to many possibilities to explore for quantum signal and information processing, and beyond.

Chapter 7
Quantum Entropy

This chapter first briefly reviews the notion of classical entropy and related informational quantities, and then it presents in detail the quantum counterparts of these informational concepts and their properties, starting with the quantum entropy. Both classical and quantum entropy-based quantities addressed in this chapter will then be useful in later chapters to assess informational performance of quantum signals and processes.

7.1 Classical Shannon Entropy

7.1.1 Definition and Some Properties

The notion of information can be assigned an empirical origin. To undertake a theoretical formalization for this notion, in a physical perspective, the **entropy** is introduced, as a quantity offering a ground for a quantitative assessment of (certain aspects of) information and information content of signals. Such quantitative assessment in particular enables the formulation of quantitative laws and relations having informational meaning and significance, and standing as important elements to compose a (physical) theory of information. To introduce a useful definition for such a (classical) entropy, we consider a (classical) random variable X that can assume some discrete value x_j with the prior probabilities $p_j = \Pr\{X = x_j\}$. It is convenient to define as $-\log(p_j)$ a quantity that can receive a dual interpretation: it can represent, before observing a realization of the random variable X, the prior uncertainty for observing the event $X = x_j$; it can also represent, after observing the realization $X = x_j$, the posterior information revealed by this observation. This dual physical interpretation matches the mathematical behavior that directly follows form the quantitative definition. A probability $p_j \in [0, 1]$ implies a non-negative $-\log(p_j) \geq 0$ to quantify uncertainty or information. When p_j monotonically increases from 0 to 1,

F. Chapeau-Blondeau, *Quantum Information and Quantum Signal Processing*,
https://doi.org/10.1007/978-3-032-27499-1_7

then $-\log(p_j)$ monotonically decreases from $+\infty$ to 0, reflecting a decreasing prior uncertainty or posterior information. When $p_j \to 0$, the prior uncertainty or posterior information $-\log(p_j) \to +\infty$ is very high, which makes sense for an event $X = x_j$ very unlikely to occur; conversely, when $p_j \to 1$, the prior uncertainty or posterior information $-\log(p_j) \to 0$ is very low, which makes sense for an event $X = x_j$ quasi certain to occur. At this (early) stage in the elaboration of an information theory, we observe that the elementary mathematical properties match the physical or empirical expectations or intuition that may preexist about the notion of information that is being formalized. Such matching between mathematical behavior and physical expectation for a quantitative information, will be observed repeatedly in the course of the elaboration of the theory, to stand both as a guide and as a mark of consistency of the theory. At some further stage of the elaboration, in sufficiently involved informational situations, prior physical intuition will reach a limit, while the mathematical working of the theory will substitute and relay to significant insights. This is what can be expected from a physical theory, and a theory of information makes no exception.

By enumerating the set of its feasible values $\{x_j\}$, the random variable X is endowed with the probability distribution $\{p_j\}$, with $p_j = \Pr\{X = x_j\}$, thus satisfying $\sum_j p_j = 1$. The **classical or Shannon entropy** $H(X)$ for the random variable X is then defined [171–173] as the average uncertainty or information

$$H(X) = -\sum_j p_j \log(p_j) \geq 0 , \tag{7.1}$$

and $H(X)$ results as a non-negative scalar quantity. Convenient alternative notations will be $H(X) \equiv H(\{p_j\})$ for the entropy of the probability distribution $\{p_j\}$, or even $H(X) \equiv H(p_j)$ for conciseness when no risk of confusion is involved. The logarithm in Eq. (7.1) can be in any valid logarithmic base $b > 1$, related to the natural logarithm $\ln(\cdot)$ by $\log_b(u) = \ln(u)/\ln(b)$, so that the entropies associated with different logarithmic bases will differ only by a constant positive multiplicative factor, usually non-critical in the present context. With the natural logarithm $\ln(\cdot)$, when $b = e$, the unit for the entropy is the "nat". In addition to the natural logarithm, a convenient choice is the binary logarithm with $b = 2$; the corresponding unit for the entropy is the "bit" (but which can also designate the plain binary digit), or alternatively the "shannon" (Sh) recommended by ISO in [174] and that we shall prefer in this book.

It is also convenient for the sequel to adopt yet another alternative notation, where we generically denote by x the values accessible to the random variable X, with probabilities $P(x) = \Pr\{X = x\}$, and the classical or Shannon entropy is defined as the average

$$\boxed{H(X) = \big\langle -\log P(X)\big\rangle = -\sum_x P(x)\log\big(P(x)\big) \geq 0 .} \tag{7.2}$$

For instance, when the random variable X has access to J discrete values $\{x_j\}_{j=1}^{J}$ with equal probabilities $p_j = 1/J$ for $j = 1$ to J, then from Eq. (7.1) we obtain the entropy $H(X) = \log(J)$. If we rely on our informational interpretation of the entropy, we can anticipate that $H(X) = \log(J)$ should be the maximum entropy for a probability distribution with J accessible values, because an equiprobable distribution corresponds to the maximum average prior uncertainty about which realization x_j is going to occur, and hence the anticipation of a maximum entropy

$$H_{\max} = \log(J) . \tag{7.3}$$

This will be proved mathematically shortly, consolidating in this way the informational interpretation of the entropy. At the other extreme, when in the set $\{p_j\}_{j=1}^{J}$ there is one probability which is 1 and the other $J-1$ probabilities are 0, then Eq. (7.1) yields $H(X) = 0$ (since by continuity $0\log(0) = 0$), that is, the minimum possible value for the non-negative entropy $H(X)$. This is again consistent as here there exists no uncertainty with only one single realization that occurs with certainty.

For example, when X is a binary random variable with two accessible values of probabilities p and $1-p$, the entropy is

$$H(X) = H_{\text{bin}}(p) = -p\log(p) - (1-p)\log(1-p) , \tag{7.4}$$

and is represented in Fig. 7.1. It illustrates the above general properties that $H_{\text{bin}}(p) = 0$ for $p = 1$ and $p = 0$, and $H_{\text{bin}}(p) = H_{\max} = \log_2(2) = 1$ Sh at equiprobability when $p = 1/2$.

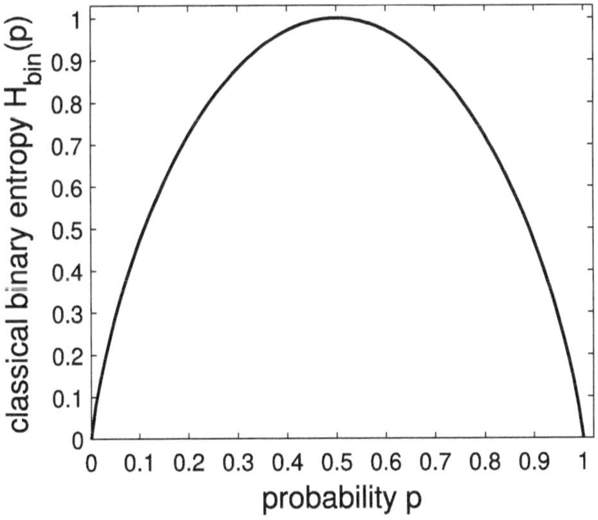

Fig. 7.1 The classical entropy $H_{\text{bin}}(p)$ of Eq. (7.4) in shannons for a binary random variable, as a function of the probability p

It is useful to consider jointly two random variables X and Y, having access respectively to discrete values generically denoted x and y, and endowed with a joint probability distribution $\Pr\{X = x\,;\, Y = y\} = P(x, y)$ lending itself to the common Bayes relations $P(x, y) = P(x|y)P(y) = P(y|x)P(x)$ with the marginals $P(x) = \sum_y P(x, y)$ and $P(y) = \sum_x P(x, y)$. For the pair (X, Y) of random variables, a **joint entropy** $H(X, Y)$ follows from the same generic notion of entropy of Eq. (7.2) as

$$H(X, Y) = -\sum_{x,y} P(x, y) \log\big(P(x, y)\big) \geq 0\,, \tag{7.5}$$

representing the prior uncertainty or posterior information, on average, associated with the joint realization of the pair (X, Y). When X and Y are statistically independent, then $P(x, y) = P(x)P(y)$ and Eq. (7.5) leads to $H(X, Y) = H(X) + H(Y)$. This additivity of the entropy which mathematically follows from the princeps definition of Eq. (7.2) in case of independence, is again well in tune with an empirical expectation for a quantitative information associated with two independent events. We can even move a little farther: additivity of the entropy as $H(X, Y) = H(X) + H(Y)$ is a necessary and sufficient condition for independence of X and Y, as we shall prove later in this Sect. 7.1.

A notion of conditional entropy also exists. From any particular realization $X = x$, a conditional distribution $P(Y = y|X = x) = P(y|x)$ follows as a valid probability distribution over Y, entailing from Eq. (7.2) a conditional entropy for Y as $H(Y|X = x) = \langle -\log P(Y|X = x)\rangle$ where the average is over the conditional distribution $P(Y = y|X = x) = P(y|x)$ of Y at fixed $X = x$. Next, by averaging over x we obtain the **conditional entropy** $H(Y|X) = \sum_x H(Y|X = x)P(x)$ for the pair of random variables (X, Y), which is also

$$H(Y|X) = -\sum_{x,y} P(x)P(y|x) \log\big(P(y|x)\big) = -\sum_{x,y} P(x, y) \log\big(P(y|x)\big) \geq 0\,, \tag{7.6}$$

interpretable as the average information revealed by Y when X is known. By symmetry, a similar definition holds for the conditional entropy $H(X|Y)$. When X and Y are independent, then from $P(y|x) = P(y)$ we deduce $H(Y|X) = H(Y)$ and similarly $H(X|Y) = H(X)$, again a mathematical behavior well in tune with the empirical expectation for a quantitative information concerning independent variables X and Y when prior knowledge of X reveals no information about Y, and vice versa. A little farther, $H(Y|X) = H(Y)$ is a necessary and sufficient condition for independence of X and Y, as we shall prove later; and similarly for $H(X|Y) = H(X)$.

With the Bayes relations used in conjunction with Eqs. (7.5)–(7.6), it can then be mathematically established rather easily that in general

$$H(X, Y) = H(Y|X) + H(X) = H(X|Y) + H(Y) \geq 0\,. \tag{7.7}$$

Again, we observe a good match of the mathematical relations of Eq. (7.7) with empirical expectation for a quantitative information: the information $H(X, Y)$ in (X, Y)

is the information $H(X)$ revealed by X plus the information $H(Y|X)$ remaining to be revealed by Y when X is known. These consequences and properties consolidate the informational interpretation associated with the entropy of Eq. (7.2).

When X completely (univocally deterministically) determines Y, then $P(Y = y|X = x) = P(y|x)$ is 1 for one particular value $Y = y$ determined by $X = x$, while it is 0 for the other values of Y; this leads to the entropy $H(Y|X = x) = 0$. This is the same type (with all values at 0 but one single 1) of conditional probability distribution $P(Y = y|X = x)$ for Y occurring for every given value x for X, and this is the only type of probability distribution $P(Y = y|X = x)$ yielding $H(Y|X = x) = 0$ as easily seen from Eq. (7.2). Consequently, on average $H(Y|X) = 0$, and $H(X, Y) = H(X)$, if and only if X completely determines Y. Furthermore, if and only if X and Y are in bijective relation, i.e., completely determine one another, we have $H(Y|X) = H(X|Y) = 0$ and $H(X, Y) = H(X) = H(Y)$. For our quantitative notions of information, again a good match is observed between the mathematical behaviors and the empirical expectations in this situation of complete determination among the variables X and Y.

7.1.2 Concavity and the Jensen Inequality

Important properties related to the entropy of Eq. (7.2) can be deduced from concavity properties. A function $f(u)$ is **concave** ($\cap$) over some interval $[u_{\min}, u_{\max}]$ of $\mathbb{R}$ when, for any u_1 and u_2 in this interval [173, 175],

$$f\big(pu_1 + (1-p)u_2\big) \ge pf(u_1) + (1-p)f(u_2)\,, \tag{7.8}$$

for any $p \in [0, 1]$. Equality for any u_1 and u_2 in $[u_{\min}, u_{\max}]$ holds in Eq. (7.8) *if* $p = 1$ or $p = 0$. When equality for any u_1 and u_2 in $[u_{\min}, u_{\max}]$ holds in Eq. (7.8) *only if* $p = 1$ or $p = 0$, the function $f(u)$ is said strictly concave. When a function $f(u)$ has a second derivative $f''(u)$ that is (strictly) negative, it is (strictly) concave. The function $h(u) = -u\ln(u)$ has a second derivative $h''(u) = -1/u$ that is strictly negative over $[0, +\infty[$, so that $h(u)$ is strictly concave over this interval. The sum of two concave functions over $[u_{\min}, u_{\max}]$ is a concave function over $[u_{\min}, u_{\max}]$. A function $f(u)$ is **convex** ($\cup$) when $-f(u)$ is concave.

A function $f(u)$ concave over $[u_{\min}, u_{\max}]$, by induction over the number of points, satisfies [173]

$$f\Big(\sum_j p_j u_j\Big) \ge \sum_j p_j f(u_j)\,, \tag{7.9}$$

with the set $\{p_j\}$ forming a probability distribution. For a function $f(u)$ strictly concave, equality for any set of u_j in $[u_{\min}, u_{\max}]$ holds in Eq. (7.9) if and only if one p_j equals 1 and all the others equal 0.

It follows that, for a concave function $f(\cdot)$, any random variable X has averages satisfying

$$f\big(\langle X\rangle\big) \geq \big\langle f(X)\big\rangle\,, \tag{7.10}$$

which is the **Jensen inequality** [173, 175].

The notion of concavity extends to functions $f(\vec{u}\,)$ of multiple real variables to satisfy the inequality

$$f\big(p\vec{u}_1 + (1-p)\vec{u}_2\big) \geq pf(\vec{u}_1) + (1-p)f(\vec{u}_2)\,. \tag{7.11}$$

The entropy $H(X)$ of Eq. (7.1) is a **concave** function in the variables $\{p_j\}$, as exemplified by Fig. 7.1.

7.1.3 *Relative Entropy*

A related quantity is the relative entropy, also known as the Kullback–Leibler divergence. It is defined [173, 175] for two probability distributions $\{p_j\}$ and $\{q_j\}$ over the same set of events indexed by j, as

$$H(p_j\|q_j) = \sum_j p_j\big[\log(p_j) - \log(q_j)\big]\,, \tag{7.12}$$

as a concise notation for the extended alternative $H\big(\{p_j\}\parallel\{q_j\}\big)$.

An important property is the **Gibbs inequality**, expressing the positivity

$$H\big(\{p_j\}\parallel\{q_j\}\big) = H(p_j\|q_j) \geq 0\,, \tag{7.13}$$

with equality if and only if $p_j = q_j$ for all j, that is, when the two probability distributions $\{p_j\}$ and $\{q_j\}$ coincide.

We have $H(p_j\|q_j) = -\sum_j p_j \log(q_j/p_j) = -\big\langle\log(q_j/p_j)\big\rangle \geq -\log\big(\langle q_j/p_j\rangle\big) = -\log\big(\sum_j q_j\big) = -\log(1) = 0$ by concavity ($\cap$) of the logarithm and the Jensen inequality of Eq. (7.10), proving that $H(p_j\|q_j)$ is non-negative. To include also the condition when $H(p_j\|q_j)$ is zero, we have the alternative and complete proof.

Proof For a proof of the Gibbs inequality, we use the natural logarithm without loss of generality, and start with $H(p_j\|q_j) = \sum_j\big[p_j\ln(p_j) - p_j\ln(q_j)\big]$. We then use the Taylor–Lagrange expansion $\ln(q_j) = \ln(p_j) + (q_j - p_j)\ln'(p_j) + (1/2)(q_j - p_j)^2\ln''(\xi_j)$ with $\xi_j \in \big[\min(p_j, q_j), \max(p_j, q_j)\big]$, involving the first and second derivatives of the natural logarithm. We then deduce $p_j\ln(p_j) - p_j\ln(q_j) = -(q_j - p_j) + (1/2)(q_j - p_j)^2 p_j/\xi_j^2$. By summing over j we finally obtain $H(p_j\|q_j) = \sum_j\big[p_j\ln(p_j) - p_j\ln(q_j)\big] = (1/2)\sum_j(q_j - p_j)^2 p_j/\xi_j^2$. Since p_j/ξ_j^2 is positive (non-negative) for all j, this is enough to prove the Gibbs inequality and its equality condition. □

After the Gibbs inequality, the relative entropy $H(p_j \| q_j)$ can be interpreted as a non-negative (asymmetric) index of departure or mismatch of two probability distributions. It is also connected with the entropy of Eq. (7.2) in the following way. With a uniform, or equiprobable, distribution $q_j = 1/J$, for $j = 1$ to J, any other probability distribution $\{p_j\}$ over the J events satisfies

$$H\big(p_j \| 1/J\big) = \sum_j p_j \log(p_j) + \sum_j p_j \log(J) , \tag{7.14}$$

and by referring to $H_{\max} = \log(J)$ as in Eq. (7.3), we can write

$$H\big(p_j \| 1/J\big) = H_{\max} - H(p_j) . \tag{7.15}$$

By the Gibbs inequality, since $H\big(p_j \| 1/J\big) \geq 0$, this proves that the uniform probability distribution over J states is, as anticipated from the informational interpretation of the entropy, the distribution achieving the maximum of the entropy, as given by Eq. (7.3), and that this distribution is unique.

Beyond two, we can consider a family of $K \geq 2$ probability distributions $\mathcal{P}_k \equiv \{p_j^{(k)}\}$ over the same set of events indexed by j, for $k = 1$ to K, and a set of K positive weights $\{\mu_k\}_{k=1}^K$ satisfying $\sum_{k=1}^K \mu_k = 1$. The average probability distribution is obtained as the convex sum $\overline{\mathcal{P}} = \sum_{k=1}^K \mu_k \mathcal{P}_k$. For the K probability distributions $\mathcal{P}_k$, one then defines the **Jensen–Shannon information divergence** as

$$J_{\mathrm{JS}}(\mathcal{P}_1, \mathcal{P}_2, \ldots, \mathcal{P}_K) = H\big(\overline{\mathcal{P}}\big) - \sum_{k=1}^K \mu_k H(\mathcal{P}_k) , \tag{7.16}$$

which is equivalently

$$J_{\mathrm{JS}}(\mathcal{P}_1, \mathcal{P}_2, \ldots, \mathcal{P}_K) = \sum_{k=1}^K \mu_k H\big(\mathcal{P}_k \| \overline{\mathcal{P}}\big) . \tag{7.17}$$

By the Gibbs inequality, the Jensen–Shannon information divergence $J_{\mathrm{JS}}(\mathcal{P}_1, \mathcal{P}_2, \ldots, \mathcal{P}_K)$ is non-negative, and it vanishes if and only if the K probability distributions $\mathcal{P}_k$ all coincide and coincide with the average distribution $\overline{\mathcal{P}}$ (which requires uniform weights $\mu_k = 1/K$ for all k). The Jensen–Shannon information divergence $J_{\mathrm{JS}}(\mathcal{P}_1, \mathcal{P}_2, \ldots, \mathcal{P}_K)$ offers a useful non-negative (symmetric) index of dissimilarity between $K \geq 2$ probability distributions.

Also, from $J_{\mathrm{JS}}(\mathcal{P}_1, \mathcal{P}_2, \ldots, \mathcal{P}_K) \geq 0$ and Eq. (7.16) follows the inequality

$$H\left(\sum_{k=1}^K \mu_k \mathcal{P}_k\right) \geq \sum_{k=1}^K \mu_k H\big(\mathcal{P}_k\big) , \tag{7.18}$$

again expressing concavity (∩) of the Shannon entropy $H(\cdot)$. In addition, the relative entropy satisfies the inequality

$$H\left(\sum_{k=1}^{K} \mu_k \mathcal{P}_k \,\Big\|\, \sum_{k=1}^{K} \mu_k \mathcal{P}'_k\right) \le \sum_{k=1}^{K} \mu_k H\left(\mathcal{P}_k \| \mathcal{P}'_k\right), \tag{7.19}$$

for two sets of probability distributions $\mathcal{P}_k$ and $\mathcal{P}'_k$, for $k = 1$ to K, expressing joint convexity (∪) of the relative entropy $H(\cdot\|\cdot)$.

7.1.4 Mutual Information

Another important informational quantity related to the entropy is the mutual information $H(X : Y)$ between two random variables X and Y, defined [171, 173] as

$$H(X : Y) = H\big(P(x, y) \,\|\, P(x)P(y)\big) \ge 0\,. \tag{7.20}$$

The mutual information $H(X : Y)$ is symmetric in the exchange of X and Y. From the definition of the relative entropy of Eq. (7.12) and the Bayes relations, alternative forms can be easily derived for the mutual information $H(X : Y)$, as

$$H(X : Y) = H(X) + H(Y) - H(X, Y) \tag{7.21}$$
$$= H(X) - H(X|Y) = H(Y) - H(Y|X) \tag{7.22}$$
$$= H(X, Y) - H(X|Y) - H(Y|X)\,. \tag{7.23}$$

By the Gibbs inequality of Eq. (7.13) applied to Eq. (7.20), it follows that the mutual information $H(X : Y)$ vanishes if and only if $P(x, y) = P(x)P(y)$. By definition, two random variables X and Y are independent if and only if their joint probability distribution factorizes as $P(x, y) = P(x)P(y)$. This may suggest that to establish independence of X and Y may require to examine all possible joint configurations (x, y) and in every case test whether the factorization $P(x, y) = P(x)P(y)$ holds. Now we have an alternative criterion: two random variables X and Y are independent if and only if their mutual information $H(X : Y)$ vanishes. This is a powerful criterion in the sense that it requires to test only one scalar quantity, the mutual information $H(X : Y)$, to decide independence. Moreover, operationally or physically, independence can be decided when $H(X : Y)$ is found negligible compared with $H(X, Y)$, which by Eq. (7.23) represents an achievable upper bound to $H(X : Y)$. The ratio $H(X : Y)/H(X, Y)$ defines a coefficient of dependence, lying in $[0, 1]$, which is 0 if and only if X and Y are independent, and is 1 if and only if X and Y are in bijective relation.

The mutual information $H(X : Y)$ in this way represents a non-negative index of departure from independence of a joint probability distribution $P(x, y)$. The alternative forms of Eqs. (7.21)–(7.23) provide additional interpretations for the mutual information $H(X : Y)$, with informational meanings which match the empirical expectations for a quantitative information theory. The mutual information $H(X : Y) = H(Y : X)$, as suggested by its name, offers a quantification of the information common to the two random variables X and Y.

The marginal entropy $H(Y)$ is a concave ($\cap$) function of the marginal probabilities $P(y) = \sum_x P(y|x)P(x)$, which are themselves linear functions of the marginal probabilities $P(x)$ at fixed conditional probabilities $P(y|x)$, so that $H(Y)$ is also a concave function of $P(x)$. The conditional entropy $H(Y|X)$ from Eq. (7.6) is a linear function of the marginal probabilities $P(x)$. As a result, the mutual information $H(X : Y) = H(Y) - H(Y|X)$, at fixed conditional probabilities $P(y|x)$, is in general a **concave** ($\cap$) function of the probability distribution $P(x)$. Meanwhile, at fixed probabilities $P(x)$, the mutual information $H(X : Y)$ is in general a **convex** ($\cup$) function of the conditional probabilities $P(y|x)$. Such properties [173] similarly hold when exchanging the roles of X and Y.

In case of independence of two random variables X and Y we have seen in Sect. 7.1.1 that $H(Y) = H(Y|X)$, but we can now state that in any case $H(Y) \geq H(Y|X)$. This follows from Eq. (7.22) and the positivity of the mutual information $H(X : Y)$. This inequality $H(Y) \geq H(Y|X)$ that is now mathematically proved can be casually expressed, based on the informational interpretation of the entropy, by stating that conditioning cannot increase information; this is again a natural expectation for a quantitative information, consolidating the informational interpretation of the entropy. A little farther, in Sect. 7.1.1 we had that X and Y independent implies $H(Y) = H(Y|X)$, from Eq. (7.22) we can now state that $H(Y) = H(Y|X)$ if and only if X and Y are independent. In a similar way, by Eq. (7.21), we have the **subadditivity** property of the entropy $H(X, Y) \leq H(X) + H(Y)$, with equality if and only if X and Y are independent, enabling to state that dependence reduces the joint entropy.

Figure 7.2 shows a graphical representation [175], sometimes known as a Venn diagram, of the relations between the informational quantities derived from the entropy of Eq. (7.2).

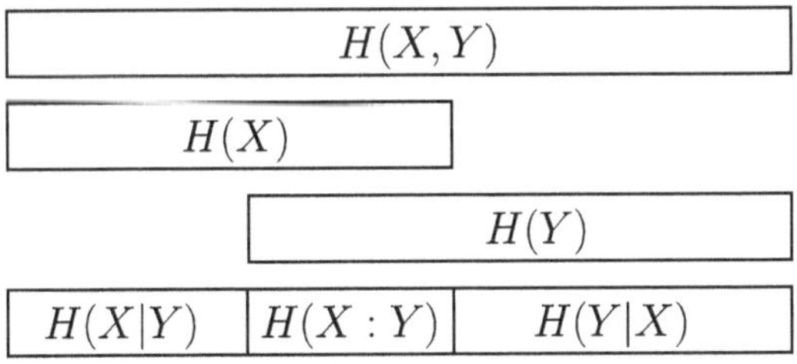

Fig. 7.2 A Venn diagram for graphical representation of the relations between the informational quantities derived from the entropy of Eq. (7.2) for two random variables X and Y, with the joint $H(X, Y)$, the marginal $H(X)$ and $H(Y)$, the conditional $H(X|Y)$ and $H(Y|X)$, entropies; the mutual information $H(X : Y) = H(Y : X)$

Three random variables X, Y, and Z form a Markov chain, denoted $X \to Y \to Z$, when their conditional probability distributions satisfy $P(z|y,x) = P(z|y)$. A consequence is that $X \to Y \to Z$ a Markov chain implies $Z \to Y \to X$ is a Markov chain, and $P(x|y,z) = P(x|y)$; whence the conditional entropies $H(X|Y,Z) = H(X|Y)$. Also for a Markov chain, from Eq. (7.22) we have the mutual informations $H(X : Y, Z) = H(X) - H(X|Y,Z) = H(X) - H(X|Y) = H(X : Y)$. But also, from $H(X : Y, Z) = H(X) - H(X|Y,Z)$, since conditioning does not increase the entropy we have $H(X|Y,Z) \le H(X|Z)$, whence $H(X : Y, Z) \ge H(X) - H(X|Z) = H(X : Z)$. We then deduce

$$H(X) \ge H(X : Y) \ge H(X : Z) , \tag{7.24}$$

representing what is known as the **data processing inequality** [173]. It expresses that the information about X that may be contained in Y cannot be increased by further processing of Y yielding Z, when the sequence $X \to Y \to Z$ forms a Markov chain.

Here our starting point was to introduce the definition of Eq. (7.1) for the entropy, and then to progressively verify by deduction that it offers a legitimate quantitative assessment of information, since in particular mathematical consequences derived from Eq. (7.1) were found to satisfy empirical or physical properties expected from a quantitative theory of information. Alternatively, a set of axioms could be imposed as a starting point, to express the properties that we think must be satisfied by a suitable quantitative assessment of information [171, 173]. With a natural set of properties (axioms), as those reflected in the informational behaviors we observed above, it is commonly the definition of Eq. (7.1) that comes out as a suitable information quantification. The possibility of such an axiomatic approach offers an additional, a priori, legitimation to the definition of the entropy in Eq. (7.1) with its informational interpretation and consequences. A strong motivation for us here, for reviewing these classical entropy-based informational quantities, is that they can all receive a quantum counterpart, in an appropriate form, and with informational interpretations that carry over to the quantum domain, as we shall soon see.

The entropy-based informational quantities introduced above can be exploited, in particular, to analyze two essential informational tasks having great fundamental and practical relevance, namely information compression and information transmission over noisy communication channels, and to characterize the efficiency and best achievable performance thereof. This leads to nontrivial informational results connected to two central theorems of (classical) information theory [171, 173]. These two essential informational tasks will be examined, especially in their quantum formulation, respectively in Chaps. 8 and 13 to come. Yet before, it is suitable to extend to the quantum domain, the entropy-based informational quantities defined above, so as to enlarge their scope and applicability to information supported by quantum signals, and obtain efficient tools to assess informational processes at the quantum level.

7.2 Quantum Information for Quantum State

7.2.1 *Information in a Quantum State*

From the quantitative notions of information presented in Sect. 7.1, to make a connection with quantum signals and systems, one could start with the question: How much information can be stored in a quantum state? With an information-carrying signal realized by a pure quantum state $|\psi\rangle = \sum_{n=1}^{N} \alpha_n |n\rangle$ of an N-dimensional Hilbert space $\mathcal{H}_N$ as in Eq. (2.1), the coordinates α_n are continuously-valued complex numbers. If one considers a set of J discrete values $\{x_j\}_{j=1}^{J}$ as in Sect. 7.1.1, one can certainly arrange J different configurations of the coordinates α_n in such a way that the information-carrying quantum signal $|\psi\rangle$ is made capable of encoding or storing the J different values $\{x_j\}_{j=1}^{J}$. For example, with a qubit state $|\psi\rangle$ in the two-dimensional Hilbert space $\mathcal{H}_2$, with the parameterization $|\psi\rangle = \cos(\theta/2)\,|0\rangle + e^{i\varphi}\sin(\theta/2)\,|1\rangle$ of Eq. (2.6), one can arrange J distinct configurations $|\psi_j\rangle$ defined by $\theta = \theta_j = (j-1)\pi/J$ for $j = 1$ to J, and a fixed value for φ, so as to encode or index J different values $\{x_j\}_{j=1}^{J}$. And this remains feasible for J arbitrarily large. When a probability distribution $\{p_j\}_{j=1}^{J}$ is introduced over the set $\{x_j\}_{j=1}^{J}$, the entropy or average information $H(X)$ that follows in Eq. (7.1) can grow arbitrarily large, with J arbitrarily large. For instance, for a uniform probability distribution $\{p_j = 1/J\}_{j=1}^{J}$, the entropy $H(X) = \log(J)$ in Eq. (7.1) goes to infinity when $J \to +\infty$. This suggests that an arbitrary large amount of information, when assessed by the classical entropy, can be stored in a quantum state. And this remains true even for a two-dimensional qubit.

This however leads to another meaningful informational question to ask concerning quantum states as information-carrying signals.

7.2.2 *Information Out of a Quantum State*

For many practical purposes, information, when stored in some physical carrier, should be retrievable in some way or another. For quantum states as information-carrying signals, this naturally leads to the question: How much information can be retrieved out of a quantum state? This especially (inevitably) when performing a quantum measurement on a quantum state.

On an N-dimensional quantum state with Hilbert space $\mathcal{H}_N$, when performing a von Neumann projective measurement, we know from Sect. 4.1.1 that in general N distinct outcomes can occur, at random, with probabilities ruled by Eq. (4.1). Such a probabilistic measurement result, according to the informational perspective of Sect. 7.1.1, can convey an entropy or average information of at most $\log(N)$, when the N measurement outcomes are equiprobable.

However, from Sect. 4.2, we also know that a generalized measurement can be performed on an N-dimensional quantum state, and such a generalized measurement can in general exhibit M distinct outcomes, with probabilities ruled by Eqs. (4.25) or (4.31). Especially one can have $M > N$, with M possibly arbitrarily large, and possibly associated with an arbitrarily large entropy in Eq. (7.1). For example, on a qubit, one can consider performing the "star" generalized measurement defined by Eqs. (4.48)–(4.49), having M possible outcomes. For a qubit prepared in the maximally mixed state $\rho = \mathrm{I}_2/2$, it follows from Eq. (4.50) that the M measurement outcomes are all equiprobable with probability $1/M$. Such measurement results, according to Sect. 7.1.1, can be associated in Eq. (7.1) with an entropy or average information of $\log(M)$, which can go arbitrarily large for a generalized measurement with $M \to +\infty$. An information similarly diverging as $\log(M)$ can even be obtained with a preparation of the qubit in a pure quantum state. This suggests that an arbitrary large amount of information can be retrieved out of a quantum state, and this is true even with a two-dimensional qubit.

However, in such scenarios, the entropy associated with the quantum measurement may be attributed essentially to the intrinsically random character of quantum measurement, and not to a controlled informational scheme where some "coherent" information is purposely stored in, and then retrieved out of, a quantum information-carrying signal. So we are led to yet another meaningful informational question concerning quantum states, and this will lead us to specific properties with important conceptual and practical significance for quantum information supported by quantum signals.

7.2.3 Input–Output Information. The Holevo Information

At this point of our informational exploration of the capabilities of quantum signals or quantum states as information carriers, we are led to the question: How much of the information purposely stored into a quantum signal can be retrieved out by quantum measurement?

For a general formulation, we return to the random variable X as examined previously, and we now consider that each accessible value x_j is encoded into a quantum state represented by the density operator $\rho_j \in \mathcal{L}(\mathcal{H}_N)$, for $j = 1$ to J. A generalized measurement associated with K POVM elements $\{\mathsf{E}_k\}_{k=1}^{K}$ is implemented, following Sect. 4.2.4, with the measurement outcome k denoted $y_k \equiv k$, for $k = 1$ to K, defining the realizations of an output random variable Y. When measuring the pre-measurement state ρ_j, the outcome y_k is obtained with the conditional probability

$$\Pr\{Y = y_k \mid X = x_j\} = \mathrm{tr}(\rho_j \mathsf{E}_k) \tag{7.25}$$

given by Eq. (4.39).

From the prior probabilities $p_j = \Pr\{X = x_j\}$, and the conditional probabilities of Eq. (7.25), an input–output mutual information $H(X : Y)$ can be deduced. It can then be proved [25, 176] that, for any POVM at the output, the input–output mutual information $H(X : Y)$ is upper bounded as

$$H(X:Y) \leq \chi\big(\{p_j, \rho_j\}\big) . \tag{7.26}$$

The quantity $\chi\big(\{p_j, \rho_j\}\big)$ is the Holevo information; it is a specifically quantum quantity, reflecting the quantum nature of the information carriers forming the statistical ensemble $\{(p_j, \rho_j)\}_{j=1}^{J}$ used for information encoding at the input. The **Holevo information** of the statistical ensemble $\{(p_j, \rho_j)\}_{j=1}^{J}$ is defined [9, 25, 176] as

$$\boxed{\chi\big(\{p_j, \rho_j\}\big) = S\left(\sum_{j=1}^{J} p_j \rho_j\right) - \sum_{j=1}^{J} p_j\, S(\rho_j) ,} \tag{7.27}$$

where

$$S(\rho) = -\operatorname{tr}\big(\rho \log(\rho)\big) \leq \log(N) , \tag{7.28}$$

is the quantum von Neumann entropy associated with a generic quantum state $\rho \in \mathcal{L}(\mathcal{H}_N)$, entailing $\chi\big(\{p_j, \rho_j\}\big) \leq \log(N)$. For our task of assessing the amount of classical information associated (that can be stored and retrieved) with a quantum state or signal, the inequality of Eq. (7.26) establishes the Holevo information $\chi\big(\{p_j, \rho_j\}\big)$ of Eq. (7.27) as a fundamental upper bound. More elements of justification for this upper bound $\chi\big(\{p_j, \rho_j\}\big)$ will become accessible in Sect. 7.6 and Eq. (7.121), after we have journeyed across the present chapter. As a first step to proceed, the von Neumann entropy $S(\rho)$ of Eq. (7.28), essential to the definition the Holevo information $\chi\big(\{p_j, \rho_j\}\big)$ of Eq. (7.27), is a quantity of major importance for quantum information theory, much like the classical Shannon entropy $H(X)$ of Eq. (7.1) is for classical information theory. And from these entropies, a similarity of the Holevo information $\chi\big(\{p_j, \rho_j\}\big)$ of Eq. (7.27) can be found with the Jensen–Shannon information of Eq. (7.16). So in the next section we will examine in more detail the quantum von Neumann entropy.

When applied to an N-dimensional quantum system of Hilbert space $\mathcal{H}_N$, the Holevo information $\chi\big(\{p_j, \rho_j\}\big)$ of Eq. (7.27) is maximized at $\log(N)$. This maximum can be achieved by any ensemble of N equiprobable and mutually orthogonal pure states of $\mathcal{H}_N$. This points to a maximum of $\log_2(N)$ shannons or bits of information that can be coherently stored and retrieved by means of an N-dimensional quantum system. For the two-dimensional qubit in $\mathcal{H}_2$, this gives a maximum of 1 shannon or 1 bit of information that can be coherently stored and retrieved. This limitation is however specific to the definite task of storing and retrieving information supported by quantum states. Although very important, this task is by no means the only concern of information processing at large. On the contrary, it rests in some sense at the very elementary ground level where the concern is merely reproducing some information that we have at disposal, but where processing per se of such information has not yet started. Beyond such reproducing of information, when actual processing starts, the point of view of Sect. 7.2.1, where an arbitrary large amount of information can be stored in a quantum state, acquires special relevance. In this vein can be placed the image processing by the Deutsch–Jozsa quantum algorithm of Sect. 5.4. As we have seen in Sect. 5.4, a whole binary image of $256 \times 256 = 2^{16}$ pixels as in Fig. 5.9 can be stored in a single quantum register of $L = 16$ qubits placed in an appropriate superposed state (while comparatively $2^{16} = 65536$ one-bit classical registers would be required instead). The concern in Sect. 5.4 is then not to retrieve such quantumly stored image, but to process it so as to classify it into one of two predefined classes. It is a matter of quantumly storing the 2^{16} bits of information of the binary image, so as to process them, in order to extract one bit of information of the class. The quantum algorithm of Sect. 5.4 solves this task by parallel processing of the 16-qubit quantum register containing the (256×256)-pixel binary image. We also find here an illustration of various possibilities for encoding information in quantum states, for storage or processing. The encoding scheme of Sect. 7.2.1 rather encodes information in the analog coordinates α_n of a quantum state $|\psi\rangle = \sum_{n=1}^{N} \alpha_n |n\rangle$, while the scheme of Sect. 5.4 rather encodes information in the basis states $\{|n\rangle\}$. Both can serve various purposes and exhibit various properties.

7.3 Quantum Von Neumann Entropy

7.3.1 Definition and Some Properties

In an N-dimensional Hilbert space $\mathcal{H}_N$, a quantum state or signal represented by the density operator $\rho \in \mathcal{L}(\mathcal{H}_N)$, has a **von Neumann or quantum entropy** defined [172, 177] as the scalar quantity

$$\boxed{S(\rho) = -\operatorname{tr}\big(\rho \log(\rho)\big) .} \tag{7.29}$$

It is convenient for the density operator ρ to introduce the diagonal or eigendecomposition $\rho = \sum_{n=1}^{N} \lambda_n |\lambda_n\rangle \langle\lambda_n|$ as in Eq. (2.37). Then $\log(\rho)$ and $-\rho\log(\rho)$ are operators of $\mathcal{L}(\mathcal{H}_N)$ having the same N (mutually orthogonal) eigenstates $\{|\lambda_n\rangle\}$ in $\mathcal{H}_N$, associated with the N eigenvalues, respectively, $\{\log(\lambda_n)\}$ and $\{-\lambda_n \log(\lambda_n)\}$. It results for Eq. (7.29) the alternative expression

$$S(\rho) = -\sum_{n=1}^{N} \lambda_n \log(\lambda_n) = H\big(\{\lambda_n\}\big) , \tag{7.30}$$

showing that the quantum or von Neumann entropy $S(\rho)$ is equivalent to the classical or Shannon entropy $H\big(\{\lambda_n\}\big) = S(\rho)$ of the probability distribution formed by the N eigenvalues $\{\lambda_n\}$ of ρ. The alternative expression of Eq. (7.30) proves that the von Neumann entropy $S(\rho)$ is a real non-negative scalar, upper bounded as $S(\rho) \le \log(N)$, and reaching the maximum $S(\rho) = S_{\max} = \log(N)$ for the uniform probability distribution $\{\lambda_n = 1/N\}$, or equivalently if and only if ρ is the maximally mixed state $\rho = \mathrm{I}_N/N$ of $\mathcal{L}(\mathcal{H}_N)$ with quantum entropy $S\big(\mathrm{I}_N/N\big) = \log(N)$. At the other extreme, the minimum entropy $S(\rho) = 0$ is obtained if and only if ρ is a pure state $\rho = |\lambda\rangle\langle\lambda|$, or equivalently ρ is a rank-one operator with a single eigenvalue at 1 and the $N-1$ others at 0. As a result, the quantum entropy $S(\rho)$, vanishing for pure states and maximized for the maximally mixed state, can be interpreted as a non-negative index assessing the mixed character or uncertainty or decoherence of a density operator ρ.

As a consequence of the upper and lower bounds to the entropy, the Holevo information of Eq. (7.27) is upper bounded as $\chi\big(\{p_j, \rho_j\}\big) \le \log(N)$, with the maximum $\log(N)$ achieved by any statistical ensemble of the form $\{(1/N, |n\rangle\langle n|)\}_{n=1}^{N}$, of N equiprobable mutually orthogonal pure states $|n\rangle$ of $\mathcal{H}_N$, each individually with minimum entropy $S\big(|n\rangle\langle n|\big) = 0$, and together with the average state $\sum_{n=1}^{N} |n\rangle\langle n|/N = \mathrm{I}_N/N$ of maximum entropy $S\big(\mathrm{I}_N/N\big) = \log(N)$.

The quantum entropy $S(\rho)$ of Eq. (7.29) is a **strictly concave** ($\cap$) function over the convex set of density operators ρ of $\mathcal{L}(\mathcal{H}_N)$. This follows from the lower bound $0 \le \chi\big(\{p_j, \rho_j\}\big)$ and the Klein inequality which are to come in Sect. 7.4.

For a product state $\rho_A \otimes \rho_B$ of two density operators $\rho_A \in \mathcal{L}(\mathcal{H}^A)$ and $\rho_B \in \mathcal{L}(\mathcal{H}^B)$, there exists an **additivity** property of the quantum entropy, as

$$S(\rho_A \otimes \rho_B) = S(\rho_A) + S(\rho_B) \; . \tag{7.31}$$

The justification of Eq. (7.31) is an occasion to introduce a useful logarithmic identity holding for two positive operators $\rho_A \in \mathcal{L}(\mathcal{H}^A)$ and $\rho_B \in \mathcal{L}(\mathcal{H}^B)$, for which we have

$$\log(\rho_A \otimes \rho_B) = \log(\rho_A) \otimes \mathrm{I}^B + \mathrm{I}^A \otimes \log(\rho_B) \; , \tag{7.32}$$

where I^A and I^B are the identity operators respectively on $\mathcal{H}^A$ and $\mathcal{H}^B$. This holds because $\rho_A \otimes \rho_B = (\rho_A \otimes \mathrm{I}^B)(\mathrm{I}^A \otimes \rho_B) = (\mathrm{I}^A \otimes \rho_B)(\rho_A \otimes \mathrm{I}^B)$, and for the logarithm of the product of these two positive operators that commute we have $\log(\rho_A \otimes \rho_B) = \log(\rho_A \otimes \mathrm{I}^B) + \log(\mathrm{I}^A \otimes \rho_B)$. Then, it is easily found by using the diagonal form of the positive operators that $\log(\rho_A \otimes \mathrm{I}^B) = \log(\rho_A) \otimes \mathrm{I}^B$ and $\log(\mathrm{I}^A \otimes \rho_B) = \mathrm{I}^A \otimes \log(\rho_B)$, and the logarithmic identity for operators of Eq. (7.32) follows.

Using Eq. (7.32), and superfluous subscripts in the trace for more clarity, we have the entropy

$$\begin{aligned}
S(\rho_A \otimes \rho_B) &= -\operatorname{tr}_{AB}\Big((\rho_A \otimes \rho_B)\log(\rho_A \otimes \rho_B)\Big) && (7.33)\\
&= -\operatorname{tr}_{AB}\Big((\rho_A \otimes \rho_B)\big[\log(\rho_A) \otimes \mathrm{I}^B + \mathrm{I}^A \otimes \log(\rho_B)\big]\Big) && (7.34)\\
&= -\operatorname{tr}_{AB}\Big(\rho_A \log(\rho_A) \otimes \rho_B + \rho_A \otimes \rho_B \log(\rho_B)\Big) && (7.35)\\
&= -\operatorname{tr}_A\big(\rho_A \log(\rho_A)\big)\operatorname{tr}_B(\rho_B) - \operatorname{tr}_A(\rho_A)\operatorname{tr}_B\big(\rho_B \log(\rho_B)\big) && (7.36)\\
&= S(\rho_A) + S(\rho_B) \; , && (7.37)
\end{aligned}$$

finally establishing the additivity of Eq. (7.31).

An alternative proof of Eq. (7.31) could have been obtained by leaning on the multiplicative structure of the eigendecomposition of a product state, with especially the eigenvalues of $\rho_A \otimes \rho_B$ given as the product of the eigenvalues of ρ_A with those of ρ_B, from which, by Eq. (7.30), the additivity of Eq. (7.31) would follow. But the less common logarithmic identity for operators introduced in Eq. (7.32) will be useful on other occasions.

An isometric operator V mapping $\mathcal{H}_N$ into a larger Hilbert space $\mathcal{H}_{N'}$ with dimension $N' \geq N$ and $\mathsf{V}^\dagger \mathsf{V} = \mathrm{I}_N$, transforms the density operator $\rho \in \mathcal{L}(\mathcal{H}_N)$ into the density operator $\rho' = \mathsf{V}\rho\mathsf{V}^\dagger \in \mathcal{L}(\mathcal{H}_{N'})$. The quantum entropy, by such a transformation, exhibits an **isometric invariance**, with

$$S\big(\mathsf{V}\rho\mathsf{V}^\dagger\big) = S(\rho) \,. \tag{7.38}$$

This follows because $S(\rho) = \mathrm{tr}\big(h(\rho)\big)$, with the useful auxiliary function

$$h(u) = -u \log(u) \,. \tag{7.39}$$

Then, since as indicated in Sect. A.3 of the Appendix we have $h\big(\mathsf{V}\rho\mathsf{V}^\dagger\big) = \mathsf{V}h(\rho)\mathsf{V}^\dagger$, we obtain $S\big(\mathsf{V}\rho\mathsf{V}^\dagger\big) = \mathrm{tr}\big(h(\mathsf{V}\rho\mathsf{V}^\dagger)\big) = \mathrm{tr}\big(\mathsf{V}h(\rho)\mathsf{V}^\dagger\big) = \mathrm{tr}\big(h(\rho)\big) = S(\rho)$ by circular invariance of the trace, as targeted. An interpretation is that an isometric evolution (including a unitary evolution) does not change the uncertainty of a density operator as assessed by the quantum entropy.

For example, a generic qubit density operator ρ with Bloch vector $\vec{r}$ in Eq. (2.61) receives the two eigenvalues $\lambda_\pm = \big(1 \pm \|\vec{r}\,\|\big)/2$ of Eq. (2.66). The quantum entropy of Eq. (7.30) for the qubit results as

$$S(\rho) = h\left(\frac{1+\|\vec{r}\,\|}{2}\right) + h\left(\frac{1-\|\vec{r}\,\|}{2}\right) = H_{\mathrm{bin}}\left(\frac{1+\|\vec{r}\,\|}{2}\right) , \tag{7.40}$$

and is represented in Fig. 7.3 as a function of the norm $\|\vec{r}\,\|$ of the Bloch vector, with the auxiliary function $h(u)$ from Eq. (7.39) and $H_{\mathrm{bin}}(\cdot)$ the classical binary entropy from Eq. (7.4).

The von Neumann entropy $S(\rho)$ of the qubit state ρ in Eq. (7.40) is a monotonically decreasing function of the norm $\|\vec{r}\,\|$ of its Bloch vector, as visible in Fig. 7.3. For pure qubit states, $\|\vec{r}\,\| = 1$ and the entropy $S(\rho)$ vanishes. When $\|\vec{r}\,\|$ decreases below 1, the state ρ becomes mixed as the qubit gradually depolarizes, and the entropy $S(\rho)$ starts to rise above 0. At $\|\vec{r}\,\| = 0$, the qubit is in the maximally mixed fully depolarized state $\rho = \mathrm{I}_2/2$, and it displays the maximum entropy $S(\rho) = \log(2) =$ 1 Sh. When passing from a pure state to the maximally mixed state, as $\|\vec{r}\,\|$ decreases from 1 to 0, gradual depolarization of the qubit state ρ increases its quantum entropy $S(\rho)$, as a mark of the increased uncertainty or decoherence of the quantum state.

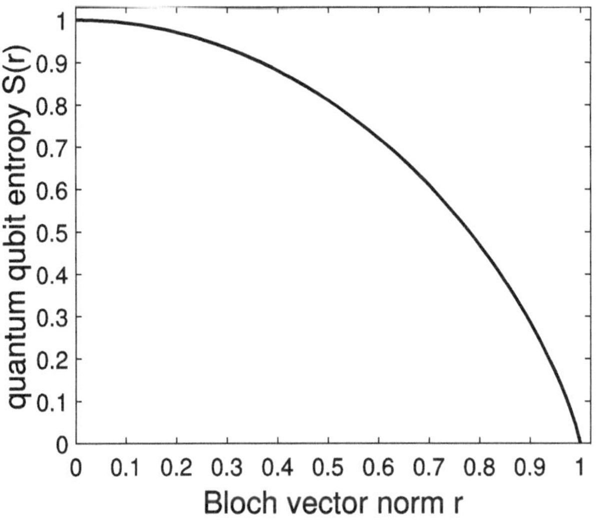

Fig. 7.3 The quantum or von Neumann entropy $S(\rho) = S(r)$ of Eq. (7.40) in shannons for a qubit state ρ, as a function of the norm $\|\vec{r}\,\| = r$ of the qubit Bloch vector $\vec{r}$

7.3.2 *Quantum Entropy and Measurement*

For an additional meaningful characterization of the quantum von Neumann entropy, we consider, as in Sect. 7.2.3 above, that the quantum signal $\rho \in \mathcal{L}(\mathcal{H}_N)$ is measured by means of a generalized measurement associated with K POVM elements $\{\mathsf{E}_k\}_{k=1}^K$, which are K positive operators of $\mathcal{L}(\mathcal{H}_N)$ satisfying the completeness relation $\sum_{k=1}^K \mathsf{E}_k = \mathrm{I}_N$ of Eq. (4.38). The K possible measurement outcomes, denoted $y_k \equiv k$ for $k = 1$ to K, define the K possible realizations of a classical random variable Y, that therefore occur with the probabilities $P_k = \Pr\{Y = y_k\} = \mathrm{tr}(\rho\mathsf{E}_k)$ according to Eq. (4.39). The classical entropy $H(Y) = -\sum_{k=1}^K P_k \log(P_k)$ associated with the probability distribution of the measurement outcomes happens to be lower bounded by the quantum entropy $S(\rho)$, for some meaningful class of generalized measurements, as we will now show. The property is established in [7] when the measuring POVM $\{\mathsf{E}_k\}$ corresponds to a standard von Neumann projective measurement as in Sect. 4.1, and in [12] when each POVM element E_k is taken as a rank-one unnormalized projector acting in $\mathcal{H}_N$.

By resorting to the diagonal decomposition of ρ as in Eq. (2.37), we have the probability

$$P_k = \mathrm{tr}\left(\sum_{n=1}^N \lambda_n \,|\lambda_n\rangle\,\langle\lambda_n|\,\mathsf{E}_k\right) = \sum_{n=1}^N \lambda_n\,\langle\lambda_n|\mathsf{E}_k|\lambda_n\rangle = \sum_{n=1}^N \lambda_n E_{kn}\,. \tag{7.41}$$

Each matrix element $E_{kn} = \langle\lambda_n|\mathsf{E}_k|\lambda_n\rangle$ is a non-negative scalar in [0, 1] since each POVM element E_k is a positive (non-negative) operator satisfying $0 \le \mathsf{E}_k \le \mathrm{I}_N$. The sum $\sum_{k=1}^{K} E_{kn} = \langle\lambda_n|\lambda_n\rangle = 1$ for every n, indicates that the set $\{E_{kn}\}_{k=1}^{K}$ forms a probability distribution over k, for every n. This is the probability distribution of the K measurement outcomes $\{y_k \equiv k\}_{k=1}^{K}$ when ρ happens to be prepared as $\rho = |\lambda_n\rangle\langle\lambda_n|$. In addition $\sum_{n=1}^{N} E_{kn} = \mathrm{tr}(\mathsf{E}_k)$, since the eigenbasis $\{|\lambda_n\rangle\}_{n=1}^{N}$ is an orthonormal basis of $\mathcal{H}_N$.

We therefore have for the measurement entropy

$$H(Y) = -\sum_{k=1}^{K} P_k \log(P_k) = -\sum_{k=1}^{K}\left(\sum_{n=1}^{N} E_{kn}\lambda_n\right)\log\left(\sum_{n=1}^{N} E_{kn}\lambda_n\right), \tag{7.42}$$

which is also

$$H(Y) = \sum_{k=1}^{K} h\left(\sum_{n=1}^{N} E_{kn}\lambda_n\right), \tag{7.43}$$

with the concave ($\cap$) function $h(u) = -u\log(u)$ of Eq. (7.39). In order to apply the Jensen inequality of Eq. (7.10) so as to deduce a meaningful lower bound to the measurement entropy $H(Y)$, we need to restrict the class of the quantum measurements involved. If we were to consider for instance a degenerate quantum measurement associated with the two POVM elements $\mathsf{E}_1 = \mathrm{I}_N$ and $\mathsf{E}_2 = \mathsf{0}$ the null operator on $\mathcal{H}_N$, we would have a perfectly valid quantum measurement, with the two probabilities $P_1 = 1$ and $P_2 = 0$, entailing a vanishing measurement entropy $H(Y) = 0$. So we restrict to quantum measurements having POVM elements E_k of rank one. Any quantum measurement in this class necessarily has a number of outcomes $K \ge N$. Each E_k is a positive, hence Hermitian, operator of $\mathcal{L}(\mathcal{H}_N)$, and since it is of rank one, it receives the diagonal representation $\mathsf{E}_k = e_k|e_k\rangle\langle e_k|$ with a normalized eigenvector $|e_k\rangle \in \mathcal{H}_N$. The single nonzero eigenvalue $e_k = \mathrm{tr}(\mathsf{E}_k) = \langle e_k|\mathsf{E}_k|e_k\rangle$ is necessarily in [0, 1] since $0 \le \mathsf{E}_k \le \mathrm{I}_N$. This class of quantum measurements includes all the standard von Neumann projective measurements of Sect. 4.1 when $K = N$, as well as generalized measurements with an arbitrarily large number of outcomes $K > N$, as in Sect. 4.2.5 for the qubit; meanwhile, it excludes the direction toward degenerate measurements with $K < N$, as evoked above with $K = 2$.

For this class of quantum measurements, featuring $\mathrm{tr}(\mathsf{E}_k) = \sum_{n=1}^{N} E_{kn} = e_k \in [0, 1]$, for each fixed k we can construct a virtual probability distribution by introducing the scalar $E_{k,N+1} = 1 - e_k \in [0, 1]$ associated with the virtual eigenvalue $\lambda_{N+1} = 0$, so that $\sum_{n=1}^{N+1} E_{kn} = 1$ and the set $\{E_{kn}\}_{n=1}^{N+1}$ forms a probability distribution over n, for every k. We can now write

$$\sum_{n=1}^{N} E_{kn}\lambda_n = \sum_{n=1}^{N+1} E_{kn}\lambda_n = \langle\lambda_n\rangle_k \tag{7.44}$$

with $\langle\lambda_n\rangle_k$ a proper statistical average of the set $\{\lambda_n\}_{n=1}^{N+1}$ according to the probability distribution $\{E_{kn}\}_{n=1}^{N+1}$ at any fixed k. This leads for the measurement entropy of Eq. (7.43) to

$$H(Y) = \sum_{k=1}^{K} h\big(\langle\lambda_n\rangle_k\big) . \tag{7.45}$$

By the Jensen inequality of Eq. (7.10), we now have

$$h\big(\langle\lambda_n\rangle_k\big) \geq \big\langle h(\lambda_n)\big\rangle_k = \sum_{n=1}^{N+1} E_{kn} h(\lambda_n) = \sum_{n=1}^{N} E_{kn} h(\lambda_n) , \tag{7.46}$$

since $h\big(\lambda_{N+1} = 0\big) = 0$. This inserted in Eq. (7.45) leads to

$$H(Y) \geq \sum_{n=1}^{N}\sum_{k=1}^{K} E_{kn} h(\lambda_n) = \sum_{n=1}^{N} h(\lambda_n) = -\sum_{n=1}^{N} \lambda_n \log(\lambda_n) = S(\rho) . \tag{7.47}$$

This finally gives the targeted inequality

$$H(Y) \geq S(\rho) , \tag{7.48}$$

stating that the classical entropy $H(Y)$ of the probability distribution arising from a generalized measurement of a quantum signal ρ is never smaller than the quantum von Neumann entropy $S(\rho)$ of Eq. (7.29). This holds for generalized measurements in the class of rank-one (unnormalized) projectors with $K \geq N$ outcomes, which includes standard von Neumann projective measurements. In this respect, such a von Neumann measurement can be obtained by measuring a quantum observable Ω performing a projective measurement in an eigenbasis of Ω; the probability distribution that results can serve to define a classical entropy $H(\Omega)$ for the observable Ω on the state ρ, therefore verifying $H(\Omega) \geq S(\rho)$. In general, the lower bound $S(\rho)$ for the measurement entropy $H(Y)$ is always achievable by measuring the quantum signal ρ by means of a standard von Neumann projective measurement in the eigenbasis $\big\{|\lambda_n\rangle\big\}$ of ρ. This provides an alternative (variational) characterization of the quantum von Neumann entropy $S(\rho)$ of Eq. (7.29), as the achievable lower bound to the classical entropy $H(Y)$ of the probability distribution arising from any measurement of ρ with rank-one operators.

7.3.3 *Joint and Conditional Quantum Entropies*

For a bipartite quantum system AB of Hilbert space $\mathcal{H}^A \otimes \mathcal{H}^B$ described by the joint density operator $\rho_{AB} \in \mathcal{L}\big(\mathcal{H}^A \otimes \mathcal{H}^B\big)$, the notion of quantum entropy of

Eq. (7.29) applies equally and leads to a joint quantum entropy $S(A, B) = S(\rho_{AB}) = -\operatorname{tr}\big(\rho_{AB}\log(\rho_{AB})\big)$.

The subsystem A alone is described by the reduced density operator $\rho_A = \operatorname{tr}_B(\rho_{AB})$ of $\mathcal{L}(\mathcal{H}^A)$ and is endowed with the (marginal) quantum entropy $S(A) = S(\rho_A) = -\operatorname{tr}\big(\rho_A\log(\rho_A)\big)$. Similarly, the subsystem B is described by the reduced density operator $\rho_B = \operatorname{tr}_A(\rho_{AB})$ of $\mathcal{L}(\mathcal{H}^B)$, with the marginal quantum entropy $S(B) = S(\rho_B) = -\operatorname{tr}\big(\rho_B\log(\rho_B)\big)$. By analogy with the classical entropy and Eq. (7.7), conditional quantum entropies can be defined as

$$S(A|B) = S(A, B) - S(B)\ , \tag{7.49}$$

$$S(B|A) = S(A, B) - S(A)\ . \tag{7.50}$$

When the joint state is separable with a joint density operator in product form $\rho_{AB} = \rho_A \otimes \rho_B$, by the additivity property of Eq. (7.31) we obtain an additive joint quantum entropy $S(A, B) = S(A) + S(B)$. The conditional quantum entropies are then given as $S(A|B) = S(A)$ and $S(B|A) = S(B)$. These properties of the quantum entropies associated with a separable or product state of a bipartite system AB are much similar to the properties of the classical entropies associated with two independent random variables X and Y in Sect. 7.1.

However, in general conditions, there exists a significant property of the conditional classical entropy that is not shared by its quantum counterpart. The conditional classical entropy $H(Y|X)$ is defined in Eq. (7.6) as the average of the non-negative classical Shannon entropy of the probability distribution $P(Y|X = x)$, and as such $H(Y|X)$ in Eq. (7.6) arrives as non-negative, by its definition. Then, from some elementary properties involving $H(Y|X)$ we deduce Eq. (7.7), which in turn entails the alternative expression $H(Y|X) = H(X, Y) - H(X)$. This is rather this last expression, extended to the quantum domain, that serves in Eqs. (7.49)–(7.50) to define the conditional quantum entropies $S(A|B)$ and $S(B|A)$. Meanwhile, the conditional quantum entropies $S(A|B)$ and $S(B|A)$ are not defined as averages of non-negative quantum von Neumann entropies. As a consequence, the conditional quantum entropies $S(A|B)$ and $S(B|A)$ defined by Eqs. (7.49)–(7.50) are not necessarily always non-negative, and significant examples thereof will be given shortly. This lack of non-negativity of the quantum conditional entropies $S(A|B)$ and $S(B|A)$ in a bipartite system AB, can be viewed as another manifestation of the distinctive quantum correlations associated with entanglement and having no classical analogue. Such distinctive properties attached to the quantum entropy are examined in more detail in the next two sections, via the important property for a bipartite state constituted by the Schmidt decomposition.

7.3.4 The Schmidt Decomposition

We consider a bipartite quantum system AB of Hilbert space $\mathcal{H}^A \otimes \mathcal{H}^B$, with the spaces $\mathcal{H}^A$ of dimension N_A and $\mathcal{H}^B$ of dimension N_B, placed in a generic pure state $|AB\rangle$ receiving the general form

$$|AB\rangle = \sum_{j=1}^{N_A} \sum_{k=1}^{N_B} \alpha_{jk} \, |j_A\rangle \otimes |k_B\rangle \ , \tag{7.51}$$

with $\{|j_A\rangle\}_{j=1}^{N_A}$ and $\{|k_B\rangle\}_{k=1}^{N_B}$ two orthonormal bases respectively of $\mathcal{H}^A$ and $\mathcal{H}^B$. The corresponding density operator is

$$\rho_{AB} = |AB\rangle \langle AB| = \sum_{j,j'=1}^{N_A} \sum_{k,k'=1}^{N_B} \alpha_{jk}\alpha^*_{j'k'} \, |j_A\rangle \langle j'_A| \otimes |k_B\rangle \langle k'_B| \ . \tag{7.52}$$

The reduced density operator for system B is

$$\rho_B = \mathrm{tr}_A\big(\rho_{AB}\big) = \sum_{j=1}^{N_A} \sum_{k,k'=1}^{N_B} \alpha_{jk}\alpha^*_{jk'} \, |k_B\rangle \langle k'_B| \ . \tag{7.53}$$

In addition, it is possible to rewrite the generic bipartite pure state of Eq. (7.51) as

$$|AB\rangle = \sum_{k=1}^{N_B} |\widetilde{k}_A\rangle \otimes |k_B\rangle \ , \tag{7.54}$$

where we have introduced the N_B vectors of $\mathcal{H}^A$ defined as

$$|\widetilde{k}_A\rangle = \sum_{j=1}^{N_A} \alpha_{jk} \, |j_A\rangle \ , \tag{7.55}$$

labeled by $k = 1$ to N_B via $\widetilde{k}_A$, which can be understood also as $\widetilde{k}_A \equiv \widetilde{k}_A(k)$.

Now we choose the orthonormal basis $\{|k_B\rangle\}_{k=1}^{N_B}$ of $\mathcal{H}^B$ as an eigenbasis of the reduced density operator ρ_B, in which ρ_B has diagonal form. With this choice of basis, the coefficients α_{jk} introduced in Eq. (7.51) satisfy

$$\sum_{j=1}^{N_A} \alpha_{jk}\alpha^*_{jk'} = p_k \delta_{kk'} \ , \tag{7.56}$$

to make Eq. (7.53) diagonal, with N_B non-negative scalars p_k, forming a probability distribution, and representing the eigenvalues of $\rho_B = \sum_{k=1}^{N_B} p_k \, |k_B\rangle \langle k_B|$.

With this choice of $\{|k_B\rangle\}_{k=1}^{N_B}$ as an eigenbasis of ρ_B, two vectors $|\widetilde{k}_A\rangle$ and $|\widetilde{k}'_A\rangle$ from Eq. (7.55) have inner product

$$\langle\widetilde{k}'_A\,|\,\widetilde{k}_A\rangle = \left(\sum_{j'=1}^{N_A} \alpha^*_{j'k'}\,\langle j'_A|\right)\left(\sum_{j=1}^{N_A} \alpha_{jk}\,|j_A\rangle\right) \tag{7.57}$$

$$= \sum_{j,j'=1}^{N_A} \alpha_{jk}\alpha^*_{j'k'}\,\langle j'_A|j_A\rangle \tag{7.58}$$

$$= \sum_{j=1}^{N_A} \alpha_{jk}\alpha^*_{jk'} = p_k\delta_{kk'}\ , \tag{7.59}$$

since $\langle j'_A|j_A\rangle = \delta_{jj'}$, showing that these N_B vectors of $\mathcal{H}^A$ are mutually orthogonal with squared norms $\langle\widetilde{k}_A\,|\,\widetilde{k}_A\rangle$ given by the p_k. For some k the vector $|\widetilde{k}_A\rangle$ may be the null vector, signaled by a squared norm $\langle\widetilde{k}_A\,|\,\widetilde{k}_A\rangle = p_k$ that vanishes. We introduce their normalized version

$$|\widehat{k}_A\rangle = \frac{|\widetilde{k}_A\rangle}{\sqrt{p_k}}\ , \tag{7.60}$$

unless $p_k = 0$ for some k in which case $|\widehat{k}_A\rangle$ is set to the null vector. The set $\{|\widehat{k}_A\rangle\}_{k=1}^{N_B}$ now forms an orthonormal set of N_B vectors of $\mathcal{H}^A$ labeled by $k = 1$ to N_B via $\widehat{k}_A$, with also possible occurrences of the null vector.

The generic bipartite pure state of Eq. (7.54) can now be written as

$$|AB\rangle = \sum_{k=1}^{N_B} \sqrt{p_k}\,|\widehat{k}_A\rangle \otimes |k_B\rangle\ . \tag{7.61}$$

From Eq. (7.61) and $\rho_{AB} = |AB\rangle\langle AB|$ we recover the reduced density operator ρ_B for system B as

$$\rho_B = \mathrm{tr}_A\big(\rho_{AB}\big) = \sum_{k=1}^{N_B} p_k\,|k_B\rangle\langle k_B| \tag{7.62}$$

in diagonal form, as expected; we also obtain the reduced density operator ρ_A for system A as

$$\rho_A = \mathrm{tr}_B\big(\rho_{AB}\big) = \sum_{k=1}^{N_B} p_k\,|\widehat{k}_A\rangle\langle\widehat{k}_A|\ . \tag{7.63}$$

The diagonal form of Eq. (7.63) represents an eigendecomposition for ρ_A. This shows that the two reduced density operators ρ_A and ρ_B derived from a bipartite pure state $|AB\rangle$ have the same set of (nonzero) eigenvalues $\{p_k\}$; the two reduced density operators ρ_A and ρ_B have the same rank (number of nonzero eigenvalues)

$N_r \le \min(N_A, N_B)$. The states $|\widehat{k_A}\rangle$ in Eq. (7.63) are eigenstates of ρ_A, while the states $|k_B\rangle$ are eigenstates of ρ_B.

The sum of Eq. (7.61) can be limited to the number $N_r \le \min(N_A, N_B)$ of nonzero eigenvalues p_k, which is the common rank of ρ_A and ρ_B, so as to give for the generic bipartite pure state of Eq. (7.51) the decomposition

$$\boxed{|AB\rangle = \sum_{k=1}^{N_r} \sqrt{p_k}\, |\widehat{k_A}\rangle \otimes |k_B\rangle\ .} \tag{7.64}$$

The form of Eq. (7.64) represents the **Schmidt decomposition** of the bipartite pure quantum state $|AB\rangle$. For the bipartite pure state $|AB\rangle$, the expansion of Eq. (7.51), referring to two arbitrary orthonormal bases for $\mathcal{H}^A$ and $\mathcal{H}^B$, may contain up to $N_A N_B$ terms, with complex coefficients α_{jk}. By contrast, by referring to the two special orthonormal bases formed from the eigenstates of the reduced density operators ρ_A and ρ_B, the Schmidt decomposition of Eq. (7.64) contains at most only $\min(N_A, N_B)$ terms, with real coefficients p_k. The rank $N_r \le \min(N_A, N_B)$ common to ρ_A and ρ_B is also known as the Schmidt number, or Schmidt rank, of the bipartite pure state $|AB\rangle$, and the N_r nonzero scalars $\sqrt{p_k}$ occurring in Eq. (7.64) are the Schmidt coefficients.

When the bipartite pure state is separable, it can be written as $|AB\rangle = |\widehat{1_A}\rangle \otimes |1_B\rangle$, which is a Schmidt decomposition with only one nonzero eigenvalue $p_1 = 1$ for each reduced state $\rho_A = |\widehat{1_A}\rangle\langle\widehat{1_A}|$ and $\rho_B = |1_B\rangle\langle 1_B|$, all other eigenvalues being zero, this corresponding to a Schmidt number of 1. In this way, a bipartite pure state is separable if and only if it has a Schmidt number of 1. Any larger Schmidt number will quantify the amount of entanglement in the bipartite pure state.

By contrast, for tripartite quantum systems ABC (or beyond), no comparable decomposition generally exists, breaking any joint tripartite pure state $|ABC\rangle$ into a sum of product states of the three subsystems that would extend the Schmidt decomposition. The reason is that if A is projected onto one of its eigenstates by a local measurement, the remaining subsystem BC may not necessarily be in a separable bipartite pure state, but it can well be in an entangled bipartite state so that no definite states exist separately for each subpart B and C.

7.3.5 *Bipartite Quantum Entropies and Entanglement*

Returning to the joint quantum entropy of Sect. 7.3.3, a bipartite quantum system AB in a pure state $\rho_{AB} = |AB\rangle\langle AB|$ exhibits a vanishing joint quantum entropy $S(A, B) = 0$. As shown by the Schmidt decomposition, the two reduced states $\rho_A = \mathrm{tr}_B(\rho_{AB}) = \mathrm{tr}_B\big(|AB\rangle\langle AB|\big)$ and $\rho_B = \mathrm{tr}_A(\rho_{AB}) = \mathrm{tr}_A\big(|AB\rangle\langle AB|\big)$ have the same rank and the same (nonzero) eigenvalues, and have therefore the same quantum entropy $S(A) = S(B)$. This is an important and frequently useful property: **two**

reduced states obtained by reduction of a common bipartite pure state share the same quantum entropy.

When the bipartite pure state $|AB\rangle$ is not separable (i.e., is entangled), the reduced states ρ_A and ρ_B are usually mixed states, with non-vanishing quantum entropies $S(A) \geq 0$ and $S(B) \geq 0$. So for a pure bipartite state $\rho_{AB} = |AB\rangle\langle AB|$ we have in general $0 = S(A, B) \leq S(A) = S(B)$, and the resulting conditional quantum entropies, as defined in Eqs. (7.49)–(7.50), verify $S(A|B) = S(A, B) - S(B) \leq 0$ and $S(B|A) = S(A, B) - S(A) \leq 0$. In this way, entangled pure bipartite states in general are associated with negative conditional quantum entropies $S(A|B) = S(B|A)$. This offers a criterion for quantum entanglement: a bipartite quantum system AB in a pure state is entangled if and only if the conditional quantum entropies $S(A|B) = S(B|A)$ are negative.

For example, for a two-qubit system AB in the entangled pure state $|AB\rangle = \big(|00\rangle + |11\rangle\big)/\sqrt{2}$ (this is the maximally entangled Bell state $|AB\rangle = |\beta_{00}\rangle$ of Eq. (2.21) above), the corresponding bipartite density operator is $\rho_{AB} = |AB\rangle\langle AB| = \big(|00\rangle\langle 00| + |00\rangle\langle 11| + |11\rangle\langle 00| + |11\rangle\langle 11|\big)/2$, and the joint quantum entropy of the pure state is $S(A, B) = 0$. The two reduced density operators, for each subsystem A and B, are $\rho_A = \mathrm{tr}_B(\rho_{AB}) = \big(|0\rangle\langle 0| + |1\rangle\langle 1|\big)/2 = \mathrm{I}_2/2$ and $\rho_B = \mathrm{tr}_A(\rho_{AB}) = \mathrm{I}_2/2$. These are two maximally mixed qubit states, each having quantum entropy $S(A) = S(B) = \log(2)$, therefore exhibiting $S(A) = \log(2) > S(A, B) = 0$ and $S(B) = \log(2) > S(A, B) = 0$, leading to negative conditional quantum entropies $S(A|B) = -S(B) = -\log(2) < 0$ and $S(B|A) = -S(A) = -\log(2) < 0$.

In a bipartite quantum system AB in a pure state $|AB\rangle$, there is no uncertainty about the quantum state of the pair AB which is completely determined, whence the vanishing joint quantum entropy $S(A, B) = 0$. When the joint pure state $|AB\rangle$ happens to be entangled and not separable, no definite individual state separately exists for each subsystem A and B, and this uncertainty is associated with the non-vanishing marginal quantum entropies $S(A) = S(B) > 0$. The entropies $S(A) = S(B) > S(A, B) = 0$ entail negative conditional quantum entropies $S(A|B) = S(B|A) < 0$ in Eqs. (7.49)–(7.50). These are distinctive features of an entangled quantum system: although the whole is completely determined, the parts may remain uncertain.

For the bipartite system AB in the pure state $|AB\rangle$ of the product space $\mathcal{H}^A \otimes \mathcal{H}^B$ as in Sect. 7.3.3, the reduced quantum entropy $S(A) = S(B)$ offers a quantitative assessment of the entanglement strength between A and B, and is called in this circumstance the **entanglement entropy**. The entanglement entropy $S(A) = S(B)$ is zero if and only if $|AB\rangle$ is a separable (a minimally entangled) state; while $S(A) = S(B)$ is maximized at $\log\big(\min(N_A, N_B)\big) > 0$. For example, for an entangled qubit pair AB in a Bell state $|AB\rangle$, the reduced states are $\rho_A = \rho_B = \mathrm{I}_2/2$, with entanglement entropy $S(A) = S(B) = \log(2)$ reaching its maximum, and identifying the Bell states as maximally entangled states for two qubits.

7.3.6 Entropy of a Mixture of Quantum States

We consider a quantum system of interest Q, with N-dimensional Hilbert space $\mathcal{H}^Q \equiv \mathcal{H}_N$, that can be in a state $\rho_j \in \mathcal{L}(\mathcal{H}_N)$, with the probability p_j, for $j = 1$ to J. The average state is defined as

$$\overline{\rho} = \sum_{j=1}^{J} p_j \, \rho_j \;, \tag{7.65}$$

and its quantum entropy satisfies the inequality

$$\boxed{S(\overline{\rho}) \leq H\big(\{p_j\}\big) + \sum_{j=1}^{J} p_j \, S(\rho_j) \,.} \tag{7.66}$$

This can be proved by introducing the eigendecomposition of each state $\rho_j \in \mathcal{L}(\mathcal{H}_N)$ as

$$\rho_j = \sum_{n=1}^{N} \lambda_n^j \, |\lambda_n^j\rangle \, \langle\lambda_n^j| \;, \tag{7.67}$$

for $j = 1$ to J. In this way, every state $\rho_j \in \mathcal{L}(\mathcal{H}_N)$ can be realized by the statistical ensemble $\{(\lambda_n^j, |\lambda_n^j\rangle)\}_{n=1}^N$ of its N mutually orthogonal eigenstates $|\lambda_n^j\rangle \in \mathcal{H}_N$, each occurring with the probability λ_n^j. The average state $\overline{\rho} \in \mathcal{L}(\mathcal{H}_N)$ of Eq. (7.65) for system Q can be realized by a statistical ensemble of the JN pure states $|\lambda_n^j\rangle \in \mathcal{H}_N$, each occurring with the probability $p_j\lambda_n^j$, for $j = 1$ to J and $n = 1$ to N.

To obtain a purification for $\overline{\rho}$, as in Sect. 2.5.3, we introduce a reference system R with dimension JN and a purifying bipartite pure state $|QR\rangle$ such that $\overline{\rho} = \mathrm{tr}_R\big(|QR\rangle\,\langle QR|\big)$. This also gives access to the reduced state for the reference system R as $\rho_R = \mathrm{tr}_Q\big(|QR\rangle\,\langle QR|\big)$. The two reduced states $\overline{\rho}$ and ρ_R are obtained by reduction of a common bipartite pure state $|QR\rangle$, consequently they share the same quantum entropy $S(\overline{\rho}) = S(\rho_R)$, as seen at the beginning of Sect. 7.3.5. The system R with dimension JN has an orthonormal reference basis we denote $\big\{|jn\rangle\big\}$, for $j = 1$ to J and $n = 1$ to N. We envisage a von Neumann projective measurement of R in this basis $\big\{|jn\rangle\big\}$. The measurement outcome is to project R on a definite state $|jn\rangle$ with the probability $p_j\lambda_n^j$, for $j = 1$ to J and $n = 1$ to N, as it results from the purification analysis done with Eq. (2.48) in Sect. 2.5.3. This probability distribution $\big\{p_j\lambda_n^j\big\}$ associated with the JN outcomes of the von Neumann measurement of R in state ρ_R, therefore exhibits a classical entropy $H\big(\big\{p_j\lambda_n^j\big\}\big)$, which by Eq. (7.48) satisfies the inequality

$$H\big(\big\{p_j\lambda_n^j\big\}\big) \geq S(\rho_R) = S(\overline{\rho}) \,. \tag{7.68}$$

We have also

$$H\big(\{p_j\lambda_n^j\}\big) = \sum_{j=1}^{J}\sum_{n=1}^{N} -p_j\lambda_n^j \log\big(p_j\lambda_n^j\big) \tag{7.69}$$

$$= \sum_{j=1}^{J}\sum_{n=1}^{N} -p_j\lambda_n^j \Big[\log(p_j) + \log\big(\lambda_n^j\big)\Big] \tag{7.70}$$

$$= \sum_{j=1}^{J} -p_j \log(p_j) \sum_{n=1}^{N} \lambda_n^j + \sum_{j=1}^{J} p_j \sum_{n=1}^{N} -\lambda_n^j \log\big(\lambda_n^j\big) \tag{7.71}$$

$$= H\big(\{p_j\}\big) + \sum_{j=1}^{J} p_j S(\rho_j)\,, \tag{7.72}$$

since $\sum_{n=1}^{N} \lambda_n^j = 1$ for every $j = 1$ to J. By combining Eq. (7.72) with Eq. (7.68) we obtain the targeted inequality of Eq. (7.66).

Equality holds in Eq. (7.66) if and only if the J density operators ρ_j have mutually orthogonal supports in $\mathcal{H}_N$. In this circumstance $S(\overline{\rho}) = H\big(\{p_j\lambda_n^j\}\big)$, and from Eq. (7.72) equality follows in Eq. (7.66).

As a consequence of Eq. (7.66), the Holevo information $\chi\big(\{p_j, \rho_j\}\big)$ defined in Eq. (7.27) is upper bounded as

$$\chi\big(\{p_j, \rho_j\}\big) = S(\overline{\rho}) - \sum_{j=1}^{J} p_j\, S(\rho_j) \le H\big(\{p_j\}\big)\,. \tag{7.73}$$

7.4 Quantum Relative Entropy

Another useful quantity is the quantum relative entropy or relative von Neumann entropy. It is defined [172, 177, 178] for two density operators ρ and ρ' of $\mathcal{L}(\mathcal{H}_N)$ as

$$\boxed{S(\rho \,\|\, \rho') = \mathrm{tr}\Big(\rho\big[\log(\rho) - \log(\rho')\big]\Big)\,.} \tag{7.74}$$

An important property is the **Klein inequality**, stipulating that

$$S(\rho \,\|\, \rho') \ge 0\,, \tag{7.75}$$

with equality if and only if $\rho = \rho'$, that is, when the two density operators ρ and ρ' coincide.

Proof To prove the Klein inequality, we start with the eigendecompositions $\rho = \sum_{n=1}^{N} \lambda_n \,|\lambda_n\rangle\,\langle\lambda_n|$ and $\rho' = \sum_{\ell=1}^{N} \lambda'_\ell \,|\lambda'_\ell\rangle\,\langle\lambda'_\ell|$ of the two density operators in $\mathcal{L}(\mathcal{H}_N)$. From the definition of the quantum relative entropy of Eq. (7.74), we obtain

$$S(\rho \parallel \rho') = \sum_{n=1}^{N} \Big[\lambda_n \log(\lambda_n) - \langle \lambda_n | \rho \log(\rho') | \lambda_n \rangle \Big] . \tag{7.76}$$

Next, we substitute into Eq. (7.76) the eigenrelations $\langle \lambda_n | \, \rho = \lambda_n \, \langle \lambda_n |$, and after this substitution we proceed with

$$\langle \lambda_n | \log(\rho') | \lambda_n \rangle = \left\langle \lambda_n \middle| \left(\sum_{\ell=1}^{N} \log(\lambda'_\ell) \, |\lambda'_\ell\rangle \, \langle \lambda'_\ell | \right) \middle| \lambda_n \right\rangle \tag{7.77}$$

$$= \sum_{\ell=1}^{N} \log(\lambda'_\ell) \, \langle \lambda_n | \lambda'_\ell \rangle \, \langle \lambda'_\ell | \lambda_n \rangle \tag{7.78}$$

$$= \sum_{\ell=1}^{N} P_{n\ell} \log(\lambda'_\ell) \, , \tag{7.79}$$

where we have introduced the non-negative real quantities $P_{n\ell} = \langle \lambda_n | \lambda'_\ell \rangle \, \langle \lambda'_\ell | \lambda_n \rangle = |\langle \lambda_n | \lambda'_\ell \rangle|^2$. We have $\sum_{\ell=1}^{N} P_{n\ell} = \langle \lambda_n | \left(\sum_{\ell=1}^{N} |\lambda'_\ell\rangle \, \langle \lambda'_\ell | \right) |\lambda_n\rangle = \langle \lambda_n | \mathrm{I}_N | \lambda_n \rangle = \langle \lambda_n | \lambda_n \rangle$ $= 1$, so that the $P_{n\ell}$ form a probability distribution over ℓ for every n. In a similar way, since also $P_{n\ell} = \langle \lambda'_\ell | \lambda_n \rangle \, \langle \lambda_n | \lambda'_\ell \rangle$, we have $\sum_{n=1}^{N} P_{n\ell} = 1$ for every ℓ. The square matrix formed as $\big[P_{n\ell} \big]$ results as an $N \times N$ real matrix of non-negative entries, with each of its N rows summing to 1 and each of its N columns summing to 1, also called a doubly stochastic matrix. Now by the strict concavity ($\cap$) of the logarithm and the Jensen inequality of Eq. (7.10), we have for every n in Eq. (7.79)

$$\sum_{\ell=1}^{N} P_{n\ell} \log(\lambda'_\ell) \leq \log\left(\sum_{\ell=1}^{N} P_{n\ell} \lambda'_\ell \right) , \tag{7.80}$$

with equality if and only if for some ℓ one $P_{n\ell}$ equals 1 while the other $P_{n\ell}$ vanish. When only one $P_{n\ell} = 1$, the sum $\sum_{\ell=1}^{N} P_{n\ell} \lambda'_\ell$ reduces to one eigenvalue λ'_ℓ for every n. This sampled eigenvalue λ'_ℓ is a function of n that we may denote q_n. For every $n = 1$ to N, when equality holds in Eq. (7.80), the N numbers q_n sample all the N eigenvalues λ'_ℓ, this being due to the double stochasticity of the matrix $\big[P_{n\ell} \big]$ that may contain only one entry 1 in each row and in each column, or in other words when for all n equality holds in Eq. (7.80) the matrix $\big[P_{n\ell} \big]$ is a permutation matrix. So when equality holds in Eq. (7.80), for all $n = 1$ to N, the set $\{q_n\}_{n=1}^{N}$ coincides with the set $\{\lambda'_\ell\}_{\ell=1}^{N}$ of the N eigenvalues of ρ', up to some renumbering. In addition, when equality holds in Eq. (7.80), for every n and its one associated ℓ, the condition $1 = P_{n\ell} = \langle \lambda_n | \lambda'_\ell \rangle \, \langle \lambda'_\ell | \lambda_n \rangle$ implies that the two so associated eigenstates $|\lambda_n\rangle$ and $|\lambda'_\ell\rangle$ coincide, while for this n the $N - 1$ remaining eigenstates of ρ' are orthogonal to $|\lambda_n\rangle$. Again, when n varies from 1 to N, the matrix $\big[P_{n\ell} \big]$ being a permutation matrix implies that each eigenstate $|\lambda_n\rangle$ comes to coincide with one distinct eigenstate $|\lambda'_\ell\rangle$, for $\ell = 1$ to N. It results that when equality holds in Eq. (7.80), for all $n = 1$ to N, the eigenset $\{|\lambda_n\rangle\}_{n=1}^N$ coincides with the eigenset $\{|\lambda'_\ell\rangle\}_{\ell=1}^N$, up to some renumbering (the same as before for the coinciding eigenvalues).

Returning to Eq. (7.76), we now have the inequality

$$S(\rho \,\|\, \rho') \geq \sum_{n=1}^{N} \big[\lambda_n \log(\lambda_n) - \lambda_n \log(q_n)\big] , \tag{7.81}$$

with equality in Eq. (7.81) in the configuration where equality holds in Eq. (7.80) for all $n = 1$ to N, that is, in the configuration where the set $\{q_n\}_{n=1}^N$ coincides with the set of eigenvalues $\{\lambda'_\ell\}_{\ell=1}^N$ and also the two sets of eigenstates $\{|\lambda_n\rangle\}_{n=1}^N$ and $\{|\lambda'_\ell\rangle\}_{\ell=1}^N$ coincide. In such configuration, the set $\{q_n\}_{n=1}^N$ forms a probability distribution as well as the set $\{\lambda_n\}_{n=1}^N$. So we recognize that Eq. (7.81) is $S(\rho\|\rho') \geq H(\lambda_n\|q_n)$, and by the Gibbs inequality of Eq. (7.13) we deduce that $S(\rho\|\rho') \geq 0$ with equality if and only if the two probability distributions $\{q_n\}_{n=1}^N$ and $\{\lambda_n\}_{n=1}^N$ coincide, which means that the two sets of eigenvalues $\{\lambda'_\ell\}_{\ell=1}^N$ and $\{\lambda_n\}_{n=1}^N$ coincide. And since also the two sets of eigenstates $\{|\lambda'_\ell\rangle\}_{\ell=1}^N$ and $\{|\lambda_n\rangle\}_{n=1}^N$ coincide, we conclude that equality holds in $S(\rho\|\rho') \geq 0$ if an only if the two density operators ρ and ρ' have the same set of eigenvalues and the same set of eigenstates, that is, the two density operators ρ and ρ' coincide, proving the Klein inequality of Eq. (7.75). □

From the Klein inequality, the quantum relative entropy $S(\rho\|\rho')$ can be interpreted as a non-negative (asymmetric) index of distinguishability of two density operators.

The maximally mixed quantum state I_N/N of $\mathcal{L}(\mathcal{H}_N)$, with quantum entropy $S\big(\mathrm{I}_N/N\big) = \log(N)$, leads for any state ρ of $\mathcal{L}(\mathcal{H}_N)$ to

$$S\left(\rho \,\middle\|\, \frac{\mathrm{I}_N}{N}\right) = \log(N) - S(\rho) . \tag{7.82}$$

The Klein inequality then proves that the quantum entropy $S(\rho)$ is maximized, at the maximum value $\log(N)$, if and only if ρ is the maximally mixed state I_N/N.

Like the quantum entropy in Eq. (7.31), the quantum relative entropy exhibits a property of additivity. For two density operators ρ_A and ρ'_A of $\mathcal{L}(\mathcal{H}^A)$, and two density operators ρ_B and ρ'_B of $\mathcal{L}(\mathcal{H}^B)$, we have the **additivity** of the quantum relative entropy as

$$S\big(\rho_A \otimes \rho_B \,\|\, \rho'_A \otimes \rho'_B\big) = S(\rho_A \,\|\, \rho'_A) + S(\rho_B \,\|\, \rho'_B) . \tag{7.83}$$

For a justification, we have from the definition Eq. (7.74) of the quantum relative entropy,

$$S\big(\rho_A \otimes \rho_B \,\|\, \rho'_A \otimes \rho'_B\big) = -S(\rho_A \otimes \rho_B) - \mathrm{tr}\Big((\rho_A \otimes \rho_B) \log(\rho'_A \otimes \rho'_B)\Big). \tag{7.84}$$

Next, we use the logarithmic identity for operators of Eq. (7.32) applied to $\log(\rho'_A \otimes \rho'_B)$ to obtain

$$S\big(\rho_A \otimes \rho_B \,\|\, \rho_A' \otimes \rho_B'\big) = -S(\rho_A \otimes \rho_B) - \mathrm{tr}\Big((\rho_A \otimes \rho_B)\big[\log(\rho_A') \otimes \mathrm{I}^B + \mathrm{I}^A \otimes \log(\rho_B')\big]\Big) \tag{7.85}$$

$$= -S(\rho_A \otimes \rho_B) - \mathrm{tr}\Big(\rho_A \log(\rho_A') \otimes \rho_B + \rho_A \otimes \rho_B \log(\rho_B')\Big) \tag{7.86}$$

$$= -S(\rho_A) - S(\rho_B) - \mathrm{tr}\big(\rho_A \log(\rho_A')\big) - \mathrm{tr}\big(\rho_B \log(\rho_B')\big) \tag{7.87}$$

$$= S(\rho_A \,\|\, \rho_A') + S(\rho_B \,\|\, \rho_B') \,, \tag{7.88}$$

finally establishing the additivity of Eq. (7.83).

Also like the quantum entropy in Eq. (7.38), the quantum relative entropy exhibits a property of isometric invariance. For two density operators ρ_1 and ρ_2 of $\mathcal{L}(\mathcal{H}_N)$, and an isometric operator V mapping $\mathcal{H}_N$ into a larger Hilbert space $\mathcal{H}_{N'}$ with dimension $N' \geq N$ and $\mathsf{V}^\dagger\mathsf{V} = \mathrm{I}_N$, we have the **isometric invariance** of the quantum relative entropy as

$$S\big(\mathsf{V}\rho_1\mathsf{V}^\dagger \,\|\, \mathsf{V}\rho_2\mathsf{V}^\dagger\big) = S(\rho_1 \,\|\, \rho_2) \,. \tag{7.89}$$

For a justification, we have from the definition Eq. (7.74) of the quantum relative entropy,

$$S\big(\mathsf{V}\rho_1\mathsf{V}^\dagger \,\|\, \mathsf{V}\rho_2\mathsf{V}^\dagger\big) = -S\big(\mathsf{V}\rho_1\mathsf{V}^\dagger\big) - \mathrm{tr}\big(\mathsf{V}\rho_1\mathsf{V}^\dagger \log(\mathsf{V}\rho_2\mathsf{V}^\dagger)\big). \tag{7.90}$$

From Sect. A.3 of the Appendix we have $\log(\mathsf{V}\rho_2\mathsf{V}^\dagger) = \mathsf{V}\log(\rho_2)\mathsf{V}^\dagger$, so that in Eq. (7.90) we have $\mathrm{tr}\big(\mathsf{V}\rho_1\mathsf{V}^\dagger \log(\mathsf{V}\rho_2\mathsf{V}^\dagger)\big) = \mathrm{tr}\big(\mathsf{V}\rho_1 \log(\rho_2)\mathsf{V}^\dagger\big) = \mathrm{tr}\big(\rho_1 \log(\rho_2)\big)$. In addition, by the isometric invariance of the quantum entropy in Eq. (7.38) we have $S\big(\mathsf{V}\rho_1\mathsf{V}^\dagger\big) = S(\rho_1)$, so that Eq. (7.90) is

$$S\big(\mathsf{V}\rho_1\mathsf{V}^\dagger \,\|\, \mathsf{V}\rho_2\mathsf{V}^\dagger\big) = -S(\rho_1) - \mathrm{tr}\big(\rho_1 \log(\rho_2)\big) = S(\rho_1 \,\|\, \rho_2) \,, \tag{7.91}$$

establishing the isometric invariance of the quantum relative entropy in Eq. (7.89). An interpretation is that an isometric evolution (including a unitary evolution) does not change the distinguishability of two density operators.

The quantum relative entropy provides an alternative characterization [178, 179] of the Holevo information $\chi\big(\{p_j, \rho_j\}\big)$ of Eq. (7.27). By appropriate rewriting and linearity of the trace, it is easily found that

$$\chi\big(\{p_j, \rho_j\}\big) = S(\overline{\rho}) - \sum_{j=1}^{J} p_j\, S(\rho_j) = \sum_{j=1}^{J} p_j\, S\big(\rho_j \,\|\, \overline{\rho}\big) \,, \tag{7.92}$$

with the average state $\overline{\rho}$ from Eq. (7.65). The Holevo information $\chi\big(\{p_j, \rho_j\}\big)$ of Eq. (7.92) and its expression from the quantum relative entropy consolidate its interpretation as a quantum version [180] of the classical Jensen–Shannon information divergence of Eqs. (7.16)–(7.17). By the Klein inequality, the Holevo information $\chi\big(\{p_j, \rho_j\}\big)$ is non-negative, and $\chi\big(\{p_j, \rho_j\}\big) = 0$ if and only if the J states ρ_j all coincide and coincide with the average state $\overline{\rho}$ (which requires uniform probabilities $p_j = 1/J$ for all j).

Along with Eq. (7.73) we obtain for the Holevo information the lower and upper bounds

$$0 \leq \chi\big(\{p_j, \rho_j\}\big) \leq H\big(\{p_j\}\big) \,, \tag{7.93}$$

supplementing $0 \leq \chi\big(\{p_j, \rho_j\}\big) \leq \log(N)$ indicated previously just after Eq. (7.28). It can be shown that the Holevo information $\chi\big(\{p_j, \rho_j\}\big)$ can be either convex (∪) or concave (∩) depending on whether the signal states $\{\rho_j\}$ or their probabilities $\{p_j\}$ are kept fixed [12]. The Holevo information $\chi\big(\{p_j, \rho_j\}\big)$ is a convex (∪) function of the states $\{\rho_j\}$ for fixed probabilities $\{p_j\}$, while it is a concave (∩) function of the probabilities $\{p_j\}$ for fixed signal states $\{\rho_j\}$.

The lower bound $0 \leq \chi\big(\{p_j, \rho_j\}\big)$ and the Klein inequality prove the strict concavity (∩) of the quantum entropy $S(\rho)$, as mentioned in Sect. 7.3.1.

7.5 Quantum Mutual Information

We consider as in Sect. 7.3.3 a bipartite quantum system AB of Hilbert space $\mathcal{H}^A \otimes \mathcal{H}^B$ described by the joint density operator $\rho_{AB} \in \mathcal{L}\big(\mathcal{H}^A \otimes \mathcal{H}^B\big)$, endowed with the joint quantum entropy $S(A, B) = S(\rho_{AB}) = -\operatorname{tr}\big(\rho_{AB} \log(\rho_{AB})\big)$. The reduced state for subsystem A is $\rho_A = \operatorname{tr}_B(\rho_{AB}) \in \mathcal{L}\big(\mathcal{H}^A\big)$ and is endowed with the (marginal) quantum entropy $S(A) = S(\rho_A) = -\operatorname{tr}\big(\rho_A \log(\rho_A)\big)$. The reduced state for subsystem B is $\rho_B = \operatorname{tr}_A(\rho_{AB}) \in \mathcal{L}\big(\mathcal{H}^B\big)$ and is endowed with the (marginal) quantum entropy $S(B) = S(\rho_B) = -\operatorname{tr}\big(\rho_B \log(\rho_B)\big)$. We examine the quantum relative entropy

$$S\big(\rho_{AB} \,\|\, \rho_A \otimes \rho_B\big) = \operatorname{tr}\big(\rho_{AB} \log(\rho_{AB})\big) - \operatorname{tr}\big(\rho_{AB} \log(\rho_A \otimes \rho_B)\big) \,. \tag{7.94}$$

From the logarithmic identity for operators of Eq. (7.32), for use in Eq. (7.94) we have

$$\begin{aligned}
\operatorname{tr}\big(\rho_{AB} \log(\rho_A \otimes \rho_B)\big) &= \operatorname{tr}\Big(\rho_{AB}\big[\log(\rho_A) \otimes \mathrm{I}^B\big]\Big) + \operatorname{tr}\Big(\rho_{AB}\big[\mathrm{I}^A \otimes \log(\rho_B)\big]\Big) && (7.95)\\
&= \operatorname{tr}_A\Big(\operatorname{tr}_B\Big(\rho_{AB}\big[\log(\rho_A) \otimes \mathrm{I}^B\big]\Big)\Big) + \operatorname{tr}_B\Big(\operatorname{tr}_A\Big(\rho_{AB}\big[\mathrm{I}^A \otimes \log(\rho_B)\big]\Big)\Big) && (7.96)\\
&= \operatorname{tr}_A\big(\rho_A \log(\rho_A)\big) + \operatorname{tr}_B\big(\rho_B \log(\rho_B)\big) && (7.97)\\
&= -S(\rho_A) - S(\rho_B) \,. && (7.98)
\end{aligned}$$

Substituting Eq. (7.98) in Eq. (7.94) we finally obtain

$$\boxed{S\big(\rho_{AB} \,\|\, \rho_A \otimes \rho_B\big) = S(\rho_A) + S(\rho_B) - S(\rho_{AB}) = S(A : B) \geq 0 \,,} \tag{7.99}$$

which provides a definition for a **quantum mutual information** $S(A:B) = S(A) + S(B) - S(A, B)$. With the conditional quantum entropies $S(A|B) = S(A, B) - S(B)$ and $S(B|A) = S(A, B) - S(A)$ defined in Eqs. (7.49)–(7.50), we also have for the quantum mutual information,

$$S(A:B) = S(A) - S(A|B) = S(B) - S(B|A) \ . \tag{7.100}$$

By the Klein inequality of Eq. (7.75), the quantum mutual information $S(A:B)$ defined in Eq. (7.99) is non-negative. From $0 \leq S(A:B) = S(A) - S(A|B) = S(B) - S(B|A)$ we deduce $S(A) \geq S(A|B)$ and $S(B) \geq S(B|A)$, indicating that conditioning cannot increase the quantum entropy, much as for the classical entropy in Sect. 7.1.4. The quantum mutual information $S(A:B)$ vanishes as $S(A:B) = 0$ if and only if $\rho_{AB} = \rho_A \otimes \rho_B$, that is, the bipartite system AB is in a separable product state with two independent subparts A and B yielding $S(A, B) = S(A) + S(B)$. At the other extreme, we have the maximum mutual information $S(A:B) = S(A) + S(B) = 2S(A) = 2S(B)$ if and only if AB is in a pure bipartite state yielding $S(A, B) = 0$.

The non-negative quantum mutual information $S(A:B) \geq 0$ shows that in general, for two correlated (entangled) systems A and B, the quantum entropy is **subadditive**, that is,

$$S(A, B) \leq S(A) + S(B) \ , \tag{7.101}$$

with equality if and only if A and B are two independent systems, when $S(A:B) = 0$. The quantum mutual information $S(A:B)$ can be seen as a non-negative index of departure from independence or separability of the two quantum systems A and B, or an index of correlation or entanglement between A and B.

The quantum mutual information provides an interesting alternative expression for the Holevo information $\chi\big(\{p_j, \rho_j\}\big)$ of Eq. (7.27). For J states $\rho_j \in \mathcal{L}(\mathcal{H}^B)$ accessible to system B with the probabilities p_j, we consider the bipartite quantum state

$$\rho_{AB} = \sum_{j=1}^{J} p_j \, |j\rangle \langle j| \otimes \rho_j \ , \tag{7.102}$$

where $\{|j\rangle\}_{j=1}^{J}$ is an orthonormal basis for the J-dimensional space $\mathcal{H}^A$ of A. The bipartite state ρ_{AB} from Eq. (7.102) represents what is called the classical–quantum state associated with the statistical ensemble $\{(p_j, \rho_j)\}_{j=1}^{J}$. We then obtain

$$\chi\big(\{p_j, \rho_j\}\big) = S(A:B) = S(\rho_A) + S(\rho_B) - S(\rho_{AB}) \ . \tag{7.103}$$

This follows since we have $\rho_A = \mathrm{tr}_B(\rho_{AB}) = \sum_j p_j \, |j\rangle \langle j|$, giving the marginal quantum entropy $S(\rho_A) = \mathrm{tr}\big(h(\rho_A)\big) = \sum_j h(p_j) = -\sum_j p_j \log(p_j)$. Also, $\rho_B = \mathrm{tr}_A(\rho_{AB}) = \sum_j p_j \rho_j = \overline{\rho}$ gives the marginal quantum entropy $S(\rho_B) = S\big(\overline{\rho}\big)$. The joint quantum entropy $S(\rho_{AB})$ can be expressed as $S(\rho_{AB}) = \mathrm{tr}\big(h(\rho_{AB})\big) = \mathrm{tr}\big(-\rho_{AB} \log(\rho_{AB})\big)$. By the identity of Eq. (A.32) we have

$$h(\rho_{AB}) = \sum_{j=1}^{J} |j\rangle\langle j| \otimes h\big(p_j\rho_j\big) , \qquad (7.104)$$

so that

$$S\big(\rho_{AB}\big) = \mathrm{tr}\big(h(\rho_{AB})\big) = \sum_{j=1}^{J} \mathrm{tr}\big(h(p_j\rho_j)\big) \qquad (7.105)$$

$$= \sum_{j=1}^{J} \mathrm{tr}\big(-p_j\rho_j \log(p_j\rho_j)\big) . \qquad (7.106)$$

We next have

$$-\, p_j\rho_j \log\big(p_j\rho_j\big) = -p_j \log(p_j)\rho_j - p_j\rho_j \log(\rho_j) , \qquad (7.107)$$

we take the trace and sum over j in Eq. (7.106) to obtain the joint quantum entropy

$$S\big(\rho_{AB}\big) = -\sum_{j=1}^{J} p_j \log(p_j) + \sum_{j=1}^{J} p_j S(\rho_j) . \qquad (7.108)$$

We therefore finally obtain $S(\rho_A) + S(\rho_B) - S(\rho_{AB}) = S(A:B)$ matching $\chi\big(\{p_j, \rho_j\}\big)$ and completing the proof of Eq. (7.103).

In addition, for the bipartite classical–quantum state ρ_{AB} of Eq. (7.102), we can obtain a supplementary expression for the conditional quantum entropy previously defined as $S(B|A) = S(A, B) - S(A)$. From $S(A) = S(\rho_A) = -\sum_j p_j \log(p_j)$ and from Eq. (7.108) for $S(A, B) = S(\rho_{AB})$, we obtain here $S(B|A) = \sum_j p_j S(\rho_j)$. Moreover, this suggests measuring A with a von Neumann measurement in the orthonormal basis $\{|j\rangle\}_{j=1}^{J}$, to obtain the result j with the probability p_j while projecting B in state ρ_j, and further define the conditional quantum entropy $S(B|A = j) = S(\rho_j)$, with average $\sum_j S(B|A = j)p_j = S(B|A)$, as in the classical case of Eq. (7.6). These supplementary expressions accessible for the conditional quantum entropy beyond $S(B|A) = S(A, B) - S(A)$, and referring to an explicit measurement of A, are specific to the bipartite classical–quantum state of Eq. (7.102).

7.6 Strong Subadditivity and Related Properties

For a tripartite quantum system ABC, there exists another important inequality stating the **strong subadditivity of the quantum entropy**, as

$$S(A, B, C) + S(B) \leq S(A, B) + S(B, C) , \qquad (7.109)$$

requiring a rather involved proof as reported in [181, 182].

Strong subadditivity of the quantum entropy is equivalent [183] to another important inequality formed by the **monotonicity of the quantum relative entropy**, under a quantum operation or process $\mathcal{E}(\cdot)$, stipulating that

$$S\big(\mathcal{E}(\rho) \parallel \mathcal{E}(\rho')\big) \leq S(\rho \parallel \rho') , \qquad (7.110)$$

for two states ρ and ρ' accessible to a quantum system with Hilbert space $\mathcal{H}_N$. This monotonicity of the quantum relative entropy is proved for instance in [182–185]. The monotonicity under quantum operation of the quantum relative entropy expresses that no processing can increase the distinguishability of two quantum states, and accordingly Eq. (7.110) is sometimes interpreted as a quantum data processing inequality, although another one more akin to the classical counterpart of Eq. (7.24) will be given in Eq. (13.60) upon cascading noisy quantum channels.

Equation (7.110) includes monotonicity of the quantum relative entropy under partial tracing, as

$$S(\rho_Q \parallel \rho'_Q) \leq S\big(\rho_{QE} \parallel \rho'_{QE}\big) , \qquad (7.111)$$

for two states ρ_{QE} and ρ'_{QE} accessible to a bipartite quantum system QE of Hilbert space $\mathcal{H}^Q \otimes \mathcal{H}^E$, with reduced states $\rho_Q = \mathrm{tr}_E(\rho_{QE})$ and $\rho'_Q = \mathrm{tr}_E(\rho'_{QE})$. This is granted by Eq. (7.110) since, as we have seen in Sect. 3.2.9 and Eq. (3.127), partial tracing of a density operator is an instance of a quantum operation or process $\mathcal{E}(\cdot)$.

For the reciprocal implication, we can introduce the Stinespring isometric extension operator Λ for a dilation of the quantum operation $\mathcal{E}(\cdot)$ on system Q with an environment space $\mathcal{H}^E$, as in Sect. 3.2.6. By the isometric invariance of Eq. (7.89) for the quantum relative entropy, we have $S\big(\Lambda\rho_Q\Lambda^\dagger \parallel \Lambda\rho'_Q\Lambda^\dagger\big) = S(\rho_Q \| \rho'_Q)$. By the monotonicity under partial tracing of Eq. (7.111), we obtain $S(\rho_Q \| \rho'_Q) = S\big(\Lambda\rho_Q\Lambda^\dagger \parallel \Lambda\rho'_Q\Lambda^\dagger\big) \geq S\big(\mathrm{tr}_E(\Lambda\rho_Q\Lambda^\dagger) \parallel \mathrm{tr}_E(\Lambda\rho'_Q\Lambda^\dagger)\big)$. And since by Eq. (3.115) we have $\mathrm{tr}_E\big(\Lambda\rho_Q\Lambda^\dagger\big) = \mathcal{E}(\rho_Q)$, we have deduced from Eq. (7.111) the monotonicity under quantum operation of Eq. (7.110). Monotonicity of the quantum relative entropy under quantum operation in Eq. (7.110) or under partial tracing in Eq. (7.111) are therefore equivalent.

When the quantum operation $\mathcal{E}(\cdot)$ is unital, so as to satisfy $\mathcal{E}(\mathrm{I}_N) = \mathrm{I}_N$, monotonicity of the quantum relative entropy in Eq. (7.110) leads to

$$S\left(\mathcal{E}(\rho) \,\middle\|\, \frac{\mathrm{I}_N}{N}\right) \leq S\left(\rho \,\middle\|\, \frac{\mathrm{I}_N}{N}\right) , \qquad (7.112)$$

and by Eq. (7.82) this is equivalent to

$$S\big(\mathcal{E}(\rho)\big) \geq S(\rho) . \qquad (7.113)$$

Therefore, unital quantum operations never decrease the quantum entropy of quantum states. But this is no longer true in general for non-unital quantum operations, with an example to be given in Sect. 9.10 just after Eq. (9.96).

The monotonicity of the quantum relative entropy is related in [182–184] to the strong subadditivity of the quantum entropy proved in [181, 183]. The strong subadditivity of the quantum entropy in Eq. (7.109) can be derived from the monotonicity of the quantum relative entropy under partial tracing in Eq. (7.111). For a tripartite quantum system ABC in state $\rho_{ABC} \in \mathcal{L}(\mathcal{H}^A \otimes \mathcal{H}^B \otimes \mathcal{H}^C)$, with the standard state reductions such as $\rho_{AB} = \mathrm{tr}_C(\rho_{ABC})$ and $\rho_B = \mathrm{tr}_{AC}(\rho_{ABC})$, the strong subadditivity of Eq. (7.109) is $S(\rho_{ABC}) + S(\rho_B) \leq S(\rho_{AB}) + S(\rho_{BC})$, equivalent to $S(\rho_{ABC}) - S(\rho_{BC}) \leq S(\rho_{AB}) - S(\rho_B)$, and it can be restated in terms of the relative entropy as

$$S\left(\rho_{AB} \,\middle\|\, \frac{\mathrm{I}^A}{N_A} \otimes \rho_B\right) \leq S\left(\rho_{ABC} \,\middle\|\, \frac{\mathrm{I}^A}{N_A} \otimes \rho_{BC}\right) \tag{7.114}$$

through the use of Eq. (7.99), with N_A the dimension of $\mathcal{H}^A$ and I^A/N_A the fully depolarized density operator of $\mathcal{L}(\mathcal{H}^A)$. The inequality in Eq. (7.114) is granted by the monotonicity of the quantum relative entropy under partial tracing over C, which in this way implies the strong subadditivity of Eq. (7.109).

Monotonicity of the quantum relative entropy therefore implies strong subadditivity of the quantum entropy. The reciprocal is also true [181–183], so that both properties are equivalent in this respect, but none of them is reputed easy to establish in the first place.

The **Holevo upper bound** stated in Eqs. (7.26)–(7.27) to limit the classical input–output mutual information $H(X : Y)$ arising when measuring a statistical ensemble $\{(p_j, \rho_j)\}_{j=1}^J$ with an arbitrary POVM $\{\mathsf{E}_k\}_{k=1}^K$, as discussed in Sect. 7.2.3, can also be derived from the monotonicity of the quantum relative entropy [186, 187]. For this purpose, an arbitrary POVM $\{\mathsf{E}_k\}_{k=1}^K$ is represented by the measurement channel

$$\mathcal{M}(\rho) = \sum_{k=1}^{K} \mathrm{tr}(\rho \mathsf{E}_k)\, |k\rangle\, \langle k| \tag{7.115}$$

as in Eq. (4.40) of Sect. 4.2.4. With the conditional measurement probabilities $\Pr\{Y = y_k \mid X = x_j\} = \mathrm{tr}(\rho_j \mathsf{E}_k)$ as in Eq. (7.25), this gives us the quantum entropy

$$S\big(\mathcal{M}(\rho_j)\big) = H(Y \mid X = x_j) = \sum_{k=1}^{K} h\big(\mathrm{tr}(\rho_j \mathsf{E}_k)\big)\,, \tag{7.116}$$

with the auxiliary function $h(u) = -u \log(u)$ of Eq. (7.39). This follows since $S\big(\mathcal{M}(\rho_j)\big) = \mathrm{tr}\, h\big(\mathcal{M}(\rho_j)\big)$, and $\mathcal{M}(\rho)$ in Eq. (7.115) is a density operator in diagonal form for which application of the nonlinear function $h(\cdot)$ operates according to

Sect. A.3 of the Appendix. We deduce the classical input–output conditional entropy $H(Y|X)$ as

$$H(Y|X) = \sum_{j=1}^{J} p_j H(Y \mid X = x_j) = \sum_{j=1}^{J} p_j S\big(\mathcal{M}(\rho_j)\big) . \tag{7.117}$$

The output probabilities are $\Pr\{Y = y_k\} = \sum_{j=1}^{J} p_j \operatorname{tr}(\rho_j \mathsf{E}_k) = \operatorname{tr}\big(\overline{\rho}\, \mathsf{E}_k\big)$, so that the classical output entropy is

$$H(Y) = \sum_{k=1}^{K} h\big(\Pr\{Y = y_k\}\big) = \sum_{k=1}^{K} h\big(\operatorname{tr}\big(\overline{\rho}\, \mathsf{E}_k\big)\big) = S\big(\mathcal{M}(\overline{\rho})\big) . \tag{7.118}$$

The classical input–output mutual information follows as

$$H(X:Y) = H(Y) - H(Y|X) = S\big(\mathcal{M}(\overline{\rho})\big) - \sum_{j=1}^{J} p_j S\big(\mathcal{M}(\rho_j)\big) . \tag{7.119}$$

The map $\mathcal{M}(\cdot)$ is linear, so that, as in Eq. (7.92), we have the alternative expression

$$H(X:Y) = \sum_{j=1}^{J} p_j S\big(\mathcal{M}(\rho_j) \parallel \mathcal{M}(\overline{\rho})\big) . \tag{7.120}$$

By the monotonicity of the quantum relative entropy in Eq. (7.110), for every j we have $S\big(\mathcal{M}(\rho_j) \parallel \mathcal{M}(\overline{\rho})\big) \leq S\big(\rho_j \parallel \overline{\rho}\big)$, so that from Eq. (7.120) we deduce the Holevo inequality

$$H(X:Y) \leq \sum_{j=1}^{J} p_j S\big(\rho_j \parallel \overline{\rho}\big) = \chi\big(\{p_j, \rho_j\}\big) , \tag{7.121}$$

which is the inequality of Eq. (7.26) establishing the Holevo information $\chi\big(\{p_j, \rho_j\}\big)$ of Eq. (7.27) as an upper bound to classical input–output mutual information $H(X:Y)$.

From Eq. (7.121), by the monotonicity of the quantum relative entropy in Eq. (7.110), we deduce

$$\chi\big(\{p_j, \mathcal{E}(\rho_j)\}\big) \leq \chi\big(\{p_j, \rho_j\}\big) , \tag{7.122}$$

for any quantum operation $\mathcal{E}(\cdot)$.

A property of **joint convexity of the quantum relative entropy** can also be derived from its monotonicity [182, 185]. For a quantum system Q with N-dimensional Hilbert space $\mathcal{H}^Q \equiv \mathcal{H}_N$, having access to two statistical ensembles

$\{(p_j, \rho_j)\}_{j=1}^J$ and $\{(p'_j, \rho'_j)\}_{j=1}^J$ of same size J, we have [9] the inequality

$$S\Big(\sum_{j=1}^J p_j\rho_j \,\Big\|\, \sum_{j=1}^J p'_j\rho'_j\Big) \le H(p_j\|p'_j) + \sum_{j=1}^J p_j S(\rho_j\|\rho'_j)\,. \tag{7.123}$$

If for the two ensembles the size J were not the same, rather obvious adaptation could be made by appropriately adding irrelevant states with zero probabilities to equalize the sizes. A proof of Eq. (7.123) might be a bit long but is another interesting exercise of computing with functions of operators, that we now undertake. We introduce an auxiliary quantum system R with a J-dimensional Hilbert space $\mathcal{H}^R \equiv \mathcal{H}_J$, endowed with the orthonormal basis $\{|j\rangle\}_{j=1}^J$. We define as in [12] the two bipartite density operators

$$\rho_{RQ} = \sum_{j=1}^J |j\rangle\langle j| \otimes p_j\rho_j\,, \tag{7.124}$$

$$\rho'_{RQ} = \sum_{j=1}^J |j\rangle\langle j| \otimes p'_j\rho'_j\,, \tag{7.125}$$

whose partial trace gives

$$\mathrm{tr}_R\big(\rho_{RQ}\big) = \sum_{j=1}^J p_j\rho_j\,, \tag{7.126}$$

$$\mathrm{tr}_R\big(\rho'_{RQ}\big) = \sum_{j=1}^J p'_j\rho'_j\,, \tag{7.127}$$

while they exhibit the quantum relative entropy

$$S\big(\rho_{RQ}\,\|\,\rho'_{RQ}\big) = \mathrm{tr}\big(\rho_{RQ}\log(\rho_{RQ})\big) - \mathrm{tr}\big(\rho_{RQ}\log(\rho'_{RQ})\big)\,. \tag{7.128}$$

To transform Eq. (7.128) we have

$$\log\big(\rho'_{RQ}\big) = \sum_{j=1}^J |j\rangle\langle j| \otimes \log\big(p'_j\rho'_j\big)\,, \tag{7.129}$$

since ρ'_{RQ} of Eq. (7.125) is a classical–quantum state benefiting from the identity in Eqs. (A.31)–(A.32) of the Appendix. For use in Eq. (7.128) we therefore have

$$\rho_{RQ}\log(\rho'_{RQ}) = \left(\sum_{j=1}^{J} |j\rangle\langle j| \otimes p_j\rho_j\right)\left(\sum_{j'=1}^{J} |j'\rangle\langle j'| \otimes \log\big(p'_{j'}\rho'_{j'}\big)\right) \tag{7.130}$$

$$= \sum_{j=1}^{J} |j\rangle\langle j| \otimes p_j\rho_j\log\big(p'_j\rho'_j\big)\,, \tag{7.131}$$

since $\langle j|j'\rangle = \delta_{jj'}$. Taking the (linear) trace acting in Eq. (7.128) we obtain

$$\mathrm{tr}\big(\rho_{RQ}\log(\rho'_{RQ})\big) = \sum_{j=1}^{J} \mathrm{tr}\big(p_j\rho_j\log(p'_j\rho'_j)\big)\,. \tag{7.132}$$

To handle each term $\mathrm{tr}\big(p_j\rho_j\log(p'_j\rho'_j)\big)$ for given j, we introduce the eigenvalues $\{\lambda_n^j\}_{n=1}^N$ of ρ'_j and refer the operator representation to the orthonormal eigenbasis of ρ'_j where holds the diagonal form $\rho'_j = \mathrm{diag}\big[\lambda_1^j, \dots, \lambda_N^j\big]$ and $p'_j\rho'_j = \mathrm{diag}\big[p'_j\lambda_1^j, \dots, p'_j\lambda_N^j\big]$, so that $\log\big(p'_j\rho'_j\big) = \log(p'_j)\mathrm{I}_N + \mathrm{diag}\big[\log(\lambda_1^j), \dots, \log(\lambda_N^j)\big] = \log(p'_j)\mathrm{I}_N + \log(\rho'_j)$. We therefore have the operator $p_j\rho_j\log(p'_j\rho'_j) = p_j\log(p'_j)\rho_j + p_j\rho_j\log(\rho'_j)$, and its trace $\mathrm{tr}\big(p_j\rho_j\log(p'_j\rho'_j)\big) = p_j\log(p'_j) + p_j\,\mathrm{tr}\big(\rho_j\log(\rho'_j)\big)$, for every $j = 1$ to J. This leads for Eq. (7.132) to

$$\mathrm{tr}\big(\rho_{RQ}\log(\rho'_{RQ})\big) = \sum_{j=1}^{J} p_j\log(p'_j) + \sum_{j=1}^{J} p_j\,\mathrm{tr}\big(\rho_j\log(\rho'_j)\big)\,, \tag{7.133}$$

and similarly

$$\mathrm{tr}\big(\rho_{RQ}\log(\rho_{RQ})\big) = \sum_{j=1}^{J} p_j\log(p_j) + \sum_{j=1}^{J} p_j\,\mathrm{tr}\big(\rho_j\log(\rho_j)\big)\,. \tag{7.134}$$

We can also observe that Eq. (7.134) is a form of Eq. (7.66) with equality, since the J operators $|j\rangle\langle j| \otimes \rho_j$ (of probability p_j) have mutually orthogonal supports in $\mathcal{H}_J \otimes \mathcal{H}_N$ and Eq. (7.31) holds with $S\big(|j\rangle\langle j|\big) = 0$ for every j. Equations (7.133) and (7.134) are the two pieces forming Eq. (7.128), which now becomes

$$S\big(\rho_{RQ}\,\|\,\rho'_{RQ}\big) = \sum_{j=1}^{J} p_j\Big[\log(p_j) - \log(p'_j)\Big] + \sum_{j=1}^{J} p_j\Big[\mathrm{tr}\big(\rho_j\log(\rho_j)\big) - \mathrm{tr}\big(\rho_j\log(\rho'_j)\big)\Big]\,, \tag{7.135}$$

where we recognize

$$S\big(\rho_{RQ}\,\|\,\rho'_{RQ}\big) = H(p_j\|p'_j) + \sum_{j=1}^{J} p_j S(\rho_j\|\rho'_j)\,. \tag{7.136}$$

By the monotonicity of the quantum relative entropy under partial tracing in Eq. (7.111), we have $S\big(\rho_{RQ} \,\|\, \rho'_{RQ}\big) \geq S\big(\mathrm{tr}_R(\rho_{RQ}) \,\|\, \mathrm{tr}_R(\rho'_{RQ})\big)$, and by exploiting Eqs. (7.126)–(7.127) we obtain the inequality

$$S\left(\sum_{j=1}^{J} p_j \rho_j \,\Big\|\, \sum_{j=1}^{J} p'_j \rho'_j\right) \leq H(p_j \| p'_j) + \sum_{j=1}^{J} p_j S(\rho_j \| \rho'_j) \tag{7.137}$$

announced in Eq. (7.123).

When the two probability distributions $\{p_j\}_{j=1}^J$ and $\{p'_j\}_{j=1}^J$ coincide, their classical relative entropy is $H(p_j \| p'_j) = 0$ and Eq. (7.137) yields

$$S\left(\sum_{j=1}^{J} p_j \rho_j \,\Big\|\, \sum_{j=1}^{J} p_j \rho'_j\right) \leq \sum_{j=1}^{J} p_j S(\rho_j \| \rho'_j) \,, \tag{7.138}$$

expressing the **joint convexity** ($\cup$) of the quantum relative entropy, with a classical counterpart in Eq. (7.19).

7.7 Quantum Discord

7.7.1 *Measurement-Based Conditional Quantum Entropy*

For the bipartite quantum system AB of Hilbert space $\mathcal{H}^A \otimes \mathcal{H}^B$, the conditional quantum entropies $S(A|B)$ and $S(B|A)$ are generally defined in Eqs. (7.49)–(7.50) from a quantum extension of the classical relation in Eq. (7.7). This gives access to a definition for instance for $S(A|B) = S(A, B) - S(B)$, with an interpretation as the quantum entropy of A knowing B, although no effective measurement to let us know B has been explicitly envisaged. This leaves room to an alternative definition of the conditional quantum entropy $S(A|B)$, referring to an explicit measurement on B. For this purpose, a generalized measurement with operators $\{\mathsf{M}_m\}$ and POVM $\{\mathsf{E}_m = \mathsf{M}_m^\dagger \mathsf{M}_m\}$ is performed on B. In the measurement of B from the bipartite system AB with pre-measurement state ρ_{AB}, the operator M_m acts on B with the probability

$$P(m) = \mathrm{tr}\big((\mathrm{I}^A \otimes \mathsf{M}_m)\rho_{AB}(\mathrm{I}^A \otimes \mathsf{M}_m^\dagger)\big) = \mathrm{tr}\big((\mathrm{I}^A \otimes \mathsf{E}_m)\rho_{AB}\big) \,, \tag{7.139}$$

to deliver for the compound AB the joint bipartite post-measurement state

$$\rho_{AB}^m = \frac{1}{P(m)}\big(\mathrm{I}^A \otimes \mathsf{M}_m\big)\rho_{AB}\big(\mathrm{I}^A \otimes \mathsf{M}_m^\dagger\big) \,. \tag{7.140}$$

The subsystem A alone terminates with the reduced density operator

$$\rho_A^m = \mathrm{tr}_B\big(\rho_{AB}^m\big) = \frac{1}{P(m)}\,\mathrm{tr}_B\big((\mathrm{I}^A \otimes \mathsf{E}_m)\rho_{AB}\big)\,, \tag{7.141}$$

exhibiting a quantum von Neumann entropy $S(\rho_A^m)$.

This leads to an alternative definition for a conditional quantum entropy $\overline{S}(A|B)$ for A, knowing the result m upon measuring B and occurring with the probability distribution $P(m)$, as the average quantum entropy

$$\overline{S}(A|B) = \sum_m S\big(\rho_A^m\big)P(m)\,. \tag{7.142}$$

This alternative, measurement-based, conditional quantum entropy $\overline{S}(A|B)$ quantifies the average quantum entropy remaining for A when the classical result m of measuring B with the POVM $\{\mathsf{E}_m\}$ is known, or equivalently when a statistical ensemble of states $\big\{\big(P(m), \rho_A^m\big)\big\}$ is specified for A. This measurement-based conditional quantum entropy $\overline{S}(A|B)$ in Eq. (7.142) is a genuine average quantum von Neumann entropy, which therefore cannot go negative, in contrast with the standard conditional quantum entropy $S(A|B) = S(A, B) - S(B)$ from Eqs. (7.49)–(7.50). The discrepancy or discord between these two conceivable conditional quantum entropies $S(A|B)$ and $\overline{S}(A|B)$ provides a useful ground to contribute to the characterization of various aspects of the correlations that may exist in a bipartite quantum state ρ_{AB}.

The average density operator for A is $\sum_m \rho_A^m P(m) = \sum_m \mathrm{tr}_B\big((\mathrm{I}^A \otimes \mathsf{E}_m)\rho_{AB}\big) = \mathrm{tr}_B\big((\mathrm{I}^A \otimes \sum_m \mathsf{E}_m)\rho_{AB}\big) = \mathrm{tr}_B\big((\mathrm{I}^A \otimes \mathrm{I}^B)\rho_{AB}\big) = \mathrm{tr}_B\big(\rho_{AB}\big) = \rho_A$. It describes the state of A resting in the statistical ensemble $\big\{\big(P(m), \rho_A^m\big)\big\}$ with the measurement result m unknown. This reduced state ρ_A for A is unaffected by the measurement performed on B, as also expressed by the no-signaling theorem of Sect. 6.3. From the concavity ($\cap$) of the quantum von Neumann entropy, we therefore deduce the inequality

$$\overline{S}(A|B) = \sum_m S\big(\rho_A^m\big)P(m) \le S\Big(\sum_m \rho_A^m P(m)\Big) = S\big(\rho_A\big) = S(A)\,. \tag{7.143}$$

Therefore, for any measuring POVM $\{\mathsf{E}_m\}$ on B, the measurement-based conditional quantum entropy $\overline{S}(A|B)$ from Eq. (7.142) is always bounded as

$$0 \le \overline{S}(A|B) \le S(A)\,, \tag{7.144}$$

which sensibly expresses that measuring B generally reduces the uncertainty on A.

When the bipartite quantum state is a separable product state $\rho_{AB} = \rho_A \otimes \rho_B$, it follows from Eq. (7.139) a probability $P(m) = \mathrm{tr}\big(\mathsf{E}_m\rho_B\big)$ and from Eq. (7.141) a post-measurement state $\rho_A^m = \rho_A$ for A whatever the result m upon measuring B. In this circumstance, the measurement-based conditional quantum entropy in Eq. (7.142) reaches its maximum $\overline{S}(A|B) = S(A)$, consistent with the fact that for a separable product state $\rho_{AB} = \rho_A \otimes \rho_B$, however B is measured, brings no information about A. In the same circumstance $\rho_{AB} = \rho_A \otimes \rho_B$, the standard conditional quantum entropy $S(A|B) = S(A)$ reaches the same maximum $S(A)$, as seen in Sect. 7.3.3.

7.7.2 Minimizing the Measurement-Based Conditional Quantum Entropy

For a given bipartite state ρ_{AB}, the measurement-based conditional quantum entropy $\overline{S}(A|B)$ of Eq. (7.142) generally depends on the choice of the measuring POVM $\{\mathsf{E}_m\}$. To make $\overline{S}(A|B)$ intrinsic to the bipartite state ρ_{AB}, it can be envisaged to determine the POVM $\{\mathsf{E}_m\}$ minimizing $\overline{S}(A|B)$, identifying the optimal POVM able to extract the maximum information from measuring B to reduce the uncertainty remaining on A at the minimum $\overline{S}_{\min}(A|B) \geq 0$. This minimization is however difficult to accomplish in general [188], moreover as the dimension grows, or when seeking a closed-form expression for $\overline{S}_{\min}(A|B)$. It is nevertheless known that the optimal POVM has to be extremal, and an extremal POVM contains at most $[\dim(B)]^2$ elements [188, 189]. When B is a (two-dimensional) qubit, an extremal POVM is formed with rank-one operators (each necessarily proportional to a rank-one projector), but not necessarily as a von Neumann measurement limited to two rank-one projectors; for B of higher dimension, an extremal POVM may contain operators with rank larger than one [189, 190].

Instead of minimizing $\overline{S}(A|B)$ over general POVMs $\{\mathsf{E}_m\}$, an interesting approach is to minimize $\overline{S}(A|B)$ over von Neumann measurements formed with a complete set of mutually orthogonal projectors $\big\{|m\rangle\langle m| = \mathsf{E}_m = \mathsf{M}_m\big\}$ spanning the space $\mathcal{H}^B$. In this circumstance, upon measuring B with such a von Neumann measurement, the projector $|m\rangle\langle m|$ acts with in Eq. (7.139) the probability

$$P(m) = \mathrm{tr}\big((\mathrm{I}^A \otimes |m\rangle\langle m|)\rho_{AB}\big) . \tag{7.145}$$

The subsystem A alone terminates with in Eq. (7.141) the reduced density operator

$$\rho_A^m = \frac{1}{P(m)} \mathrm{tr}_B\big((\mathrm{I}^A \otimes |m\rangle\langle m|)\rho_{AB}\big) , \tag{7.146}$$

while the compound AB is placed in the joint bipartite post-measurement state

$$\rho_{AB}^m = \frac{1}{P(m)}\big(\mathrm{I}^A \otimes |m\rangle\langle m|\big)\rho_{AB}\big(\mathrm{I}^A \otimes |m\rangle\langle m|\big) = \rho_A^m \otimes |m\rangle\langle m| , \tag{7.147}$$

since B has been projected onto $|m\rangle$. For the compound AB, the average post-measurement state

$$\overline{\rho}'_{AB} = \sum_m P(m)\rho^m_{AB} = \sum_m P(m)\rho^m_A \otimes |m\rangle\langle m| \tag{7.148}$$

has then the structure of a classical–quantum state, as encountered in Eq. (7.102) and verifying the identity from Eqs. (A.31)–(A.32). By reproducing on $\overline{\rho}'_{AB}$ the same derivation as in Eqs. (7.102)–(7.108), we obtain in place of Eq. (7.108) the quantum entropy

$$S(\overline{\rho}'_{AB}) = -\sum_m P(m)\log\big(P(m)\big) + \sum_m P(m)S(\rho^m_A) , \tag{7.149}$$

which is also

$$S(\overline{\rho}'_{AB}) = S\Big(\mathrm{tr}_A(\overline{\rho}'_{AB})\Big) + \overline{S}(A|B) . \tag{7.150}$$

This leads, for the conditional quantum entropy $\overline{S}(A|B)$ of Eq. (7.142), upon a von Neumann measurement $\big\{|m\rangle\langle m|\big\}$ on B and from the average state $\overline{\rho}'_{AB}$ of Eq. (7.148), to the supplementary expression

$$\overline{S}(A|B) = S(\overline{\rho}'_{AB}) - S\Big(\mathrm{tr}_A(\overline{\rho}'_{AB})\Big) . \tag{7.151}$$

By invoking from [172] the concavity ($\cap$) of the generic conditional quantum entropy function $S_{\mathrm{cond}}(\rho_{AB}) = S(\rho_{AB}) - S(\mathrm{tr}_A(\rho_{AB}))$ with respect to a bipartite state ρ_{AB}, Ref. [191] reports the inequality

$$S(\overline{\rho}'_{AB}) - S\Big(\mathrm{tr}_A(\overline{\rho}'_{AB})\Big) \geq S(\rho_{AB}) - S\big(\mathrm{tr}_A(\rho_{AB})\big) , \tag{7.152}$$

which is therefore equivalent to

$$\overline{S}(A|B) \geq S(A|B) . \tag{7.153}$$

Equations (7.151) and (7.152)–(7.153) hold for any von Neumann measurement $\big\{|m\rangle\langle m|\big\}$ on B, especially for the optimal von Neumann measurement minimizing $\overline{S}(A|B)$ at $\overline{S}_{\min}(A|B)$. The inequality of Eq. (7.153) can receive a sensible interpretation, expressing that the uncertainty remaining on A after any von Neumann measurement on B, is never smaller than the (initial) uncertainty on A with knowledge of a complete and intact quantum system B before B has been collapsed into a classical outcome by a von Neumann measurement. Furthermore, this important inequality of Eq. (7.153) remains true for a broader class of measurements beyond von Neumann ones [191]. Although Eq. (7.151) which served to obtain Eq. (7.153) is established here only for von Neumann measurements on B when $\overline{\rho}'_{AB}$ is the classical–quantum state of Eq. (7.148), Ref. [192], based on the strong subadditivity of Eq. (7.109) for the quantum von Neumann entropy, reports that Eq. (7.153) remains true for a broader class of measurements on B.

A meaningful reference of bipartite state is when ρ_{AB} is the pure state $\rho_{AB} = |\psi_{AB}\rangle\langle\psi_{AB}|$ with the state vector

$$|\psi_{AB}\rangle = \sum_{a,b} \alpha_{ab}\,|a\rangle \otimes |b\rangle \;\in \mathcal{H}^A \otimes \mathcal{H}^B\;, \tag{7.154}$$

with $\{|a\rangle\}$ an orthonormal basis for the Hilbert space $\mathcal{H}^A$ and $\{|b\rangle\}$ one for $\mathcal{H}^B$.

Upon measuring B with an arbitrary POVM $\{\mathsf{E}_m = \mathsf{M}_m^\dagger \mathsf{M}_m\}$, the probability in Eq. (7.139) for obtaining m is

$$P(m) = \langle\psi_{AB}|\,\mathrm{I}^A \otimes \mathsf{E}_m|\psi_{AB}\rangle = \sum_{a,b,b'} \alpha_{ab}\alpha^*_{ab'}\,\langle b'|\mathsf{E}_m|b\rangle\;, \tag{7.155}$$

while the compound AB terminates in Eq. (7.140) in the pure state $\rho^m_{AB} = |\psi^m_{AB}\rangle\langle\psi^m_{AB}|$ with the state vector

$$|\psi^m_{AB}\rangle = \frac{1}{\sqrt{P(m)}}(\mathrm{I}^A \otimes \mathsf{M}_m)\,|\psi_{AB}\rangle = \frac{1}{\sqrt{P(m)}}\sum_{a,b} \alpha_{ab}\,|a\rangle \otimes \mathsf{M}_m\,|b\rangle \;\in \mathcal{H}^A \otimes \mathcal{H}^B\;. \tag{7.156}$$

The subsystem A alone terminates with in Eq. (7.141) the reduced density operator

$$\rho^m_A = \mathrm{tr}_B\big(\rho^m_{AB}\big) = \frac{1}{P(m)}\sum_{a,a',b,b'} \alpha_{ab}\alpha^*_{a'b'}\,\langle b'|\mathsf{E}_m|b\rangle\,|a\rangle\langle a'| \;\in \mathcal{L}\big(\mathcal{H}^A\big)\,, \tag{7.157}$$

which in general remains a mixed state, with a non-vanishing quantum entropy $S(\rho^m_A)$, leading in Eq. (7.142) to a non-vanishing conditional quantum entropy $\overline{S}(A|B)$.

There is however the possibility of measuring B with a von Neumann measurement in the orthonormal basis $\{|m\rangle \equiv |b\rangle\}$ of $\mathcal{H}^B$, to give in Eq. (7.155) the probability

$$P(m) = \sum_a |\alpha_{am}|^2\,, \tag{7.158}$$

and in Eq. (7.156) the pure separable product state

$$|\psi^m_{AB}\rangle = \frac{1}{\sqrt{P(m)}}\sum_a \alpha_{am}\,|a\rangle \otimes |m\rangle\;. \tag{7.159}$$

The subsystem A alone then terminates in Eq. (7.157) with the reduced density operator $\rho^m_A = |\psi^m_A\rangle\langle\psi^m_A|$ representing the pure state

$$|\psi^m_A\rangle = \frac{1}{\sqrt{P(m)}}\sum_a \alpha_{am}\,|a\rangle \tag{7.160}$$

of the space $\mathcal{H}^A$, endowed with a vanishing quantum entropy $S(\rho_A^m) = 0$, for any measurement result m, leading in Eq. (7.142) to a vanishing conditional quantum entropy $\overline{S}(A|B) = 0$. Moreover, since always $0 \le \overline{S}(A|B)$, the von Neumann measurement of B achieving the minimum $\overline{S}(A|B) = 0$ is optimal for minimizing conditional quantum entropy $\overline{S}(A|B)$ at $\overline{S}_{\min}(A|B) = 0$.

When the bipartite system AB is in a pure state $|\psi_{AB}\rangle$, by measuring B with a (any) von Neumann measurement, knowing the measurement result m places A in a well defined pure state, leaving no uncertainty on A as quantified by a vanishing conditional quantum entropy $\overline{S}(A|B) = \overline{S}_{\min}(A|B) = 0$. In addition, as we have seen earlier in Sect. 7.3.5, for a pure bipartite state $|\psi_{AB}\rangle$ the standard conditional quantum entropy $S(A|B)$ is generally negative, so that the inequality $\overline{S}(A|B) \ge S(A|B)$ of Eq. (7.153) holds as it should. In this respect, the non-negative conditional quantum entropy $\overline{S}(A|B)$ exhibits more of a classic behavior, while the possibly negative conditional entropy $S(A|B)$ is more specifically quantum, offering two distinct points of view on the conditional uncertainty in a bipartite state.

7.7.3 Measurement-Based Quantum Mutual Information

The alternative, measurement-based, conditional quantum entropy $\overline{S}(A|B)$ of Eq. (7.142) leads to an alternative quantum mutual information

$$\overline{S}(A:B) = S(A) - \overline{S}(A|B) \ge 0\,, \tag{7.161}$$

instead of $S(A:B) = S(A) - S(A|B)$ from Eq. (7.100). This alternative, measurement-based, mutual information $\overline{S}(A:B)$ is asymmetric in the exchange of A and B, as a result of the asymmetry of the conditional quantum entropy $\overline{S}(A|B)$ involving a measurement on B in $\mathcal{H}^B$, not generally symmetric with a measurement on A in $\mathcal{H}^A$. By means of the quantum relative entropy defined in Eq. (7.74), it is readily established that also

$$\overline{S}(A:B) = \sum_m P(m)\, S\big(\rho_A^m \,\|\, \rho_A\big)\,. \tag{7.162}$$

Equation (7.162) provides, beyond the inequality of Eq. (7.143), another demonstration that $\overline{S}(A:B) \ge 0$. Equation (7.162) also indicates that the measurement-based mutual information $\overline{S}(A:B) = 0$ if and only if $\rho_A^m = \rho_A$ for any result m upon measuring B. When the bipartite quantum state is a separable product state $\rho_{AB} = \rho_A \otimes \rho_B$, we therefore have such a vanishing mutual information $\overline{S}(A:B) = 0$ for any measuring POVM $\{\mathsf{E}_m\}$ on B. In the same circumstance $\rho_{AB} = \rho_A \otimes \rho_B$, the standard quantum mutual information $S(A:B) = 0$ also vanishes, as seen in Sect. 7.5.

By the inequality of Eq. (7.66) for a mixture of quantum states applied to the statistical ensemble $\big\{\big(P(m), \rho_A^m\big)\big\}$, we obtain $S(\rho_A) \le H\big(\{P(m)\}\big) + \overline{S}(A|B)$, to deduce the inequality $H\big(\{P(m)\}\big) \geq S(\rho_A) - \overline{S}(A|B) = \overline{S}(A:B)$. The measurement-based quantum mutual information $\overline{S}(A:B)$ is therefore always bounded as

$$0 \leq \overline{S}(A:B) \leq H\Big(\big\{P(m)\big\}\Big) , \tag{7.163}$$

for any measuring POVM $\{\mathsf{E}_m\}$ on B.

For any von Neumann measurements on B or any generalized measurements on B for which the inequality of Eq. (7.153) holds, follows the inequality on the mutual informations

$$\overline{S}(A:B) \leq S(A:B) . \tag{7.164}$$

When the measurement-based conditional quantum entropy $\overline{S}(A|B)$ is minimized at $\overline{S}_{\min}(A|B)$ over the measurements of B as discussed in Sect. 7.7.2, a maximum measurement-based mutual information $\overline{S}_{\max}(A:B) \geq 0$ results in Eq. (7.161), assessing the correlation between A and B captured by the most efficient measurement of B.

7.7.4 Quantum Discord

The **quantum discord** is defined [191, 193, 194] as the difference

$$D(A:B) = S(A:B) - \overline{S}(A:B) \tag{7.165}$$

$$= \overline{S}(A|B) - S(A|B) = \overline{S}(A|B) + S(B) - S(A,B) . \tag{7.166}$$

Usually, but not necessarily [193], the measurement-based conditional quantum entropy $\overline{S}(A|B)$ is taken at its minimum $\overline{S}_{\min}(A|B)$ over the measurements of B, associated with the measurement-based quantum mutual information $\overline{S}(A:B)$ at the maximum $\overline{S}_{\max}(A:B) = S(A) - \overline{S}_{\min}(A|B)$.

For any von Neumann measurements on B or any generalized measurements on B for which the inequality of Eq. (7.153) holds, by the inequality of Eq. (7.164), follows a non-negative quantum discord $D(A:B) \geq 0$.

The quantum discord $D(A:B)$ is intended to assess the quantum character, as opposed to the classical character, of the correlation in a bipartite state ρ_{AB}. Equation (7.165) can be rewritten as

$$\underbrace{S(A:B)}_{\text{total}} = \underbrace{D(A:B)}_{\text{quantum}} + \underbrace{\overline{S}(A:B)}_{\text{classical}} , \tag{7.167}$$

so as to distribute the total correlation quantified by the mutual information $S(A:B)$, into a purely quantum correlation quantified by the quantum discord $D(A:B)$, and a classical correlation quantified by the mutual information $\overline{S}(A:B)$ issued from a classical result upon measuring B. The quantum discord $D(A:B)$ quantifies the departure from classicality of the bipartite state ρ_{AB} or the mismatch between the quantum mutual information $S(A:B)$ and a classical mutual information $\overline{S}(A:B)$, a mismatch that is more pronounced when the bipartite state ρ_{AB} carries more quantum correlations. The quantum correlations here can be understood as those that cannot be completely characterized by means of local operations separately on A and B supplemented by classical communication between A and B, as for instance the correlations observed with the entangled pure states analyzed in Sect. 6.2.

For the separable product state $\rho_{AB} = \rho_A \otimes \rho_B$, when there is no correlation whatsoever, both $\overline{S}(A:B) = 0$ and $S(A:B) = 0$ as we have just seen in Sect. 7.7.3, leading to a vanishing quantum discord $D(A:B) = 0$.

For the important reference of a pure bipartite state $\rho_{AB} = |\psi_{AB}\rangle\langle\psi_{AB}|$, we have seen at the end of Sect. 7.7.2 that the measurement-based conditional quantum entropy $\overline{S}(A|B)$ is minimized at $\overline{S}_{\min}(A|B) = 0$ by any von Neumann measurement on B. For a pure state $\rho_{AB} = |\psi_{AB}\rangle\langle\psi_{AB}|$, we have the joint quantum entropy $S(A, B) = 0$. In the generic situation where $|\psi_{AB}\rangle$ is entangled between A and B, the reduced states ρ_A and ρ_B are mixed states, with marginal quantum entropies $S(A) = S(B)$. The standard quantum mutual information is $S(A:B) = 2S(A)$, while the measurement-based quantum mutual information is $\overline{S}(A:B) = S(A)$, and the quantum discord is $D(A:B) = S(A)$. Thus, in a pure (entangled) bipartite state ρ_{AB}, of the total correlation $S(A:B) = 2S(A)$, half is of classical character as quantified by $\overline{S}(A:B) = S(A)$, and half is of quantum character as quantified by $D(A:B) = S(A)$. This accords with the situation examined in Sect. 6.2, where only half of the information stored in a bipartite pure entangled state can be recovered by local operations and classical communication.

The bipartite state

$$\rho_{AB} = \sum_b p_b \rho_A^b \otimes |b\rangle\langle b| \; , \tag{7.168}$$

with $\{p_b\}$ a probability distribution, $\{\rho_A^b\}$ a set of density operators for A and $\{|b\rangle\}$ a set of orthonormal states of $\mathcal{H}^B$, is again a classical–quantum state as in Eq. (7.148), so that it exhibits, as in Eq. (7.149), the joint quantum entropy $S(A, B) = H(\{p_b\}) + \sum_b p_b S(\rho_A^b)$. For A the reduced state $\rho_A = \mathrm{tr}_B(\rho_{AB}) = \sum_b p_b \rho_A^b$ has the marginal quantum entropy $S(A) = S(\sum_b p_b \rho_A^b)$, while for B the reduced state $\rho_B = \mathrm{tr}_A(\rho_{AB}) = \sum_b p_b |b\rangle\langle b|$ has the marginal quantum entropy $S(B) = H(\{p_b\})$. Follows the quantum mutual information $S(A:B) = S(A) + S(B) - S(A, B) = S(\sum_b p_b \rho_A^b) - \sum_b p_b S(\rho_A^b) \geq 0$. When B is measured by a von Neumann measurement in the basis $\{|m\rangle \equiv |b\rangle\}$, the probability in Eq. (7.139) is $P(m) = p_m$, the compound AB terminates in Eq. (7.140) in the state $\rho_{AB}^m = \rho_A^m \otimes |m\rangle\langle m|$, while subsystem A alone terminates with the reduced density operator ρ_A^m, leading to the conditional quantum entropy $\overline{S}(A|B) = \sum_m p_m S(\rho_A^m)$. We then obtain the measurement-based quantum mutual information $\overline{S}(A:B) = S(A) - \overline{S}(A|B) = S(\sum_b p_b \rho_A^b) - \sum_m p_m S(\rho_A^m)$, coinciding with $S(A:B)$ above, to yield the vanishing quantum discord $D(A:B) = S(A:B) - \overline{S}(A:B) = 0$. This indicates that in the bipartite state ρ_{AB} of the classical–quantum form of Eq. (7.168), of the total correlation quantified by $S(A:B) \geq 0$, there is only classical correlation quantified by $\overline{S}(A:B) = S(A:B)$, and no quantum correlation since $D(A:B) = 0$.

More generally, bipartite states under the form of the statistical mixture

$$\rho_{AB} = \sum_{j,k} p_{jk} \rho_A^j \otimes \rho_B^k \tag{7.169}$$

are considered for instance in [191, 193, 194]. These states are unentangled in the sense that they can be generated by separable density operators $\rho_A^j \otimes \rho_B^k$ selected locally for A and B according to the joint classical probability distribution p_{jk} as a source of correlation [170, 195]. Depending on the conditions, such states can nevertheless give rise to nonzero quantum discord $D(A:B)$, revealing that statistical mixtures of separable density operators can induce quantum correlation as quantified by the quantum discord, in absence of entanglement [191, 193, 194]. In this respect, quantum discord can point to quantum correlations going beyond entanglement and available as a resource for useful performance not classically accessible [196, 197]. And beyond bipartite states, the quantum discord can be generalized to progressively characterize the correlations among tripartite quantum states or even multipartite quantum states [198, 199].

7.7.5 An Example of Quantum Discord for Bipartite Qubit States

To further illustrate the capability of the quantum discord to contribute to characterizing the correlations that may exist in a bipartite quantum state, we consider for a qubit pair AB the family of bipartite states defined by the density operator

$$\rho_{AB} = p \left|\beta_{00}\right\rangle \left\langle\beta_{00}\right| + (1-p) \left|\beta_{10}\right\rangle \left\langle\beta_{10}\right| , \tag{7.170}$$

interpolating between the two Bell states $|\beta_{00}\rangle = \big(|00\rangle + |11\rangle\big)/\sqrt{2}$ and $|\beta_{10}\rangle = \big(|00\rangle - |11\rangle\big)/\sqrt{2}$ from Eqs. (2.21) and (2.23), with a probability $p \in [0, 1]$ controlling the statistical ensemble $\big\{\big(p, |\beta_{00}\rangle\big), \big(1-p, |\beta_{10}\rangle\big)\big\}$ realizing the state ρ_{AB}.

When $p = 0$ or $p = 1$, the bipartite density operator ρ_{AB} in Eq. (7.170) represents a maximally entangled pure state, respectively the Bell state $|\beta_{00}\rangle$ or $|\beta_{10}\rangle$. These are also the two Bell states $|\phi_\pm\rangle$ considered in Eq. (6.12) in Sect. 6.2, where we have seen that such Bell states contain specific quantum correlations, associated with quantum entanglement here, that cannot be resolved by local measurement and classical communication.

By contrast, when $p = 1/2$, the bipartite density operator ρ_{AB} in Eq. (7.170) reduces to the mixed quantum state $\big(|00\rangle\langle 00| + |11\rangle\langle 11|\big)/2 = \big(|0\rangle\langle 0| \otimes |0\rangle\langle 0| + |1\rangle\langle 1| \otimes |1\rangle\langle 1|\big)/2$, representing the statistical ensemble $\big\{\big(1/2, |00\rangle\big) ; \big(1/2, |11\rangle\big)\big\}$

of two equiprobable separable states. The bipartite density operator ρ_{AB} in this case contains only classical correlations between A and B, amenable to local processing and classical communication, in the sense that the state represented by ρ_{AB} can be produced by first locally preparing qubit A in the state $|0\rangle$ or $|1\rangle$ equiprobably at random, and then classically informing B of the preparation of A to prepare B similarly.

The quantum discord that we are going to work out for ρ_{AB}, as we shall see, offers a quantitative assessment for such quantum or classical character of the correlations in the bipartite state, according to p.

In the pair AB we consider measuring the qubit B with a von Neumann measurement in the orthonormal basis $\big\{ |m\rangle\,, |m^\perp\rangle \big\}$ of the plane $\mathcal{H}_2$, with the parameterization $|m\rangle = \cos\big(\theta_m/2\big)\,|0\rangle + e^{i\varphi_m} \sin\big(\theta_m/2\big)\,|1\rangle$. With the symmetry in the bipartite state ρ_{AB} of Eq. (7.170), the measurement probability from Eq. (7.145) comes out as $P(m) = 1/2$ for anyone of the two possible results of the von Neumann measurement on B, and for any tuning of the measurement vector $|m\rangle$ in $\mathcal{H}_2$. Next, after projecting B onto $|m\rangle$, the reduced density operator for A in Eq. (7.146) follows as

$$\rho_A^m = \begin{bmatrix} |\langle 0|m\rangle|^2 & (2p-1)\,\langle m|0\rangle\,\langle 1|m\rangle \\ (2p-1)\,\langle 0|m\rangle\,\langle m|1\rangle & |\langle 1|m\rangle|^2 \end{bmatrix}, \tag{7.171}$$

endowed with the two eigenvalues

$$\lambda_A^{\pm} = \frac{1}{2}\left(1 \pm \sqrt{1 - 4(1-p)p\sin^2(\theta_m)}\right), \tag{7.172}$$

yielding the quantum entropy

$$S\big(\rho_A^m\big) = H_{\text{bin}}\big(\lambda_A^+\big)\,, \tag{7.173}$$

with $H_{\text{bin}}(\cdot)$ the classical binary entropy from Eq. (7.4) and Fig. 7.1. In the other alternative, with same probability $1/2$, where the von Neumann measurement projects the qubit B onto $|m^\perp\rangle$, the qubit A terminates in a reduced state having the same quantum entropy as in Eq. (7.173). We therefore obtain in Eq. (7.142) the conditional quantum entropy

$$\overline{S}(A|B) = H_{\text{bin}}\big(\lambda_A^+\big)\,. \tag{7.174}$$

The conditional quantum entropy $\overline{S}(A|B)$ in Eq. (7.174) is then minimized at $\overline{S}_{\min}(A|B) = 0$, by the optimal von Neumann measurement of B in the basis $\big\{ |0\rangle\,, |1\rangle \big\}$ at the angle $\theta_m = 0$, giving in Eq. (7.172) the eigenvalue $\lambda_A^+ = 1$ for any $p \in [0, 1]$. By this optimal measurement, from the bipartite state ρ_{AB} of Eq. (7.170), when B is measured in state $|0\rangle$ then it is certain that A is also in state $|0\rangle$, and similarly when B is measured in $|1\rangle$ it is certain that A is also in $|1\rangle$, this leaving no uncertainty on A, whence $\overline{S}(A|B) = \overline{S}_{\min}(A|B) = 0$. Moreover, since always $0 \le \overline{S}(A|B)$, there exists no generalized measurement of the qubit B to further reduce the conditional quantum entropy $\overline{S}(A|B)$ below $\overline{S}_{\min}(A|B) = 0$.

At the other extreme, the conditional quantum entropy $\overline{S}(A|B)$ in Eq. (7.174) is maximized by the von Neumann measurement of B in the basis $\{\,|+\rangle\,, |-\rangle\}$ at the angle $\theta_m = \pi/2$, giving in Eq. (7.172) the eigenvalue $\lambda_A^+ = p$, leaving the qubit A in the statistical ensemble $\{(p, |+\rangle), (1-p, |-\rangle)\}$ of two orthogonal eigenstates in $\mathcal{H}_2$, and in Eq. (7.174) a conditional quantum entropy maximized at $\overline{S}(A|B) = H_{\text{bin}}(p)$. The maximum remaining uncertainty is when $p = 1/2$ and $\lambda_A^{\pm} = 1/2$ for $\overline{S}(A|B) = \log(2)$, which is the overall maximum entropy for a qubit.

To proceed to the quantum discord, we note that the bipartite density operator $\rho_{AB} \in \mathcal{L}(\mathcal{H}_2^{\otimes 2})$ in Eq. (7.170) is in diagonal form in the Bell basis of $\mathcal{H}_2^{\otimes 2}$; it therefore has the joint quantum entropy $S(\rho_{AB}) = S(A, B) = H_{\text{bin}}(p)$. From $\text{tr}_B(\rho_{AB}) = \rho_A$ we obtain $\rho_A = \text{I}_2/2$ as the maximally mixed qubit state for the reduced density operator of qubit A, having therefore the quantum entropy $S(\rho_A) = S(A) = \log(2)$. Similarly, $\text{tr}_A(\rho_{AB}) = \rho_B = \text{I}_2/2$ has the quantum entropy $S(\rho_B) = S(B) = \log(2)$. So we deduce the standard quantum mutual information

$$S(A:B) = S(A) + S(B) - S(A, B) = 2\log(2) - H_{\text{bin}}(p)\,. \tag{7.175}$$

As p varies in $[0, 1]$, the quantum mutual information $S(A:B)$ in Eq. (7.175) remains strictly positive and varies in $[\log(2), 2\log(2)]$, manifesting strong correlation which persists between A and B in the bipartite state ρ_{AB} for any p. The strongest correlation manifested by $S(A:B)$ is at $p = 0$ or $p = 1$, when $S(A:B) = 2\log(2)$, and ρ_{AB} represents a maximally entangled pure state. The weakest (although significantly nonzero) correlation is at $p = 1/2$, when $S(A:B) = \log(2)$, and ρ_{AB} represents a classically correlated mixed state. The quantum or classical character of the correlations manifested globally by $S(A:B)$ of Eq. (7.175) can be assessed by the quantum discord.

Following Eq. (7.161), a measurement-based quantum mutual information at its maximum can be obtained as

$$\overline{S}_{\max}(A:B) = S(A) - \overline{S}_{\min}(A|B) = \log(2)\,, \tag{7.176}$$

valid for any $p \in [0, 1]$ and applying when the qubit B is measured with the optimal von Neumann measurement in the basis $\{\,|0\rangle\,, |1\rangle\}$. And following Eq. (7.165), the quantum discord can finally be obtained as

$$D(A:B) = S(A:B) - \overline{S}_{\max}(A:B) = \log(2) - H_{\text{bin}}(p)\,. \tag{7.177}$$

The quantum discord $D(A:B)$ is non-negative for any $p \in [0, 1]$ as expected. It reaches its minimum $D(A:B) = 0$ at $p = 1/2$, when the correlations between A and B are of purely classical character, with ρ_{AB} a classically correlated mixed state. Meanwhile, the maximum discord $D(A:B) = \log(2)$ is reached at $p = 0$ or $p = 1$, when the correlations between A and B are of strongly quantum character, with

ρ_{AB} a maximally entangled pure state. For p in between, following Eq. (7.167), of the total correlation quantified by $S(A:B) = 2\log(2) - H_{\text{bin}}(p)$, the quantum discord $D(A:B) = \log(2) - H_{\text{bin}}(p)$ quantifies the quantum part, while $\overline{S}_{\max}(A:B) = \log(2)$ quantifies the classical part. Here, only the quantum part $D(A:B)$ of the correlation varies with p, which controls the statistical mixing of the two entangled Bell states in the bipartite state ρ_{AB} of Eq. (7.170).

7.8 Summary on Quantum Entropy

In this section we summarize important quantum entropy-based quantities presented in this chapter, along with their most essential properties, providing a ground for a quantitative assessment of information supported by quantum signals. These results will especially be useful to us in the sequel, most notably in Chap. 8 on quantum information compression, and in Chap. 13 on quantum information communication.

A quantum state represented by the density operator $\rho \in \mathcal{L}(\mathcal{H}_N)$ has a **quantum or von Neumann entropy** defined in Sect. 7.3.1 as

$$S(\rho) = -\operatorname{tr}\big(\rho\log(\rho)\big) = -\sum_{n=1}^{N} \lambda_n \log(\lambda_n) = H\big(\{\lambda_n\}\big) \le \log(N) , \tag{7.178}$$

with λ_n the eigenvalues of ρ.

The quantum entropy $S(\rho)$ of Eq. (7.178) is a strictly concave ($\cap$) function over the convex set of density operators ρ of $\mathcal{L}(\mathcal{H}_N)$, it satisfies the additivity property $S(\rho_A \otimes \rho_B) = S(\rho_A) + S(\rho_B)$ and an isometric invariance $S\big(\mathsf{V}\rho\mathsf{V}^\dagger\big) = S(\rho)$. It provides a lower bound as $S(\rho) \le H(Y)$ to the classical entropy $H(Y)$ from any generalized measurement of ρ with rank-one operators.

For a statistical ensemble of J quantum states $\{(p_j, \rho_j)\}_{j=1}^J$ with average state $\overline{\rho} = \sum_{j=1}^J p_j\rho_j$ we have, from Eq. (7.66) in Sect. 7.3.6, the inequality

$$S\big(\overline{\rho}\big) \le H\big(\{p_j\}\big) + \sum_{j=1}^{J} p_j\, S(\rho_j) . \tag{7.179}$$

The **quantum relative entropy** of two density operators ρ and ρ' of $\mathcal{L}(\mathcal{H}_N)$, from Eq. (7.74) in Sect. 7.4, is

$$S(\rho \,\|\, \rho') = \operatorname{tr}\Big(\rho\big[\log(\rho) - \log(\rho')\big]\Big) , \tag{7.180}$$

which verifies the Klein inequality $S(\rho\|\rho') \ge 0$ to vanish if and only if $\rho = \rho'$.

It satisfies the additivity property $S\big(\rho_A \otimes \rho_B \,\|\, \rho'_A \otimes \rho'_B\big) = S(\rho_A \| \rho'_A) + S(\rho_B \| \rho'_B)$ of Eq. (7.83), and an isometric invariance $S\big(\mathsf{V}\rho_1\mathsf{V}^\dagger \,\|\, \mathsf{V}\rho_2\mathsf{V}^\dagger\big) = S(\rho_1 \| \rho_2)$ in Eq. (7.89).

It is **monotonic**, under any quantum operation or process $\mathcal{E}(\cdot)$ with $S\big(\mathcal{E}(\rho) \,\|\, \mathcal{E}(\rho')\big) \leq S(\rho \| \rho')$ from Eq. (7.110), and under partial tracing with $S(\rho_Q \| \rho'_Q) \leq S\big(\rho_{QE} \,\|\, \rho'_{QE}\big)$ from Eq. (7.111), both being equivalent.

It relates to the quantum entropy through $S\big(\rho \,\|\, \mathrm{I}_N/N\big) = \log(N) - S(\rho)$, so that when $\mathcal{E}(\cdot)$ is unital we have $S\big(\mathcal{E}(\rho)\big) \geq S(\rho)$.

For two statistical ensembles $\{(p_j, \rho_j)\}_{j=1}^J$ and $\{(p'_j, \rho'_j)\}_{j=1}^J$ we have from Eq. (7.123) the inequality

$$S\Big(\sum_{j=1}^J p_j\rho_j \,\Big\|\, \sum_{j=1}^J p'_j\rho'_j\Big) \leq H(p_j \| p'_j) + \sum_{j=1}^J p_j S(\rho_j \| \rho'_j)\,, \tag{7.181}$$

leading for two identical probability distributions $\{p_j\}_{j=1}^J$ and $\{p'_j\}_{j=1}^J$ to the **joint convexity** ($\cup$) of Eq. (7.138) as

$$S\Big(\sum_{j=1}^J p_j\rho_j \,\Big\|\, \sum_{j=1}^J p_j\rho'_j\Big) \leq \sum_{j=1}^J p_j S(\rho_j \| \rho'_j)\,. \tag{7.182}$$

The **Holevo information** of the statistical ensemble $\{(p_j, \rho_j)\}_{j=1}^J$ of J quantum states of $\mathcal{L}(\mathcal{H}_N)$ is defined in Eq. (7.27) or in Eq. (7.92) as

$$\chi\big(\{p_j, \rho_j\}\big) = S(\overline{\rho}) - \sum_{j=1}^J p_j\, S(\rho_j) = \sum_{j=1}^J p_j\, S\big(\rho_j \,\|\, \overline{\rho}\big) \leq \log(N)\,. \tag{7.183}$$

It is bounded as $0 \leq \chi\big(\{p_j, \rho_j\}\big) \leq H\big(\{p_j\}\big)$, and $\chi\big(\{p_j, \rho_j\}\big) = 0$ if and only if the J states ρ_j all coincide and coincide with the average state $\overline{\rho}$.

The Holevo information provides the upper bound $\chi\big(\{p_j, \rho_j\}\big) \geq H(X:Y)$ of Eq. (7.26) or Eq. (7.121) limiting by the Holevo inequality the classical input–output mutual information $H(X:Y)$ that can be stored and retrieved with the quantum ensemble $\{(p_j, \rho_j)\}_{j=1}^J$ or that can be communicated over a quantum channel measuring $\{(p_j, \rho_j)\}_{j=1}^J$ as output.

The Holevo information $\chi\big(\{p_j, \rho_j\}\big)$ is convex ($\cup$) in the states $\{\rho_j\}$ for fixed probabilities $\{p_j\}$, while it is concave ($\cap$) in the probabilities $\{p_j\}$ for fixed signal states $\{\rho_j\}$.

For a bipartite quantum system AB, from Sect. 7.3.3 we have the **conditional quantum entropies** $S(A|B) = S(A,B) - S(B)$ and $S(B|A) = S(A,B) - S(A)$ of Eqs. (7.49)–(7.50). By the Schmidt decomposition of Eq. (7.64) in Sect. 7.3.4, two reduced states $\rho_A = \mathrm{tr}_B\big(|AB\rangle\,\langle AB|\big)$ and $\rho_B = \mathrm{tr}_A\big(|AB\rangle\,\langle AB|\big)$ obtained by reduction from a common bipartite pure state $|AB\rangle$ share the same quantum entropy $S(A) = S(B)$. And as seen in Sect. 7.3.5, a bipartite quantum system AB in a pure state $|AB\rangle$ is entangled if and only if the conditional quantum entropies $S(A|B) = S(B|A)$ are negative.

The **quantum mutual information** of a bipartite quantum system AB is defined in Sect. 7.5 by Eq. (7.99) as

$$S(A:B) = S\big(\rho_{AB} \,\|\, \rho_A \otimes \rho_B\big) = S(A) + S(B) - S(A,B) , \tag{7.184}$$

and is also $S(A:B) = S(A) + S(B) - S(A,B) = S(A) - S(A|B) = S(B) - S(B|A)$. It is non-negative, and $S(A:B) = 0$ if and only if the two subsystems A and B are separable with $\rho_{AB} = \rho_A \otimes \rho_B$, while $S(A:B) = S(A) + S(B)$ if and only if AB is in a pure bipartite state yielding $S(A,B) = 0$.

It follows that the quantum entropy is generally subadditive with $S(A,B) \leq S(A) + S(B)$, and for a tripartite quantum system ABC it is strongly subadditive with $S(A,B,C) + S(B) \leq S(A,B) + S(B,C)$ as stated in Eq. (7.109). Strong subadditivity of the quantum entropy is equivalent to monotonicity of the quantum relative entropy. Also, monotonicity of the quantum relative entropy implies the Holevo inequality, as shown with the measurement channel of Eq. (7.115).

Finally, the **quantum discord** $D(A:B)$ presented in Sect. 7.7 offers a quantity to assess the quantumness of a bipartite quantum state ρ_{AB}, or the quantum or classical character of the correlations that may exist in the bipartite state ρ_{AB}.

Chapter 8
Quantum Information Compression

Information compression is a central topic of information processing, with both fundamental and practical relevance, in the classical and in the quantum realms alike, and specially within our signal-processing perspective. This chapter first briefly reviews the topic of classical information compression to arrive at the central result of classical information theory provided by the Shannon theorem for classical source coding. The chapter then gives a detailed presentation of the quantum counterpart, with quantum information compression, with a comparable result provided by the Schumacher theorem for quantum source coding, standing as a central result of quantum information theory.

8.1 Classical Information Compression

A classical information source is represented as a process that can emit information-carrying signals under the form of random symbols X drawn among the J characters of a source alphabet $\{s_1, s_2, \dots, s_J\} = \mathcal{S}$, with the prior probabilities $\Pr\{X = s_j\} = p_j$ for $j = 1$ to J, so as to form the statistical ensemble $\big\{(p_j, s_j)\big\}_{j=1}^{J}$. Such a classical information source $\big\{(p_j, s_j)\big\}_{j=1}^{J}$ can be assigned the classical or Shannon entropy

$$H(X) = -\sum_{j=1}^{J} p_j \log(p_j) \, . \tag{8.1}$$

We consider such a classical information source emitting a random sequence $\vec{X}^L = X_1 X_2 \cdots X_L$ of L consecutive random symbols, each drawn independently from the source alphabet $\mathcal{S}$ with the probabilities $\{p_j\}_{j=1}^{J}$. A generic realization $\vec{x}^L$ of the random sequence $\vec{X}^L$ can be assigned the probability

F. Chapeau-Blondeau, *Quantum Information and Quantum Signal Processing*,
https://doi.org/10.1007/978-3-032-27499-1_8

$$\Pr\{\vec{X}^L = \vec{x}^L\} = P(\vec{x}^L) = \prod_{j=1}^{J} p_j^{L_j}, \tag{8.2}$$

where each L_j is the number of occurrences of the character s_j in the sequence $\vec{x}^L$, with $\sum_{j=1}^{J} L_j = L$. From Eq. (8.2) we have

$$-\frac{1}{L}\log P(\vec{x}^L) = -\sum_{j=1}^{J}\frac{L_j}{L}\log(p_j). \tag{8.3}$$

At large length L, by the law of large numbers (when it holds in this way the source is said to be ergodic), the relative frequencies L_j/L tend to the corresponding probabilities p_j, for $j = 1$ to J, so that we have for Eq. (8.3) the limit

$$-\frac{1}{L}\log P(\vec{x}^L) \xrightarrow[L\to\infty]{} -\sum_{j=1}^{J} p_j\log(p_j) = H(X). \tag{8.4}$$

This is a convergence in probability, which can also be stated as

$$\lim_{L\to\infty}\Pr\left\{\left|-\frac{1}{L}\log P(\vec{X}^L) - H(X)\right| \le \varepsilon\right\} = 1, \tag{8.5}$$

for any $\varepsilon > 0$. In particular for any $\varepsilon > 0$ as small as we wish, there always exists a sufficiently large length L, above which the probability in Eq. (8.5) is as close to 1 as we wish.

For the realization of the random sequence $\vec{X}^L$, among the complete set $\mathcal{S}^L$ of the J^L feasible sequences $\vec{x}^L$ of length L, we define the **typical subset** $\mathcal{T}_\varepsilon^L$ as the subset of all sequences $\vec{x}^L$ with a probability (of being emitted by the source) $P(\vec{x}^L)$ satisfying

$$\left|-\frac{1}{L}\log P(\vec{x}^L) - H(X)\right| \le \varepsilon. \tag{8.6}$$

A sequence $\vec{x}^L$ belonging to the typical subset $\mathcal{T}_\varepsilon^L$ is called a typical sequence. The typical sequences are the sequences $\vec{x}^L$ for which $-\log P(\vec{x}^L)/L$ is very close to the entropy $H(X)$, with a departure ε as small as we wish provided L is large enough, by Eq. (8.5); by Eq. (8.3) they can be understood as the sequences for which the relative frequencies L_j/L are very close to the probabilities p_j. So a typical sequence can be seen a sequence $\vec{x}^L$ containing nearly Lp_j characters s_j, for $j = 1$ to J.

By the convergence in probability of Eq. (8.5) following from the law of large numbers, for a length L sufficiently large, we have for the probability of the typical subset

$$\Pr\{\mathcal{T}_\varepsilon^L\} > 1 - \varepsilon, \tag{8.7}$$

quantifying the probability that a sequence $\vec{x}^L$ emitted by the source belongs to the typical subset $\mathcal{T}_\varepsilon^L$. When we consider the entropy $H(X) \equiv H_2(X)$ in Eq. (8.1) with the logarithm of base 2 that can be inverted by exponentiation to 2, we also obtain from Eq. (8.6) a characterization of the probability $P(\vec{x}^L)$ of a typical sequence $\vec{x}^L \in \mathcal{T}_\varepsilon^L$ as

$$2^{-(H_2+\varepsilon)L} \leq P(\vec{x}^L) \leq 2^{-(H_2-\varepsilon)L} . \tag{8.8}$$

The number, denoted $|\mathcal{T}_\varepsilon^L|$, of sequences in the typical subset $\mathcal{T}_\varepsilon^L$ can be sharply bounded by

$$(1-\varepsilon)2^{(H_2-\varepsilon)L} \leq |\mathcal{T}_\varepsilon^L| \leq 2^{(H_2+\varepsilon)L} . \tag{8.9}$$

The lower bound in Eq. (8.9) follows from

$$1-\varepsilon < \Pr\{\mathcal{T}_\varepsilon^L\} = \sum_{\vec{x}^L \in \mathcal{T}_\varepsilon^L} P(\vec{x}^L) \leq |\mathcal{T}_\varepsilon^L| 2^{-(H_2-\varepsilon)L} , \tag{8.10}$$

implying $(1-\varepsilon)2^{(H_2-\varepsilon)L} \leq |\mathcal{T}_\varepsilon^L|$. The upper bound in Eq. (8.9) follows from

$$1 = \sum_{\vec{x}^L \in \mathcal{S}^L} P(\vec{x}^L) \geq \sum_{\vec{x}^L \in \mathcal{T}_\varepsilon^L} P(\vec{x}^L) \geq |\mathcal{T}_\varepsilon^L| 2^{-(H_2+\varepsilon)L} , \tag{8.11}$$

implying $|\mathcal{T}_\varepsilon^L| \leq 2^{(H_2+\varepsilon)L}$.

In this way, by Eq. (8.7) the typical subset $\mathcal{T}_\varepsilon^L$ has a probability $\Pr\{\mathcal{T}_\varepsilon^L\}$ of nearly 1, all the sequences $\vec{x}^L$ of the typical subset being nearly equiprobable each with a common probability of $2^{-LH_2(X)}$ by Eq. (8.8), and the number $|\mathcal{T}_\varepsilon^L|$ of sequences in the typical subset is nearly $2^{LH_2(X)}$ by Eq. (8.9), "nearly" here meaning "as much as we wish provided the length L is large enough". This property is known as **asymptotic equipartition**.

These properties of sequences $\vec{x}^L$ emitted by the information source, especially the properties emerging at large length L with typical sequences, allow us to devise a **compression** scheme for lossless compression or recoding of the source. We envisage recoding each sequence $\vec{x}^L$ emitted by the source with a binary code. For the typical sequences of $\mathcal{T}_\varepsilon^L$, their number $|\mathcal{T}_\varepsilon^L|$ being no larger than $2^{(H_2+\varepsilon)L}$ by Eq. (8.9), we can arrange binary words of size no larger than $(H_2+\varepsilon)L$ bits for their coding. We add an extra bit in case $(H_2+\varepsilon)L$ is not an integer, and another extra bit to serve as a binary prefix indicating that a typical sequence is being coded. So binary words with no more than $(H_2+\varepsilon)L+2$ bits can be used for coding the typical sequences of $\mathcal{T}_\varepsilon^L$. For untypical sequences $\vec{x}^L$ outside $\mathcal{T}_\varepsilon^L$, their number being no larger than J^L, they can be coded with binary words of no more than $L\log_2(J)+2$ bits, including a binary prefix. A binary prefix at 0 initiates a codeword for a typical sequence, while a binary prefix at 1 initiates a codeword for an untypical sequence. In this way we have a one-to-one univocally decodable code for lossless recoding of each sequence $\vec{x}^L$ of length L emitted by the information source.

By weighting each of the two coding lengths by the probabilistic weights, respectively $\Pr\{\mathcal{T}_\varepsilon^L\}$ and $1 - \Pr\{\mathcal{T}_\varepsilon^L\}$ for the typical and untypical sequences, we obtain an average number of binary characters for the coding of a source sequence of length L, as

$$\overline{\ell}_L = \big[(H_2 + \varepsilon)L + 2\big]\Pr\{\mathcal{T}_\varepsilon^L\} + \big[L\log_2(J) + 2\big]\Big[1 - \Pr\{\mathcal{T}_\varepsilon^L\}\Big]. \tag{8.12}$$

With $\overline{\ell} = \overline{\ell}_L/L$ we obtain an average number of binary characters for the coding of a single source symbol, as

$$\overline{\ell} = \left(H_2 + \varepsilon + \frac{2}{L}\right)\Pr\{\mathcal{T}_\varepsilon^L\} + \left[\log_2(J) + \frac{2}{L}\right]\Big[1 - \Pr\{\mathcal{T}_\varepsilon^L\}\Big]. \tag{8.13}$$

We can take $\varepsilon \to 0$ as $L \to \infty$, when $\Pr\{\mathcal{T}_\varepsilon^L\} \to 1$ in Eq. (8.7), and in this limit Eq. (8.13) gives

$$\overline{\ell} \longrightarrow H_2(X). \tag{8.14}$$

This indicates that by operating on longer and longer source sequences as $L \to \infty$, we have access to an average coding length $\overline{\ell}$ tending to the entropy $H(X)$. Instead of a binary alphabet with two binary characters 0 and 1 for recoding, we can envisage a J-ary coding alphabet with J characters matching the cardinal of the source alphabet $\mathcal{S}$. The previous coding argument would apply equally with J-ary characters instead of binary characters, and it would lead in Eq. (8.14) to an average coding length of $\overline{\ell} \to H_J(X)$ of J-ary characters per source symbol, with the entropy $H_J(X)$ in logarithm of base J for Eq. (8.1). For a source drawing from an alphabet with J characters, the entropy in base J is $H_J(X) \le \log_J(J) = 1$. So our lossless recoding scheme leads to $\overline{\ell} \to H_J(X) \le 1$, that is, a coding rate or average number $\overline{\ell}$ of J-ary characters of less than 1 for each symbol emitted by the source drawing from a J-ary alphabet $\mathcal{S}$. Our procedure is indeed in general a compression scheme, unless the source is at maximum entropy $H_J(X) = \log_J(J) = 1$ when it cannot be further compressed.

So a classical information source can be compressed with no loss at a compression rate given by its classical entropy. This is the **Shannon theorem for classical source coding** or classical information compression [171, 173]. The theorem can be supplemented to include that no lossless compression is possible at a rate smaller than the entropy. This is suggested by the count we made of the source sequences exhibiting asymptotic equipartition, and can be proved [175] for instance by showing that (instead of $\mathcal{T}_\varepsilon^L$) any subset $\mathcal{S}_0$ of $\mathcal{S}^L$ with cardinal strictly below $2^{LH_2(X)} = J^{LH_J(X)}$ has a probability $\Pr\{\mathcal{S}_0\}$ that necessarily goes to zero as $L \to \infty$, so that it is worthless to seek to concentrate the compression capability on such subset. Another possibility for the proof [173] is to invoke the Kraft inequality, which fixes a minimum to the length of any uniquely decodable recoding sequences, to prescribe an average coding length $\overline{\ell}$ no smaller than the source entropy. The source coding theorem can also be associated with practical compression techniques to efficiently approach the optimal performance set by the source entropy, such as Huffman coding, or arithmetic coding, or Lempel–Ziv coding [173, 175]. The source coding theorem further consolidates the status of the entropy as a legitimate quantitative assessment for the information content of an information source, and it provides helpful guidance for practice. The source coding theorem also stands as a non-trivial result with strong informational significance, concretizing the expectation we expressed at the end of the first paragraph of Sect. 7.1.1, for the profit from the mathematical working of a (physical) information theory.

The Shannon source coding theorem can be extended for compression of information sources exhibiting statistical dependence between consecutive symbols $X_1, X_2, \dots, X_L$. In this circumstance, in place of Eq. (8.4), the convergence of $-\log P(\vec{x}^{\,L})/L$ at large length L is to the average entropy or entropy rate $\overline{H}(X) = \lim_{L\to\infty} H(X_1, X_2, \dots, X_L)/L$. Then, in a similar way, the average entropy $\overline{H}(X)$ fixes the minimal (best) rate for lossless compression of the information source with dependence. For the entropy, to handle the complication of the asymptotic limit $L \to \infty$, some short-term structuration in the statistical dependence is commonly assumed, through a Markovian assumption, very often quite realistic physically, where the dependence occurs via conditioning by the most recent past symbols. The average entropy $\overline{H}(X)$ then follows from the transition probabilities of the underlying Markov chain.

We can note that in the asymptotic regime $L \to \infty$ of the optimal lossless compression, each source sequence $\overrightarrow{X}^{L}$ comprising L source symbols, is univocally encoded into a coding sequence comprising $LH(X)$ (or $L\overline{H}(X)$ with dependence) coding symbols, or more precisely $LH_D(X)$ coding symbols from a D-ary coding alphabet (we took $D = 2$ in the binary coding above). The source sequences of length L have an entropy of $LH(X)$, they are in one-to-one correspondence with the coding sequences, so that the coding sequences have the same entropy $LH(X)$. This is an entropy of $LH_D(X)$ for the D-ary coding sequences comprising $LH_D(X)$ coding symbols, which gives an entropy per coding symbol of 1, that is, an entropy of one D-ary unit of information per D-ary coding symbol. This is the maximum entropy $\log_D(D) = 1$ in D-ary unit for the D-ary source realized by the coder. This means that the stream of coding symbols output by the coder is at maximum entropy on its alphabet. This could be anticipated from (or alternatively this can serve as a consolidation of) the consistency of the informational reasoning, because if the coding stream were not at maximum entropy, it could be further compressed, providing a more compact compression for the original source, but this is not possible since the compression is optimal in the first place. This is consistent, and maybe a little paradoxical, in that the stream of coding symbols performing the optimal lossless compression of the information from the source, despite its advanced elaboration and sophistication, displays the same statistics, and same maximum entropy, as a stream of random symbols independently and equiprobably drawn from the coding alphabet.

We now use these elements of classical information compression as a guideline to address the compression of a quantum information source.

8.2 Quantum Information Compression

8.2.1 *Schumacher Theorem for Quantum Source Coding*

A quantum information source is represented as a process that can emit information-carrying signals under the form of pure (non-necessarily orthogonal) quantum states or symbols $|\phi_j\rangle$ from an N-dimensional Hilbert space $\mathcal{H}_N$, with prior probabilities p_j, for $j = 1$ to J, so as to form the statistical ensemble of quantum states $\big\{(p_j, |\phi_j\rangle)\big\}_{j=1}^{J}$. Such a quantum information source is represented by the density operator

$$\rho = \sum_{j=1}^{J} p_j \, |\phi_j\rangle \, \langle\phi_j| \, , \tag{8.15}$$

having the quantum or von Neumann entropy $S(\rho) = -\operatorname{tr}(\rho \log \rho)$. The density operator $\rho \in \mathcal{L}(\mathcal{H}_N)$, with N eigenstates $|\lambda_n\rangle$ and N eigenvalues λ_n, receives the diagonal representation

$$\rho = \sum_{n=1}^{N} \lambda_n \, |\lambda_n\rangle \, \langle\lambda_n| \, , \tag{8.16}$$

associated with the equivalent statistical ensemble $\big\{(\lambda_n, |\lambda_n\rangle)\big\}_{n=1}^{N}$ for ρ. The probability distribution formed by the N eigenvalues λ_n has a classical or Shannon entropy

$$H\big(\{\lambda_n\}\big) = -\sum_{n=1}^{N} \lambda_n \log(\lambda_n) = S(\rho) \, , \tag{8.17}$$

coinciding with the quantum or von Neumann entropy $S(\rho)$. Meanwhile, the N eigenstates provide an orthonormal (eigen)basis $\{|\lambda_n\rangle\}_{n=1}^{N}$ for the space $\mathcal{H}_N$.

We consider the quantum information source emitting a sequence $|\psi^L\rangle$ of L consecutive quantum symbols, each of these L symbols being drawn independently from the source alphabet $\{|\phi_j\rangle\}_{j=1}^{J}$ with the probabilities $\{p_j\}_{j=1}^{J}$. When compression of the quantum source is envisaged, significant differences exist with the compression of a classical source in Sect. 8.1. We could say that, in some sense, the classical coder in Sect. 8.1 has complete knowledge of the sequences of symbols being compressed, while the quantum coder has more limited knowledge. To begin with, the quantum symbols $|\phi_j\rangle$ emitted by the quantum source may not be all perfectly distinguishable by quantum measurement, while by contrast the classical symbols s_j for the classical source in Sect. 8.1 are perfectly distinguished. A plain quantum measurement would generally incur some destructive effect without enabling to (exactly) identify the symbols $|\phi_j\rangle$ in a sequence. So the quantum coder cannot get to know by quantum measurement the sequences to be compressed. Also, once a lossless compression is completed, the classical coder in Sect. 8.1 can keep with it a complete knowledge or record of the sequence that has been compressed; meanwhile, the quantum coder cannot, by an effect of the no-cloning theorem. In short, the classical coder compresses perfectly known sequences drawn from the known source ensemble $\{(p_j, s_j)\}_{j=1}^{J}$, while the quantum coder has to compress unknown sequences drawn from the known source ensemble $\{(p_j, |\phi_j\rangle)\}_{j=1}^{J}$. In the face of such limited knowledge, the quantum coder will not aim to recode more efficiently all the J^L length-L source sequences from $\mathcal{H}_N^{\otimes L}$, that in fact cannot be known or distinguished when they show up. Instead, the quantum coder will compress by recoding a basis of representation focusing on some subspace of $\mathcal{H}_N^{\otimes L}$ where a (typical) length-L source sequence lies with high probability, as we are going to see.

Each source symbol $|\phi_j\rangle \in \mathcal{H}_N$ can be expressed as a superposition of states from the eigenbasis $\{|\lambda_n\rangle\}_{n=1}^{N}$ of $\mathcal{H}_N$. The two statistical ensembles $\{(p_j, |\phi_j\rangle)\}_{j=1}^{J}$ and $\{(\lambda_n, |\lambda_n\rangle)\}_{n=1}^{N}$ are equivalent ensembles represented by the same density operator ρ. In this respect, a sequence $|\psi^L\rangle$ of L source symbols $|\phi_j\rangle$ drawn among J with the probabilities p_j, and a sequence of L eigenstates $|\lambda_n\rangle$ drawn among N with the probabilities λ_n, are represented by the common density operator $\rho^{\otimes L} \in \mathcal{L}(\mathcal{H}_N^{\otimes L})$, and they exhibit the same statistics upon any quantum measurement. Anyone of the J^L sequences $|\psi^L\rangle$ of L symbols $|\phi_j\rangle$ can be decomposed in the eigenbasis $\{|\lambda_n\rangle\}^{\otimes L}$ of $\rho^{\otimes L}$ forming an orthonormal basis of the product space $\mathcal{H}_N^{\otimes L}$ with dimension N^L. An eigenstate of $\rho^{\otimes L}$, generically denoted $|\lambda^{(L)}\rangle$, is formed as a tensor product of L eigenstates each taken from $\{|\lambda_n\rangle\}_{n=1}^{N}$ among the N^L possible products; the associated eigenvalue $\lambda^{(L)}$ is formed as the corresponding product of L eigenvalues each taken from $\{\lambda_n\}_{n=1}^{N}$. We have the two equivalent expressions,

$$\rho^{\otimes L} = \sum_{|\psi^L\rangle \in \mathcal{H}_N^{\otimes L}} P(|\psi^L\rangle)\, |\psi^L\rangle\langle\psi^L| = \sum_{\lambda^{(L)} \in \{\lambda_n\}^L} \lambda^{(L)}\, |\lambda^{(L)}\rangle\langle\lambda^{(L)}| \,, \tag{8.18}$$

with $P(|\psi^L\rangle)$ the probability that the sequence $|\psi^L\rangle$ is emitted by the quantum source. In Eq. (8.18), the first sum runs over a set of J^L sequences $|\psi^L\rangle$ of $\mathcal{H}_N^{\otimes L}$, while the second sum generically runs over the set of the N^L orthonormal basis sequences $|\lambda^{(L)}\rangle$ of $\mathcal{H}_N^{\otimes L}$.

We define the **typical subspace** $\mathcal{H}_{\text{typ}}$ of $\mathcal{H}_N^{\otimes L}$ as the subspace spanned by the sequences $|\lambda^{(L)}\rangle$ taken from the eigenbasis $\{|\lambda_n\rangle\}^{\otimes L}$ of $\rho^{\otimes L}$, associated with eigenvalues $\lambda^{(L)}$ satisfying

$$\left| -\frac{1}{L}\log\big(\lambda^{(L)}\big) - S(\rho) \right| \le \varepsilon \,. \tag{8.19}$$

Such sequences $|\lambda^{(L)}\rangle$ from $\{|\lambda_n\rangle\}^{\otimes L}$ are all mutually orthogonal. The quantum states or sequences belonging to the subspace $\mathcal{H}_{\text{typ}}$ are called typical sequences. The (mutually orthogonal) typical sequences $|\lambda^{(L)}\rangle$ satisfying Eq. (8.19) form an orthonormal basis for the typical subspace $\mathcal{H}_{\text{typ}}$. Or in other words, the typical subspace $\mathcal{H}_{\text{typ}}$ is the subspace of $\mathcal{H}_N^{\otimes L}$ spanned by the typical basis sequences $|\lambda^{(L)}\rangle$ satisfying

Eq. (8.19). Among the orthonormal basis $\{|\lambda_n\rangle\}^{\otimes L}$ for $\mathcal{H}_N^{\otimes L}$, the typicality condition of Eq. (8.19) selects a subset of (typical) basis states to define $\mathcal{H}_{\text{typ}}$.

The quantum typical subspace $\mathcal{H}_{\text{typ}}$ associated with the probability distribution $\{\lambda_n\}_{n=1}^N$, plays a role similar to that of the classical typical subset $\mathcal{T}_\varepsilon^L$ associated with the probability distribution $\{p_j\}_{j=1}^J$ and previously defined in Sect. 8.1 via Eq. (8.6). Similar properties following from the law of large numbers at large length L can be obtained (and proved) as in Sect. 8.1. In particular, the typical basis sequences defined via Eq. (8.19) can be understood as the sequences of $\mathcal{H}_N^{\otimes L}$ tensoring L eigenstates from $\{|\lambda_n\rangle\}_{n=1}^N$ and comprising (nearly) a number $L\lambda_n$ of state $|\lambda_n\rangle$, for $n = 1$ to N. The number of such typical sequences, denoted $|\mathcal{H}_{\text{typ}}|$, at large L verifies

$$\log\big(|\mathcal{H}_{\text{typ}}|\big) \approx LH\big(\{\lambda_n\}\big) = LS(\rho) \,, \tag{8.20}$$

for reasons similar to those following from Eq. (8.9). And $|\mathcal{H}_{\text{typ}}|$ is also the dimension of the typical subspace $\mathcal{H}_{\text{typ}} \subseteq \mathcal{H}_N^{\otimes L}$ so that $|\mathcal{H}_{\text{typ}}| \leq N^L$. We introduce the subspace $\mathcal{H}_{\text{typ}}^\perp$, the orthogonal complement to $\mathcal{H}_{\text{typ}}$ in $\mathcal{H}_N^{\otimes L}$. We obtain the direct sum $\mathcal{H}_N^{\otimes L} = \mathcal{H}_{\text{typ}} \oplus \mathcal{H}_{\text{typ}}^\perp$, indicating that any state or sequence $|\psi^L\rangle$ of $\mathcal{H}_N^{\otimes L}$ is uniquely decomposable as the sum of a vector from $\mathcal{H}_{\text{typ}}$ and a vector from $\mathcal{H}_{\text{typ}}^\perp$.

For the **compression** of a sequence $|\psi^L\rangle \in \mathcal{H}_N^{\otimes L}$ emitted by the quantum information source, a quantum measurement of $|\psi^L\rangle$ can be performed by means of the complete set of two projectors $\big\{\Pi_{\text{typ}}, \Pi_{\text{typ}}^\perp\big\}$ acting in $\mathcal{H}_N^{\otimes L}$, where Π_{typ} is the projector onto the typical subspace $\mathcal{H}_{\text{typ}}$, and $\Pi_{\text{typ}}^\perp$ the projector onto the orthogonal subspace $\mathcal{H}_{\text{typ}}^\perp$. The (conditional) probability of projecting a given $|\psi^L\rangle$ onto $\mathcal{H}_{\text{typ}}$ is

$$P\big(\mathcal{H}_{\text{typ}}\big|\,|\psi^L\rangle\big) = \langle\psi^L|\,\Pi_{\text{typ}}\,|\psi^L\rangle = \text{tr}\Big(\Pi_{\text{typ}}\,|\psi^L\rangle\,\langle\psi^L|\Big) \,. \tag{8.21}$$

The (conditional) probability of projecting $|\psi^L\rangle$ onto $\mathcal{H}_{\text{typ}}^\perp$ is the complementary probability

$$P\big(\mathcal{H}_{\text{typ}}^\perp\big|\,|\psi^L\rangle\big) = 1 - P\big(\mathcal{H}_{\text{typ}}\big|\,|\psi^L\rangle\big) = \langle\psi^L|\,\Pi_{\text{typ}}^\perp\,|\psi^L\rangle = \text{tr}\Big(\Pi_{\text{typ}}^\perp\,|\psi^L\rangle\,\langle\psi^L|\Big) \,. \tag{8.22}$$

As we indicated, upon any quantum measurement, $|\psi^L\rangle$ exhibits the same statistics as L independent draws from the statistical ensemble $\{(\lambda_n, |\lambda_n\rangle)\}_{n=1}^N$. At large length L, by the law of large numbers, L independent draws from $\{(\lambda_n, |\lambda_n\rangle)\}_{n=1}^N$ produce a sequence of the typical subspace $\mathcal{H}_{\text{typ}}$ with a probability tending to 1, for reasons similar to those following from Eq. (8.7). This implies the same probability tending to 1, that the sequence $|\psi^L\rangle$ is projected onto the typical subspace $\mathcal{H}_{\text{typ}}$. Accordingly, at large length L, a sequence $|\psi^L\rangle$ emitted by the source displays in Eq. (8.21) a high probability $P\big(\mathcal{H}_{\text{typ}}\big|\,|\psi^L\rangle\big)$ of being projected onto $\mathcal{H}_{\text{typ}}$, with $P\big(\mathcal{H}_{\text{typ}}\big|\,|\psi^L\rangle\big)$ tending to 1 as $L \to \infty$, and this equivalently represents a probability tending to 1 for the source sequence $|\psi^L\rangle$ to belong to the typical subspace $\mathcal{H}_{\text{typ}}$ in the first place. Long sequences $|\psi^L\rangle$ emitted by the quantum source tend to belong to the typical subspace $\mathcal{H}_{\text{typ}}$ with high probability.

By averaging Eq. (8.21) over the prior probability $P(|\psi^L\rangle)$ that the sequence $|\psi^L\rangle$ is emitted by the source, and using Eq. (8.18), we obtain the overall probability $P(\mathcal{H}_{\rm typ})$ that the measurement by $\{\Pi_{\rm typ}, \Pi_{\rm typ}^{\perp}\}$ projects onto $\mathcal{H}_{\rm typ}$, as

$$P(\mathcal{H}_{\rm typ}) = \sum_{|\psi^L\rangle \in \mathcal{H}_N^{\otimes L}} P(\mathcal{H}_{\rm typ}\big|\,|\psi^L\rangle)\, P(|\psi^L\rangle) = {\rm tr}\Big(\Pi_{\rm typ}\rho^{\otimes L}\Big)\,, \tag{8.23}$$

which in the same way tends to 1 at large length L.

So upon measuring a given sequence $|\psi^L\rangle$ with $\{\Pi_{\rm typ}, \Pi_{\rm typ}^{\perp}\}$, two eventualities ensue. In the first eventuality, $|\psi^L\rangle$ gets projected onto $\mathcal{H}_{\rm typ}$; this occurs with the conditional probability $P(\mathcal{H}_{\rm typ}\big|\,|\psi^L\rangle)$ of Eq. (8.21), and the post-measurement state is

$$|\psi^L_{\rm post}\rangle = \frac{\Pi_{\rm typ}\,|\psi^L\rangle}{\sqrt{P(\mathcal{H}_{\rm typ}\big|\,|\psi^L\rangle)}} \in \mathcal{H}_{\rm typ}\ . \tag{8.24}$$

In particular, if $|\psi^L\rangle$ was originally in $\mathcal{H}_{\rm typ}$, then $\Pi_{\rm typ}\,|\psi^L\rangle = |\psi^L\rangle$ and $P(\mathcal{H}_{\rm typ}\big|\,|\psi^L\rangle) = 1$, so that $|\psi^L_{\rm post}\rangle = |\psi^L\rangle$ and such a state $|\psi^L\rangle \in \mathcal{H}_{\rm typ}$ is invariant in the measurement by $\{\Pi_{\rm typ}, \Pi_{\rm typ}^{\perp}\}$. And a state $|\psi^L\rangle$ in $\mathcal{H}_{\rm typ}$ is an eventuality with a probability that tends to 1 as $L \to \infty$, as we indicated above. In general, the post-measurement state $|\psi^L_{\rm post}\rangle$ obtained in Eq. (8.24) belongs to the subspace $\mathcal{H}_{\rm typ}$ having a dimension $|\mathcal{H}_{\rm typ}| \le N^L$. So $|\psi^L_{\rm post}\rangle \in \mathcal{H}_{\rm typ}$ can be unitarily mapped into a state $|\psi_{\rm cod}\rangle$ belonging to a coding space $\mathcal{H}_{\rm cod}$ with dimension $|\mathcal{H}_{\rm typ}|$, supported by a coding quantum system with this dimension $|\mathcal{H}_{\rm typ}|$. This mapping can be accomplished by a coding unitary operator $\mathsf{U}_{\rm cod}$ transforming each of the $|\mathcal{H}_{\rm typ}|$ basis states $\{|\lambda_n\rangle\}^{\otimes L}$ from $\mathcal{H}_N^{\otimes L}$ that happen to be typical so as to form a basis for the typical subspace $\mathcal{H}_{\rm typ}$, into the $|\mathcal{H}_{\rm typ}|$ standard basis states of $\mathcal{H}_{\rm cod}$. To match the dimensionality, we may add that $\mathsf{U}_{\rm cod}$ will unitarily map the entire space $\mathcal{H}_N^{\otimes L}$ into a space of the form $\mathcal{H}_{\rm cod} \otimes \mathcal{H}_{\rm ext}$, with the extra space $\mathcal{H}_{\rm ext}$ chosen with the adequate dimension, and supported by another extra quantum system with this dimension. As a unitary operator, $\mathsf{U}_{\rm cod}$ maps the space $\mathcal{H}_N^{\otimes L}$ into a space (or possibly a subspace of) $\mathcal{H}_{\rm cod} \otimes \mathcal{H}_{\rm ext}$ with same dimension. The unitary $\mathsf{U}_{\rm cod}$ is designed so that the $|\mathcal{H}_{\rm typ}|$ basis states of the typical subspace $\mathcal{H}_{\rm typ}$ are mapped into $|\mathcal{H}_{\rm typ}|$ orthonormal states of $\mathcal{H}_{\rm cod} \otimes |0_{\rm ext}\rangle$, where $|0_{\rm ext}\rangle$ is an unimportant fixed state of $\mathcal{H}_{\rm ext}$ acting as a flag to mark the image by $\mathsf{U}_{\rm cod}$ of a (typical) basis state of $\mathcal{H}_{\rm typ}$. The remaining (untypical) basis states of $\mathcal{H}_N^{\otimes L}$ are unitarily mapped by $\mathsf{U}_{\rm cod}$ into the same number of orthonormal states belonging to the orthogonal complement of $\mathcal{H}_{\rm cod} \otimes |0_{\rm ext}\rangle$. In this way, the coding transformation $\mathsf{U}_{\rm cod}$ maps any state $|\psi^L_{\rm post}\rangle \in \mathcal{H}_{\rm typ} \subseteq \mathcal{H}_N^{\otimes L}$ into a state of the form

$$\mathsf{U}_{\rm cod}\,|\psi^L_{\rm post}\rangle = |\psi_{\rm cod}\rangle \otimes |0_{\rm ext}\rangle\ , \tag{8.25}$$

where $|\psi_{\rm cod}\rangle$ is a pure state belonging to the $|\mathcal{H}_{\rm typ}|$-dimensional space $\mathcal{H}_{\rm cod}$. Then $|\psi_{\rm cod}\rangle$ alone is retained as the compressed state, while $|0_{\rm ext}\rangle$ is discarded; that is, only the coding quantum system with dimension $|\mathcal{H}_{\text{typ}}|$ is retained, while the extra quantum system is discarded.

In the second eventuality upon measuring $|\psi^L\rangle$ with $\{\Pi_{\text{typ}}, \Pi_{\text{typ}}^{\perp}\}$, the state $|\psi^L\rangle$ gets projected onto the complementary subspace $\mathcal{H}_{\text{typ}}^{\perp}$. This occurs with the complementary probability $P(\mathcal{H}_{\text{typ}}^{\perp}|\,|\psi^L\rangle)$ of Eq. (8.22). In this eventuality, on the projected state lying in $\mathcal{H}_{\text{typ}}^{\perp}$, applying U_{cod} delivers an unimportant pure state $|\Psi\rangle$ of $\mathcal{H}_{\text{cod}} \otimes \mathcal{H}_{\text{ext}}$, usually entangled between $\mathcal{H}_{\text{cod}}$ and $\mathcal{H}_{\text{ext}}$, yet this $|\Psi\rangle$ has no component in $\mathcal{H}_{\text{cod}} \otimes |0_{\text{ext}}\rangle$ as it belongs to the orthogonal complement of $\mathcal{H}_{\text{cod}} \otimes |0_{\text{ext}}\rangle$. For compression, when the extra quantum system with space $\mathcal{H}_{\text{ext}}$ is discarded, the coding quantum system with space $\mathcal{H}_{\text{cod}}$ is in the (usually mixed) unimportant state $\rho_{\text{uni}} = \text{tr}_{\text{ext}}(|\Psi\rangle\langle\Psi|)$.

So the compression of a state $|\psi^L\rangle \in \mathcal{H}_N^{\otimes L}$ emitted by the source includes performing a measurement of $|\psi^L\rangle$ with $\{\Pi_{\text{typ}}, \Pi_{\text{typ}}^{\perp}\}$, on the post-measurement state in $\mathcal{H}_N^{\otimes L}$ apply the unitary U_{cod} to obtain a state in $\mathcal{H}_{\text{cod}} \otimes \mathcal{H}_{\text{ext}}$, finally discard the extra quantum system with space $\mathcal{H}_{\text{ext}}$ and retain only the state of the coding system with space $\mathcal{H}_{\text{cod}}$ as the compressed state.

For **decompression**, the coding quantum system of space $\mathcal{H}_{\text{cod}}$ is augmented by tensoring with an extra system of space (isomorphic to) $\mathcal{H}_{\text{ext}}$ placed in the fixed pure state $|0_{\text{ext}}\rangle$. This restores a quantum state in a space matching the initial dimension N^L, and to which the inverse coding transformation $\mathsf{U}_{\text{cod}}^{-1}$ is applied. In the eventuality occurring with the probability $P(\mathcal{H}_{\text{typ}}|\,|\psi^L\rangle)$ where the measurement by $\{\Pi_{\text{typ}}, \Pi_{\text{typ}}^{\perp}\}$ projected $|\psi^L\rangle$ onto $\mathcal{H}_{\text{typ}}$, the augmented state is the pure state $|\psi_{\text{cod}}\rangle \otimes |0_{\text{ext}}\rangle$ of Eq. (8.25), from which $\mathsf{U}_{\text{cod}}^{-1}$ reconstructs the projected state $|\psi_{\text{post}}^L\rangle$ of Eq. (8.24). Alternatively, in the eventuality occurring with the complementary probability $P(\mathcal{H}_{\text{typ}}^{\perp}|\,|\psi^L\rangle)$ where the projection was onto $\mathcal{H}_{\text{typ}}^{\perp}$, the augmented state is the mixed state $\rho_{\text{uni}} \otimes |0_{\text{ext}}\rangle\langle 0_{\text{ext}}|$, from which $\mathsf{U}_{\text{cod}}^{-1}$ reconstructs the unimportant state $\rho_{\text{uni}}' = \mathsf{U}_{\text{cod}}^{-1}(\rho_{\text{uni}} \otimes |0_{\text{ext}}\rangle\langle 0_{\text{ext}}|)\mathsf{U}_{\text{cod}}^{-1\dagger}$. The decompressed state thus arrives as a mixed state represented by the density operator

$$\rho_{\text{dec}} = P(\mathcal{H}_{\text{typ}}|\,|\psi^L\rangle)\,|\psi_{\text{post}}^L\rangle\langle\psi_{\text{post}}^L| + P(\mathcal{H}_{\text{typ}}^{\perp}|\,|\psi^L\rangle)\rho_{\text{uni}}' , \tag{8.26}$$

or equivalently, by using Eqs. (8.21) and (8.22),

$$\rho_{\text{dec}} = \Pi_{\text{typ}}\,|\psi^L\rangle\langle\psi^L|\,\Pi_{\text{typ}} + \langle\psi^L|\Pi_{\text{typ}}^{\perp}|\psi^L\rangle\,\rho_{\text{uni}}' . \tag{8.27}$$

This compression–decompression scheme, when operated on sequences $|\psi^L\rangle$ with large length L, by the law of large numbers, will deliver a decompressed state, being with high probability, very close or similar to the initial state $|\psi^L\rangle$. This similarity can be assessed by means of the fidelity F between the pure state $|\psi^L\rangle$ emitted by the source and the mixed state ρ_{dec} of Eq. (8.27) reconstructed by the decompression. The fidelity is defined as

$$F(|\psi^L\rangle, \rho_{\text{dec}}) = \langle\psi^L|\,\rho_{\text{dec}}\,|\psi^L\rangle , \tag{8.28}$$

and it offers a similarity index $F \in [0, 1]$, which is 1 if and only if ρ_{dec} coincides with the pure state $|\psi^L\rangle\langle\psi^L|$, and is 0 when ρ_{dec} represents a pure state orthogonal to $|\psi^L\rangle$. From Eq. (8.27), we obtain for the fidelity

$$F\big(|\psi^L\rangle, \rho_{\text{dec}}\big) = \langle\psi^L|\,\Pi_{\text{typ}}\,|\psi^L\rangle\,\langle\psi^L|\,\Pi_{\text{typ}}\,|\psi^L\rangle + \langle\psi^L|\Pi_{\text{typ}}^{\perp}|\psi^L\rangle\,\langle\psi^L|\rho'_{\text{uni}}|\psi^L\rangle \tag{8.29}$$

$$= \Big[P\big(\mathcal{H}_{\text{typ}}\big|\,|\psi^L\rangle\big)\Big]^2 + \Big[1 - P\big(\mathcal{H}_{\text{typ}}\big|\,|\psi^L\rangle\big)\Big]\langle\psi^L|\rho'_{\text{uni}}|\psi^L\rangle\ . \tag{8.30}$$

The scalar $\langle\psi^L|\rho'_{\text{uni}}|\psi^L\rangle$ of Eq. (8.30) is a real in [0, 1] for any length L. At large length L, the probability $P\big(\mathcal{H}_{\text{typ}}\big|\,|\psi^L\rangle\big)$ tends to 1, so that in Eq. (8.30) the fidelity $F\big(|\psi^L\rangle, \rho_{\text{dec}}\big)$ between the initial sequence $|\psi^L\rangle$ and its reconstructed version also tends to 1. This expresses the possibility, asymptotically at large length L, of perfectly faithful reconstruction with a probability tending to 1, when compressing an arbitrary sequence $|\psi^L\rangle$ emitted by the quantum information source.

The initial uncompressed sequence $|\psi^L\rangle$ belongs to the space $\mathcal{H}_N^{\otimes L}$ of dimension N^L, obtained by tensoring L symbol-carrying quantum systems each with dimension N. If the coding space $\mathcal{H}_{\text{cod}}$ of dimension $|\mathcal{H}_{\text{typ}}|$ is realized by tensoring similar N-dimensional quantum systems, a number $\log_N\big(|\mathcal{H}_{\text{typ}}|\big) \approx LS_N(\rho)$ is required, according to Eq. (8.20), with $S_N(\rho)$ the quantum entropy of the information source with a logarithm of base N. We know that $S_N(\rho) \le \log_N(N) = 1$. A number of $LS_N(\rho) \le L$ instead of L quantum systems with dimension N expresses indeed a compression, when the coding space $\mathcal{H}_{\text{cod}}$ of dimension $|\mathcal{H}_{\text{typ}}|$ replaces the initial space $\mathcal{H}_N^{\otimes L}$ of dimension N^L, with a compression rate per quantum symbol given by the quantum entropy $S(\rho)$.

The fidelity $F\big(|\psi^L\rangle, \rho_{\text{dec}}\big)$ of Eq. (8.30) characterizes, in probabilistic terms, the compression–decompression of an arbitrary sequence $|\psi^L\rangle$, and this fidelity F goes to 1 as the probability $P\big(\mathcal{H}_{\text{typ}}\big|\,|\psi^L\rangle\big)$ goes to 1 at large L. This characterization in probability of the fidelity can be supplemented by a characterization on average. We consider averaging F of Eq. (8.30) according to the prior probability $P\big(|\psi^L\rangle\big)$ of each possible sequence $|\psi^L\rangle$ emitted by the source, so as to yield the average fidelity

$$\overline{F} = \sum_{|\psi^L\rangle \in \mathcal{H}_N^{\otimes L}} P\big(|\psi^L\rangle\big)\, F\big(|\psi^L\rangle, \rho_{\text{dec}}\big)\ . \tag{8.31}$$

Since for any real x from $(x-1)^2 \ge 0$ one has $x^2 \ge 2x - 1$, we obtain from Eq. (8.30) the inequality

$$F\big(|\psi^L\rangle, \rho_{\text{dec}}\big)\ \ge\ 2P\big(\mathcal{H}_{\text{typ}}\big|\,|\psi^L\rangle\big) - 1\ , \tag{8.32}$$

and upon averaging via Eq. (8.31),

$$\overline{F} \ge 2P\big(\mathcal{H}_{\text{typ}}\big) - 1\ , \tag{8.33}$$

where we made use of Eq. (8.23). At large length L, the probability $P(\mathcal{H}_{\text{typ}})$ from Eq. (8.23) tends to 1, leading from Eq. (8.33) to $\overline{F} \geq 1$ for the average fidelity. Since the fidelity F can never exceed 1, this means that the average fidelity $\overline{F}$ goes to 1 at large L. This expresses that on average, asymptotically at large length L, perfect reconstruction can be achieved.

So a quantum information source can be compressed asymptotically with no loss at a compression rate given by its quantum entropy $S(\rho)$. This is the **Schumacher theorem for quantum source coding** or quantum information compression [4, 200]. The compression rate $S(\rho)$ expresses the accessible reduction of dimensionality of the Hilbert space for representing long sequences of quantum states emitted by the source and with asymptotically no information loss. A strong parallel can be appreciated with the Shannon theorem for classical source coding presented in Sect. 8.1. In a similar way, the Schumacher quantum coding theorem can be supplemented to include that no lossless quantum compression is possible at a rate smaller than the quantum entropy $S(\rho)$. This can be justified [4, 200] by showing that a projector Π_0 (instead of Π_{typ}) onto any subspace of $\mathcal{H}_N^{\otimes L}$ with dimension strictly below $N^{LS_N(\rho)}$ leads to a probability $\text{tr}(\Pi_0 \rho^{\otimes L})$ that necessarily goes to zero as $L \to \infty$, so that it is worthless to seek to concentrate the compression capability on such subspace. No improvement over the optimal compression rate $S(\rho)$ can be obtained, even when resorting to non-unitary quantum processes for coding or decoding to extend the unitary quantum process U_{cod} considered above for the Schumacher theorem [12, 201].

Instead of the fidelity of Eq. (8.28), the distortion between the initial and reconstructed states can be evaluated by a distance between two density operators ρ_1 and ρ_2 like the trace distance $D_1(\rho_1, \rho_2)$ defined in Sect. 3.2.3 with Eq. (3.51). We have seen in Sect. 3.2.3 that the trace norm or ℓ_1-norm of an operator $\mathsf{A} \in \mathcal{L}(\mathcal{H}_N)$ is defined as $\|\mathsf{A}\|_1 = \text{tr}(|\mathsf{A}|)$, with the absolute value $|\mathsf{A}| = \sqrt{\mathsf{A}^\dagger \mathsf{A}}$. The trace distance of two density operators ρ_1 and ρ_2 of $\mathcal{L}(\mathcal{H}_N)$ is defined in Eq. (3.51) as

$$D_1(\rho_1, \rho_2) = \frac{1}{2}\|\rho_1 - \rho_2\|_1 = \frac{1}{2}\,\text{tr}\big(|\rho_1 - \rho_2|\big) , \tag{8.34}$$

with $0 \leq D_1(\rho_1, \rho_2) \leq 1$, and $D_1(\rho_1, \rho_2) = 0$ if and only if $\rho_1 = \rho_2$. The fidelity $F(\rho_1, \rho_2)$ between two arbitrary density operators ρ_1 and ρ_2 satisfies [12, 202] the inequalities

$$0 \leq 1 - \sqrt{F(\rho_1, \rho_2)} \leq D_1(\rho_1, \rho_2) \leq \sqrt{1 - F(\rho_1, \rho_2)} \leq 1 , \tag{8.35}$$

this in particular being verified by the fidelity $F\big(\rho_1 = |\psi_1\rangle\langle\psi_1|,\, \rho_2\big) = \langle\psi_1|\rho_2|\psi_1\rangle$ between a pure state $\rho_1 = |\psi_1\rangle\langle\psi_1|$ and an arbitrary mixed state ρ_2 as used in Eq. (8.28). So by Eq. (8.35), in the source coding procedure, the fidelity $F\big(|\psi^L\rangle,\, \rho_{\text{dec}}\big)$ going to 1 in Eq. (8.28) is equivalent to a trace distance $D_1\big(|\psi^L\rangle\langle\psi^L|,\, \rho_{\text{dec}}\big)$ going to 0, between the initial pure state $|\psi^L\rangle$ and the reconstructed state ρ_{dec}, expressing similarly perfect (asymptotic) reconstruction. The

Schumacher compression theorem therefore applies equivalently when evaluating distortion with the fidelity or with the trace distance.

Another distortion criterion can rely on the Hilbert–Schmidt norm or ℓ_2-norm of an operator $\mathsf{A} \in \mathcal{L}(\mathcal{H}_N)$ defined as the square root $\|\mathsf{A}\|_2 = \sqrt{\operatorname{tr}\left(|\mathsf{A}|^2\right)}$. The Hilbert–Schmidt distance of two density operators ρ_1 and ρ_2 of $\mathcal{L}(\mathcal{H}_N)$ is defined as

$$D_2(\rho_1, \rho_2) = \frac{1}{2}\|\rho_1 - \rho_2\|_2 = \frac{1}{2}\sqrt{\operatorname{tr}\left(|\rho_1 - \rho_2|^2\right)} \,. \tag{8.36}$$

The operator $|\rho_1 - \rho_2|$ is a positive operator of $\mathcal{L}(\mathcal{H}_N)$, with N positive eigenvalues μ_n, so that the Hilbert–Schmidt distance of Eq. (8.36) gives $2D_2(\rho_1, \rho_2) = \sqrt{\sum_n \mu_n^2}$, while the trace distance of Eq. (8.34) gives $2D_1(\rho_1, \rho_2) = \sum_n \mu_n \geq 2D_2(\rho_1, \rho_2)$, therefore entailing the inequalities

$$0 \leq D_2(\rho_1, \rho_2) \leq D_1(\rho_1, \rho_2) \leq 1 \,. \tag{8.37}$$

So by Eq. (8.37), in the source coding procedure a trace distance D_1 going to 0 is equivalent to a Hilbert–Schmidt distance D_2 going to 0. The Schumacher compression theorem therefore applies equivalently when evaluating distortion with the Hilbert–Schmidt distance.

The compression procedure does not assign univocal codes to the sequences of quantum symbols $|\phi_j\rangle$ emitted by the source, which in some sense remain unknown to the coder as indicated earlier. For quantum compression, only the source ensemble $\left\{\left(p_j, |\phi_j\rangle\right)\right\}$ needs to be known, or even more strictly, the density operator ρ of Eq. (8.15) is enough, to determine its eigendecomposition defining the typical subspace $\mathcal{H}_{\text{typ}}$ and then work out, via the projectors $\left\{\Pi_{\text{typ}}, \Pi_{\text{typ}}^{\perp}\right\}$ and the unitary U_{cod}, the compression (and decompression) of the source. The quantum compression procedure essentially focuses (via the unitary U_{cod}) on compressing a basis of representation for length-L source sequences, with a privileged (shorter) encoding for the typical basis states $\{|\lambda_n\rangle\}^{\otimes L}$ defined via Eq. (8.19) and spanning the typical subspace $\mathcal{H}_{\text{typ}}$ where all the probabilistic weight of the source sequences tends to concentrate. In this respect, the coding procedure rather constructs a compression of the typical subspace, and from it, indirectly, follows a compression of the input sequences (that anyway remain unknown). The quantum signals $|\phi_j\rangle$ are emitted by the source sequentially as tensor-product states over the space $\mathcal{H}_N^{\otimes L}$, they are encoded via U_{cod} over a compressed space $\mathcal{H}_{\text{cod}}$ with a dimension reduced optimally by the factor $S(\rho)$ per symbol, and they are finally reconstructed via $\mathsf{U}_{\text{cod}}^{-1}$ over the initial space $\mathcal{H}_N^{\otimes L}$, but the sequences of emitted signals $|\phi_j\rangle$ and their reconstructions are nowhere known or observed in the procedure. The information they contain is compressed and restored, but it is nowhere accessed. In particular, if instead of the original ensemble $\left\{\left(p_j, |\phi_j\rangle\right)\right\}_{j=1}^{J}$ the quantum source were to change to emit from an equivalent ensemble, such as the eigenset $\left\{\left(\lambda_n, |\lambda_n\rangle\right)\right\}_{n=1}^{N}$ or any isometrically equivalent ensemble as in Eq. (2.38) represented by the same density operator ρ, the optimal quantum coder would not change and would remain equally optimal.

Compared to classical compression, in the more stringent conditions of quantum compression, we note that perfect reconstruction only occurs asymptotically, at large length $L \to \infty$. At finite L, some distortion in the reconstruction may remain, quantified by a fidelity in Eq. (8.30) or Eq. (8.31) reaching below 1, to go to 1 asymptotically as $L \to \infty$. By contrast, classical compression in Sect. 8.1, in less stringent conditions, arranges univocal codes for all source sequences of length L, to ensure no loss at any L, and it achieves the optimal minimum compression rate only asymptotically as $L \to \infty$. Classical compression is lossless at any length L and optimal at $L \to \infty$, while quantum compression is lossless and optimal only at $L \to \infty$.

A special case of quantum compression is when the source ensemble $\{(p_j, |\phi_j\rangle)\}_{j=1}^{J}$ is made of $J \leq N$ quantum symbols $|\phi_j\rangle$ that are mutually orthogonal in $\mathcal{H}_N$ and therefore perfectly distinguishable and identifiable by a von Neumann projective measurement. By measuring the quantum symbols in this way, a quantum sequence from the source can be converted with no loss into the classical sequence of the measurement results, which can then be classically compressed at the minimal rate given by the Shannon classical entropy $H(\{p_j\})$, which here coincides with the quantum entropy $S(\rho)$ of the Schumacher coding theorem.

In addition to the quantum block-coding procedure considered above, while targeting the optimal performance of the Schumacher theorem, practical alternative techniques inspired from their classical counterparts have been proposed for quantum compression, with quantum Huffman coding [203], arithmetic coding [204] or Lempel–Ziv type universal compression [205].

8.2.2 An Example of Quantum Compression

To illustrate the procedure of quantum compression presented in Sect. 8.2.1, we consider the compression of a quantum information source drawing from the statistical ensemble comprising $J = 3$ qubit states

$$\Big\{|\phi_1\rangle = |0\rangle\,, |\phi_2\rangle = |1\rangle\,, |\phi_3\rangle = |+\rangle\Big\}\,, \tag{8.38}$$

with respective probabilities $\{p_1 = 1/7, p_2 = 2/7, p_3 = 4/7\}$. This is the same ensemble considered in Sect. 2.5.5 and defined in Eq. (2.71). The corresponding density operator ρ representing the source is given in Eq. (2.72), and its eigendecomposition is given along with Eq. (2.73) and associated eigenset

$$\Big\{\big(\lambda_+ \approx 0.795, |\lambda_+\rangle\big)\,;\,\big(\lambda_- \approx 0.205, |\lambda_-\rangle\big)\Big\}\,. \tag{8.39}$$

The 3 qubit states of the source alphabet of Eq. (8.38) can be expressed in the eigenbasis $\{|\lambda_+\rangle\,, |\lambda_-\rangle\}$ of the qubit Hilbert space $\mathcal{H}_2$ as

$$\begin{cases} |\phi_1\rangle \approx 0.615\,|\lambda_+\rangle + 0.788\,|\lambda_-\rangle\ , & (8.40)\\ |\phi_2\rangle \approx 0.788\,|\lambda_+\rangle - 0.615\,|\lambda_-\rangle\ , & (8.41)\\ |\phi_3\rangle \approx 0.993\,|\lambda_+\rangle + 0.122\,|\lambda_-\rangle\ . & (8.42) \end{cases}$$

The quantum entropy $S(\rho)$ of the quantum source, following Eq. (8.17), results as $S(\rho) \approx 0.733$ Sh.

If we consider long sequences $|\psi^L\rangle$ of L consecutive qubit states drawn from the quantum ensemble of Eq. (8.38), by proper coding as explained in Sect. 8.2.1, it should therefore be possible to reduce the number of qubits by a factor of $S(\rho) \approx 0.733$ with vanishing loss of information from the source. The optimal performance is usually obtained asymptotically when $L \to \infty$, but it is possible to observe at work the essential elements enabling quantum compression as soon as small L. So as an instructive illustration we address the compression of sequences $|\psi^{L=3}\rangle$ of $L = 3$ consecutive qubit states emitted by the quantum source of Eq. (8.38). At $L = 3$, the typical sequences from the eigenbasis of $\mathcal{H}_2^{\otimes 3}$ defined via Eq. (8.19) would comprise $L\lambda_+ \approx 2.4$ states $|\lambda_+\rangle$ and $L\lambda_- \approx 0.6$ state $|\lambda_-\rangle$. Crudely, we arrange the typical subspace $\mathcal{H}_{\rm typ}$ as the subspace of $\mathcal{H}_2^{\otimes 3}$ spanned by the 4 eigenstates

$$\Big\{|\lambda_+\rangle\,|\lambda_+\rangle\,|\lambda_-\rangle\,,\ |\lambda_+\rangle\,|\lambda_-\rangle\,|\lambda_+\rangle\,,\ |\lambda_-\rangle\,|\lambda_+\rangle\,|\lambda_+\rangle\,,\ |\lambda_+\rangle\,|\lambda_+\rangle\,|\lambda_+\rangle\Big\}\,, \tag{8.43}$$

comprising 2 or 3 states $|\lambda_+\rangle$. By using Eqs. (8.40)–(8.42) to develop the tensor products, one can visualize how any of the $J^L = 3^3 = 27$ distinct 3-qubit sequences $|\psi^{L=3}\rangle$ the source can emit, decomposes as the sum of a vector of the typical subspace $\mathcal{H}_{\rm typ}$ and a vector of its orthogonal complement $\mathcal{H}_{\rm typ}^{\perp}$.

For the coding procedure, the measurement by $\big\{\Pi_{\rm typ},\, \Pi_{\rm typ}^{\perp}\big\}$ employs two complementary projectors acting in $\mathcal{H}_2^{\otimes 3}$. The conditional probabilities $P\big(\mathcal{H}_{\rm typ}\big|\,|\psi^{L=3}\rangle\big)$ can be computed according to Eq. (8.21) for the 27 possible configurations of the 3-qubit source sequence $|\psi^{L=3}\rangle$. The overall probability $P\big(\mathcal{H}_{\rm typ}\big)$ of Eq. (8.23) then follows as $P\big(\mathcal{H}_{\rm typ}\big) \approx 0.891$. To approach the optimal compression rate $S(\rho) \approx 0.733$ Sh, we choose to compress each 3-qubit sequence $|\psi^{L=3}\rangle$ emitted by the source, with only 2 qubits instead of 3.

A direct approach, with no preliminary coding, could consist, in order to compress each 3-qubit source sequence $|\psi^{L=3}\rangle = |q_1\rangle\,|q_2\rangle\,|q_3\rangle$, in issuing a 2-qubit compressed version, say $|q_1\rangle\,|q_2\rangle$, that would retain only the first two qubits and simply discard the third qubit. From $|q_1\rangle\,|q_2\rangle$, for decompression, the 3-qubit reconstructed sequence could then be taken as $|q_1\rangle\,|q_2\rangle\,|\phi_3\rangle$, since it is the state $|\phi_3\rangle$ of the source alphabet of Eq. (8.38) that exhibits the maximum prior probability. This direct compression–decompression protocol, with a rate 2/3, achieves, between the initial and reconstructed 3-qubit sequences, an average fidelity of $\overline{F}_{\rm nocod} \approx 0.786$, over the 27 distinct sequences the source can emit. A larger fidelity can be obtained, by following the approach of Sect. 8.2.1, in particular exploiting the typical subspace $\mathcal{H}_{\rm typ}$, as we now show.

To follow the approach of Sect. 8.2.1, we devise the coding operator $\mathsf{U}_{\rm cod}$ as a unitary transformation from $\mathcal{H}_2^{\otimes 3}$ into $\mathcal{H}_2^{\otimes 3}$, mapping any of the 4 basis states of

Eq. (8.43) spanning the typical subspace $\mathcal{H}_{\rm typ}$, and therefore also any of the states of $\mathcal{H}_{\rm typ}$, into a state of the form $|\psi_{\rm cod}\rangle \otimes |0\rangle$, adhering to Eq. (8.25), where $|\psi_{\rm cod}\rangle$ is a 2-qubit state of $\mathcal{H}_2^{\otimes 2} \equiv \mathcal{H}_{\rm cod}$, complemented with a "flag" qubit state $|0\rangle \equiv |0_{\rm ext}\rangle$ of $\mathcal{H}_2 \equiv \mathcal{H}_{\rm ext}$. Meanwhile, any source sequence $|\psi^{L=3}\rangle$ projected by $\{\Pi_{\rm typ}, \Pi_{\rm typ}^{\perp}\}$ onto the untypical subspace $\mathcal{H}_{\rm typ}^{\perp}$, would receive by $\mathsf{U}_{\rm cod}$, an unimportant code with a flag qubit at $|1\rangle \in \mathcal{H}_2$. For compression here, when the flag qubit is discarded, the two coding qubits that are retained are always in a pure 2-qubit state $|\psi_{\rm cod}\rangle$.

The decompression operates as described in Sect. 8.2.1. The 2-qubit compressed state $|\psi_{\rm cod}\rangle$ is augmented to the 3-qubit state $|\psi_{\rm cod}\rangle \otimes |0\rangle$. The inversion $\mathsf{U}_{\rm cod}^{-1}\, |\psi_{\rm cod}\rangle \otimes |0\rangle$ delivers a 3-qubit reconstructed state as in Eqs. (8.26)–(8.27). Via Eq. (8.30), we then have access to the average fidelity $\overline{F}$ of Eq. (8.31), between the initial and reconstructed 3-qubit sequences. We obtained a lower bound to the average fidelity of Eq. (8.31) as $\overline{F} > 0.814$; the lower bound occurs because, in the small-probability eventuality of a projection onto the untypical subspace $\mathcal{H}_{\rm typ}^{\perp}$, we set to zero the positive term $\langle\psi^L|\rho'_{\rm uni}|\psi^L\rangle$ in each fidelity $F\big(|\psi^L\rangle\,,\rho_{\rm dec}\big)$ of Eq. (8.30), to dispense ourselves from a complete explicit specification of the coding unitary $\mathsf{U}_{\rm cod}$ that would determine the unimportant (pre)coding 3-qubit state $|\Psi\rangle$ for each $|\psi^{L=3}\rangle$ projected onto $\mathcal{H}_{\rm typ}^{\perp}$. So we find that this compression–decompression protocol exploiting the typical subspace achieves a fidelity $\overline{F}$ larger than that of the direct protocol considered above, with a similar compression rate of 2/3. In this way, exploiting the typical subspace enables better compression efficiency. By the Schumacher compression theorem of Sect. 8.2.1, perfect fidelity of reconstruction tending to 1 can be expected, at a compression rate going to $S(\rho) \approx 0.733$ Sh, by operating on long source sequences $|\psi^L\rangle$ of length $L \to \infty$.

The statistical ensemble $\{(p_j, |\phi_j\rangle)\}_{j=1}^3$ from Eq. (8.38) can be associated with a classical entropy $H\big(\{p_j\}\big) \approx 1.379$ Sh, larger than the quantum entropy $S(\rho) \approx 0.733$ Sh. This classical entropy would provide the optimal rate of $H\big(\{p_j\}\big) \approx 1.379$ coding bits per source symbol, for lossless compression of a 3-state classical source with the same probability distribution $\{p_j\}_{j=1}^3$. But since the 3 quantum states $|\phi_j\rangle$ of Eq. (8.38) are non-orthogonal and therefore non-perfectly distinguishable, they carry less information and accordingly they require less resource, with a smaller optimal rate of $S(\rho) \approx 0.733$ qubit per source symbol, for their lossless compression.

8.2.3 Compression of Mixed Quantum States

The statistical ensemble of J pure states $\big\{(p_j, |\phi_j\rangle)\big\}_{j=1}^J$ introduced at the beginning of Sect. 8.2.1 is a useful model for representing a quantum source of controlled information. However, one could naturally ask about extending the problem of compression to a quantum source emitting from a statistical ensemble of J mixed states $\big\{(p_j, \rho_j)\big\}_{j=1}^J$. The average density operator associated with such a mixed-state source is

$$\rho = \sum_{j=1}^{J} p_j \rho_j \ ; \tag{8.44}$$

it can receive an eigendecomposition $\{(\lambda_n, |\lambda_n\rangle)\}$ as in Eq. (8.16), and a quantum entropy $S(\rho)$ as in Eq. (8.17).

For mixed-state sources [206] proves that lossless compression can be achieved at a rate given by the quantum entropy $S(\rho)$ of the source, in the same (asymptotic) regime of long sequences as in the Schumacher compression theorem of Sect. 8.2.1 with pure states. However, although the entropy $S(\rho)$ is an achievable rate, $S(\rho)$ is not an adequate lower bound for lossless compression of a mixed-state quantum source. This can be seen with a source emitting a single mixed state ρ_1 with a probability $p_1 = 1$. Any (non-degenerate) mixed state ρ_1 has a nonzero quantum entropy $S(\rho) = S(\rho_1) > 0$, but asymptotically long sequences of the same source state ρ_1 can be transmitted (compressed and reconstructed) with an average rate going to 0, that is, a rate below the entropy $S(\rho) = S(\rho_1) > 0$.

As an alternative, it can be proved [207, 208] that the Holevo information

$$\chi\left(\{p_j, \rho_j\}\right) = S(\rho) - \sum_{j=1}^{J} p_j \, S(\rho_j) \le H\left(\{p_j\}\right) , \tag{8.45}$$

as defined in Eq. (7.27), constitutes a lower bound to the rate of lossless compression of a general mixed-state quantum source. The reason, as argued in [208], is that since a mixed-state ensemble $\{(p_j, \rho_j)\}$ can always be used to encode and losslessly transmit a quantity $\chi\left(\{p_j, \rho_j\}\right)$ of classical information per quantum symbol (see Sect. 13.2.2 to come on quantum communication), if such an ensemble could be compressed at a rate lower than $\chi\left(\{p_j, \rho_j\}\right)$, this would imply an error-free transmission of classical information at a rate larger than $\chi\left(\{p_j, \rho_j\}\right)$, which is not permitted. However, this lower bound $\chi\left(\{p_j, \rho_j\}\right)$ to the rate of lossless compression of a general quantum source, is not known to be generally achievable.

The lower bound $\chi\left(\{p_j, \rho_j\}\right)$ is achievable in some special cases. In the case of the above elementary example of a source emitting a single mixed state ρ_1, the lower bound $\chi\left(\{p_1, \rho_1\}\right) = 0$ coincides with the achievable vanishing rate. In the case of J pure states $\rho_j = |\phi_j\rangle\langle\phi_j|$ having entropy $S(\rho_j) = 0$, the lower bound $\chi\left(\{p_j, \rho_j\}\right) = S(\rho)$ coincides with the achievable lower bound prescribed by the Schumacher compression theorem of Sect. 8.2.1 for a pure-state information source.

Also, with an ensemble of J (necessarily $J \le N$) mixed states $\rho_j \in \mathcal{L}(\mathcal{H}_N)$ having mutually orthogonal supports supp(ρ_j) in $\mathcal{H}_N$, the Holevo information $\chi\left(\{p_j, \rho_j\}\right) = H\left(\{p_j\}\right) \le \log(N)$ is an achievable lower bound for the compression rate. The J states ρ_j remaining perfectly distinguishable can be measured, and the measurement results among J can be losslessly compressed as a classical source at the optimal compression rate $H\left(\{p_j\}\right) = \chi\left(\{p_j, \rho_j\}\right)$.

In addition, [209, 210] provide optimal compression rates for mixed-state sources with a fidelity criterion extended to mixed states. The optimal compression rate is obtained in [209, 210] as the von Neumann entropy of a reduced source resulting from a decomposition of the mixed-state source into essential and redundant parts. Reference [12] discusses various distortion criteria for mixed-state compression, showing that they may be associated with different optimal rates and compression–decompression schemes.

8.3 Quantum Storage of Classical Information

For practical applications, it is natural to envisage to encode a classical information source $\{(p_j, s_j)\}_{j=1}^{J}$ as considered in Sect. 8.1, into a quantum information source as considered in Sect. 8.2, for instance to have access to reduced physical volume of storage at the quantum level, as for example when storing one binary digit into one electron spin. Some aspects of this issue have been discussed at the end of Sect. 7.2.3 introducing the Holevo information.

If the J classical symbols s_j are univocally encoded into $J \leq N$ quantum states $|\phi_j\rangle \in \mathcal{H}_N$ that are mutually orthogonal in $\mathcal{H}_N$, then the coding states $|\phi_j\rangle$ remain perfectly distinguishable through quantum measurement, and the classical information can be exactly recovered from its quantum encoding. There is no loss of information in the quantum encoding. The pure-state quantum ensemble $\{(p_j, |\phi_j\rangle)\}$ has the von Neumann quantum entropy $S(\rho) = H(\{p_j\}) \leq \log(J) \leq \log(N)$. An optimal lossless compression can therefore be accomplished on the pure-state quantum source $\{(p_j, |\phi_j\rangle)\}$ at a rate $S(\rho) = H(\{p_j\})$ that is no different from the rate $H(\{p_j\})$ of an optimal compression directly on the classical source. This was already noted at the end of Sect. 8.2.1 as a special case of the Schumacher quantum compression.

Similarly, if the J classical symbols s_j are univocally encoded into $J \leq N$ mixed states $\rho_j \in \mathcal{L}(\mathcal{H}_N)$ that have mutually orthogonal supports $\mathrm{supp}(\rho_j)$ in $\mathcal{H}_N$, then the coding states ρ_j also remain perfectly distinguishable and the classical information can be exactly recovered from its quantum encoding. Again, there is no loss of information in the quantum encoding. The mixed-state quantum ensemble $\{(p_j, \rho_j)\}_{j=1}^{J}$ has the Holevo information $\chi(\{p_j, \rho_j\}) = H(\{p_j\}) \leq \log(J) \leq \log(N)$. As noted in Sect. 8.2.3, an optimal lossless compression can therefore be accomplished on such mixed-state quantum source $\{(p_j, \rho_j)\}_{j=1}^{J}$ at a rate $\chi(\{p_j, \rho_j\}) = H(\{p_j\})$ that is again no different from the rate $H(\{p_j\})$ of an optimal compression directly on the classical source.

By contrast, if the J classical symbols s_j are encoded into J quantum states that are not perfectly distinguishable through quantum measurement, then some information is lost in such non-invertible encoding protocol. In such situation $\chi(\{p_j, \rho_j\}) \leq S(\rho) \leq H(\{p_j\})$. The quantum source with a smaller information content can be optimally compressed with no loss at a minimum rate between $\chi(\{p_j, \rho_j\})$ and $S(\rho)$, while the classical source with a larger information content requires a larger rate $H(\{p_j\})$ to be optimally compressed with no loss. The quantum encoding here is an example of a lossy information channel, and optimal transmission of classical information over such lossy quantum channel is a topic of major importance, which we will examine in detail in Chap. 13.

8.4 Entanglement Dilution and Concentration

Entanglement dilution has already been evoked at the end of Sect. 6.6 as an application of quantum teleportation. For a bipartite quantum system AB, entanglement dilution consists in converting stereotyped maximally entangled states into entangled states of a selected arbitrary type, and this through only local operations and classical communication [211, 212].

Entanglement concentration, known also as entanglement distillation or purification (possibly with some minute variants), can be seen as the inverse process, consisting, for a bipartite system AB, in upgrading partially entangled states into maximally entangled states, also through only local operations and classical communication [211, 213]. Entanglement distillation is important to restore the quality of diluted entanglement, by concentrating it into maximal entanglement, representing an essential resource for quantum information processing (exploited for instance, as we have seen, in superdense coding, teleportation, or cryptography). Different methods have been proposed to perform entanglement concentration [211, 213–215]. For an illustration, we sketch below the Schmidt projection method introduced in [211]. We describe the method, as in [211], to distill qubit pairs, but the generalization to bipartite quantum states of arbitrary dimension is straightforward, as indicated in [211].

We consider a qubit pair AB shared between Alice and Bob, as in the situation of bipartite entanglement examined in Sect. 7.3.5 for instance. The qubit pair AB is in a (partially) entangled pure state $|AB\rangle$ of the product space $\mathcal{H}_2 \otimes \mathcal{H}_2$, that we can generally express in the Schmidt decomposition of Eq. (7.64) as

$$|AB\rangle = \sqrt{p_0}\,|0_A\rangle\,|0_B\rangle + \sqrt{p_1}\,|1_A\rangle\,|1_B\rangle = \sum_{k=0}^{1} \sqrt{p_k}\,|k_A\rangle\,|k_B\rangle \ , \tag{8.46}$$

where the set $\{p_0, p_1\}$ forms a probability distribution. When the state $|AB\rangle$ is measured in the orthonormal eigenbasis $\{|k_A\rangle\,|k_B\rangle\}$ of $\mathcal{H}_2 \otimes \mathcal{H}_2$ formed from the eigenstates of the reduced density operators $\rho_A = \mathrm{tr}_B\big(|AB\rangle\,\langle AB|\big)$ and $\rho_B = \mathrm{tr}_A\big(|AB\rangle\,\langle AB|\big)$, there is a probability p_k of obtaining a given $k = 0$ or 1, projecting $|AB\rangle$ onto the corresponding basis state $|k_A\rangle\,|k_B\rangle$ of $\mathcal{H}_2 \otimes \mathcal{H}_2$. This is a separable measurement between A and B. To obtain the measurement result $|k_A\rangle\,|k_B\rangle$, Alice can measure her qubit A locally in the eigenbasis $\{|0_A\rangle\,, |1_A\rangle\}$ of ρ_A to get $|k_A\rangle$, while Bob measures his qubit B in the eigenbasis $\{|0_B\rangle\,, |1_B\rangle\}$ of ρ_B to get $|k_B\rangle$. Moreover, due to the entanglement in $|AB\rangle$ of Eq. (8.46), Alice and Bob always measure the same value of k. Accordingly, Alice and Bob can choose each to perform its own local measurement, or otherwise it is enough for Alice to locally measure A and classically communicate her value of k to Bob, or reciprocally.

With L shared qubit pairs, we consider repeating $|AB\rangle$ in a sequence of L consecutive identical states, which jointly realize the L-fold product state

$$|AB\rangle^{\otimes L} = \Big(\sqrt{p_0}\,|0_A\rangle\,|0_B\rangle + \sqrt{p_1}\,|1_A\rangle\,|1_B\rangle\Big)^{\otimes L}, \tag{8.47}$$

of the space $(\mathcal{H}_2 \otimes \mathcal{H}_2)^{\otimes L}$. The expansion of the L-fold tensor product of Eq. (8.47) produces a total of 2^L terms, each of these terms arriving with a scalar multiplicative coefficient made from products of the factors $\sqrt{p_k}$. These 2^L multiplicative coefficients are not all different, but they are the same over $L+1$ definite classes of terms, corresponding to a given statistical type. Each of these uniform classes comes with a cardinal given by the binomial coefficient $C_\ell^L = L!/\big(\ell!(L-\ell)!\big)$, associated with the multiplicative coefficient $\sqrt{p_0}^{\,\ell}\sqrt{p_1}^{\,L-\ell}$ shared in common by these equiweighted basis states of $(\mathcal{H}_2 \otimes \mathcal{H}_2)^{\otimes L}$ sequencing tensor products of ℓ states $|0_A\rangle\,|0_B\rangle$ with $L-\ell$ states $|1_A\rangle\,|1_B\rangle$, for every integer $\ell \in \{0, 1, \dots, L\}$ labeling the uniform class. A generalized measurement of $|AB\rangle^{\otimes L}$ can be envisaged, made of the $L+1$ mutually orthogonal projectors Π_ℓ onto each of the $L+1$ uniform classes over the space $(\mathcal{H}_2 \otimes \mathcal{H}_2)^{\otimes L}$, so that the measurement result discloses only a statistical type or class ℓ. This is again a separable measurement that can be performed locally by Alice and Bob. Each projector Π_ℓ acts with the probability $C_\ell^L p_0^\ell p_1^{L-\ell}$ of the binomial distribution, and it produces a post-measurement state belonging to a subspace of $(\mathcal{H}_2 \otimes \mathcal{H}_2)^{\otimes L}$ associated with the class ℓ and spanned by the C_ℓ^L equiweighted basis states composing the class ℓ. The post-measurement state, from its equiweighted constitution over the class, is a maximally entangled state of this subspace. This maximally entangled state is shared and obtained by Alice and Bob through local measurement and classical communication. The equiweighted character in the post-measurement state is an essential feature to upgrade partially entangled states as in Eq. (8.46) into maximally entangled states. As indicated in [211], such maximally entangled states obtained in a subspace conditioned on the measurement outcome can be exploited without further ado for various applications such as teleportation over their subspace; or they can be further transformed into standard forms such as definite Bell states.

In the limit of long sequences, the probabilistic post-selection by the measurement outcome will tend to vanish. In the asymptotic limit of long sequences $L \to \infty$, with a probability tending to 1, the measurement will select the class of typical sequences, associated with the typical subspace as defined in Sect. 8.2.1 via Eq. (8.19). This typical class is composed of basis sequences comprising approximately Lp_0 states $|0_A\rangle\,|0_B\rangle$ in tensor products with Lp_1 states $|1_A\rangle\,|1_B\rangle$, all equiweighted in the maximally entangled post-measurement state. The cardinal of the typical class, or number of typical basis sequences it contains, has its logarithm given by the entropy $LH\big(\{p_k\}\big) = LS(\rho_A) = LS(\rho_B)$. This fixes the best performance for the distillation protocol [211, 213]: sequences of L partially entangled states $|AB\rangle$ can be distilled into sequences of $LS(\rho_A) = LS(\rho_B) \leq L$ maximally entangled states, in the limit of long sequences $L \to \infty$. In other words, the entanglement entropy $S(\rho_A) = S(\rho_B)$, as defined at the end of Sect. 7.3.5, is the maximum rate for converting partially entangled states $|AB\rangle$ into maximally entangled states. This remains true, beyond qubit pairs, for bipartite states of arbitrary dimension. To project onto the typical subspace and obtain the equiweighted constitution, a measurement that can be used by Alice and Bob is similar to the measurement $\{\Pi_{\text{typ}}, \Pi_{\text{typ}}^{\perp}\}$ involved in Eqs. (8.21)–(8.22) for the compression protocol of Sect. 8.2.1. And a unitary encoding by U_{cod} as in Eq. (8.25) can be performed to reconfigure the maximally entangled states, for instance as sequences of selected Bell states.

Entanglement dilution [211, 212] can also be performed at the asymptotic rate given by the entanglement entropy $S(\rho_A) = S(\rho_B)$. Alice can locally prepare both parts of the entangled state $|AB\rangle$ of Eq. (8.46), assemble the L-fold sequence $|AB\rangle^{\otimes L}$ of Eq. (8.47), and perform the generalized measurement of $|AB\rangle^{\otimes L}$ via the set of projectors $\{\Pi_\ell\}$ or alternatively via the pair of projectors $\{\Pi_{\text{typ}}, \Pi_{\text{typ}}^{\perp}\}$. In the asymptotic limit of long sequences $L \to \infty$, after measuring the L-fold sequence $|AB\rangle^{\otimes L}$, the measurement by Alice delivers (with a probability $\to 1$) a typical sequence drawn equiprobably from the typical subspace spanned by the class of L-fold typical basis sequences, whose cardinal has its logarithm given by $LH(\{p_k\}) = LS(\rho_A) = LS(\rho_B)$. This typical subspace can be optimally coded (compressed) into a space supported by $LS(\rho_A) \leq L$ qubit pairs instead of the L initial qubit pairs. The L-fold basis sequences are tensor products of L terms $|0_A\rangle\,|0_B\rangle$ and $|1_A\rangle\,|1_B\rangle$ that are separable between A and B, and similarly the coding unitary transformation like U_{cod} in Eq. (8.25) can be applied by Alice separately on A and B. Of a compressed sequence of length $LS(\rho_A) \leq L$, the part belonging to Bob, by the teleportation protocol of Sect. 6.6, can be teleported to Bob by consuming $LS(\rho_A)$ maximally entangled Bell states shared between Alice and Bob, while conserving the prior entanglement with Alice as indicated at the end of Sect. 6.6. Then, on his part in the received sequence, Bob can locally apply his part of the inverse coding unitary transformation so as to reconstruct (decompress) his part of the sequence. Alice locally performs a similar decompression on her part of the sequence. This finally reconstructs the whole initial bipartite sequence of length L, now shared between Alice and Bob. In this way, L qubit pairs in an arbitrary (partially) entangled state $|AB\rangle$ owned by Alice, can be established (diluted) in shared entanglement with Bob, by consuming $LS(\rho_A) \leq L$ maximally entangled Bell states, corresponding to an entanglement dilution rate [211, 216] also controlled by the entanglement entropy $S(\rho_A) = S(\rho_B)$.

In addition, it can be shown that, for bipartite pure states, in the asymptotic limit $L \to \infty$, entanglement dilution and concentration can be performed with vanishing classical communication cost [217]. Entanglement dilution and concentration are feasible for pure states and for mixed states as well [170, 213]. They can be extended, beyond bipartite quantum systems, to multipartite systems [170, 218]. The methods themselves, and their practical implementations, today can still benefit from further improvement, toward enhanced mastery of the important resource of quantum entanglement, its quantitative assessment and the cost of its establishment between distant parties, for relevance to signal and information processing [170, 219].

8.5 Summary on Quantum Information Compression

The lossless compression of an information source is a central structuring topic of information theory, conceptually and practically, for the classical and quantum domains alike. Especially, it consolidates the status of the entropy for a meaningful quantitative assessment of information.

A standard situation of quantum compression considers a quantum information source emitting sequences of signals or symbols of the Hilbert space $\mathcal{H}_N$ independently drawn from the statistical ensemble $\{(p_j, |\phi_j\rangle)\}_{j=1}^J$ introduced at the beginning of Sect. 8.2.1. The quantum information source is represented by the density operator $\rho = \sum_{j=1}^J p_j |\phi_j\rangle \langle\phi_j|$ from Eq. (8.15), coming with the eigendecomposition of Eq. (8.16). The density operator ρ alone, via its eigenset $\{(\lambda_n, |\lambda_n\rangle)\}_{n=1}^N$, determines the **quantum entropy** $S(\rho) = -\operatorname{tr}(\rho \log \rho)$ of the information source via Eq. (8.17), and also the **typical subspace** $\mathcal{H}_{\text{typ}}$ of $\mathcal{H}_N^{\otimes L}$ defined by Eq. (8.19) when emitting sequences $|\psi^L\rangle$ of L consecutive independent symbols drawn from $\{(p_j, |\phi_j\rangle)\}_{j=1}^J$; finally, ρ also determines a whole optimal procedure for lossless coding and decoding of the information source.

A number L of independent draws from the source ensemble $\{(p_j, |\phi_j\rangle)\}_{j=1}^J$ exhibit the same measurement statistics as L independent draws from the eigenset $\{(\lambda_n, |\lambda_n\rangle)\}_{n=1}^N$. As a consequence, in the asymptotic regime of long source sequences of length $L \to \infty$, the probability goes to 1 that the sequence $|\psi^L\rangle$ emitted by the source belongs to the typical subspace $\mathcal{H}_{\text{typ}} \subseteq \mathcal{H}_N^{\otimes L}$ having the dimension $|\mathcal{H}_{\text{typ}}| \leq N^L$ governed by Eq. (8.20) and tending to $N^{LS_N(\rho)}$.

A **coding procedure** of the source sequences $|\psi^L\rangle$ can start by measuring $|\psi^L\rangle$ with the two projectors $\left\{\Pi_{\text{typ}}, \Pi_{\text{typ}}^{\perp}\right\}$ involved in Eqs. (8.21)–(8.22), with the effect of projecting $|\psi^L\rangle$ onto the typical subspace $\mathcal{H}_{\text{typ}}$ with a probability tending to 1 in the limit of long sequences $L \to \infty$. The typical subspace $\mathcal{H}_{\text{typ}}$ can be unitarily mapped into a coding space $\mathcal{H}_{\text{cod}}$ with dimension essentially $|\mathcal{H}_{\text{typ}}| \approx N^{LS_N(\rho)} \leq N^L$ supplemented by an unimportant extra space $\mathcal{H}_{\text{ext}}$ to match the dimension N^L of $\mathcal{H}_N^{\otimes L}$ globally unitarily mapped into the product space $\mathcal{H}_{\text{cod}} \otimes \mathcal{H}_{\text{ext}}$. For every source sequence $|\psi^L\rangle \in \mathcal{H}_N^{\otimes L}$, compression is achieved by retaining only the part resulting in $\mathcal{H}_{\text{cod}}$ while discarding the part in $\mathcal{H}_{\text{ext}}$. At large length $L \to \infty$, a source sequence $|\psi^L\rangle$ is in $\mathcal{H}_{\text{typ}}$ with a probability going to 1, so that no loss is incurred in its compression. The compressed states from $\mathcal{H}_{\text{cod}}$ enable to recover the uncompressed initial states $|\psi^L\rangle \in \mathcal{H}_N^{\otimes L}$ with asymptotically perfect fidelity.

In this way, a number of $LS_N(\rho) \leq L$ instead of L quantum systems of $\mathcal{H}_N$ with dimension N can be used to compress with asymptotically no loss the information source, achieving an optimal compression rate per quantum symbol given by the quantum entropy $S(\rho)$, while no smaller rate can maintain a lossless compression, this essentially forming the **Schumacher theorem for quantum source coding** of Sect. 8.2.1.

Extension to compression of mixed-state information sources was addressed in Sect. 8.2.3. Related to information compression, the important operations of entanglement dilution and entanglement concentration between two distant parties, can also be achieved with an optimal rate controlled by the quantum entropy, as discussed in Sect. 8.4.

Chapter 9
Quantum Noise

Noise is a major concern in signal processing, classical and quantum alike. Quantum noise can be understood as any disturbance affecting a quantum system in an uncontrolled fashion, so that a well defined state vector or pure state can no longer be assigned to the quantum system. Quantum noise in this way is assimilable to the phenomenon of decoherence at large, when quantum states lose distinctive quantum properties attached to pure states [6, 7]. Such uncertainty in the state or loss of coherence may originate in a random state preparation from a statistical ensemble, or in a rather similar fashion it may originate in random fluctuations of state parameters, such as relative phases in a superposition, or such as the noisy preparation of the qubit previously described in Sect. 2.5.6. The loss of coherence very often originates in a quantum system of interest getting entangled to an uncontrolled, unobserved, environment, with as a result its state becoming uncertain. In all such scenarios inducing uncertainty on the state, the affected or noisy quantum state uniformly can be described with the common notion of density operator, representing a mixed state.

Quantum noise affecting quantum states in this way matches the conception of noise prominent in standard signal processing, when uncontrolled nuisance factors cause the alteration of some signal of interest. Accordingly, the generic terminology of "quantum noise" is preferred in this book, as in [9] for instance, to encompass the notion of quantum decoherence. Quantum noise, or quantum decoherence, is a ubiquitous factor with a major impact on the performance of quantum processing and quantum technologies. It is therefore realistic and crucial to take it into account in the framework of quantum information and quantum signal processing. In this chapter, several approaches to the modeling of quantum noise will be presented, with an emphasis on the Kraus representation, for its generality and also for its being well suited to our signals-and-systems orientation. The chapter will then concentrate on quantum noise models relevant to qubit or multiple-qubit states or signals. Theoretical noise models will be described. Circuit models will also be introduced,

F. Chapeau-Blondeau, *Quantum Information and Quantum Signal Processing*,
https://doi.org/10.1007/978-3-032-27499-1_9

enabling experimental noise simulation with controlled noise parameters on quantum hardware, in particular offering a test bed for quantum signal processing with noise.

9.1 Modeling Quantum Noise

A quantum noise in general can be modeled as a quantum process $\mathcal{N}(\cdot)$ transforming an input density operator $\rho \in \mathcal{L}(\mathcal{H}_N)$ into an output (noisy) density operator $\mathcal{N}(\rho) \in \mathcal{L}(\mathcal{H}_N)$, while forming the quantum channel depicted in Fig. 9.1.

The modeling of the quantum noise process or quantum noise channel $\mathcal{N}(\cdot)$ resorts to the general description of the evolution of a quantum state as developed in Sect. 3.2, particularly when it is a non-unitary (non-invertible) open-system evolution resulting from interaction with an uncontrolled environment. Accordingly, the quantum noise is modeled by the completely positive trace-preserving (linear) superoperator $\mathcal{N}(\cdot)$ endowed with the properties presented in Sect. 3.2. A non-unitary quantum process $\mathcal{N}(\cdot)$ typically transforms a pure state into a mixed state, in this way introducing some uncertainty in the quantum state and motivating the noise interpretation.

To proceed to a more explicit characterization, a possible approach, as explained in Sect. 3.2, is to define the quantum noise process $\mathcal{N}(\cdot)$ via an operator-sum or Kraus representation as in Eq. (3.45), acting on an input density operator $\rho \in \mathcal{L}(\mathcal{H}_N)$ according to the quantum operation

$$\rho \longmapsto \mathcal{N}(\rho) = \sum_{k=1}^{K} \Lambda_k \rho \Lambda_k^\dagger , \tag{9.1}$$

with the Kraus operators $\Lambda_k \in \mathcal{L}(\mathcal{H}_N)$ satisfying the completeness relation

$$\sum_{k=1}^{K} \Lambda_k^\dagger \Lambda_k = \mathrm{I}_N \tag{9.2}$$

of Eq. (3.47). A definite noise process or decohering process can then be modeled by specifying a set of Kraus operators $\{\Lambda_k\}_{k=1}^K$ satisfying Eq. (9.2) and acting according to Eq. (9.1). An interesting feature, as we have seen in Sect. 3.2.4, is that the number K of Kraus operators can be matched to the dimension N of the quantum system experiencing the noise and does not need to exceed N^2, irrespective of the dimen-

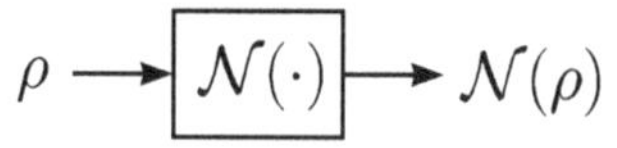

Fig. 9.1 A quantum noise process or channel performing the input–output quantum operation $\rho \mapsto \mathcal{N}(\rho)$

sionality or number of degrees of freedom of the uncontrolled environment causing the noise.

An alternative approach for specifying a noise model can be to rely on the Stinespring dilated unitary representation presented in Sect. 3.2.1. To describe the effect of a noise on a principal quantum system Q, the approach would proceed by explicitly specifying a model for the environment E causing the noise or decohering effect. This would include specification of an initial state $|e_0\rangle$ for the environment E, specification of the joint evolution by the unitary U_{QE} of the system–environment compound QE, and finally the partial tracing over the environment E, as dictated by Eq. (3.37). The initial density operator ρ of the principal quantum system Q would then undergo the transformation

$$\rho \longmapsto \mathcal{N}(\rho) = \mathrm{tr}_E\Big(\mathsf{U}_{QE}\big(\rho \otimes |e_0\rangle\langle e_0|\big)\mathsf{U}_{QE}^\dagger\Big) . \tag{9.3}$$

This approach may involve some substantial modeling, depending on the assumptions and characteristics taken into account for describing the environment. However, there exists a canonical procedure, presented in Sect. 3.2.8, to design an environment model matched to a given noise process $\mathcal{N}(\cdot)$ defined by K Kraus operators Λ_k, and resorting to a modeled environment E with same dimension K. This canonical procedure is based on the joint unitary evolution of Eq. (3.121), reading

$$\mathsf{U}_{QE}\,|\psi_{QE}\rangle = \mathsf{U}_{QE}\,|Q\rangle \otimes |e_0\rangle = \sum_{k=1}^{K} \Lambda_k\,|Q\rangle \otimes |e_k\rangle , \tag{9.4}$$

when acting on a principal system Q in the pure state $\rho = |Q\rangle\langle Q|$, to be followed on the transformed joint density operator for QE, by the partial trace $\mathrm{tr}_E(\cdot)$ of Eq. (9.3) to obtain $\mathcal{N}(\rho)$. The noise model is then explicitly determined by specifying the initial state $|e_0\rangle$ of the environment E, along with the joint unitary evolution U_{QE} to satisfy Eq. (9.4). This approach offers an environment model acting as a simulator for the quantum noise $\mathcal{N}(\cdot)$.

These two approaches (Kraus representation or dilated unitary representation) for the modeling of a quantum noise will now be further specified for the fundamental case of the qubit, and for multiple-qubit systems. We will present various useful quantum noise models or noise processes relevant to the qubit, and essential to quantum information and quantum signal processing. Moreover, quantum noise affecting multiple qubits, from its mere nature, can often be assumed to act independently on each individual qubit of a multiple-qubit state, and accordingly be modeled separately on each individual qubit. Quantum noise models for the qubit are therefore also meaningful for quantum states of arbitrary dimension obtained by assembling qubits.

9.2 Modeling Qubit Noise

As presented in Sect. 2.5.4, the quantum state of a qubit is generally represented by a density operator $\rho \in \mathcal{L}(\mathcal{H}_2)$, associated with a Bloch vector $\vec{r} \in \mathbb{R}^3$ in Eq. (2.61). A qubit noise $\mathcal{N}(\cdot)$, according to Eq. (9.1), can be described by the quantum operation

$$\rho \longmapsto \rho' = \mathcal{N}(\rho) = \sum_{k=1}^{K} \Lambda_k \rho \Lambda_k^\dagger . \tag{9.5}$$

Equation (9.5) defines a qubit noise model under the form of a quantum process or channel receiving the input state $\rho \in \mathcal{L}(\mathcal{H}_2)$ and delivering the output (noisy) state $\rho' \in \mathcal{L}(\mathcal{H}_2)$. The noise model of Eq. (9.5) is completely characterized by the K Kraus operators $\Lambda_k \in \mathcal{L}(\mathcal{H}_2)$ satisfying, according to Eq. (9.2), the completeness relation

$$\sum_{k=1}^{K} \Lambda_k^\dagger \Lambda_k = \mathrm{I}_2 . \tag{9.6}$$

For the qubit, as indicated at the end of in Sect. 3.2.4, the Kraus operators Λ_k need not be more than $K = 4$, to represent all the feasible decohering evolutions or noise models in Eq. (9.5).

In addition, according to Sect. 3.2.11 and Eq. (3.146), the evolution of Eq. (9.5) is equivalent to the transformation of the qubit Bloch vector by the affine map of the Bloch ball into itself,

$$\vec{r} \longmapsto \vec{r}\,' = A\vec{r} + \vec{c} , \tag{9.7}$$

with A a 3×3 real matrix and $\vec{c}$ a real vector in $\mathbb{R}^3$, and delivering the Bloch vector $\vec{r}\,' \in \mathbb{R}^3$ characterizing the output noisy density operator ρ' of the qubit.

A noise model for the qubit can then be specified by a set of Kraus operators $\{\Lambda_k\}$ for Eq. (9.5) or equivalently by a pair $(A, \vec{c})$ for Eq. (9.7). In the sequel, we will present some definite choices thereof corresponding to specific and important quantum noise models meaningful to the qubit.

Another approach for specifying a noise model for the qubit relies on the dilated unitary representation of Eq. (9.3). It requires to specify a joint qubit–environment unitary evolution operator U_{QE}, along with an initial state $|e_0\rangle$ for the environment. In the canonical procedure of Eq. (9.4), the model for the environment E will operate on a principal qubit Q taken in the generic pure state

$$|Q\rangle = \alpha_0 |0\rangle + \alpha_1 |1\rangle . \tag{9.8}$$

The initial state $|e_0\rangle$ of the environment E and the joint unitary evolution U_{QE} are then chosen so as to satisfy Eq. (9.4) giving here

$$|\psi'_{QE}\rangle = \mathsf{U}_{QE}\,|Q\rangle \otimes |e_0\rangle = \sum_{k=1}^{K} \Lambda_k\,|Q\rangle \otimes |e_k\rangle \tag{9.9}$$

$$= \sum_{k=1}^{K} \Big(\alpha_0 \Lambda_k\,|0\rangle \otimes |e_k\rangle + \alpha_1 \Lambda_k\,|1\rangle \otimes |e_k\rangle \Big)\,, \tag{9.10}$$

and determined by the K Kraus operators Λ_k of the qubit noise model.

This approach of specifying an environment model for simulating a quantum noise, since $K = 4$ Kraus operators are enough to represent any qubit noise, requires at most a 4-dimensional environment E, realizable by at most a 2-qubit environment model. This will give us the opportunity in the sequel, to also introduce simple noise simulators, with two or sometimes one qubit for the environment, to simulate by means of quantum circuits important qubit noise models presented in this chapter.

When specifying a qubit noise model with Kraus operators Λ_k of $\mathcal{L}(\mathcal{H}_2)$, it will often be convenient to exploit the Pauli basis $\{\mathrm{I}_2, \sigma_x, \sigma_y, \sigma_z\}$ of $\mathcal{L}(\mathcal{H}_2)$. In this perspective, it is useful to refer to the standard parameterization $\rho = \big[\rho_{00}, \rho_{01}\,;\, \rho_{10}, \rho_{11}\big]$ of Eq. (2.59) and introduce the three transformed states

$$\sigma_x \rho \sigma_x^\dagger = \begin{bmatrix} \rho_{11} & \rho_{10} \\ \rho_{01} & \rho_{00} \end{bmatrix}, \tag{9.11}$$

which is the Bloch vector transformation $\vec{r} = [r_x, r_y, r_z]^\top \mapsto [r_x, -r_y, -r_z]^\top$ or in $\mathbb{R}^3$ the rotation by π around the Ox axis,

$$\sigma_y \rho \sigma_y^\dagger = \begin{bmatrix} \rho_{11} & -\rho_{10} \\ -\rho_{01} & \rho_{00} \end{bmatrix}, \tag{9.12}$$

which is the Bloch vector transformation $\vec{r} \mapsto [-r_x, r_y, -r_z]^\top$ or in $\mathbb{R}^3$ the rotation by π around the Oy axis, and

$$\sigma_z \rho \sigma_z^\dagger = \begin{bmatrix} \rho_{00} & -\rho_{01} \\ -\rho_{10} & \rho_{11} \end{bmatrix}, \tag{9.13}$$

which is the Bloch vector transformation $\vec{r} \mapsto [-r_x, -r_y, r_z]^\top$ or in $\mathbb{R}^3$ the rotation by π around the Oz axis. This especially leads, for any density operator $\rho \in \mathcal{L}(\mathcal{H}_2)$, to the following identity

$$\rho + \sigma_x \rho \sigma_x^\dagger + \sigma_y \rho \sigma_y^\dagger + \sigma_z \rho \sigma_z^\dagger = 2\,\mathrm{tr}(\rho)\mathrm{I}_2 = 2\mathrm{I}_2\,. \tag{9.14}$$

For handling a qubit state in the Bloch representation of Eq. (2.61), one has therefore the useful relationships

$$\sigma_x (\vec{r} \cdot \vec{\sigma}\,) \sigma_x^\dagger = r_x \sigma_x - r_y \sigma_y - r_z \sigma_z\,, \tag{9.15}$$

$$\sigma_y(\vec{r}\cdot\vec{\sigma})\sigma_y^\dagger = -r_x\sigma_x + r_y\sigma_y - r_z\sigma_z\,, \tag{9.16}$$

$$\sigma_z(\vec{r}\cdot\vec{\sigma})\sigma_z^\dagger = -r_x\sigma_x - r_y\sigma_y + r_z\sigma_z\,, \tag{9.17}$$

leading to the related identity

$$\sigma_x(\vec{r}\cdot\vec{\sigma})\sigma_x^\dagger + \sigma_y(\vec{r}\cdot\vec{\sigma})\sigma_y^\dagger + \sigma_z(\vec{r}\cdot\vec{\sigma})\sigma_z^\dagger = -\vec{r}\cdot\vec{\sigma}\,. \tag{9.18}$$

9.3 Bit-Flip Noise

The first important qubit noise model is the bit-flip noise. It is defined by the two Kraus operators $\Lambda_1 = \sqrt{1-p}\,\mathrm{I}_2$ and $\Lambda_2 = \sqrt{p}\sigma_x$, to implement in Eq. (9.5) the quantum operation

$$\rho' = \mathcal{N}(\rho) = (1-p)\rho + p\sigma_x\rho\sigma_x^\dagger\,, \tag{9.19}$$

with p a real parameter in $[0, 1]$. Leaning on the probabilistic interpretation of the Kraus representation discussed in Sect. 3.2.5, the action of the bit-flip noise can be viewed as a random alteration that flips the qubit state by applying the Pauli operator σ_x with a probability p, while leaving the qubit unchanged with a probability $1-p$.

From the standard parameterization $\rho = \big[\rho_{00}, \rho_{01}\,; \rho_{10}, \rho_{11}\big]$ of Eq. (2.59), via Eq. (9.11) the transformed state is

$$\rho' = \mathcal{N}(\rho) = \begin{bmatrix} (1-p)\rho_{00} + p\rho_{11} & (1-p)\rho_{01} + p\rho_{10} \\ p\rho_{01} + (1-p)\rho_{10} & p\rho_{00} + (1-p)\rho_{11} \end{bmatrix}. \tag{9.20}$$

Equivalently, via Eq. (9.15), we have in Eq. (9.7) the transformed Bloch vector

$$\vec{r}\,' = A\vec{r} = \begin{bmatrix} 1 & 0 & 0 \\ 0 & 1-2p & 0 \\ 0 & 0 & 1-2p \end{bmatrix} \vec{r}\,. \tag{9.21}$$

It results that the action of the bit-flip noise is to contract the Bloch ball in $\mathbb{R}^3$ by the factor $1-2p$ along the Oy and Oz directions, while it leaves it unaffected along the Ox direction. A real parameter p in $[0, 1]$ is therefore required for a valid noise model contracting the Bloch ball into itself.

In this contraction of the Bloch ball, the norm $\|\vec{r}\,\|$ of the input Bloch vector never increases, with an output norm $\|\vec{r}\,'\| \le \|\vec{r}\,\|$. As a consequence, the quantum von Neumann entropy $S(\rho')$, as it results from Eq. (7.40) for the output noisy state of the qubit, is never smaller than the entropy $S(\rho)$ of the input state, that is $S(\rho') \ge S(\rho)$. Such quantum entropies, at the input or at the output, will be especially useful to us in Chap. 13 on quantum communication, for assessing the performance of a noisy quantum channel for transmission of information.

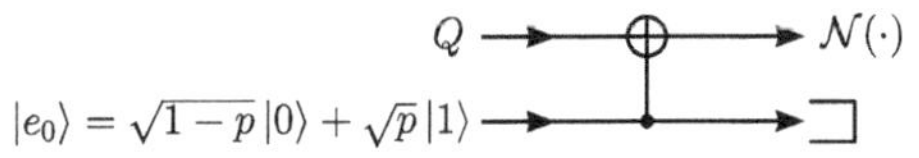

Fig. 9.2 Quantum circuit using a Cnot gate equivalent to a Pauli σ_x controlled gate, to simulate the bit-flip noise by realizing the quantum operation $\mathcal{N}(\cdot)$ of Eq. (9.19) on the principal qubit Q. The one-qubit environment E prepared in the pure state $|e_0\rangle = \sqrt{1-p}\,|0\rangle + \sqrt{p}\,|1\rangle$ sets the noise parameter p in the presence of a fixed hardware circuit and is discarded at the output. The uncertainty resulting for the output state of the principal qubit Q induces the targeted bit-flip noise

The set of Kraus operators defines a model for the environment E via the canonical procedure of Sect. 3.2.8. The $K = 2$ Kraus operators of the bit-flip noise in Eq. (9.19) determine a two-dimensional environment, realizable with a single qubit. We can resort to a one-qubit environment E referred to the orthonormal basis $\{|e_k\rangle\} = \{|e_1\rangle \equiv |0\rangle\,, |e_2\rangle \equiv |1\rangle\}$. For the evolution of the composite system QE according to Eqs. (9.9)–(9.10), we have the right-hand side

$$\sum_{k=1}^{2} \Lambda_k\,|Q\rangle \otimes |e_k\rangle = \sqrt{1-p}\,|Q\rangle \otimes |0\rangle + \sqrt{p}\sigma_x\,|Q\rangle \otimes |1\rangle \tag{9.22}$$

$$= \sqrt{1-p}\big(\alpha_0\,|00\rangle + \alpha_1\,|10\rangle\big) + \sqrt{p}\big(\alpha_0\,|11\rangle + \alpha_1\,|01\rangle\big)\,. \tag{9.23}$$

We then need to devise, for Eq. (9.9), the joint unitary evolution U_{QE} of the qubit pair QE together with the initial state $|e_0\rangle$ of the environment qubit E, so that $\mathsf{U}_{QE}\,|Q\rangle \otimes |e_0\rangle$ matches Eqs. (9.22)–(9.23). This can be accomplished with the two-qubit circuit of Fig. 9.2.

The quantum circuit in Fig. 9.2 uses a Cnot gate which is equivalently a Pauli σ_x controlled gate. The control is achieved by the environment qubit E in the pure state $|e_0\rangle = \sqrt{1-p}\,|0\rangle + \sqrt{p}\,|1\rangle$ which enables to set the probability p of the bit-flip noise. In this way, the noise simulator of Fig. 9.2 presents the interesting property of involving a fixed hardware (the Cnot gate) and a noise level p adjustable by the external signal $|e_0\rangle$. In turn, the control qubit state $|e_0\rangle$ can be easily obtained from a state $|0\rangle$ subjected to the rotation gate $\mathsf{R}_y(\xi)$ from Eq. (3.26) at an angle $\xi = 2\arcsin\big(\sqrt{p}\big)$.

When the principal qubit Q in Fig. 9.2 is in the generic state $|Q\rangle$ of Eq. (9.8) then the two-qubit input state in Fig. 9.2 is

$$|Q\rangle \otimes |e_0\rangle = \alpha_0\sqrt{1-p}\,|00\rangle + \alpha_1\sqrt{1-p}\,|10\rangle + \alpha_0\sqrt{p}\,|01\rangle + \alpha_1\sqrt{p}\,|11\rangle\;, \tag{9.24}$$

and, by the action of the Cnot gate, a two-qubit output state follows as

$$|\psi'_{QE}\rangle = \mathsf{U}_{QE}\,|Q\rangle \otimes |e_0\rangle = \alpha_0\sqrt{1-p}\,|00\rangle + \alpha_1\sqrt{1-p}\,|10\rangle + \alpha_0\sqrt{p}\,|11\rangle + \alpha_1\sqrt{p}\,|01\rangle\;, \tag{9.25}$$

which is exactly the targeted behavior of Eqs. (9.22)–(9.23).

In general the output state $|\psi'_{QE}\rangle$ in Eq. (9.25) describes a principal qubit Q that has become entangled with the environment (control) qubit E. When the control qubit in Fig. 9.2 is discarded (unaccessed) at the output—as it simulates an uncontrolled environment E—then the resulting uncertainty on the state of the principal qubit Q is described by the operation of partial tracing over the environment E, as it occurs in Eq. (9.3). The general procedure of Sect. 3.2.8 ensures that the targeted noise model $\rho \mapsto \mathcal{N}(\rho)$ is obtained in this way. Yet, this can be verified by explicitly calculating from Eq. (9.25) the partial trace of $\rho'_{QE} = |\psi'_{QE}\rangle\langle\psi'_{QE}|$ over the environment E. The final state of the principal qubit Q results as

$$\mathrm{tr}_E\big(|\psi'_{QE}\rangle\langle\psi'_{QE}|\big) = \begin{bmatrix} (1-p)|\alpha_0|^2 + p|\alpha_1|^2 & (1-p)\alpha_0\alpha_1^* + p\alpha_0^*\alpha_1 \\ (1-p)\alpha_0^*\alpha_1 + p\alpha_0\alpha_1^* & (1-p)|\alpha_1|^2 + p|\alpha_0|^2 \end{bmatrix} = \mathcal{N}\big(|Q\rangle\langle Q|\big)\,, \tag{9.26}$$

which is exactly the noisy state $\mathcal{N}\big(|Q\rangle\langle Q|\big)$ of Eq. (9.20) for a qubit initially in the pure state $|Q\rangle$ with density operator $\rho = |Q\rangle\langle Q|$ as in Eq. (2.58). By linearity, the circuit of Fig. 9.2 will also transform according to Eq. (9.19) mixed input states ρ of Q as convex sums of pure states.

In addition, the joint output state $\rho'_{QE} = |\psi'_{QE}\rangle\langle\psi'_{QE}|$ from Eq. (9.25), can alternatively be traced over the principal qubit Q. This provides a characterization for the final reduced state $\rho'_E = \mathrm{tr}_Q\big(\rho'_{QE}\big)$ of the environment qubit E after its interaction with Q, and adhering to the general picture of Sect. 3.2.7 for the evolution of the environment. Equivalently, the final state ρ'_E of the environment, as generally expressed in Eq. (3.119), can be characterized via its matrix elements of Eq. (3.120), to yield here

$$\rho'_E = \begin{bmatrix} 1-p & \sqrt{(1-p)p}\,r_x \\ \sqrt{(1-p)p}\,r_x & p \end{bmatrix}, \tag{9.27}$$

valid for any input state ρ, pure or mixed, of the principal qubit Q, characterized by the input Bloch vector $\vec{r} = [r_x, r_y, r_z]^\top$. By Eqs. (2.63)–(2.65) we find that the state ρ'_E is a qubit density operator characterized by the Bloch vector $\big[2\sqrt{(1-p)p}\,r_x,\ 0,\ 1-2p\big]^\top$. The state ρ'_E of Eq. (9.27) has the two eigenvalues

$$\lambda_\pm = \frac{1}{2} \pm \frac{1}{2}\Big[4(1-p)p\,r_x^2 + (1-2p)^2\Big]^{1/2}. \tag{9.28}$$

They can especially be used to compute the final quantum entropy $S(\rho'_E) = H\big(\{\lambda_\pm\}\big)$ of the environment qubit E. In the noise model, the environment E always starts in the pure state $|e_0\rangle$ with a zero initial quantum entropy $S\big(|e_0\rangle\langle e_0|\big) = 0$, and E terminates with a final entropy $S(\rho'_E)$, which represents the entropy generated in the environment E during the interaction with the principal qubit Q resulting in the bit-flip noise here. As we shall see in Chap. 13 on quantum communication, the quantum entropy $S(\rho'_E)$ generated in the environment will stand as a key quantity to assess (via the entropy exchange of Sect. 13.3.1) the performance of a noisy quantum communication channel for transmission of quantum information.

The final state ρ'_E of the environment, as it directly follows from the Kraus operators via Eq. (3.119), is not affected by the constituent details of the environment model serving for the Stinespring dilated representation of the noise process. This was generally noticed at the end of Sect. 3.2.7, and is attested here by Eq. (9.27) and Eq. (9.28) which are not directly affected by the initial state $|e_0\rangle$ of the environment qubit E, and are essentially influenced by the input state ρ of the principal qubit Q and by the level p of the bit-flip noise in Eq. (9.19). The same is true for the resulting quantum entropy $S(\rho'_E)$ generated in the environment, which is determined essentially by the input state of the principal system Q and by the quantum noise $\mathcal{N}(\cdot)$, but not by the specific details of alternative equivalent modelings of the environment, as it will also be further verified in general conditions in Sect. 13.3.1.

A bit-flip noise, in practice, can occur in many situations, when the interaction of the qubit with the environment results in a possibility of random flip in its state as described by Eq. (9.19). This is the case for instance in the example of the electronic spin in the earth magnetic field evoked in Sect. 3.2.5 in relation to Eq. (3.102). The example of the noisy preparation of the qubit described in Sect. 2.5.6 can also be viewed as the action of a bit-flip noise process. As we argued in the penultimate paragraph of Sect. 3.2.5, the noisy preparation that flips $|0\rangle$ into $|1\rangle = \sigma_x |0\rangle$ with the flip probability $p = \langle \sin^2(\xi) \rangle$, can be modeled with the two Kraus operators $\{\Lambda_1 = \sqrt{1-p}\, \mathrm{I}_2,\ \Lambda_2 = \sqrt{p}\sigma_x\}$, which is a bit-flip noise process according to Eq. (9.19). This illustrates how various detailed scenarios where a qubit interacts with an uncontrolled environment can be modeled by the generic and concise methodology of quantum operation with Kraus operators, with here the bit-flip noise process or channel of Eq. (9.19).

9.4 Phase-Flip Noise

Another, rather comparable, noise model for the qubit is the phase-flip noise. This quantum noise randomly flips the qubit phase by applying the Pauli operator σ_z with probability p, while it leaves the qubit unchanged with probability $1-p$. It is defined by the two Kraus operators $\Lambda_1 = \sqrt{1-p}\, \mathrm{I}_2$ and $\Lambda_2 = \sqrt{p}\sigma_z$, to implement in Eq. (9.5) the quantum operation

$$\rho' = \mathcal{N}(\rho) = (1-p)\rho + p\sigma_z \rho \sigma_z^\dagger \,. \tag{9.29}$$

From the standard parameterization $\rho = [\rho_{00}, \rho_{01}\, ; \rho_{10}, \rho_{11}]$ of Eq. (2.59), via Eq. (9.13) the transformed state is

$$\rho' = \mathcal{N}(\rho) = \begin{bmatrix} \rho_{00} & (1-2p)\rho_{01} \\ (1-2p)\rho_{10} & \rho_{11} \end{bmatrix} . \tag{9.30}$$

Equivalently, via Eq. (9.17), we have in Eq. (9.7) the transformed Bloch vector

$$\vec{r}\,' = A\vec{r} = \begin{bmatrix} 1-2p & 0 & 0 \\ 0 & 1-2p & 0 \\ 0 & 0 & 1 \end{bmatrix} \vec{r}\,. \tag{9.31}$$

It results that the action of the phase-flip noise is to contract the Bloch ball in $\mathbb{R}^3$ by the factor $1-2p$ along the Ox and Oy directions, while it leaves it unaffected along the Oz direction. In this contraction, the norm $\|\vec{r}\,\|$ of the input Bloch vector never increases, with an output norm $\|\vec{r}\,'\| \le \|\vec{r}\,\|$. As a consequence again, the quantum entropy $S(\rho')$, as it results from Eq. (7.40) for the output noisy state of the qubit, is never smaller than the entropy $S(\rho)$ of the input state, that is $S(\rho') \ge S(\rho)$.

The $K=2$ Kraus operators of the phase-flip noise in Eq. (9.29) again determine, by the canonical procedure of Sect. 3.2.8, a two-dimensional environment E, realizable with a single qubit. To simulate the phase-flip noise of Eq. (9.29) on the principal qubit Q by the procedure of Sect. 3.2.8, as we have done for the bit-flip noise, a quantum circuit can be devised acting on a one-qubit environment E referred to the orthonormal basis $\{|e_k\rangle\} = \{|0\rangle\,,|1\rangle\}$, so that the composite system QE according to Eq. (9.9) satisfies

$$\mathsf{U}_{QE}\,|Q\rangle \otimes |e_0\rangle = \sqrt{1-p}\,|Q\rangle \otimes |0\rangle + \sqrt{p}\sigma_z\,|Q\rangle \otimes |1\rangle\;, \tag{9.32}$$

as determined by the two Kraus operators of the phase-flip noise. This can be accomplished with the two-qubit circuit of Fig. 9.3, which uses a Pauli σ_z controlled gate. The practical implementation of the controlled qubit gate adheres to the general considerations of Sect. 5.6, and more specifically here to Fig. 5.15. In the quantum circuit of Fig. 9.3, the control is achieved by an environment qubit E placed in the pure state $|e_0\rangle = \sqrt{1-p}\,|0\rangle + \sqrt{p}\,|1\rangle$ which again provides the advantage of adjusting the probability p of the phase-flip noise with the external signal $|e_0\rangle$ in the presence of a fixed hardware for the phase-flip noise simulator.

The operation of the quantum circuit of Fig. 9.3 can be analyzed as done with Eqs. (9.24)–(9.25) for the bit-flip noise, and it verifies that the targeted behavior according to Eq. (9.32) is indeed obtained, so as to simulate the phase-flip noise $\mathcal{N}(\cdot)$ of Eq. (9.29).

In addition, after its interaction with the principal qubit Q, the environment qubit E terminates in the final reduced state ρ'_E from Eq. (3.119) characterized, via its

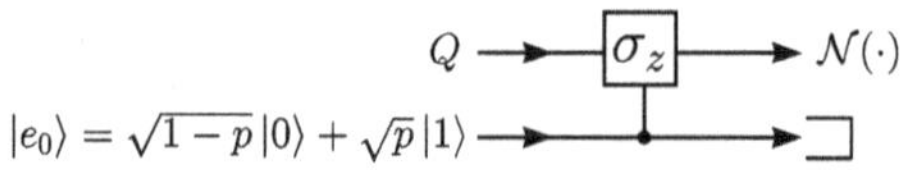

Fig. 9.3 Quantum circuit using a Pauli σ_z controlled gate, to simulate the phase-flip noise process by realizing the quantum operation $\mathcal{N}(\cdot)$ of Eq. (9.29) on the principal qubit Q. The one-qubit environment E prepared in the pure state $|e_0\rangle = \sqrt{1-p}\,|0\rangle + \sqrt{p}\,|1\rangle$ sets the noise parameter p in the presence of a fixed hardware circuit and is discarded at the output

matrix elements of Eq. (3.120), as

$$\rho'_E = \begin{bmatrix} 1-p & \sqrt{(1-p)p}\, r_z \\ \sqrt{(1-p)p}\, r_z & p \end{bmatrix} . \tag{9.33}$$

By Eqs. (2.63)–(2.65), the state ρ'_E is a qubit density operator characterized by the Bloch vector $\left[2\sqrt{(1-p)p}\, r_z,\ 0,\ 1-2p\right]^\top$. The state ρ'_E of Eq. (9.33) has the two eigenvalues

$$\lambda_\pm = \frac{1}{2} \pm \frac{1}{2}\left[4(1-p)p\, r_z^2 + (1-2p)^2\right]^{1/2} . \tag{9.34}$$

They can be used to compute the final quantum entropy $S(\rho'_E) = H\big(\{\lambda_\pm\}\big)$ of the environment qubit E, which is also the entropy generated in the environment by the action of the phase-flip noise.

An example of occurrence of a phase-flip noise can be obtained with the interferometer of Fig. 3.4. We consider that this interferometer is governed by a qubit unitary operator U_ξ as in Eq. (3.18) defined by the matrix of Eq. (3.31) referred to the basis $\{|0\rangle\,, |1\rangle\}$. Accordingly, when it receives as input a qubit (photon) in the generic pure state $|Q\rangle = \alpha_0\,|0\rangle + \alpha_1\,|1\rangle$, the interferometer delivers the output state $|Q'_\xi\rangle = \mathsf{U}_\xi\,|Q\rangle = \alpha_0\,|0\rangle + \alpha_1 e^{i\xi}\,|1\rangle$. We also consider that some random fluctuations give rise to a phase shift ξ which is now a random variable, distributed according to the probability density function $p_\xi(\xi)$ as in Sect. 2.5.6 (as for instance mechanical vibrations in the two arms of the interferometer inducing random variations in the difference in length of the two optical paths). This is an effect of randomization of the relative phase between the two components of the qubit state, giving rise to a random state $|Q'_\xi\rangle$ associated with the probability distribution $p_\xi(\xi)$, to form at the output a statistical ensemble of pure states described by the density operator

$$\rho' = \int_\xi p_\xi(\xi)\,|Q'_\xi\rangle\,\langle Q'_\xi|\,d\xi \tag{9.35}$$

similar to Eq. (2.85), and having the matrix representation

$$\rho' = \begin{bmatrix} |\alpha_0|^2 & \alpha_0\alpha_1^*\langle e^{-i\xi}\rangle \\ \alpha_0^*\alpha_1\langle e^{i\xi}\rangle & |\alpha_1|^2 \end{bmatrix} , \tag{9.36}$$

where the statistical averages like $\langle e^{i\xi}\rangle$ are according to the probability density $p_\xi(\xi)$.

From the input state $\rho = |Q\rangle\,\langle Q|$ as in Eq. (2.58), and via Eqs. (2.63)–(2.65), it is found that the input–output transformation $\rho \mapsto \rho'$ implemented by the noisy interferometer is equivalent to the transformation $\vec{r} \mapsto \vec{r}\,' = A\vec{r}$ of the Bloch vectors, with the matrix

$$A = \begin{bmatrix} \langle\cos(\xi)\rangle & -\langle\sin(\xi)\rangle & 0 \\ \langle\sin(\xi)\rangle & \langle\cos(\xi)\rangle & 0 \\ 0 & 0 & 1 \end{bmatrix}. \tag{9.37}$$

In the simple case where $\langle\sin(\xi)\rangle = 0$ (this occurs for instance with an even probability density $p_\xi(\xi)$ as in Fig. 2.4, describing a phase ξ with symmetric fluctuations around 0), by comparing A with Eq. (9.31) we see that the noisy interferometer implements a phase-flip noise process with flip probability $p = \big(1 - \langle\cos(\xi)\rangle\big)/2 = \langle\sin^2(\xi/2)\rangle$. In the basis $\{|0\rangle, |1\rangle\}$, between the two components of the qubit state, a phase fluctuation ξ symmetric about zero is equivalent to a phase-flip noise. Although the phase fluctuation is continuous in its detail, its effect on the two-dimensional qubit is equivalent to the discrete phase flips according to Eq. (9.29).

More generally, the phase ξ will fluctuate with an arbitrary distribution $p_\xi(\xi)$ around a constant value $\overline{\xi} \neq 0$, enabling for A of Eq. (9.37) the factorization

$$A = \begin{bmatrix} \langle\cos(\xi - \overline{\xi})\rangle & -\langle\sin(\xi - \overline{\xi})\rangle & 0 \\ \langle\sin(\xi - \overline{\xi})\rangle & \langle\cos(\xi - \overline{\xi})\rangle & 0 \\ 0 & 0 & 1 \end{bmatrix} \begin{bmatrix} \cos(\overline{\xi}) & -\sin(\overline{\xi}) & 0 \\ \sin(\overline{\xi}) & \cos(\overline{\xi}) & 0 \\ 0 & 0 & 1 \end{bmatrix}. \tag{9.38}$$

The constant phase $\overline{\xi}$ is defined so that the average $\langle\sin(\xi - \overline{\xi})\rangle = 0$ in Eq. (9.38), which is obtained by $\tan(\overline{\xi}) = \langle\sin(\xi)\rangle / \langle\cos(\xi)\rangle$. Then, the first matrix in the right-hand side of Eq. (9.38) becomes exactly the matrix of a phase-flip noise as in Eq. (9.31) with flip probability p given by $1 - 2p = \langle\cos(\xi - \overline{\xi})\rangle \Leftrightarrow p = \langle\sin^2[(\xi - \overline{\xi})/2]\rangle$. Meanwhile, the second matrix in the right-hand side of Eq. (9.38) represents a rotation matrix of the Bloch vector by the angle $\overline{\xi}$ around the Oz axis in $\mathbb{R}^3$. This represents the effect of a unitary qubit operator $\mathsf{U}_{\overline{\xi}}$, an instance of the generic form of Eq. (3.18). So the noisy interferometer can be analyzed as a noise-free interferometer implementing on the incoming qubit a phase shift by the constant angle $\overline{\xi}$ via the unitary $\mathsf{U}_{\overline{\xi}}$, followed by a phase-flip noise $\mathcal{N}(\cdot)$ as in Eq. (9.31) on the qubit with flip probability $p = \langle\sin^2[(\xi - \overline{\xi})/2]\rangle$. Globally, this is the quantum process $\rho \mapsto \mathsf{U}_{\overline{\xi}}\rho\mathsf{U}_{\overline{\xi}}^\dagger \mapsto \mathcal{N}(\mathsf{U}_{\overline{\xi}}\rho\mathsf{U}_{\overline{\xi}}^\dagger) = \mathcal{E}(\rho) = \rho'$, with the quantum process $\mathcal{E}(\cdot)$ characterized by the two Kraus operators $\sqrt{1-p}\,\mathsf{U}_{\overline{\xi}}$ and $\sqrt{p}\,\sigma_z\mathsf{U}_{\overline{\xi}}$. Alternatively, the noisy interferometer can be analyzed as the phase-flip noise applied first on the incoming qubit, followed by the coherent phase shift by $\mathsf{U}_{\overline{\xi}}$, since these two processes commute here (although this is not always the rule), giving the equivalent quantum process $\rho \mapsto \mathcal{N}(\rho) \mapsto \mathsf{U}_{\overline{\xi}}\mathcal{N}(\rho)\mathsf{U}_{\overline{\xi}}^\dagger = \mathcal{E}(\rho)$, with the two Kraus operators $\sqrt{1-p}\,\mathsf{U}_{\overline{\xi}}$ and $\sqrt{p}\,\mathsf{U}_{\overline{\xi}}\sigma_z$.

At this stage, important properties can be emphasized for a quantum process or channel.

- First, as already mentioned, a noise process like Eq. (9.29) or Eq. (9.19), although it describes discrete random flips, offers, for a finite-dimensional quantum system, the appropriate modeling of the effect of analog or continuously-distributed fluctuations. This is observed with the noisy interferometer above, with analog fluctuations in the optical paths which, when taken into account according to the principles of quantum evolution as set in Chap. 3, give rise to a discrete phase-flip noise. This was also observed with the noisy preparation of a qubit in Sect. 2.5.6, where analog fluctuations in some part alignment give rise to a discrete bit-flip noise. Phrased in a terminology in use in signal processing, the physical reality might be analog, its operational formalization is digital.

- Second, any conceivable physical evolution of a quantum system—closed or open, unitary or non—can be understood as a process transforming a density operator into another density operator, which can be described via the operator-sum of Eq. (3.45) by a finite set of Kraus operators $\{\Lambda_k\}$. Although such Kraus operators $\{\Lambda_k\}$ are thus conferred a fundamental status in the quantum theory, they may nevertheless be derived, in given physical situations, from empirical or phenomenological considerations, yet framed by the principles of quantum evolution of Chap. 3. Such phenomenological considerations may be recognized in the handling of the random fluctuations present with the noisy interferometer above or the noisy qubit preparation of Sect. 2.5.6, nevertheless leading to a valid quantum evolution.
- Third, a quantum process or channel as the noisy interferometer above, when analyzed according to the principles of Chap. 3 on quantum evolution, can naturally appear as an association of two more elementary and meaningful processes. In this way, the noisy interferometer comes up as a noise-free interferometer performing the unitary phase shift by $\mathsf{U}_{\overline{\xi}}$, associated with a reference quantum noise model (phase-flip here) among those presented in this chapter. The different associated processes combine in a natural way, based on the principles of quantum evolution of Chap. 3 and the rules of linear algebra. This matches well with the conceptualization common in signals-and-systems theory, where a given system can be assigned a well-defined operational role or functionality in transforming/processing information-carrying signals, and where such operative systems can be associated/cascaded to perform, stage by stage, some more elaborate signal processing task. This picture can be carried over to the quantum realm, with the quantum states (density operators) materializing the signals, and the quantum processes materializing the operative systems, which can be combined or cascaded for transforming, stage by stage, the density operators, performing in this way some relevant quantum signal processing.

Finally, we mention that another useful set of two Kraus operators can also give rise to a phase-flip noise as in Eq. (9.29). Consider the alternative set of two Kraus operators defined in the computational basis by

$$\Lambda_1' = \begin{bmatrix} 1 & 0 \\ 0 & \sqrt{1-\gamma} \end{bmatrix} = |0\rangle\langle 0| + \sqrt{1-\gamma}\,|1\rangle\langle 1| \ , \tag{9.39}$$

$$\Lambda_2' = \begin{bmatrix} 0 & 0 \\ 0 & \sqrt{\gamma} \end{bmatrix} = \sqrt{\gamma}\,|1\rangle\langle 1| \ , \tag{9.40}$$

with a parameter $\gamma \in [0, 1]$. This alternative Kraus set defines a qubit noise process transforming the qubit Bloch vector by $\vec{r} = [r_x, r_y, r_z]^\top \mapsto \big[\sqrt{1-\gamma}\, r_x, \sqrt{1-\gamma}\, r_y, r_z\big]^\top$. The two alternative Kraus operators of Eqs. (9.39)–(9.40) are related to the two former Kraus operators $\Lambda_1 = \sqrt{1-p}\, \mathrm{I}_2$ and $\Lambda_2 = \sqrt{p}\sigma_z$ of Eq. (9.29) by

$$\begin{cases} \Lambda_1' = \sqrt{1-p}\Lambda_1 + \sqrt{p}\Lambda_2\,, & (9.41) \\ \Lambda_2' = \sqrt{p}\Lambda_1 - \sqrt{1-p}\Lambda_2\,, & (9.42) \end{cases}$$

when $p = \big(1 - \sqrt{1-\gamma}\big)/2$. The relations of Eqs. (9.41)–(9.42) form an instance of the isometric (unitary here) relation of Eq. (3.72), so that the alternative Kraus set $\{\Lambda_1', \Lambda_2'\}$ equally represents a phase-flip noise, also ruled by the former Kraus set $\{\Lambda_1, \Lambda_2\}$. The noise model based on the alternative Kraus set $\{\Lambda_1', \Lambda_2'\}$ of Eqs. (9.39)–(9.40) is referred to as a *phase damping noise*. It is also a phase-flip noise, yet it makes possible an alternative interpretation of the random action by the noise, replacing the random flips to the phase, and which can be meaningful in some specific physical contexts. The operator Λ_1' in Eq. (9.39) leaves the (ground) state $|0\rangle$ unaffected, and it reduces or damps by a factor $\sqrt{1-\gamma}$ the probability amplitude of the (excited) state $|1\rangle$, so that after the action of Λ_1' there would be less probability of finding the qubit in the (excited) state $|1\rangle$. The operator Λ_2' in Eq. (9.40) with $\Lambda_2'\,|0\rangle = 0$ destroys the state $|0\rangle$, like when a photon is absorbed, and reduces by a factor $\sqrt{\gamma}$ the probability amplitude of the state $|1\rangle$.

Accordingly, the qubit noise represented by the quantum process $\mathcal{N}(\cdot)$ of this Sect. 9.4, which can describe a phase damping noise as well as a phase-flip noise, is sometimes referred to simply as a *quantum phase noise*.

9.5 Bit-Phase-Flip Noise

Another related qubit noise is the bit-phase-flip noise. This quantum noise performs both a bit flip and phase flip to the qubit by applying the Pauli operator $\sigma_y = i\sigma_x\sigma_z$ with probability p, while it leaves the qubit unchanged with probability $1-p$. It is defined by the two Kraus operators $\Lambda_1 = \sqrt{1-p}\, \mathrm{I}_2$ and $\Lambda_2 = \sqrt{p}\sigma_y$, to implement in Eq. (9.5) the quantum operation

$$\rho' = \mathcal{N}(\rho) = (1-p)\rho + p\sigma_y\rho\sigma_y^\dagger\,. \tag{9.43}$$

From the parameterization $\rho = \big[\rho_{00}, \rho_{01}\,;\, \rho_{10}, \rho_{11}\big]$ of Eq. (2.59), via Eq. (9.12) the transformed state is

$$\rho' = \mathcal{N}(\rho) = \begin{bmatrix} (1-p)\rho_{00} + p\rho_{11} & (1-p)\rho_{01} - p\rho_{10} \\ -\,p\rho_{01} + (1-p)\rho_{10} & p\rho_{00} + (1-p)\rho_{11} \end{bmatrix}. \tag{9.44}$$

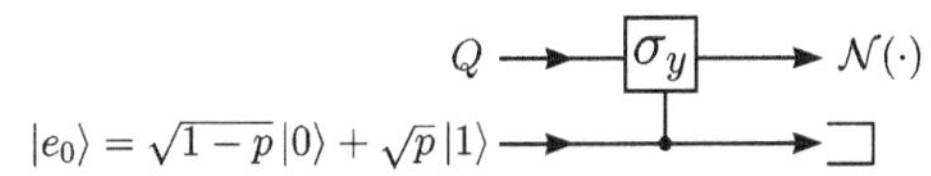

Fig. 9.4 Quantum circuit using a Pauli σ_y controlled gate, to simulate the bit-phase-flip noise by realizing the quantum operation $\mathcal{N}(\cdot)$ of Eq. (9.43) on the principal qubit Q. The one-qubit environment E prepared in the pure state $|e_0\rangle = \sqrt{1-p}\,|0\rangle + \sqrt{p}\,|1\rangle$ sets the noise parameter p in the presence of a fixed hardware circuit and is discarded at the output

The transformation of the Bloch vector in Eq. (9.7) amounts to

$$\vec{r}\,' = A\vec{r} = \begin{bmatrix} 1-2p & 0 & 0 \\ 0 & 1 & 0 \\ 0 & 0 & 1-2p \end{bmatrix} \vec{r}\,. \tag{9.45}$$

It results that the action of the bit-phase-flip noise is to contract the Bloch ball in $\mathbb{R}^3$ by the factor $1-2p$ along the Ox and Oz directions, while it leaves it unaffected along the Oy direction. In this contraction, the norm $\|\vec{r}\,\|$ of the input Bloch vector never increases, with an output norm $\|\vec{r}\,'\| \le \|\vec{r}\,\|$. As a consequence again, the quantum entropy $S(\rho')$, as it results from Eq. (7.40) for the output noisy state of the qubit, is never smaller than the entropy $S(\rho)$ of the input state, that is $S(\rho') \ge S(\rho)$.

The $K = 2$ Kraus operators of the bit-phase-flip noise in Eq. (9.43) again determine, by the canonical procedure of Sect. 3.2.8, a two-dimensional environment E, realizable with a single qubit. To simulate the bit-phase-flip noise of Eq. (9.43) on the principal qubit Q by the procedure of Sect. 3.2.8, a quantum circuit can be devised acting on a one-qubit environment E referred to the orthonormal basis $\{|e_k\rangle\} = \{|0\rangle, |1\rangle\}$, so that the composite system QE according to Eq. (9.9) satisfies

$$\mathsf{U}_{QE}\,|Q\rangle \otimes |e_0\rangle = \sqrt{1-p}\,|Q\rangle \otimes |0\rangle + \sqrt{p}\sigma_y\,|Q\rangle \otimes |1\rangle\ . \tag{9.46}$$

This can be accomplished by the two-qubit circuit of Fig. 9.4 using a Pauli σ_y controlled gate, implementable as in Fig. 5.16. The control is achieved by an environment qubit E placed in the pure state $|e_0\rangle = \sqrt{1-p}\,|0\rangle + \sqrt{p}\,|1\rangle$ with again the advantage of adjusting the probability p of the bit-phase-flip noise through the external signal $|e_0\rangle$ in the presence of a fixed hardware for the noise simulator.

For the bit-phase-flip noise, the two Kraus operators from Eq. (9.43) provide direct access, via Eq. (3.119), to the final state ρ'_E of the two-dimensional environment E, characterized, via its matrix elements of Eq. (3.120), as

$$\rho'_E = \begin{bmatrix} 1-p & \sqrt{(1-p)p}\,r_y \\ \sqrt{(1-p)p}\,r_y & p \end{bmatrix}. \tag{9.47}$$

By Eqs. (2.63)–(2.65), the state ρ'_E is a qubit density operator with Bloch vector $\left[2\sqrt{(1-p)p}\,r_y,\ 0,\ 1-2p\right]^{\mathsf{T}}$. The state ρ'_E of Eq. (9.47) has the two eigenvalues

$$\lambda_{\pm} = \frac{1}{2} \pm \frac{1}{2}\Big[4(1-p)p\,r_y^2 + (1-2p)^2\Big]^{1/2} . \tag{9.48}$$

They can be used to compute the quantum entropy $S(\rho'_E)$ of the environment, which is also the entropy generated in the environment by the action of the bit-phase-flip noise on the input state ρ.

9.6 Depolarizing Noise

The depolarizing noise leaves the qubit unchanged with probability $1-p$, or it applies one of the three Pauli operators randomly with probability $p/3$. It is defined by the four Kraus operators $\sqrt{1-p}\,\mathrm{I}_2$ and $\sqrt{p/3}\sigma_k$ with $k = x, y, z$, to implement in Eq. (9.5) the quantum operation

$$\rho' = \mathcal{N}(\rho) = (1-p)\rho + \frac{p}{3}\big(\sigma_x\rho\sigma_x^\dagger + \sigma_y\rho\sigma_y^\dagger + \sigma_z\rho\sigma_z^\dagger\big) , \tag{9.49}$$

already met in Eq. (3.103). Equation (9.49) can also be written as

$$\rho' = \mathcal{N}(\rho) = (1-p')\rho + \frac{p'}{4}\big(\mathrm{I}_2\rho\mathrm{I}_2^\dagger + \sigma_x\rho\sigma_x^\dagger + \sigma_y\rho\sigma_y^\dagger + \sigma_z\rho\sigma_z^\dagger\big) , \tag{9.50}$$

with $p' = 4p/3$, offering the alternative interpretation of the action of the depolarizing noise, as to leave the qubit state ρ unchanged with probability $1-p'$, or to apply one of the four operators of the Pauli basis $\{\mathrm{I}_2, \sigma_x, \sigma_y, \sigma_z\}$ of $\mathcal{L}(\mathcal{H}_2)$ randomly with probability $p'/4$. Obviously, applying the trivial Pauli operator I_2 is again an unchanged qubit state, accumulating the total probability $1 - p' + p'/4 = 1 - p$. Finally, by using the identity of Eq. (9.14), it is also possible to obtain for the quantum operation of Eq. (9.49) the equivalent (non-Kraus) form

$$\rho' = \mathcal{N}(\rho) = \alpha\rho + (1-\alpha)\frac{\mathrm{I}_2}{2} , \tag{9.51}$$

with $\alpha = 1 - 4p/3 = 1 - p'$. This offers the alternative interpretation of the action of the depolarizing noise, as to leave the qubit state ρ unchanged with probability α, or to replace it with probability $1-\alpha$ by the maximally mixed state $\mathrm{I}_2/2$ where the qubit is completely depolarized. The fully depolarizing noise is when $\alpha = 0$ in Eq. (9.51), and the qubit state ρ is forced into the completely depolarized state $\mathrm{I}_2/2$; this corresponds in Eq. (9.49) to $1 - p = p/3 = 1/4$, that is, an equiprobable random application to ρ of any of the four Pauli operators.

In addition, Eq. (9.51) offers a natural extension of the depolarizing noise to the qudit with Hilbert space $\mathcal{H}_N$ of arbitrary dimension N, via the transformation

$$\mathcal{N}(\rho) = \alpha\rho + (1-\alpha)\frac{\mathrm{I}_N}{N} , \tag{9.52}$$

where the qudit state $\rho \in \mathcal{L}(\mathcal{H}_N)$ is replaced with probability $1-\alpha$ by the maximally mixed qudit state I_N/N.

For the qubit, with the parameterization $\rho = \left[\rho_{00}, \rho_{01} ; \rho_{10}, \rho_{11}\right]$ of Eq. (2.59) used in Eq. (9.49) or (9.51), the transformed state results as

$$\rho' = \mathcal{N}(\rho) = \begin{bmatrix} \left(1-\frac{2}{3}p\right)\rho_{00} + \frac{2}{3}p\rho_{11} & \left(1-\frac{4}{3}p\right)\rho_{01} \\ \left(1-\frac{4}{3}p\right)\rho_{10} & \frac{2}{3}p\rho_{00} + \left(1-\frac{2}{3}p\right)\rho_{11} \end{bmatrix} , \tag{9.53}$$

which is also

$$\rho' = \mathcal{N}(\rho) = \begin{bmatrix} \frac{1+\alpha}{2}\rho_{00} + \frac{1-\alpha}{2}\rho_{11} & \alpha\rho_{01} \\ \alpha\rho_{10} & \frac{1-\alpha}{2}\rho_{00} + \frac{1+\alpha}{2}\rho_{11} \end{bmatrix} . \tag{9.54}$$

The transformation of the Bloch vector in Eq. (9.7) amounts to

$$\vec{r}\,' = A\vec{r} = \begin{bmatrix} \alpha & 0 & 0 \\ 0 & \alpha & 0 \\ 0 & 0 & \alpha \end{bmatrix} \vec{r} = \alpha\vec{r} . \tag{9.55}$$

It results that the action of the depolarizing noise is to contract the Bloch ball by the factor α uniformly in the three directions of $\mathbb{R}^3$. In this contraction, the norm $\|\vec{r}\,\|$ of the input Bloch vector never increases, since $|\alpha| \in [0, 1]$, with an output norm $\|\vec{r}\,'\| \le \|\vec{r}\,\|$. As a consequence again, the quantum entropy $S(\rho')$, as it results from Eq. (7.40) for the output noisy state of the qubit, is never smaller than the entropy $S(\rho)$ of the input state, that is $S(\rho') \ge S(\rho)$.

The $K = 4$ Kraus operators of the depolarizing noise in Eq. (9.49) determine, by the canonical procedure of Sect. 3.2.8, a four-dimensional environment E, realizable with a pair of qubits. To simulate the depolarizing noise of Eq. (9.49) on the principal qubit Q, it is possible to design a quantum circuit adhering to the generic procedure of Sect. 3.2.8, as we did for the bit-flip, phase-flip and bit-phase-flip noises. The two-qubit environment E referred to the orthonormal basis $\left\{|e_k\rangle\right\}_{k=1}^{4} = \left\{|00\rangle , |01\rangle , |10\rangle , |11\rangle\right\}$, should be devised so that the composite system QE according to Eq. (9.9) satisfies

$$\mathsf{U}_{QE}\,|Q\rangle \otimes |e_0\rangle = \sqrt{1-p}\,|Q\rangle \otimes |00\rangle + \sqrt{\frac{p}{3}}\Big(\sigma_x\,|Q\rangle \otimes |01\rangle + \sigma_y\,|Q\rangle \otimes |10\rangle + \sigma_z\,|Q\rangle \otimes |11\rangle\Big) . \tag{9.56}$$

In the left-hand side of Eq. (9.56), any pure state $|e_0\rangle \in \mathcal{H}_2^{\otimes 2}$ can be inserted to specify the initial state of the qubit pair representing the environment E; and then Eq. (9.56) takes part in the definition of the unitary evolution U_{QE} evolving the joint state $|Q\rangle \otimes |e_0\rangle$ of the composite system QE for any state $|Q\rangle$ of the principal qubit Q. There are (infinitely) many choices of the couple $\big(|e_0\rangle\,, \mathsf{U}_{QE}\big)$ that can accomplish the evolution of Eq. (9.56). An interesting one is, as before with the simulating circuits of Figs. 9.2, 9.3 and 9.4, to seek to have the parameter p of the depolarizing noise of Eq. (9.49) adjustable (programmable) via the initial state $|e_0\rangle$ of the environment E acting on a fixed (p-independent) hardware of the simulating circuit. As we are going to see, this can even be achieved with a parameter p carried by only one of the two qubits representing the environment E, prepared in the simple two-qubit separable pure state

$$|e_0\rangle = |p\rangle \otimes |0\rangle = \Big(\sqrt{1-p}\,|0\rangle + \sqrt{p}\,|1\rangle\Big) \otimes |0\rangle = \sqrt{1-p}\,|00\rangle + \sqrt{p}\,|10\rangle\ . \tag{9.57}$$

As part of the simulating circuit, it is then convenient to have the state $|e_0\rangle$ of Eq. (9.57) undergo the two-qubit unitary transformation U_0 realizing

$$\mathsf{U}_0\,|e_0\rangle = \sqrt{1-p}\,|00\rangle + \sqrt{\frac{p}{3}}\Big(|01\rangle + |10\rangle + |11\rangle\Big)\ . \tag{9.58}$$

This can be accomplished by the p-independent unitary transformation U_0 defined by its action on the computational basis of $\mathcal{H}_2^{\otimes 2}$ as

$$\mathsf{U}_0 \begin{cases} |00\rangle \longmapsto |00\rangle\ , \\ |01\rangle \longmapsto \sqrt{\dfrac{1}{2}}\Big(|01\rangle - |11\rangle\Big)\ , \\ |10\rangle \longmapsto \sqrt{\dfrac{1}{3}}\Big(|01\rangle + |10\rangle + |11\rangle\Big)\ , \\ |11\rangle \longmapsto \sqrt{\dfrac{1}{6}}\Big(|01\rangle - 2\,|10\rangle + |11\rangle\Big)\ . \end{cases} \tag{9.59}$$

In principle, as we have seen in Sect. 3.1.5, the two-qubit unitary gate U_0 defined by Eq. (9.59) can always be realized by combining single-qubit gates and two-qubit Cnot gates. The two-qubit transformed state $\mathsf{U}_0\,|e_0\rangle$ resulting from Eq. (9.58) then provides two control qubits acting in the simulating circuit of Fig. 9.5, to finalize the transformation of Eq. (9.56) delivered on the circuit output.

We note that in the quantum operation of Eq. (9.49) defining the depolarizing noise, in the fourth Kraus operator σ_z can equivalently be replaced by $i\sigma_z$ which defines the same depolarizing noise. It is rather this equivalent form of the quantum operation that is simulated by the circuit of Fig. 9.5, since instead of σ_z we have $\sigma_x\sigma_y = i\sigma_z$ that is actually implemented on the principal qubit Q when the control $\mathsf{U}_0\,|e_0\rangle$ is at $|11\rangle$.

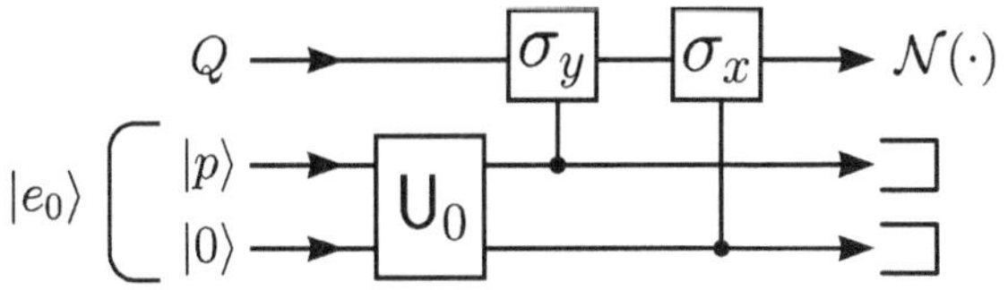

Fig. 9.5 Quantum circuit simulating the depolarizing noise by realizing the quantum operation $\mathcal{N}(\cdot)$ of Eq. (9.49) on the principal qubit Q. The two-qubit environment E prepared in the separable pure state $|e_0\rangle = |p\rangle \otimes |0\rangle$ of Eq. (9.57) with $|p\rangle = \sqrt{1-p}\,|0\rangle + \sqrt{p}\,|1\rangle$, is used to set the noise parameter p in the presence of a fixed hardware circuit. The unitary gate U_0 is defined by Eq. (9.59) and provides the control qubits to the Pauli gates σ_x and σ_y to implement a transformation equivalent to Eq. (9.56)

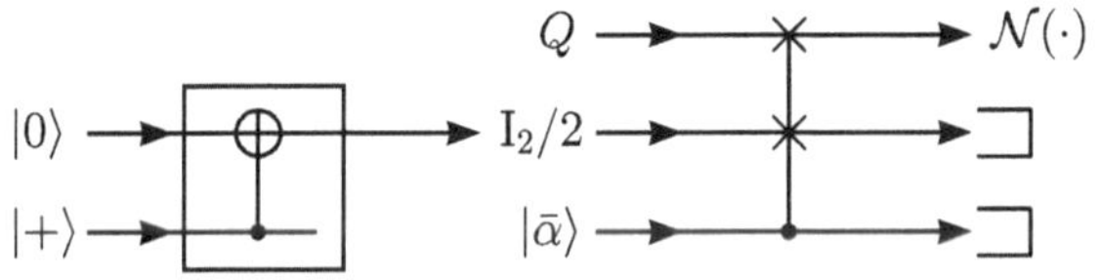

Fig. 9.6 Quantum circuit using a controlled-swap gate to simulate the depolarizing noise by realizing the transformation of Eq. (9.51) on the principal qubit Q. A control qubit prepared in the pure state $|\bar{\alpha}\rangle = \sqrt{\alpha}\,|0\rangle + \sqrt{1-\alpha}\,|1\rangle$ sets the noise parameter α in the presence of a fixed hardware circuit. A front Cnot gate is shown to produce the maximally mixed qubit state $\mathrm{I}_2/2$ from two input qubits $|0\rangle$ and $|+\rangle$ followed by partial tracing when the qubit $|+\rangle$ is discarded

In Fig. 9.5, the two control qubits in the state $\mathsf{U}_0\,|e_0\rangle$ of Eq. (9.58), trigger with the appropriate probability amplitudes the transformations to the principal qubit state $|Q\rangle$ equivalent to Eq. (9.56). In this way, the circuit of Fig. 9.5 materializes the definition of a valid couple $\left(|e_0\rangle, \mathsf{U}_{QE}\right)$ that simulates for the composite system QE an evolution as in Eq. (9.56). Since the simulating solution of Fig. 9.5 adheres to the generic approach of Sect. 3.2.8, it is then granted (and readily verifiable explicitly) that partial tracing over the two-qubit environment E of the composite density operator ρ'_{QE} resulting from the quantum state of Eq. (9.56) leads for the principal qubit Q to the state transformation of Eq. (9.49) defining the depolarizing noise, as targeted. Accordingly, in the circuit of Fig. 9.5, when the two control qubits are discarded (unobserved) at the output, the principal qubit Q with input state ρ terminates in the noisy state $\mathcal{N}(\rho)$ from Eq. (9.49).

There exists an alternative for the simulation of the depolarizing noise, that, although it does not follow the generic procedure of Sect. 3.2.8, offers another interesting solution. It is based on the alternative form of Eq. (9.51) for the state transformation realized by the depolarizing noise, which with probability $1-\alpha$ forces the output state at the maximally mixed qubit state $\mathrm{I}_2/2$ while it changes nothing with probability α. This can be simulated by the quantum circuit of Fig. 9.6, which uses a controlled-swap gate from Fig. 5.17.

For the operation of the controlled-swap in Fig. 9.6, the maximally mixed input state $\mathrm{I}_2/2$ represents an incoherent qubit which is at $|0\rangle$ with probability $1/2$ and at $|1\rangle$ with probability $1/2$. In the circuit of Fig. 9.6, the three-qubit output state is then $|\phi_1\rangle = \sqrt{\alpha}\,|Q\rangle\,|0\rangle\,|0\rangle + \sqrt{1-\alpha}\,|0\rangle\,|Q\rangle\,|1\rangle$ with probability $1/2$, and $|\phi_2\rangle = \sqrt{\alpha}\,|Q\rangle\,|1\rangle\,|0\rangle + \sqrt{1-\alpha}\,|1\rangle\,|Q\rangle\,|1\rangle$ with probability $1/2$, such a statistical ensemble being represented by the output density operator $\rho_{\text{out}} = \big(|\phi_1\rangle\langle\phi_1| + |\phi_2\rangle\langle\phi_2|\big)/2$. The principal qubit Q at the output, by tracing over the two other qubits identified with the environment E, has thus the reduced state $\text{tr}_E(\rho_{\text{out}}) = \alpha\,|Q\rangle\langle Q| + (1-\alpha)\text{I}_2/2 = \mathcal{N}\big(|Q\rangle\langle Q|\big)$ as targeted to implement Eq. (9.51). For a mixed input state ρ, by the property of convexity-linearity already invoked several times before, the transmission follows as in Eq. (9.51).

After the action of the depolarizing noise, the four-dimensional environment E terminates in the final state ρ'_E of Eq. (3.119) characterized, via its matrix elements of Eq. (3.120), as

$$\rho'_E = \begin{bmatrix} 1-p & \dfrac{\sqrt{(1-p)p}}{\sqrt{3}}r_x & \dfrac{\sqrt{(1-p)p}}{\sqrt{3}}r_y & \dfrac{\sqrt{(1-p)p}}{\sqrt{3}}r_z \\ \mathtt{cc} & \dfrac{p}{3} & -i\dfrac{p}{3}r_z & i\dfrac{p}{3}r_y \\ \mathtt{cc} & i\dfrac{p}{3}r_z & \dfrac{p}{3} & -i\dfrac{p}{3}r_x \\ \mathtt{cc} & -i\dfrac{p}{3}r_y & i\dfrac{p}{3}r_x & \dfrac{p}{3} \end{bmatrix}, \tag{9.60}$$

where the entries marked "cc" follow as complex conjugate from the Hermitian symmetry of the matrix. Then this density operator ρ'_E has the four eigenvalues

$$\left\{\begin{aligned} \lambda_\pm^{(1)} &= \frac{1}{6}(3-2p) \pm \frac{1}{6}\Big[12(1-p)p\,\|\vec{r}\,\|^2 + (4p-3)^2\Big]^{1/2}, && (9.61)\\ \lambda_\pm^{(2)} &= \frac{1}{3}\big(1 \pm \|\vec{r}\,\|\big)p\,. && (9.62)\end{aligned}\right.$$

They can be used to compute the quantum entropy $S(\rho'_E)$ of the environment which is also the entropy generated in the environment by the action of the depolarizing noise on the input state ρ.

The depolarizing noise is entanglement breaking [25] for $p \geq 1/2 \Leftrightarrow \alpha \leq 1/3$.

9.7 General Pauli Noise

A general Pauli noise [12, 220] acts through random applications of the four Pauli operators $\{\sigma_0 \equiv \text{I}_2,\ \sigma_x,\ \sigma_y,\ \sigma_z\}$ which form an orthogonal basis for operators on $\mathcal{H}_2$. A general Pauli noise implements in Eq. (9.5) the quantum operation

$$\rho' = \mathcal{N}(\rho) = \sum_{k=0,x,y,z} p_k \sigma_k \rho \sigma_k^\dagger\,, \tag{9.63}$$

with the $\{p_k\}$ a probability distribution, and the four Kraus operators $\Lambda_k = \sqrt{p_k}\sigma_k$. From the parameterization $\rho = [\rho_{00}, \rho_{01}\,;\,\rho_{10}, \rho_{11}]$ of Eq. (2.59), by using Eqs. (9.11)–(9.13) in Eq. (9.63), the transformed state follows as

$$\rho' = \mathcal{N}(\rho) = \begin{bmatrix} (p_0+p_z)\rho_{00} + (p_x+p_y)\rho_{11} & (p_0-p_z)\rho_{01} + (p_x-p_y)\rho_{10} \\ (p_x-p_y)\rho_{01} + (p_0-p_z)\rho_{10} & (p_x+p_y)\rho_{00} + (p_0+p_z)\rho_{11} \end{bmatrix}. \tag{9.64}$$

The transformation of the Bloch vector in Eq. (9.7) amounts to

$$\vec{r}\,' = A\vec{r} = \begin{bmatrix} a_x & 0 & 0 \\ 0 & a_y & 0 \\ 0 & 0 & a_z \end{bmatrix} \vec{r}\,, \tag{9.65}$$

with a diagonal matrix A defined by the three real scalar entries

$$\begin{cases} a_x = p_0 + p_x - p_y - p_z = 1 - 2(p_y + p_z)\,, & (9.66) \\ a_y = p_0 - p_x + p_y - p_z = 1 - 2(p_x + p_z)\,, & (9.67) \\ a_z = p_0 - p_x - p_y + p_z = 1 - 2(p_x + p_y)\,. & (9.68) \end{cases}$$

For the qubit, the class of Eq. (9.63) in particular contains important Pauli noises such as the bit-flip, the phase-flip, the bit-phase-flip, the depolarizing noises. The Pauli noises are all characterized by a vector $\vec{c} = \vec{0}$ in Eq. (9.7), and are therefore unital noises, satisfying $\mathcal{N}(\mathrm{I}_2) = \mathrm{I}_2$ as indicated in Sect. 3.2.11 on non-unitary qubit processes. The three real parameters a_ℓ in Eqs. (9.66)–(9.68) are contraction factors satisfying $0 \le |a_\ell| \le 1$ for all $\ell \in \{x, y, z\}$, to guarantee that the Bloch ball of valid Bloch vectors is mapped into itself. In this way, the transformation of Eq. (9.63) by the noise contracts the Bloch vectors in a generally nonuniform way along the three directions Ox, Oy and Oz of $\mathbb{R}^3$. An increasing level of noise corresponds to a more pronounced contraction. The maximum level of contraction would occur as $a_x = a_y = a_z = 0$, corresponding to a noisy state $\mathcal{N}(\rho)$ with a null Bloch vector $\vec{0}$ characterizing the maximally mixed qubit state $\mathrm{I}_2/2$ identifiable with the maximally noisy state.

Generically, the $K = 4$ Kraus operators of the general Pauli noise in Eq. (9.63) determine, by the canonical procedure of Sect. 3.2.8, a four-dimensional environment E, realizable with a pair of qubits. To simulate the general Pauli noise of Eq. (9.63) on the principal qubit Q by the procedure of Sect. 3.2.8, a quantum circuit can be devised acting on a two-qubit environment E referred to the orthonormal basis $\{|e_k\rangle\}_{k=1}^4 = \{|00\rangle, |01\rangle, |10\rangle, |11\rangle\}$, so that the composite system QE according to Eq. (9.9) satisfies

$$\mathsf{U}_{QE}\,|Q\rangle \otimes |e_0\rangle = \sqrt{p_0}\,|Q\rangle \otimes |00\rangle + \sqrt{p_x}\sigma_x\,|Q\rangle \otimes |01\rangle + \sqrt{p_y}\sigma_y\,|Q\rangle \otimes |10\rangle + \sqrt{p_z}\sigma_z\,|Q\rangle \otimes |11\rangle\,. \tag{9.69}$$

Again we can seek the interesting possibility of adjusting (programming) the noise parameters p_k of Eq. (9.63) via the initial state $|e_0\rangle \in \mathcal{H}_2^{\otimes 2}$ of the environment E

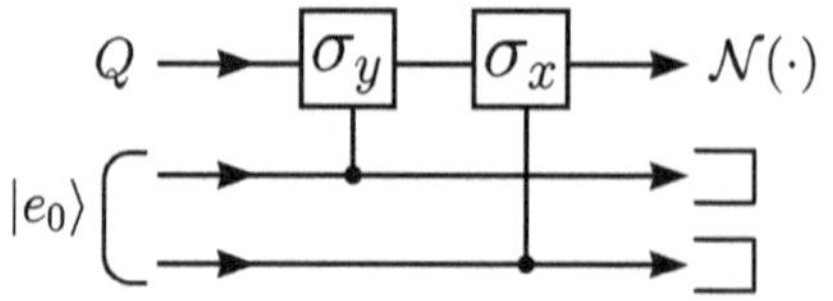

Fig. 9.7 Quantum circuit simulating a general Pauli noise by realizing the quantum operation $\mathcal{N}(\cdot)$ of Eq. (9.63) on the principal qubit Q. The two-qubit environment E prepared in the pure entangled state $|e_0\rangle$ of Eq. (9.70) sets the four noise parameters p_k in the presence of a fixed hardware circuit and provides the control qubits to the Pauli gates σ_x and σ_y to implement a transformation equivalent to Eq. (9.69)

to act on a fixed hardware of the simulating circuit. For this purpose, a two-qubit separable product state $|e_0\rangle$ does not contain enough free parameters to generally represent the four parameters p_k in Eq. (9.63). However, the two-qubit entangled state

$$|e_0\rangle = \sqrt{p_0}\,|00\rangle + \sqrt{p_x}\,|01\rangle + \sqrt{p_y}\,|10\rangle + \sqrt{p_z}\,|11\rangle \tag{9.70}$$

is generally suited, and can be used, in a way comparable to Fig. 9.5, to provide two control qubits acting on Pauli gates as shown in Fig. 9.7.

The circuit of Fig. 9.7, as in Fig. 9.5, implements in Eq. (9.63) the equivalent transformation $i\sigma_z = \sigma_x\sigma_y$ instead of σ_z, with the probability amplitude $\sqrt{p_z}$.

Instead of the three-qubit circuit of Fig. 9.7 with a two-qubit environment following the generic procedure of Sect. 3.2.8, for simulating quantum Pauli noises another solution is proposed in [220]. The quantum circuit of [220] uses three environment qubits (instead of two in Fig. 9.7), each enabling control on one of the three independent noise parameters (p_x, p_y, p_z) in Eq. (9.63), to simulate a large (infinite) subclass of interesting Pauli noises. The alternative circuit of [220] provides a more direct selection of the noise parameters, in particular exempting one from fabricating the two-qubit entangled state $|e_0\rangle$ of Eq. (9.70) required in Fig. 9.7.

After the action of the Pauli noise, the four-dimensional environment terminates in the final state ρ'_E of Eq. (3.119) characterized, via its matrix elements of Eq. (3.120), as

$$\rho'_E = \begin{bmatrix} p_0 & \sqrt{p_0 p_x} r_x & \sqrt{p_0 p_y} r_y & \sqrt{p_0 p_z} r_z \\ \sqrt{p_0 p_x} r_x & p_x & -i\sqrt{p_x p_y} r_z & i\sqrt{p_x p_z} r_y \\ \sqrt{p_0 p_y} r_y & i\sqrt{p_x p_y} r_z & p_y & -i\sqrt{p_y p_z} r_x \\ \sqrt{p_0 p_z} r_z & -i\sqrt{p_x p_z} r_y & i\sqrt{p_y p_z} r_x & p_z \end{bmatrix}. \tag{9.71}$$

The four eigenvalues of ρ'_E here are difficult to express analytically, in general conditions. Depending on the prospect, their numerical evaluation can also be envisaged.

9.8 Completely Decohering Qubit Process

An interesting qubit noise model that is unital but not in the class of Pauli noises of Sect. 9.7 is the noise process defined by the two Hermitian non-unitary Kraus operators $\Lambda_0 = |0\rangle\langle 0|$ and $\Lambda_1 = |1\rangle\langle 1|$, implementing the quantum operation

$$\rho' = \mathcal{N}(\rho) = |0\rangle\langle 0|\,\rho\,|0\rangle\langle 0| + |1\rangle\langle 1|\,\rho\,|1\rangle\langle 1| \ , \tag{9.72}$$

which is also

$$\rho' = \mathcal{N}(\rho) = \rho_{00}\,|0\rangle\langle 0| + \rho_{11}\,|1\rangle\langle 1| = \begin{bmatrix} \rho_{00} & 0 \\ 0 & \rho_{11} \end{bmatrix} , \tag{9.73}$$

for an input qubit with density operator $\rho = \left[\rho_{00}, \rho_{01} \,;\, \rho_{10}, \rho_{11}\right]$ in the standard parameterization of Eq. (2.59). The operation of the process $\mathcal{N}(\cdot)$ is equivalent in Eq. (9.7) to the transformation of the Bloch vector

$$\vec{r}\,' = A\vec{r} = \begin{bmatrix} 0 & 0 & 0 \\ 0 & 0 & 0 \\ 0 & 0 & 1 \end{bmatrix} \vec{r} \ . \tag{9.74}$$

As visible from Eq. (9.73), the action of this qubit noise process is to completely cancel the coherences or off-diagonal terms of the input density operator ρ, while preserving the populations or diagonal terms of ρ. In this regard, this is a completely decohering process or channel. Only the two pure states $|0\rangle$ and $|1\rangle$ are transmitted with no alteration, since $\mathcal{N}\big(|0\rangle\langle 0|\big) = |0\rangle\langle 0|$ and $\mathcal{N}\big(|1\rangle\langle 1|\big) = |1\rangle\langle 1|$. Any other pure input state $|Q\rangle$ formed as the coherent superposition $|Q\rangle = \alpha_0\,|0\rangle + \alpha_1\,|1\rangle$, with α_0 and α_1 nonzero, gets transformed into the mixed state $\mathcal{N}\big(|Q\rangle\langle Q|\big) = |\alpha_0|^2\,|0\rangle\langle 0| + |\alpha_1|^2\,|1\rangle\langle 1|$, representing the incoherent statistical mixture $\big\{\big(|\alpha_0|^2, |0\rangle\big), \big(|\alpha_1|^2, |1\rangle\big)\big\}$ of the two individually invariant pure states $|0\rangle$ and $|1\rangle$. Alternatively, the quantum process $\mathcal{N}(\cdot)$ of Eqs. (9.72)–(9.73) can be interpreted as an *unread* von Neumann measurement of the qubit in the orthonormal basis $\big\{|0\rangle\,, |1\rangle\big\}$, as discussed at the end of Sect. 4.2.3.

The $K = 2$ (non-unitary) Kraus operators of the completely decohering process in Eq. (9.72) determine, by the canonical procedure of Sect. 3.2.8, a two-dimensional environment E, realizable with a single qubit. To simulate the completely decohering process of Eq. (9.72) on the principal qubit Q by the procedure of Sect. 3.2.8, a quantum circuit can be devised acting on a one-qubit environment E referred to the orthonormal basis $\big\{|e_k\rangle\big\} = \big\{|0\rangle\,, |1\rangle\big\}$. For an input qubit Q in the generic pure state $|Q\rangle = \alpha_0\,|0\rangle + \alpha_1\,|1\rangle$, the joint unitary evolution of the composite system QE, according to Eqs. (9.9)–(9.10), should be

$$|\psi'_{QE}\rangle = \mathsf{U}_{QE}\,|Q\rangle \otimes |e_0\rangle = \alpha_0\,|0\rangle \otimes |0\rangle + \alpha_1\,|1\rangle \otimes |1\rangle \ . \tag{9.75}$$

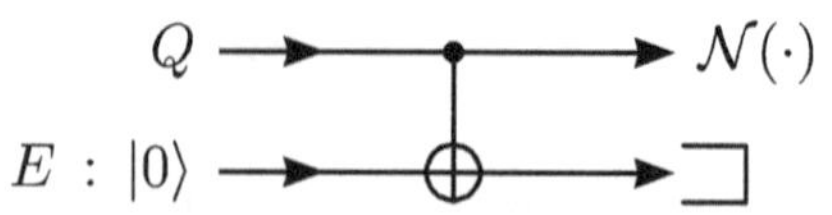

Fig. 9.8 Quantum circuit using a Cnot gate to simulate the completely decohering noise process or channel $\mathcal{N}(\cdot)$ of Eq. (9.72) on the principal qubit Q. The one-qubit environment E is initialized in the pure state $|e_0\rangle = |0\rangle$. The quantum circuit realizes the joint two-qubit unitary evolution U_{QE} of Eq. (9.75)

This two-qubit joint evolution can be accomplished with a Cnot gate arranged as in Fig. 9.8.

When the environment qubit E is discarded (unobserved) at the circuit output in Fig. 9.8, by partial tracing over the environment E we obtain $\mathrm{tr}_E\big(|\psi'_{QE}\rangle\langle\psi'_{QE}|\big) = \mathcal{N}\big(|Q\rangle\langle Q|\big) = |\alpha_0|^2\,|0\rangle\langle 0| + |\alpha_1|^2\,|1\rangle\langle 1|$, as targeted in Eq. (9.73). Then, at the output when the principal qubit Q described by the reduced quantum state $\rho' = \mathrm{tr}_E\big(|\psi'_{QE}\rangle\langle\psi'_{QE}|\big) = \mathcal{N}\big(|Q\rangle\langle Q|\big)$ is measured in the orthonormal basis $\big\{|0\rangle, |1\rangle\big\}$, it has the probability $\Pr\big\{|0\rangle\big\} = \mathrm{tr}\big(\rho'\Pi_0\big) = \langle 0|\rho'|0\rangle = |\alpha_0|^2$ of being found in state $|0\rangle$, with the projector $\Pi_0 = |0\rangle\langle 0|$. The same probability $\Pr\big\{|0\rangle\big\}$ can be computed equivalently from the full two-qubit pure state $|\psi'_{QE}\rangle$ of Eq. (9.75), subjected to the two measuring projectors $\big\{|0\rangle\langle 0| \otimes \mathrm{I}_2, |1\rangle\langle 1| \otimes \mathrm{I}_2\big\}$ to represent an unobserved environment qubit E, and to yield $\Pr\big\{|0\rangle\big\} = \langle\psi'_{QE}|\Pi_0 \otimes \mathrm{I}_2|\psi'_{QE}\rangle = |\alpha_0|^2$. It results that when the input preparation of the qubit Q is $|Q\rangle = |0\rangle$ with $(\alpha_0, \alpha_1) = (1, 0)$, at the output Q is always measured in state $|0\rangle$, and similarly when Q is prepared as $|Q\rangle = |1\rangle$ with $(\alpha_0, \alpha_1) = (0, 1)$ it is always measured as $|1\rangle$, this manifesting the invariance individually of the two basis states $|0\rangle$ and $|1\rangle$ as already noted above. In this situation, the principal qubit Q and the environment qubit E, by the joint unitary evolution U_{QE} of Eq. (9.75), terminate at the output in a joint state $|\psi'_{QE}\rangle$ that remains a two-qubit *separable* state. The principal qubit Q initialized in state $|0\rangle$ or $|1\rangle$ at the input, in the channel transmission, does not entangle with the unobserved environment E, and the state of Q remains unaltered.

By contrast, when the input qubit Q is initialized in the coherent superposition $|Q\rangle = \alpha_0\,|0\rangle + \alpha_1\,|1\rangle$, to check that the invariant transmission disappears, we can envisage measuring Q at the output in the orthonormal basis $\big\{|Q\rangle, |Q^\perp\rangle\big\}$ of $\mathcal{H}_2$, with the state $|Q^\perp\rangle$ orthogonal to $|Q\rangle$ in the plane $\mathcal{H}_2$. The probability $\Pr\big\{|Q\rangle\big\}$ of measuring Q at the output in the same state $|Q\rangle$ in which it was prepared at the input, is $\Pr\big\{|Q\rangle\big\} = \mathrm{tr}\big(\rho'\Pi_Q\big) = \langle Q|\rho'|Q\rangle = |\alpha_0|^4 + |\alpha_1|^4$, with the projector $\Pi_Q = |Q\rangle\langle Q|$. Equivalently, when the environment qubit E is unobserved in Fig. 9.8, the same probability is obtained as $\Pr\big\{|Q\rangle\big\} = \langle\psi'_{QE}|\Pi_Q \otimes \mathrm{I}_2|\psi'_{QE}\rangle = |\alpha_0|^4 + |\alpha_1|^4$. This probability $\Pr\big\{|Q\rangle\big\} = |\alpha_0|^4 + |\alpha_1|^4$ in general differs from 1, except in the two special cases examined above when $(\alpha_0, \alpha_1) = (1, 0)$ or $(0, 1)$. This expresses that, although the two basis states $|0\rangle$ and $|1\rangle$ are individually invariant by the completely decohering process $\mathcal{N}(\cdot)$ of Eqs. (9.72)–(9.73), any non-degenerate coherent superposition of the two basis states $|0\rangle$ and $|1\rangle$ is generally non-invariant in the transmission by $\mathcal{N}(\cdot)$, as a mark of the lost coherence. With an input superposition, the principal qubit Q and the environment qubit E, by the joint unitary evolution U_{QE} of Eq. (9.75), generally terminate at the output in a joint state $|\psi'_{QE}\rangle$ that has become a two-qubit *entangled* state. In the transmission by $\mathcal{N}(\cdot)$, the principal qubit Q gets entangled with an unobserved environment E, as a source of the loss of coherence, and the entangled qubit Q at the output no longer has a well defined pure state of its own. The coherent input superposition described by the pure qubit state $|Q\rangle = \alpha_0 |0\rangle + \alpha_1 |1\rangle$, is transformed by the channel into the incoherent output mixture described by the statistical ensemble $\{(|\alpha_0|^2, |0\rangle), (|\alpha_1|^2, |1\rangle)\}$ representing the mixed state with density operator $\mathcal{N}(|Q\rangle\langle Q|) = |\alpha_0|^2 |0\rangle\langle 0| + |\alpha_1|^2 |1\rangle\langle 1|$, with the corresponding uncertainty caused by Q becoming entangled to the environment E. This is a fundamental mechanism of loss of coherence or of acquired uncertainty or of degradation by noise, for a quantum state, by becoming entangled with an uncontrolled environment.

Furthermore, this situation of the qubit Q serving as control in Fig. 9.8, is generally representative of the situation of any control qubit materialized by a straight arrow traversing end-to-end a quantum circuit without interruption by a quantum gate, as it appears in Fig. 9.8 and also on many other occasions for instance in Chap. 5 on quantum circuits. Such a control qubit is usually prepared at the input in a coherent one-qubit separable pure state. Such control qubit will remain in the same invariant one-qubit pure state, as rightly suggested by the uninterrupted input–output arrow, only when it does not entangle with the rest of the circuit. Otherwise, if the control qubit entangles with the rest of the circuit, it will no longer keep a well defined coherent separable pure state of its own, a fortiori no longer reproducing the one-qubit separable input preparation, despite the uninterrupted input–output arrow. On an N-dimensional quantum state $|\psi\rangle$ of $\mathcal{H}_N$, generically a control qubit will use its two computational-basis states $\{|0\rangle, |1\rangle\}$ to trigger one or the other of two unitary transformations $\{\mathsf{U}_0, \mathsf{U}_1\}$ in $\mathcal{H}_N$, so as to realize in $\mathcal{H}_2 \otimes \mathcal{H}_N$ the unitary transformation

$$\begin{cases} |0\rangle \otimes |\psi\rangle \longmapsto |0\rangle \otimes \mathsf{U}_0 |\psi\rangle \ , & (9.76) \\ |1\rangle \otimes |\psi\rangle \longmapsto |1\rangle \otimes \mathsf{U}_1 |\psi\rangle \ . & (9.77) \end{cases}$$

Accordingly, when the control qubit is prepared in one of its two computational-basis states $|0\rangle$ or $|1\rangle$, it does not entangle with the rest of the circuit and its preparation remains invariant. Otherwise, when the control qubit is prepared in the non-degenerate superposition $\alpha_0 |0\rangle + \alpha_1 |1\rangle$, it terminates in the generally entangled state $\alpha_0 |0\rangle \otimes \mathsf{U}_0 |\psi\rangle + \alpha_1 |1\rangle \otimes \mathsf{U}_1 |\psi\rangle$, replacing its well-defined one-qubit pure state by the mixed qubit (reduced) state

$$\begin{bmatrix} |\alpha_0|^2 & \alpha_0\alpha_1^* \langle\psi|\mathsf{U}_1^\dagger \mathsf{U}_0|\psi\rangle \\ \alpha_0^*\alpha_1 \langle\psi|\mathsf{U}_0^\dagger \mathsf{U}_1|\psi\rangle & |\alpha_1|^2 \end{bmatrix}, \tag{9.78}$$

with only the populations $|\alpha_0|^2$ and $|\alpha_1|^2$ remaining invariant in the input–output transformation, and usually a decay of the coherences.

Returning to the completely decohering process of Eq. (9.72), after it has acted the two-dimensional environment E terminates in a final state ρ'_E given by its matrix elements of Eq. (3.120), and resulting as $\rho'_E = \rho_{00}\,|0\rangle\langle 0| + \rho_{11}\,|1\rangle\langle 1|$. This density operator ρ'_E has the two eigenvalues ρ_{00} and $\rho_{11} = 1 - \rho_{00}$, which can be used in particular to compute the final quantum entropy $S(\rho'_E)$ generated in the environment E by the decohering interaction.

The completely decohering process can also be found to arise through a mechanism of **randomization in the coordinates** of a quantum state. We consider the quantum process that completely randomizes the relative phase between the two components of a qubit state, by the input–output unitary transformation $|Q\rangle = \alpha_0\,|0\rangle + \alpha_1\,|1\rangle \mapsto |Q_\xi\rangle = \alpha_0\,|0\rangle + e^{i\xi}\alpha_1\,|1\rangle$ where ξ is a random phase uniformly distributed over the interval $[0, 2\pi[$. By averaging the random state $|Q_\xi\rangle\langle Q_\xi|$ over the phase distribution, we obtain the (average) density operator $\overline{\rho} = \big\langle\, |Q_\xi\rangle\langle Q_\xi|\,\big\rangle = |\alpha_0|^2\,|0\rangle\langle 0| + |\alpha_1|^2\,|1\rangle\langle 1|$ to represent the statistical ensemble generated at the output in response to the input $|Q\rangle\langle Q|$. The linear map $|Q\rangle\langle Q| \mapsto \overline{\rho} = \big\langle\, |Q_\xi\rangle\langle Q_\xi|\,\big\rangle$ resulting from the randomization and averaging, is precisely the characterization of the completely decohering qubit process we had earlier, manifesting that phase randomization is a mechanism that can give rise to the completely decohering process.

In addition, the completely decohering qubit process of Eqs. (9.72)–(9.73) can also be obtained from the phase-flip noise of Eq. (9.31) at a flip probability $p = 1/2$. In this respect, phase randomization by a continuously-distributed random phase ξ, can equivalently be modeled as a discrete random phase flip as with the phase-flip noise of Sect. 9.4.

More broadly, a quantum state $|Q\rangle$ can experience some random alteration in its defining coordinates or parameters, described by the input–output unitary transformation $|Q\rangle \mapsto |Q_\xi\rangle$, where $|Q_\xi\rangle$ is a random state parameterized by the random ξ that can generally be a vector of multiple parameters. Averaging over ξ the density operator $|Q_\xi\rangle\langle Q_\xi|$ leads to the (average) density operator $\overline{\rho} = \big\langle\, |Q_\xi\rangle\langle Q_\xi|\,\big\rangle$ to represent the output statistical ensemble. Through the (linear) averaging, the input–output transformation yields $|Q\rangle\langle Q| \mapsto \overline{\rho} = \mathcal{N}\big(|Q\rangle\langle Q|\big)$ standing as a linear map $\mathcal{N}(\cdot)$ mapping an input density operator into an output density operator. Such a map $\mathcal{N}(\cdot)$ is a quantum process as those reported in Sect. 3.2, and, in addition to the present random parameterization model, $\mathcal{N}(\cdot)$ can receive both an equivalent Kraus representation or a Stinespring dilated unitary representation. In the other direction, these general representations (Kraus or Stinespring) reported in Sect. 3.2 for non-unitary quantum processes, can as well encompass scenarios of randomization in the parameters defining a quantum state. The noisy preparation of the qubit in Sect. 2.5.6, the noisy interferometer in Sect. 9.4, the phase randomization for the completely decohering process above, can be seen as examples of this type.

9.9 Amplitude Damping Noise

The amplitude damping noise [9, 12] is a qubit noise process defined in Eq. (9.5) by the $K = 2$ Kraus operators

$$\Lambda_1 = \begin{bmatrix} 1 & 0 \\ 0 & \sqrt{1-\gamma} \end{bmatrix} = |0\rangle\langle 0| + \sqrt{1-\gamma}\,|1\rangle\langle 1| \ , \tag{9.79}$$

$$\Lambda_2 = \begin{bmatrix} 0 & \sqrt{\gamma} \\ 0 & 0 \end{bmatrix} = \sqrt{\gamma}\,|0\rangle\langle 1| \ , \tag{9.80}$$

with the damping factor $\gamma \in [0, 1]$. For an input qubit with density operator $\rho = [\rho_{00}, \rho_{01}\,;\, \rho_{10}, \rho_{11}]$ in the standard parameterization of Eq. (2.59), the quantum operation resulting in Eq. (9.5) leads to the transformed state

$$\rho' = \mathcal{N}(\rho) = \begin{bmatrix} (1-\gamma)\rho_{00} + \gamma & \sqrt{1-\gamma}\,\rho_{01} \\ \sqrt{1-\gamma}\,\rho_{10} & (1-\gamma)\rho_{11} \end{bmatrix} . \tag{9.81}$$

The action of the amplitude damping noise can be interpreted as follows. The operator Λ_1 in Eq. (9.79) leaves the (ground) state $|0\rangle$ unaffected, and it reduces or damps by a factor $\sqrt{1-\gamma}$ the probability amplitude of the (excited) state $|1\rangle$, so that after the action of Λ_1 there would be less probability of finding the qubit in the (excited) state $|1\rangle$. The operator Λ_2 in Eq. (9.80) transforms the excited state $|1\rangle$ into the ground state $|0\rangle$ with a probability amplitude $\sqrt{\gamma}$, while it nullifies the ground state $|0\rangle$ with $\Lambda_2\,|0\rangle = 0$ the null vector of the space $\mathcal{H}_2$. In this way, the amplitude damping noise offers a description of the relaxation of an excited state $|1\rangle$ into a ground state $|0\rangle$ via a dissipation of energy into an environment at zero temperature [9, 12]

The transformation by the amplitude damping noise of the qubit Bloch vector in Eq. (9.7) can be obtained from Eq. (9.81), via Eqs. (2.63)–(2.65), and comes out as

$$\vec{r}\,' = A\vec{r} + \vec{c} = \begin{bmatrix} \sqrt{1-\gamma} & 0 & 0 \\ 0 & \sqrt{1-\gamma} & 0 \\ 0 & 0 & 1-\gamma \end{bmatrix} \vec{r} + \begin{bmatrix} 0 \\ 0 \\ \gamma \end{bmatrix} . \tag{9.82}$$

With the vector $\vec{c} \neq \vec{0}$ in Eq. (9.82), the amplitude damping noise is a non-unital qubit noise, by contrast with the Pauli noises examined previously which, with $\vec{c} = \vec{0}$, were unital.

The $K = 2$ Kraus operators of the amplitude damping noise, by the canonical procedure of Sect. 3.2.8, determine a two-dimensional environment E, realizable with a single qubit. The simulation of the amplitude damping noise will be addressed in the next Sect. 9.10, under its generalized form, and is finally achievable by means of the quantum circuit of Fig. 9.14 at $p = 1$.

After the action of the amplitude damping noise, the two-dimensional environment terminates in the final state ρ'_E of Eq. (3.119) characterized, via its matrix elements of Eq. (3.120), as

$$\rho'_E = \begin{bmatrix} 1-\gamma\rho_{11} & \sqrt{\gamma}\,\rho_{01} \\ \sqrt{\gamma}\,\rho_{10} & \gamma\rho_{11} \end{bmatrix}, \tag{9.83}$$

or equivalently ρ'_E is a qubit density operator with Bloch vector $\left[\sqrt{\gamma}r_x,\ \sqrt{\gamma}r_y,\ \gamma r_z + 1-\gamma\right]^\top$. The state ρ'_E of Eq. (9.83) has the two eigenvalues

$$\lambda_\pm = \frac{1}{2} \pm \frac{1}{2}\left[4\gamma|\rho_{01}|^2 + (1-2\gamma\rho_{11})^2\right]^{1/2} \tag{9.84}$$

$$= \frac{1}{2} \pm \frac{1}{2}\left[\gamma\left(r_x^2+r_y^2\right) + \left(\gamma r_z + 1-\gamma\right)^2\right]^{1/2}. \tag{9.85}$$

They can be used to compute the quantum entropy $S(\rho'_E)$ of the environment, which is also the entropy generated in the environment by the action of the amplitude damping noise on the input state ρ.

9.10 Generalized Amplitude Damping or Thermal Noise

The generalized amplitude damping noise [9, 221–223] is a qubit noise defined in Eq. (9.5) by the $K=4$ Kraus operators

$$\Lambda_1 = \sqrt{p}\begin{bmatrix} 1 & 0 \\ 0 & \sqrt{1-\gamma} \end{bmatrix} = \sqrt{p}\left(|0\rangle\langle 0| + \sqrt{1-\gamma}\,|1\rangle\langle 1|\right), \tag{9.86}$$

$$\Lambda_2 = \sqrt{p}\begin{bmatrix} 0 & \sqrt{\gamma} \\ 0 & 0 \end{bmatrix} = \sqrt{p}\sqrt{\gamma}\,|0\rangle\langle 1|\,, \tag{9.87}$$

$$\Lambda_3 = \sqrt{1-p}\begin{bmatrix} \sqrt{1-\gamma} & 0 \\ 0 & 1 \end{bmatrix} = \sqrt{1-p}\left(\sqrt{1-\gamma}\,|0\rangle\langle 0| + |1\rangle\langle 1|\right), \tag{9.88}$$

$$\Lambda_4 = \sqrt{1-p}\begin{bmatrix} 0 & 0 \\ \sqrt{\gamma} & 0 \end{bmatrix} = \sqrt{1-p}\sqrt{\gamma}\,|1\rangle\langle 0|\,. \tag{9.89}$$

For an input qubit with density operator $\rho = \left[\rho_{00}, \rho_{01}\,;\, \rho_{10}, \rho_{11}\right]$ in the standard parameterization of Eq. (2.59), the quantum operation resulting in Eq. (9.5) leads to the transformed state

$$\rho' = \mathcal{N}(\rho) = \begin{bmatrix} (1-\gamma)\rho_{00} + p\gamma & \sqrt{1-\gamma}\,\rho_{01} \\ \sqrt{1-\gamma}\,\rho_{10} & (1-\gamma)\rho_{11} + (1-p)\gamma \end{bmatrix}. \tag{9.90}$$

The transformation of the qubit Bloch vector in Eq. (9.7), from Eq. (9.90) via Eqs. (2.63)–(2.65), follows as

$$\vec{r}\,' = A\vec{r} + \vec{c} = \begin{bmatrix} \sqrt{1-\gamma} & 0 & 0 \\ 0 & \sqrt{1-\gamma} & 0 \\ 0 & 0 & 1-\gamma \end{bmatrix} \vec{r} + \begin{bmatrix} 0 \\ 0 \\ (2p-1)\gamma \end{bmatrix}, \tag{9.91}$$

identifying with $\vec{c} \neq \vec{0}$ a non-unital qubit noise.

The generalized amplitude damping noise can also be identified as a **quantum thermal noise**. This noise offers a model to describe the interaction of a qubit with an uncontrolled environment representing a thermal bath at temperature T. The parameter $\gamma \in [0, 1]$ stands as a damping or coupling factor, which often can be related [9, 222, 223] to the time duration t of the interaction of the qubit with the thermal bath by

$$\gamma = 1 - e^{-t/\tau}, \tag{9.92}$$

where τ is a time constant for the interaction (such as the spin-lattice relaxation time τ in magnetic resonance). At long interaction time $t \gg \tau$, then the damping or coupling $\gamma \to 1$, and the qubit relaxes to the equilibrium or thermalized mixed state $\rho_\infty = p\,|0\rangle\langle 0| + (1-p)\,|1\rangle\langle 1|$ of Bloch vector $\vec{r}_\infty = [0, 0, 2p-1]^\top$. Accordingly, at equilibrium the qubit rests in a statistical ensemble with probabilities p for the ground state $|0\rangle$ and $1-p$ for the excited state $|1\rangle$. With the energies E_0 and $E_1 > E_0$ assigned respectively to the ground state $|0\rangle$ and excited state $|1\rangle$, the equilibrium probabilities are determined by the Boltzmann distribution, yielding

$$p = \frac{\exp\big(-E_0/(k_B T)\big)}{\exp\big(-E_0/(k_B T)\big) + \exp\big(-E_1/(k_B T)\big)} = \frac{1}{1 + \exp\big(-(E_1 - E_0)/(k_B T)\big)}, \tag{9.93}$$

with $k_B \approx 1.381 \times 10^{-23}$ J/K the Boltzmann constant. The equilibrium state can also be written

$$\rho_\infty = \frac{1}{Z}\exp\left(-\frac{\mathsf{H}}{k_B T}\right) = \frac{1}{Z}\left[\exp\left(-\frac{E_0}{k_B T}\right)|0\rangle\langle 0| + \exp\left(-\frac{E_1}{k_B T}\right)|1\rangle\langle 1|\right], \tag{9.94}$$

with the qubit Hamiltonian or energy operator having the diagonal form $\mathsf{H} = E_0\,|0\rangle\langle 0| + E_1\,|1\rangle\langle 1|$, and the partition function $Z = \operatorname{tr}\exp\big(-\mathsf{H}/(k_B T)\big) = \exp\big(-E_0/(k_B T)\big) + \exp\big(-E_1/(k_B T)\big)$. In this way, in the quantum thermal noise of Eqs. (9.86)–(9.91), the probability p is determined via Eq. (9.93) by the temperature T of the bath. From Eq. (9.93), the probability p is a decreasing function of the temperature T. At $T = 0$ the probability is $p = 1$ for the ground state $|0\rangle$, while at $T \to +\infty$ the ground state $|0\rangle$ and excited state $|1\rangle$ are equiprobable with $p = 1/2$. Therefore, from Eq. (9.93), when the temperature T monotonically increases from 0 to $+\infty$, the probability p monotonically decreases from 1 to $1/2$. Typical evolutions of p with T are shown in Fig. 9.9, for various energy gaps $E_1 - E_0$.

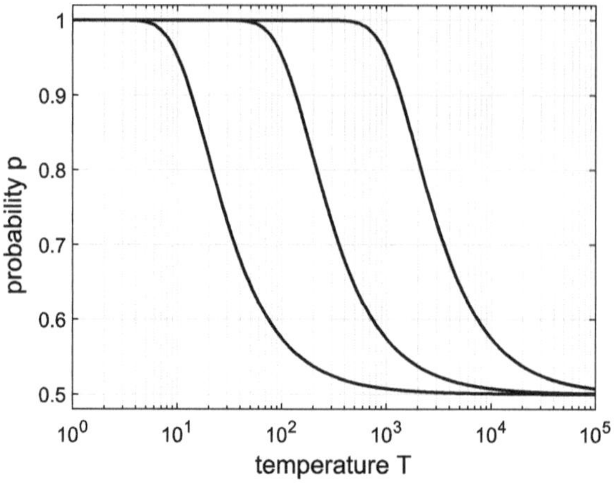

Fig. 9.9 For the quantum thermal noise, the probability p according to Eq. (9.93) of the ground state $|0\rangle$ at equilibrium, as a function of the temperature T in kelvins of the thermal bath. The energy gap $E_1 - E_0 = f k_B T_{300}$ is some fraction f of the thermal energy at room temperature $k_B T_{300}$ with $T_{300} = 300$ kelvins. For the three curves from left to right is $f = 0.1$, $f = 1$ and $f = 10$

In Eqs. (9.86)–(9.89), the two Kraus operators Λ_1 and Λ_2 describe the relaxation of the excited state $|1\rangle$ into the ground state $|0\rangle$ with the qubit dissipating energy into the thermal bath, much like in the amplitude damping noise of Eqs. (9.79)–(9.80), yet with the influence of the probability p here for the relaxation. The two Kraus operators Λ_3 and Λ_4 describe the opposite process, influenced by the probability $1 - p$, of an excitation of the ground states $|0\rangle$ into the excited state $|1\rangle$ while the qubit absorbs energy from the bath.

The output noisy state $\rho' = \mathcal{N}(\rho)$ determined by the Bloch vector $\vec{r}\,'$ of Eq. (9.91) is influenced by the noise temperature T only through the probability p. The special case $p = 1$ recovers the amplitude damping noise of Sect. 9.9 describing the interaction of a qubit with a thermal bath at temperature $T = 0$.

The input Bloch vector $\vec{r} = [r_x, r_y, r_z]^\top$ transformed by Eq. (9.91) yields the output Bloch vector $\vec{r}\,'$ having the squared norm

$$\|\vec{r}\,'\|^2 = (1 - \gamma)(r_x^2 + r_y^2) + r_z'^2 \,, \tag{9.95}$$

with the squared z-component

$$r_z'^2 = \big[(1 - \gamma) r_z + (2p - 1)\gamma\big]^2 \tag{9.96}$$

especially carrying via p the dependence of $\|\vec{r}\,'\|^2$ with the noise temperature T.

As we have seen, at long interaction time $t \to +\infty$ with the thermal bath, the damping or coupling factor $\gamma \to 1$ and the output state relaxes to the equilibrium mixed state $\rho_\infty = p\,|0\rangle\,\langle 0| + (1 - p)\,|1\rangle\,\langle 1|$ determined via p by the temperature T of the bath, and irrespective of the initial state ρ of the qubit. The equilibrium state ρ_∞ is characterized by the quantum von Neumann entropy $S(\rho_\infty) = H_{\rm bin}(p)$ in Eq. (7.40), and at any finite temperature T the probability $p > 1/2$ so that $S(\rho_\infty) < \log(2)$ is strictly below the maximum qubit entropy $\log(2)$. When the qubit is initialized in the maximally mixed state $\rho = \mathrm{I}_2/2$ with maximum entropy $S(\rho = \mathrm{I}_2/2) = \log(2)$, it relaxes equally to ρ_∞ with the smaller entropy $S(\rho_\infty) < \log(2)$. This is an example of a non-unital evolution that may decrease the quantum entropy, as here in the configuration where a high-entropy initial state $\rho = \mathrm{I}_2/2$ relaxes to an equilibrium state ρ_∞ with lower entropy. By contrast, unital evolutions never decrease the quantum entropy, as stated by Eq. (7.113), and as we have explicitly verified with the unital qubit noises previously presented in this chapter.

The $K = 4$ Kraus operators of Eqs. (9.86)–(9.89) for the quantum thermal noise determine, by the canonical procedure of Sect. 3.2.8, a four-dimensional environment E, realizable with a pair of qubits. To simulate the quantum thermal noise from Eqs. (9.86)–(9.89) on the principal qubit Q by the procedure of Sect. 3.2.8, a quantum circuit can be devised acting on a two-qubit environment E referred to the orthonormal basis $\big\{|e_k\rangle\big\}_{k=1}^4 = \big\{|00\rangle, |01\rangle, |10\rangle, |11\rangle\big\}$. The joint unitary evolution of the composite system QE, according to Eqs. (9.9)–(9.10), should be

$$\mathsf{U}_{QE}\,|Q\rangle \otimes |e_0\rangle = \Lambda_1\,|Q\rangle \otimes |00\rangle + \Lambda_2\,|Q\rangle \otimes |01\rangle + \Lambda_3\,|Q\rangle \otimes |10\rangle + \Lambda_4\,|Q\rangle \otimes |11\rangle \tag{9.97}$$

$$= \sqrt{p}\Big[\alpha_0\,|000\rangle + \alpha_1\sqrt{\gamma}\,|001\rangle + \alpha_1\sqrt{1-\gamma}\,|100\rangle\Big] + \sqrt{1-p}\Big[\alpha_0\sqrt{1-\gamma}\,|010\rangle + \alpha_1\,|110\rangle + \alpha_0\sqrt{\gamma}\,|111\rangle\Big], \tag{9.98}$$

with again the preparation $|Q\rangle = \alpha_0\,|0\rangle + \alpha_1\,|1\rangle$ for qubit Q, as in Eq. (9.8).

It is seen that the state of Eq. (9.98), in the 8-dimensional space $\mathcal{H}_2^{\otimes 3}$ of the 3-qubit compound QE, assumes no component in the subspace spanned by $\big\{|011\rangle, |101\rangle\big\}$, but instead remains in a 6-dimensional subspace of $\mathcal{H}_2^{\otimes 3}$. With the choice of fixed unitary U_{QE} independent of (p, γ), if one seeks to adjust the two parameters $(p, \gamma) \in [0, 1]^2$ of the thermal noise via the initial state $|e_0\rangle \in \mathcal{H}_2^{\otimes 2}$ of the environment E, for instance via the separable state $|e_0\rangle = \big(\sqrt{1-p}\,|0\rangle + \sqrt{p}\,|1\rangle\big) \otimes \big(\sqrt{1-\gamma}\,|0\rangle + \sqrt{\gamma}\,|1\rangle\big)$ or a more involved entangled state of $\mathcal{H}_2^{\otimes 2}$, then usually $|e_0\rangle$ will vary in the 4-dimensional space $\mathcal{H}_2^{\otimes 2}$ when (p, γ) cover the domain $[0, 1]^2$. When the qubit state $|Q\rangle$ varies in $\mathcal{H}_2$ via Eq. (9.8) and $|e_0\rangle$ varies in $\mathcal{H}_2^{\otimes 4}$, then the joint product state $|Q\rangle \otimes |e_0\rangle$ will in general vary in the 8-dimensional space $\mathcal{H}_2^{\otimes 3}$. Since the 8-dimensional space $\mathcal{H}_2^{\otimes 3}$ cannot be unitarily mapped onto the 6-dimensional subspace defined by Eq. (9.98), we find in this direction no (p, γ)-independent unitary U_{QE} to satisfy Eqs. (9.97)–(9.98).

As an alternative, to arrive at a suitable choice of the constituents $\big(|e_0\rangle, \mathsf{U}_{QE}\big)$ of an environment model to satisfy Eqs. (9.97)–(9.98), especially affording efficient control of the two noise parameters (p, γ), we will follow the approach of [223], exploiting a model known as the thermal attenuator [224, 225] exhibiting some essential equivalence with the qubit thermal noise considered here. In the thermal attenuator, the environment E is represented by a two-dimensional quantum system

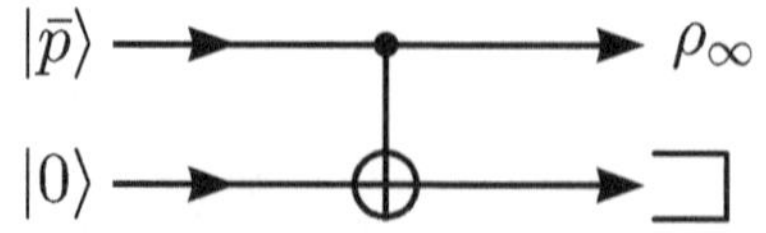

Fig. 9.10 A Cnot gate fed by a target qubit in state $|0\rangle$ and a control qubit in state $|\bar{p}\rangle = \sqrt{p}\,|0\rangle + \sqrt{1-p}\,|1\rangle$. On the two-qubit output state $\sqrt{p}\,|00\rangle + \sqrt{1-p}\,|11\rangle$, when the target qubit is discarded, the control qubit is placed in the mixed state $\rho_\infty = p\,|0\rangle\langle 0| + (1-p)\,|1\rangle\langle 1|$

prepared in the mixed state $\rho_\infty = p\,|0\rangle\langle 0| + (1-p)\,|1\rangle\langle 1|$ we introduced above as the equilibrium or thermalized state imposed by the thermal bath. The input qubit Q and the two-dimensional environment E interact through the unitary operator

$$\widetilde{\mathsf{U}}_{QE} = \begin{bmatrix} 1 & 0 & 0 & 0 \\ 0 & \sqrt{1-\gamma} & \sqrt{\gamma} & 0 \\ 0 & -\sqrt{\gamma} & \sqrt{1-\gamma} & 0 \\ 0 & 0 & 0 & 1 \end{bmatrix}, \tag{9.99}$$

transforming the bipartite state according to $\rho \otimes \rho_\infty \mapsto \widetilde{\mathsf{U}}_{QE}(\rho \otimes \rho_\infty)\widetilde{\mathsf{U}}_{QE}^\dagger = \rho'_{QE}$. On this transformed state ρ'_{QE}, the partial trace over the environment E is easily computed and leads to the state $\mathrm{tr}_E\big(\rho'_{QE}\big) = \mathcal{N}(\rho)$ as targeted in Eq. (9.90). Strictly speaking, for the qubit thermal noise, we are not dealing here with a proper Stinespring dilated unitary representation, as in Sect. 3.2.8, because the environment starts in a mixed qubit state ρ_∞ instead of the pure state $|e_0\rangle$. It is nevertheless possible to introduce a purification of the initial state ρ_∞ of the environment. This can be realized by resorting to an additional auxiliary qubit, which is entangled with the environment qubit by a preparation in the pure two-qubit state $\sqrt{p}\,|00\rangle + \sqrt{1-p}\,|11\rangle$; then when partial tracing over the auxiliary qubit is performed, the environment qubit gets placed in the targeted mixed state $\rho_\infty = p\,|0\rangle\langle 0| + (1-p)\,|1\rangle\langle 1|$. Moreover, in practice this operation is readily accomplished by means of a Cnot gate fed with a target qubit in state $|0\rangle$ and a control qubit in state $|\bar{p}\rangle = \sqrt{p}\,|0\rangle + \sqrt{1-p}\,|1\rangle$, as depicted in Fig. 9.10. In turn, the qubit state $|\bar{p}\rangle$ is easily obtained from a state $|0\rangle$ subjected to the rotation gate $\mathsf{R}_\mathsf{y}(\xi)$ from Eq. (3.26) at an angle $\xi = 2\arccos\big(\sqrt{p}\big)$. In Fig. 9.10, the two qubits at the Cnot output jointly terminate in the bipartite entangled pure state $\sqrt{p}\,|00\rangle + \sqrt{1-p}\,|11\rangle$. When the target qubit is then discarded (non-measured), the control qubit is described by the mixed state $\rho_\infty = p\,|0\rangle\langle 0| + (1-p)\,|1\rangle\langle 1|$ as expected.

The process represented in Fig. 9.10 restores a proper Stinespring dilated unitary representation for the qubit thermal noise, with a two-qubit environment model prepared in the pure state $\sqrt{p}\,|00\rangle + \sqrt{1-p}\,|11\rangle$ and a joint dilated unitary evolution obtained by combining the unitary operation of Fig. 9.10 and that of Eq. (9.99). An advantage in constructing in this way a simulator model for the qubit thermal noise is that we obtain a convenient separated control of the two noise parameters (p, γ): the probability p is adjusted via an auxiliary qubit in the initial state

QE	T_0	QE	$\widetilde{\mathsf{U}}\mathsf{c}$	QE	$\mathsf{T}_0^{-1} \equiv \mathsf{T}_0$	QE
$\|00\rangle$	$\longmapsto$	$\|00\rangle$	$\longmapsto$	$\|00\rangle$	$\longmapsto$	$\|00\rangle$
$\|01\rangle$	$\longmapsto$	$\|01\rangle$	$\boldsymbol{\longmapsto}$	$\left(\sqrt{1-\gamma}\,\|0\rangle - \sqrt{\gamma}\,\|1\rangle\right)\|1\rangle$	$\boldsymbol{\longmapsto}$	$\sqrt{1-\gamma}\,\|01\rangle - \sqrt{\gamma}\,\|10\rangle$
$\|10\rangle$	$\boldsymbol{\longmapsto}$	$\|11\rangle$	$\boldsymbol{\longmapsto}$	$\left(\sqrt{\gamma}\,\|0\rangle + \sqrt{1-\gamma}\,\|1\rangle\right)\|1\rangle$	$\boldsymbol{\longmapsto}$	$\sqrt{\gamma}\,\|01\rangle + \sqrt{1-\gamma}\,\|10\rangle$
$\|11\rangle$	$\boldsymbol{\longmapsto}$	$\|10\rangle$	$\longmapsto$	$\|10\rangle$	$\boldsymbol{\longmapsto}$	$\|11\rangle$

Fig. 9.11 Acting on the qubit pair QE successively transformed, and as also given in [223], the sequence of three elementary unitary transformations: T_0, then the controlled $\widetilde{\mathsf{U}}\mathsf{c}$ obtained from Eq. (9.100), and finally $\mathsf{T}_0^{-1} \equiv \mathsf{T}_0$, to accomplish the unitary transformation $\widetilde{\mathsf{U}}_{QE}$ of Eq. (9.99). Bold arrows indicate nontrivial actions of the corresponding transformations

$|\bar{p}\rangle = \sqrt{p}\,|0\rangle + \sqrt{1-p}\,|1\rangle$ of Fig. 9.10, while the coupling factor γ is adjusted by the unitary $\widetilde{\mathsf{U}}_{QE}$ of Eq. (9.99).

It is also convenient to look for a simple implementation of the two-qubit unitary $\widetilde{\mathsf{U}}_{QE}$ of Eq. (9.99), acting on the principal qubit Q and on an environment qubit E prepared in the mixed state ρ_∞ by the circuit of Fig. 9.10. Relying on their universality property for the synthesis of quantum circuits, as indicated in Sect. 3.1.5, we know that the two-qubit Cnot gate supplemented by one-qubit gates, can supply an implementation for $\widetilde{\mathsf{U}}_{QE}$ as well as for any unitary operator. An approach for the synthesis, based on [9, 18–20], would start by decomposing the unitary operator into two-level unitary matrices, which are unitary matrices acting non-trivially only on two or fewer vector components. Here, our unitary matrix $\widetilde{\mathsf{U}}_{QE}$ we consider in Eq. (9.99) is already a two-level unitary matrix, acting non-trivially only on the two vector components along $|01\rangle$ and $|10\rangle$. The two-level part of the matrix $\widetilde{\mathsf{U}}_{QE}$ can be extracted as

$$\widetilde{\mathsf{U}} = \begin{bmatrix} \sqrt{1-\gamma} & \sqrt{\gamma} \\ -\sqrt{\gamma} & \sqrt{1-\gamma} \end{bmatrix}. \tag{9.100}$$

The compound QE of the principal qubit Q and environment qubit E undergoes a two-level unitary transformation by $\widetilde{\mathsf{U}}_{QE}$ of Eq. (9.99), which can be decomposed into the sequence of three elementary unitary transformations defined in Fig. 9.11 by their action on the computational basis of the bipartite space $\mathcal{H}_2^{\otimes 2}$.

In the three-step sequence of Fig. 9.11, the first transformation T_0 rearranges the vector components for the qubit pair QE so that the two-level transformation of Eq. (9.99) gets applied only to qubit Q in the qubit pair. This transformation T_0 amounts to a controlled-Not on qubit E controlled by qubit Q, and it can therefore receive a simple circuit implementation with a Cnot gate. The second transformation $\widetilde{\mathsf{U}}\mathsf{c}$ in the sequence of Fig. 9.11 is the one-qubit unitary operator $\widetilde{\mathsf{U}}$ from Eq. (9.100), acting only on qubit Q, yet under the control of qubit E. We will therefore have to arrange a controlled version of the one-qubit gate corresponding to $\widetilde{\mathsf{U}}$ in Eq. (9.100). The third transformation in the sequence of Fig. 9.11 restores the initial arrangement of the vector components for the qubit pair QE, through the inverse transformation

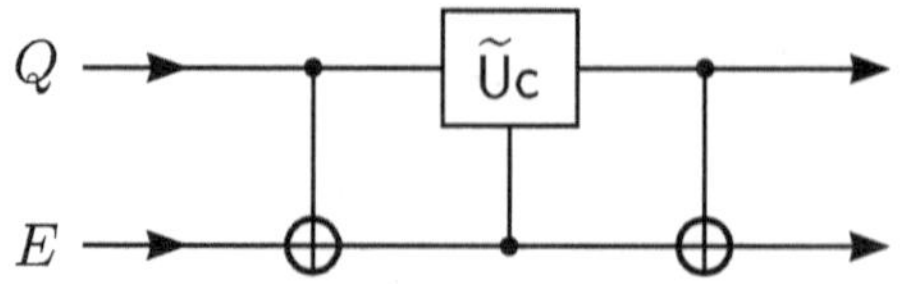

Fig. 9.12 Quantum circuit realizing on the qubit pair QE the unitary transformation $\widetilde{\mathsf{U}}_{QE}$ of Eq. (9.99) via the three-step sequence of Fig. 9.11, and employing a controlled version $\widetilde{\mathsf{U}}\mathsf{c}$ of the gate $\widetilde{\mathsf{U}}$ from Eq. (9.100)

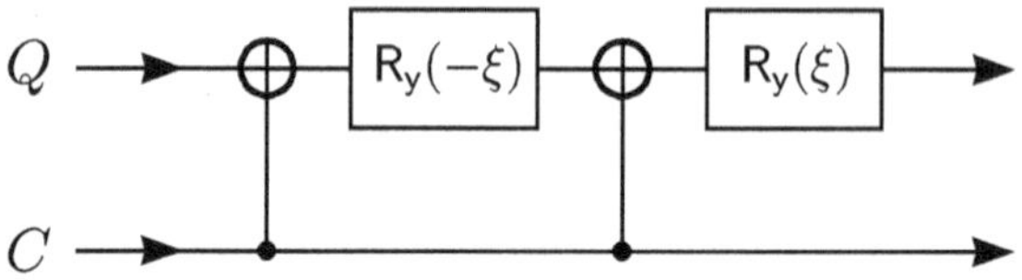

Fig. 9.13 Quantum circuit implementation of the controlled gate $\widetilde{\mathsf{U}}\mathsf{c}$ used in Fig. 9.12, featuring two one-qubit standard $\mathsf{R_y}$ rotation gates from Eq. (3.26) and two Cnot gates, in order to apply the transformation $\widetilde{\mathsf{U}}$ to the target qubit Q under the control of the qubit C

$\mathsf{T}_0^{-1} \equiv \mathsf{T}_0$ coinciding with T_0. After the three-step sequence of Fig. 9.11, the state of the qubit pair QE has experienced the targeted unitary transformation $\widetilde{\mathsf{U}}_{QE}$ in Eq. (9.99). The sequence of Fig. 9.11 can receive a simple circuit implementation represented in Fig. 9.12.

The quantum circuit of Fig. 9.12 uses a controlled gate $\widetilde{\mathsf{U}}\mathsf{c}$ realizing a controlled version of the one-qubit gate $\widetilde{\mathsf{U}}$ from Eq. (9.100). An implementation of the controlled gate $\widetilde{\mathsf{U}}\mathsf{c}$ is materialized by a circuit where a control qubit at $|1\rangle$ applies the gate $\widetilde{\mathsf{U}}$ to the target qubit, while a control qubit at $|0\rangle$ leaves the target qubit unaffected. For an explicit synthesis of such a controlled gate $\widetilde{\mathsf{U}}\mathsf{c}$, we follow the generic procedure described in Sect. 5.6 to construct a controlled version of an arbitrary one-qubit gate. The unitary transformation $\widetilde{\mathsf{U}}$ of Eq. (9.100) is a rotation in the complex plane $\mathcal{H}_2$ by the angle $\xi \in [-\pi/2, 0]$ defined by $\cos(\xi) = \sqrt{1-\gamma}$ and $\sin(\xi) = -\sqrt{\gamma}$, giving $\xi = -\arcsin(\sqrt{\gamma})$. It can be realized as $\widetilde{\mathsf{U}} = \mathsf{R_y}(2\xi)$, with the qubit rotation gate $\mathsf{R_y}(\cdot)$ from Eq. (3.26). For our synthesis we also have $\widetilde{\mathsf{U}} = \mathsf{R_y}(\xi)\sigma_x\mathsf{R_y}(-\xi)\sigma_x$, with the standard inversion or "Not" Pauli operator σ_x. This decomposition for $\widetilde{\mathsf{U}}$ follows from Sect. 5.6 and it can be verified here by directly calculating the corresponding matrix products. We also have $\mathsf{R_y}(\xi)\mathsf{R_y}(-\xi) = \mathrm{I}_2$. It results that when the control is applied on σ_x with a standard Cnot gate, the targeted controlled gate $\widetilde{\mathsf{U}}\mathsf{c}$ is obtained. The circuit implementation of this procedure is represented in Fig. 9.13.

Now the three quantum circuits of Figs. 9.10, 9.12, and 9.13 can be associated, in order to obtain the simulator circuit represented in Fig. 9.14, capable of simulating on the qubit Q an arbitrary quantum thermal noise with controlled parameters (p, γ).

The quantum circuit of Fig. 9.14 realizes a Stinespring dilated unitary representation according to the considerations of Sect. 3.2.8, to simulate the qubit thermal noise $\mathcal{N}(\cdot)$ of Eq. (9.90) with two adjustable parameters (p, γ). The 4 Kraus oper-

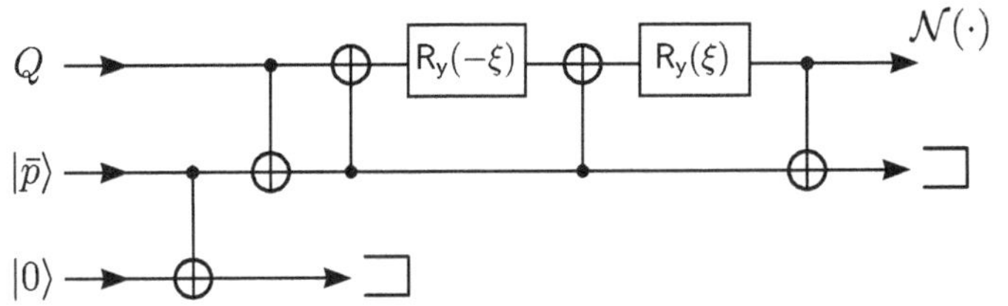

Fig. 9.14 Quantum circuit simulating the quantum thermal noise $\mathcal{N}(\cdot)$ of Eq. (9.90) on the principal qubit Q, with two adjustable parameters (p, γ). The input auxiliary qubit in state $|\bar{p}\rangle = \sqrt{p}\,|0\rangle + \sqrt{1-p}\,|1\rangle \in \mathcal{H}_2$ sets the probability p, while the two one-qubit rotation gates $\mathsf{R_y}(\cdot)$ with the angle $\xi = -\arcsin(\sqrt{\gamma})$ set the damping or coupling factor γ for the noise

ators in Eqs. (9.86)–(9.89) of the thermal noise determine a 4-dimensional environment implemented by two auxiliary (environment) qubits in Fig. 9.14. Including the principal qubit Q experiencing the thermal noise, this leads to a total of 3 qubits employed by the noise simulator. The two environment qubits are initialized in the pure separable state $|e_0\rangle = |\bar{p}\rangle \otimes |0\rangle \in \mathcal{H}_2^{\otimes 2}$. The environment qubit prepared in the state $|\bar{p}\rangle = \sqrt{p}\,|0\rangle + \sqrt{1-p}\,|1\rangle \in \mathcal{H}_2$ allows one to adjust the probability p of the thermal noise. The angle $\xi = -\arcsin(\sqrt{\gamma})$ for the rotation gates allows one to adjust the damping or coupling factor γ of the thermal noise. The quantum circuit of Fig. 9.14 to simulate the qubit thermal noise relies only on elementary quantum gates: 5 two-qubit Cnot gates and 2 one-qubit rotation gates $\mathsf{R_y}(\cdot)$. The two environment qubits in the circuit of Fig. 9.14 are finally discarded (non-measured), and the circuit output marked $\mathcal{N}(\cdot)$ delivers the noisy qubit Q affected by a quantum thermal noise of Eq. (9.90) endowed with two controlled parameters (p, γ).

The synthesis arriving at the circuit of Fig. 9.14 for the simulation of a qubit thermal noise was also addressed in [223], and here it provides a worked-out example of a non-trivial synthesis for a specific-purpose quantum circuit (a thermal noise simulator here). Along with the other noise simulator circuits presented earlier in this chapter, it contributes here to obtain a systematic catalog of essential qubit noise models with their simulation circuits, with broad relevance to quantum technology. The approach also illustrates the universal capability for quantum synthesis of the set of one-qubit gates supplemented by the two-qubit Cnot gate, as indicated in Sect. 3.1.5. Moreover, this universal set for quantum synthesis, although it is formed with unitary gates, is not restricted to the synthesis of unitary quantum processes, but it can as well manage the synthesis of non-unitary quantum processes, as the thermal noise in Fig. 9.14 or all the noise simulator circuits presented earlier in this chapter. The canonical procedure of Sect. 3.2.8 is a generic approach for such synthesis or simulation of non-unitary quantum processes, as already indicated in Sect. 3.2.8 and as now illustrated several times in the present chapter. Finally we observe that the benefit to quantum information processing does not only reside in the synthesis of unitary processes for coherent quantum computation, but more broadly, synthesis of non-unitary quantum processes is also relevant.

In the model based on Eqs. (9.97)–(9.98), after the action of the qubit thermal noise, the four-dimensional environment E terminates in the final state ρ'_E of Eq. (3.119) characterized, via its matrix elements of Eq. (3.120), as

$$\rho'_E = \begin{bmatrix} p(1-\gamma\rho_{11}) & p\sqrt{\gamma}\rho_{01} & \sqrt{p(1-p)}\sqrt{1-\gamma} & \sqrt{p(1-p)}\sqrt{\gamma(1-\gamma)}\rho_{10} \\ p\sqrt{\gamma}\rho_{10} & p\gamma\rho_{11} & \sqrt{p(1-p)}\sqrt{\gamma(1-\gamma)}\rho_{10} & 0 \\ \mathtt{cc} & \mathtt{cc} & (1-p)(1-\gamma\rho_{00}) & (1-p)\sqrt{\gamma}\rho_{10} \\ \mathtt{cc} & 0 & (1-p)\sqrt{\gamma}\rho_{01} & (1-p)\gamma\rho_{00} \end{bmatrix}, \tag{9.101}$$

where the entries marked "cc" follow as complex conjugate from the Hermitian symmetry of the matrix. In addition, the final state ρ'_E is given in [226] as a function of the Bloch vector $\vec{r} = [r_x, r_y, r_z]^\top$ characterizing the input state ρ of the principal qubit Q. For the four eigenvalues of ρ'_E in Eq. (9.101), exact analytical expressions can be found, but they are too bulky to be written down here in general form.

For an interesting special case, we can consider the restricted class where the input signal ρ is characterized by the Bloch vector $\vec{r} = [0, 0, r_z]^\top$ with $r_z \in [-1, 1]$, leading in Eq. (9.101) to a final state of the environment expressible as

$$\rho'_E = \frac{1}{2}\begin{bmatrix} p[2-\gamma(1-r_z)] & 0 & 2\sqrt{p(1-p)}\sqrt{1-\gamma} & 0 \\ 0 & p\gamma(1-r_z) & 0 & 0 \\ 2\sqrt{p(1-p)}\sqrt{1-\gamma} & 0 & (1-p)[2-\gamma(1+r_z)] & 0 \\ 0 & 0 & 0 & (1-p)\gamma(1+r_z) \end{bmatrix}. \tag{9.102}$$

To evaluate the quantum entropy $S(\rho'_E)$, the four eigenvalues of ρ'_E from Eq. (9.102) follow as

$$\lambda_1 = \frac{1}{4}\left[(2p-1)\gamma r_z + 2 - \gamma - \sqrt{\Gamma}\right], \tag{9.103}$$

$$\lambda_2 = \frac{1}{4}\left[(2p-1)\gamma r_z + 2 - \gamma + \sqrt{\Gamma}\right], \tag{9.104}$$

$$\lambda_3 = \frac{1}{2}(1-p)\gamma(1+r_z), \tag{9.105}$$

$$\lambda_4 = \frac{1}{2}p\gamma(1-r_z), \tag{9.106}$$

with

$$\Gamma = \gamma^2 r_z^2 + 2(2p-1)\gamma(2-\gamma)r_z + (2p-1)^2\gamma^2 + 4(1-\gamma). \tag{9.107}$$

We obtain in this way a characterization of the quantum entropy $S(\rho'_E)$ generated in the thermal bath E by the action of the thermal noise, as a function of the input signal

ρ via the coordinate r_z, and of the properties of the noise process via the coupling γ to the bath and the probability p conveying the temperature T of the bath.

9.11 Squeezed Generalized Amplitude Damping Noise

The squeezed generalized amplitude damping (SGAD) noise is a sophisticated noise model for the qubit, which generalizes or extends both the amplitude damping noise of Sect. 9.9 and the generalized amplitude damping noise of Sect. 9.10. The SGAD noise [221, 227–230] describes the interaction of a qubit with an environment represented by a squeezed thermal bath. Squeezing of a thermal bath is obtained by a specific coupling operation capable of introducing correlations between the modes or thermal bosons (photons, phonons) of the bath, with possibilities of reducing the dispersion or uncertainty attached to quantum states via Heisenberg inequalities. For a qubit interacting with a squeezed thermal bath, the SGAD noise can be modeled in Eq. (9.5) by the $K = 4$ Kraus operators

$$\Lambda_1 = \sqrt{p}\begin{bmatrix} \sqrt{1-\mu} & 0 \\ 0 & \sqrt{1-\nu} \end{bmatrix}, \tag{9.108}$$

$$\Lambda_2 = \sqrt{p}\begin{bmatrix} 0 & \sqrt{\nu} \\ \sqrt{\mu}e^{-i\Phi} & 0 \end{bmatrix}, \tag{9.109}$$

$$\Lambda_3 = \sqrt{1-p}\begin{bmatrix} \sqrt{1-\gamma} & 0 \\ 0 & 1 \end{bmatrix}, \tag{9.110}$$

$$\Lambda_4 = \sqrt{1-p}\begin{bmatrix} 0 & 0 \\ \sqrt{\gamma} & 0 \end{bmatrix}, \tag{9.111}$$

with the real parameters μ, ν, γ and p all in $[0, 1]$, and Φ a phase angle in $[0, 2\pi[$. From $\rho = \big[\rho_{00}, \rho_{01}\,;\, \rho_{10}, \rho_{11}\big]$ as in Eq. (2.59), the quantum operation resulting in Eq. (9.5) leads to the transformed state

$$\mathcal{N}(\rho) = \begin{bmatrix} \big[p(1-\mu) + (1-p)(1-\gamma)\big]\rho_{00} + p\nu\rho_{11} & \big[p\sqrt{1-\mu}\sqrt{1-\nu} + (1-p)\sqrt{1-\gamma}\big]\rho_{01} + p\sqrt{\mu\nu}e^{i\Phi}\rho_{10} \\ \mathrm{cc} & \big[p\mu + (1-p)\gamma\big]\rho_{00} + \big[p(1-\nu) + 1 - p\big]\rho_{11} \end{bmatrix}. \tag{9.112}$$

with again "cc" for the complex conjugate.

The transformation of the Bloch vector in Eq. (9.7) is

$$\vec{r}\,' = A\vec{r} + \vec{c} = \begin{bmatrix} a_{xx} & a_{xy} & 0 \\ a_{xy} & a_{yy} & 0 \\ 0 & 0 & a_{zz} \end{bmatrix}\vec{r} + \begin{bmatrix} 0 \\ 0 \\ c_z \end{bmatrix}, \tag{9.113}$$

with the components

$$a_{xx} = p\big[\sqrt{1-\mu}\sqrt{1-\nu} + \cos(\Phi)\sqrt{\mu\nu}\,\big] + (1-p)\sqrt{1-\gamma}\ , \tag{9.114}$$

$$a_{yy} = p\big[\sqrt{1-\mu}\sqrt{1-\nu} - \cos(\Phi)\sqrt{\mu\nu}\,\big] + (1-p)\sqrt{1-\gamma}\ , \tag{9.115}$$

$$a_{xy} = -p\sin(\Phi)\sqrt{\mu\nu}\ , \tag{9.116}$$

$$a_{zz} = 1 - p(\mu+\nu) - (1-p)\gamma\ , \tag{9.117}$$

$$c_z = -p(\mu-\nu) - (1-p)\gamma\ . \tag{9.118}$$

More elements can be found in [221, 227–230] concerning the detailed physical significance of the parameters characterizing the SGAD noise in Eqs. (9.108)–(9.111). With no squeezing, $\mu = 0$ and $\nu = \gamma$ in Eqs. (9.108)–(9.111) and the SGAD noise reduces to the generalized amplitude damping noise of Sect. 9.10.

The $K = 4$ Kraus operators of the SGAD noise, by the procedure of Sect. 3.2.8, determine a four-dimensional environment E, which terminates in the final state ρ'_E of Eq. (3.119) represented in the computational basis by the 4×4 matrix $W = \big[w_{k\ell}\big]$ with elements $w_{k\ell} = \mathrm{tr}\big(\Lambda_k \rho \Lambda_\ell^\dagger\big)$ from Eq. (3.120). The four matrix elements $(w_{33}, w_{34}, w_{43}, w_{44})$ are the same as those for the thermal noise in Eq. (9.101), since both noises share the same Kraus operators Λ_3 and Λ_4. The other matrix elements for the SGAD noise are

$$w_{11} = p\big[(1-\mu)\rho_{00} + (1-\nu)\rho_{11}\big]\ , \tag{9.119}$$

$$w_{12} = p\Big[\sqrt{(1-\mu)\nu}\,\rho_{01} + \sqrt{\mu(1-\nu)}\,e^{i\Phi}\rho_{10}\Big]\ , \tag{9.120}$$

$$w_{13} = \sqrt{(1-p)p}\,\Big[\sqrt{(1-\gamma)(1-\mu)}\,\rho_{00} + \sqrt{1-\nu}\,\rho_{11}\Big]\ , \tag{9.121}$$

$$w_{14} = \sqrt{(1-p)p}\sqrt{\gamma(1-\nu)}\,\rho_{10}\ , \tag{9.122}$$

$$w_{22} = p\big(\mu\rho_{00} + \nu\rho_{11}\big)\ , \tag{9.123}$$

$$w_{23} = \sqrt{(1-p)p}\,\Big[\sqrt{\mu}e^{-i\Phi}\rho_{01} + \sqrt{(1-\gamma)\nu}\,\rho_{10}\Big]\ , \tag{9.124}$$

$$w_{24} = \sqrt{(1-p)p}\sqrt{\gamma\mu}\,e^{-i\Phi}\rho_{00}\ , \tag{9.125}$$

supplemented by their complex conjugate as dictated by the hermiticity of W. The four eigenvalues of ρ'_E are difficult to characterize analytically, but can be computed numerically in definite conditions.

9.12 Commuting Qubit Noise

We will now describe a useful class of qubit noise models, which represents the class of qubit noises that commute with an arbitrarily given unitary operator acting on the qubit. An instance of such situation was encountered with the example of the noisy interferometer addressed in Sect. 9.4, featuring a phase-flip noise which was found to commute with the unitary (dephasing) action of the interferometer. This type of commuting behavior of a noise model is important, because it represents a noise process that can equivalently take place before or after the action of the unitary operator on the qubit. Moreover, it can even represent two separate noise processes that could take place, one before and one after, the action of the unitary operator, and could altogether be lumped into a single, localized, noise process. Given a unitary qubit operator, we want to characterize the class of qubit noise processes that are lumpable in this way, or equivalently that commute with the unitary.

We consider an arbitrary qubit unitary operator, under the form of U_ξ of Eq. (3.18) in Sect. 3.1.4. Such a unitary U_ξ is characterized by a unit vector $\vec{n} \in \mathbb{R}^3$ which determines the two orthogonal eigenstates of U_ξ in $\mathcal{H}_2$ that were denoted $|\pm\vec{n}\,\rangle$ in Sect. 3.1.4 (just before Eq. (3.30)). The qubit Hilbert space $\mathcal{H}_2$ can be conveniently referred to the orthonormal basis $\{\,|0'\rangle = |\vec{n}\,\rangle\,,\ |1'\rangle = |-\vec{n}\,\rangle\}$ substituting to the standard computational basis $\{|0\rangle\,,|1\rangle\}$. Any noise model on the qubit can be completely defined by specifying the action of the superoperator $\mathcal{N}(\cdot)$ on a basis for operators on $\mathcal{H}_2$. As in [231], we define the alternative Pauli operators as $\sigma_x' = |0'\rangle\langle 1'| + |1'\rangle\langle 0'|$, $\sigma_y' = i\,|1'\rangle\langle 0'| - i\,|0'\rangle\langle 1'|$ and $\sigma_z' = |0'\rangle\langle 0'| - |1'\rangle\langle 1'|$. Associated with I_2 the identity on $\mathcal{H}_2$, the alternative set $\{\mathrm{I}_2, \sigma_x', \sigma_y', \sigma_z'\}$ as well as the standard set $\{\mathrm{I}_2, \sigma_x, \sigma_y, \sigma_z\}$ form two orthogonal bases for operators on $\mathcal{H}_2$. We then consider the class of quantum noise models defined by

$$\mathcal{N}(\mathrm{I}_2) = \mathrm{I}_2 + \mu_0\sigma_z' \,, \tag{9.126}$$

$$\mathcal{N}(\sigma_z') = \mu_1\sigma_z' \,, \tag{9.127}$$

$$\mathcal{N}(\sigma_x' \pm i\sigma_y') = \mu_2 e^{\mp i\phi}(\sigma_x' \pm i\sigma_y') \,, \tag{9.128}$$

providing a complete characterization of the noise superoperator $\mathcal{N}(\cdot)$ in the alternative Pauli basis $\{\mathrm{I}_2, \sigma_x', \sigma_y', \sigma_z'\}$ of $\mathcal{L}(\mathcal{H}_2)$. From Eqs. (9.126)–(9.128) we deduce

$$\mathcal{N}\big(|0'\rangle\langle 0'|\big) = \frac{1}{2}\big(1 + \mu_0 + \mu_1\big)\,|0'\rangle\langle 0'| + \frac{1}{2}\big(1 - \mu_0 - \mu_1\big)\,|1'\rangle\langle 1'| \,, \tag{9.129}$$

$$\mathcal{N}\big(|0'\rangle\langle 1'|\big) = \mu_2\,|0'\rangle\langle 1'|\,e^{-i\phi} \,, \tag{9.130}$$

$$\mathcal{N}\big(|1'\rangle\langle 0'|\big) = \mu_2\,|1'\rangle\langle 0'|\,e^{i\phi} \,, \tag{9.131}$$

$$\mathcal{N}\big(|1'\rangle\langle 1'|\big) = \frac{1}{2}\big(1 + \mu_0 - \mu_1\big)\,|0'\rangle\langle 0'| + \frac{1}{2}\big(1 - \mu_0 + \mu_1\big)\,|1'\rangle\langle 1'| \,, \tag{9.132}$$

providing a complete characterization of $\mathcal{N}(\cdot)$ in the alternative computational basis $\{|0'\rangle\langle 0'|\,, |0'\rangle\langle 1'|\,, |1'\rangle\langle 0'|\,, |1'\rangle\langle 1'|\}$ of $\mathcal{L}(\mathcal{H}_2)$. Then the noisy state $\rho' = \mathcal{N}(\rho)$ follows, for any qubit density operator $\rho \in \mathcal{L}(\mathcal{H}_2)$, by linearity based on the parameterization of ρ in Eq. (2.60), as

$$\mathcal{N}(\rho) = \frac{1}{2}\begin{bmatrix} 1 + \mu_0 + (\rho_{00} - \rho_{11})\mu_1 & 2\rho_{01}\mu_2 e^{-i\phi} \\ 2\rho_{10}\mu_2 e^{i\phi} & 1 - \mu_0 - (\rho_{00} - \rho_{11})\mu_1 \end{bmatrix} . \tag{9.133}$$

The two noise parameters μ_1 and μ_2 are reals in [0, 1] and satisfy $\mu_1 \geq \mu_2^2$ to ensure complete positivity of the map $\mathcal{N}(\cdot)$; also $\mu_0 = (1-\mu_1)\mu$ with the real constant $\mu \in [-1, 1]$. The real parameter ϕ is a phase angle. Based on [232], the noise parameters can be further modeled via time-resolved dynamics specifying $\mu_1 = e^{-t/\tau_1}$ and $\mu_2 = e^{-t/\tau_2}$, with a longitudinal relaxation time τ_1 and a transverse dephasing time τ_2 satisfying $\tau_2 \leq 2\tau_1$ for complete positivity, and $\phi = \omega t$, with t the time interval where the qubit is exposed to the noise interaction. The noisy state of Eq. (9.133) therefore takes the form

$$\mathcal{N}(\rho) = \frac{1}{2}\begin{bmatrix} 1+\mu+(\rho_{00}-\rho_{11}-\mu)e^{-t/\tau_1} & 2\rho_{01}e^{-i\phi}e^{-t/\tau_2} \\ 2\rho_{10}e^{i\phi}e^{-t/\tau_2} & 1-\mu-(\rho_{00}-\rho_{11}-\mu)e^{-t/\tau_1} \end{bmatrix}. \tag{9.134}$$

Following [231, 232], the quantum noise model of Eqs. (9.126)–(9.128) represents the most general noise model independently acting on single qubits, with the underlying temporal dynamics stationary and continuously differentiable, and commuting with the unitary U_ξ of axis $\vec{n}$. At long interaction time $t \to +\infty$, the noisy qubit of Eq. (9.134) relaxes to the equilibrium mixed state

$$\mathcal{N}(\rho) = \frac{1}{2}\begin{bmatrix} 1+\mu & 0 \\ 0 & 1-\mu \end{bmatrix}, \tag{9.135}$$

independent of the initial state ρ, with populations (probabilities) respectively $(1 \pm \mu)/2$ for the two basis states $|\pm\vec{n}\,\rangle$, and with the coherences (the off-diagonal terms) reaching zero. This is a typical effect of quantum noise or decoherence, to transform pure or coherent states into mixed states or incoherent statistical mixtures by gradually destroying the coherences. Coherent quantum states by so losing their coherence gradually become like classical states. The time scale of this evolution characterizes the decoherence time of the quantum states and is often a crucial duration above which specific quantum properties may be lost.

For the Bloch representation associated with the alternative Pauli basis $\{\mathrm{I}_2, \sigma_x', \sigma_y', \sigma_z'\}$, we define in $\mathcal{H}_2$ the pure state $|+'\rangle = \big(|0'\rangle + |1'\rangle\big)/\sqrt{2}$ having the (unit) Bloch vector $\vec{n}_\perp$ in $\mathbb{R}^3$ orthogonal to $\vec{n}$. For the (alternative) Bloch representation, $\mathbb{R}^3$ is referred to the orthonormal basis $\big\{\vec{n}_\perp, \vec{n}_\perp' = \vec{n} \times \vec{n}_\perp, \vec{n}\,\big\}$. Any density operator ρ of $\mathcal{L}(\mathcal{H}_2)$ has then the Bloch representation $\rho = (\mathrm{I}_2 + \vec{r}\cdot\vec{\sigma}\,')/2$, with the formal vector $\vec{\sigma}\,' = [\sigma_x', \sigma_y', \sigma_z']$, and a Bloch vector $\vec{r} = [r_x, r_y, r_z]^\top$ referred to the orthonormal basis $\big\{\vec{n}_\perp, \vec{n}_\perp', \vec{n}\,\big\}$ of $\mathbb{R}^3$. The unit Bloch vector with $r_z = 1$ is $\vec{r} = \vec{n}$ representing the pure state $\rho = |0'\rangle\langle 0'|$; the unit Bloch vector with $r_x = 1$ is $\vec{r} = \vec{n}_\perp$ representing the pure state $\rho = |+'\rangle\langle +'|$; while the unit Bloch vector with $r_y = 1$ is $\vec{r} = \vec{n}_\perp'$ representing the pure state $\rho = |+i'\rangle\langle +i'|$ with $|+i'\rangle = \big(|0'\rangle + i\,|1'\rangle\big)/\sqrt{2}$.

The transformation $\rho' = \mathcal{N}(\rho)$ by the noise can then be worked out in this Bloch representation via Eqs. (9.126)–(9.128), and it leads to a transformed state $\rho' = (\mathrm{I}_2 + \vec{r}\,'\cdot\vec{\sigma}\,')/2$ characterized by the Bloch vector resulting from the affine transformation

$$\vec{r}\,' = A\vec{r} + \vec{c} = \begin{bmatrix} \mu_2\cos(\phi) & -\mu_2\sin(\phi) & 0 \\ \mu_2\sin(\phi) & \mu_2\cos(\phi) & 0 \\ 0 & 0 & \mu_1 \end{bmatrix} \vec{r} + \begin{bmatrix} 0 \\ 0 \\ \mu_0 \end{bmatrix}. \tag{9.136}$$

The action of the matrix A in Eq. (9.136) is to contract the Bloch vector by μ_1 in the direction $\vec{n}$ and by μ_2 in the plane $(\vec{n}_\perp, \vec{n}_\perp')$ orthogonal to $\vec{n}$ in $\mathbb{R}^3$, with larger contraction as a larger level of noise. The matrix A also induces a (coherent) rotation by the angle ϕ around the axis $\vec{n}$, which by itself does not reduce the norm of the Bloch vectors and so does not increase the entropy of the qubit states from Eq. (7.40), and is thus a coherent rather than a genuine noisy action on the qubit. Finally, there is in Eq. (9.136) a translation by $\vec{c} = [0, 0, \mu_0]^\top$ parallel to the axis $\vec{n}$.

The transformation by the noise described in Bloch representation in Eq. (9.136) renders transparent its commutation property with the arbitrary unitary U_ξ. In Bloch representation in $\mathbb{R}^3$, the action of U_ξ is a rotation around the axis $\vec{n}$ by the angle ξ. The noise of Eqs. (9.126)–(9.128) is matched to the axis $\vec{n}$ via the basis $\big\{|0'\rangle = |\vec{n}\,\rangle\,, |1'\rangle = |-\vec{n}\,\rangle\big\}$, so that the resulting affine transformation of $\mathbb{R}^3$ in Eq. (9.136) commutes with the rotation by U_ξ, and is the most general acceptable affine transformation to do so.

We have characterized here, via Eqs. (9.126)–(9.128) or Eq. (9.136), the whole class of qubit noises that commute with a given unitary U_ξ. In particular, this allows us to deduce that the standard phase-flip, depolarizing, thermal noises presented earlier, belong to the class of noises commuting with the unitary operators U_ξ characterized by $\vec{n} = \vec{e}_z$. The standard bit-flip noise is in the class with $\vec{n} = \vec{e}_x$, and the bit-phase-flip with $\vec{n} = \vec{e}_y$. Meanwhile, general Pauli noises and the SGAD noise will usually never commute with any qubit unitary operator.

9.13 Noise on Multiple Qubits

A quantum noise on quantum states of arbitrary dimension can generally be described as a non-unitary process according to Sect. 9.1. As we have indicated, any finite-dimensional quantum states can be materialized as multiple-qubit states. For multiple-qubit states, it can often be assumed that the noise acts independently on each qubit. In this circumstance, there is no need to define higher-dimensional noise models, and the noise model defined on each qubit by the quantum process $\mathcal{N}(\cdot)$ will extend to higher dimension according to the tensor-product structure set by the joint quantum states.

For instance, a qubit pair AB with Hilbert space $\mathcal{H}_2^{\otimes 2}$ has a state characterized by the density operator $\rho_{AB} \in \mathcal{L}(\mathcal{H}_2^{\otimes 2})$. On the qubit A acts a quantum noise implementing in Eq. (9.5) the quantum process $\mathcal{N}_A(\cdot)$ defined by the K Kraus operators $\Lambda_k \in \mathcal{L}(\mathcal{H}_2)$. On the qubit B independently acts a quantum noise implementing in Eq. (9.5) the quantum process $\mathcal{N}_B(\cdot)$ defined by the K' Kraus operators $\Lambda'_{k'} \in \mathcal{L}(\mathcal{H}_2)$. The qubit pair AB then experiences a noise process defined by the KK' Kraus operators $\Lambda_k \otimes \Lambda'_{k'}$, acting to produce the noisy qubit pair with density operator

$\rho'_{AB} \in \mathcal{L}(\mathcal{H}_2^{\otimes 2})$ given by

$$\rho'_{AB} = \mathcal{N}_{AB}(\rho_{AB}) = \mathcal{N}_A \otimes \mathcal{N}_B(\rho_{AB}) = \sum_{k=1}^{K}\sum_{k'=1}^{K'}(\Lambda_k \otimes \Lambda'_{k'})\rho_{AB}(\Lambda_k^\dagger \otimes \Lambda'^\dagger_{k'}) . \tag{9.137}$$

Only when the joint state is separable as the product $\rho_{AB} = \rho_A \otimes \rho_B$ will the noisy state follow as the separable product state $\rho'_{AB} = \mathcal{N}_{AB}(\rho_{AB}) = \mathcal{N}_A(\rho_A) \otimes \mathcal{N}_B(\rho_B)$. Otherwise, with ρ_{AB} entangled, the effect of the superoperator $\mathcal{N}_{AB}(\cdot)$ from $\mathcal{L}(\mathcal{H}_2^{\otimes 2})$ onto $\mathcal{L}(\mathcal{H}_2^{\otimes 2})$ in Eq. (9.137) will have to be worked out in a joint fashion. This can be done for instance by first establishing the transformation by $\mathcal{N}_{AB}(\cdot)$ of the (separable) computational basis of $\mathcal{L}(\mathcal{H}_2^{\otimes 2})$ according to the tensor-product structure. For example the (separable) basis operator $|00\rangle\langle 01| = |0\rangle\langle 0| \otimes |0\rangle\langle 1|$ transforms as $\mathcal{N}_{AB}(|00\rangle\langle 01|) = \mathcal{N}_{AB}(|0\rangle\langle 0| \otimes |0\rangle\langle 1|) = \mathcal{N}_A(|0\rangle\langle 0|) \otimes \mathcal{N}_B(|0\rangle\langle 1|)$, with transformations like $\mathcal{N}_B(|0\rangle\langle 1|)$ handled as in Eq. (3.49), and in a similar way for the 16 (separable) basis operators of $\mathcal{L}(\mathcal{H}_2^{\otimes 2})$. Then, from the decomposition of ρ_{AB} in the computational basis of $\mathcal{L}(\mathcal{H}_2^{\otimes 2})$, by linearity, follows the transformed state $\mathcal{N}_{AB}(\rho_{AB})$ for any $\rho_{AB} \in \mathcal{L}(\mathcal{H}_2^{\otimes 2})$.

Alternatively, if the quantum noise cannot be assumed to act independently on the two qubits A and B, a joint model $\mathcal{N}_{AB}(\cdot)$ has to be introduced, for instance through Kraus operators acting jointly in the space $\mathcal{H}_2^{\otimes 2}$ and not factorizable as tensor products of Kraus operators acting separately in $\mathcal{H}_2$. And similarly for correlated noise on multiple qubits in larger number.

9.14 Temporal Dynamics for Quantum Noise

The considerations to be presented in this section will refer to a quantum noise process $\rho \mapsto \mathcal{N}(\rho)$, but they also equally apply to any quantum open evolution $\rho \mapsto \mathcal{E}(\rho)$ as in Sect. 3.2, not necessarily interpreted as a quantum noise.

9.14.1 Time-Dependent Kraus Operators

As we already indicated in Sect. 3.2.12, when resorting to a quantum operation $\rho \mapsto \mathcal{N}(\rho)$ as in Fig. 9.1 for the modeling of a quantum noise, the temporal dynamics remain unresolved. The action of the noise is described by the quantum process $\mathcal{N}(\cdot)$, acting as the functional block of Fig. 9.1, transforming the input state ρ into the output state $\mathcal{N}(\rho)$, without explicitly specifying the time course of the input–output transformation. Such a functional-block transformation is well suited for many applications in quantum information processing with a signals-and-systems theoretic orientation.

A possibility to restore explicit continuous-time dynamics is by considering that the input state $\rho(t_0)$ is applied from some initial time t_0 to experience the noise process, which at each time $t \geq t_0$ produces the transformed state $\rho(t)$ by the quantum operation

$$\rho(t_0) \longmapsto \rho(t) = \mathcal{N}\big(\rho(t_0)\big) = \sum_{k=1}^{K} \Lambda_k(t)\rho(t_0)\Lambda_k^\dagger(t) . \tag{9.138}$$

Time-dependent Kraus operators $\Lambda_k(t)$ are introduced to generate the continuous-time evolution of the output state $\rho(t)$ at each time $t \geq t_0$. The appropriate time dependence of the Kraus operators usually has its origin in their derivation, whether issued from an underlying dilated unitary representation or from a more empirical derivation, as discussed in Sect. 3.2.

This possibility of time-dependent Kraus operators can be materialized with the qubit noise models of this chapter by introducing time dependence in their parameters. For example, in Sect. 9.10 for the qubit thermal noise, a temporal dependence was envisaged in Eq. (9.92) for the coupling factor γ with the thermal bath. Also, the derivation of Sect. 9.12 on commuting noises contains temporal dynamics appropriate for parameters of the models in the class of Eq. (9.136). By adapting the choice of the symmetry axis represented by the vector $\vec{n}$ in Sect. 9.12, this class of noise models of Eq. (9.136) contains the bit-flip, phase-flip, bit-phase-flip noises, as well as the generalized amplitude damping noise, which would all come in this way with time-dependent parameters. Typically, the time dependence can take place via temporal relaxations as $e^{-t/\tau}$ in some model parameters.

For example, a decohering probability p in the Kraus operators of the qubit noise models of this chapter, can be considered as describing the decoherence of the qubit via interacting with its environment over a very short time interval Δt, with a proportionality $p = \Delta t/\tau$ when $\Delta t \ll \tau$ and $p \ll 1$. The interaction over a time interval $t = n\Delta t$ can be described by applying n times the noise model of the evolution over Δt. For instance for the phase damping of Eq. (9.31), this would lead for the qubit Bloch vector to the global evolution

$$\vec{r} \longmapsto \begin{bmatrix} (1-2p)^n & 0 & 0 \\ 0 & (1-2p)^n & 0 \\ 0 & 0 & 1 \end{bmatrix} \vec{r} . \tag{9.139}$$

At long interaction time $t \gg \Delta t$, the term $(1-2p)^n$ goes to $e^{-2t/\tau}$. The same global effect is captured by a time-varying probability $p(t) = \big(1 - e^{-2t/\tau}\big)/2$ to act in the initial noise model of Eq. (9.31) and accordingly to be placed in the Kraus operators of the corresponding phase-flip noise process. With this modeling, at long interaction time $t \gg \tau$, the decohering probability goes to $p(t) = 1/2$, describing the completely decohering process of Sect. 9.8. The same time-dependent decohering probability $p(t) = \big(1 - e^{-2t/\tau}\big)/2$ holds equally, in similar conditions, for the bit-flip noise of Sect. 9.3 and for the bit-phase-flip noise of Sect. 9.5. The time constant

τ here determines the decoherence time of the quantum states, above which specific quantum properties may be lost.

More involved time dependence can be introduced, for instance for the parameters of the squeezed generalized amplitude damping noise of Sect. 9.11, as discussed in [227–229]. If one wants to dig deeper into the physics underlying quantum noise, for instance to identify the mechanisms and values governing the parameters of a noise model presented here, like for example the time constants τ or τ_1 and τ_2 appearing in Eq. (9.134), then [69, 233, 234] can be consulted.

9.14.2 Lindblad Differential Equation

An alternative to time-dependent Kraus operators for continuous-time modeling of open quantum dynamics is via a temporal differential equation, the Lindblad equation [44, 45] mentioned previously in Sect. 3.2.12. Before examining such continuous-time non-unitary evolution of the state of an open quantum system, it is helpful to start with the unitary evolution of the state of a closed quantum system, where we anticipate to encounter again the Schrödinger equation of Sect. 3.1.2.

A closed quantum system with Hilbert space $\mathcal{H}_N$, in an infinitesimal time interval dt, undergoes an evolution of its state vector from $|\psi(t)\rangle$ to $|\psi(t+dt)\rangle$. According to the principle of unitary evolution of Sect. 3.1.1 and its Eq. (3.1), this evolution occurs as $|\psi(t+dt)\rangle = \mathsf{U}(t+dt,t)\,|\psi(t)\rangle$ where $\mathsf{U}(t+dt,t)$ is a unitary (evolution) operator of $\mathcal{L}(\mathcal{H}_N)$. Assuming a continuous change in time, the evolution operator from t to $t+dt$ can be modeled as $\mathsf{U}(t+dt,t) = \mathrm{I}_N + \mathsf{L}dt$, since at the limit $dt \to 0$, when $|\psi(t+dt)\rangle = |\psi(t)\rangle$, we have the trivial evolution by the identity operator $\mathrm{I}_N = \lim_{dt\to 0} \mathsf{U}(t+dt,t)$ of $\mathcal{L}(\mathcal{H}_N)$. The unitarity requirement $\mathsf{U}(t+dt,t)\mathsf{U}^\dagger(t+dt,t) = \mathrm{I}_N$, when applied up to first order in dt, imposes that $\mathsf{L}^\dagger = -\mathsf{L}$, i.e. our "rate" operator L modeling the infinitesimal change of the evolution operator over dt should be anti-Hermitian. We then obtain to first order in dt the evolution $|\psi(t+dt)\rangle = |\psi(t)\rangle + \mathsf{L}\,|\psi(t)\rangle\,dt$. And via $|\psi(t+dt)\rangle = |\psi(t)\rangle + \frac{d}{dt}\,|\psi(t)\rangle\,dt$, we obtain the rate of change of the quantum state as $\frac{d}{dt}\,|\psi(t)\rangle = \mathsf{L}\,|\psi(t)\rangle$. By comparing with the Schrödinger equation (3.3), we find that our rate operator is $\mathsf{L} = -(i/\hbar)\mathsf{H}$, what we somehow already knew from Eq. (3.7) applied to the infinitesimal time interval from t to $t+dt$. When the state of the closed quantum system is represented by a density operator, we have by Eq. (3.2) the evolution $\rho(t+dt) = \mathsf{U}(t+dt,t)\rho(t)\mathsf{U}^\dagger(t+dt,t)$, which is up to first order in dt the evolution $\rho(t+dt) = \rho(t) + \big(\mathsf{L}\rho(t) + \rho(t)\mathsf{L}^\dagger\big)dt = \rho(t) + [\mathsf{L},\rho(t)]dt$. The rate of change of the density operator follows as $\frac{d}{dt}\rho(t) = [\mathsf{L},\rho(t)]$ which is similar to Eq. (3.8) when $\mathsf{L} = -(i/\hbar)\mathsf{H}$. We will now reproduce this line of reasoning for the non-unitary evolution of the state of an open quantum system.

An open quantum system with Hilbert space $\mathcal{H}_N$, in an infinitesimal time interval dt, undergoes a non-unitary evolution of its density operator from $\rho(t)$ to $\rho(t+dt)$. According to Eq. (3.45) in Sect. 3.2.2, this non-unitary evolution can be represented by a quantum operation that we write here

$$\rho(t+dt) = \sum_{k=0}^{K} \Lambda_k(t+dt,t)\rho(t)\Lambda_k^\dagger(t+dt,t) , \tag{9.140}$$

with $K+1$ Kraus operators $\Lambda_k(t+dt,t) \in \mathcal{L}(\mathcal{H}_N)$ to implement the non-unitary evolution in the infinitesimal time interval from t to $t+dt$. We assume a continuous change in time and a variation $\rho(t+dt) - \rho(t) = \frac{d}{dt}\rho(t)dt$ that is of first order in dt at a rate $\frac{d}{dt}\rho(t)$ we want to characterize. Among the $K+1$ Kraus operators in Eq. (9.140), one (say Λ_0) is assumed to vary according to

$$\Lambda_0(t+dt,t) = \mathrm{I}_N + \mathsf{L}_0 dt . \tag{9.141}$$

For $k \neq 0$, the other Kraus operators are taken to vary as

$$\Lambda_k(t+dt,t) = \mathsf{L}_k\sqrt{dt} , \tag{9.142}$$

so that the multiplicative contribution in Eq. (9.140) of each of these $\Lambda_k(t+dt,t)$ be in dt, as $\Lambda_k(t+dt,t)\rho(t)\Lambda_k^\dagger(t+dt,t) = \mathsf{L}_k\rho(t)\mathsf{L}_k^\dagger dt$. The choice in Eq. (9.141) ensures that at the limit $dt \to 0$, when $\rho(t+dt) = \rho(t)$, the evolution of Eq. (9.140) is trivially governed by the identity operator I_N of $\mathcal{L}(\mathcal{H}_N)$. At this stage, L_0 in Eq. (9.141) (like the other operators L_k) is an arbitrary operator of $\mathcal{L}(\mathcal{H}_N)$, and it can always be decomposed as the sum $\mathsf{L}_0 = \mathsf{L}_0^{(\mathrm{h})} + \mathsf{L}_0^{(\mathrm{ah})}$ of a Hermitian operator $\mathsf{L}_0^{(\mathrm{h})} = \big(\mathsf{L}_0 + \mathsf{L}_0^\dagger\big)/2$ plus an anti-Hermitian operator $\mathsf{L}_0^{(\mathrm{ah})} = \big(\mathsf{L}_0 - \mathsf{L}_0^\dagger\big)/2$, with $\mathsf{L}_0^{(\mathrm{h})\dagger} = \mathsf{L}_0^{(\mathrm{h})}$ and $\mathsf{L}_0^{(\mathrm{ah})\dagger} = -\mathsf{L}_0^{(\mathrm{ah})}$. This decomposition of L_0 is convenient to incorporate the completeness relation of Eq. (3.47) for the Kraus operators $\Lambda_k(t+dt,t)$. For this purpose, to first order in dt we have $\Lambda_0^\dagger\Lambda_0 = \mathrm{I}_N + 2\mathsf{L}_0^{(\mathrm{h})}dt$; and $\Lambda_k^\dagger\Lambda_k = \mathsf{L}_k^\dagger\mathsf{L}_k dt$ for $k \neq 0$. The completeness relation of Eq. (3.47) for the Kraus operators $\Lambda_k(t+dt,t)$ therefore imposes to satisfy, to first order in dt,

$$\mathrm{I}_N = \sum_{k=0}^{K} \Lambda_k^\dagger\Lambda_k = \mathrm{I}_N + \left(2\mathsf{L}_0^{(\mathrm{h})} + \sum_{k=1}^{K} \mathsf{L}_k^\dagger\mathsf{L}_k\right)dt . \tag{9.143}$$

This indicates that the Hermitian part of L_0 is constrained to be

$$\mathsf{L}_0^{(\mathrm{h})} = -\frac{1}{2}\sum_{k=1}^{K} \mathsf{L}_k^\dagger\mathsf{L}_k , \tag{9.144}$$

and this is enough to satisfy the completeness relation of Eq. (3.47), along with completely arbitrary operators L_k on $\mathcal{H}_N$ for $k = 1$ to K.

The Kraus operators of Eqs. (9.141)–(9.142) can now be inserted in Eq. (9.140), and to first order in dt this provides the rate of change of the density operator as

$$\frac{d}{dt}\rho(t) = \left[\mathsf{L}_0^{(\mathrm{ah})}, \rho(t)\right] + \sum_{k=1}^{K}\left(\mathsf{L}_k\rho(t)\mathsf{L}_k^\dagger - \frac{1}{2}\left\{\mathsf{L}_k^\dagger\mathsf{L}_k, \rho(t)\right\}\right), \tag{9.145}$$

which is also

$$\frac{d}{dt}\rho(t) = \left[\mathsf{L}_0^{(\mathrm{ah})}, \rho(t)\right] + \frac{1}{2}\sum_{k=1}^{K}\left(\left[\mathsf{L}_k, \rho(t)\mathsf{L}_k^\dagger\right] + \left[\mathsf{L}_k\rho(t), \mathsf{L}_k^\dagger\right]\right), \tag{9.146}$$

involving the commutator $[\cdot\,,\cdot]$ and anticommutator $\{\cdot\,,\cdot\}$ from Eqs. (A.15)–(A.16) for a pair of operators. By referring to Eq. (3.8) for the evolution of a closed system, we have the identification for the anti-Hermitian operator $\mathsf{L}_0^{(\mathrm{ah})} = -(i/\hbar)\mathsf{H}$, with H a Hamiltonian describing a unitary evolution of the system as if it were isolated. The other operators L_k in Eq. (9.145) describe the coupling of the system to the environment, each representing a change the state of the system may experience with a probability of order dt. Equation (9.145) is a master equation for the rate of change of the density operator, and it is known as the **Lindblad equation**, while the operators L_k in Eq. (9.145) are known as Lindblad operators.

The Lindblad equation (9.145) assumes a continuous variation in time for the density operator, and in this respect it is less general than a global Kraus transformation from time t_0 to time t as in the quantum operation of Eq. (9.138), which may include discontinuous jumps in the evolution. The Lindblad operators L_k, much like the Kraus operators Λ_k, can be determined based on empirical or phenomenological considerations for the evolution, or alternatively on more fundamental mechanisms. For example, for the qubit, dephasing can be described by a Lindblad operators $\mathsf{L}_1 \propto \sigma_z$ flipping the phase, spontaneous emission by a Lindblad operators $\mathsf{L}_2 \propto |0\rangle\langle 1|$ taking the excited state $|1\rangle$ to the ground state $|0\rangle$, spontaneous absorption by a Lindblad operators $\mathsf{L}_3 \propto |1\rangle\langle 0|$ taking the ground state $|0\rangle$ to the excited state $|1\rangle$.

The Lindblad equation (9.145) also involves some Markovian approximation [6, 7, 45]. The Kraus form of the evolution, as obtained in Sect. 3.2.2 with Eq. (3.45), is deduced when the system and environment start in a separable state, while the continuous-time evolution may gradually entangle the system and the environment, so that applying Eq. (9.140) at each time t involves some approximation. The separation assumption represents an acceptable approximation when the environment contains a large number of degrees of freedom so that its state is little affected in the interaction with the system of interest, or when the environment rapidly dissipates the correlations it may have with the system. The environment can be assumed in a stationary state $\overline{\rho}_E$ and the joint state for the system–environment compound is approximated as the product $\rho(t) \otimes \overline{\rho}_E$. Such approximation proves adequate in many physical situations of interest. The continuous-time evolution of Eq. (9.145) implies some coarse graining over time, sampled at a time scale dt not too small so that neglecting the system–environment correlations at time t does not significantly change the value of $\rho(t + dt)$. With such sufficiently large time scale, the future evolution of the state $\rho(t)$ of the system is conditioned essentially by the present state $\rho(t)$ and not by its past history with the environment, as a Markovian approximation.

By contrast, the functional-block evolution of $\rho(t)$ by a process $\mathcal{N}(\cdot)$, between an initial time and a final time, is not limited by comparable assumptions.

More on the Lindblad equation and its physical significance can be found for instance in [6, 7, 45]. In the present book, for quantum information, rather than the Lindblad differential equation, we preferentially favor the functional-block transformation of Eq. (9.138) or Eq. (3.45) with Kraus operators, for the non-unitary decohering or noisy evolution of quantum states, in a signals-and-systems theoretic perspective.

9.15 Summary on Quantum Noise

Quantum noise or decoherence can be understood as any uncontrolled alteration of a quantum state. The alteration can result from randomized coordinates or random preparation out of a statistical ensemble, or very often from interaction with an uncontrolled environment. A quantum noise can be modeled as a non-unitary open-system evolution according to Sect. 3.2, materialized by a quantum process or quantum operation $\mathcal{N}(\cdot)$, transforming the density operator representing the quantum system of interest, from an input density operator ρ into an output (noisy) density operator $\mathcal{N}(\rho)$, as depicted in Fig. 9.1.

The quantum process $\rho \mapsto \mathcal{N}(\rho)$ can be described by specifying a set of **Kraus operators** $\{\Lambda_k\}$, characterizing the quantum noise, and acting in Eq. (9.1). The Kraus representation in Eq. (9.1) is specially suited for a functional-block modeling of the quantum noise, with no more than N^2 Kraus operators needed to represent any quantum noise altering a density operator ρ on an N-dimensional Hilbert space $\mathcal{H}_N$.

Alternatively, the quantum process $\rho \mapsto \mathcal{N}(\rho)$ can be described via a **Stinespring dilated unitary representation,** involving an explicit model for the environment altering the quantum state of interest, and resting essentially on Eqs. (9.3)–(9.4). The dilated unitary representation from Eqs. (9.3)–(9.4) is specially suited for constructing **simulation circuits** for quantum noises, with control of the noise parameters, particularly for a test bed in quantum signal processing. Especially, the canonical procedure of Sect. 3.2.8 and Eq. (9.4) simulates a noise process defined by K Kraus operators, by means of a K-dimensional environment model coupled to the quantum system of interest, as exemplified many times in this chapter.

These approaches for the modeling of quantum noise have then been explicitly specified for the fundamental case of the qubit. The main quantum noise models relevant to the qubit have been described, via their Kraus operators, and via quantum circuits for their controlled simulation. Beyond, for multiple qubits, the noise can often be assumed independent between qubits and be treated via tensor products of single-qubit noise models; otherwise specific multi-qubit noise models have to be introduced to handle the noise correlation between qubits. The **qubit noise models** presented in this chapter will be essential to us repeatedly in the following chapters, to analyze various important quantum signal-processing tasks, in the presence of noise.

Chapter 10
Quantum Detection

In this chapter, we examine the task of quantum state detection, as it is addressed for instance in the seminal monograph [235]. This task can also be referred to as quantum state discrimination. It represents a fundamental task of quantum information processing, aiming at distinguishing quantum states based on their measurement, so as to efficiently extract useful information. This is especially relevant for distinguishing signaling quantum states altered by noise, as they result from the considerations presented in Chap. 9 on quantum noise. Detection from noisy signals is also an essential topic of classical signal processing. Although one may expect somewhere significant differences between classical and quantum detection, nevertheless at a sufficiently high level a close parallelism can be drawn between the two, as we shall see in this chapter, substantiating our perspective of quantum signal processing.

In this chapter, we will start with a concise survey of classical detection from noisy (random) signals, as it is common in standard signal processing. These references of classical detection, as it will then be visible, can be carried over to quantum detection, and they will serve us as guidelines to the presentation of quantum detection forming the central topic of this chapter. Detection or discrimination of two states or signals can be solved rather thoroughly, with explicit theoretical characterizations of optimal solutions, in the classical and in the quantum domains alike. Accordingly it represents an essential reference, for the theoretical design and the practical application of detection strategies. For its importance, we will first concentrate on such two-state detection, with an in-depth analysis, and then extend the approach to multiple-state detection or discrimination, to arrive at the square-root quantum detector or great practical value.

F. Chapeau-Blondeau, *Quantum Information and Quantum Signal Processing*,
https://doi.org/10.1007/978-3-032-27499-1_10

10.1 Classical Detection

In classical signal detection theory [236–239], one considers a dataset $\vec{x} = (x_1, x_2, \dots, x_N)^\top$ of N real values forming one realization of N samples taken from a random (noisy) signal. The dataset $\vec{x}$ can have been produced either under hypothesis H_0 with prior probability P_0, or under hypothesis H_1 with prior probability $P_1 = 1 - P_0$. For example, these two hypotheses H_0 and H_1 may represent two scenarios where an analog waveform encoding a binary digit 0 or another analog waveform encoding a binary digit 1 is emitted on a communication channel over which they get corrupted by some noise before becoming accessible at a receiver measuring $\vec{x}$ for decoding. From one realization or one measurement of the dataset $\vec{x}$, one wants to decide (detect) whether this realization was produced under hypothesis H_0 or H_1 (e.g., whether a bit 0 or a bit 1 is buried in the noisy measurements). A measured dataset $\vec{x}$ that serves to decide between hypotheses H_0 and H_1 can be viewed as a point $\vec{x}$ in the real measurement space $\mathbb{R}^N$. Any detection strategy or detector can be viewed as a bipartition of the measurement space $\mathbb{R}^N$ into two regions $\mathcal{R}_0$ and $\mathcal{R}_1$, such that $\mathcal{R}_0 \cup \mathcal{R}_1 = \mathbb{R}^N$ and $\mathcal{R}_0 \cap \mathcal{R}_1 = \emptyset$, defined in such a way that if the measured $\vec{x}$ falls in $\mathcal{R}_0$ then hypothesis H_0 is decided, and if $\vec{x}$ falls in $\mathcal{R}_1$ then H_1 is decided. It follows that the conditional probability for the detector to decide H_k when H_j is true, for j and k in $\{0, 1\}$, can be expressed via the N-dimensional integral

$$\Pr\{\vec{x} \in \mathcal{R}_k \,|\, \mathrm{H}_j\} = \int_{\mathcal{R}_k} p(\vec{x}\,|\mathrm{H}_j) d\vec{x} \,, \tag{10.1}$$

with $d\vec{x} = dx_1 \cdots dx_N$, and the probability density $p(\vec{x}\,|\mathrm{H}_j)$ that the measurement $\vec{x}$ is obtained when hypothesis H_j holds, for any $\vec{x} \in \mathbb{R}^N$. Each probability density $p(\vec{x}\,|\mathrm{H}_j)$ is determined by the manner the underlying signal (e.g., the waveform for 0 or 1) gets altered by the noise and therefrom induces the measurable data. These two probability densities $p(\vec{x}\,|\mathrm{H}_j)$ preexist as common but unrelated to any detection strategy that may be chosen to devise a decision from a measurement $\vec{x}$. The performance of a detection strategy or detector can be assessed by means of various meaningful criteria, so as to subsequently target the optimal detector maximizing a selected performance criterion.

10.1.1 Minimal-Error Detection

In this context of random (noisy) measurements, a meaningful way of evaluating the performance of a detector is by resorting to the overall probability P_{er} of making a detection error. Related to it is the probability of a successful detection $P_{\mathrm{suc}} = 1 - P_{\mathrm{er}}$. The probability of detection error can be evaluated as

$$P_{\mathrm{er}} = \Pr\{\vec{x} \in \mathcal{R}_1 \,|\, \mathrm{H}_0\} P_0 + \Pr\{\vec{x} \in \mathcal{R}_0 \,|\, \mathrm{H}_1\} P_1 \,, \tag{10.2}$$

which, via Eq. (10.1), is also

$$P_{\text{er}} = P_0 \int_{\mathcal{R}_1} p(\vec{x}\,|\text{H}_0)d\vec{x} + P_1 \int_{\mathcal{R}_0} p(\vec{x}\,|\text{H}_1)d\vec{x}\ . \tag{10.3}$$

Since $\mathcal{R}_0$ and $\mathcal{R}_1$ are complementary in $\mathbb{R}^N$, one has

$$\Pr\{\vec{x} \in \mathcal{R}_0 \,|\, \text{H}_j\} = 1 - \Pr\{\vec{x} \in \mathcal{R}_1 \,|\, \text{H}_j\}\ , \quad \text{for } j = 0, 1, \tag{10.4}$$

and the probability of error P_{er} of Eq. (10.3) can be expressed with a single integral over $\mathcal{R}_1$ as

$$P_{\text{er}} = P_1 - \int_{\mathcal{R}_1} \big[P_1 p(\vec{x}\,|\text{H}_1) - P_0 p(\vec{x}\,|\text{H}_0)\big]d\vec{x}\ . \tag{10.5}$$

In Eq. (10.5), the handle that varies with the detector is restricted to the region $\mathcal{R}_1$ of $\mathbb{R}^N$, the other quantities being fixed by the detection problem itself. For efficient detection, it is then appropriate to devise the region $\mathcal{R}_1$ in order to minimize the probability of error P_{er} in Eq. (10.5). This minimization is accomplished by making as large as possible the integral over $\mathcal{R}_1$ in the right-hand side of Eq. (10.5), and in turn this is accomplished by placing into $\mathcal{R}_1$ all and only the points $\vec{x}$ of $\mathbb{R}^N$ at which the integrand is positive. The optimal setting of the region $\mathcal{R}_1$ minimizing P_{er} is thus defined as the set $\mathcal{R}_1^{\text{opt}} = \big\{\vec{x} \in \mathbb{R}^N \,\big|\, P_1 p(\vec{x}\,|\text{H}_1) - P_0 p(\vec{x}\,|\text{H}_0) > 0\big\}$, and $\mathcal{R}_0^{\text{opt}}$ is the complementary set in $\mathbb{R}^N$. The points $\vec{x}$ on the boundary where $P_1 p(\vec{x}\,|\text{H}_1) - P_0 p(\vec{x}\,|\text{H}_0) = 0$ can indifferently be placed in $\mathcal{R}_1^{\text{opt}}$ or $\mathcal{R}_0^{\text{opt}}$ and will not affect the value of P_{er} in Eq. (10.5). In a more operational way, the functioning of the optimal detector minimizing P_{er} can be based on the scalar function

$$T(\vec{x}\,) = P_1 p(\vec{x}\,|\text{H}_1) - P_0 p(\vec{x}\,|\text{H}_0)\ , \tag{10.6}$$

known as a test statistic, in such a way that, at the measured point $\vec{x}$, if $T(\vec{x}\,)$ is found > 0 then H_1 is decided, otherwise H_0 is decided. Commonly, in classical detection, the test statistic is rather chosen here as the (likelihood) ratio $p(\vec{x}\,|\text{H}_1)/p(\vec{x}\,|\text{H}_0)$ which is compared with the decision threshold P_0/P_1 for an equivalent decision; yet the test statistic $T(\vec{x}\,)$ of Eq. (10.6) carries over more adequately to quantum detection.

The optimal detector achieves in Eq. (10.5) the minimal probability of error expressible as

$$P_{\text{er}}^{\min} = P_1 - \int_{\mathcal{R}_1^{\text{opt}}} \big|T(\vec{x}\,)\big|\,d\vec{x} = P_0 - \int_{\mathcal{R}_0^{\text{opt}}} \big|T(\vec{x}\,)\big|\,d\vec{x}\ . \tag{10.7}$$

This is so because over $\mathcal{R}_1^{\text{opt}}$ the test statistic $T(\vec{x}\,)$ is positive so that $T(\vec{x}\,) = |T(\vec{x}\,)|$, while over $\mathcal{R}_0^{\text{opt}}$ it is negative so that $T(\vec{x}\,) = -|T(\vec{x}\,)|$; also the integral of $T(\vec{x}\,)$ over $\mathbb{R}^N = \mathcal{R}_0^{\text{opt}} \cup \mathcal{R}_1^{\text{opt}}$ is $P_1 - P_0$ and is also the sum of the integral of $T(\vec{x}\,)$ over

$\mathcal{R}_0^{\rm opt}$ plus the integral of $T(\vec{x})$ over $\mathcal{R}_1^{\rm opt}$. By combining the two expressions for $P_{\rm er}^{\min}$ in Eq. (10.7), one finally obtains the minimal probability of error achieved by the optimal detector as

$$P_{\rm er}^{\min} = \frac{1}{2} - \frac{1}{2}\int_{\mathbb{R}^N} \big|T(\vec{x})\big| d\vec{x} \,. \tag{10.8}$$

The form of Eq. (10.8), in particular, guarantees that $P_{\rm er}^{\min}$ never exceeds $1/2$.

This completes the characterization of the minimal-error optimal detector, which operates at each measured data point $\vec{x}$ by evaluating the test statistic $T(\vec{x})$ of Eq. (10.6) and comparing it with 0 to return its binary decision. In doing so, the detector achieves the probability of error $P_{\rm er}^{\min}$ of Eq. (10.8) which is minimal over all conceivable detection strategies exploiting the same data $\vec{x}$.

10.1.2 Minimal-Cost Detection

Minimizing the probability of detection error $P_{\rm er}$ of Eq. (10.2) requires the knowledge of the prior probabilities P_0 and $P_1 = 1 - P_0$ of the two hypotheses H_0 and H_1. Relying on prior knowledge, such an approach is referred to as a Bayesian approach. There exists an extended Bayesian formulation to optimal detection, which introduces specific costs C_{kj} for each of the four possible decision scenarios, where C_{kj} is the cost for the detector to decide H_k when H_j is true, for j and k in $\{0, 1\}$. A natural performance criterion, extending the probability of detection error $P_{\rm er}$ of Eq. (10.2), is the average detection cost

$$\begin{aligned}\overline{C} &= C_{10}\Pr\{\vec{x} \in \mathcal{R}_1 \,|\, \mathrm{H}_0\}P_0 + C_{00}\Pr\{\vec{x} \in \mathcal{R}_0 \,|\, \mathrm{H}_0\}P_0 \\ &\quad + C_{01}\Pr\{\vec{x} \in \mathcal{R}_0 \,|\, \mathrm{H}_1\}P_1 + C_{11}\Pr\{\vec{x} \in \mathcal{R}_1 \,|\, \mathrm{H}_1\}P_1 \,,\end{aligned} \tag{10.9}$$

which, by exploiting Eqs. (10.1) and (10.4), is also

$$\begin{aligned}\overline{C} &= C_{00}P_0 + C_{01}P_1 \\ &\quad - \int_{\mathcal{R}_1} \big[(C_{01} - C_{11})P_1\, p(\vec{x}\,|\mathrm{H}_1) - (C_{10} - C_{00})P_0\, p(\vec{x}\,|\mathrm{H}_0)\big] d\vec{x} \,.\end{aligned} \tag{10.10}$$

The probability of error $P_{\rm er}$ of Eq. (10.2) is the average cost $\overline{C}$ in the special case where correct decisions have no cost with $C_{00} = C_{11} = 0$ and wrong decisions have unit cost with $C_{01} = C_{10} = 1$. In general, penalizing wrong decisions requires $C_{10} \geq C_{00}$ and $C_{01} \geq C_{11}$ for consistency. Minimizing the average detection cost of Eq. (10.10) amounts to making as large as possible the integral over $\mathcal{R}_1$ in the right-hand side of Eq. (10.10). This is accomplished by the detector using the test statistic

$$T(\vec{x}) = (C_{01} - C_{11})P_1\, p(\vec{x}\,|\mathrm{H}_1) - (C_{10} - C_{00})P_0\, p(\vec{x}\,|\mathrm{H}_0) \,, \tag{10.11}$$

to decide H_1 if $T(\vec{x}) > 0$, and H_0 otherwise. And this optimal Bayesian detector achieves in Eq. (10.10) the minimum detection cost

$$\overline{C}_{\min} = \frac{1}{2}\big[(C_{01} + C_{11})P_1 + (C_{10} + C_{00})P_0\big] - \frac{1}{2}\int_{\mathbb{R}^N} \big|T(\vec{x})\big| d\vec{x} . \qquad (10.12)$$

10.1.3 Neyman–Pearson Detection

In some situations, it is difficult to assign meaningful values to the prior probabilities P_0 and P_1. An approach then to devising a useful optimal detector is to resort to the Neyman–Pearson criterion, involving solely conditional probabilities. Under hypothesis H_0 is defined the conditional probability of deciding H_1 and known as the probability of false alarm $P_f = \Pr\{\vec{x} \in \mathcal{R}_1 \mid H_0\} = \int_{\mathcal{R}_1} p(\vec{x}\,|H_0)d\vec{x}$; meanwhile, under hypothesis H_1 is defined the conditional probability of deciding H_1 and known as the probability of detection $P_d = \Pr\{\vec{x} \in \mathcal{R}_1 \mid H_1\} = \int_{\mathcal{R}_1} p(\vec{x}\,|H_1)d\vec{x}$. The Neyman–Pearson criterion consists in selecting the detector maximizing the probability of detection P_d, in the presence of a probability of false alarm no larger than a prescribed level P_f. If at the optimum $P_f = \int_{\mathcal{R}_1} p(\vec{x}\,|H_0)d\vec{x}$ were below the prescribed level, an enlargement of the region $\mathcal{R}_1$—the integration domain—could be envisaged, which would also lead to an increase of $P_d = \int_{\mathcal{R}_1} p(\vec{x}\,|H_1)d\vec{x}$, which is not permitted at the optimum. So the optimum necessarily saturates the prescribed level for P_f. The Neyman–Pearson optimal detector therefore maximizes the probability of detection P_d with a probability of false alarm at the prescribed level P_f, achieving

$$\max_{\mathcal{R}_1} P_d = \int_{\mathcal{R}_1} p(\vec{x}\,|H_1)d\vec{x} , \qquad (10.13)$$

under the constraint $\int_{\mathcal{R}_1} p(\vec{x}\,|H_0)d\vec{x} = P_f$ prescribed. This represents a constrained maximization which can be solved with the Lagrangian $\mathcal{L} = \int_{\mathcal{R}_1} p(\vec{x}\,|H_1)d\vec{x} - \mu\big[\int_{\mathcal{R}_1} p(\vec{x}\,|H_0)d\vec{x} - P_f\big]$ and Lagrange multiplier μ, so as to find $\mathcal{R}_1$ maximizing

$$\int_{\mathcal{R}_1} \big[p(\vec{x}\,|H_1) - \mu p(\vec{x}\,|H_0)\big]d\vec{x} . \qquad (10.14)$$

The maximization of Eq. (10.14) is accomplished by the detector using the test statistic

$$T(\vec{x}) = p(\vec{x}\,|H_1) - \mu p(\vec{x}\,|H_0) , \qquad (10.15)$$

to decide H_1 at each data point $\vec{x}$ belonging to the region $\mathcal{R}_1^{\text{opt}}$ where $T(\vec{x}) > 0$, and H_0 otherwise. This determines the procedure of the Neyman–Pearson optimal detector, via the optimal region $\mathcal{R}_1^{\text{opt}}(\mu)$ defined by $T(\vec{x}) > 0$, except for the still undetermined Lagrange multiplier μ. This procedure achieves a probability of false alarm which is forced at the prescribed level P_f via

$$\int_{\mathcal{R}_1^{\text{opt}}(\mu)} p(\vec{x}\,|\text{H}_0)d\vec{x} = P_f \ , \tag{10.16}$$

providing an equation which is inverted to fix the scalar value $\mu = \mu(P_f)$ of the Lagrange multiplier. This $\mu(P_f)$ placed in the test statistic of Eq. (10.15) completely determines the procedure of the Neyman–Pearson optimal detector, which is then guaranteed to operate at the prescribed probability of false alarm P_f, while achieving the maximal probability of detection:

$$P_d^{\max}(P_f) = \int_{\mathcal{R}_1^{\text{opt}}[\mu(P_f)]} p(\vec{x}\,|\text{H}_1)d\vec{x} \ . \tag{10.17}$$

The curve of the two conditional probabilities $P_d = P_d(P_f)$ giving P_d as a function of P_f is known as the receiver operating characteristic (ROC) of a detector, and it stands as a useful tool to assess and compare the performance of different detectors. Equation (10.17) provides a closed-form expression for the ROC curve characterizing the Neyman–Pearson optimal detector. In practice, the inversion $\mu = \mu(P_f)$ from Eq. (10.16) may be uneasy to perform analytically; yet Eqs. (10.16) and (10.17) together will at least always provide a useful parametric form as a function of the Lagrange multiplier μ for the ROC curve $\big(P_f(\mu), P_d(\mu)\big)$ characterizing in the plane (P_f, P_d) the Neyman–Pearson optimal detector.

10.1.4 Minimax Detection

When the prior probabilities $(P_0, P_1 = 1 - P_0)$ are unknown, another reasonable approach to devise an optimal detector is to turn to minimax detection [236, 237]. It consists in considering the probability P_{er} of detection error as in Eq. (10.2), and the maximum value $P_{\text{er}}^{\max}$ it can assume for the worse-case P_1, and then select the detector that minimizes this maximum probability $P_{\text{er}}^{\max}$, defining in this way the minimax optimal detector. By referring to the decision region $\mathcal{R}_1$ of $\mathbb{R}^N$ which defines a detection scheme, the minimax optimal detector achieves

$$\min_{\mathcal{R}_1} \left(\max_{P_1} P_{\text{er}} \right) . \tag{10.18}$$

Alternatively, in terms of the probability of detection success $P_{\text{suc}} = 1 - P_{\text{er}}$, the minimax optimal detector equivalently achieves

$$\max_{\mathcal{R}_1} \left(\min_{P_1} P_{\text{suc}} \right) . \tag{10.19}$$

For a more explicit characterization of the minimax optimal detector, the probability of detection error P_{er} of Eq. (10.2), since $P_0 = 1 - P_1$, can be expressed as

$$P_{\text{er}} = \Pr\{\vec{x} \in \mathcal{R}_1 \,|\, \mathrm{H}_0\} + P_1\Big[\Pr\{\vec{x} \in \mathcal{R}_0 \,|\, \mathrm{H}_1\} - \Pr\{\vec{x} \in \mathcal{R}_1 \,|\, \mathrm{H}_0\}\Big]\,, \tag{10.20}$$

showing it is an affine function of P_1. This affine function $P_{\text{er}}(P_1)$ in Eq. (10.20) assumes the maximum

$$P_{\text{er}}^{\max} = \Pr\{\vec{x} \in \mathcal{R}_1 \,|\, \mathrm{H}_0\} = \int_{\mathcal{R}_1} p(\vec{x}\,|\mathrm{H}_0)d\vec{x} \tag{10.21}$$

at $P_1 = 0$ when $\Pr\{\vec{x} \in \mathcal{R}_0 \,|\, \mathrm{H}_1\} \le \Pr\{\vec{x} \in \mathcal{R}_1 \,|\, \mathrm{H}_0\}$ (in this condition the slope of the affine function $P_{\text{er}}(P_1)$ is negative), or else $P_{\text{er}}(P_1)$ assumes the maximum

$$P_{\text{er}}^{\max} = \Pr\{\vec{x} \in \mathcal{R}_0 \,|\, \mathrm{H}_1\} = 1 - \int_{\mathcal{R}_1} p(\vec{x}\,|\mathrm{H}_1)d\vec{x} \tag{10.22}$$

at $P_1 = 1$ otherwise (when the slope of the affine function $P_{\text{er}}(P_1)$ is positive). As the decision region $\mathcal{R}_1$ is increased from the empty set $\emptyset$ to the whole $\mathbb{R}^N$, the probability of Eq. (10.21) increases from 0 while the probability of Eq. (10.22) decreases from 1. The maximum of the two, as $\mathcal{R}_1$ increases, reaches a minimum for a region $\mathcal{R}_1 = \overline{\mathcal{R}}_1$ associated with the cross-over of the two probabilities in Eqs. (10.21)–(10.22) occurring when the affine function $P_{\text{er}}(P_1)$ is flat with zero slope, or equivalently when

$$\Pr\{\vec{x} \in \mathcal{R}_1 \,|\, \mathrm{H}_0\} = \Pr\{\vec{x} \in \mathcal{R}_0 \,|\, \mathrm{H}_1\}\,, \tag{10.23}$$

and realizing a minimum of the maximum probability $P_{\text{er}}^{\max}$ that can be characterized as

$$\big(P_{\text{er}}^{\max}\big)_{\min} = \int_{\overline{\mathcal{R}}_1} p(\vec{x}\,|\mathrm{H}_0)d\vec{x} = 1 - \int_{\overline{\mathcal{R}}_1} p(\vec{x}\,|\mathrm{H}_1)d\vec{x}\,, \tag{10.24}$$

which gives also the equivalent condition

$$\int_{\overline{\mathcal{R}}_1} \big[p(\vec{x}\,|\mathrm{H}_0) + p(\vec{x}\,|\mathrm{H}_1)\big]d\vec{x} = 1\,. \tag{10.25}$$

Equation (10.23), or (10.24) or (10.25), stands as a necessary condition that must be satisfied by the minimax optimal detector defined by the decision region $\overline{\mathcal{R}}_1$. The positive integrand in Eq. (10.25) would integrate to 2 over $\mathbb{R}^N$, so there usually exists an infinite number of subregions $\overline{\mathcal{R}}_1$ of $\mathbb{R}^N$ that could enclose a volume of 1 under the integrand so as to satisfy Eq. (10.25). To complete the characterization of the minimax optimal detector, we must therefore find the region $\overline{\mathcal{R}}_1$ satisfying Eq. (10.25) and achieving the minimum of $\big(P_{\text{er}}^{\max}\big)_{\min}$ in Eq. (10.24). This is equivalent to finding $\overline{\mathcal{R}}_1$ solving the constrained maximization

$$\max_{\overline{\mathcal{R}}_1} \int_{\overline{\mathcal{R}}_1} p(\vec{x}\,|\mathrm{H}_1)d\vec{x}\ , \tag{10.26}$$

under the constraint of Eq. (10.25).

Before we come to this resolution, it can be noted that, via Eq. (10.4), the condition of Eq. (10.23) is equivalent to

$$\Pr\{\vec{x} \in \mathcal{R}_0 \mid \mathrm{H}_0\} = \Pr\{\vec{x} \in \mathcal{R}_1 \mid \mathrm{H}_1\}\ . \tag{10.27}$$

Also, for the equivalent definition of Eq. (10.19), the probability of detection success $P_{\text{suc}} = 1 - P_{\text{er}} = \Pr\{\vec{x} \in \mathcal{R}_0 \mid \mathrm{H}_0\} + P_1\big[\Pr\{\vec{x} \in \mathcal{R}_1 \mid \mathrm{H}_1\} - \Pr\{\vec{x} \in \mathcal{R}_0 \mid \mathrm{H}_0\}\big]$ is also an affine function of P_1, and the cross-over in the two maximum probabilities of error $P_{\text{er}}^{\max}$ of Eqs. (10.21)–(10.22) is associated with the cross-over in the two minimum probabilities of success

$$P_{\text{suc}}^{\min} = \Pr\{\vec{x} \in \mathcal{R}_0 \mid \mathrm{H}_0\} = \int_{\mathcal{R}_0} p(\vec{x}\,|\mathrm{H}_0)d\vec{x} \tag{10.28}$$

at $P_1 = 0$, and

$$P_{\text{suc}}^{\min} = \Pr\{\vec{x} \in \mathcal{R}_1 \mid \mathrm{H}_1\} = \int_{\mathcal{R}_1} p(\vec{x}\,|\mathrm{H}_1)d\vec{x} \tag{10.29}$$

at $P_1 = 1$, occurring similarly for a region $\mathcal{R}_1 = \overline{\mathcal{R}}_1$ satisfying Eqs. (10.24)–(10.25) to realize the maximum in the minimal success probability $P_{\text{suc}}^{\min}$. This leads to the equivalent definition of the minimax optimal detector

$$\max_{\mathcal{R}_1}\Big(\min_j \Pr\{\vec{x} \in \mathcal{R}_j \mid \mathrm{H}_j\}\Big)\ . \tag{10.30}$$

This equivalent definition presents the advantage of not referring to the unknown prior probabilities (P_0, P_1) to envisage their worst-case configuration. It defines the minimax optimal detector as the detector for which the minimal conditional probability $\Pr\{\vec{x} \in \mathcal{R}_j \mid \mathrm{H}_j\}$ of correct detection gets maximized. As a necessary condition, we now know from Eq. (10.27), that at the optimum, the two conditional probabilities $\Pr\{\vec{x} \in \mathcal{R}_j \mid \mathrm{H}_j\}$, for $j = 0, 1$, are equal, and they also get jointly maximized by the minimax optimal detector. So we have with Eqs. (10.18), (10.19), and (10.30), three equivalent definitions of the minimax optimal detector. We now return to completing its explicit characterization by solving the constrained maximization of Eq. (10.26).

The constrained maximization of Eq. (10.26) exhibits the same form as the constrained maximization of Eq. (10.13) for the Neyman–Pearson detector. In a similar way, it can be solved with the Lagrangian $\mathcal{L} = \int_{\overline{\mathcal{R}}_1} p(\vec{x}\,|\mathrm{H}_1)d\vec{x} - \mu\Big(\int_{\overline{\mathcal{R}}_1}\big[p(\vec{x}\,|\mathrm{H}_0) + p(\vec{x}\,|\mathrm{H}_1)\big]d\vec{x} - 1\Big)$ and Lagrange multiplier μ, so as to find $\overline{\mathcal{R}}_1$ maximizing

$$\int_{\overline{\mathcal{R}}_1} \big[p(\vec{x}\,|\mathrm{H}_1) - \mu' p(\vec{x}\,|\mathrm{H}_0)\big] d\vec{x}\ , \tag{10.31}$$

with $\mu' = \mu/(1-\mu)$ having, as μ, the status of a free parameter covering $\mathbb{R}$. As before, the maximization of Eq. (10.31) is accomplished by the detector employing the test statistic

$$T(\vec{x}\,) = p(\vec{x}\,|\mathrm{H}_1) - \mu' p(\vec{x}\,|\mathrm{H}_0)\ , \tag{10.32}$$

to decide H_1 at each data point $\vec{x}$ belonging to the region $\overline{\mathcal{R}}_1^{\mathrm{opt}}$ where $T(\vec{x}\,) > 0$, and H_0 otherwise. This determines the procedure of the minimax optimal detector, via the optimal region $\overline{\mathcal{R}}_1^{\mathrm{opt}}(\mu')$ defined by $T(\vec{x}\,) > 0$, except for the still undetermined parameter μ'. When this parametric region $\overline{\mathcal{R}}_1^{\mathrm{opt}}(\mu')$ is used in the constraint of Eq. (10.25), an equation is obtained whose inversion fixes the value of μ'. This μ' inserted in the test statistic of Eq. (10.32) completely determines the procedure of the minimax optimal detector.

By definition, the minimax optimal detector coincides with the minimal-error optimal detector of Sect. 10.1.1 acting at the prior probability $P_1 = P_1^{\max}$ standing as the maximizer of the minimal probability of error $P_{\mathrm{er}}^{\min}(P_1)$ of Eq. (10.8). Yet, the characterization we have obtained here for the minimax optimal detector is independent and dispenses us from obtaining the characterization of $P_{\mathrm{er}}^{\min}$ and $P_1^{\max}$ from Eq. (10.8). The minimax optimal detector can also be extended to the criterion of minimal Bayesian cost as considered in Sect. 10.1.2. The Bayesian cost $\overline{C}$ of Eq. (10.9) is also an affine function of the prior probability P_1, and considering the maximum cost $\overline{C}^{\max}$ for the worse-case P_1, one can seek the optimal detector minimizing this maximum cost $\overline{C}^{\max}$. This extended minimax optimal detector also coincides with the minimal-cost optimal detector of Sect. 10.1.2 acting at the prior probability P_1 maximizing the minimal cost of Eq. (10.12).

The various optimal detectors we have characterized in Sect. 10.1 constitute standard references of statistical signal processing. They formalize important and significant approaches and properties relevant to the solution of the detection task optimizing different meaningful performance criteria. A remarkable property is that the optimal detectors, to return their optimal decision, only need to consider a scalar quantity—the test statistic $T(\vec{x}\,)$—function of the dataset $\vec{x}$. To optimally decide from an N-point dataset $\vec{x}$, there is no need to devise composite decision schemes assigning special roles to every particular point x_n from the dataset $\vec{x}$, for $n = 1$ to N; by contrast, a single globalizing scalar $T(\vec{x}\,)$ is enough (and optimal). Furthermore, this scalar test statistic $T(\vec{x}\,)$ to determine the optimal decision is involved simply in a dichotomic test on its sign. And it is found that the optimal test statistic $T(\vec{x}\,)$, for the different performance criteria, receives the same generic form $T(\vec{x}\,) = p(\vec{x}\,|\mathrm{H}_1) - \mu p(\vec{x}\,|\mathrm{H}_0)$. This is true for the minimal-error detector in Eq. (10.6), for the minimal-cost detector in Eq. (10.11), for the Neyman–Pearson detector in Eq. (10.15), for the minimax detector in Eq. (10.32), with in each case, a parameter μ adapted to the criterion being optimized. Finally, the procedure of the optimal detectors and their optimal performances are essentially governed by the two conditional probabilities $p(\vec{x}\,|\mathrm{H}_j)$ of the data $\vec{x}$ in each of the two hypotheses H_j (possibly supplemented by the two priors P_j), which encapsulate all the relevant features of the detection task to be solved optimally. For the sequel, as we are going to see, these references from classical detection can be carried over, in appropriate forms, to serve as helpful guidelines to the study of quantum detection, which we now undertake.

10.2 Optimal Detection of Two Quantum States

In quantum detection [235, 240–242], a quantum system with N-dimensional Hilbert space $\mathcal{H}_N$ has a state, pure or mixed, represented by the density operator $\rho \in \mathcal{L}(\mathcal{H}_N)$. The state ρ can be prepared either as the density operator ρ_0 with prior probability P_0, or as the density operator ρ_1 with prior probability $P_1 = 1 - P_0$. The detection task is to determine, from a measurement of the state (the signal) ρ, whether the quantum system is in state ρ_0 or ρ_1. To obtain a solution, since a binary decision is targeted, a quantum detection strategy or quantum detector considers measuring ρ by means of a two-outcome generalized measurement, defined by two measurement operators adhering to Sect. 4.2, associated with the positive operator-valued measure (POVM) having the two elements $\{\mathsf{E}_0, \mathsf{E}_1\}$ following Sect. 4.2.4. Each of the two POVM elements E_k, for $k = 0, 1$, is a positive (Hermitian) operator of $\mathcal{L}(\mathcal{H}_N)$ satisfying $0 \leq \mathsf{E}_k \leq \mathrm{I}_N$, and together summing to the identity operator $\mathrm{I}_N = \mathsf{E}_0 + \mathsf{E}_1$ according to Eq. (4.38); the two operators E_k are also called detection operators [235]. When the measurement outcome corresponding to E_k is obtained, then the quantum detector decides that the preparation of the quantum system was the state ρ_k, for $k = 0, 1$. Except in the special case where the supports of ρ_0 and ρ_1 span two orthogonal subspaces of $\mathcal{H}_N$, the two quantum states ρ_0 and ρ_1 in general cannot be perfectly distinguished (especially when they are noisy), and some detection errors are therefore to be expected. A meaningful strategy for a quantum detector is then to optimally select the pair of detection operators $\{\mathsf{E}_0, \mathsf{E}_1\}$ in order to minimize the cost of such detection errors. A characterization of optimal strategies for quantum detection under various criteria can be found, for instance, in [235]. Optimal strategies for quantum detection can be presented following the same line as in Sect. 10.1 for classical detection.

The conditional probability for obtaining the outcome k upon measuring the state ρ_j, for j and k in $\{0, 1\}$, is according to Eq. (4.39) given by the operator trace

$$\Pr\{\mathsf{E}_k|\rho_j\} = \mathrm{tr}(\mathsf{E}_k\rho_j) , \tag{10.33}$$

paralleling Eq. (10.1). The four conditional probabilities following from Eq. (10.33) then stand as essential ingredients for evaluating the performance of various quantum detectors or detection strategies.

10.2.1 Minimal-Error Detection

For quantum detection, we have the overall probability of detection error $P_{\rm er} = \Pr\{\mathsf{E}_1|\rho_0\}P_0 + \Pr\{\mathsf{E}_0|\rho_1\}P_1$ resulting as

$$P_{\rm er} = \mathrm{tr}(\mathsf{E}_1\rho_0)P_0 + \mathrm{tr}(\mathsf{E}_0\rho_1)P_1 \tag{10.34}$$

$$= \mathrm{tr}\big(\mathsf{E}_1\rho_0P_0 + (\mathrm{I}_N - \mathsf{E}_1)\rho_1P_1\big) \tag{10.35}$$

$$= P_1 - \mathrm{tr}\big((P_1\rho_1 - P_0\rho_0)\mathsf{E}_1\big)\,, \tag{10.36}$$

since $\mathsf{E}_0 = \mathrm{I}_N - \mathsf{E}_1$ and the ρ_j are with unit trace, and displaying a parallel with Eq. (10.5). From Eq. (10.36), the probability of detection error can also be expressed as

$$P_{\rm er} = P_1 - \mathrm{tr}(\mathsf{T}\mathsf{E}_1)\,, \tag{10.37}$$

with the Hermitian test operator

$$\mathsf{T} = P_1\rho_1 - P_0\rho_0\,, \tag{10.38}$$

which is a counterpart to the classical test statistic of Eq. (10.6).

We then seek the optimal choice for the pair of detection operators $\big\{\mathsf{E}_0 = \mathrm{I}_N - \mathsf{E}_1, \mathsf{E}_1\big\}$ that minimizes the probability of detection error $P_{\rm er}$ from Eq. (10.37). This is achieved by finding the detection operator E_1 that maximizes the term $\mathrm{tr}(\mathsf{T}\mathsf{E}_1)$ in the right-hand side of Eq. (10.37). To characterize this optimal detection operator, we introduce the eigen- or diagonal decomposition of the Hermitian test operator T as

$$\mathsf{T} = \sum_{n=1}^{N} \lambda_n\,|\lambda_n\rangle\,\langle\lambda_n|\,, \tag{10.39}$$

with the N (real) eigenvalues λ_n associated with the N eigenstates $|\lambda_n\rangle$ forming an orthonormal basis of the space $\mathcal{H}_N$. We then obtain

$$\mathrm{tr}(\mathsf{T}\mathsf{E}_1) = \sum_{n=1}^{N} \lambda_n\,\mathrm{tr}\big(|\lambda_n\rangle\,\langle\lambda_n|\,\mathsf{E}_1\big) = \sum_{n=1}^{N} \lambda_n\,\langle\lambda_n|\mathsf{E}_1|\lambda_n\rangle\,. \tag{10.40}$$

Each scalar $\langle\lambda_n|\mathsf{E}_1|\lambda_n\rangle$ in Eq. (10.40) is a real confined between 0 and 1, since E_1 is a positive operator verifying $0 \le \mathsf{E}_1 \le \mathrm{I}_N$. For each n, the maximum value of 1 is reached by $\langle\lambda_n|\mathsf{E}_1|\lambda_n\rangle$ when E_1 is the projector $|\lambda_n\rangle\,\langle\lambda_n|$ onto the eigensubspace spanned by $|\lambda_n\rangle$. The detection operator E_1 maximizing the sum in the right-hand side of Eq. (10.40) is thus realized by summing the rank-one projectors $|\lambda_n\rangle\,\langle\lambda_n|$ for all (and only) the eigenstates $|\lambda_n\rangle$ associated with a positive eigenvalue λ_n, i.e.:

$$\mathsf{E}_1^{\rm opt} = \sum_{\lambda_n>0} |\lambda_n\rangle\,\langle\lambda_n|\,, \tag{10.41}$$

achieving for $\mathrm{tr}(\mathsf{T}\mathsf{E}_1)$ in Eq. (10.40) the maximum $\sum_{\lambda_n>0}\lambda_n$. The optimal detection operator $\mathsf{E}_1^{\rm opt}$ defined by Eq. (10.41) is thus the projector onto the subspace of $\mathcal{H}_N$ spanned by the eigenstates $|\lambda_n\rangle$ associated with all the positive eigenvalues λ_n of the test operator T. The complementary detection operator $\mathsf{E}_0^{\rm opt} = \mathrm{I}_N - \mathsf{E}_1^{\rm opt}$ is the projector onto the orthogonal subspace in $\mathcal{H}_N$. If zero happens to be an eigenvalue of T, the associated projector can indifferently be incorporated into $\mathsf{E}_1^{\rm opt}$ or $\mathsf{E}_0^{\rm opt}$, as it always contributes for zero in the probability of error $P_{\rm er}$ from Eq. (10.37).

The optimal detection POVM defined by Eq. (10.41) achieves in Eq. (10.37) the minimal probability of error:

$$P_{\rm er}^{\rm min} = P_1 - \sum_{\lambda_n>0}\lambda_n \,, \tag{10.42}$$

which subtracts from the prior P_1 all the positive eigenvalues of the test operator T. Since the eigenvalues λ_n sum to $\mathrm{tr}(\mathsf{T}) = P_1 - P_0$, one has equivalently $P_{\rm er}^{\rm min} = P_0 + \sum_{\lambda_n<0}\lambda_n$ summing over the negative eigenvalues of T, or by combining with Eq. (10.42),

$$P_{\rm er}^{\rm min} = \frac{1}{2} - \frac{1}{2}\sum_{n=1}^{N}|\lambda_n| = \frac{1}{2} - \frac{1}{2}\,\mathrm{tr}\big(|\mathsf{T}|\big) \,, \tag{10.43}$$

which is a counterpart to Eq. (10.8).

With the notion of trace distance introduced at the beginning of Sect. 3.2.3, we have $\mathrm{tr}\big(|\mathsf{T}|\big)/2 = \mathrm{tr}\big(|P_1\rho_1 - P_0\rho_0|\big)/2 = D_1\big(P_1\rho_1, P_0\rho_0\big)$, so that also

$$P_{\rm er}^{\rm min} = \frac{1}{2} - D_1\big(P_1\rho_1, P_0\rho_0\big) \,, \tag{10.44}$$

with $D_1\big(P_1\rho_1, P_0\rho_0\big) = \|P_1\rho_1 - P_0\rho_0\|_1/2 \le \big(\|P_1\rho_1\|_1 + \|P_0\rho_0\|_1\big)/2 = (P_1 + P_0)/2 = 1/2$ ensuring that $P_{\rm er}^{\rm min}$ stays between 0 and 1/2.

In this way, we have obtained a complete characterization of the quantum detector minimizing the probability of detection error, along with its optimal performance $P_{\rm er}^{\rm min}$ of Eq. (10.43). As anticipated in Sect. 4.2.4, the requirement on the statistical performance $P_{\rm er}$ only constrains the POVM, with subsequently in principle an infinite number of ways of settling an associated pair of measurement operators to implement the POVM. Since the optimal POVM from Eq. (10.41) is made of two projectors acting in $\mathcal{H}_N$, the two measurement operators coinciding with the two POVM elements $\{\mathsf{E}_0^{\rm opt}, \mathsf{E}_1^{\rm opt}\}$ form a valid set of measurement operators associated with the POVM and can be used for a practical implementation of the optimal quantum detector.

10.2.2 Minimal-Cost Detection

For quantum detection optimizing a Bayesian cost, the average cost paralleling Eq. (10.9) is

$$\overline{C} = C_{10}\,\mathrm{tr}(\mathsf{E}_1\rho_0)P_0 + C_{00}\,\mathrm{tr}(\mathsf{E}_0\rho_0)P_0 + C_{01}\,\mathrm{tr}(\mathsf{E}_0\rho_1)P_1 + C_{11}\,\mathrm{tr}(\mathsf{E}_1\rho_1)P_1\;. \tag{10.45}$$

It can be put under the form

$$\overline{C} = C_{00}P_0 + C_{01}P_1 - \mathrm{tr}(\mathsf{T}\mathsf{E}_1)\;, \tag{10.46}$$

with the Hermitian test operator

$$\mathsf{T} = (C_{01} - C_{11})P_1\rho_1 - (C_{10} - C_{00})P_0\rho_0\;. \tag{10.47}$$

The minimization of the average detection cost $\overline{C}$ in Eq. (10.46) is again realized by finding the detection operator E_1 maximizing $\mathrm{tr}(\mathsf{T}\mathsf{E}_1)$. This is achieved by the optimal detection operator $\mathsf{E}_1^{\mathrm{opt}}$ as in Eq. (10.41), representing the projector onto the eigensubspace of T in Eq. (10.47) associated with the positive eigenvalues. The complementary detection operator is $\mathsf{E}_0^{\mathrm{opt}} = \mathrm{I}_N - \mathsf{E}_1^{\mathrm{opt}}$. This optimal quantum detector achieves in Eq. (10.46) the minimal detection cost

$$\overline{C}_{\mathrm{min}} = \frac{1}{2}\Big[(C_{01} + C_{11})P_1 + (C_{00} + C_{10})P_0\Big] - \frac{1}{2}\,\mathrm{tr}\big(|\mathsf{T}|\big)\;, \tag{10.48}$$

paralleling Eq. (10.12).

10.2.3 Neyman–Pearson Detection

The quantum Neyman–Pearson detector seeks to maximize the probability of detection $P_d = \Pr\{\mathsf{E}_1|\rho_1\} = \mathrm{tr}(\mathsf{E}_1\rho_1)$ with a probability of false alarm $\Pr\{\mathsf{E}_1|\rho_0\} = \mathrm{tr}(\mathsf{E}_1\rho_0)$ no larger than a prescribed level P_f. This is equivalent to saturating P_f, as for the classical detector. This constrained maximization can be solved with the Lagrangian $\mathcal{L} = \mathrm{tr}(\mathsf{E}_1\rho_1) - \mu\big[\mathrm{tr}(\mathsf{E}_1\rho_0) - P_f\big]$ and Lagrange multiplier μ, so as to determine the detection operator E_1 maximizing

$$\mathrm{tr}(\mathsf{E}_1\rho_1) - \mu\,\mathrm{tr}(\mathsf{E}_1\rho_0) = \mathrm{tr}(\mathsf{T}\mathsf{E}_1)\;, \tag{10.49}$$

with the Hermitian test operator

$$\mathsf{T} = \rho_1 - \mu\rho_0\;. \tag{10.50}$$

The maximization of $\text{tr}(\mathsf{T}\mathsf{E}_1)$ in Eq. (10.49) is achieved by the μ-dependent detection operator $\mathsf{E}_1^{\text{opt}}(\mu)$ as in Eq. (10.41), representing the projector onto the eigensubspace of T in Eq. (10.50) associated with the positive eigenvalues. The undetermined Lagrange multiplier μ is settled by forcing the probability of false alarm at the prescribed level P_f via

$$\text{tr}\big(\mathsf{E}_1^{\text{opt}}(\mu)\rho_0\big) = P_f \ , \tag{10.51}$$

providing an equation which is inverted to fix the scalar value $\mu = \mu(P_f)$ of the Lagrange multiplier. This $\mu(P_f)$ inserted in the test operator of Eq. (10.50) completely determines the quantum Neyman–Pearson optimal detector, which is then guaranteed to operate at the prescribed probability of false alarm P_f, while achieving the maximal probability of detection:

$$P_d^{\max}(P_f) = \text{tr}\Big(\mathsf{E}_1^{\text{opt}}\big(\mu(P_f)\big)\rho_1\Big) \ . \tag{10.52}$$

Equation (10.52), delivering $P_d^{\max}$ as a function of P_f, provides the ROC curve characterizing the quantum Neyman–Pearson optimal detector.

10.2.4 Minimax Detection

The minimax quantum detector considers, as in Eq. (10.34), the probability of detection error

$$P_{\text{er}} = \text{tr}(\mathsf{E}_1\rho_0) + P_1\big[\text{tr}(\mathsf{E}_0\rho_1) - \text{tr}(\mathsf{E}_1\rho_0)\big] \ , \tag{10.53}$$

which is an affine function of P_1 similar to Eq. (10.20), and it adjusts the detection POVM $\{\mathsf{E}_0 = \text{I}_N - \mathsf{E}_1, \mathsf{E}_1\}$ to achieve

$$\min_{\mathsf{E}_1}\left(\max_{P_1} P_{\text{er}}\right) , \tag{10.54}$$

in similarity with Eq. (10.18). By the same arguments as for the classical minimax detector in Sect. 10.1.4, the solution has to satisfy the necessary condition:

$$\text{tr}(\mathsf{E}_1\rho_0) + \text{tr}(\mathsf{E}_1\rho_1) = 1 \ , \tag{10.55}$$

paralleling the condition receiving the three equivalent forms of Eqs. (10.23)–(10.25). The constrained maximization of Eq. (10.26) becomes here

$$\max_{\mathsf{E}_1} \text{tr}(\mathsf{E}_1\rho_1) \ , \tag{10.56}$$

under the constraint of Eq. (10.55). This constrained maximization, by the Lagrange multiplier method as used previously, proceeds through finding the detection operator E_1 maximizing $\mathrm{tr}(\mathsf{T}\mathsf{E}_1)$, with the Hermitian test operator:

$$\mathsf{T} = \rho_1 - \mu' \rho_0 \;, \tag{10.57}$$

paralleling Eq. (10.32). This maximization is solved by the μ'-dependent detection operator $\mathsf{E}_1^{\mathrm{opt}}(\mu')$ representing the projector onto the eigensubspace of T associated with the positive eigenvalues. The undetermined scalar parameter μ' is settled via the constraint of Eq. (10.55) by solving

$$\mathrm{tr}\big(\mathsf{E}_1^{\mathrm{opt}}(\mu')(\rho_0 + \rho_1)\big) = 1 \;, \tag{10.58}$$

and this in principle completely determines, via the optimal POVM $\big\{\mathsf{E}_0^{\mathrm{opt}} = \mathrm{I}_N - \mathsf{E}_1^{\mathrm{opt}}, \mathsf{E}_1^{\mathrm{opt}}\big\}$, the quantum minimax optimal detector.

10.3 Detection of Two Pure Quantum States

The optimal quantum detectors characterized in Sect. 10.2 apply generally with two mixed states ρ_0 and ρ_1. An important application is to the reference situation of the detection of two pure states $|\psi_0\rangle$ and $|\psi_1\rangle$ in the N-dimensional space $\mathcal{H}_N$. For illustration, we will first work out explicitly the characterization of the minimal-error detection of two pure quantum states $|\psi_0\rangle$ and $|\psi_1\rangle$.

Two pure quantum states $|\psi_0\rangle$ and $|\psi_1\rangle$ of $\mathcal{H}_N$ have density operators $\rho_0 = |\psi_0\rangle\langle\psi_0|$ and $\rho_1 = |\psi_1\rangle\langle\psi_1|$. For their minimal-error detection, the Hermitian test operator of Eq. (10.38) is

$$\mathsf{T} = P_1 |\psi_1\rangle\langle\psi_1| - P_0 |\psi_0\rangle\langle\psi_0| \;. \tag{10.59}$$

To carry out the eigendecomposition of T, it can be noted that any state $|\psi_\perp\rangle$ orthogonal to the vector plane $\big(|\psi_0\rangle, |\psi_1\rangle\big)$ is an eigenstate of T associated with the eigenvalue 0. Such eigenstates generically span an $(N-2)$-dimensional subspace of $\mathcal{H}_N$. The two remaining eigenstates of T are to be found in the vector plane $\big(|\psi_0\rangle, |\psi_1\rangle\big)$ since altogether the N eigenstates span $\mathcal{H}_N$. For their determination, it is possible, without loss of generality, to use the parameterization

$$|\psi_1\rangle = \alpha_0 |\psi_0\rangle + \alpha_1 |\psi_0^\perp\rangle \;, \tag{10.60}$$

with the state $|\psi_0^\perp\rangle$ orthogonal to $|\psi_0\rangle$ in the vector plane $\big(|\psi_0\rangle, |\psi_1\rangle\big)$, and the two complex coordinates satisfying $|\alpha_0|^2 + |\alpha_1|^2 = 1$. The inner product $\langle\psi_0|\psi_1\rangle = \alpha_0$ is an essential parameter reflecting the relative configuration of $|\psi_0\rangle$ and $|\psi_1\rangle$ in $\mathcal{H}_N$.

The determination of the two eigenstates lying in the plane $\big(|\psi_0\rangle\,,|\psi_1\rangle\big)$ can be performed by considering the restriction of the test operator T to the plane $\big(|\psi_0\rangle\,,|\psi_1\rangle\big)$ where it is represented by the matrix

$$\mathsf{T} = \begin{bmatrix} P_1|\alpha_0|^2 - P_0 & P_1\alpha_0\alpha_1^* \\ P_1\alpha_0^*\alpha_1 & P_1|\alpha_1|^2 \end{bmatrix} \tag{10.61}$$

referred to the orthonormal basis $\big\{|\psi_0\rangle\,,|\psi_0^\perp\rangle\big\}$ of the plane $\big(|\psi_0\rangle\,,|\psi_1\rangle\big)$. The characteristic equation of T is $\det(\mathsf{T}-\lambda \mathrm{I}_2) = \lambda^2 - \mathrm{tr}(\mathsf{T})\lambda + \det(\mathsf{T}) = 0$, with $\mathrm{tr}(\mathsf{T}) = P_1 - P_0$ and $\det(\mathsf{T}) = -P_0P_1|\alpha_1|^2$. With the determinant $\det(\mathsf{T}) \le 0$ generically the two eigenvalues of T have opposite signs (except in the degenerate case $\alpha_1 = 0$ when $|\psi_1\rangle \equiv |\psi_0\rangle$). The discriminant of the characteristic equation is $\Delta = \mathrm{tr}^2(\mathsf{T}) - 4\det(\mathsf{T}) = 1 - 4P_0P_1|\alpha_0|^2$ and is generically positive, so that the two eigenvalues of T follow as

$$\lambda_\pm = \frac{1}{2}\Big[\mathrm{tr}(\mathsf{T}) \pm \sqrt{\Delta}\Big] = \frac{1}{2}\Big[P_1 - P_0 \pm \sqrt{1-4P_0P_1|\alpha_0|^2}\Big]. \tag{10.62}$$

From the eigenvalues, Eq. (10.43) leads to

$$P_{\mathrm{er}}^{\mathrm{min}} = \frac{1}{2}\Big[1 - \sqrt{1-4P_0P_1|\langle\psi_0|\psi_1\rangle|^2}\Big], \tag{10.63}$$

which is generally the probability of error achieved by the minimal-error detector of two pure quantum states $|\psi_0\rangle$ and $|\psi_1\rangle$ of $\mathcal{H}_N$. This Eq. (10.63) is sometimes referred to as the Helstrom (lower) bound [235]. It is essentially controlled by the inner product $\langle\psi_0|\psi_1\rangle$ of the two states to be detected. When the two pure states $|\psi_0\rangle$ and $|\psi_1\rangle$ are orthogonal, then $\langle\psi_0|\psi_1\rangle = 0$ and $P_{\mathrm{er}}^{\mathrm{min}} = 0$ since $|\psi_0\rangle$ and $|\psi_1\rangle$ are perfectly distinguishable. Otherwise they are not, and the best that can be done in terms of probability of detection error is Eq. (10.63). Some typical evolutions of the probability of detection error of Eq. (10.63) are depicted in Fig. 10.1.

Associated with the two eigenvalues $\lambda_\pm$ of Eq. (10.62) are the two eigenstates $|\lambda_\pm\rangle$. In the basis $\big\{|\psi_0\rangle\,,|\psi_0^\perp\rangle\big\}$, before normalization, they can be characterized by the coordinates

$$|\lambda_\pm\rangle = \begin{bmatrix} \big(|\alpha_0|^2 - |\alpha_1|^2\big)P_1 - P_0 \pm \sqrt{\Delta} \\ 2\alpha_0^*\alpha_1 P_1 \end{bmatrix}. \tag{10.64}$$

After normalization, $|\lambda_+\rangle$ defines the optimal detection operator $\mathsf{E}_1^{\mathrm{opt}} = |\lambda_+\rangle\langle\lambda_+|$ according to Eq. (10.41) as a rank-one projector acting in $\mathcal{H}_N$. The other optimal detection operator can be taken as the complementary projector $\mathsf{E}_0^{\mathrm{opt}} = \mathrm{I}_N - \mathsf{E}_1^{\mathrm{opt}}$ in $\mathcal{H}_N$ acting as $|\lambda_-\rangle\langle\lambda_-|$ since the detection is essentially decided from the projection in the plane $\big(|\psi_0\rangle\,,|\psi_1\rangle\big)$ coinciding with the plane $\big(|\lambda_+\rangle\,,|\lambda_-\rangle\big)$, and the states to be measured have no components outside this plane.

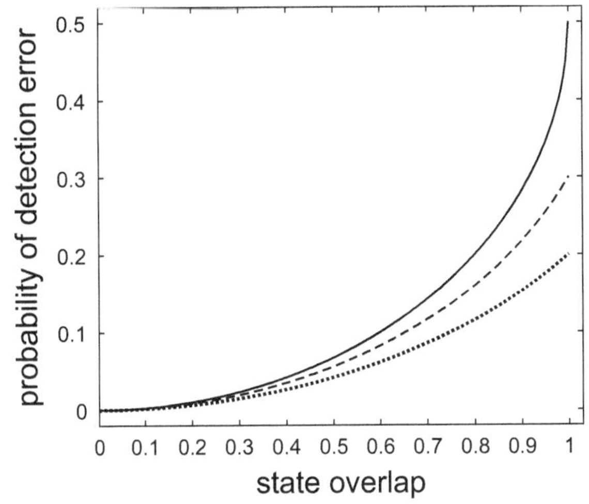

Fig. 10.1 For minimal-error detection of two pure quantum states $|\psi_0\rangle$ and $|\psi_1\rangle$ in $\mathcal{H}_N$, as a function of their overlap $|\langle\psi_0|\psi_1\rangle|$, the minimal probability of error $P_{\text{er}}^{\min}$ of Eq. (10.63) with the prior probability $P_0 = 0.5$ (solid line), $P_0 = 0.7$ (dashed line), $P_0 = 0.8$ (dotted line)

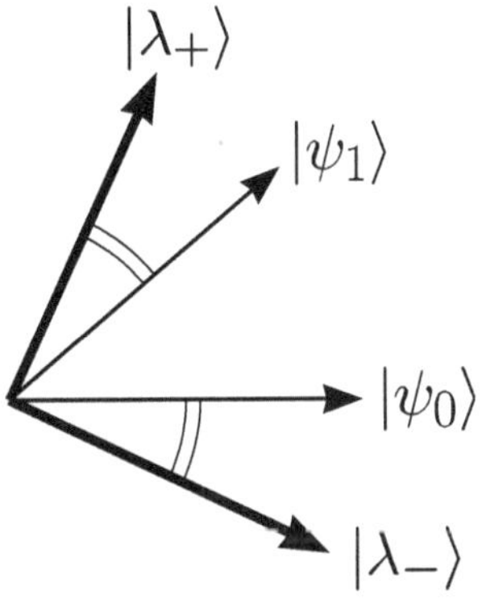

Fig. 10.2 In the vector plane formed by the two pure quantum states $|\psi_0\rangle$ and $|\psi_1\rangle$ to be distinguished in $\mathcal{H}_N$, the two orthogonal eigenstates $|\lambda_+\rangle$ and $|\lambda_-\rangle$ from Eq. (10.64) after normalization and defining the two projectors of the optimal detector achieving the minimal probability of error $P_{\text{er}}^{\min}$ of Eq. (10.63), with the prior probabilities $P_0 = P_1 = 1/2$

There exists a nice geometrical picture for the action of the optimal projectors $|\lambda_+\rangle\langle\lambda_+|$ and $|\lambda_-\rangle\langle\lambda_-|$ in the plane $\big(|\psi_0\rangle, |\psi_1\rangle\big)$, as illustrated in Fig. 10.2. The two orthogonal eigenstates $|\lambda_+\rangle$ and $|\lambda_-\rangle$ form a sector at right angle in the plane $\big(|\psi_0\rangle, |\psi_1\rangle\big)$. This right-angle sector sits, in relation to the pair $\big(|\psi_0\rangle, |\psi_1\rangle\big)$, in a symmetrical position when $P_0 = P_1 = 1/2$. As P_1 increases above P_0, the right-angle sector rotates in the plane with $|\lambda_+\rangle$ approaching $|\psi_1\rangle$, and conversely, when P_1 decreases below P_0, in the rotation $|\lambda_-\rangle$ approaches $|\psi_0\rangle$.

Before addressing the Neyman–Pearson and minimax detections of two pure quantum states in the N-dimensional space $\mathcal{H}_N$, we will proceed with additional elements relevant to the detection of qubit states in the two-dimensional space $\mathcal{H}_2$. This is consistent with the fundamental status of the qubit for quantum information, but also, as we just witnessed in Sect. 10.3, the detection of two pure states $|\psi_0\rangle$ and $|\psi_1\rangle$ in $\mathcal{H}_N$ is essentially a two-dimensional problem, solvable in the vector plane $\big(|\psi_0\rangle, |\psi_1\rangle\big)$, and further elements on the detection for qubits in $\mathcal{H}_2$ will offer an easier path to the Neyman–Pearson and minimax detections of two pure states in $\mathcal{H}_N$.

10.4 Detection of Two Qubit States

Efficient detection strategies are specially useful for mixed quantum states, typically representing signaling states altered by some quantum noise. Such situation is typical and frequently encountered in quantum communication, with, for instance, a qubit, formed by a photon encoding some binary information by means of its two basis states of polarization, which gets altered by some quantum noise as it travels across an optical fiber and is received as a noisy photon on which a detection task is to be performed so as to recover the correct signaling state intended at the emitter. So it is useful now, by applying the theory of Sect. 10.2, to show how the optimal detection of general mixed states of a qubit can be conveniently characterized, especially by means of the Bloch representation.

For the detection task, the two qubit states ρ_0 and ρ_1 of $\mathcal{L}(\mathcal{H}_2)$ can be represented, according to Eq. (2.61), in the Bloch representation:

$$\rho_j = \frac{1}{2}\big(\mathrm{I}_2 + \vec{r}_j \cdot \vec{\sigma}\big), \qquad j = 0, 1, \tag{10.65}$$

with the two Bloch vectors $\vec{r}_j \in \mathbb{R}^3$ of Euclidean norm $\|\vec{r}_j\| = 1$ for a pure state and $\|\vec{r}_j\| < 1$ for a mixed state. As we have seen in Sect. 10.2, the optimal quantum detectors, their structure and performance, are essentially controlled by the (Hermitian) test operator T and its eigendecomposition, according to its various expressions of Eq. (10.38) or (10.47) or (10.50) or (10.57) depending on the criterion to be optimized. For the qubit, the test operator T and its eigendecomposition can be efficiently handled in Bloch representation.

10.4.1 Minimal-Error Detection

We consider first the optimal quantum detector achieving minimal probability of error, and controlled by the test operator T in Eq. (10.38). In Bloch representation, the test operator $\mathsf{T} = P_1\rho_1 - P_0\rho_0$ of Eq. (10.38) is

$$\mathsf{T} = \frac{1}{2}\big[(P_1 - P_0)\mathrm{I}_2 + \vec{\tau}\cdot\vec{\sigma}\big]\,, \tag{10.66}$$

characterized by the test Bloch vector of $\mathbb{R}^3$,

$$\vec{\tau} = P_1\vec{r}_1 - P_0\vec{r}_0\,. \tag{10.67}$$

For the eigendecomposition of T, which is a Hermitian operator of the qubit, according to Sect. 4.3.4 we have the two eigenvalues

$$\lambda_\pm = \frac{1}{2}\Big(P_1 - P_0 \pm \|\vec{\tau}\,\|\Big)\,, \tag{10.68}$$

and the two projectors onto the two orthogonal eigenstates

$$|\lambda_\pm\rangle\,\langle\lambda_\pm| = \frac{1}{2}\left(\mathrm{I}_2 \pm \frac{\vec{\tau}}{\|\vec{\tau}\,\|}\cdot\vec{\sigma}\right). \tag{10.69}$$

The signs of the two eigenvalues $\lambda_\pm$ can also be discussed by means of the determinant $\det(\mathsf{T}) = \big[(P_1 - P_0)^2 - \|\vec{\tau}\,\|^2\big]/4 = \lambda_+\lambda_-$. The sign of $\det(\mathsf{T})$ changes with the situation of $\|\vec{\tau}\,\|^2 = P_1^2\|\vec{r}_1\|^2 + P_0^2\|\vec{r}_0\|^2 - 2P_0P_1\vec{r}_0\vec{r}_1$ in relation to $(P_1 - P_0)^2 = P_1^2 + P_0^2 - 2P_0P_1$.

When $\|\vec{\tau}\,\| > |P_1 - P_0|$ then $\det(\mathsf{T}) < 0$ and the two eigenvalues $\lambda_\pm$ have opposite signs, with λ_+ the positive eigenvalue in Eq. (10.68) and the pair of optimal detection operators $\mathsf{E}_1^{\mathrm{opt}} = |\lambda_+\rangle\,\langle\lambda_+|$ and $\mathsf{E}_0^{\mathrm{opt}} = |\lambda_-\rangle\,\langle\lambda_-|$ in Eq. (10.69); the performance from Eqs. (10.42) or (10.43) is then $P_{\mathrm{er}}^{\min} = \big(1 - \|\vec{\tau}\,\|\big)/2$.

When $\|\vec{\tau}\,\| \le |P_1 - P_0|$ then $\det(\mathsf{T}) \ge 0$ and the two eigenvalues $\lambda_\pm$ have the same sign. Both eigenvalues in Eq. (10.68) are then positive when $P_1 \ge P_0$, so that the optimal detection operator $\mathsf{E}_1^{\mathrm{opt}} = \mathrm{I}_2$ always occurs in the measurement and ρ_1 is always decided, while $\mathsf{E}_0^{\mathrm{opt}}$ is the null operator of $\mathcal{L}(\mathcal{H}_2)$ and it never occurs; this incurs an error with probability P_0, when ρ_0 happens to hold. The situation is reversed when $P_0 \ge P_1$.

Summarizing, the performance of the minimal-error optimal detector is therefore

$$\begin{cases} \dfrac{1}{2}\Big(1 - \|\vec{\tau}\,\|\Big)\,, & \text{when } \|\vec{\tau}\,\| > |P_1 - P_0|\,, \qquad (10.70)\\[2mm] \min(P_0, P_1)\,, & \text{when } \|\vec{\tau}\,\| < |P_1 - P_0|\,, \qquad (10.71)\end{cases}$$

with continuity at the boundary of the two regimes. The norm $\|\vec{\tau}\,\|$ of the test Bloch vector $\vec{\tau}$ from Eq. (10.67) acts as an index of contrast between the two qubit states ρ_0 and ρ_1 to be detected. When $\|\vec{\tau}\,\|$ is too small, as in Eq. (10.71), the quantum measurement is trivial and of no use and the optimal detector always decides the state with the largest prior probability, but this cannot occur when $P_0 = P_1$, or with two distinct pure states ρ_0 and ρ_1.

When $P_0 \neq P_1$, there may exist two strictly distinct qubit states ρ_0 and ρ_1, yet close enough in their contrast $\|\vec{\tau}\| < |P_1 - P_0|$ so that Eq. (10.71) holds. For discriminating between two such distinct states, no approach can therefore be more efficient, in terms of probability of error, than the trivial detector that always decides the state with largest prior probability. Any conceivable quantum measurement is completely non-informative about the two states [243], and cannot aid to discriminate between them, although the two quantum states are strictly distinct. There hardly exists an analogous classical situation, where two strictly distinct classical signals could not benefit from any measurement to improve their discrimination.

With two pure qubit states $\rho_0 = |\psi_0\rangle\langle\psi_0|$ and $\rho_1 = |\psi_1\rangle\langle\psi_1|$, it is the regime of Eq. (10.70) that always holds, since $\|\vec{r}_0\| = 1 = \|\vec{r}_1\|$ implies $\|\vec{\tau}\|^2 = P_1^2 + P_0^2 - 2P_0P_1\vec{r}_0\vec{r}_1 \geq |P_1 - P_0|^2 = P_1^2 + P_0^2 - 2P_0P_1$. Furthermore, it is possible to relate the inner product $|\langle\psi_0|\psi_1\rangle|^2 = \mathrm{tr}(\rho_0\rho_1)$ in $\mathcal{H}_2$ to the inner product $\vec{r}_0\vec{r}_1$ in $\mathbb{R}^3$. In Bloch representation, by the relation of Eq. (A.45) in the Appendix, one also has $\mathrm{tr}(\rho_0\rho_1) = \mathrm{tr}\big[(1 + \vec{r}_0\vec{r}_1)\mathrm{I}_2/4\big] = \big(1 + \vec{r}_0\vec{r}_1\big)/2$, leading to

$$\vec{r}_0\vec{r}_1 = 2|\langle\psi_0|\psi_1\rangle|^2 - 1 . \tag{10.72}$$

It follows that for two pure qubit states $\|\vec{\tau}\|^2 = P_1^2 + P_0^2 - 2P_0P_1\vec{r}_0\vec{r}_1 = 1 - 4P_0 P_1|\langle\psi_0|\psi_1\rangle|^2$, and Eq. (10.70) coincides with Eq. (10.63). This shows, as we announced, that the treatment of the two-dimensional qubit in $\mathcal{H}_2$ contains the solution to the general detection of two pure quantum states in $\mathcal{H}_N$.

The qubit case in $\mathcal{H}_2$ can also serve us to work out the Neyman–Pearson and the minimax detections of two pure quantum states in $\mathcal{H}_N$, which were left pending at the end of Sect. 10.3.

10.4.2 Neyman–Pearson Detection

For the two qubit states ρ_0 and ρ_1 from Eq. (10.65), the quantum Neyman–Pearson optimal detector is governed by the test operator $\mathsf{T} = \rho_1 - \mu\rho_0$ of Eq. (10.50) which is in Bloch representation:

$$\mathsf{T} = \frac{1}{2}\left[(1-\mu)\mathrm{I}_2 + \vec{\tau}\cdot\vec{\sigma}\right] , \tag{10.73}$$

characterized by the test Bloch vector:

$$\vec{\tau} = \vec{r}_1 - \mu\vec{r}_0 . \tag{10.74}$$

The two eigenvalues and two eigenstates of the test operator T can then be obtained as a function of the test Bloch vector $\vec{\tau}$, so as to accomplish the characterization of the quantum Neyman–Pearson optimal detector according to the program of Sect. 10.2.3.

The two eigenvalues of T in Eq. (10.73) are $\lambda_\pm = \big(1 - \mu \pm \|\vec{\tau}\,\|\big)/2$, with $\|\vec{\tau}\,\|^2 = (\vec{r}_1 - \mu\vec{r}_0)^2 = \|\vec{r}_1\|^2 + \mu^2\|\vec{r}_0\|^2 - 2\mu\vec{r}_0\vec{r}_1$. The signs of the two eigenvalues $\lambda_\pm$ can be discussed by means of the determinant $\det(\mathsf{T}) = \big[(1-\mu)^2 - \|\vec{\tau}\,\|^2\big]/4 = \lambda_+\lambda_-$. There is a nontrivial (not involving the identity operator I_2) measurement when the two eigenvalues $\lambda_\pm$ have opposite signs, when $\det(\mathsf{T}) < 0$ or equivalently $\|\vec{\tau}\,\|^2 > (1-\mu)^2$. This condition essentially depends on the two Bloch vectors $\vec{r}_0$ and $\vec{r}_1$ and their relative configuration in $\mathbb{R}^3$.

A situation of interest that can be further characterized in general analytical terms is the detection of two pure states $\rho_0 = |\psi_0\rangle\langle\psi_0|$ and $\rho_1 = |\psi_1\rangle\langle\psi_1|$, when $\|\vec{r}_0\| = 1 = \|\vec{r}_1\|$ and $\|\vec{\tau}\,\|^2 = 1 + \mu^2 - 2\mu\vec{r}_0\vec{r}_1$. For $\mu \leq 0$ then $\det(\mathsf{T}) \geq 0$ and the two eigenvalues $\lambda_\pm \geq 0$; the optimal detection operator $\mathsf{E}_1^{\text{opt}} = \mathrm{I}_2$ so that ρ_1 is always decided and $P_d = 1$; this is a trivial detector that makes no use of any quantum measurement to return its decision, as in Eq. (10.71). A nontrivial detector exists for $\mu > 0$, leading to $\det(\mathsf{T}) < 0$ and only one positive eigenvalue λ_+, associated with the optimal detection operator $\mathsf{E}_1^{\text{opt}} = |\lambda_+\rangle\langle\lambda_+| = \big(\mathrm{I}_2 + \vec{\tau}\cdot\vec{\sigma}/\|\vec{\tau}\,\|\big)/2$. By the relation of Eq. (A.45), this leads to the probability of false alarm of Eq. (10.51) as

$$P_f = \operatorname{tr}\big(\mathsf{E}_1^{\text{opt}}\rho_0\big) = \frac{1}{2}\left(1 + \vec{r}_0\frac{\vec{\tau}}{\|\vec{\tau}\,\|}\right) \tag{10.75}$$

$$= \frac{1}{2}\left(1 + \frac{\vec{r}_0\vec{r}_1 - \mu}{\sqrt{1 + \mu^2 - 2\mu\vec{r}_0\vec{r}_1}}\right) \tag{10.76}$$

since $\vec{r}_0\vec{\tau} = \vec{r}_0\vec{r}_1 - \mu$, and to the probability of detection of Eq. (10.52) as

$$P_d^{\max} = \operatorname{tr}\big(\mathsf{E}_1^{\text{opt}}\rho_1\big) = \frac{1}{2}\left(1 + \vec{r}_1\frac{\vec{\tau}}{\|\vec{\tau}\,\|}\right) \tag{10.77}$$

$$= \frac{1}{2}\left(1 + \frac{1 - \mu\vec{r}_0\vec{r}_1}{\sqrt{1 + \mu^2 - 2\mu\vec{r}_0\vec{r}_1}}\right) \tag{10.78}$$

since $\vec{r}_1\vec{\tau} = 1 - \mu\vec{r}_0\vec{r}_1$.

Together Eqs. (10.76) and (10.78) delineate the ROC curve $\big(P_f(\mu), P_d^{\max}(\mu)\big)$ of the Neyman–Pearson optimal detector of two pure qubit states $|\psi_0\rangle$ and $|\psi_1\rangle$ in $\mathcal{H}_2$, in a parametric form as a function of the Lagrange multiplier μ. From Eq. (10.72) for the inner product, Eqs. (10.76) and (10.78) also contain the characterization of the performance of the Neyman–Pearson optimal detector for two pure quantum states $|\psi_0\rangle$ and $|\psi_1\rangle$ from a Hilbert space $\mathcal{H}_N$ with arbitrary dimension N, which is essentially a two-dimensional problem solvable in the vector plane $(|\psi_0\rangle, |\psi_1\rangle)$ as executed with two qubit states in $\mathcal{H}_2$.

Finally, it is possible to eliminate μ between Eqs. (10.76) and (10.78), to obtain as in [235]:

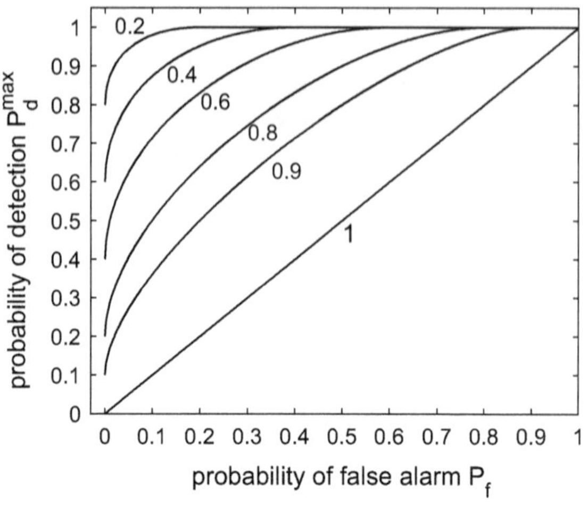

Fig. 10.3 ROC curves characterizing the performance of the Neyman–Pearson optimal detector of two pure quantum states $|\psi_0\rangle$ and $|\psi_1\rangle$ of the N-dimensional space $\mathcal{H}_N$, showing the probability of detection $P_d^{\max}$ as a function of the probability of false alarm P_f, according to Eqs. (10.79)–(10.80), for six values of the squared overlap $|\langle\psi_0|\psi_1\rangle|^2$ from 0.2 to 1

$$P_d^{\max} = \begin{cases} \left[\sqrt{|\langle\psi_0|\psi_1\rangle|^2 P_f} + \sqrt{(1-|\langle\psi_0|\psi_1\rangle|^2)(1-P_f)}\right]^2, & \text{when } 0 \le P_f \le |\langle\psi_0|\psi_1\rangle|^2, \quad (10.79) \\ 1, & \text{when } |\langle\psi_0|\psi_1\rangle|^2 \le P_f \le 1, \quad (10.80) \end{cases}$$

which represents a closed-form expression of the ROC curve characterizing the performance of the Neyman–Pearson optimal detector of two pure quantum states $|\psi_0\rangle$ and $|\psi_1\rangle$ of $\mathcal{H}_N$.

From Eqs. (10.76) and (10.78), we can also obtain the ratio

$$\frac{2P_f - 1}{2P_d^{\max} - 1} = \frac{\vec{r}_0\vec{r}_1 - \mu}{1 - \mu\vec{r}_0\vec{r}_1} = g\,. \tag{10.81}$$

This ratio g is accessible, via Eqs. (10.79)–(10.80), as soon as P_f is specified, and it leads to a characterization of the value of the Lagrange multiplier μ at the optimum as

$$\mu = \mu_{\text{opt}} = \frac{\vec{r}_0\vec{r}_1 - g}{1 - g\vec{r}_0\vec{r}_1}\,. \tag{10.82}$$

This value of μ_{opt} completes, via Eq. (10.74), the determination of the Neyman–Pearson optimal detection POVM $\{\mathsf{E}_0^{\text{opt}}, \mathsf{E}_1^{\text{opt}}\}$.

This characterization of the Neyman–Pearson optimal detector generally applies, via Eq. (10.72), to the detection of two pure quantum states $|\psi_0\rangle$ and $|\psi_1\rangle$ in the N-dimensional space $\mathcal{H}_N$. For illustration of the performance, typical ROC curves according to Eqs. (10.79)–(10.80) are represented in Fig. 10.3, for different values of $|\langle\psi_0|\psi_1\rangle|^2$ characterizing the overlap of the two pure states $|\psi_0\rangle$ and $|\psi_1\rangle$ to be detected.

In Eqs. (10.79)–(10.80) and Fig. 10.3, the growth of P_f from 0 to 1 is covered with a Lagrange multiplier μ decreasing from $+\infty$ to $-\infty$ in Eqs. (10.76) and (10.78). The point $P_f = 0$ corresponds to $\mu = +\infty$ in Eq. (10.76), and to $P_d^{\max} = (1 - \vec{r}_0\vec{r}_1)/2 = 1 - |\langle\psi_0|\psi_1\rangle|^2$ in Eq. (10.78). As P_f grows above 0, the corresponding μ decreases in Eq. (10.76), and $P_d^{\max}$ increases in Eq. (10.78). The point where $P_d^{\max}$ reaches the maximum of 1 corresponds in Eq. (10.78) to $\mu = 0$, and to $P_f = (1 + \vec{r}_0\vec{r}_1)/2 = |\langle\psi_0|\psi_1\rangle|^2$ in Eq. (10.76). Beyond is the range of Eq. (10.80), where $P_d^{\max}$ saturates at 1, when P_f continues to grow to 1, while μ in Eq. (10.76) decreases from 0 to $-\infty$. Finally, $P_f = 1$ is reached when $\mu = -\infty$ in Eq. (10.76).

The ROC curves of Fig. 10.3 illustrate the expected property of an improved detection performance as the overlap $|\langle\psi_0|\psi_1\rangle|$ between the two quantum states decreases.

10.4.3 Minimax Detection

For the two qubit states ρ_0 and ρ_1 of Eq. (10.65), the quantum minimax optimal detector is governed by the test operator $\mathsf{T} = \rho_1 - \mu'\rho_0$ of Eq. (10.57) which is in Bloch representation:

$$\mathsf{T} = \frac{1}{2}\big[(1 - \mu')\mathrm{I}_2 + \vec{\tau}\cdot\vec{\sigma}\big], \tag{10.83}$$

exhibiting the same form as in the Neyman–Pearson detection of Eq. (10.73), with here the test Bloch vector $\vec{\tau} = \vec{r}_1 - \mu'\vec{r}_0$. The μ'-dependent optimal detection operator $\mathsf{E}_1^{\mathrm{opt}}(\mu')$ follows as the projector onto the eigensubspace of T associated with the positive eigenvalues, and it rests on the eigendecomposition of T which can be handled in Bloch representation as discussed just after Eqs. (10.73)–(10.74) for the Neyman–Pearson detection. From the μ'-dependent optimal POVM $\{\mathsf{E}_0^{\mathrm{opt}} = \mathrm{I}_N - \mathsf{E}_1^{\mathrm{opt}}, \mathsf{E}_1^{\mathrm{opt}}\}$, the final determination of the parameter μ' is realized by solving Eq. (10.58).

For two pure states $\rho_0 = |\psi_0\rangle\langle\psi_0|$ and $\rho_1 = |\psi_1\rangle\langle\psi_1|$, when $\|\vec{r}_0\| = 1 = \|\vec{r}_1\|$, the eigendecomposition of T performed in Sect. 10.4.2 for the Neyman–Pearson detection carries over to the minimax detection here, with μ becoming μ'. For solving Eq. (10.58), by the similarity of form with the Neyman–Pearson detection, we have $\mathrm{tr}(\mathsf{E}_1^{\mathrm{opt}}\rho_0)$ given by Eqs. (10.75)–(10.76), and $\mathrm{tr}(\mathsf{E}_1^{\mathrm{opt}}\rho_1)$ given by Eqs. (10.77)–(10.78), so that from the constraint of Eq. (10.58) we obtain $\vec{r}_0\vec{r}_1 - \mu' + 1 - \mu'\vec{r}_0\vec{r}_1 = 0$, giving $\mu' = \mu'_{\mathrm{opt}} = 1$.

Then $\mu'_{\mathrm{opt}} = 1$ inserted in Eq. (10.83) yields the test operator $\mathsf{T} = (\vec{r}_1 - \vec{r}_0)\cdot\vec{\sigma}/2$ with the two eigenvalues $\lambda_\pm = \pm\|\vec{r}_1 - \vec{r}_0\|/2$ and the projectors $|\lambda_\pm\rangle\langle\lambda_\pm| = \big[\mathrm{I}_2 \pm (\vec{r}_1 - \vec{r}_0)\cdot\vec{\sigma}/\|\vec{r}_1 - \vec{r}_0\|\big]/2$ onto its two eigenstates, fixing the two optimal detection operators $\mathsf{E}_0^{\mathrm{opt}} = |\lambda_-\rangle\langle\lambda_-|$ and $\mathsf{E}_1^{\mathrm{opt}} = |\lambda_+\rangle\langle\lambda_+|$ of the minimax optimal detector of two pure qubit states.

The performance of this minimax optimal detector can be characterized by the two conditional probabilities of correct detection $\mathrm{tr}(\mathsf{E}_1^{\mathrm{opt}}\rho_1) = \big[1 + \vec{r}_1(\vec{r}_1 - \vec{r}_0)/\|\vec{r}_1 - \vec{r}_0\|\big]/2$ and $\mathrm{tr}(\mathsf{E}_0^{\mathrm{opt}}\rho_0) = \big[1 - \vec{r}_0(\vec{r}_1 - \vec{r}_0)/\|\vec{r}_1 - \vec{r}_0\|\big]/2$, which both coincide since

$\text{tr}(\mathsf{E}_1^{\text{opt}}\rho_1) = \text{tr}(\mathsf{E}_0^{\text{opt}}\rho_0) = \big(1+\sqrt{1-\vec{r}_0\vec{r}_1}/\sqrt{2}\big)/2$ as it should from Eq. (10.55). From Eq. (10.72) for the inner product, we also obtain $\text{tr}(\mathsf{E}_1^{\text{opt}}\rho_1) = \text{tr}(\mathsf{E}_0^{\text{opt}}\rho_0) = \big(1+\sqrt{1-|\langle\psi_0|\psi_1\rangle|^2}\big)/2$.

As before, this resolution also contains the characterization of the minimax optimal detector for two pure quantum states $|\psi_0\rangle$ and $|\psi_1\rangle$ from a Hilbert space $\mathcal{H}_N$ with arbitrary dimension N, which is essentially a two-dimensional problem solvable in the vector plane $(|\psi_0\rangle, |\psi_1\rangle)$ as executed with two qubit states in $\mathcal{H}_2$. Also, by comparing with the results of Sect. 10.4.1, it can be observed that for two pure states $|\psi_0\rangle$ and $|\psi_1\rangle$, the optimal detection POVM $\{\mathsf{E}_0^{\text{opt}}, \mathsf{E}_1^{\text{opt}}\}$ of the minimax optimal detector coincides with that of the minimal-error detector with equal priors $P_0 = P_1 = 1/2 = P_1^{\max}$ (the maximizer of $P_{\text{er}}^{\min}$), as characterized in Fig. 10.2, and both sharing the same probability of error P_{er} from Eq. (10.63).

10.5 Detection from a Noisy Qubit

We now want to focus on another aspect, with great relevance to quantum signal and information processing, where a qubit has its state ρ prepared in either states ρ_0 or ρ_1 as in Sect. 10.4, to serve as two signaling states in some communication protocol. Prior to detection, the information-carrying qubit is affected by some quantum noise or quantum non-unitary process, modeled by the superoperator $\mathcal{N}(\cdot)$ as addressed in Chap. 9, so that the noisy qubit has a state described by the density operator $\rho' = \mathcal{N}(\rho)$. Only the noisy quantum state ρ' is accessible for detection, as it could be received at the output of a noisy communication channel. It is then to be decided whether ρ' is in state $\rho_0' = \mathcal{N}(\rho_0)$ or $\rho_1' = \mathcal{N}(\rho_1)$, so as to deduce whether the initial signaling state emitted at the channel input was ρ_0 or ρ_1. From Sect. 10.4, we now know the optimal quantum detectors, and the optimal efficiency they achieve, according to various performance criteria. This optimal efficiency can now be subjected to a further optimization step, specially meaningful in the context of quantum communication, and consisting in determining the optimal pair of input signaling states $\{\rho_0, \rho_1\}$ to maximize the efficiency of the optimal detector from the noisy output $\rho' = \mathcal{N}(\rho)$, in the presence of a definite noise model $\mathcal{N}(\cdot)$ on the communication channel. We now examine this issue, which is also addressed, for instance, in [221].

In this context of quantum communication where the input signaling pair $\{\rho_0, \rho_1\}$ is purposely devised for efficient transmission over a noisy channel, known prior probabilities $(P_0, P_1 = 1 - P_0)$ represent a realistic reference situation, and we will therefore consider the situation of minimal-error optimal detection. Alternative performance criteria could be treated as well from the general considerations on quantum detection presented above. According to Sect. 10.4, minimal-error detection from the noisy qubit state $\rho' = \mathcal{N}(\rho)$ is governed by the test operator T from Eq. (10.38), taking here the form:

$$\mathsf{T} = P_1\rho_1' - P_0\rho_0' = P_1\mathcal{N}(\rho_1) - P_0\mathcal{N}(\rho_0)\,, \tag{10.84}$$

and expressible in Bloch representation as

$$\mathsf{T} = \frac{1}{2}\big[(P_1 - P_0)\mathrm{I}_2 + \vec{\tau}\,' \cdot \vec{\sigma}\big]\,, \tag{10.85}$$

which is analogous to Eq. (10.66). The action of the noise $\rho \mapsto \rho' = \mathcal{N}(\rho)$ on qubit states transforms their Bloch vector in $\mathbb{R}^3$ by the affine map of Eq. (9.7):

$$\vec{r} \longmapsto \vec{r}\,' = A\vec{r} + \vec{c}\,, \tag{10.86}$$

where the real pair $\{A, \vec{c}\,\}$ characterizes the noise model. In Eq. (10.85), we have the output test Bloch vector

$$\vec{\tau}\,' = P_1\vec{r}_1' - P_0\vec{r}_0' = A\vec{\tau} + \vec{c}\,'\,, \tag{10.87}$$

with $\vec{c}\,' = (P_1 - P_0)\vec{c}$ and the input test Bloch vector

$$\vec{\tau} = P_1\vec{r}_1 - P_0\vec{r}_0\,, \tag{10.88}$$

as in Eq. (10.67) prior to the action of the noise.

The performance of the minimal-error detector now follows from Eqs. (10.70)–(10.71) yet applying to the noisy qubit, so as to give

$$P_{\rm er}^{\rm min} = \begin{cases} \dfrac{1}{2}\Big(1 - \|\vec{\tau}\,'\|\Big)\,, & \text{when } \|\vec{\tau}\,'\| \geq |P_1 - P_0|\,, \quad (10.89)\\ \min(P_0, P_1)\,, & \text{when } \|\vec{\tau}\,'\| < |P_1 - P_0|\,. \quad (10.90) \end{cases}$$

This probability of error $P_{\rm er}^{\rm min}$ of the optimal detector is small when the index of contrast $\|\vec{\tau}\,'\|$ between the two noisy output states is large. It is then meaningful to optimize the two input signaling states $\big\{\rho_0, \rho_1\big\}$ so as to minimize the probability of error $P_{\rm er}^{\rm min}$ of Eqs. (10.89)–(10.90). This amounts to selecting the two input signaling states $\big\{\rho_0, \rho_1\big\}$, or equivalently their two Bloch vectors $\big\{\vec{r}_0, \vec{r}_1\big\}$, in order to maximize the norm $\|\vec{\tau}\,'\|$ of the output test vector of Eq. (10.87). We now address this maximization in the presence of a given noise model characterized by a specified real pair $\{A, \vec{c}\,\}$.

The output test vector $\vec{\tau}\,'$ of Eq. (10.87) has a squared norm expressible as

$$\|\vec{\tau}\,'\|^2 = \|A\vec{\tau}\,\|^2 + \|\vec{c}\,'\|^2 + 2\vec{v}\,\vec{\tau}\,, \tag{10.91}$$

with the vector $\vec{v} = A^\top\vec{c}\,'$ and $\vec{v}\,\vec{\tau}$ which is the standard real inner product of two column vectors of $\mathbb{R}^3$. Minimizing the probability of error $P_{\rm er}^{\rm min}$ of Eqs. (10.89)–(10.90) is equivalent to maximizing the squared normed $\|\vec{\tau}\,'\|^2$ in Eq. (10.91), and this by a proper choice of the two signaling states $\big\{\rho_0, \rho_1\big\}$ or equivalently of their two Bloch vectors $\big\{\vec{r}_0, \vec{r}_1\big\}$. Since $\|\vec{\tau}\,'\|^2$ in Eq. (10.91) depends on $\big\{\vec{r}_0, \vec{r}_1\big\}$ only through the input test vector $\vec{\tau}$ of Eq. (10.88), maximization of $\|\vec{\tau}\,'\|^2$ needs to be performed simply according to the components of $\vec{\tau} = [\tau_x, \tau_y, \tau_z]^\top$, and not separately $\vec{r}_0$ and $\vec{r}_1$. Also, since the input Bloch vectors $\vec{r}_0$ and $\vec{r}_1$ have at most unit norm, the input test vector $\vec{\tau}$ of Eq. (10.88) is constrained by $\|\vec{\tau}\|^2 \le 1$. Minimizing the probability of error $P_{\rm er}^{\rm min}$ of Eqs. (10.89)–(10.90) then amounts to maximizing $\|\vec{\tau}\,'\|^2$ of Eq. (10.91) according to the components of $\vec{\tau}$ of Eq. (10.88) subject to the constraint $\|\vec{\tau}\|^2 \le 1$.

An important property is that the maximum of $\|\vec{\tau}\,'\|^2$ in Eq. (10.91) always occurs at the saturated constraint $\|\vec{\tau}\| = 1$, indicating two orthogonal pure signaling states $\{\rho_0, \rho_1\}$ at the optimum. This holds since in Eq. (10.91) the two terms $\|A\vec{\tau}\|^2$ and $\|\vec{c}\,'\|^2$ are non-negative, so necessarily at the maximum of $\|\vec{\tau}\,'\|^2$ the third term $2\vec{v}\,\vec{\tau}$ is also non-negative; if it were negative, the change $\vec{\tau} \to -\vec{\tau}$ would increase $\|\vec{\tau}\,'\|^2$, which is not feasible at the maximum. Consequently, at the maximum of $\|\vec{\tau}\,'\|^2$ in Eq. (10.91), necessarily $\|\vec{\tau}\| = 1$; if it were not so, a uniform scaling of $\vec{\tau}$ to reach $\|\vec{\tau}\| = 1$ would also increase $\|\vec{\tau}\,'\|^2$, which is not feasible at the maximum. For $\vec{\tau}$ in Eq. (10.88), this necessary condition $\|\vec{\tau}\| = 1$ can only be obtained by the choice $\vec{r}_0 = -\vec{r}_1$ and $\|\vec{r}_0\| = \|\vec{r}_1\| = 1$, indicating necessarily two orthogonal pure signaling states $\{\rho_0 = |\psi_0\rangle\langle\psi_0|, \rho_1 = |\psi_1\rangle\langle\psi_1|\}$ to minimize the probability of detection error $P_{\rm er}^{\rm min}$ of Eqs. (10.89)–(10.90) achieved by the maximum of $\|\vec{\tau}\,'\|^2$ in Eq. (10.91). The same conclusion could be reached by a more direct convexity argument, with $P_{\rm er}^{\rm min}$ a concave ($\cap$) function of $\vec{\tau}$, its minimum occurs on the frontier of the permitted domain where $\|\vec{\tau}\| = 1$.

To further characterize the optimal $\vec{\tau}$ maximizing $\|\vec{\tau}\,'\|^2$ in Eq. (10.91), we refer to a property mentioned at the end of Sect. 3.2.11. By polar decomposition we have $A = UG$, with U a real unitary matrix, and G a real symmetric matrix with three real eigenvalues $\{g_1, g_2, g_3\}$ and three eigenvectors $\{\vec{g}_1, \vec{g}_2, \vec{g}_3\}$ forming an orthonormal basis of $\mathbb{R}^3$. The transformation of the test vector $\vec{\tau}$ in Eq. (10.87) is thus a contraction by G along the axes $\{\vec{g}_1, \vec{g}_2, \vec{g}_3\}$ followed by a real unitary transformation U and a translation by $\vec{c}\,'$. It is convenient to express the two vectors $\vec{\tau} = [\tau_1, \tau_2, \tau_3]^\top$ and $\vec{v} = [v_1, v_2, v_3]^\top$ with their coordinates referring to the orthonormal eigenbasis $\{\vec{g}_1, \vec{g}_2, \vec{g}_3\}$ of G. Then it follows that $\|A\vec{\tau}\|^2 = \|G\vec{\tau}\|^2 = \sum_{\ell=1}^3 g_\ell^2 \tau_\ell^2$. As a result, the constrained maximization of $\|\vec{\tau}\,'\|^2$ in Eq. (10.91) can be expressed in the orthonormal eigenbasis $\{\vec{g}_1, \vec{g}_2, \vec{g}_3\}$ of G as

$$\left\{ \begin{aligned} &\max_{\vec{\tau}} \left[C(\vec{\tau}) = \sum_{\ell=1}^{3} \left(g_\ell^2 \tau_\ell^2 + 2 v_\ell \tau_\ell \right) \right], && (10.92) \\ &\text{subject to } \ \|\vec{\tau}\|^2 = \sum_{\ell=1}^{3} \tau_\ell^2 = 1 \,. && (10.93) \end{aligned} \right.$$

A solution $\vec{\tau} = \vec{\tau}^{\,\rm opt}$ solving the constrained maximization of Eqs. (10.92)–(10.93) defines the optimal signaling pair of pure states $\{\rho_0^{\rm opt}, \rho_1^{\rm opt}\}$ by $\{\vec{r}_0 = -\vec{\tau}^{\,\rm opt}, \vec{r}_1 = \vec{\tau}^{\,\rm opt}\}$. The output test vector $\vec{\tau}\,' = A\vec{\tau}^{\,\rm opt} + \vec{c}\,'$ resulting in Eq. (10.87) in turn determines the optimal detection operators $\{\mathsf{E}_0^{\rm opt}, \mathsf{E}_1^{\rm opt}\}$ and associated performance $P_{\rm er}^{\rm min}$ in Eqs. (10.89)–(10.90), maximizing the efficiency for detection from the noisy qubit. Closed-form analytical solutions for the maximizer $\vec{\tau}^{\,\rm opt}$ of Eqs. (10.92)–(10.93) are however not easily accessible in general. For the resolution, one can, for instance, use Eq. (10.93) to obtain $\tau_3^2 = 1 - \tau_1^2 - \tau_2^2$ so as to eliminate τ_3 in Eq. (10.92) and then perform a maximization of Eq. (10.92) in (τ_1, τ_2) over the unit disk $\tau_1^2 + \tau_2^2 \leq 1$. Alternatively, one can define a Lagrangian $\mathcal{L}$ for the system of Eqs. (10.92)–(10.93) as $\mathcal{L} = C(\vec{\tau}\,) + \mu\big(\|\vec{\tau}\,\|^2 - 1\big)$ with Lagrange multiplier μ. The stationarity condition $\partial\mathcal{L}/\partial\tau_\ell = 0$ yields $\tau_\ell = -v_\ell/(g_\ell^2 + \mu)$, for $\ell = 1, 2, 3$; this inserted in the constraint $\|\vec{\tau}\,\|^2 = 1$ then yields an equation in the remaining sole unknown μ. Yet, none of these two approaches allows to reach closed-form analytical solutions in general. However, special conditions or specific noise models important to the qubit lead to closed-form analytical solutions as we are going to see. In addition, numerical resolution of Eqs. (10.92)–(10.93) can always be performed for any given noise model $\{A, \vec{c}\,\}$.

10.5.1 Detection with Unital Qubit Noises

Special configurations where the solution to the maximization of Eqs. (10.92)–(10.93) can be further specified in general analytical terms occur when in Eq. (10.87) the vector $\vec{c}\,' = (P_1 - P_0)\vec{c}$ vanishes. This condition $\vec{c}\,' = \vec{0}$ holds for any noise with $\vec{c} = \vec{0}$ in Eq. (10.86). This is the class of unital qubit noises, as defined in Sect. 3.2.11, and in this respect we encounter here an interesting property of unital qubit noises as to afford an analytical resolution to the maximization of Eqs. (10.92)–(10.93). In particular, the class of unital qubit noises contains the whole class of Pauli noises of Sect. 9.7. The condition $\vec{c}\,' = \vec{0}$ also holds, for any qubit noise, when $P_0 = P_1 = 1/2$ with two equal prior probabilities.

When $\vec{c}\,' = \vec{0}$ in Eq. (10.87), the optimal $\vec{\tau}$ maximizing $\|\vec{\tau}\,'\|^2$ in Eq. (10.91), or equivalently solving Eqs. (10.92)–(10.93), is a unit vector $\vec{\tau}^{\,\mathrm{opt}}$ matching the eigenvector $\vec{g}_\ell$ having the eigenvalue with maximum modulus $|g_\ell|$. This dominant eigenvector identifies the direction in $\mathbb{R}^3$ of least contraction by the noise. It achieves the maximum $\|\vec{\tau}\,'\|_{\max} = |g_\ell|$ to be used in Eqs. (10.89)–(10.90) to express the minimal probability of detection error $P_{\mathrm{er}}^{\min}$. This $\vec{\tau}^{\,\mathrm{opt}}$ defines the optimal input signaling pair $\{\rho_0^{\mathrm{opt}}, \rho_1^{\mathrm{opt}}\}$ by $\{\vec{r}_0 = -\vec{\tau}^{\,\mathrm{opt}}, \vec{r}_1 = \vec{\tau}^{\,\mathrm{opt}}\}$, modulo the change $\vec{\tau}^{\,\mathrm{opt}} \to -\vec{\tau}^{\,\mathrm{opt}}$ for an equivalent solution. The associated optimal detection operators $\{\mathsf{E}_0^{\mathrm{opt}}, \mathsf{E}_1^{\mathrm{opt}}\}$ from the noisy output are formed by the two projectors $\big(\mathrm{I}_2 \pm \vec{\tau}^{\,\mathrm{opt}} \cdot \vec{\sigma}\,\big)/2$ as long as $|g_\ell| \geq |P_1 - P_0|$ or otherwise by the identity I_2 for always deciding for the maximum prior P_0 or P_1. This maximizes the performance in minimal-error detection from the noisy qubit.

For instance, with the general Pauli noises of Sect. 9.7, the matrix A of Eq. (9.65) is already in diagonal form, so that the dominant eigenvector $\vec{\tau}^{\,\mathrm{opt}} = \vec{g}_\ell$ solving Eqs. (10.92)–(10.93) is one of $\{\vec{e}_x, \vec{e}_y, \vec{e}_z\}$, the standard orthonormal basis of $\mathbb{R}^3$, with the corresponding eigenvalue g_ℓ determined in Eqs. (9.66)–(9.68) by the largest $|g_k| = |a_k|$ for $k = x, y$ or z.

With the phase-flip (Pauli) noise of Sect. 9.4, the dominant eigenvector solving Eqs. (10.92)–(10.93) is $\vec{\tau}^{\,\text{opt}} = \vec{e}_z$ corresponding to the maximum eigenvalue 1 of A in Eq. (9.31). The solution $\vec{\tau}^{\,\text{opt}} = \vec{e}_z$ identifies an optimal input signaling pair $\left\{\rho_0^{\text{opt}} = |0\rangle\langle 0|\,,\ \rho_1^{\text{opt}} = |1\rangle\langle 1|\right\}$, and the maximum eigenvalue 1 leads in Eqs. (10.89)–(10.90) to the minimal probability of error $P_{\text{er}}^{\min} = 0$, achieved by the optimal pair of operators $\left\{\mathsf{E}_0^{\text{opt}} = |0\rangle\langle 0|\,,\ \mathsf{E}_1^{\text{opt}} = |1\rangle\langle 1|\right\}$ for detection from the noisy qubit, whatever the level p of the phase-flip noise in Eq. (9.29). This indicates that using the optimal input signaling pair $\left\{\rho_0^{\text{opt}} = |0\rangle\langle 0|\,,\ \rho_1^{\text{opt}} = |1\rangle\langle 1|\right\}$ along with the optimal minimal-error detector $\left\{\mathsf{E}_0^{\text{opt}},\ \mathsf{E}_1^{\text{opt}}\right\}$ from the noisy output affords the possibility of transmission without error over a noisy communication channel corrupted by phase-flip noise. This possibility of signaling without error is due to the existence of an invariant axis or noise-free axis in $\mathbb{R}^3$ (the Oz axis for the phase-flip noise), along which the Bloch vector $\vec{r}$ experiences no contraction by the noise matrix A in Eq. (9.31). The two antipodal unit Bloch vectors along this invariant axis ($\pm\vec{e}_z$ for the phase-flip noise) are associated with two orthogonal pure qubit states in $\mathcal{H}_2$ (the states $|\psi_0\rangle = |0\rangle$ and $|\psi_1\rangle = |1\rangle$ for the phase-flip noise) enabling an optimal signaling pair $\left\{\rho_0^{\text{opt}} = |\psi_0\rangle\langle\psi_0|\,,\ \rho_1^{\text{opt}} = |\psi_1\rangle\langle\psi_1|\right\}$ that is both unaffected by the noise and perfectly distinguishable by the optimal detector $\left\{\mathsf{E}_0^{\text{opt}} = |\psi_0\rangle\langle\psi_0|\,,\ \mathsf{E}_1^{\text{opt}} = |\psi_1\rangle\langle\psi_1|\right\}$.

A similar possibility of error-free transmission exists with the bit-flip noise of Eq. (9.21) of invariant axis Ox and optimal signaling pair $\left\{|\psi_0\rangle\,, |\psi_1\rangle\right\} = \left\{|+\rangle\,, |-\rangle\right\}$, and with the bit-phase-flip noise of Eq. (9.45) of invariant axis Oy and optimal signaling pair $\left\{|\psi_0\rangle\,, |\psi_1\rangle\right\} = \left\{|+i\rangle\,, |-i\rangle\right\}$ from Eqs. (2.13)–(2.14). Accessing such possibility of error-free transmission requires to be able both to signal with the optimal input pair and to detect with the optimal detector. On the contrary, no possibility of error-free transmission exists with the depolarizing noise of Eq. (9.55), having no invariant axis in $\mathbb{R}^3$. Due to the isotropic contraction by A of Eq. (9.55), any pair of orthogonal pure states $\left\{|\psi_0\rangle\,, |\psi_1\rangle\right\}$ is an optimal input signaling pair, achieving $\|\vec{\tau}\,'\|_{\max} = |\alpha|$ for the minimal probability of error $P_{\text{er}}^{\min}$ in Eqs. (10.89)–(10.90). In addition, as noted after Eqs. (10.70)–(10.71), with sufficiently strong noise at small contraction factor α, the contrast of the two noisy states may diminish to satisfy $\|\vec{\tau}\,'\|_{\max} = |\alpha| < |P_1 - P_0|$ so that Eq. (10.90) holds; then no measurement of the optimally prepared qubit affected by depolarizing noise will aid to improve the discrimination performance over the trivial detector that always decides the state with the largest prior probability.

10.5.2 Detection with Qubit Thermal Noise

With the non-unital generalized amplitude damping noise or quantum thermal noise of Sect. 9.10, the transformation of the test Bloch vector in Eq. (10.87) is governed by the pair $\{A, \vec{c}\}$ defined in Eq. (9.91), with $\vec{c} = [0, 0, c_z]^\top \neq \vec{0}$ and a matrix A diagonal in the standard orthonormal basis $\left\{\vec{e}_x, \vec{e}_y, \vec{e}_z\right\}$ of $\mathbb{R}^3$. The scalar criterion of Eq. (10.92) is

$$C(\vec{\tau}\,) = (1-\gamma)\big(\tau_x^2+\tau_y^2\big) + (1-\gamma)^2\tau_z^2 + 2(1-\gamma)(P_1-P_0)c_z\tau_z\,, \tag{10.94}$$

and its constrained maximization is accomplished by

$$\tau_z = \tau_z^{\rm opt} = (P_1-P_0)(2p-1), \quad \tau_x^2+\tau_y^2 = 1-\tau_z^2\,. \tag{10.95}$$

Due to the symmetry in the plane (Ox, Oy) of the thermal noise model of Eq. (9.91), some freedom remains in the two components τ_x and τ_y for a valid solution in Eq. (10.95). The solution from Eq. (10.95) achieves the maximum

$$\|\vec{\tau}\,'\|^2_{\max} = 1-\gamma\Big[1-\big(\tau_z^{\rm opt}\big)^2\Big] \tag{10.96}$$

to be used in Eqs. (10.89)–(10.90) to obtain the minimal probability of error $P_{\rm er}^{\min}$ achievable with the optimal input signaling pair of pure states defined by $\vec{\tau}^{\,\rm opt}$.

As we have seen in Sect. 9.10, in the quantum thermal noise the temperature T of the thermal bath is conveyed by the probability p present in Eq. (9.91). Meanwhile, the coupling factor γ reflects the interaction time of the qubit with the thermal bath. The optimal input signaling pair $\{\rho_0^{\rm opt}, \rho_1^{\rm opt}\}$, which is determined by $\vec{\tau}^{\,\rm opt}$, is found by Eq. (10.95) independent of the coupling γ, and dependent on the temperature T except when $P_0 = P_1$. The optimal detection operators $\{\mathsf{E}_0^{\rm opt}, \mathsf{E}_1^{\rm opt}\}$ and associated performance $P_{\rm er}^{\min}$, which are determined by $\vec{\tau}\,' = A\vec{\tau}^{\,\rm opt} + \vec{c}\,'$, are dependent on the coupling γ and are dependent on the temperature T except when $P_0 = P_1$.

In the important special case where $P_0 = P_1$, we have $\tau_z^{\rm opt} = 0$ and $\|\vec{\tau}\,'\|^2_{\max} = 1-\gamma$. The minimal probability of error follows as $P_{\rm er}^{\min} = \big(1-\sqrt{1-\gamma}\big)/2$ and is independent of the temperature T of the thermal noise. Any pair $\big\{|\psi_0\rangle, |\psi_1\rangle\big\}$ of two orthogonal pure states with Bloch vectors in the plane (Ox, Oy) is an optimal input signaling pair, for instance, $\big\{|\psi_0\rangle = |+\rangle, |\psi_1\rangle = |-\rangle\big\}$, and the associated optimal detection operators are the projectors $\big\{\mathsf{E}_0^{\rm opt} = |\psi_0\rangle\langle\psi_0|, \mathsf{E}_1^{\rm opt} = |\psi_1\rangle\langle\psi_1|\big\}$. At long interaction time with the thermal bath, the coupling $\gamma \to 1$ and $P_{\rm er}^{\min} \to 1/2$ since the two pure signaling states (orthogonal at the input) terminate at the output in the same thermalized mixed state ρ_∞ and become useless for the detection task.

In the broader situation of the squeezed generalized amplitude damping noise of Sect. 9.11, the optimal quantum detector and its performance are characterized in [221].

10.6 Quantum Discrimination Between Multiple States

The notions we have analyzed, concerning quantum detection or discrimination of two quantum states, set an important reference and can also be extended to a larger number of states among which to discriminate. We can consider an N-dimensional quantum system of Hilbert space $\mathcal{H}_N$, which can be in one of $J \ge 2$ alternative states $\rho_j \in \mathcal{L}(\mathcal{H}_N)$, for $j = 1$ to J, with prior probabilities P_j with $\sum_{j=1}^{J} P_j = 1$.

Discrimination is performed by measuring the system via a POVM with J elements $\{\mathsf{E}_k\}_{k=1}^J$, in such a way that if the operator E_k is measured, then the state ρ_k is decided. On occasion, E_k can even be a null operator, if it is more or optimally efficient to never decide for some definite state ρ_k. According to the preparation ρ_j, the various measurement outcomes are obtained with conditional probabilities similar to Eq. (10.33). The performance can be evaluated with the probability of successful discrimination:

$$P_{\text{suc}} = \sum_{j=1}^{J} \text{tr}\big(\mathsf{E}_j\, \rho_j\big) P_j \,. \tag{10.97}$$

It is then appropriate to seek the optimal quantum measurement characterized by the J discrimination operators E_k, subject to $0 \leq \mathsf{E}_k \leq \mathrm{I}_N$ and $\sum_{k=1}^J \mathsf{E}_k = \mathrm{I}_N$, in order to maximize the probability of success P_{suc} of Eq. (10.97). This equivalently minimizes the probability of a discrimination error $P_{\text{er}} = 1 - P_{\text{suc}}$.

For multiple-state discrimination, the constrained optimization of the performance, such as P_{suc} of Eq. (10.97), is in general difficult to perform analytically, as reflected in [244, 245]. As a general property, it is known [235, 244, 246] that the optimal measurement maximizing the probability of success P_{suc} of Eq. (10.97) must satisfy the necessary and sufficient conditions

$$\sum_{j=1}^{J} P_j\, \rho_j \mathsf{E}_j \geq P_k\, \rho_k, \quad \forall k \,, \tag{10.98}$$

from which another set of necessary and sufficient conditions can be derived as

$$\mathsf{E}_j\big(P_j\, \rho_j - P_k\, \rho_k\big)\mathsf{E}_k = 0, \quad \forall j, k \,. \tag{10.99}$$

The optimal measurement is explicitly known in certain special cases [244, 245], especially for qubit states [247–249]. For example, [250] addresses the minimal-error discrimination of $J = 4$ mixed qubit states with arbitrary prior probabilities P_j, it characterizes the optimal measuring POVM with J (non-null) elements, but also reports the possibility of conditions with a degenerate optimal POVM incorporating null elements or equivalently a number $< J$ of non-null elements. In addition, general lower and upper bounds have been found for the optimal performance in discriminating arbitrary states [244, 251]. Alternatively, numerical methods have been developed to compute optimal solutions to multiple-state discrimination [252, 253], mainly resorting to semidefinite programming. The subject of discrimination among multiple quantum states is further investigated, for instance, in [254–260] but is today not yet fully solved in all relevant aspects.

There exists a slightly different approach to quantum discrimination with multiple (at least three) possible decisions, known as unambiguous or error-free discrimination, and which has attracted much attention. We will now examine this problem in more detail, as an example of a meaningful quantum discrimination task with multiple possible decisions.

10.7 Error-Free or Unambiguous Quantum Discrimination

Error-free or unambiguous discrimination applies to $J \geq 2$ quantum states $\rho_j \in \mathcal{L}(\mathcal{H}_N)$ that typically are not of full rank, except possibly one at most that may be of full rank, or equivalently each support $\text{supp}(\rho_j)$ is strictly contained in $\mathcal{H}_N$, i.e., $\text{supp}(\rho_j) \subset \mathcal{H}_N$, except possibly for one. When in addition the J supports are sufficiently distinct or non-overlapping, it becomes possible to envisage a measurement set via a POVM comprising discrimination operators that, when measured (when they act to produce the measurement outcome), identify a state ρ_j unambiguously or with no error, and supplemented by an inconclusive POVM operator indicating the circumstance where the measurement is unable to identify unambiguously a given ρ_j. The measurement set or POVM typically contains $J + 1$ operators (or sometimes less, but with always one inconclusive operator), among which J operators (or less) that when measured identify a given state ρ_j with no error, and one inconclusive operator that when measured signals that no error-free decision can be taken.

For illustration we will consider the important reference task of error-free or unambiguous discrimination of two pure quantum states in $\mathcal{H}_N$, which especially can be worked out analytically. Such task was first considered for two pure states with equal prior probabilities in [261–263], and extended to arbitrary priors in [264].

10.7.1 Error-Free Discrimination of Two Pure States

Two pure quantum states $|\psi_0\rangle$ and $|\psi_1\rangle$ of $\mathcal{H}_N$ have (rank-one) density operators $\rho_0 = |\psi_0\rangle\langle\psi_0|$ and $\rho_1 = |\psi_1\rangle\langle\psi_1|$ and prior probabilities, respectively, P_0 and $P_1 = 1 - P_0$. In Sect. 10.3, we have treated the detection of two such quantum states, with a binary measurement protocol which is always conclusive but prone to some error, and we have characterized the optimal measurement protocol having minimal probability of error. For error-free detection or discrimination, we now envisage a ternary measurement protocol, which is not always conclusive, but never makes error. The measurement operates via a POVM with three (positive) elements or discrimination operators $\{\mathsf{E}_0, \mathsf{E}_1, \mathsf{E}_{\text{inc}}\}$ summing to the identity I_N. The discrimination operators are devised in such a way that when the outcome corresponding to E_j is obtained, then it is certain that the quantum system is in state ρ_j, for $j = 0, 1$; in conjunction, the outcome corresponding to E_{inc} identifies the circumstance where no ascertained decision can be returned.

As for the binary detection of Sect. 10.3, the ternary error-free discrimination can essentially be solved in the vector plane spanned by $|\psi_0\rangle$ and $|\psi_1\rangle$. Also, as in Eq. (10.60), it is possible, without loss of generality, to use the parameterization

$$|\psi_1\rangle = \alpha_0 |\psi_0\rangle + \alpha_1 |\psi_0^\perp\rangle , \tag{10.100}$$

with the state $|\psi_0^\perp\rangle$ orthogonal to $|\psi_0\rangle$ in the plane $\big(|\psi_0\rangle , |\psi_1\rangle\big)$, and the two complex coordinates satisfying $|\alpha_0|^2 + |\alpha_1|^2 = 1$. The inner product $\langle\psi_0|\psi_1\rangle = \alpha_0$ reflecting the relative configuration of $|\psi_0\rangle$ and $|\psi_1\rangle$ in $\mathcal{H}_N$ is again an essential parameter for expressing the solution. We also introduce

$$|\psi_1^\perp\rangle = \alpha_1^* |\psi_0\rangle - \alpha_0^* |\psi_0^\perp\rangle , \tag{10.101}$$

as the state orthogonal to $|\psi_1\rangle$ in the plane $\big(|\psi_0\rangle , |\psi_1\rangle\big)$.

The two orthogonal projectors $|\psi_0^\perp\rangle\langle\psi_0^\perp|$ and $|\psi_1^\perp\rangle\langle\psi_1^\perp|$ can now be used to construct the two discrimination operators as

$$\begin{cases} \mathsf{E}_0 = a_0 |\psi_1^\perp\rangle\langle\psi_1^\perp| , & (10.102) \\ \mathsf{E}_1 = a_1 |\psi_0^\perp\rangle\langle\psi_0^\perp| , & (10.103) \end{cases}$$

with real coefficient $a_j \in [0, 1]$ to ensure $0 \le \mathsf{E}_j \le \mathrm{I}_N$, for $j = 0, 1$. We then obtain $\mathsf{E}_0 |\psi_1\rangle = 0$, and $\langle\psi_1|\mathsf{E}_0|\psi_1\rangle = 0$ for the probability of measuring E_0 when the state is $|\psi_1\rangle$. Thus, upon measurement of the quantum system with the POVM $\{\mathsf{E}_0, \mathsf{E}_1, \mathsf{E}_{\text{inc}}\}$, when the outcome corresponding to E_0 is obtained, we are sure that the system is not in the state $|\psi_1\rangle$, i.e., that it is in the state $|\psi_0\rangle$. In a similar way, since $\mathsf{E}_1 |\psi_0\rangle = 0$ and $\langle\psi_0|\mathsf{E}_1|\psi_0\rangle = 0$, when the measurement outcome corresponding to E_1 is obtained, we are sure that the system is in the state $|\psi_1\rangle$, since it cannot be in the state $|\psi_0\rangle$. In this way, for any a_0, a_1 in $[0, 1]$, the two operators E_0, E_1 of Eqs. (10.102)–(10.103) are two positive discrimination operators returning unambiguous or error-free decisions. The three discrimination operators $\{\mathsf{E}_0, \mathsf{E}_1, \mathsf{E}_{\text{inc}}\}$ to constitute a valid POVM need to be three positive operators summing to the identity operator I_N, so that we deduce the third, inconclusive, discrimination operator as

$$\mathsf{E}_{\text{inc}} = \mathrm{I}_N - \mathsf{E}_0 - \mathsf{E}_1 . \tag{10.104}$$

We must select the proportionality coefficients (a_0, a_1) to ensure that E_{inc} of Eq. (10.104) remains a positive operator satisfying $0 \le \mathsf{E}_{\text{inc}} \le \mathrm{I}_N$. For the complete determination of the solution, we then look for the optimum POVM $\{\mathsf{E}_0, \mathsf{E}_1, \mathsf{E}_{\text{inc}}\}$ maximizing the discrimination performance.

The probability of obtaining a successful discrimination is

$$\begin{aligned} P_{\text{suc}} &= \mathrm{tr}(\mathsf{E}_0\rho_0) P_0 + \mathrm{tr}(\mathsf{E}_1\rho_1) P_1 & (10.105) \\ &= \langle\psi_0|\mathsf{E}_0|\psi_0\rangle\, P_0 + \langle\psi_1|\mathsf{E}_1|\psi_1\rangle\, P_1 , & (10.106) \end{aligned}$$

or, alternatively, the probability of obtaining a inconclusive result is

$$P_{\text{inc}} = 1 - P_{\text{suc}} = \langle\psi_0|\mathsf{E}_{\text{inc}}|\psi_0\rangle\, P_0 + \langle\psi_1|\mathsf{E}_{\text{inc}}|\psi_1\rangle\, P_1 . \tag{10.107}$$

By Eqs. (10.102)–(10.103), one has $\langle\psi_0|\mathsf{E}_0|\psi_0\rangle = a_0|\langle\psi_0|\psi_1^\perp\rangle|^2$ and $\langle\psi_1|\mathsf{E}_1|\psi_1\rangle = a_1|\langle\psi_1|\psi_0^\perp\rangle|^2$, so that $P_{\rm suc}$ of Eq. (10.106) is also

$$P_{\rm suc} = a_0|\langle\psi_0|\psi_1^\perp\rangle|^2 P_0 + a_1|\langle\psi_1|\psi_0^\perp\rangle|^2 P_1 \ . \tag{10.108}$$

The inner products $\langle\psi_0|\psi_1^\perp\rangle = \alpha_1^*$ and $\langle\psi_1|\psi_0^\perp\rangle = \alpha_1$ lead then to

$$P_{\rm suc} = a_0|\alpha_1|^2 P_0 + a_1|\alpha_1|^2 P_1 \ . \tag{10.109}$$

We then determine the coefficients (a_0, a_1) in order to maximize the probability of success $P_{\rm suc}$ or equivalently minimize the probability of inconclusive result $P_{\rm inc} = 1 - P_{\rm suc}$, while ensuring $\mathsf{E}_{\rm inc}$ of Eq. (10.104) satisfies $0 \le \mathsf{E}_{\rm inc} \le \mathrm{I}_N$.

The condition $0 \le \mathsf{E}_j \le \mathrm{I}_N$ satisfied by both E_0 and E_1 in Eq. (10.104) guarantees that $\mathsf{E}_{\rm inc}$ in Eq. (10.104) is a Hermitian operator satisfying $\mathsf{E}_{\rm inc} \le \mathrm{I}_N$. It remains to ensure that $\mathsf{E}_{\rm inc}$ is positive, i.e., $\mathsf{E}_{\rm inc} \ge 0$. For this purpose, we consider the restriction of the operator $\mathsf{E}_{\rm inc}$ to the vector plane $\big(|\psi_0\rangle\,,|\psi_1\rangle\big)$ and represented by the matrix

$$\mathsf{E}_{\rm inc} = \begin{bmatrix} 1 - a_0|\alpha_1|^2 & a_0\alpha_0\alpha_1^* \\ a_0\alpha_0^*\alpha_1 & 1 - a_0|\alpha_0|^2 - a_1 \end{bmatrix} \tag{10.110}$$

referred to the orthonormal basis $\{|\psi_0\rangle\,,|\psi_0^\perp\rangle\}$ of the plane $\big(|\psi_0\rangle\,,|\psi_1\rangle\big)$. The positivity of the operator $\mathsf{E}_{\rm inc}$ in Eq. (10.104) is granted by the positivity of the matrix in Eq. (10.110). The 2×2 Hermitian matrix $\mathsf{E}_{\rm inc}$ in Eq. (10.110) is positive if and only if it has two positive eigenvalues, or equivalently $\mathrm{tr}(\mathsf{E}_{\rm inc}) \ge 0$ and $\det(\mathsf{E}_{\rm inc}) \ge 0$. These last two conditions translate into

$$\begin{cases} \mathrm{tr}(\mathsf{E}_{\rm inc}) = 2 - a_0 - a_1 \ge 0\ , & (10.111)\\ \det(\mathsf{E}_{\rm inc}) = 1 - (a_0 + a_1) + a_0a_1|\alpha_1|^2 \ge 0\ . & (10.112)\end{cases}$$

While satisfying the conditions of Eqs. (10.111)–(10.112), the coefficients (a_0, a_1) are settled to maximize the probability of success $P_{\rm suc}$ of Eq. (10.109).

The criterion $P_{\rm suc}$ of Eq. (10.109) is an increasing function of both variables a_0 and a_1 separately. Therefore, in the permitted domain for (a_0, a_1) defined over $[0, 1]^2$ by the constraints of Eqs. (10.111)–(10.112), the maximum of $P_{\rm suc}$ is necessarily located on a boundary, where both a_0 and a_1 can no longer increase separately. If it were not so, taking a step in a direction where a_0 or a_1 alone could be increased would raise $P_{\rm suc}$. Moreover, when the constraint of Eq. (10.112) is satisfied, it can be verified that Eq. (10.111) is also satisfied for any $(a_0, a_1) \in [0, 1]^2$. Therefore Eq. (10.112) is sufficient to delimit the permitted domain for $(a_0, a_1) \in [0, 1]^2$, and the maximum of $P_{\rm suc}$ can be searched on the boundary when this constraint (10.112) is set to equality. This enables to eliminate a_1 as a function of a_0 and substitute in $P_{\rm suc}$ of Eq. (10.109) which can then be readily maximized as a single-variable function in a_0. This leads to the solution maximizing $P_{\rm suc}$ in Eq. (10.109), under the constraints of Eqs. (10.111)–(10.112), similar to [244], as

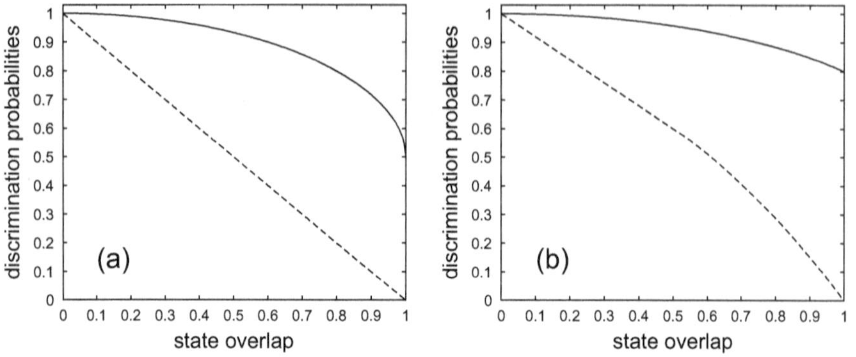

Fig. 10.4 For optimal discrimination of two pure quantum states $|\psi_0\rangle$ and $|\psi_1\rangle$ in the N-dimensional space $\mathcal{H}_N$, as a function of their overlap $|\langle\psi_0|\psi_1\rangle|$, the probability of success $P_{\text{suc}}^{\max}$ from Eq. (10.115) for error-free or unambiguous discrimination with a three-outcome measurement (dashed line), and $P_{\text{suc}}^{\max} = 1 - P_{\text{er}}^{\min}$ from Eq. (10.63) for the minimal-error detector of Sect. 10.3 with an always conclusive two-outcome measurement (solid line). The prior probability is $P_0 = 0.5$ in (**a**), and $P_0 = 0.8$ in (**b**)

$$\begin{cases} a_0^{\text{opt}} = \dfrac{1 - \sqrt{P_1/P_0}\,|\alpha_0|}{|\alpha_1|^2}\,, & (10.113)\\[2ex] a_1^{\text{opt}} = \dfrac{1 - \sqrt{P_0/P_1}\,|\alpha_0|}{|\alpha_1|^2}\,, & (10.114)\end{cases}$$

achieving the maximum probability of successful discrimination:

$$P_{\text{suc}}^{\max} = 1 - 2\sqrt{P_0 P_1}\,|\alpha_0| = 1 - 2\sqrt{P_0 P_1}\,|\langle\psi_0|\psi_1\rangle|\,. \qquad (10.115)$$

This solution ceases to hold when $\sqrt{P_{\max}/P_{\min}}\,|\alpha_0| > 1$ with $P_{\max} = \max(P_0, P_1)$ and $P_{\min} = \min(P_0, P_1)$, in which case the optimal measurement is defined by $\big(a_0^{\text{opt}} = 1, a_1^{\text{opt}} = 0\big)$ for $P_0 = P_{\max}$ or by $\big(a_0^{\text{opt}} = 0, a_1^{\text{opt}} = 1\big)$ for $P_1 = P_{\max}$, and it achieves $P_{\text{suc}}^{\max} = P_{\max}|\alpha_1|^2$.

As anticipated, the solution defined by Eqs. (10.113)–(10.115) bears a general significance independent of the specific planar parameterization of Eq. (10.100) and conveyed via the inner product $\langle\psi_0|\psi_1\rangle$. In Eqs. (10.113)–(10.115), expressing $\alpha_0 = \langle\psi_0|\psi_1\rangle$ and $|\alpha_1|^2 = 1 - |\langle\psi_0|\psi_1\rangle|^2$ restores their general status as the solution to the error-free or unambiguous discrimination of two pure quantum states $|\psi_0\rangle$ and $|\psi_1\rangle$ in the N-dimensional Hilbert space $\mathcal{H}_N$.

It is interesting to compare the performance of Eq. (10.115) with the probability of discrimination success $P_{\text{suc}}^{\max} = 1 - P_{\text{er}}^{\min}$ as it results from Eq. (10.63) for the (always conclusive) minimal-error detector of two pure states of Sect. 10.3. This is accomplished in Fig. 10.4.

As illustrated in Fig. 10.4, the probability of performing a successful discrimination is in general smaller with the ternary error-free detector of Eq. (10.115) than with the binary minimal-error detector of Sect. 10.3. For every measurement performed, the minimal-error detector of Sect. 10.3 always returns a conclusive decision about the state; as a result, it is more often successful for the discrimination, but on occasion it commits errors, by deciding the wrong state. By contrast, the ternary detector of Eq. (10.115) for some performed measurements may return an inconclusive outcome; it is less often successful for the discrimination, but it never commits errors, any conclusive decision it returns about one state or the other is always correct.

10.7.2 Error-Free Discrimination of Two Mixed States

Instead of pure states, we now consider the more general situation where the quantum system with N-dimensional Hilbert space $\mathcal{H}_N$ can be in one of two mixed quantum states represented by two density operators ρ_0 and ρ_1, respectively, with prior probabilities P_0 and $P_1 = 1 - P_0$. We consider the generic situation where ρ_0 and ρ_1 are not of full rank and have nonidentical supports in $\mathcal{H}_N$. One ρ_j with full rank (but not two) would still allow error-free discrimination and is included as a degenerate case of the generic treatment below.

We devise a three-element POVM $\{\mathsf{E}_0, \mathsf{E}_1, \mathsf{E}_{\text{inc}}\}$ for error-free discrimination of the two mixed states ρ_0 and ρ_1 of $\mathcal{L}(\mathcal{H}_N)$. For $j = 0, 1$, we consider the support $\text{supp}(\rho_j)$ of ρ_j which is a subspace of $\mathcal{H}_N$. Its orthogonal complement $[\text{supp}(\rho_j)]^\perp = \ker(\rho_j)$ is the kernel or null space of ρ_j, forming another subspace of $\mathcal{H}_N$. The orthogonal projector onto the subspace $\ker(\rho_j)$ is denoted Π_j, for $j = 0, 1$. We can then select the discrimination operator $\mathsf{E}_0 = a_0\Pi_1$ with $a_0 \in [0, 1]$ to ensure $0 \leq \mathsf{E}_0 \leq \mathrm{I}_N$. The probability of measuring E_0 when the state is ρ_1 is the trace $\text{tr}(\mathsf{E}_0\rho_1) = a_0 \,\text{tr}(\Pi_1\rho_1) = 0$, this is because $\rho_1\Pi_1$ and $\Pi_1\rho_1$ are the null operator mapping any vector of $\mathcal{H}_N$ into the null vector. Thus, upon measurement of the quantum system with the POVM $\{\mathsf{E}_0, \mathsf{E}_1, \mathsf{E}_{\text{inc}}\}$, when the outcome corresponding to E_0 is obtained, we are sure that the system is not in state ρ_1, i.e., that it is in state ρ_0. In a similar way, we select the discrimination operator $\mathsf{E}_1 = a_1\Pi_0$ with $a_1 \in [0, 1]$, and when E_1 is measured we are sure that the system is in state ρ_1. The remaining, inconclusive, discrimination operator is $\mathsf{E}_{\text{inc}} = \mathrm{I}_N - \mathsf{E}_0 - \mathsf{E}_1$ as in Eq. (10.104). The two proportionality coefficients $(a_0, a_1) \in [0, 1]^2$ are chosen to maximize as in Eq. (10.105) the probability of successful discrimination:

$$P_{\text{suc}} = \text{tr}(\mathsf{E}_0\rho_0)P_0 + \text{tr}(\mathsf{E}_1\rho_1)P_1 \;, \tag{10.116}$$

or equivalently minimize the probability of an inconclusive result $P_{\text{inc}} = 1 - P_{\text{suc}}$, while maintaining $0 \leq \mathsf{E}_{\text{inc}} \leq \mathrm{I}_N$.

Unambiguous or error-free discrimination for two mixed states was first considered in [265]. Reduction theorems given in [266, 267] show that the general problem can always be reduced to discriminating two states each of rank r jointly spanning a $2r$-dimensional space. The simplest case that is not reducible to pure state discrimination is the problem of two rank-2 density operators in a four-dimensional space, which is analyzed in [268, 269]. Also [244] offers an overview. According to the vision emerging from [244, 269], the explicit characterization of the general solution to unambiguous discrimination of two mixed states is not fully completed, but solutions are known for some special cases, as the one we now examine.

In the degenerate case where one ρ_j, say ρ_0, is of full rank, then the null space $\ker(\rho_0)$ is reduced to the null vector, the projector Π_0 and POVM element $\mathsf{E}_1 = a_1\Pi_0$ are both the null operator, and consequently E_1 is never measured and ρ_1 is never decided, since it could be confused with ρ_0 by $\mathrm{supp}(\rho_1) \subset \mathrm{supp}(\rho_0) \equiv \mathcal{H}_N$. The POVM reduces to $\{\mathsf{E}_0, \mathsf{E}_{\mathrm{inc}}\}$. The probability of success in Eq. (10.116) becomes $P_{\mathrm{suc}} = \mathrm{tr}(\mathsf{E}_0\rho_0)P_0$. It is maximized with $\mathsf{E}_0 = \Pi_1$ projecting onto $\ker(\rho_1)$, and $\mathsf{E}_{\mathrm{inc}} = \Pi_1^{\perp}$ the projector onto $\mathrm{supp}(\rho_1)$. Consequently, each time the preparation is ρ_1 the inconclusive $\mathsf{E}_{\mathrm{inc}}$ is measured. On some occasions, $\mathsf{E}_{\mathrm{inc}}$ can be measured also when the preparation is ρ_0. But when E_0 is measured, it is certain that the preparation is ρ_0.

Two mixed states with one-dimensional null spaces.
A special case is when the two mixed states ρ_0 and ρ_1 have one-dimensional (distinct) null spaces in $\mathcal{H}_N$, amenable to a solution formally similar to the discrimination of two pure states of Sect. 10.7.1, as we now report based on [265].

For the one-dimensional null spaces $\ker(\rho_0)$ and $\ker(\rho_1)$, we introduce $|u_0\rangle$ and $|u_1\rangle$ as two (non-collinear) unit vectors of $\mathcal{H}_N$ spanning $\ker(\rho_0)$ and $\ker(\rho_1)$, respectively. Without loss of generally, analogously with Eq. (10.100), we can use in the vector plane $\big(|u_0\rangle, |u_1\rangle\big)$ of $\mathcal{H}_N$ the parameterization:

$$|u_1\rangle = \alpha_0 |u_0\rangle + \alpha_1 |u_0^{\perp}\rangle \ , \tag{10.117}$$

with the unit vector $|u_0^{\perp}\rangle$ orthogonal to $|u_0\rangle$ in the plane $\big(|u_0\rangle, |u_1\rangle\big)$, and the two complex coordinates satisfying $|\alpha_0|^2 + |\alpha_1|^2 = 1$. The inner product $\langle u_0|u_1\rangle = \alpha_0$ is analogously the essential parameter reflecting the relative configuration of $|u_0\rangle$ and $|u_1\rangle$ in $\mathcal{H}_N$, and therefrom of ρ_0 and ρ_1.

The discrimination operator E_0 is chosen proportional to the projector $|u_1\rangle\langle u_1| = \Pi_1$ onto $\ker(\rho_1)$, i.e., $\mathsf{E}_0 = a_0|u_1\rangle\langle u_1|$ with $a_0 \in [0, 1]$, and similarly $\mathsf{E}_1 = a_1|u_0\rangle\langle u_0|$ with $a_1 \in [0, 1]$. For the measurement probabilities occurring in P_{suc} of Eq. (10.116), one has $\mathrm{tr}(\mathsf{E}_0\rho_0) = a_0\,\mathrm{tr}\big(|u_1\rangle\langle u_1|\rho_0\big) = a_0\langle u_1|\rho_0|u_1\rangle$ and $\mathrm{tr}(\mathsf{E}_1\rho_1) = a_1\langle u_0|\rho_1|u_0\rangle$, yielding

$$\begin{aligned} P_{\mathrm{suc}} &= a_0\langle u_1|\rho_0|u_1\rangle P_0 + a_1\langle u_0|\rho_1|u_0\rangle P_1 &\quad (10.118)\\ &= a_0 Q_0 + a_1 Q_1 \ , &\quad (10.119)\end{aligned}$$

with the two real scalars

$$\begin{cases} Q_0 = \langle u_1|\rho_0|u_1\rangle P_0 \ , & (10.120)\\ Q_1 = \langle u_0|\rho_1|u_0\rangle P_1 \ . & (10.121)\end{cases}$$

Equations (10.118)–(10.119) are similar in form to Eqs. (10.108)–(10.109), with $Q_0 \in [0, 1]$ similar to $|\alpha_1|^2 P_0$ and $Q_1 \in [0, 1]$ similar to $|\alpha_1|^2 P_1$.

We now seek the two coefficients $(a_0, a_1) \in [0, 1]^2$ maximizing $P_{\rm suc}$ in Eqs. (10.118)–(10.119) while maintaining $0 \le \mathsf{E}_{\rm inc} \le \mathrm{I}_N$. The situation of $\mathsf{E}_{\rm inc} = \mathrm{I}_N - \mathsf{E}_0 - \mathsf{E}_1$ is again controlled by its restriction to the vector plane $\big(|u_0\rangle, |u_1\rangle\big)$ represented by the matrix

$$\mathsf{E}_{\rm inc} = \begin{bmatrix} 1 - a_0|\alpha_0|^2 - a_1 & -a_0\alpha_0\alpha_1^* \\ -a_0\alpha_0^*\alpha_1 & 1 - a_0|\alpha_1|^2 \end{bmatrix} \tag{10.122}$$

referred to the orthonormal basis $\{|u_0\rangle, |u_0^\perp\rangle\}$ of the plane $\big(|u_0\rangle, |u_1\rangle\big)$. Equation (10.122) is analogous to Eq. (10.110). The requirements on the operator $\mathsf{E}_{\rm inc}$ boil down again to requiring the positivity of the matrix in Eq. (10.122), which is obtained by exactly the same conditions as in Eqs. (10.111)–(10.112).

Maximizing $P_{\rm suc}$ of Eq. (10.118)–(10.119) subject to the constraints of Eqs. (10.111)–(10.112) is a problem formally similar to the constrained maximization of Eq. (10.109), and it therefore receives a solution similar to Eqs. (10.113)–(10.115). Mutatis mutandis, the solution maximizing $P_{\rm suc}$ of Eqs. (10.118)–(10.119) subject to the constraints of Eqs. (10.111)–(10.112) is

$$\begin{cases} a_0^{\rm opt} = \dfrac{1 - \sqrt{Q_1/Q_0}\,|\alpha_0|}{|\alpha_1|^2}, & (10.123) \\[2ex] a_1^{\rm opt} = \dfrac{1 - \sqrt{Q_0/Q_1}\,|\alpha_0|}{|\alpha_1|^2}, & (10.124) \end{cases}$$

achieving the maximum probability of successful discrimination

$$P_{\rm suc}^{\max} = 1 - 2\frac{\sqrt{Q_0 Q_1}}{|\alpha_1|^2}|\alpha_0| \tag{10.125}$$

$$= 1 - 2\frac{\sqrt{P_0 P_1 \langle u_1|\rho_0|u_1\rangle \langle u_0|\rho_1|u_0\rangle}}{1 - |\langle u_0|u_1\rangle|^2} |\langle u_0|u_1\rangle|, \tag{10.126}$$

and paralleling Eqs. (10.113)–(10.115). This solution ceases to hold when $\sqrt{Q_{\max}/Q_{\min}}\,|\alpha_0| > 1$ with $Q_{\max} = \max(Q_0, Q_1)$ and $Q_{\min} = \min(Q_0, Q_1)$, in which case the optimal measurement is defined by $\big(a_0^{\rm opt} = 1, a_1^{\rm opt} = 0\big)$ for $Q_0 = Q_{\max}$ or by $\big(a_0^{\rm opt} = 0, a_1^{\rm opt} = 1\big)$ for $Q_1 = Q_{\max}$, and it achieves $P_{\rm suc}^{\max} = Q_{\max}$.

We can view the solution of Eqs. (10.123)–(10.125) for optimal error-free or unambiguous discrimination of two mixed states ρ_0, ρ_1 with one-dimensional null spaces, as a general case from which the solution of Eqs. (10.113)–(10.115) for two pure states $|\psi_0\rangle$, $|\psi_1\rangle$ can be recovered, when $\rho_j = |\psi_j\rangle\langle\psi_j|$ and $|u_j\rangle = |\psi_j^\perp\rangle$, for $j = 0, 1$.

10.7.3 Error-Free Discrimination of More Than Two States

We can consider error-free or unambiguous discrimination of a number $J > 2$ of quantum states, represented in general by J density operators $\rho_j \in \mathcal{L}(\mathcal{H}_N)$, with respective prior probabilities P_j, for $j = 1$ to J. For error-free discrimination to be possible, typically every ρ_j is expected to be with a defective rank $< N$, except at most one that may possibly be of full rank. If more than one ρ_j were of full rank, none of the states ρ_j could be ascertained with no risk of error. Error-free discrimination can be envisaged by measuring via orthogonal projectors in $\mathcal{H}_N$. A quantum state ρ_j can well be measured (projected) outside its support $\text{supp}(\rho_j)$; the place of $\mathcal{H}_N$ where a state ρ_j cannot be measured (projected) is its null space $\ker(\rho_j)$. The feasibility of error-free discrimination of states ρ_j will therefore be determined by the relative situation of their null spaces $\ker(\rho_j)$; typically they should be nontrivial and sufficiently distinct. Error-free discrimination of a given state ρ_j is feasible when ρ_j can be measured in a nontrivial subspace of the intersection of all the $J - 1$ null spaces of all the other $J - 1$ states. A discrimination operator (a POVM element) E_j accomplishing error-free discrimination of such a state ρ_j can be chosen as proportional to the orthogonal projector onto this subspace intersection of the $J - 1$ null spaces. In this way, when E_j is the projector found to act in the measurement, it is sure that the pre-measurement is ρ_j and cannot be any of the $J - 1$ other states. As before, the probability of successful discrimination to be maximized is

$$P_{\text{suc}} = \sum_{j=1}^{J} \text{tr}(\mathsf{E}_j \rho_j) P_j \,, \tag{10.127}$$

equivalently minimizing the probability of an inconclusive result $P_{\text{inc}} = 1 - P_{\text{suc}}$.

The explicit characterization of the solution to such error-free or unambiguous discrimination of multiple quantum states is not today fully achieved in general, but solutions are known for some special cases [244, 246, 269–272], and also general lower and upper bounds on the probability of success have been found [244, 246, 272]. Alternatively, numerical methods have been developed to compute optimal solutions to unambiguous discrimination [270, 273–275], mainly resorting to semidefinite programming.

10.7.4 Entropy Assessment of the Performance

Error-free discrimination is reminiscent of soft-decision decoding in communication. In this respect, for an assessment of the performance in quantum state discrimination, in place of the probability of success from measuring the output, it can be interesting to examine, between the input preparation and the output decision, the input–output classical mutual information as defined in Sect. 7.1.4. For fixed given input states, if one seeks the optimal quantum measurement maximizing the input–output mutual

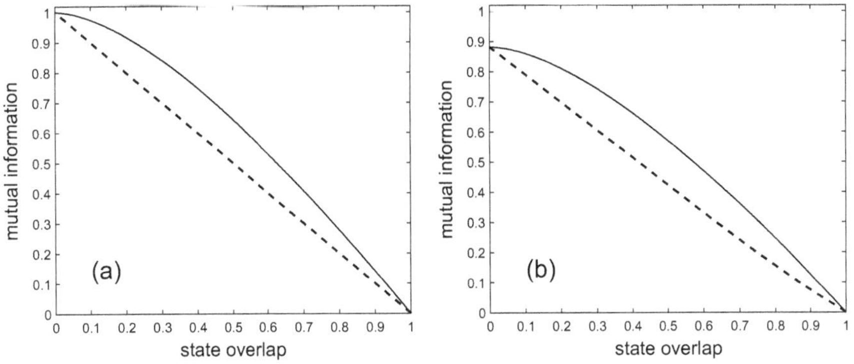

Fig. 10.5 Input–output classical mutual information (in shannons) between the two input states $|\psi_0\rangle$ and $|\psi_1\rangle$ and the output decisions resulting from the quantum measurement, as a function of the state overlap $|\langle\psi_0|\psi_1\rangle|$. Dashed line: with the three-outcome measurement from Eqs. (10.113)–(10.114) achieving optimal unambiguous or error-free discrimination. Solid line: with the two-outcome measurement from Sect. 10.3 achieving minimal-error conclusive discrimination, and standing also as the optimal quantum measurement maximizing the input–output mutual information. The prior probability is $P_0 = 0.5$ in (**a**), and $P_0 = 0.7$ in (**b**)

information, this amounts to computing what is known as the "accessible information" [49, 244, 276, 277], and this will be examined in more generality in Chap. 13 on quantum communication and more specifically its Sect. 13.2.2.

For two pure quantum states, it is known that the mutual information is maximized when the states are prepared with equal probability and when the minimal-error measurement is employed [244]. The input–output mutual information is chiefly determined by the input–output conditional probabilities as those from Eq. (10.33). An illustration is provided in Fig. 10.5, comparing the input–output mutual information for the discrimination of two pure states $|\psi_0\rangle$ and $|\psi_1\rangle$ of the N-dimensional space $\mathcal{H}_N$, by means of the two-outcome minimal-error detector from Sect. 10.3 and by means of the three-outcome error-free detector from Sect. 10.7.1.

It can be seen in Fig. 10.5 that the three outcomes afforded by the measurement optimal for unambiguous or error-free discrimination do not allow an increased mutual information compared with the two-outcome measurement of Sect. 10.3 achieving minimal-error conclusive discrimination, and that is optimal here to maximize the mutual information.

10.8 Square-Root Quantum Detector

10.8.1 General Characterization

For detection or discrimination of multiple quantum states, as we have indicated, there are generally no closed-form analytical solutions known for the optimal measurement associated with an arbitrary set of states to discriminate. As an alternative, there exists an approach that, although usually suboptimal, offers an explicit characterization of a useful quantum measurement from any given set of states to discriminate, as we are going to see now.

We consider as before, a quantum system that can be in one of $J \geq 2$ quantum states represented by the density operators ρ_j, for $j = 1$ to J, with prior probabilities P_j satisfying $\sum_{j=1}^{J} P_j = 1$. Jointly, the supports of these J feasible states ρ_j span a vector space with dimension N, and it is possible to take this N-dimensional space as the working Hilbert space $\mathcal{H}_N$, so that $\rho_j \in \mathcal{L}(\mathcal{H}_N)$ for $j = 1$ to J. This assumption preserves the generality of the treatment, much as in Sect. 10.3 for the detection of two pure states, which was handled in a vector plane without loss of generality.

The average density operator

$$\rho = \sum_{j=1}^{J} P_j \, \rho_j \tag{10.128}$$

is a full-rank operator of $\mathcal{L}(\mathcal{H}_N)$ having $\mathcal{H}_N$ for support. It is a strictly positive operator of $\mathcal{L}(\mathcal{H}_N)$; it therefore has a unique positive square root $\rho^{1/2}$ which is also a full-rank strictly positive operator of $\mathcal{L}(\mathcal{H}_N)$, which is therefore invertible as $\rho^{-1/2}$. A characterization of the square root of the density operator ρ (and its inverse) follows from the considerations of Sect. 4.2.4 on POVM.

To detect the state of the quantum system, a quantum measurement is considered by means of a POVM consisting of J elements taken as the detection or discrimination operators $\mathsf{E}_j \in \mathcal{L}(\mathcal{H}_N)$ defined as

$$\boxed{\mathsf{E}_j = \rho^{-1/2} P_j \, \rho_j \, \rho^{-1/2},} \tag{10.129}$$

for $j = 1$ to J. These are J positive operators defining a valid POVM, especially satisfying

$$\sum_{j=1}^{J} \mathsf{E}_j = \mathrm{I}_N \, , \tag{10.130}$$

matching the identity operator I_N on $\mathcal{H}_N$ as required by the completeness relation of Eq. (4.38). This POVM from Eq. (10.129) is known as the **square-root measurement** [244, 278, 279], and the associated quantum detector deciding ρ_j when E_j is measured is known as the **square-root detector**. A significant advantage is that

Eq. (10.129) stands as an explicitly characterized quantum measurement valid for any set of states $\{(P_j, \rho_j)\}$ to discriminate. Beyond this advantage, the square-root detector from Eq. (10.129) owns no general optimality properties, but it is often "pretty good" in situations of practical interest (and Eq. (10.129) is sometimes referred to also as the "pretty good measurement") [244, 278, 279]. Moreover, the square-root detector is sometimes optimal, coinciding with the minimal-error detector [280].

When the quantum system is in state ρ_j, the conditional probability of measuring the discrimination operator E_k is

$$\Pr\{\mathsf{E}_k|\rho_j\} = \mathrm{tr}(\mathsf{E}_k\,\rho_j)\,, \tag{10.131}$$

for integers j and k in $\{1, 2, \ldots, J\}$, similar to Eq. (10.33) when $J = 2$. If the native Hilbert space of the quantum system was initially larger than $\mathcal{H}_N$, the POVM operators could be supplemented by an additional projector projecting onto the complementary subspace $\mathcal{H}_N^\perp$ orthogonal to $\mathcal{H}_N$ in the larger space. However, the additional measurement outcome of projecting onto $\mathcal{H}_N^\perp$ would never occur, since the state vectors to be measured have no component in $\mathcal{H}_N^\perp$. So this supplementary projector can be omitted, and the working Hilbert space taken as $\mathcal{H}_N$.

From Eq. (10.131), the probability P_{suc} of successful discrimination follows according to Eq. (10.97) as

$$P_{\text{suc}} = \sum_{j=1}^{J} P_j^2\,\mathrm{tr}\Big(\rho^{-1/2}\,\rho_j\,\rho^{-1/2}\,\rho_j\Big)\,. \tag{10.132}$$

One remarkable property of the square-root detector from Eq. (10.129) occurs for discriminating J pure quantum states $|\psi_j\rangle \in \mathcal{H}_N$, spanning the Hilbert space $\mathcal{H}_N$ with dimension $N \le J$. Since $\rho_j = |\psi_j\rangle\langle\psi_j|$, each discrimination operator E_j in Eq. (10.129) follows as a rank-one operator $\mathsf{E}_j = |m_j\rangle\langle m_j|$ of $\mathcal{L}(\mathcal{H}_N)$, defined by the measurement vectors (generally unnormalized)

$$\boxed{|m_j\rangle = \rho^{-1/2}\sqrt{P_j}\,|\psi_j\rangle\,,} \tag{10.133}$$

for $j = 1$ to J, with here

$$\rho = \sum_{j=1}^{J} P_j\,|\psi_j\rangle\langle\psi_j|\,. \tag{10.134}$$

It can then be shown [279] that the square-root detector from Eq. (10.133) minimizes the (weighted) squared error

$$e^2 = \sum_{j=1}^{J}\Big\|\sqrt{P_j}\,|\psi_j\rangle - |m_j\rangle\Big\|^2\,, \tag{10.135}$$

between the weighted states $\sqrt{P_j}\,|\psi_j\rangle$ to be discriminated and the corresponding measurement vectors $|m_j\rangle$ of Eq. (10.133). Accordingly, the square-root detector is also known as the least-squares detector [279, 281, 282].

The squared error of Eq. (10.135) can always be expanded as

$$e^2 = \sum_{j=1}^{J} \Big[P_j \langle \psi_j|\psi_j\rangle + \langle m_j|m_j\rangle - 2\sqrt{P_j}\,\langle m_j|\psi_j\rangle \Big], \tag{10.136}$$

since $\rho^{-1/2}$ is positive and hence $\langle m_j|\psi_j\rangle = \langle \psi_j|m_j\rangle = \sqrt{P_j}\,\langle \psi_j|\rho^{-1/2}|\psi_j\rangle$ is real. Any discrimination operators $\mathsf{E}_j = |m_j\rangle\langle m_j|$ forming a valid POVM satisfy Eq. (10.130), whose trace yields $\sum_{j=1}^{J} \langle m_j|m_j\rangle = \mathrm{tr}(\mathrm{I}_N) = N$, so that Eq. (10.136) is also

$$e^2 = 1 + N - 2\sum_{j=1}^{J} \sqrt{P_j}\,\langle m_j|\psi_j\rangle = 1 + N - 2\,\mathrm{tr}\big(\rho^{1/2}\big)\,. \tag{10.137}$$

Meanwhile, Eq. (10.131) gives the conditional probabilities $\Pr\{\mathsf{E}_k|\rho_j\} = |\langle m_k|\psi_j\rangle|^2$, so that the probability P_{suc} of successful discrimination from Eq. (10.97) follows as

$$P_{\mathrm{suc}} = \sum_{j=1}^{J} P_j \left|\langle m_j|\psi_j\rangle\right|^2 = \sum_{j=1}^{J} P_j^2 \left|\langle \psi_j|\rho^{-1/2}|\psi_j\rangle\right|^2 . \tag{10.138}$$

It is visible that determining the J measurement vectors $|m_j\rangle$ so as to minimize the squared error e^2 of Eq. (10.137) is in general not equivalent to maximizing the probability P_{suc} of Eq. (10.138), even for equiprobable states when $P_j = 1/J$ for all j. The J measurement vectors $|m_j\rangle$ of the square-root detector from Eq. (10.133) always minimize the weighted squared error of Eq. (10.135); yet they do not in general maximize the success probability P_{suc} of Eq. (10.138). In addition, for equiprobable states with $P_j = 1/J$ for all j, the square-root detector from Eq. (10.133) also minimizes the unweighted squared error $\sum_{j=1}^{J} \|\, |\psi_j\rangle - |m_j\rangle \,\|^2$, as well as minimizing the weighted squared error $\sum_{j=1}^{J} \|\, |\psi_j\rangle/\sqrt{J} - |m_j\rangle \,\|^2$ of Eq. (10.135), as it can be verified from Eq. (10.137). Moreover, there exist situations of practical interest where the square-root detector from Eq. (10.133) also maximizes the success probability P_{suc} of Eq. (10.138).

For example, we can consider as in [244, 279, 283] the set of J symmetric pure states $|\psi_j\rangle \in \mathcal{H}_N$ with equal prior probabilities $P_j = 1/J$ and given by

$$|\psi_j\rangle = \mathsf{U}^j\,|\psi_0\rangle \quad \text{for } j = 0, 1, \ldots, J-1, \tag{10.139}$$

where U is a unitary operator on $\mathcal{H}_N$ satisfying $\mathsf{U}^J = \mathrm{I}_N$ and $|\psi_0\rangle$ is any pure state of $\mathcal{H}_N$. A structure of algebraic group can be assigned to the set of operators U^j, and for the square-root detector the resulting measurement vectors $|m_j\rangle$ in Eq. (10.133) exhibit the same symmetries and satisfy $|m_j\rangle = \mathsf{U}^j\,|m_0\rangle$, so that $\langle m_j|\psi_j\rangle = \langle m_0|\psi_0\rangle$ for all j, giving $P_{\text{suc}} = |\langle m_0|\psi_0\rangle|^2 = |\langle\psi_0|\rho^{-1/2}|\psi_0\rangle|^2/J$ in Eq. (10.138). In addition, it is found [244, 279, 283] that here the square-root detector both minimizes the squared error e^2 of Eq. (10.137) and maximizes the success probability P_{suc} of Eq. (10.138). So for such equiprobable set of J symmetric pure states, the square-root detector from Eq. (10.133) coincides with the minimal-error optimal detector.

For instance, the set of J pure qubit states

$$|\psi_j\rangle = \cos\left(\frac{2\pi j}{J}\right)|0\rangle + \sin\left(\frac{2\pi j}{J}\right)|1\rangle\ , \tag{10.140}$$

for $j = 0$ to $J-1$, is generated from $|\psi_0\rangle = |0\rangle$ by the qubit unitary operator

$$\mathsf{U} = \begin{bmatrix} \cos(2\pi/J) & -\sin(2\pi/J) \\ \sin(2\pi/J) & \cos(2\pi/J) \end{bmatrix}, \tag{10.141}$$

which is the rotation by the angle $2\pi/J$ in the vector plane $\mathcal{H}_2$. With the prior probabilities $P_j = 1/J$, the average density operator of Eq. (10.134) follows as $\rho = \mathrm{I}_2/2$, so that $\rho^{-1/2} = \sqrt{2}\,\mathrm{I}_2$. As a result, the square-root measurement of Eq. (10.133) is defined by the J measurement vectors $|m_j\rangle = \sqrt{2/J}\,|\psi_j\rangle$. This is the POVM associated with the qubit generalized measurement of Sect. 4.2.5 with $\mathsf{E}_j = |m_j\rangle\langle m_j| = \mathsf{M}_j^\dagger \mathsf{M}_j$. This is also the optimal POVM achieving minimal-error detection or discrimination of the J pure qubit states of Eq. (10.140) with equal prior probabilities. It achieves in Eq. (10.138) the success probability $P_{\text{suc}} = 2/J$.

Finally, for $J = N$ states $|\psi_j\rangle$ spanning the Hilbert space $\mathcal{H}_N$, the set of N vectors $\{|m_j\rangle\}$ from Eq. (10.133) form an orthonormal basis of $\mathcal{H}_N$ defining a von Neumann projective measurement. Since they satisfy $\sum_{j=1}^N |m_j\rangle\langle m_j| = \mathrm{I}_N$, the property follows from the argument given just after Eq. (A.10) in the Appendix. In this circumstance, Eq. (10.133) provides an explicit analytical expression for an orthonormal basis of $\mathcal{H}_N$ constructed from a set of N linearly independent vectors $|\psi_j\rangle$ of $\mathcal{H}_N$, offering in this way an analytical solution to the orthogonalization problem otherwise tackled by the Gram–Schmidt algorithm.

10.8.2 Square-Root Detector for the Qubit

For generally mixed qubit states, the square-root detector can be rather thoroughly characterized analytically in Bloch representation. For this purpose, we consider that each of the qubit states $\rho_j = \big(\mathrm{I}_2 + \vec{r}_j\cdot\vec{\sigma}\big)/2$ of $\mathcal{L}(\mathcal{H}_2)$ is characterized, as in Eq. (10.65), by a Bloch vector $\vec{r}_j$, for $j = 1$ to J. The average state ρ of Eq. (10.128) is then the qubit density operator of $\mathcal{L}(\mathcal{H}_2)$ characterized by the averaged Bloch vector

$$\vec{r} = \sum_{j=1}^{J} P_j \vec{r}_j \,. \tag{10.142}$$

According to Eq. (2.66), the average density operator ρ has the two positive (except in degenerate cases) eigenvalues $\lambda_\pm = \big(1 \pm \|\vec{r}\,\|\big)/2$ associated with the two projectors $|\lambda_\pm\rangle\langle\lambda_\pm|$ of Eq. (2.70). The diagonal form $\rho = \lambda_+ |\lambda_+\rangle\langle\lambda_+| + \lambda_- |\lambda_-\rangle\langle\lambda_-|$ then enables the characterization of the square-root operator $\rho^{1/2}$ and its inverse as

$$\rho^{-1/2} = \lambda_+^{-1/2} |\lambda_+\rangle\langle\lambda_+| + \lambda_-^{-1/2} |\lambda_-\rangle\langle\lambda_-| \,, \tag{10.143}$$

which is also in Bloch representation:

$$\rho^{-1/2} = \frac{1}{2}\Big[\Big(\lambda_+^{-1/2} + \lambda_-^{-1/2}\Big)\mathrm{I}_2 + \Big(\lambda_+^{-1/2} - \lambda_-^{-1/2}\Big)\frac{\vec{r}}{\|\vec{r}\,\|}\cdot\vec{\sigma}\Big] \tag{10.144}$$

$$= a_0\mathrm{I}_2 + \vec{a}\cdot\vec{\sigma} \,. \tag{10.145}$$

The J discrimination operators $\mathsf{E}_j \in \mathcal{L}(\mathcal{H}_2)$ from Eq. (10.129) can then be characterized in Bloch representation by the product $\mathsf{E}_j = \big(a_0\mathrm{I}_2 + \vec{a}\cdot\vec{\sigma}\big)\big(\mathrm{I}_2 + \vec{r}_j\cdot\vec{\sigma}\big)\big(a_0\mathrm{I}_2 + \vec{a}\cdot\vec{\sigma}\big)P_j/2$, which can be evaluated by means of the identity of Eq. (A.45), so as to yield

$$\mathsf{E}_j = b_j\mathrm{I}_2 + \vec{c}_j\cdot\vec{\sigma} \,, \tag{10.146}$$

with

$$b_j = \big(a_0^2 + 2a_0\vec{a}\,\vec{r}_j + \vec{a}^{\,2}\big)\frac{P_j}{2} \,, \tag{10.147}$$

$$\vec{c}_j = \Big[2a_0\vec{a} + (a_0^2 - \vec{a}^{\,2})\vec{r}_j + 2(\vec{a}\,\vec{r}_j)\vec{a}\,\Big]\frac{P_j}{2} \,. \tag{10.148}$$

The J conditional probabilities $\mathrm{tr}(\mathsf{E}_j\,\rho_j)$ then follow as $\mathrm{tr}(\mathsf{E}_j\,\rho_j) = b_j + \vec{c}_j\vec{r}_j$ or also

$$\mathrm{tr}(\mathsf{E}_j\,\rho_j) = \Big[a_0^2 + \vec{a}^{\,2} + 4a_0\vec{a}\,\vec{r}_j + (a_0^2 - \vec{a}^{\,2})\vec{r}_j^{\,2} + 2\big(\vec{a}\,\vec{r}_j\big)^2\Big]\frac{P_j}{2} \,, \tag{10.149}$$

to determine the probability of successful discrimination P_{suc} in Eq. (10.97).

As an application, we can consider the detection of two pure qubit states $|\psi_0\rangle$ and $|\psi_1\rangle$, with prior probabilities P_0 and $P_1 = 1 - P_0$, respectively, characterized by the two unit Bloch vectors $\vec{r}_0$ and $\vec{r}_1$. The minimal-error detector is characterized in Sect. 10.4.1 and also in Fig. 10.2, and it achieves the minimal probability of error $P_{\mathrm{er}}^{\min}$ of Eqs. (10.70)–(10.71). The square-root detector is governed from Eq. (10.142) by the averaged Bloch vector $\vec{r} = P_0\vec{r}_0 + P_1\vec{r}_1$, and it achieves the probability of error $P_{\mathrm{er}}^{\mathrm{sqr}} = 1 - P_{\mathrm{suc}}$ resulting from Eq. (10.97) and Eq. (10.149). The performances $P_{\mathrm{er}}^{\min}$ and $P_{\mathrm{er}}^{\mathrm{sqr}}$ are compared in Fig. 10.6 for detection between $|\psi_0\rangle = |0\rangle$ and $|\psi_1\rangle = |+\rangle$, as a function of the prior probability P_0.

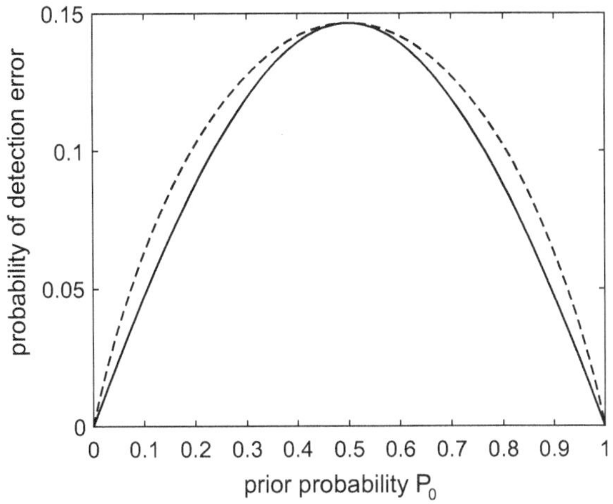

Fig. 10.6 For the detection of two pure qubit states $|\psi_0\rangle = |0\rangle$ and $|\psi_1\rangle = |+\rangle$, as a function of the prior probability P_0, the probability of detection error $P_{\rm er}^{\rm min}$ of the minimal-error optimal detector (solid line), and $P_{\rm er}^{\rm sqr}$ of the square-root detector (dashed line)

In Fig. 10.6, as expected, the probability of detection error $P_{\rm er}^{\rm sqr}$ of the square-root detector is always (slightly) larger than the probability $P_{\rm er}^{\rm min}$ of the minimal-error optimal detector. This is so for any prior probability P_0 in Fig. 10.6, except in the degenerate cases $P_0 = 0$ and $P_0 = 1$ where exact detection is feasible with no error, and in the case $P_0 = 1/2$ where the performances $P_{\rm er}^{\rm min}$ and $P_{\rm er}^{\rm sqr}$ of the two detectors coincide, as visible in Fig. 10.6. Beyond the conditions of Fig. 10.6 for the qubit, it can be checked that at $P_0 = 1/2 = P_1$, for any pair of pure quantum states $|\psi_0\rangle$ and $|\psi_1\rangle$ in the N-dimensional Hilbert space $\mathcal{H}_N$, the square-root detector resulting from Eq. (10.133) coincides with the minimal-error detector characterized in Sect. 10.3 and Fig. 10.2.

As another application, we examine the detection of two orthogonal pure qubit states $|0\rangle$ and $|1\rangle$, altered by a depolarizing noise with contraction factor α acting as in Eq. (9.55), the two states occurring with equal prior probabilities $P_0 = P_1 = 1/2$. This could represent a binary communication protocol over a noisy quantum channel. The two noisy states ρ_j to be detected are characterized in $\mathbb{R}^3$ by the Bloch vector $\vec{r}_j = (-1)^j \alpha \vec{e}_z$, for $j = 0$ or 1. For the minimal-error quantum detector, the test operator in Eq. (10.66) is $\mathsf{T} = -\alpha\sigma_z$, having the two eigenvalues $\mp\alpha$ and eigenstates $|0\rangle$ and $|1\rangle$. As a result, the minimal-error detector, according to Sect. 10.4.1, implements the optimal POVM with the two detection operators $\mathsf{E}_j^{\rm opt} = \big[\mathrm{I}_2 + (-1)^j \sigma_z\big]/2$ or equivalently $\big\{\mathsf{E}_0^{\rm opt} = |0\rangle\langle 0|\,,\ \mathsf{E}_1^{\rm opt} = |1\rangle\langle 1|\big\}$ and achieves the minimal probability of detection error $P_{\rm er}^{\rm min} = (1-\alpha)/2$. Meanwhile, the square-root detector is governed in Eq. (10.142) by the averaged Bloch vector $\vec{r} = \vec{0}$, the average density operator of Eq. (10.134) follows as $\rho = \mathrm{I}_2/2$, so that $\rho^{-1/2} = \sqrt{2}\,\mathrm{I}_2$ leads to the detection operators $\mathsf{E}_j^{\rm sqr} = \big[\mathrm{I}_2 + (-1)^j \alpha\sigma_z\big]/2 = \rho_j$ achieving the probability of detection error $P_{\rm er}^{\rm sqr} = (1-\alpha^2)/2$. The performances $P_{\rm er}^{\rm min} = (1-\alpha)/2$ and $P_{\rm er}^{\rm sqr} = (1-\alpha^2)/2$ are compared in Fig. 10.7, as a function of the level of the depolarizing noise quantified by the contraction factor α.

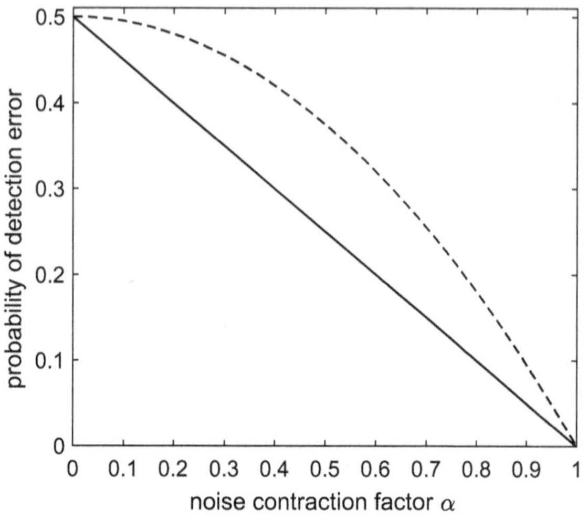

Fig. 10.7 For the detection of the two qubit states $|0\rangle$ and $|1\rangle$ altered by a quantum depolarizing noise with contraction factor α in abscissa, the probability of detection error $P_{\rm er}^{\rm min} = (1-\alpha)/2$ of the minimal-error optimal detector (solid line), and $P_{\rm er}^{\rm sqr} = (1-\alpha^2)/2$ of the square-root detector (dashed line)

In Fig. 10.7, as expected again, the probability of detection error $P_{\rm er}^{\rm sqr}$ of the square-root detector is always larger than the probability $P_{\rm er}^{\rm min}$ of the minimal-error optimal detector. This is so for any contraction factor α of the depolarizing noise, except in the degenerate cases $\alpha = 1$ with no noise, and $\alpha = 0$ at maximum noise when the two states to be detected both collapse into the maximally mixed state $\mathrm{I}_2/2$.

As a third application, we examine the detection of the three qubit states $\rho_1 = |0\rangle\langle 0|$, $\rho_2 = \mathrm{I}_2/2$ and $\rho_3 = |1\rangle\langle 1|$, with, respectively, the prior probabilities $P_1 = p/2$, $P_2 = 1-p$, and $P_3 = p/2$. These states are characterized, respectively, by the three Bloch vectors $\vec{r}_1 = \vec{e}_z$, $\vec{r}_2 = \vec{0}$, and $\vec{r}_3 = -\vec{e}_z$.

By direct reasoning here, it is found that the detection POVM $\big\{\mathsf{E}_1 = |0\rangle\langle 0|\,,\ \mathsf{E}_2 = 0,\ \mathsf{E}_3 = |1\rangle\langle 1|\big\}$ when $p \ge 1-p$, or $\big\{\mathsf{E}_1 = 0,\ \mathsf{E}_2 = \mathrm{I}_2,\ \mathsf{E}_3 = 0\big\}$ when $p \le 1-p$, achieves the probability of detection success $P_{\rm suc} = \max(p, 1-p)$ or equivalently the probability of detection error $P_{\rm er} = 1 - P_{\rm suc} = \min(p, 1-p)$. It can be known that this detection scheme is the optimal minimal-error detector because its probability of detection error $P_{\rm er} = \min(p, 1-p) = P_{\rm er}^{\rm min}$ reaches the lower bound prescribed in [251].

For the square-root detector, the average density operator of Eq. (10.128) follows as $\rho = \mathrm{I}_2/2$, so that $\rho^{-1/2} = \sqrt{2}\,\mathrm{I}_2$ leads in Eq. (10.129) to the detection POVM $\big\{\mathsf{E}_1^{\rm sqr} = p\,|0\rangle\langle 0|\,,\ \mathsf{E}_2^{\rm sqr} = (1-p)\mathrm{I}_2,\ \mathsf{E}_3^{\rm sqr} = p\,|1\rangle\langle 1|\big\}$. This square-root detector achieves the probability of detection success $P_{\rm suc}^{\rm sqr} = p^2 + (1-p)^2$ or equivalently the probability of detection error $P_{\rm er}^{\rm sqr} = 1 - P_{\rm suc}^{\rm sqr} = 1 - p^2 - (1-p)^2$.

The probabilities of detection error, $P_{\rm er}^{\rm min}$ for the optimal minimal-error detector and $P_{\rm er}^{\rm sqr}$ for the square-root detector, are compared in Fig. 10.8 as a function of the prior probability p.

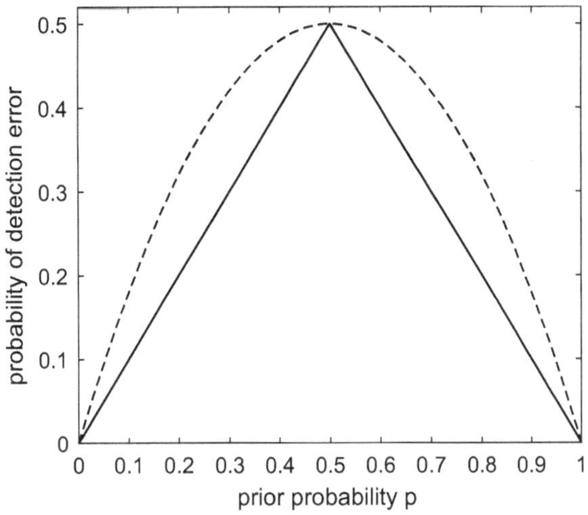

Fig. 10.8 For the detection of the three qubit states (see text) characterized by the prior probability p in abscissa, the probability of detection error $P_{\text{er}}^{\min} = \min(p, 1-p)$ of the minimal-error optimal detector (solid line), and $P_{\text{er}}^{\text{sqr}} = 1 - p^2 - (1-p)^2$ of the square-root detector (dashed line)

Again, the results of Fig. 10.8 illustrate how the probability of detection error $P_{\text{er}}^{\text{sqr}}$ of the square-root detector is, as expected, generally larger than the probability $P_{\text{er}}^{\min}$ of the minimal-error optimal detector. Nevertheless, as we have argued, the square-root detector offers an interesting alternative, through its usually more direct and explicit characterization.

10.9 Summary on Quantum Detection

Quantum detection following [235] or **quantum state discrimination** is an important task of signal and information processing. It involves an information-carrying quantum system with N-dimensional Hilbert space $\mathcal{H}_N$, which can be prepared in one among $J \geq 2$ possible states ρ_j with prior probabilities P_j. The preparation then is to be decided (efficiently) by measuring the quantum system with a set of detection operators (POVM elements) $\{\mathsf{E}_k\}$. Various meaningful criteria can be selected to assess the efficiency, but all are crucially governed by the conditional probabilities $\Pr\{\mathsf{E}_k|\rho_j\} = \operatorname{tr}(\mathsf{E}_k\rho_j)$ of deciding for k in place of j when the state ρ_j holds, as they are found for instance in Eq. (10.33) or Eq. (10.131).

The case of **binary detection**, with $J = 2$ possible quantum states in $\mathcal{H}_N$, is specially important. It can be solved analytically to a large extent and accordingly provides a useful reference, and it has great practical relevance for applications such as binary signaling for quantum communication. A Hermitian **test operator** T characterizes the dissimilarity of the two quantum states $\{\rho_0, \rho_1\}$ to be detected, with slightly differing expressions for T according to the criterion to be optimized, as seen from Eqs. (10.38), (10.47), (10.50), or (10.57). The eigendecomposition of T

determines both the **optimal detection** operators $\{\mathsf{E}_k\}$ and the associated maximized efficiency.

The optimal discrimination of more than two quantum states ρ_j in $\mathcal{H}_N$ in general has to be solved numerically, or possibly analytically in some special cases. As an alternative, the **square-root detector** presented in Sect. 10.8, although suboptimal in general, stands as an explicit, convenient, and often (very) efficient detection strategy, to generally discriminate among any given set of quantum states ρ_j with their arbitrary prior probabilities P_j.

Quantum state detection or discrimination is specially useful in the context of quantum signaling in the presence of **noise**, as addressed for example in Sect. 10.5. In such context, some signal states $\{\rho_j\}$ are purposely prepared, but get altered by some definite quantum noise process $\mathcal{N}(\cdot)$ realizing the input–output transformation $\rho_j \mapsto \mathcal{N}(\rho_j)$, and the detection or discrimination has to be performed from the noisy output quantum states $\{\mathcal{N}(\rho_j)\}$. From measuring the noisy output states $\{\mathcal{N}(\rho_j)\}$, an optimal detection strategy, as it follows from the characterizations worked out in this chapter, will come with some maximized efficiency, which may then be further optimized by seeking the optimal input signaling states $\{\rho_j\}$ so as to maximize the detection efficiency of the optimal quantum detector from the noisy output.

Chapter 11
Quantum Estimation

In this chapter we examine the task of parameter estimation from a quantum state. It stands as another fundamental task of quantum information processing, central to sensing and metrology, aiming at (precisely) ascribing a value to some parameter of interest by means of quantum measurement on some parameter-dependent quantum state or signal. The parameter can equally characterize a quantum process when the parameter-dependent quantum state being measured serves as an excitation signal to probe the process. We start the chapter with a brief review of the classical theory of parameter estimation in a statistical framework. This will subsequently serve us as a guideline for developing the quantum theory of estimation, while appreciating some close parallel between the two. Quantum estimation is in particular assessed with the fundamental notion of Fisher information, relevant in its quantum and classical versions, which are both to be presented. Classical estimation is also needed to complete a quantum estimation task by a classical estimation step from the result of a quantum measurement, as we shall see.

11.1 Classical Estimation

Parameter estimation from observed data is an important task of classical signal processing [236, 239, 284]. We briefly review the principles thereof, starting from estimation of a deterministic parameter, and then extend this basis to a random parameter in Bayesian estimation.

11.1.1 Deterministic Parameter

In classical estimation theory [236, 239, 284] one considers a dataset $\vec{x} = (x_1, x_2, \ldots, x_N)$ of N real values representing N samples taken from a random

F. Chapeau-Blondeau, *Quantum Information and Quantum Signal Processing*,
https://doi.org/10.1007/978-3-032-27499-1_11

(noisy) signal. The dataset $\vec{x} \in \mathbb{R}^N$ is distributed according to the (multidimensional) probability density function denoted $p(\vec{x}\,; \xi)$. The probability distribution $p(\vec{x}\,; \xi)$ is dependent upon a deterministic scalar parameter ξ, with a definite but unknown value that is to be estimated from observed realizations[1] of the dataset $\vec{x}$. From any realization of $\vec{x}$, one can conceive a function $\widehat{\xi}(\vec{x}\,)$, known as an estimator, assigning the value $\widehat{\xi}$ to the unknown parameter ξ. A meaningful way of assessing the performance of an estimator $\widehat{\xi}(\vec{x}\,)$ is by the mean-squared error defined as the average

$$e^2\big(\widehat{\xi}\,\big) = \big\langle [\,\widehat{\xi}(\vec{x}\,) - \xi]^2 \big\rangle = \int_{\mathbb{R}^N} \big[\,\widehat{\xi}(\vec{x}\,) - \xi\big]^2 p(\vec{x}\,; \xi) d\vec{x}\;, \tag{11.1}$$

with $d\vec{x} = dx_1 dx_2 \cdots dx_N$ an infinitesimal domain of $\mathbb{R}^N$. One can also introduce the bias $b(\widehat{\xi}\,) = \big\langle \widehat{\xi}(\vec{x}\,) - \xi \big\rangle = \big\langle \widehat{\xi}(\vec{x}\,) \big\rangle - \xi$ and variance $\mathrm{var}(\widehat{\xi}\,)$ of the estimator $\widehat{\xi}(\vec{x}\,)$, which are related by $e^2\big(\widehat{\xi}\,\big) = b^2(\widehat{\xi}\,) + \mathrm{var}(\widehat{\xi}\,)$.

For any conceivable estimators $\widehat{\xi}(\vec{x}\,)$, the mean-squared error $e^2\big(\widehat{\xi}\,\big)$ can be lower bounded in the following way. The so-called **statistical score** is the scalar random variable $V(\vec{x}\,; \xi)$, function of the vector random variable $\vec{x}$, defined by

$$V(\vec{x}\,; \xi) = \partial_\xi \ln p(\vec{x}\,; \xi) = \frac{\partial_\xi p(\vec{x}\,; \xi)}{p(\vec{x}\,; \xi)}\;, \tag{11.2}$$

with $\partial_\xi \cdot \equiv \partial \cdot / \partial \xi$ as the derivative relative to ξ. The score has vanishing mean $\big\langle V(\vec{x}\,; \xi) \big\rangle = 0$, since

$$\big\langle V(\vec{x}\,; \xi) \big\rangle = \int_{\mathbb{R}^N} \frac{\partial_\xi p(\vec{x}\,; \xi)}{p(\vec{x}\,; \xi)} p(\vec{x}\,; \xi) d\vec{x} = \partial_\xi \int_{\mathbb{R}^N} p(\vec{x}\,; \xi) d\vec{x} = \partial_\xi 1 = 0\;. \tag{11.3}$$

The variance of the score $\mathrm{var}[V(\vec{x}\,; \xi)] = \big\langle V^2(\vec{x}\,; \xi) \big\rangle$ defines the **classical Fisher information** about the parameter ξ contained in the data $\vec{x}$,

$$F_c(\xi\,; \vec{x}\,) = \big\langle V^2(\vec{x}\,; \xi) \big\rangle = \int_{\mathbb{R}^N} \frac{1}{p(\vec{x}\,; \xi)} \big[\partial_\xi p(\vec{x}\,; \xi)\big]^2 d\vec{x}\;. \tag{11.4}$$

The Fisher information $F_c(\xi\,; \vec{x}\,)$ is a non-negative real scalar; it can be expected to depend on ξ in general; it is independent of $\vec{x}$ but the second argument $\vec{x}$ is left in $F_c(\xi\,; \vec{x}\,)$ to indicate that the Fisher information is contained in the data $\vec{x}$ (about the parameter ξ). It is easily shown that the classical Fisher information $F_c(\xi\,; \vec{x}\,)$ defined by Eq. (11.4) is equivalently

$$F_c(\xi\,; \vec{x}\,) = -\big\langle \partial_\xi V(\vec{x}\,; \xi) \big\rangle\;. \tag{11.5}$$

---

[1] In some places of this book, as in [284] for instance, we sometimes use the same notation for the random signal $\vec{x}$ per se and for a generic realization of it, to avoid duplicated notations for closely related quantities, still distinguishable from the context and usage.

Generally, the Cauchy–Schwarz inequality on random variables gives

$$\mathrm{var}(\widehat{\xi}\,)\,\mathrm{var}(V) \geq \Big\langle \big(\widehat{\xi} - \langle\widehat{\xi}\rangle\big)\big(V - \langle V\rangle\big)\Big\rangle^2 . \tag{11.6}$$

The left side of Eq. (11.6) is $\mathrm{var}(\widehat{\xi})F_c(\xi\,;\vec{x}\,)$; and since $\langle V\rangle = 0$, the right side of Eq. (11.6) is $\langle\widehat{\xi}\,V\rangle^2$. By Eq. (11.2), we have

$$\langle\widehat{\xi}\,V\rangle = \left\langle \widehat{\xi}(\vec{x}\,)\frac{\partial_\xi p(\vec{x}\,;\xi)}{p(\vec{x}\,;\xi)}\right\rangle = \int_{\mathbb{R}^N} \widehat{\xi}(\vec{x}\,)\,\partial_\xi p(\vec{x}\,;\xi)d\vec{x} = \partial_\xi \int_{\mathbb{R}^N} \widehat{\xi}(\vec{x}\,)\,p(\vec{x}\,;\xi)d\vec{x}\,, \tag{11.7}$$

since the estimator $\widehat{\xi}(\vec{x}\,)$ is independent of the unknown parameter ξ, by consistency of the approach where $\widehat{\xi}(\vec{x}\,)$ is constructed to estimate an unknown ξ. It follows from Eq. (11.7) that $\langle\widehat{\xi}\,V\rangle = \partial_\xi\langle\widehat{\xi}\rangle = 1 + \partial_\xi b(\widehat{\xi}\,)$ involving the derivative of the bias $b(\widehat{\xi}\,)$. Then, Eq. (11.6) leads to the **Cramér–Rao inequality**

$$\mathrm{var}(\widehat{\xi}\,) \geq \frac{\big[1 + \partial_\xi b(\widehat{\xi}\,)\big]^2}{F_c(\xi\,;\vec{x}\,)}\,, \tag{11.8}$$

establishing a lower bound for the variance of any estimator $\widehat{\xi}(\vec{x}\,)$. Equivalently, the mean-squared error $e^2\big(\widehat{\xi}\,\big)$ is lower bounded by the alternative Cramér–Rao inequality reading

$$e^2\big(\widehat{\xi}\,\big) \geq b^2(\widehat{\xi}\,) + \frac{\big[1 + \partial_\xi b(\widehat{\xi}\,)\big]^2}{F_c(\xi\,;\vec{x}\,)}\,. \tag{11.9}$$

The right-hand side of the Cramér–Rao inequality, in Eq. (11.8) or Eq. (11.9), is known as the Cramér–Rao (lower) bound. The Cramér–Rao bound involves the (reciprocal of the) Fisher information $F_c(\xi\,;\vec{x}\,)$. The Fisher information $F_c(\xi\,;\vec{x}\,)$ is intrinsic to the relation of the unknown parameter ξ with the dataset $\vec{x}$ used for its estimation, and in this respect it is independent of, and common to, all estimators $\widehat{\xi}(\vec{x}\,)$ envisaged for the estimation. So in general, a higher Fisher information $F_c(\xi\,;\vec{x}\,)$ can be expected to translate into a lower Cramér–Rao bound and therefrom higher efficiency for estimation. In addition, the Cramér–Rao bound in the right-hand side of Eq. (11.8) or Eq. (11.9) can depend on the estimator via its bias $b(\widehat{\xi}\,)$; while for the important class of unbiased estimators with $b(\widehat{\xi}\,) = 0$, the bound also becomes intrinsic as the Fisher information $F_c(\xi\,;\vec{x}\,)$, and completely determined by it.

Three Properties: We have the three useful properties for classical estimation.

(1) When the dataset $\vec{x} = (x_1, x_2, \ldots, x_N)$ is formed by N independent random variables x_n, the joint probability density factorizes as $p(\vec{x}\,) = p(x_1)p(x_2)\cdots p(x_N)$ so that the score is additive and writes $V(\vec{x}\,;\xi) = \sum_{n=1}^N V(x_n\,;\xi)$, with mutually independent individual scores $V(x_n\,;\xi) = \partial_\xi \ln p(x_n\,;\xi)$. As a consequence, the Fisher information (the variance of the score) is also **additive** and writes

$F_c(\xi\,;\vec{x}\,) = \sum_{n=1}^{N} F_c(\xi\,;x_n)$, with $F_c(\xi\,;x_n) = \text{var}\big[V(x_n\,;\xi)\big]$ the Fisher information about the parameter ξ contained in the datapoint x_n.

(2) When an (N-dimensional) ξ-independent invertible transformation $f(\cdot)$ is applied to the dataset as $f(\vec{x}\,) = \vec{y}$, the corresponding probability densities are related by $p_y(\vec{y}\,;\xi)d\vec{y} = p_x(\vec{x}\,;\xi)d\vec{x}$ when in addition $d\vec{y} = |\det(J_f)|d\vec{x}$, with J_f the Jacobian matrix of the transformation $f(\cdot)$. It then results from Eq. (11.2) that the scores of the initial data $\vec{x}$ and transformed data $\vec{y}$ coincide, i.e., $V(\vec{y}\,;\xi) = V(\vec{x}\,;\xi)$. From Eq. (11.4), the Fisher informations also coincide, i.e., $F_c(\xi\,;\vec{y}\,) = F_c(\xi\,;\vec{x}\,)$. More generally, for a non-invertible ξ-independent transformation $\vec{y} = f(\vec{x}\,)$, the Fisher information never increases, satisfying [285] the **monotonicity** relation $F_c(\xi\,;\vec{y}\,) \le F_c(\xi\,;\vec{x}\,)$.

(3) Let $p_1(\vec{x}\,;\xi)$ and $p_2(\vec{x}\,;\xi)$ two elements of the convex set of N-dimensional probability densities, and a scalar $\mu \in [0, 1]$. If we denote by $F_c[p(\vec{x}\,;\xi)]$ the Fisher information in Eq. (11.4) associated with the generic density $p(\vec{x}\,;\xi)$, we then have the **convexity** property $F_c[\mu p_1(\vec{x}\,;\xi) + (1-\mu)p_2(\vec{x}\,;\xi)] \le \mu F_c[p_1(\vec{x}\,;\xi)] + (1-\mu)F_c[p_2(\vec{x}\,;\xi)]$, as proved for instance in [286], and which naturally extends to a convex superposition of an arbitrary number of probability densities.

There exists a general methodology enabling to construct a useful estimator, known as the **maximum-likelihood estimator**. It considers the likelihood function, which is nothing else than the probability density $p(\vec{x}\,;\xi)$, but where now the data $\vec{x}$ is considered as a fixed realization given by some definite observation while the parameter ξ is considered as a undetermined variable. The maximum-likelihood estimator then returns that value of the variable ξ achieving the maximization

$$\widehat{\xi}_{\text{ML}}(\vec{x}\,) = \arg\max_{\xi}\; p(\vec{x}\,;\xi)\,, \tag{11.10}$$

for any realization of the dataset $\vec{x}$ delivered by the measurement. Equivalently, the maximizer $\widehat{\xi}_{\text{ML}}(\vec{x}\,)$ also maximizes the loglikelihood $\ln\big(p(\vec{x}\,;\xi)\big)$, which on occasion provides a criterion easier to handle.

The maximum-likelihood estimator $\widehat{\xi}_{\text{ML}}(\vec{x}\,)$ of Eq. (11.10) a priori displays appealing properties. The maximum-likelihood procedure is in principle always applicable, since the probability density $p(\vec{x}\,;\xi)$ of the measured data can in general always be conceived; and it is an a priori reasonable approach, as it consists in assigning to the unknown parameter ξ that value that makes the most probable the dataset $\vec{x}$ we have in hand. Also, a posteriori, optimality properties can be proved for the maximum-likelihood estimator $\widehat{\xi}_{\text{ML}}(\vec{x}\,)$ of Eq. (11.10). In the limit of a large number N of independent data points x_n, by the law of large numbers and the central limit theorem, it can be proved [239, 284] that $\widehat{\xi}_{\text{ML}}(\vec{x}\,)$ becomes unbiased, achieves the minimum mean-squared error $1/F_c(\xi\,;\vec{x}\,)$, and acquires a Gaussian or normal distribution $\widehat{\xi}_{\text{ML}}(\vec{x}\,) \sim \mathcal{N}\big(\xi, 1/F_c(\xi\,;\vec{x}\,)\big)$ with mean ξ and variance $1/F_c(\xi\,;\vec{x}\,)$. It is in this respect the best conceivable estimator. For these reasons, the maximum-likelihood estimator is very often employed in practice, even for small size N of the dataset, and it stands as an essential reference for estimation.

Upon a change of parameter $\xi = \xi(\nu) = g(\nu)$, the score functions transform as $V(\vec{x}\,;\nu) = (\partial\xi/\partial\nu)V(\vec{x}\,;\xi)$, the Fisher informations relative to the parameters ξ and ν are related by $F_c(\nu) = (\partial\xi/\partial\nu)^2 F_c(\xi)$, and the maximum-likelihood estimators by $\widehat{\xi}_{\rm ML} = g\big(\widehat{\nu}_{\rm ML}\big)$.

11.1.2 Multiple-Parameter Estimation

There exists a natural extension, when from the measured data $\vec{x}$, the task is to estimate a set of multiple unknown parameters [236, 239, 287] that we denote by the vector $\vec{\xi} = [\xi_1, \xi_2, \dots]^\top$. From the measured data $\vec{x}$, a vector estimator $\widehat{\vec{\xi}}(\vec{x}\,) = \big[\widehat{\xi}_1(\vec{x}\,),\ \widehat{\xi}_2(\vec{x}\,),\ \dots\big]^\top$ is formed with each component offering a scalar estimator for the corresponding scalar parameter. Any such vector estimator is assigned a vector bias $\vec{b}(\widehat{\vec{\xi}}\,) = \big\langle\widehat{\vec{\xi}}(\vec{x}\,)\big\rangle - \vec{\xi} = \big[\widehat{\xi}_1 - \xi_1,\ \widehat{\xi}_2 - \xi_2,\ \dots\big]^\top$, and a covariance matrix of the estimation error

$$C(\widehat{\vec{\xi}}\,) = \Big\langle\big[\widehat{\vec{\xi}}(\vec{x}\,) - \vec{\xi}\,\big]\big[\widehat{\vec{\xi}}(\vec{x}\,) - \vec{\xi}\,\big]^\top\Big\rangle\,, \tag{11.11}$$

where the statistical average $\langle\cdot\rangle$ is calculated according to the probability distribution $p(\vec{x}\,;\vec{\xi}\,)$ of the data $\vec{x}$ function of the vector parameter $\vec{\xi}$. The covariance matrix of the error is lower bounded by the Cramér–Rao matrix inequality

$$C(\widehat{\vec{\xi}}\,) \ge \vec{b}\,\vec{b}^\top + \Big(I + \vec{\nabla}\vec{b}^\top\Big)\, F_c^{-1}(\vec{\xi}\,;\vec{x}\,)\,\Big(I + \vec{\nabla}\vec{b}^\top\Big)^\top\,. \tag{11.12}$$

In Eq. (11.12), we have the nabla derivative vector $\vec{\nabla}\cdot = [\partial\cdot/\partial\xi_1,\ \partial\cdot/\partial\xi_2,\ \dots]^\top$ and the associated identity matrix I; also $F_c^{-1}\big(\vec{\xi}\,;\vec{x}\,\big)$ is the inverse[2] of the Fisher information matrix $F_c\big(\vec{\xi}\,;\vec{x}\,\big)$. The classical Fisher information matrix $F_c\big(\vec{\xi}\,;\vec{x}\,\big)$ is the square symmetric non-negative definite matrix defined as the covariance matrix $F_c\big(\vec{\xi}\,;\vec{x}\,\big) = \big\langle\vec{V}\vec{V}^\top\big\rangle$ of the score vector $\vec{V}(\vec{x}\,;\vec{\xi}\,) = \vec{\nabla}\ln p(\vec{x}\,;\vec{\xi}\,)$ having the vanishing mean $\big\langle\vec{V}\big\rangle = \vec{0}$. Equivalently, the classical Fisher information matrix $F_c\big(\vec{\xi}\,;\vec{x}\,\big) = \big[F_{jk}\big]$ is defined by the matrix elements

$$F_{jk} = \big\langle V_j V_k\big\rangle = \left\langle \frac{1}{p^2(\vec{x}\,;\vec{\xi}\,)}\,\frac{\partial p(\vec{x}\,;\vec{\xi}\,)}{\partial\xi_j}\,\frac{\partial p(\vec{x}\,;\vec{\xi}\,)}{\partial\xi_k}\right\rangle \tag{11.13}$$

$$= -\left\langle\frac{\partial^2}{\partial\xi_j\partial\xi_k}\Big[\ln p(\vec{x}\,;\vec{\xi}\,)\Big]\right\rangle\,. \tag{11.14}$$

The vector maximum-likelihood estimator is defined as

---

[2] More generally, one must resort to the Moore–Penrose pseudoinverse when $F_c\big(\vec{\xi}\,;\vec{x}\,\big)$ is not invertible [287, 288].

$$\widehat{\vec{\xi}}_{\mathrm{ML}}(\vec{x}\,) = \arg\max_{\vec{\xi}}\, p(\vec{x}\,;\vec{\xi}\,)\,. \tag{11.15}$$

Asymptotically, for a large number N of independent data points $(x_1, x_2, \ldots, x_N) = \vec{x}$, the vector maximum-likelihood estimator becomes unbiased, saturates the Cramér–Rao inequality of Eq. (11.12), and acquires a (multidimensional) Gaussian distribution $\widehat{\vec{\xi}}_{\mathrm{ML}}(\vec{x}\,) \sim \mathcal{N}\big(\vec{\xi}, F_c^{-1}(\vec{\xi}\,;\vec{x}\,)\big)$ with mean $\vec{\xi}$ and covariance matrix $F_c^{-1}\big(\vec{\xi}\,;\vec{x}\,\big)$.

Upon a change of parameter $\vec{\xi} = \vec{\xi}\big(\vec{\nu}\,\big) = g\big(\vec{\nu}\,\big)$ via a multidimensional function $g(\cdot)$, the components of the score vector transform as $V_\ell\big(\vec{x}\,;\vec{\nu}\,\big) = \sum_k J_{\ell k} V_k\big(\vec{x}\,;\vec{\xi}\,\big)$, with the Jacobian matrix $J = \big[J_{\ell k} = \partial\xi_k/\partial\nu_\ell\big]$. The Fisher information matrices relative to the parameters $\vec{\xi}$ and $\vec{\nu}$ are related by $F_c\big(\vec{\nu}\,\big) = J\,F_c\big(\vec{\xi}\,\big)J^\top$, and the maximum-likelihood estimators by $\widehat{\vec{\xi}}_{\mathrm{ML}} = g\big(\widehat{\vec{\nu}}_{\mathrm{ML}}\big)$.

The three properties indicated in Sect. 11.1.1 for the scalar Fisher information, concerning additivity, monotonicity and convexity, naturally extend to the Fisher information matrix.

11.1.3 Random Parameter: Bayesian Estimation

A related but somewhat distinct approach to estimation consists in the following Bayesian approach [236, 239, 284], which we introduce for a single scalar parameter, while it naturally extends to multiple parameters. Instead of considering an unknown parameter with a fixed deterministic value ξ, Bayesian estimation considers a random parameter ξ distributed according to the prior probability density $p_\xi(\xi)$. The data $\vec{x} = (x_1, x_2, \ldots, x_N)$ produced under a given realization of ξ is governed by the conditional probability density $p(\vec{x}\,|\xi)$. The Bayes rule provides access to a posterior probability density for the parameter ξ, given the data $\vec{x}$, as

$$p(\xi\,|\vec{x}\,) = \frac{p(\vec{x}\,|\xi)p_\xi(\xi)}{p(\vec{x}\,)}\,, \tag{11.16}$$

with the total probability density

$$p(\vec{x}\,) = \int_{-\infty}^{\infty} p(\vec{x}\,|\xi)p_\xi(\xi)d\xi\,. \tag{11.17}$$

Usually in a Bayesian framework, the density $p_\xi(\xi)$ is known as describing the prior situation, or our prior ignorance, of the parameter ξ; meanwhile, the conditional distribution $p(\vec{x}\,|\xi)$ is known from a model of how the data $\vec{x}$ is produced in the presence of a value ξ of the parameter. The posterior distribution $p(\xi\,|\vec{x}\,)$ then follows from Eq. (11.16). The posterior distribution $p(\xi\,|\vec{x}\,)$ will then determine the optimal Bayesian estimator, for any observed data $\vec{x}$ we have in hand, in the following way.

From a given realization of the observed data $\vec{x}$, an estimator $\widehat{\xi}(\vec{x})$ for the unknown parameter ξ is now characterized by the mean-squared error

$$e^2\big(\widehat{\xi}\,|\vec{x}\big) = \big\langle[\widehat{\xi}(\vec{x}) - \xi]^2\big\rangle_{\xi|\vec{x}} = \int_{-\infty}^{\infty} \big[\widehat{\xi}(\vec{x}) - \xi\big]^2 p(\xi\,|\vec{x})d\xi\,, \tag{11.18}$$

with the average denoted $\langle\cdot\rangle_{\xi|\vec{x}}$ taken according to the posterior probability density $p(\xi\,|\vec{x})$. We have equivalently

$$e^2\big(\widehat{\xi}\,|\vec{x}\big) = \big[\widehat{\xi}(\vec{x}) - \langle\xi\rangle_{\xi|\vec{x}}\big]^2 + \langle\xi^2\rangle_{\xi|\vec{x}} - \langle\xi\rangle^2_{\xi|\vec{x}}\,, \tag{11.19}$$

or

$$e^2\big(\widehat{\xi}\,|\vec{x}\big) = \big[\widehat{\xi}(\vec{x}) - \langle\xi\rangle_{\xi|\vec{x}}\big]^2 + \mathrm{var}(\xi\,|\vec{x})\,, \tag{11.20}$$

with the posterior variance of the parameter

$$\mathrm{var}(\xi\,|\vec{x}) = \langle\xi^2\rangle_{\xi|\vec{x}} - \langle\xi\rangle^2_{\xi|\vec{x}} = \int_{-\infty}^{\infty} \big[\xi - \langle\xi\rangle_{\xi|\vec{x}}\big]^2 p(\xi\,|\vec{x})d\xi\,. \tag{11.21}$$

In the right side of Eq. (11.20), the variance $\mathrm{var}(\xi\,|\vec{x})$ is independent of the estimator $\widehat{\xi}(\vec{x})$, so that minimization of the mean-squared error $e^2\big(\widehat{\xi}\,|\vec{x}\big)$ is realized by the **optimal Bayesian estimator**

$$\widehat{\xi}_{\mathrm{B}}(\vec{x}) = \langle\xi\rangle_{\xi|\vec{x}} = \int_{-\infty}^{\infty} \xi\; p(\xi\,|\vec{x})d\xi\,, \tag{11.22}$$

which is the posterior mean of the parameter, achieving the minimal mean-squared error $e^2\big(\widehat{\xi}_{\mathrm{B}}|\vec{x}\big) = \mathrm{var}(\xi\,|\vec{x})$ of Eq. (11.21). For each observed data $\vec{x}$, an estimate $\widehat{\xi}_{\mathrm{B}}(\vec{x})$ follows for ξ via Eq. (11.22) from the knowledge of the posterior distribution $p(\xi\,|\vec{x})$ we have from Eq. (11.16).

There exists an equivalent to the Cramér–Rao inequality of Eqs. (11.8)–(11.9) for lower bounding the mean-squared error of a Bayesian estimator [236, 289]. By the Cauchy–Schwarz inequality we have

$$\Big\langle\big[\widehat{\xi}(\vec{x}) - \xi\big]^2\Big\rangle_{\vec{x},\xi} \Big\langle\big[\partial_\xi \ln p(\vec{x},\xi)\big]^2\Big\rangle_{\vec{x},\xi} \geq \Big\langle\big[\widehat{\xi}(\vec{x}) - \xi\big]\big[\partial_\xi \ln p(\vec{x},\xi)\big]\Big\rangle^2_{\vec{x},\xi}\,, \tag{11.23}$$

with the average denoted $\big\langle\cdot\big\rangle_{\vec{x},\xi}$ taken according to the joint probability density $p(\vec{x},\xi) = p(\vec{x}\,|\xi)p_\xi(\xi)$.

For handling the right side of Eq. (11.23) we have

$$\Big\langle [\widehat{\xi}(\vec{x}\,) - \xi][\partial_\xi \ln p(\vec{x}, \xi)]\Big\rangle_{\vec{x},\xi} = \Big\langle [\widehat{\xi}(\vec{x}\,) - \xi]\frac{\partial_\xi p(\vec{x}, \xi)}{p(\vec{x}, \xi)}\Big\rangle_{\vec{x},\xi} \tag{11.24}$$

$$= \int_{\mathbb{R}^N}\int_{-\infty}^{\infty} [\widehat{\xi}(\vec{x}\,) - \xi]\partial_\xi p(\vec{x}, \xi) d\xi\, d\vec{x}\,. \tag{11.25}$$

For Eq. (11.25) we have

$$\int_{-\infty}^{\infty} \widehat{\xi}(\vec{x}\,)\partial_\xi p(\vec{x}, \xi) d\xi = \Big[\widehat{\xi}(\vec{x}\,)p(\vec{x}, \xi)\Big]_{\xi=-\infty}^{\xi=+\infty} = 0 - 0 = 0\,, \tag{11.26}$$

since the estimator $\widehat{\xi}(\vec{x}\,)$ is independent of the parameter ξ and the density $p(\vec{x}, \xi)$ integrates to 1 over its support so that $p(\vec{x}, \xi) \to 0$ when $\xi \to \pm\infty$. In addition for Eq. (11.25) by integration by parts we also have

$$-\int_{-\infty}^{\infty} \xi\,\partial_\xi p(\vec{x}, \xi) d\xi = -\Big[\xi p(\vec{x}, \xi)\Big]_{\xi=-\infty}^{\xi=+\infty} + \int_{-\infty}^{\infty} p(\vec{x}, \xi) d\xi\,. \tag{11.27}$$

For a well-behaved density $p(\vec{x}, \xi)$ having a finite first statistical moment, $\xi p(\vec{x}, \xi)$ is finitely integrable with $\xi p(\vec{x}, \xi)$ returning to zero at $\xi \to \pm\infty$, so that $\big[\xi p(\vec{x}, \xi)\big]_{\xi=-\infty}^{\xi=+\infty} = 0 - 0 = 0$. Then by completing with the integration over $\vec{x}$, Eq. (11.27) integrates to 1, and since there is zero contribution from Eq. (11.26), the right side of the Cauchy–Schwarz inequality of Eq. (11.23) also evaluates to 1.

For handling the left side of Eq. (11.23) we have

$$\Big\langle \big[\partial_\xi \ln p(\vec{x}, \xi)\big]^2\Big\rangle_{\vec{x},\xi} = \Big\langle \big[\partial_\xi \ln p(\vec{x}\,|\xi) + \partial_\xi \ln p_\xi(\xi)\big]^2\Big\rangle_{\vec{x},\xi}\,. \tag{11.28}$$

When the squared bracket $[\cdot]^2$ in the right side of Eq. (11.28) is expanded, it introduces a crossed term of the form

$$\Big\langle \big[\partial_\xi \ln p(\vec{x}\,|\xi)\big]\big[\partial_\xi \ln p_\xi(\xi)\big]\Big\rangle_{\vec{x},\xi} = \int_{\mathbb{R}^N}\int_{-\infty}^{\infty} \big[\partial_\xi p(\vec{x}\,|\xi)\big]\big[\partial_\xi p_\xi(\xi)\big] d\xi\, d\vec{x} \tag{11.29}$$

$$= \int_{-\infty}^{\infty} \big[\partial_\xi p_\xi(\xi)\big]\left[\partial_\xi \int_{\mathbb{R}^N} p(\vec{x}\,|\xi) d\vec{x}\right] d\xi \tag{11.30}$$

$$= \int_{-\infty}^{\infty} \big[\partial_\xi p_\xi(\xi)\big]\big[\partial_\xi 1\big]\, d\xi = 0\,, \tag{11.31}$$

since the conditional density $p(\vec{x}\,|\xi)$ integrates over $\vec{x}$ to 1 for any ξ. The crossed term therefore vanishes, and Eq. (11.28) reduces to

$$
\begin{aligned}
&\left\langle\left[\partial_\xi \ln p(\vec{x},\xi)\right]^2\right\rangle_{\vec{x},\xi} \\
&= \left\langle\left[\frac{\partial_\xi p(\vec{x}\,|\xi)}{p(\vec{x}\,|\xi)}\right]^2\right\rangle_{\vec{x},\xi} + \left\langle\left[\frac{\partial_\xi p_\xi(\xi)}{p_\xi(\xi)}\right]^2\right\rangle_{\vec{x},\xi} &(11.32)\\
&= \int_{-\infty}^{\infty}\int_{\mathbb{R}^N}\frac{[\partial_\xi p(\vec{x}\,|\xi)]^2}{p(\vec{x}\,|\xi)}d\vec{x}\; p_\xi(\xi)d\xi + \int_{-\infty}^{\infty}\frac{[\partial_\xi p_\xi(\xi)]^2}{p_\xi(\xi)}d\xi \int_{\mathbb{R}^N} p(\vec{x}\,|\xi)d\vec{x} &(11.33)\\
&= \int_{-\infty}^{\infty} F_c(\xi\,;\vec{x}\,)p_\xi(\xi)d\xi + F_c^{\text{pri}}(\xi) &(11.34)\\
&= \left\langle F_c(\xi\,;\vec{x}\,)\right\rangle_\xi + F_c^{\text{pri}}(\xi)\,. &(11.35)
\end{aligned}
$$

In Eq. (11.34) we have introduced

$$
\int_{-\infty}^{\infty} F_c(\xi\,;\vec{x}\,)p_\xi(\xi)d\xi = \left\langle F_c(\xi\,;\vec{x}\,)\right\rangle_\xi \tag{11.36}
$$

which is the classical Fisher information $F_c(\xi\,;\vec{x}\,)$ of Eq. (11.4) averaged over the prior density $p_\xi(\xi)$ of the parameter, and

$$
F_c^{\text{pri}}(\xi) = \int_{-\infty}^{\infty}\frac{[\partial_\xi p_\xi(\xi)]^2}{p_\xi(\xi)}d\xi \tag{11.37}
$$

is the Fisher information that can be attached to the prior density $p_\xi(\xi)$ of the parameter ξ.

Finally, returning to the Cauchy–Schwarz inequality of Eq. (11.23), we deduce

$$
\left\langle\left[\widehat{\xi}(\vec{x}\,) - \xi\right]^2\right\rangle_{\vec{x},\xi} \ge \frac{1}{\left\langle F_c(\xi\,;\vec{x}\,)\right\rangle_\xi + F_c^{\text{pri}}(\xi)}\,, \tag{11.38}
$$

which is the **Van Trees inequality** [236], placing a lower bound to the mean-squared error of any Bayesian estimator $\widehat{\xi}(\vec{x}\,)$. This is the Bayesian counterpart of the Cramér–Rao inequality of Eqs. (11.8)–(11.9).

Alternatively, from $p(\vec{x},\xi) = p(\xi\,|\vec{x}\,)p(\vec{x}\,)$, since $\partial_\xi \ln p(\vec{x}\,) = 0$, the left side of Eq. (11.32) is also

$$
\begin{aligned}
\left\langle\left[\partial_\xi \ln p(\xi\,|\vec{x}\,)\right]^2\right\rangle_{\vec{x},\xi} &= \int_{\mathbb{R}^N}\int_{-\infty}^{\infty}\frac{[\partial_\xi p(\xi\,|\vec{x}\,)]^2}{p(\xi\,|\vec{x}\,)}d\xi\; p(\vec{x}\,)d\vec{x} &(11.39)\\
&= \int_{\mathbb{R}^N} F_c^{\text{pos}}(\vec{x}\,;\xi)\; p(\vec{x}\,)d\vec{x} &(11.40)\\
&= \left\langle F_c^{\text{pos}}(\vec{x}\,;\xi)\right\rangle_{\vec{x}}\,, &(11.41)
\end{aligned}
$$

where we have introduced an $\vec{x}$-dependent posterior Fisher information

$$F_c^{\text{pos}}(\vec{x}\,;\xi) = \int_{-\infty}^{\infty} \frac{[\partial_\xi p(\xi\,|\vec{x}\,)]^2}{p(\xi\,|\vec{x}\,)} d\xi \tag{11.42}$$

therefore satisfying

$$\big\langle F_c^{\text{pos}}(\vec{x}\,;\xi)\big\rangle_{\vec{x}} = \big\langle F_c(\xi\,;\vec{x}\,)\big\rangle_\xi + F_c^{\text{pri}}(\xi)\,, \tag{11.43}$$

so that the lower bound in the right side of the Van Trees inequality of Eq. (11.38) is also $1/\big\langle F_c^{\text{pos}}(\vec{x}\,;\xi)\big\rangle_{\vec{x}}$.

The notion of bias, as defined previously in Sect. 11.1.1, does not apply as such to a Bayesian estimator as for instance $\widehat{\xi}_{\text{B}}(\vec{x}\,)$ of Eq. (11.22), since the bias represents the average departure of an estimator from the true value of the parameter ξ and in Bayesian estimation there is no such true value of the parameter, but rather a distribution of values. Nevertheless, if a Bayesian bias were defined as $\big\langle\widehat{\xi}(\vec{x}\,) - \xi\big\rangle_{\xi|\vec{x}}$, then the optimal Bayesian estimator $\widehat{\xi}_{\text{B}}(\vec{x}\,)$ of Eq. (11.22) would be unbiased in this sense for any observed data $\vec{x}$.

When the dataset $\vec{x} = (x_1, x_2, \ldots, x_N)$ is formed by N independent and identically distributed random variables x_n, then $\big\langle F_c(\xi\,;\vec{x}\,)\big\rangle_\xi = N\big\langle F_c(\xi\,;x_n)\big\rangle_\xi$ dominates at large N over $F_c^{\text{pri}}(\xi)$ and the lower bound in the right side of Eq. (11.38) reduces to $1/\big\langle F_c(\xi\,;\vec{x}\,)\big\rangle_\xi$. Moreover, the optimal Bayesian estimator $\widehat{\xi}_{\text{B}}(\vec{x}\,)$ of Eq. (11.22) can be shown, asymptotically, to saturate this bound in Eq. (11.38) and to exhibit a Gaussian distribution [289, 290], much like the maximum-likelihood estimator for the deterministic parameter of Sect. 11.1.1.

The notions of estimators, mean-squared estimation error, governed by the Fisher information, for deterministic as well as for random parameters, as presented in this Section, will now serve as guidelines to address quantum parameter estimation.

11.2 Parameter Estimation on a Quantum State

We now come to the task of quantum parameter estimation [235, 291, 292]. As for classical estimation, the main considerations useful to quantum estimation will be naturally introduced for the situation of a deterministic scalar parameter, and they will be subsequently extended in various directions.

We consider a quantum system with N-dimensional Hilbert space $\mathcal{H}_N$, whose state is represented by the density operator $\rho_\xi \in \mathcal{L}(\mathcal{H}_N)$. The density operator ρ_ξ is dependent upon a deterministic scalar parameter ξ, which assumes a definite but unknown value. We undertake to estimate the value of ξ by means of quantum measurement on the state ρ_ξ.

In general, a quantum measurement on ρ_ξ can be carried out by means of a POVM $\{\mathsf{E}_m\}_{m=1}^M$ with M POVM operators E_m for M measurement outcomes, as presented in Sect. 4.2.4. Here also, when devising useful estimation strategies, the meaningful performance assessments involve essentially the POVM $\{\mathsf{E}_m\}$, as anticipated in

Sect. 4.2.4, leaving subsequent latitude to choose underlying measurement operators $\{\mathsf{M}_m\}$ practically implementing a (any) selected POVM as $\{\mathsf{E}_m = \mathsf{M}_m^\dagger \mathsf{M}_m\}$.

Upon measuring the parameter-dependent quantum signal ρ_ξ, the measurement outcome m is obtained with the probability

$$P(m\,;\xi) = \mathrm{tr}\big(\mathsf{E}_m \rho_\xi\big)\,, \tag{11.44}$$

for $m = 1$ to M, as in Eq. (4.39). The measurement outcome m, with the status of a classical random variable, can be used to devise an estimator $\widehat{\xi}(m)$ for the unknown parameter ξ, much like in Sect. 11.1.1, and whose performance is now addressed.

11.2.1 Quantum Score and Quantum Fisher Information

The performance of any such estimator is governed by the classical Fisher information

$$F_c(\xi\,;m) = \sum_{m=1}^{M} \frac{1}{P(m\,;\xi)}\Big[\partial_\xi P(m\,;\xi)\Big]^2 = \sum_{m=1}^{M} \frac{1}{\mathrm{tr}(\mathsf{E}_m \rho_\xi)}\Big[\partial_\xi\, \mathrm{tr}\big(\mathsf{E}_m \rho_\xi\big)\Big]^2 \tag{11.45}$$

about the parameter ξ contained in the data m, which is similar to Eq. (11.4) for the discrete probability distribution $P(m\,;\xi) = \mathrm{tr}\big(\mathsf{E}_m \rho_\xi\big)$ of the random variable m function of the unknown parameter ξ. The derivative ∂_ξ present in the Fisher information $F_c(\xi\,;m)$ of Eq. (11.45) operates as $\partial_\xi P(m\,;\xi) = \partial_\xi\, \mathrm{tr}\big(\mathsf{E}_m \rho_\xi\big) = \mathrm{tr}\big(\mathsf{E}_m \partial_\xi \rho_\xi\big)$, since the measurement operator E_m, much like the classical estimator $\widehat{\xi}(m)$, ought to be independent of the unknown parameter ξ, and so as to yield

$$F_c(\xi\,;m) = \sum_{m=1}^{M} \frac{1}{\mathrm{tr}(\mathsf{E}_m \rho_\xi)}\Big[\mathrm{tr}\big(\mathsf{E}_m \partial_\xi \rho_\xi\big)\Big]^2 . \tag{11.46}$$

The derivative ∂_ξ applying in Eq. (11.45) to the classical scalar probability $P(m\,;\xi)$, now in Eq. (11.46) applies as $\partial_\xi \rho_\xi$ to the quantum density operator ρ_ξ, and to push further the quantum analysis we will follow the approach of [293]. A convenient way to handle the derivative (a linear operation) of a density operator is through the introduction of another (linear) operator V_ξ, known as the symmetrized logarithmic derivative, and defined by the equation (yet to be solved) [235, 294, 295]

$$\partial_\xi \rho_\xi \equiv \frac{\partial \rho_\xi}{\partial \xi} = \frac{1}{2}\Big(\mathsf{V}_\xi \rho_\xi + \rho_\xi \mathsf{V}_\xi\Big)\,. \tag{11.47}$$

Since ρ_ξ is a positive hence Hermitian operator satisfying $\rho_\xi^\dagger = \rho_\xi$, hermiticity is also transferred to its derivative to yield $(\partial_\xi \rho_\xi)^\dagger = (\rho_\xi \mathsf{V}_\xi^\dagger + \mathsf{V}_\xi^\dagger \rho_\xi)/2 = \partial_\xi \rho_\xi$, for any ρ_ξ, so that by identification with Eq. (11.47) we deduce $\mathsf{V}_\xi^\dagger = \mathsf{V}_\xi$. Therefore, Eq. (11.47) defines a Hermitian operator V_ξ, which is matched to the density operator ρ_ξ, and for this reason generally carries dependence on the unknown parameter ξ. This way of handling the derivative in Eq. (11.47) is in particular preferable to the alternatives $\partial_\xi \rho_\xi = \mathsf{V}_\xi \rho_\xi$ or $\partial_\xi \rho_\xi = \rho_\xi \mathsf{V}_\xi$; these alternatives would more directly parallel the classical approach of Sect. 11.1.1 with Eq. (11.2) for the scalar classical score $V(\vec{x}\,;\xi)$ yielding $\partial_\xi p = Vp = pV$, yet they would not set a Hermitian operator V_ξ, whence the symmetrized version of Eq. (11.47).

Furthermore, by taking the trace of Eq. (11.47) we obtain $\mathrm{tr}(\partial_\xi \rho_\xi) = \mathrm{tr}(\rho_\xi \mathsf{V}_\xi)$ by circularity of the trace; this is also $\mathrm{tr}(\rho_\xi \mathsf{V}_\xi) = \partial_\xi\, \mathrm{tr}(\rho_\xi) = \partial_\xi 1 = 0$, so that we deduce the vanishing mean $\langle \mathsf{V}_\xi \rangle = \mathrm{tr}(\rho_\xi \mathsf{V}_\xi) = 0$. The quantum observable V_ξ with zero mean is also known as the **quantum score** operator, and it can be regarded as a quantum analogue of the classical scalar score function $V(\vec{x}\,;\xi)$ of Eq. (11.2), as it will also further verify in the sequel.

From Eq. (11.47) we obtain

$$\mathrm{tr}\big(\mathsf{E}_m \partial_\xi \rho_\xi\big) = \frac{1}{2}\Big[\mathrm{tr}(\mathsf{E}_m \mathsf{V}_\xi \rho_\xi) + \mathrm{tr}(\mathsf{E}_m \rho_\xi \mathsf{V}_\xi)\Big] \tag{11.48}$$

$$= \frac{1}{2}\Big[\mathrm{tr}(\rho_\xi \mathsf{E}_m \mathsf{V}_\xi) + \mathrm{tr}\big((\rho_\xi \mathsf{E}_m \mathsf{V}_\xi)^\dagger\big)\Big] \tag{11.49}$$

$$= \mathrm{Re}\,\mathrm{tr}\big(\rho_\xi \mathsf{E}_m \mathsf{V}_\xi\big)\ . \tag{11.50}$$

This real part of the trace in Eq. (11.50) provides an expression for $\mathrm{tr}(\mathsf{E}_m \partial_\xi \rho_\xi)$, which can be inserted in Eq. (11.46) to yield the classical Fisher information

$$F_c(\xi\,;m) = \sum_{m=1}^{M} \frac{1}{\mathrm{tr}(\rho_\xi \mathsf{E}_m)} \Big[\mathrm{Re}\,\mathrm{tr}\big(\rho_\xi \mathsf{E}_m \mathsf{V}_\xi\big)\Big]^2 . \tag{11.51}$$

In the quantum domain, for efficient estimation from measurement of the quantum state ρ_ξ, one has then the faculty to choose the measuring POVM $\{\mathsf{E}_m\}_{m=1}^M$ so as to maximize the classical Fisher information $F_c(\xi\,;m)$ in Eq. (11.51). In this respect, there exists an upper bound [292, 296] limiting from above the classical Fisher information $F_c(\xi\,;m)$ of Eq. (11.51) for all feasible POVM, and that can be derived from the Cauchy–Schwarz inequality as we now show.

For $\mathrm{tr}(\rho_\xi \mathsf{E}_m)$ which is a real positive scalar, a useful factorization introduces the two operators $\mathsf{A}^\dagger = \sqrt{\rho_\xi}\sqrt{\mathsf{E}_m}/\sqrt{\mathrm{tr}(\rho_\xi \mathsf{E}_m)}$ and $\mathsf{B} = \sqrt{\mathsf{E}_m}\mathsf{V}_\xi\sqrt{\rho_\xi}$ giving $\mathrm{tr}(\mathsf{A}^\dagger \mathsf{B}) = \mathrm{tr}\big(\rho_\xi \mathsf{E}_m \mathsf{V}_\xi\big)/\sqrt{\mathrm{tr}(\rho_\xi \mathsf{E}_m)}$, so as to write for Eq. (11.51),

$$F_c(\xi\,;m) = \sum_{m=1}^{M}\Big[\mathrm{Re}\,\mathrm{tr}(\mathsf{A}^\dagger \mathsf{B})\Big]^2 \le \sum_{m=1}^{M}\big|\mathrm{tr}(\mathsf{A}^\dagger \mathsf{B})\big|^2\ , \tag{11.52}$$

since, by the Pythagorean theorem, the squared real part of a complex number is never larger than its squared modulus. The Hilbert–Schmidt inner product $\mathrm{tr}(\mathsf{A}^\dagger \mathsf{B})$ of two operators satisfies the Cauchy–Schwarz inequality $\left|\mathrm{tr}(\mathsf{A}^\dagger\mathsf{B})\right|^2 \le \mathrm{tr}(\mathsf{A}^\dagger\mathsf{A})\,\mathrm{tr}(\mathsf{B}^\dagger\mathsf{B})$ from Eq. (A.17) of the Appendix. Both operators $\sqrt{\rho_\xi}$ and $\sqrt{\mathsf{E}_m}$ are Hermitian, since the positive operators ρ_ξ and E_m are. We then deduce $\mathrm{tr}(\mathsf{A}^\dagger\mathsf{A}) = 1$ and $\mathrm{tr}(\mathsf{B}^\dagger\mathsf{B}) = \mathrm{tr}\big(\mathsf{E}_m\mathsf{V}_\xi\rho_\xi\mathsf{V}_\xi\big)$, leading from Eq. (11.52) to

$$F_c(\xi\,;m) \le \sum_{m=1}^{M} \left|\mathrm{tr}(\mathsf{A}^\dagger\mathsf{B})\right|^2 \tag{11.53}$$

$$\le \sum_{m=1}^{M} \mathrm{tr}\big(\mathsf{E}_m\mathsf{V}_\xi\rho_\xi\mathsf{V}_\xi\big) = \mathrm{tr}\big(\mathsf{V}_\xi\rho_\xi\mathsf{V}_\xi\big)\,, \tag{11.54}$$

where the last equality arises because the elements E_m of any valid POVM satisfy $\sum_{m=1}^{M}\mathsf{E}_m = \mathrm{I}_N$ summing to the identity operator I_N on $\mathcal{H}_N$. Equation (11.54) leads then to the targeted inequality

$$F_c(\xi\,;m) \le \mathrm{tr}\big(\rho_\xi\mathsf{V}_\xi^2\big) \equiv F_q(\xi)\,, \tag{11.55}$$

defining the **quantum Fisher information**

$$\boxed{F_q(\xi) = \mathrm{tr}\big(\rho_\xi\mathsf{V}_\xi^2\big) = \big\langle\mathsf{V}_\xi^2\big\rangle\,,} \tag{11.56}$$

limiting the classical Fisher information $F_c(\xi\,;m)$, and exhibiting analogy with its classical counterpart in Eq. (11.4). The relation to the score, in Eq. (11.56) and in Eq. (11.4), also reinforces the interpretation of the operator V_ξ as the quantum score. The operator V_ξ is Hermitian, so V_ξ^2 is (Hermitian) positive, and its average $F_q(\xi)$ is a non-negative real scalar. In addition, from Eq. (11.47) one has $\partial_\xi(\rho_\xi\mathsf{V}_\xi) = \rho_\xi\partial_\xi\mathsf{V}_\xi + (\partial_\xi\rho_\xi)\mathsf{V}_\xi = \rho_\xi\partial_\xi\mathsf{V}_\xi + \big(\mathsf{V}_\xi\rho_\xi\mathsf{V}_\xi + \rho_\xi\mathsf{V}_\xi^2\big)/2$; and since $\mathrm{tr}\big(\partial_\xi(\rho_\xi\mathsf{V}_\xi)\big) = \partial_\xi\,\mathrm{tr}\big(\rho_\xi\mathsf{V}_\xi\big) = \partial_\xi 0 = 0$, one deduces $0 = \mathrm{tr}\big(\rho_\xi\partial_\xi\mathsf{V}_\xi\big) + \mathrm{tr}\big(\mathsf{V}_\xi\partial_\xi\rho_\xi\big) = \mathrm{tr}\big(\rho_\xi\partial_\xi\mathsf{V}_\xi\big) + \mathrm{tr}\big(\rho_\xi\mathsf{V}_\xi^2\big)$, giving $\mathrm{tr}\big(\mathsf{V}_\xi\partial_\xi\rho_\xi\big) = -\,\mathrm{tr}\big(\rho_\xi\partial_\xi\mathsf{V}_\xi\big)$, and the alternative expression $F_q(\xi) = \mathrm{tr}\big(\mathsf{V}_\xi\partial_\xi\rho_\xi\big)$, which is also

$$F_q(\xi) = -\,\mathrm{tr}\big(\rho_\xi\partial_\xi\mathsf{V}_\xi\big) = -\big\langle\partial_\xi\mathsf{V}_\xi\big\rangle \tag{11.57}$$

exhibiting similarity with Eq. (11.5).

The classical Fisher information $F_c(\xi\,;m)$, as visible in Eq. (11.45), is in general dependent upon the measuring POVM $\{\mathsf{E}_m\}_{m=1}^{M}$, via the probabilities $P(m\,;\xi) = \mathrm{tr}(\mathsf{E}_m\rho_\xi)$ of Eq. (11.44) for the measurement outcomes m. By contrast, as we have seen in Sect. 11.1, the classical Fisher information $F_c(\xi\,;m)$ does not refer to, and is therefore independent of, any estimator $\widehat{\xi}(m)$ employed for estimating the unknown parameter ξ from the measurement outcome m. Now, the quantum Fisher information $F_q(\xi)$ defined in Eq. (11.56) does not refer to any measuring POVM. It is intrinsic to the relationship of the unknown parameter ξ with the quantum state ρ_ξ, and solely determined by it. In this respect, $F_q(\xi)$ offers a common upper bound limiting via

Eq. (11.55) any POVM-dependent classical Fisher information $F_c(\xi\,;m)$. For these reasons, the quantum Fisher information $F_q(\xi)$ represents an important reference for quantum estimation, assessing the overall maximal efficiency that can be envisaged.

Now, for efficient estimation, we would like to identify an optimal POVM $\{\mathsf{E}_m\}_{m=1}^M$ saturating the inequality of Eq. (11.55), that is, a POVM achieving $F_c(\xi\,;m) = F_q(\xi)$. The conditions for achieving $F_c(\xi\,;m) = F_q(\xi)$ are analyzed for instance in [296–299], and are also given in [293] that we follow here while adapting to the notations of this textbook. Achieving $F_c(\xi\,;m) = F_q(\xi)$ requires first to achieve equality in Eq. (11.52) or (11.53), and a necessary and sufficient condition for this is for $\mathrm{tr}(\mathsf{A}^\dagger\mathsf{B})$ to be real. The second requirement is to achieve equality also in Eq. (11.54), and a necessary and sufficient condition for this is for A and B to be proportional. These two requirements are jointly satisfied by the equivalent necessary and sufficient condition of A and B proportional through a real proportionality coefficient, that we write $\mathsf{A} \propto_{\mathbb{R}} \mathsf{B}$, so as to saturate the inequality of Eq. (11.55). In this way, equality in Eq. (11.55) is achieved if and only if each POVM element E_m realizes [296, 297]

$$\sqrt{\mathsf{E}_m}\,\mathsf{V}_\xi\sqrt{\rho_\xi} \propto_{\mathbb{R}} \sqrt{\mathsf{E}_m}\sqrt{\rho_\xi}\;. \tag{11.58}$$

A sufficient condition to satisfy Eq. (11.58) is to satisfy $\sqrt{\mathsf{E}_m}\,\mathsf{V}_\xi \propto_{\mathbb{R}} \sqrt{\mathsf{E}_m}$, which becomes a sufficient and necessary condition when ρ_ξ, and hence $\sqrt{\rho_\xi}$, are full-rank operators. Also, the complete set of POVM elements $\{\mathsf{E}_m\}_{m=1}^M$ we are seeking together must sum to the identity operator I_N on $\mathcal{H}_N$. The known operators E_m solutions to Eq. (11.58) are usually expressed in terms of the eigendecomposition of the quantum score V_ξ. A POVM $\{\mathsf{E}_m\}_{m=1}^M$ with general significance known [296–299] to satisfy Eq. (11.58) is the projective measurement realized by the projectors onto the eigensubspaces of the score operator V_ξ. This amounts to defining the measuring POVM as a measurement of the observable V_ξ. In this circumstance, V_ξ and $\sqrt{\mathsf{E}_m}$ commute in Eq. (11.58), and the proportionality coefficient is the (real) eigenvalue in the corresponding eigensubspace of V_ξ. However such a POVM, being characterized through V_ξ, generally carries explicit dependence on the unknown parameter ξ, as V_ξ does. In this context of estimation, such ξ-dependent solutions are therefore unacceptable (inaccessible) for a practically feasible quantum measurement. Also, only ξ-independent POVM and estimators are constrained by the inequalities of Eqs. (11.55) and (11.8), as it appeared in their derivations; if ξ-dependent solutions were allowed, the best estimator would always be $\widehat{\xi}(\vec{x}) = \xi$, irrespective of any measurement, which is clearly unacceptable since ξ is unknown. In general, there is no known solution, and a fortiori no known methodology, to obtain an optimal acceptable (ξ-independent) POVM $\{\mathsf{E}_m\}_{m=1}^M$ that would satisfy Eq. (11.58) and maximize the Fisher information $F_c(\xi\,;m)$ at $F_q(\xi)$ in Eq. (11.55). In this way, the upper bound formed by the quantum Fisher information $F_q(\xi)$ in Eq. (11.55), does not generally stand as an achievable upper bound. Situations may exist with no definite (parameter-independent as it should) measuring POVM able to achieve equality in Eq. (11.55).

Even if one cannot generally define a ξ-independent optimal POVM saturating Eq. (11.55), one can however resort to adaptive techniques [297, 300–306] to progressively adjust a modifiable POVM via repeated measurements on several replicas of the state ρ_ξ undergoing the estimation. For instance [297], one can start with a rough estimate of ξ from a non-optimized initial POVM, then exploit this estimate of ξ to construct a current optimal POVM based on the eigendecomposition of the score V_ξ at this current estimate of ξ, then redo an estimation with this current "locally optimal" POVM, and iterate over successive replicas of ρ_ξ. In this way, by adaptive techniques, if not with one fixed ξ-independent POVM, the maximal performance determined by $F_q(\xi)$ in Eq. (11.55) can often be reached or sufficiently approached. In any case, the quantum Fisher information $F_q(\xi)$ is always a meaningful reference to situate the performance of any quantum estimation strategy.

Upon a change of parameter $\xi = \xi(\nu)$, the classical properties mentioned at the end of Sect. 11.1.1 apply in a comparable way in the quantum case. The quantum score operators transform as $\mathsf{V}_\nu = (\partial\xi/\partial\nu)\mathsf{V}_\xi$, and the quantum Fisher informations relative to the parameters ξ and ν verify $F_q(\nu) = (\partial\xi/\partial\nu)^2 F_q(\xi)$.

Five Properties: We have the five useful properties for quantum estimation.

(1) We consider a ξ-dependent density operator $\rho_\xi^{(1)}$ associated via Eq. (11.47) with the quantum score $\mathsf{V}_\xi^{(1)}$, and another ξ-dependent density operator $\rho_\xi^{(2)}$ associated via Eq. (11.47) with the quantum score $\mathsf{V}_\xi^{(2)}$. The bipartite density operator $\rho_\xi = \rho_\xi^{(1)} \otimes \rho_\xi^{(2)}$ is associated via Eq. (11.47) with a quantum score V_ξ, which is given as $\mathsf{V}_\xi = \mathsf{V}_\xi^{(1)} \otimes \mathrm{I}^{(2)} + \mathrm{I}^{(1)} \otimes \mathsf{V}_\xi^{(2)}$ so as to satisfy Eq. (11.47), where $\mathrm{I}^{(1)}$ and $\mathrm{I}^{(2)}$ are the identity operators on the Hilbert spaces of, respectively, $\rho_\xi^{(1)}$ and $\rho_\xi^{(2)}$ (which need not be the same). We note $F_q(\xi\,;\rho_\xi)$ the quantum Fisher information defined by Eqs. (11.56) or (11.57) for the bipartite state ρ_ξ. According to Eq. (11.57), it can be computed as $F_q(\xi\,;\rho_\xi) = -\operatorname{tr}(\rho_\xi \partial_\xi \mathsf{V}_\xi) = -\operatorname{tr}\big[\big(\rho_\xi^{(1)} \otimes \rho_\xi^{(2)}\big)\big(\partial_\xi \mathsf{V}_\xi^{(1)} \otimes \mathrm{I}^{(2)} + \mathrm{I}^{(1)} \otimes \partial_\xi \mathsf{V}_\xi^{(2)}\big)\big]$. This is also $F_q(\xi\,;\rho_\xi) = -\operatorname{tr}\big(\rho_\xi^{(1)}\partial_\xi \mathsf{V}_\xi^{(1)} \otimes \rho_\xi^{(2)} + \rho_\xi^{(1)} \otimes \rho_\xi^{(2)}\partial_\xi \mathsf{V}_\xi^{(2)}\big) = -\operatorname{tr}\big(\rho_\xi^{(1)}\partial_\xi \mathsf{V}_\xi^{(1)}\big) - \operatorname{tr}\big(\rho_\xi^{(2)}\partial_\xi \mathsf{V}_\xi^{(2)}\big)$, which proves the **additivity** of the quantum Fisher information upon tensoring, $F_q\big(\xi\,;\rho_\xi = \rho_\xi^{(1)} \otimes \rho_\xi^{(2)}\big) = F_q\big(\xi\,;\rho_\xi^{(1)}\big) + F_q\big(\xi\,;\rho_\xi^{(2)}\big)$.

(2) We consider a ξ-independent unitary operator U evolving the quantum state according to $\rho_\xi \mapsto \mathsf{U}\rho_\xi\mathsf{U}^\dagger = \rho'_\xi$. The quantum score associated with the unitarily transformed state ρ'_ξ is the Hermitian operator V'_ξ solving $\partial_\xi \rho'_\xi = \big(\mathsf{V}'_\xi \rho'_\xi + \rho'_\xi \mathsf{V}'_\xi\big)/2$, according to Eq. (11.47), or equivalently solving $\mathsf{U}(\partial_\xi\rho_\xi)\mathsf{U}^\dagger = \big(\mathsf{V}'_\xi\mathsf{U}\rho_\xi\mathsf{U}^\dagger + \mathsf{U}\rho_\xi\mathsf{U}^\dagger\mathsf{V}'_\xi\big)/2$. We verify that $\mathsf{V}'_\xi = \mathsf{U}\mathsf{V}_\xi\mathsf{U}^\dagger$ solves this equation, since when substituted in this equation it gives $\mathsf{U}(\partial_\xi\rho_\xi)\mathsf{U}^\dagger = \big(\mathsf{U}\mathsf{V}_\xi\rho_\xi\mathsf{U}^\dagger + \mathsf{U}\rho_\xi\mathsf{V}_\xi\mathsf{U}^\dagger\big)/2$, which is also $\mathsf{U}(\partial_\xi\rho_\xi)\mathsf{U}^\dagger = \mathsf{U}\big[(\mathsf{V}_\xi\rho_\xi + \rho_\xi\mathsf{V}_\xi)/2\big]\mathsf{U}^\dagger$, which is granted since V_ξ solves Eq. (11.47). Now, by Eq. (11.56), the quantum Fisher information associated with the unitarily transformed state ρ'_ξ is $F'_q(\xi) = \operatorname{tr}\big(\rho'_\xi \mathsf{V}'^2_\xi\big)$, which is $F'_q(\xi) = \operatorname{tr}\big(\mathsf{U}\rho_\xi\mathsf{U}^\dagger\,\mathsf{U}\mathsf{V}_\xi^2\mathsf{U}^\dagger\big) = \operatorname{tr}\big(\mathsf{U}\rho_\xi\mathsf{V}_\xi^2\mathsf{U}^\dagger\big) = \operatorname{tr}\big(\rho_\xi\mathsf{V}_\xi^2\big) = F_q(\xi)$. So under a ξ-independent unitary evolution of the state $\rho_\xi \mapsto \mathsf{U}\rho_\xi\mathsf{U}^\dagger$, the quantum score evolves according to

$\mathsf{V}_\xi \mapsto \mathsf{U}\mathsf{V}_\xi\mathsf{U}^\dagger$, while the quantum Fisher information exhibits a **unitary invariance** with $F_q(\xi\,;\mathsf{U}\rho_\xi\mathsf{U}^\dagger) = F_q(\xi\,;\rho_\xi)$.

(3) When the quantum state ρ_ξ is evolved by the more general ξ-independent quantum operation or process $\rho_\xi \mapsto \mathcal{E}(\rho_\xi)$ as in Eq. (3.45), the quantum Fisher information obeys [307, 308] the **monotonicity** relation $F_q\big(\xi\,;\mathcal{E}(\rho_\xi)\big) \le F_q(\xi\,;\rho_\xi)$. No preprocessing by $\mathcal{E}(\cdot)$ prior to the quantum measurement can enhance the estimation efficiency assessed by the quantum Fisher information.

(4) Let $\rho_\xi^{(1)}$ and $\rho_\xi^{(2)}$ be two elements of the convex set of density operators on $\mathcal{H}_N$, and a scalar $\mu \in [0, 1]$. If we denote by $F_q[\rho_\xi]$ the quantum Fisher information of Eqs. (11.56)–(11.57) associated with the generic density operator ρ_ξ, we then have the **convexity** property $F_q\big[\mu\rho_\xi^{(1)} + (1-\mu)\rho_\xi^{(2)}\big] \le \mu F_q\big[\rho_\xi^{(1)}\big] + (1-\mu)F_q\big[\rho_\xi^{(2)}\big]$, as proved for instance in [309, 310], and which naturally extends to a convex superposition of an arbitrary number of density operators.

(5) In many useful situations, the dependence in the parameter ξ of the quantum state ρ_ξ is inherited from a ξ-dependent quantum process $\mathcal{E}_\xi(\cdot)$ —unitary as in Sect. 3.1 and Eq. (3.2), or non-unitary as in Sect. 3.2 and Eq. (3.37) or (3.45)—acting on a ξ-independent quantum state ρ_0 (an input probe), to deliver the ξ-dependent state $\rho_\xi = \mathcal{E}_\xi(\rho_0)$. Then, as a consequence of the linearity of the map $\rho_\xi = \mathcal{E}_\xi(\rho_0)$, the quantum Fisher information $F_q\big[\rho_\xi = \mathcal{E}_\xi(\rho_0)\big]$ when seen as a functional of the input density operator ρ_0 is also [311] a convex ($\cup$) functional of ρ_0. When ρ_0 is a convex combination of density operators, such as $\rho_0 = \mu\rho_0^{(1)} + (1-\mu)\rho_0^{(2)}$, then by linearity $\mathcal{E}_\xi(\rho_0) = \mu\mathcal{E}_\xi\big(\rho_0^{(1)}\big) + (1-\mu)\mathcal{E}_\xi\big(\rho_0^{(2)}\big)$, and by the convexity in ρ_ξ from property (4) above, one has

$$F_q\big[\mathcal{E}_\xi(\rho_0)\big] \le \mu F_q\big[\mathcal{E}_\xi(\rho_0^{(1)})\big] + (1-\mu)F_q\big[\mathcal{E}_\xi(\rho_0^{(2)})\big]\,, \tag{11.59}$$

expressing convexity in ρ_0.

Equation (11.59) implies that if the input probe ρ_0 can be convexly decomposed under a form $\rho_0 = \mu\rho_0^{(1)} + (1-\mu)\rho_0^{(2)}$, then $F_q\big[\rho_\xi = \mathcal{E}_\xi(\rho_0)\big]$ is not maximal at this ρ_0. Conversely, $F_q\big[\rho_\xi = \mathcal{E}_\xi(\rho_0)\big]$ is at a maximum necessarily for a ρ_0 that is not convexly decomposable [309, 311]. The states ρ_0 that are not convexly decomposable are rank-one density operators of the form $\rho_0 = |\psi_0\rangle\langle\psi_0|$. This establishes the important property that the quantum Fisher information $F_q\big[\rho_\xi = \mathcal{E}_\xi(\rho_0)\big]$ accessible from the transformed state $\rho_\xi = \mathcal{E}_\xi(\rho_0)$ is maximized by a pure input probe $\rho_0 = |\psi_0\rangle\langle\psi_0|$.

11.2.2 Computing the Quantum Fisher Information

The definition of the quantum Fisher information $F_q(\xi)$ in Eq. (11.56) rests on the quantum score V_ξ introduced by Eq. (11.47) for a given density operator $\rho_\xi \in \mathcal{L}(\mathcal{H}_N)$. To solve Eq. (11.47) and obtain an explicit characterization of the quantum score operator V_ξ, we use the diagonal eigendecomposition of ρ_ξ as

$$\rho_\xi = \sum_{n=1}^{N} \lambda_n \left|\lambda_n\right\rangle \left\langle\lambda_n\right| , \tag{11.60}$$

in its orthonormal eigenbasis $\{|\lambda_n\rangle\}_{n=1}^N$ of $\mathcal{H}_N$ associated with the non-negative eigenvalues $\{\lambda_n\}_{n=1}^N$. Then Eq. (11.47) is also

$$2\partial_\xi \rho_\xi = \sum_{n=1}^{N} \lambda_n \Big(\left|\lambda_n\right\rangle \left\langle\lambda_n\right| V_\xi + V_\xi \left|\lambda_n\right\rangle \left\langle\lambda_n\right| \Big) . \tag{11.61}$$

For an eigenstate $|\lambda_k\rangle$, from Eq. (11.61), by orthonormality of the basis $\{|\lambda_n\rangle\}_{n=1}^N$, we have

$$2\left\langle\lambda_k\right| \partial_\xi \rho_\xi = \lambda_k \left\langle\lambda_k\right| V_\xi + \sum_{n=1}^{N} \lambda_n \left\langle\lambda_k\right| V_\xi \left|\lambda_n\right\rangle \left\langle\lambda_n\right| , \tag{11.62}$$

giving for an eigenstate $|\lambda_\ell\rangle$,

$$2\left\langle\lambda_k\right| \partial_\xi \rho_\xi \left|\lambda_\ell\right\rangle = \lambda_k \left\langle\lambda_k\right| V_\xi \left|\lambda_\ell\right\rangle + \lambda_\ell \left\langle\lambda_k\right| V_\xi \left|\lambda_\ell\right\rangle . \tag{11.63}$$

This leads to a characterization of the score operator V_ξ by its matrix elements relative to the eigenbasis $\{|\lambda_n\rangle\}_{n=1}^N$, reading

$$\left\langle\lambda_k\right| V_\xi \left|\lambda_\ell\right\rangle = 2\frac{\left\langle\lambda_k\right| \partial_\xi \rho_\xi \left|\lambda_\ell\right\rangle}{\lambda_k + \lambda_\ell} , \tag{11.64}$$

when $\lambda_k + \lambda_\ell \neq 0$. The eigenvalues of ρ_ξ are all non-negative, but if for some pair (k, ℓ) both eigenvalues λ_k and λ_l are simultaneously zero, Eq. (11.63) yields $\langle\lambda_k|\partial_\xi \rho_\xi|\lambda_\ell\rangle = 0$, and the corresponding matrix element $\langle\lambda_k|V_\xi|\lambda_\ell\rangle$ can also be taken as zero, which is enough for V_ξ to satisfy Eq. (11.61). We therefore obtain the explicit expression for the quantum score

$$\boxed{V_\xi = 2\sum_{k,\ell} \frac{\left\langle\lambda_k\right| \partial_\xi \rho_\xi \left|\lambda_\ell\right\rangle}{\lambda_k + \lambda_\ell} \left|\lambda_k\right\rangle \left\langle\lambda_\ell\right| ,} \tag{11.65}$$

where the sums include every pair (k, ℓ) with $\lambda_k + \lambda_\ell \neq 0$. Equation (11.65) is also given for instance in [292, 312].

Both the eigenvalues λ_n and eigenstates $|\lambda_n\rangle$ of ρ_ξ may depend on the parameter ξ, so that by differentiating the diagonal decomposition of ρ_ξ in Eq. (11.60), we obtain

$$\partial_\xi \rho_\xi = \sum_{n=1}^{N} \Big((\partial_\xi \lambda_n) |\lambda_n\rangle \langle \lambda_n| + \lambda_n |\partial_\xi \lambda_n\rangle \langle \lambda_n| + \lambda_n |\lambda_n\rangle \langle \partial_\xi \lambda_n| \Big) , \qquad (11.66)$$

where $|\partial_\xi \lambda_n\rangle$ denotes the differentiated state vector $\partial_\xi |\lambda_n\rangle \equiv \partial |\lambda_n\rangle / \partial \xi$. Also, since $\langle \lambda_k|\lambda_\ell\rangle = \delta_{k\ell}$, one has $0 = \partial_\xi \langle \lambda_k|\lambda_\ell\rangle = \langle \partial_\xi \lambda_k|\lambda_\ell\rangle + \langle \lambda_k|\partial_\xi \lambda_\ell\rangle$, so that $\langle \partial_\xi \lambda_k|\lambda_\ell\rangle = - \langle \lambda_k|\partial_\xi \lambda_\ell\rangle$; especially $\langle \partial_\xi \lambda_k|\lambda_k\rangle = - \langle \lambda_k|\partial_\xi \lambda_k\rangle = - \langle \partial_\xi \lambda_k|\lambda_k\rangle^*$, so $\langle \partial_\xi \lambda_k|\lambda_k\rangle$ and $\langle \lambda_k|\partial_\xi \lambda_k\rangle$ are purely imaginary (if not zero). This used with Eqs. (11.65)–(11.66) leads to an alternative expression for the quantum score V_ξ, where the derivative $\partial_\xi \rho_\xi$ appearing in Eq. (11.65) is now transferred to the eigenvalues and eigenstates of ρ_ξ, and reading

$$\boxed{\mathsf{V}_\xi = \sum_k \frac{\partial_\xi \lambda_k}{\lambda_k} |\lambda_k\rangle \langle \lambda_k| + 2 \sum_{k,\ell} \frac{\lambda_k - \lambda_\ell}{\lambda_k + \lambda_\ell} \langle \partial_\xi \lambda_k|\lambda_\ell\rangle |\lambda_k\rangle \langle \lambda_\ell| ,} \qquad (11.67)$$

with the first sum running on every eigenvalue $\lambda_k \neq 0$, and the second sum on every pair (k, ℓ) with $\lambda_k + \lambda_\ell \neq 0$. Equation (11.67) is also given for instance in [292].

For the quantum Fisher information $F_q(\xi) = \mathrm{tr}(\rho_\xi \mathsf{V}_\xi^2)$ of Eq. (11.56), we can use Eq. (11.65) to obtain V_ξ^2 and then, via Eq. (11.60), deduce

$$\boxed{F_q(\xi) = 2 \sum_{k,\ell} \frac{\big|\langle \lambda_k|\partial_\xi \rho_\xi|\lambda_\ell\rangle\big|^2}{\lambda_k + \lambda_\ell} .} \qquad (11.68)$$

Equation (11.68) is also given for instance in [312]. By means of Eq. (11.66) to transfer the derivative to the eigenvalues and eigenstates of ρ_ξ, we further obtain the alternative expression

$$\boxed{F_q(\xi) = \sum_k \frac{(\partial_\xi \lambda_k)^2}{\lambda_k} + 2 \sum_{k,\ell} \frac{(\lambda_k - \lambda_\ell)^2}{\lambda_k + \lambda_\ell} \big|\langle \partial_\xi \lambda_k|\lambda_\ell\rangle\big|^2 .} \qquad (11.69)$$

Equation (11.69) is also given for instance in [292].

In Eq. (11.69), the first sum vanishes when the eigenvalues λ_n of ρ_ξ are all independent of ξ, while the second (double) sum in Eq. (11.69) vanishes when the eigenstates $|\lambda_n\rangle$ of ρ_ξ are all independent of ξ. The first sum in Eq. (11.69) is similar to a classical Fisher information, analogous to Eq. (11.4), for the discrete probability distribution $\{\lambda_n\}_{n=1}^N$. It would characterize the classical random outcomes resulting from measuring ρ_ξ by means of a von Neumann projective measurement in its orthonormal eigenbasis $\{|\lambda_n\rangle\}_{n=1}^N$ of $\mathcal{H}_N$. The second (double) sum in Eq. (11.69) has a more specific quantum character. When it vanishes with all the eigenstates $|\lambda_n\rangle$ of ρ_ξ independent of ξ, it gives $\mathsf{V}_\xi = \partial_\xi \ln(\rho_\xi)$ in Eq. (11.67). This also indicates that when the dependence on ξ of the quantum state ρ_ξ is carried only by its eigenvalues $\{\lambda_n\}$

and not by its eigenstates $\{|\lambda_n\rangle\}$, then there exists a quantum measurement of ρ_ξ characterized by a classical Fisher information $F_c(\xi)$ reaching the quantum Fisher information $F_q(\xi)$: it is the von Neumann measurement of ρ_ξ in its ξ-independent eigenbasis $\{|\lambda_n\rangle\}_{n=1}^N$, with the measurement outcomes governed by the probability distribution $\{\lambda_n\}_{n=1}^N$.

The double sum in Eq. (11.69) involves a coefficient $c_{k\ell} = (\lambda_k - \lambda_\ell)^2/(\lambda_k + \lambda_\ell) = c_{\ell k}$ symmetric or invariant in the exchange of the indices k and ℓ, and $|\langle\partial_\xi\lambda_k|\lambda_\ell\rangle|^2 = |\langle\partial_\xi\lambda_\ell|\lambda_k\rangle|^2$ is similarly symmetric. As a result, any antisymmetric coefficient $d_{k\ell} = -d_{\ell k}$ added to $c_{k\ell}$ in Eq. (11.69) will leave the double sum invariant. By this mechanism, instead of $c_{k\ell}$, for instance the coefficient $2\lambda_k(\lambda_k - \lambda_\ell)^2/(\lambda_k + \lambda_\ell)^2$ or also the coefficient $2\lambda_k(\lambda_k - \lambda_\ell)/(\lambda_k + \lambda_\ell)$, leave the double sum invariant in Eq. (11.69). Accordingly, many (infinitely many) alternative coefficients are feasible, instead of $c_{k\ell}$ in Eq. (11.69), to define the same quantum Fisher information $F_q(\xi)$, as reflected in the literature on the subject.

For the special but important case of a **pure state**, the density operator $\rho_\xi = |\lambda\rangle\langle\lambda|$ has a single nonzero eigenvalue $\lambda = 1$ (independent of ξ) associated with the ξ-dependent eigenstate $|\lambda\rangle$. As a result, more concise characterizations of V_ξ and $F_q(\xi)$ are also accessible. A pure state gives $\rho_\xi^2 = \rho_\xi$, and therefore $\rho_\xi\partial_\xi\rho_\xi + (\partial_\xi\rho_\xi)\rho_\xi = \partial_\xi\rho_\xi$, and by comparing with Eq. (11.47) we obtain for the quantum score of a pure state the expressions

$$\boxed{\mathsf{V}_\xi = 2\partial_\xi\rho_\xi = 2\Big(|\partial_\xi\lambda\rangle\langle\lambda| + |\lambda\rangle\langle\partial_\xi\lambda|\Big)\,.} \tag{11.70}$$

The quantum Fisher information $F_q(\xi) = \mathrm{tr}(\rho_\xi\mathsf{V}_\xi^2) = \mathrm{tr}(\mathsf{V}_\xi\partial_\xi\rho_\xi)$ from Eq. (11.56), for a pure state then follows as

$$\boxed{F_q(\xi) = 2\,\mathrm{tr}\big[(\partial_\xi\rho_\xi)^2\big] = 4\Big(\langle\partial_\xi\lambda|\partial_\xi\lambda\rangle + \langle\partial_\xi\lambda|\lambda\rangle^2\Big)\,.} \tag{11.71}$$

In general for quantum estimation from a quantum state ρ_ξ, pure or mixed, the present characterization of the quantum Fisher information $F_q(\xi)$ will be useful to select conditions, when adjustable, affording a larger $F_q(\xi)$ for larger efficiency. The characterization of $F_q(\xi)$ will be useful also to select a measuring POVM $\{\mathsf{E}_m\}_{m=1}^M$ achieving in Eqs. (11.45)–(11.46) a classical Fisher information

$$F_c(\xi) = \sum_{m=1}^{M} \frac{1}{\mathrm{tr}(\mathsf{E}_m\rho_\xi)}\Big[\mathrm{tr}\big(\mathsf{E}_m\partial_\xi\rho_\xi\big)\Big]^2 \tag{11.72}$$

coming close (as close as possible) to its upper bound $F_q(\xi)$. From the POVM, the measurement outcome, with its M possible values, then stands as a classical random variable, upon which will be based the estimation of the parameter ξ with classical techniques, as discussed in Sect. 11.1, among which one of choice is maximum likelihood.

With a definite POVM $\{\mathsf{E}_m\}_{m=1}^{M}$, when repeating L independent measurements on L successive replicas of ρ_ξ, a dataset $\{L_m\}$ is collected with a number L_m of outcomes m, for $m = 1$ to M, with $\sum_{m=1}^{M} L_m = L$. From the resulting multinomial probability distribution, the loglikelihood follows as $\ln P\big(\{L_m\}\,;\xi\big) = \sum_{m=1}^{M} L_m \ln\big(P(m\,;\xi)\big) + \ln(L!) - \sum_{m=1}^{M} \ln(L_m!)$. The maximum-likelihood estimator is then determined as the solution to

$$\widehat{\xi}_{\mathrm{ML}}\big(\{L_m\}\big) = \arg\max_{\xi} \sum_{m=1}^{M} L_m \ln\big(\mathrm{tr}(\mathsf{E}_m\,\rho_\xi)\big) \,, \tag{11.73}$$

with $\widehat{\xi}_{\mathrm{ML}}$ coming with its known asymptotic optimality properties at large size L, as indicated at the end of Sect. 11.1.1. The quantum estimation task in this way is finalized by a classical estimation step from the results of quantum measurement.

These principles of quantum parameter estimation will now be applied, in the next section, to the important task of quantum phase estimation, when the unknown parameter ξ to be estimated is some quantum phase of interest. Quantum phase estimation encompasses many situations of great relevance to quantum signal and information processing.

11.3 Quantum Phase Estimation

Phase estimation plays an important role in several areas of quantum signal and information processing. For instance in quantum metrology, it takes place in many applications such as interferometry, magnetometry, atomic clocks, frequency standards, and many other high-precision high-sensitivity physical measurements [293, 301, 313–316]. For quantum computation, phase estimation was encountered in Chap. 5 for the Grover search algorithm with an unknown number of solutions at the end of Sect. 5.5, and it will be encountered again in Chap. 12 for the Shor factoring algorithm of Sect. 12.6.

A phase estimation task typically arises when a quantum signal (a quantum state) acting as a probe experiences a unitary transformation U_ξ. Such a unitary process U_ξ of $\mathcal{L}(\mathcal{H}_N)$ is generically expressed as

$$\mathsf{U}_\xi = \exp(-i\xi\mathsf{G}) \,, \tag{11.74}$$

where $\mathsf{G} \in \mathcal{L}(\mathcal{H}_N)$ is the (ξ-independent) Hermitian generator of U_ξ, and ξ is the unknown phase to be estimated. The evolution operator from Eq. (3.6) can be seen as an instance of Eq. (11.74), with the generator G related to the Hamiltonian H, and a phase ξ related to the duration $t - t_0$ of the evolution. We note that G and U_ξ from

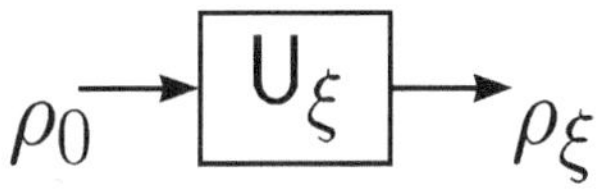

Fig. 11.1 An input probe with density operator ρ_0 acts as an excitation signal applied to the unitary process U_ξ of Eq. (11.74) under estimation, so as to deliver the ξ-dependent output density operator ρ_ξ resulting from Eq. (11.75), and standing as the ξ-dependent probing signal to be exploited for estimating the quantum phase ξ

Eq. (11.74) share the same eigenstates in $\mathcal{H}_N$ and therefore generally commute, and as well G^2 and U_ξ commute.

The probe signal is prepared in the initial state $\rho_0 \in \mathcal{L}(\mathcal{H}_N)$ to serve as an input excitation to the unitary process U_ξ so as to realize the input–output transformation

$$\rho_0 \longmapsto U_\xi \rho_0 U_\xi^\dagger = \exp(-i\xi G)\rho_0 \exp(i\xi G) = \rho_\xi \,, \tag{11.75}$$

delivering on the output the ξ-dependent probing state ρ_ξ. The task is then, from the probing state ρ_ξ at the output, to efficiently estimate the quantum phase ξ. The estimation setting is depicted in Fig. 11.1, and it can be viewed as an input–output signal transformation designed for a task of system identification concerning the process U_ξ, as in signals-and-systems theory.

11.3.1 Performance with Unitary Parameterization

When the probing state ρ_ξ receives its dependence on the parameter ξ from the unitary process of Eq. (11.75), the following properties ensue.

The average $\langle G\rangle_\xi$ of the observable G in the state ρ_ξ follows, according to Eq. (4.59), as $\langle G\rangle_\xi = \mathrm{tr}(\rho_\xi G) = \mathrm{tr}\big(U_\xi \rho_0 U_\xi^\dagger G\big) = \mathrm{tr}\big(U_\xi \rho_0 G U_\xi^\dagger\big) = \mathrm{tr}(\rho_0 G) = \langle G\rangle_0$ by commutation of G and U_ξ and circularity of the trace, indicating that the averages of G in the states ρ_ξ and ρ_0 coincide. For similar reasons, the averages $\langle G^2\rangle_\xi = \langle G^2\rangle_0$ coincide. And for the Hermitian dispersion operator $\Delta G = G - \langle G\rangle\, I_N$, the averages of its square $\langle \Delta G^2\rangle_\xi = \langle \Delta G^2\rangle_0$ also coincide in the states ρ_ξ and ρ_0.

The Hermitian generator $G \in \mathcal{L}(\mathcal{H}_N)$ has eigenstates $\big\{|u_n\rangle\big\}_{n=1}^N$ forming an orthonormal basis of $\mathcal{H}_N$, associated with the (real) eigenvalues $\{u_n\}_{n=1}^N$. Then, all four operators U_ξ, G, G^2, and ΔG share the same eigenstates $\big\{|u_n\rangle\big\}_{n=1}^N$. Each eigenstate transforms as $|u_n\rangle \mapsto U_\xi\, |u_n\rangle = \exp(-i\xi u_n)\, |u_n\rangle$.

When the output probing state ρ_ξ has eigenvalues λ_k and eigenstates $|\lambda_k\rangle$, the input probe state ρ_0 has the same eigenvalues λ_k and has eigenstates $|\lambda_k^{(0)}\rangle$ such that $|\lambda_k\rangle = U_\xi\, |\lambda_k^{(0)}\rangle$, for $k = 1$ to N. The output probing state ρ_ξ from Eq. (11.75) differentiates as

$$\partial_\xi \rho_\xi = -i\mathsf{G}\mathsf{U}_\xi \rho_0 \mathsf{U}_\xi^\dagger + \mathsf{U}_\xi \rho_0 \mathsf{U}_\xi^\dagger i\mathsf{G} = i\mathsf{U}_\xi [\rho_0, \mathsf{G}]\mathsf{U}_\xi^\dagger = i[\rho_\xi, \mathsf{G}] \ , \tag{11.76}$$

involving the commutator of Eq. (A.15) from the Appendix. The output state ρ_ξ in Eq. (11.75) is pure or mixed, respectively, if and only if the input state ρ_0 is pure or mixed.

For a mixed probing state ρ_ξ, the quantum score of Eq. (11.65) follows as

$$\mathsf{V}_\xi = 2i \sum_{k,\ell} \frac{1}{\lambda_k + \lambda_\ell} \left|\lambda_k\right\rangle \left\langle\lambda_k\right| [\rho_\xi, \mathsf{G}] \left|\lambda_\ell\right\rangle \left\langle\lambda_\ell\right| \ , \tag{11.77}$$

which is also $\mathsf{V}_\xi = \mathsf{U}_\xi \mathsf{V}_0 \mathsf{U}_\xi^\dagger$ with

$$\mathsf{V}_0 = 2i \sum_{k,\ell} \frac{1}{\lambda_k + \lambda_\ell} \left|\lambda_k^{(0)}\right\rangle \left\langle\lambda_k^{(0)}\right| [\rho_0, \mathsf{G}] \left|\lambda_\ell^{(0)}\right\rangle \left\langle\lambda_\ell^{(0)}\right| \ . \tag{11.78}$$

Also for the mixed probing state ρ_ξ, the quantum Fisher information of Eq. (11.68) follows as

$$F_q(\xi) = 2 \sum_{k,\ell} \frac{(\lambda_k - \lambda_\ell)^2}{\lambda_k + \lambda_\ell} \left|\langle\lambda_k|\mathsf{G}|\lambda_\ell\rangle\right|^2 \tag{11.79}$$

$$= 2 \sum_{k,\ell} \frac{(\lambda_k - \lambda_\ell)^2}{\lambda_k + \lambda_\ell} \left|\langle\lambda_k^{(0)}|\mathsf{G}|\lambda_\ell^{(0)}\rangle\right|^2 . \tag{11.80}$$

A pure output probing state $\rho_\xi = |\psi_\xi\rangle\langle\psi_\xi|$ is associated with the pure input probe state $\rho_0 = |\psi_0\rangle\langle\psi_0|$. Then, by Eq. (11.76), the quantum score of Eq. (11.70) directly follows as $\mathsf{V}_\xi = i2[\rho_\xi, \mathsf{G}] = \mathsf{U}_\xi \mathsf{V}_0 \mathsf{U}_\xi^\dagger$ with $\mathsf{V}_0 = i2[\rho_0, \mathsf{G}]$. For the quantum Fisher information $F_q(\xi)$ of Eq. (11.71), the transformed state vector $|\psi_\xi\rangle = \mathsf{U}_\xi |\psi_0\rangle = \exp(-i\xi\mathsf{G})|\psi_0\rangle$ yields the derivative $\partial_\xi|\psi_\xi\rangle \equiv |\partial_\xi\psi_\xi\rangle = -i\mathsf{G}|\psi_\xi\rangle$, leading in Eq. (11.71) to

$$F_q(\xi) = 4\Big(\langle\psi_\xi|\mathsf{G}^2|\psi_\xi\rangle - \langle\psi_\xi|\mathsf{G}|\psi_\xi\rangle^2\Big) \tag{11.81}$$

$$= 4\Big(\langle\psi_0|\mathsf{G}^2|\psi_0\rangle - \langle\psi_0|\mathsf{G}|\psi_0\rangle^2\Big) . \tag{11.82}$$

This is equivalently

$$F_q(\xi) = 4\,\langle\Delta\mathsf{G}^2\rangle_\xi = 4\,\langle\Delta\mathsf{G}^2\rangle_0 \ , \tag{11.83}$$

so that with a pure probe, the quantum Fisher information $F_q(\xi)$ is given by the average value of the squared dispersion of the Hermitian generator G, equivalently in the state $|\psi_\xi\rangle$ or $|\psi_0\rangle$.

For an input probe ρ_0 commuting with U_ξ or equivalently with G, the transformed output state resulting from Eq. (11.75) is $\rho_\xi \equiv \rho_0$ and is thus insensitive to the phase

ξ. Accordingly, the quantum score V_ξ and quantum Fisher information $F_q(\xi)$ vanish, since the output state ρ_ξ is of no use to estimate the phase ξ. On the contrary, we would like now to characterize the input probe ρ_0 and the whole strategy with maximum efficiency to estimate the phase ξ.

11.3.2 An Optimal Strategy for Phase Estimation

To devise an efficient strategy for estimating the phase ξ, we start by looking at the quantum Fisher information $F_q(\xi)$ of Eqs. (11.79)-(11.80) or (11.83) characterizing the best conceivable performance. From the convexity property (4) of the quantum Fisher information indicated at the end of Sect. 11.2.1, we know that $F_q(\xi)$ will be maximized by using an input probe signal formed by a pure input state $\rho_0 = |\psi_0\rangle\langle\psi_0|$.

Maximization of the quantum Fisher information $F_q(\xi)$ of Eq. (11.83) is then accomplished [313] by the input probe state $|\psi_0\rangle$ in which the generator G is maximally dispersed. This is the state

$$|\psi_0\rangle = |\psi_0^{\text{opt}}\rangle = \frac{1}{\sqrt{2}}\big(|u_{\min}\rangle + |u_{\max}\rangle\big) , \tag{11.84}$$

formed as an equally weighted superposition of the (orthogonal) eigenstates $|u_{\min}\rangle$ and $|u_{\max}\rangle$ associated respectively with the minimum $u_{\min}$ and maximum $u_{\max}$ eigenvalues of the generator G. This optimal input signal $|\psi_0^{\text{opt}}\rangle$ realizes in Eq. (11.83) the maximum Fisher information $F_q^{\max}(\xi) = 4\langle\psi_0^{\text{opt}}|\Delta\mathsf{G}^2|\psi_0^{\text{opt}}\rangle$ resulting as

$$F_q^{\max}(\xi) = \big(u_{\max} - u_{\min}\big)^2 . \tag{11.85}$$

We now examine whether a quantum measurement exists or not with a classical Fisher information $F_c(\xi)$ from Eq. (11.45)–(11.46) reaching the quantum Fisher information $F_q^{\max}(\xi)$ of Eq. (11.85). When the optimal input signal $|\psi_0^{\text{opt}}\rangle$ of Eq. (11.84) is used to probe the unitary process $\mathsf{U}_\xi = \exp(-i\xi\mathsf{G})$ under estimation, the resulting output state follows also as a pure state, given by

$$|\psi_\xi\rangle = \mathsf{U}_\xi\,|\psi_0^{\text{opt}}\rangle = \frac{1}{\sqrt{2}}\Big[\exp(-i\xi u_{\min})\,|u_{\min}\rangle + \exp(-i\xi u_{\max})\,|u_{\max}\rangle\Big] . \tag{11.86}$$

From a pure output probing state $\rho_\xi = |\psi_\xi\rangle\langle\psi_\xi|$, the phase estimation procedure can be completed as follows. For estimating the phase ξ, a conceivable quantum measurement of $|\psi_\xi\rangle$ could be to resort to a von Neumann projective measurement in the orthonormal eigenbasis $\big\{|u_n\rangle\big\}_{n=1}^{N}$ of $\mathcal{H}_N$. This would lead to the binary outcome of projecting $|\psi_\xi\rangle$ onto $|u_{\min}\rangle$ with the probability $|\langle u_{\min}|\psi_\xi\rangle|^2 = 1/2$, or onto

$|u_{\max}\rangle$ with the probability $|\langle u_{\max}|\psi_\xi\rangle|^2 = 1/2$. This unfortunately is a ξ-independent measurement outcome, affording no possibility of estimating the phase ξ.

An alternative projective measurement could be conceived with the input state $|\psi_0^{\rm opt}\rangle$ of Eq. (11.84) as part of the projective vectors. The probability of projecting $|\psi_\xi\rangle$ of Eq. (11.86) onto $|\psi_0^{\rm opt}\rangle$ follows from their inner product $\langle\psi_0^{\rm opt}|\psi_\xi\rangle = \big[\exp(-i\xi u_{\min}) + \exp(-i\xi u_{\max})\big]/2$ as the probability

$$P_0 = \left|\langle\psi_0^{\rm opt}|\psi_\xi\rangle\right|^2 = \cos^2\!\left(\frac{u_{\max}-u_{\min}}{2}\,\xi\right) = \frac{1}{2}\Big[1+\cos\big((u_{\max}-u_{\min})\xi\big)\Big], \tag{11.87}$$

which is a ξ-dependent probability, signaling we are on a good track. A possibility then would be to supplement the projective measurement with the single additional projector $\mathrm{I}_N - |\psi_0^{\rm opt}\rangle\langle\psi_0^{\rm opt}|$ projecting onto the subspace of $\mathcal{H}_N$ orthogonal to $|\psi_0^{\rm opt}\rangle$. This forms a valid quantum measurement, with only two orthogonal projectors to span $\mathcal{H}_N$, and yielding two possible outcomes. When applied to $|\psi_\xi\rangle$ of Eq. (11.86), either $|\psi_\xi\rangle$ is projected onto $|\psi_0^{\rm opt}\rangle$ with the probability P_0 of Eq. (11.87), or $|\psi_\xi\rangle$ is projected onto the orthogonal subspace with the complementary probability $1-P_0$.

According to Eq. (11.45), such a two-outcome measurement is characterized by the classical Fisher information

$$F_c(\xi) = \frac{1}{P_0}\left(\partial_\xi P_0\right)^2 + \frac{1}{1-P_0}\left[\partial_\xi(1-P_0)\right]^2 \tag{11.88}$$

$$= \frac{1}{(1-P_0)P_0}\left(\partial_\xi P_0\right)^2 . \tag{11.89}$$

With P_0 given by Eq. (11.87), we finally obtain

$$F_c(\xi) = \left(u_{\max}-u_{\min}\right)^2 . \tag{11.90}$$

We find that the classical Fisher information $F_c(\xi)$ of Eq. (11.90) exactly matches the maximal quantum Fisher information $F_q^{\max}(\xi)$ of Eq. (11.85), and therefore saturates the inequality of Eq. (11.55). This indicates that our two-outcome measurement of $|\psi_\xi\rangle$ from Eq. (11.86), is in fact an optimal measurement, achieving the maximal classical Fisher information $F_c(\xi) = F_q^{\max}(\xi)$ among all conceivable quantum measurements. In particular, there is no need to conceive a quantum measurement with a number of outcomes larger than two; it would not lead to a larger classical Fisher information.

From the outcomes of the optimal quantum measurement of the ξ-dependent state $|\psi_\xi\rangle$ of Eq. (11.86), the final step is to conceive an estimator for the quantum phase ξ. The maximum-likelihood procedure presented at the end of Sect. 11.1.1 will allow us to reach the best estimation performance determined by the maximal Fisher information $F_c(\xi) = F_q^{\max}(\xi)$ of Eq. (11.90), at least in the limit of a large number L of repeated measurements on L independent replicas of $|\psi_\xi\rangle$, obtained by repeating L times the experiment where the input signal $|\psi_0^{\rm opt}\rangle$ of Eq. (11.84) probes U_ξ to generate $|\psi_\xi\rangle$. This can be carried out as follows.

The measurement probability P_0 of Eq. (11.87) can be rewritten as the linear (affine) function

$$P_0 = \frac{1}{2}\big[1 + s(\xi)\big] \tag{11.91}$$

of the ξ-dependent auxiliary parameter

$$s(\xi) = \cos\big((u_{\max} - u_{\min})\xi\big) , \tag{11.92}$$

giving in particular for the derivative

$$\left(\frac{\partial s}{\partial \xi}\right)^2 = \big(u_{\max} - u_{\min}\big)^2\big(1 - s^2\big) . \tag{11.93}$$

From L independent measurements of $|\psi_\xi\rangle$, we denote by $L_0 \in \{0, 1, \ldots, L\}$ the number of times the projection onto $|\psi_0^{\text{opt}}\rangle$ occurs. From there, estimating s is similar to estimating the parameter of a binomial probability distribution, and with the likelihood $[(1+s)/2]^{L_0}[(1-s)/2]^{L-L_0}$ the maximum-likelihood estimator for s follows as

$$\widehat{s}_{\text{ML}} = 2\frac{L_0}{L} - 1 \in [-1, 1] . \tag{11.94}$$

For the binomial distribution, based on the expectation $\langle L_0\rangle = LP_0 = (1+s)L/2$ and variance $\text{var}(L_0) = LP_0(1-P_0) = (1-s^2)L/4$, it follows that $\langle\widehat{s}_{\text{ML}}\rangle = s$ so that $\widehat{s}_{\text{ML}}$ is an unbiased estimator of s for any size L, with the mean-squared error $\langle(\widehat{s}_{\text{ML}} - s)^2\rangle = \text{var}(\widehat{s}_{\text{ML}}) = 4\text{var}(L_0)/L^2 = (1-s^2)/L$. From Eq. (11.45), the classical Fisher information for estimating s is $F_c(s) = (\partial_s P_0)^2/P_0 + \big[\partial_s(1-P_0)\big]^2/(1-P_0) = 1/(1-s^2)$, establishing that $\langle(\widehat{s}_{\text{ML}} - s)^2\rangle = \text{var}(\widehat{s}_{\text{ML}}) = 1/[LF_c(s)] = (1-s^2)/L$, expressing that $\widehat{s}_{\text{ML}}$ of Eq. (11.94) is a maximally efficient estimator of s, for any size L. With Eq. (11.92), we also verify that $F_c(\xi) = (\partial s/\partial\xi)^2 F_c(s)$ consistently with the definition of the classical Fisher information from Eq. (11.45).

Based on the behavior of the likelihood [284] for a transformed parameter as in Eq. (11.92), the determination of the maximum-likelihood estimator $\widehat{\xi}_{\text{ML}}$ for ξ can be accomplished by inverting the nonlinear relation $\widehat{s}_{\text{ML}} = s(\widehat{\xi}_{\text{ML}})$ as

$$\widehat{\xi}_{\text{ML}} = \frac{\arccos\big(\widehat{s}_{\text{ML}}\big)}{u_{\max} - u_{\min}} , \tag{11.95}$$

or also

$$\widehat{\xi}_{\text{ML}} = \frac{1}{u_{\max} - u_{\min}} \arccos\left(2\frac{L_0}{L} - 1\right) . \tag{11.96}$$

For quantitative analysis of the properties of the maximum-likelihood estimator $\widehat{\xi}_{\text{ML}}$ in Eq. (11.96) for ξ, the theory of error propagation can be exploited when

L is large and the two estimators depart slightly from the corresponding true values. For the expectations, we have $\langle\widehat{s}_{\rm ML}\rangle \approx s\big(\langle\widehat{\xi}_{\rm ML}\rangle\big)$, implying that as L gets large, $\langle\widehat{\xi}_{\rm ML}\rangle = \xi$ and $\widehat{\xi}_{\rm ML}$ becomes an unbiased estimator of ξ, as expected for a maximum-likelihood estimator. For the variance of $\widehat{\xi}_{\rm ML}$, we have $\mathrm{var}\big(\widehat{s}_{\rm ML}\big) \approx (\partial s/\partial\xi)^2\mathrm{var}\big(\widehat{\xi}_{\rm ML}\big)$. As a result, at large L, the variance $\mathrm{var}\big(\widehat{\xi}_{\rm ML}\big)$ tends to coincide with the mean-squared error $e^2\big(\widehat{\xi}_{\rm ML}\big) = \big\langle(\widehat{\xi}_{\rm ML} - \xi)^2\big\rangle$ and yields $e^2\big(\widehat{\xi}_{\rm ML}\big) = \mathrm{var}\big(\widehat{\xi}_{\rm ML}\big) = 1/\big[L(u_{\max} - u_{\min})^2\big]$, which is also

$$e^2\big(\widehat{\xi}_{\rm ML}\big) = \mathrm{var}\big(\widehat{\xi}_{\rm ML}\big) = \frac{1}{L(u_{\max} - u_{\min})^2} = \frac{1}{L F_q^{\max}(\xi)}\,, \tag{11.97}$$

involving the Fisher information $F_q^{\max}(\xi) = F_c(\xi)$ of Eqs. (11.85) and (11.90). This is the best performance for the mean-squared error $e^2\big(\widehat{\xi}_{\rm ML}\big)$, saturating the Cramér–Rao inequality, and expected upon a large number L of repeated independent experiments for operating the maximum-likelihood estimator $\widehat{\xi}_{\rm ML}$ of Eq. (11.96) for the quantum phase ξ.

In this way we have obtained a complete characterization of an optimal strategy maximizing the efficiency for estimating the quantum phase ξ of the unitary operator U_ξ in Fig. 11.1. The unitary process U_ξ is probed by the optimal pure input signal $|\psi_0^{\rm opt}\rangle$ of Eq. (11.84), to afford the maximal quantum Fisher information $F_q^{\max}(\xi)$ of Eq. (11.85). The output probing signal $|\psi_\xi\rangle = \mathsf{U}_\xi\,|\psi_0^{\rm opt}\rangle$ is measured by the two-outcome measurement using the two projectors $\big\{|\psi_0^{\rm opt}\rangle\langle\psi_0^{\rm opt}|\,,\ \mathrm{I}_N - |\psi_0^{\rm opt}\rangle\langle\psi_0^{\rm opt}|\big\}$, to afford the maximal classical Fisher information $F_c(\xi) = F_q^{\max}(\xi)$. From the measurement outcomes, the maximum-likelihood estimator $\widehat{\xi}_{\rm ML}$ of Eq. (11.96) gives access to the minimal mean-squared estimation error $e^2\big(\widehat{\xi}_{\rm ML}\big) = 1/\big[LF_q^{\max}(\xi)\big]$ of Eq. (11.97) which is the minimal error upon a large number L of repeated independent measurements.

Several remarks can be made on these principles of quantum phase estimation:

(1) Interferometry. An important application of these principles of quantum phase estimation is for estimating the phase shift ξ of an optical interferometer as in Fig. 3.4, implementing the qubit unitary process $\mathsf{U}_\xi \in \mathcal{L}(\mathcal{H}_2)$ of Eq. (3.17) acting at the elementary level of individual photons. The corresponding Hermitian generator in Eq. (11.74) is $\mathsf{G} = \vec{n}\cdot\vec{\sigma}/2 \in \mathcal{L}(\mathcal{H}_2)$, endowed with the two eigenvalues $\pm 1/2$. The two eigenstates of G and U_ξ are the two orthogonal polarization states denoted $|\pm\vec{n}\,\rangle$ in Fig. 3.4. The optimal configuration to probe the interferometer is to use at the input a photon prepared in an equally weighted superposition of the two orthogonal eigenpolarizations $|\psi_0^{\rm opt}\rangle = \big(|\vec{n}\,\rangle + |-\vec{n}\,\rangle\big)/\sqrt{2}$ according to Eq. (11.84). The optimal measurement of the output photon is by a von Neumann projective measurement in the orthonormal basis $\big\{|\psi_0^{\rm opt}\rangle\,,\ |\psi_{0\perp}^{\rm opt}\rangle = \big(|\vec{n}\,\rangle - |-\vec{n}\,\rangle\big)/\sqrt{2}\big\}$, equivalent to measuring the spin observable $\Omega = \vec{n}\cdot\vec{\sigma}$. From L repeated measurements, the maximum-likelihood estimator of Eq. (11.96) for the phase is $\widehat{\xi}_{\rm ML} = \arccos\big(2L_0/L - 1\big)$. We shall examine further, later in this chapter, quantum parameter estimation for this fundamental case of the qubit.

(2) Local Estimation. This optimal strategy for quantum phase estimation achieves the maximal efficiency determined by Eq. (11.85) while returning for the phase ξ, via the estimator of Eq. (11.96), an estimate lying in the interval $[0, \pi]/(u_{\max} - u_{\min})$. Different phases ξ with $(u_{\max} - u_{\min})\xi$ differing by integer multiples of π will not be distinguished. For example, in the previous Remark (1) above on interferometry, the estimation always returns an estimate of the photonic phase in the interval $[0, \pi]$, whatever the true phase ξ or optical path length in the interferometer. Only the parameter $s(\xi) \in [-1, 1]$ of Eq. (11.92) is an identifiable parameter, in the terminology of signals-and-systems theory, since the measurement results only depend on the parameter $s(\xi)$ and not directly on the phase ξ itself. From a parameter $s(\xi)$ estimated in $[-1, 1]$, some determination of the phase modulo $\pi/(u_{\max} - u_{\min})$ may therefore be missed by the estimator of Eq. (11.96). Such a situation is commonly referred to as a *local estimation*, local to the estimation range for the estimate. Yet, this is not a specifically quantum feature, but rather a consequence of estimating a parameter ξ that determines the physical measurements via some periodicity invariance as in Eq. (11.92).

(3) Measuring an Observable. An interesting alternative way exists for describing the optimal quantum measurement using the pair of projectors $\{|\psi_0^{\mathrm{opt}}\rangle\langle\psi_0^{\mathrm{opt}}|, \mathrm{I}_N - |\psi_0^{\mathrm{opt}}\rangle\langle\psi_0^{\mathrm{opt}}|\}$ defined by the optimal input $|\psi_0^{\mathrm{opt}}\rangle$ from Eq. (11.84). Following Sect. 4.3, one can consider measuring the observable

$$\Omega = |u_{\min}\rangle\langle u_{\max}| + |u_{\max}\rangle\langle u_{\min}| \;, \tag{11.98}$$

acting in $\mathcal{H}_N$.

We have for this observable $\Omega\,|u_{\min}\rangle = |u_{\max}\rangle$ and $\Omega\,|u_{\max}\rangle = |u_{\min}\rangle$. The observable Ω has the eigenstate $|\psi_0^{\mathrm{opt}}\rangle = \big(|u_{\min}\rangle + |u_{\max}\rangle\big)/\sqrt{2}$ of Eq. (11.84) with eigenvalue $+1$, and the eigenstate

$$|\psi_{0\perp}^{\mathrm{opt}}\rangle = \frac{1}{\sqrt{2}}\big(|u_{\min}\rangle - |u_{\max}\rangle\big) \tag{11.99}$$

with eigenvalue -1; the rest are $N-2$ mutually orthogonal eigenstates spanning the subspace of $\mathcal{H}_N$ orthogonal to the vector plane $\big(|\psi_0^{\mathrm{opt}}\rangle, |\psi_{0\perp}^{\mathrm{opt}}\rangle\big)$ associated with the degenerate eigenvalue 0, and whose specific determination has no impact for our purpose here. Measuring the observable Ω is equivalent to performing a projective measurement by means of the three projectors onto each of the three mutually orthogonal eigensubspaces of Ω. The probing signal $|\psi_\xi\rangle$ of Eq. (11.86) always lies in the vector plane $\big(|\psi_0^{\mathrm{opt}}\rangle, |\psi_{0\perp}^{\mathrm{opt}}\rangle\big)$, which coincides with the vector plane $\big(|u_{\min}\rangle, |u_{\max}\rangle\big)$, since $|\psi_0^{\mathrm{opt}}\rangle$ and $|\psi_{0\perp}^{\mathrm{opt}}\rangle$ are linear combinations of $|u_{\min}\rangle$ and $|u_{\max}\rangle$. In the state $|\psi_\xi\rangle$, measuring the observable Ω projects $|\psi_\xi\rangle$ onto $|\psi_0^{\mathrm{opt}}\rangle$ with the probability P_0 of Eq. (11.87) and discloses the eigenvalue $+1$ for the observable; otherwise $|\psi_\xi\rangle$ is projected onto $|\psi_{0\perp}^{\mathrm{opt}}\rangle$ with the complementary probability $1 - P_0$ and discloses the eigenvalue -1 for the observable; and there is zero probability that $|\psi_\xi\rangle$ gets projected outside the vector plane $\left(|\psi_0^{\text{opt}}\rangle, |\psi_{0\perp}^{\text{opt}}\rangle\right)$, or equivalently that the eigenvalue 0 of Ω shows up as a measurement result. From Eq. (11.86) we have

$$\Omega\,|\psi_\xi\rangle = \frac{1}{\sqrt{2}}\Big[\exp(-i\xi u_{\min})\,|u_{\max}\rangle + \exp(-i\xi u_{\max})\,|u_{\min}\rangle\Big]\,, \tag{11.100}$$

so that the average value $\langle\Omega\rangle$ of the observable Ω in state $|\psi_\xi\rangle$ follows, by Eq. (4.60), as

$$\langle\Omega\rangle = \langle\psi_\xi|\Omega|\psi_\xi\rangle = \cos\big((u_{\max} - u_{\min})\xi\big)\,. \tag{11.101}$$

This is also $\langle\Omega\rangle = +1 \times P_0 - 1 \times (1 - P_0) = 2P_0 - 1$, and also $\langle\Omega\rangle$ coincides with the parameter $s(\xi)$ of Eq. (11.92). So estimating the average $\langle\Omega\rangle$ is equivalent to estimating the parameter $s(\xi)$ of Eq. (11.92). A number L of repetitions of the measurement of the observable Ω will yield L_0 values at $+1$ and $L - L_0$ values at -1, and an estimate for $\langle\Omega\rangle \equiv s$ following from Eq. (11.94), for an equivalent estimation of the phase ξ.

(4) Inoperative Probes. It can be noted that if the input probe is prepared in an eigenstate of the unitary U_ξ as $|\psi_0\rangle = |u_n\rangle$, then it transforms as $|\psi_0\rangle \mapsto \mathsf{U}_\xi\,|\psi_0\rangle = \exp(-i\xi u_n)\,|\psi_0\rangle = |\psi_\xi\rangle$, experiencing only a multiplication by the global ξ-dependent phase factor $\exp(-i\xi u_n)$. Such a global phase factor is inaccessible by measuring $|\psi_\xi\rangle$, yielding no access to the unknown phase ξ. This is also the situation of an input density operator $\rho_0 = |\psi_0\rangle\langle\psi_0|$ commuting with the unitary U_ξ, and a transformed state $\rho_\xi = \rho_0$ insensitive to ξ, as indicated at the end of Sect. 11.3.1. In addition, since $|\psi_0\rangle = |u_n\rangle$ is also an eigenstate of G, we have the average $\langle\Delta\mathsf{G}^2\rangle_0 = 0$. Follows in Eq. (11.83) a Fisher information $F_q(\xi) = 0$, consistently for an input probe state that is completely inoperative for estimating the phase ξ. By continuity, when the input probe $|\psi_0\rangle$ approaches such inoperative states eigenstates of U_ξ, one can expect a reduction of the quantum Fisher information $F_q(\xi)$, associated with a reduction of the efficiency for phase estimation. Moreover, for the preparation of the probe, these inoperative eigenstates span the whole Hilbert space $\mathcal{H}_N$. A blind uninformed preparation of the probe could well fall close to such inoperative states and entail a poor estimation efficiency. On the contrary, the characterization of the optimal input probe from Eq. (11.84) offers a quite useful guidance.

11.3.3 Phase Estimation with Noise

The optimal strategy of Sect. 11.3.2 fixes an important reference for quantum phase estimation, with the essential steps (input excitation, output measurement, estimator) which can be generally characterized to achieve optimality. The picture can however be extended to include an important feature under the form of a quantum noise hindering the estimation task. The transformed state $\mathsf{U}_\xi\rho_0\mathsf{U}_\xi^\dagger$ from Eq. (11.75), before

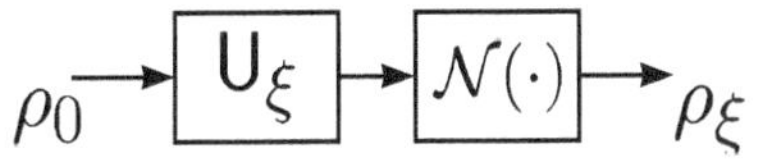

Fig. 11.2 An input probe with density operator ρ_0 acts as an excitation signal applied to the unitary process U_ξ of Eq. (11.74) under estimation. The transformed signal $\mathsf{U}_\xi \rho_0 \mathsf{U}_\xi^\dagger$ is then affected by the quantum noise $\mathcal{N}(\cdot)$ before delivering the ξ-dependent noisy output signal $\rho_\xi = \mathcal{N}(\mathsf{U}_\xi \rho_0 \mathsf{U}_\xi^\dagger)$ accessible for estimating the phase ξ

it becomes accessible for estimating the phase ξ, may be affected by some quantum noise, generally representable by the quantum process $\mathcal{N}(\cdot)$ as addressed in Sect. 9.1. Such a situation of quantum phase estimation from a noisy quantum signal is depicted in Fig. 11.2.

When estimation of the phase ξ has to be performed from the noisy signal $\rho_\xi = \mathcal{N}(\mathsf{U}_\xi \rho_0 \mathsf{U}_\xi^\dagger)$, it is in general more difficult to explicitly characterize the optimal strategy, depending on the noise $\mathcal{N}(\cdot)$. The quantum Fisher information $F_q(\xi)$ itself, the optimal input state ρ_0 replacing Eq. (11.84) in order to maximize $F_q(\xi)$, the optimal output measurement to reach the maximal $F_c(\xi) \leq F_q(\xi)$, are not always easily accessible. Situations may exist with no (valid (ξ-independent)) quantum measurement with a classical Fisher information $F_c(\xi)$ generally reaching the quantum Fisher information $F_q(\xi)$. For any state ρ_ξ it can be shown [317] that the quantum Fisher information $F_q(\xi)$ can never exceed the upper bound $4\langle \Delta\mathsf{G}^2\rangle_\xi = 4\,\mathrm{tr}(\rho_\xi \Delta\mathsf{G}^2)$ of the squared dispersion $\Delta\mathsf{G}^2$ averaged over ρ_ξ. Due to the monotonicity property (3) indicated at the end of Sect. 11.2.1, the quantum Fisher information $F_q(\xi\,;\rho_\xi)$ from the noisy signal $\rho_\xi = \mathcal{N}(\mathsf{U}_\xi \rho_0 \mathsf{U}_\xi^\dagger)$ is never larger than the quantum Fisher information $F_q(\xi\,;\mathsf{U}_\xi \rho_0 \mathsf{U}_\xi^\dagger)$ from the noise-free signal $\mathsf{U}_\xi \rho_0 \mathsf{U}_\xi^\dagger$. In addition, by the convexity property (5) of the quantum Fisher information indicated at the end of Sect. 11.2.1, maximization of $F_q(\xi\,;\rho_\xi)$ in the presence of noise always occurs with a pure input probe $\rho_0 = |\psi_0\rangle\langle\psi_0|$, which however may no longer correspond to the pure state $|\psi_0^{\mathrm{opt}}\rangle$ of Eq. (11.84) optimal in the noise free situation. And for any pure input state $\rho_0 = |\psi_0\rangle\langle\psi_0|$, the Fisher information $F_q(\xi\,;\rho_\xi)$ cannot exceed the upper bound $4\langle\psi_0|\Delta\mathsf{G}^2|\psi_0\rangle$ of Eq. (11.83). In particular, inoperative input probes $|\psi_0\rangle$ chosen as eigenstates of U_ξ, as seen in Remark (4) terminating Sect. 11.3.2, will remain inoperative in the presence of noise.

Comparable properties hold when the quantum noise $\mathcal{N}(\cdot)$ in Fig. 11.2 acts before, instead of after, the unitary U_ξ. For phase estimation from the noisy signal $\rho_\xi = \mathsf{U}_\xi \mathcal{N}(\rho_0)\mathsf{U}_\xi^\dagger$, maximization of the quantum Fisher information $F_q(\xi\,;\rho_\xi)$ occurs for a pure optimal input state $\rho_0^{\mathrm{opt}} = |\psi_0^{\mathrm{opt}}\rangle\langle\psi_0^{\mathrm{opt}}|$; for any input state ρ_0, the Fisher information $F_q(\xi\,;\rho_\xi)$ cannot exceed the upper bound $4\langle\psi_0^{\mathrm{opt}}|\Delta\mathsf{G}^2|\psi_0^{\mathrm{opt}}\rangle$ resulting from Eq. (11.83). And comparable properties still hold with two independent noises acting before and after the unitary U_ξ in Fig. 11.2.

Beyond these general properties, the strategy to efficient estimation (regarding the input excitation, output measurement, final estimator) will critically depend on the noise process $\mathcal{N}(\cdot)$ and dimensionality of the probing quantum signals. Quantum estimation in the presence of noise will be further specified for the fundamental case of the qubit later in this chapter.

11.3.4 Phase Estimation with Entanglement

Efficient estimation, with an estimator like $\widehat{\xi}_{\rm ML}$ of Eq. (11.96), envisages repeating L times the measurement, on L independent replicas of the probing state $\rho_\xi \in \mathcal{L}(\mathcal{H}_N)$, obtained from L independent replicas of the input probe $\rho_0 \in \mathcal{L}(\mathcal{H}_N)$ successively probing the unitary process U_ξ of Fig. 11.1. Instead of L independent replicas of the input probe $\rho_0 \in \mathcal{L}(\mathcal{H}_N)$, one could consider an entangled state with the L input probes prepared in a composite state with density operator $\rho_0^{(L)} \in \mathcal{L}\big(\mathcal{H}_N^{\otimes L}\big)$. The L independent probes ρ_0 correspond to the multiple-probe separable product state $\rho_0^{(L)} = \rho_0^{\otimes L}$, but as we are going to see, some interesting benefit can be gained by considering a multiple-probe prepared in an entangled state $\rho_0^{(L)}$ of $\mathcal{L}\big(\mathcal{H}_N^{\otimes L}\big)$.

Each of the L entangled probes can be applied successively, one after the other, onto the unitary process U_ξ of Fig. 11.1. When acting on the input multiple-probe state $\rho_0^{(L)} \in \mathcal{L}\big(\mathcal{H}_N^{\otimes L}\big)$, the unitary process U_ξ of Eq. (11.74) performs the transformation

$$\rho_0^{(L)} \longmapsto \mathsf{U}_\xi^{(L)} \rho_0^{(L)} \mathsf{U}_\xi^{(L)\dagger} = \rho_\xi^{(L)} , \tag{11.102}$$

with the unitary operator

$$\mathsf{U}_\xi^{(L)} = \mathsf{U}_\xi^{\otimes L} = \exp\big(-i\xi \mathsf{G}^{(L)}\big) \tag{11.103}$$

of $\mathcal{L}\big(\mathcal{H}_N^{\otimes L}\big)$. In Eq. (11.103), the Hermitian generator $\mathsf{G}^{(L)} \in \mathcal{L}\big(\mathcal{H}_N^{\otimes L}\big)$ is $\mathsf{G}^{(L)} = \mathsf{G} \otimes \mathrm{I}_N \otimes \cdots \otimes \mathrm{I}_N + \mathrm{I}_N \otimes \mathsf{G} \otimes \mathrm{I}_N \otimes \cdots \otimes \mathrm{I}_N + \cdots + \mathrm{I}_N \otimes \cdots \otimes \mathrm{I}_N \otimes \mathsf{G}$, involving a sum of L distinct factors each formed as the product of G with $L-1$ identities I_N.

We observe that the phase estimation task based on Eq. (11.103) operating in $\mathcal{H}_N^{\otimes L}$ has a structure analogue to the phase estimation task based on Eq. (11.74) operating in $\mathcal{H}_N$, with $\mathsf{G}^{(L)}$ replacing G. Consequently, the optimal estimation strategy of Sect. 11.3.2 applies as well, mutatis mutandis.

In particular, the quantum Fisher information $F_q^{(L)}(\xi)$ attached to the L-fold probing state $\rho_\xi^{(L)}$ of Eq. (11.102) is maximized by a pure L-fold multiple input probe $|\psi_0^{(L)}\rangle \in \mathcal{H}_N^{\otimes L}$ with density operator $\rho_0^{(L)} = |\psi_0^{(L)}\rangle\langle\psi_0^{(L)}|$ in Eq. (11.102). A pure input probe achieves, as in Eq. (11.83), a quantum Fisher information

$$F_q^{(L)}(\xi) = 4\Big\langle\big(\Delta\mathsf{G}^{(L)}\big)^2\Big\rangle_\xi = 4\Big\langle\big(\Delta\mathsf{G}^{(L)}\big)^2\Big\rangle_0 , \tag{11.104}$$

given by the squared dispersion operator $\Delta\mathsf{G}^{(L)}$ averaged equivalently over $|\psi_0^{(L)}\rangle$ or over $|\psi_\xi^{(L)}\rangle = \mathsf{U}_\xi^{(L)}\,|\psi_0^{(L)}\rangle$.

The optimal estimation from $\rho_\xi^{(L)}$ is governed by the eigenstates $|u_{\min}^{(L)}\rangle$ and $|u_{\max}^{(L)}\rangle$ associated respectively with the minimum $u_{\min}^{(L)}$ and maximum $u_{\max}^{(L)}$ eigenvalues of $\mathsf{G}^{(L)}$. The eigendecomposition of $\mathsf{G}^{(L)}$ or $\mathsf{U}_\xi^{(L)}$ is known from the eigendecomposition of G or U_ξ engaged in the L-fold tensor product. In this respect, for $\mathsf{G}^{(L)}$, the minimum eigenvalue is $u_{\min}^{(L)} = Lu_{\min}$ associated with the eigenstate $|u_{\min}^{(L)}\rangle = |u_{\min}\rangle^{\otimes L}$, the maximum eigenvalue is $u_{\max}^{(L)} = Lu_{\max}$ associated with the eigenstate $|u_{\max}^{(L)}\rangle = |u_{\max}\rangle^{\otimes L}$. As a result, the optimal state for the input multiple-probe is the pure state, defined analogously to Eq. (11.84), as

$$|\psi_0^{(L)}\rangle = |\psi_0^{(L)\mathrm{opt}}\rangle = \frac{1}{\sqrt{2}}\Big(|u_{\min}\rangle^{\otimes L} + |u_{\max}\rangle^{\otimes L}\Big)\,, \tag{11.105}$$

which is generally an entangled state of $\mathcal{H}_N^{\otimes L}$ for the L-fold multiple-probe. This achieves the maximum quantum Fisher information, defined analogously to Eq. (11.85), as

$$F_q^{(L)\max}(\xi) = L^2(u_{\max} - u_{\min})^2 = L^2 F_q^{\max}(\xi)\,, \tag{11.106}$$

which amplifies by a factor L^2 the single-probe Fisher information $F_q^{\max}(\xi)$ of Eq. (11.85). This scaling as L^2 of the quantum Fisher information of the L-fold multiple-probe is known as the **Heisenberg scaling** [313, 318–320]. Comparatively, by the additivity property (1) at the end of Sect. 11.2.1, a number L of independent probes lead to a scaling proportional to L for the quantum Fisher information, referred to as the standard, or shot-noise, scaling. The Heisenberg quadratic scaling in L^2 arises by entangling the L probes, taking place in this way via a specifically quantum mechanism. It offers an entanglement-based quantum enhancement, improving over the standard linear scaling in L of L independent probes which is also the classical scaling expected from L independent measurements in classical estimation.

When the optimal input multiple-probe $|\psi_0^{(L)\mathrm{opt}}\rangle$ of Eq. (11.105) is used to probe the unitary process $\mathsf{U}_\xi^{(L)}$ of Eq. (11.103), the resulting output state follows as

$$|\psi_\xi^{(L)}\rangle = \mathsf{U}_\xi^{(L)}\,|\psi_0^{(L)\mathrm{opt}}\rangle = \frac{1}{\sqrt{2}}\Big[\exp(-iL\xi u_{\min})\,|u_{\min}\rangle^{\otimes L} + \exp(-iL\xi u_{\max})\,|u_{\max}\rangle^{\otimes L}\Big]\,. \tag{11.107}$$

Then, an optimal quantum measurement of $|\psi_\xi^{(L)}\rangle$ in $\mathcal{H}_N^{\otimes L}$ uses, as before, the projector $|\psi_0^{(L)\mathrm{opt}}\rangle\langle\psi_0^{(L)\mathrm{opt}}|$ and the complementary projector $\mathrm{I}_N^{\otimes L} - |\psi_0^{(L)\mathrm{opt}}\rangle\langle\psi_0^{(L)\mathrm{opt}}|$ acting in $\mathcal{H}_N^{\otimes L}$. This two-outcome measurement in $\mathcal{H}_N^{\otimes L}$ achieves a classical Fisher information reaching the maximum quantum Fisher information $F_q^{(L)\max}(\xi)$ of Eq. (11.106), and therefore both Fisher informations benefit from the same Heisenberg scaling.

The measurement results are governed by the probability $P_0^{(L)}$ of projecting $|\psi_\xi^{(L)}\rangle$ onto $|\psi_0^{(L)\mathrm{opt}}\rangle$, defined analogously to Eq. (11.87), as

$$P_0^{(L)} = \left|\langle\psi_0^{(L)\text{opt}}|\psi_\xi^{(L)}\rangle\right|^2 = \cos^2\left(\frac{u_{\max}-u_{\min}}{2}\,L\xi\right) = \frac{1}{2}\Big[1+\cos\big((u_{\max}-u_{\min})L\xi\big)\Big]\,. \tag{11.108}$$

The measurement probability $P_0^{(L)}$ is related to the unknown phase ξ, as

$$P_0^{(L)} = \frac{1}{2}\left[1+s^{(L)}(\xi)\right], \tag{11.109}$$

via the ξ-dependent auxiliary parameter

$$s^{(L)}(\xi) = \cos\big((u_{\max}-u_{\min})L\xi\big)\,, \tag{11.110}$$

in similarity with Eqs. (11.91)–(11.92).

We can also introduce the state vector of $\mathcal{H}_N^{\otimes L}$, defined in analogy with Eq. (11.99), as

$$|\psi_{0\perp}^{(L)\text{opt}}\rangle = \frac{1}{\sqrt{2}}\Big(|u_{\min}\rangle^{\otimes L} - |u_{\max}\rangle^{\otimes L}\Big)\,, \tag{11.111}$$

and orthogonal to $|\psi_0^{(L)\text{opt}}\rangle$ of Eq. (11.105). The two states $|\psi_0^{(L)\text{opt}}\rangle$ and $|\psi_{0\perp}^{(L)\text{opt}}\rangle$ lie in the vector plane $\big(|u_{\min}\rangle^{\otimes L}, |u_{\max}\rangle^{\otimes L}\big)$ of $\mathcal{H}_N^{\otimes L}$, which coincides with the vector plane $\big(|\psi_0^{(L)\text{opt}}\rangle, |\psi_{0\perp}^{(L)\text{opt}}\rangle\big)$ of $\mathcal{H}_N^{\otimes L}$. The probing state $|\psi_\xi^{(L)}\rangle$ of Eq. (11.107) also lies in the vector plane $\big(|u_{\min}\rangle^{\otimes L}, |u_{\max}\rangle^{\otimes L}\big)$ or $\big(|\psi_0^{(L)\text{opt}}\rangle, |\psi_{0\perp}^{(L)\text{opt}}\rangle\big)$ of $\mathcal{H}_N^{\otimes L}$. So the projective measurement of $|\psi_\xi^{(L)}\rangle$ we envisage can only project onto $|\psi_0^{(L)\text{opt}}\rangle$ with the probability $P_0^{(L)}$ of Eq. (11.109) or onto $|\psi_{0\perp}^{(L)\text{opt}}\rangle$ with the complementary probability $1-P_0^{(L)}$, and there is zero probability that $|\psi_\xi^{(L)}\rangle$ gets projected outside the plane $\big(|\psi_0^{(L)\text{opt}}\rangle, |\psi_{0\perp}^{(L)\text{opt}}\rangle\big)$.

By repeating L' times the two-outcome measurement of $|\psi_\xi^{(L)}\rangle$, we obtain L'_0 times the projection onto $|\psi_0^{(L)\text{opt}}\rangle$, giving access to the maximum-likelihood estimator for $s^{(L)}(\xi)$ as

$$\widehat{s}_{\text{ML}}^{(L)} = 2\frac{L'_0}{L'} - 1 \in [-1, 1]\,, \tag{11.112}$$

which in turn gives access to the maximum-likelihood estimator $L\widehat{\xi}_{\text{ML}}$ for the unknown phase $L\xi$ as

$$L\,\widehat{\xi}_{\text{ML}} = \frac{\arccos\big(\widehat{s}_{\text{ML}}^{(L)}\big)}{u_{\max}-u_{\min}}\,, \tag{11.113}$$

finally yielding the maximum-likelihood estimator $\widehat{\xi}_{\text{ML}}$ for the unknown phase ξ as

$$\widehat{\xi}_{\text{ML}} = \frac{1}{(u_{\max}-u_{\min})L}\arccos\left(2\frac{L'_0}{L'}-1\right)\,. \tag{11.114}$$

By the same argument as in Sect. 11.3.2, which applies equally to the L-fold multiple-probe, we are sure that we have here the most efficient estimation strategy for estimating the phase ξ from the L-fold multiple probing state $|\psi_\xi^{(L)}\rangle$. In particular, the maximum-likelihood estimator $\widehat{\xi}_{\rm ML}$ of Eq. (11.114), at large size L', is unbiased and endowed with a mean-squared error $e^2(\widehat{\xi}_{\rm ML}) = \mathrm{var}(\widehat{\xi}_{\rm ML}) = 1/[L'L^2 F_q^{\max}(\xi)] = 1/[L'L^2(u_{\max} - u_{\min})^2]$, involving the Fisher information $F_q^{\max}(\xi)$ of Eq. (11.85) benefiting from the Heisenberg quadratic scaling in L^2 from Eq. (11.106) by probing with an L-fold multiple-probe. By comparison, with the single-probe strategy of Sect. 11.3.2, via a total of $L'L$ independent measurement of $|\psi_\xi\rangle$, the estimator of Eq. (11.96) reaches the mean-squared error $e^2(\widehat{\xi}_{\rm ML}) = \mathrm{var}(\widehat{\xi}_{\rm ML}) = 1/[L'L F_q^{\max}(\xi)] = 1/[L'L(u_{\max} - u_{\min})^2]$ following from Eq. (11.97). So the L-fold multiple-probe affords a division by L of the mean-squared estimation error. There exists a trade-off, between measuring $L'L$ independent probes in the simpler single-probe strategy, versus measuring L' blocks of L entangled probes in the more elaborated L-fold multiple-probe strategy. The trade-off can be handled in different ways, especially if for instance one has at disposal a fixed total budget of usable probes. If it was only a matter of maximizing the Fisher information, then all the LL' probes of the budget should be entangled into a single block. But then only one block or one multiple-probe would be available for measurement, and we would not be in the asymptotic regime of a large number of measurements where the Fisher information precisely determines the mean-squared error of the maximum-likelihood estimator in use for the phase. With one single measurement, the estimator can be expected to have a relatively large mean-squared error, which in addition remains to characterize as long as we are not in the asymptotic regime where it is determined by the Fisher information. Increasing the number of measurements, then, would reduce the mean-squared error of the estimator, but at the same time the Fisher information. In general, the multiple-probe strategy offers a possibility for improved efficiency, stemming from the Heisenberg enhanced scaling of the Fisher information. This also comes at the cost of handling entangled probes, which are more difficult to process, and more fragile to noise and decoherence as the conditions of Sect. 11.3.3 become relevant.

The four remarks terminating Sect. 11.3.2 can be updated for the multiple-probe strategy offering the possibility of entanglement-assisted estimation.

(1) Interferometry. Optical interferometry can use multiple entangled photons to benefit from the Heisenberg enhanced performance in phase estimation, as addressed for instance in [66, 314, 318, 321].

(2) Local Estimation. With the L-fold multiple-probe strategy, as before the phase ξ is not directly identifiable, but indirectly via the parameter $s^{(L)}(\xi)$ of Eqs. (11.110) and (11.112). Through Eq. (11.113), the parameter $L\xi$ is first estimated, by an estimate returned in the interval $[0, \pi]/(u_{\max} - u_{\min})$. The estimate $\widehat{\xi}_{\rm ML}$ for ξ follows from Eq. (11.114) and is returned in the interval $[0, \pi]/[(u_{\max} - u_{\min})L]$. This estimate $\widehat{\xi}_{\rm ML}$ for ξ benefits from the Heisenberg enhanced precision with $e^2(\widehat{\xi}_{\rm ML}) = \mathrm{var}(\widehat{\xi}_{\rm ML}) = 1/[L'L^2 F_q^{\max}(\xi)]$, as we have seen. But this enhanced precision of the L-fold multiple-probe strategy comes at the cost of an L-fold narrowing of the local character of the estimate for ξ, which is now confined to an interval L times reduced in range compared with the single-probe strategy.

It can also be noted that what is accomplished with the L-fold multiple-probe is equivalent to the process of a single probe that would probe the amplified phase $\xi_L = L\xi$ characterizing the unitary U_{ξ_L} instead of the elementary phase ξ characterizing the unitary U_ξ. The state vector of the single probe would transform as $|\psi\rangle \mapsto \mathsf{U}_{\xi_L}|\psi\rangle = \exp(-i\xi_L\mathsf{G})|\psi\rangle$, and the optimal input $|\psi_0^{\mathrm{opt}}\rangle$ of Eq. (11.84) would achieve the same maximum quantum Fisher information $F_q^{\max}(\xi_L) = (u_{\max} - u_{\min})^2$ as in Eq. (11.85) but this time for estimating the amplified phase $\xi_L = L\xi$, which gives $F_q^{\max}(\xi) = L^2 F_q^{\max}(\xi_L) = L^2(u_{\max} - u_{\min})^2$, as in a change of parameter similar to $\xi = \xi(\nu)$ addressed in Sect. 11.2.1 just before the final five properties (1)–(5), i.e., an L^2-times amplified Fisher information for estimating ξ as in Eq. (11.106). A maximum-likelihood estimator for ξ_L with the mean-squared error $e^2(\widehat{\xi_L}) = \mathrm{var}(\widehat{\xi_L}) = 1/F_q^{\max}(\xi_L) = 1/(u_{\max} - u_{\min})^2$ would lead for ξ to an L^2-times smaller mean-squared error, equivalent to the performance in a single use of the L-fold multiple-probe. The amplified phase $\xi_L = L\xi$ is accessible when the same probe can be made to recirculate L times in succession across the unitary process U_ξ of Fig. 11.1. This could be for instance the same single photon, recirculating L times across the interferometer of Fig. 3.4, before it gets measured, if ever the procedure can be controlled in this way without anticipated (destructive) measurement of the circulating photon.

(3) Measuring an Observable. For estimating the phase ξ via measuring the L-fold multiple-probe in the state $|\psi_\xi^{(L)}\rangle$ of Eq. (11.107), instead of the two-outcome optimal measurement using the projector $|\psi_0^{(L)\mathrm{opt}}\rangle\langle\psi_0^{(L)\mathrm{opt}}|$ and the complementary projector in $\mathcal{H}_N^{\otimes L}$, one can consider measuring on $|\psi_\xi^{(L)}\rangle$ the observable $\Omega^{\otimes L}$ acting in $\mathcal{H}_N^{\otimes L}$ and following from the observable Ω of Eq. (11.98). The eigendecomposition of $\Omega^{\otimes L}$ is related to the eigendecomposition of Ω via the L-fold tensor product, so we deduce that $\Omega^{\otimes L}$ has the three eigenvalues $\{+1, -1, 0\}$, arising from L-fold products of the three eigenvalues $\{+1, -1, 0\}$ of Ω. The observable $\Omega^{\otimes L}$ verifies $\Omega^{\otimes L}|u_{\min}\rangle^{\otimes L} = \big(\Omega|u_{\min}\rangle\big)^{\otimes L} = |u_{\max}\rangle^{\otimes L}$ and $\Omega^{\otimes L}|u_{\max}\rangle^{\otimes L} = |u_{\min}\rangle^{\otimes L}$. Consequently, the state vector $|\psi_0^{(L)\mathrm{opt}}\rangle$ of Eq. (11.105) verifies $\Omega^{\otimes L}|\psi_0^{(L)\mathrm{opt}}\rangle = |\psi_0^{(L)\mathrm{opt}}\rangle$ and is therefore an eigenstate of $\Omega^{\otimes L}$ with eigenvalue $+1$; also $|\psi_{0\perp}^{(L)\mathrm{opt}}\rangle$ of Eq. (11.111) is an eigenstate of $\Omega^{\otimes L}$ with eigenvalue -1. The other $N^L - 2$ eigenstates of $\Omega^{\otimes L}$ are to be found in the subspace of $\mathcal{H}_N^{\otimes L}$ orthogonal to the vector plane $\big(|\psi_0^{(L)\mathrm{opt}}\rangle, |\psi_{0\perp}^{(L)\mathrm{opt}}\rangle\big)$ which is the same as the vector plane $\big(|u_{\min}\rangle^{\otimes L}, |u_{\max}\rangle^{\otimes L}\big)$. Any eigenstate in this subspace will necessarily have a component along an eigenstate of Ω lying outside the plane $\big(|u_{\min}\rangle, |u_{\max}\rangle\big)$, i.e., a component along an eigenstate of Ω associated with the eigenvalue 0. As a consequence, via the L-fold product, these $N^L - 2$ remaining eigenstates of $\Omega^{\otimes L}$ are also associated with the eigenvalue 0.

From Eq. (11.107) we have

$$\Omega^{\otimes L}\,|\psi_\xi^{(L)}\rangle = \frac{1}{\sqrt{2}}\Big[\exp(-iL\xi u_{\min})\,|u_{\max}\rangle^{\otimes L} + \exp(-iL\xi u_{\max})\,|u_{\min}\rangle^{\otimes L}\Big]\,, \tag{11.115}$$

so that the average value $\langle\Omega^{\otimes L}\rangle$ of the observable $\Omega^{\otimes L}$ in state $|\psi_\xi^{(L)}\rangle$ follows as

$$\langle\Omega^{\otimes L}\rangle = \langle\psi_\xi^{(L)}|\Omega^{\otimes L}|\psi_\xi^{(L)}\rangle = \cos\big((u_{\max} - u_{\min})L\xi\big)\,, \tag{11.116}$$

showing that $\langle\Omega^{\otimes L}\rangle$ coincides with the parameter $s^{(L)}(\xi)$ of Eq. (11.110). As before, estimating the observable average $\langle\Omega^{\otimes L}\rangle$ is equivalent to estimating the parameter $s^{(L)}(\xi)$, and an estimate for $s^{(L)}(\xi) \equiv \langle\Omega^{\otimes L}\rangle$ accomplishes the estimation of the phase ξ via Eq. (11.113). Since $|\psi_\xi^{(L)}\rangle$ is in the plane $\big(|\psi_0^{(L)\mathrm{opt}}\rangle\,, |\psi_{0\perp}^{(L)\mathrm{opt}}\rangle\big)$, on state $|\psi_\xi^{(L)}\rangle$ measuring $\Omega^{\otimes L}$ can only lead to the eigenvalues $+1$ or -1 and never to the eigenvalue 0. A significant advantage of measuring the observable $\Omega^{\otimes L}$ for estimation is that $\Omega^{\otimes L}$ is a separable observable of the L probes. Consequently, the observable $\Omega^{\otimes L}$ can be measured by measuring separately the observable Ω on each individual probe of the L-fold multiple-probe in state $|\psi_\xi^{(L)}\rangle$. Each measurement of Ω on one of the L individual probes will reveal an eigenvalue of Ω. The product of the resulting L eigenvalues will reveal the eigenvalue measured for $\Omega^{\otimes L}$. This product can only be $+1$ or -1, with the respective probabilities that we (momentarily) denote $P_L(+1)$ and $P_L(-1) = 1 - P_L(+1)$. We deduce the average $\langle\Omega^{\otimes L}\rangle = 2P_L(+1) - 1$, so that $P_L(+1) = \big(1 + \langle\Omega^{\otimes L}\rangle\big)/2$. But since $\langle\Omega^{\otimes L}\rangle$ and $s^{(L)}(\xi)$ coincide, from Eq. (11.109) we deduce that $P_L(+1) = P_0^{(L)}$. The separable measurement of $\Omega^{\otimes L}$ and the joint measurement via the projector $|\psi_0^{(L)\mathrm{opt}}\rangle\langle\psi_0^{(L)\mathrm{opt}}|$ are both with the two outcomes ±1 with same statistics. Both measurements have therefore equivalent statistical performances for estimation. Measuring $\Omega^{\otimes L}$ via L separate measurements of Ω on each of the L individual probes is usually easier to perform compared with the joint measurement based on the projector $|\psi_0^{(L)\mathrm{opt}}\rangle\langle\psi_0^{(L)\mathrm{opt}}|$ acting collectively on the L probes. These are not two completely equivalent measurements, for the reason that the measured state $|\psi_\xi^{(L)}\rangle$ gets projected onto distinct post-measurement states. The two measurements operate in two distinct eigenbases of $\Omega^{\otimes L}$, one separable, one entangled; the possibility of two distinct eigenbases stemming from the degeneracy of the eigenvalues of $\Omega^{\otimes L}$. In the separable measurement, the post-measurement state is an L-fold tensor product of $|\psi_0^{\mathrm{opt}}\rangle$ and $|\psi_{0\perp}^{\mathrm{opt}}\rangle$, while in the joint collective measurement it is either $|\psi_0^{(L)\mathrm{opt}}\rangle$ or $|\psi_{0\perp}^{(L)\mathrm{opt}}\rangle$. But all are two-outcome measurements with same statistics and same performance for phase estimation. Again, measuring the observable Ω is essentially equivalent to performing a projective measurement projecting onto the two state vectors $\big\{|\psi_0^{\mathrm{opt}}\rangle\,, |\psi_{0\perp}^{\mathrm{opt}}\rangle\big\}$ associated with Eq. (11.99).

(4) Inoperative Probes. If the L-fold multiple input probe is prepared in a state $|\psi_0^{(L)}\rangle$ which is an eigenstate of the unitary process $\mathsf{U}_\xi^{(L)} = \mathsf{U}_\xi^{\otimes L}$, the transformed state $\mathsf{U}_\xi^{\otimes L}\,|\psi_0^{(L)}\rangle = |\psi_\xi^{(L)}\rangle$ amounts to the input $|\psi_0^{(L)}\rangle$ multiplied by a ξ-dependent global phase factor. Such a global phase factor is inaccessible by measuring $|\psi_\xi^{(L)}\rangle$, yielding no access to the unknown phase ξ. In addition, since such an input $|\psi_0^{(L)}\rangle$ is also an eigenstate of $\mathsf{G}^{\otimes L}$, we have the average $\langle(\Delta \mathsf{G}^{(L)})^2\rangle_0 = 0$. Follows in Eq. (11.104) a quantum Fisher information $F_q^{(L)}(\xi) = 0$, consistently for an input probe $|\psi_0^{(L)}\rangle$ that is completely inoperative for estimating the phase ξ. Such an inoperative input $|\psi_0^{(L)}\rangle$, eigenstate of $\mathsf{U}_\xi^{\otimes L}$, can be a separable state of $\mathcal{H}_N^{\otimes L}$, formed as a tensor product of L eigenstates of U_ξ. It can also be an entangled state of $\mathcal{H}_N^{\otimes L}$, because of the degeneracy of the eigenvalues of $\mathsf{U}_\xi^{\otimes L}$. For illustration, when $L = 2$, both $|u_{\min}\rangle \otimes |u_{\max}\rangle$ and $|u_{\max}\rangle \otimes |u_{\min}\rangle$ are separable eigenstates of $\mathsf{U}_\xi^{\otimes 2}$, associated with the same eigenvalue $\exp\big(-i\xi(u_{\min} + u_{\max})\big)$. As a consequence, the two states $|w_\pm\rangle = \big(|u_{\min}\rangle \otimes |u_{\max}\rangle \pm |u_{\max}\rangle \otimes |u_{\min}\rangle\big)/\sqrt{2}$ are also two eigenstates of $\mathsf{U}_\xi^{\otimes 2}$, associated with the same eigenvalue $\exp\big(-i\xi(u_{\min} + u_{\max})\big)$, but are generally two entangled states of $\mathcal{H}_N^{\otimes 2}$. Any linear combination of $|w_+\rangle$ and $|w_-\rangle$ will also (almost) generally stand as an entangled eigenstate of $\mathsf{U}_\xi^{\otimes 2}$ with the same eigenvalue $\exp\big(-i\xi(u_{\min} + u_{\max})\big)$. These linear combinations form a two-dimensional subspace of $\mathcal{H}_N^{\otimes 2}$, where the input probes are completely inoperative for estimating the phase ξ. And in a similar way at $L = 2$, assembling other pairs of eigenstates of U_ξ produces many other two-dimensional subspaces of $\mathcal{H}_N^{\otimes 2}$ inoperative for phase estimation. At any size L, when the L-fold multiple input probe $|\psi_0^{(L)}\rangle$ approaches such inoperative states eigenstates of $\mathsf{U}_\xi^{\otimes L}$, by continuity one can expect a reduction of the quantum Fisher information $F_q^{(L)}(\xi)$, associated with a reduction of the efficiency for phase estimation. In this respect, exciting blindly at the input may incur poor estimation efficiency, while the optimal input probing as characterized in Eq. (11.105) offers useful guidance.

11.4 Estimation on Qubits

In this section, the principles of quantum estimation of Sect. 11.3 are further specified to the fundamental case of the qubit. For this purpose, we specifically exploit the Bloch representation accessible for the qubit, as also realized for instance in [293, 322].

11.4.1 Quantum Fisher Information for the Qubit

A qubit with Hilbert space $\mathcal{H}_2$ has a ξ-dependent state described in the Bloch representation of Eq. (2.60) by the density operator

$$\rho_\xi = \frac{1}{2}\Big(\mathrm{I}_2 + \vec{r}_\xi \cdot \vec{\sigma}\Big), \tag{11.117}$$

with the Bloch vector $\vec{r}_\xi \in \mathbb{R}^3$ carrying the dependence of the qubit state with the unknown parameter ξ to be estimated. The state of Eq. (11.117) differentiates as $\partial_\xi \rho_\xi = \partial_\xi \vec{r}_\xi \cdot \vec{\sigma}/2$. For the case of a **mixed qubit state** ρ_ξ, the quantum score operator of Eq. (11.65) is also

$$\mathsf{V}_\xi = 2\sum_{k,\ell} \frac{1}{\lambda_k + \lambda_\ell} |\lambda_k\rangle \langle\lambda_k|\partial_\xi \rho_\xi|\lambda_\ell\rangle \langle\lambda_\ell| \ , \tag{11.118}$$

and for the qubit it can be evaluated in Bloch representation by means of the two eigenstates from Eq. (2.70) for ρ_ξ. For example, among the four terms in the sum of Eq. (11.118) for the qubit, one is

$$\frac{1}{\lambda_- + \lambda_+} |\lambda_-\rangle \langle\lambda_-|\partial_\xi \rho_\xi|\lambda_+\rangle \langle\lambda_+| = \frac{\mathrm{I}_2 - \vec{r}_\xi \cdot \vec{\sigma}/\|\vec{r}_\xi\|}{2} \frac{\partial_\xi \vec{r}_\xi \cdot \vec{\sigma}}{2} \frac{\mathrm{I}_2 + \vec{r}_\xi \cdot \vec{\sigma}/\|\vec{r}_\xi\|}{2} . \tag{11.119}$$

To evaluate expressions such as Eq. (11.119) involving products of non-commuting operators on $\mathcal{H}_2$, we use the identities of Eqs. (A.45)–(A.46) from the Appendix. Handling Eq. (11.118) in this manner for the qubit [293], we finally obtain the quantum score in Bloch representation

$$\boxed{\mathsf{V}_\xi = -\frac{\vec{r}_\xi\, \partial_\xi \vec{r}_\xi}{1 - \vec{r}_\xi^{\,2}} \mathrm{I}_2 + \left(\frac{\vec{r}_\xi\, \partial_\xi \vec{r}_\xi}{1 - \vec{r}_\xi^{\,2}} \vec{r}_\xi + \partial_\xi \vec{r}_\xi \right) \cdot \vec{\sigma} \ .} \tag{11.120}$$

By a comparable approach in Bloch representation, the quantum Fisher information $F_q(\xi) = \mathrm{tr}(\rho_\xi \mathsf{V}_\xi^2) = \mathrm{tr}(\mathsf{V}_\xi \partial_\xi \rho_\xi)$ follows for the qubit as

$$\boxed{F_q(\xi) = \frac{\left(\vec{r}_\xi\, \partial_\xi \vec{r}_\xi\right)^2}{1 - \vec{r}_\xi^{\,2}} + \left(\partial_\xi \vec{r}_\xi\right)^2 .} \tag{11.121}$$

For the case of a **pure qubit state** ρ_ξ, differentiating $\|\vec{r}_\xi\|^2 = \vec{r}_\xi\, \vec{r}_\xi = 1$ gives $2\vec{r}_\xi\, \partial_\xi \vec{r}_\xi = 0$, whence $\vec{r}_\xi\, \partial_\xi \vec{r}_\xi = 0$, indicating that the Bloch vector $\vec{r}_\xi$ is orthogonal to its derivative $\partial_\xi \vec{r}_\xi$ in $\mathbb{R}^3$. The quantum score $\mathsf{V}_\xi = 2\partial_\xi \rho_\xi$ of Eq. (11.70) in Bloch representation is

$$\boxed{\mathsf{V}_\xi = \partial_\xi \vec{r}_\xi \cdot \vec{\sigma} \ ,} \tag{11.122}$$

and the quantum Fisher information $F_q(\xi) = \mathrm{tr}(\mathsf{V}_\xi \partial_\xi \rho_\xi)$ of Eq. (11.71) is

$$\boxed{F_q(\xi) = \left(\partial_\xi \vec{r}_\xi\right)^2 .} \tag{11.123}$$

Equations (11.121) and (11.123) offer a general characterization of the quantum Fisher information $F_q(\xi)$ assessing the performance in estimating any scalar parameter ξ attached to a qubit state ρ_ξ, mixed or pure.

11.4.2 Classical Fisher Information for the Qubit

For estimation, the qubit in state ρ_ξ can generally be measured by means of a POVM $\{\mathsf{E}_m\}_{m=1}^M$ with M POVM operators $\mathsf{E}_m \in \mathcal{L}(\mathcal{H}_2)$. Each POVM element $\mathsf{E}_m = b_m \mathrm{I}_2 + \vec{a}_m \cdot \vec{\sigma}$ can be parameterized in Bloch representation according to Eq. (4.53). The resulting measurement probabilities follow from Eq. (4.54) as

$$P(m\,;\xi) = \mathrm{tr}(\mathsf{E}_m \rho_\xi) = b_m + \vec{a}_m \vec{r}_\xi \ . \tag{11.124}$$

With the derivatives $\partial_\xi P(m\,;\xi) = \vec{a}_m \partial_\xi \vec{r}_\xi$, a classical Fisher information follows from Eq. (11.45) as

$$F_c(\xi) = \sum_{m=1}^{M} \frac{\left(\vec{a}_m \partial_\xi \vec{r}_\xi\right)^2}{b_m + \vec{a}_m \vec{r}_\xi} \ . \tag{11.125}$$

A definite estimation task on a qubit will come with a definite ξ-dependent Bloch vector $\vec{r}_\xi$. It will then be meaningful to look for a POVM $\left\{(b_m, \vec{a}_m)\right\}_{m=1}^M$ with a classical Fisher information $F_c(\xi)$ in Eq. (11.125) coming as close as possible to the quantum Fisher information $F_q(\xi)$ of Eqs. (11.121) or (11.123). In this respect, as already indicated, it may not be feasible, in definite situations, to find a valid (ξ-independent) POVM achieving the equality $F_c(\xi) = F_q(\xi)$.

A special but quite useful and common POVM arises when measuring on the qubit in state ρ_ξ a spin observable of the form $\Omega = \vec{\omega} \cdot \vec{\sigma}$, as defined at the end of Sect. 4.3.4, with a unit vector $\vec{\omega} \in \mathbb{R}^3$. This gives rise to the POVM with the two elements $\mathsf{E}_\pm = \left(\mathrm{I}_2 \pm \vec{\omega} \cdot \vec{\sigma}\right)/2$. Two (eigen)values ± 1 can be measured for the observable, with the probabilities $\Pr\{\pm 1\} = \left(1 \pm \vec{\omega}\,\vec{r}_\xi\right)/2$ from Eq. (4.80). The corresponding classical Fisher information in Eq. (11.125) follows as

$$F_c(\xi) = \frac{\left(\vec{\omega}\,\partial_\xi \vec{r}_\xi\right)^2}{1 - (\vec{\omega}\,\vec{r}_\xi)^2} \ . \tag{11.126}$$

For illustration, we now examine several significant examples of parameter estimation from a qubit state.

11.4.3 Estimation of a Pure Qubit State

For the generic pure qubit state $|\psi\rangle = \cos(\theta/2)\,|0\rangle + e^{i\varphi}\sin(\theta/2)\,|1\rangle$ from Eq. (2.6), the associated Bloch vector $\vec{r} = [\sin(\theta)\cos(\varphi),\ \sin(\theta)\sin(\varphi),\ \cos(\theta)]^\top$ in $\mathbb{R}^3$ has the two derivatives $\partial_\theta\vec{r} = [\cos(\theta)\cos(\varphi),\ \cos(\theta)\sin(\varphi),\ -\sin(\theta)]^\top$ and $\partial_\varphi\vec{r} = [-\sin(\theta)\sin(\varphi),\ \sin(\theta)\cos(\varphi),\ 0]^\top$. At fixed φ, for estimating the parameter $\xi \equiv \theta$, we have from Eq. (11.123) the quantum Fisher information $F_q(\theta) = (\partial_\theta\vec{r}\,)^2 = 1$ for all φ. Meanwhile, at fixed θ, for estimating the parameter $\xi \equiv \varphi$, we have from Eq. (11.123) the quantum Fisher information $F_q(\varphi) = (\partial_\varphi\vec{r}\,)^2 = \sin^2(\theta)$, which is maximized at $F_q(\varphi) = 1$ when $\theta = \pi/2$.

A von Neumann measurement in the computational basis $\big\{|0\rangle\,, |1\rangle\big\}$ has a projection probability $\Pr\big\{|0\rangle\big\} = P_0 = |\langle 0|\psi\rangle|^2 = \cos^2(\theta/2) = \big[1+\cos(\theta)\big]/2$, for all φ. For estimation of the qubit coordinate θ, such measurement achieves, from Eq. (11.45), the classical Fisher information $F_c(\theta) = (\partial_\theta P_0)^2/\big[(1-P_0)P_0\big] = 1 = F_q(\theta)$, for all φ. Therefore, this von Neumann measurement in the basis $\big\{|0\rangle\,, |1\rangle\big\}$—equivalent to measuring the qubit observable σ_z—is optimal for estimating the qubit coordinate θ. It is however completely incapable of estimating the qubit coordinate φ, since the measurement results are insensitive to φ.

An alternative measurement of the qubit in the Hadamard basis $\big\{|+\rangle\,, |-\rangle\big\}$ has a projection probability $\Pr\big\{|+\rangle\big\} = P_+ = |\langle +|\psi\rangle|^2 = \big[1+\sin(\theta)\cos(\varphi)\big]/2$. For estimation of the qubit coordinate θ, such measurement achieves, from Eq. (11.45), the classical Fisher information

$$F_c(\theta) = \frac{\big(\partial_\theta P_+\big)^2}{(1-P_+)P_+} = \frac{\cos^2(\theta)\cos^2(\varphi)}{1-\sin^2(\theta)\cos^2(\varphi)}\,, \tag{11.127}$$

which is maximized at $F_c(\theta) = 1 = F_q(\theta)$ when $\varphi = 0$. Therefore, for estimating the qubit coordinate θ, measuring in the basis $\big\{|+\rangle\,, |-\rangle\big\}$—equivalent to measuring the qubit observable σ_x—is optimal only when $\varphi = 0$, and is less efficient, with $F_c(\theta) < F_q(\theta) = 1$, when $\varphi \neq 0$; while measuring in $\big\{|0\rangle\,, |1\rangle\big\}$ is optimal for any φ. And for estimating the qubit coordinate φ, the measurement in the Hadamard basis $\big\{|+\rangle\,, |-\rangle\big\}$ achieves the classical Fisher information

$$F_c(\varphi) = \frac{\big(\partial_\varphi P_+\big)^2}{(1-P_+)P_+} = \frac{\sin^2(\theta)\sin^2(\varphi)}{1-\sin^2(\theta)\cos^2(\varphi)}\,, \tag{11.128}$$

which is maximized at $F_c(\varphi) = 1 = F_q(\varphi)$ when $\theta = \pi/2$. Therefore, for estimating the qubit coordinate φ, measuring in the basis $\big\{|+\rangle\,, |-\rangle\big\}$ is optimal only when $\theta = \pi/2$, and is less efficient otherwise with $F_c(\theta) \le F_q(\theta) = \sin^2(\theta)$.

As illustrated by this example, different parameters usually require different quantum measurements for efficient estimation.

11.4.4 Qubit Phase Estimation

Qubit phase estimation is an important metrological task arising when a qubit acts as an excitation signal to probe a unitary process $\mathsf{U}_\xi \in \mathcal{L}(\mathcal{H}_2)$ from Eq. (11.74). The operation somehow contains a generalization of estimation of the qubit phase coordinate φ we just considered above in Sect. 11.4.3. For phase estimation, the qubit unitary operator $\mathsf{U}_\xi \in \mathcal{L}(\mathcal{H}_2)$ in Eq. (11.74) receives the general parameterization

$$\mathsf{U}_\xi = \exp\left(-i\,\frac{\xi}{2}\vec{n}\cdot\vec{\sigma}\right) = \exp(-i\xi\mathsf{G})\,, \tag{11.129}$$

as in Eq. (3.18), characterized by the unit vector $\vec{n}$ of $\mathbb{R}^3$, and ξ is the unknown phase parameter to be estimated. The Hermitian generator in Eq. (11.129) is $\mathsf{G} = \vec{n}\cdot\vec{\sigma}/2 \in \mathcal{L}(\mathcal{H}_2)$, endowed with the two eigenvalues $u_\pm = \pm 1/2$. The two eigenstates of G and U_ξ are the two orthogonal state vectors denoted $|\pm\vec{n}\,\rangle$ associated with the two density operators $|\pm\vec{n}\,\rangle\langle\pm\vec{n}\,|$ of unit Bloch vectors $\pm\vec{n}$.

When the input density operator $\rho_0 \in \mathcal{L}(\mathcal{H}_2)$ in Fig. 11.1 has Bloch vector $\vec{r}_0 \in \mathbb{R}^3$, the transformation of Eq. (11.75) produces the output state ρ_ξ having Bloch vector $\vec{r}_\xi \in \mathbb{R}^3$ formed by $\vec{r}_0$ rotated around the axis $\vec{n}$ by the angle ξ. In this way, the phase angle ξ originally belonging to the qubit process U_ξ of Eq. (11.129) is transferred to the qubit state for its estimation. Exploiting the Bloch representation as in [323], the identity of Eq. (A.45) from the Appendix can be used to deduce the commutator $[\rho_0, \mathsf{G}] = -i(\vec{n}\times\vec{r}_0)\cdot\vec{\sigma}/2$. Then, by using the eigendecomposition of Eqs. (2.66) and (2.70) for the qubit, the quantum score in Eq. (11.78) can be evaluated as the compact expression

$$\mathsf{V}_0 = \left(\vec{n}\times\vec{r}_0\right)\cdot\vec{\sigma}\,, \tag{11.130}$$

which also holds when ρ_0 is a pure qubit state ruled by $\mathsf{V}_0 = i2[\rho_0, \mathsf{G}]$. Next, the quantum Fisher information of Eq. (11.80) evaluates to

$$F_q(\xi) = \left(\vec{n}\times\vec{r}_0\right)^2\,, \tag{11.131}$$

which also holds when ρ_0 is a pure qubit state ruled by Eq. (11.82).

Alternatively, the Bloch vector $\vec{r}_0 \in \mathbb{R}^3$ can always be written as $\vec{r}_0 = r_{0\parallel}\vec{n} + r_{0\perp}\vec{n}_\perp$ defining the unit vector $\vec{n}_\perp$ orthogonal to $\vec{n}$ in the plane $\left(\vec{n}, \vec{r}_0\right)$ of $\mathbb{R}^3$. The rotated Bloch vector becomes $\vec{r}_\xi = r_{0\parallel}\vec{n} + r_{0\perp}\cos(\xi)\vec{n}_\perp + r_{0\perp}\sin(\xi)\vec{n}\,'_\perp$ with a third orthogonal unit vector $\vec{n}\,'_\perp = \vec{n}\times\vec{n}_\perp$ of $\mathbb{R}^3$. The derivative ensues as $\partial_\xi\vec{r}_\xi = -r_{0\perp}\sin(\xi)\vec{n}_\perp + r_{0\perp}\cos(\xi)\vec{n}\,'_\perp = \vec{n}\times\vec{r}_\xi$, providing an intrinsic geometric characterization of the derivative

$$\partial_\xi\vec{r}_\xi = \vec{n}\times\vec{r}_\xi\,, \tag{11.132}$$

which leads to the vanishing inner product $\vec{r}_\xi\, \partial_\xi \vec{r}_\xi = 0$. It then follows in Eqs. (11.120) and (11.122) equally that the quantum score is always $\mathsf{V}_\xi = \partial_\xi \vec{r}_\xi \cdot \vec{\sigma}$, equivalent to

$$\mathsf{V}_\xi = \left(\vec{n} \times \vec{r}_\xi\right) \cdot \vec{\sigma} \,, \tag{11.133}$$

matching $\mathsf{U}_\xi \mathsf{V}_0 \mathsf{U}_\xi^\dagger = \mathsf{V}_\xi$ from Eq. (11.130). It also follows in Eqs. (11.121) and (11.123) equally that the quantum Fisher information is always $F_q(\xi) = \left(\partial_\xi \vec{r}_\xi\right)^2$, equivalent to $F_q(\xi) = \left(\vec{n} \times \vec{r}_\xi\right)^2$, which is similar to Eq. (11.131) due to the geometry of $\vec{r}_0$ and $\vec{r}_\xi$ fixing $\|\vec{n} \times \vec{r}_0\| = \|\vec{n} \times \vec{r}_\xi\|$.

It can now be observed, from Eq. (11.131), that any pure input probe ρ_0 with a unit Bloch vector $\vec{r}_0$ orthogonal[3] to the axis $\vec{n}$ maximizes the quantum Fisher information $F_q(\xi)$ at the maximum $F_q^{\max}(\xi) = 1$, for any phase ξ. Consistently, one such optimal input state is $|\psi_0^{\text{opt}}\rangle$ of Eq. (11.84), which for the qubit here takes the form

$$|\psi_0^{\text{opt}}\rangle = \frac{1}{\sqrt{2}}\big(|\vec{n}\,\rangle + |{-\vec{n}}\,\rangle\big) \,, \tag{11.134}$$

associated with the density operator

$$\rho_0^{\text{opt}} = |\psi_0^{\text{opt}}\rangle\, \langle\psi_0^{\text{opt}}| = \frac{1}{2}\big(\mathrm{I}_2 + \vec{r}_0^{\,\text{opt}} \cdot \vec{\sigma}\,\big) \,, \tag{11.135}$$

with an optimal unit Bloch vector $\vec{r}_0^{\,\text{opt}}$ orthogonal to the rotation axis $\vec{n}$ in $\mathbb{R}^3$. When the unit vector $\vec{n}$ in $\mathbb{R}^3$ is characterized by the polar coordinates (θ_n, φ_n) or the Cartesian coordinates $\vec{n} = [\sin(\theta_n)\cos(\varphi_n),\ \sin(\theta_n)\sin(\varphi_n),\ \cos(\theta_n)]^\top$, the unit vector $\vec{r}_0^{\,\text{opt}} \in \mathbb{R}^3$ of Eq. (11.135) follows with the coordinates $\vec{r}_0^{\,\text{opt}} = [-\cos(\theta_n)\cos(\varphi_n),\ -\cos(\theta_n)\sin(\varphi_n),\ \sin(\theta_n)]^\top = -\partial_{\theta_n}\vec{n}$, verifying that $\vec{r}_0^{\,\text{opt}}$ is orthogonal to $\vec{n}$.

For qubit phase estimation, the derivative of Eq. (11.132) determines the classical Fisher information $F_c(\xi)$ in Eq. (11.125) or (11.126). In particular, upon measuring on ρ_ξ the spin observable $\Omega = \vec{\omega} \cdot \vec{\sigma}$, the classical Fisher information of Eq. (11.126) follows as

$$F_c(\xi) = \frac{\left[\vec{\omega}\,(\vec{n} \times \vec{r}_\xi)\right]^2}{1 - (\vec{\omega}\,\vec{r}_\xi)^2} \,. \tag{11.136}$$

With an optimal pure input probe of unit $\vec{r}_0$ orthogonal to the rotation axis $\vec{n}$, the rotated Bloch vector $\vec{r}_\xi$ is also orthogonal to $\vec{n}$ and keeps unit norm. In another alternative orthonormal basis $\{\vec{n}, \vec{n}_\perp = \vec{r}_\xi, \vec{n}'_\perp = \vec{n} \times \vec{n}_\perp\}$ of $\mathbb{R}^3$, we introduce for the real unit vector $\vec{\omega}$ the three Cartesian coordinates $\vec{\omega} = [\omega_1, \omega_2, \omega_3]^\top$, with $\omega_1^2 + \omega_2^2 + \omega_3^2 = 1$, to express Eq. (11.136) as

[3] Conversely, a unit Bloch vector $\vec{r}_0$ parallel to $\vec{n}$ characterizes an eigenstate of U_ξ and a pure input probe $|\psi_0\rangle$ inoperative for estimation, endowed with a vanishing Fisher information $F_q(\xi) = 0$ in Eq. (11.131), as seen in Remark (4) terminating Sect. 11.3.2.

$$F_c(\xi) = \frac{\omega_3^2}{\omega_1^2 + \omega_3^2} \, . \tag{11.137}$$

By fixing $\omega_1 = 0$, the classical Fisher information $F_c(\xi)$ of Eq. (11.137) is thus maximized at $F_c^{\max}(\xi) = F_q^{\max}(\xi) = 1$, for any (ω_2, ω_3). Therefore, any spin observable $\Omega = \vec{\omega} \cdot \vec{\sigma}$, defined by a unit vector $\vec{\omega} \in \mathbb{R}^3$ taken in the plane orthogonal to the rotation axis $\vec{n}$, provides an optimal quantum measurement capable of achieving the maximal Fisher information $F_c(\xi) = F_c^{\max}(\xi) = F_q^{\max}(\xi) = 1$ uniformly for any unknown phase ξ.

Consistently, one such optimal observable is the observable Ω defined in Eq. (11.98), which for the qubit here takes the form

$$\Omega = |\vec{n}\,\rangle\langle -\vec{n}\,| + |-\vec{n}\,\rangle\langle \vec{n}\,| \, , \tag{11.138}$$

which is equivalently the spin observable $\Omega = \vec{r}_0^{\,\text{opt}} \cdot \vec{\sigma}$ defined by the unit vector $\vec{r}_0^{\,\text{opt}} \perp \vec{n}$ also occurring in Eq. (11.135). The resulting measurement probabilities are $\Pr\{\pm 1\} = \big[1 \pm \cos(\xi)\big]/2$ associated with the average value $\langle \Omega \rangle = \cos(\xi)$.

The quantum score V_ξ itself in Eq. (11.133) is another optimal observable $\mathsf{V}_\xi = \vec{\omega}_\xi \cdot \vec{\sigma}$ with $\vec{\omega}_\xi = \vec{n} \times \vec{r}_\xi$ a unit vector orthogonal to $\vec{n}$. This observable V_ξ achieves the same maximal Fisher information $F_c(\xi) = F_c^{\max}(\xi) = F_q^{\max}(\xi) = 1$; however it is a ξ-dependent observable, and therefore not practically realizable, as it was also discussed in Sect. 11.2.1 after Eq. (11.58).

From an optimal qubit observable like Ω of Eq. (11.138), estimation of the average value $\langle \Omega \rangle = \cos(\xi)$ here provides access as in Sect. 11.3.2, to an optimal estimation of the qubit phase ξ in the conditions of Sect. 11.3.2. In particular, the maximum-likelihood estimator of Sect. 11.3.2 for the qubit phase ξ achieves in Eq. (11.97) a minimal mean-squared error $e^2\big(\widehat{\xi}_{\text{ML}}\big) = 1/L$, while returning for ξ an estimate in the range $[0, \pi]$.

This optimal estimation strategy uses independent qubits to probe the unitary process U_ξ. Alternatively, L-fold multiple-probes can be exploited for qubit phase estimation, following Sect. 11.3.4 in a straightforward manner. In particular, the L-fold multiple-probe strategy of Sect. 11.3.4, for the qubit phase ξ will achieve in Eq. (11.106) the maximal quantum Fisher information $F_q^{(L)\max}(\xi) = L^2$ exhibiting the entanglement-enhanced Heisenberg scaling, while returning for ξ a local estimate in the range $[0, \pi/L]$.

11.4.5 Qubit Phase Estimation with Noise

The considerations of Sect. 11.3.3 concerning the impact of noise on the phase estimation procedure can be further developed for the qubit phase estimation [323–325]. When estimation of the qubit phase ξ takes place in the presence of noise, in the configuration of Fig. 11.2, the input probe signal ρ_0 is first transformed by the unitary process U_ξ into the intermediate ξ-dependent density operator $\rho_1(\xi) = \mathsf{U}_\xi \rho_0 \mathsf{U}_\xi^\dagger$, which is subsequently affected by the quantum noise $\mathcal{N}(\cdot)$ so as to produce the output noisy density operator $\rho_\xi = \mathcal{N}\big(\mathsf{U}_\xi \rho_0 \mathsf{U}_\xi^\dagger\big)$ accessible for estimating the phase ξ.

The procedure can be conveniently described in the Bloch representation of qubit states, as for instance exploited in [323]. The input state ρ_0, with Bloch vector $\vec{r}_0$, gives rise to an intermediate state $\rho_1(\xi) = \mathsf{U}_\xi \rho_0 \mathsf{U}_\xi^\dagger$, with Bloch vector $\vec{r}_1(\xi) = U_\xi \vec{r}_0$, with the 3×3 real unitary rotation matrix U_ξ defined in Eq. (3.21). The noise then implements in $\mathbb{R}^3$ the affine transformation of Eq. (9.7) as

$$\vec{r}_1(\xi) \longmapsto \vec{r}_\xi = A\vec{r}_1(\xi) + \vec{c}\,, \tag{11.139}$$

with $(A, \vec{c}\,)$ characterizing the noise $\mathcal{N}(\cdot)$. The derivative of the output Bloch vector follows as $\partial_\xi \vec{r}_\xi = A\partial_\xi \vec{r}_1(\xi)$, with also $\partial_\xi \vec{r}_1(\xi) = \vec{n} \times \vec{r}_1(\xi)$ as in Eq. (11.132). The quantum Fisher information of Eq. (11.121) receives its dependence on the phase ξ via $\vec{r}_1(\xi)$ and follows as

$$F_q(\xi) = \frac{\big[(A\vec{r}_1 + \vec{c}\,)A\partial_\xi \vec{r}_1\big]^2}{1 - (A\vec{r}_1 + \vec{c}\,)^2} + \big(A\partial_\xi \vec{r}_1\big)^2 \tag{11.140}$$

$$= \frac{\big[(A\vec{r}_1 + \vec{c}\,)A(\vec{n} \times \vec{r}_1)\big]^2}{1 - (A\vec{r}_1 + \vec{c}\,)^2} + \big[A(\vec{n} \times \vec{r}_1)\big]^2 . \tag{11.141}$$

A meaningful task then is to look for the Bloch vector $\vec{r}_1$ maximizing the quantum Fisher information $F_q(\xi)$ of Eq. (11.141), which in turn will determine via $\vec{r}_0 = U_\xi^\top \vec{r}_1$ the input probe signal ρ_0 maximizing $F_q(\xi)$. The maximum of $F_q(\xi)$ in Eq. (11.141) will in general depend on the geometric configuration in $\mathbb{R}^3$ of the axis $\vec{n}$ of the unitary U_ξ in relation to the parameters $(A, \vec{c}\,)$ characterizing the noise. By the convexity property (5) of the quantum Fisher information indicated at the end of Sect. 11.2.1, we know that $F_q(\xi)$ will be maximized by a pure input state ρ_0, and therefore a pure state ρ_1, associated with a unit Bloch vector $\vec{r}_1$; in addition this maximum of $F_q(\xi)$ will be ≤ 1, the maximum Fisher information with no noise. For the maximization of $F_q(\xi)$ in Eqs. (11.140)–(11.141) it is convenient to refer the Bloch vectors of $\mathbb{R}^3$ to the orthonormal basis $\{\vec{n}, \vec{n}_\perp, \vec{n}'_\perp = \vec{n} \times \vec{n}_\perp\}$, where here $\vec{n}_\perp$ is any unit vector of $\mathbb{R}^3$ orthogonal to the axis $\vec{n}$ of U_ξ. In this basis, the unit vector $\vec{r}_1$ has the polar coordinates (θ_1, φ_1), with φ_1 the azimuth angle around $\vec{n}$ and θ_1 the coelevation angle with $\vec{n}$, and the Cartesian coordinates

$$\vec{r}_1 = \big[\cos(\theta_1), \sin(\theta_1)\cos(\varphi_1), \sin(\theta_1)\sin(\varphi_1)\big]^\top . \tag{11.142}$$

And since $\partial_\xi \vec{r}_1 = \vec{n} \times \vec{r}_1$, we have

$$\partial_\xi \vec{r}_1 = \big[0, -\sin(\theta_1)\sin(\varphi_1), \sin(\theta_1)\cos(\varphi_1)\big]^\top . \tag{11.143}$$

By inserting Eqs. (11.142) and (11.143) in Eq. (11.140), we obtain $F_q(\theta_1, \varphi_1; \xi)$ specifying the quantum Fisher information as a function of the two angles (θ_1, φ_1). For given axis $\vec{n}$ and noise $(A, \vec{c})$, it is then in principle straightforward, at least numerically, to find the optimal configuration $(\theta_1^{\rm opt}, \varphi_1^{\rm opt})$ determining $\vec{r}_1 = \vec{r}_1^{\rm\ opt}$ maximizing $F_q(\theta_1, \varphi_1; \xi)$ in Eqs. (11.140)–(11.141). The corresponding optimal input state $\rho_0^{\rm opt}$, before the rotation by ξ, is then defined by the unit Bloch vector $\vec{r}_0^{\rm\ opt}$ with polar coordinates $(\theta_0^{\rm opt} = \theta_1^{\rm opt}, \varphi_0^{\rm opt} = \varphi_1^{\rm opt} - \xi)$, by virtue of referring to the basis $\{\vec{n}, \vec{n}_\perp, \vec{n}'_\perp\}$.

Such maximization of the quantum Fisher information $F_q(\xi)$ in Eqs. (11.140)–(11.141) is explicitly worked out in [293], for specific noise models $(A, \vec{c})$ relevant to the qubit, as presented in Chap. 9. These resolutions of [293] show, as it could be anticipated, that the optimal input probe state $\rho_0^{\rm opt}$ generally comes out as a ξ-dependent solution, dependent on the unknown phase ξ under estimation. The maximizer $(\theta_1^{\rm opt}, \varphi_1^{\rm opt})$ is intrinsic to the criterion function $F_q(\theta_1, \varphi_1)$ in Eq. (11.141), and both are independent of the phase ξ. The dependence on the phase ξ is induced in the subsequent solution for the input probe $\rho_0^{\rm opt}$ with coordinates $(\theta_0^{\rm opt} = \theta_1^{\rm opt}, \varphi_0^{\rm opt} = \varphi_1^{\rm opt} - \xi)$. For Pauli noises or more general unital noises, [293] reports optimal configurations generally occurring at $\theta_0^{\rm opt} = \theta_1^{\rm opt} = \pi/2$, i.e., an optimal input probe $\rho_0^{\rm opt}$ orthogonal to the rotation axis $\vec{n}$, however supplemented by a ξ-dependent azimuth angle $\varphi_0^{\rm opt} = \varphi_1^{\rm opt} - \xi$ around $\vec{n}$. For non-unital noises, the optimal input configurations are found in [293] as $(\theta_0^{\rm opt} = \theta_1^{\rm opt} \neq \pi/2, \varphi_0^{\rm opt} = \varphi_1^{\rm opt} - \xi)$, i.e., as ξ-dependent probes $\rho_0^{\rm opt}$ nonorthogonal to $\vec{n}$. Instead of such inaccessible ξ-dependent input probes $\rho_0^{\rm opt}$, the maximum of the Fisher information $F_q(\xi)$ in Eqs. (11.140)–(11.141) may alternatively be reached with adaptive approaches, iteratively adjusting the input probe ρ_0 as the estimate for the phase ξ is progressively refined, and comparable to the adaptive techniques evoked in Sect. 11.2.1 just before the final five properties (1)–(5). Otherwise, with a fixed input probe ρ_0, one will have to admit to operate with a Fisher information $F_q(\xi)$ at a level below its maximum, yet completely predictable from Eqs. (11.140)–(11.141), and also in general dependent on the unknown phase ξ under estimation.

Once the input probe signal ρ_0 is settled, fixing in Eqs. (11.140)–(11.141) a quantum Fisher information $F_q(\xi)$ at a given level for each ξ, the next step for estimation is to devise a measurement protocol of the noisy output state ρ_ξ depicted in Fig. 11.2. For this it is appropriate to look for a measurement protocol affording a classical Fisher information $F_c(\xi)$ coming as close as possible to the upper bound prescribed by the quantum Fisher information $F_q(\xi)$. When measuring the noisy qubit state ρ_ξ with the general POVM of Sect. 11.4.2, the classical Fisher information of Eq. (11.125) follows as

$$F_c(\xi) = \sum_{m=1}^{M} \frac{\left(\vec{a}_m A \partial_\xi \vec{r}_1\right)^2}{b_m + \vec{a}_m (A\vec{r}_1 + \vec{c}\,)} \tag{11.144}$$

$$= \sum_{m=1}^{M} \frac{\left[\vec{a}_m A (\vec{n} \times \vec{r}_1)\right]^2}{b_m + \vec{a}_m (A\vec{r}_1 + \vec{c}\,)} \,. \tag{11.145}$$

Alternatively, as in Sect. 11.4.2, when measuring on the noisy qubit state ρ_ξ the spin observable $\Omega = \vec{\omega} \cdot \vec{\sigma}$, the classical Fisher information of Eq. (11.126) follows as

$$F_c(\xi) = \frac{\left(\vec{\omega}\, A \partial_\xi \vec{r}_1\right)^2}{1 - \left[\vec{\omega}\,(A\vec{r}_1 + \vec{c}\,)\right]^2} \tag{11.146}$$

$$= \frac{\left[\vec{\omega}\, A (\vec{n} \times \vec{r}_1)\right]^2}{1 - \left[\vec{\omega}\,(A\vec{r}_1 + \vec{c}\,)\right]^2} \,. \tag{11.147}$$

For given axis $\vec{n}$ and noise $(A, \vec{c}\,)$, the classical Fisher information $F_c(\xi)$ of Eqs. (11.145) or (11.147) is dependent on the ξ-dependent rotated Bloch vector $\vec{r}_1(\xi) = U_\xi \vec{r}_0$. In general, maximization of $F_c(\xi)$ will have to take place in the presence of a specified input probe ρ_0 and $\vec{r}_0$, so as to optimally match the output measurement to this specific input excitation.

The choice of the output measurement then fixes in Eqs. (11.145) or (11.147) a classical Fisher information $F_c(\xi)$ at a given level. From the measurement results, the maximum-likelihood estimator $\widehat{\xi}_{\mathrm{ML}}$ for the qubit phase ξ will allow one to reach a mean-squared estimation error $e^2\big(\widehat{\xi}_{\mathrm{ML}}\big)$ controlled by the reciprocal Fisher information $1/F_c(\xi)$.

11.4.6 Qubit Phase Estimation with Depolarizing Noise

For illustration, we now explicitly work out the important reference case of the depolarizing noise affecting the qubit phase estimation, as analyzed for instance in [323]. According to Sect. 9.6, the depolarizing noise on the qubit is characterized by a zero translation $\vec{c} = \vec{0}$ and an isotropic matrix $A = \alpha I_3$ with α the contraction factor introduced in Eq. (9.55). As a result, the action of the noise in Eq. (11.139) reduces to the scalar contraction $\vec{r}_1(\xi) \mapsto \vec{r}_\xi = \alpha \vec{r}_1(\xi)$. The derivative $\partial_\xi \vec{r}_\xi = \vec{n} \times \vec{r}_\xi \perp \vec{r}_\xi$ so that the quantum Fisher information of Eqs. (11.141) or (11.121) reduces to

$$F_q(\xi) = \left(\vec{n} \times \vec{r}_\xi\right)^2 = \alpha^2 \left(\vec{n} \times \vec{r}_0\right)^2 \,, \tag{11.148}$$

coinciding with Eq. (11.131) at no noise when $\alpha = 1$.

Maximization of the quantum Fisher information $F_q(\xi)$ of Eq. (11.148) is accomplished by an optimal pure input probe ρ_0^{opt} chosen with any unit Bloch vector $\vec{r}_0$

orthogonal to the axis $\vec{n}$, just as in the noise-free case of Sect. 11.4.4; and this optimal ρ_0^{opt} from Eqs. (11.134)–(11.135) achieves in Eq. (11.148) the maximum

$$F_q^{\max}(\xi) = \alpha^2 \leq 1 \,. \tag{11.149}$$

This especially demonstrates the possibility of an optimal input probe independent of the unknown phase ξ, a rather exceptional property in the presence of noise, as explained above, and occurring here thanks to the high symmetry (isotropy) of the depolarizing noise. By contrast, for less symmetric qubit noises, such as the bit-flip or the phase-flip or the bit-phase-flip noises, the optimal input probe maximizing $F_q(\xi)$ in general is dependent on ξ, except in some special configurations where the rotation axis $\vec{n}$ is parallel to an invariant axis Ox, Oy, or Oz in $\mathbb{R}^3$ of the noise [293].

When measuring on the noisy output ρ_ξ the spin observable $\Omega = \vec{\omega} \cdot \vec{\sigma}$, the classical Fisher information of Eq. (11.147) follows as

$$F_c(\xi) = \alpha^2 \frac{\left[\vec{\omega}\left(\vec{n} \times \vec{r}_1\right)\right]^2}{1 - \alpha^2 (\vec{\omega}\vec{r}_1)^2} \,, \tag{11.150}$$

matching the noise-free case of Eq. (11.136) when $\alpha = 1$.

With an optimal pure input probe of unit $\vec{r}_0$ orthogonal to the rotation axis $\vec{n}$, the rotated Bloch vector $\vec{r}_1(\xi)$ is also orthogonal to $\vec{n}$ and keeps unit norm. In the orthonormal basis $\left\{\vec{n}, \vec{n}_\perp = \vec{r}_1, \vec{n}'_\perp = \vec{n} \times \vec{n}_\perp\right\}$ of $\mathbb{R}^3$, we again introduce for the real unit vector $\vec{\omega}$ the three Cartesian coordinates $\vec{\omega} = [\omega_1, \omega_2, \omega_3]^\top$, with $\omega_1^2 + \omega_2^2 + \omega_3^2 = 1$, to express Eq. (11.150) as

$$F_c(\xi) = \alpha^2 \frac{\omega_3^2}{1 - \alpha^2 \omega_2^2} \,. \tag{11.151}$$

To maximize $F_c(\xi)$ in Eq. (11.151) it is necessary to fix $\omega_1 = 0$, then the reciprocal $1/F_c(\xi) = 1 + (1 - \alpha^2)/(\alpha^2 \omega_3^2)$ is minimized by $\omega_3^2 = 1$, therefore $F_c(\xi)$ is maximized by $(\omega_1 = 0, \omega_2 = 0, \omega_3 = \pm 1)$ at $F_c^{\max}(\xi) = F_q^{\max}(\xi) = \alpha^2$. This stands as a necessary and sufficient condition, in order to maximize $F_c(\xi)$ of Eq. (11.150) at $F_c^{\max}(\xi) = F_q^{\max}(\xi) = \alpha^2$, to take $\vec{\omega} \in \mathbb{R}^3$ in the plane orthogonal to the rotation axis $\vec{n}$ and also orthogonal to the rotated Bloch vector $\vec{r}_1(\xi)$. The optimal configuration for estimating the qubit phase ξ is represented in Fig. 11.3.

The optimal configuration of Fig. 11.3, requiring for the spin observable $\Omega = \vec{\omega} \cdot \vec{\sigma}$ at the output, a measurement vector $\vec{\omega}$ orthogonal to the rotated Bloch vector $\vec{r}_1(\xi) = U_\xi \vec{r}_0$, is in general a ξ-dependent solution. This means that, contrary to the noise-free case of Sect. 11.4.4, with the depolarizing noise there is no ξ-independent optimal spin observable $\Omega = \vec{\omega} \cdot \vec{\sigma}$ whose measurement would realize the overall maximal Fisher information $F_c(\xi) = F_q^{\max}(\xi) = \alpha^2$ uniformly for any ξ.

We can also explicitly exhibit, as done for the noise-free case of Sect. 11.4.4 after Eq. (11.138), the observable formed by the quantum score V_ξ. From Eq. (11.120), with the depolarizing noise, we have the score

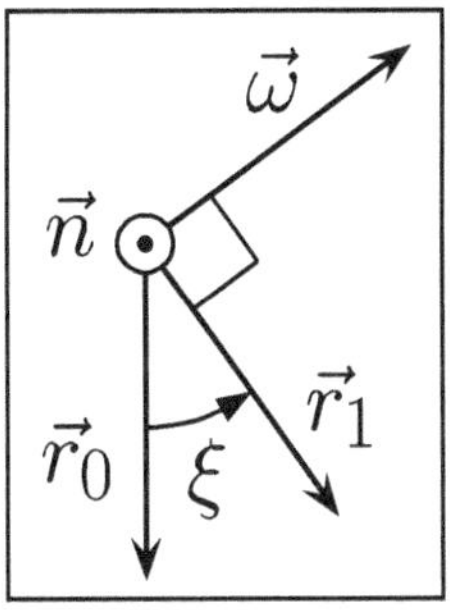

Fig. 11.3 Optimal configuration for estimating the qubit phase ξ with depolarizing noise: The graph is in the plane $(\vec{r}_0, \vec{r}_1)$ of $\mathbb{R}^3$ orthogonal to the rotation axis $\vec{n}$ of the unitary U_ξ which is also the plane $(\vec{n}_\perp, \vec{n}'_\perp)$. A unit Bloch vector $\vec{r}_0$ orthogonal to the axis $\vec{n}$ forms the optimal input excitation maximizing the quantum Fisher information $F_q(\xi)$ at $F_q^{\max}(\xi) = \alpha^2$. In addition, the optimal output spin observable requires a measurement vector $\vec{\omega}$ in this plane $(\vec{r}_0, \vec{r}_1)$, and orthogonal to the rotated Bloch vector $\vec{r}_1(\xi) = U_\xi \vec{r}_0$, so as to reach a classical Fisher information $F_c(\xi)$ matching its upper bound $F_c^{\max}(\xi) = F_q^{\max}(\xi) = \alpha^2$

$$\mathsf{V}_\xi = \partial_\xi \vec{r}_\xi \cdot \vec{\sigma} = \big(\vec{n} \times \vec{r}_\xi\big) \cdot \vec{\sigma} = \alpha\big(\vec{n} \times \vec{r}_1\big) \cdot \vec{\sigma} \; . \tag{11.152}$$

This score V_ξ stands as the optimal observable $\mathsf{V}_\xi = \alpha\Omega_\xi$ proportional to the spin observable $\Omega_\xi = \vec{\omega}_\xi \cdot \vec{\sigma}$ with $\vec{\omega}_\xi = \vec{n} \times \vec{r}_1(\xi)$ a unit vector orthogonal to $\vec{n}$ and to the rotated Bloch vector $\vec{r}_1(\xi)$, which is the optimal spin observable we just characterized before. From Sect. 4.3.4 on observables of the qubit, we know that both observables Ω_ξ and $\mathsf{V}_\xi = \alpha\Omega_\xi$ have the same measurement statistics and are therefore equally optimal for estimating the phase ξ. The optimal observable $\mathsf{V}_\xi = \big(\vec{n} \times \vec{r}_\xi\big) \cdot \vec{\sigma}$ from Eq. (11.152) achieves the maximal Fisher information $F_c(\xi) = F_c^{\max}(\xi) = F_q^{\max}(\xi) = \alpha^2$; however it is a ξ-dependent observable, and therefore not practically accessible, since the phase ξ is unknown in the first place.

To cope with such ξ-dependent optimal observable, it is possible to turn to adaptive approaches, as evoked in Sect. 11.2.1, to iteratively adjust the measurement vector $\vec{\omega}$ of a spin observable in the plane orthogonal to $\vec{n}$ as the estimate of the phase ξ is progressively refined.

Alternatively, to circumvent the ξ-dependent optimal observable, a feasible option is to turn to a measurement of the output signal ρ_ξ by means of a more general POVM, extending the spin observable equivalent to a two-element POVM, as done in [323]. In addition here, the approach will also stand as a worked-out example illustrating a generic benefit we evoked toward the end of Sect. 4.2.2 on generalized measurement, of measuring a quantum state with a more general POVM rather than by means of an observable or a standard von Neumann measurement. A general M-element POVM $\{\mathsf{E}_m\}_{m=1}^M$ as in Sect. 11.4.2, according to Eq. (11.125), achieves in presence of the depolarizing noise, the classical Fisher information

$$F_c(\xi) = \alpha^2 \sum_{m=1}^{M} \frac{\left[\vec{a}_m(\vec{n} \times \vec{r}_1)\right]^2}{b_m + \alpha \vec{a}_m \vec{r}_1} . \tag{11.153}$$

From the characterization we performed of the optimal two-element POVM or optimal observable requiring a unit measurement vector $\vec{\omega}$ in the plane orthogonal to $\vec{n}$, we turn for the more general POVM to $M \geq 2$ operators $\mathsf{E}_m = \left(\mathrm{I}_2 + \vec{a}_m \cdot \vec{\sigma}\right)/M$, with $\|\vec{a}_m\| = 1$, each E_m proportional to a projector in $\mathcal{H}_2$, and with each measurement vector $\vec{a}_m$ of $\mathbb{R}^3$ placed in the plane orthogonal to the axis $\vec{n}$. This family of POVM operators encompasses the ξ-dependent optimal POVM at $M = 2$. With the planar structure of the measurement vectors $\vec{a}_m$, these POVM also are comparable to the qubit "star" POVM presented in Sect. 4.2.5. This uniform planar structure for the POVM is motivated because, with an unknown phase ξ, no information is available on the probing vector $\alpha\vec{r}_1(\xi)$ that is tracked, except that it lies in the plane orthogonal to $\vec{n}$ as shown in Fig. 11.3. The optimal pure input probe $\rho_0^{\rm opt}$ has also its unit Bloch vector $\vec{r}_0$ in the plane orthogonal to $\vec{n}$. In the orthonormal basis $\{\vec{r}_0^{\,\rm opt}, \vec{n} \times \vec{r}_0^{\,\rm opt}\}$ of this plane, each unit vector $\vec{a}_m$ is taken with Cartesian coordinates $\vec{a}_m = [\cos(\phi_m), \sin(\phi_m)]^\top$, with the azimuth angle ϕ_m around $\vec{n}$, for $m = 1$ to M. This leads for the classical Fisher information of Eq. (11.153) to

$$F_c(\xi) = \frac{\alpha^2}{M} \sum_{m=1}^{M} \frac{\sin^2(\phi_m - \xi)}{1 + \alpha\cos(\phi_m - \xi)} . \tag{11.154}$$

With an unknown phase angle ξ for the rotated Bloch vector $\vec{r}_1(\xi)$ in the plane orthogonal to $\vec{n}$, we have no motivation for a choice other than uniformly distributing the M measurement vectors $\vec{a}_m$ in this plane, with $\phi_m = 2\pi(m-1)/M$ for $m = 1$ to M. With this choice, Fig. 11.4 represents the classical Fisher information $F_c(\xi)$ from Eq. (11.154), as a function of the number M of POVM elements, for various values of the noise level α and of the phase ξ to be estimated, as also reported in [323] in other configurations.

In Fig. 11.4, for a small number $M \geq 2$ of POVM elements, a clear dependence of the performance $F_c(\xi)$ on the phase ξ is observed. This can be expected with a small number M of measurement vectors $\vec{a}_m$ to track the rotated Bloch vector $\vec{r}_1(\xi)$ at an angle ξ in the plane orthogonal to $\vec{n}$. However, for sufficiently large M, around $M = 8$ to 9 in Fig. 11.4, the uniform distribution in the plane of a larger number M of measurement vectors $\vec{a}_m$, reduces the dependence on ξ, with the Fisher information $F_c(\xi)$ tending to stabilize at an asymptotic limiting value. At large number M, when the sum in Eq. (11.154) tends to an integral over $[0, 2\pi[$, the Fisher information $F_c(\xi)$ tends to a limit that can be expressed as

$$F_c^\infty(\xi) = \frac{\alpha^2}{2\pi} \int_0^{2\pi} \frac{\sin^2(\phi)}{1 + \alpha\cos(\phi)} d\phi = 1 - \sqrt{1 - \alpha^2} , \tag{11.155}$$

independent of the phase ξ. In practice, $M = 8$ to 10 allows $F_c(\xi)$ of Eq. (11.154) to come close to the limit $F_c^\infty(\xi)$ of Eq. (11.155) for any phase ξ and noise level α,

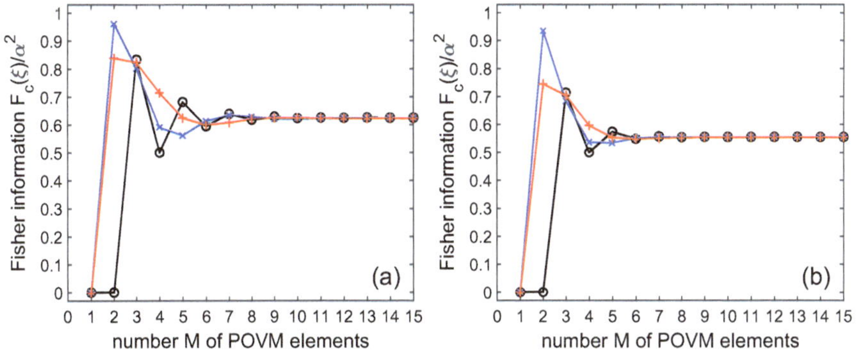

Fig. 11.4 Classical Fisher information $F_c(\xi)/\alpha^2$ from Eq. (11.154) normalized by the overall maximum prescribed by the quantum Fisher information $F_q^{\max}(\xi) = \alpha^2$, as a function of the number M of POVM elements for the quantum measurement, for various values of the phase angle ξ to be estimated: (∘) $\xi = 0$, (×) $\xi = 0.6\pi$, (+) $\xi = 1.3\pi$. The qubit depolarizing noise has a contraction factor $\alpha = 0.8$ in (a), and $\alpha = 0.6$ in (b)

Fig. 11.5 Classical Fisher information $F_c^{\infty}(\xi)$ of Eq. (11.155) at large number M of POVM elements (solid line), and overall maximum prescribed by the quantum Fisher information $F_q^{\max}(\xi) = \alpha^2$ from Eq. (11.149) (dashed line), as a function of the contraction factor α of the qubit depolarizing noise

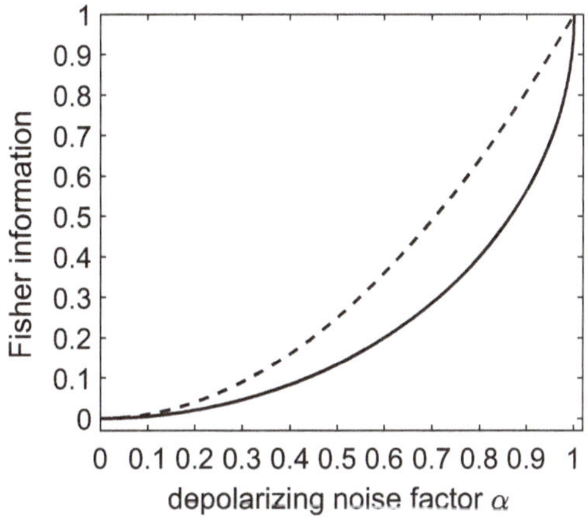

since numerically, beginning at $M = 8$, it is found that $F_c(\xi)$ of Eq. (11.154) never drops below $0.9F_c^{\infty}(\xi)$. The classical Fisher information $F_c^{\infty}(\xi)$ of Eq. (11.155) at large number M is compared in Fig. 11.5 with the overall maximum prescribed by the quantum Fisher information $F_q^{\max}(\xi) = \alpha^2$ from Eq. (11.149).

In Fig. 11.5, for any noise contraction factor α, the performance $F_c^{\infty}(\xi)$ remains below the maximum $F_q^{\max}(\xi) = \alpha^2$, demonstrating that, with depolarizing noise, even with a large number M of POVM elements, it is not possible to obtain a ξ-independent measurement exhibiting a classical Fisher information $F_c(\xi)$ reaching the maximum $F_q^{\max}(\xi)$ uniformly for all phase ξ, although this was possible without noise as seen in Sect. 11.4.4. Further analysis on this topic of estimation from a noisy qubit can be found in [323], especially including an implementation of the maximum-likelihood phase estimator $\widehat{\xi}_{\mathrm{ML}}$ and its test against the optimal performance characterized here.

11.4.7 Phase Estimation from Entangled Qubits

For estimating the phase ξ of the qubit unitary U_ξ of Eq. (11.129), it is possible, as in Sect. 11.3.4, to probe U_ξ with multiple entangled qubits. When there is no noise, phase estimation from entangled qubits can be handled by a straightforward application of the approach of Sect. 11.3.4. The picture is more versatile in the presence of noise, as we now examine.

To handle phase estimation form multiple entangled noisy qubits, entangled qubit states cannot be described in the (helpful) Bloch representation, as we have done so far in this Sect. 11.4. The Bloch representation of Eq. (11.117) applies to individual qubit states, or separable tensor-product states of qubits. Entangled states of multiple qubits have to be handled, for phase estimation, as the general quantum states of Sect. 11.3 and more specifically of Sect. 11.3.4.

A qubit has its state vector in the Hilbert space $\mathcal{H}_2$ referred to the computational orthonormal basis $\big\{|0\rangle\,,|1\rangle\big\}$; its density operator is in the space $\mathcal{L}(\mathcal{H}_2)$ referred to the standard orthonormal basis $\big\{|0\rangle\langle 0|\,,|0\rangle\langle 1|\,,|1\rangle\langle 0|\,,|1\rangle\langle 1|\big\}$. A composite system of L qubits has its state vector in the tensor-product Hilbert space $\mathcal{H}_2^{\otimes L}$ referred to the computational orthonormal basis $\big\{|0\rangle\,,|1\rangle\big\}^{\otimes L}$; its density operator is in the space $\mathcal{L}\big(\mathcal{H}_2^{\otimes L}\big)$ referred to the standard orthonormal basis $\big\{|0\rangle\langle 0|\,,|0\rangle\langle 1|\,,|1\rangle\langle 0|\,,|1\rangle\langle 1|\big\}^{\otimes L}$.

As a worked-out example here to illustrate the benefit of entanglement for quantum estimation, as generally presented in Sect. 11.3.4, we rely on the situation also considered in [311]. For phase estimation, we consider a qubit unitary operator $\mathsf{U}_\xi \in \mathcal{L}(\mathcal{H}_2)$ under the form

$$\mathsf{U}_\xi = \begin{bmatrix} 1 & 0 \\ 0 & e^{i\xi} \end{bmatrix}, \tag{11.156}$$

which is the form of Eq. (3.31). This amounts, with an appropriate origin for the phases, to refer U_ξ to its orthonormal eigenbasis, or equivalently here to select the computational basis $\big\{|0\rangle\,,|1\rangle\big\}$ as the eigenbasis of U_ξ. Intrinsically, the phase ξ to be estimated remains the phase difference between the two eigenvalues of U_ξ, which is an intrinsic property of U_ξ independent of its representation basis. The choice of Eq. (11.156) will ease the notations without affecting the faculty of conveying useful ideas relevant to estimation from multiple noisy qubits. The action of U_ξ on the standard basis of $\mathcal{L}(\mathcal{H}_2)$, through $|j\rangle\langle k| \mapsto \mathsf{U}_\xi\,|j\rangle\langle k|\,\mathsf{U}_\xi^\dagger = \mathcal{U}_\xi\big(|j\rangle\langle k|\big)$, can then be expressed as

$$\begin{cases} \mathcal{U}_\xi\big(|0\rangle\langle 0|\big) = |0\rangle\langle 0| \ , & (11.157) \\ \mathcal{U}_\xi\big(|0\rangle\langle 1|\big) = e^{-i\xi}\,|0\rangle\langle 1| \ , & (11.158) \\ \mathcal{U}_\xi\big(|1\rangle\langle 0|\big) = e^{i\xi}\,|1\rangle\langle 0| \ , & (11.159) \\ \mathcal{U}_\xi\big(|1\rangle\langle 1|\big) = |1\rangle\langle 1| \ . & (11.160) \end{cases}$$

Any qubit noise $\mathcal{N}(\cdot)$, via its Kraus representation of Eq. (9.5) and via Eq. (3.49), can also be completely characterized by its action on the standard basis of $\mathcal{L}(\mathcal{H}_2)$. The cascade as in Fig. 11.2 of the two quantum processes $\mathcal{U}_\xi(\cdot)$ and $\mathcal{N}(\cdot)$ results in the ξ-dependent quantum process $\mathcal{N}\circ\mathcal{U}_\xi(\cdot) = \mathcal{N}\big(\mathcal{U}_\xi(\cdot)\big) = \mathcal{T}_\xi(\cdot)$, with $\mathcal{T}_\xi(\cdot)$ also readily characterized by its action on the standard basis of $\mathcal{L}(\mathcal{H}_2)$. Then, $\mathcal{T}_\xi(\cdot)$ engaged in the L-fold tensor product allows one to characterize the transformation by the cascade of Fig. 11.2 of a composite input probe $\rho_0 \in \mathcal{L}(\mathcal{H}_2^{\otimes L})$ of L entangled qubits, typically as the map $\rho_0 \mapsto \mathcal{T}_\xi^{\otimes L}(\rho_0) = \rho_\xi$. The resulting ξ-dependent output state of the L qubits can then be treated for estimation as the generic state ρ_ξ of Sect. 11.3. Accordingly, the eigendecomposition $\big\{\lambda_k, |\lambda_k\rangle\big\}_{k=1}^{2^L}$ of $\rho_\xi \in \mathcal{L}(\mathcal{H}_2^{\otimes L})$ has to be established, and the derivative $\partial_\xi \rho_\xi$ can be used to deduce the quantum Fisher information $F_q(\xi)$ of Eq. (11.68). The L-qubit input signal $\rho_0 \in \mathcal{L}(\mathcal{H}_2^{\otimes L})$ in Fig. 11.2 can then be optimized in order to maximize $F_q(\xi)$. For measuring ρ_ξ a POVM $\{\mathsf{E}_m\}_{m=1}^M$ is introduced, characterized by means of its classical Fisher information $F_c(\xi)$ of Eq. (11.72), and optimized to come as close as possible to its upper bound $F_q(\xi)$. From the measurement results delivered by such an optimized POVM, the maximum-likelihood estimator will efficiently complete the estimation task of the phase ξ from the noisy L-qubit signal.

This is the program that was explicitly carried out in Sect. 11.3.4, with a complete analytical characterization of the optimal phase estimation strategy with entanglement, and applicable to an L-qubit entangled probe, yet in the absence of noise. Here in the presence of noise, the optimal estimation strategy cannot be as straightforwardly characterized, especially analytically, and the explicit characterization may be expected to grow in complication as the number L of noisy entangled qubits increases. For illustration, we will work out here the situation of a qubit pair in the presence of depolarizing[4] noise, as also addressed for instance in [311].

The depolarizing noise of Sect. 9.6 transforms a qubit density operator according to Eq. (9.49), and, as a function of the contraction factor α introduced in Eq. (9.55), its action on the standard basis of $\mathcal{L}(\mathcal{H}_2)$ follows as

[4] Especially, the depolarizing noise displays an isotropic character, conveyed for instance by Eq. (9.55), which is compatible and does not compete with the prior choice of the reference basis $\big\{|0\rangle, |1\rangle\big\}$ for $\mathcal{H}_2$ as the eigenstates of U_ξ. Otherwise, a nonisotropic qubit noise would also impose a requirement on the reference basis for $\mathcal{H}_2$, which would have to be accommodated with the representation of the qubit unitary U_ξ. For instance, the introduction of a phase-flip noise as in Sect. 9.4 would fix the states $|0\rangle$ and $|1\rangle$ (as its two eigenstates), making them unavailable to also represent the two eigenstates of a general U_ξ, which would then have to be handled in the general form of Eq. (11.129) instead of Eq. (11.156).

$$\begin{cases} \mathcal{N}\big(|0\rangle\langle 0|\big) = \dfrac{1+\alpha}{2}\,|0\rangle\langle 0| + \dfrac{1-\alpha}{2}\,|1\rangle\langle 1|\ , & (11.161) \\ \mathcal{N}\big(|0\rangle\langle 1|\big) = \alpha\,|0\rangle\langle 1|\ , & (11.162) \\ \mathcal{N}\big(|1\rangle\langle 0|\big) = \alpha\,|1\rangle\langle 0|\ , & (11.163) \\ \mathcal{N}\big(|1\rangle\langle 1|\big) = \dfrac{1-\alpha}{2}\,|0\rangle\langle 0| + \dfrac{1+\alpha}{2}\,|1\rangle\langle 1|\ . & (11.164) \end{cases}$$

By coupling with Eqs. (11.157)–(11.160), the cascaded process $\mathcal{T}_\xi(\cdot) = \mathcal{N} \circ \mathcal{U}_\xi(\cdot)$ results as

$$\begin{cases} \mathcal{T}_\xi\big(|0\rangle\langle 0|\big) = \dfrac{1+\alpha}{2}\,|0\rangle\langle 0| + \dfrac{1-\alpha}{2}\,|1\rangle\langle 1|\ , & (11.165) \\ \mathcal{T}_\xi\big(|0\rangle\langle 1|\big) = \alpha e^{-i\xi}\,|0\rangle\langle 1|\ , & (11.166) \\ \mathcal{T}_\xi\big(|1\rangle\langle 0|\big) = \alpha e^{i\xi}\,|1\rangle\langle 0|\ , & (11.167) \\ \mathcal{T}_\xi\big(|1\rangle\langle 1|\big) = \dfrac{1-\alpha}{2}\,|0\rangle\langle 0| + \dfrac{1+\alpha}{2}\,|1\rangle\langle 1|\ . & (11.168) \end{cases}$$

In particular, it can be verified that here $\mathcal{N}(\cdot)$ and $\mathcal{U}_\xi(\cdot)$ commute on the basis states $\big\{|j\rangle\langle k|\,;\, j,k = 0,1\big\}$, so that they commute everywhere with $\mathcal{N} \circ \mathcal{U}_\xi(\cdot) = \mathcal{U}_\xi \circ \mathcal{N}(\cdot) = \mathcal{T}_\xi(\cdot)$; therefore the depolarizing noise $\mathcal{N}(\cdot)$ applied after or before the unitary operator U_ξ has the same effect. This commutation property here is due to the specific, isotropic, character of the depolarizing noise, and is not generally the rule with other quantum noises.

A two-qubit entangled input probe is represented in the standard basis of $\mathcal{L}\big(\mathcal{H}_2^{\otimes 2}\big)$ by the density operator

$$\rho_0 = \begin{bmatrix} \rho_{00} & \rho_{01} & \rho_{02} & \rho_{03} \\ \rho_{10} & \rho_{11} & \rho_{12} & \rho_{13} \\ \rho_{20} & \rho_{21} & \rho_{22} & \rho_{23} \\ \rho_{30} & \rho_{31} & \rho_{32} & \rho_{33} \end{bmatrix} . \tag{11.169}$$

Transformation of the Two Qubits: For probing with two qubits, each of the two qubits entangled in the input pair can be sent successively across the noisy unitary process of Fig. 11.2. This realizes the input–output transformation $\rho_0 \mapsto \mathcal{T}_\xi \otimes \mathcal{T}_\xi(\rho_0) = \rho_\xi$, leading via Eqs. (11.165)–(11.168) to the output density operator

$$\rho_\xi = \begin{bmatrix} d_{00} & \alpha e^{-i\xi}(b\rho_{01} + c\rho_{23}) & \alpha e^{-i\xi}(b\rho_{02} + c\rho_{13}) & \alpha^2 e^{-i2\xi}\rho_{03} \\ \alpha e^{i\xi}(b\rho_{10} + c\rho_{32}) & d_{11} & \alpha^2 \rho_{12} & \alpha e^{-i\xi}(c\rho_{02} + b\rho_{13}) \\ \alpha e^{i\xi}(b\rho_{20} + c\rho_{31}) & \alpha^2 \rho_{21} & d_{22} & \alpha e^{-i\xi}(c\rho_{01} + b\rho_{23}) \\ \alpha^2 e^{i2\xi}\rho_{30} & \alpha e^{i\xi}(c\rho_{20} + b\rho_{31}) & \alpha e^{i\xi}(c\rho_{10} + b\rho_{32}) & d_{33} \end{bmatrix} , \tag{11.170}$$

with the four diagonal terms

$$\begin{cases} d_{00} = b^2\rho_{00} + bc(\rho_{11} + \rho_{22}) + c^2\rho_{33} , & (11.171) \\ d_{11} = b^2\rho_{11} + bc(\rho_{00} + \rho_{33}) + c^2\rho_{22} , & (11.172) \\ d_{22} = b^2\rho_{22} + bc(\rho_{00} + \rho_{33}) + c^2\rho_{11} , & (11.173) \\ d_{33} = b^2\rho_{33} + bc(\rho_{11} + \rho_{22}) + c^2\rho_{00} , & (11.174) \end{cases}$$

and the shorthand notation $b = (1+\alpha)/2$ and $c = (1-\alpha)/2$. Follows the derivative

$$\partial_\xi \rho_\xi = \begin{bmatrix} 0 & -i\alpha e^{-i\xi}(b\rho_{01} + c\rho_{23}) & -i\alpha e^{-i\xi}(b\rho_{02} + c\rho_{13}) & -i2\alpha^2 e^{-i2\xi}\rho_{03} \\ i\alpha e^{i\xi}(b\rho_{10} + c\rho_{32}) & 0 & 0 & -i\alpha e^{-i\xi}(c\rho_{02} + b\rho_{13}) \\ i\alpha e^{i\xi}(b\rho_{20} + c\rho_{31}) & 0 & 0 & -i\alpha e^{-i\xi}(c\rho_{01} + b\rho_{23}) \\ i2\alpha^2 e^{i2\xi}\rho_{30} & i\alpha e^{i\xi}(c\rho_{20} + b\rho_{31}) & i\alpha e^{i\xi}(c\rho_{10} + b\rho_{32}) & 0 \end{bmatrix} . \quad (11.175)$$

Instead of tackling the (relatively involved) optimization of the general two-qubit input probe ρ_0 of Eq. (11.169) to maximize $F_q(\xi)$, one can alternatively consider the input signal realized by the two-qubit entangled state

$$|\psi_0\rangle = \frac{1}{\sqrt{2}}\Big(|00\rangle + |11\rangle\Big) . \quad (11.176)$$

This is, for our two-qubit input probe, the input state of Eq. (11.105), which is optimal in the absence of noise, and is a maximally entangled state of the qubit pair. It is associated in Eq. (11.169) with the (much simplified) density operator $\rho_0 = |\psi_0\rangle\langle\psi_0|$ reading

$$\rho_0 = \frac{1}{2}\begin{bmatrix} 1&0&0&1 \\ 0&0&0&0 \\ 0&0&0&0 \\ 1&0&0&1 \end{bmatrix} . \quad (11.177)$$

The resulting two-qubit noisy output state $\rho_\xi = \mathcal{T}_\xi \otimes \mathcal{T}_\xi(\rho_0)$ from Eq. (11.170) follows as

$$\rho_\xi = \frac{1}{2}\begin{bmatrix} b^2+c^2 & 0 & 0 & \alpha^2 e^{-i2\xi} \\ 0 & 2bc & 0 & 0 \\ 0 & 0 & 2bc & 0 \\ \alpha^2 e^{i2\xi} & 0 & 0 & b^2+c^2 \end{bmatrix} , \quad (11.178)$$

and its derivative from Eq. (11.175) as

$$\partial_\xi \rho_\xi = \frac{1}{2}\begin{bmatrix} 0 & 0 & 0 & -i2\alpha^2 e^{-i2\xi} \\ 0 & 0 & 0 & 0 \\ 0 & 0 & 0 & 0 \\ i2\alpha^2 e^{i2\xi} & 0 & 0 & 0 \end{bmatrix} . \quad (11.179)$$

For the eigendecomposition of ρ_ξ from Eq. (11.178), both $|01\rangle = [0, 1, 0, 0]^\top$ and $|10\rangle = [0, 0, 1, 0]^\top$ are eigenstates of ρ_ξ with the same eigenvalue $bc = (1-\alpha^2)/4$.

Since ρ_ξ is Hermitian with four mutually orthogonal eigenstates in $\mathcal{H}_2^{\otimes 2}$, the two remaining eigenstates of ρ_ξ are to be found in the two-dimensional subspace spanned by $\{|00\rangle, |11\rangle\}$. They can be searched as the solutions to the two-dimensional eigen-decomposition of the matrix

$$\rho_\xi^{(2)} = \frac{1}{2}\begin{bmatrix} b^2+c^2 & \alpha^2 e^{-i2\xi} \\ \alpha^2 e^{i2\xi} & b^2+c^2 \end{bmatrix} \tag{11.180}$$

representing the restriction of ρ_ξ to the vector plane $(|00\rangle, |11\rangle)$. The resulting degree-two characteristic equation $\det(\rho_\xi^{(2)} - \lambda I_2) = 0$ has two real roots $\lambda_\pm$ summing to $\lambda_+ + \lambda_- = b^2 + c^2 = (1+\alpha^2)/2$. The two associated eigenstates $|\lambda_\pm\rangle$ come under the form $|\lambda_+\rangle = e_0|00\rangle + e_1 e^{i2\xi}|11\rangle$ and $|\lambda_-\rangle = e_1 e^{-i2\xi}|00\rangle - e_0|11\rangle$, with e_0 and e_1 real satisfying $e_0^2 + e_1^2 = 1$. Due to the form of Eq. (11.179) for $\partial_\xi \rho_\xi$, only the two eigenstates $|\lambda_\pm\rangle$ lying in the plane $(|00\rangle, |11\rangle)$ contribute to the quantum Fisher information $F_q(\xi)$ of Eq. (11.68), moreover only through the two crossed terms

$$\langle\lambda_+|\partial_\xi \rho_\xi|\lambda_-\rangle = \langle\lambda_-|\partial_\xi \rho_\xi|\lambda_+\rangle^* = i\alpha^2 e^{-i2\xi} , \tag{11.181}$$

while each of the two rectangular terms $\langle\lambda_+|\partial_\xi \rho_\xi|\lambda_+\rangle$ and $\langle\lambda_-|\partial_\xi \rho_\xi|\lambda_-\rangle$ vanishes.

Finally, the quantum Fisher information $F_q(\xi)$ of Eq. (11.68) evaluates to

$$F_q(\xi) = \frac{8\alpha^4}{1+\alpha^2} . \tag{11.182}$$

This represents the quantum Fisher information $F_q(\xi)$ accessible from a noisy entangled qubit pair prepared in the state $|\psi_0\rangle$ of Eq. (11.176). It is interesting to compare it with $F_q(\xi) = 2\alpha^2$ deduced from Eq. (11.149) as the maximal Fisher information accessible from two separable noisy qubits optimally prepared in the state $|+\rangle \otimes |+\rangle$ here. This comparison is accomplished in Fig. 11.6.

It is observed in Fig. 11.6 that without noise, at $\alpha = 1$, the entangled qubit pair of Eq. (11.176) is much more efficient for the phase estimation since it yields a quantum Fisher information $F_q(\xi) = 4$ in Eq. (11.182) which is twice the maximal Fisher information $F_q(\xi) = 2\alpha^2 = 2$ of two separable qubits deduced from Eq. (11.149). This is the Heisenberg enhancement announced in Sect. 11.3.4 on the occasion of Eq. (11.106): with an L-qubit separable probe the quantum Fisher information scales as L, while there exists a possibility of scaling as L^2 with an L-qubit entangled probe, as observed in Fig. 11.6 for $L = 2$. However, as also indicated in Sect. 11.3.4, this entanglement-based enhancement is fragile to noise. In Fig. 11.6, when the level of noise increases as the contraction factor α decreases below 1, the superiority of the Fisher information $F_q(\xi)$ of the entangled pair gradually diminishes. A crossover occurs at $\alpha = \alpha_c = 1/\sqrt{3} \approx 0.577$ in Fig. 11.6. At higher noise levels, when $\alpha < \alpha_c$, the separable qubit pair exhibits a Fisher information $F_q(\xi) = 2\alpha^2$ becoming superior to that of the entangled pair in Eq. (11.182). The entangled probe is more sensitive to the phase ξ been estimated (as observed in Fig. 11.6 with no noise at

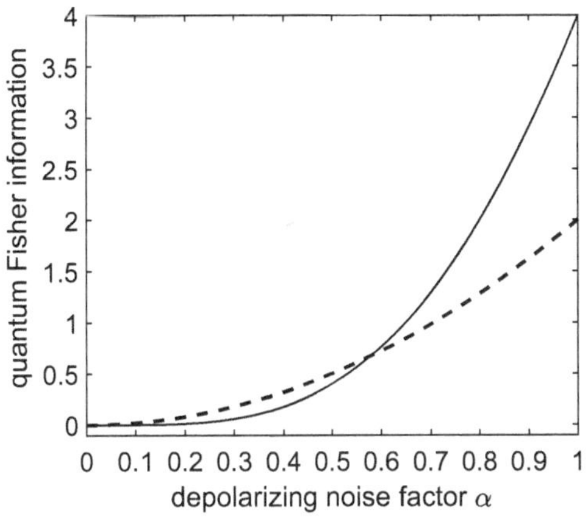

Fig. 11.6 For phase estimation in the setting of Fig. 11.2, the quantum Fisher information $F_q(\xi)$ as a function of the contraction factor α of the depolarizing noise. Solid line: $F_q(\xi)$ is from Eq. (11.182) when probing with an entangled qubit pair prepared in the state $|\psi_0\rangle$ of Eq. (11.176) and transformed as $\mathcal{T}_\xi \otimes \mathcal{T}_\xi(\rho_0) = \rho_\xi$. Dashed line: $F_q(\xi) = 2\alpha^2$, the maximal Fisher information accessible from two separable qubits resulting from Eq. (11.149). A crossover occurs at $\alpha = \alpha_c = 1/\sqrt{3} \approx 0.577$

$\alpha = 1$), but at the same time it is more fragile to noise. This is rather typical of the status of entangled signals for quantum information processing in the presence of noise; whether they can bring useful enhancement over separable signals depends on the conditions, which need be examined closely. More elements can be found on entangled qubit states for phase estimation in [231, 311] for instance, with other entangled preparations, differing from $|\psi_0\rangle$ of Eq. (11.176), since we proved that this $|\psi_0\rangle$ of Eq. (11.176) is the optimal two-qubit input probe only with no noise, and in the presence of noise it gradually ceases to be so (although the optimal one is less easily characterized). The impact of the number L of entangled qubits in the probe, beyond $L = 2$, is also considered in [231], where, depending on the conditions, finite optimal entanglement sizes L are reported, with maximum efficiency at partial instead of complete entanglement. In addition, there exists another interesting (intriguing) way of exploiting an entangled qubit pair for probing the unitary process for phase estimation, illustrative of the distinctive and non-intuitive capabilities of quantum entanglement, that we now examine.

One Passive Qubit Entangled in the Probe: Instead of letting the two qubits of the entangled pair interact with the unitary U_ξ to be estimated, an interesting alternative is to let only one. The approach follows the setting of Fig. 11.7, also considered for instance in [311], so as to realize the input–output two-qubit transformation $\rho_0 \mapsto \mathcal{T}_\xi \otimes \mathcal{I}_2(\rho_0) = \rho_\xi$, where $\mathcal{I}_2(\cdot)$ is the one-qubit identity superoperator defined by $\mathcal{I}_2\big(|j\rangle\langle k|\big) = |j\rangle\langle k|$ over the standard basis of $\mathcal{L}(\mathcal{H}_2)$.

With the generic input $\rho_0 \in \mathcal{L}(\mathcal{H}_2^{\otimes 2})$ from Eq. (11.169), the setting of Fig. 11.7 leads to the output density operator

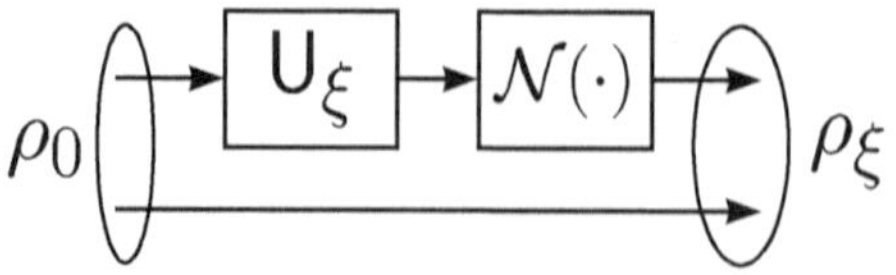

Fig. 11.7 A pair of qubits prepared in the entangled input state $\rho_0 \in \mathcal{L}(\mathcal{H}_2^{\otimes 2})$ where only the first qubit of the pair interacts with the unitary U_ξ and noise $\mathcal{N}(\cdot)$, while the second qubit is left untouched and is passive in this respect for probing U_ξ

$$\rho_\xi = \begin{bmatrix} b\rho_{00} + c\rho_{22} & b\rho_{01} + c\rho_{23} & \alpha e^{-i\xi}\rho_{02} & \alpha e^{-i\xi}\rho_{03} \\ b\rho_{10} + c\rho_{32} & b\rho_{11} + c\rho_{33} & \alpha e^{-i\xi}\rho_{12} & \alpha e^{-i\xi}\rho_{13} \\ \alpha e^{i\xi}\rho_{20} & \alpha e^{i\xi}\rho_{21} & c\rho_{00} + b\rho_{22} & c\rho_{01} + b\rho_{23} \\ \alpha e^{i\xi}\rho_{30} & \alpha e^{i\xi}\rho_{31} & c\rho_{10} + b\rho_{32} & c\rho_{11} + b\rho_{33} \end{bmatrix}, \tag{11.183}$$

having the derivative

$$\partial_\xi \rho_\xi = \begin{bmatrix} 0 & 0 & -i\alpha e^{-i\xi}\rho_{02} & -i\alpha e^{-i\xi}\rho_{03} \\ 0 & 0 & -i\alpha e^{-i\xi}\rho_{12} & -i\alpha e^{-i\xi}\rho_{13} \\ i\alpha e^{i\xi}\rho_{20} & i\alpha e^{i\xi}\rho_{21} & 0 & 0 \\ i\alpha e^{i\xi}\rho_{30} & i\alpha e^{i\xi}\rho_{31} & 0 & 0 \end{bmatrix}. \tag{11.184}$$

With the conservative choice of Eqs. (11.176)–(11.177) for the input probe $\rho_0 = |\psi_0\rangle\langle\psi_0|$, the two-qubit noisy output state $\rho_\xi = \mathcal{T}_\xi \otimes \mathcal{I}_2(\rho_0)$ from Eq. (11.183) follows as

$$\rho_\xi = \frac{1}{2}\begin{bmatrix} b & 0 & 0 & \alpha e^{-i\xi} \\ 0 & c & 0 & 0 \\ 0 & 0 & c & 0 \\ \alpha e^{i\xi} & 0 & 0 & b \end{bmatrix}, \tag{11.185}$$

and its derivative from Eq. (11.184) as

$$\partial_\xi \rho_\xi = \frac{1}{2}\begin{bmatrix} 0 & 0 & 0 & -i\alpha e^{-i\xi} \\ 0 & 0 & 0 & 0 \\ 0 & 0 & 0 & 0 \\ i\alpha e^{i\xi} & 0 & 0 & 0 \end{bmatrix}. \tag{11.186}$$

For the eigendecomposition of ρ_ξ from Eq. (11.185), both $|01\rangle = [0, 1, 0, 0]^\top$ and $|10\rangle = [0, 0, 1, 0]^\top$ are eigenstates of ρ_ξ with the same eigenvalue $c/2 = (1-\alpha)/4$. The two remaining eigenstates of ρ_ξ are to be found in the two-dimensional subspace spanned by $\{|00\rangle, |11\rangle\}$, as the solutions to the two-dimensional eigendecomposition of the matrix

$$\rho_\xi^{(2)} = \frac{1}{2}\begin{bmatrix} b & \alpha e^{-i\xi} \\ \alpha e^{i\xi} & b \end{bmatrix} \tag{11.187}$$

representing the restriction of ρ_ξ to the vector plane $\big(|00\rangle\,, |11\rangle\big)$. The resulting degree-two characteristic equation $\det\big(\rho_\xi^{(2)} - \lambda \mathrm{I}_2\big) = 0$ has two real roots $\lambda_\pm$ summing to $b = (1+\alpha)/2$. The two associated eigenstates $|\lambda_\pm\rangle$ come under the form $|\lambda_+\rangle = e_0\,|00\rangle + e_1 e^{i\xi}\,|11\rangle$ and $|\lambda_-\rangle = e_1 e^{-i\xi}\,|00\rangle - e_0\,|11\rangle$, with e_0 and e_1 real satisfying $e_0^2 + e_1^2 = 1$. Due to the form of Eq. (11.186) for $\partial_\xi \rho_\xi$, only the two eigenstates $|\lambda_\pm\rangle$ lying in the plane $\big(|00\rangle\,, |11\rangle\big)$ contribute to the quantum Fisher information $F_q(\xi)$ of Eq. (11.68), moreover only through the two crossed terms

$$\langle\lambda_+|\partial_\xi \rho_\xi|\lambda_-\rangle = \langle\lambda_-|\partial_\xi \rho_\xi|\lambda_+\rangle^* = i\frac{1}{2}\alpha e^{-i\xi}\,, \tag{11.188}$$

while each of the two rectangular terms $\langle\lambda_+|\partial_\xi \rho_\xi|\lambda_+\rangle$ and $\langle\lambda_-|\partial_\xi \rho_\xi|\lambda_-\rangle$ vanishes.

Finally, the quantum Fisher information $F_q(\xi)$ of Eq. (11.68) evaluates to

$$F_q(\xi) = \frac{2\alpha^2}{1+\alpha}\,. \tag{11.189}$$

This represents the quantum Fisher information $F_q(\xi)$ accessible from a noisy entangled qubit pair prepared in the state $|\psi_0\rangle$ of Eq. (11.176), but when only one qubit of the pair is active to probe the unitary U_ξ according to the setting of Fig. 11.7. It is interesting to compare it with $F_q(\xi) = F_q^{\max}(\xi) = \alpha^2$ of Eq. (11.149) as the maximal Fisher information accessible from a single probing qubit optimally prepared in the state $|\psi_0^{\mathrm{opt}}\rangle = |+\rangle$ from Eq. (11.134). This comparison is accomplished in Fig. 11.8.

A remarkable feature in Fig. 11.8 is that, except at the extremes $\alpha = 0$ and 1, the quantum Fisher information $F_q(\xi)$ of the entangled qubit pair is always larger than that of the optimal single qubit. In this respect, an entangled qubit pair with only one active qubit as in Fig. 11.7 is always more efficient for estimation than one optimal qubit in the setting of Fig. 11.2. This is remarkable because the additional passive qubit entangled in the pair of Fig. 11.7 never interacts with the unitary U_ξ under estimation. Yet its presence induces a net improvement in the estimation efficiency. This is a rather counterintuitive benefit afforded by quantum entanglement. It can be noted in Fig. 11.8 that when there is no noise, at $\alpha = 1$, both performances coincide; and it is only in the presence of noise, at $\alpha < 1$, that the entangled pair is strictly superior for estimation. In Fig. 11.8, the gain or ratio of the two Fisher informations $2/(1+\alpha)$ goes to 2 as the level of noise increases when the contraction factor $\alpha \to 0$. This indicates that at very strong depolarizing noise when $\alpha \to 0$, the setting of Fig. 11.7 with a passive entangled qubit in the pair, tends to be twice more efficient for estimation than the one-qubit optimal setting of Fig. 11.2. By contrast, in the distinct situation of Fig. 11.6, we rather found that entanglement was best beneficial at small noise. These examples illustrate various benefits obtainable from entanglement for enhanced performance in quantum signal processing in the presence of noise, yet occurring under variable and versatile forms according to the conditions and tasks involved. In addition, for estimation in the presence of noise, the optimal entanglement maximizing the benefit is not necessarily easy to characterize, depending on the conditions, especially when the size of the entangled

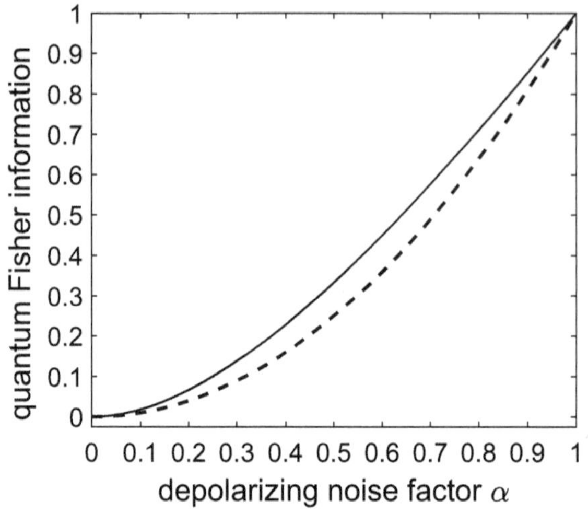

Fig. 11.8 For phase estimation in the setting of Fig. 11.7, the quantum Fisher information $F_q(\xi)$ as a function of the contraction factor α of the depolarizing noise. Solid line: $F_q(\xi)$ is from Eq. (11.189) when probing with an entangled qubit pair prepared in the state $|\psi_0\rangle$ of Eq. (11.176) and transformed as $\mathcal{T}_\xi \otimes \mathcal{I}_2(\rho_0) = \rho_\xi$ with only one qubit active for probing the unitary U_ξ. Dashed line: $F_q(\xi) = F_q^{\max}(\xi) = \alpha^2$, the maximal Fisher information of Eq. (11.149) achieved by a single probing qubit

probing signal increases. The two-qubit entangled probe $|\psi_0\rangle = \big(|00\rangle + |11\rangle\big)/\sqrt{2}$ from Eq. (11.176) was shown capable of improving over the separable probe, in various conditions, but with noise this $|\psi_0\rangle$ was not proved to be the optimal two-qubit entangled probe; this is similar (and even more challenging) when the size of the entangled probe increases. For phase estimation involving entanglement of active and passive qubits, [231, 311, 326] explore broader excitation conditions, beyond $|\psi_0\rangle$ of Eq. (11.176), and report optimal entanglement configurations that differ from maximal entanglement, depending on the noise conditions.

It can be noted that if in the ξ-dependent two-qubit probing state ρ_ξ of Eq. (11.185) a partial trace is taken, relative to either the active or the passive qubit, then in both cases the reduced state results as $\mathrm{I}_2/2$. This is a ξ-independent qubit state, which therefore cannot serve to estimate the phase ξ. This indicates that with the entangled pair of Eq. (11.185), the active or the passive qubit, when measured alone, are equally useless for estimating ξ. The two qubits of the entangled pair of Eq. (11.185) must be measured coherently, for enabling phase estimation, although they need not be measured with a joint two-qubit measurement, and a separable measurement of each qubit of the pair is possible for estimation—although presumably non-optimal. By contrast, with two active probing qubits in the state ρ_ξ of Eq. (11.178), a similar partial tracing always leads to a ξ-dependent reduced state, which can be measured alone to estimate ξ.

More generally, for the setting of Fig. 11.7, if instead of the two-qubit state of Eq. (11.185) the more general two-qubit state ρ_ξ of Eq. (11.183) is traced over the first, active, qubit, then a ξ-independent reduced state always results for the passive qubit, useless for phase estimation. This indicates that in the setting of Fig. 11.7, the passive qubit alone can never be measured to estimate the phase ξ. By contrast, when ρ_ξ of Eq. (11.183) is traced over the second, passive, qubit, then a ξ-dependent reduced state may result, provided the two-qubit input probe ρ_0 in Eq. (11.169) is appropriate, so as to enable estimation by measuring the active qubit alone.

The possibility of enhanced performance by entangling an active probing signal with a passive auxiliary quantum signal as in Fig. 11.7, can be exploited in various areas of quantum sensing and metrology, for example in quantum illumination radar. Quantum illumination consists in an electromagnetic beam active to illuminate a target, where the photons of the active beam are entangled with passive photons held at the receiver without interaction with the target [327]. In quantum illumination radar, for detecting a weakly reflecting target immersed in background noise, joint processing at the receiver of the active and passive entangled signals can lead to enhanced performance [328–332]. This illustrates again the possible benefits and potentialities to be explored of quantum properties for improved sensing, metrology, and signal processing [316, 328, 333].

Beyond the characterization of the phase estimation from noisy entangled qubits by means of the quantum Fisher information $F_q(\xi)$ undertaken in this Sect. 11.4.7, we will not go further into the characterization of an explicit measurement protocol through its classical Fisher information $F_c(\xi)$. This can be accomplished from the general considerations of Sect. 11.2. The measurement strategies feasible for phase estimation from noisy entangled qubits will in general exhibit similar properties as those encountered in Sects. 11.4.5 and 11.4.6 for phase estimation from noisy separable qubits. The same circumstances will typically be met, such as a ξ-dependent optimal measurement that need be circumvented, the possibility of resorting to standard von Neumann measurements or alternatively to generalized POVM, as separable or as joint measurements. In each scenario, the performance of the measurement will be characterizable by means of the classical Fisher information $F_c(\xi)$ and its comparison with its upper bound $F_q(\xi)$. There is no one definite optimal measurement protocol that singles out in this situation of phase estimation from noisy entangled qubits, in contrary to the case of phase estimation from noise-free entangled qubits solved in Sect. 11.3.4.

11.4.8 Stochastic Resonance in Phase Estimation

For further illustration we address another application of the principles of quantum estimation. We consider phase estimation from a noisy qubit, for which we have seen in Sect. 11.4.5 that in general both the optimal input probe signal and the optimal measurement, may be expected to depend on the unknown phase ξ to be estimated. In this context, we are led to consider phase estimation with a fixed given input probe signal and a fixed given measurement. In these conditions, for given ranges of the phase ξ to be estimated, we will show the (counterintuitive) possibility of increasing the level of noise to obtain an enhancement of the efficiency for phase estimation. This can be seen as an instance of the phenomenon of stochastic resonance [334, 335], which characterizes situations where maximum efficiency for information processing is obtained at a nonzero level of noise, which has been reported in classical settings [336–340], and which in the quantum framework assigns a beneficial role to noise and decoherence [33, 341–344].

We consider a qubit involved in a task of phase estimation in the presence of noise, in the configuration of Fig. 11.2. The unitary operator U_ξ in Eq. (11.129) is characterized by the axis $\vec{n} = [0, 1, 0]^\top = \vec{e}_y$. The noise $\mathcal{N}(\cdot)$ acting on the qubit in Fig. 11.2, is taken to be a generalized amplitude damping noise or quantum thermal noise as defined in Sect. 9.10. This describes a probing qubit coupled to an environment represented as a thermal bath. The temperature T of the bath, as it appears in Eq. (9.93), will be the natural parameter chosen to represent the level of the thermal noise. For the alteration of the probing qubit in Eq. (11.139), the matrix A and translation vector $\vec{c}$ of the quantum thermal noise, are determined, according to Eq. (9.91), by the damping or coupling factor γ and the equilibrium probability p conveying the influence of the temperature T via Eq. (9.93).

The unitary process U_ξ of Eq. (11.129) is probed by a pure input state ρ_0 prepared with the Bloch vector $\vec{r}_0 = [1, 0, 0]^\top = \vec{e}_x$. This is at least an input probe $\vec{r}_0$ orthogonal to the rotation axis $\vec{n} = \vec{e}_y$, which is, according to [293], a necessary (but not sufficient) condition for optimality of the probe with (non-unital) thermal noise, much like with unital qubit noises as we indicated in Sect. 11.4.5 before Eq. (11.144). Measurement of the probing qubit is performed via a spin observable $\Omega = \vec{\omega} \cdot \vec{\sigma}$, characterized by the vector $\vec{\omega} = [0, 0, -1]^\top = -\vec{e}_z$ also orthogonal to the axis $\vec{n} = \vec{e}_y$. The efficiency of this measurement for estimating the phase ξ is assessed by the classical Fisher information $F_c(\xi)$ from Eq. (11.147).

Figure 11.9 shows two evolutions of the Fisher information $F_c(\xi)$ of Eq. (11.147) as a function of the level of the thermal noise quantified by the bath temperature T. For more direct significance, in Fig. 11.9 the influence of the temperature is quantitatively displayed as a function of the auxiliary variable $T_p = 2(1 - p)$ interpreted as an effective or equivalent noise temperature. In this way, from Eq. (9.93), this T_p is a monotonically increasing function $T_p = T_p(T)$ of the bath temperature T, for any value of the energy gap $E_1 - E_0 > 0$. When T is 0 then p is 1 and T_p is 0, while when T is $+\infty$ then p is $1/2$ and T_p is 1. So a bath temperature T increasing from 0 to $+\infty$ is monotonically mapped into an effective temperature T_p increasing from 0 to 1. This offers the convenience of a finite range in terms of $T_p \in [0, 1]$ to convey the influence of the bath temperature $T \in [0, +\infty[$, and also releases the quantitative analysis from the unimportant particular value of the energy gap $E_1 - E_0 > 0$.

For estimation with fixed probe $\vec{r}_0$ and measurement vector $\vec{\omega}$, Fig. 11.9 shows, as the noise temperature T_p increases and depending on the conditions, the possibility of a Fisher information $F_c(\xi)$ that increases or that antiresonates at a minimum for a critical temperature, instead of a monotonic decay of $F_c(\xi)$ with increasing T_p. The Fisher information $F_c(\xi)$ in Eq. (11.147) is controlled by the geometric configuration in $\mathbb{R}^3$ of the rotated probe $\vec{r}_1(\xi) = U_\xi \vec{r}_0$ and measurement vector $\vec{\omega}$, in

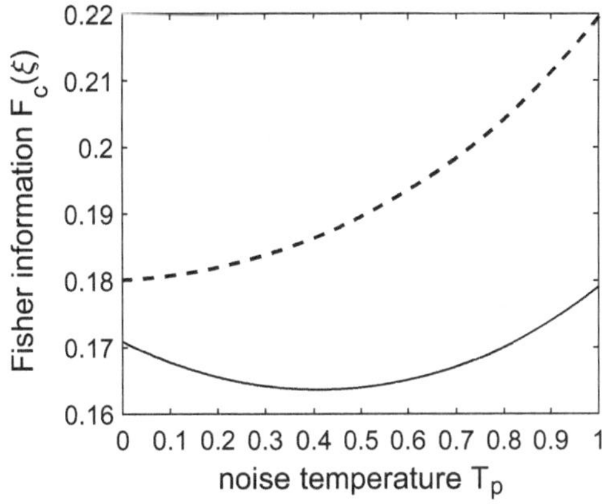

Fig. 11.9 Classical Fisher information $F_c(\xi)$ of Eq. (11.147) as a function of the effective noise temperature $T_p = 2(1-p)$, for phase estimation from a qubit affected by a quantum thermal noise. The noise coupling factor is $\gamma = 0.4$ with a phase $\xi = 0.25\pi$ (dashed line); and $\gamma = 0.5$ with $\xi = 0.2\pi$ (solid line)

relation to the rotation axis $\vec{n}$ and noise parameters $(A, \vec{c})$. An increasing temperature changes the noise parameters and therefrom the geometry controlling $F_c(\xi)$. As a result, as visible in Fig. 11.9, an increasing noise temperature T_p can cause a more favorable geometry inducing an increasing Fisher information $F_c(\xi)$, and therefrom a more efficient phase estimation. Some other conditions with an increasing noise temperature would lead to a decreasing $F_c(\xi)$, more in line with the conventional expectation.

More elements on noise-assisted quantum phase estimation can be found in [344], also in [33] for the effect with unital quantum noises, while other parameters for noise-enhanced quantum estimation are addressed in [343], and other tasks of noise-enhanced quantum signal processing are explored in [226, 341, 342, 345–349].

11.4.9 Frequency Estimation

When a quantum system is governed by a Hamiltonian operator H constant in time, according to Eq. (3.6) the corresponding unitary evolution operator is $\mathsf{U}_t = \exp\bigl(-i(\mathsf{H}/\hbar)t\bigr)$. It exhibits the same form as U_ξ of Eq. (11.74) when the generator G relates to the reduced Hamiltonian $\mathsf{H}/\hbar$, and estimation of the phase ξ coincides with estimation of the time duration t the probing system experiences the evolution by $\mathsf{U}_t = \exp\bigl(-i(\mathsf{H}/\hbar)t\bigr)$. In this circumstance, the analyses we developed here for phase estimation are also relevant for estimating such an interaction time.

In addition, upon a change of parameter $\xi = \xi(\nu)$, as indicated in Sect. 11.2.1 just before the final five properties (1)–(5), the quantum Fisher informations relative to the parameters ξ and ν verify $F_q(\nu) = (\partial\xi/\partial\nu)^2 F_q(\xi)$. In this respect, in particular, for a phase ξ building up over the time t as $\xi = \nu t$ with ν an (angular) frequency, estimation of the phase ξ is connected and relevant to estimation of the frequency ν. Frequency estimation is encountered in many areas of quantum metrology [350, 351], such as frequency standards, quantum clocks, magnetometry, atomic spectroscopy, and determination of atomic transition frequencies. For example, in quantum magnetometry [7], a two-state magnetic spin (a qubit) is used to probe a static magnetic field of magnitude B_0 parallel to the Oz axis. The spin is governed by the time-constant Hamiltonian operator $\mathsf{H} = \hbar\nu\sigma_z/2$ with ν the Larmor frequency proportional to the magnetic field B_0. The associated unitary evolution operator is $\mathsf{U}_\nu = \exp\big(-i(\mathsf{H}/\hbar)t\big) = \exp\big(-i\nu t\sigma_z/2\big)$, similar to the generic U_ξ of Eq. (11.129) with an axis $\vec{n} = \vec{e}_z$, a phase $\xi = \nu t$ and a generator $\mathsf{G} = \sigma_z/2$. Estimating the frequency ν therefore estimates the magnetic field B_0.

For frequency estimation, the classical Fisher informations $F_c(\nu)$ for the parameter ν and $F_c(\xi)$ for the parameter $\xi = \nu t$ are related by $F_c(\nu) = t^2 F_c(\xi)$, and similarly $F_q(\nu) = t^2 F_q(\xi)$. In this respect, the estimation strategies guided by the Fisher informations, for the phase ξ or the frequency ν, bear close relationship.

For instance, for qubit frequency estimation, based on Eq. (11.131), for the noise-free situation we deduce a quantum Fisher information

$$F_q(\nu) = t^2\big(\vec{n} \times \vec{r}_0\big)^2 . \tag{11.190}$$

Estimation of the frequency ν is therefore optimized by a pure input probe with a unit Bloch vector $\vec{r}_0$ orthogonal to the rotation axis $\vec{n}$, just like estimation of the phase $\xi = \nu t$. The resulting maximum is $F_q^{\max}(\nu) = t^2$. For noise-free frequency estimation, we can thus expect an accuracy (the reciprocal of the root-mean-squared error) increasing linearly with the probing time t (the time the probe is exposed to the magnetic field in magnetometry).

The picture is different for frequency estimation in the presence of quantum noise. For a qubit affected by depolarizing noise, based on Eq. (11.149), we deduce the maximal achievable quantum Fisher information as

$$F_q^{\max}(\nu) = t^2\alpha^2 . \tag{11.191}$$

It is reasonable here to assume that the contraction factor α assessing the level of quantum noise is also dependent on the time the qubit is exposed to the noise. These temporal aspects relate to the considerations of Sect. 9.14. A typical temporal variation for the contraction factor α of a depolarizing noise is $\alpha = e^{-t/\tau}$, with a relaxation time constant τ or decoherence time, function of the coupling of the qubit to its environment. The Fisher information $F_q^{\max}(\nu) = t^2 e^{-2t/\tau}$ following in Eq. (11.191) is now found to be maximized by a probing time $t = \tau$. This indicates that in the presence of noise, there exists an optimal probing time $t = \tau$ to maximize the efficiency in frequency estimation. A sufficiently prolonged interaction time t is favorable to frequency estimation as in the noise-free case above, but now with longer probing time t the qubit is also more exposed to the noise and becomes less efficient for estimation, whence the compromise at an intermediate optimal probing time $t = \tau$.

Many more aspects concerning frequency estimation can be addressed from the considerations on phase estimation we developed above.

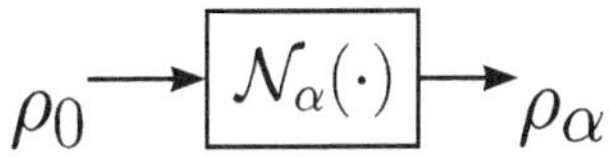

Fig. 11.10 An input qubit with density operator ρ_0 acts as an excitation signal applied to the depolarizing noise process or channel $\mathcal{N}_\alpha(\cdot)$, to deliver the output density operator ρ_α to be exploited for estimating the noise contraction factor α

11.4.10 Noise Level Estimation

For another example of quantum parameter estimation, further illustrating the general considerations of Sect. 11.2, but now with a parameter not directly related to the phase of a unitary process, we turn to a task of noise level estimation, to estimate a parameter characterizing a non-unitary process, as for instance considered in [309, 323, 352]. We will treat the estimation of the contraction factor α of a qubit depolarizing noise process $\mathcal{N}_\alpha(\cdot)$ as presented in Sect. 9.6. This factor could represent the level of noise over a noisy qubit channel, for instance an optical fiber transmitting photons. The noise factor α now plays the role of the generic unknown parameter $\xi \equiv \alpha$. A probe qubit in state ρ_0 is applied at the input of the depolarizing noise process or channel $\mathcal{N}_\alpha(\cdot)$, which delivers the output signal ρ_α to be measured to estimate the noise level α, as depicted in Fig. 11.10.

The input state ρ_0 with Bloch vector $\vec{r}_0$ is transformed via Eq. (9.55) into the output state ρ_α with Bloch vector $\vec{r}_\alpha = \alpha\vec{r}_0$. From Eq. (11.121), the quantum Fisher information relative to the noise factor α is

$$F_q(\alpha) = \frac{\left(\vec{r}_\alpha\, \partial_\alpha \vec{r}_\alpha\right)^2}{1 - \vec{r}_\alpha^{\,2}} + \left(\partial_\alpha \vec{r}_\alpha\right)^2 . \tag{11.192}$$

The output Bloch vector $\vec{r}_\alpha = \alpha\vec{r}_0$ has the derivative $\partial_\alpha \vec{r}_\alpha = \vec{r}_0$, inserted in Eq. (11.192) to give

$$F_q(\alpha) = \frac{\vec{r}_0^{\,2}}{1 - \alpha^2 \vec{r}_0^{\,2}} . \tag{11.193}$$

Maximization of the quantum Fisher information $F_q(\alpha)$ of Eq. (11.193) is realized by $\|\vec{r}_0\| = 1$, that is, by a pure input probe ρ_0 with an arbitrary orientation for $\vec{r}_0$ in $\mathbb{R}^3$, so as to achieve the maximum

$$F_q^{\max}(\alpha) = \frac{1}{1 - \alpha^2} \tag{11.194}$$

in Eq. (11.193). Any pure input probe ρ_0 is therefore optimal for estimation.

Next, to determine an optimal measurement of the output signal ρ_α, we consider measuring a spin observable $\Omega = \vec{\omega} \cdot \vec{\sigma}$. The resulting two measurement outcomes ± 1 have, according to Eq. (4.80), the probabilities

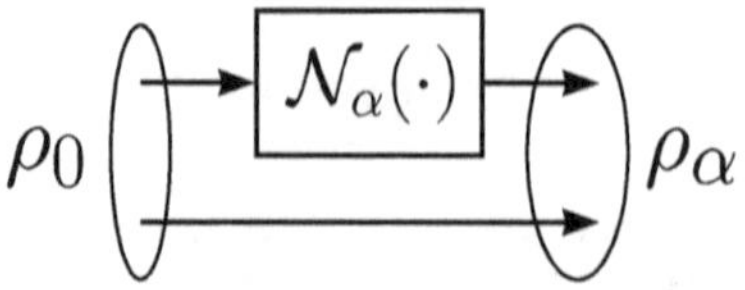

Fig. 11.11 A pair of qubits prepared in the entangled input state ρ_0, where only the first qubit of the pair interacts with the noise process $\mathcal{N}_\alpha(\cdot)$ for estimating the noise level α. The second qubit is left untouched and is passive in this respect for probing $\mathcal{N}_\alpha(\cdot)$

$$\Pr\{\pm 1\} = \frac{1}{2}\big(1 \pm \vec{r}_\alpha \vec{\omega}\big) = \frac{1}{2}\big(1 \pm \alpha \vec{r}_0 \vec{\omega}\big) . \tag{11.195}$$

From Eq. (11.126), the corresponding classical Fisher information follows as

$$F_c(\alpha) = \frac{(\vec{r}_0\vec{\omega})^2}{1 - \alpha^2(\vec{r}_0\vec{\omega})^2} . \tag{11.196}$$

Maximization of the classical Fisher information $F_c(\alpha)$ of Eq. (11.196) is realized by $(\vec{r}_0\vec{\omega})^2 = 1$, that is, by a measurement vector matched to the input by $\vec{\omega} = \pm\vec{r}_0$, to achieve for $F_c(\alpha)$ the maximum $F_q^{\max}(\alpha)$ of Eq. (11.194). With this optimal choice, the observable $\Omega = \vec{r}_0 \cdot \vec{\sigma}$ (or $\Omega = -\vec{r}_0 \cdot \vec{\sigma}$) defines an optimal measurement, uniformly for any α. The optimal choice $\Omega = \vec{r}_0 \cdot \vec{\sigma}$ gives in Eq. (11.195) the two probabilities $\Pr\{\pm 1\} = (1 \pm \alpha)/2$.

Upon repeating L times the preparation of the pure input probe ρ_0 and on the output ρ_α measuring the observable $\Omega = \vec{r}_0 \cdot \vec{\sigma}$, a number L_+ of outcomes $+1$ is obtained. The maximum-likelihood estimator for the noise (unconstrained) contraction factor α follows as

$$\widehat{\alpha}_{\rm ML} = 2\frac{L_+}{L} - 1 . \tag{11.197}$$

It is unbiased and with mean-squared error $\big\langle(\widehat{\alpha}_{\rm ML} - \alpha)^2\big\rangle = (1-\alpha^2)/L = 1/[LF_c(\alpha)] = 1/[LF_q^{\max}(\alpha)]$, establishing $\widehat{\alpha}_{\rm ML}$ of Eq. (11.197) as an optimal estimator from any number L of independent measurements. This defines the optimal input state for the probe qubit, the optimal output measurement and optimal estimator, affording maximal efficiency to estimate the contraction factor α assessing the level of a qubit depolarizing noise.

It is also possible to investigate the exploitation of entanglement to contribute to the task of noise level estimation. For illustration, we analyze the benefit of adding an auxiliary qubit as in Fig. 11.11. This is a setting where the probe qubit, for estimating the noise level, is entangled to an auxiliary passive qubit, as in Fig. 11.7 where it was found beneficial to phase estimation.

The qubit pair at the input in Fig. 11.11 is prepared in the pure state $|\psi_0\rangle = \big(|00\rangle + |11\rangle\big)/\sqrt{2}$ of Eq. (11.176) with $\rho_0 = |\psi_0\rangle\langle\psi_0|$. We already know that maximizing the quantum Fisher information $F_q(\alpha)$ requires a pure input state; in addition, due to the isotropy of the qubit depolarizing noise, [309] reports that the maximally entangled state of Eq. (11.176) is optimal for maximizing $F_q(\alpha)$. The transformed state at the output in Fig. 11.11 is $\rho_\alpha = \mathcal{N}_\alpha \otimes \mathcal{I}_2(\rho_0)$ which is in the computational basis

$$\rho_\alpha = \frac{1}{4}\begin{bmatrix} 1+\alpha & 0 & 0 & 2\alpha \\ 0 & 1-\alpha & 0 & 0 \\ 0 & 0 & 1-\alpha & 0 \\ 2\alpha & 0 & 0 & 1+\alpha \end{bmatrix}, \tag{11.198}$$

with derivative

$$\partial_\alpha \rho_\alpha = \frac{1}{4}\begin{bmatrix} 1 & 0 & 0 & 2 \\ 0 & -1 & 0 & 0 \\ 0 & 0 & -1 & 0 \\ 2 & 0 & 0 & 1 \end{bmatrix}. \tag{11.199}$$

For the evaluation of the quantum Fisher information from Eq. (11.68), we note that ρ_α has the four (α-independent) eigenstates $\{|\lambda_1\rangle = |\psi_0\rangle, |\lambda_2\rangle = |01\rangle, |\lambda_3\rangle = |10\rangle, |\lambda_4\rangle = (|00\rangle - |11\rangle)/\sqrt{2}\}$ respectively with the four eigenvalues $\{\lambda_1 = (1+3\alpha)/4,\ \lambda_2 = \lambda_3 = \lambda_4 = (1-\alpha)/4\}$. Follow the matrix elements $\langle\lambda_1|\partial_\alpha\rho_\alpha|\lambda_1\rangle = 3/4$ and $\langle\lambda_2|\partial_\alpha\rho_\alpha|\lambda_2\rangle = \langle\lambda_3|\partial_\alpha\rho_\alpha|\lambda_3\rangle = \langle\lambda_4|\partial_\alpha\rho_\alpha|\lambda_4\rangle = -1/4$, all other crossed matrix elements $\langle\lambda_k|\partial_\alpha\rho_\alpha|\lambda_\ell\rangle$ being zero. Finally, from Eq. (11.68) we obtain the quantum Fisher information for the entangled qubit pair of Fig. 11.11, as

$$F_q^{(1)}(\alpha) = \frac{3}{(1+3\alpha)(1-\alpha)}, \tag{11.200}$$

as also found in [309]. The quantum Fisher information $F_q^{(1)}(\alpha)$ of Eq. (11.200) for the entangled qubit pair is compared in Fig. 11.12 with the maximal Fisher information $F_q^{\max}(\alpha)$ of Eq. (11.194) from a single probe qubit, via the ratio or gain

$$\frac{F_q^{(1)}(\alpha)}{F_q^{\max}(\alpha)} = 3\frac{1+\alpha}{1+3\alpha}, \tag{11.201}$$

as a function of the depolarizing noise contraction factor α to be estimated.

Figure 11.12 shows a Fisher information gain $F_q^{(1)}(\alpha)/F_q^{\max}(\alpha)$ ranging between 1.5 and 3, i.e., a gain always larger than 1 for any noise level α. In this respect, the presence of the passive qubit entangled in the probe always induces a net enhancement of the quantum Fisher information, and therefrom of the estimation efficiency. For noise level estimation here, this is a similar entanglement-based enhancement as it was observed with the setting of Fig. 11.7 for phase estimation. This is again a somewhat counterintuitive and remarkable property of quantum entanglement, because similarly in Fig. 11.11, the entangled auxiliary qubit never interacts with the process $\mathcal{N}_\alpha(\cdot)$ to be estimated, yet its presence produces a net enhancement of the estimation efficiency.

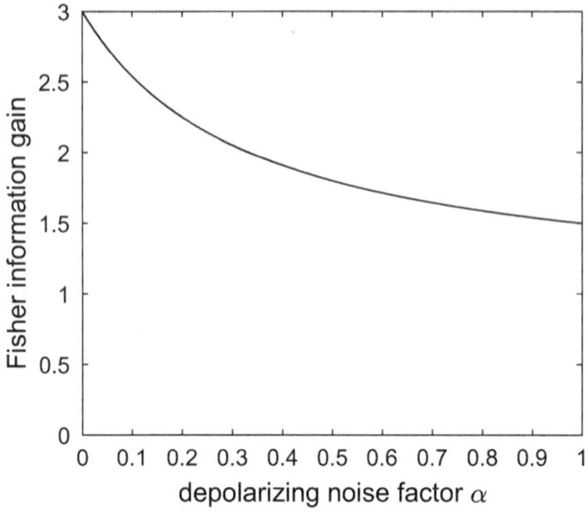

Fig. 11.12 As a function of the depolarizing noise contraction factor α to be estimated, the Fisher information gain as the ratio $F_q^{(1)}(\alpha)/F_q^{\max}(\alpha)$ from Eq. (11.201) where $F_q^{\max}(\alpha)$ is the maximal quantum Fisher information of Eq. (11.194) from a single optimal probe qubit in the setting of Fig. 11.10, and $F_q^{(1)}(\alpha)$ is the quantum Fisher information of Eq. (11.200) from an entangled qubit pair in the setting of Fig. 11.11 with only one active probing qubit in the pair

For measuring the output qubit pair in state ρ_α in Fig. 11.11, we recognize that the four eigenstates $\{|\lambda_n\rangle\}$ of ρ_α are independent of the unknown parameter α, and only its four eigenvalues $\{\lambda_n\}$ are dependent on α. Based on the remark made just after Eq. (11.69), we therefore know that there exists a quantum measurement of ρ_α characterized by a classical Fisher information $F_c(\alpha)$ reaching the quantum Fisher information $F_q(\alpha)$: it is the von Neumann measurement of ρ_α in its eigenbasis $\{|\lambda_1\rangle, |\lambda_2\rangle, |\lambda_3\rangle, |\lambda_4\rangle\}$. Moreover, because of the degeneracy of the three eigenvalues $\lambda_2 = \lambda_3 = \lambda_4$, it is possible to sum the three associated rank-one projectors into a single projector with rank three. So it is enough and optimal to consider the two-outcome measurement defined by the projector $|\lambda_1\rangle\langle\lambda_1| = |\psi_0\rangle\langle\psi_0|$ and the complementary projector $\mathrm{I}_2^{\otimes 2} - |\psi_0\rangle\langle\psi_0|$ acting in $\mathcal{H}_2^{\otimes 2}$. The probability that the measurement of the output qubit pair in state ρ_α projects onto $|\psi_0\rangle$ is $P_0 = \langle\psi_0|\rho_\alpha|\psi_0\rangle = \lambda_1 = (1+3\alpha)/4$, with $\partial_\alpha P_0 = 3/4$. This two-outcome measurement has a classical Fisher information similar to Eq. (11.89), and following as

$$F_c(\alpha) = \frac{3}{(1+3\alpha)(1-\alpha)} . \tag{11.202}$$

As expected, this $F_c(\alpha)$ exactly matches the quantum Fisher information $F_q^{(1)}(\alpha)$ of Eq. (11.200), indicating that in this respect our two-outcome measurement of the output qubit pair in state ρ_α is optimal for estimating the noise level α.

Other issues can be addressed, such as the possibility of an efficient separable measurement of the entangled qubit pair in state ρ_α, and its performance in relation to the quantum Fisher information of Eq. (11.200). This can be treated with the considerations on quantum estimation developed and illustrated in this chapter. We note that here also, as in the setting of Fig. 11.7, the two entangled qubits output in Fig. 11.11 must be measured coherently. If one qubit of the pair, either the passive or the active qubit, is discarded (non-measured), the effect is deduced from the reduced density operator obtained by tracing ρ_α of Eq. (11.198) over one qubit; in both cases this leads to the reduced density operator $\mathrm{I}_2/2$ inoperative for estimating α.

Still many other issues can be investigated from the considerations on quantum estimation developed and illustrated in this chapter. For example, with two active qubits in the probing pair, with again the preparation $|\psi_0\rangle = \big(|00\rangle + |11\rangle\big)/\sqrt{2}$ of Eq. (11.176) and $\rho_0 = |\psi_0\rangle\langle\psi_0|$, when the two entangled qubits are sent, one after the other, across the noise channel $\mathcal{N}_\alpha(\cdot)$ of Fig. 11.10, the transformed state at the output is $\rho_\alpha = \mathcal{N}_\alpha \otimes \mathcal{N}_\alpha(\rho_0)$ which is in the computational basis

$$\rho_\alpha = \frac{1}{4}\begin{bmatrix} 1+\alpha^2 & 0 & 0 & 2\alpha^2 \\ 0 & 1-\alpha^2 & 0 & 0 \\ 0 & 0 & 1-\alpha^2 & 0 \\ 2\alpha^2 & 0 & 0 & 1+\alpha^2 \end{bmatrix} . \tag{11.203}$$

The derivative $\partial_\alpha \rho_\alpha$ follows, and the eigendecomposition of ρ_α can be accomplished explicitly as for the setting of Fig. 11.11 where only one qubit of the pair was active for probing. Finally, the resulting quantum Fisher information of Eq. (11.68) is obtained, and is also reported in [309] to be

$$F_q^{(2)}(\alpha) = \frac{12\alpha^2}{(1+3\alpha^2)(1-\alpha^2)} . \tag{11.204}$$

The quantum Fisher information $F_q^{(2)}(\alpha)$ of Eq. (11.204) for this entangled qubit pair is compared in Fig. 11.13 with the maximal Fisher information $2F_q^{\max}(\alpha)$ resulting from Eq. (11.194) for two independent probe qubits successively applied and optimally processed in the setting of Fig. 11.10, via the gain

$$\frac{F_q^{(2)}(\alpha)}{2F_q^{\max}(\alpha)} = \frac{6\alpha^2}{1+3\alpha^2} , \tag{11.205}$$

as a function of the depolarizing noise contraction factor α to be estimated.

Figure 11.13 shows a Fisher information gain $F_q^{(2)}(\alpha)/2F_q^{\max}(\alpha)$ larger than 1 only for $\alpha > \alpha_c = 1/\sqrt{3} \approx 0.578$. In this respect, the entangled qubit pair is more efficient for estimation only when $\alpha > \alpha_c$, i.e., at small noise levels. Otherwise, at large noise, two independent qubits each optimally processed are more efficient. This again illustrates the versatile role of quantum entanglement for quantum signal processing in the presence of noise, which can bring useful benefit over more conven-

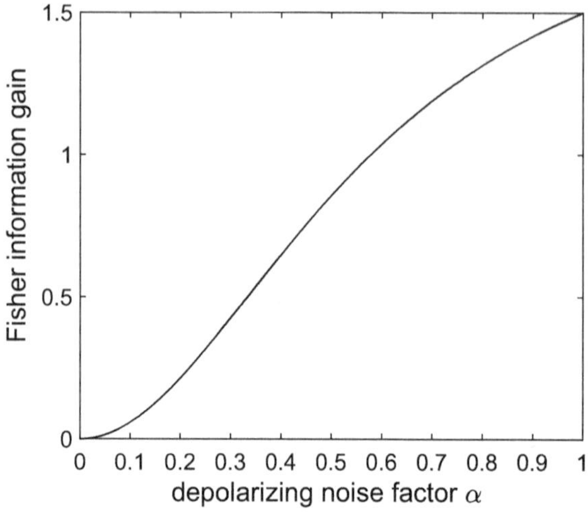

Fig. 11.13 As a function of the depolarizing noise contraction factor α to be estimated, the Fisher information gain as the ratio $F_q^{(2)}(\alpha)/2F_q^{\max}(\alpha)$ from Eq. (11.205) where $2F_q^{\max}(\alpha)$ is the maximal quantum Fisher information from Eq. (11.194) for two independent probe qubits successively applied in the setting of Fig. 11.10, and $F_q^{(2)}(\alpha)$ is the quantum Fisher information of Eq. (11.204) from an entangled qubit pair with two active probing qubits in the pair

tional techniques, depending on the conditions, which need to be analyzed explicitly. Meanwhile, to characterize the optimal entanglement maximizing the benefit usually remains a challenging task.

11.5 Multiple Quantum Parameters

11.5.1 Quantum Fisher Information Matrix

As in the classical case addressed in Sect. 11.1.2, a natural extension is to consider a quantum state $\rho_{\vec{\xi}} \in \mathcal{L}(\mathcal{H}_N)$ depending on a vector of multiple unknown parameters $\vec{\xi} = [\xi_1, \xi_2, \ldots]^\top$ to be estimated [292, 308, 353].

A quantum score Hermitian operator V_j is associated via Eq. (11.47) with each derivative $\partial\rho_{\vec{\xi}}/\partial\xi_j \equiv \partial_j\rho_{\vec{\xi}}$, which comes with the zero mean $\langle \mathsf{V}_j\rangle = \mathrm{tr}\big(\rho_{\vec{\xi}}\mathsf{V}_j\big) = 0$. They can be assembled into a column vector of operators $\vec{\mathsf{V}} = [\mathsf{V}_1, \mathsf{V}_2, \ldots]^\top$, having the zero mean $\langle\vec{\mathsf{V}}\rangle = \big[\langle \mathsf{V}_1\rangle, \langle \mathsf{V}_2\rangle, \ldots\big]^\top = \vec{0}$. The quantum Fisher information matrix is the symmetrized covariance $F_q\big(\vec{\xi}\,\big) = \big\langle \vec{\mathsf{V}}\vec{\mathsf{V}}^\top + (\vec{\mathsf{V}}\vec{\mathsf{V}}^\top)^\top\big\rangle/2$. It is a square real symmetric non-negative definite matrix $F_q\big(\vec{\xi}\,\big) = \big[F_{jk}\big(\vec{\xi}\,\big)\big]$ equivalently defined by the matrix elements

$$F_{jk}(\vec{\xi}\,) = \frac{1}{2}\operatorname{tr}\big(\rho_{\vec{\xi}}\,(\mathsf{V}_j\mathsf{V}_k + \mathsf{V}_k\mathsf{V}_j)\big) \tag{11.206}$$

$$= \operatorname{tr}\big((\partial_j\rho_{\vec{\xi}})\mathsf{V}_k\big) = \operatorname{tr}\big((\partial_k\rho_{\vec{\xi}})\mathsf{V}_j\big) \tag{11.207}$$

$$= -\operatorname{tr}\big(\rho_{\vec{\xi}}\,\partial_j\mathsf{V}_k\big) = -\big\langle\partial_j\mathsf{V}_k\big\rangle \tag{11.208}$$

$$= -\operatorname{tr}\big(\rho_{\vec{\xi}}\,\partial_k\mathsf{V}_j\big) = -\big\langle\partial_k\mathsf{V}_j\big\rangle\,. \tag{11.209}$$

When referring to a diagonal decomposition like Eq. (11.60) for the density operator $\rho_{\vec{\xi}}$, then for each component V_j of the quantum score, the expressions of Eqs. (11.65) and (11.67) apply with every derivative ∂_j; meanwhile, for the quantum Fisher information matrix we have the elements

$$F_{jk}(\vec{\xi}\,) = 2\sum_{\ell,m}\frac{\langle\lambda_\ell|\partial_j\rho_{\vec{\xi}}\,|\lambda_m\rangle\,\langle\lambda_m|\partial_k\rho_{\vec{\xi}}\,|\lambda_\ell\rangle}{\lambda_\ell+\lambda_m}\,, \tag{11.210}$$

or equivalently

$$F_{jk}(\vec{\xi}\,) = \sum_{\ell}\frac{(\partial_j\lambda_\ell)(\partial_k\lambda_\ell)}{\lambda_\ell} + \sum_{\ell,m}\frac{(\lambda_\ell-\lambda_m)^2}{\lambda_\ell+\lambda_m}\Big(\langle\lambda_\ell|\partial_j\lambda_m\rangle\,\langle\partial_k\lambda_m|\lambda_\ell\rangle + \langle\lambda_\ell|\partial_k\lambda_m\rangle\,\langle\partial_j\lambda_m|\lambda_\ell\rangle\Big)\,, \tag{11.211}$$

where the sum involving ℓ alone runs on every eigenvalue $\lambda_\ell \neq 0$, and the two sums on the pair (ℓ, m) run on eigenvalues $\lambda_\ell + \lambda_m \neq 0$. Equation (11.211) is also given for instance in [292, 308].

Measuring $\rho_{\vec{\xi}}$ by means of an arbitrary POVM $\{\mathsf{E}_m\}_{m=1}^M$ leads to the probability distribution $P\big(m\,;\vec{\xi}\,\big) = \operatorname{tr}\big(\mathsf{E}_m\,\rho_{\vec{\xi}}\big)$, for $m = 1$ to M. The resulting classical Fisher information matrix $F_c\big(\vec{\xi}\,\big)$ of Sect. 11.1.2, as for the single-parameter case, is upper bounded [308, 354] by the quantum Fisher information matrix $F_q\big(\vec{\xi}\,\big)$, fixing the matrix inequality $F_c\big(\vec{\xi}\,\big) \leq F_q\big(\vec{\xi}\,\big)$.

To optimize the estimation, one may exploit any flexibility existing on $\rho_{\vec{\xi}}$ to maximize in some meaningful sense (which may not be univocal in this multi-parametric context) the quantum Fisher information $F_q\big(\vec{\xi}\,\big)$, and then select the POVM $\{\mathsf{E}_m\}_{m=1}^M$ also to judiciously maximize the classical Fisher information $F_c\big(\vec{\xi}\,\big)$ toward its upper bound $F_q\big(\vec{\xi}\,\big)$. With a definite POVM $\{\mathsf{E}_m\}_{m=1}^M$, when repeating L independent measurements on L successive replicas of $\rho_{\vec{\xi}}$, a dataset $\{L_m\}$ is collected with a number L_m of outcomes m, for $m = 1$ to M, with $\sum_{m=1}^M L_m = L$. From the resulting multinomial probability distribution, the loglikelihood follows as $\ln P\big(\{L_m\}\,;\vec{\xi}\,\big) = \sum_{m=1}^M L_m \ln\big(P(m\,;\vec{\xi}\,)\big) + \ln(L!) - \sum_{m=1}^M \ln(L_m!)$. The vector maximum-likelihood estimator is then determined as the solution to

$$\widehat{\vec{\xi}}_{\mathrm{ML}}\big(\{L_m\}\big) = \arg\max_{\vec{\xi}}\sum_{m=1}^{M} L_m \ln\big(\operatorname{tr}(\mathsf{E}_m\,\rho_{\vec{\xi}})\big)\,, \tag{11.212}$$

with $\widehat{\vec{\xi}}_{\text{ML}}$ coming with its known asymptotic optimality properties at large size L, as indicated in Sect. 11.1.2, much like in the single-parameter case of Eq. (11.73).

Upon a change of parameter $\vec{\xi} = \vec{\xi}(\vec{\nu}) = g(\vec{\nu})$ via a multidimensional function $g(\cdot)$, the operator components of the quantum score vector transform as $\mathsf{V}_\ell(\vec{\nu}) = \sum_k J_{\ell k}\mathsf{V}_k(\vec{\xi})$, with the Jacobian matrix $J = \left[J_{\ell k} = \partial\xi_k/\partial\nu_\ell\right]$. The quantum Fisher information matrices relative to the parameters $\vec{\xi}$ and $\vec{\nu}$ are related by $F_q(\vec{\nu}) = JF_q(\vec{\xi})J^\top$.

The five properties indicated at the end of Sect. 11.2.1 for the scalar quantum Fisher information, concerning additivity, unitary invariance, monotonicity and convexity, naturally extend [308] to the quantum Fisher information matrix.

11.5.2 Multiple Qubit Parameters

For a qubit state $\rho_{\vec{\xi}}$ in the Bloch representation of Eq. (11.117), with a Bloch vector $\vec{r}_{\vec{\xi}}$ carrying the dependence on the vector parameter $\vec{\xi}$, we have [308] for the quantum Fisher information matrix $F_q(\vec{\xi}) = \left[F_{jk}(\vec{\xi})\right]$ the elements

$$F_{jk}(\vec{\xi}) = \frac{\left(\vec{r}_{\vec{\xi}}\,\partial_j\vec{r}_{\vec{\xi}}\right)\left(\vec{r}_{\vec{\xi}}\,\partial_k\vec{r}_{\vec{\xi}}\right)}{1-\vec{r}_{\vec{\xi}}^{\,2}} + \left(\partial_j\vec{r}_{\vec{\xi}}\right)\left(\partial_k\vec{r}_{\vec{\xi}}\right) \tag{11.213}$$

for the general case of a mixed qubit state $\rho_{\vec{\xi}}$, while they reduce to $F_{jk}(\vec{\xi}) = \left(\partial_j\vec{r}_{\vec{\xi}}\right)\left(\partial_k\vec{r}_{\vec{\xi}}\right)$ for the special case of a pure qubit state $\rho_{\vec{\xi}}$.

For illustration, an example of multiple-parameter estimation is provided by the estimation of the two coordinates $\vec{\xi} \equiv [\theta, \varphi]^\top$ of the generic pure qubit state $|\psi\rangle = \cos(\theta/2)\,|0\rangle + e^{i\varphi}\sin(\theta/2)\,|1\rangle$ also considered in Sect. 11.4.3. From the two derivatives $\partial_\theta\vec{r}$ and $\partial_\varphi\vec{r}$ given in Sect. 11.4.3, we obtain $\left(\partial_\theta\vec{r}\right)\left(\partial_\varphi\vec{r}\right) = 0$, and finally the quantum Fisher information matrix as the 2×2 diagonal matrix $F_q(\theta,\varphi) = \left[1,\ 0;0,\ \sin^2(\theta)\right]$. No measurement of the qubit is maximally efficient simultaneously for estimating θ and φ, and to achieve a classical Fisher information matrix $F_c(\theta,\varphi)$ matching the quantum Fisher information matrix $F_q(\theta,\varphi)$, as reflected in Sect. 11.4.3. Typically some compromise may be found, by admitting a reduced performance with a given quantum measurement, or by using two distinct measurements for estimating θ and φ. We shall come back with more detail below in Sect. 11.6.1 dealing with quantum state tomography, to such multiple-parameter estimation of a quantum state through combined use of several quantum measurements.

Another example of multiple-parameter estimation is provided by the estimation of the two parameters $\vec{\xi} \equiv [\gamma, p]^\top$ characterizing a qubit thermal noise as described in Sect. 9.10. The thermal noise process is probed by an input probe qubit prepared in the state with Bloch vector $\vec{r} = [r_x, r_y, r_z]^\top$. After alteration by the noise process, the probe qubit terminates in the state $\rho_{\vec{\xi}}$ with the Bloch vector $\vec{r}_{\vec{\xi}}$ resulting from Eq. (9.91) as

$$\vec{r}_{\vec{\xi}} = \begin{bmatrix} \sqrt{1-\gamma}\, r_x \\ \sqrt{1-\gamma}\, r_y \\ (1-\gamma)r_z + (2p-1)\gamma \end{bmatrix} . \tag{11.214}$$

The state $\rho_{\vec{\xi}}$ is then used to estimate the two noise parameters $[\gamma, p]^\top \equiv \vec{\xi}$ via $\vec{r}_{\vec{\xi}} \equiv \vec{r}_{\vec{\xi}}\,(\gamma, p)$. For this multiparameter estimation task, the quantum Fisher information matrix $F_q(\gamma, p)$ is the 2×2 matrix with the four elements of Eq. (11.213) determined by the derivatives

$$\partial_\gamma \vec{r}_{\vec{\xi}} = \begin{bmatrix} \dfrac{-1/2}{\sqrt{1-\gamma}} r_x \\ \dfrac{-1/2}{\sqrt{1-\gamma}} r_y \\ -r_z + 2p - 1 \end{bmatrix} \quad \text{and} \quad \partial_p \vec{r}_{\vec{\xi}} = \begin{bmatrix} 0 \\ 0 \\ 2\gamma \end{bmatrix} . \tag{11.215}$$

From the derivatives of Eq. (11.215), the 2×2 quantum Fisher information matrix $F_q(\gamma, p)$ of Eq. (11.213) is fully accessible, although we do not write its explicit expression which is rather bulky. A natural guide then is to seek to optimize the input probe $\vec{r} = [r_x, r_y, r_z]^\top$ so as to maximize, as much as possible, in some useful sense, the quantum Fisher information $F_q(\gamma, p)$. By the convexity of the Fisher information, we already know that $F_q(\gamma, p)$ will be maximized by a pure input probe, i.e., a unit-norm input Bloch vector $\vec{r}$. Further optimization of the input probe will usually depend on the conditions. In particular, it is typically found that the optimal input probe maximizing the diagonal element $F_{\gamma\gamma}$ of the Fisher information matrix (characterizing the separate estimation of γ at fixed p), is not the same as the optimal input probe maximizing the diagonal element F_{pp} of the Fisher information matrix (characterizing the separate estimation of p at fixed γ). For joint estimation of the pair (γ, p) a compromise is to be found for the optimal input probe, for instance by maximizing the trace $F_{\gamma\gamma} + F_{pp}$ of the 2×2 quantum Fisher information matrix $F_q(\gamma, p)$.

Next, for estimation, the qubit state $\rho_{\vec{\xi}}$ can generally be measured by the M-element POVM $\{\mathsf{E}_m\}_{m=1}^M$ defined in Eq. (4.53), producing M possible measurement outcomes governed by the probability distribution

$$P_m(\gamma, p) = \mathrm{tr}\big(\mathsf{E}_m\, \rho_{\vec{\xi}}\big) = b_m + \vec{a}_m \vec{r}_{\vec{\xi}} \ , \quad \text{for} \ \ m = 1, 2, \ldots M \ . \tag{11.216}$$

For such a measuring POVM, the classical Fisher information matrix $F_c(\gamma, p)$ follows from Eq. (11.14) with the four matrix elements

$$F_{jk}^{(c)}(\gamma, p) = \sum_{m=1}^{M} \frac{\big(\vec{a}_m \partial_j \vec{r}_{\vec{\xi}}\big)\big(\vec{a}_m \partial_k \vec{r}_{\vec{\xi}}\big)}{b_m + \vec{a}_m \vec{r}_{\vec{\xi}}} , \tag{11.217}$$

which are fully known from the derivatives of Eq. (11.215). A natural guide then is to seek to optimize the measuring POVM $\{(\vec{a}_m, b_m)\}_{m=1}^M$ so as to maximize, as much as possible, in some useful sense, the classical Fisher information $F_c(\gamma, p)$ toward its upper bound $F_q(\gamma, p)$.

Finally, from the measurement outcomes, the maximum-likelihood estimator $\widehat{\vec{\xi}}_{\rm ML} = \left[\widehat{\gamma}_{\rm ML}, \widehat{p}_{\rm ML}\right]^\top$ can be implemented. When repeating L independent measurements on L independent qubits successively probing the thermal noise process, a number L_m of outcomes m are obtained, for $m = 1$ to M, with $\sum_{m=1}^M L_m = L$. From the resulting multinomial probability distribution, the loglikelihood follows as $\ln P\big(\{L_m\}; (\gamma, p)\big) = \sum_{m=1}^M L_m \ln(P_m) + \ln(L!) - \sum_{m=1}^M \ln(L_m!)$. As an instance of Eq. (11.212), the maximum-likelihood estimator is then determined as the solution to

$$\big(\widehat{\gamma}_{\rm ML}, \widehat{p}_{\rm ML}\big) = \underset{(\gamma, p)}{\arg\max} \sum_{m=1}^{M} L_m \ln\big(b_m + \vec{a}_m \vec{r}_{\vec{\xi}}\,(\gamma, p)\big)\,, \tag{11.218}$$

which usually here needs to be solved numerically, from any measurement dataset $\{L_m\}_{m=1}^M$.

Instead of successive *independent* probing qubits, *entangled* qubits can be applied at the input to probe the thermal noise process, as done for instance in Sect. 11.4.7. Enhancement of the Fisher informations, both quantum and classical, and therefore of the estimation performance, can result. The situation is however more complicated to handle, with higher-dimensional quantum states, whose quantum Fisher information resorts to Eqs. (11.210)–(11.211) instead of Eq. (11.213). The issue of the optimal entanglement structure is also non-necessarily simple to resolve. Yet substantial benefit for estimation can result from entanglement. Even passive auxiliary qubits, entangled in the input probe but not interacting with the thermal noise process to estimate, as the passive qubit of Figs. 11.7 or 11.11, can bring a net improvement of the Fisher information and of the estimation performance, in definite conditions.

As we have seen in Sect. 9.10, the action of the quantum thermal noise is characterized in Eqs. (9.86)–(9.91) by the pair of parameters (γ, p), with γ the coupling factor of Eq. (9.92) and p the equilibrium probability of Eq. (9.93). These two model parameters (γ, p) can be related to more underlying physical parameters. The coupling factor γ can be related to a relaxation time constant τ as $\gamma = \gamma(\tau) = 1 - e^{-t/\tau}$, for a fixed interaction time t with the thermal bath. The equilibrium probability p can be related to the temperature T of the bath as $p = p(T) = 1/\big[1 + \exp\big(-\Delta E/(k_B T)\big)\big]$, for a fixed energy gap $\Delta E = E_1 - E_0$ between the excited and ground basis states of the qubit. This defines a transformation $g(\cdot)$, as evoked above at the end of Sect. 11.5.1, between the parameters (γ, p) and (τ, T). The associated Jacobian matrix involves the derivatives $\partial\gamma/\partial\tau = -(1-\gamma)t/\tau^2$ and $\partial p/\partial T = -(1-p)p\Delta E/(k_B T^2)$. This gives access to the Fisher informations in (τ, T), governing the joint estimation of the relaxation time constant τ and noise temperature T.

11.6 Quantum State and Process Tomographies

11.6.1 Quantum State Tomography

Quantum state tomography can be seen as a special and important instance of a task of multiple-parameter quantum estimation. The problem involved in quantum state tomography is to estimate, via a series of measurements, the complete state (generally a density operator) of a quantum system. This can be accomplished, for a quantum system with N-dimensional Hilbert space $\mathcal{H}_N$, by the estimation of the coefficients or coordinates of its density operator $\rho \in \mathcal{L}(\mathcal{H}_N)$ expressed as a linear combination in an appropriate orthonormal basis of $\mathcal{L}(\mathcal{H}_N)$. This confers in this respect a specific (linear) parametric structure to this task of multiple-parameter estimation involved in quantum state tomography, which can be handled in various ways.

It is always possible to consider a *Hermitian* orthonormal basis $\{\mathsf{B}_j\}$ of $\mathcal{L}(\mathcal{H}_N)$, i.e., formed by Hermitian operators of $\mathcal{L}(\mathcal{H}_N)$. This is possible because the set of Hermitian operators of $\mathcal{L}(\mathcal{H}_N)$ forms a vector space over the scalar field $\mathbb{R}$. This vector space of Hermitian operators always admits an orthonormal basis (and in fact an infinite number, unitarily related). Any operator $\mathsf{A} \in \mathcal{L}(\mathcal{H}_N)$ can always be decomposed as $\mathsf{A} = \mathsf{A}' + i\mathsf{A}''$ with the two Hermitian operators $\mathsf{A}' = (\mathsf{A} + \mathsf{A}^\dagger)/2$ and $\mathsf{A}'' = -i(\mathsf{A} - \mathsf{A}^\dagger)/2$ of $\mathcal{L}(\mathcal{H}_N)$. In the Hermitian basis $\{\mathsf{B}_j\}$, they can be decomposed as $\mathsf{A}' = \sum_j \alpha'_j \mathsf{B}_j$ and $\mathsf{A}'' = \sum_j \alpha''_j \mathsf{B}_j$ with $\{\alpha'_j\}$ and $\{\alpha''_j\}$ real coordinates. The arbitrary operator A can then be decomposed as $\mathsf{A} = \sum_j (\alpha'_j + i\alpha''_j)\mathsf{B}_j$ establishing $\{\mathsf{B}_j\}$ also as a (Hermitian) basis of $\mathcal{L}(\mathcal{H}_N)$.

An effective construction of a Hermitian orthonormal basis of $\mathcal{L}(\mathcal{H}_N)$ can be described as follows. The standard orthonormal basis of the N^2-dimensional space $\mathcal{L}(\mathcal{H}_N)$ is formed by the set of the N^2 square $N \times N$ matrices with a single 1 at each intersection of a row and a column and 0 in all the other $N^2 - 1$ positions. This is not a Hermitian basis, but the non-Hermitian matrices can be made Hermitian by pairwise combinations. A Hermitian basis is obtained with the N diagonal (Hermitian) matrices with a single 1 on the diagonal and 0 elsewhere, supplemented by the $(N-1)N/2$ Hermitian matrices with a pair of 1 at symmetric off-diagonal positions and 0 elsewhere, and supplemented by the $(N-1)N/2$ Hermitian matrices with a pair $(i, -i)$ at symmetric off-diagonal positions and 0 elsewhere. We obtain in this way a set of N^2 linearly independent Hermitian matrices forming an orthogonal Hermitian basis (the "Hermitianized" standard basis) for $\mathcal{L}(\mathcal{H}_N)$, and finally each matrix can be suitably normalized, leading to the Hermitian orthonormal basis $\{\mathsf{B}_j^{\text{stand}}\}$ for operators of $\mathcal{L}(\mathcal{H}_N)$. From the set $\{\mathsf{B}_j^{\text{stand}}\}$, with an arbitrary unitary operator U of $\mathcal{L}(\mathcal{H}_N)$, the set $\{\mathsf{U}\mathsf{B}_j^{\text{stand}}\mathsf{U}^\dagger\}$ offers another Hermitian orthonormal basis of $\mathcal{L}(\mathcal{H}_N)$.

For quantum state tomography, the density operator $\rho \in \mathcal{L}(\mathcal{H}_N)$ can be decomposed into a Hermitian orthonormal basis $\{\mathsf{B}_j\}$ of $\mathcal{L}(\mathcal{H}_N)$, as

$$\rho = \sum_{j=1}^{N^2} \alpha_j \mathsf{B}_j \,. \tag{11.219}$$

Estimation of the (real) coordinates α_j, for $j = 1$ to N^2, thus provides a complete characterization of the density operator ρ, as targeted in quantum state tomography. An interesting feature is that each real coordinate α_j can be estimated separately, as in a single-parameter estimation task. This is achieved by measuring, in the state ρ, each of the observables B_j, separately.

According to Eq. (4.59), the average value of the observable B_j in the state ρ is $\langle \mathsf{B}_j \rangle = \mathrm{tr}(\rho \mathsf{B}_j)$. The operators $\{\mathsf{B}_j\}$ forming an orthonormal basis have a Hilbert–Schmidt inner product $\mathrm{tr}\big(\mathsf{B}_j^\dagger \mathsf{B}_k\big)$ satisfying the orthonormality relations $\mathrm{tr}\big(\mathsf{B}_j^\dagger \mathsf{B}_k\big) = \delta_{jk}$. As a consequence, from Eq. (11.219) we obtain $\langle \mathsf{B}_j \rangle = \alpha_j$. So each coordinate α_j can be obtained by estimating the average $\langle \mathsf{B}_j \rangle$ of the observable B_j in the state ρ. This can be performed experimentally by repeating the measurement of the observable B_j over a large number of identical replicas of the state ρ, and by an empirical average of the values returned by the measurements. The same estimation is done separately, by measuring each observable B_j to estimate the corresponding coordinate $\alpha_j = \langle \mathsf{B}_j \rangle$, for $j = 1$ to N^2. In this way, the multiparametric task of estimating N^2 coordinates α_j can be turned into N^2 single-parameter estimation tasks (one for each α_j) performed independently. Such decoupling is not always feasible for more general (nonlinear) settings of multiparametric estimation. Each single-parameter estimation is a priori easier to handle, especially to control its situation in relation to the best performance specified by the (scalar) Fisher informations. Yet, the observables B_j are usually incompatible (non-commuting) and therefore have to be measured separately, successively, each through a separate campaign of repeated preparations and measurements of ρ, making this separable approach more costly in terms of implied measurements.

For example, for estimation of a general qubit state $\rho \in \mathcal{L}(\mathcal{H}_2)$, it is convenient to refer to the Hermitian orthogonal Pauli basis of $\mathcal{L}(\mathcal{H}_2)$ as introduced in Sect. A.5 of the Appendix. A general qubit state ρ is fully characterized by its three coordinates $[r_x, r_y, r_z]^\top$ of its Bloch representation of Eq. (2.61). They can be estimated as the three averages $r_x = \langle \sigma_x \rangle$, $r_y = \langle \sigma_y \rangle$ and $r_z = \langle \sigma_z \rangle$, via three separate campaigns of repeated measurements for each of the three (non-commuting) Pauli operators on the state ρ. For tomography of multiple-qubit states, the corresponding tensor products of one-qubit Pauli operators offer a convenient Hermitian orthogonal basis.

Another method, more closely adhering to a direct multiparametric approach, is by measuring the state ρ with a general M-element POVM $\{\mathsf{E}_m\}_{m=1}^M$, producing M possible measurement outcomes m governed by the probability distribution $P_m = \mathrm{tr}(\mathsf{E}_m \rho)$, for $m = 1$ to M. A campaign repeating L independent measurements on L independent replicas of the state ρ, collects a dataset $\{L_m\}$ with a number L_m of outcomes m, for $m = 1$ to M, with $\sum_{m=1}^M L_m = L$. From the resulting multinomial probability distribution, the loglikelihood follows as $\ln P\big(\{L_m\}\,;\rho\big) = \sum_{m=1}^M L_m \ln\big(\mathrm{tr}(\mathsf{E}_m \rho)\big) + \ln(L!) - \sum_{m=1}^M \ln(L_m!)$. By means of a suitable multiparametric representation for the unknown state ρ, a maximum-likelihood estimator is then determined as the solution to

$$\widehat{\rho}_{\rm ML}\big(\{L_m\}\big) = \arg\max_{\rho} \sum_{m=1}^{M} L_m \ln\big(\mathrm{tr}(\mathsf{E}_m\,\rho)\big)\,. \tag{11.220}$$

This is a global solution to the multiparametric estimation, where a single measurement campaign is used to globally estimate at once the whole set of unknown parameters characterizing the state ρ. An additional constraint may be imposed to ensure that the returned estimate remains a positive (density) operator.

The Fisher informations, classical and quantum, scalar or matrix, can be used to evaluate the performance of a quantum state tomography, depending on the conditions and approaches. The choice of the observables B_j in Eq. (11.219) may have a significant impact on the performance, as well as the choice of the POVM $\{\mathsf{E}_m\}$ in Eq. (11.220) and associated parameterization of the unknown density operator ρ under estimation. To limit the number of required measurements while controlling the accuracy of the estimation, in practice a tomographic approach is often further adapted to some special classes of density operators ρ. Relevant conditions often met in practice include pure or quasi pure states, with ρ a low-rank operator, or product states, or states with particular symmetries or restricted forms. Fewer, well devised, measurements may then be sufficient to reach a given estimation accuracy. Selecting a basis $\{\mathsf{B}_j\}$ in Eq. (11.219) where ρ can be correctly approximated parsimoniously with only a few dominant coordinates α_j can stand as an interesting feature. Or, instead of referring to a complete set of observables $\{\mathsf{B}_j\}$ as in Eq. (11.219) forming a basis of $\mathcal{L}(\mathcal{H}_N)$, one may turn to a reduced incomplete set of observables, and from the measured averaged values of the observables, an underdetermined inverse problem can be solved where a regularized (minimum-norm, maximum-entropy) solution $\widehat{\rho}$ can be sought that matches the measured constraints (the averages). In these directions, generic signal processing techniques, such as maximum entropy reconstruction, or compressive sensing, or more generally signal reconstruction and approximation methods, have been studied to contribute to efficient quantum state tomography [355–362].

11.6.2 Quantum Process Tomography

Another important task related to multiparameter estimation, and closely connected to quantum state tomography, is quantum process tomography. It consists in estimating a complete characterization of a quantum process or operation $\mathcal{E}(\cdot)$ as presented in Sect. 3.2. We can consider an input density operator $\rho \in \mathcal{L}(\mathcal{H})$ experiencing the quantum process $\rho \mapsto \mathcal{E}(\rho) = \rho'$ delivering an output density operator $\rho' \in \mathcal{L}(\mathcal{H}')$, possibly with two different input and output Hilbert spaces $\mathcal{H}$ and $\mathcal{H}'$ as in Sect. 3.2.9.

A complete characterization of the quantum process $\mathcal{E}(\cdot)$ from $\mathcal{L}(\mathcal{H})$ onto $\mathcal{L}(\mathcal{H}')$, can be obtained by specifying how $\mathcal{E}(\cdot)$ transforms an orthonormal basis $\big\{|j\rangle\langle k|\big\}$ of $\mathcal{L}(\mathcal{H})$, with integers j, k from 1 to $\dim(\mathcal{H})$. To obtain $\mathcal{E}\big(|j\rangle\langle k|\big)$, a basis vector

$|j\rangle\langle k|$ of the vector space $\mathcal{L}(\mathcal{H})$ is however not generally a density operator, when $j \neq k$, and thus not a valid state of $\mathcal{L}(\mathcal{H})$ directly accessible to prepare an input quantum state to act as an excitation signal to the process $\mathcal{E}(\cdot)$. Such basis vector $|j\rangle\langle k|$ is nevertheless accessible as a linear combination of valid density operators of $\mathcal{L}(\mathcal{H})$, by the identity of Eq. (3.48). Through such linear combinations, $\mathcal{E}\big(|j\rangle\langle k|\big)$ is subsequently accessible with only evaluations of terms as $\mathcal{E}\big(|\ell\rangle\langle\ell|\big)$, which form valid density operators of $\mathcal{L}(\mathcal{H}')$. Each such density operator $\mathcal{E}\big(|\ell\rangle\langle\ell|\big)$ can be estimated by a procedure of quantum state tomography, as in Sect. 11.6.1. This provides a feasible solution to quantum process tomography.

An alternative useful basis of $\mathcal{L}(\mathcal{H}) \equiv \mathcal{L}(\mathcal{H}_N)$ can be obtained by exploiting the Hermitianized standard basis of $\mathcal{L}(\mathcal{H}_N)$ constructed in Sect. 11.6.1 above, which is a basis of Hermitian matrices, from which we shall deduce a basis of positive (Hermitian) matrices, that is, a basis of density operators for $\mathcal{L}(\mathcal{H}_N)$. In the Hermitianized standard basis, the N diagonal matrices are positive unit-trace, and hence represent density operators of $\mathcal{L}(\mathcal{H}_N)$, in fact the N density operators $|j\rangle\langle j|$ of $\mathcal{L}(\mathcal{H}_N)$. The other $(N-1)N$ non-diagonal matrices are Hermitian matrices with eigenvalues -1, 0, and 1. They are turned into positive matrices by adding the $N \times N$ identity matrix I_N to each of them. This displaces the eigenvalues to 0, 1 and 2, while the resulting $(N-1)N$ matrices are now positive matrices. The identity matrix I_N is a linear combination of the N diagonal matrices of the standard basis of $\mathcal{L}(\mathcal{H}_N)$. Adding such a linear combination to each of the $(N-1)N$ non-diagonal matrices does not change the linear independence among the N^2 matrices. After such a transformation of the Hermitianized standard basis, we have obtained a set of N^2 linearly independent positive matrices, therefore forming a basis for the N^2-dimensional space $\mathcal{L}(\mathcal{H}_N)$. However, this positive basis is no longer orthogonal. The addition of I_N makes the vectors oblique. Each of the vectors (matrices) of this positive basis can be normalized to unit trace, so as to deliver a (non-orthogonal) basis of density operators for $\mathcal{L}(\mathcal{H}_N)$. Alternatively, a closely related set of N^2 density operators providing a non-orthogonal basis for $\mathcal{L}(\mathcal{H}_N)$ is formed by the N projectors onto the N basis states $|j\rangle$ of $\mathcal{H}_N$, supplemented by the $(N-1)N$ projectors onto the $(N-1)N$ states of the form $\big(|j\rangle + |k\rangle\big)/\sqrt{2}$ and $\big(|j\rangle + i\,|k\rangle\big)/\sqrt{2}$ for $j = 1$ to $N-1$ and $k = j+1$ to N, as given in [363]. Beyond, there is no possibility of an orthogonal basis of density operators for $\mathcal{L}(\mathcal{H}_N)$; this would provide for an N-dimensional quantum system, N^2 perfectly distinguishable quantum states, although only N are possible. In this vein, for the two-dimensional qubit with Hilbert space $\mathcal{H}_2$, the set $\big\{\mathrm{I}_2/2,\ (\sigma_x + \mathrm{I}_2)/2,\ (\sigma_y + \mathrm{I}_2)/2,\ (\sigma_z + \mathrm{I}_2)/2\big\}$ forms a non-orthogonal basis of density operators for $\mathcal{L}(\mathcal{H}_2)$. Alternatively, the N^2 projectors evoked above are onto the $N^2 = 4$ qubit states $\big\{|0\rangle,\ |1\rangle,\ \big(|0\rangle + |1\rangle\big)/\sqrt{2},\ \big(|0\rangle + i\,|1\rangle\big)/\sqrt{2}\,\big\}$. For multiple-qubit states, the corresponding tensor products of the single-qubit basis operators offer a basis of multiple-qubit density operators. For quantum process tomography, the image of such a basis of density operators by a quantum process $\mathcal{E}(\cdot)$ provides a complete characterization of $\mathcal{E}(\cdot)$. The implementation is facilitated when the density operators to be applied to $\mathcal{E}(\cdot)$ represent pure states, as with the N^2 projectors evoked above for $\mathcal{L}(\mathcal{H}_N)$ or with the procedure exploiting Eq. (3.48).

When a parametric model exists for the quantum process $\mathcal{E}(\cdot)$ to be characterized, its tomography can be cast into a multiparametric estimation task, as previously considered, and approached with corresponding methodologies such as maximum likelihood. When a Kraus representation $\mathcal{E}(\rho) = \sum_\ell \Lambda_\ell \rho \Lambda_\ell^\dagger$ as in Eq. (3.45) is accessible for the process, an orthonormal basis for operators from $\mathcal{H}$ onto $\mathcal{H}'$ can be selected as $\{\mathsf{F}_n\}$ for the integer index $n = 1$ to $\dim(\mathcal{H}) \times \dim(\mathcal{H}')$, so as to express each Kraus operator as $\Lambda_\ell = \sum_n f_{\ell n} \mathsf{F}_n$. The process is therefore $\mathcal{E}(\rho) = \sum_n \sum_{n'} \chi_{nn'} \mathsf{F}_n \rho \mathsf{F}_{n'}^\dagger$ with the coefficients $\chi_{nn'} = \sum_\ell f_{\ell n} f_{\ell n'}^*$, and the completeness relation of the Kraus operators as in Eq. (3.47) reading $\sum_n \sum_{n'} \chi_{nn'} \mathsf{F}_{n'}^\dagger \mathsf{F}_n = \mathrm{I}^{\mathcal{H}}$. The process $\mathcal{E}(\cdot)$ is here completely characterized by the square Hermitian positive matrix $[\chi_{nn'}]$ known as the **process matrix**, offering for $\mathcal{E}(\cdot)$ a definite parameterization for estimation, where the choice of the Kraus representation $\{\Lambda_\ell\}$ and operator basis $\{\mathsf{F}_n\}$ may play crucial roles. Many variants and elaborations are feasible for quantum process tomography, in similarity with the situation of quantum state tomography [363–366]. Also quantum process tomography remains today a rather formidable effort, as the dimensionality grows, in terms of number of parameters and number of measurements required for their (accurate) estimation.

Quantum process tomography, so as to estimate a complete characterization of a quantum process $\mathcal{E}(\cdot)$, can also be handled through an estimation of the associated Choi–Jamiołkowski state presented in Sect. 3.2.10, casting it into a task of state tomography.

11.7 Quantum Bayesian Estimation

11.7.1 Bayesian Methodology of Quantum Estimation

As in the classical situation of Sect. 11.1.3, quantum Bayesian estimation [367–369] arises with random parameters. We introduce it with a scalar parameter ξ, while quantum Bayesian estimation naturally extends to a vector $\vec{\xi}$ of multiple parameters. The unknown parameter to be estimated is considered to be a random parameter ξ distributed according to the prior probability density $p_\xi(\xi)$. Measurement of the parameter-dependent quantum state ρ_ξ, following Eq. (11.44), induces the conditional probability distribution

$$P(m|\xi) = \mathrm{tr}(\mathsf{E}_m \rho_\xi), \tag{11.221}$$

for the outcomes $m = 1$ to M. These conditional probabilities determine, as in Eq. (11.16), the posterior probability density for the unknown parameter ξ as

$$p(\xi|m) = \frac{\mathrm{tr}(\mathsf{E}_m \rho_\xi) p_\xi(\xi)}{\int_{-\infty}^{\infty} \mathrm{tr}(\mathsf{E}_m \rho_\xi) p_\xi(\xi) d\xi}. \tag{11.222}$$

From the measurement outcome m, the posterior probability density of Eq. (11.222) determines the optimal Bayesian estimator $\widehat{\xi}_B(m) = \langle\xi\rangle_{\xi|m}$ as the posterior mean as in Eq. (11.22), and its mean-squared error $e^2\big(\widehat{\xi}_B|m\big) = \mathrm{var}(\xi|m)$ as in Eq. (11.21).

A sensible task then is to seek to optimize the choice of the measuring POVM $\{\mathsf{E}_m\}_{m=1}^M$, and possibly the design of the probing signal ρ_ξ, in order to minimize the mean-squared error $e^2\big(\widehat{\xi}_B|m\big) = \mathrm{var}(\xi|m)$ of the optimal Bayesian estimator $\widehat{\xi}_B(m)$. This optimization task will generally depend on the structure of the quantum state ρ_ξ and its dependence on ξ.

A useful guide is offered by the Van Trees inequality of Eq. (11.38) controlling the maximal efficiency in Bayesian estimation. The prior Fisher information $F_c^{\mathrm{pri}}(\xi)$ in Eq. (11.37) is fixed by the prior density $p_\xi(\xi)$, while the average Fisher information $\big\langle F_c(\xi\,;m)\big\rangle_\xi$ according to $p_\xi(\xi)$, with $F_c(\xi\,;m) = F_c(\xi)$ given by Eq. (11.72), can be maximized by selecting the POVM. The classical Fisher information $F_c(\xi)$ is upper bounded, for any ξ, by the quantum Fisher information $F_q(\xi)$ which is independent of the POVM. The average $\big\langle F_c(\xi\,;m)\big\rangle_\xi$ is therefore upper bounded by the average $\big\langle F_q(\xi)\big\rangle_\xi$. The upper bound $\big\langle F_q(\xi)\big\rangle_\xi$ is independent of the POVM, but can be maximized by selecting the other adjustable estimation conditions (such as the input probe when estimating the phase of a unitary process as in Sect. 11.3). When the upper bound $\big\langle F_q(\xi)\big\rangle_\xi$ is settled, the POVM can be selected to obtain an average $\big\langle F_c(\xi\,;m)\big\rangle_\xi$ coming as close as possible to its upper bound $\big\langle F_q(\xi)\big\rangle_\xi$. And once the POVM is settled, the optimal Bayesian estimator $\widehat{\xi}_B(m)$ follows as the posterior mean from Eq. (11.222).

The Fisher information, in its quantum and classical versions, is thus confirmed as an essential tool for guiding the elaboration of efficient strategies for estimation of random as well as of deterministic quantum parameters.

11.7.2 Bayesian Qubit Phase Estimation

For illustration, we treat the qubit phase estimation of Sect. 11.4.4 according to a Bayesian approach.

In the setting of Fig. 11.1, at the input the probe qubit is prepared in an optimal state ρ_0 having a unit Bloch vector $\vec{r}_0 = \vec{r}_0^{\,\mathrm{opt}}$ orthogonal to the axis $\vec{n}$ of the qubit unitary U_ξ under estimation. As we deduced from Eq. (11.131), such an input probe signal ensures a maximal quantum Fisher information $F_q^{\max}(\xi) = 1$ uniformly for any phase ξ. Accordingly, the average $\big\langle F_q(\xi)\big\rangle_\xi$ is also maximized at $\big\langle F_q^{\max}(\xi)\big\rangle_\xi = F_q^{\max}(\xi) = 1$ for any prior probability density $p_\xi(\xi)$ for the phase ξ.

Next, measurement of the ξ-dependent output signal ρ_ξ can be accomplished by measuring the optimal spin observable $\Omega = \vec{r}_0^{\,\mathrm{opt}} \cdot \vec{\sigma}$ from Eq. (11.138), with $M = 2$ possible outcomes $m = \pm 1$. This ensures, as seen in Sect. 11.4.4, a maximized classical Fisher information $F_c(\xi) = F_q^{\max}(\xi)$ uniformly for any phase ξ. Accordingly, the average $\langle F_c(\xi)\rangle_\xi$ is also maximized at $F_q^{\max}(\xi)$ for any prior density $p_\xi(\xi)$ of the phase ξ.

The conditional measurement probabilities for Eq. (11.221) result as

$$P(m = \pm 1|\xi) = \frac{1}{2}\big[1 \pm \cos(\xi)\big] . \tag{11.223}$$

The posterior probability density of Eq. (11.222) for the phase ξ follows as

$$p(\xi|m = \pm 1) = \frac{1 \pm \cos(\xi)}{1 \pm \langle\cos(\xi)\rangle_\xi} p_\xi(\xi) , \tag{11.224}$$

and determines the optimal Bayesian estimator $\widehat{\xi}_{\mathrm{B}}(m = \pm 1) = \langle\xi\rangle_{\xi|m}$ as the posterior mean.

For further explicit characterization, we consider the prior density $p_\xi(\xi) = 1/\pi$ uniform over the support $[0, \pi]$. Follow the average $\langle\cos(\xi)\rangle_\xi = 0$, and the posterior density $p(\xi|m = \pm 1) = [1 \pm \cos(\xi)]/\pi$ over the support $[0, \pi]$. The optimal Bayesian estimator is then $\widehat{\xi}_{\mathrm{B}}(m = \pm 1) = \pi/2 \mp 2/\pi$, and it achieves uniformly the root-mean-squared error $\sqrt{\mathrm{var}(\xi|m = \pm 1)} = (\pi^2/12 - 4/\pi^2)^{1/2} \approx 0.206\pi$.

This is the scenario upon performing one single measurement on a single qubit for Bayesian phase estimation. A more efficient scenario, when accessible, would consist in repeating L times the measurement from L independent replicas of the probing state ρ_ξ, while assuming the whole batch of L replicas is produced under the same value ξ of the random parameter. When using the same two-outcome qubit measurement as in Eq. (11.223), we collect a number L_+ of outcomes $+1$ and $L - L_+$ outcomes -1. A given configuration $(L_+, L - L_+)$ of the measurement outcomes is thus obtained with the conditional probability

$$P(L_+|\xi) = \left[\frac{1 + \cos(\xi)}{2}\right]^{L_+}\left[\frac{1 - \cos(\xi)}{2}\right]^{L-L_+} = \big[\cos(\xi/2)\big]^{2L_+}\big[\sin(\xi/2)\big]^{2(L-L_+)} . \tag{11.225}$$

The difficult task then, as is common in a Bayesian framework, is to analytically perform the integration of $P(L_+|\xi)$ over the prior $p_\xi(\xi)$ to deduce the posterior distribution $p(\xi|L_+)$ of the parameter and to characterize the optimal Bayesian estimator $\widehat{\xi}_{\mathrm{B}}(L_+)$ as the posterior mean. Alternatively, this can be performed numerically. Or also, in the large L limit, as exploited in [368] or as evoked at the end of Sect. 11.1.3, a Gaussian approximation [290] becomes accessible for the posterior distribution $p(\xi|L_+)$, by the Laplace Bernstein von Mises theorem.

These specificities of Bayesian estimation are not by themselves of quantum nature, but rather are characteristic of standard Bayesian approaches, and so we will not go further here in this direction. More specifically for our quantum signal processing perspective, this shows how similar considerations apply for classical or quantum signals equally.

11.8 Summary on Quantum Estimation

Estimation of a parameter ξ characterizing a quantum state $\rho_\xi \in \mathcal{L}(\mathcal{H}_N)$ is crucially governed by the **quantum Fisher information** $F_q(\xi)$, which quantifies the information about the parameter ξ contained in the quantum state ρ_ξ. The quantum Fisher information $F_q(\xi)$ is defined in Eq. (11.56) from the quantum score V_ξ as $F_q(\xi) = \mathrm{tr}\big(\rho_\xi \mathsf{V}_\xi^2\big) = \big\langle \mathsf{V}_\xi^2 \big\rangle$. A more explicit expression for $F_q(\xi)$ follows from the eigendecomposition of the density operator ρ_ξ, and via the dependence on ξ of its eigenvalues $\big\{\lambda_n\big\}_{n=1}^N$ and eigenstates $\big\{|\lambda_n\rangle\big\}_{n=1}^N$, in Eq. (11.69) for a mixed state ρ_ξ and in Eq. (11.71) for a pure state ρ_ξ.

For estimation, the ξ-dependent quantum state ρ_ξ is measured, in general by a POVM $\{\mathsf{E}_m\}_{m=1}^M$ exhibiting M measurement outcomes with the probability distribution $P(m\,;\xi) = \mathrm{tr}\big(\mathsf{E}_m \rho_\xi\big)$ of Eq. (11.44). From the measurement results, any estimator $\widehat{\xi}$ for ξ exhibits a mean-squared error $e^2\big(\widehat{\xi}\big) = \big\langle\big(\widehat{\xi} - \xi\big)^2\big\rangle$ lower bounded by the Cramér–Rao inequality of Eq. (11.9), controlled by the **classical Fisher information** $F_c(\xi)$ from Eq. (11.72). In turn, the classical Fisher information $F_c(\xi)$ is upper bounded by the quantum Fisher information $F_q(\xi)$, with $F_c(\xi) \le F_q(\xi)$ according to Eq. (11.55).

To optimize the estimation, one may exploit any flexibility existing on the probing quantum signal ρ_ξ to maximize the quantum Fisher information $F_q(\xi)$, then select the POVM $\{\mathsf{E}_m\}_{m=1}^M$ measuring ρ_ξ so as to maximize the classical Fisher information $F_c(\xi)$ toward its upper bound $F_q(\xi)$, and finally from the (classical) measurement results implement a trusted estimation technique such as the **maximum-likelihood estimator**.

An essential application is for **estimation of the phase** ξ of a unitary quantum process $\mathsf{U}_\xi = \exp(-i\xi\mathsf{G})$ with given Hermitian generator $\mathsf{G} \in \mathcal{L}(\mathcal{H}_N)$ as in Eq. (11.74), through the transformation of an input probe signal ρ_0 according to $\rho_0 \mapsto \mathsf{U}_\xi \rho_0 \mathsf{U}_\xi^\dagger = \rho_\xi$ as in Eq. (11.75). The optimal input probe ρ_0 maximizing the quantum Fisher information $F_q(\xi)$ is the pure input signal $\rho_0 = |\psi_0^{\mathrm{opt}}\rangle\langle\psi_0^{\mathrm{opt}}|$ with $|\psi_0^{\mathrm{opt}}\rangle = \big(|u_{\min}\rangle + |u_{\max}\rangle\big)/\sqrt{2}$ and $|u_{\min}\rangle$, $|u_{\max}\rangle$ the two eigenstates of G (and U_ξ) respectively with minimum and maximum eigenvalues $u_{\min}$, $u_{\max}$, to achieve the maximum $F_q^{\max}(\xi) = (u_{\max} - u_{\min})^2$ of Eq. (11.85). The optimal quantum measurement is formed by the pair of projectors $\big\{|\psi_0^{\mathrm{opt}}\rangle\langle\psi_0^{\mathrm{opt}}|\,,\, \mathrm{I}_N - |\psi_0^{\mathrm{opt}}\rangle\langle\psi_0^{\mathrm{opt}}|\big\}$ achieving the maximum feasible for the classical Fisher information $F_c(\xi) = F_q^{\max}(\xi)$ in Eq. (11.90). This is equivalent to measuring the observable $\Omega = |u_{\min}\rangle\langle u_{\max}| + |u_{\max}\rangle\langle u_{\min}|$ of Eq. (11.98). From the measurement results, in the limit of a large number L of independent repetitions, the maximum-likelihood estimator $\widehat{\xi}_{\mathrm{ML}}$ of Eq. (11.96) achieves the minimal mean-squared estimation error $e^2\big(\widehat{\xi}_{\mathrm{ML}}\big)$ in Eq. (11.97).

With L independent probes, the Fisher informations, quantum and classical, usually scale linearly with L, entailing a mean-squared estimation error decaying as $1/L$, this conforming with the shot-noise classical scaling. Alternatively, as seen in Sect. 11.3.4, optimally entangling the L probes can lead to a quadratic scaling in

L^2 for the quantum and the classical Fisher informations, associated with a mean-squared estimation error decaying as $1/L^2$, a performance enhancement known as the **Heisenberg quantum scaling**.

The analysis can be further specified for parameter estimation from a qubit state ρ_ξ, pure or mixed, especially by means of the Bloch representation, as accomplished in Sect. 11.4. This served us to address the important situation of **phase estimation from a qubit altered by quantum noise**, in various scenarios, especially showing possibilities to exploit entanglement and counter the noise.

The considerations can be extended to estimation of **multiple parameters** (assembled in a vector $\vec{\xi}$) characterizing a quantum state $\rho_{\vec{\xi}} \in \mathcal{L}(\mathcal{H}_N)$, as examined in Sect. 11.5, leading especially to the quantum Fisher information matrix $F_q(\vec{\xi}) = \left[F_{jk}(\vec{\xi})\right]$ of Sect. 11.5.1.

An important instance of multiple-parameter estimation is **quantum tomography**, for states or processes. Quantum tomography consists in estimating, through appropriate (repeated) quantum measurements, a whole set of parameters characterizing a quantum state or a quantum process. The parameterization can be linear through their coordinates in an appropriates basis, or otherwise through another (nonlinear) parameterization. Various approaches can be exploited for quantum state or process tomographies, as discussed in Sect. 11.6. In particular, quantum process tomography can be cast into a quantum state tomography through estimation of the Choi–Jamiołkowski state.

For quantum estimation, beyond deterministic parameters, random parameters can be handled as well, following a **Bayesian estimation** framework presented in Sect. 11.7, and involving a prior probability distribution $p_\xi(\xi)$ for the parameter ξ characterizing the quantum state ρ_ξ to be measured for estimation.

Chapter 12
Quantum Fourier Transform

The Fourier transform is a cardinal tool of signal processing, classical, and quantum as well. Classical and quantum Fourier transforms are closely related, but each also shows specificities of its own. Quantum theory in finite-dimensional Hilbert spaces, which is our main focus in this book, naturally connects with the *discrete* Fourier transform. Accordingly, in this chapter, we will first briefly review the classical (discrete) Fourier transform, especially when seen as a change of coordinates (arguably a common view in classical signal processing), but also as a change of basis (arguably more akin to quantum signal processing). We will then present in more detail the quantum Fourier transform, followed by some of its prominent applications for quantum signal processing.

For classical signal processing, the Fourier transform is often exploited with an orientation of amplitude processing, as for instance with linear filtering to selectively amplify or attenuate the spectral components of a signal according to their frequencies. For quantum signal processing, the phase or relative phases in quantum states or signals play an essential role in determining their properties of coherence, interference and interaction, and therefrom their capabilities for information processing and computation. Accordingly, the quantum Fourier transform finds prominent application for phase processing, in turn broadly useful to quantum sensing and metrology and to quantum computation, as we shall see.

12.1 Classical (Discrete) Fourier Transform

12.1.1 Change of Basis

The set of discrete-time signals under the form of a sequence of N scalar values numbered by an integer temporal index $t = 0$ to $N - 1$, can receive the structure of an N-dimensional (complex) vector space by introducing a natural orthonormal basis, called the temporal basis, defined by the set of N discrete-time Dirac signals

F. Chapeau-Blondeau, *Quantum Information and Quantum Signal Processing*,
https://doi.org/10.1007/978-3-032-27499-1_12

$$b_\tau(t) = \delta_{\tau t}\ , \quad \text{for } \tau \in \{0, 1, \ldots, N-1\}\ , \tag{12.1}$$

with $\delta_{\tau t}$ the Kronecker delta or discrete-time Dirac impulse. Each basis signal $b_\tau(t)$ from Eq. (12.1) is a temporal sequence of N scalar values at integer times $t = 0$ to $N - 1$, with $b_\tau(t) = 1$ for $t = \tau$ and $b_\tau(t) = 0$ otherwise; and each basis signal $b_\tau(t)$ is labeled by the integer $\tau \in \{0, 1, \ldots, N-1\}$. Two basis signals from Eq. (12.1) are orthogonal according to the scalar inner product $\langle b_{\tau'}(t), b_\tau(t)\rangle = \sum_{t=0}^{N-1} b_{\tau'}^*(t) b_\tau(t) = \delta_{\tau\tau'}$.

For this vector space of signals, there also exists another orthonormal basis, called the frequency basis or Fourier basis, defined by the set of N discrete-time harmonic signals

$$\widetilde{b}_\nu(t) = \frac{1}{\sqrt{N}} \exp\left(i2\pi \frac{\nu t}{N}\right), \quad \text{for } \nu \in \{0, 1, \ldots, N-1\}\ . \tag{12.2}$$

In a similar way, each basis signal $\widetilde{b}_\nu(t)$ from Eq. (12.2) is a temporal sequence of N scalar values at integer times $t = 0$ to $N - 1$, and it is labeled by an integer $\nu \in \{0, 1, \ldots, N-1\}$. Two basis signals from Eq. (12.2) are orthogonal according to the inner product $\langle \widetilde{b}_{\nu'}(t), \widetilde{b}_\nu(t)\rangle = \sum_{t=0}^{N-1} \widetilde{b}_{\nu'}^*(t) \widetilde{b}_\nu(t) = \delta_{\nu\nu'}$, so that the N vectors $\{\widetilde{b}_\nu(t)\}_{\nu=0}^{N-1}$ indeed form an orthonormal basis for the N-dimensional vector space of discrete-time signals of length N.

The temporal basis $\{b_\tau(t)\}_{\tau=0}^{N-1}$ and the frequency basis $\{\widetilde{b}_\nu(t)\}_{\nu=0}^{N-1}$, as two orthonormal bases of the same N-dimensional vector space, can be related by a unitary transformation, expressible in matrix form as

$$\begin{matrix}\uparrow\\ \nu\\ \downarrow\end{matrix} \begin{bmatrix} \vdots \\ \widetilde{b}_\nu(t) \\ \vdots \end{bmatrix} = \mathsf{U_F} \begin{bmatrix} \vdots \\ b_\tau(t) \\ \vdots \end{bmatrix} \begin{matrix}\uparrow\\ \tau\\ \downarrow\end{matrix}\ , \tag{12.3}$$

or $\left[\widetilde{b}_\nu(t)\right] = \mathsf{U_F}\left[b_\tau(t)\right]$, where $\left[b_\tau(t)\right]$ is a column vector of N entries $b_\tau(t)$ labeled by the integer $\tau \in \{0, 1, \ldots, N-1\}$, and similarly for the column vector $\left[\widetilde{b}_\nu(t)\right]$ labeled by $\nu \in \{0, 1, \ldots, N-1\}$, at each time t. Meanwhile in Eq. (12.3), operates $\mathsf{U_F}$ which is an $N \times N$ symmetric unitary matrix, having rows labeled by ν and columns by τ, defined as

$$\mathsf{U_F} = \frac{1}{\sqrt{N}} \begin{bmatrix} \cdots & \cdots & \cdots \\ \vdots & \vdots & \vdots \\ \cdots & \exp\left(i2\pi \frac{\nu\tau}{N}\right) & \cdots \\ \vdots & \vdots & \vdots \\ \cdots & \cdots & \cdots \end{bmatrix} \begin{matrix}\uparrow\\ \nu\\ \downarrow\end{matrix}\ . \tag{12.4}$$

The matrix relationship of Eq. (12.3) is equivalent to

$$\widetilde{b}_\nu(t) = \frac{1}{\sqrt{N}} \sum_{\tau=0}^{N-1} \exp\left(i2\pi \frac{\nu\tau}{N}\right) b_\tau(t) \,, \quad \text{for } \nu \in \{0, 1, \ldots, N-1\} \,, \tag{12.5}$$

which in turn is equivalent to Eq. (12.2), since $b_\tau(t) = \delta_{\tau t}$ by Eq. (12.1).

The N columns of the matrix $\mathsf{U_F}$ from Eq. (12.4) form N column vectors

$$\frac{1}{\sqrt{N}}\left[1, \exp\left(i2\pi\frac{1}{N}\tau\right), \exp\left(i2\pi\frac{2}{N}\tau\right), \ldots, \exp\left(i2\pi\frac{\nu}{N}\tau\right), \ldots, \exp\left(i2\pi\frac{N-1}{N}\tau\right)\right]^\top \tag{12.6}$$

labeled by $\tau \in \{0, 1, \cdots, N-1\}$ which are orthonormal vectors, with inner products $\delta_{\tau\tau'}$, which is the same condition as $\langle b_{\tau'}(t), b_\tau(t)\rangle = \delta_{\tau\tau'}$ mentioned previously. This orthonormality alone ensures that $\mathsf{U_F}$ is a unitary matrix. A similar orthonormality condition holds for the N rows of the matrix $\mathsf{U_F}$ from Eq. (12.4), labeled by $\nu \in \{0, 1, \ldots, N-1\}$, which alone equivalently ensures that $\mathsf{U_F}$ is a unitary matrix.

Alternatively, the general term of the unitary matrix $\mathsf{U_F}$ from Eq. (12.4) can be written as $N^{-1/2} z^{\nu\tau \ (\mathrm{mod}\, N)}$ with $z = \exp(i2\pi/N)$ the N-th complex root of unity, to give

$$\mathsf{U_F} = \frac{1}{\sqrt{N}} \begin{bmatrix} 1 & 1 & 1 & 1 & \cdots & 1 \\ 1 & z & z^2 & z^3 & \cdots & z^{N-1} \\ 1 & z^2 & z^4 & z^6 & \cdots & \cdots \\ \vdots & & & \vdots & & \vdots \\ 1 & \cdots & & z^{\nu\tau \ (\mathrm{mod}\, N)} & & \cdots \\ \vdots & & & \vdots & & \vdots \\ 1 & z^{N-1} & & \cdots & & \cdots \end{bmatrix} \begin{matrix} \uparrow \\ \nu \\ \downarrow \end{matrix} . \tag{12.7}$$

The introduction of two sets of discrete-time signals, $\{b_\tau(t)\}_{\tau=0}^{N-1}$ in Eq. (12.1) and $\{\widetilde{b}_\nu(t)\}_{\nu=0}^{N-1}$ in Eq. (12.2), is a natural approach in classical digital signal processing for defining two useful orthonormal bases, the temporal basis and the frequency or Fourier basis. Beyond, the matrix relationship of Eq. (12.3), although it may seem a bit redundant once Eqs. (12.1) and (12.2) have been introduced, is nevertheless useful as it explicitly exhibits the unitary transformation $\mathsf{U_F}$ connecting the two bases, which will be central to the extension of the Fourier transform to the quantum domain.

The matrix $\mathsf{U_F}$ introduced in Eq. (12.3) and specified in Eq. (12.4) or (12.7) defines a unitary transformation, or Fourier transformation, to pass from the temporal basis to the frequency or Fourier basis. The change of basis $\left[\widetilde{b}_\nu(t)\right] = \mathsf{U_F}\left[b_\tau(t)\right]$ from Eq. (12.3) is invertible as $\left[b_\tau(t)\right] = \mathsf{U_F^\dagger}\left[\widetilde{b}_\nu(t)\right]$, with the unitary matrix $\mathsf{U_F}$ of Eq. (12.4) satisfying $\mathsf{U_F^{-1}} = \mathsf{U_F^\dagger}$. Equivalently, the inverse transformation reversing Eq. (12.5) reads

$$b_\tau(t) = \frac{1}{\sqrt{N}} \sum_{\nu=0}^{N-1} \exp\left(-i2\pi \frac{\nu\tau}{N}\right) \widetilde{b}_\nu(t) \,, \quad \text{for } \tau \in \{0, 1, \ldots, N-1\} \,. \tag{12.8}$$

The frequency basis $\{\widetilde{b}_\nu(t)\}_{\nu=0}^{N-1}$ represents the Fourier transform of the temporal basis $\{b_\tau(t)\}_{\tau=0}^{N-1}$, as defined by Eq. (12.5) or equivalently by the unitary transformation U_F in Eq. (12.3). Conversely, the temporal basis $\{b_\tau(t)\}_{\tau=0}^{N-1}$ represents the inverse Fourier transform of the frequency basis $\{\widetilde{b}_\nu(t)\}_{\nu=0}^{N-1}$, as defined by Eq. (12.8) or equivalently by the adjoint unitary transformation $\mathsf{U}_\mathsf{F}^\dagger$ deduced from Eq. (12.4). Its action to transform an orthonormal basis via U_F completely defines the Fourier transform, which can however receive an alternative presentation as a change of coordinates.

12.1.2 Change of Coordinates

An arbitrary discrete-time signal $x(t)$ with integer $t = 0$ to $N-1$, can be expressed in the temporal basis $\{b_\tau(t)\}_{\tau=0}^{N-1}$ as the linear superposition

$$x(t) = \sum_{\tau=0}^{N-1} x_\tau b_\tau(t) , \tag{12.9}$$

with the coordinate x_τ of $x(t)$ on $b_\tau(t)$ given by the inner product

$$x_\tau = \langle b_\tau(t), x(t)\rangle = \sum_{t=0}^{N-1} x(t) b_\tau^*(t) = \sum_{t=0}^{N-1} x(t)\delta_{\tau t} = x(\tau) . \tag{12.10}$$

This same signal $x(t)$ can also be expressed in the frequency basis $\{\widetilde{b}_\nu(t)\}_{\nu=0}^{N-1}$ as the linear superposition

$$x(t) = \sum_{\nu=0}^{N-1} \widetilde{x}_\nu \widetilde{b}_\nu(t) , \tag{12.11}$$

with the coordinate $\widetilde{x}_\nu$ of $x(t)$ on $\widetilde{b}_\nu(t)$ given by the inner product

$$\widetilde{x}_\nu = \langle \widetilde{b}_\nu(t), x(t)\rangle = \sum_{t=0}^{N-1} x(t)\widetilde{b}_\nu^*(t) = \frac{1}{\sqrt{N}} \sum_{t=0}^{N-1} x(t) \exp\left(-i2\pi \frac{\nu t}{N}\right) . \tag{12.12}$$

The change from the time coordinate x_τ of Eq. (12.10) to the frequency coordinate $\widetilde{x}_\nu$ of Eq. (12.12) can be formally derived by replacing in Eq. (12.9) each vector $b_\tau(t)$ of the time basis by its expression in terms of the vectors $\widetilde{b}_\nu(t)$ of the frequency basis as given by Eq. (12.8). Then by permuting the sums in τ and in ν one identifies the weight of each basis vector $\widetilde{b}_\nu(t)$ which according to Eq. (12.11) defines the frequency coordinates as a function of the time coordinates by

$$\widetilde{x}_\nu = \frac{1}{\sqrt{N}} \sum_{\tau=0}^{N-1} x_\tau \exp\left(-i2\pi \frac{\nu\tau}{N}\right) = \mathrm{FT}\big[x_\tau\big], \quad \text{for } \nu \in \{0, 1, \dots, N-1\}, \tag{12.13}$$

or alternatively in matrix form $\big[\widetilde{x}_\nu\big] = \mathsf{U}_\mathsf{F}^\dagger \big[x_\tau\big]$. This is a rather formal derivation, since Eq. (12.13) can be found similar to Eq. (12.12) if one is willing to confuse the indices t and τ, that otherwise have different statuses. Equation (12.13) defines the Fourier transform $\mathrm{FT}\big[x_\tau\big]$ of the sequence of N time coordinates x_τ defining the discrete-time signal in the temporal basis $\{b_\tau(t)\}_{\tau=0}^{N-1}$.

A similar process from Eq. (12.11) via Eq. (12.5) leads to the reverse change of coordinate

$$x_\tau = \frac{1}{\sqrt{N}} \sum_{\nu=0}^{N-1} \widetilde{x}_\nu \exp\left(i2\pi \frac{\nu\tau}{N}\right) = \mathrm{FT}^{-1}\big[\widetilde{x}_\nu\big], \quad \text{for } \tau \in \{0, 1, \dots, N-1\}, \tag{12.14}$$

or alternatively in matrix form $\big[x_\tau\big] = \mathsf{U}_\mathsf{F} \big[\widetilde{x}_\nu\big]$, defining the inverse Fourier transform $\mathrm{FT}^{-1}\big[\widetilde{x}_\nu\big]$ of the sequence of N frequency coordinates $\widetilde{x}_\nu$.

This is a well-known result in linear algebra: from the temporal representation to the frequency representation, when the change of basis $\big[\widetilde{b}_\nu(t)\big] = \mathsf{U}_\mathsf{F} \big[b_\tau(t)\big]$ is governed by the unitary matrix U_F, the change of coordinates $\big[\widetilde{x}_\nu\big] = \mathsf{U}_\mathsf{F}^\dagger \big[x_\tau\big]$ is governed by the adjoint matrix $\mathsf{U}_\mathsf{F}^\dagger$. The frequency coordinates $\{\widetilde{x}_\nu\}_{\nu=0}^{N-1}$ in Eq. (12.13) are considered as the Fourier transform of the temporal coordinates $\{x_\tau\}_{\tau=0}^{N-1}$, while the temporal coordinates $\{x_\tau\}_{\tau=0}^{N-1}$ in Eq. (12.14) are considered as the inverse Fourier transform of the frequency coordinates $\{\widetilde{x}_\nu\}_{\nu=0}^{N-1}$, as is common in digital signal processing. Very crudely, the change of coordinates involves a "$-i$" in the Fourier transformation of Eq. (12.13), while the change of basis involves a "$+i$" in the Fourier transformation of Eq. (12.5). Arguably, for the Fourier transform, the change of coordinates with "$-i$" is more common to classical signal processing, while the change of basis with "$+i$" is more common to quantum signal processing.

Practically, the computation of the Fourier transform, according to Eq. (12.13) for instance, for each discrete frequency ν requires summing in time τ over N values, and this for each of the N frequencies ν. This amounts to a total number of operations or algorithmic complexity of order $O\big(N^2\big)$ for a direct computation of the Fourier transform according to Eq. (12.13), for arbitrary size N. There however exists a more efficient alternative, exploiting the factorization properties of the exponential function, when the size N is a power of two $N = 2^L$ and the fast Fourier transform algorithm [370] takes a smaller number of operations of order $O\big(N \log N\big) \equiv O\big(L2^L\big)$ to compute the Fourier transform. Still, to anticipate, we will see shortly that quantum circuits, in the style of those presented in Chap. 5, by parallelism via quantum superposition, afford yet a smaller algorithmic complexity $O\big((\log N)^2\big) \equiv O\big(L^2\big)$ for computing the quantum Fourier transform.

To summarize these classical references of signal processing, in a nutshell, the Fourier-transformed basis is $\big[\widetilde{b}_\nu(t)\big] = \mathsf{U}_\mathsf{F} \big[b_\tau(t)\big]$, while the Fourier-transformed coordinate is $\big[\widetilde{x}_\nu\big] = \mathsf{U}_\mathsf{F}^\dagger \big[x_\tau\big]$. The unitary transformation U_F defined in Eq. (12.4) will be the central notion to carry over the Fourier transform to the quantum domain, as we now consider.

12.2 Quantum Fourier Transform

12.2.1 Change of Basis

A quantum system has its state vector belonging to an N-dimensional Hilbert space $\mathcal{H}_N$ referred to the computational orthonormal basis formed by the set of N vectors $|j\rangle \in \{|0\rangle, |1\rangle, \dots, |N-1\rangle\}$. The **quantum Fourier transform** can be defined as the unitary operator U_F of $\mathcal{L}(\mathcal{H}_N)$ specified by the transformation of the N vectors of the computational basis under the form

$$\boxed{|j\rangle \xmapsto{\;\mathsf{U}_\mathsf{F}\;} |\widetilde{j}\rangle = \frac{1}{\sqrt{N}} \sum_{k=0}^{N-1} \exp\left(i2\pi \frac{jk}{N}\right) |k\rangle ,} \tag{12.15}$$

when $j = 0$ to $N-1$. The unitary transformation of Eq. (12.15) delivers a set of N vectors $\{|\widetilde{j}\rangle\} = \{|\widetilde{0}\rangle, |\widetilde{1}\rangle, \dots, |\widetilde{N-1}\rangle\}$ obtained as $|\widetilde{j}\rangle = \mathsf{U}_\mathsf{F}|j\rangle$. The N vectors $|\widetilde{j}\rangle$ being with unit norm and mutually orthogonal establish U_F as unitary. The unitary operator U_F of Eq. (12.15) is represented in the computational basis of $\mathcal{H}_N$ by the $N \times N$ symmetric unitary matrix

$$\mathsf{U}_\mathsf{F} = \frac{1}{\sqrt{N}} \begin{bmatrix} \cdots & \cdots & \cdots \\ \vdots & \vdots & \vdots \\ \cdots & \exp\left(i2\pi \dfrac{jk}{N}\right) & \cdots \\ \vdots & \vdots & \vdots \\ \cdots & \cdots & \cdots \end{bmatrix} \begin{matrix} \\ \uparrow \\ k \\ \downarrow \\ \end{matrix} , \tag{12.16}$$

similar to the change-of-basis unitary matrix of Eq. (12.4). The unitary character of the representation matrix U_F in Eqs. (12.4) and (12.16) ensures that the operator U_F defined in Eq. (12.15) is indeed a unitary operator of $\mathcal{L}(\mathcal{H}_N)$. Alternatively, the general term of the unitary matrix U_F can be written as $N^{-1/2} z^{jk \pmod N}$ with $z = \exp(i2\pi/N)$ the N-th complex root of unity, to give

$$\mathsf{U_F} = \frac{1}{\sqrt{N}} \begin{bmatrix} 1 & 1 & 1 & 1 & \cdots & 1 \\ 1 & z & z^2 & z^3 & \cdots & z^{N-1} \\ 1 & z^2 & z^4 & z^6 & \cdots & \cdots \\ \vdots & & & \vdots & & \vdots \\ 1 & \cdots & & z^{jk \,(\mathrm{mod}\, N)} & & \cdots \\ \vdots & & & \vdots & & \vdots \\ 1 & z^{N-1} & & \cdots & & \cdots \end{bmatrix} \begin{matrix} \uparrow \\ k \\ \downarrow \end{matrix} , \tag{12.17}$$

similar to the matrix of Eq. (12.7). The set of the N vectors $\{|\widetilde{\jmath}\rangle\} = \{|\widetilde{0}\rangle, |\widetilde{1}\rangle, \ldots, |\widetilde{N-1}\rangle\}$ resulting from the unitary transformation of Eq. (12.15), when $j = 0$ to $N-1$, forms another orthonormal basis of $\mathcal{H}_N$, and $\{|\widetilde{\jmath}\rangle\}$ represents the Fourier basis or Fourier transform of the computational basis $\{|j\rangle\} = \{|0\rangle, |1\rangle, \ldots, |N-1\rangle\}$.

For example, when $N = 3$, the original computational basis is $\{|0\rangle, |1\rangle, |2\rangle\}$ with vectors $|0\rangle = [1, 0, 0]^\top$, $|1\rangle = [0, 1, 0]^\top$ and $|2\rangle = [0, 0, 1]^\top$. The corresponding Fourier basis is $\{|\widetilde{0}\rangle, |\widetilde{1}\rangle, |\widetilde{2}\rangle\}$ with vectors

$$|\widetilde{0}\rangle = \frac{1}{\sqrt{3}} \begin{bmatrix} 1 \\ 1 \\ 1 \end{bmatrix} \begin{matrix} \uparrow \\ k \\ \downarrow \end{matrix} , \quad |\widetilde{1}\rangle = \frac{1}{\sqrt{3}} \begin{bmatrix} 1 \\ \exp(i2\pi/3) \\ \exp(i4\pi/3) \end{bmatrix} \begin{matrix} \uparrow \\ k \\ \downarrow \end{matrix} , \quad |\widetilde{2}\rangle = \frac{1}{\sqrt{3}} \begin{bmatrix} 1 \\ \exp(i4\pi/3) \\ \exp(i8\pi/3) \end{bmatrix} \begin{matrix} \uparrow \\ k \\ \downarrow \end{matrix} . \tag{12.18}$$

Generally, the reverse change of basis inverting the transformation of Eq. (12.15) and transforming the Fourier basis $\{|\widetilde{\jmath}\rangle\} = \{|\widetilde{0}\rangle, |\widetilde{1}\rangle, \ldots, |\widetilde{N-1}\rangle\}$ back into the original computational basis $\{|j\rangle\} = \{|0\rangle, |1\rangle, \ldots, |N-1\rangle\}$, operates as

$$|\widetilde{\jmath}\rangle \xmapsto{\ \mathsf{U_F^\dagger}\ } |j\rangle = \frac{1}{\sqrt{N}} \sum_{k=0}^{N-1} \exp\left(-i2\pi \frac{jk}{N}\right) |\widetilde{k}\rangle , \tag{12.19}$$

or equivalently $|j\rangle = \mathsf{U_F^\dagger} |\widetilde{\jmath}\rangle$, defining the inverse quantum Fourier transform as the unitary operator $\mathsf{U_F^{-1}} = \mathsf{U_F^\dagger}$.

Any state vector $|\psi\rangle$ expressed in the computational basis as

$$|\psi\rangle = \sum_{j=0}^{N\ 1} x_j \, |j\rangle , \tag{12.20}$$

can be assigned a Fourier transform $\mathsf{U_F} |\psi\rangle$ which, by linearity and according to Eq. (12.15) follows as

$$\mathsf{U}_\mathsf{F} |\psi\rangle = \sum_{j=0}^{N-1} x_j \mathsf{U}_\mathsf{F} |j\rangle , \tag{12.21}$$

$$= \sum_{j=0}^{N-1} x_j \frac{1}{\sqrt{N}} \sum_{k=0}^{N-1} \exp\left(i2\pi \frac{jk}{N} \right) |k\rangle , \tag{12.22}$$

$$= \sum_{k=0}^{N-1} y_k |k\rangle , \tag{12.23}$$

with the scalar coordinate

$$y_k = \frac{1}{\sqrt{N}} \sum_{j=0}^{N-1} x_j \exp\left(i2\pi \frac{jk}{N} \right) , \tag{12.24}$$

also expressible in matrix form with the column vector $\big[y_k\big] = \mathsf{U}_\mathsf{F}\big[x_j\big]$, or $\big[y_k\big] = \mathrm{FT}^{-1}\big[x_j\big]$ with the notion of Fourier transform from Eqs. (12.13)–(12.14) for a scalar sequence. Here, this is not quite the notion of change of coordinates as in Sect. 12.1.2, because the column vectors of coordinates $\big[x_j\big]$ and $\big[y_k\big]$ belong to two distinct state vectors $|\psi\rangle$ and $\mathsf{U}_\mathsf{F} |\psi\rangle$ of $\mathcal{H}_N$.

The initial state vector $|\psi\rangle$ of Eq. (12.20), when measured in the computational basis $\{|j\rangle\}_{j=0}^{N-1}$, has the probability $|x_j|^2$ of being found in state $|j\rangle$. After the Fourier transform of Eqs. (12.21)–(12.23), the transformed state vector $\mathsf{U}_\mathsf{F} |\psi\rangle$, when measured in the same computational basis $\{|k\rangle\}_{k=0}^{N-1}$, has the probability $|y_k|^2$ of being found in state $|k\rangle$. Each coordinate y_k of the Fourier-transformed state $\mathsf{U}_\mathsf{F} |\psi\rangle$ is a superposition of all coordinates x_j of the initial state $|\psi\rangle$, weighted by special phase patterns conveyed by the Fourier factors $\exp\big(i2\pi jk/N\big)$ in Eq. (12.24), to act in a summation interpretable as accomplishing some phase interferences, and altogether ensuring reversibility in the representation. The phases or relative phases in the coordinates of quantum states play an important role in determining their properties, their behavior upon measurement, and accordingly their capabilities for information processing and computation. As a consequence, such (phase) operation realized by obtaining the Fourier-transformed state $\mathsf{U}_\mathsf{F} |\psi\rangle$, is often a useful step when processing (with) quantum states. This is essentially what is accomplished in the application to be described shortly in Sect. 12.3 of the Fourier transform for quantum phase estimation. And similarly in the processing described in Sect. 5.4 of the Deutsch–Jozsa algorithm (with rather a Hadamard instead of a Fourier transform), where the final quantum measurement on the transformed state is key to the resolution.

12.2.2 Change of Coordinates

The notion of change of coordinates rather applies to a same state vector $|\psi\rangle = \sum_{j=0}^{N-1} x_j\,|j\rangle = \sum_{k=0}^{N-1} \widetilde{x}_k\,|\widetilde{k}\,\rangle$ having coordinates $[x_j]$ in the original computational basis $\{|j\rangle\}_{j=0}^{N-1}$ of $\mathcal{H}_N$ and coordinates $[\widetilde{x}_k]$ in the Fourier basis $\{|\widetilde{k}\,\rangle\}_{k=0}^{N-1}$ of $\mathcal{H}_N$. From the change of basis of Eq. (12.19) one obtains

$$|\psi\rangle = \sum_{j=0}^{N-1} x_j\,|j\rangle\;, \tag{12.25}$$

$$= \sum_{j=0}^{N-1} x_j \frac{1}{\sqrt{N}} \sum_{k=0}^{N-1} \exp\!\left(-i2\pi\frac{jk}{N}\right)|\widetilde{k}\rangle\;, \tag{12.26}$$

$$= \sum_{k=0}^{N-1} \widetilde{x}_k\,|\widetilde{k}\rangle\;, \tag{12.27}$$

with the scalar coordinate

$$\widetilde{x}_k = \frac{1}{\sqrt{N}} \sum_{j=0}^{N-1} x_j \exp\!\left(-i2\pi\frac{jk}{N}\right), \tag{12.28}$$

also expressible in matrix form as $[\widetilde{x}_k] = \mathsf{U}_\mathsf{F}^\dagger[x_j]$, or $[\widetilde{x}_k] = \mathrm{FT}[x_j]$ with the notion of Fourier transform of Eqs. (12.13)–(12.14) for a scalar sequence. This matches, for a quantum state vector here, the notion of change of coordinates between the initial and Fourier bases as in Eq. (12.13).

12.2.3 Product Form

For the quantum Fourier transform in Eq. (12.15) of the computational basis of $\mathcal{H}_N$, there exists a useful alternative product form stemming from the convenient identities, for any scalar α,

$$\sum_{k=0}^{2^L-1} \exp\!\left(\alpha\frac{k}{2^L}\right)|k\rangle = \bigotimes_{\ell=1}^{L}\Big[|0\rangle + \exp(\alpha 2^{-\ell})\,|1\rangle\Big] \tag{12.29}$$

$$= \bigotimes_{\ell=1}^{L}\left[|0\rangle + \exp\!\left(\alpha\frac{2^{L-\ell}}{2^L}\right)|1\rangle\right], \tag{12.30}$$

accommodating for a possible binary representation for the integer k if convenient. This is easily established by expanding the L-fold tensor product in Eq. (12.30). For example, for $L = 3$ one has

$$\bigotimes_{\ell=1}^{3}\left[\,|0\rangle + \exp\left(\alpha\frac{2^{3-\ell}}{2^3}\right)|1\rangle\right]$$

$$= \left[\,|0\rangle + \exp\left(\alpha\frac{2^2}{2^3}\right)|1\rangle\right]\otimes\left[\,|0\rangle + \exp\left(\alpha\frac{2^1}{2^3}\right)|1\rangle\right]\otimes\left[\,|0\rangle + \exp\left(\alpha\frac{2^0}{2^3}\right)|1\rangle\right] \tag{12.31}$$

$$\begin{aligned}= &\ |000\rangle + \exp\left(\alpha\frac{2^0}{2^3}\right)|001\rangle + \exp\left(\alpha\frac{2^1}{2^3}\right)|010\rangle + \\ &\exp\left(\alpha\frac{2^1+2^0}{2^3}\right)|011\rangle + \exp\left(\alpha\frac{2^2}{2^3}\right)|100\rangle + \exp\left(\alpha\frac{2^2+2^0}{2^3}\right)|101\rangle + \\ &\exp\left(\alpha\frac{2^2+2^1}{2^3}\right)|110\rangle + \exp\left(\alpha\frac{2^2+2^1+2^0}{2^3}\right)|111\rangle \end{aligned} \tag{12.32}$$

$$\begin{aligned}= &\ |000\rangle + \exp(\alpha/2^3)\,|001\rangle + \exp(\alpha 2/2^3)\,|010\rangle + \\ &\exp(\alpha 3/2^3)\,|011\rangle + \exp(\alpha 4/2^3)\,|100\rangle + \exp(\alpha 5/2^3)\,|101\rangle + \\ &\exp(\alpha 6/2^3)\,|110\rangle + \exp(\alpha 7/2^3)\,|111\rangle \end{aligned} \tag{12.33}$$

$$= \sum_{k=0}^{7}\exp\left(\alpha\frac{k}{2^3}\right)|k\rangle\ . \tag{12.34}$$

It results that when realizing N-dimensional quantum states like $|k\rangle$ of $\mathcal{H}_N$ by assembling qubits, when $N = 2^L$ for L-qubit quantum states, the Fourier transform of Eq. (12.15) receives the equivalent product form

$$|j\rangle \longmapsto |\widetilde{j}\rangle = \frac{1}{2^{L/2}}\bigotimes_{\ell=1}^{L}\left[\,|0\rangle + \exp(i2\pi j 2^{-\ell})\,|1\rangle\right] = \frac{1}{2^{L/2}}\bigotimes_{\ell=1}^{L}\left[\,|0\rangle + \exp\left(i2\pi j\frac{2^{L-\ell}}{2^L}\right)|1\rangle\right], \tag{12.35}$$

which follows from the identities of Eqs. (12.29)–(12.30) when $\alpha = i2\pi j$. This product form for the Fourier transform of Eq. (12.15) indicates in particular that the L qubits in state $|\widetilde{j}\rangle$ of Eq. (12.35) are unentangled.

Additional relations exist, that are specially useful for circuit implementation with elementary quantum gates of the quantum Fourier transform. With the binary representation $j = j_1 j_2 \cdots j_L$, or more formally $j = j_1 2^{L-1} + j_2 2^{L-2} + \cdots + j_L 2^0$, one has $j/2^L = 0.j_1 j_2 \cdots j_L$, or more formally $j/2^L = j_1 2^{-1} + j_2 2^{-2} + \cdots + j_L 2^{-L}$. Also $j2^{-\ell} \equiv j_{L-\ell+1}2^{-1} + j_{L-\ell+2}2^{-2} + \cdots + j_L 2^{-\ell}$ (mod 1), the congruence mod-

ulo 1 manifesting equal values resulting from the two arguments acting in the exponential $\exp(i2\pi j2^{-\ell})$. This gives
$\ell = 1 \Longrightarrow j2^{-\ell} \equiv j_L 2^{-1} = 0.j_L \pmod 1$,
$\ell = 2 \Longrightarrow j2^{-\ell} \equiv j_{L-1}2^{-1} + j_L 2^{-2} = 0.j_{L-1}j_L \pmod 1$,
$\ell = 3 \Longrightarrow j2^{-\ell} \equiv j_{L-2}2^{-1} + j_{L-1}2^{-2} + j_L 2^{-3} = 0.j_{L-2}j_{L-1}j_L \pmod 1$,
$\ell = 4 \Longrightarrow j2^{-\ell} \equiv j_{L-3}2^{-1} + j_{L-2}2^{-2} + j_{L-1}2^{-3} + j_L 2^{-4} = 0.j_{L-3}j_{L-2}j_{L-1}j_L$ (mod 1),
...
$\ell = L \Longrightarrow j2^{-\ell} \equiv j_1 2^{-1} + j_2 2^{-2} + j_3 2^{-3} + \cdots + j_L 2^{-L} = 0.j_1 j_2 \cdots j_L \pmod 1$.

In this way, when j_L is the least significant bit of j, the L-fold tensor product in Eq. (12.35) can then be written as (omitting the prefactor $1/2^{L/2}$)

$$\Big[|0\rangle + |1\rangle \exp(i2\pi 0.j_L)\Big] \otimes \Big[|0\rangle + |1\rangle \exp(i2\pi 0.j_{L-1}j_L)\Big] \otimes \cdots \otimes \Big[|0\rangle + |1\rangle \exp(i2\pi 0.j_1 j_2 \cdots j_L)\Big] . \tag{12.36}$$

Alternatively, the indices can be made to run in the other direction. With the binary representation $j = j_{L-1} \cdots j_1 j_0$, or more formally $j = j_{L-1}2^{L-1} + \cdots + j_1 2^1 + j_0 2^0$, one has $j/2^L = 0.j_{L-1} \cdots j_1 j_0$, or more formally $j/2^L = j_{L-1}2^{-1} + \cdots + j_1 2^{-L+1} + j_0 2^{-L}$.
Also $j2^{-\ell} \equiv j_{\ell-1}2^{-1} + j_{\ell-2}2^{-2} + \cdots + j_0 2^{-\ell} \pmod 1$, giving
$\ell = 1 \Longrightarrow j2^{-\ell} \equiv j_0 2^{-1} = 0.j_0 \pmod 1$,
$\ell = 2 \Longrightarrow j2^{-\ell} \equiv j_1 2^{-1} + j_0 2^{-2} = 0.j_1 j_0 \pmod 1$,
$\ell = 3 \Longrightarrow j2^{-\ell} \equiv j_2 2^{-1} + j_1 2^{-2} + j_0 2^{-3} = 0.j_2 j_1 j_0 \pmod 1$,
$\ell = 4 \Longrightarrow j2^{-\ell} \equiv j_3 2^{-1} + j_2 2^{-2} + j_1 2^{-3} + j_0 2^{-4} = 0.j_3 j_2 j_1 j_0 \pmod 1$,
...
$\ell = L \Longrightarrow j2^{-\ell} \equiv j_{L-1}2^{-1} + j_{L-2}2^{-2} + j_{L-3}2^{-3} + \cdots + j_0 2^{-L} = 0.j_{L-1} \cdots j_1 j_0 \pmod 1$.

In this way, when j_0 is the least significant bit of j, the L-fold tensor product in Eq. (12.35) can then be written as (omitting the prefactor $1/2^{L/2}$)

$$\Big[|0\rangle + |1\rangle \exp(i2\pi 0.j_0)\Big] \otimes \Big[|0\rangle + |1\rangle \exp(i2\pi 0.j_1 j_0)\Big] \otimes \cdots \otimes \Big[|0\rangle + |1\rangle \exp(i2\pi 0.j_{L-1} \cdots j_1 j_0)\Big] . \tag{12.37}$$

12.2.4 Quantum Fourier Circuit

The product form of Eq. (12.35) holds the key to an efficient circuit for computing the quantum Fourier transform. For this quantum circuit, an important gate will be the qubit phase gate similar to Eq. (3.31), at various phases 2^{-k}, under the form

$$\mathsf{U}_{2^{-k}} \equiv \mathsf{R}_k = \begin{bmatrix} 1 & 0 \\ 0 & \exp(i2\pi 2^{-k}) \end{bmatrix} . \tag{12.38}$$

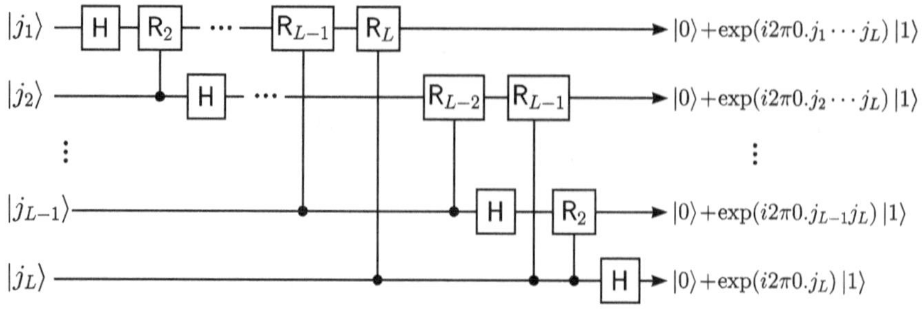

Fig. 12.1 Quantum circuit involving L input qubits so as to compute the quantum Fourier transform by implementing the product form of Eqs. (12.35)–(12.36). In the binary numbering of the basis states of the L qubits, j_1 is the most significant bit while j_L is the least significant. The normalization factor $1/\sqrt{2}$ is omitted in the output states

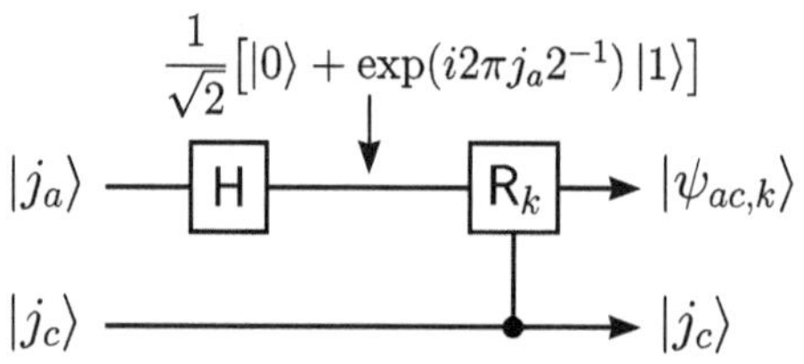

Fig. 12.2 The action on the computational basis when the phase gate R_k from Eq. (12.38) is controlled by the qubit $|j_c\rangle$ with the binary digit $j_c = 0$ or 1, to produce the output state $|\psi_{ac,k}\rangle$ from Eqs. (12.40)–(12.41)

Each qubit gate R_k will be used in a controlled form, following the methodology of Sect. 5.6. The quantum circuit will input L qubits, numbered from $\ell = 1$ to L. According to their binary weights $j_\ell \in \{0, 1\}$, the qubits will experience cascaded phase gates R_k and also serve as control to the phase gates acting on qubits with higher significance in the binary representation of j, in order to implement the products as in Eq. (12.36). A quantum circuit resulting for the quantum Fourier transform is shown in Fig. 12.1, and is also found for instance in [9, 107].

A basic operation in the circuit of Fig. 12.1 is when the phase gate R_k from Eq. (12.38) is controlled by the qubit $|j_c\rangle$ with the binary digit $j_c = 0$ or 1, as depicted in Fig. 12.2.

In Fig. 12.2, from the input $|j_a\rangle$ with $j_a \in \{0, 1\}$, the Hadamard gate H delivers the output state $\big(|0\rangle \pm |1\rangle\big)/\sqrt{2}$ expressible as

$$\mathsf{H}\,|j_a\rangle = \frac{1}{\sqrt{2}}\Big[|0\rangle + \exp\big(i2\pi j_a 2^{-1}\big)\,|1\rangle\Big] = \frac{1}{\sqrt{2}}\Big[|0\rangle + \exp\big(i2\pi 0.j_a\big)\,|1\rangle\Big], \tag{12.39}$$

since $\exp(i2\pi 0.1) = \exp\big(i2\pi 2^{-1}\big) = \exp(i\pi) = -1$. Also in Fig. 12.2, when the state $\big(|0\rangle \pm |1\rangle\big)/\sqrt{2}$ is input to the phase gate R_k, the output is $\big(|0\rangle \pm |1\rangle\big)/\sqrt{2}$ if $j_c = 0$, and it is $\big[|0\rangle \pm \exp\big(i2\pi 2^{-k}\big)\,|1\rangle\big]/\sqrt{2}$ if $j_c = 1$. These two output configurations for the phase gate R_k can be combined in the single expression $\big[|0\rangle \pm \exp\big(i2\pi j_c 2^{-k}\big)\,|1\rangle\big]/\sqrt{2}$ for $j_c = 0$ or 1. Combining with the action of the Hadamard gate H in the form of Eq. (12.39), we obtain in Fig. 12.2 the output state

$$|\psi_{ac,k}\rangle = \frac{1}{\sqrt{2}}\Big[|0\rangle + \exp(i2\pi j_a 2^{-1})\exp(i2\pi j_c 2^{-k})\,|1\rangle\Big] \tag{12.40}$$

$$= \frac{1}{\sqrt{2}}\Big[|0\rangle + \exp\big(i2\pi(j_a 2^{-1} + j_c 2^{-k})\big)\,|1\rangle\Big]. \tag{12.41}$$

By such operation, the bit j_c can be shifted at position k in the fractional part of $0.j$ as it occurs on the circuit output in Fig. 12.1. The Hadamard gates H and phase gates R_k combine their actions in the circuit of Fig. 12.1, to compute the product form of Eq. (12.36) for the quantum Fourier transform when the quantum states are represented on L qubits.

The total number of gates in the quantum circuit of Fig. 12.1, summed from top to bottom across the L horizontal pipelines associated with the L qubits j_ℓ, amounts to $L + (L-1) + \cdots + 2 + 1 = (L+1)L/2$. This represents a number of gates or operations of order $O\big(L^2\big) \equiv O\big((\log N)^2\big)$ to compute the quantum Fourier transform on L qubits representing the quantum states from a Hilbert space with dimension $N = 2^L$. By comparison, as we indicated at the end of Sect. 12.1, a classical Fourier transform on L bits, by the fast Fourier transform algorithm, would take $O\big(L2^L\big) \equiv O\big(N \log N\big)$ operations. An exponential speedup is afforded by the quantum computation of the Fourier transform.

Any joint quantum state on L qubits can be applied on the input in Fig. 12.1 as a linear superposition of the computational-basis states, like the state $|\psi\rangle$ in Eq. (12.20). The corresponding Fourier-transformed state $\mathsf{U}_\mathsf{F}\,|\psi\rangle$ according to Eqs. (12.23)–(12.24) is then computed in $O\big(L^2\big)$ complexity by the circuit of Fig. 12.1 by linearity, benefiting again from a form of parallelism via quantum superposition. While the $N = 2^L$ Fourier coefficients like the quantities y_k in Eqs. (12.23)–(12.24) are present in the superposed state resulting in the circuit output, direct access to these coefficients y_k however remains limited by the bottleneck of quantum measurement. A direct quantum measurement on the Fourier-transformed state $\mathsf{U}_\mathsf{F}\,|\psi\rangle$ at the output in Fig. 12.1, would lead to a random projection onto one basis state $|k\rangle$ of the L qubits, selected at random according to the Born rule, with a probability given by the squared modulus $|y_k|^2$ of the corresponding Fourier coefficient y_k in Eq. (12.23). Some prior quantum processing usefully exploiting the Fourier-transformed state is likely to take place before the final quantum measurement. We now come to such useful processing exploiting the quantum Fourier transform.

12.3 Quantum Fourier Transform for Phase Estimation

A prominent application of the quantum Fourier transform consists in quantum phase estimation, with a specific contribution that usefully supplements the approaches to quantum phase estimation reported in Sect. 11.3, for reasons that we shall discuss in Sect. 12.3.6. In a nutshell, to anticipate, the Fourier-based estimation delivers a global phase estimate via a one-shot almost deterministic result; while the approaches of Sect. 11.3, relying on statistical averaging from repeated measurements, in comparison deliver a local phase estimate, yet with usually higher accuracy and less circuit complexity [371].

12.3.1 A Fourier-Based Phase Estimator

Closed evolutions of quantum states are realized by unitary operators. Unitary operators come with unit-modulus eigenvalues. Such unit-modulus complex numbers stand as important characteristics of the state transformation, and they can be characterized by their phase, which we now seek to estimate. For this purpose, we consider a quantum process represented by the unitary operator U_ξ, endowed with an eigenvalue denoted $\exp(i2\pi\xi)$ associated with the eigenstate $|u\rangle$, so that $\mathsf{U}_\xi\,|u\rangle = \exp(i2\pi\xi)\,|u\rangle$. The task is then, when the eigenstate $|u\rangle$ is specified, for the associated eigenvalue $\exp(i2\pi\xi)$, to estimate the unknown value of the phase represented by $\xi \in [0, 1[$. A convenient solution based on the quantum Fourier transform, commonly originated in the reference [372] by Kitaev and further analyzed in [107], proceeds as follows.

Relying on the possibility of controlled quantum operations presented in Sect. 5.6, we introduce a qubit in state $|c\rangle$ to control the application of the unitary U_ξ, so as to realize the controlled-U_ξ process, denoted cU_ξ, and operating as $\mathsf{cU}_\xi\,|c=0\rangle\,|v\rangle = |c=0\rangle\,|v\rangle$ and $\mathsf{cU}_\xi\,|c=1\rangle\,|v\rangle = |c=1\rangle\,\mathsf{U}_\xi\,|v\rangle$, that is, the arbitrary state $|v\rangle$ is transformed by U_ξ when the control qubit is in state $|c\rangle = |1\rangle$, while the state $|v\rangle$ remains unchanged with a control qubit in state $|c\rangle = |0\rangle$. The control qubit placed in the superposed state $|c\rangle = |+\rangle = \big(|0\rangle + |1\rangle\big)/\sqrt{2}$ leads to the transformation $\mathsf{cU}_\xi\,|+\rangle\,|v\rangle = \big(|0\rangle\,|v\rangle + |1\rangle\,\mathsf{U}_\xi\,|v\rangle\big)/\sqrt{2}$ which, for an arbitrary $|v\rangle$, generally entangles the control qubit with the controlled system. However, the eigenstate $|u\rangle$ of U_ξ transforms as $\mathsf{cU}_\xi\,|0\rangle\,|u\rangle = |0\rangle\,|u\rangle$ and $\mathsf{cU}_\xi\,|1\rangle\,|u\rangle = \exp(i2\pi\xi)\,|1\rangle\,|u\rangle$, and when the control qubit is placed in the superposed state $|c\rangle = |+\rangle$, the controlled-U_ξ process realizes the transformation

$$\frac{1}{\sqrt{2}}\big(|0\rangle + |1\rangle\big)\,|u\rangle \longmapsto \frac{1}{\sqrt{2}}\Big(|0\rangle\,|u\rangle + |1\rangle\,\mathsf{U}_\xi\,|u\rangle\Big) = \frac{1}{\sqrt{2}}\Big[|0\rangle + \exp(i2\pi\xi)\,|1\rangle\Big]\,|u\rangle\,, \quad (12.42)$$

where the control qubit terminates in the transformed state $\big[|0\rangle + \exp(i2\pi\xi)\,|1\rangle\big]/\sqrt{2}$ while the controlled system remains in the separable state $|u\rangle$. In this way, by the

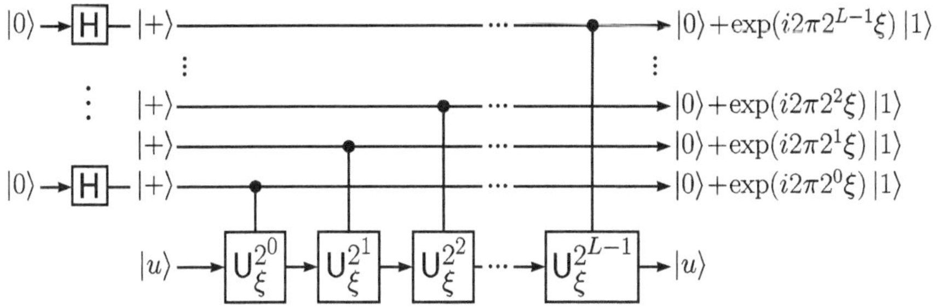

Fig. 12.3 Quantum circuit involving L input qubits initialized in the state $|+\rangle = \mathsf{H}\,|0\rangle$ to act as control qubits to the L controlled-$\mathsf{U}_\xi^{2^\ell}$ gates, for $\ell = 0$ to $L-1$. The $\mathsf{U}_\xi^{2^\ell}$ gates are excited by their common eigenstate $|u\rangle$ which remains unchanged all through the transformation. Each of the L control qubits terminates in a transformed state from Eq. (12.43) where the normalization factors $1/\sqrt{2}$ have been omitted in the figure. We note that the unitary operator U_ξ does not need to process qubits, i.e., the state $|u\rangle$ is not necessarily a qubit state. Qubits are introduced as the L control qubits, and it is these L qubits that are further processed to estimate the phase ξ

operation of the controlled process U_ξ, the information of the phase ξ is transferred to the control qubit. This is a significant property, by which the control qubit prepared in the superposed state $|+\rangle$ can now be processed to serve in the estimation of the phase ξ.

We further assume that the process U_ξ is accessible when raised to an arbitrary power of the form 2^ℓ for appropriate values of the integer $\ell \geq 0$. This is achievable by cascading 2^ℓ copies of the process U_ξ to realize the process $\mathsf{U}_\xi^{2^\ell}$. Especially, application to the eigenstate $|u\rangle$ yields $\mathsf{U}_\xi^{2^\ell}\,|u\rangle = \exp(i2\pi 2^\ell \xi)\,|u\rangle$. With access also to a controlled version of the process $\mathsf{U}_\xi^{2^\ell}$, which follows by cascading the controlled-U_ξ process, we are able as in Eq. (12.42) to perform the state transformation

$$\frac{1}{\sqrt{2}}(|0\rangle + |1\rangle)\,|u\rangle \longmapsto \frac{1}{\sqrt{2}}\Big(|0\rangle\,|u\rangle + |1\rangle\,\mathsf{U}_\xi^{2^\ell}\,|u\rangle\Big) = \frac{1}{\sqrt{2}}\Big[|0\rangle + \exp\big(i2\pi 2^\ell \xi\big)\,|1\rangle\Big]\,|u\rangle\ . \tag{12.43}$$

The output of this transformation delivers, in the right-hand size of Eq. (12.43), a qubit state $\big[|0\rangle + \exp(i2\pi 2^\ell \xi)\,|1\rangle\big]/\sqrt{2}$ standing as the separable state in which the control qubit to $\mathsf{U}_\xi^{2^\ell}$ terminates, while the eigenstate $|u\rangle$ remains unaffected.

It is then possible to assemble by tensoring L such controlled processes, with L control qubits prepared in the superposed state $|+\rangle^{\otimes L}$, according to the circuit architecture depicted in Fig. 12.3 and resulting from [107, 372]. As it will appear in the analysis, the value of L determines the precision or resolution in estimating the phase ξ, with finer resolution or higher precision as L increases.

In the circuit of Fig. 12.3, the L input qubits, each prepared in state $|+\rangle = \mathsf{H}\,|0\rangle$, act as control qubits to the L controlled-$\mathsf{U}_\xi^{2^\ell}$ gates, for $\ell = 0$ to $L-1$, so as to perform the input–output transformation expressible as

$$|+\rangle^{\otimes L}|u\rangle = \frac{1}{2^{L/2}}\sum_{k=0}^{2^L-1}|k\rangle|u\rangle \longmapsto \frac{1}{2^{L/2}}\sum_{k=0}^{2^L-1}|k\rangle\,\mathsf{U}_\xi^k|u\rangle = \frac{1}{2^{L/2}}\sum_{k=0}^{2^L-1}|k\rangle\exp(i2\pi k\xi)|u\rangle\,, \tag{12.44}$$

with $\mathsf{U}_\xi^0 \equiv \mathsf{Id}$ the identity operator. At the output in Eq. (12.44), we obtain the separable product state $|\widetilde{\psi}_\xi\rangle|u\rangle$, where the L control qubits terminate in the joint quantum state $|\widetilde{\psi}_\xi\rangle$. This state $|\widetilde{\psi}_\xi\rangle$ is an L-qubit separable state whose constitution can also be obtained from Fig. 12.3 as the product form

$$|\widetilde{\psi}_\xi\rangle = \frac{1}{2^{L/2}}\Big[|0\rangle+\exp\big(i2\pi 2^{L-1}\xi\big)|1\rangle\Big]\Big[|0\rangle+\exp\big(i2\pi 2^{L-2}\xi\big)|1\rangle\Big]\cdots$$
$$\cdots\Big[|0\rangle+\exp\big(i2\pi 2^{0}\xi\big)|1\rangle\Big] \tag{12.45}$$
$$= \frac{1}{2^{L/2}}\bigotimes_{\ell=1}^{L}\left[|0\rangle+\exp\left(i2\pi j_\xi\frac{2^{L-\ell}}{2^L}\right)|1\rangle\right] \tag{12.46}$$
$$= \frac{1}{2^{L/2}}\sum_{k=0}^{2^L-1}\exp\left(i2\pi\frac{j_\xi k}{2^L}\right)|k\rangle\,, \tag{12.47}$$

with the new variable $j_\xi/2^L = \xi \Leftrightarrow j_\xi = \xi 2^L$, and the equality of Eq. (12.47) which results from the identities of Eqs. (12.29)–(12.30) when $\alpha = i2\pi j_\xi$. To this L-qubit quantum state $|\widetilde{\psi}_\xi\rangle$ existing in a Hilbert space with dimension 2^L, it is then convenient to apply an inverse Fourier transform, as defined via Eq. (12.19), to obtain the state

$$|\psi_\xi\rangle = \mathsf{U}_\mathsf{F}^\dagger|\widetilde{\psi}_\xi\rangle\,. \tag{12.48}$$

With an unknown phase $\xi \in [0, 1[$, one has $j_\xi = \xi 2^L \in [0, 2^L[$. When j_ξ is precisely an integer $j \in [0, 2^L[$, then the state $|\widetilde{\psi}_\xi\rangle$ of Eqs. (12.45)–(12.47) is precisely $|\widetilde{j}\rangle$, the Fourier transform of the basis state $|j\rangle$ as in Eq. (12.15), and therefore $|\psi_\xi\rangle = |j\rangle$ in Eq. (12.48). So when the L-qubit state $|\psi_\xi\rangle$ of Eq. (12.48) is measured in the computational basis, the outcome j is obtained with probability 1, and from j the estimate $\widehat{\xi} = j/2^L$ delivers the true value ξ of the phase.

In the more general case when $j_\xi = \xi 2^L \in [0, 2^L[$ is not an integer, measuring the L-qubit state $|\psi_\xi\rangle$ of Eq. (12.48) in the computational basis, will project onto a basis state $|j\rangle$, from which the estimate $\widehat{\xi} = j/2^L$ is with high probability very close to the true value ξ of the phase.

We now work out a quantitative analysis of the performance of the estimator $\widehat{\xi} = j/2^L$. The inverse Fourier transform of the L-qubit state $|\widetilde{\psi}_\xi\rangle$ from Eq. (12.47) is

$$|\psi_\xi\rangle = \mathsf{U}_\mathsf{F}^\dagger|\widetilde{\psi}_\xi\rangle = \frac{1}{2^{L/2}}\sum_{k=0}^{2^L-1}\exp\left(i2\pi\frac{j_\xi k}{2^L}\right)\mathsf{U}_\mathsf{F}^\dagger|k\rangle\,. \tag{12.49}$$

From Eq. (12.19) we obtain

$$\mathsf{U}_\mathsf{F}^\dagger \left|k\right\rangle = \frac{1}{2^{L/2}} \sum_{j=0}^{2^L-1} \exp\left(-i2\pi \frac{jk}{2^L}\right) \left|j\right\rangle , \tag{12.50}$$

giving for Eq. (12.49),

$$\left|\psi_\xi\right\rangle = \frac{1}{2^L} \sum_{j=0}^{2^L-1} \sum_{k=0}^{2^L-1} \exp\left(i2\pi \frac{(j_\xi - j)k}{2^L}\right) \left|j\right\rangle = \sum_{j=0}^{2^L-1} S_j \left|j\right\rangle . \tag{12.51}$$

The sum

$$S_j = \frac{1}{2^L} \sum_{k=0}^{2^L-1} \exp\left(i2\pi \frac{(j_\xi - j)k}{2^L}\right) \tag{12.52}$$

contained in Eq. (12.51), when j_ξ is some integer $j_0 \in [0, 2^L[$, amounts to

$$S_j = \delta_{jj_0} , \tag{12.53}$$

with δ_{jj_0} the Kronecker delta. Meanwhile, for more generic j_ξ, the geometric series in S_j of Eq. (12.52) sums to

$$S_j = \exp\left(i\pi \frac{2^L - 1}{2^L}(j_\xi - j)\right) \frac{1}{2^L} \frac{\sin\left[\pi(j_\xi - j)\right]}{\sin\left[\pi(j_\xi - j)/2^L\right]} . \tag{12.54}$$

We note that when j_ξ tends to an integer $j_0 \in [0, 2^L[$, the limit of Eq. (12.54) when $j_\xi \to j_0$ recovers Eq. (12.53). In general, S_j of Eq. (12.54) can be seen as a function of j sharply peaked around the integer closest to j_ξ, and even more peaked as the size L gets larger.

We then obtain a complete characterization in the computational basis of the L-qubit state $\left|\psi_\xi\right\rangle$ expressed in the concise form

$$\left|\psi_\xi\right\rangle = \sum_{j=0}^{2^L-1} S_j \left|j\right\rangle \tag{12.55}$$

stemming from Eq. (12.51), where the coordinates S_j are explicitly known from Eq. (12.54). Upon measuring the state $\left|\psi_\xi\right\rangle$ of Eq. (12.55) in the computational basis, projection onto the particular basis state $\left|j\right\rangle$ occurs with the probability $\Pr(j;\xi) = |S_j|^2$ which, by Eq. (12.54), is

$$\Pr(j;\xi) = \frac{1}{2^{2L}} \frac{\sin^2\left[2^L\pi(\xi - j/2^L)\right]}{\sin^2\left[\pi(\xi - j/2^L)\right]} . \tag{12.56}$$

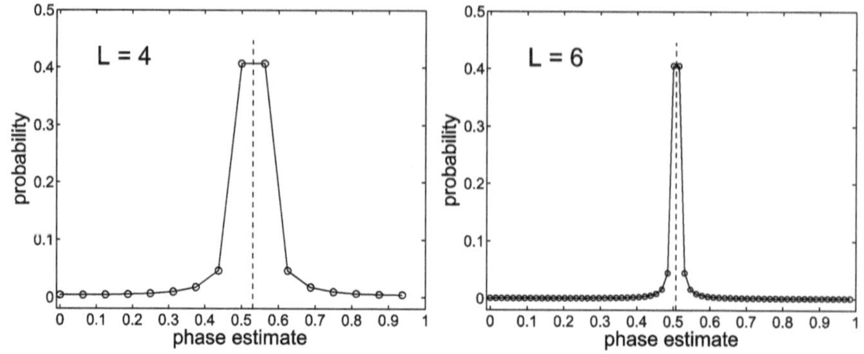

Fig. 12.4 The probability $\Pr(\widehat{\xi} = j/2^L) = \Pr(j; \xi)$ from Eq. (12.56) for the estimate $\widehat{\xi}$ of the phase in abscissa, at two distinct values of the number L of qubits fixing the resolution 2^{-L} of the estimation. In abscissa, the values accessible to the estimate $\widehat{\xi} = j/2^L \in [0, 1[$ are discrete, materialized by open circles (∘) and occur indexed by the integer $j \in [0, 2^L[$. For each L, the true value of the phase to be estimated is taken as $\xi = 0.5 + 2^{-L}/2$ (with $j_\xi = \xi 2^L = 0.5 + 2^{L-1}$ maximally distant from an integer) and is shown by the vertical dotted line

For a fixed measurement outcome j, the probability $\Pr(j; \xi)$ of Eq. (12.56), when seen as a function of ξ, is maximized by $\widehat{\xi} = j/2^L$, which therefore stands as the maximum-likelihood estimator for the phase ξ from the outcome of measuring $|\psi_\xi\rangle$ of Eq. (12.55) in the computational basis. For a given phase ξ, the probability $\Pr(\widehat{\xi} = j/2^L) = \Pr(j; \xi)$ following from Eq. (12.56) for the phase estimate $\widehat{\xi} = j/2^L$ is represented in Fig. 12.4 for two different values of the number L of qubits fixing the resolution 2^{-L} of the estimation.

Figure 12.4 shows, as anticipated, how the probability of the phase estimate $\widehat{\xi} = j/2^L$ gets more and more peaked around the true phase $\xi \in [0, 1[$, as the resolution 2^{-L} of the estimation gets finer.

For another viewpoint, Fig. 12.5 shows the probability $\Pr(j; \xi)$ of Eq. (12.56), as a function of the phase ξ, for a fixed given value of the measurement outcome taken to be $j = 2^L/2$.

In the example of Fig. 12.5, when the unknown phase ξ is such that $j_\xi = \xi 2^L$ exactly matches the integer $j = 2^L/2$, then the probability $\Pr(j = 2^L/2; \xi)$ is 1 for this fixed estimate $\widehat{\xi} = 1/2$. And this peak of the probability $\Pr(j = 2^L/2; \xi)$ with magnitude 1 gets narrower as the resolution 2^{-L} of the estimation gets finer.

For the estimator $\widehat{\xi} = j/2^L$, we can obtain a mean-squared error $e^2(\widehat{\xi}) = \langle(\widehat{\xi} - \xi)^2\rangle$, as

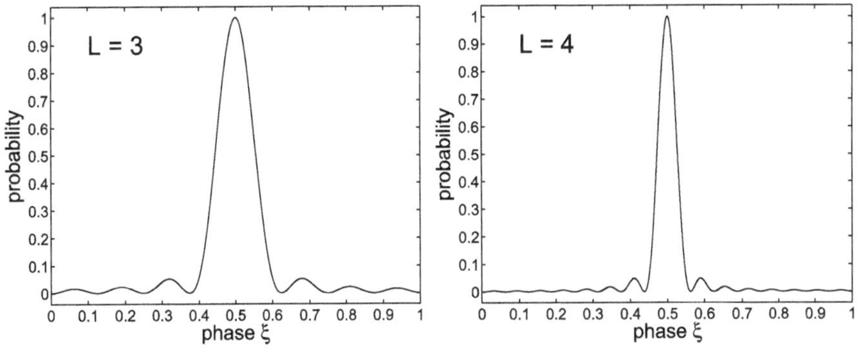

Fig. 12.5 The probability $\Pr(j = 2^L/2; \xi)$ from Eq. (12.56) as a function of the phase ξ for a fixed given value of the measurement outcome $j = 2^L/2$, and at two distinct values of the number L of qubits fixing the resolution 2^{-L} of the estimation

$$e^2(\widehat{\xi}\,) = \sum_{j=0}^{2^L-1} \big(\widehat{\xi} - \xi\big)^2 \Pr(j; \xi) \tag{12.57}$$

$$= \frac{1}{2^{2L}} \sum_{j=0}^{2^L-1} \big(\xi - j/2^L\big)^2 \frac{\sin^2\big[2^L\pi(\xi - j/2^L)\big]}{\sin^2\big[\pi(\xi - j/2^L)\big]} \,. \tag{12.58}$$

The mean-squared error of Eq. (12.58) is in general dependent on the parameter ξ. At a given resolution size L of the estimation, as we have seen, the error $e^2(\widehat{\xi}\,)$ is minimized at zero when $j_\xi = \xi 2^L$ is an integer $j_0 \in [0, 2^L[$. Conversely, the error $e^2(\widehat{\xi}\,)$ of Eq. (12.58) is maximized when j_ξ is maximally distant from an integer $j_0 \in [0, 2^L[$, i.e., when j_ξ is of the form $j_\xi = j_0 \pm 1/2$, as in the conditions of Fig. 12.4 for instance. Such maximum mean-squared error $e^2(\widehat{\xi}\,)$ is represented (by its square root) in Fig. 12.6, as a function of the size L fixing the resolution of the estimation.

Figure 12.6 shows a root-mean-squared estimation error which decays as $1/\sqrt{N}$ with $N = 2^L$, which is a common trend for statistical estimation from N independent data points. The performance of the Fourier-based estimator in Fig. 12.6 is achieved via a total of $N - 1 = 2^L - 1 = 2^{L-1} + 2^{L-2} + \cdots + 2^0$ evaluations of the unitary process U_ξ in the circuit of Fig. 12.3 according to Eq. (12.45), this to yield a root-mean-squared estimation error $\sim 1/\sqrt{N}$.

To push further the analytical characterization, we can also compute the alternative estimation error as the average

$$e_s^2(\widehat{\xi}\,) = \frac{1}{\pi^2}\big\langle \sin^2[\pi(\widehat{\xi} - \xi)]\big\rangle = \frac{1}{2\pi^2}\Big[1 - \big\langle \cos[2\pi(\widehat{\xi} - \xi)]\big\rangle\Big] = \frac{1}{4\pi^2}\Big\langle \big|e^{i2\pi\widehat{\xi}} - e^{i2\pi\xi}\big|^2\Big\rangle, \tag{12.59}$$

also considered for instance in [294, 301, 302, 304, 373]. With the probabilistic weight $\Pr(j; \xi)$ from Eq. (12.56) for the average $\langle\cdot\rangle$, the estimation error of

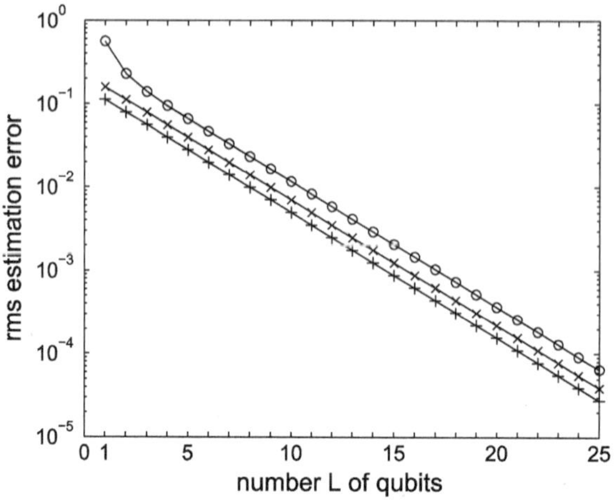

Fig. 12.6 Root-mean-squared estimation error as a function of the number L of qubits fixing the resolution 2^{-L} of the estimation. With ($\circ$) is the error $e(\widehat{\xi})$ evaluated at each L from Eq. (12.58) in conditions where it is maximal, i.e., when $j_\xi = \xi 2^L$ is maximally distant from an integer $j_0 \in [0, 2^L - 1[$, obtained here for estimating the phase $\xi = 0.5 + 2^{-L}/2$, as also considered in Fig. 12.4. With ($\times$) is the average root-mean-squared error $2^{-L/2}/(\sqrt{2}\pi)$ following from Eq. (12.62). With (+) is the root-mean-squared error $2^{-L/2}/(2\pi)$ of the direct maximum-likelihood estimator (with no Fourier transform) for ξ deduced from Eq. (11.97)

Eq. (12.59) amounts to

$$e_s^2(\widehat{\xi}) = \sum_{j=0}^{2^L-1} \frac{1}{\pi^2} \sin^2[\pi(\widehat{\xi} - \xi)] \Pr(j; \xi) \tag{12.60}$$

$$= \frac{2^{-2L}}{\pi^2} \sum_{j=0}^{2^L-1} \sin^2\big[2^L \pi(\xi - j/2^L)\big] . \tag{12.61}$$

When the estimation error $\widehat{\xi} - \xi$ gets small, as it occurs as the size L gets large, then $\sin[\pi(\widehat{\xi} - \xi)] \approx \pi(\widehat{\xi} - \xi)$ so that the alternative estimation error $e_s^2(\widehat{\xi})$ of Eqs. (12.59)–(12.61) tends to coincide with the conventional mean-squared error $e^2(\widehat{\xi})$ of Eq. (12.58). The main motivation for $e_s^2(\widehat{\xi})$ is that it is more easily handled analytically. By linearizing the squared sine and via a series of complex exponentials, Eq. (12.61) can be explicitly summed to yield

$$e^2(\widehat{\xi}) \approx e_s^2(\widehat{\xi}) = \frac{2^{-L}}{2\pi^2}\big[1 - \cos(2\pi j_\xi)\big] = \frac{2^{-L}}{\pi^2} \sin^2(\pi j_\xi) . \tag{12.62}$$

We observe that, as $e^2(\widehat{\xi})$ of Eq. (12.58), the estimation error $e_s^2(\widehat{\xi})$ of Eq. (12.62) vanishes when $j_\xi = \xi 2^L$ is an integer $j_0 \in [0, 2^L[$, and it is maximized at $e_s^2(\widehat{\xi}) = 2^{-L}/\pi^2$ when j_ξ is a half-integer $j_0 \pm 1/2$. In such circumstances, it is meaningful to average the estimation error when the true value $\xi = j_\xi/2^L$ of the phase to be estimated, uniformly covers a resolution cell of width 2^{-L}, i.e., when j_ξ is uniformly distributed over an interval of the form $[j_0 - 1/2, j_0 + 1/2[$. Averaging in this way $e_s^2(\widehat{\xi}) \approx e^2(\widehat{\xi})$ of Eq. (12.62) leads to the average mean-squared error $2^{-L}/(2\pi^2)$. The corresponding average root-mean-squared error $2^{-L/2}/(\sqrt{2}\,\pi)$ is also depicted in Fig. 12.6, and as expected it is observed below the maximal error resulting from Eq. (12.58) in Fig. 12.6.

This average mean-squared error of $1/(2\pi^2 N)$ is achieved by the Fourier-based estimator via a total of $N - 1 = 2^L - 1$ evaluations of the unitary process U_ξ in the circuit of Fig. 12.3 according to Eq. (12.45). A possible alternative is to consider the direct maximum-likelihood estimator, requiring no Fourier transform, examined in Sect. 11.3.2, with its mean-squared error of Eq. (11.97). The direct maximum-likelihood estimator for $2\pi\xi$ with N evaluations of U_ξ achieves a mean-squared error of $1/N$; this is a reference obtained from Eq. (11.97) when $(u_{\max} - u_{\min})^2 = 1$, as is the case for instance for qubit phase estimation. This is equivalent for the parameter ξ considered here, to a mean-squared error of $1/(4\pi^2 N)$, which is two times smaller than the average mean-squared error of the Fourier-based estimator. The corresponding root-mean-squared error $1/(2\pi\sqrt{N})$ of the direct maximum-likelihood estimator for ξ is also depicted in Fig. 12.6. We will come back later, in Sect. 12.3.6, on this comparison of the Fourier-based phase estimation with direct maximum-likelihood estimation, articulated through the distinction between global and local phase estimation.

12.3.2 Another Fourier-Based Phase Estimator

For Fourier-based phase estimation, there exist other ways of probing the dephasing process U_ξ so as to obtain a ξ-dependent quantum state or signal whose measurement will provide access to an estimate for the unknown phase ξ.

An interesting way, also exploited in [371, 374], is to take $N - 1$ qubits, for which we consider the joint state, denoted $|\overline{k}\rangle$, for integer $k = 0$ to $N - 1$, formed with the k first qubits placed in state $|1\rangle$ while the remaining qubits are placed in state $|0\rangle$, according to

$$|\overline{k}\rangle = \big|\overbrace{0\cdots 0\,\underbrace{1\cdots 1}_{k}}^{N-1}\big\rangle\,. \tag{12.63}$$

The set of N states $\{|\overline{k}\rangle\}$, for $k = 0$ to $N - 1$, forms an orthonormal basis for the N-dimensional subspace $\overline{\mathcal{H}}_N$ of the 2^{N-1}-dimensional Hilbert space of the $N - 1$ qubits. The state of the $N - 1$ qubits will remain in this subspace $\overline{\mathcal{H}}_N$ over the whole processing for phase estimation. For instance, for $N = 4$ with $N - 1 = 3$ qubits, we have for $\overline{\mathcal{H}}_4$ the four-dimensional basis $\{|\overline{0}\rangle = |000\rangle\,, |\overline{1}\rangle = |001\rangle\,, |\overline{2}\rangle = |011\rangle\,, |\overline{3}\rangle = |111\rangle\}$.

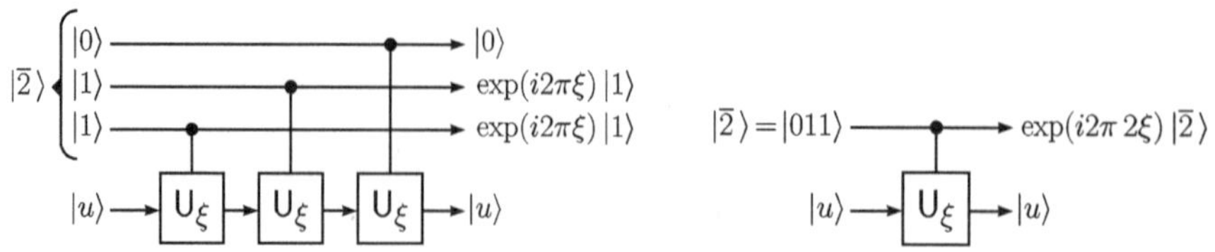

Fig. 12.7 Quantum circuit involving $N-1=3$ input qubits prepared in the state $|\overline{2}\,\rangle = |011\rangle$ and acting as control qubits to the controlled-U_ξ gate, to operate either in parallel (left) or sequentially (right). The U_ξ gate is fed with its eigenstate $|u\rangle$ which remains unchanged all through the transformation. The $N-1$ control qubits terminate in the joint state $\exp(i2\pi\,2\xi)\,|011\rangle = \exp(i2\pi\,2\xi)\,|\overline{2}\,\rangle$ separable from $|u\rangle$, as in Eq. (12.64)

As in Sect. 12.3.1, the $N-1$ qubits in state $|\overline{k}\,\rangle$ are used as control qubits to the controlled-U_ξ process. The state transformation is realized according to the circuit setting depicted in Fig. 12.7, which illustrates the operation when $N=4$ and $N-1=3$ control qubits are involved. The state transformation, as shown by Fig. 12.7, can be carried out either in parallel, by applying the $N-1$ control qubits simultaneously to $N-1$ identical copies of the controlled-U_ξ gate, or sequentially, by applying the $N-1$ control qubits one after the other to the same controlled-U_ξ gate.

The $N-1$ control qubits starting in the joint state $|\overline{k}\,\rangle$ terminate in the joint state $\exp(i2\pi k\xi)\,|\overline{k}\,\rangle$ which remains separable from the eigenstate $|u\rangle$ of U_ξ, and therefore implements the input–output transformation

$$|\overline{k}\,\rangle \longmapsto \exp(i2\pi k\xi)\,|\overline{k}\,\rangle \tag{12.64}$$

of the $(N-1)$-qubit state. We note that as in Sect. 12.3.1, the unitary operator U_ξ shown in Fig. 12.7 does not need to process qubits, i.e., the state $|u\rangle$ is not necessarily a qubit state. Yet the control qubits in state $|\overline{k}\,\rangle$ interact with U_ξ in such a way that they get influenced by U_ξ and terminate in a ξ-dependent $(N-1)$-qubit state that remains separable from $|u\rangle$, as described by Eq. (12.64).

The $N-1$ input control qubits can be prepared in a uniform superposition of the N basis states $\{|\overline{k}\,\rangle\}$ of $\overline{\mathcal{H}}_N$, and by linearity applying on Eq. (12.64), they experience the input–output transformation

$$\frac{1}{\sqrt{N}}\sum_{k=0}^{N-1}|\overline{k}\,\rangle \longmapsto \frac{1}{\sqrt{N}}\sum_{k=0}^{N-1}\exp(i2\pi k\xi)\,|\overline{k}\,\rangle = |\widetilde{\psi}_\xi\rangle\ . \tag{12.65}$$

The $(N-1)$-qubit output state $|\widetilde{\psi}_\xi\rangle \in \overline{\mathcal{H}}_N$ in Eq. (12.65) can also be put under the form

$$|\widetilde{\psi}_\xi\rangle = \frac{1}{\sqrt{N}}\sum_{k=0}^{N-1}\exp\left(i2\pi\frac{j_\xi k}{N}\right)|\overline{k}\,\rangle\ , \tag{12.66}$$

with the variable $j_\xi/N = \xi$. When $N = 2^L$, the state $|\widetilde{\psi}_\xi\rangle$ of Eq. (12.66) is similar in form to the state $|\widetilde{\psi}_\xi\rangle$ of Eq. (12.47). The two approaches, of Eq. (12.47) and Eq. (12.66), to reach the output state $|\widetilde{\psi}_\xi\rangle$ that probes the unknown phase ξ, rely on two distinct trade-offs between circuit complexity and number of probing qubits. The approach of Eq. (12.47) uses a reduced number L of probing qubits, but it requires implementation of the more complicated processes $\mathsf{U}_\xi^{2^\ell}$ at various powers 2^ℓ as in Fig. 12.3. Conversely, for an equivalent operation, the approach of Eq. (12.66) uses a larger number $2^L - 1$ of probing qubits, but it requires implementation only of the elemental process U_ξ, moreover in a single copy in the sequential implementation of Fig. 12.7. In spite of these differences, both approaches resort to a similar number $N - 1 = 2^L - 1$ evaluations of the process U_ξ to obtain the output state $|\widetilde{\psi}_\xi\rangle$.

Based on the similarity of form with Eq. (12.47), the same processing ensues for the $(N-1)$-qubit output state $|\widetilde{\psi}_\xi\rangle \in \overline{\mathcal{H}}_N$ of Eq. (12.66), which undergoes an inverse Fourier transform as in Eq. (12.48), defined as in Eq. (12.19) but now operating in the N-dimensional subspace $\overline{\mathcal{H}}_N$ spanned by the orthonormal basis $\{|\overline{k}\rangle\}$ according to

$$\mathsf{U}_\mathsf{F}^\dagger |\overline{k}\rangle = \frac{1}{\sqrt{N}} \sum_{j=0}^{N-1} \exp\left(-i2\pi \frac{jk}{N}\right) |\overline{j}\rangle \tag{12.67}$$

which maintains the state in this subspace $\overline{\mathcal{H}}_N$.

After such an inverse Fourier transform on $|\widetilde{\psi}_\xi\rangle \in \overline{\mathcal{H}}_N$ of Eq. (12.66), we obtain analogously to Eq. (12.55), the $(N-1)$-qubit state

$$\mathsf{U}_\mathsf{F}^\dagger |\widetilde{\psi}_\xi\rangle = |\psi_\xi\rangle = \sum_{j=0}^{N-1} S_j |\overline{j}\rangle \in \overline{\mathcal{H}}_N \tag{12.68}$$

with the coefficients

$$S_j = \frac{1}{N} \sum_{k=0}^{N-1} \exp\left(i2\pi \frac{(j_\xi - j)k}{N}\right) \tag{12.69}$$

$$= \exp\left(i\pi \frac{N-1}{N}(j_\xi - j)\right) \frac{1}{N} \frac{\sin[\pi(j_\xi - j)]}{\sin[\pi(j_\xi - j)/N]}, \tag{12.70}$$

where here $j_\xi = \xi N$, and paralleling Eq. (12.54).

When the $N-1$ output qubits are measured in the N-dimensional orthonormal basis $\{|\overline{j}\rangle\}$, the state $|\psi_\xi\rangle$ of Eq. (12.68) gets projected with the probability $|S_j|^2$ onto one particular basis state $|\overline{j}\rangle$ that identifies an integer $j \in [0, N-1]$ from which the maximum-likelihood phase estimate $\widehat{\xi} = j/N$ follows. We note that equivalently, it is enough to measure each of the $N-1$ qubits separately in the computational basis $\{|0\rangle, |1\rangle\}$, and count the number j of qubits found in state $|1\rangle$.

The phase estimator $\widehat{\xi} = j/N$ shares the same properties as those characterized in Sect. 12.3.1 with Eqs. (12.56)–(12.62), when $N = 2^L$. In particular, it exhibits a mean-squared estimation error $e^2(\widehat{\xi})$ with a similar decay as N^{-1}, as reflected in Fig. 12.6, for a similar number $N-1=2^L-1$ of evaluations of the unitary process U_ξ.

12.3.3 Optimized Input Superposition

The approach of Sect. 12.3.2 also affords a possibility for improving the estimation performance. Instead of the uniform input superposition of Eq. (12.65), the $N-1$ control qubits can be prepared in an arbitrary nonuniform superposition of the N basis states $\{|\overline{k}\rangle\}$ of $\overline{\mathcal{H}}_N$, so as to generalize the input–output transformation of Eqs. (12.65)–(12.66) by

$$\sum_{k=0}^{N-1} a_k\,|\overline{k}\,\rangle \longmapsto \sum_{k=0}^{N-1} a_k \exp\!\left(i2\pi\frac{j_\xi k}{N}\right)|\overline{k}\,\rangle = |\widetilde{\psi}_\xi\rangle\;, \tag{12.71}$$

with the complex coefficients $a_k \in \mathbb{C}$ verifying $\sum_{k=0}^{N-1}|a_k|^2=1$, as also exploited in [371, 374]. By comparison, Eq. (12.65) represents an excitation by the uniform input superposition with $a_k=1/\sqrt{N}$ for all $k=0$ to $N-1$.

The state $|\widetilde{\psi}_\xi\rangle \in \overline{\mathcal{H}}_N$ of Eq. (12.71), generalizing that of Eq. (12.66), then undergoes an inverse Fourier transform according to Eq. (12.67) to give

$$\mathsf{U}_\mathsf{F}^\dagger\,|\widetilde{\psi}_\xi\rangle = |\psi_\xi\rangle = \sum_{j=0}^{N-1} S'_j\,|\overline{j}\,\rangle \in \overline{\mathcal{H}}_N\;, \tag{12.72}$$

with now the coefficients

$$S'_j = \frac{1}{\sqrt{N}}\sum_{k=0}^{N-1} a_k \exp\!\left(i2\pi\frac{(j_\xi - j)k}{N}\right), \tag{12.73}$$

in place of Eq. (12.69). As before in Sect. 12.3.2, upon measuring the state $|\psi_\xi\rangle \in \overline{\mathcal{H}}_N$ of Eq. (12.72) in the orthonormal basis $\{|\overline{j}\,\rangle\}$ of $\overline{\mathcal{H}}_N$, the outcome delivers the integer $j\in[0,N-1]$ with the probability $|S'_j|^2$ and produces the maximum-likelihood phase estimate $\widehat{\xi}=j/N$; and this is equivalent to measuring each of the $N-1$ qubits separately in the computational basis $\{|0\rangle\,,|1\rangle\}$, and count the number j of qubits found in state $|1\rangle$.

From Eq. (12.73), we then have the squared modulus

$$|S'_j|^2 = S'_j S'^{*}_j = \frac{1}{N}\left[\sum_{k=0}^{N-1} a_k \exp\left(i2\pi\frac{(j_\xi - j)k}{N}\right)\right]\left[\sum_{\ell=0}^{N-1} a_\ell^* \exp\left(-i2\pi\frac{(j_\xi - j)\ell}{N}\right)\right] \tag{12.74}$$

$$= \frac{1}{N}\sum_{k=0}^{N-1}\sum_{\ell=0}^{N-1} a_k a_\ell^* \exp\left(i2\pi\frac{(j_\xi - j)(k-\ell)}{N}\right) \tag{12.75}$$

$$= \frac{1}{N}\sum_{k=0}^{N-1}\sum_{\ell=0}^{N-1} |a_k||a_\ell| \cos\left[2\pi\left(\xi - \frac{j}{N}\right)(k-\ell) + \phi_k - \phi_\ell\right], \tag{12.76}$$

since $|S'_j|^2$ is real, and with for each complex coefficient $a_k = |a_k|\exp(i\phi_k)$.

To characterize the performance of this nonuniform excitation, it is convenient to compute the estimation error $e_s^2(\widehat{\xi})$ from Eq. (12.59) which coincides with the conventional mean-squared error at large size N when the error is small. For this purpose, following [374], we have the average

$$\left\langle \cos[2\pi(\xi - \widehat{\xi})]\right\rangle = \sum_{j=0}^{N-1} |S'_j|^2 \cos\left[2\pi\left(\xi - \frac{j}{N}\right)\right] \tag{12.77}$$

$$= \sum_{k=0}^{N-1}\sum_{\ell=0}^{N-1} |a_k||a_\ell|\, W_{k\ell}, \tag{12.78}$$

with the coefficients

$$W_{k\ell} = \frac{1}{N}\sum_{j=0}^{N-1} \cos\left[2\pi\left(\xi - \frac{j}{N}\right)(k-\ell) + \phi_k - \phi_\ell\right]\cos\left[2\pi\left(\xi - \frac{j}{N}\right)\right]. \tag{12.79}$$

The product of two cosines is then linearized to give

$$W_{k\ell} = \frac{1}{2N}\sum_{j=0}^{N-1}\left\{\cos\left[2\pi\left(\xi - \frac{j}{N}\right)(k-\ell+1) + \phi_k - \phi_\ell\right] + \cos\left[2\pi\left(\xi - \frac{j}{N}\right)(k-\ell-1) + \phi_k - \phi_\ell\right]\right\}. \tag{12.80}$$

The two sums of cosines in Eq. (12.80) can be explicitly evaluated by means of a geometric series of complex exponentials, and for the different integers $k, \ell \in [0, N-1]$. We then arrive for the average of Eqs. (12.77)–(12.78) at

$$\left\langle \cos[2\pi(\xi - \widehat{\xi})]\right\rangle = \sum_{k=0}^{N-2} |a_{k+1}||a_k| \cos(\phi_{k+1} - \phi_k) + |a_{N-1}||a_0| \cos(2\pi N\xi + \phi_{N-1} - \phi_0). \tag{12.81}$$

Based on Eq. (12.81), the estimation error $e_s^2(\widehat{\xi})$ from Eq. (12.59) can be made independent of the parameter ξ, which is a valuable property ensuring for the estimator $\widehat{\xi} = j/N$ an intrinsic performance independent of the unknown phase ξ being estimated. This can be obtained by selecting for Eq. (12.81) an input superposition with the coefficient $a_0 = 0$. This entails an average in Eq. (12.81) reducing to

$$\big\langle \cos[2\pi(\xi - \widehat{\xi}\,)] \big\rangle = \sum_{k=0}^{N-2} |a_{k+1}||a_k| \cos(\phi_{k+1} - \phi_k) , \tag{12.82}$$

and affording in Eq. (12.59) the ξ-independent estimation error

$$e_s^2(\widehat{\xi}\,) = \frac{1}{2\pi^2}\left[1 - \sum_{k=0}^{N-2} |a_{k+1}||a_k| \cos(\phi_{k+1} - \phi_k)\right] . \tag{12.83}$$

The next step for a good performance is to select the remaining coefficients a_k of the input superposition $\sum_{k=0}^{N-1} a_k\, |\overline{k}\,\rangle$ from Eq. (12.71), so as to minimize the estimation error $e_s^2(\widehat{\xi}\,)$ in Eq. (12.83), or equivalently maximize the sum in the right side of Eq. (12.82), subject to the normalization constraint $\sum_{k=0}^{N-1} |a_k|^2 = 1$. The arguments ϕ_k appearing in Eq. (12.82) are completely free angles not restricted by the normalization constraint. To maximize Eq. (12.82) these ϕ_k should therefore all be selected so that $\cos(\phi_{k+1} - \phi_k) = 1$ for all k; and this can be obtained by $\phi_k = 0$ for all k in Eq. (12.82). The maximum of Eq. (12.82) is thus achievable by a set of real coefficients a_k in the input superposition; and relative phase differences between them would generally degrade the performance. For real coefficients a_k then the constrained maximization of Eq. (12.82) is accomplished by

$$a_k = \sqrt{\frac{2}{N}}\, \sin\!\left(\pi \frac{k}{N}\right) , \quad k = 0, 1, \ldots, N-1, \tag{12.84}$$

which verify the canonical equations of the Lagrange multiplier method. The same optimal coefficients as those of Eq. (12.84) are also found in [301] (its Eq. (12)) or in [302] (its Eq. (40)), yet in somewhat distinct conditions of Fourier-based quantum phase estimation [301, 302, 374]. With Eq. (12.84), the coefficients S'_j in Eq. (12.73) then acquire the form

$$S'_j = \frac{\sqrt{2}}{N} \sum_{k=0}^{N-1} \sin\!\left(\pi \frac{k}{N}\right) \exp\!\left(i2\pi \frac{(j_\xi - j)k}{N}\right) , \tag{12.85}$$

and the sum in Eq. (12.85) can be explicitly evaluated to give

$$S'_j = \frac{1}{N\sqrt{2}} \frac{\sin(\pi/N)}{\cos\big[2\pi(j_\xi - j)/N\big] - \cos(\pi/N)} \Big(1 + \exp\big[i2\pi(j_\xi - j)\big]\Big) . \tag{12.86}$$

For the phase estimate $\widehat{\xi}$, the probability

$$\Pr\big(\widehat{\xi} = j/N\big) = |S'_j|^2 \tag{12.87}$$

readily follows, and is represented in Fig. 12.8.

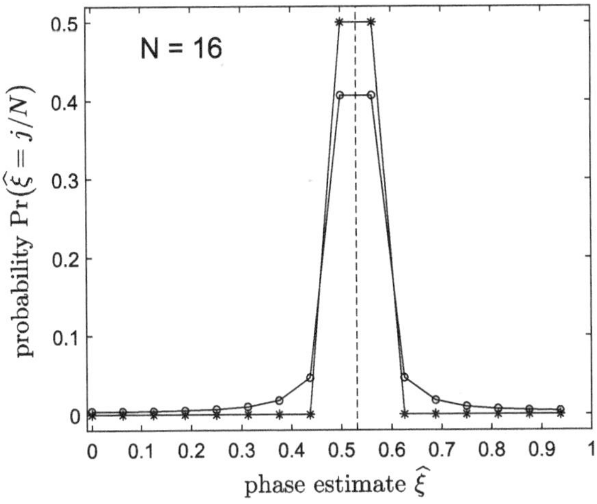

Fig. 12.8 The probability $\Pr(\widehat{\xi} = j/N)$ for the estimate $\widehat{\xi}$ of the phase in abscissa, as it results: ($*$) from Eq. (12.87) with the optimal nonuniform input superposition of Eq. (12.84), or ($\circ$) from Eq. (12.56) with the non-optimal uniform input superposition, when $N = 2^L = 16$ (with ($\circ$) the same curve as at $L = 4$ in Fig. 12.4). The true value of the phase to be estimated here is $\xi = 0.5 + 0.5/N$ indicated by the vertical dotted line

It is consistently observed in Fig. 12.8 that the distribution of probabilities $\Pr(\widehat{\xi} = j/N)$ of Eq. (12.87) resulting from the optimal nonuniform input superposition of Eq. (12.84) is, in generic conditions, more peaked around the true phase ξ than the distribution of Eq. (12.56) resulting from a uniform input superposition instead. This naturally leads to a reduced estimation error as expected when optimizing the input superposition.

Moreover, the optimal nonuniform coefficients of Eq. (12.84) achieve for $\langle \cos[2\pi(\xi - \widehat{\xi})] \rangle$ in Eq. (12.82) the maximum $\cos(\pi/N)$, and this leads in Eq. (12.83) or Eq. (12.59) to the minimal estimation error

$$e_s^2(\widehat{\xi}) = \frac{1}{2\pi^2}\left[1 - \cos\left(\frac{\pi}{N}\right)\right] = \frac{1}{\pi^2}\sin^2\left(\frac{\pi}{2N}\right) . \tag{12.88}$$

Especially, at large size N, the estimation error $e_s^2(\widehat{\xi})$ of Eq. (12.88) coincides with the conventional mean-squared error $e^2(\widehat{\xi})$, which results as $e^2(\widehat{\xi}) \approx 1/(4N^2)$.

We observe here a specific benefit that can be obtained from an optimized nonuniform superposition in the transformation of Eq. (12.71) for phase estimation on the unitary process U_ξ. With N evaluations of the unitary U_ξ, the protocols of Sects. 12.3.1 and 12.3.2 achieve in general a mean-squared estimation error evolving as $1/N$. This is a standard limit also achieved in classical parameter estimation from N independent measurements. By contrast, the protocol of Sect. 12.3.3 with the optimized quantum superposition of Eq. (12.84), is able to achieve a reduced mean-squared estimation error evolving as $1/N^2$. Such improvement in the scaling of the error for quantum estimation, is identifiable with the Heisenberg enhancement presented in Sect. 11.3.4.

To supplement the mean-squared estimation error, we can also turn to the Fisher informations, classical and quantum, as presented in Chap. 11, in order to characterize the situation of the Fourier-based phase estimation protocols, in relation to the best performance that can be envisaged. For the derivations of the Fisher informations of the Fourier-based protocols, we shall follow [371] which considered phase estimation on a qubit unitary process U_ξ in place of the more general quantum process U_ξ we introduced in Sect. 12.3.1; and in addition here we shall present a comparison of the situations of the uniform and of the optimized input superpositions as probing signal. This will provide interesting worked-out examples illustrating the usefulness of the Fisher information in quantum signal processing.

12.3.4 Classical Fisher Information

In the assessment of the performance of the Fourier-based estimators for the quantum phase, it is meaningful to examine the classical Fisher information introduced in Sect. 11.1. The Fourier-based estimators operate from measurement of the ξ-dependent Fourier-transformed signal $|\psi_\xi\rangle = \mathsf{U}_\mathsf{F}^\dagger |\widetilde{\psi}_\xi\rangle$, delivering the outcome j governed by the probability distribution $\Pr(j;\xi)$, from Eq. (12.56) with the uniform input superposition, or from Eq. (12.87) with the optimized nonuniform input superposition. The classical Fisher information contained in the data j about the unknown phase ξ results, according to Eq. (11.45), as the expectation

$$F_c(\xi) = \Big\langle \big[\partial_\xi \ln \Pr(j;\xi)\big]^2 \Big\rangle = \sum_{j=0}^{N-1} \frac{\big[\partial_\xi \Pr(j;\xi)\big]^2}{\Pr(j;\xi)} . \tag{12.89}$$

When $\Pr(j;\xi) = |S'_j|^2$ as in Eq. (12.87) for the optimized input superposition from Eq. (12.84), we equivalently have

$$F_c(\xi) = 4\sum_{j=0}^{N-1} \big(\partial_\xi |S'_j|\big)^2 ; \tag{12.90}$$

and analogously when $\Pr(j;\xi) = |S_j|^2$, with the derivative $\partial_\xi |S_j|$, from Eq. (12.56) or (12.70) for the uniform input superposition of Eq. (12.65). The derivative $\partial_\xi \cdot = \partial \cdot / \partial_\xi$ relative to the phase ξ can then be analytically performed, from Eq. (12.86) for S'_j or from Eq. (12.54) for S_j, to yield the classical Fisher information $F_c(\xi)$. Figure 12.9 compares the classical Fisher information $F_c(\xi)$ resulting from Eqs. (12.89)–(12.90), when using the uniform input superposition of Eq. (12.65) or the optimized input superposition of Eq. (12.84).

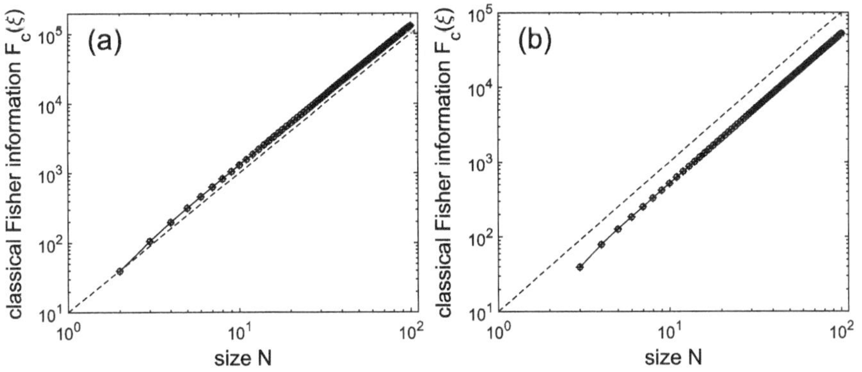

Fig. 12.9 Upon measuring the ξ-dependent Fourier-transformed signal $|\psi_\xi\rangle = \mathsf{U}_\mathsf{F}^\dagger |\widetilde{\psi}_\xi\rangle$, the classical Fisher information $F_c(\xi)$ from Eqs. (12.89)–(12.90), as a function of the size N dimensioning the $(N-1)$-qubit probing signal, for estimation of the phase ξ, with (○) when $\xi = 0.5 + 0.5/N$ and (+) when $\xi = \sqrt{3}/3 \approx 0.577$. The dashed diagonal with slope 2 represents the growth as N^2. (**a**) is with the uniform input superposition $a_k = 1/\sqrt{N}$ in Eq. (12.71) or (12.65), and (**b**) is with the optimized coefficients a_k of Eq. (12.84)

The evolutions of Fig. 12.9 show in particular a classical Fisher information $F_c(\xi)$ exhibiting no significant dependence on the value of the unknown phase ξ being estimated. Such a parameter-independent Fisher information $F_c(\xi)$ in Fig. 12.9 is a valuable feature, not always observed in parameter estimation, indicating here a uniform performance assessed by $F_c(\xi)$ that remains the same for any range of the phase ξ being estimated. As discussed in [371], such an invariance of $F_c(\xi)$ with ξ here stems from a periodic character in $\upsilon = \xi - j/N$ of $\left(\partial_\xi |S'_j|\right)^2$ or $\left(\partial_\xi |S_j|\right)^2$ involved in the sum of Eq. (12.90), where ξ, at increasing N, tends to act as an irrelevant origin as in an integration over one period.

Another significant property observed in Fig. 12.9 is that the Fisher information $F_c(\xi)$ resulting from Eqs. (12.89)–(12.90), increases as N^2, with the size N dimensioning the probing signal and the number of measured qubits. This scaling as N^2 of $F_c(\xi)$ applies equally for the uniform input superposition of Eq. (12.65) and for the optimized nonuniform superposition from Eq. (12.84), as visible in Fig. 12.9.

However, there exists an important difference between the uniform and the optimized nonuniform input superpositions, and concerning the performance of the maximum-likelihood estimator $\widehat{\xi} = j/N$ from the result of measuring the ξ-dependent Fourier-transformed signal $|\psi_\xi\rangle = \mathsf{U}_\mathsf{F}^\dagger |\widetilde{\psi}_\xi\rangle$. With the optimized nonuniform superposition of Eq. (12.84), the Fourier-based estimator $\widehat{\xi} = j/N$ achieves a mean-squared error $e^2(\widehat{\xi}\,) \approx e_s^2(\widehat{\xi}\,)$ in Eq. (12.88) displaying a $1/N^2$ decay matching the N^2 scaling of the Fisher information $F_c(\xi)$. By contrast, with the uniform input superposition of Eq. (12.65), the estimator $\widehat{\xi} = j/N$ achieves the mean-squared error $e^2(\widehat{\xi}\,)$ of Eq. (12.62) displaying a less favorable $1/N$ decay, as reflected in Fig. 12.6, unable to match the N^2 scaling of the Fisher information $F_c(\xi)$. We note here that the large N asymptotics is not the limit of a large number of independent measure-

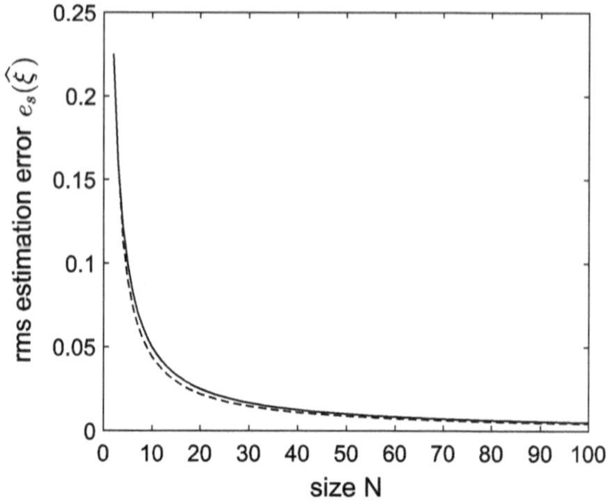

Fig. 12.10 For the Fourier-based phase estimator $\widehat{\xi}$ from Sect. 12.3.3 with the optimized input superposition, the root-mean-squared estimation error $e_s(\widehat{\xi}) \approx e(\widehat{\xi})$ from Eq. (12.88) (solid line) compared with $1/\sqrt{F_c(\xi)}$ (dashed line) with $F_c(\xi)$ the classical Fisher information of Fig. 12.9(b) (+), as a function of the size N dimensioning the probing signal

ments where the mean-squared error of the maximum-likelihood estimator matches the reciprocal of the classical Fisher information. The Fourier-based estimators considered here are one-shot estimators, usually performing a single measurement of the ξ-dependent signal $|\psi_\xi\rangle$ to estimate the phase ξ. They come with a bias decaying with N and a mean-squared error not bound to match the Cramér–Rao bound controlled by reciprocal of the classical Fisher information.

In the estimation protocol using the optimized input superposition, it is also interesting to quantitatively compare, beyond their variations controlled by N^2, the Fisher information $F_c(\xi)$ of Fig. 12.9 and the mean-squared estimation error $e^2(\widehat{\xi}) \approx e_s^2(\widehat{\xi})$ obtained from Eq. (12.88). Such a comparison is realized in Fig. 12.10, in the conditions of Fig. 12.9.

In Fig. 12.10, the root-mean-squared (rms) estimation error $e(\widehat{\xi}) \approx e_s(\widehat{\xi})$ from Eq. (12.88) is found just slightly above the bound $1/\sqrt{F_c(\xi)}$, which is the asymptotic lower bound that would limit the maximum-likelihood estimator from a large number L_{rep} of independent measurements repeated on L_{rep} independent replicas obtained for $|\psi_\xi\rangle \in \overline{\mathcal{H}}_N$ of Eq. (12.72). By comparison, the rms estimation error $e(\widehat{\xi})$ in Fig. 12.10 characterizes one single measurement of the $(N-1)$-qubit state $|\psi_\xi\rangle$. So it is observed that as soon as $L_{\text{rep}} = 1$ measurement, the rms estimation error $e(\widehat{\xi})$ in Fig. 12.10 comes very close to the best asymptotic performance at large L_{rep}. The results of Fig. 12.10 show that the one-shot Fourier-based estimator $\widehat{\xi}$ optimized according to Eqs. (12.84)–(12.88) achieves an rms estimation error $e(\widehat{\xi})$ very close to (barely above) the asymptotic lower bound $1/\sqrt{F_c(\xi)}$ for any signal size N, and with a decay as $\sim 1/N$.

This performance assessment characterizes the phase estimation protocol where the ξ-dependent probing signal $|\widetilde{\psi}_\xi\rangle$ is inverse Fourier transformed to deliver the signal $|\psi_\xi\rangle = \mathsf{U}_\mathsf{F}^\dagger |\widetilde{\psi}_\xi\rangle$ which is measured via a standard von Neumann quantum measurement in an orthonormal projective basis of $\overline{\mathcal{H}}_N$. This estimation protocol is effectively capable of returning a phase estimate in the interval $[0, 2\pi[$, as we have seen, with an arbitrary precision increasing with the number of qubits (sized by N) exploited. Yet, in terms of quantum signal processing for quantum phase estimation, one could wonder whether some more efficient processing could still be conceived, for instance, with a substitute to the inverse Fourier transform or with some generalized quantum measurement. The quantum Fisher information is a valuable tool to address such issue.

12.3.5 Quantum Fisher Information

For further assessment of the performance of the Fourier-based approach to quantum phase estimation, we now examine the quantum Fisher information introduced in Sect. 11.2.1.

When estimation of the phase ξ is based on a ξ-dependent pure state like $|\widetilde{\psi}_\xi\rangle = \sum_{k=0}^{N-1} a_k \exp(i2\pi k\xi)\,|\overline{k}\,\rangle$ of Eq. (12.71), the corresponding quantum Fisher information is expressible from Eq. (11.71) as

$$F_q(\xi) = 4\Big(\langle \partial_\xi \widetilde{\psi}_\xi | \partial_\xi \widetilde{\psi}_\xi \rangle + \langle \partial_\xi \widetilde{\psi}_\xi | \widetilde{\psi}_\xi \rangle^2\Big) . \tag{12.91}$$

The differentiated state

$$|\partial_\xi \widetilde{\psi}_\xi \rangle = i2\pi \sum_{k=0}^{N-1} k a_k \exp(i2\pi k\xi)\,|\overline{k}\,\rangle \tag{12.92}$$

gives access to the inner products

$$\langle \partial_\xi \widetilde{\psi}_\xi | \partial_\xi \widetilde{\psi}_\xi \rangle = 4\pi^2 \sum_{k=0}^{N-1} k^2 |a_k|^2 \tag{12.93}$$

and

$$\langle \partial_\xi \widetilde{\psi}_\xi | \widetilde{\psi}_\xi \rangle = -i2\pi \sum_{k=0}^{N-1} k |a_k|^2 , \tag{12.94}$$

so that the quantum Fisher information of Eq. (12.91) follows as

$$F_q(\xi) = 16\pi^2 \left[\sum_{k=0}^{N-1} k^2 |a_k|^2 - \left(\sum_{k=0}^{N-1} k |a_k|^2 \right)^2 \right] . \tag{12.95}$$

We obtain in Eq. (12.95) a quantum Fisher information $F_q(\xi)$ which is independent of the phase ξ to be estimated. This property is not necessarily granted for every parametric dependence of a quantum state, and it ensures here a uniform performance $F_q(\xi)$ for any phase ξ. We also note that if the estimation were based on another ξ-dependent pure state $\mathsf{U}\,|\widetilde{\psi}_\xi\rangle$, obtained by applying any ξ-independent unitary transformation U to the state $|\widetilde{\psi}_\xi\rangle$, then the quantum Fisher information $F_q(\xi)$ of Eq. (12.91) would remain unchanged, by virtue of $\mathsf{U}^\dagger\mathsf{U} = \mathsf{Id}$, or by property (2) at the end of Sect. 11.2.1. In this respect, the same quantum Fisher information $F_q(\xi)$ of Eq. (12.95) applies when the inverse-Fourier-transformed state $|\psi_\xi\rangle = \mathsf{U}_\mathsf{F}^\dagger\,|\widetilde{\psi}_\xi\rangle$ from Eqs. (12.72), (12.68) or (12.49) is measured for estimating the phase ξ.

In the input superposition for $|\widetilde{\psi}_\xi\rangle$ in Eq. (12.71), when the coefficients a_k are optimized according to Eq. (12.84) so as to minimize the estimation error $e_s^2(\widehat{\xi})$ of Eqs. (12.59) and (12.83), then the quantum Fisher information of Eq. (12.95) can be evaluated to be

$$F_q(\xi) = 4\pi^2\left[\frac{1}{3}N^2 + \frac{2}{3} - \frac{2}{\sin^2(\pi/N)}\right], \tag{12.96}$$

exhibiting at large N the limit $F_q(\xi) \to (4\pi^2/3 - 8)N^2 \approx 5.16N^2$. This $F_q(\xi)$ of Eq. (12.96) in particular fixes an upper bound to the classical Fisher information $F_c(\xi)$ of Eq. (12.90) and Fig. 12.9(b), with both $F_q(\xi)$ and $F_c(\xi)$ sharing the same scaling as $\sim N^2$.

Otherwise, with the uniform input superposition of Eq. (12.65), when $a_k = 1/\sqrt{N}$ for $k = 0$ to $N-1$, we obtain in Eq. (12.95) the quantum Fisher information

$$F_q(\xi) = 4\pi^2\left(\frac{1}{3}N^2 - \frac{1}{3}\right), \tag{12.97}$$

exhibiting at large N the limit $F_q(\xi) \to (4\pi^2/3)N^2 \approx 13.16N^2$. This quantum Fisher information $F_q(\xi)$ is an upper bound to the classical Fisher information $F_c(\xi)$ of Fig. 12.9(a), with both $F_q(\xi)$ and $F_c(\xi)$ sharing the same scaling as $\sim N^2$.

Furthermore, the quantum Fisher information $F_q(\xi)$ is quantitatively compared with the corresponding classical Fisher information $F_c(\xi)$ of Sect. 12.3.4, for the optimized nonuniform and for the uniform input superpositions, in the graphs of Fig. 12.11.

Figure 12.11 shows that the quantum Fisher information $F_q(\xi)$ of this Sect. 12.3.5 and the classical Fisher information $F_c(\xi)$ of Sect. 12.3.4 precisely match one another. The Fisher informations of the uniform input superposition are still larger than the Fisher informations of the optimized superposition from Eq. (12.84). But we remember that with the uniform input superposition the Fourier-based estimator displays a mean-squared error unable to match the (reciprocal) Fisher information $F_c(\xi)$ (especially its N^2 scaling), while with the optimized superposition it does match $F_c(\xi)$ and its N^2 scaling (see Fig. 12.10).

The analysis of the quantum Fisher information also suggests an alternative approach for efficient phase estimation. Instead of optimizing the input coefficients a_k in Eq. (12.71) in order to minimize the estimation error $e_s^2(\widehat{\xi})$ of Eqs. (12.59) and (12.83), one could consider selecting the input coefficients a_k so as to maximize the quantum Fisher information $F_q(\xi)$ of Eq. (12.95). Examining this possibility

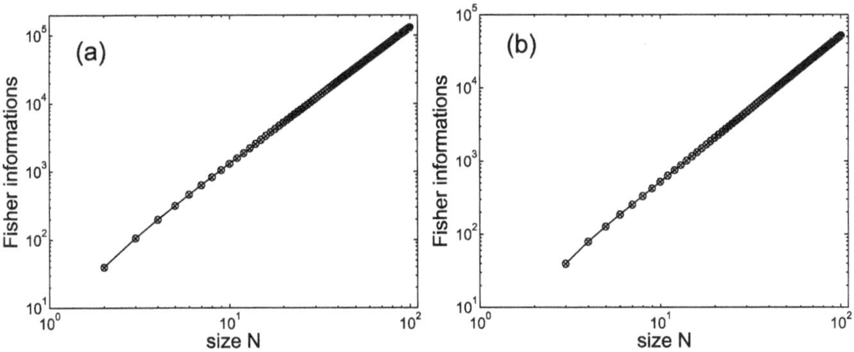

Fig. 12.11 Quantum and classical Fisher informations as a function of the size N dimensioning the $(N-1)$-qubit probing signal for estimation of the phase ξ. (**a**) is with the uniform input superposition $a_k = 1/\sqrt{N}$: ($\circ$) is the quantum $F_q(\xi)$ of Eq. (12.97), ($\times$) is the classical $F_c(\xi)$ from Eq. (12.89) as in Fig. 12.9(a). (**b**) is with the optimized coefficients a_k of Eq. (12.84): ($\circ$) is the quantum $F_q(\xi)$ of Eq. (12.96), ($\times$) is the classical $F_c(\xi)$ from Eq. (12.89) as in Fig. 12.9(b)

will lead us to an important distinction between local and global quantum phase estimation, that we now address.

12.3.6 Local Versus Global Quantum Phase Estimation

So we seek to determine the input coefficients a_k maximizing the quantum Fisher information $F_q(\xi)$ of Eq. (12.95). This constrained maximization of Eq. (12.95) is solved for instance in [371] via the Lagrange multiplier method. It is found that the maximal quantum Fisher information $F_q(\xi)$ in Eq. (12.95) is obtained by the coefficients $a_0 = a_{N-1} = 1/\sqrt{2}$ and $a_k = 0$ for $k = 1$ to $N-2$. This corresponds in Eq. (12.71) to the $(N-1)$-qubit optimal input signal $\big(|\overline{0}\rangle + |\overline{N-1}\rangle\big)/\sqrt{2} = \big(|0\cdots 0\rangle + |1\cdots 1\rangle\big)/\sqrt{2}$. Such an optimal signal is comparable to the so-called photonic NOON states representing an extreme configuration for populating optical modes with photons [375, 376]. This optimal input is also similar to the quantum state maximizing the dispersion of the generator of the unitary process on $\overline{\mathcal{H}}_N$ defined by Eq. (12.71), as discussed in Sect. 11.3.2. At the optimum, the quantum Fisher information of Eq. (12.95) is maximized at

$$F_q(\xi) = 4\pi^2(N-1)^2 , \tag{12.98}$$

exhibiting at large N the limit $F_q(\xi) \to 4\pi^2 N^2 \approx 39.48N^2$. It can be observed that the quantum Fisher information $F_q(\xi)$ of Eq. (12.98) is strictly larger than that of Eq. (12.96) and also that of Eq. (12.97), for any size $N \geq 2$. However, it remains to be verified whether, and how, the estimation performance associated with $F_q(\xi)$

of Eq. (12.98) can actually be achieved by an effective estimation strategy. It can indeed, but in a slightly restrictive manner, as we now explain.

The $(N-1)$-qubit optimal input signal

$$|\psi_{\rm in}\rangle = \frac{1}{\sqrt{2}}\Big(|\overline{0}\,\rangle + |\overline{N-1}\,\rangle\Big) \in \overline{\mathcal{H}}_N \tag{12.99}$$

achieving the maximal $F_q(\xi)$ of Eq. (12.98), produced by the transformation of Eq. (12.71), the ξ-dependent output signal

$$|\widetilde{\psi}_\xi\rangle = \frac{1}{\sqrt{2}}\Big(|\overline{0}\,\rangle + \exp\big[i2\pi(N-1)\xi\big]\,|\overline{N-1}\,\rangle\Big)\,. \tag{12.100}$$

To estimate the phase ξ, a direct measurement of the ξ-dependent state $|\widetilde{\psi}_\xi\rangle \in \overline{\mathcal{H}}_N$ of Eq. (12.100) can be envisaged. A projective measurement in the Hilbert space $\overline{\mathcal{H}}_N$ can be designed, with one of the projective vectors of $\overline{\mathcal{H}}_N$ provided by the input state $|\psi_{\rm in}\rangle \in \overline{\mathcal{H}}_N$ of Eq. (12.99). From the inner product $\langle\psi_{\rm in}|\widetilde{\psi}_\xi\rangle = \big(1+\exp[i2\pi(N-1)\xi]\big)/2$, the probability in the measurement of projecting $|\widetilde{\psi}_\xi\rangle$ onto $|\psi_{\rm in}\rangle$ results as

$$P_0 = \big|\langle\psi_{\rm in}|\widetilde{\psi}_\xi\rangle\big|^2 = \frac{1}{4}\big|1+\exp[i2\pi(N-1)\xi]\,\big|^2 \tag{12.101}$$

$$= \frac{1}{2} + \frac{1}{2}\cos\big[2\pi(N-1)\xi\big] \tag{12.102}$$

$$= \cos^2\big[\pi(N-1)\xi\big]\,. \tag{12.103}$$

It is enough to supplement the quantum measurement in $\overline{\mathcal{H}}_N$ by a single additional projector projecting onto the subspace of $\overline{\mathcal{H}}_N$ orthogonal to the state vector $|\psi_{\rm in}\rangle$. This realizes a licit quantum measurement for $|\widetilde{\psi}_\xi\rangle \in \overline{\mathcal{H}}_N$, with only two orthogonal projectors to span $\overline{\mathcal{H}}_N$, delivering two possible outcomes: either $|\widetilde{\psi}_\xi\rangle$ is projected onto $|\psi_{\rm in}\rangle$ with the probability P_0 of Eq. (12.103), or $|\widetilde{\psi}_\xi\rangle$ is projected onto the orthogonal subspace with the complementary probability $1-P_0$.

In conformity with Eq. (11.45) or Eq. (12.89), such a two-outcome measurement leads to the classical Fisher information

$$F_c(\xi) = \frac{1}{P_0}\big(\partial_\xi P_0\big)^2 + \frac{1}{1-P_0}\big(\partial_\xi(1-P_0)\big)^2 = \frac{\big(\partial_\xi P_0\big)^2}{(1-P_0)P_0}\,. \tag{12.104}$$

By inserting the probability P_0 from Eqs. (12.102)–(12.103) and exploiting some elementary trigonometric identities, from Eq. (12.104) we finally obtain

$$F_c(\xi) = 4\pi^2(N-1)^2\,. \tag{12.105}$$

We therefore find that the two-outcome measurement of the multiple-qubit optimized output state $|\widetilde{\psi}_\xi\rangle$ of Eq. (12.100) achieves a classical Fisher information

$F_c(\xi)$ in Eq. (12.105) that exactly matches the quantum Fisher information $F_q(\xi)$ of Eq. (12.98), which is itself maximized over all possible input superpositions in Eq. (12.71). No other estimation protocol exploiting the ξ-dependent output state $|\widetilde{\psi}_\xi\rangle$ of Eq. (12.71) can therefore be expected to achieve a higher efficiency in terms of the performance assessed by the Fisher informations, classical and quantum. For effective estimation of the phase ξ, the data delivered by the two-outcome measurement of the multiple-qubit state $|\widetilde{\psi}_\xi\rangle$ of Eq. (12.100), can then be employed to implement the maximum-likelihood estimator, following an approach much similar to the multiple-qubit approach of Eqs. (11.109)–(11.114). The maximum-likelihood estimator will then achieve the best performance specified by the two coinciding Fisher informations $F_q(\xi)$ in Eq. (12.98) and $F_c(\xi)$ in Eq. (12.105), at least asymptotically in the limit of a large data record obtained by measuring a large number $L_{\rm rep}$ of independent repetitions of $|\widetilde{\psi}_\xi\rangle$. So with its superior Fisher informations in Eqs. (12.98) and (12.105), this direct maximum-likelihood estimation protocol, directly measuring $|\widetilde{\psi}_\xi\rangle$ of Eq. (12.100) with no Fourier transform, seems to be strictly more efficient than the alternative Fourier-based estimation protocols. Nevertheless, there exists substantial motivation for also considering the Fourier-based estimation protocol, as we now explain.

A major reason is that the probing signal $|\widetilde{\psi}_\xi\rangle$ of Eq. (12.100), on which is grounded the estimation with its superior Fisher informations of Eqs. (12.98) and (12.105), probes the unknown phase ξ only through the multiple $(N-1)\xi$, performing what is known as parameter amplification [302]. The two-outcome measurement is governed by the probability distribution $(P_0, 1-P_0)$, with P_0 of Eq. (12.102) which is formally similar to Eq. (11.109) and can be expressed as $P_0 = [1+s(\xi)]/2$ with $s(\xi) = \cos[2\pi(N-1)\xi]$. Over $L_{\rm rep}$ repeated measurements of $|\widetilde{\psi}_\xi\rangle$ of Eq. (12.100) delivering L_0 projections onto $|\psi_{\rm in}\rangle$, the maximum-likelihood procedure first provides access to an estimate $\widehat{s}_{\rm ML} = 2L_0/L_{\rm rep} - 1 \in [-1, 1]$ for $s(\xi)$ as in Eq. (11.112). An estimate is deduced for the amplified angle $2\pi(N-1)\xi$ as $\arccos(\widehat{s}_{\rm ML}) \in [0, 2\pi]$ mod 2π. Moreover, since the measurement probability P_0 is affected only by $\cos[2\pi(N-1)\xi]$, for the amplified angle $2\pi(N-1)\xi$ the estimate is returned in $[0,\pi]$, and the possibility of an angle in $[\pi, 2\pi]$ is unseen by the estimator, unless some additional information exists enabling to locate in $[0,\pi]$ or $[\pi, 2\pi]$ the targeted amplified angle. Equivalently, the sign of $2\pi(N-1)\xi$ in $[0,\pi]$ or $[-\pi, 0]$ is unseen. An amplified angle $2\pi(N-1)\xi$ known (estimated) over a range of extension π mod 2π, leads for the targeted phase ξ to several distinct determinations, over range cells of extension $1/\big(2(N-1)\big)$ mod $1/(N-1)$, that are compatible with the measured data, and cannot be distinguished by using these data. This ambiguity can be lifted only with prior or additional information limiting the unknown phase ξ to one specific determination or one definite range cell. In this respect, this estimation protocol involving parameter amplification realizes what can be called a **local estimation**, when one seeks to estimate, with high accuracy, a small change around a priorly known phase.

By contrast, the Fourier-based estimation, by exploiting instead of Eq. (12.100) the ξ-dependent signal $|\widetilde{\psi}_\xi\rangle$ of Eq. (12.71), simultaneously probes a series of N multiples $k\xi$, with $k=0$ to $N-1$, of the unknown phase ξ. By comparison, Eq. (12.100) only probes the two multiples at $k = 0$ and $k = N - 1$. Then, from $|\widetilde{\psi}_\xi\rangle$ of Eq. (12.71), the measurement of the Fourier-transformed state $|\psi_\xi\rangle = \mathsf{U}_\mathsf{F}^\dagger\, |\widetilde{\psi}_\xi\rangle$ of Eq. (12.72) with its N possible outcomes is able to exploit this richer information originating in $|\widetilde{\psi}_\xi\rangle$ to estimate the phase ξ. The resulting estimator $\widehat{\xi} = j/N$ returns an estimate over the whole range $[0, 1[$ for ξ, with no ambiguity from multiple unresolved determinations. In this respect, the Fourier-based estimation approach realizes what can be called a **global estimation**, over the whole feasible range of the unknown phase ξ.

So both approaches, local and global, to quantum phase estimation, process a signal of $N - 1$ qubits while offering complementary properties. Both approaches benefit from the favorable Heisenberg-enhanced rms error decreasing as $1/N$, in place of the more traditional $1/\sqrt{N}$. The local approach with performance in Eq. (12.98) is the most accurate, with an rms error decreasing as $1/(2\pi N) \approx 0.16/N$; yet only to precisely estimate a local change around a priorly known phase. The global Fourier-based approach with performance in Eq. (12.96) is a little less accurate (around three times less), with an rms error decreasing as $1/\big(\sqrt{4\pi^2/3 - 8}\, N\big) \approx 0.44/N$; but for estimation over the whole feasible range of the unknown phase. Both approaches have their precision which in principle can be made arbitrarily high by increasing the size N dimensioning the probing signal or the resolution $1/N$.

In addition, the global Fourier-based approach of Eq. (12.96) is specifically useful as it involves a single N-outcome von Neumann measurement in $\mathcal{H}_N$ of the $(N - 1)$-qubit Fourier-transformed signal $|\psi_\xi\rangle$ from Eq. (12.72), to deliver a one-shot global estimate directly for the phase ξ itself. Meanwhile, the more accurate local approach of Eq. (12.98) involves a direct (no prior Fourier transform) two-outcome measurement of the $(N - 1)$-qubit signal $|\widetilde{\psi}_\xi\rangle$ from Eq. (12.100), usually repeated L_{rep} times, to estimate first the amplified phase $(N - 1)\xi$ via statistical averaging, so as to achieve its superior accuracy for estimating the phase ξ. The local approach also avoids the circuit complexity inherent to the Fourier transform.

Quantum estimation can be referred to statistical estimation theory, because a quantum measurement is essentially a random process, delivering a random outcome at each realization. To estimate a deterministic parameter having a fixed value, for a statistical estimate from N independent random measurements (realizations), statistical estimation theory dictates an rms error or rms departure from the true value, decaying as $\sim 1/\sqrt{N}$. By comparison, a classical deterministic measurement operating with a resolution of $1/N$, would estimate a deterministic parameter with an error or precision as $\sim 1/N$. In this respect, the Fourier-based estimation with the uniform input superposition, measuring an N-dimensional quantum state to achieve an rms error $\sim 1/\sqrt{N}$, is therefore dominated by the statistical rms error $\sim 1/\sqrt{N}$ dictated by statistical estimation theory from random measurements. By contrast, the Fourier-based estimation with the optimized nonuniform input superposition achieves an rms error $\sim 1/N$, much like a deterministic measurement with a resolution $1/N$. In some sense, the random character of quantum measurement, imposing the $\sim 1/\sqrt{N}$ rms error, is like suppressed by exploiting the optimized input superposition.

Still other quantum phase estimation techniques have been reported, offering other properties and trade-offs, for instance for lifting the ambiguity of multiple phase determinations, while also preserving the valuable Heisenberg-enhanced efficiency. Adaptive techniques have been reported to gradually adjust the measurement from successive estimates, or techniques switching between several excitation signals or measurements according to some predefined schedules, or techniques resorting to more than one copy of the unitary gate U_ξ under estimation [300–306]. By comparison, as we indicated earlier, the original Fourier-based approach of Sect. 12.3.1 needs several copies of the gate U_ξ and less probing qubits, while comparatively the Fourier-based approach introduced in Sect. 12.3.2 needs more qubits but a single copy of the gate U_ξ and each qubit (although entangled to the others) can be applied and measured separately.

The Fourier-based phase estimation of this Sect. 12.3 has been analyzed in the presence of **quantum noise** in [374], when the output signal $|\widetilde{\psi}_\xi\rangle$ before it can be processed (by Fourier transform) for phase estimation, is degraded by some additional phase noise. The results of [374] show that the conditions optimal for estimation are not the same with noise as without. Also, in the presence of noise, increasing the size N of the probing signal is no longer uniformly beneficial. This increases the resolution of the estimation, but also in some sense the exposition to noise, so that to estimate a phase with noise it turns out that an optimum exists in the size N of the probing signal, to maximize the performance. The optimal entanglement of the probing signal, as specified by Eq. (12.84) with no noise, also changes with the noise. These results of [374] are illustrative of the rather general situation that the optimal processings of quantum signals with no noise, usually change in the presence of noise. They must be characterized anew, and are often dependent on the type and also on the level of quantum noise. And they are not always straightforward to characterize.

12.3.7 Phase Estimation for a Qubit Unitary Process

In this Sect. 12.3, the unitary process U_ξ involved in the estimation task of the phase ξ acts on quantum states of arbitrary dimension; U_ξ is not required to be an operator acting on qubit states in the two-dimensional Hilbert space $\mathcal{H}_2$. Accordingly, the eigenstate $|u\rangle$ involved in Figs. 12.3 and 12.7 is not necessarily a qubit state. Qubits are introduced as control qubits in Figs. 12.3 and 12.7. These control qubits interact with the unitary U_ξ in such a way that they get influenced by U_ξ and terminate in a ξ-dependent multiple-qubit state, as in Eqs. (12.45)–(12.47) or in Eqs. (12.65)–(12.66), that is further processed to estimate the phase ξ.

For a unitary process U_ξ acting directly on qubit states in the two-dimensional Hilbert space $\mathcal{H}_2$, as explained in Sect. 3.1.4, it is always possible to refer $\mathcal{H}_2$ to the orthonormal basis formed by the two eigenstates of U_ξ which can be denoted $\{|0\rangle, |1\rangle\}$, and to consider for U_ξ the generic form

$$\mathsf{U}_\xi = \begin{bmatrix} 1 & 0 \\ 0 & \exp(i2\pi\xi) \end{bmatrix}, \qquad (12.106)$$

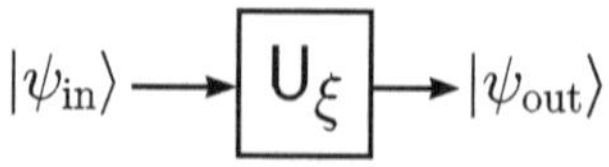

Fig. 12.12 A quantum gate materializing the generic qubit unitary process U_ξ from Eq. (12.106) acting on individual qubits which can be applied sequentially so as to jointly form an N-qubit state $|\psi_{\rm in}\rangle$ as input. Then, by sequentially transforming each qubit, the gate produces the N-qubit state $|\psi_{\rm out}\rangle$ as output

similar to Eq. (3.31). To undertake estimation of the phase ξ in Eq. (12.106), since U_ξ directly acts on qubits, no auxiliary control qubits need be introduced and no controlled version of the process U_ξ is required. One can apply multiple-qubit states sequentially to a single process U_ξ, as depicted in Fig. 12.12.

For instance, when the $(N-1)$-qubit states $|\overline{k}\,\rangle$ of Eq. (12.63) are used, we obtain the input–output transformation

$$|\overline{k}\,\rangle \longmapsto \mathsf{U}_\xi^{\otimes(N-1)}\,|\overline{k}\,\rangle = \exp(i2\pi k\xi)\,|\overline{k}\,\rangle \tag{12.107}$$

similar to Eq. (12.64). And by linearity, for an arbitrary input superposition, we obtain the input–output transformation

$$|\psi_{\rm in}\rangle = \sum_{k=0}^{N-1} a_k\,|\overline{k}\,\rangle \longmapsto \mathsf{U}_\xi^{\otimes(N-1)}\,|\psi_{\rm in}\rangle = \sum_{k=0}^{N-1} a_k \exp\!\left(i2\pi\frac{j_\xi k}{N}\right)|\overline{k}\,\rangle = |\psi_{\rm out}\rangle\ , \tag{12.108}$$

with $j_\xi = \xi N$. Then $|\psi_{\rm out}\rangle$ of Eq. (12.108) is similar to $|\widetilde{\psi}_\xi\rangle$ of Eq. (12.71). This state $|\psi_{\rm out}\rangle \equiv |\widetilde{\psi}_\xi\rangle$ is therefore equally amenable to the processings and analyses developed in this Sect. 12.3 to estimate the phase ξ.

From the $(N-1)$-qubit signal $|\psi_{\rm out}\rangle \equiv |\widetilde{\psi}_\xi\rangle$, the optimal estimation of Eqs. (12.98)–(12.105) coincides with the optimal estimation of Eqs. (11.102)–(11.110) in Sect. 11.3.4. The generator corresponding to the unitary process in Eq. (12.106) is $\mathsf{G} = \pi\sigma_z$, with two eigenvalues $u_{\max} = \pi$ and $u_{\min} = -\pi$ and two eigenstates $|u_{\max}\rangle = |0\rangle$ and $|u_{\min}\rangle = |1\rangle$. The $(N-1)$-qubit optimal input probe of Eq. (12.99) coincides with Eq. (11.105), to achieve the quantum Fisher information of Eq. (12.98) which is the maximum $F_q^{(L)\max}(\xi)$ of Eq. (11.106) when $F_q^{\max}(\xi) = 4\pi^2$ and $L = N-1$ qubits, benefiting from the Heisenberg-enhanced scaling.

12.4 Period Finding

Another important application of the quantum Fourier transform is for period finding, as exploited for instance in the celebrated Shor quantum algorithm [377–379] performing prime factorization of integers in polynomial complexity in the length of the integer (while comparatively all known classical algorithms exhibit exponential complexity). We consider a periodic function $f(\cdot)$ defined over the integers with an integer period $r \in \{1, \dots, N-1\}$ such that $f(x) = f(x+r)$ for all x integer. For a generic function $f(\cdot)$ with no specified properties, finding the period r by classical means usually requires a number of at least $O(N)$ evaluations of $f(\cdot)$. With a quantum approach, this can be accomplished with a single evaluation of $f(\cdot)$, on an appropriate superposition of input arguments, followed by an appropriate processing, as we are going to see.

An N-dimensional Hilbert space $\mathcal{H}_N$ of quantum states is referred to the orthonormal basis $\big\{|x\rangle\big\}$ for integer $x \in \{0, 1, \dots, N-1\}$. Leaning on the results of Chap. 5, we envisage a quantum circuit capable of producing the superposed quantum state, comparable to the right-hand side of Eq. (5.13), having here the form

$$|\psi\rangle = \frac{1}{\sqrt{N}} \sum_{x=0}^{N-1} |x\rangle\, |f(x)\rangle \ . \tag{12.109}$$

For all integer x, each state vector $|f(x)\rangle$ is a basis state from an auxiliary Hilbert space $\mathcal{H}$ chosen with the appropriate dimension so as to represent by orthogonal basis states all feasible values $f(x)$ of the periodic function $f(\cdot)$. The bipartite state $|\psi\rangle \in \mathcal{H}_N \otimes \mathcal{H}$ of Eq. (12.109) is obtained through a single evaluation of the function $f(\cdot)$, on a superposed input as in Eq. (5.13), to afford some form of parallel computation. At the output, the state $|\psi\rangle \in \mathcal{H}_N \otimes \mathcal{H}$ of Eq. (12.109) contains all the N values $|f(x)\rangle$ of the function $f(\cdot)$ for all integers x over at least one period or more, placed in a uniform quantum superposition, with each value $|f(x)\rangle \in \mathcal{H}$ individually indexed via the set of the corresponding N orthogonal addresses $|x\rangle \in \mathcal{H}_N$. In this way, the bipartite state $|\psi\rangle \in \mathcal{H}_N \otimes \mathcal{H}$ of Eq. (12.109) contains the repeated values of $|f(x)\rangle$ separated by the period r, the task being to exploit such constitutive information to determine r.

An inverse quantum Fourier transform is performed via Eq. (12.19) on the first part of $|\psi\rangle$ or register containing $|x\rangle \in \mathcal{H}_N$, so as to produce

$$|\widetilde{\psi}\rangle = \mathsf{U}_\mathsf{F}^\dagger \otimes \mathsf{Id}\, |\psi\rangle = \frac{1}{\sqrt{N}} \sum_{x=0}^{N-1} \mathsf{U}_\mathsf{F}^\dagger\, |x\rangle\, |f(x)\rangle \tag{12.110}$$

$$= \frac{1}{\sqrt{N}} \sum_{x=0}^{N-1} \frac{1}{\sqrt{N}} \sum_{k=0}^{N-1} \exp\!\left(-i2\pi \frac{xk}{N}\right) |k\rangle\, |f(x)\rangle \tag{12.111}$$

$$= \frac{1}{\sqrt{N}} \sum_{k=0}^{N-1} |k\rangle\, |\widetilde{f}(k)\rangle \ , \tag{12.112}$$

with the vector $|\widetilde{f}(k)\rangle$ of the Hilbert space $\mathcal{H}$ defined as

$$|\widetilde{f}(k)\rangle = \frac{1}{\sqrt{N}} \sum_{x=0}^{N-1} \exp\!\left(-i2\pi \frac{xk}{N}\right) |f(x)\rangle \ , \tag{12.113}$$

which is not necessarily with unit norm. The state $|\widetilde{\psi}\,\rangle$ is a valid normalized state of the tensor product Hilbert space $\mathcal{H}_N \otimes \mathcal{H}$, so that its unit squared norm $1 = \langle \widetilde{\psi}\,|\widetilde{\psi}\,\rangle$ gives

$$1 = \langle \widetilde{\psi}\,|\widetilde{\psi}\,\rangle = \frac{1}{N}\sum_{k=0}^{N-1} \langle \widetilde{f}(k)|\widetilde{f}(k)\rangle\ , \tag{12.114}$$

leading to the upper bound $N \geq \langle \widetilde{f}(k)|\widetilde{f}(k)\rangle$ on each squared norm for every k. The sequence of $|\widetilde{f}(k)\rangle$ for $k \in \{0, 1, \ldots, N-1\}$ can be interpreted as a Fourier transform for the quantum function $|f(x)\rangle$ or sequence of N quantum states $|f(x)\rangle$ for $x \in \{0, 1, \ldots, N-1\}$, as follows from Eq. (12.28). When $|f(x)\rangle$ is a periodic sequence with period r, its Fourier transform $|\widetilde{f}(k)\rangle$ consists of a line spectrum, made of sharp spectral lines occurring at integer multiple of the fundamental frequency $1/r$. To see this more formally and quantitatively, we first analyze the simpler situation where the period r is an exact submultiple of the length N, and later we will consider the situation where this condition does not hold.

When the period r is an exact submultiple of the length N, we write $nr = N$ for some integer $n \geq 1$. A given $x \in \{0, 1, \ldots, r-1\}$ fixes an initial or reference phase over the period. For each such $x \in \{0, 1, \ldots, r-1\}$, the sum in Eq. (12.113) runs over the arguments $x + \ell r$ for $\ell = 0$ to $n-1$, which all share the same value $f(x) = f(x+\ell r)$, so that their collective contribution to the summand in Eq. (12.113) amounts to

$$|F(x)\rangle = \exp\!\left(-i2\pi\frac{xk}{N}\right)|f(x)\rangle \sum_{\ell=0}^{n-1}\exp\!\left(-i2\pi\frac{\ell rk}{N}\right)\,, \tag{12.115}$$

The sum in Eq. (12.115) evaluates to

$$S(k) = \sum_{\ell=0}^{n-1}\exp\!\left(-i2\pi\frac{\ell rk}{N}\right) = n\delta_{k,mn}\ , \tag{12.116}$$

being nonzero only at $k = mn$, for integer $m \in \{0, 1, \ldots, r-1\}$, i.e., when k is an integer multiple of $n = N/r$. The sum in Eq. (12.113) is then $|\widetilde{f}(k)\rangle = \sum_{x=0}^{r-1}|F(x)\rangle/\sqrt{N}$ and it evaluates to

$$|\widetilde{f}(k)\rangle = \begin{cases} \dfrac{n}{\sqrt{N}}\displaystyle\sum_{x=0}^{r-1}\exp\!\left(-i2\pi\frac{xk}{N}\right)|f(x)\rangle & \text{for } k = m\dfrac{N}{r},\ \text{with } m \in \{0, 1, \ldots, r-1\}\,, \\[2ex] 0 & \text{for other } k \in \{0, 1, \ldots, N-1\}\,. \end{cases} \tag{12.117}$$

Such sequence $|\widetilde{f}(k)\rangle$ represents the line spectrum announced above, with spectral lines at integer multiples of N/r.

Then, on the superposed state $|\widetilde{\psi}\rangle \in \mathcal{H}_N \otimes \mathcal{H}$ of Eq. (12.112), the first register containing $|k\rangle \in \mathcal{H}_N$ is measured in the computational basis of the N-dimensional space $\mathcal{H}_N$. This outputs at random a value k multiple of N/r, since according to Eq. (12.117) other k, associated with a zero weight $|\widetilde{f}(k)\rangle$, are absent in the superposition of Eq. (12.112). From such a multiple of N/r, the (classical) continued fraction algorithm [378, 379] usually allows one to deduce the integer period $r \in \{1, \ldots, N-1\}$. Alternatively, the process can be repeated several times to collect several multiples of N/r whose greatest common divisor retrieves N/r and hence r.

When the period r does not divide the length N, the integer period r is now such that $(n-1)r + n_0 = N$ with integer $n_0 \in \{1, \ldots, r-1\}$. For the vector $|\widetilde{f}(k)\rangle$ of Eq. (12.113), the sharp spectral lines defined by Eq. (12.117) will tend to broaden for values of k around the integer multiples mN/r where now N/r is no longer an integer. The regime of interest in practice is when $n \gg 1$ or $N \gg r$, with a sufficiently large number of periods r covered by the integers $x \in \{0, 1, \ldots, N-1\}$, so as to allow an accurate determination of r. Or the other way around, for a given period r, a larger range N increases the accuracy in determining r.

In the sum over x in Eq. (12.113), we again factorize the repeated value of $|f(x)\rangle$ for each $x \in \{1, \ldots, r-1\}$, leading this time to

$$|\widetilde{f}(k)\rangle = \frac{1}{\sqrt{N}}\left[S_0(k)\sum_{x=0}^{n_0-1}\exp\left(-i2\pi\frac{xk}{N}\right)|f(x)\rangle + S_1(k)\sum_{x=n_0}^{r-1}\exp\left(-i2\pi\frac{xk}{N}\right)|f(x)\rangle\right], \tag{12.118}$$

with the two sums

$$S_0(k) = 1 + \exp\left(-i2\pi\frac{kr}{N}\right) + \cdots + \exp\left(-i2\pi k\frac{(n-1)r}{N}\right) \tag{12.119}$$

$$= \exp\left(-i\pi k\frac{(n-1)r}{N}\right)\frac{\sin\left(\pi k\frac{nr}{N}\right)}{\sin\left(\pi k\frac{r}{N}\right)} \tag{12.120}$$

and

$$S_1(k) = 1 + \exp\left(-i2\pi\frac{kr}{N}\right) + \cdots + \exp\left(-i2\pi k\frac{(n-2)r}{N}\right) \tag{12.121}$$

$$= \exp\left(-i\pi k\frac{(n-2)r}{N}\right)\frac{\sin\left(\pi k\frac{n-1}{N}r\right)}{\sin\left(\pi k\frac{r}{N}\right)}. \tag{12.122}$$

In addition, when $n_0 \to r$ so that $nr = N$, one finds for $S_0(k)$ in Eq. (12.120) the limit

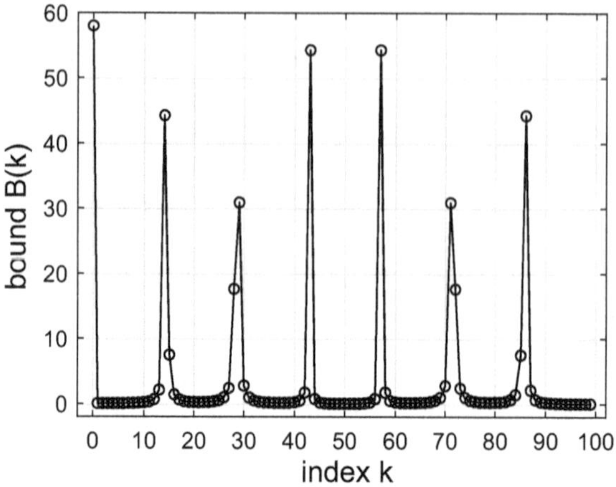

Fig. 12.13 The upper bound $B(k)$ of Eq. (12.124) limiting the squared norm $\langle \widetilde{f}(k)|\widetilde{f}(k)\rangle$, as a function of the integer $k \in \{0, \ldots, N-1\}$ materialized at $\circ$, and reflecting the structure of line spectrum of the vector sequence $|\widetilde{f}(k)\rangle$ of Eq. (12.118) with r sharp spectral lines occurring around integer multiples of the noninteger value N/r. Here the period is $r = 7$ and the length $N = 100$, giving $n_0 = 2$ and $N/r \approx 14.286$

$$S_0(k) = \frac{\sin(\pi k)}{\sin(\pi k/n)} = \begin{cases} n & \text{for } k = mn = m\dfrac{N}{r}\,, \\ 0 & \text{for other } k \in \{0, 1, \ldots, N-1\}\,, \end{cases} \tag{12.123}$$

which is identical to $S(k)$ of Eq. (12.116), and no term associated with $S_1(k)$ subsists in Eq. (12.118) which then coincides with Eq. (12.117).

The vector $|\widetilde{f}(k)\rangle \in \mathcal{H}$ of Eq. (12.118) in squared norm can be upper bounded as

$$\langle \widetilde{f}(k)|\widetilde{f}(k)\rangle \le \frac{1}{N}\Big[|S_0(k)|^2 n_0^2 + |S_1(k)|^2 (r-n_0)^2\Big] = B(k)\,. \tag{12.124}$$

The bound $B(k)$ of Eq. (12.124) reflects the structure of line spectrum of the vector $|\widetilde{f}(k)\rangle$ of Eq. (12.118) as a function of $k \in \{0, \ldots, N-1\}$, featuring r sharp spectral lines occurring around integer multiples of the noninteger value N/r. This is illustrated in Fig. 12.13 in typical conditions.

Then, when on the superposed state $|\widetilde{\psi}\rangle \in \mathcal{H}_N \otimes \mathcal{H}$ of Eq. (12.112) the first register containing $|k\rangle \in \mathcal{H}_N$ is measured in the computational basis of the N-dimensional space $\mathcal{H}_N$, with high probability an integer k is obtained that is very close to a multiple of the noninteger value N/r where the weight $|\widetilde{f}(k)\rangle$ of Eq. (12.118) is large. Again, from such result, the (classical) continued fraction algorithm [378, 379] usually allows one to deduce the integer period r, or several repetitions can be done to collect several (approximate) multiples to help.

As we mentioned, this quantum algorithm for finding the period $r \in \{1, \ldots, N-1\}$ of the function $f(\cdot)$ basically requires only one evaluation of $f(\cdot)$, on a superposed quantum state affording some parallel processing conveyed by Eqs. (12.109)–(12.112). When the integers are represented on L bits and the quantum states like $|x\rangle$ are materialized on L qubits with $2^L \geq N$, the (inverse) quantum Fourier transform in Eq. (12.110) can be computed with the algorithmic complexity $O(L^2)$ indicated at the end of Sect. 12.2. The (classical) continued fraction algorithm operates [9] with a polynomial algorithmic complexity $O(L^3)$. As a result, the overall algorithmic complexity of the period-finding quantum algorithm is ultimately determined by the complexity of computing (one evaluation of) the function $f(\cdot)$, in relation to this complexity $O(L^3)$. In many situations of interest, a polynomial complexity in the length L of the representation of integers is obtained by the quantum period-finding algorithm. Meanwhile, classical approaches requiring at least $O(N) \equiv O(2^L)$ evaluations of $f(\cdot)$ comparatively remain with exponential complexity.

12.5 Order Finding

Order finding is a problem of number theory or arithmetic offering another situation where a quantum resolution in polynomial algorithmic complexity exists when all known classical resolutions comparatively exhibit exponential complexity. Order finding also stands as an important step in the factorization of integers into prime numbers in polynomial complexity by the Shor quantum algorithm [377–379]. The quantum resolution of order finding exploits the quantum Fourier transform, under two possible alternative approaches, either as a period-finding task [378, 379] or as a phase estimation task [9]. This we are going to see, especially to further illustrate significant applications of the quantum Fourier transform.

For two coprime integers $N > a > 0$, the order of a modulo N is the smallest positive integer $r > 0$ such that $a^r = 1 \pmod N$. Finding the order r with classical methods (known today) requires a number of operations or algorithmic complexity of order $O(N)$, or an exponential complexity $O(2^L)$ in the length $L \gtrsim \log_2(N)$ or number of bits L to represent the integer N. Comparatively, with quantum means, the order-finding problem can be solved [378, 379] with an algorithmic complexity polynomial in the length L.

We note that for any two integer exponents x_1 and x_2 in the range $0 < x_1 < x_2 \leq r$, we necessarily have $a^{x_1} \neq a^{x_2} \pmod N$. Otherwise we would have $a^{x_1} = a^{x_2} = a^{x_1} a^{x_2 - x_1} \pmod N$, and since a is coprime to N, also a has an inverse modulo N, and so has a^{x_1}, and we would deduce $a^{x_2 - x_1} = 1 \pmod N$, so that the order of a modulo N would be $\leq x_2 - x_1 < r$ instead of r, which is not permitted by definition of the order r. Consequently, the sequence $(a, a^2, \ldots, a^r)$ is a sequence of r distinct integers modulo N with values in the set $\{0, 1, 2, \ldots, N-1\}$ of cardinality N. It follows that the order r satisfies $r \leq N$.

12.5.1 Order Finding by Period Finding

We also have $1 = a^0 = a^r \pmod N$, and the order r stands as the period of the function defined for an integer argument x as

$$f_a(x) = a^x \pmod N\,. \tag{12.125}$$

Consequently, the problem of finding the order r of a modulo N is equivalent [378, 379] to finding the period r of the modular-exponentiation function $f_a(x)$ from Eq. (12.125). For finding its period r, the function $f_a(x)$ from Eq. (12.125) can be handled by the quantum method of Sect. 12.4. Implementing the computation of $f_a(x)$ can be accomplished by sequential squaring $a \mapsto a^2 \pmod N$ performed no more than L times to obtain the factors a^{2^j} up to $2^j = 2^L \geq N$, which are then selectively multiplied according to the corresponding bits in the representation of the L-bit integer argument x. Each multiplication involving two L-bit integers can be performed with no more than $O(L^2)$ elementary operations. This leads [378] to an algorithmic complexity of $O(L^3)$ for evaluating the modular-exponentiation function $f_a(\cdot)$ from Eq. (12.125) with integers represented on L bits.

Period finding on $f_a(\cdot)$ by the quantum method of Sect. 12.4 requires only one evaluation of $f_a(\cdot)$, followed by the processing exploiting the quantum Fourier transform and the continued fraction algorithm with an overall $O(L^3)$ algorithmic complexity. As a result, via period-finding on $f_a(\cdot)$, the order of a modulo N can be obtained in a polynomial $O(L^3)$ complexity.

12.5.2 Order Finding by Phase Estimation

For solving the order-finding problem, the quantum Fourier transform can be exploited in an alternative manner [9, 372], performing a quantum phase estimation as in Sect. 12.3. Moreover, this application of quantum phase estimation for order finding, is specifically interesting to examine, as it faces a complication that is not taken into account in the generic methodology presented in Sect. 12.3. To anticipate, it is the complication where the eigenstate denoted $|u\rangle$ in Fig. 12.3 or 12.7 and used as an excitation to estimate the phase ξ, is itself dependent on the unknown phase ξ, and therefore not directly accessible as an excitation signal.

For order finding here, the unitary process, similar to U_ξ in Sect. 12.3, and undergoing the phase estimation, acts in the N-dimensional Hilbert space $\mathcal{H}_N$ with computational orthonormal basis $\{|0\rangle, |1\rangle, |2\rangle, \ldots, |N-1\rangle\}$. It takes the form of the unitary operator U_a from $\mathcal{H}_N$ onto $\mathcal{H}_N$ defined on the computational basis of $\mathcal{H}_N$ by

$$\mathsf{U}_a\,|x\rangle = |ax \pmod N\rangle\,, \tag{12.126}$$

for the integer $x \subset \{0, 1, 2, \ldots, N-1\}$. If x is materialized as an L-bit word, on the overhead where $N \le x \le 2^L - 1$ the unitary U_a of Eq. (12.126) can be extended by acting trivially on the corresponding supernumerary states $|x\rangle$. An integer a coprime to N ensures that a is invertible modulo N, and as a consequence, when x covers the N integers $\{0, 1, 2, \ldots, N-1\}$, the modular product $ax \pmod N$ generates N distinct integers in $\{0, 1, 2, \ldots, N-1\}$, and Eq. (12.126) is a bijective mapping establishing U_a as unitary. If two distinct integers $x_1 \neq x_2$ were to give $ax_1 = ax_2 \pmod N$, we would have $a(x_2 - x_1) = 0 \pmod N$, so a would be a divider of zero, which cannot be the case since a is invertible modulo N.

For any integer $s \in \{0, 1, \ldots, r-1\}$, the r (normalized) states defined as

$$|u_s\rangle = \frac{1}{\sqrt{r}} \sum_{k=0}^{r-1} \exp\left(-i2\pi \frac{sk}{r}\right) \left|a^k \pmod N\right\rangle, \tag{12.127}$$

are r eigenstates of the unitary U_a of Eq. (12.126). We have

$$\mathsf{U}_a |u_s\rangle = \frac{1}{\sqrt{r}} \sum_{k=0}^{r-1} \exp\left(-i2\pi \frac{sk}{r}\right) \left|a^{k+1} \pmod N\right\rangle \tag{12.128}$$

$$= \exp\left(i2\pi \frac{s}{r}\right) |u_s\rangle , \tag{12.129}$$

since the value of the summand is the same in $k = 0$ and $k = r$, identifying $\exp(i2\pi s/r)$ as the eigenvalue associated with the eigenstate $|u_s\rangle$ of the unitary U_a.

For a given integer $s \in \{0, 1, \ldots, r-1\}$, the methodology of quantum phase estimation of Sect. 12.3 could be invoked to estimate the phase s/r, and from here deduce the order r. This however would require an excitation signal to the unitary process U_a under estimation prepared as the associated eigenstate $|u_s\rangle$ from Eq. (12.127), and this we cannot do, since we do not know (yet) the order r. Alternatively, it is possible to excite the unitary process U_a with a superposition of its eigenstates $|u_s\rangle$ from Eq. (12.127). A similar step was taken in Eqs. (5.57)–(5.58) for the Grover quantum search algorithm of Sect. 5.5. From the geometric series $\sum_{s=0}^{r-1} \exp(-i2\pi s\ell/r) = r\delta_{\ell r}$ with periodicity r in ℓ, we get the superposition of eigenstates

$$\frac{1}{\sqrt{r}} \sum_{s=0}^{r-1} \exp\left(i2\pi \frac{s\ell}{r}\right) |u_s\rangle = \left|a^\ell \pmod N\right\rangle, \tag{12.130}$$

leading in $\ell = 0$ or in $\ell = r$ to the simple superposition

$$\frac{1}{\sqrt{r}} \sum_{s=0}^{r-1} |u_s\rangle = |1\rangle \tag{12.131}$$

amounting to a superposed quantum state reducing to the basis state $|1\rangle$, which is readily accessible as an input excitation. The same phase-estimation procedure as in Sect. 12.3.1 can then be applied, with now the superposed state $|1\rangle$ of Eq. (12.131) as an input excitation to the unitary process U_a under estimation. This phase-estimation procedure, including the final inverse quantum Fourier transform of Eq. (12.48), is end-to-end a linear processing, ruled by the superposition principle of linear algebra, for its action on the superposed input from Eq. (12.131). Loosely speaking, at the output each peaked pattern similar to the distribution of coefficients S_j occurring in Eq. (12.55), for instance, will appear in linear superposition, each contributed by an eigenstate $|u_s\rangle$ with phase s/r present in the input superposition of Eq. (12.131). As a result, on the multiple-qubit signal serving to probe the unknown phase, the final quantum measurement at the output after the inverse quantum Fourier transform, delivers an estimate $\widehat{s/r}$ for one of the phases s/r present in the eigenvalues $\exp\bigl(i2\pi s/r\bigr)$ associated with the superposition of Eq. (12.131), for a given integer s selected at random by the Born rule, at the location of a peak in the probability amplitude. In this respect, a significant role is conferred to a sufficiently fine resolution 2^{-L_q} afforded by L_q probing qubits for the phase estimation in relation to the step $1/r$.

From an estimate $\widehat{s/r}$ for the phase s/r, the last step is to deduce the integer order r, in the presence of an unknown integer s. To circumvent the unknown s, we remember that we are dealing with two integers s and r known to satisfy $0 \le s < r \le N$, with known modulus N. With these constraints on s and r, the rational number s/r closest to the estimate $\widehat{s/r}$ can be found by the (classical) continued fraction algorithm, as also exploited in the period-finding quantum procedure of Sect. 12.4. The integer $\widehat{r}$ returned by the continued fraction algorithm represents the correct order r with high probability. As a verification it is easily checked that indeed $a^{\widehat{r}} = 1 \pmod{N}$. If it were not the case, the same order-finding procedure would have to be run anew, providing another estimate $\widehat{s'/r}$ of a phase s'/r with the same order r and another integer s', which alone can serve to estimate the order r. Additionally, the two estimates $\widehat{s/r}$ and $\widehat{s'/r}$ can be combined, jointly offering a more stringent set of constraints narrowing the values admissible for the integer order r. After one or a few such phase estimation steps, the order r can be obtained with quasi certainty.

When the quantum states for representing the integers $x \in \{0, 1, 2, \dots, N-1\}$ are materialized on L qubits with $2^L \ge N$, the implementation of the unitary process U_a of Eq. (12.126) can be realized through the multiplication of two L-bit integers, which can be performed with no more than $O\bigl(L^2\bigr)$ elementary operations. The phase estimation protocol, in the implementation of Fig. 12.3, in addition introduces L_q auxiliary qubits to probe the phase of U_a. The quantum circuit as in Fig. 12.3 for phase estimation involves the process U_a at various powers $\mathsf{U}_a^{2^j}$ for $j = 0$ to $L_\text{q} - 1$. As for the computation of the function $f_a(\cdot)$ from Eq. (12.125), a number L_q of the above multiplications can be performed to obtain an algorithmic complexity $O\bigl(L^2 L_\text{q}\bigr)$ for the circuit for phase estimation. For a sufficiently fine resolution 2^{-L_q} to estimate the phase, the appropriate number L_q of probing qubits scales linearly with the length L of the representation of integers for order finding. We therefore deduce a polynomial algorithmic complexity $O(L^3)$ for solving the order-finding problem via quantum phase estimation, similar to the complexity of the approach via period-finding of Sect. 12.5.1, both exploiting the quantum Fourier transform.

12.6 Shor Quantum Factoring Algorithm

The Shor quantum factoring algorithm [377–379] is a randomized algorithm that factors an integer of length L into primes in polynomial complexity, while all classical algorithms known today comparatively exhibit exponential complexity in L. As a rule, factoring can be converted into an order-finding task, that the Shor algorithm solves efficiently via the Fourier-based quantum methodology of Sect. 12.5, as the crucial speedup step. The Shor factoring algorithm also exploits the (classical) Euclidean algorithm, enabling for two integers N_1 and N_2 represented on L bits, to find their greatest common divisor $\mathrm{GCD}(N_1, N_2)$ in a step of polynomial complexity no larger than $O(L^3)$.

To find a nontrivial factor of a composite (odd) integer N, the Shor algorithm essentially addresses the following steps.

- Start by randomly choosing an integer x in the range 2 to $N - 1$. If x happens to divide N we are done, and this can be checked by computing $\mathrm{GCD}(x, N)$ with the Euclidean algorithm.
- Otherwise x is coprime to N, and the order r of x modulo N can be found efficiently by the quantum methodology of Sect. 12.5. We therefore have $x^r - 1 = 0 \pmod{N}$.
- Then if the order r is even, $r/2$ is an integer, enabling the factorization $(x^{r/2} - 1)(x^{r/2} + 1) = x^r - 1 = 0 \pmod{N}$. Therefore, N divides the product $(x^{r/2} - 1)(x^{r/2} + 1)$. But N does not divide $x^{r/2} - 1$, since we would have $x^{r/2} - 1 = 0 \pmod{N}$ and the order of x modulo N would be $\leq r/2$ instead of r.
- If in addition N does not divide $x^{r/2} + 1$, that is $x^{r/2} + 1 \neq 0 \pmod{N}$, then necessarily the factors of N are distributed between the factors of $x^{r/2} + 1$ and of $x^{r/2} - 1$, and necessarily N has a nontrivial common factor[1] with $x^{r/2} + 1$, and another one with $x^{r/2} - 1$. As a result, each one of $\mathrm{GCD}(x^{r/2} + 1, N)$ and $\mathrm{GCD}(x^{r/2} - 1, N)$ provides a nontrivial factor of N. For any x coprime to N, this algorithm succeeds in returning a nontrivial factor of N, unless either (1) the order r is odd, or (2) r is even and $x^{r/2} = -1 = N - 1 \pmod{N}$. In this last case (2), the algorithm obtains the trivial factors N and 1.

As a useful supplement, it can be proved [378–380] (using the Chinese remainder theorem and the group structure under multiplication modulo N of the integers coprime to N) that, when m is the number of distinct prime factors of N, we have the probability

[1] For example, $N = 6$ divides the product $4 \times 3 = 12$, and N does not divide neither 4 nor 3, so N has a common factor with 4 (which is 2) and a common factor with 3 (which is 3).

$$\Pr\{r \text{ is even and } x^{r/2} \not\equiv -1 \pmod N\} \geq 1 - \frac{1}{2^{m-1}} \,. \tag{12.132}$$

This guarantees that the Shor factoring algorithm in general succeeds with high probability. In the most stringent case when $m = 2$, the probability of Eq. (12.132) is larger than $1/2$, so that two repetitions of the algorithm on average suffice. A composite integer N with $m = 2$ prime factors is the situation of the cryptographic key $k_{\mathrm{RSA}} \equiv N$ of the RSA cryptosystem evoked at the end of Sect. 5.5, which can therefore be broken in polynomial complexity by the Shor quantum factoring algorithm. The special case with $m = 1$ when N is a power of a single prime can be treated separately by classical means in $O(L^3)$ complexity for factoring N.

All the steps in the Shor factoring algorithm can be performed by classical means with polynomial algorithmic complexity in the length L of the integer N, except the order-finding step which today remains (classically) with exponential complexity. The efficient quantum resolution of the order-finding step, by means of the quantum methodology of Sect. 12.5, leads for the Shor factoring algorithm to an overall polynomial algorithmic complexity in $O(L^3)$. The algorithm may then be repeated (no more than $O(L)$ times) to obtain a complete factorization of N. Incidently, the performance of the Shor quantum algorithm does not solve the P-versus-NP conjecture from computational complexity theory [381], since the factorization of integers (and a fortiori order finding) is not proved to be an NP-complete problem.

The Shor quantum algorithm for factoring integers in polynomial complexity stands as a prominent landmark, demonstrating a specific quantum contribution to information processing, to achieve a performance unattainable by classical means. The Shor algorithm was introduced in 1994 by [377] for its principle. Experimental implementations were then reported, first in 2001 by [382] for factoring $N = 15 = 3 \times 5$, then in 2012 by [383] for factoring $N = 21 = 3 \times 7$, and in 2019 by [384] for $N = 35 = 5 \times 7$.

For a further illustration of the principle, we consider factoring $N = 55$. We select $x = 4$ coprime to N, with $\mathrm{GCD}(x = 4, N = 55) = 1$. The order of x modulo N is $r = 10$, with $4^{10} = 1\,048\,576 = 19\,065 \times 55 + 1$. It follows that $x^5 + 1 = 1025 = 5^2 \times 41$ with $\mathrm{GCD}(x^5 + 1, N) = 5$ providing a factor of N, and $x^5 - 1 = 1023 = 3 \times 11 \times 31$ with $\mathrm{GCD}(x^5 - 1, N) = 11$ providing another factor of N. We obtain the product $N = 55 = 5 \times 11$ completing the factorization of N.

12.7 HHL Algorithm for Linear Systems of Equations

Another important application of the quantum Fourier transform is the HHL quantum algorithm introduced in [385] to solve a linear system of algebraic equations. The HHL algorithm again takes advantage of the Fourier-based phase estimation from Sect. 12.3. The task is to find $\vec{x}$ solving the linear system of equations $A\vec{x} = \vec{b}$, where A is an $N \times N$ matrix of $\mathbb{C}^N \times \mathbb{C}^N$ and $\vec{b}$ a unit-norm column vector of $\mathbb{C}^N$. The matrix A is supposed to be Hermitian, otherwise one would have to solve the enlarged system

$$\begin{bmatrix} [0] & A \\ A^\dagger & [0] \end{bmatrix} \begin{bmatrix} \vec{0} \\ \vec{x} \end{bmatrix} = \begin{bmatrix} \vec{b} \\ \vec{0} \end{bmatrix} , \tag{12.133}$$

involving a Hermitian matrix. For a quantum resolution by the HHL algorithm [385], the eigendecomposition of A is invoked as

$$A = \sum_{n=1}^{N} a_n \, |a_n\rangle \, \langle a_n| \; , \tag{12.134}$$

introducing an orthonormal eigenbasis $\{|a_n\rangle\}$ of the N-dimensional Hilbert space $\mathcal{H}_N$ and associated (real) eigenvalues $\{a_n\}$. The vector $\vec{b}$ is represented by the state vector $|\vec{b}\,\rangle$ of $\mathcal{H}_N$ expressed in the eigenbasis as

$$|\vec{b}\,\rangle = \sum_{n=1}^{N} b_n \, |a_n\rangle \; . \tag{12.135}$$

The matrix A is readily inverted in its eigenbasis as

$$A^{-1} = \sum_{n=1}^{N} a_n^{-1} \, |a_n\rangle \, \langle a_n| \; . \tag{12.136}$$

The solution $\vec{x} = A^{-1}\vec{b}$ of the linear system will be supplied by the HHL quantum algorithm in the eigenbasis under the form

$$|\vec{x}\,\rangle = C \sum_{n=1}^{N} a_n^{-1} b_n \, |a_n\rangle \; , \tag{12.137}$$

where C is a constant ensuring that $|\vec{x}\,\rangle$ is a valid (normalized) state vector of $\mathcal{H}_N$.

To accomplish this resolution in the eigenbasis, the HHL algorithm exploits the Fourier-based phase estimation procedure from Sect. 12.3.1 in the following way. The matrix A Hermitian leads to a matrix $\exp(iA)$ which is unitary, and being unitary, $\exp(iA)$ can in principle receive a quantum implementation by means of a quantum circuit. For more flexibility, the algorithm implements the unitary matrix $\exp(iAt)$, where t is a real scalar. Various techniques exist [385–387] for efficiently implementing such a unitary transformation $\exp(iAt)$. The unitary matrix $\exp(iAt)$ is endowed with the same N eigenstates $|a_n\rangle$ and N eigenvalues $\exp(ia_nt)$, and t serves to control the range covered by the N phases a_nt in relation to $[0, 2\pi[$. The Fourier-based phase estimation aims at extracting from $\exp(iAt)$ the information contained in the phase a_nt, so that from the eigenvalue a_n the solution in Eq. (12.137) can be finally obtained (constructed). In the phase-estimation circuit of Fig. 12.3, the unitary matrix

$\exp(iAt)$ plays the role of the unitary operator U_ξ. As in Fig. 12.3, a set of L auxiliary qubits is introduced, prepared in the state $|+\rangle^{\otimes L}$, to serve similarly as control qubits to the unitary $\exp(iAt)$ raised at different integer powers. By the same transfer mechanism as in Eqs. (12.42)–(12.44), the L control qubits act to extract the value of the phase $a_n t$, and their number L fixes the precision or resolution in estimating or representing the phase $a_n t$. When in the circuit of Fig. 12.3 the eigenstate $|a_n\rangle$ (instead of $|u\rangle$) is applied at the input, along with the L control qubits in state $|+\rangle^{\otimes L}$, the same transformation as in Eq. (12.44) takes place, receiving here the form

$$|+\rangle^{\otimes L}\,|a_n\rangle \longmapsto \frac{1}{2^{L/2}}\sum_{k=0}^{2^L-1}|k\rangle\exp(ikAt)\,|a_n\rangle = \frac{1}{2^{L/2}}\sum_{k=0}^{2^L-1}|k\rangle\exp(ika_nt)\,|a_n\rangle\ . \tag{12.138}$$

An inverse Fourier transform by the operator $\mathsf{U}_\mathsf{F}^\dagger$ is then performed on the L control qubits, as in Eq. (12.49), to yield as in Eq. (12.51) a transformed state taking here the form

$$|\Psi_n\rangle = \sum_{j=0}^{2^L-1} S_j(n)\,|j\rangle\,|a_n\rangle\ . \tag{12.139}$$

This state $|\Psi_n\rangle$ is similar to the state $|\psi_\xi\rangle$ of Eq. (12.55) for what concerns the L control qubits, and in addition $|\Psi_n\rangle$ also carries the eigenstate $|a_n\rangle$ that served as the input excitation. Accordingly, the corresponding amplitude $S_j(n)$ in Eq. (12.139) is a function of n keeping track of the eigenstate $|a_n\rangle$ that elicited it. As we have seen on the occasion of Eq. (12.54), the amplitude $S_j(n)$ as a function of j is sharply peaked around an integer $\tilde{j}_n \in [0, 2^L[$ such that $a_n t \approx 2\pi\,\tilde{j}_n/2^L$. When $2^L a_n t/(2\pi)$ is precisely an integer j_0, the amplitude $S_j(n)$ for a given n as a function of j is the Kronecker delta $S_j(n) = \delta_{jj_0}$ of Eq. (12.53), giving $\tilde{j}_n = j_0$; otherwise $S_j(n)$ is a narrow peak around an integer $\tilde{j}_n \approx 2^L a_n t/(2\pi)$ described by the shape of Eq. (12.54), and for which the Kronecker delta $S_j(n) = \delta_{j\tilde{j}_n}$ offers a suitable approximation. Accordingly, to more easily grasp the mode of operation of the HHL algorithm, a helpful approximation is to take $S_j(n) \approx \delta_{j\tilde{j}_n}$ at this integer $\tilde{j}_n \approx 2^L a_n t/(2\pi)$, as also done for instance in [388], leading for Eq. (12.139) to

$$|\Psi_n\rangle \approx |\tilde{j}_n\rangle\,|a_n\rangle\ , \tag{12.140}$$

as a suitable approximation for the state $|\Psi_n\rangle$. Otherwise, for a more accurate quantitative analysis, a distributed or spread peaked function of j according to Eq. (12.54) should be used instead for $S_j(n)$, which would introduce no insurmountable difficulty but tends to obscure the essential mechanism of the algorithm on which we want to concentrate in this report. In the end, when it all comes down to measuring quantum states, the approximation resulting from Eq. (12.140) will provide a quite adequate description, as much as the phase estimation is more finely resolved at sufficiently large size L.

In this way, the state $|\Psi_n\rangle$ of Eq. (12.140) provides a quantum representation, via a computational-basis state $|\tilde{j}_n\rangle$ of the L control qubits, where the integer $\tilde{j}_n \in [0, 2^L[$ can be seen as an L-bit code identifying the eigenvalue a_n via $a_n t \approx 2\pi\, \tilde{j}_n/2^L$. The Fourier-based phase estimation enabled a quantum encoding of the eigenvalue a_n, at an L-bit resolution, by means of a computational-basis state $|\tilde{j}_n\rangle$. In addition, in the state $|\Psi_n\rangle$ constructed in Eq. (12.140), the encoding $|\tilde{j}_n\rangle$ of the eigenvalue a_n is accompanied by the corresponding eigenstate $|a_n\rangle$, where for the sequel $|a_n\rangle$ can be seen as an address to refer the coded eigenvalue $|\tilde{j}_n\rangle$. If we were interested, as in Sect. 12.3.1, only in estimating the phase a_n associated with (elicited by) the eigenvector $|a_n\rangle$, a quantum measurement of $|\Psi_n\rangle$ in the computational basis of the L control qubits, would follow, delivering via $\tilde{j}_n$ an estimate of the targeted phase a_n. The goal is different here.

For the HHL algorithm, in the phase-estimation circuit of Fig. 12.3, instead of exciting at the input by a single eigenstate $|a_n\rangle$, the excitation is simultaneously by all the eigenstates $|a_n\rangle$ contained in the linear superposition of Eq. (12.135) realizing the quantum state $|\vec{b}\,\rangle$. This is performed simply by applying $|\vec{b}\,\rangle$ on the input marked $|u\rangle$ in the circuit of Fig. 12.3. The state $|\vec{b}\,\rangle$ can be materialized in the computational basis of $\mathcal{H}_N$, and there is no need to explicitly work out the decomposition of $|\vec{b}\,\rangle$ in the eigenbasis $\{|a_n\rangle\}$, which anyway is a priori unknown. The decomposition in Eq. (12.135) is built-in with $|\vec{b}\,\rangle$, whatever the basis for its materialization, and it is this built-in decomposition of $|\vec{b}\,\rangle$ in Eq. (12.135) that controls the action of $|\vec{b}\,\rangle$ in the phase-estimation circuit of Fig. 12.3. All the steps involved with the circuit of Fig. 12.3, including the final inverse Fourier transform $\mathsf{U}_\mathsf{F}^\dagger$ that follows, are linear operations, so that when all the eigenvalues a_n are excited in superposition by $|\vec{b}\,\rangle$ in Eq. (12.135), we obtain in place of Eq. (12.140) the superposed analogue state

$$|\Psi\rangle \approx \sum_{n=1}^{N} b_n\, |\tilde{j}_n\rangle\, |a_n\rangle\ . \tag{12.141}$$

A comparable situation where the Fourier-based phase estimation operates while exciting with multiple eigenstates in superposition at the input, was previously encountered for instance with Eqs. (5.57)–(5.58) for the Grover search algorithm with an unknown number of solutions, or with Eqs. (12.130)–(12.131) for order finding. In the superposed state $|\Psi\rangle$ of Eq. (12.141), all the N eigenvalues a_n are represented, as $a_n \approx 2\pi\, \tilde{j}_n/(t2^L) = \tilde{a}_n$, each by means of an L-bit binary code for the integer $\tilde{j}_n \in [0, 2^L[$ carried by the L-qubit basis states $|\tilde{j}_n\rangle$, and each eigenvalue representation $|\tilde{j}_n\rangle$ is separately identified by a set of N mutually orthogonal addresses $|a_n\rangle$. In this respect, all the N (approximated) eigenvalues $\tilde{a}_n$ have been imprinted under a quantum form in the superposed state $|\Psi\rangle$, where they remain perfectly and separately identified by their mutually orthogonal addresses $|a_n\rangle$. In $|\Psi\rangle$ the N eigenvalues fit in superposition on a resolution grid allowing 2^L distinct configurations, standing as N sharp peaks (approximated as Kronecker peaks), and the size L fixing the resolution is supposed large enough so that the N peaks do not inconveniently overlap. The parameters L and t help on this issue.

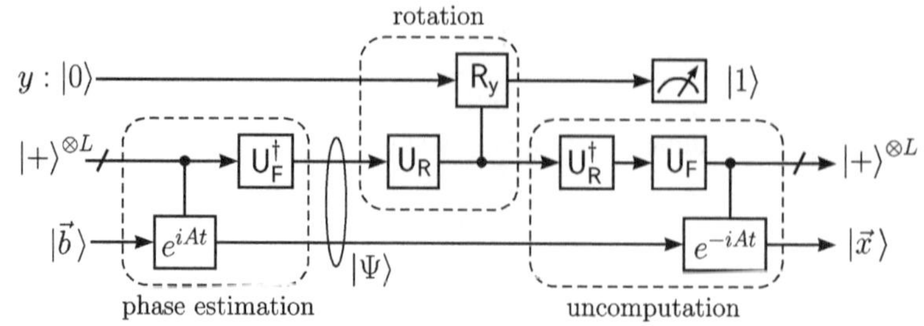

Fig. 12.14 Quantum circuit sketching the operation of the HHL algorithm for solving a linear system of equations. The Fourier-based phase estimation on the unitary process e^{iAt} delivers the state $|\Psi\rangle$ of Eq. (12.141), serving to prepare via the unitary $\mathsf{U_R}$, the control to the qubit rotation $\mathsf{R_y}$ implementing the transformation of Eq. (12.142) on the auxiliary qubit y. An uncomputation is then performed to disentangle on the output the solution $|\vec{x}\,\rangle$ of Eq. (12.137) obtained, conditioned on an auxiliary qubit y measured as $|1\rangle$, for the linear system

A quantum circuit implementing the whole HHL algorithm is sketched in Fig. 12.14.

The highly structured quantum state $|\Psi\rangle$ of Eq. (12.141), also shown in Fig. 12.14, then will serve to construct the targeted solution under the form of Eq. (12.137). The approach especially can rely on the ability of quantum (unitary) circuits to implement any conceivable Boolean functions on any number of bits, as reported in Sect. 5.2. The eigenvalue $\tilde{a}_n$ from its quantum representation $|\tilde{j}_n\rangle$ contained in $|\Psi\rangle$, can be used to obtain the reciprocal $\tilde{a}_n^{-1}$, and also a qubit rotation gate $\mathsf{R_y}(\theta_n)$ as in Eq. (3.26) capable of performing in the complex plane $\mathcal{H}_2$ the qubit rotation

$$\mathsf{R_y}(\theta_n)\,|0\rangle = \cos\big(\theta_n/2\big)\,|0\rangle + \sin\big(\theta_n/2\big)\,|1\rangle = \sqrt{1-\big(c\tilde{a}_n^{-1}\big)^2}\;|0\rangle + c\tilde{a}_n^{-1}\,|1\rangle\;, \tag{12.142}$$

with the angle $\theta_n = 2\arcsin\big(c\tilde{a}_n^{-1}\big) \Leftrightarrow \sin\big(\theta_n/2\big) = c\tilde{a}_n^{-1}$, and c a constant ensuring that the rotation angle θ_n remains in $[0, \pi]$. Beyond the feasibility in principle of the controlled rotation in Eq. (12.142), for its implementation more elements and explicit techniques can be traced, for instance, from [386, 387, 389]. The binary representation $|\tilde{j}_n\rangle$ of the eigenvalue $\tilde{a}_n$ can be seen as an address pointing to the corresponding rotation gate $\mathsf{R_y}(\theta_n)$. The targeted rotation gate $\mathsf{R_y}(\theta_n)$ may be implemented by a pipeline cascading successive rotation gates $\mathsf{R_y}(\cdot)$ each controlled by a single bit of $|\tilde{j}_n\rangle$ to rotate by appropriate increments decomposing the angle θ_n, as for example shown in [386, 387, 389]. This realizes a qubit rotation gate $\mathsf{R_y}(\theta_n)$ conditioned on the state $|\tilde{j}_n\rangle$. The unitary circuit $\mathsf{U_R}$ in Fig. 12.14 encapsulates the stage preparing the control to the qubit rotation $\mathsf{R_y}(\cdot)$. Such operation takes place in linear superposition for every n present in Eq. (12.141). An auxiliary qubit y in state $|y\rangle$ initialized at $|y = 0\rangle$ is introduced to experience the rotation $\mathsf{R_y}(\theta_n)$ of Eq. (12.142) conditioned on $|\tilde{j}_n\rangle$, so that the state from Eq. (12.141) is further transformed as

$$|\Psi\rangle\,|y=0\rangle \longmapsto \sum_{n=1}^{N} b_n\,|\tilde{j}_n\rangle\,|a_n\rangle \left(\sqrt{1-\left(c\tilde{a}_n^{-1}\right)^2}\;|y=0\rangle + c\tilde{a}_n^{-1}\,|y=1\rangle\right) . \tag{12.143}$$

Next, from the state in Eq. (12.143), a von Neumann measurement of the auxiliary qubit y is performed in the computational basis $\{|0\rangle\,,|1\rangle\}$. When as a result the auxiliary qubit y is found in the basis state $|1\rangle$, the pre-measurement state from Eq. (12.143) gets projected onto the state

$$C\sum_{n=1}^{N}\tilde{a}_n^{-1} b_n\,|\tilde{j}_n\rangle\,|a_n\rangle\,|y=1\rangle\ , \tag{12.144}$$

with the constant

$$C=\left(\sum_{n=1}^{N}\left(\tilde{a}_n^{-1}|b_n|\right)^2\right)^{-1/2} \tag{12.145}$$

normalizing the post-measurement state. In the post-measurement state of Eq. (12.144), the auxiliary qubit y is in the separable state $|1\rangle$, and the rest comes close to the targeted solution state of Eq. (12.137), except that each component addressed by $|a_n\rangle$ remains entangled with the associated eigenvalue $|\tilde{j}_n\rangle$ which is no longer needed.

In the HHL algorithm, the quantum state

$$C\sum_{n=1}^{N}\tilde{a}_n^{-1} b_n\,|\tilde{j}_n\rangle\,|a_n\rangle \tag{12.146}$$

obtained after measuring the qubit y in state $|1\rangle$ from Eq. (12.144), is then disentangled by undoing the operation of phase estimation and preparation of the controlled rotation. This is performed by the stage of uncomputation indicated in Fig. 12.14, where the L auxiliary qubits encoding the phase $|\tilde{j}_n\rangle$ in Eq. (12.146) go through the inverse unitary $\mathsf{U}_\mathsf{R}^\dagger$, then through the Fourier transform gate U_F, to serve as controls to the inverse process $\exp(-iAt)$. As a result of this uncomputation, the state of Eq. (12.146) gets transformed into the output state

$$|+\rangle^{\otimes L}\,C\sum_{n=1}^{N}\tilde{a}_n^{-1} b_n\,|a_n\rangle = |+\rangle^{\otimes L}\,|\vec{x}\,\rangle\ , \tag{12.147}$$

where the L control qubits are returned to their initial state $|+\rangle^{\otimes L}$ and (an approximation of) the solution $|\vec{x}\,\rangle$ of Eq. (12.137) is separately accessible. In this way, on the output, conditioned on the postselection rule of an auxiliary qubit y measured as $|1\rangle$, the quantum register that at the input contained $|\vec{b}\,\rangle$, now contains on the output the solution $|\vec{x}\,\rangle$ of Eq. (12.137) obtained for the linear system.

The solution $\vec{x}$ to the linear system is made accessible by the HHL algorithm under the form of the quantum state $|\vec{x}\rangle$, and an explicit readout from $|\vec{x}\rangle$ of all the N scalar components of $\vec{x}$, if it was attempted, would be restricted by the bottleneck of quantum measurement. More skillful exploitation of the solution $|\vec{x}\rangle$ can be conceived. An interesting exploitation of the solution $|\vec{x}\rangle$, very often relevant in practice, is to evaluate some functional $\vec{x}^\dagger \Omega \vec{x}$ of the solution $\vec{x}$ under the form $\langle \vec{x} \,|\, \Omega \,|\, \vec{x} \rangle$, where Ω is an observable representing a physical quantity of interest. Successive repetitions of the HHL algorithm on the same input data A and $\vec{b}$, terminated by a measurement on $|\vec{x}\rangle$ of the observable Ω, will estimate the expectation value $\langle \vec{x} \,|\, \Omega \,|\, \vec{x} \rangle$.

An important aspect is that all the steps of the HHL algorithm can be implemented efficiently, especially the quantum process $\exp(iAt)$ and the controlled rotation via U_R, and their inverse [385, 386]. For solving an $N \times N$ linear system, a classical approach with best efficiency is the conjugate gradient method [385, 388] affording an $O(N)$ algorithmic complexity. Meanwhile, the HHL quantum algorithm, especially by computing on all the eigenvalues in parallel, allows one to achieve an $O(\log N)$ complexity, affording in this way an exponential speedup. Detailed elements on how to obtain such computational advantage from the HHL algorithm can be found for instance in [385, 388, 389], along with discussion of runtime and success performance.

The HHL quantum algorithm for solving linear systems of equations, with its possibility of exponential speedup over classical solutions, stands as an important quantum contribution to a ubiquitous computational task, and accordingly it is applied and extended in many areas and directions [388, 389].

12.8 Summary on Quantum Fourier Transform

The **quantum Fourier transform** is the unitary operator U_F of $\mathcal{L}(\mathcal{H}_N)$ defined by the transformation of the computational basis of $\mathcal{H}_N$ given in Eq. (12.15) or equivalently represented by the $N \times N$ symmetric unitary matrix U_F from Eqs. (12.16)–(12.17). When the quantum states are realized as L-qubit states, the quantum Fourier transform receives the equivalent product form of Eq. (12.35), and it can be computed by the quantum circuit of Fig. 12.1 with a polynomial $O(L^2)$ algorithmic complexity.

A prominent application of the quantum Fourier transform examined in Sect. 12.3 is for **quantum phase estimation**, to estimate the phase ξ from an eigenvalue $\exp(i2\pi\xi)$ of a unitary process U_ξ. In essence, a quantum state is constructed as a superposition of harmonic components at integer multiples $k\xi$ of the unknown phase ξ (as in Eqs. (12.44), (12.65) or (12.71)) and similar to $\sum_k \exp(i2\pi k\xi)\,|k\rangle$, and this state is converted by Fourier transform into a single-peak distribution of the probability amplitude (like the corresponding coefficients S_j or S'_j) of an output state similar to $\sum_j S_j\,|j\rangle$, which is measured to reveal the phase ξ from the peak location. Various schemes are possible according to the number of probing qubits and the number of copies of the quantum gate for U_ξ. Typically, with an N-dimensional Fourier transform associated with N evaluations of the process U_ξ, an rms estimation error scaling as $1/\sqrt{N}$ is accessible that conforms with standard statistical estimation theory; while with an optimally entangled probing signal a more efficient rms error in $1/N$ can be obtained. As discussed in Sect. 12.3.6, the Fourier transform methodology enables a one-shot global estimation of the quantum phase ξ over the whole angular range $2\pi\xi \in [0, 2\pi[$; while comparatively, the methodology of Sect. 11.3, by averaging from repeated measurements, performs a local phase estimation, yet with higher accuracy and less circuit complexity.

Another important application of the quantum Fourier transform is for **finding the period** r of a function $f(x)$ defined on integer arguments x, as examined in Sect. 12.4. The procedure performs a single evaluation of $f(\cdot)$ in parallel on a quantum superposition, to construct a superposed quantum state $|\psi\rangle = N^{-1/2}\sum_x |x\rangle\,|f(x)\rangle$ in Eq. (12.109) that contains the values of $f(x)$ for all (relevant) arguments x to reflect its periodicity r. The periodic structure in $|\psi\rangle$, by the quantum Fourier transform is turned into a multiple-peak structure $|\widetilde{f}(k)\rangle$ carried by the quantum state $|\widetilde{\psi}\rangle = N^{-1/2}\sum_k |k\rangle\,|\widetilde{f}(k)\rangle$ of Eq. (12.112), whose measurement outputs a peak location revealing the period r (by the continued fraction algorithm).

At the root of both applications – phase estimation and period finding – lies the behavior, quite common to signal processing, that the Fourier transform turns periodic structures into peaked structures, with periodicities connected to peak locations; the main quantum aspect, arguably, being to devise how to imprint and read out such structures on quantum states or signals. For quantum signal processing, a convenient behavior at work is that, from peaked structures imprinted across the probability amplitudes of a quantum state, high probabilities sit in the peaks, and so quantum measurement, by the Born rule, delivers the location of a peak with high probability. Here with the Fourier processing, after the action of the unitary $\mathsf{U}_\mathsf{F}^\dagger$ in Eq. (12.48) or (12.68), each peak revealed by quantum measurement has its origin (before $\mathsf{U}_\mathsf{F}^\dagger$) in a harmonic component with frequency j forming a column of the Fourier matrix U_F in Eqs. (12.16)–(12.17). Each (harmonic) column of U_F is the image by U_F of a basis state (a peak). Reciprocally, after processing by $\mathsf{U}_\mathsf{F}^\dagger$, each peak revealed by quantum measurement detects the presence, in the input, of the image by U_F of a basis state (a harmonic component with the Fourier processing). The scheme can be extended to detect input components other than harmonic components, other components or waveforms of interest that would be defined by the columns of another unitary matrix U_W substituting the Fourier unitary U_F, and processing by the inverse $\mathsf{U}_\mathsf{W}^\dagger$ would reveal the presence, in the input, of such waveforms, as output peaks to be measured. The Deutsch–Jozsa quantum algorithm does more or less this, as we argued in Sect. 5.4, with a Hadamard transform for U_W, to test the presence or not of a Hadamard component at zero frequency (a DC component) in input images as in Figs. 5.8–5.9.

Finally in this chapter the quantum Fourier transform is invoked to solve the problem of **finding the order** r of an integer a modulo N, as examined in Sect. 12.5, and the problem of factoring an integer into primes by the **Shor quantum algorithm** of Sect. 12.6. Both problems by classical means remain today with exponential algorithmic complexity in the length L of the representation for integers, while the quantum resolutions based on the quantum Fourier transform achieve a polynomial $O(L^3)$ complexity. The **HHL quantum algorithm** of Sect. 12.7, via Fourier-based phase estimation, can solve an $N \times N$ linear system of equations in $O(\log N)$ complexity instead of $O(N)$ classically.

Period finding, order finding, the Shor factoring of integers, along with other comparable problems like discrete logarithms or Simon's problem, are instances of the broader (Abelian) hidden subgroup problem [390–395], consisting, for an algebraic group G, in finding a (hidden) subgroup over which a criterion function $f(\cdot)$ has some invariance (constant by cosets) properties. With an essentially similar quantum approach using parallel evaluation of $f(\cdot)$ and the quantum Fourier transform of $f(\cdot)$ over a finite Abelian group G, an efficient resolution for the hidden subgroup problem can be obtained [390, 391, 394, 395] with an algorithmic complexity polynomial in the size $\log |G|$. Beyond, efficiently finding hidden subgroups of non-Abelian groups G remains today more challenging and holds the key to improved solutions to other practical problems of interest [392, 393, 396].

Chapter 13
Quantum Communication

In this chapter, we will first briefly review the transmission of classical information over a classical (noisy) communication channel, establishing the possibility of error-free communication at a maximum rate given by the classical information capacity of the channel. We will then transpose this framework to the transmission of information over a quantum communication channel. Over a quantum channel the transmission is supported by quantum signals, but the information communicated end-to-end across the channel can be taken as classical or quantum. Various meaningful performance criteria and information capacities ensue, that will be presented. We will also study important examples of quantum communication channels, with their optimal conditions of operation, and the derivation of their classical and quantum information capacities.

13.1 Classical Channel Transmitting Classical Information

A classical information channel at its input emits information-carrying signals under the form of random symbols X drawn among the J characters of an input alphabet $\{s_1, s_2, \dots, s_J\}$, with the prior probabilities $\Pr\{X = s_j\} = p_j$ for $j = 1$ to J. Each input symbol X transmitted over the channel produces an output symbol Y which takes values in the output alphabet $\{s'_1, s'_2, \dots, s'_{J'}\}$ with a size J' possibly different from the input size J. The constitution of the communication channel fixes the input–output conditional probabilities $p_{k|j} = \Pr\{Y = s'_k \mid X = s_j\}$ for $(j, k) \in \{1, \dots, J\} \times \{1, \dots, J'\}$. Follow the input–output joint probabilities $p_{kj} = p_{k|j} p_j$ and the output probabilities $p'_k = \sum_j p_{k|j} p_j$. Most frequently, the communication channel will be a noisy transmission device, precluding a one-to-one exact input–output transmission of the information-carrying symbols.

A central result of classical information theory, originally established in [171] and with great impact on communication technologies, is that, over such a generally noisy

F. Chapeau-Blondeau, *Quantum Information and Quantum Signal Processing*,
https://doi.org/10.1007/978-3-032-27499-1_13

transmission channel, it is nevertheless possible to transmit information with no error. This central result [171, 173] incorporates three essential aspects, showing that (i) it is indeed possible to transmit information with no error over a noisy communication channel, (ii) at a non-vanishing rate that can be explicitly quantified and maximized, (iii) by means of a (channel) coding procedure whose general principle is exhibited.

The challenge and achievement can be appreciated by means of a simple binary communication channel transmitting binary symbols 0 and 1. The noisy channel, for example, can randomly flip each binary symbol with a probability f of, say, $f = 0.1$. Since the channel introduces errors in this way, how can it be possible to transmit information with no error over such a noisy channel? One could think of robustifying against noise each input binary symbol with a repetition code, where each binary (logical) symbol would be repeated a given number of, say L, times, as $000\cdots$ or $111\cdots$, prior to transmission over the noisy channel, followed by a majority decoding over each L-bit block at the output to restore a binary (logical) symbol. With such a repetition code, it is easy to show that, as the repetition length $L \to \infty$, the probability of failure of the majority decoding decays as $f^{L/2}$, so that vanishing error in transmitting each input binary (logical) symbol can be obtained, yet at a transmission rate $1/L$ also going to zero. Obviously, this is not the appropriate procedure to obtain vanishing transmission error at a non-vanishing rate. An appropriate way, as established by the theory in [171, 173], consists, instead of encoding separately each logical binary symbol at the input, in considering long sequences of L consecutive input binary symbols, much like it was considered for source coding in Sect. 8.1. Each individual input binary symbol can be randomly flipped with a probability $f = 0.1$, so that it can be randomly transformed into an erroneous binary symbol to incur an irreparable transmission error with a nonzero probability. However, a long sequence of L binary symbols cannot be transformed by the noise (with significant probability) into any other possible sequence of L binary symbols. At large length L, by the law of large numbers, a sequence of L binary symbols can only be transformed by the noise (with significant probability) into another sequence of L binary symbols differing by about $f = 10\,\%$ of its binary symbols. When such "error clouds" sized as $10\,\%$ do not overlap at the output, the input L-bit binary sequences that generated them can be perfectly distinguished and recovered from the output L-bit (noisy) binary sequences. To obtain non-overlapping error clouds at the output, only a fraction of the 2^L possible binary sequences of length L chosen sufficiently spaced apart, should be used at the input for signaling, each envisaged as a centroid for an error cloud at the output. This is in essence the principle enabling error-free transmission over a noisy communication channel at a non-vanishing rate. We will now present these elements more formally and in general conditions.

We consider the general communication channel introduced above when emitting random sequences $\overrightarrow{X}^L = X_1X_2\cdots X_L$ of L consecutive independent input symbols. The input symbols are taken independent here because we want to characterize the maximum possible rate for information transmission over the channel, and as for the subadditivity of entropy seen in Sect. 7.1.4, dependence here would reduce the entropy or information content of the stream of symbols and the transmission rate,[1] so that maximum rate is obtained with independent symbols [12, 397]. For a given realization $\vec{x}^L = x_1 x_2 \cdots x_L$ of the input sequence $\vec{X}^L$, the channel associates a random output sequence $\vec{Y}^L$ of length L occurring with the conditional probability

$$P\big(\vec{Y}^L \,|\, \vec{X}^L = \vec{x}^L\big) = \prod_{\ell=1}^{L} P(Y_\ell \,|\, X_\ell = x_\ell) \,, \tag{13.1}$$

leading to

$$-\frac{1}{L} \log P\big(\vec{Y}^L \,|\, \vec{X}^L = \vec{x}^L\big) = \frac{1}{L} \sum_{\ell=1}^{L} -\log P(Y_\ell \,|\, X_\ell = x_\ell) \,. \tag{13.2}$$

We consider the regime at large length L, where we examine the channel transmitting long sequences of L consecutive symbols. At large length $L \to \infty$, by the law of large numbers, with arguments similar to those employed in Sect. 8.1 for source compression, one can prove the following properties [173], taking place with a probability tending to 1:

(1) The channel emits an input sequence $\vec{x}^L$ that is typical, relative to the input probability distribution $\{p_j\}$, with symbols x_ℓ in Eqs. (13.1) and (13.2) visiting the possible values s_j with relative frequencies matching the input probabilities p_j. By Eq. (8.9) such typical input sequences are in number $2^{LH_2(X)}$, and they are all equiprobable each with probability $2^{-LH_2(X)}$ by Eq. (8.8).

(2) The couples (Y_ℓ, x_ℓ) in Eq. (13.2) also occur with relative frequencies matching the input–output joint probabilities $p_{kj} = p_{k|j} p_j$, and in such circumstance the output sequence $\vec{Y}^L$ is said to be jointly typical with the input sequence $\vec{x}^L$.

(3) The output sequence $\vec{Y}^L$ is also typical, relative to the output probability distribution $\big\{p'_k = \sum_j p_{k|j} p_j\big\}$. In their totality the typical output sequences are in number $2^{LH_2(Y)}$ and are all equiprobable each with probability $2^{-LH_2(Y)}$.

(4) The sample average in the right-hand side of Eq. (13.2) therefore tends to the statistical expectation according to the input–output joint probability distribution $\{p_{kj} = p_{k|j} p_j\}$.

So we have for Eq. (13.2) the limit

$$-\frac{1}{L} \log P\big(\vec{Y}^L \,|\, \vec{X}^L = \vec{x}^L\big) \xrightarrow[L \to \infty]{} \big\langle -\log P(Y|X)\big\rangle = H(Y|X) \,, \tag{13.3}$$

introducing the input–output conditional entropy $H(Y|X)$ of Eq. (7.6). We use for the entropies the logarithm of base 2 that can be inverted by exponentiation to 2, so, with $H(Y|X) \equiv H_2(Y|X)$, from Eq. (13.3), we obtain at the limit

[1] At least for stationary memoryless channels. See also [12] its Theorem 13.1.1.

$$P\left(\vec{Y}^{L} \mid \vec{X}^{L}=\vec{x}^{L}\right)=2^{-L H_2(Y|X)}\,. \tag{13.4}$$

At large length $L \to \infty$, by Property (2) above, the sequence $\vec{Y}^{L}$ that is obtained at the output is a sequence that is jointly typical with the particular sequence $\vec{x}^{L}$ emitted at the input (that happens to be typical by Property (1)). By Eq. (13.4), each such feasible output sequence $\vec{Y}^{L}$ has a similar probability of $2^{-LH_2(Y|X)}$, irrespective of the detailed constitutions of the (typical) input sequence $\vec{x}^{L}$ that appeared at the input, and of the output sequence (jointly typical with $\vec{x}^{L}$) induced at the output. To match the overall probability 1 of obtaining an (any) output sequence $\vec{Y}^{L}$ jointly typical with $\vec{x}^{L}$, we deduce that the number of output sequences jointly typical with $\vec{x}^{L}$ is $2^{LH_2(Y|X)}$ (so that this number $2^{LH_2(Y|X)}$ multiplied by the common probability $2^{-LH_2(Y|X)}$ matches the overall probability 1). So each emitted input sequence $\vec{x}^{L}$ can be altered to terminate into a jointly typical output subset containing $2^{LH_2(Y|X)}$ sequences. Also, by Property (3) we know that any sequence $\vec{Y}^{L}$ that can appear at the output is a typical output sequence, and that the total number of such typical output sequences is $2^{LH_2(Y)}$. So we evaluate at

$$\frac{2^{LH_2(Y)}}{2^{LH_2(Y|X)}}=2^{L[H_2(Y)-H_2(Y|X)]}=2^{LH_2(X:Y)} \tag{13.5}$$

the maximal number of jointly typical output subsets of cardinal $2^{LH_2(Y|X)}$ that can fit with no overlap inside the output set of $2^{LH_2(Y)}$ typical output sequences, as illustrated in Fig. 13.1. The maximum number in Eq. (13.5) happens when the jointly typical output subsets are optimally spaced in the output set of $2^{LH_2(Y)}$ typical output sequences. And when there is no overlap at the output, the input sequence $\vec{x}^{L}$ can be univocally recovered from its noisy transmission at the output. So Eq. (13.5) also represents the maximal number of input sequences that can be distinguished and recovered with no error[2] from the noisy channel output; and this number is directly determined in Eq. (13.5) by the input–output mutual information $H_2(X:Y)=H_2(Y)-H_2(Y|X)$ of the channel, as defined in Eqs. (7.20)–(7.23).

Alternatively, the maximal number of error-free transmissible sequences in Eq. (13.5) can be expressed equivalently by means of the entropies in base J as $J^{LH_J(X:Y)}$, with the J-ary mutual information $H_J(X:Y)=H_J(Y)-H_J(Y|X)$. With a J-ary input alphabet, the total number of possible sequences of length L is J^L, and the input–output mutual information $H_J(X:Y)$ is upper bounded by the input entropy $H_J(X)$ as $H_J(X:Y)\le H_J(X)\le \log_J(J)=1$. So all the J^L sequences of L input symbols cannot be transmitted error-free, but only a subset of at most $J^{LH_J(X:Y)}\le J^L$ of them, those mapped by the noisy channel into disjoint subsets at the output. This corresponds to a maximal transmission rate of $\log_J\left(J^{LH_J(X:Y)}\right)/L=H_J(X:Y)\le 1$, expressing that, on average, less than one J-ary unit of information can be transmitted error-free per J-ary symbol sent across

[2] In the sense of arbitrarily small error associated with the asymptotic character of the law of large numbers at large length $L \to \infty$.

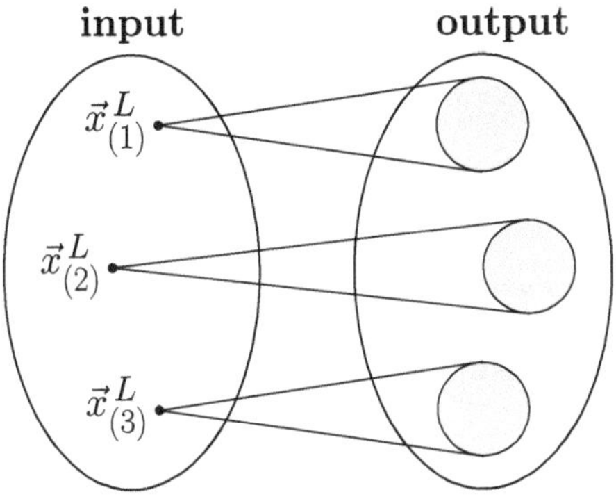

Fig. 13.1 Operation of the classical communication channel when transmitting long sequences of L consecutive symbols of length $L \to \infty$. In the output set, each gray disk contains $2^{LH_2(Y|X)}$ sequences $\overrightarrow{Y}^L$ that can be associated (that are jointly typical) with each particular typical sequence $\vec{x}^L$ that can appear at the channel input. The output set contains a total of $2^{LH_2(Y)}$ typical sequences $\overrightarrow{Y}^L$ that can appear at the channel output. Provided they are optimally spaced, a maximum number $2^{LH_2(Y)}/2^{LH_2(Y|X)} = 2^{L[H_2(Y)-H_2(Y|X)]} = 2^{LH_2(X:Y)}$ of gray disks can fit with no overlap inside the output set

the noisy channel. The other way round, this expresses that the input–output mutual information $H(X:Y)$ of a channel quantifies the maximal rate for error-free transmission of information over this noisy channel.

To further maximize the transmission rate $H(X:Y)$, it is possible to select the probability distribution $\{p_j\} \equiv \{P(x)\}$ at the input so as to solve the maximization

$$C(X,Y) = \max_{\{p_j\}} H(X:Y) \,, \tag{13.6}$$

and defining the **information capacity** $C(X,Y)$ of the communication channel. As indicated in Sect. 7.1.4, given the input–output conditional probabilities $\{P(y|x)\}$ fixed by the channel, the input–output mutual information $H(X:Y)$ is in general a concave ($\cap$) function of the input probability distribution $\{p_j \equiv P(x)\}$, usually with a unique optimum $\{p_j^{\text{opt}}\}$ for the input distribution $\{p_j\}$ solving the maximization of Eq. (13.6). The information capacity $C(X,Y)$ resulting from the maximization of Eq. (13.6) represents the overall maximal rate for error-free transmission of information over the noisy channel.

These results form the **Shannon theorem for classical channel coding** [171, 173], stating that error-free communication is possible over a noisy channel, at a maximum rate given by the information capacity of Eq. (13.6). It can be supplemented by proving that no error-free communication is possible at a rate larger than the information capacity of the channel. This theorem stands as a central result of classical information theory, especially useful as a guide for the development of communication technologies. It also consolidates the status of the entropy and entropy-based quantities like the mutual information, as legitimate concepts for a quantitative assessment of information. As the source coding theorem in Sect. 8.1, the channel coding theorem here stands as another non-trivial result with strong informational significance, again concretizing the expectation we expressed at the end of the first paragraph of Sect. 7.1.1, for the profit from the mathematical working of a (physical) information theory.

With similar considerations, the transmission can be analyzed in the other direction, establishing that for each particular sequence $\vec{y}^{\,L}$ of length $L \to \infty$ received at the output, the input sequences that may have produced this output $\vec{y}^{\,L}$ form a subset of $2^{LH_2(X|Y)}$ input sequences $\overrightarrow{X}^L$ of length L as those that are jointly typical with $\vec{y}^{\,L}$, among the $2^{LH_2(X)}$ typical sequences that can appear at the input. A maximum number of $2^{LH_2(X)}/2^{LH_2(X|Y)} = 2^{L[H_2(X)-H_2(X|Y)]} = 2^{LH_2(X:Y)}$ of such input subsets can be arranged with no overlap in the input set, so that no two input subsets can reach the same output $\vec{y}^{\,L}$, to ensure a univocal recovery of the channel input from a noisy output $\vec{y}^{\,L}$. This equivalently leads to the same maximum rate $H(X:Y)$ for error-free transmission over the noisy channel.

The above reasoning establishing the Shannon theorem prescribes that error-free transmission can be achieved by optimal spacing of the joint-typicality subsets. This is the role of the operation of **channel coding**, to employ the capacity-achieving input probability distribution $\{p_j^{\text{opt}}\}$ solving Eq. (13.6), to generate codewords to optimally recode the stream of input symbols before they are sent across the channel. A direct protocol for channel coding is **random coding**, which operates in the following way. The input to the coding stage is a source that emits J-ary symbols, of which we consider long sequences of L_{in} consecutive symbols. Their recoding is envisaged with a J-ary coding alphabet, with a same cardinal J as the source alphabet, but source and coding alphabets of different sizes could do as well. To transmit over the noisy channel at any rate $R \le C(X, Y) \le \log_J(J) = 1$, the J-ary codewords are selected as sequences with a length of $L_{\text{cod}} = L_{\text{in}}/R \ge L_{\text{in}}$ of J-ary characters. For coding the $J^{L_{\text{in}}}$ input sequences of length L_{in}, an identical number $J^{L_{\text{in}}}$ of J-ary codewords of length L_{cod} are selected at random, with each of their J-ary character drawn independently according to the capacity-achieving input probability distribution $\{p_j^{\text{opt}}\}$. The (long) codewords that will show up in the channel operation will result as coding sequences typical for the distribution $\{p_j^{\text{opt}}\}$. Such random selection is enough to ensure optimal spacing of the joint-typicality subsets. By operating on long input sequences with $L_{\text{in}} \to \infty$, error-free transmission can be achieved at any rate R no larger than the channel capacity $C(X, Y)$.

This random coding protocol in practice implies nearest neighbor decoding, where each noisy output sequence has to be compared with all codewords so as to restore the nearest one. This usually induces a decoding cost exponential in the length L_{in}. To avoid such costly decoding, more sophisticated coding techniques have been developed, which introduce structures in the codewords helpful for their recovery from degraded versions, with, for instance, parity checks or syndrome processing to more directly and efficiently recover an adequate codeword while avoiding exhaustive search. The performance and distance to the asymptotic optimum of the Shannon theorem when operating at finite length L_{in}, which is required in practice, are also meaningful criteria to devise an efficient coding protocol. These aspects are addressed by the field of channel coding [398].

For illustration, the noisy binary channel envisaged as an example above with a probability of bit flip f, is known as a **binary symmetric channel**. Its information capacity can be evaluated as

$$C(X, Y) = \log(2) - H_{\text{bin}}(f) , \tag{13.7}$$

with $H_{\text{bin}}(\cdot)$ the binary entropy of Eq. (7.4) and Fig. 7.1. For $f = 0.1$ the capacity is $C \approx 0.531$ Sh, or 0.531 binary unit of information per binary symbol sent over the channel. So with a channel coding operation, as explained above, inflating the number of binary symbols by a factor of 2, slightly above $1/C \approx 1.883$, it is in principle possible to transmit with no error over such a noisy channel, by encoding long input sequences of binary symbols.

13.2 Quantum Channel Transmitting Classical Information

13.2.1 Model of Quantum Communication Channel

To proceed to the quantum domain, in this section we will examine the communication of classical information yet taking place over a quantum channel. This represents situations where the end-to-end information to be transmitted is classical, but its transmission is supported by quantum states or signals over a quantum channel. These are important situations for practical applications and technologies, for instance, to have access to reduced physical size of information carriers at the quantum level, as when transmitting individual photons in optical fibers or individual electrons in electric wires [399]. Compared with the classical communication channels examined in Sect. 13.1, such quantum communication channels exhibit specificities of their own, stemming from the quantum nature of the information carriers. One can think of the fact that distinct quantum states or signals are not necessarily distinguishable by quantum measurement, or the fact that quantum measurement is intrinsically random, or the impact of quantum noise or decoherence on the communication channels. An important difference is that the quantum hardware alone does not completely fix the communication channel, and the choice of an output measurement by the receiver is a crucial option with no classical analogue. So we now concentrate on such quantum channels for communicating classical information, as important devices of quantum information and signal processing.

We start as in Sect. 13.1 with a classical input emitting information-carrying signals under the form of random symbols X drawn among the J characters of an input alphabet $\{s_1, s_2, \ldots, s_J\}$, with the prior probabilities $\Pr\{X = s_j\} = p_j$ for $j = 1$ to J. For communication over a quantum channel, each input character s_j is encoded into a quantum state represented by the density operator $\rho_j \in \mathcal{L}(\mathcal{H}_N)$ on the N-dimensional Hilbert space $\mathcal{H}_N$.

The quantum channel usually implements a noisy input–output transmission of the quantum states, which can be described, as discussed in Sect. 3.2, by a (non-unitary) quantum process under the operator-sum form of Eq. (3.45), that we here write

$$\rho \longmapsto \mathcal{N}(\rho) = \sum_{\ell} \Lambda_\ell \rho \Lambda_\ell^\dagger \ , \tag{13.8}$$

where the set of Kraus operators $\{\Lambda_\ell\}$ defines the action of the channel $\mathcal{N}(\cdot)$. The operators Λ_ℓ in general are operators from the N-dimensional input Hilbert space $\mathcal{H}_N$ onto the N'-dimensional output Hilbert space $\mathcal{H}_{N'}$, with an $N' \times N$ representation matrix. Very commonly the input and output spaces have the same dimension $N = N'$, but in some circumstances these may differ, as for instance with the quantum erasure channel [25] to be examined in Sect. 13.2.8. The Kraus operators $\{\Lambda_\ell\}$ are bound to satisfy the completeness relation

$$\sum_{\ell} \Lambda_\ell^\dagger \Lambda_\ell = \mathrm{I}_N \ , \tag{13.9}$$

as in Eq. (3.47), where here I_N is the identity operator on the input space $\mathcal{H}_N$. The quantum process $\rho \mapsto \mathcal{N}(\rho)$ of Eq. (13.8) realizes a completely positive trace-preserving linear map from $\mathcal{L}(\mathcal{H}_N)$ onto $\mathcal{L}(\mathcal{H}_{N'})$, transforming an input density operator $\rho \in \mathcal{L}(\mathcal{H}_N)$ into an output (noisy) density operator $\rho' = \mathcal{N}(\rho) \in \mathcal{L}(\mathcal{H}_{N'})$. The number of Kraus operators need not be larger than NN' to represent all feasible maps.

At the channel output, a generalized quantum measurement, as discussed in Sect. 4.2, is performed by means of a K-element POVM defined by the K positive operators $\mathsf{E}_k \in \mathcal{L}(\mathcal{H}_{N'})$, for $k = 1$ to K, bound to satisfy the completeness relation

$$\sum_{k=1}^{K} \mathsf{E}_k = \mathrm{I}_{N'} \ , \tag{13.10}$$

as in Eq. (4.38), where here $\mathrm{I}_{N'}$ is the identity operator on the output space $\mathcal{H}_{N'}$. The result of the quantum measurement occurs at random and is represented by the classical random variable Y, taking the value s'_k when the POVM operator E_k operates, for $k = 1$ to K. With the noisy output states denoted $\rho'_j = \mathcal{N}(\rho_j)$, the input–output transmission (conditional) probabilities follow as

$$\Pr\{Y = s'_k \mid X = s_j\} = \mathrm{tr}(\rho'_j \mathsf{E}_k) \ . \tag{13.11}$$

Here also, for a meaningful assessment of the performance of the quantum channel for classical information communication, essentially the measuring POVM $\{\mathsf{E}_k\}$ matters,

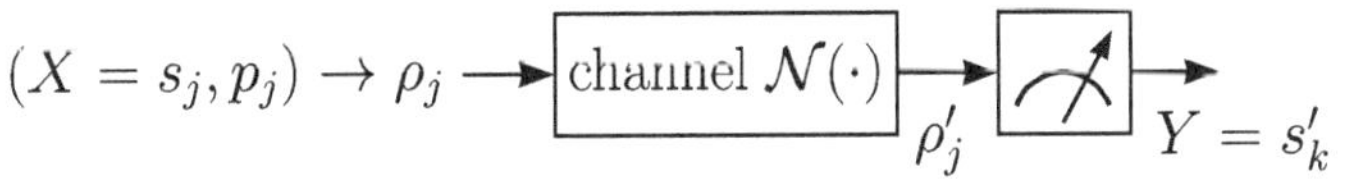

Fig. 13.2 The quantum communication channel: At the input the classical symbol $X = s_j$ occurring with prior probability p_j is encoded into the quantum state or symbol $\rho_j \in \mathcal{L}(\mathcal{H}_N)$, which is transmitted over the (noisy) quantum channel $\mathcal{N}(\cdot)$ to deliver the output quantum state $\rho'_j = \mathcal{N}(\rho_j) \in \mathcal{L}(\mathcal{H}_{N'})$, which is measured by a K-element POVM $\{\mathsf{E}_k\}$ yielding at the output the classical symbol $Y = s'_k$ according to the input–output conditional probabilities of Eq. (13.11)

to determine the output labels Y, and not directly the underlying measuring operators $\{\mathsf{M}_k\}$ which follow to realize the POVM as $\left\{\mathsf{E}_k = \mathsf{M}_k^\dagger \mathsf{M}_k\right\}$.

The whole communication process maps the input classical information X into the output classical information Y, mediated by transmission over the quantum channel $\mathcal{N}(\cdot)$, and is depicted in Fig. 13.2.

A fundamental quantity to assess the input–output transmission of classical information by the quantum channel of Fig. 13.2, is the input–output mutual information analyzed in Sect. 13.1,

$$H(X:Y) = H(Y) - H(Y|X) = H(X) - H(X|Y)\,. \tag{13.12}$$

This $H(X:Y)$ is determined by the input probability distribution $\{p_j\}$ (which is essentially classical), and by the input–output transmission probabilities of Eq. (13.11) fixed by the output measurement $\{\mathsf{E}_k\}$ (which is essentially quantum). From the results of Sect. 13.1 and the Shannon theorem for classical channel coding, we know that the input–output mutual information $H(X:Y)$ represents an achievable rate for error-free transmission of classical information, end-to-end over the quantum communication channel here. For a given quantum channel $\mathcal{N}(\cdot)$, the input–output mutual information $H(X:Y)$ is dependent on the input statistical ensemble $\left\{(p_j, \rho_j)\right\}$, and on the output measuring POVM $\{\mathsf{E}_k\}$. Accordingly, we will now examine various strategies that are accessible for optimizing the mutual information $H(X:Y)$ of the communication channel.

13.2.2 Optimizing the Output Measurement. The Holevo Information

For a quantum channel $\mathcal{N}(\cdot)$ with a given statistical ensemble $\{(p_j, \rho_j)\}_{j=1}^J$ of input signaling states, each choice of an output POVM $\{\mathsf{E}_k\}_{k=1}^K$ fixes a classical channel in the sense of Sect. 13.1. For the resulting communication channel, a meaningful issue is to determine the output POVM $\{\mathsf{E}_k\}_{k=1}^K$ maximizing the input–output mutual information $H(X:Y)$ of Eq. (13.12). The **accessible information** $I_{\text{acc}}(X,Y)$ is defined [9, 12] as the maximum of $H(X:Y)$ over all feasible output POVM $\{\mathsf{E}_k\}$, that is,

$$\boxed{I_{\rm acc}(X, Y) = \max_{\{\mathsf{E}_k\}} H(X:Y)\,.} \tag{13.13}$$

It is in general not straightforward to characterize the optimal POVM $\{\mathsf{E}_k\}$ solving the maximization of Eq. (13.13) and fixing the level $I_{\rm acc}(X, Y)$ of the accessible information for a given output ensemble $\{(p_j, \rho'_j)\}_{j=1}^J$ resulting by $\rho'_j = \mathcal{N}(\rho_j)$ from a given input ensemble $\{(p_j, \rho_j)\}_{j=1}^J$. Similar to the convexity (∪) of the mutual information $H(X : Y)$ in the input–output conditional probabilities $P(y|x)$ at fixed input probabilities $\{p_j\} \equiv \{P(x)\}$, at fixed input ensemble $\{(p_j, \rho_j)\}_{j=1}^J$ the mutual information $H(X : Y)$ is convex (∪) in the output POVM $\{\mathsf{E}_k\}$ over the convex set of output POVMs as described in Sect. 4.2.4. In this respect $H(X : Y)$ would exhibit a unique minimum, but Eq. (13.13) is concerned with a maximum. This maximum in Eq. (13.13) is therefore to be found on the boundary of the convex set of output POVMs, at an extreme POVM formed with only rank-one POVM operators, those that are not decomposable as a (non-degenerate) convex sum of operators [49, 276]. It is known [276] that in general the maximum of $H(X : Y)$ in Eq. (13.13) can be achieved by $K \in \left[N', N'^2\right]$ rank-one operators $\mathsf{E}_k = |m_k\rangle\langle m_k|$ with the K vectors $|m_k\rangle \in \mathcal{H}_{N'}$ of squared norm $\langle m_k|m_k\rangle \leq 1$ (not necessarily easy to characterize). Further characterization exists of the optimal POVM $\{\mathsf{E}_k\}$ when the output ensemble $\{(p_j, \rho'_j)\}_{j=1}^J$ to be measured has special symmetries [276].

In addition, there exists a general upper bound [12, 176] establishing

$$I_{\rm acc}(X, Y) \leq \chi\big(\{p_j, \rho'_j\}\big)\,, \tag{13.14}$$

with the **Holevo information** $\chi\big(\{p_j, \rho'_j\}\big)$ associated with the output statistical ensemble $\{(p_j, \rho'_j)\}_{j=1}^J$, defined according to Eq. (7.27) as

$$\chi\big(\{p_j, \rho'_j\}\big) = S\big(\overline{\rho}\,'\big) - \sum_{j=1}^J p_j\, S(\rho'_j) \leq \log(N')\,, \tag{13.15}$$

or alternatively, by Eq. (7.92), as

$$\chi\big(\{p_j, \rho'_j\}\big) = \sum_{j=1}^J p_j\, S\big(\rho'_j \,\|\, \overline{\rho}\,'\big)\,, \tag{13.16}$$

with the average output state

$$\overline{\rho}\,' = \sum_{j=1}^J p_j\, \rho'_j\,. \tag{13.17}$$

By linearity of the map $\mathcal{N}(\cdot)$, the average output state in Eq. (13.17) is also $\overline{\rho}\,' = \mathcal{N}(\overline{\rho})$ with the average input state $\overline{\rho} = \sum_{j=1}^J p_j \rho_j$.

A POVM $\{\mathsf{E}_k\}_{k=1}^{K}$ raising in Eq. (13.13) the accessible information $I_{\mathrm{acc}}(X, Y)$ at the level of the Holevo upper bound $\chi\left(\{p_j, \rho'_j\}\right)$ may not always exist [176, 400]. Nevertheless, the Holevo information $\chi\left(\{p_j, \rho'_j\}\right)$ forms a **maximal achievable rate** for error-free classical information transmission over the quantum channel. This can be obtained by means of a suitable **channel coding and decoding** protocol, relying on long sequences of consecutive quantum symbols collectively measured at the output. Inspired by the classical random coding of Sect. 13.1, such a protocol can operate in the following way, serving both for a proof of achievability and an operational procedure [12, 400–402]. The signaling ensemble $\{(p_j, \rho_j)\}_{j=1}^{J}$ can be employed to generate codewords randomly drawn as long sequences of L_{cod} consecutive independent input symbols ρ_j. At large length L_{cod}, by the law of large numbers, as with the classical channel of Sect. 13.1 or with the quantum compression of Sect. 8.2, the performance is determined essentially by the behavior on typical sequences. To communicate at a rate $R \le \chi_J\left(\{p_j, \rho'_j\}\right) \le \log_J(J) = 1$, among the $J^{L_{\mathrm{cod}}}$ feasible coding sequences, pruning is performed at random so as to select $J^{RL_{\mathrm{cod}}} \le J^{L_{\mathrm{cod}}}$ codewords sufficiently distinguishable from their noisy version at the channel output. At large length L_{cod}, the selected codewords tend to be typical for the statistical ensemble $\{(p_j, \rho_j)\}_{j=1}^{J}$. At the output the noisy codewords are collectively measured by a joint quantum measurement over L_{cod}-fold blocks and collectively decoded. This collective output decoding is an essential ingredient in the protocol, because if instead individual measurements for decoding were performed separately on each consecutive independent symbol, the transmission rate would be ruled by the input–output mutual information $H(X : Y)$ and its maximum $I_{\mathrm{acc}}(X, Y)$, with in general no guarantee of reaching as high as $\chi\left(\{p_j, \rho'_j\}\right)$. But for long input sequences of L_{cod} consecutive independent symbols, measuring at the output with a joint POVM acting collectively over the L_{cod}-fold blocks, can maximize the error-free transmission rate at $\chi\left(\{p_j, \rho'_j\}\right)$. The collective POVM is somehow a specifically quantum feature, so that the end-to-end transmission $X \mapsto Y$ mediated by the quantum channel differs from a purely classical transmission. In particular, with a collective POVM the output (classical) probabilities do not follow as product probabilities from individual symbols. Moreover, a notable property is that on sequences asymptotically at $L_{\mathrm{cod}} \to \infty$, an optimal POVM, such as in Eq. (13.13), is not required. It is enough to resort to the square-root measurement of Sect. 10.8 associated with the codewords in use (alternatively, sequential collective measurements can also do [12, 403–405]). In the limit $L_{\mathrm{cod}} \to \infty$ it is established that a transmission rate R up to $\chi\left(\{p_j, \rho'_j\}\right)$ can be achieved per input symbol with asymptotically vanishing error, and that $\chi\left(\{p_j, \rho'_j\}\right)$ is the maximum rate permitting an error-free transmission [12, 400–402]. In this way, the Holevo information $\chi\left(\{p_j, \rho'_j\}\right)$ represents the maximum rate achievable for error-free transmission of classical information, via consecutive independent uses of the quantum channel $\mathcal{N}(\cdot)$ operating with a given statistical ensemble $\{(p_j, \rho_j)\}_{j=1}^{J}$ of input signaling states.

The Holevo information presents the advantage of characterizing the maximum achievable rate of error-free transmission of classical information over the quantum channel using a given input signaling ensemble $\{(p_j, \rho_j)\}_{j=1}^{J}$, yet without the need to refer to any specific measuring POVM and to any $L_{\rm cod}$-long blocks explicitly. It provides the useful reference of a maximal efficiency, which is always achievable in known asymptotic conditions at $L_{\rm cod} \to \infty$. However, for practical implementation, the asymptotic regime $L_{\rm cod} \to \infty$ remains difficult to handle, and at finite length $L_{\rm cod} \geq 2$, efficient collective POVMs acting in $\mathcal{H}_{N'}^{\otimes L_{\rm cod}}$ to maximize the input–output transmission rate, generally remain difficult to characterize, depending on the conditions, a priori no less difficult than the optimal separable POVM (at $L_{\rm cod} = 1$) acting in $\mathcal{H}_{N'}$ and solving the maximization of Eq. (13.13).

13.2.3 Optimizing the Input Ensemble. The Holevo Capacity

Transmission over a given quantum channel $\mathcal{N}(\cdot)$ can be further optimized by adjusting the statistical ensemble $\{(p_j, \rho_j)\}_{j=1}^{J}$ of input signaling states. It is then meaningful to look for the input ensemble $\{(p_j, \rho_j)\}$ maximizing the Holevo information $\chi\big(\{p_j, \rho'_j\}\big)$ of Eq. (13.15), and achieving the **Holevo capacity** of the channel

$$\boxed{\chi_{\max}(\mathcal{N}) = \max_{\{p_j, \rho_j\}} \chi\big(\{p_j, \rho'_j\}\big)\ ,} \tag{13.18}$$

with $\rho'_j = \mathcal{N}(\rho_j)$. The Holevo capacity $\chi_{\max}(\mathcal{N})$ of Eq. (13.18) represents the overall maximal rate achievable for error-free transmission of classical information over the quantum channel $\mathcal{N}(\cdot)$, when resorting to product states or consecutive independent uses of the channel [12, 400–402].

To communicate with no error at this maximum rate $\chi_{\max}(\mathcal{N})$, the block coding and decoding protocol presented above can be employed with the optimal signaling ensemble $\big\{(p_j, \rho_j)\big\}$ solving the maximization of Eq. (13.18). The maximum rate $\chi_{\max}(\mathcal{N})$ can be achieved in general in the asymptotic regime of long blocks of size $L_{\rm cod} \to \infty$ of consecutive independent input symbols collectively measured and decoded at the output [12, 401, 402]. In practice however, to achieve the Holevo capacity $\chi_{\max}(\mathcal{N})$ in this way, the computational cost of the decoding scheme grows exponentially with the length $L_{\rm cod}$, as with the classical random coding, so that the design of alternative coding–decoding techniques for quantum channels remains today an active area of research.

Upon maximizing the Holevo information $\chi\big(\{p_j, \rho'_j\}\big)$ of Eq. (13.15) in order to solve Eq. (13.18), the second term of $\chi\big(\{p_j, \rho'_j\}\big)$ in Eq. (13.15)—the average output entropy $\sum_j p_j S(\rho'_j)$—would go to zero in the favorable configuration where all the output states ρ'_j would approach pure states (each having zero entropy); the first term—the entropy $S(\overline{\rho}\,')$ of the average output state—would go to its maximum $\log(N')$ in the favorable configuration where all the output states ρ'_j would average to the maximally mixed state $\mathrm{I}_{N'}/N'$, expressing in some sense that the output states ρ'_j collectively contain sufficient variability so as to distribute uniformly over the entire output space. So maximizing the Holevo information $\chi\big(\{p_j, \rho'_j\}\big)$ of Eq. (13.15) to solve Eq. (13.18), loosely speaking, leads us to select the input ensemble so as to approach, at the output, maximally distinguishable pure states spanning the entire output space, as much as possible within the constraints imposed by the channel $\mathcal{N}(\cdot)$.

From what we indicated at the end of Sect. 7.4, the Holevo information $\chi\left(\{p_j, \rho_j'\}\right)$ is a convex ($\cup$) function of the output states $\{\rho_j'\}$, for a fixed probability distribution $\{p_j\}$. The output states $\rho_j' = \mathcal{N}(\rho_j)$ are linear functions of the input states ρ_j, so that $\chi\left(\{p_j, \rho_j'\}\right)$ is also a convex ($\cup$) function of the input states $\{\rho_j\}$, for a fixed probability distribution $\{p_j\}$. As a consequence, a maximum of $\chi\left(\{p_j, \rho_j'\}\right)$ necessarily occurs on the boundary of the convex set of density operators ρ_j, i.e., for pure input states $\rho_j = |\psi_j\rangle\langle\psi_j|$ (that are not decomposable as a convex sum of several pure states). Meanwhile, the Holevo information $\chi\left(\{p_j, \rho_j'\}\right)$ is a concave ($\cap$) function of the probability distribution $\{p_j\}$, for fixed signal states $\{\rho_j\}$ and $\left\{\rho_j' = \mathcal{N}(\rho_j)\right\}$. Consequently, a local maximum of $\chi\left(\{p_j, \rho_j'\}\right)$ is not necessarily a global maximum.

For a given quantum channel $\mathcal{N}(\cdot)$, the maximization of Eq. (13.18) is not necessarily easy to solve. However, it is known [406, 407] that the maximum of $\chi\left(\{p_j, \rho_j'\}\right)$ can be achieved by $J \in \left[N, N^2\right]$ pure input states $\rho_j = |\psi_j\rangle\langle\psi_j|$ with $|\psi_j\rangle \in \mathcal{H}_N$ (not necessarily easy to characterize).

An alternative characterization [179, 408] of the Holevo capacity of Eq. (13.18) is

$$\chi_{\max}(\mathcal{N}) = \min_{\rho_2} \max_{\rho_1} S\big(\mathcal{N}(\rho_1) \| \mathcal{N}(\rho_2)\big) , \tag{13.19}$$

making use of the quantum relative entropy of Eq. (7.74). The minimax condition of Eq. (13.19) fixes the value of $\chi_{\max}(\mathcal{N})$ but does not completely determine an optimal ensemble $\left\{\left(p_j, \rho_j = |\psi_j\rangle\langle\psi_j|\right)\right\}$ of input states achieving $\chi_{\max}(\mathcal{N})$ in Eq. (13.18). Several properties are established in [179] for an optimal input ensemble $\left\{\left(p_j, \rho_j = |\psi_j\rangle\langle\psi_j|\right)\right\}$ achieving the maximum Holevo information $\chi_{\max}(\mathcal{N})$ in Eq. (13.18):

- There might be different optimal input ensembles $\left\{(p_j, \rho_j)\right\}$ achieving the maximum $\chi_{\max}(\mathcal{N})$ in Eq. (13.18), but they all share [409] the same common average input state $\overline{\rho} = \sum_j p_j \rho_j = \overline{\rho}_{\text{opt}}$, and consequently the same average output state $\overline{\rho}'_{\text{opt}} = \mathcal{N}\left(\overline{\rho}_{\text{opt}}\right)$.
- Any feasible output state $\mathcal{N}(\rho)$ satisfies $S\big(\mathcal{N}(\rho) \| \mathcal{N}(\overline{\rho}_{\text{opt}})\big) \le \chi_{\max}(\mathcal{N})$, otherwise such a ρ included in the optimal input ensemble would produce a strict increase of χ and alter $\overline{\rho}_{\text{opt}}$.
- In an optimal input ensemble $\left\{(p_j, \rho_j)\right\}$, each optimal input state ρ_j of probability $p_j \neq 0$ realizes the (same) maximum $S\big(\mathcal{N}(\rho_j) \| \mathcal{N}(\overline{\rho}_{\text{opt}})\big) = \chi_{\max}(\mathcal{N})$, otherwise the average relative entropy of Eq. (13.16) could not reach $\chi_{\max}(\mathcal{N})$.
- An optimal input ensemble has an image by $\mathcal{N}(\cdot)$ whose support coincides with the support of the image by $\mathcal{N}(\cdot)$ of the whole input space, otherwise a state $\mathcal{N}(\rho)$ outside this support would increase χ by the infinite amount $S\big(\mathcal{N}(\rho) \| \mathcal{N}(\overline{\rho}_{\text{opt}})\big) = \infty$ in Eq. (13.16).

• For any $\overline{\rho}$ that solves the minimization $\min_{\overline{\rho}} \max_{\rho} S\big(\mathcal{N}(\rho)\|\mathcal{N}(\overline{\rho})\big)$, there exists an optimal input ensemble $\big\{(p_j, \rho_j)\big\}$ with average state $\sum_j p_j \rho_j = \overline{\rho}$.

In practice, such a $\overline{\rho}$ solving the minimization $\min_{\overline{\rho}} \max_{\rho} S\big(\mathcal{N}(\rho)\|\mathcal{N}(\overline{\rho})\big)$ will be associated with several $\rho = \rho_j$ solutions to the minimax problem, for a given unique $\overline{\rho}$. Then several of these solutions $\rho = \rho_j$ should be selected and weighted by appropriate probabilities p_j so that they match the prescribed average state $\sum_j p_j \rho_j = \overline{\rho}$ and form an optimal input ensemble $\big\{(p_j, \rho_j)\big\}$.

A simpler optimization [25], meaningful when one has to operate with a given quantum channel $\mathcal{N}(\cdot)$ imposing the (quantum) signaling states $\{\rho_j\}_{j=1}^J$, is, instead of the maximization of Eq. (13.18), to maximize the Holevo information $\chi\big(\{p_j, \rho_j'\}\big)$ only with respect to the prior (classical) probabilities $\{p_j\}_{j=1}^J$, with fixed $\{\rho_j\}$. In addition, with the practical necessity of referring to an explicit POVM, instead of $\chi\big(\{p_j, \rho_j'\}\big)$ one is rather faced with the problem of maximizing the information $I_{\mathrm{acc}}(X, Y)$ or even $H(X : Y)$, with respect to $\big\{p_j, |\psi_j\rangle\big\}$ or to $\{p_j\}$ only. The solutions thereof are also not generally characterized.

13.2.4 Exploiting Entanglement

Two quantum channels $\mathcal{N}_1$ and $\mathcal{N}_2$ involved in a tensor product $\mathcal{N}_1 \otimes \mathcal{N}_2$ provide a representation of two distinct uses of channel $\mathcal{N}_1$ and channel $\mathcal{N}_2$. For independent uses of the channels with product quantum states, the Holevo capacity of Eq. (13.18) is additive [406, 407], that is

$$\chi_{\max}(\mathcal{N}_1 \otimes \mathcal{N}_2) = \chi_{\max}(\mathcal{N}_1) + \chi_{\max}(\mathcal{N}_2) . \tag{13.20}$$

In general, especially for non-independent but entangled uses of the channels with non-product entangled states, the Holevo capacity of Eq. (13.18) is always **superadditive** [406, 407], that is

$$\chi_{\max}(\mathcal{N}_1 \otimes \mathcal{N}_2) \geq \chi_{\max}(\mathcal{N}_1) + \chi_{\max}(\mathcal{N}_2) . \tag{13.21}$$

This is natural since $\chi_{\max}(\mathcal{N}_1 \otimes \mathcal{N}_2)$ can at least be as large as $\chi_{\max}(\mathcal{N}_1) + \chi_{\max}(\mathcal{N}_2)$ by using independent product states as in Eq. (13.20). The truly important issue is whether, by using non-independent entangled uses of the channels, one can reach $\chi_{\max}(\mathcal{N}_1 \otimes \mathcal{N}_2) > \chi_{\max}(\mathcal{N}_1) + \chi_{\max}(\mathcal{N}_2)$, i.e., a strictly superadditive Holevo capacity. This is a crucial issue since strict superadditivity of the Holevo capacity would enable to increase the error-free transmission rate by using entangled inputs to a channel.

Additivity in Eq. (13.21), excluding the possibility of strict superadditivity in Eq. (13.21), has been proved for a significant range of quantum channels. Additivity of $\chi_{\max}(\mathcal{N}_1 \otimes \mathcal{N}_2)$ has been proved when one channel is the identity channel, or a unital qubit channel, or a classical–quantum or a quantum–classical channel, or an entanglement-breaking channel [406]. Additivity of $\chi_{\max}(\mathcal{N} \otimes \mathcal{N})$ has been proved for unital qubit channels, the depolarizing channel, the erasure channel, the purely lossy bosonic channel, the whole class of entanglement-breaking channels [25]. For a while it has even been conjectured that all quantum channels might exhibit an additive Holevo capacity. But it is now known that this is not so.

An example of strict superadditivity in Eq. (13.21) has been reported in [410]. This has been found in [410] to occur with channels of the form $\mathcal{N}(\rho) = \sum_{m=1}^{M} \Lambda_m \rho \Lambda_m^\dagger$, of M Kraus operators $\Lambda_m = \sqrt{P_m}\mathsf{U}_m$ with M random unitary operators U_m from $\mathcal{H}_N$ onto $\mathcal{H}_N$, random probabilities P_m, and $1 \ll M \ll N$, requiring a (high) dimension $N \geq 183$, as shown in [411].

Additivity of the Holevo capacity in Eq. (13.20) has been proved in [407] equivalent to additivity of the minimal output entropy $S_{\min}(\mathcal{N}_1 \otimes \mathcal{N}_2) = S_{\min}(\mathcal{N}_1) + S_{\min}(\mathcal{N}_2)$, with $S_{\min}(\mathcal{N}) = \min_\rho S\big(\mathcal{N}(\rho)\big)$ this minimum being achievable over *pure* input states $\rho = |\psi\rangle\langle\psi|$ on $\mathcal{H}_N$.

The work of [410] reports a violation of the additivity of the minimal output entropy, under the form of subadditivity $S_{\min}(\mathcal{N}_1 \otimes \mathcal{N}_2) < S_{\min}(\mathcal{N}_1) + S_{\min}(\mathcal{N}_2)$, where an entangled input state accross the two channels exhibits a lower output entropy, and hence greater resistance to noise, than any unentangled state. This is equivalent to strict superadditivity of the Holevo capacity in Eq. (13.21), for entangled states more resistant to noise.

After the example of superadditivity in [410], today it remains largely an open issue to identify whether there exist other classes of quantum channels with a superadditive Holevo capacity, simpler (non-random) classes, more generic, physically motivated, superadditive channels. This superadditivity issue is quite important and significant for quantum communication, because, as we indicated in Sect. 13.1, in classical communication channels, maximum rate of information transmission occurs with independent consecutive symbols, or independent consecutive uses of the channel, and correlation or dependence between consecutive uses do not increase the transmission rate [12, 397]. This is no longer generally true with quantum channels, as demonstrated by [410], where quantum correlation by entanglement between consecutive uses of a quantum channel, may increase the maximum rate of error-free information transmission. This stands as another remarkable manifestation of the distinctive character of quantum correlation by entanglement, offering unparalleled properties useful to quantum signal and information processing. Complete understanding of the potentialities of entanglement for quantum communication, of their conditions of applicability, of the performance they entail, is however not yet achieved today and requires further theoretical efforts, let aside the technological opportunities that may result.

13.2.5 Classical Capacity of a Quantum Channel. The HSW Theorem

The classical information capacity $C(\mathcal{N})$ of a quantum channel $\mathcal{N}(\cdot)$, in similarity with the case of the classical channel considered in Sect. 13.1 and Eq. (13.6), is defined as the maximum achievable rate (per channel use) for error-free transmission of classical information over $\mathcal{N}(\cdot)$. The classical information is supported by the classical random variables (X, Y) occurring in Fig. 13.2, and the quantum channel $\mathcal{N}(\cdot)$ can receive consecutive inputs chosen independent or correlated by entanglement, whatever may be more efficient to maximize the transmission rate. In view of the properties of the Holevo capacity $\chi_{\max}(\mathcal{N})$, and especially its superadditive character in general as expressed by Eq. (13.21), the **classical capacity** $C(\mathcal{N})$ of a quantum channel $\mathcal{N}$ is generally given by the so-called "regularized" Holevo capacity as

$$\boxed{C(\mathcal{N}) = \lim_{L\to\infty} \frac{1}{L}\, \chi_{\max}\left(\mathcal{N}^{\otimes L}\right) .} \tag{13.22}$$

This definition in general involves maximization of the Holevo information over L-fold entangled input states, conveyed by the L-fold channel $\mathcal{N}^{\otimes L}$ representing L consecutive uses of the physical channel $\mathcal{N}$. This amounts to signaling with macrosymbols from $\mathcal{L}(\mathcal{H}_N^{\otimes L})$ realized as sequences of L possibly entangled symbols from $\mathcal{L}(\mathcal{H}_N)$. Achievability is in general obtained by encoding followed by collective measurement and decoding on asymptotically long sequences of macrosymbols. Moreover, with $\chi_{\max}\left(\mathcal{N}^{\otimes L}\right)$ increasing with L in case of strict superadditivity, the asymptotic regime of large length $L \to \infty$ has to be considered in general in Eq. (13.22) to characterize the quantum capacity $C(\mathcal{N})$. This central result in Eq. (13.22) establishing the classical capacity of a quantum channel by means of the Holevo information, is commonly referred to as the **Holevo–Schumacher–Westmoreland (HSW) theorem** of quantum information theory [12, 183, 401, 402].

For the special case of quantum channels with an additive Holevo capacity $\chi_{\max}$ in Eq. (13.22), for any size L one has $\chi_{\max}\left(\mathcal{N}^{\otimes L}\right) = L\, \chi_{\max}(\mathcal{N})$, so that the general definition of Eq. (13.22) reduces to a single-use or single-letter formula

$$C(\mathcal{N}) = \chi_{\max}(\mathcal{N}) , \tag{13.23}$$

and the classical capacity $C(\mathcal{N})$ of the quantum channel $\mathcal{N}$ coincides with its Holevo capacity $\chi_{\max}(\mathcal{N})$ of Eq. (13.18). Otherwise, the channel Holevo capacity $\chi_{\max}(\mathcal{N})$ of Eq. (13.18) forms a lower bound to the classical capacity $C(\mathcal{N})$, when $\chi_{\max}$ is non-additive, that is, strictly superadditive.

Today we do not have a complete characterization of the quantum channels with a non-additive (strictly superadditive) or an additive Holevo capacity $\chi_{\max}(\mathcal{N})$, when for the classical capacity $C(\mathcal{N})$ the regularized formula of Eq. (13.22) can be replaced by the single-use formula of Eq. (13.23), and of the conditions in which communication of classical information over a quantum channel may benefit from entangled

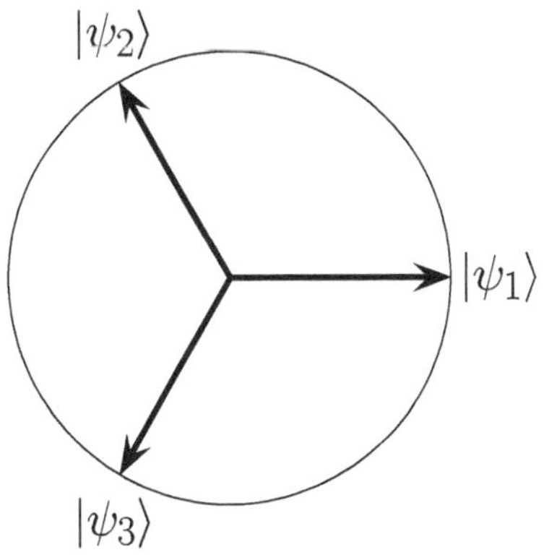

Fig. 13.3 The trine ensemble composed in the vector plane $\mathcal{H}_2$ of the three pure qubit states $|\psi_1\rangle = |0\rangle$, $|\psi_2\rangle = |\psi_+\rangle$ and $|\psi_3\rangle = |\psi_-\rangle$ from Eq. (13.24)

inputs to enhance the transmission rate. Due to its asymptotic character, the general definition of Eq. (13.22) is in practice difficult to handle whenever necessary. However, as indicated above in Sect. 13.2.4, many quantum channels commonly encountered exhibit an additive Holevo capacity, so that their performance for classical information communication can be characterized from the single-use capacity of Eq. (13.23). Now in the sequel, in the rest of this Sect. 13.2, we will work out several significant examples regarding communication of classical information over quantum channels, with explicit derivations of classical capacities $C(\mathcal{N})$.

13.2.6 An Example of Quantum Communication

As an example, we consider quantum communication by means of qubit states. We examine the situation considered in [400, 412, 413] of a statistical ensemble of three signaling states, known as the trine ensemble, formed by the three pure qubit states $|\psi_1\rangle = |0\rangle$, $|\psi_2\rangle = |\psi_+\rangle$ and $|\psi_3\rangle = |\psi_-\rangle$ with

$$|\psi_\pm\rangle = -\frac{1}{2}\,|0\rangle \pm \frac{\sqrt{3}}{2}\,|1\rangle \ , \tag{13.24}$$

as depicted in Fig. 13.3, with prior probabilities $p_j = 1/3$ for $j = 1$ to 3.

The quantum transmission channel $\mathcal{N}(\cdot)$ is taken as the ideal noise-free identity channel, i.e., $\mathcal{N}(\rho) = \rho$ for any qubit density operator ρ on the two-dimensional Hilbert space $\mathcal{H}_2$. Nevertheless, since the three signaling symbols $\{|\psi_1\rangle\,, |\psi_2\rangle\,, |\psi_3\rangle\}$ are not completely distinguishable, quantum measurement at the output will usually lead to some unavoidable (statistical) error in recognizing each individual symbol. A given level of such error is associated with a given input–output classical mutual information for the channel, and this mutual information fixes an achievable rate for error-free transmission of classical information over the channel, provided we resort to an appropriate coding and decoding protocol, as we explained above. When signaling with $\{|\psi_1\rangle, |\psi_2\rangle, |\psi_3\rangle\}$ is imposed over the channel, we therefore seek to optimize the output measurement so as to maximize the input–output classical mutual information.

At the channel output, the average state of Eq. (13.17) follows as

$$\overline{\rho}' = \sum_{j=1}^{3} \frac{1}{3} |\psi_j\rangle \langle\psi_j| = \frac{\mathrm{I}_2}{2}, \tag{13.25}$$

which is the maximally mixed qubit state, with maximal quantum entropy $S(\mathrm{I}_2/2) = \log(2) = 1$ Sh in Eq. (7.29). Each of the three pure density operators $\rho'_j = \mathcal{N}(\rho_j) = \rho_j = |\psi_j\rangle \langle\psi_j|$ has zero quantum entropy $S(|\psi_j\rangle \langle\psi_j|) = 0$ in Eq. (7.29). The Holevo information of the output ensemble (the equiprobable trine ensemble) results in Eq. (13.15) as $\chi = \log(2) = 1$ Sh.

At the output, different measuring POVM can be envisaged.

• Measurement can be performed by a **standard von Neumann measurement** in the orthonormal projective basis $\{|0\rangle, |1\rangle\}$ of $\mathcal{H}_2$. This represents an output measurement with two outcomes, that can be noted $Y = 0$ when the measurement projects onto $|0\rangle$, and $Y = 1$ when it projects onto $|1\rangle$; meanwhile, the input $X = 1$ is encoded by $|\psi_1\rangle = |0\rangle$, the input $X = 2$ by $|\psi_2\rangle = |\psi_+\rangle$ and $X = 3$ by $|\psi_3\rangle = |\psi_-\rangle$. From Eq. (13.11) we deduce the six input–output conditional probabilities $\Pr\{Y = 0|X = 1\} = 1 = 1 - \Pr\{Y = 1|X = 1\}$ and $\Pr\{Y = 0|X = 2\} = \Pr\{Y = 0|X = 3\} = 1/4 = 1 - \Pr\{Y = 1|X = 2\} = 1 - \Pr\{Y = 1|X = 3\}$. The input–output classical conditional entropy follows as $H(Y|X) = (4/3)\log(2) - (1/2)\log(3)$. Also, the two output probabilities are $\Pr\{Y = 0\} = \Pr\{Y = 1\} = 1/2$, whence the output classical entropy $H(Y) = \log(2)$. The input–output mutual information of Eq. (13.12) results as $H(X : Y) = (1/2)\log(3) - (1/3)\log(2) \approx 0.459$ Sh. It is easy to check that this represents the maximal mutual information $H(X : Y)$ that can be extracted by measuring with a standard von Neumann measurement in $\mathcal{H}_2$. This is rather far below the Holevo information $\chi = \log(2) = 1$ Sh fixing an upper bound to $H(X : Y)$ via Eq. (13.14). To try to do better, one as the resource to turn to a generalized measurement, especially a three-outcome measurement at the output.

• A generalized measurement useful to discriminate quantum states like the three pure qubit states $|\psi_j\rangle$ of the trine ensemble, is the **square-root measurement** that we presented in Sect. 10.8, which, although not necessarily optimal, is often efficient in practical situations of interest. In general, for discriminating among J pure states $|\psi_j\rangle \in \mathcal{H}_N$ with prior probabilities p_j, the square-root measurement is defined by the J rank-one operators $\mathsf{E}_j = |m_j\rangle \langle m_j|$ with the J vectors $|m_j\rangle = \rho^{-1/2}\sqrt{p_j}\,|\psi_j\rangle$ of $\mathcal{H}_N$ also given in Eq. (10.133). For the qubit equiprobable trine ensemble, having the average density operator $\rho = \overline{\rho}' = \mathrm{I}_2/2$ in Eq. (13.25), the square-root measurement is characterized by the three vectors $|m_j\rangle = \sqrt{2/3}\,|\psi_j\rangle$ and three POVM elements $\mathsf{E}_j = (2/3)\,|\psi_j\rangle \langle\psi_j|$, for $j = 1$ to 3, exhibiting the same symmetry as the trine ensemble. Respectively, when each E_j operates, the measurement outcome is

$Y = j$, for $j = 1$ to 3. From Eq. (13.11), we have the input–output conditional probabilities $\Pr\{Y = 1|X = 1\} = 2/3$ and $\Pr\{Y = 2|X = 1\} = \Pr\{Y = 3|X = 1\} = 1/6$. The other 3 + 3 conditional probabilities follow by symmetry. The input–output classical conditional entropy follows as $H(Y|X) = \log(3) - (1/3)\log(2)$. Also, by symmetry, the three output probabilities are $\Pr\{Y = 1\} = \Pr\{Y = 2\} = \Pr\{Y = 3\} = 1/3$, whence the output classical entropy $H(Y) = \log(3)$. The input–output mutual information of Eq. (13.12) results as $H(X : Y) = (1/3)\log(2) \approx 0.333$ Sh. It is below the Holevo information $\chi = 1$ Sh of the equiprobable trine ensemble, and also below the maximal mutual information $H(X : Y) \approx 0.459$ Sh of the von Neumann measurement treated earlier.

• The **optimal measurement** maximizing the mutual information $H(X : Y)$ in Eq. (13.13) can be obtained as follows. For the equiprobable trine ensemble, the results of [276] serve to establish that the optimal measurement is the POVM determined by the three elements

$$\mathsf{E}_j = \frac{2}{3}\Big(\mathrm{I}_2 - |\psi_j\rangle\langle\psi_j|\Big), \quad \text{for } j = 1 \text{ to } 3, \tag{13.26}$$

also exhibiting the symmetry of the trine ensemble. Each of the three operators E_j in Eq. (13.26) is proportional [413] to an orthogonal projector on a direction orthogonal to each state $|\psi_j\rangle$ in the qubit plane $\mathcal{H}_2$. From Eq. (13.11), we have the input–output conditional probabilities $\Pr\{Y = 1|X = 1\} = 0$ and $\Pr\{Y = 2|X = 1\} = \Pr\{Y = 3|X = 1\} = 1/2$. The other 3 + 3 conditional probabilities follow by symmetry. We see that E_j being measured excludes that the state was $|\psi_j\rangle$, while it places equal probabilities on the two other states. The input–output classical conditional entropy follows as $H(Y|X) = \log(2)$. Also, by symmetry, the three output probabilities are $\Pr\{Y = 1\} = \Pr\{Y = 2\} = \Pr\{Y = 3\} = 1/3$, whence the output classical entropy $H(Y) = \log(3)$. The input–output mutual information of Eq. (13.12) results as $H(X : Y) = \log(3) - \log(2) \approx 0.585$ Sh. Based on [276], we know this is the maximum information that can be extracted from the equiprobable trine ensemble by any measurement acting on individual qubits, and accordingly it fixes $I_{\text{acc}}(X, Y) = \log(3) - \log(2) \approx 0.585$ Sh in Eq. (13.13). This accessible information $I_{\text{acc}}(X, Y) \approx 0.585$ Sh is still below the Holevo information $\chi = 1$ Sh of the equiprobable trine ensemble, but no quantum measurement $\{\mathsf{E}_k\}_{k=1}^{K}$ acting on individual qubits can extract more information in Eq. (13.13) from the equiprobable trine ensemble.

We note that such an explicit characterization of the optimal measuring POVM, and an assessment of its performance $I_{\text{acc}}(X, Y)$ relative to its upper bound χ, is far from being equally accessible in other conditions. For instance, when the three states of the trine ensemble are no longer equiprobable, or when some quantum noise $\mathcal{N}(\cdot)$ intervenes in the communication process, the optimal strategy is to be determined anew. An example is reported in [25], with a non-equiprobable trine and the noise-free channel, showing that when signaling is performed at the input with $p_1 = 0$ and $p_2 = p_3 = 1/2$, the square-root measurement at the output is a standard two-outcome von Neumann measurement, capable of extracting an information of

$H(X:Y) \approx 0.645$ Sh, which is above the performance $I_{\text{acc}}(X, Y) \approx 0.585$ Sh of the equiprobable trine. However, the optimal input probability distribution $\{p_j\}_{j=1}^3$ for the trine ensemble and the optimal output measurement to maximize the mutual information $H(X:Y)$ remain unknown, as it was noted in [25], "as for most of the Shannon theory" of quantum communication channels.

As suggested by the block coding and decoding protocol presented in Sect. 13.2.2, more information may be extracted by moving to collective processing of blocks of qubits. To gain insight in this direction, we can consider qubit pairs, as realized in [400, 412]. Among the nine possible blocks of two qubits, by pruning we select as codewords the three two-qubit states $|\psi_j\rangle \otimes |\psi_j\rangle$ constructed from the trine ensemble (and forming the so-called double-trine ensemble), with prior probabilities $p_j = 1/3$, for $j = 1$ to 3. This concatenation and pruning to generate the equiprobable double-trine ensemble for signaling, maintains the same statistics as the equiprobable trine ensemble $\{(p_j = 1/3, |\psi_j\rangle)\}$ over the channel. The equiprobable double-trine ensemble, in place of Eq. (13.25), is associated with the average density operator $\overline{\rho}'$ having the nonzero eigenvalues 1/2, 1/4 and 1/4, and the quantum entropy $S(\overline{\rho}') = 3\log(2)/2 = 3/2$ Sh, leading to a Holevo information $\chi^{(2)} = 3/4$ Sh per qubit. Compared with the Holevo information $\chi = 1$ Sh per qubit of the one-qubit signaling, this could suggest a lesser efficiency of the two-qubit signaling, which exploits for signaling only three among the nine feasible states. But this is not so. The three two-qubit signaling states $|\psi_j\rangle \otimes |\psi_j\rangle$ constitute three linearly independent states in the four-dimensional space $\mathcal{H}_2^{\otimes 2}$, while the three one-qubit signaling states $|\psi_j\rangle$ do not in $\mathcal{H}_2$, and in this respect the signals have become more distinguishable. Altogether, this leads to a superior efficiency for the two-qubit signaling, that can be assessed as follows. With the two-qubit signaling, at the output the square-root measurement can be implemented in $\mathcal{H}_2^{\otimes 2}$. This in general takes the form of a joint global measurement in $\mathcal{H}_2^{\otimes 2}$, acting collectively on the two signaling qubits, not reducible to two separate measurements on each qubit, and projecting onto entangled post-measurement states in $\mathcal{H}_2^{\otimes 2}$. The interesting result is that an information per qubit of $H(X:Y) \approx 0.685$ Sh can be extracted at the output, at the cost of processing blocks of two qubits collectively measured [400, 412, 413]. This comes closer to the Holevo upper bound $\chi^{(2)} = 3/4$ Sh per qubit of the double-trine ensemble, and also improves over the maximal accessible information $I_{\text{acc}}(X, Y) \approx 0.585$ Sh per qubit of the one-qubit signaling, and approaches the ultimate limit of $\chi = 1$ Sh per qubit. For the two-qubit signaling, the information per qubit of $H(X:Y) \approx 0.685$ Sh in [400, 412, 413] is not proved to correspond to the optimal two-qubit strategy reaching (and determining) the maximum accessible information with a two-qubit POVM. Ultimately, a maximum mutual information per qubit reaching the Holevo upper bound $\chi = 1$ Sh, can eventually be obtained, in principle, by collective processing of longer and longer blocks of qubits; but the path to this, in practice, is today poorly charted.

To proceed with other illustrations, we now study important examples of quantum communication channels, that especially lend themselves to an explicit characterization of their classical information capacity $C(\mathcal{N})$, or at least their Holevo capacity $\chi_{\max}(\mathcal{N})$, that we carry out.

13.2.7 Classical Capacity of the Qudit Depolarizing Channel

An important communication channel is the qudit depolarizing channel, acting on N-dimensional quantum systems or qudits, experiencing during their transmission the noisy input–output transformation

$$\mathcal{N}(\rho) = \alpha\rho + (1-\alpha)\frac{\mathrm{I}_N}{N} = \rho' , \tag{13.27}$$

also representing the effect of the qudit depolarizing noise of Eq. (9.52). In the noisy transmission by Eq. (13.27), an input qudit state $\rho \in \mathcal{L}(\mathcal{H}_N)$ is unaffected with probability α, while it is replaced by the maximally mixed qudit state I_N/N with probability $1-\alpha$.

For the qudit depolarizing channel $\mathcal{N}(\cdot)$, the Holevo capacity $\chi_{\max}$ is known to be additive [414], so that its classical information capacity $C(\mathcal{N})$ coincides with the channel Holevo capacity $\chi_{\max}(\mathcal{N})$ as in Eq. (13.23). So we seek to characterize the capacity $C(\mathcal{N}) = \chi_{\max}(\mathcal{N})$ from Eq. (13.18), as accomplished in [414] or in [12]. We know that the maximization in Eq. (13.18) can be achieved by pure input states $|\psi_j\rangle$. For any pure state $|\psi\rangle$ of the qudit Hilbert space $\mathcal{H}_N$, the input–output transformation of Eq. (13.27) can be rewritten as

$$\mathcal{N}\big(|\psi\rangle\langle\psi|\big) = \left(\alpha + \frac{1-\alpha}{N}\right)|\psi\rangle\langle\psi| + \frac{1-\alpha}{N}\Big(\mathrm{I}_N - |\psi\rangle\langle\psi|\Big) . \tag{13.28}$$

In order to compute its quantum entropy, the output density operator $\mathcal{N}\big(|\psi\rangle\langle\psi|\big) \in \mathcal{L}(\mathcal{H}_N)$ can receive an eigendecomposition from Eq. (13.28). The density operator $\mathcal{N}\big(|\psi\rangle\langle\psi|\big)$ in Eq. (13.28) has $|\psi\rangle$ as an eigenstate, associated with the eigenvalue $\alpha + (1-\alpha)/N$. The $N-1$ remaining eigenstates of $\mathcal{N}\big(|\psi\rangle\langle\psi|\big)$ can be taken as $N-1$ states of $\mathcal{H}_N$, mutually orthogonal and each orthogonal to $|\psi\rangle$; they are all associated with the eigenvalue $(1-\alpha)/N$ occurring with multiplicity $N-1$. As a consequence, the quantum entropy of Eq. (7.29) for the output density operator $\mathcal{N}\big(|\psi\rangle\langle\psi|\big)$ in Eq. (13.28) follows as

$$S\big(\mathcal{N}(|\psi\rangle\langle\psi|)\big) = h\left(\alpha + \frac{1-\alpha}{N}\right) + (N-1)h\left(\frac{1-\alpha}{N}\right) , \tag{13.29}$$

with the auxiliary function $h(u) = -u\log(u)$ from Eq. (7.39). And this output entropy is the same for any pure input state $|\psi\rangle$. Any statistical ensemble $\big\{\big(p_j, |\psi_j\rangle\big)\big\}$ of pure input states will therefore lead in Eq. (13.15) to the same average output entropy

$$\sum_j p_j\, S(\rho_j') = h\left(\alpha + \frac{1-\alpha}{N}\right) + (N-1)h\left(\frac{1-\alpha}{N}\right). \tag{13.30}$$

To maximize the Holevo information $\chi\left(\{p_j, \rho_j'\}\right)$ in Eq. (13.15) over input ensembles of pure states, the task then reduces to maximizing the entropy $S(\overline{\rho}\,')$ of the average output state $\overline{\rho}\,'$ of Eq. (13.17). The overall maximum of $S\left(\overline{\rho}\,'\right)$ is $\log(N)$, achieved by the maximally mixed average output state $\overline{\rho}\,' = \mathrm{I}_N/N$. The channel of Eq. (13.27) is a unital channel satisfying $\mathcal{N}\left(\mathrm{I}_N/N\right) = \mathrm{I}_N/N$, and by linearity of the map $\mathcal{N}(\cdot)$ the average output state $\overline{\rho}\,' = \mathrm{I}_N/N = \mathcal{N}(\overline{\rho})$ is achieved by the average input state $\overline{\rho} = \mathrm{I}_N/N$. In turn, the average input state $\overline{\rho} = \mathrm{I}_N/N$ is achieved by any input ensemble formed by N mutually orthogonal pure states $|\psi_j\rangle$ with equal probability $p_j = 1/N$. Such an optimal input ensemble achieves in Eq. (13.18) the maximum $\chi_{\max}(\mathcal{N}) = C(\mathcal{N})$ that determines the classical information capacity of the qudit depolarizing channel as

$$\boxed{C(\mathcal{N}) = \log(N) - h\left(\alpha + \frac{1-\alpha}{N}\right) - (N-1)h\left(\frac{1-\alpha}{N}\right).} \tag{13.31}$$

With the HSW theorem of Sect. 13.2.5 we know that the rate $C(\mathcal{N}) = \chi_{\max}(\mathcal{N})$ can be achieved for error-free communication by transmitting long blocks of $L_{\text{cod}} \to \infty$ independent input states collectively measured at the output. But this collective processing is not necessary for the qudit depolarizing channel, since we will see that there exists a quantum measurement acting on individual qudits capable of reaching an input–output mutual information $H(X:Y)$ in Eq. (13.13) matching the capacity $C(\mathcal{N})$ of Eq. (13.31). An optimal input ensemble can be taken as the computational basis $\{|j\rangle\}_{j=1}^N$ of $\mathcal{H}_N$ where the N signaling states $|\psi_j\rangle = |j\rangle$ are equiprobable with probabilities $p_j = 1/N$. The output of the channel is measured by a standard von Neumann measurement in the same computational basis, defining the N measuring POVM operators $\mathsf{E}_k = |k\rangle\langle k|$ for $k = 1$ to N. The input–output transmission (conditional) probabilities result in Eq. (13.11) as $\Pr\{Y = k \mid X = j\} = (1-\alpha)/N$ when $k \neq j$, and $\Pr\{Y = j \mid X = j\} = \alpha + (1-\alpha)/N$. This entails an input–output conditional Shannon entropy

$$H(Y|X) = h\left(\alpha + \frac{1-\alpha}{N}\right) + (N-1)h\left(\frac{1-\alpha}{N}\right), \tag{13.32}$$

which happens here to coincide with the average output quantum entropy of Eq. (13.30). Also, the output probabilities follow as $\Pr\{Y = k\} = 1/N$ for $k = 1$ to N, entailing the output classical entropy $H(Y) = \log(N)$. The input–output mutual information in Eq. (13.12) results as

$$H(X:Y) = \log(N) - h\left(\alpha + \frac{1-\alpha}{N}\right) - (N-1)h\left(\frac{1-\alpha}{N}\right), \tag{13.33}$$

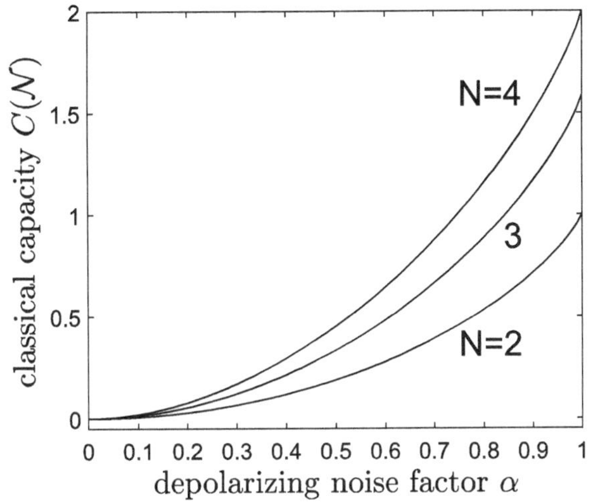

Fig. 13.4 For the qudit depolarizing channel of Eq. (13.27), the classical information capacity $C(\mathcal{N})$ (in shannons) of Eq. (13.31), as a function of the noise factor α, for three values of the dimension N of the qudit

which exactly matches the classical capacity $C(\mathcal{N})$ of Eq. (13.31). Figure 13.4 represents the classical capacity $C(\mathcal{N})$ of Eq. (13.31), as a function of the factor α assessing the noisiness of the depolarizing channel in Eq. (13.27), for different values of the dimension N of the qudit.

In this way, with a qudit depolarizing channel, each qudit can be encoded and measured separately, to provide a classical channel $X \mapsto Y$ with an input–output mutual information $H(X : Y)$ matching the classical capacity $C(\mathcal{N})$ of the quantum channel.

We have identified the optimal input signaling ensemble of N orthogonal pure states, and the optimal output von Neumann measurement, enabling to operate an N-dimensional qudit depolarizing channel at its classical capacity $C(\mathcal{N}) = \chi_{\max}(\mathcal{N})$ of Eq. (13.31), by processing individual qudits for input encoding and output measurement. Since for the qudit depolarizing channel the Holevo capacity $\chi_{\max}$ is additive [414], signaling with entangled blocks of qudits cannot increase the error-free transmission rate above the single-qudit capacity $\chi_{\max}(\mathcal{N}) = C(\mathcal{N})$ of Eq. (13.31).

13.2.8 Classical Capacity of the Qudit Erasure Channel

A quantum erasure channel was evoked at the beginning of Sect. 3.2.9 as an example of channel with different input and output spaces. The erasure channel also represents another instance of a quantum noise process $\mathcal{N}(\cdot)$ degrading a quantum state or signal, when the information carrier gets lost or absorbed. For N-dimensional qudits, the quantum erasure channel [12, 25] maps input quantum states of the space $\mathcal{H}_N$ into output quantum states of the $(N+1)$-dimensional space $\mathcal{H}_{N+1} = \mathcal{H}_N \oplus \{|\times\rangle\} \supset \mathcal{H}_N$ obtained by augmenting $\mathcal{H}_N$ with the one-dimensional space spanned by a quantum state $|\times\rangle$ orthogonal to $\mathcal{H}_N$ informing the receiver that the qudit has been erased at the output. Such a qudit erasure channel, on an input density operator $\rho \in \mathcal{L}(\mathcal{H}_N)$, applies the completely positive trace-preserving linear map (in non-Kraus form)

$$\mathcal{N}(\rho) = (1-p)\rho + p\,|\times\rangle\langle\times| \ , \tag{13.34}$$

delivering an output density operator $\mathcal{N}(\rho) = \rho' \in \mathcal{L}(\mathcal{H}_{N+1})$, with $p \in [0,1]$ the probability that the qudit gets erased in the transmission.

For the qudit erasure channel $\mathcal{N}(\cdot)$, the Holevo capacity $\chi_{\max}$ is known to be additive [25, 415], so that its classical information capacity $C(\mathcal{N})$ coincides with the channel Holevo capacity $\chi_{\max}(\mathcal{N})$ as in Eq. (13.23). To characterize the capacity $C(\mathcal{N}) = \chi_{\max}(\mathcal{N})$, we need to accomplish the maximization of Eq. (13.18) on pure input states $|\psi_j\rangle$ of $\mathcal{H}_N$. For any pure input state $|\psi\rangle \in \mathcal{H}_N$, the resulting output state $\mathcal{N}\big(|\psi\rangle\langle\psi|\big) = (1-p)\,|\psi\rangle\langle\psi| + p\,|\times\rangle\langle\times| \in \mathcal{L}(\mathcal{H}_{N+1})$ receives the following eigendecomposition in $\mathcal{H}_{N+1}$. The density operator $\mathcal{N}\big(|\psi\rangle\langle\psi|\big)$ has the eigenstate $|\psi\rangle \in \mathcal{H}_N \subset \mathcal{H}_{N+1}$ with eigenvalue $1-p$; it also has as eigenstate any of the $N-1$ pure states of $\mathcal{H}_N \subset \mathcal{H}_{N+1}$ mutually orthogonal and orthogonal to $|\psi\rangle$, each with eigenvalue 0 receiving therefore the multiplicity $N-1$; finally it has the eigenstate $|\times\rangle \in \mathcal{H}_{N+1}$ with eigenvalue p. As a result, the output quantum entropy $S\big(\mathcal{N}(|\psi\rangle\langle\psi|)\big) = h(1-p) + h(p)$ is the same for any pure input state $|\psi\rangle$, and will also represent the average output entropy for any input ensemble of pure states $\big\{\big(p_j, |\psi_j\rangle\big)\big\}$, to serve in Eq. (13.15).

To maximize the Holevo information $\chi\big(\{p_j, \rho'_j\}\big)$ in Eq. (13.15) we now need only to maximize the entropy $S(\overline{\rho}\,')$ of the average output state $\overline{\rho}\,' = \mathcal{N}(\overline{\rho})$. This is accomplished when the average input state $\overline{\rho}$ is the maximum-entropy state $\overline{\rho} = \mathrm{I}_N/N$ on the input space $\mathcal{H}_N$. This average input state $\overline{\rho} = \mathrm{I}_N/N$ is achieved by an input ensemble formed by N mutually orthogonal pure states $|\psi_j\rangle \in \mathcal{H}_N$ with equal probability $p_j = 1/N$. The corresponding average output state results as $\overline{\rho}\,' = \mathcal{N}\big(\mathrm{I}_N/N\big) = (1-p)\mathrm{I}_N/N + p\,|\times\rangle\langle\times|$. Any N mutually orthogonal pure states of $\mathcal{H}_N \subset \mathcal{H}_{N+1}$ form N eigenstates for $\overline{\rho}\,'$, each with the same eigenvalue $(1-p)/N$ therefore receiving the multiplicity N; the remaining eigenstate of $\overline{\rho}\,'$ is $|\times\rangle$ with eigenvalue p. The entropy of the average output state follows as $S\big(\overline{\rho}\,' = \mathcal{N}(\mathrm{I}_N/N)\big) = Nh\big((1-p)/N\big) + h(p)$. This is less than the maximal entropy $S\big(\mathrm{I}_{N+1}/(N+1)\big) = (N+1)h\big(1/(N+1)\big) = \log(N+1)$ of the maximally mixed state $\mathrm{I}_{N+1}/(N+1)$ on the output space $\mathcal{H}_{N+1}$. But due to the constraint imposed by the erasure channel of Eq. (13.34), the input provides no control on the output subspace spanned by $|\times\rangle$; there is in general no input in $\mathcal{H}_N$ enabling to reach the average output $\overline{\rho}\,' = \mathrm{I}_{N+1}/(N+1)$ on $\mathcal{H}_{N+1}$ (except in the special case $p = 1/(N+1)$ where the average input $\overline{\rho} = \mathrm{I}_N/N$ obtains the average $\overline{\rho}\,' = \mathrm{I}_{N+1}/(N+1)$ at the output), and it is the maximimum-entropy input $\overline{\rho} = \mathrm{I}_N/N$ that produces the maximum entropy $S\big(\overline{\rho}\,' = \mathcal{N}(\mathrm{I}_N/N)\big) = Nh\big((1-p)/N\big) + h(p)$ achievable at the output, for any probability p imposed by the channel. The maximum of Eq. (13.18) determines the classical information capacity $\chi_{\max}(\mathcal{N}) = C(\mathcal{N})$ of the qudit erasure channel, resulting as

$$\boxed{C(\mathcal{N}) = (1 - p)\log(N)\,,} \tag{13.35}$$

as also given in [12, 415].

The classical capacity $\chi_{\max}(\mathcal{N}) = C(\mathcal{N})$ of Eq. (13.35) is achieved in Eq. (13.18) by any input ensemble $\{(p_j = 1/N, |\psi_j\rangle)\}$ of N mutually orthogonal and equiprobable pure states $|\psi_j\rangle \in \mathcal{H}_N$. In addition, the transmission rate $C(\mathcal{N})$ can be obtained by single-qudit measurement at the channel output. A von Neumann measurement is performed at the output in the projective orthonormal basis $\{|\psi_j\rangle, |\times\rangle\}$ of $\mathcal{H}_{N+1}$ formed by augmenting with $|\times\rangle$ the input signaling basis $\{|\psi_j\rangle\}$ of $\mathcal{H}_N$. The input–output transmission (conditional) probabilities follow from Eq. (13.11), and they lead to an input–output classical conditional entropy $H(Y|X) = h(1-p) + h(p)$ and to an output classical entropy $H(Y) = Nh\big((1-p)/N\big) + h(p)$. The input–output mutual information of Eq. (13.12) results as $H(X:Y) = (1-p)\log(N)$, and exactly matches the classical capacity $C(\mathcal{N})$ of Eq. (13.35). In addition, since for the qudit erasure channel the Holevo capacity $\chi_{\max}$ is additive [25, 415], signaling with entangled blocks of qudits cannot increase the error-free transmission rate above the single-qudit capacity $\chi_{\max}(\mathcal{N}) = C(\mathcal{N})$ of Eq. (13.35).

13.2.9 *Classical Capacity of Unital Qubit Channels*

A unital qubit channel $\mathcal{N}(\cdot)$ has the unitary operator I_2 on $\mathcal{H}_2$ as a fixed point, and equivalently satisfies $\mathcal{N}(I_2/2) = I_2/2$. Unital channels acting on qubits implement an input–output transformation $\rho \mapsto \mathcal{N}(\rho) = \rho'$ of the qubit density operator in $\mathcal{L}(\mathcal{H}_2)$, equivalent to a linear transformation

$$\vec{r} \longmapsto A\vec{r} = \vec{r}\,' \tag{13.36}$$

of the Bloch vector in $\mathbb{R}^3$, where A as a 3×3 real matrix. This also represents the effect of a qubit unital noise in Eq. (3.146) when $\vec{c} = \vec{0}$.

For unital qubit channels the Holevo capacity $\chi_{\max}$ is known to be additive [31], so that its classical information capacity $C(\mathcal{N})$ coincides with the channel Holevo capacity $\chi_{\max}(\mathcal{N})$ as in Eq. (13.23). To characterize the capacity $C(\mathcal{N}) = \chi_{\max}(\mathcal{N})$ from Eq. (13.18), we need to accomplish the maximization of Eq. (13.18) on pure input qubit states $|\psi_j\rangle \subset \mathcal{H}_2$. The pure qubit states in $\mathcal{H}_2$ have Bloch vectors lying on the Bloch sphere (its surface) in $\mathbb{R}^3$. The Bloch sphere, by the linear map of Eq. (13.36), is transformed into a tilted ellipsoid in $\mathbb{R}^3$, centered at the origin and fitted inside the (initial) Bloch sphere, as we explained at the end of Sect. 3.2.11. If we write the polar decomposition $A = UG$, the real positive symmetric matrix $G = \sqrt{A^\top A}$ has the three eigenvalues $\{g_1, g_2, g_3\}$ in $[0, 1]$ arranged in decreasing order, associated with the three eigenvectors $\{\vec{g}_1, \vec{g}_2, \vec{g}_3\}$ forming an orthonormal basis of $\mathbb{R}^3$. The major eigenvector $\vec{g}_1$ associated with the largest eigenvalue g_1, provides two normalized antipodal Bloch vectors $\pm\vec{g}_1$ representing two orthogonal pure input states $|\psi_1\rangle$ and $|\psi_2\rangle = |\psi_1^\perp\rangle$ in $\mathcal{H}_2$. These two input Bloch vectors transform as $\pm A\vec{g}_1 = \pm g_1 U\vec{g}_1 = \pm\vec{r}_1{}'$ and provide two antipodal output Bloch vectors $\pm\vec{r}_1{}'$ with norm $\|\vec{r}_1{}'\| = g_1 \leq 1$. The two vectors $\pm\vec{r}_1{}'$ define the major axis of the tilted ellipsoid mentioned above, and also stand as the output Bloch vectors with maximal norm $\|\vec{r}_1{}'\| = g_1 \leq 1$ in $\mathbb{R}^3$. These two Bloch vectors $\pm\vec{r}_1{}'$ also represent two output density operators ρ_1' and ρ_2' with minimal quantum entropy $S(\rho_1') = S(\rho_2') = S_{\min} = h\big((1+g_1)/2\big) + h\big((1-g_1)/2\big) = H_{\text{bin}}\big((1+g_1)/2\big)$, where we made use of the qubit entropy from Eq. (7.40).

When the two orthogonal pure input signals $\big\{|\psi_1\rangle, |\psi_2\rangle\big\}$ defined by $\pm\vec{g}_1$ are applied with any probability distribution $\{p_1, p_2\}$, they induce an average output entropy (to serve in Eq. (13.15)) that is at its minimum $S_{\min}$. When these two input states are equiprobable, at $p_1 = p_2 = 1/2$, the average input density operator is $\overline{\rho} = \mathrm{I}_2/2$, the maximally mixed qubit state with maximal entropy $S(\mathrm{I}_2/2) = \log(2)$. Since we are dealing with a unital (linear) channel, satisfying $\mathcal{N}(\mathrm{I}_2) = \mathrm{I}_2$, we have the average output density operator $\overline{\rho}\,' = \mathcal{N}\big(\overline{\rho}\big) = \mathrm{I}_2/2$. This is the average output state with maximum entropy $S\big(\overline{\rho}\,'\big) = \log(2)$. This identifies the conditions achieving in Eq. (13.18) the maximum $\chi_{\max}(\mathcal{N}) = C(\mathcal{N})$ determining the classical information capacity of the (any) qubit unital channel as

$$\boxed{C(\mathcal{N}) = \log(2) - H_{\text{bin}}\left(\frac{1+g_1}{2}\right).} \tag{13.37}$$

So for a given unital qubit channel defined by the real matrix A in Eq. (13.36), an optimal input signaling ensemble is formed by the two equiprobable orthogonal states of $\mathcal{H}_2$ determined by the major eigenvector $\vec{g}_1$ of the matrix G given as the positive square root $G = \sqrt{A^\top A}$. It achieves the classical capacity $C(\mathcal{N})$ of Eq. (13.37) determined by the corresponding major eigenvalue g_1 of G.

As before, with the HSW theorem of Sect. 13.2.5 we know that the rate $C(\mathcal{N})$ can be achieved for error-free communication by transmitting long blocks of $L_{\text{cod}} \to \infty$ independent input qubit states collectively measured at the output. But this collective processing is not necessary for unital qubit channels, since we are going to see that there exists a quantum measurement acting on individual qubits capable of reaching an input–output mutual information $H(X:Y)$ in Eq. (13.13) matching the classical capacity $C(\mathcal{N})$ of Eq. (13.37). When signaling with the optimal input ensemble, we consider measuring on an output qubit the spin observable $\Omega = \vec{\omega}\cdot\vec{\sigma}$ with the (unit) measurement vector $\vec{\omega} = \vec{r}_1{}'/\|\vec{r}_1{}'\| = A\vec{g}_1/g_1 = U\vec{g}_1$ in $\mathbb{R}^3$. As we have seen in Sect. 4.3.4, this is equivalent to performing a von Neumann measurement in the orthonormal projective basis of $\mathcal{H}_2$ formed by the two orthogonal pure qubit states with Bloch vectors $\pm\vec{r}_1{}'/\|\vec{r}_1{}'\| = \pm\vec{\omega}$.

The input encoding of $X = 1$ is with the signaling state $|\psi_1\rangle$ of Bloch vector $\vec{g}_1$. It induces the output state $\mathcal{N}\big(|\psi_1\rangle\langle\psi_1|\big) = \rho_1'$ with Bloch vector $\vec{r}_1{}' = A\vec{g}_1 = g_1 U\vec{g}_1$, that when measured projects onto the pure state corresponding to $\vec{\omega}$ and decoded as $Y = 1$ with the conditional probability

$$\Pr\{Y = 1 \mid X = 1\} = \frac{1}{2}\Big(1 + \vec{r}_1{}'\vec{\omega}\Big) = \frac{1}{2}(1 + g_1)\,, \tag{13.38}$$

according to Eq. (4.80). The input encoding of $X = 2$ is with the signaling state $|\psi_2\rangle$ of Bloch vector $-\vec{g}_1$. It induces the output state $\mathcal{N}\big(|\psi_2\rangle\langle\psi_2|\big) = \rho_2'$ with Bloch vector $-\vec{r}_1{}' = -A\vec{g}_1 = -g_1 U\vec{g}_1$, that when measured projects onto the pure state corresponding to $-\vec{\omega}$ and decoded as $Y = 2$ with the conditional probability

$$\Pr\{Y = 2 \mid X = 2\} = \frac{1}{2}\Big(1 - (-\vec{r}_1{}')\vec{\omega}\Big) = \frac{1}{2}(1 + g_1)\,. \tag{13.39}$$

The other two input–output conditional probabilities follow by complementation to 1. This entails an input–output classical conditional entropy

$$H(Y|X) = h\left(\frac{1+g_1}{2}\right) + h\left(\frac{1-g_1}{2}\right) = H_{\mathrm{bin}}\left(\frac{1+g_1}{2}\right). \tag{13.40}$$

With two equiprobable states in the optimal input ensemble, the output probabilities follow as $\Pr\{Y = 1\} = \Pr\{Y = 2\} = 1/2$, entailing the output classical entropy $H(Y) = \log(2)$. The input–output mutual information in Eq. (13.12) results as

$$H(X:Y) = \log(2) - H_{\mathrm{bin}}\left(\frac{1+g_1}{2}\right), \tag{13.41}$$

which exactly matches the classical capacity $C(\mathcal{N})$ of Eq. (13.37).

Here also, with a unital qubit channel, each qubit can be encoded and measured separately, to provide a classical channel $X \mapsto Y$ with an input–output mutual information $H(X:Y)$ matching the classical capacity $C(\mathcal{N})$ of the quantum channel.

We have identified the optimal input signaling pair of orthogonal pure states defined by the Bloch vectors $\pm\vec{g}_1$, and the optimal output von Neumann measurement, enabling to operate any unital qubit channel at its classical capacity $C(\mathcal{N})$ of Eq. (13.37), by processing individual qubits for input encoding and output measurement. In addition, since for unital qubit channels the Holevo capacity $\chi_{\max}$ is additive [31], signaling with entangled blocks of qubits cannot increase the error-free transmission rate above the single-qubit capacity $\chi_{\max}(\mathcal{N}) = C(\mathcal{N})$ of Eq. (13.37).

Important instances of a unital qubit channel are the **bit-flip**, the **phase-flip**, and the **bit-phase-flip channels** following, respectively, from the noise models of Sects. 9.3, 9.4 and 9.5. These three noisy qubit channels all exhibit an invariant axis in $\mathbb{R}^3$, respectively, Ox, Oz, and Oy, that experiences no contraction in the transformation of Eq. (13.36), and is associated with a major eigenvalue $g_1 = 1$. This implies in Eq. (13.37) a classical capacity $C(\mathcal{N}) = \log(2)$ that is the overall maximal information capacity for a channel transmitting classical binary information, as it would occur for a channel with no noise, for instance, the identity channel. This invariant axis in $\mathbb{R}^3$ is associated with an orthonormal basis in $\mathcal{H}_2$ immune to noise for signaling, and given by the eigenbasis of the corresponding Pauli operator defining the qubit noise model. For instance, for the phase-flip channel the immune basis is $\{|0\rangle, |1\rangle\}$, for the bit-flip channel it is $\{|+\rangle, |-\rangle\}$, and for the bit-phase-flip channel it is $\{|+i\rangle, |-i\rangle\}$, as it results from Sect. A.5 of the Appendix. This immune basis can be used for direct encoding of the two values of a classical binary information, that will then be transmitted with no alteration over the quantum channel, at any noise level p. Signaling on the immune basis with two equiprobable states enables error-free transmission at the maximum rate given by the classical capacity $C(\mathcal{N}) = \log(2)$, which is the performance of a perfect classical binary channel. So although we have genuine noisy quantum channels with the bit-flip, the phase-flip, and the bit-phase-flip qubit channels, by means of an optimal two-state signaling along their invariant axis, it is possible to obtain perfect communication of classical information, at any noise level p over these quantum channels.

Another important unital qubit channel is the **completely decohering qubit channel** as it results from Sect. 9.8. This noisy qubit channel $\mathcal{N}$ leaves the axis Oz invariant in $\mathbb{R}^3$, that experiences no contraction in the transformation of Eq. (13.36) realized by Eq. (9.74), and is associated with a major eigenvalue $g_1 = 1$. Accordingly, as seen in Sect. 9.8, the two pure basis states $|0\rangle$ and $|1\rangle$ are invariant by $\mathcal{N}$, enabling a signaling immune to noise for classical binary information that can then be transmitted with no error. It follows that this completely decohering qubit channel $\mathcal{N}$ has a classical capacity in Eq. (13.37) of $C(\mathcal{N}) = \log(2)$, representing the overall maximum achievable for a qubit channel, and obtained by the input statistical ensemble of two equiprobable pure states $\{(1/2, |0\rangle), (1/2, |1\rangle)\}$.

As for the three qubit flip channels above, with the completely decohering qubit channel also, it is possible, by signaling along the invariant axis, to obtain perfect communication of *classical* information over such noisy quantum channels. This will no longer hold true when we get interested in communicating *quantum* information over such channels, as it will be addressed in Sect. 13.3.

13.2.10 Classical Capacity of the Qubit Thermal Noise Channel

The qubit thermal noise channel is also the generalized amplitude damping noise channel as it results from the quantum noise model of Sect. 9.10. It represents an important instance of a non-unital qubit channel. Compared with unital qubit channels treated in Sect. 13.2.9, for non-unital qubit channels the characterization of the Holevo capacity $\chi_{\max}(\mathcal{N})$ and a fortiori of the classical capacity $C(\mathcal{N})$ are more difficult to accomplish explicitly. The nonzero translation vector $\vec{c}$ of Eq. (3.146) in the qubit transformation $\vec{r} \mapsto A\vec{r} + \vec{c}$ in $\mathbb{R}^3$ by the channel complicates the geometry of the quantum states and hinder the maximization of Eq. (13.18), to begin with. It is known in general, for non-unital qubit channels, that the Holevo capacity $\chi_{\max}(\mathcal{N})$ can be achieved with an optimal input ensemble containing at most four pure states [416]. However, as argued in [416], a maximum of four optimal input states concern specific qubit channels with low symmetry. In [408, 416], non-unital qubit channels are shown with an optimal input ensemble having three or two states.

The qubit thermal noise channel is a non-unital qubit channel that exhibits some symmetry relative to the Oz axis in $\mathbb{R}^3$, as reflected by its Bloch representation of Eq. (9.91). However, for this important non-unital qubit channel, it is not known in general, except in some special cases, that the Holevo capacity $\chi_{\max}(\mathcal{N})$ as it results from Eq. (13.18), is additive, and that the classical capacity $C(\mathcal{N})$ of Eq. (13.22) reduces to the single-use formula of Eq. (13.23).

To characterize the Holevo capacity $\chi_{\max}(\mathcal{N})$ from Eq. (13.18) of the qubit thermal noise channel $\mathcal{N}(\cdot)$, Refs. [409, 417] address the special case of a zero temperature $T = 0$ at a probability $p = 1$ in the model of Eqs. (9.86)–(9.89), that comes to coincide with the model of amplitude damping noise of Sect. 9.9. For the general case of an arbitrary temperature T in the qubit thermal noise channel $\mathcal{N}(\cdot)$, to characterize the Holevo capacity $\chi_{\max}(\mathcal{N})$ it is found in [408, 409] that an optimal input ensemble achieving the maximum in Eq. (13.18), is generally formed by two nonorthogonal and equiprobable pure qubit states, equidistant from the Oz axis in Bloch representation.

For two qubit density operators ρ_1 and ρ_2, respectively with Bloch vectors $\vec{r}_1$ and $\vec{r}_2$, the quantum relative entropy $S(\rho_1\|\rho_2) = \mathrm{tr}\big(\rho_1 \log(\rho_1) - \rho_1 \log(\rho_2)\big)$ of Eq. (7.74) can be evaluated [409] as

$$S(\rho_1\|\rho_2) = \frac{1}{2}\log(1-r_1^2) + \frac{r_1}{2}\log\left(\frac{1+r_1}{1-r_1}\right) - \frac{1}{2}\log(1-r_2^2) - \frac{r_1}{2}\cos(\xi)\log\left(\frac{1+r_2}{1-r_2}\right), \tag{13.42}$$

with the norms $r_1 = \|\vec{r}_1\|$ and $r_2 = \|\vec{r}_2\|$, and cosine $\cos(\xi) = \vec{r}_1\vec{r}_2/(r_1 r_2)$.

Based on this relation, [408, 418] obtain a characterization of the Holevo capacity of the qubit thermal noise channel, by solving Eq. (13.19), under the form (see also [225])

$$\boxed{\chi_{\max}(\mathcal{N}) = \frac{1}{2}\Big[f\big(r(q)\big) - f(q)\Big]}, \tag{13.43}$$

with

$$f(u) = (1+u)\log(1+u) + (1-u)\log(1-u), \tag{13.44}$$

$$f'(u) = \frac{d}{du}f(u) = \log\left(\frac{1+u}{1-u}\right), \tag{13.45}$$

the function

$$r(q) = \sqrt{1-\gamma - \frac{[q-(2p-1)\gamma]^2}{1-\gamma} + q^2}, \tag{13.46}$$

and q determined as the solution in $[-1, 1]$ to

$$(q-2p+1)\gamma f'\big(r(q)\big) + (1-\gamma)r(q)f'(q) = 0. \tag{13.47}$$

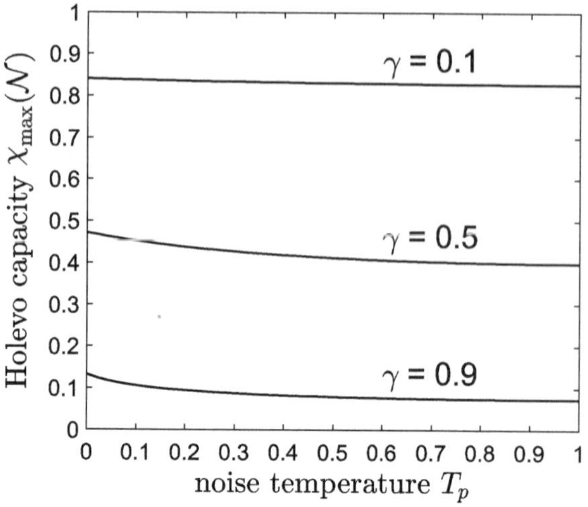

Fig. 13.5 For the qubit thermal noise channel, the Holevo capacity $\chi_{\max}(\mathcal{N})$ (in shannons) of Eq. (13.43), as a function of the effective noise temperature $T_p = 2(1-p)$ as in Fig. 11.9, for three values of the coupling γ

The optimal signaling ensemble achieves the average output state $\overline{\rho}'_{\rm opt}$ having the Bloch vector $[0, 0, q]^\top$ according to [418]. The two optimal input states can be chosen with the Bloch vectors $\vec{r}_\pm^{\rm\,opt} = \big[\pm\sin(\theta_{\rm opt}),\ 0,\ \cos(\theta_{\rm opt})\big]^\top$ and equal prior probabilities 1/2, according to [418]. The corresponding output states follow from Eq. (9.91) as $\vec{r}_\pm^{\,\prime} = \big[\pm\sqrt{1-\gamma}\sin(\theta_{\rm opt}),\ 0,\ (1-\gamma)\cos(\theta_{\rm opt}) + (2p-1)\gamma\big]^\top$, so that to match the average output state $\overline{\rho}'_{\rm opt}$ we have $q = (1-\gamma)\cos(\theta_{\rm opt}) + (2p-1)\gamma$, which serves to determine $\theta_{\rm opt}$ from q solving Eq. (13.47). In addition, we have $r(q) = \|\vec{r}_\pm^{\,\prime}\| = \sqrt{(1-\gamma)\sin^2(\theta_{\rm opt}) + q^2}$, which is similar to Eq. (13.46).

Figures 13.5 and 13.6 display the evolution of the Holevo capacity $\chi_{\max}(\mathcal{N})$ from Eq. (13.43), obtained via numerical resolution of Eq. (13.47), in various configurations of the qubit thermal noise channel. For convenience, as in Fig. 11.9, the temperature $T \in [0, +\infty[$ of the thermal noise is represented by the equivalent effective temperature $T_p = 2(1-p) \in [0, 1]$, as introduced in Sect. 11.4.8.

At long interaction times $t \gg \tau$ of the qubit with the thermal bath, the coupling $\gamma = 1 - e^{-t/\tau}$ increases to 1 while the Holevo capacity $\chi_{\max}(\mathcal{N})$ decreases and tends to vanish, as visible in Fig. 13.5. This is the thermalization regime of the quantum channel, where all signaling states terminate in the same thermalized state $\rho_\infty = p\,|0\rangle\langle 0| + (1-p)\,|1\rangle\langle 1|$ determined by the noise temperature via the probability p, but with no input information subsisting at the output, entailing a vanishing capacity $\chi_{\max}(\mathcal{N}) \to 0$ as $\gamma \to 1$ in Fig. 13.5. On the contrary, at very short interaction times $t \ll \tau$ of the qubit with the thermal bath, the coupling $\gamma = 1 - e^{-t/\tau}$ goes to zero, the thermal noise channel in Eq. (9.91) tends to the identity channel, and the Holevo capacity $\chi_{\max}(\mathcal{N})$ tends to its maximum of 1 shannon per transmitted qubit, as visible in Fig. 13.5.

With the probability $p = 1/2$ at an infinite temperature T, then $\vec{c} = \vec{0}$ and in Eq. (9.91) the qubit thermal noise channel reduces to a unital qubit channel, with the

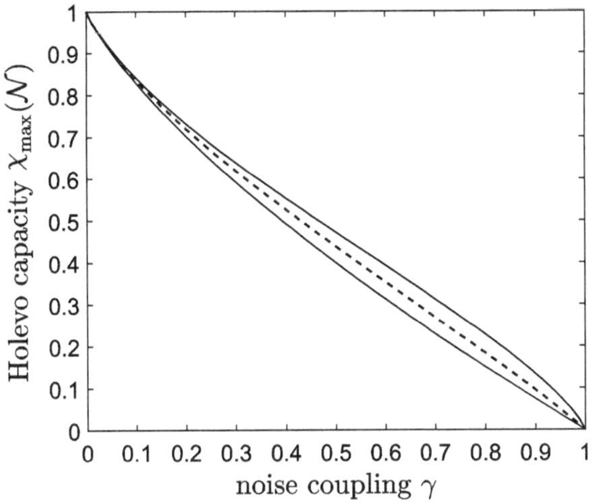

Fig. 13.6 For the qubit thermal noise channel, the Holevo capacity $\chi_{\max}(\mathcal{N})$ (in shannons) of Eq. (13.43), as a function of the noise coupling γ, for three values of the noise probability p. The right-most solid curve is $p = 1$ corresponding to a zero temperature T as well as T_p, the left-most solid curve is $p = 1/2$ corresponding to an infinite temperature T equivalent to $T_p = 1$, the intermediate dashed curve is $p = 0.9$. At $p = 1/2$ the channel becomes unital and its Holevo capacity $\chi_{\max}(\mathcal{N})$ coincides with its classical information capacity $C(\mathcal{N}) = \chi_{\max}(\mathcal{N})$ also characterized in Eq. (13.37)

channel classical capacity $C(\mathcal{N})$ governed by Eq. (13.37). The solution to Eq. (13.47) is $q = 0$, leading to $r = \sqrt{1-\gamma}$ in Eq. (13.46), and $\chi_{\max}(\mathcal{N}) = f\big(\sqrt{1-\gamma}\big)/2$ in Eq. (13.43). For the channel the major eigenvalue is $g_1 = \sqrt{1-\gamma}$, so that $\chi_{\max}(\mathcal{N}) = f\big(\sqrt{1-\gamma}\big)/2$ coincides with the classical capacity $C(\mathcal{N}) = \chi_{\max}(\mathcal{N})$ as given by Eq. (13.37) when $\chi_{\max}(\mathcal{N})$ is additive for the unital qubit channel, as displayed in Fig. 13.6 at $p = 1/2$. An optimal input signaling pair achieving the capacity $C(\mathcal{N}) = \chi_{\max}(\mathcal{N})$ is $\big\{|+\rangle, |-\rangle\big\}$ at $\theta_{\text{opt}} = \pi/2$ with $\vec{r}_{\pm}^{\,\text{opt}} = \pm\vec{e}_x$. We observe in this way in Fig. 13.6 that at an infinite temperature T, the classical capacity $C(\mathcal{N}) = \chi_{\max}(\mathcal{N}) = f\big(\sqrt{1-\gamma}\big)/2$ of the thermal noise channel does not generally vanish, provided the signaling qubits are not exposed to the thermal noise for a too long time $t \gg \tau$, so that the coupling $\gamma = 1 - e^{-t/\tau}$ remains below 1.

In general, for any temperature T, when signaling with the optimal input pair defined by the equiprobable Bloch vectors $\vec{r}_{\pm}^{\,\text{opt}}$ indicated above, a useful quantum measurement at the output is the von Neumann measurement projecting in the orthonormal basis $\big\{|+\rangle, |-\rangle\big\}$, equivalent to measuring the spin observable σ_x on the output qubit. This measurement is the minimal-error detector of Sect. 10.4.1 for discriminating between the two noisy states $\vec{r}_{\pm}^{\,\prime}$ at the channel output. This optimal detector is associated in Eq. (10.67) with the test Bloch vector $\vec{\tau} = \sqrt{1-\gamma}\,\sin(\theta_{\text{opt}})\vec{e}_x$, and in Eq. (10.69) with the two measuring projectors $|\pm\rangle\langle\pm| = \big(\mathrm{I}_2 \pm \sigma_x\big)/2$. It achieves the minimal probability of detection error given by Eq. (10.70) as $P_{\text{er}}^{\min} = \big[1 - \sqrt{1-\gamma}\,|\sin(\theta_{\text{opt}})|\big]/2$. Due to the symmetry of the channel, this also repre-

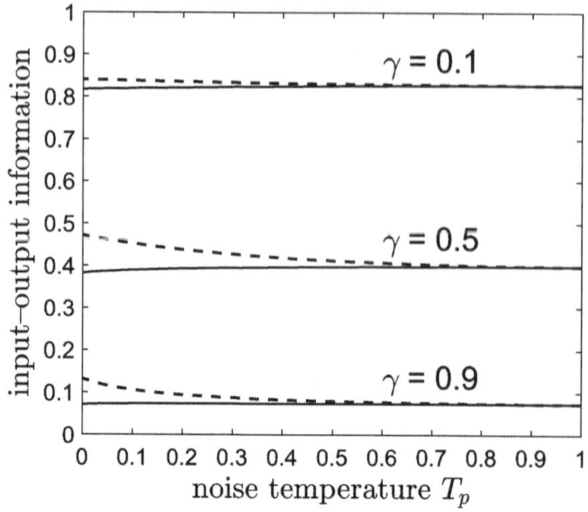

Fig. 13.7 For the qubit thermal noise channel, when signaling with the optimal input pair $\vec{r}_{\pm}^{\,\text{opt}}$ and measuring the output in the basis $\{|+\rangle, |-\rangle\}$, the input–output mutual information $H(X:Y)$ of Eq. (13.49) (solid line) compared with the Holevo capacity $\chi_{\max}(\mathcal{N})$ of Eq. (13.43) (dashed line) in shannons, as a function of the effective noise temperature $T_p = 2(1-p)$, for three values of the coupling γ

sents each of the two individual (conditional) probabilities of transmission error $\Pr\{|-\rangle \,|\, \vec{r}_{+}^{\,\text{opt}}\} = \Pr\{|+\rangle \,|\, \vec{r}_{-}^{\,\text{opt}}\} = P_{\text{er}}^{\min}$. From these two conditional probabilities, we deduce the input–output classical conditional entropy

$$H(Y|X) = h\left(\frac{1}{2}\left[1+\sqrt{1-\gamma}\,\big|\sin(\theta_{\text{opt}})\big|\right]\right) + h\left(\frac{1}{2}\left[1-\sqrt{1-\gamma}\,\big|\sin(\theta_{\text{opt}})\big|\right]\right); \tag{13.48}$$

and with the output entropy $H(Y) = \log(2)$, the input–output mutual information follows as

$$H(X:Y) = \log(2) - h\left(\frac{1}{2}\left[1+\sqrt{1-\gamma}\,\big|\sin(\theta_{\text{opt}})\big|\right]\right) - h\left(\frac{1}{2}\left[1-\sqrt{1-\gamma}\,\big|\sin(\theta_{\text{opt}})\big|\right]\right). \tag{13.49}$$

This measurement in the basis $\{|+\rangle, |-\rangle\}$, achieving the minimal probability of detection error at the output, is also the von Neumann measurement of the output qubit that maximizes the input–output mutual information $H(X:Y)$ in Eq. (13.49). No von Neumann measurement or qubit observable at the output can achieve a larger mutual information $H(X:Y)$. It solves the maximization of Eq. (13.13) when the measurement operators are restricted to a qubit observable. It is meaningful to compare this maximum mutual information $H(X:Y)$ of Eq. (13.49) extractable by a qubit observable at the output, with its upper bound provided by the Holevo capacity $\chi_{\max}(\mathcal{N})$ of Eq. (13.43); this is accomplished in Fig. 13.7.

We observe in Fig. 13.7 that the mutual information $H(X:Y)$ remains below the Holevo capacity $\chi_{\max}(\mathcal{N})$, for any noise temperature. The gap reduces as the temperature increases, with $H(X:Y)$ tending to $\chi_{\max}(\mathcal{N})$ at large temperature. This is in contrast with the unital qubit channels studied earlier in Sect. 13.2.9, where

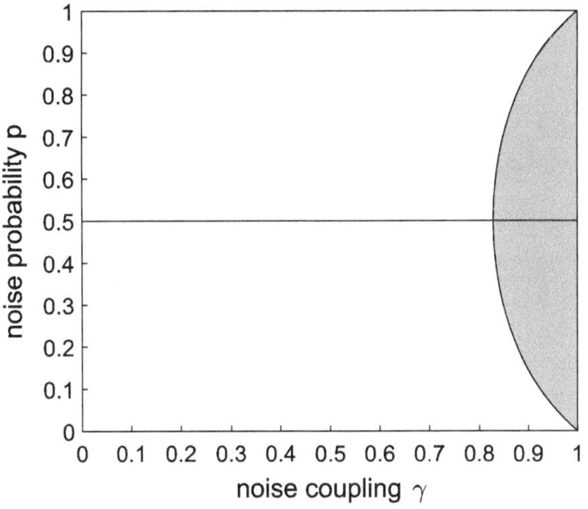

Fig. 13.8 In the plane (γ, p) of the parameters of the qubit thermal noise channel, the region in gray is where the conditions of Eqs. (13.50) and (13.51) are satisfied and the channel is entanglement breaking, implying an additive Holevo capacity $\chi_{\max}(\mathcal{N})$ matching the channel classical capacity $C(\mathcal{N})$

a qubit observable at the output always exists to achieve an input–output mutual information $H(X : Y)$ matching the Holevo capacity $\chi_{\max}(\mathcal{N})$.

As indicated in Sect. 3.2.10, a quantum channel is entanglement breaking if and only if its Choi–Jamiołkowski state is separable [36]. Entanglement-breaking channels have an additive Holevo capacity [25]. The qubit thermal noise channel is entanglement breaking if and only if its two parameters (γ, p) satisfy [225] the conditions

$$\begin{cases} 0.8284 \approx 2\left(\sqrt{2}-1\right) \leq \gamma \leq 1\,, & (13.50) \\ \dfrac{1}{2}\left(1-\sqrt{\dfrac{\gamma^2+4\gamma-4}{\gamma^2}}\right) \leq p \leq \dfrac{1}{2}\left(1+\sqrt{\dfrac{\gamma^2+4\gamma-4}{\gamma^2}}\right), & (13.51) \end{cases}$$

with the corresponding region in the plane (γ, p) displayed in Fig. 13.8. When the thermal noise channel $\mathcal{N}(\cdot)$ is entanglement breaking, we therefore have the channel Holevo capacity $\chi_{\max}(\mathcal{N})$ of Eq. (13.43) which matches the channel classical capacity $C(\mathcal{N})$, as provided by the single-use formula of Eq. (13.23).

As we have just seen, when $p = 1/2$ at an infinite temperature T the qubit thermal noise channel reduces, at any $\gamma \in [0, 1]$, to a unital qubit channel also exhibiting an additive Holevo capacity $\chi_{\max}$, so that here also $\chi_{\max}(\mathcal{N})$ of Eq. (13.43) matches the classical capacity $C(\mathcal{N})$, as in Eq. (13.23).

Outside these conditions on (γ, p), the additivity or non-additivity of the Holevo capacity $\chi_{\max}$ for the qubit thermal noise channel is presented in [225] as remaining an important open problem, (even at $p = 1$ at vanishing temperature $T = 0$ when it reduces to the (non-unital) amplitude damping noise channel of Sect. 9.9). The classical capacity $C(\mathcal{N})$ of the qubit thermal noise channel therefore remains today unknown in these conditions.

13.3 Quantum Channel Transmitting Quantum Information

With the model of quantum communication channel defined in Sect. 13.2.1 and Fig. 13.2, the concern throughout Sect. 13.2 was about the faithful transmission $X \mapsto Y$ of classical information carried by the classical variables (X, Y) and mediated by the quantum channel via the quantum signals $\big(\rho_j, \mathcal{N}(\rho_j) = \rho'_j\big)$. Assessment of the performance was grounded on the input–output classical mutual information $H(X : Y)$, and the conditions for its maximization leading to the classical information capacity $C(\mathcal{N})$ of Eq. (13.22) characterizing the maximal rate achievable for error-free transmission of classical information over the quantum channel $\mathcal{N}$. This perspective encompasses, as we indicated, important situations for practical application of quantum technologies, useful to classical information in the classical world, yet assisted by quantum resources.

Nevertheless, with this model of quantum communication channel of Sect. 13.2.1 and Fig. 13.2, one can choose instead to focus on faithful transmission of the quantum signals themselves, such as $\rho_j \mapsto \mathcal{N}(\rho_j) = \rho'_j$ or any unknown sequences thereof. Accordingly, performance evaluation should substitute criteria applying to the quantum signals themselves across the quantum channel $\mathcal{N}$. This will require distinctive informational quantities specifically meaningful for assessing the performance in this quantum perspective, and that we now introduce.

13.3.1 Entropy Exchange

A useful informational quantity concerning the quantum signals involved in the operation of a quantum channel is the entropy exchange $S_{\rm ex}(\rho, \mathcal{N})$, which represents [9, 419] the amount of entropy generated in the environment by the action of a quantum channel $\mathcal{N}(\cdot)$ implementing the noisy transmission of the input state ρ. As we have seen in Sect. 3.2.7, the action of a (primal) channel $\mathcal{N}(\cdot)$ on the environment can be described by means of a dual complementary or conjugate channel $\widetilde{\mathcal{N}}(\cdot)$ from the input space $\mathcal{H}_N$ into the environment space $\mathcal{H}^E$. When the environment starts in a pure state (this is always feasible, possibly via a purification step as in Sect. 2.5.3), it starts with a zero quantum entropy, and after the action of the quantum channel $\mathcal{N}(\cdot)$ on the input state ρ the environment terminates, as we have shown in Sect. 3.2.7 and Eqs. (3.119) and (3.120), generally in a mixed state $\rho'_E = \widetilde{\mathcal{N}}(\rho)$ represented by the $K \times K$ matrix W with elements

$$w_{k\ell} = \mathrm{tr}\big(\Lambda_k \rho \Lambda_\ell^\dagger\big) , \tag{13.52}$$

for $k, \ell = 1$ to K. The **entropy exchange** is then equivalent to the final quantum entropy of the environment, that is,

$$S_{\rm ex}(\rho,\mathcal{N}) = S(\rho'_E) = -\,{\rm tr}\big(W\log(W)\big)\;. \tag{13.53}$$

The final state $\rho'_E = \widetilde{\mathcal{N}}(\rho)$ of the environment is represented by the $K \times K$ matrix $W = [w_{k\ell}]$ from Eq. (13.52) when the quantum channel $\mathcal{N}(\cdot)$ is modeled by the set $\{\Lambda_k\}_{k=1}^K$ of K Kraus operators. However, as we have seen in Sect. 3.2.4 with Eq. (3.72), there exists an isometric freedom in the choice of the set of Kraus operators representing a given quantum channel $\mathcal{N}(\cdot)$. The entropy exchange $S_{\rm ex}(\rho,\mathcal{N}) = S(\rho'_E)$ is nevertheless invariant in this process, and this can be justified as follows.

For a given quantum channel $\mathcal{N}(\cdot)$ from $\mathcal{L}(\mathcal{H}_N)$ onto $\mathcal{L}(\mathcal{H}_{N'})$, the smallest size K for the set $\{\Lambda_k\}_{k=1}^K$ of Kraus operators is given by the Choi rank of $\mathcal{N}(\cdot)$, no larger than NN', as indicated in Sect. 3.2.10. The minimal-size Kraus set $\{\Lambda_k\}_{k=1}^K$ is formed by $K \le NN'$ Kraus operators defined via Eq. (3.135) from the eigendecomposition of the Choi–Jamiołkowski density operator of Sect. 3.2.10. With this minimal-size Kraus set $\{\Lambda_k\}_{k=1}^K$ of size $K \le NN'$, the $K \times K$ matrix $W = [w_{k\ell}]$ of Eq. (13.52) representing the density operator $\rho'_E = \widetilde{\mathcal{N}}(\rho)$ of the environment, is a positive unit-trace matrix with K eigenstates $|g_k\rangle$ associated with K eigenvalues g_k forming a probability distribution and determining the entropy exchange as

$$S_{\rm ex}(\rho,\mathcal{N}) = S(\rho'_E) = -\,{\rm tr}\big(W\log(W)\big) = -\sum_{k=1}^{K} g_k\log(g_k) \le \log(NN')\;. \tag{13.54}$$

For the quantum channel $\mathcal{N}(\cdot)$, any alternative set of Kraus operators $\{\Lambda'_\ell\}_{\ell=1}^{K'}$ with larger size $K' \ge K$, can be isometrically deduced from the minimal-size Kraus set $\{\Lambda_k\}_{k=1}^K$ of size $K \le NN'$, by the isometry of Eq. (3.72) reading

$$\Lambda'_\ell = \sum_{k=1}^{K} v_{\ell k}\Lambda_k\;, \tag{13.55}$$

where the complex numbers $[v_{\ell k}] = V$ form the $K' \times K$ matrix V of an isometry, and satisfy $\sum_{\ell=1}^{K'} v^*_{\ell k}v_{\ell k'} = \delta_{kk'}$, or equivalently $V^\dagger V = I_K$ with I_K the $K \times K$ identity matrix. The set of Kraus operators $\{\Lambda'_\ell\}_{\ell=1}^{K'}$ defines an alternative model for the environment, which after interaction with the principal quantum system in state ρ terminates, according to Eq. (13.52), in a mixed state of density operator represented by the $K' \times K'$ matrix W' with elements

$$w'_{k\ell} = {\rm tr}\big(\Lambda'_k\rho\Lambda'^\dagger_\ell\big)\;, \tag{13.56}$$

for $k, \ell = 1$ to K'. By inserting Eq. (13.55) into Eq. (13.56), we deduce the matrix relation

$$W' = VWV^\dagger\;. \tag{13.57}$$

The $K' \times K'$ matrix $W' = [w'_{k\ell}]$ of Eq. (13.57) representing the density operator of the alternative environment model, is a positive unit-trace matrix. The eigenrelation $W|g_k\rangle = g_k|g_k\rangle$ gives $VW|g_k\rangle = g_kV|g_k\rangle$, and since $VW = W'V$, we obtain $W'V|g_k\rangle = g_kV|g_k\rangle$, showing that W' possesses the K eigenstates $V|g_k\rangle$ associated with the K eigenvalues g_k. The K (positive) eigenvalues g_k sum to 1, the $K' \times K'$ matrix W' is positive with unit trace just as W, so the $K' - K$ remaining (positive) eigenvalues of W' are 0. Therefore, the alternative environment model with a different density operator W' sharing the same nonzero eigenvalues, has a same quantum entropy $-\operatorname{tr}\big(W'\log(W')\big) = -\operatorname{tr}\big(W\log(W)\big)$ corresponding to the same entropy exchange $S_{\rm ex}(\rho, \mathcal{N}) = S(\rho'_E)$ as in Eq. (13.54). This applies equally for any environment model for the quantum channel $\mathcal{N}(\cdot)$: they all lead to the same entropy exchange $S_{\rm ex}(\rho, \mathcal{N}) = S(\rho'_E)$ as in Eq. (13.54), or same entropy generated in the environment by the operation of the quantum channel $\mathcal{N}(\cdot)$ transmitting the input ρ.

If we do not want to explicitly exhibit the eigendecomposition of W' in relation to that of W for two isometrically related environment models, a shorter path invokes the isometric invariance of the quantum entropy in Eq. (7.38), to write $\operatorname{tr}\big(-W'\log(W')\big) = S(W') = S\big(VWV^\dagger\big) = S(W) = \operatorname{tr}\big(-W\log(W)\big)$, establishing the entropy invariance of the two alternative environment models.

When $\mathcal{N}(\cdot)$ is a unitary channel with a minimal-size Kraus set containing a single unitary operator $\Lambda_1 = \mathsf{U}$ implementing a reversible evolution of the input state ρ, with $K = 1$ the resulting entropy exchange vanishes in Eq. (13.54). No entropy is generated in the environment by a reversible unitary evolution. By contrast, a non-unitary evolution of ρ by $K > 1$ Kraus operators will in general produce a non-vanishing entropy exchange in Eq. (13.54), with entropy generated into the environment by a non-unitary non-reversible evolution.

13.3.2 Coherent Information

From the output entropy $S(\rho')$ and the entropy exchange $S_{\rm ex}(\rho, \mathcal{N}) = S(\rho'_E)$ one defines for the quantum channel the **coherent information**

$$I_{\rm co}(\rho, \mathcal{N}) = S(\rho') - S(\rho'_E) \tag{13.58}$$

$$= S\big(\mathcal{N}(\rho)\big) - S_{\rm ex}(\rho, \mathcal{N})\ , \tag{13.59}$$

as introduced in [419, 420]. The coherent information $I_{\rm co}(\rho, \mathcal{N})$ is not necessarily non-negative. The coherent information $I_{\rm co}(\rho, \mathcal{N})$ is a function of both the channel $\mathcal{N}$ and its input ρ, and it somehow quantifies how much of the information from the input ρ remains after it has traveled across the channel $\mathcal{N}$ and leaked into the environment. We will see that the coherent information is closely related to the quantum information capacity of the channel.

As seen in Sect. 3.2.1, the quantum channel $\mathcal{N}(\cdot)$ can always receive a Stinespring dilated unitary representation associated with a joint unitary evolution $|\psi'_{QE}\rangle = \mathsf{U}_{QE}|Q\rangle \otimes |e_0\rangle$ written here for a pure state $\rho = |Q\rangle\langle Q|$ at the channel input, as in Eq. (3.34). The state at the channel output follows as the reduced state

$\rho' = \mathcal{N}(\rho) = \mathrm{tr}_E\big(|\psi'_{QE}\rangle\langle\psi'_{QE}|\big)$, while the final state of the environment is $\rho'_E = \mathrm{tr}_Q\big(|\psi'_{QE}\rangle\langle\psi'_{QE}|\big)$. From Sect. 7.3.5 concerning quantum entropies from a bipartite state, with a pure input state $\rho = |Q\rangle\langle Q|$ the bipartite output state $|\psi'_{QE}\rangle\langle\psi'_{QE}|$ is pure, and we deduce that the two reduced states have the same quantum entropy $S(\rho') = S(\rho'_E)$, implying in Eq. (13.58) a vanishing coherent information $I_{\rm co}(\rho, \mathcal{N}) = 0$.

Meanwhile, for a mixed input state ρ, the coherent information $I_{\rm co}(\rho, \mathcal{N})$ of Eq. (13.58) is in general non-vanishing.

An additivity property holds [421] for the coherent information of Eqs. (13.58)–(13.59), giving, for two independent channels with separable inputs, $I_{\rm co}(\rho_1 \otimes \rho_2, \mathcal{N}_1 \otimes \mathcal{N}_2) = I_{\rm co}(\rho_1, \mathcal{N}_1) + I_{\rm co}(\rho_2, \mathcal{N}_2)$, as a result of the additivity of quantum entropies for product states in Eq. (7.31).

When two quantum channels $\mathcal{N}_1(\cdot)$ and $\mathcal{N}_2(\cdot)$ are cascaded to realize the transformation $\rho \mapsto \mathcal{N}_1(\rho) \mapsto \mathcal{N}_2 \circ \mathcal{N}_1(\rho)$, a **quantum data processing inequality** analogue to Eq. (7.24) can be proved [419, 421], taking the form

$$S(\rho) \geq I_{\rm co}(\rho, \mathcal{N}_1) \geq I_{\rm co}(\rho, \mathcal{N}_2 \circ \mathcal{N}_1) \,, \qquad (13.60)$$

where the first inequality follows as a special case of the second inequality, since the input entropy $S(\rho)$ is also the coherent information for the trivial identity channel. In this respect, the quantum data processing inequality of Eq. (13.60) assigns to the quantum coherent information a role resembling that of the classical mutual information.

One can define a coherent information $I_{\rm co}(\mathcal{N})$ intrinsic to a quantum channel $\mathcal{N}$, by the maximization over the input signal $\rho \in \mathcal{L}(\mathcal{H}_N)$ reading

$$I_{\rm co}(\mathcal{N}) = \max_{\rho} I_{\rm co}\big(\rho, \mathcal{N}\big) \,. \qquad (13.61)$$

Although for some input state ρ the coherent information $I_{\rm co}(\rho, \mathcal{N})$ can be negative, the channel coherent information $I_{\rm co}(\mathcal{N})$ of Eq. (13.61) is non-negative [12, 422], with always $I_{\rm co}(\mathcal{N}) \geq 0$. This follows because, since $I_{\rm co}(\rho, \mathcal{N})$ vanishes for pure states ρ, enlarging the maximization to mixed states ρ can only produce a non-negative maximum in Eq. (13.61). Moreover, a nonzero channel coherent information $I_{\rm co}(\mathcal{N})$ in Eq. (13.61) is always achieved by a mixed input state ρ. Yet, the maximization of Eq. (13.61) is generally a **non-concave** ($\not\cap$) optimization problem [422], with the possibility of local maxima that are not the global maximum.

The channel coherent information $I_{\rm co}(\mathcal{N})$ of Eq. (13.61) is in general non-additive. The additivity property mentioned above for the coherent information $I_{\rm co}\big(\rho, \mathcal{N}\big)$ in general does not translate into additivity of the channel coherent information $I_{\rm co}(\mathcal{N})$. In general, $I_{\rm co}(\mathcal{N} \otimes \mathcal{N}) \neq 2I_{\rm co}(\mathcal{N})$, because the maximization resulting from Eq. (13.61) for defining $I_{\rm co}(\mathcal{N} \otimes \mathcal{N})$ generally has to consider entangled states over $\mathcal{L}(\mathcal{H}_N \otimes \mathcal{H}_N)$, not limited to product states as in the additivity property of $I_{\rm co}\big(\rho, \mathcal{N}\big)$ mentioned above. In general, the channel coherent information $I_{\rm co}(\mathcal{N})$ is

superadditive, satisfying $I_{co}(\mathcal{N} \otimes \mathcal{N}) \geq 2I_{co}(\mathcal{N})$, since $I_{co}(\mathcal{N} \otimes \mathcal{N})$ can trivially be at least as large as $2I_{co}(\mathcal{N})$ by using separable product states.

In the convex set of quantum channels $\mathcal{N}$ over $\mathcal{L}(\mathcal{H}_N)$ as envisaged at the end of Sect. 3.2.9, the channel coherent information $I_{co}(\mathcal{N})$ is not known to be convex ($\cup$) in general, so that the inequality $I_{co}\big(p\mathcal{N}_1 + (1-p)\mathcal{N}_2\big) \leq pI_{co}(\mathcal{N}_1) + (1-p)I_{co}(\mathcal{N}_2)$ is not generally satisfied.

13.3.3 Quantum Capacity of a Quantum Channel. The LSD Theorem

Over a quantum channel, as for classical information in Sect. 13.2, it is possible to faithfully transmit quantum information, mainly also by benefiting from the law of large numbers and the dominant behavior on typical subsets of asymptotically long signaling sequences [12, 420, 423, 424]. The maximum rate of faithful transmission of quantum information over a quantum channel $\mathcal{N}(\cdot)$ is given by its quantum capacity $Q(\mathcal{N})$. The role of the quantum capacity $Q(\mathcal{N})$ for faithful quantum communication can be envisioned by referring to what has been accomplished for faithful quantum source coding in Sect. 8.2.1.

In Sect. 8.2.1, for lossless compression of a quantum information source, the source sequences $|\psi^{L_{sou}}\rangle$ of length L_{sou} from the space $\mathcal{H}_N^{\otimes L_{sou}}$ with dimension $N^{L_{sou}}$, by the optimal compression procedure at $L_{sou} \to \infty$, are encoded into a space $\mathcal{H}_{cod}$ with smaller dimension $N^{L_{sou} S_N(\rho)}$, with asymptotically vanishing error, and in this way it achieves the minimal compression rate given by the source entropy $S_N(\rho) \leq \log_N(N) = 1$.

Now for error-free quantum communication over a channel $\mathcal{N}(\cdot)$, at a rate $R \leq Q_N(\mathcal{N}) \leq \log_N(N) = 1$, a **coding and decoding** procedure [420] can be envisaged as follows. The input sequences $|\psi^{L_{in}}\rangle$ of length L_{in} that are to be encoded, belong to the input space $\mathcal{H}_N^{\otimes L_{in}}$ with dimension $N^{L_{in}}$. Each basis state from the input space $\mathcal{H}_N^{\otimes L_{in}}$ is encoded by a codeword taken from a larger space $\mathcal{H}_N^{\otimes L_{in}/R}$ with larger dimension $N^{L_{in}/R}$. In this way, a number L_{in} of N-dimensional quantum systems, get encoded into a larger number $L_{in}/R \geq L_{in}$ of N-dimensional quantum systems. The codewords are sparsely distributed in $\mathcal{H}_N^{\otimes L_{in}/R}$, and span a coding subspace in $\mathcal{H}_N^{\otimes L_{in}/R}$, constructed in such a way that, from a corrupted encoding at the channel output, faithful recovery of the input sequences can be obtained at the decoding stage. Encoding and decoding employ quantum processes similar to those presented in Sect. 8.2.1 for quantum source compression, relying on unitary processes or subspace projections by measurement. The input states $|\psi^{L_{in}}\rangle \in \mathcal{H}_N^{\otimes L_{in}}$, undergoing encoding into $\mathcal{H}_N^{\otimes L_{in}/R}$, transmission across the quantum channel, and decoding, lead to decoded output states in $\mathcal{H}_N^{\otimes L_{in}}$ devised to exhibit perfect fidelity with the input states or sequences, in the sense of the quantitative fidelity criteria also used in Sect. 8.2.1 to assess quantum source compression, where perfect fidelity is obtained with a probability going to 1 asymptotically as $L_{in} \to \infty$. Perfect transmission in this sense is targeted for any quantum input sequences $|\psi^{L_{\text{in}}}\rangle$, and not only for a restricted set of quantum signaling states enabling encoding of classical information over the quantum channel (for instance, $\{|\psi_1\rangle, |\psi_2\rangle, |\psi_3\rangle\}$ in the example of Sect. 13.2.6). Quantum communication is rather concerned with faithful transmission of a continuously infinite set of sequences $|\psi^{L_{\text{in}}}\rangle$ instead of a finite set of discrete signaling states in Sect. 13.2, or in other words any superposition of states from the input alphabet must be faithfully transmitted. The distinction between transmitting classical versus quantum information will be vividly exemplified in the sequel with the two-dimensional qubit channels to be examined in Sect. 13.3.4, or with the completely decohering qubit channel in Sect. 13.3.6. For quantum communication, by the encoding and decoding procedures envisaged above, in the asymptotic regime of large length $L_{\text{in}} \to \infty$, transmission with perfect input–output fidelity can be achieved at any rate R no larger than the quantum capacity $Q_N(\mathcal{N}) \leq \log_N(N) = 1$. In simple terms, the quantum capacity $Q(\mathcal{N})$ of a quantum channel $\mathcal{N}$, is the maximum rate $R \leq 1$ at which L_{in} input qudits (more properly, their quantum state) can be encoded into $L_{\text{in}}/R \geq L_{\text{in}}$ qudits of the same dimension, in such a way that the L_{in} input qudits can be recovered from the L_{in}/R qudits corrupted at the output, with perfect fidelity in the asymptotic limit $L_{\text{in}} \to \infty$.

It can then be shown [420, 423, 424] that the **quantum capacity** $Q(\mathcal{N})$ can be derived from the channel coherent information, by the regularized expression,

$$\boxed{Q(\mathcal{N}) = \lim_{L\to\infty} \frac{1}{L} I_{\text{co}}\big(\mathcal{N}^{\otimes L}\big)\,.} \tag{13.62}$$

This central result establishing the quantum information capacity of a quantum channel, has been obtained through a series of successive arguments [26, 419, 421, 425], and has come to be known as the **Lloyd–Shor–Devetak (LSD) theorem** of quantum information theory [12, 183, 426, 427]. This characterization in Eq. (13.62) of the quantum capacity $Q(\mathcal{N})$ of a quantum channel $\mathcal{N}$ in general requires a "regularized" version of the channel coherent information $I_{\text{co}}\big(\mathcal{N}^{\otimes L}\big)$, involving the L-fold channel $\mathcal{N}^{\otimes L}$ representing L consecutive uses of the physical channel $\mathcal{N}$ conveying L generally entangled N-dimensional quantum systems, and considered in the asymptotic limit of large size $L \to \infty$, as for the classical capacity $C(\mathcal{N})$ of Eq. (13.22). Such regularization in general cannot be avoided, since the channel coherent information $I_{\text{co}}(\mathcal{N})$ in Eq. (13.61) is generally non-additive, but **superadditive**, as indicated at the end of Sect. 13.3.2. Superadditivity holds the promise of enhanced transmission rate over the quantum channel by using entangled inputs. Yet it also implies, with the regularization at $L \to \infty$ in Eq. (13.62), that an explicit calculation of a quantum capacity $Q(\mathcal{N})$ remains in general a difficult task.

As for the channel coherent information $I_{\text{co}}(\mathcal{N})$ of Eq. (13.61), in the convex set of quantum channels $\mathcal{N}$ over $\mathcal{L}(\mathcal{H}_N)$, the quantum capacity $Q(\mathcal{N})$ is not known to be convex (∪) in general, so that the inequality $Q\big(p\mathcal{N}_1 + (1-p)\mathcal{N}_2\big) \leq pQ(\mathcal{N}_1) + (1-p)Q(\mathcal{N}_2)$ is not generally satisfied.

For the special case of quantum channels with an additive channel coherent information $I_{\rm co}(\mathcal{N})$ in Eq. (13.61), for any size L one has $I_{\rm co}\big(\mathcal{N}^{\otimes L}\big) = L I_{\rm co}(\mathcal{N})$, so that the definition of Eq. (13.62) reduces to a single-use or single-letter formula

$$Q(\mathcal{N}) = I_{\rm co}\big(\mathcal{N}\big)\,, \tag{13.63}$$

and the quantum capacity $Q(\mathcal{N})$ of the channel coincides with the channel coherent information $I_{\rm co}(\mathcal{N})$ of Eq. (13.61). Otherwise, the channel coherent information $I_{\rm co}(\mathcal{N})$ of Eq. (13.61) forms a lower bound to the quantum capacity $Q(\mathcal{N})$, when $I_{\rm co}(\mathcal{N})$ is non-additive, that is, strictly superadditive.

Today we do not have a complete characterization of the quantum channels with a non-additive (strictly superadditive) or an additive channel coherent information $I_{\rm co}(\mathcal{N})$, when for the quantum capacity $Q(\mathcal{N})$ the regularized formula of Eq. (13.62) can be replaced by the single-use formula of Eq. (13.63), and of the conditions in which communication of quantum information over a quantum channel can benefit from entangled inputs to enhance the transmission rate. This is the same issue as encountered in Sect. 13.2.4 for communication of classical information over a quantum channel. Entanglement between consecutive uses of a quantum channel stands as a resource capable of enhancing the communication rates, classical and quantum, yet in conditions and with efficiencies that today are far from being completely characterized.

As indicated above, to achieve the quantum capacity $Q(\mathcal{N})$ of Eq. (13.62) requires an appropriate coding and decoding protocol, applied on input sequences of length $L_{\rm in}$, and usually operating in the asymptotic regime $L_{\rm in} \to \infty$ [12, 424, 428, 429]. The determination of an optimal code subspace for a given quantum channel $\mathcal{N}$ remains a difficult task, and the asymptotic regime of long sequences remains difficult to practically handle. Accordingly, effective techniques for quantum channel coding represent today an active area of research.

A class of quantum channels is known with additive channel coherent information $I_{\rm co}(\mathcal{N})$. These are the **degradable quantum channels** [25, 430]. A (primal) channel $\mathcal{N}$ from an input space $\mathcal{H}^A$ into an output space $\mathcal{H}^B$, can be assigned a dual complementary or conjugate channel $\widetilde{\mathcal{N}}$ from the input space $\mathcal{H}^A$ into the environment space $\mathcal{H}^E$, as defined in Sect. 3.2.7. The channel $\mathcal{N}$ from $\mathcal{H}^A$ into $\mathcal{H}^B$ is degradable [25, 430] when it can be degraded into its complementary channel $\widetilde{\mathcal{N}}$, that is, there exists a (degrading) channel $\mathcal{T}$ from $\mathcal{H}^B$ into $\mathcal{H}^E$ such that $\widetilde{\mathcal{N}} = \mathcal{T} \circ \mathcal{N}$. Roughly speaking, a degradable channel $\mathcal{N}$ is a channel that is less noisy than its complementary channel $\widetilde{\mathcal{N}}$, so that following $\mathcal{N}$ an extra noisy transformation $\mathcal{T}$ can produce $\widetilde{\mathcal{N}}$. In other words, the final state of the environment of a degradable channel can be deduced from the output state of the channel.

For degradable quantum channels, the channel coherent information $I_{\rm co}(\mathcal{N})$ is additive [430, 431], and the quantum capacity $Q(\mathcal{N})$ is given by the single-use formula of Eq. (13.63) and it coincides with the channel coherent information $I_{\rm co}(\mathcal{N})$ of Eq. (13.61). Moreover, for a degradable quantum channel $\mathcal{N}$, the coherent information $I_{\rm co}(\rho, \mathcal{N})$ of Eqs. (13.58) and (13.59) is a concave ($\cap$) function [432] of the input density operator ρ, which greatly facilitates the maximization of Eq. (13.61) determining the quantum capacity $Q(\mathcal{N})$ of Eq. (13.63).

A channel $\mathcal{N}$ is **antidegradable** [25] when there exists a channel $\mathcal{T}'$ from $\mathcal{H}^E$ into $\mathcal{H}^B$ such that $\mathcal{N} = \mathcal{T}' \circ \widetilde{\mathcal{N}}$. A channel $\mathcal{N}$ degradable implies $\widetilde{\mathcal{N}}$ antidegradable, and $\mathcal{N}$ antidegradable implies $\widetilde{\mathcal{N}}$ degradable. Every entanglement-breaking quantum channel is antidegradable [431]. It can be proved [12, 25] that any antidegradable channel $\mathcal{N}$ has zero quantum capacity $Q(\mathcal{N}) = 0$. This can be understood from the no-cloning theorem of Sect. 3.1.3. For an antidegradable channel $\mathcal{N}$, if the quantum capacity $Q(\mathcal{N})$ were nonzero, for any pure input state $|\psi\rangle$, after proper encoding and transmission over the channel $\mathcal{N}$, the channel output, after proper decoding, would be able to recover the pure input state $|\psi\rangle$ with no error. Yet at the same time, the information leaked through $\widetilde{\mathcal{N}}$ into the environment during the transmission over $\mathcal{N}$, when fed into the degrading channel $\mathcal{T}'$, would enable the environment to reconstruct the channel output, and from it by the same decoding, to recover the pure input state $|\psi\rangle$ with no error. For any pure input state $|\psi\rangle$, the channel output and the environment would together possess two copies of $|\psi\rangle$, which is not permitted by the no-cloning theorem, and consequently precludes an antidegradable channel from having a nonzero quantum capacity.

Examples of degradable quantum channels are the generalized dephasing channels discussed in [430], the qubit amplitude damping channel discussed in [417], the quantum Hadamard channels discussed in [12]. Yet many (most) quantum channels are neither degradable nor antidegradable [422, 431].

13.3.4 Quantum Capacity of Two-Dimensional Qubit Channels

An interesting class of quantum channels for which the quantum capacity can be completely characterized is formed by the qubit channels with a two-dimensional environment, that is, an environment that can be modeled by a single qubit [433]. These are also qubit channels that can be represented by two Kraus operators acting in the two-dimensional Hilbert space $\mathcal{H}_2$. As shown in [431, 433], every such quantum channel is either degradable or antidegradable. This class contains unital channels important to the qubit, such as the bit-flip, the phase-flip, the bit-phase-flip channels, as they follow from the corresponding qubit noise models presented in Chap. 9. For these important unital qubit channels, the quantum capacity can be explicitly derived via a uniform approach as follows.

We consider unital qubit channels characterized by the two Kraus operators $\sqrt{1-p}\,\mathsf{U}_0$ and $\sqrt{p}\,\mathsf{U}_1$, implementing the input–output quantum operation

$$\rho \longmapsto \mathcal{N}(\rho) = (1-p)\mathsf{U}_0\rho\mathsf{U}_0^\dagger + p\mathsf{U}_1\rho\mathsf{U}_1^\dagger\,, \qquad (13.64)$$

where U_0 and U_1 are two unitary operators acting in the qubit Hilbert space $\mathcal{H}_2$, and $p \in [0, 1]$ is a probability. Such quantum channels are degradable [433], so that their quantum capacity $Q(\mathcal{N})$ is given by the single-use expression of Eq. (13.63) and coincides with the channel coherent information $I_{co}(\mathcal{N})$ of Eq. (13.61). We also notice that the qubit channels of Eq. (13.64) are unital, satisfying $\mathcal{N}(\mathrm{I}_2) = \mathrm{I}_2$, so that their classical capacity $C(\mathcal{N})$ is ruled by Sect. 13.2.9. But here we are interested in their quantum capacity $Q(\mathcal{N})$.

The channel transformation in Eq. (13.64) can be factored as

$$\mathcal{N}(\rho) = \mathsf{U}_0\Big[(1-p)\rho + p\mathsf{U}_0^\dagger\mathsf{U}_1\rho\mathsf{U}_1^\dagger\mathsf{U}_0\Big]\mathsf{U}_0^\dagger\,, \tag{13.65}$$

so that we are led to consider the reduced channel

$$\rho \longmapsto \mathcal{N}_{\mathrm{red}}(\rho) = (1-p)\rho + p\mathsf{U}_0^\dagger\mathsf{U}_1\rho\mathsf{U}_1^\dagger\mathsf{U}_0\,, \tag{13.66}$$

with two Kraus operators $\sqrt{1-p}\,\mathrm{I}_2$ and $\sqrt{p}\,\mathsf{U}_0^\dagger\mathsf{U}_1$ constructed from the two unitary operators I_2 and $\mathsf{U}_0^\dagger\mathsf{U}_1$. The output states $\mathcal{N}_{\mathrm{red}}(\rho)$ and $\mathcal{N}(\rho) = \mathsf{U}_0\mathcal{N}_{\mathrm{red}}(\rho)\mathsf{U}_0^\dagger$ are related via the unitary transformation U_0 on the channel output. They have therefore the same quantum entropy $S\big(\mathcal{N}(\rho)\big) = S\big(\mathcal{N}_{\mathrm{red}}(\rho)\big)$.

In addition, both channels $\mathcal{N}(\cdot)$ and $\mathcal{N}_{\mathrm{red}}(\cdot)$, with two Kraus operators, can be represented by a two-dimensional environment model, by the canonical the procedure of Sect. 3.2.8. The final state ρ_E' of the environment from Eq. (3.119) can be worked out via its matrix elements of Eq. (3.120), and it comes out that both channels $\mathcal{N}(\cdot)$ and $\mathcal{N}_{\mathrm{red}}(\cdot)$ produce the same final state of the environment, representable by the matrix

$$\rho_E' = \begin{bmatrix} 1-p & \sqrt{(1-p)p}\;\mathrm{tr}\big(\mathsf{U}_0\rho\mathsf{U}_1^\dagger\big) \\ \sqrt{(1-p)p}\;\mathrm{tr}\big(\mathsf{U}_1\rho\mathsf{U}_0^\dagger\big) & p \end{bmatrix}, \tag{13.67}$$

by circularity of the trace. As a result, both channels $\mathcal{N}(\cdot)$ and $\mathcal{N}_{\mathrm{red}}(\cdot)$ generate the same quantum entropy $S(\rho_E')$ into the environment. With same output entropies and same environment entropies, both channels $\mathcal{N}(\cdot)$ and $\mathcal{N}_{\mathrm{red}}(\cdot)$ have therefore the same coherent information $I_{co}\big(\rho, \mathcal{N}\big) = I_{co}\big(\rho, \mathcal{N}_{\mathrm{red}}\big)$, and also the same quantum capacity $Q(\mathcal{N}) = Q(\mathcal{N}_{\mathrm{red}})$ from Eq. (13.63).

To characterize this quantum capacity, instead of the extended channel form of Eq. (13.64), it is thus enough, without loss of generality, to consider a quantum channel of the generic form

$$\rho \longmapsto \mathcal{N}(\rho) = (1-p)\rho + p\mathsf{U}_\xi\rho\mathsf{U}_\xi^\dagger\,, \tag{13.68}$$

these two channels of Eqs. (13.64) and (13.68) sharing the same quantum capacity when $\mathsf{U}_\xi = \mathsf{U}_0^\dagger\mathsf{U}_1$, as it follows from Eqs. (13.65) and (13.66). In addition, the unitary operator U_ξ of Eq. (13.68) can be represented by a 2×2 matrix of the generic diagonal form

$$\mathsf{U}_\xi = \begin{bmatrix} 1 & 0 \\ 0 & e^{i\xi} \end{bmatrix}, \tag{13.69}$$

as in Eq. (3.31), by defining the computational orthonormal basis $\{|0\rangle, |1\rangle\}$ of $\mathcal{H}_2$ with the two eigenstates of U_ξ and by selecting the origin of the phases so that one of the two complex unit-modulus eigenvalues of U_ξ is 1. This parameterization for the unitary U_ξ in the qubit channel of Eq. (13.68) entails no loss of generality for the calculation of its quantum capacity $Q(\mathcal{N})$.

From an input qubit with density operator ρ in the standard parameterization $\rho = [\rho_{00}, \rho_{01} ; \rho_{01}^*, \rho_{11}]$ of Eq. (2.59), the channel of Eq. (13.68) delivers an output qubit with density operator $\rho' = \mathcal{N}(\rho)$ given by

$$\rho' = \mathcal{N}(\rho) = \begin{bmatrix} \rho_{00} & (1 - p + pe^{-i\xi})\rho_{01} \\ (1 - p + pe^{i\xi})\rho_{01}^* & \rho_{11} \end{bmatrix}. \tag{13.70}$$

From an input density operator ρ of Bloch vector $\vec{r} = [r_x, r_y, r_z]^\top$, by using Eqs. (2.63)–(2.65) to go from the Hilbert space representation to the Bloch representation, the output density operator $\rho' = \mathcal{N}(\rho)$ of Eq. (13.70) can be endowed with a Bloch vector $\vec{r}\,' = [r'_x, r'_y, r'_z]^\top$ deduced as

$$\vec{r}\,' \left| \begin{aligned} r'_x &= [1 - p + p\cos(\xi)]r_x - p\sin(\xi)r_y && (13.71) \\ r'_y &= p\sin(\xi)r_x + [1 - p + p\cos(\xi)]r_y && (13.72) \\ r'_z &= r_z\,. && (13.73) \end{aligned} \right.$$

The Bloch transformation $\vec{r}\,' = A\vec{r}$ in $\mathbb{R}^3$ has an invariant axis with the Oz axis, and the matrix A is endowed with the eigenvector $\vec{g}_1 = [0, 0, 1]^\top$ associated with the major (largest) eigenvalue $g_1 = 1$. As a result, by referring to Sect. 13.2.9, it follows that the (unital) qubit channels of Eq. (13.68), and consequently of Eq. (13.64), have a classical capacity $C(\mathcal{N}) = \log(2)$ in Eq. (13.37), the maximum classical capacity for qubit channels, allowing perfect error-free transmission of classical information through binary signaling along the invariant axis in $\mathbb{R}^3$ via the eigenbasis of $\mathsf{U}_\xi = \mathsf{U}_0^\dagger \mathsf{U}_1$ immune to noise.

Focusing here on the quantum capacity $Q(\mathcal{N})$, we have for the output signal ρ' the two eigenvalues

$$\lambda'_\pm = \frac{1}{2}\big(1 \pm \|\vec{r}\,'\|\big), \tag{13.74}$$

and the quantum entropy $S(\rho')$ of the channel output is then obtained, according to Eq. (7.40), as

$$S(\rho') = H_{\text{bin}}\left(\frac{1 + \|\vec{r}\,'\|}{2}\right), \tag{13.75}$$

with $H_{\text{bin}}(\cdot)$ the classical entropy of a binary variable defined by Eq. (7.4) and represented in Fig. 7.1.

For the task of determining the quantum capacity $Q(\mathcal{N})$ of the channel, a useful invariance property exists with the output entropy $S(\rho')$ of Eq. (13.75). When the input state is transformed as $\vec{r} \to [-r_x, -r_y, r_z]^\top$ by reversing the sign of the x and y components of the input Bloch vector $\vec{r}$, the output in Eqs. (13.71)–(13.73) undergoes a similar transformation $\vec{r}\,' \to [-r_x', -r_y', r_z']^\top$, so that we observe an invariance in the output norm $\|\vec{r}\,'\|$, and hence the same invariance in the output entropy $S(\rho')$ of Eq. (13.75). This transformation of the input Bloch vector $\vec{r} \to [-r_x, -r_y, r_z]^\top$ is equivalent to the transformation of the input density operator $\rho \to \sigma_z \rho \sigma_z^\dagger$. This is expressed by Eq. (9.13) or can be easily verified by considering the transformation of Eqs. (9.29) and (9.31) at $p = 1$. Consequently, the two input states ρ and $\sigma_z \rho \sigma_z^\dagger$ produce the same quantum entropy of Eq. (13.75) at the channel output.

For the environment now, its final state $\rho_E' = \widetilde{\mathcal{N}}(\rho)$, defining the complementary channel $\widetilde{\mathcal{N}}$, can be worked out via its matrix elements as in Eq. (13.67), to yield

$$\rho_E' = \widetilde{\mathcal{N}}(\rho) = \begin{bmatrix} 1-p & \sqrt{(1-p)p}\left(\rho_{00} + e^{-i\xi}\rho_{11}\right) \\ \sqrt{(1-p)p}\left(\rho_{00} + e^{i\xi}\rho_{11}\right) & p \end{bmatrix}. \tag{13.76}$$

And via Eqs. (2.63)–(2.65), the final state $\rho_E' = \widetilde{\mathcal{N}}(\rho)$ can be described by a Bloch vector $\vec{s}\,' = [s_x', s_y', s_z']^\top$ following as

$$\vec{s}\,' \left| \begin{aligned} s_x' &= \sqrt{(1-p)p}\left[1 + r_z + \cos(\xi)(1 - r_z)\right] && (13.77) \\ s_y' &= \sqrt{(1-p)p}\ \sin(\xi)(1 - r_z) && (13.78) \\ s_z' &= 1 - 2p\,. && (13.79) \end{aligned} \right.$$

We deduce, as in Eq. (13.75), the final quantum entropy of the environment as

$$S(\rho_E') = H_{\text{bin}}\left(\frac{1 + \|\vec{s}\,'\|}{2}\right). \tag{13.80}$$

We can now explicitly exhibit the degrading channel $\mathcal{T}$ degrading the primal channel $\mathcal{N}$ into its complementary channel $\widetilde{\mathcal{N}}$ according to $\widetilde{\mathcal{N}} = \mathcal{T} \circ \mathcal{N}$, and establishing $\mathcal{N}$ as a degradable quantum channel. From the structure of $\mathcal{N}$ in Eqs. (13.71)–(13.73) that leaves the input component $r_z' = r_z$ invariant, and the structure of $\widetilde{\mathcal{N}}$ in Eqs. (13.77)–(13.79) with the final components $[s_x', s_y', s_z']$ depending solely on $r_z = r_z'$, we deduce that one has $\widetilde{\mathcal{N}} = \widetilde{\mathcal{N}} \circ \mathcal{N}$, establishing $\mathcal{T} \equiv \widetilde{\mathcal{N}}$ as a valid degrading channel for $\mathcal{N}$. In this way, the qubit channel of Eq. (13.68), and therefore also the qubit channel[3] of Eq. (13.64), are degradable quantum channels, with an additive channel coherent information $I_{\rm co}(\mathcal{N})$, and with a quantum capacity $Q(\mathcal{N})$ given by the single-use formula of Eq. (13.63).

We also observe with Eqs. (13.77)–(13.79) that, in the input transformation $\vec{r} \to [-r_x, -r_y, r_z]^\top$, the norm $\|\vec{s}\,'\|$ and hence the environment entropy $S(\rho'_E)$ in Eq. (13.80) remain invariant, as $S(\rho')$ in Eq. (13.75). So the input states ρ and $\sigma_z\rho\sigma_z^\dagger$ produce the same output entropy in Eq. (13.75) and same final entropy of the environment in Eq. (13.80).

By combining the output entropy $S(\rho')$ of Eq. (13.75) and the environment entropy $S(\rho'_E)$ of Eq. (13.80), we now have access to the coherent information $I_{\rm co}(\rho, \mathcal{N}) = S(\rho') - S(\rho'_E)$ from Eq. (13.58), associated with the input state ρ across the channel $\mathcal{N}$, which comes under the form

$$I_{\rm co}(\rho, \mathcal{N}) = S(\rho') - S(\rho'_E) = H_{\rm bin}\left(\frac{1 + \|\vec{r}\,'\|}{2}\right) - H_{\rm bin}\left(\frac{1 + \|\vec{s}\,'\|}{2}\right) . \tag{13.81}$$

And since the two entropies $S(\rho')$ and $S(\rho'_E)$ are invariant in the input transformation $\rho \to \sigma_z\rho\sigma_z^\dagger$, also invariant is the coherent information, to give

$$I_{\rm co}(\rho, \mathcal{N}) = I_{\rm co}\big(\sigma_z\rho\sigma_z^\dagger, \mathcal{N}\big) . \tag{13.82}$$

From the coherent information $I_{\rm co}(\rho, \mathcal{N})$ of Eq. (13.81), we are led to the maximization

$$Q(\mathcal{N}) = I_{\rm co}(\mathcal{N}) = \max_{\rho \in \mathcal{L}(\mathcal{H}_2)} I_{\rm co}(\rho, \mathcal{N}) \tag{13.83}$$

$$= \max_{\rho \in \mathcal{L}(\mathcal{H}_2)} H_{\rm bin}\left(\frac{1 + \|\vec{r}\,'\|}{2}\right) - H_{\rm bin}\left(\frac{1 + \|\vec{s}\,'\|}{2}\right) . \tag{13.84}$$

For an input signal $\rho \in \mathcal{L}(\mathcal{H}_2)$ parameterized by a Bloch vector $\vec{r} = [r_x, r_y, r_z]^\top$, with the norms $\|\vec{r}\,'\|$ and $\|\vec{s}\,'\|$ resulting from Eqs. (13.71) to (13.73) and (13.77)–(13.79), the maximization in Eq. (13.84) arrives as a maximization in the three variables (r_x, r_y, r_z). There however exists a possibility to reduce to one variable, by exploiting the invariance property of Eq. (13.82). As we indicated in Sect. 13.3.3, for a degradable channel $\mathcal{N}$, the coherent information $I_{\rm co}(\rho, \mathcal{N})$ is a concave ($\cap$) function of the input density operator ρ. So with the channel $\mathcal{N}$ degradable, by the concavity ($\cap$) of $I_{\rm co}(\rho, \mathcal{N})$ we have

$$\frac{1}{2} I_{\rm co}(\rho, \mathcal{N}) + \frac{1}{2} I_{\rm co}\big(\sigma_z\rho\sigma_z^\dagger, \mathcal{N}\big) \le I_{\rm co}\left(\frac{1}{2}\rho + \frac{1}{2}\sigma_z\rho\sigma_z^\dagger, \mathcal{N}\right) , \tag{13.85}$$

[3] When the primal channel changes from $\rho \mapsto \mathcal{N}(\rho)$ to $\rho \mapsto \mathsf{U}_0\mathcal{N}(\rho)\mathsf{U}_0^\dagger$, the degrading channel changes from $\rho \mapsto \mathcal{T}(\rho)$ to $\rho \mapsto \mathcal{T}\big(\mathsf{U}_0^\dagger\rho\mathsf{U}_0\big)$, with an invariant complementary channel $\rho \mapsto \widetilde{\mathcal{N}}(\rho) \equiv \mathcal{T}(\rho)$.

which by the invariance of Eq. (13.82) is also

$$I_{co}(\rho, \mathcal{N}) \leq I_{co}\left(\frac{1}{2}\rho + \frac{1}{2}\sigma_z\rho\sigma_z^\dagger, \mathcal{N}\right). \tag{13.86}$$

The density operator $(\rho + \sigma_z\rho\sigma_z^\dagger)/2$ appearing in the right-hand side of Eq. (13.86) is the density operator with Bloch vector $[0, 0, r_z]^\top$. This is easily verified since we recognize the transformation of Eqs. (9.29) and (9.31) acting $p = 1/2$. The inequality of Eq. (13.86) establishes that the maximum of $I_{co}(\rho, \mathcal{N})$ in Eqs. (13.83) and (13.84) is achieved by a density operator ρ with a Bloch vector restricted to the form $[0, 0, r_z]^\top$, that is, a density operator ρ diagonal in the computational basis.

With the input $\vec{r} = [0, 0, r_z]^\top$, we have from Eqs. (13.71) to (13.73) the squared norm $\|\vec{r}\,'\|^2 = r_z^2$, and from Eqs. (13.77) to (13.79) the squared norm

$$\|\vec{s}\,'\|^2 = (1-p)^2 + p^2 + 2(1-p)p\big[r_z^2 + (1-r_z^2)\cos(\xi)\big]. \tag{13.87}$$

The quantum capacity $Q(\mathcal{N})$ is therefore obtained from Eq. (13.84) by the one-variable maximization

$$Q(\mathcal{N}) = \max_{r_z \in [0,1]} H_{\text{bin}}\left(\frac{1+r_z}{2}\right) - H_{\text{bin}}\left(\frac{1+\sqrt{(1-p)^2 + p^2 + 2(1-p)p\big[r_z^2 + (1-r_z^2)\cos(\xi)\big]}}{2}\right). \tag{13.88}$$

The maximum in Eq. (13.88), for any p and $\cos(\xi)$, is always achieved by $r_z = 0$, that is, by the maximally mixed state $\rho = \mathrm{I}_2/2$ at the channel input. Consequently, the quantum capacity $Q(\mathcal{N})$ of the qubit channel of Eq. (13.68), and of Eq. (13.64), results as

$$\boxed{Q(\mathcal{N}) = \log(2) - H_{\text{bin}}\left(\frac{1+\sqrt{(1-p)^2 + p^2 + 2(1-p)p\cos(\xi)}}{2}\right).} \tag{13.89}$$

Figure 13.9 represents the quantum capacity $Q(\mathcal{N})$ of Eq. (13.89), as a function of the noise probability p assessing the noisy transmission by the quantum channel, for several values of the phase angle ξ specifying the channel model via Eq. (13.69).

The channel model of Eq. (13.64) with its quantum capacity in Eq. (13.89), encompasses as special cases the **bit-flip**, the **phase-flip** and the **bit-phase-flip qubit channels**. These three important qubit channels are obtained with the three Pauli operators, respectively, σ_x, σ_z and σ_y, for the unitary U_ξ in Eq. (13.68). Each Pauli operator has ± 1 for its two eigenvalues, and when each is referred to its orthonormal eigenbasis, all three Pauli operators lead to a same phase angle $\xi = \pi$ in Eq. (13.69). With $\xi = \pi$ in Eq. (13.89), we obtain the quantum capacity

$$\boxed{Q(\mathcal{N}) = \log(2) - H_{\text{bin}}(p),} \tag{13.90}$$

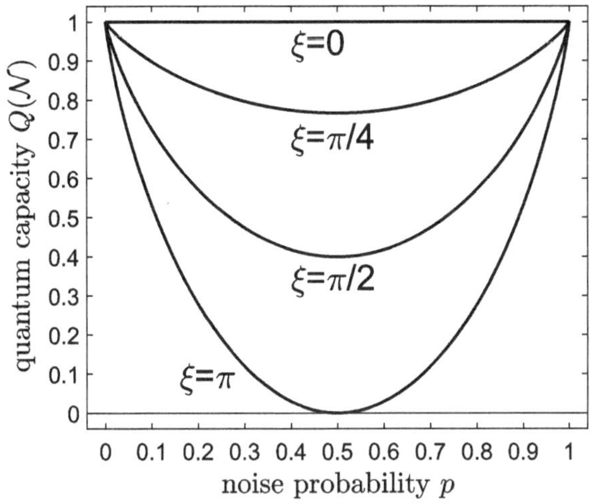

Fig. 13.9 For the two-dimensional qubit channel of Eq. (13.64), and of Eq. (13.68), as a function of the noise probability p over the channel, the quantum capacity $Q(\mathcal{N})$ in shannons of Eq. (13.89), for four values of the phase angle ξ specifying the channel model via Eq. (13.69)

which is therefore the same, for the bit-flip, the phase-flip, and the bit-phase-flip qubit channels. This capacity of Eq. (13.90) is shown in Fig. 13.9 at $\xi = \pi$. With the quantum capacity of Eq. (13.90), for the three qubit "flip" channels, we recognize also the classical capacity of the classical binary symmetric channel with a flip probability p, as in Eq. (13.7). The value $p = 1/2$ is the most detrimental probability, with a vanishing capacity, both classical and quantum, as visible in Fig. 13.9 at $\xi = \pi$.

The quantum capacity of Eq. (13.90) also gives us an occasion to visualize an important distinction, for a quantum channel $\mathcal{N}$, between its classical capacity $C(\mathcal{N})$ and its quantum capacity $Q(\mathcal{N})$. For a quantum channel $\mathcal{N}$, the classical capacity $C(\mathcal{N})$ quantifies the maximum rate for error-free transmission of *classical* information, while the quantum capacity $Q(\mathcal{N})$ quantifies the maximum rate for error-free transmission of *quantum* information. The distinction can be exemplified as follows.

We have seen at the end of Sect. 13.2.9, that each of the three qubit flip channels has an invariant axis in $\mathbb{R}^3$ associated with an orthonormal basis in $\mathcal{H}_2$ immune to noise, and given by the eigenbasis of the corresponding Pauli operator. For instance, the phase-flip channel has the immune basis $\{|0\rangle, |1\rangle\}$, given that the channel transformation $|0\rangle \to |0\rangle$ and $|1\rangle \to -|1\rangle \equiv |1\rangle$ preserves perfect identification of the two basis states at the output; the bit-flip channel has the immune basis $\{|+\rangle, |-\rangle\}$, and $\{|+i\rangle, |-i\rangle\}$ for the bit-phase-flip channel. Such an immune basis can be used for direct encoding of the two values of a classical binary information, that will then be transmitted across the quantum channel with no alteration. For each of the three qubit flip channels, this possibility of perfect transmission of every classical binary information translates into a classical capacity $C(\mathcal{N}) = \log(2)$ reaching the overall maximal capacity afforded by a binary channel, and is unaffected by the noise level p over the quantum channel.

By contrast, for the three qubit flip channels, the quantum capacity $Q(\mathcal{N})$ of Eq. (13.90) is affected by the noise level p, as visible in Fig. 13.9 at $\xi = \pi$. The reason is that, while the classical capacity $C(\mathcal{N})$ is concerned "only" with error-free transmission of an orthonormal basis allowing classical signaling, the quantum capacity $Q(\mathcal{N})$, more stringent, is concerned with error-free transmission of *any coherent superposition* of the (basis) states. For instance, the phase-flip channel transmits with no error the two basis states $|0\rangle$ and $|1\rangle$, but it alters the superposition $|+\rangle = \big(|0\rangle + |1\rangle\big)/\sqrt{2}$, that is changed to $|-\rangle$ with probability p. This translates into a quantum capacity $Q(\mathcal{N})$ affected by the noise level p over the quantum channel, as visible in Fig. 13.9.

Also, as a rule for any quantum channel [183] we always have $Q(\mathcal{N}) \leq C(\mathcal{N})$, since the quantum capacity $Q(\mathcal{N})$ is more stringent than the classical capacity $C(\mathcal{N})$, in the sense that error-free transmission of any coherent quantum states a fortiori enables error-free transmission of a set of basis states, but not conversely.

The phase angle ξ specifying the channel model in Eq. (13.69) and in Fig. 13.9, can be related to a notion of proximity between the two unitary operators U_0 and U_1 that define the general channel model of Eq. (13.64). In relation to the general channel model of Eq. (13.64), the generic form of Eq. (13.69) involves the unitary operator $\mathsf{U}_\xi = \mathsf{U}_0^\dagger \mathsf{U}_1$. Its trace $\mathrm{tr}\big(\mathsf{U}_\xi\big) = \mathrm{tr}\big(\mathsf{U}_0^\dagger \mathsf{U}_1\big) = 1 + e^{i\xi}$ introduces the Hilbert–Schmidt inner product $\mathrm{tr}\big(\mathsf{U}_0^\dagger \mathsf{U}_1\big)$ of the two unitary operators U_0 and U_1, so as to give $|\mathrm{tr}(\mathsf{U}_0^\dagger \mathsf{U}_1)|^2 = 2\big[1 + \cos(\xi)\big] = 4\cos^2(\xi/2)$. By normalizing the inner product with the Hilbert–Schmidt norms, we have access via its cosine to an angle $(\widehat{\mathsf{U}_0, \mathsf{U}_1})$ between the two unitary operators U_0 and U_1, amounting to $(\widehat{\mathsf{U}_0, \mathsf{U}_1}) = \xi/2$. So the phase angle ξ controlling the quantum capacity $Q(\mathcal{N})$ in Eq. (13.89) is determined by the overlap $\big|\mathrm{tr}(\mathsf{U}_0^\dagger \mathsf{U}_1)\big|$ between the two unitary operators U_0 and U_1 of the channel model of Eq. (13.64). At $\xi = \pi$ in Fig. 13.9, the two unitaries U_0 and U_1 are orthogonal or maximally distant; the random flips between U_0 and U_1 implemented by the channel model in Eq. (13.64) are the most deleterious; the channel capacity $Q(\mathcal{N})$ assumes the smallest values. Otherwise, as $\xi \to 0$ in Fig. 13.9, the two unitaries U_0 and U_1 tend to align; the random flips between U_0 and U_1 represent less and less harmful alterations to the qubit states; the channel capacity $Q(\mathcal{N})$ increases to larger values, up to the maximal capacity $Q(\mathcal{N}) = \log(2)$ at $\xi = 0$ when the two unitaries U_0 and U_1 coincide.

The class of quantum channels of Eq. (13.64) encompasses the three flip channels important to the qubit, and enables a complete characterization of the quantum capacity $Q(\mathcal{N})$ in Eq. (13.89). This class of Eq. (13.64) nevertheless does not contain all the qubit channels with a two-dimensional environment. The amplitude damping channel resulting from Sect. 9.9 is also with a two-dimensional environment, but is a non-unital qubit channel outside the class of Eq. (13.64). Its quantum capacity should be treated separately, however with some similarities in the derivation with the methodology developed in this Section for the channels of Eq. (13.64), as we are going to see in the next section.

13.3.5 Quantum Capacity of the Amplitude Damping Channel

The quantum capacity of the amplitude damping channel is characterized, for instance, in [417] or also in [433]. The amplitude damping channel is realized by the qubit noise channel $\mathcal{N}(\cdot)$ presented in Sect. 9.9. From an input qubit with density operator $\rho = \left[\rho_{00}, \rho_{01} \,;\, \rho_{01}^*, \rho_{11}\right]$ in the standard parameterization of Eq. (2.59), the channel delivers an output qubit with density operator $\rho' = \mathcal{N}(\rho)$ given by Eq. (9.81), rewritten here as

$$\rho' = \mathcal{N}(\rho) = \begin{bmatrix} 1-(1-\gamma)\rho_{11} & \sqrt{1-\gamma}\,\rho_{01} \\ \sqrt{1-\gamma}\,\rho_{01}^* & (1-\gamma)\rho_{11} \end{bmatrix}, \tag{13.91}$$

with the damping factor $\gamma \in [0, 1]$. Equivalently, when the input ρ has a Bloch vector $\vec{r} = [r_x, r_y, r_z]^\top$, the channel output ρ' is endowed with a Bloch vector $\vec{r}\,' = [r_x', r_y', r_z']^\top$ following from Eq. (9.82) as

$$\vec{r}\,' \left| \begin{aligned} r_x' &= \sqrt{1-\gamma}\, r_x & (13.92) \\ r_y' &= \sqrt{1-\gamma}\, r_y & (13.93) \\ r_z' &= (1-\gamma) r_z + \gamma\,. & (13.94) \end{aligned} \right.$$

Two eigenvalues for the output signal ρ' result as

$$\lambda_\pm' = \frac{1}{2}\left(1 \pm \|\vec{r}\,'\|\right), \tag{13.95}$$

with the output norm $\|\vec{r}\,'\|$ given via

$$\|\vec{r}\,'\|^2 = (1-\gamma)\left(r_x^2 + r_y^2\right) + \left[(1-\gamma)r_z + \gamma\right]^2. \tag{13.96}$$

The quantum entropy $S(\rho')$ of the channel output is then obtained, according to Eq. (7.40), as

$$S(\rho') = H_{\text{bin}}\left(\frac{1 + \|\vec{r}\,'\|}{2}\right), \tag{13.97}$$

with $H_{\text{bin}}(\cdot)$ again the classical entropy of a binary variable defined by Eq. (7.4).

We can observe the same useful invariance as in Sect. 13.3.4. In the input transformation $\vec{r} \to [-r_x, -r_y, r_z]^\top$ associated with $\rho \to \sigma_z \rho \sigma_z^\dagger$, the output norm $\|\vec{r}\,'\|$ of Eq. (13.96) is invariant, and so is the output entropy from Eq. (13.97). Therefore, the two input states ρ and $\sigma_z \rho \sigma_z^\dagger$ lead to the same output entropy $S\big(\mathcal{N}(\rho)\big) = S\big(\mathcal{N}(\sigma_z \rho \sigma_z^\dagger)\big)$ in Eq. (13.97).

For modeling the environment now, the two Kraus operators of Eqs. (9.79) and (9.80) representing the amplitude damping channel determine, by the procedure of Sect. 3.2.8, a two-dimensional model for the environment E. When transmitting an input state ρ, the action of the channel $\mathcal{N}(\cdot)$ places the two-dimensional environment E in the final state $\rho'_E = \widetilde{\mathcal{N}}(\rho)$ resulting from Eq. (9.83) as

$$\rho'_E = \widetilde{\mathcal{N}}(\rho) = \begin{bmatrix} 1-\gamma\rho_{11} & \sqrt{\gamma}\,\rho_{01} \\ \sqrt{\gamma}\,\rho_{01}^* & \gamma\rho_{11} \end{bmatrix}, \tag{13.98}$$

and defining the complementary channel $\widetilde{\mathcal{N}}(\cdot)$ from the input into the environment E.

By comparing Eqs. (13.91) and (13.98), we realize that the complementary channel $\widetilde{\mathcal{N}}(\cdot)$ can be seen as a qubit amplitude damping channel as well, characterized by the damping factor $\widetilde{\gamma} = 1-\gamma \in [0, 1]$. The final state ρ'_E of the environment E can be described by a Bloch vector $\vec{s}\,' = [s'_x, s'_y, s'_z]^\top$ therefore following as

$$\vec{s}\,' \left| \begin{aligned} s'_x &= \sqrt{\gamma}\,r_x && (13.99)\\ s'_y &= \sqrt{\gamma}\,r_y && (13.100)\\ s'_z &= \gamma r_z + 1 - \gamma\,, && (13.101) \end{aligned} \right.$$

having a form similar to Eqs. (13.92)–(13.94). From the squared norm

$$\|\vec{s}\,'\|^2 = \gamma\left(r_x^2 + r_y^2\right) + \left(\gamma r_z + 1 - \gamma\right)^2, \tag{13.102}$$

we deduce, as in Eq. (13.97), the final quantum entropy of the environment E as

$$S(\rho'_E) = H_{\text{bin}}\left(\frac{1 + \|\vec{s}\,'\|}{2}\right). \tag{13.103}$$

We also observe that, in the input transformation $\vec{r} \to [-r_x, -r_y, r_z]^\top$, the norm $\|\vec{s}\,'\|$ and hence the entropy $S(\rho'_E)$ exhibit the same invariance as $\|\vec{r}\,'\|$ and $S(\rho')$, so that the input states ρ and $\sigma_z\rho\sigma_z^\dagger$ produce the same final entropy of Eq. (13.103) in the environment of the amplitude damping channel.

By combining the output entropy $S(\rho')$ of Eq. (13.97) and the environment entropy $S(\rho'_E)$ of Eq. (13.103), we now have access to the coherent information $I_{\text{co}}(\rho, \mathcal{N}) = S(\rho') - S(\rho'_E)$ from Eq. (13.58), associated with the input state ρ across the amplitude damping channel $\mathcal{N}(\cdot)$, which comes under the form

$$I_{\text{co}}(\rho, \mathcal{N}) = S(\rho') - S(\rho'_E) = H_{\text{bin}}\left(\frac{1 + \|\vec{r}\,'\|}{2}\right) - H_{\text{bin}}\left(\frac{1 + \|\vec{s}\,'\|}{2}\right). \tag{13.104}$$

And since the two entropies $S(\rho')$ and $S(\rho'_E)$ are invariant in the input transformation $\rho \to \sigma_z\rho\sigma_z^\dagger$, also invariant is the coherent information, to give

$$I_{\rm co}(\rho, \mathcal{N}) = I_{\rm co}\big(\sigma_z\rho\sigma_z^\dagger, \mathcal{N}\big) . \tag{13.105}$$

We now show that the amplitude damping channel is degradable, by explicitly exhibiting the possibility of a degrading channel $\mathcal{T}$ from the channel output into the environment E, such that $\widetilde{\mathcal{N}} = \mathcal{T} \circ \mathcal{N}$. By considering the actions of the channels $\mathcal{N}$ and $\widetilde{\mathcal{N}}$ in Bloch representation, as expressed respectively by Eqs. (13.92)–(13.94) and (13.99)–(13.101), we are led to consider a quantum superoperator $\mathcal{T}(\cdot)$ acting on qubit density operators, and defined by its action on a generic Bloch vector $\vec{r}$ according to

$$\vec{r} \longmapsto \begin{bmatrix} \sqrt{\dfrac{\gamma}{1-\gamma}} & 0 & 0 \\ 0 & \sqrt{\dfrac{\gamma}{1-\gamma}} & 0 \\ 0 & 0 & \dfrac{\gamma}{1-\gamma} \end{bmatrix} \vec{r} + \begin{bmatrix} 0 \\ 0 \\ 1 - \dfrac{\gamma}{1-\gamma} \end{bmatrix} . \tag{13.106}$$

It is easy to verify by their actions in Bloch representation that, by cascading the transformation $\mathcal{N}$ of Eqs. (13.92)–(13.94) and then the transformation $\mathcal{T}$ of Eq. (13.106), the transformation $\widetilde{\mathcal{N}}$ of Eqs. (13.99)–(13.101) is obtained. The remaining concern is to identify the conditions where the superoperator $\mathcal{T}(\cdot)$ defined by Eq. (13.106) is a valid quantum operation or process as required.

By comparing with Eq. (9.82) defining the action of a valid amplitude damping channel, we observe that Eq. (13.106) itself defines an amplitude damping channel, with damping factor $\gamma_{\mathcal{T}} = 1 - \gamma/(1-\gamma)$, provided $\gamma_{\mathcal{T}}$ remains in the validity domain $[0, 1]$, which occurs for $\gamma \in [0, 1/2]$. So we conclude that for $\gamma \in [0, 1/2]$, the superoperator $\mathcal{T}(\cdot)$ defined by Eq. (13.106) is a valid quantum process, representing a degrading channel for $\mathcal{N}$. Therefore, for $\gamma \in [0, 1/2]$, the damping channel $\mathcal{N}$ is a degradable quantum channel. The degradation $\widetilde{\mathcal{N}} = \mathcal{T} \circ \mathcal{N}$ that is obtained occurs as a form of a multiplicative chain rule taking place when cascading two amplitude damping channels. As a rule, it is easily verified that when cascading an amplitude damping channel with damping factor γ_1, followed by an amplitude damping channel with damping factor γ_2, one obtains an amplitude damping channel with damping factor γ_3, with the multiplicative connection $1 - \gamma_3 = (1-\gamma_2)(1-\gamma_1)$.

Otherwise for Eq. (13.106), when $\gamma \in]1/2, 1]$, the parameter $\gamma_{\mathcal{T}} = 1 - \gamma/(1-\gamma)$ moves outside the interval $[0, 1]$; the transformation of Eq. (13.106) is no longer sufficiently contractive to map the Bloch ball into itself, so that it does not represent a valid quantum process. However, in this range $\gamma \in]1/2, 1]$, by the multiplicative chain rule for amplitude damping channels, we have the alternative degradation $\mathcal{N} = \mathcal{T}' \circ \widetilde{\mathcal{N}}$, with $\mathcal{T}'$ an amplitude damping channel with damping factor $\gamma'_{\mathcal{T}} = 1 - \widetilde{\gamma}/(1-\widetilde{\gamma}) = 1 - (1-\gamma)/\gamma \in [0, 1]$. So, for $\gamma \in]1/2, 1]$, we have with $\mathcal{N} = \mathcal{T}' \circ \widetilde{\mathcal{N}}$ a degradation for the complementary channel $\widetilde{\mathcal{N}}$, which is here degradable, establishing that in this range the amplitude damping channel $\mathcal{N}$ is antidegradable, with therefore a zero quantum capacity $Q(\mathcal{N})$.

The final task is to obtain the quantum capacity $Q(\mathcal{N})$ in the range $\gamma \in [0, 1/2]$ where the amplitude damping channel $\mathcal{N}$ is degradable. When the channel is degradable, its quantum capacity $Q(\mathcal{N})$ is defined by the single-use maximization of Eq. (13.63) and it coincides with the channel coherent information $I_{\mathrm{co}}(\mathcal{N})$ of Eq. (13.61). So, from the coherent information $I_{\mathrm{co}}(\rho, \mathcal{N})$ of Eq. (13.104), we are led to the maximization

$$Q(\mathcal{N}) = I_{\mathrm{co}}(\mathcal{N}) = \max_{\rho \in \mathcal{L}(\mathcal{H}_2)} I_{\mathrm{co}}(\rho, \mathcal{N}) \tag{13.107}$$

$$= \max_{\rho \in \mathcal{L}(\mathcal{H}_2)} H_{\mathrm{bin}}\left(\frac{1 + \|\vec{r}\,'\|}{2}\right) - H_{\mathrm{bin}}\left(\frac{1 + \|\vec{s}\,'\|}{2}\right). \tag{13.108}$$

To accomplish the maximization of Eqs. (13.107) and (13.108), the same argument as in Sect. 13.3.4 applies. By concavity ($\cap$) of $I_{\mathrm{co}}(\rho, \mathcal{N})$ for a degradable channel $\mathcal{N}$, the inequalities of Eqs. (13.85) and (13.86) hold, and the maximum of $I_{\mathrm{co}}(\rho, \mathcal{N})$ is achieved by an input density operator ρ with a Bloch vector of the form $[0, 0, r_z]^\top$ diagonal in the computational basis. Returning to the norms of Eqs. (13.96) and (13.102) with the input $\vec{r} = [0, 0, r_z]^\top$, we obtain in place of Eq. (13.108) the one-variable maximization

$$Q(\mathcal{N}) = \max_{r_z \in [-1,1]} H_{\mathrm{bin}}\left(\frac{1 + (1-\gamma)r_z + \gamma}{2}\right) - H_{\mathrm{bin}}\left(\frac{1 + \gamma r_z + 1 - \gamma}{2}\right). \tag{13.109}$$

It can be convenient to change the parameterization via $r_z = 1 - 2\rho_{11}$ to obtain equivalently

$$Q(\mathcal{N}) = \max_{\rho_{11} \in [0,1]} H_{\mathrm{bin}}\big((1-\gamma)\rho_{11}\big) - H_{\mathrm{bin}}\big(\gamma\rho_{11}\big), \tag{13.110}$$

as also obtained in [417]. The maximization of Eq. (13.110) is then easily solved numerically.

To summarize, the quantum capacity $Q(\mathcal{N})$ of the amplitude damping channel $\mathcal{N}$ is given by

$$Q(\mathcal{N}) = \begin{cases} \max\limits_{\rho_{11} \in [0,1]} H_{\mathrm{bin}}\big((1-\gamma)\rho_{11}\big) - H_{\mathrm{bin}}\big(\gamma\rho_{11}\big), & \text{for } \gamma \in [0, 1/2], \quad (13.111) \\ 0, & \text{for } \gamma \in [1/2, 1]. \quad (13.112) \end{cases}$$

Figure 13.10 represents the quantum capacity $Q(\mathcal{N})$ of Eqs. (13.111) and (13.112) for the amplitude damping channel, along with the optimal input coordinate $\rho_{11} = \rho_{11}^{\mathrm{opt}}$ achieving the quantum capacity in Eq. (13.111). As visible in Fig. 13.10, the optimal input defined by ρ_{11}^{opt} observed in [0.4, 0.5] corresponds to a mixed state as expected, for signaling. As the noise damping factor γ goes to 0, the amplitude damping

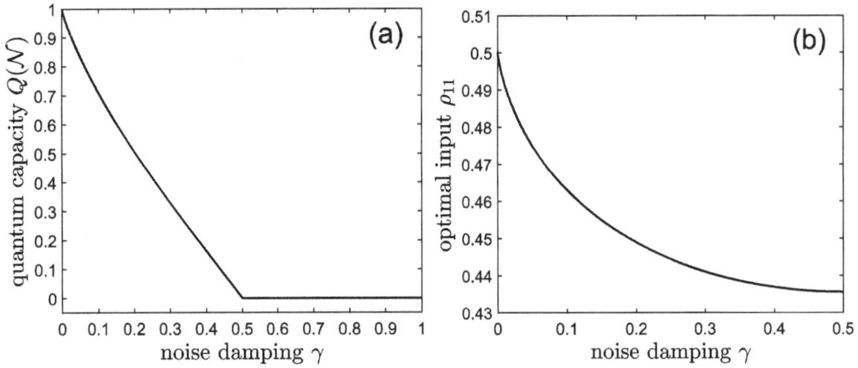

Fig. 13.10 For the qubit amplitude damping channel of Eq. (13.91), as a function of the noise damping γ, Panel (a) is the quantum capacity $Q(\mathcal{N})$ in shannons of Eqs. (13.111) and (13.112), Panel (b) is the optimal input coordinate $\rho_{11} = \rho_{11}^{\text{opt}}$ achieving the quantum capacity $Q(\mathcal{N})$ in Eq. (13.111)

channel evolves to the identity channel, its quantum capacity $Q(\mathcal{N})$ going to the overall maximum $\log(2)$ is achieved by the optimal input $\rho_{11}^{\text{opt}} = 1/2$, corresponding to the maximally mixed input state $I_2/2$.

13.3.6 *Quantum Capacity of the Completely Decohering Qubit Channel*

Beyond these important qubit channels, unital in Sect. 13.3.4 and non-unital in Sect. 13.3.5, still other qubit channels with a two-dimensional environment exist. For instance, outside the class of Sect. 13.3.4 is the completely decohering qubit channel presented in Sect. 9.8, which is the unital qubit channel $\mathcal{N}$ defined by Eqs. (9.72) and (9.73). From an input qubit with density operator $\rho = \left[\rho_{00}, \rho_{01}\,;\, \rho_{01}^*, \rho_{11}\right]$ in the standard parameterization of Eq. (2.59), this channel produces the output state $\rho' = \mathcal{N}(\rho) = \rho_{00}\,|0\rangle\langle 0| + \rho_{11}\,|1\rangle\langle 1|$ of Eq. (9.73). As we have seen in Sect. 9.8, the two (non-unitary) Kraus operators $\Lambda_0 = |0\rangle\langle 0|$ and $\Lambda_1 = |1\rangle\langle 1|$ for the channel determine a two-dimensional environment E, terminating in a final state $\rho_E' = \widetilde{\mathcal{N}}(\rho)$ given by its matrix elements of Eq. (3.120), and resulting as $\rho_E' = \widetilde{\mathcal{N}}(\rho) = \rho_{00}\,|0\rangle\langle 0| + \rho_{11}\,|1\rangle\langle 1|$. The channel output $\rho' = \mathcal{N}(\rho)$ and environment final state $\rho_E' = \widetilde{\mathcal{N}}(\rho)$ therefore coincide. Both qubit channels $\mathcal{N}$ and $\widetilde{\mathcal{N}}$ are at the same time degradable and antidegradable, with in each case the identity channel as degrading channel. The output and environment quantum entropies $S(\rho') = S(\rho_E') = H_{\text{bin}}(\rho_{00})$ coincide. The coherent information $I_{\text{co}}(\rho, \mathcal{N}) = S(\rho') - S(\rho_E') = 0$ vanishes for any input ρ, and so do both quantum capacities $Q(\mathcal{N}) = Q(\widetilde{\mathcal{N}}) = 0$.

In addition, as we have seen at the end of Sect. 13.2.9, the completely decohering qubit channel $\mathcal{N}$ has a classical capacity of $C(\mathcal{N}) = \log(2)$, representing the overall maximum achievable for a qubit channel, and obtained by the input statistical ensemble of two equiprobable pure states $\{(1/2, |0\rangle), (1/2, |1\rangle)\}$.

We have here with this completely decohering qubit channel $\mathcal{N}$, a quantum channel with maximum classical capacity $C(\mathcal{N})$ and zero quantum capacity $Q(\mathcal{N})$, a vivid illustration of the difference between the classical and quantum capacities of a quantum channel. This quantum channel $\mathcal{N}$ is able to transmit with no alteration the two basis states $|0\rangle$ and $|1\rangle$, that can be used for classical signaling, this translating into a maximum classical capacity $C(\mathcal{N})$; meanwhile, $\mathcal{N}$ is unable to transmit with no alteration any (non-degenerate) coherent superposition $\alpha_0 |0\rangle + \alpha_1 |1\rangle$ but completely cancels the coherence of the superposition, this translating into a zero quantum capacity $Q(\mathcal{N})$.

As we noted in Sect. 9.8, during the transmission of the two invariant basis states $|0\rangle$ and $|1\rangle$, the principal qubit Q does not get entangled to the environment E. There is no alteration of the transmitted state, and also no quantum entropy $S(\rho'_E) = H_{\text{bin}}(\rho_{00}) = 0$ generated into the environment, when $\rho_{00} = 1$ or 0. By contrast, during the transmission of a non-degenerate coherent superposition $|Q\rangle = \alpha_0 |0\rangle + \alpha_1 |1\rangle$, the principal qubit Q gets entangled to the environment E. The transmitted state is altered and transformed into an incoherent mixture, and a nonzero entropy $S(\rho'_E) = H_{\text{bin}}(\rho_{00}) > 0$ is generated into the environment. This is a generic mechanism taking place in noisy or decohering quantum channels: coupling by entanglement to the environment generates entropy into the environment while transmitted pure states get degraded into mixed states.

Still other qubit channels with a two-dimensional environment are treated in [433]. All the qubit channels with a two-dimensional environment are degradable or antidegradable quantum channels [431, 433], and consequently their quantum capacity is determined by the single-use formula of Eq. (13.63) and can be characterized explicitly [433]. Yet, qubit channels can also have a larger environment, up to four-dimensional, that are not known to be generally degradable or antidegradable, and whose quantum capacity today is not fully characterized.

13.3.7 Quantum Capacity of the Depolarizing Channel

The depolarizing channel is realized by the depolarizing noise model $\mathcal{N}(\cdot)$ presented in Sect. 9.6. On a qubit, this channel implements the quantum process given in Eq. (9.49) as

$$\rho' = \mathcal{N}(\rho) = (1-p)\rho + \frac{p}{3}\left(\sigma_x\rho\sigma_x^\dagger + \sigma_y\rho\sigma_y^\dagger + \sigma_z\rho\sigma_z^\dagger\right), \tag{13.113}$$

equivalent to the uniform contraction of the Bloch vector $\vec{r} \mapsto \vec{r}\,' = \alpha\vec{r}$ given in Eq. (9.55) with the contraction factor $\alpha = 1 - 4p/3$. The quantum entropy of the output state follows from Eq. (7.40) as

$$S(\rho') = H_{\text{bin}}\left(\frac{1+\alpha\|\vec{r}\,\|}{2}\right) . \tag{13.114}$$

The four Kraus operators of the qubit depolarizing channel, by the procedure of Sect. 3.2.8, determine a four-dimensional environment E. With an input density operator ρ of Bloch vector $\vec{r} = [r_x, r_y, r_z]^\top$, the environment E terminates in the final state $\rho'_E = \widetilde{\mathcal{N}}(\rho)$ given in Eq. (9.60). This density operator ρ'_E has the four eigenvalues given in Eqs. (9.61)–(9.62). From these four eigenvalues, the final quantum entropy $S(\rho'_E)$ of the environment E is obtained by Eq. (7.30). The coherent information follows as $I_{\text{co}}(\rho, \mathcal{N}) = S(\rho') - S(\rho'_E)$. We observe from Eq. (13.114) and Eqs. (9.61)–(9.62) that the two quantum entropies $S(\rho')$ and $S(\rho'_E)$, and therefore the coherent information $I_{\text{co}}(\rho, \mathcal{N})$, are determined by the norm $\|\vec{r}\,\|$ of the Bloch vector $\vec{r}$ characterizing the input density operator ρ. Consequently, the maximization of $I_{\text{co}}(\rho, \mathcal{N})$ addressed in Eq. (13.61) reduces to a one-variable maximization according to $\|\vec{r}\,\|$, so as to determine the channel coherent information $I_{\text{co}}(\mathcal{N})$. It is then found numerically that the maximum of $I_{\text{co}}(\rho, \mathcal{N})$ occurs at $\|\vec{r}\,\| = 0$ for a channel probability $p \in [0, p_c]$ with the critical value $p_c \approx 0.1893$; and for $p > p_c$ the maximum of $I_{\text{co}}(\rho, \mathcal{N})$ occurs at $\|\vec{r}\,\| = 1$ and delivers a zero channel coherent information $I_{\text{co}}(\mathcal{N}) = 0$. To summarize, the channel coherent information $I_{\text{co}}(\mathcal{N})$ of the qubit depolarizing channel can be expressed as

$$\boxed{I_{\text{co}}(\mathcal{N}) = \begin{cases} \log(2) - H(\vec{p}\,) , & \text{for } 0 \le p \le p_c \approx 0.1893 , \\ 0 , & \text{for } p_c \le p \le 1 , \end{cases}} \tag{13.115–13.116}$$

with, for the probability distribution $\vec{p} = [1-p, p/3, p/3, p/3]^\top$ of the depolarizing channel, the classical entropy $H(\vec{p}\,) = -(1-p)\log(1-p) - p\log\big(p/3\big) = H_{\text{bin}}(p) + p\log(3)$, crossing $\log(2)$ for $p = p_c \approx 0.1893$ and culminating at $2\log(2)$ for $p = 3/4$. This $I_{\text{co}}(\mathcal{N})$ for the depolarizing channel is also reported in [434] for instance.

The depolarizing channel is generally not degradable and its channel coherent information $I_{\text{co}}(\mathcal{N})$ of Eqs. (13.115) and (13.116) is strictly superadditive [12, 435]. Therefore $I_{\text{co}}(\mathcal{N})$ of Eqs. (13.115) and (13.116) does not coincide with the quantum capacity $Q(\mathcal{N})$ of Eq. (13.62), but stands only as a lower bound to it. The quantum capacity $Q(\mathcal{N})$ of the qubit depolarizing channel of Eq. (13.113) remains generally unknown today, or so to speak, we do not know that it is known. For the qubit depolarizing channel of Eq. (13.113), from an argument of quantum cloning, [15] proves that the quantum capacity $Q(\mathcal{N})$ vanishes for a depolarizing probability $p \ge 1/4 \Leftrightarrow \alpha \le 2/3$. Otherwise, a simple upper bound is known as

$$Q(\mathcal{N}) \begin{cases} \le \log(2) - H_{\text{bin}}(p) , & \text{for } 0 \le p < 1/4 , \qquad (13.117) \\ = 0 , & \text{for } 1/4 \le p \le 1 ; \qquad (13.118) \end{cases}$$

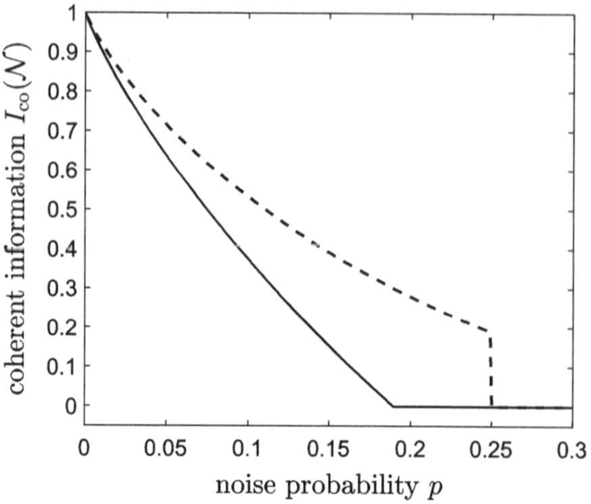

Fig. 13.11 For the qubit depolarizing channel of Eq. (13.113), as a function of the noise probability p over the channel, the channel coherent information $I_{co}(\mathcal{N})$ of Eqs. (13.115) and (13.116) (solid line) forming a lower bound to the quantum capacity $Q(\mathcal{N})$, and the upper bound of Eqs. (13.117) and (13.118) (dashed line)

and more elaborated upper bounds are given, for instance, in [436]. In addition from [25], the qubit depolarizing channel is entanglement breaking, and therefore antidegradable, for $p \geq 1/2 \Leftrightarrow \alpha \leq 1/3$; more generally, for the qudit with dimension N, the depolarizing channel, acting according to Eq. (9.52), is entanglement breaking, and thus with zero quantum capacity, for $\alpha \leq 1/(N+1)$.

Figure 13.11 shows the channel coherent information $I_{co}(\mathcal{N})$ of Eqs. (13.115) and (13.116) forming a lower bound to the quantum capacity $Q(\mathcal{N})$, along with the upper bound of Eqs. (13.117) and (13.118). The quantum capacity $Q(\mathcal{N})$ of the qubit depolarizing channel is somewhere in between, but today it is not known precisely where. Similar ignorance holds, a fortiori, for the quantum capacity of the qudit depolarizing channel in arbitrary dimension N.

13.3.8 Quantum Capacity of the Qudit Erasure Channel

The qudit erasure channel has been considered in Sect. 13.2.8 for its classical capacity. To work out its quantum capacity now, it is convenient to describe this quantum channel as connecting an N-dimensional input Hilbert space $\mathcal{H}^A \equiv \mathcal{H}_N$ to an $(N+1)$-dimensional output Hilbert space $\mathcal{H}^B \equiv \mathcal{H}_{N+1}$. The qudit erasure channel takes as input a density operator $\rho \in \mathcal{L}(\mathcal{H}_N)$, and delivers an output density operator $\rho' = \mathcal{N}(\rho) \in \mathcal{L}(\mathcal{H}_{N+1})$ given as in Eq. (13.34) by

$$\rho' = \mathcal{N}(\rho) = (1-p)\rho + p\,|\times\rangle\langle\times|\ . \tag{13.119}$$

The standard orthonormal basis $\{|1\rangle, |2\rangle, \ldots, |N\rangle\}$ is introduced for the N-dimensional space $\mathcal{H}_N$, and the $(N+1)$-dimensional space $\mathcal{H}_{N+1} = \mathcal{H}_N \oplus \{|\times\rangle\} \supset \mathcal{H}_N$ is referred to the orthonormal basis $\{|1\rangle, |2\rangle, \ldots, |N\rangle, |\times\rangle\}$. The quantum process of Eq. (13.119) can equivalently be described by the set of $N+1$ Kraus operators from $\mathcal{H}_N$ into $\mathcal{H}_{N+1}$ given by $\Lambda_0 = \sqrt{1-p}\sum_{n=1}^{N} |n_B\rangle\langle n_A|$ and $\Lambda_n = \sqrt{p}\,|\times\rangle\langle n_A|$, for $n = 1$ to N. The output density operator $\rho' = \mathcal{N}(\rho) \in \mathcal{L}(\mathcal{H}_{N+1})$ can be described by the $(N+1)\times(N+1)$ block matrix

$$\rho' = \mathcal{N}(\rho) = \begin{bmatrix} \begin{bmatrix} & & \\ & (1-p)\rho & \\ & & \end{bmatrix} & \begin{matrix} 0 \\ \vdots \\ 0 \end{matrix} \\ 0 \cdots 0 & p \end{bmatrix} . \tag{13.120}$$

By the canonical procedure of Sect. 3.2.8, the set of $N+1$ Kraus operators determines an $(N+1)$-dimensional environment E, that terminates in the final state $\rho'_E = \widetilde{\mathcal{N}}(\rho)$ characterized via its matrix elements of Eq. (13.52), yielding the representation by the $(N+1)\times(N+1)$ block matrix

$$\rho'_E = \widetilde{\mathcal{N}}(\rho) = \begin{bmatrix} 1-p & 0 \cdots 0 \\ \begin{matrix} 0 \\ \vdots \\ 0 \end{matrix} & \begin{bmatrix} & & \\ & p\rho & \\ & & \end{bmatrix} \end{bmatrix} . \tag{13.121}$$

The qudit erasure channel is [25, 415] degradable for an erasure probability $p \in [0, 1/2]$ and antidegradable for $p \in [1/2, 1]$. Its quantum capacity $Q(\mathcal{N})$, when it is nonzero for $p \in [0, 1/2]$, can therefore be obtained from the single-use formula of Eq. (13.63). For the maximization of the coherent information in Eq. (13.61), it turns out that in the range $p \in [0, 1/2]$, the maximum of $I_{\text{co}}(\rho, \mathcal{N}) = S(\rho') - S(\rho'_E)$ is achieved by the maximally mixed input state $\rho = \mathrm{I}_N/N$. The corresponding output state is represented in Eq. (13.120) by the $(N+1)\times(N+1)$ diagonal matrix $\rho' = \mathcal{N}\big(\mathrm{I}_N/N\big) = \mathrm{diag}\big[(1-p)/N, \ldots, (1-p)/N, p\big]$, having the quantum entropy $S(\rho') = Nh\big((1-p)/N\big) + h(p)$. Meanwhile, the corresponding final state of the environment is represented in Eq. (13.121) by the $(N+1)\times(N+1)$ diagonal matrix $\rho'_E = \widetilde{\mathcal{N}}\big(\mathrm{I}_N/N\big) = \mathrm{diag}\big[1-p, p/N, \ldots, p/N\big]$, having the quantum entropy $S(\rho'_E) = h(1-p) + Nh\big(p/N\big)$. The difference of these two entropies yields the channel coherent information $I_{\text{co}}(\mathcal{N}) = S(\rho') - S(\rho'_E) = (1-2p)\log(N)$, leading to the quantum capacity of the qudit erasure channel as

$$\boxed{Q(\mathcal{N}) = \begin{cases} (1-2p)\log(N)\,, & \text{for } \ p \in [0, 1/2]\,, \\ 0\,, & \text{for } \ p \in [1/2, 1]\,, \end{cases}} \qquad \begin{matrix} (13.122) \\ (13.123) \end{matrix}$$

also given in [25, 415]. As expected, the quantum capacity $Q(\mathcal{N})$ in Eqs. (13.122) and (13.123) is smaller than the classical capacity $C(\mathcal{N})$ in Eq. (13.35).

The complementary quantum channel to the erasure channel is again [25] an erasure channel where the erasure probability p is changed to $1-p$.

13.3.9 Superactivation and Other Quantum Enhancements

Many quantum channels of interest remain today with a quantum capacity $Q(\mathcal{N})$ that is not explicitly characterized. The channel coherent information $I_{co}(\mathcal{N})$ that determines the quantum capacity $Q(\mathcal{N})$ in Eq. (13.62), when it is non-additive (superadditive) or non-convex under convex combination of quantum channels, may induce non-intuitive uncommon behaviors of the quantum capacity.

Non-convexity produces the remarkable phenomenon of **superactivation** [437, 438], where two quantum channels $\mathcal{N}_1$ and $\mathcal{N}_2$, each with zero quantum capacity, when used together in a convex combination, realize a combined channel that may show a nonzero quantum capacity. This is established in [437] with a zero-capacity entanglement-binding channel $\mathcal{N}_2$ associated with a zero-capacity erasure channel $\mathcal{N}_1$ as in Sect. 13.3.8 at $p = 1/2$; or in [438] with a zero-capacity depolarizing channel $\mathcal{N}_1$ as in Sect. 13.3.7 with $p \in [1/2, 1]$. Each time, a combined channel with nonzero quantum capacity is reported [437, 438].

Remarkable effects also result from **superadditivity** in the channel coherent information $I_{co}(\mathcal{N})$. Quantum channels are reported in [439] with a zero coherent information $I_{co}(\mathcal{N}) = 0$, but a nonzero quantum capacity $Q(\mathcal{N}) > 0$. In [439] a channel $\mathcal{N}$ with zero coherent information $I_{co}(\mathcal{N}) = 0$, when employed in several ($L > 1$) entangled uses, realizes a composite channel $\mathcal{N}^{\otimes L}$ exhibiting a positive coherent information $I_{co}\big(\mathcal{N}^{\otimes L}\big) > 0$, and consequently a positive quantum capacity $Q(\mathcal{N}) > 0$ in the regularization of Eq. (13.62). Additionally in [439], quantum channels are shown for which an arbitrary large number L of (entangled) uses are necessary to reach a nonzero coherent information $I_{co}\big(\mathcal{N}^{\otimes L}\big) > 0$. Also, in [440], quantum channels are shown to exist that keep a strictly superadditive coherent information for an unbounded number L of uses.

Coherent superposition, which applies to quantum states, can be extended to quantum channels. Quantum states can be realized by placing different basis states in a coherent superposition. In a comparable way, quantum processes or channels can be associated in coherent superpositions of various types [441–445]. The resulting superposed quantum process can exhibit properties that are not present in the individual processes assembled in the superposition. Quantum channels coherently superposed in this fashion have been shown to exhibit enhanced capabilities for information communication. For example, [446] considers two completely depolarizing qudit channels $\mathcal{N}_1$ and $\mathcal{N}_2$, each with zero classical (and a fortiori quantum) information capacity, for instance, at $\alpha = 0$ the fully depolarizing channel of Fig. 13.4. When such two zero-capacity channels are cascaded in a standard pipeline, either in the causal order $\mathcal{N}_2 \circ \mathcal{N}_1$ or $\mathcal{N}_1 \circ \mathcal{N}_2$, a cascaded channel with zero capacity results. An arrangement can be designed [442, 447, 448] where one causal order or the other can be switched by means of a control qubit placed, respectively, in the basis state $|0\rangle$ or $|1\rangle$. Placing the control qubit in the coherent superposition $\big(|0\rangle + |1\rangle\big)/\sqrt{2}$ associates the two channels in a quantum superposition of simultaneously the two causal orders $\mathcal{N}_2 \circ \mathcal{N}_1$ and $\mathcal{N}_1 \circ \mathcal{N}_2$. The information-carrying signal then traverses the superposed channel simultaneously in the two causal orders $\mathcal{N}_2 \circ \mathcal{N}_1$ and $\mathcal{N}_1 \circ \mathcal{N}_2$, or equivalently in indefinite causal order. An unexpected outcome is that a nonzero capacity may result for the superposed channel [446]. Two quantum channels with zero capacity, when associated in a **coherent superposition of channels with indefinite causal order**, may yield a quantum channel with nonzero capacity. Partially or completely depolarizing channels and other types of quantum channels also, can benefit from coherent superposition for enhanced information capacity, classical or quantum [444, 449–451]. Beyond information communication, such coherently superposed quantum processes have been shown to enable enhanced performance in various quantum signal and information processing tasks, for quantum metrology [452–456], quantum parameter estimation [457–460] or process discrimination [461, 462]. Such coherently superposed quantum processes represent specifically quantum devices, with distinctive properties that are not today fully explored or exploited.

The presence of the asymptotic regularization of Eq. (13.62) and the uncommon behaviors reported above in this section, reinforce the idea that today, an explicit characterization of the quantum capacity $Q(\mathcal{N})$ of a quantum channel, often remains a difficult task. Even to decide whether a channel has a nonzero quantum capacity $Q(\mathcal{N})$ may be difficult in general. Moreover, the capacity in itself may not be sufficient for a complete account of the capability of a quantum channel for information communication, since, as indicated by the uncommon behaviors or non-conventional effects reported above in this Sect. 13.3.9, this may depend also on the external conditions in which the channel is operated or associated. Further investigations are still needed today in this direction [427].

13.4 Other Useful Informational Quantities

13.4.1 Private Classical Capacity

Another informational quantity that is found meaningful for communication over a quantum channel is the private classical capacity [12, 424, 463]. This can be connected with the classical wiretap channel [464]. As we have seen in Sect. 13.2.2 with Eq. (13.14), the Holevo information $\chi\big(\{p_j, \mathcal{N}(\rho_j)\}\big)$ of the output ensemble $\big\{(p_j, \mathcal{N}(\rho_j))\big\}$ represents an achievable upper bound to the input–output classical mutual information $H(X:Y)$ characterizing the quantum channel $\mathcal{N}(\cdot)$. There also exists a complementary quantum channel $\widetilde{\mathcal{N}}(\cdot)$ from the input into the environment E, and the Holevo information $\chi\big(\{p_j, \widetilde{\mathcal{N}}(\rho_j)\}\big)$ in a similar way quantifies the classical information from the input sent (leaked) into the environment. One then defines the **private classical information** of the quantum channel $\mathcal{N}(\cdot)$ transmitting the input ensemble $\{(p_j, \rho_j)\}_{j=1}^J$, as

$$I_p\Big(\{p_j, \mathcal{N}(\rho_j)\}\Big) = \chi\Big(\{p_j, \mathcal{N}(\rho_j)\}\Big) - \chi\Big(\{p_j, \widetilde{\mathcal{N}}(\rho_j)\}\Big) . \tag{13.124}$$

A maximization can be envisaged over the input ensemble $\{(p_j, \rho_j)\}_{j=1}^{J}$, as in Eq. (13.18), to give

$$I_p^{\max}(\mathcal{N}) = \max_{\{p_j,\rho_j\}} I_p\Big(\big\{p_j, \mathcal{N}(\rho_j)\big\}\Big) , \tag{13.125}$$

as a private classical information intrinsic to the quantum channel $\mathcal{N}(\cdot)$.

As the Holevo capacity $\chi_{\max}(\mathcal{N})$ of Eq. (13.18), and the channel coherent information $I_{co}(\mathcal{N})$ of Eq. (13.61), the channel private classical information $I_p^{\max}(\mathcal{N})$ of Eq. (13.125), is generally not additive, but superadditive, satisfying $I_p^{\max}(\mathcal{N}\otimes\mathcal{N}) \geq 2I_p^{\max}(\mathcal{N})$. One is then led to define the **private classical capacity** of the quantum channel $\mathcal{N}(\cdot)$, by the regularized formula

$$\boxed{C_p(\mathcal{N}) = \lim_{L\to\infty} \frac{1}{L}\, I_p^{\max}\big(\mathcal{N}^{\otimes L}\big) .} \tag{13.126}$$

Returning to the classical capacity $C(X, Y)$ of a classical channel, although it comes from its definition of Eq. (13.6) with a prior interpretation as a measure of information in common between the input and output, it is nevertheless not directly obvious from Eq. (13.6) that $C(X, Y)$ represents the maximum achievable rate for error-free transmission of classical information over the noisy channel (with appropriate encoding at the input)—but this should be proved separately. In a comparable way, the quantity $C_p(\mathcal{N})$ of Eq. (13.126) comes with a prior interpretation as a measure of private classical information between the input and output, since with Eq. (13.124) it is defined from the maximum amount $\chi\big(\{p_j, \mathcal{N}(\rho_j)\}\big)$ of classical information sent by the input to the output, diminished by the maximum classical information $\chi\big(\{p_j, \widetilde{\mathcal{N}}(\rho_j)\}\big)$ leaked into the environment. In addition, it can be proved [12, 424, 463] that $C_p(\mathcal{N})$ of Eq. (13.126) is indeed an information capacity, as it represents the maximum achievable rate for error-free transmission of private classical information over the quantum channel $\mathcal{N}$ (with appropriate encoding at the input). "Private" here means that classical information is transmitted from the input to the output of the channel $\mathcal{N}$, while the surrounding environment E extracts no information about the transmission.

By expressing the Holevo information in terms of quantum entropies, as in Eq. (13.15), we obtain for Eq. (13.124) the alternative expression

$$I_p\Big(\big\{p_j, \mathcal{N}(\rho_j)\big\}\Big) = S(\mathcal{N}(\overline{\rho})) - \sum_j p_j\, S(\mathcal{N}(\rho_j)) - \Big[S(\widetilde{\mathcal{N}}(\overline{\rho})) - \sum_j p_j\, S(\widetilde{\mathcal{N}}(\rho_j))\Big] . \tag{13.127}$$

For a pure input signaling state ρ_j, the states $\mathcal{N}(\rho_j)$ and $\widetilde{\mathcal{N}}(\rho_j)$ are two reduced states obtained from a pure bipartite state, and therefore they share the same quantum entropy $S\big(\mathcal{N}(\rho_j)\big) = S\big(\widetilde{\mathcal{N}}(\rho_j)\big)$, as seen in Sect. 7.3.5. When all the input signaling states ρ_j are pure, we deduce

$$I_p\Big(\{p_j, \mathcal{N}(\rho_j)\}\Big) = S(\mathcal{N}(\overline{\rho})) - S(\widetilde{\mathcal{N}}(\overline{\rho})) \tag{13.128}$$

$$= I_{\text{co}}(\overline{\rho}, \mathcal{N}) , \tag{13.129}$$

expressing that the private classical information $I_p\Big(\{p_j, \mathcal{N}(\rho_j)\}\Big)$ of a pure input ensemble coincides with the coherent information $I_{\text{co}}(\overline{\rho}, \mathcal{N})$ realized by the average input state $\overline{\rho} = \sum_j p_j \rho_j$.

The optimal density operator $\overline{\rho} = \overline{\rho}_{\text{opt}}$ maximizing $I_{\text{co}}(\overline{\rho}, \mathcal{N})$ in Eq. (13.129) so as to achieve the channel coherent information $I_{\text{co}}(\mathcal{N})$ of Eq. (13.61), can always be realized by an ensemble of pure input states (e.g., the eigenstates of $\overline{\rho}_{\text{opt}}$). As a consequence of Eq. (13.129), when maximizing $I_p\Big(\{p_j, \mathcal{N}(\rho_j)\}\Big)$ over all (pure or mixed) input ensembles $\{(p_j, \rho_j)\}$, the maximum yielding the channel private information $I_p^{\max}(\mathcal{N})$ will never be smaller than the channel coherent information $I_{\text{co}}(\mathcal{N})$, that is, $I_p^{\max}(\mathcal{N}) \geq I_{\text{co}}(\mathcal{N})$ in general for any quantum channel. Implementing the regularization step in the limit $L \to \infty$, we obtain between the information capacities of any quantum channel $\mathcal{N}$, the inequalities

$$C(\mathcal{N}) \geq C_p(\mathcal{N}) \geq Q(\mathcal{N}) , \tag{13.130}$$

with the first inequality following rather directly from the difference of Holevo informations in Eq. (13.124) in relation to Eq. (13.22).

In general, for an arbitrary quantum channel, the private information capacity $C_p(\mathcal{N})$ of Eq. (13.126) is a quantity that may be difficult to explicitly characterize, especially due to the regularization with the limit $L \to \infty$ in Eq. (13.126). Yet, for degradable quantum channels, the regularization in Eq. (13.126) is not necessary, and the private classical capacity $C_p(\mathcal{N})$ of Eq. (13.126) coincides with the channel private classical information $I_p^{\max}(\mathcal{N})$ of Eq. (13.125), and in addition $C_p(\mathcal{N})$ is also equal to the quantum capacity $Q(\mathcal{N})$ of the channel [12]. On the other hand, for antidegradable quantum channels, as the quantum capacity $Q(\mathcal{N})$, the private classical capacity $C_p(\mathcal{N})$ vanishes. In addition, [465] shows that $C_p(\mathcal{N}) = Q(\mathcal{N})$ when the quantum capacity of the complementary channel $\widetilde{\mathcal{N}}$ vanishes, with $Q(\widetilde{\mathcal{N}}) = 0$; and when $C_p(\widetilde{\mathcal{N}}) = 0$, the relevant information quantities are additive and give $C_p(\mathcal{N}) = I_p^{\max}(\mathcal{N}) = I_{\text{co}}(\mathcal{N}) = Q(\mathcal{N})$.

These results on the private classical information capacity are meaningful [466] in particular to characterize the best achievable performance for quantum distribution of a cryptographic key between two distant parties, as addressed in Sect. 6.7.

Beyond classical information, private communication of quantum information is similar to plain communication of quantum information as addressed in Sect. 13.3. The communication of quantum information in the capacity-achieving error-free conditions accomplished in Sect. 13.3 is necessarily private between the sender and receiver of the quantum channel. A third party (an eavesdropper) cannot obtain some copy of the coherent quantum information received by the receiver, otherwise this would violate the no-cloning theorem [12, 424, 467].

13.4.2 Input–Output Quantum Mutual Information

One could envisage for a quantum channel an input–output quantum mutual information as $S(A:B) = S(A) + S(B) - S(A,B)$ presented in Sect. 7.5 with Eq. (7.99). A limitation is that this notion of quantum mutual information $S(A:B)$ of Sect. 7.5 cannot be applied to relate the input and output of a quantum channel. The reason for this limitation is that the input and output of a quantum channel do not stand as two distinct quantum systems that would form the two parts of a bipartite composite system similar to AB in Sect. 7.5. The input and output of a quantum channel generally are realized by the same unique quantum system, successively placed in the input and output states while traveling across the channel, like a photon across an optical fiber for instance. In this way, between the input and output of a quantum communication channel, there is no direct possibility of a quantum mutual information defined as $S(A:B)$ of Sect. 7.5. There however exists a possibility, less direct, to define an input–output quantum mutual information meaningful to a quantum channel [12, 25], as we now explain.

We consider a quantum channel $\mathcal{N}$ connecting an input A to an output B. In general, with an input space $\mathcal{H}^A$ and an input state of density operator ρ_A, it is always possible to introduce a reference system R with Hilbert space $\mathcal{H}^R \equiv \mathcal{H}^A$ and a pure bipartite state $|RA\rangle \in \mathcal{H}^R \otimes \mathcal{H}^A$ realizing a purification for the input system A, as we explained in Sect. 2.5.3. Since they are obtained by partial tracing on a pure bipartite state $|RA\rangle$, the two reduced states $\rho_A = \mathrm{tr}_R\big(|RA\rangle\langle RA|\big)$ and $\rho_R = \mathrm{tr}_A\big(|RA\rangle\langle RA|\big)$ share the same eigenvalues and same quantum entropy $S(\rho_A) = S(\rho_R)$, as seen in Sect. 7.3.5. When the input system A is sent across the quantum channel $\mathcal{N}$ while the reference system R is left untouched, their bipartite state evolves as

$$|RA\rangle\langle RA| \longmapsto (\mathcal{I}_R \otimes \mathcal{N})\big(|RA\rangle\langle RA|\big) = \rho_{RB}\,, \tag{13.131}$$

where $\mathcal{I}_R$ is the identity superoperator for the reference space. The bipartite output state ρ_{RB} can be reduced to obtain $\rho_B = \mathrm{tr}_R\big(\rho_{RB}\big) = \mathcal{N}(\rho_A)$, thanks to Eq. (A.25) and the Kraus decomposition for $\mathcal{N}$. It is then possible to introduce the informational quantity

$$I_{\mathrm{mut}}(\rho_A, \mathcal{N}) = S(\rho_R : \rho_B) = S(\rho_R) + S(\rho_B) - S(\rho_{RB}) \tag{13.132}$$

representing the quantum mutual information of Eq. (7.99) between the reference R and channel output B. As we have seen, $S(\rho_R)$ is the same as the input entropy $S(\rho_A)$. In this respect, the reference system R can be considered as a replica of the input A which remains intact during the transmission by the channel. The quantity $S(\rho_R : \rho_B)$ is a legitimate quantum mutual information, as in Sect. 7.5, between two distinct quantum systems, the intact reference R similar to input A and the channel output B. As a result, $I_{\mathrm{mut}}(\rho_A, \mathcal{N})$ of Eq. (13.132) can be interpreted as an acceptable input–output mutual information for the quantum channel $\mathcal{N}$ transmitting the input signal ρ_A.

For an alternative expression of the joint quantum entropy $S(\rho_{RB})$ in Eq. (13.132), it is possible to refer to a Stinespring isometric representation for the channel $\mathcal{N}(\cdot)$. We introduce a Stinespring isometric extension operator Λ as in Sect. 3.2.6, from $\mathcal{H}^A$ onto $\mathcal{H}^A \otimes \mathcal{H}^E$, where $\mathcal{H}^E$ is the Hilbert space for the model environment E. We obtain the Stinespring isometric evolution associated with the quantum channel as

$$|RA\rangle \longmapsto \left(\mathrm{I}^R \otimes \Lambda\right)|RA\rangle = |RBE\rangle \;, \tag{13.133}$$

where I^R is the identity operator on the reference space $\mathcal{H}^R$. We have the reduced state $\rho_{RB} = \mathrm{tr}_E\big(|RBE\rangle\langle RBE|\big)$, and the final state of the environment $\rho'_E = \widetilde{\mathcal{N}}(\rho) = \mathrm{tr}_{RB}\big(|RBE\rangle\langle RBE|\big)$. In the split $RB|E$, the two reduced states ρ_{RB} and ρ'_E are obtained by partial tracing on a pure bipartite state $|RBE\rangle$, so that they share the same quantum entropy $S(\rho_{RB}) = S(\rho'_E)$. We therefore obtain for Eq. (13.132) the alternative expression

$$I_{\mathrm{mut}}(\rho_A, \mathcal{N}) = S(\rho_R : \rho_B) = S(\rho_A) + S(\rho_B) - S(\rho'_E) \,. \tag{13.134}$$

With the input state $\rho_A = \rho$, the output state $\rho_B = \mathcal{N}(\rho)$, and the entropy exchange $S_{\mathrm{ex}}(\rho, \mathcal{N}) = S(\rho'_E)$ from Eq. (13.53), we finally obtain the **channel input–output quantum mutual information** as

$$I_{\mathrm{mut}}(\rho, \mathcal{N}) = S(\rho) + S\big(\mathcal{N}(\rho)\big) - S_{\mathrm{ex}}(\rho, \mathcal{N}) \tag{13.135}$$

$$= S(\rho) + I_{\mathrm{co}}(\rho, \mathcal{N}) \,. \tag{13.136}$$

The input–output quantum mutual information $I_{\mathrm{mut}}(\rho, \mathcal{N})$ of Eq. (13.136) is non-negative, and it vanishes for pure input states ρ, just like the entropy $S(\rho)$ and coherent information $I_{\mathrm{co}}(\rho, \mathcal{N})$ do.

As with the coherent information $I_{\mathrm{co}}(\mathcal{N})$ of Sect. 13.3.2, it is possible to define [12] an input–output quantum mutual information intrinsic to the quantum channel $\mathcal{N}$, via a maximization over the input state $\rho \in \mathcal{L}(\mathcal{H}_N)$, as

$$I_{\mathrm{mut}}(\mathcal{N}) = \max_{\rho} I_{\mathrm{mut}}(\rho, \mathcal{N}) \,. \tag{13.137}$$

Contrary to the situation of the channel coherent information $I_{\mathrm{co}}(\mathcal{N})$ in Eq. (13.61), the channel quantum mutual information $I_{\mathrm{mut}}(\mathcal{N})$ in Eq. (13.137) is additive [12]. For two distinct quantum channels $\mathcal{N}_1$ and $\mathcal{N}_2$, the tensor-product channel satisfies the additivity property $I_{\mathrm{mut}}(\mathcal{N}_1 \otimes \mathcal{N}_2) = I_{\mathrm{mut}}(\mathcal{N}_1) + I_{\mathrm{mut}}(\mathcal{N}_2)$.

It is visible with Eq. (13.136) that the input–output quantum mutual information $I_{\mathrm{mut}}(\rho, \mathcal{N})$ differs from the coherent information $I_{\mathrm{co}}(\rho, \mathcal{N})$. The quantum capacity $Q(\mathcal{N})$ is determined via a maximization of the coherent information $I_{\mathrm{co}}(\rho, \mathcal{N})$, as we explained in Sect. 13.3.3, and therefore, contrary to the classical situation, maximization of the input–output quantum mutual information $I_{\mathrm{mut}}(\rho, \mathcal{N})$ in general does not determine the quantum capacity $Q(\mathcal{N})$ of a quantum channel $\mathcal{N}$.

For example, for the qubit depolarizing channel, from the results of Sect. 13.3.7 providing a characterization of the coherent information $I_{co}(\rho, \mathcal{N})$, the maximization in Eq. (13.137) reduces to a one-variable maximization according to the norm $\|\vec{r}\,\|$ of the input Bloch vector, as for obtaining Eqs. (13.115) and (13.116). This one-variable maximization in Eq. (13.137) can be performed numerically, with the maximum occurring at $\|\vec{r}\,\| = 0$ for the maximally mixed input state $\rho = \mathrm{I}_2/2$, to yield the quantum mutual information of the qubit depolarizing channel as

$$I_{mut}(\mathcal{N}) = 2\log(2) - H(\vec{p}\,) = 2\log(2) - H_{bin}(p) - p\log(3) \tag{13.138}$$

$$= 2\log(2) + (1-p)\log(1-p) + p\log\big(p/3\big)\,. \tag{13.139}$$

This $I_{mut}(\mathcal{N})$ vanishes for a channel probability $p = 3/4$.

In general, for the tripartite state $|RBE\rangle$ produced by the channel operation in Eq. (13.133), it is possible to consider the alternative split $B|RE$, and we obtain the reduced state $\rho_{RE} = \mathrm{tr}_B\big(|RBE\rangle\,\langle RBE|\big)$ and the quantum entropies $S(\rho_{RE}) = S(\rho_B)$. We can then define

$$I_{mut}\big(\rho, \widetilde{\mathcal{N}}\big) = S(\rho_R : \rho'_E) \tag{13.140}$$

$$= S\big(\rho_R\big) + S\big(\rho'_E\big) - S\big(\rho_{RE}\big) \tag{13.141}$$

$$= S\big(\rho_A\big) + S\big(\rho'_E\big) - S\big(\rho_B\big) \tag{13.142}$$

$$= S\big(\rho\big) - I_{co}\big(\rho, \mathcal{N}\big)\,, \tag{13.143}$$

which represents the quantum mutual information for the conjugate channel $\widetilde{\mathcal{N}}$ between the input A and the environment E, also known as the **information loss** [25]. The information loss $I_{mut}\big(\rho, \widetilde{\mathcal{N}}\big)$ is non-negative and vanishes for pure input states ρ, just like $S(\rho)$ and $I_{co}(\rho, \mathcal{N})$.

For the tripartite state $|RBE\rangle$, by considering the split $R|BE$ we obtain the reduced state $\rho_{BE} = \mathrm{tr}_R\big(|RBE\rangle\,\langle RBE|\big)$ and the quantum entropies $S(\rho_{BE}) = S(\rho_R) = S(\rho_A)$. We can then define

$$S\big(\mathcal{N}(\rho) : \widetilde{\mathcal{N}}(\rho)\big) = S\big(\rho_B\big) + S\big(\rho'_E\big) - S\big(\rho_{BE}\big) \tag{13.144}$$

$$= S\big(\rho_B\big) + S\big(\rho'_E\big) - S\big(\rho_A\big) \tag{13.145}$$

$$= S\big(\mathcal{N}(\rho)\big) + S\big(\widetilde{\mathcal{N}}(\rho)\big) - S\big(\rho\big)\,, \tag{13.146}$$

which represents the quantum mutual information between the output B and the final state of the environment E, also known as the **information noise** [25]. The information noise $S\big(\mathcal{N}(\rho) : \widetilde{\mathcal{N}}(\rho)\big)$ is non-negative and does not generally vanish with pure input states ρ.

Another important significance to the input–output quantum mutual information arises for entanglement-assisted communication, as we now explain.

13.4.3 Entanglement-Assisted Capacity

Entanglement-assisted communication is exemplified by the superdense coding protocol presented in Sect. 6.5, that can serve as a guide for a generalization as follows. For entanglement-assisted communication, before transmission over some quantum channel $\mathcal{N}$, a bipartite quantum system AB prepared in the entangled pure state $|AB\rangle$ is distributed between the sender obtaining subsystem A and the receiver obtaining subsystem B. At the channel input, J classical symbols $\{s_j\}_{j=1}^{J}$ as introduced in Fig. 13.2, are encoded by means of a set of J unitary operators $\{\mathsf{U}_j\}_{j=1}^{J}$. The coding operator U_j applied to the subsystem A accessible at the input, produces the transformed bipartite state $\big(\mathsf{U}_j \otimes \mathrm{I}^B\big)\,|AB\rangle = |\Psi_j\rangle$, with I^B the identity operator for the subsystem B which is left untouched in the encoding. The subsystem A after the coding step is sent across the quantum channel $\mathcal{N}$, to produce the transformed bipartite state $\big(\mathcal{N} \otimes \mathcal{I}_B\big)\big(|\Psi_j\rangle\,\langle\Psi_j|\big) = \rho'_{AB}$, with $\mathcal{I}_B$ the identity superoperator for subsystem B. The bipartite quantum state ρ'_{AB} accessible at the channel output is then measured by the receiver, in order to recover (as efficiently as possible) the classical symbol s_j encoded by the sender at the input.

The entanglement-assisted classical capacity $C_{\mathrm{ea}}(\mathcal{N})$ can then be defined as usual as the maximum rate of error-free transmission of classical information over the quantum channel, with at disposal an unlimited amount of prior entanglement. The characterization of this capacity $C_{\mathrm{ea}}(\mathcal{N})$ was first accomplished in [468–470]. The optimal input encoding resorts to a generalization of the Pauli operators used in the superdense coding protocol of Sect. 6.5, optimization according to the output measurement can be handled by an approach as in Sect. 13.2.2 for the accessible information. The protocol consumes prior entanglement at a rate of $S(\rho_A)$ entangled pair per channel use, with $\rho_A = \mathrm{tr}_B\big(|AB\rangle\,\langle AB|\big)$. It is then established [468–470] that the **entanglement-assisted classical capacity** $C_{\mathrm{ea}}(\mathcal{N})$ of the quantum channel $\mathcal{N}$ can be characterized as

$$\boxed{C_{\mathrm{ea}}(\mathcal{N}) = I_{\mathrm{mut}}(\mathcal{N})\,.} \tag{13.147}$$

The entanglement-assisted classical capacity $C_{\mathrm{ea}}(\mathcal{N})$ is therefore given by the input–output quantum mutual information maximized over the input state as in Eq. (13.137), much like the classical capacity $C(X, Y)$ of a classical channel in Eq. (13.6). Especially, there is no need for a regularization step since the channel quantum mutual information is additive for entangled uses. The entanglement-assisted classical capacity $C_{\mathrm{ea}}(\mathcal{N})$ is a fortiori larger than its unassisted counterpart $C(\mathcal{N})$, so that we generally have $C_{\mathrm{ea}}(\mathcal{N}) \geq C(\mathcal{N}) \geq Q(\mathcal{N})$.

For example, when $\mathcal{N}$ is the noise-free identity channel realizing $\mathcal{N}(\rho) = \rho$ for every input $\rho \in \mathcal{L}(\mathcal{H}_N)$, no entropy is generated into the environment, so that $S\big(\widetilde{\mathcal{N}}(\rho)\big) = 0$, and the coherent information of Eqs. (13.58) and (13.59) reduces to $I_{\mathrm{co}}(\rho, \mathcal{N}) = S\big(\mathcal{N}(\rho)\big) = S(\rho)$. The input–output quantum mutual information in Eq. (13.136) follows as $I_{\mathrm{mut}}(\rho, \mathcal{N}) = 2S(\rho)$, which is maximized by the maximally mixed input state $\rho = \mathrm{I}_N/N = \rho_A$ to yield the entanglement-assisted classical capacity $C_{\mathrm{ea}}(\mathcal{N}) = 2\log(N)$. For a qubit channel with dimension $N = 2$, we have the capacity $C_{\mathrm{ea}}(\mathcal{N}) = 2\log(2) = 2\,\mathrm{Sh}$, which is the transmission rate observed in the superdense coding protocol of Sect. 6.5.

One may also be interested in entanglement-assisted communication of quantum information, that is, as in Sect. 13.3, error-free transmission of any coherent superposition of quantum input states. The teleportation protocol presented in Sect. 6.6 can serve as a basic resource for accomplishing entanglement-assisted communication of quantum states, between two distant parties sharing prior entanglement. Different resource arrangements are possible, with associated performance, as discussed in [12, 471, 472]. By combining teleportation and superdense coding, a proof is given in [473] that the entanglement-assisted quantum capacity $Q_{\mathrm{ea}}(\mathcal{N})$ is always one half of its classical counterpart, that is $Q_{\mathrm{ea}}(\mathcal{N}) = C_{\mathrm{ea}}(\mathcal{N})/2$.

13.5 Summary on Quantum Communication

With quantum signals as information carriers, quantum channels can be employed to communicate end-to-end classical or quantum information. The communication is commonly hindered by quantum noise over the channel. The best performance can be assessed by means of entropy-based information capacities, standardly defined as a maximum rate of reliable (error-free) information transmission that can be achieved over the channel, usually in an asymptotic limit of long and appropriately encoded and decoded sequences of signaling symbols or macrosymbols. The type of information, classical or quantum, the constraints or resources with the quantum channel, and the reliability criterion, determine several meaningful information capacities.

We briefly summarize below important informational quantities that were presented in this chapter, meaningful to communication over quantum channels, and especially relevant to the characterization of their information capacities. In addition, the chapter also carried out the explicit calculation of significant information capacities of important quantum channels. Nevertheless, today many situations remain where significant information capacities—classical and even more so quantum—of important quantum channels are not completely known or characterized. A major difficulty lies in the regularizing step often required to characterize an information capacity for a quantum channel, when maximum efficiency for communication may be obtained by arbitrarily long sequences of entangled channel uses. Entanglement here again stands out as a specifically quantum resource that may bring enhanced performance, since by contrast for communication across classical channels correlation between consecutive uses of a channel cannot improve the best efficiency [12, 397]. For quantum channels, the (uneasy) asymptotic regime handling long sequences of quantum signals may be involved both in the derivation of information capacities (which is not the case for classical channels) and in the capacity-achieving coding–decoding communication protocol (which is usually also the case with classical channels). In this respect, practical coding–decoding techniques for quantum channels, to efficiently achieve or approach the (asymptotic) best performance stipulated by the available capacities, also remain today a direction open for further research and technological developments.

As a meaningful reference, the **classical capacity of a classical channel** is

$$\boxed{C(X, Y) = \max_{\{p_j\}} H(X : Y) ,} \tag{13.148}$$

as given by Eq. (13.6). Now come different informational quantities relevant to quantum channels.

For a classical information transmission $X \mapsto Y$ by means of an input signaling ensemble $\{(p_j, \rho_j)\}_{j=1}^J$ across a quantum channel $\mathcal{N}$ producing the output states $\rho'_j = \mathcal{N}(\rho_j)$ measured by the POVM $\{\mathsf{E}_k\}$, the **accessible information** at the output according to Eq. (13.13) is

$$I_{\text{acc}}(X, Y) = \max_{\{\mathsf{E}_k\}} H(X : Y) \leq \chi\big(\{p_j, \rho'_j\}\big) , \tag{13.149}$$

with the **Holevo information**

$$\chi\big(\{p_j, \rho'_j\}\big) = S(\overline{\rho}') - \sum_{j=1}^J p_j\, S(\rho'_j) = \sum_{j=1}^J p_j\, S\big(\rho'_j \,\|\, \overline{\rho}'\big) \tag{13.150}$$

of Eqs. (13.15) and (13.16), and average output state

$$\overline{\rho}' = \sum_{j=1}^J p_j\, \rho'_j . \tag{13.151}$$

For a given input ensemble $\{(p_j, \rho_j)\}_{j=1}^J$ and a quantum channel $\mathcal{N}(\cdot)$, the Holevo information $\chi\big(\{p_j, \rho'_j\}\big)$ from Eq. (13.150) is a **maximal rate achievable** for error-free transmission of classical information, usually by inputting independent consecutive symbols collectively measured at the output over asymptotically long sequences.

The **Holevo capacity** of the quantum channel $\mathcal{N}$ from Eqs. (13.18) and (13.19) is

$$\chi_{\max}(\mathcal{N}) = \max_{\{p_j, \rho_j\}} \chi\big(\{p_j, \rho'_j\}\big) = \min_{\rho_2} \max_{\rho_1} S\big(\mathcal{N}(\rho_1) \| \mathcal{N}(\rho_2)\big) , \tag{13.152}$$

with the maximum of $\chi\big(\{p_j, \rho'_j\}\big)$ that can be achieved by $J \in \big[N, N^2\big]$ pure input states $\rho_j = |\psi_j\rangle\langle\psi_j|$ with $|\psi_j\rangle \in \mathcal{H}_N$. The Holevo capacity $\chi_{\max}(\mathcal{N})$ stands as the maximal achievable rate for error-free transmission of classical information over the quantum channel $\mathcal{N}(\cdot)$ when using product states or independent consecutive uses of the channel. In general $\chi_{\max}(\mathcal{N})$ is **superadditive**, with $\chi_{\max}\big(\mathcal{N} \otimes \mathcal{N}\big) \geq 2\chi_{\max}(\mathcal{N})$. Entangled consecutive uses can bring improvement for channels where $\chi_{\max}(\mathcal{N})$ is strictly superadditive.

The **classical capacity** $C(\mathcal{N})$ of a quantum channel $\mathcal{N}$ is generally given by the regularized Holevo capacity according to Eq. (13.22) as

$$\boxed{C(\mathcal{N}) = \lim_{L\to\infty} \frac{1}{L}\, \chi_{\max}\big(\mathcal{N}^{\otimes L}\big)\,.} \tag{13.153}$$

By the **HSW theorem** of Sect. 13.2.5 the classical information capacity $C(\mathcal{N})$ from Eq. (13.153) is the maximum achievable rate for error-free transmission of classical information over the quantum channel $\mathcal{N}$.

An additive Holevo capacity $\chi_{\max}(\mathcal{N})$ yielding $C(\mathcal{N}) = \chi_{\max}(\mathcal{N})$ is exhibited by all entanglement-breaking channels, by the qudit depolarizing channel of Sect. 13.2.7, by the qudit erasure channel of Sect. 13.2.8, by all unital qubit channels in Sect. 13.2.9. By contrast, the (non-unital) qubit thermal noise channel of Sect. 13.2.10 in general is not known to exhibit an additive Holevo capacity $\chi_{\max}(\mathcal{N})$, hindering the characterization of its classical capacity $C(\mathcal{N})$ when it requires asymptotic regularization at length $L \to \infty$ in Eq. (13.153).

With the input state $\rho \in \mathcal{L}(\mathcal{H}_N)$ transmitted by the quantum channel $\mathcal{N}$, the **entropy exchange** $S_{\text{ex}}(\rho, \mathcal{N})$ is the quantum entropy of the final state $\rho'_E = \tilde{\mathcal{N}}(\rho)$ of the environment, given in Eq. (13.53) as

$$S_{\text{ex}}(\rho, \mathcal{N}) = S(\rho'_E) = S\big(\tilde{\mathcal{N}}(\rho)\big)\,. \tag{13.154}$$

The **coherent information** from Eqs. (13.58) and (13.59) is

$$I_{\text{co}}(\rho, \mathcal{N}) = S(\rho') - S(\rho'_E) = S\big(\mathcal{N}(\rho)\big) - S\big(\tilde{\mathcal{N}}(\rho)\big)\,. \tag{13.155}$$

The coherent information $I_{\text{co}}(\mathcal{N})$ intrinsic to the quantum channel $\mathcal{N}$ from Eq. (13.61) is

$$I_{\text{co}}(\mathcal{N}) = \max_{\rho} I_{\text{co}}\big(\rho, \mathcal{N}\big)\,. \tag{13.156}$$

In general $I_{\text{co}}(\mathcal{N})$ is **superadditive**, with $I_{\text{co}}\big(\mathcal{N} \otimes \mathcal{N}\big) \geq 2 I_{\text{co}}(\mathcal{N})$.

The **quantum capacity** $Q(\mathcal{N})$ of the quantum channel $\mathcal{N}$ is generally given in Eq. (13.62) by the regularized channel coherent information as

$$\boxed{Q(\mathcal{N}) = \lim_{L\to\infty} \frac{1}{L}\, I_{\text{co}}\big(\mathcal{N}^{\otimes L}\big)\,.} \tag{13.157}$$

By the **LSD theorem** of Sect. 13.3.3 the quantum information capacity $Q(\mathcal{N})$ from Eq. (13.157) is the maximum achievable rate for error-free transmission of quantum information over the quantum channel $\mathcal{N}$.

An additive channel coherent information $I_{\text{co}}(\mathcal{N})$ yielding $Q(\mathcal{N}) = I_{\text{co}}(\mathcal{N})$ is exhibited by all degradable quantum channels, such as the two-dimensional channels of Sect. 13.3.4 and the amplitude damping channel of Sect. 13.3.5, while the depolarizing channel of Sect. 13.3.7 is generally not degradable. Non-additive $I_{co}(\mathcal{N})$ usually requires asymptotic regularization at length $L \to \infty$ in Eq. (13.157), and remains today difficult to handle to characterize the quantum capacity $Q(\mathcal{N})$.

From Sect. 13.4.1, the private classical information of the quantum channel $\mathcal{N}$ transmitting the input ensemble $\{(p_j, \rho_j)\}_{j=1}^J$ is given in Eq. (13.124) as

$$I_p\Big(\{p_j, \mathcal{N}(\rho_j)\}\Big) = \chi\Big(\{p_j, \mathcal{N}(\rho_j)\}\Big) - \chi\Big(\{p_j, \widetilde{\mathcal{N}}(\rho_j)\}\Big) . \tag{13.158}$$

The private classical information intrinsic to the quantum channel $\mathcal{N}$ is given in Eq. (13.125) as

$$I_p^{\max}(\mathcal{N}) = \max_{\{p_j, \rho_j\}} I_p\Big(\{p_j, \mathcal{N}(\rho_j)\}\Big) . \tag{13.159}$$

The **private classical capacity** of the quantum channel $\mathcal{N}$ is given by the regularized channel private information in Eq. (13.126) as

$$\boxed{C_p(\mathcal{N}) = \lim_{L\to\infty} \frac{1}{L} I_p^{\max}\left(\mathcal{N}^{\otimes L}\right) .} \tag{13.160}$$

From Sect. 13.4.2, the **input–output quantum mutual information** with the input state $\rho \in \mathcal{L}(\mathcal{H}_N)$ transmitted by the quantum channel $\mathcal{N}$ is

$$I_{\text{mut}}\big(\rho, \mathcal{N}\big) = S(\rho) + I_{co}(\rho, \mathcal{N}) . \tag{13.161}$$

The quantum mutual information $I_{\text{mut}}(\mathcal{N})$ intrinsic to the quantum channel $\mathcal{N}$ is

$$I_{\text{mut}}(\mathcal{N}) = \max_{\rho} I_{\text{mut}}\big(\rho, \mathcal{N}\big) , \tag{13.162}$$

which also gives in Eq. (13.147) the **entanglement-assisted classical capacity** $C_{\text{ea}}(\mathcal{N})$ of the quantum channel $\mathcal{N}$ as

$$\boxed{C_{\text{ea}}(\mathcal{N}) = I_{\text{mut}}(\mathcal{N}) ,} \tag{13.163}$$

and the **entanglement-assisted quantum capacity** $Q_{\text{ea}}(\mathcal{N}) = C_{\text{ea}}(\mathcal{N})/2$ from [473].

Chapter 14
Quantum Coding and Error Correction

As we have seen in Chap. 13, over a noisy quantum channel error-free communication is feasible in principle by transmitting asymptotically long sequences of quantum signals properly encoded and decoded. This possibility essentially relies on highly probable average or typical behaviors asymptotically arising from the law of large numbers—a fundamental tenet of statistical information theory, classical and quantum alike. Such asymptotic regime is however difficult to handle in practice. Alternatively, this statistical asymptotic condition can be circumvented by deterministic schemes for error correction acting on individual or finite blocks of quantum states. In the present chapter essentially we will show the possibility and explain the basic mechanisms, of quantum coding with error-correction capabilities, in order to protect quantum states or signals against degradations or errors, wherever they may occur, for communication, processing or for fault-tolerant computation.

14.1 Quantum Error-Correcting Code

Classical signals supporting classical information may be redundantly encoded in some appropriate forms so as to protect the information they carry against noise or errors [474–476]. In a comparable way, it is desirable to have access to such protection for quantum information-carrying signals. Quantum information-carrying signals are constrained by the no-cloning theorem of Sect. 3.1.3, generally precluding direct duplication of unknown quantum states. Quantum coding seeking to encode quantum information in some redundant way so as to obtain protection against errors will have to cope with this quantum no-cloning constraint. In addition, to acquire information about an error that may have occurred on an information-carrying quantum state, quantum error correction will have to cope with a destructive or disturbing character of quantum measurement, which in general collapses the state and may destroy encoded information. We shall show in this chapter the feasibility of quantum coding for error correction, while coping with these quantum constraints (no

F. Chapeau-Blondeau, *Quantum Information and Quantum Signal Processing*,
https://doi.org/10.1007/978-3-032-27499-1_14

cloning, destructive measurement), to protect quantum states against quantum noise and errors. Surveys specialized on this topic can be found, for instance, in [477–480].

A useful approach for protecting quantum information by means of error-correcting codes employs qubits as information carriers and arranges an encoding of each of the two basis states of the qubit according to

$$\begin{cases} |0\rangle \longrightarrow |0_\mathrm{L}\rangle \ , & (14.1) \\ |1\rangle \longrightarrow |1_\mathrm{L}\rangle \ . & (14.2) \end{cases}$$

The two-qubit basis states are encoded as two logical states $|0_\mathrm{L}\rangle$ and $|1_\mathrm{L}\rangle$, usually selected as two orthogonal states of a block of K physical qubits, with a size $K > 1$ to introduce redundancy for more robustness in the representation.

The encoding of an arbitrary input qubit state $|\psi\rangle = \alpha_0\,|0\rangle + \alpha_1\,|1\rangle$ is performed as

$$|\psi\rangle = \alpha_0\,|0\rangle + \alpha_1\,|1\rangle \longrightarrow \alpha_0\,|0_\mathrm{L}\rangle + \alpha_1\,|1_\mathrm{L}\rangle \equiv |\psi_\mathrm{L}\rangle \ . \tag{14.3}$$

We are dealing in this chapter with *quantum* coding for error correction, capable of protecting any superposed quantum states like $|\psi\rangle$ above, while by contrast *classical* coding would rather protect the basis or signaling states like $|0\rangle$ and $|1\rangle$ alone. When classical coding protects the two signaling states of an information-carrying binary variable, quantum coding seeks to protect the continuously infinite number of feasible coherent states like $|\psi\rangle$ of Eq. (14.3) for a qubit. We note that the generic K-qubit coding state $|\psi_\mathrm{L}\rangle$ of Eq. (14.3) is generally an entangled state of the K coding qubits. As a way of protecting the information supported by the coherent superposed state $|\psi\rangle$ of a logical qubit, the information is distributed across a block of K entangled physical qubits.

We envisaged in Sect. 13.1 a repetition code to protect classical information. A similar coding approach may be exploited to protect quantum information, as we now examine.

14.2 Correcting a Quantum Bit Flip

14.2.1 A Quantum Repetition Code

For the qubit, a simple error-correcting code is the repetition code which encodes every logical qubit into a block of three physical qubits. The two basis states of the qubit are encoded as

$$\begin{cases} |0\rangle \longrightarrow |000\rangle \equiv |0_\mathrm{L}\rangle \ , & (14.4) \\ |1\rangle \longrightarrow |111\rangle \equiv |1_\mathrm{L}\rangle \ . & (14.5) \end{cases}$$

This three-qubit encoding can be implemented by the quantum circuit of Fig. 14.1, exploiting two auxiliary qubits initialized in state $|0\rangle$ and two Cnot gates.

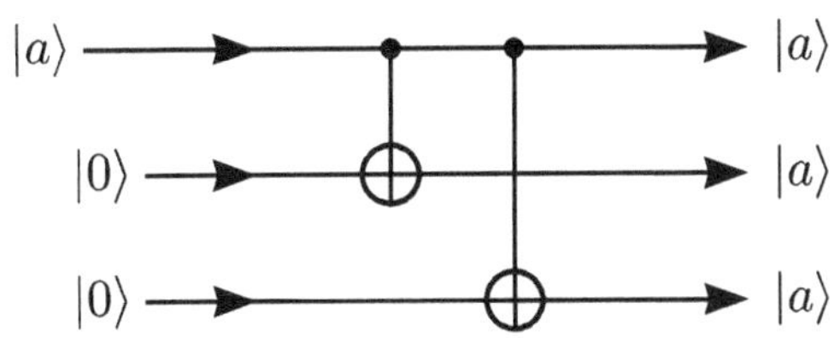

Fig. 14.1 Quantum circuit implementing the three-qubit repetition encoding of Eqs. (14.4)–(14.5). With $a = 0$ or 1 at the input, the circuit delivers the output quantum state $|a\rangle \otimes |a\rangle \otimes |a\rangle = |aaa\rangle$

The encoding of an arbitrary input superposition $|\psi\rangle = \alpha_0 |0\rangle + \alpha_1 |1\rangle$ is realized as

$$|\psi\rangle = \alpha_0 |0\rangle + \alpha_1 |1\rangle \longrightarrow |\psi_\mathrm{L}\rangle = \alpha_0 |0_\mathrm{L}\rangle + \alpha_1 |1_\mathrm{L}\rangle = \alpha_0 |000\rangle + \alpha_1 |111\rangle \ . \tag{14.6}$$

By linearity in the circuit of Fig. 14.1, the encoding $|\psi\rangle \to |\psi_\mathrm{L}\rangle$ is accomplished by applying the quantum signal $|\psi\rangle$ to the upper input marked $|a\rangle$. We note that the repetition coding, while it operates according to Eq. (14.6) for any arbitrary input state $|\psi\rangle$, does not conflict with the no-cloning theorem of Sect. 3.1.3, since it does not generally produce the (prohibited) state $|\psi\rangle \otimes |\psi\rangle \otimes |\psi\rangle$ for all $|\psi\rangle$. By this repetition code, every input quantum signal $|\psi\rangle$ is protected against a bit-flip error occurring on a single qubit of the three-qubit coding state. This can be understood from the standard notion of minimum distance d of a code [474–476], which is the smallest Hamming distance between any two codewords, and which in principle allows $\lfloor (d-1)/2 \rfloor$ errors to be corrected. The two quantum coding states of Eqs. (14.4)–(14.5) set a minimum distance of $d = 3$ for the three-qubit repetition code, allowing to detect and correct one qubit flipped in error, as we are going to see. To this purpose, it is convenient to represent a bit-flip error on a qubit in the following way. For instance, when a bit flip occurs on the second qubit, the three-qubit coding state gets altered as

$$|\psi_\mathrm{L}\rangle = \alpha_0 |000\rangle + \alpha_1 |111\rangle \longmapsto \mathsf{X}_2 |\psi_\mathrm{L}\rangle = \alpha_0 |010\rangle + \alpha_1 |101\rangle \ , \tag{14.7}$$

where we use the operator X_2 to indicate that the bit-flip Pauli operator $\mathsf{X} \equiv \sigma_x$ acts on the second qubit. It is equivalently $\mathsf{X}_2 = \mathsf{Id} \otimes \mathsf{X} \otimes \mathsf{Id}$, with here Id the identity operator on the qubit.

In the eight-dimensional space $\mathcal{H}_2^{\otimes 3}$ of the three coding qubits, the encoding of Eq. (14.6) defines a two-dimensional subspace $\overline{\mathcal{H}}_0 \equiv \mathrm{span}\big(|000\rangle\,, |111\rangle\big)$, which contains any valid coding state $|\psi_\mathrm{L}\rangle$. A bit-flip error occurring on the second qubit of $|\psi_\mathrm{L}\rangle$ as in Eq. (14.7) takes the coding state $|\psi_\mathrm{L}\rangle$ to the subspace $\overline{\mathcal{H}}_2 \equiv \mathrm{span}\big(|010\rangle\,, |101\rangle\big)$. An error on the first qubit takes the state $|\psi_\mathrm{L}\rangle$ to the subspace $\overline{\mathcal{H}}_1 \equiv \mathrm{span}\big(|100\rangle\,, |011\rangle\big)$, while an error on the third qubit takes the state $|\psi_\mathrm{L}\rangle$ to the subspace $\overline{\mathcal{H}}_3 \equiv \mathrm{span}\big(|001\rangle\,, |110\rangle\big)$. In this way, every single bit-flip error that may occur sends the code subspace $\overline{\mathcal{H}}_0$ into an orthogonal subspace, either $\overline{\mathcal{H}}_1$, $\overline{\mathcal{H}}_2$ or $\overline{\mathcal{H}}_3$ depending on the error, from where the error can be univocally identified and then corrected, as we shall see. This is a general feature that can serve as a guideline, in broader conditions, to devise an error-correcting code (a code subspace) to protect against a given set of errors: ensure that the coding states when corrupted by an error are sent into orthogonal subspaces where they remain perfectly distinguishable [481].

14.2.2 Error Detection by Projective Measurement

With the three-qubit repetition code of Eq. (14.6), to detect (and then correct) a single bit-flip error, a collective measurement can be performed of the three-qubit coding state by means of the four projectors onto the four mutually orthogonal subspaces $(\overline{\mathcal{H}}_0, \overline{\mathcal{H}}_1, \overline{\mathcal{H}}_2, \overline{\mathcal{H}}_3)$, reading, respectively:

$$\begin{cases} \mathsf{M}_0 = |000\rangle\langle 000| + |111\rangle\langle 111| \,, & (14.8) \\ \mathsf{M}_1 = |100\rangle\langle 100| + |011\rangle\langle 011| \,, & (14.9) \\ \mathsf{M}_2 = |010\rangle\langle 010| + |101\rangle\langle 101| \,, & (14.10) \\ \mathsf{M}_3 = |001\rangle\langle 001| + |110\rangle\langle 110| \,. & (14.11) \end{cases}$$

These four projectors together decompose the identity $\mathsf{Id}^{\otimes 3}$ of the three-qubit space $\mathcal{H}_2^{\otimes 3}$ as $\sum_{m=0}^{3} \mathsf{M}_m^\dagger \mathsf{M}_m = \mathsf{Id}^{\otimes 3}$ and constitute a licit generalized measurement as introduced in Sect. 4.2. When the errors are modeled as a bit-flip quantum noise with flip probability p as in Sect. 9.3 and Eq. (9.19) acting independently on each qubit, upon measuring the three-qubit coding state with the measurement $\{\mathsf{M}_m\}_{m=0}^{3}$, one of four results can be obtained as follows.

- The measurement operator M_0 operates when no error took place on the three coding qubits, this occurring with probability $(1-p)^3$, or alternatively when all three coding qubits were corrupted by an independent bit-flip error, this occurring with probability p^3.
- M_1 operates when only one bit-flip error took place, on the first qubit, this occurring with probability $p(1-p)^2$, or alternatively when two errors took place on qubits 2 and 3, this occurring with probability $(1-p)p^2$.
- M_2 operates when only one bit-flip error took place, on the second qubit, this occurring with probability $p(1-p)^2$, or alternatively when two errors took place on qubits 1 and 3, this occurring with probability $(1-p)p^2$.
- M_3 operates when only one bit-flip error took place, on the third qubit, this occurring with probability $p(1-p)^2$, or alternatively when two errors took place on qubits 1 and 2, this occurring with probability $(1-p)p^2$.

The configurations with zero or one error are more probable than the configurations with two or three errors. In the limit of small $p \to 0$, a configuration with one error has a probability of order p or $O(p)$, in comparison the configurations with two errors of probability $O(p^2)$ and with three errors of probability $O(p^3)$ become negligible. Accordingly, error correction, to begin with, concentrates on correcting one error. In the less probable circumstance where more than one error occur, error correction devised for one error will generally fail to restore the original uncorrupted coding state.

The measurement result that is obtained is known as the **error syndrome**; it indicates which one of the four operators M_m acted in the measurement. Under the assumption of zero or one bit-flip error, the error syndrome perfectly identifies the error configuration. In addition, an important feature is that the measurement by $\{\mathsf{M}_m\}_{m=0}^3$ leaves unaltered the measured quantum state. When (and only then) the measured quantum state lies entirely in one of the subspaces $(\overline{\mathcal{H}}_0, \overline{\mathcal{H}}_1, \overline{\mathcal{H}}_2, \overline{\mathcal{H}}_3)$, it is left invariant in the measurement by $\{\mathsf{M}_m\}_{m=0}^3$, and this is the situation expected with the repetition code of Eq. (14.6) possibly corrupted by a single bit-flip error. Correlatively, the measurement reveals nothing about the information encoded in the constitution of the input superposition $|\psi\rangle = \alpha_0 |0\rangle + \alpha_1 |1\rangle$. This is another general feature expected from a quantum error-correction protocol: to extract information about the errors that may have occurred in order to correct them, but to reveal nothing about the encoded quantum state so as to leave it unaltered [9].

The information in the error syndrome can be exploited for error correction. This is accomplished by implementing a bit-flip Pauli operator X on the qubit designated by the syndrome as being in error, so as to restore its correct state as it was initially before corruption by the bit-flip noise, yet without the need to know which was this initial state of the qubit in order to restore it.

14.2.3 Error-Detection Quantum Circuit

For the three-qubit repetition code from Eq. (14.6), it is possible to envisage the error-correction protocol of a single bit-flip error, without explicitly resorting to a direct measurement of the three-qubit coding state. As an alternative, a single bit-flip error can be detected by performing parity checks on the coding qubits, as we now explain.

A bit-flip mismatch between two coding qubits can be detected by a parity check that consists in testing whether these two coding qubits have the same state (the case of even parity) or not (the case of odd parity), as done by the quantum circuit of Fig. 14.2. The circuit in Fig. 14.2 uses two Cnot gates, and it outputs an auxiliary qubit placed in state $|0\rangle$ or $|1\rangle$ according to the even or odd parity of the two coding qubits q_1 and q_2 acting as Cnot controls. The result of the parity check is obtained by measuring the auxiliary qubit in its computational basis $\{|0\rangle, |1\rangle\}$, to collect the binary syndrome denoted syn in Fig. 14.2, and this is achieved without revealing or disturbing the state of the two coding qubits.

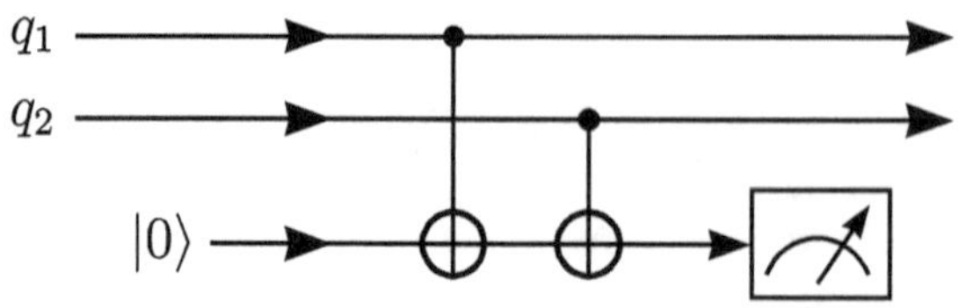

$\|q_1q_2\rangle$	syn
$\|00\rangle$	0
$\|01\rangle$	1
$\|10\rangle$	1
$\|11\rangle$	0

Fig. 14.2 Elementary parity-check cell using two Cnot gates to detect a bit-flip mismatch between two coding qubits q_1 and q_2. The auxiliary qubit initialized in state $|0\rangle$ is measured in the basis $\{|0\rangle, |1\rangle\}$ to deliver the binary syndrome denoted syn in the table describing the transformation for the two coding qubits. The parity check is achieved without altering the two coding qubits

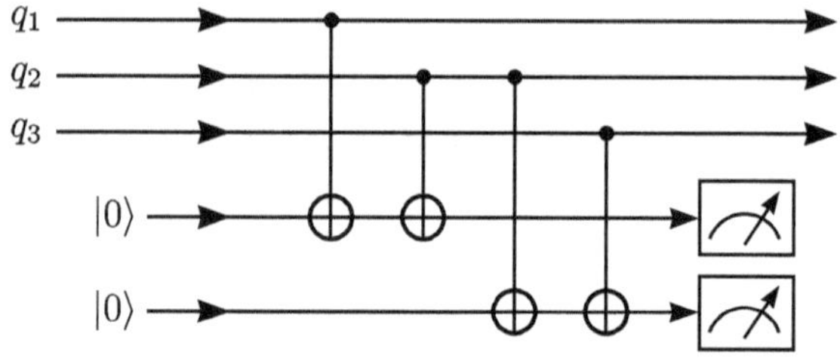

Fig. 14.3 Detection circuit for a single bit-flip error in the three-qubit repetition code from Eq. (14.6): The error-detection circuit combines two elementary parity-check cells as in Fig. 14.2, to perform two pairwise parity checks on the three input coding qubits q_1, q_2, q_3, without altering their state. Measuring the two auxiliary qubits at the output delivers the error syndrome which can be subsequently exploited for error correction on the coding qubits

We note that if an initial phase-flip error by the Pauli operator $\mathsf{Z} \equiv \sigma_z$ as $|0\rangle \mapsto \mathsf{Z}|0\rangle = |0\rangle$ and $|1\rangle \mapsto \mathsf{Z}|1\rangle = -|1\rangle$ is present on the coding qubits at the input of the elementary bit-flip detection cell of Fig. 14.2, the parity-check operation is not affected, and the coding qubits are output in the same state as at the input. In this respect, the circuit of Fig. 14.2 detects a bit-flip mismatch between the two coding qubits and is insensitive to a phase-flip error.

Two elementary parity-check cells as in Fig. 14.2 can be combined according to Fig. 14.3, to form a circuit to detect a single bit-flip error in the three-qubit repetition code from Eq. (14.6). The two auxiliary qubits output in Fig. 14.3 are then measured in the computational basis, to deliver an error syndrome identifying the coding qubit in error.

In the error-detection circuit of Fig. 14.3, the three coding qubits q_1, q_2, q_3 control four Cnot gates acting on two auxiliary qubits initialized in state $|0\rangle$, in order to perform two pairwise parity checks on the three coding qubits. On its $3 + 2$ input qubits, the circuit of Fig. 14.3 realizes the input–output transformations

$$|\psi_L\rangle\,|00\rangle = \big(\alpha_0\,|000\rangle + \alpha_1\,|111\rangle\big)\,|00\rangle \longmapsto \big(\alpha_0\,|000\rangle + \alpha_1\,|111\rangle\big)\,|00\rangle = |\psi_L\rangle\,|00\rangle \tag{14.12}$$

$$\big(\mathsf{X}_1\,|\psi_L\rangle\big)\,|00\rangle = \big(\alpha_0\,|100\rangle + \alpha_1\,|011\rangle\big)\,|00\rangle \longmapsto \big(\alpha_0\,|100\rangle + \alpha_1\,|011\rangle\big)\,|10\rangle = \big(\mathsf{X}_1\,|\psi_L\rangle\big)\,|10\rangle \tag{14.13}$$

$$\big(\mathsf{X}_2\,|\psi_L\rangle\big)\,|00\rangle = \big(\alpha_0\,|010\rangle + \alpha_1\,|101\rangle\big)\,|00\rangle \longmapsto \big(\alpha_0\,|010\rangle + \alpha_1\,|101\rangle\big)\,|11\rangle = \big(\mathsf{X}_2\,|\psi_L\rangle\big)\,|11\rangle \tag{14.14}$$

$$\big(\mathsf{X}_3\,|\psi_L\rangle\big)\,|00\rangle = \big(\alpha_0\,|001\rangle + \alpha_1\,|110\rangle\big)\,|00\rangle \longmapsto \big(\alpha_0\,|001\rangle + \alpha_1\,|110\rangle\big)\,|01\rangle = \big(\mathsf{X}_3\,|\psi_L\rangle\big)\,|01\rangle \tag{14.15}$$

successively in case of no error, or a bit-flip error on qubit q_1, q_2, or q_3. The four transformations of Eqs. (14.12)–(14.15) define the unitary action of the circuit of Fig. 14.3 over the full space $\mathcal{H}_2^{\otimes 3}$ of the three coding qubits or over the larger space $\mathcal{H}_2^{\otimes 3} \otimes \mathrm{span}\big(|00\rangle\big)$ of the $3+2$ qubits.

The two auxiliary qubits at the output in Fig. 14.3 are measured separately in the computational basis $\big\{|0\rangle\,,|1\rangle\big\}$, to deliver an error syndrome under the form of a two-digit binary code indicating the occurrence of zero or one bit-flip error and its location on qubit q_1, q_2, or q_3 of the coding state. The error-detection circuit capable of detecting and locating a bit-flip error in the three-qubit coding state does not reveal anything about the encoded information, and it leaves unaltered the three-qubit coding state, as visible on the four outputs from Eqs. (14.12)–(14.15). The error syndrome can subsequently be used to classically trigger a bit-flip Pauli operator X applied to the corrupted coding qubit for its correction.

We noted that the elementary parity-check cell which detects a bit-flip mismatch in Fig. 14.2 is insensitive to a phase-flip error that may affect the two qubits q_1 or q_2. Accordingly, for the three-qubit repetition code from Eq. (14.6), if in addition, before or after the bit-flip error by X, a phase-flip error by Z had occurred, such Z error would remain unseen in the error-detection protocol of a bit-flip error X by the circuit of Fig. 14.3. A phase-flip error Z occurring somewhere on one qubit of the three-qubit coding state flips its phase, for instance, according to $|\psi_L\rangle = \alpha_0\,|000\rangle + \alpha_1\,|111\rangle \mapsto \alpha_0\,|000\rangle - \alpha_1\,|111\rangle$ if the Z error occurs alone. It is easily verified that the parity checks taking place in Eqs. (14.12)–(14.15) to detect a bit-flip error are not sensitive and not affected by a phase-flip error (a change of sign) that would occur somewhere in the three-qubit coding state (or even two or three phase-flip errors). The parity checks in Eqs. (14.12)–(14.15) operate in the same way and the detection of the bit-flip error proceeds unaffected and leaves in place the phase-flip error. So a phase-flip error, if it may occur, must be corrected separately from a bit-flip error, as we shall soon examine in Sect. 14.3.

The error-detection circuit of Fig. 14.3 can also be analyzed as a materialization of an auxiliary system intended to realize the generalized measurement by $\{\mathsf{M}_m\}_{m=0}^{3}$ of Eqs. (14.8)–(14.11), and exemplifying the procedure of Sect. 4.2.2 for performing a generalized measurement. The two auxiliary qubits in Fig. 14.3 form the auxiliary system A of the procedure of Sect. 4.2.2. The four transformations of Eqs. (14.12)–(14.15) can be used to define the unitary action of the circuit of Fig. 14.3 on the computational basis of the full space $\mathcal{H}_2^{\otimes 3}$ of the three coding qubits, and this enables us to verify that this action can equivalently be expressed, for any three-qubit coding state $|\Psi_{\mathrm{cod}}\rangle = |q_1q_2q_3\rangle \in \mathcal{H}_2^{\otimes 3}$, as the unitary transformation

$$|\Psi_{\rm cod}\rangle \otimes |00\rangle \overset{\mathsf{U}}{\longmapsto} \mathsf{U}\,|\Psi_{\rm cod}\rangle \otimes |00\rangle = \sum_{m=0}^{3} \mathsf{M}_m\,|\Psi_{\rm cod}\rangle \otimes |m\rangle \ . \tag{14.16}$$

Equation (14.16) is an instance of Eq. (4.21), with the two auxiliary qubits initialized in the state $|00\rangle = |0\rangle \otimes |0\rangle \equiv |e_0\rangle$, and the unitary operator U from $\mathcal{H}_2^{\otimes 3} \otimes \mathcal{H}_2^{\otimes 2}$ onto $\mathcal{H}_2^{\otimes 3} \otimes \mathcal{H}_2^{\otimes 2}$ describing the action of the error-detection circuit in Fig. 14.3 and Eqs. (14.12)–(14.15), bringing into play the four measurement operators $\{\mathsf{M}_m\}_{m=0}^{3}$ of Eqs. (14.8)–(14.11). Each projection $\mathsf{M}_m\,|\Psi_{\rm cod}\rangle$ is triggered when measuring the two auxiliary qubits in the computational-basis state $|m\rangle$. This demonstrates an equivalence in this respect between the generalized measurement by $\{\mathsf{M}_m\}_{m=0}^{3}$ of Eqs. (14.8)–(14.11), and the error-detection circuit of Fig. 14.3.

14.2.4 Error Detection by Observables

A third equivalent approach is possible to envisage the correction of a single bit-flip error in the three-qubit repetition code of Eq. (14.6). From the three coding qubits in state $|q_1q_2q_3\rangle$, the error syndrome can equivalently be obtained by measuring the two observables $\mathsf{Z}_1\mathsf{Z}_2$ and $\mathsf{Z}_2\mathsf{Z}_3$, standing for $\mathsf{Z} \otimes \mathsf{Z} \otimes \mathsf{Id}$ and $\mathsf{Id} \otimes \mathsf{Z} \otimes \mathsf{Z}$, with the Pauli operator $\mathsf{Z} \equiv \sigma_z = |0\rangle\langle 0| - |1\rangle\langle 1|$ of Eq. (A.38) in the Appendix. This can be seen more explicitly as follows. Equivalently, the two observables are

$$\mathsf{Z}_1\mathsf{Z}_2 = \Big(|00\rangle\langle 00| + |11\rangle\langle 11| - |01\rangle\langle 01| - |10\rangle\langle 10|\Big) \otimes \mathsf{Id}\ , \tag{14.17}$$

$$\mathsf{Z}_2\mathsf{Z}_3 = \mathsf{Id} \otimes \Big(|00\rangle\langle 00| + |11\rangle\langle 11| - |01\rangle\langle 01| - |10\rangle\langle 10|\Big)\ . \tag{14.18}$$

Equations (14.17)–(14.18) show the two observables $\mathsf{Z}_1\mathsf{Z}_2$ and $\mathsf{Z}_2\mathsf{Z}_3$ in diagonal form. These two observables $\mathsf{Z}_1\mathsf{Z}_2$ and $\mathsf{Z}_2\mathsf{Z}_3$ have eigenvalues ± 1, they commute and can be measured one after the other in whatever order.

Measuring the observable $\mathsf{Z}_1\mathsf{Z}_2$ is equivalent to performing a projective measurement onto the mutually orthogonal $+1$ and -1 eigensubspaces of this observable, subjecting the three-qubit coding state to the two projectors

$$\Pi_{12} = \big(|00\rangle\langle 00| + |11\rangle\langle 11|\big) \otimes \mathsf{Id}\ , \tag{14.19}$$

$$\Pi_{12}^{\perp} = \big(|10\rangle\langle 10| + |01\rangle\langle 01|\big) \otimes \mathsf{Id}\ , \tag{14.20}$$

to determine the two measurement results, respectively, $Z_1Z_2 = +1$ or $Z_1Z_2 = -1$. By using the diagonal or eigendecomposition of the observable $\mathsf{Z}_1\mathsf{Z}_2$ with eigenvalues ± 1, we also have

$$\Pi_{12} = \frac{1}{2}\left(\mathsf{Id}^{\otimes 3} + \mathsf{Z}_1\mathsf{Z}_2\right), \tag{14.21}$$

$$\Pi_{12}^{\perp} = \frac{1}{2}\left(\mathsf{Id}^{\otimes 3} - \mathsf{Z}_1\mathsf{Z}_2\right). \tag{14.22}$$

When Π_{12} acts, the corresponding measurement result $Z_1Z_2 = +1$ indicates that the qubits 1 and 2 have the same (yet undisclosed) value, while $Z_1Z_2 = -1$ when $\Pi_{12}^{\perp}$ acts indicates that their value is different.

In a similar way, measuring the observable $\mathsf{Z}_2\mathsf{Z}_3$ is equivalent to performing a projective measurement onto the mutually orthogonal $+1$ and -1 eigensubspaces of this observable, subjecting the three-qubit coding state to the two projectors:

$$\Pi_{23} = \mathsf{Id} \otimes \left(|00\rangle\langle 00| + |11\rangle\langle 11|\right), \tag{14.23}$$

$$\Pi_{23}^{\perp} = \mathsf{Id} \otimes \left(|10\rangle\langle 10| + |01\rangle\langle 01|\right), \tag{14.24}$$

to determine the two measurement results, respectively, $Z_2Z_3 = +1$ or $Z_2Z_3 = -1$. Again, by using the diagonal or eigendecomposition of the observable $\mathsf{Z}_2\mathsf{Z}_3$ with eigenvalues ± 1, we also have

$$\Pi_{23} = \frac{1}{2}\left(\mathsf{Id}^{\otimes 3} + \mathsf{Z}_2\mathsf{Z}_3\right), \tag{14.25}$$

$$\Pi_{23}^{\perp} = \frac{1}{2}\left(\mathsf{Id}^{\otimes 3} - \mathsf{Z}_2\mathsf{Z}_3\right). \tag{14.26}$$

When Π_{23} acts, the corresponding measurement result $Z_2Z_3 = +1$ indicates that the qubits 2 and 3 have the same (yet undisclosed) value, while $Z_2Z_3 = -1$ when $\Pi_{23}^{\perp}$ acts indicates that their value is different.

Together the two measurement results Z_1Z_2 and Z_2Z_3 deliver an error syndrome under the form of a two-digit binary code identifying the error configuration with zero or one flipped qubit among three. For example, with no error, the intact coding state $|q_1q_2q_3\rangle = \alpha_0|000\rangle + \alpha_1|111\rangle = |\psi_L\rangle$ gives the probability $\langle q_1q_2q_3|\Pi_{12}|q_1q_2q_3\rangle = |\alpha_0|^2 + |\alpha_1|^2 = 1$ of measuring $Z_1Z_2 = +1$, and the probability $\langle q_1q_2q_3|\Pi_{23}|q_1q_2q_3\rangle = |\alpha_0|^2 + |\alpha_1|^2 = 1$ of measuring $Z_2Z_3 = +1$, with a post-measurement state remaining $|q_1q_2q_3\rangle = |\psi_L\rangle$. With the first qubit flipped in error, $|q_1q_2q_3\rangle = \alpha_0|100\rangle + \alpha_1|011\rangle = \mathsf{X}_1|\psi_L\rangle$ gives the probability $\langle q_1q_2q_3|\Pi_{12}^{\perp}|q_1q_2q_3\rangle = |\alpha_0|^2 + |\alpha_1|^2 = 1$ of measuring $Z_1Z_2 = -1$, and the probability $\langle q_1q_2q_3|\Pi_{23}|q_1q_2q_3\rangle = |\alpha_0|^2 + |\alpha_1|^2 = 1$ of measuring $Z_2Z_3 = +1$, with also a post-measurement state that remains $|q_1q_2q_3\rangle = \mathsf{X}_1|\psi_L\rangle$ and that can be corrected for the bit-flip error to restore $|\psi_L\rangle$. More concisely, these behaviors reflect the property that the code subspace $\overline{\mathcal{H}}_0$ and the three single-error subspaces $\left(\overline{\mathcal{H}}_1, \overline{\mathcal{H}}_2, \overline{\mathcal{H}}_3\right)$, are eigenspaces of the observables $\mathsf{Z}_1\mathsf{Z}_2$ and $\mathsf{Z}_2\mathsf{Z}_3$, associated with the eigenvalue $+1$ for $\overline{\mathcal{H}}_0$ and -1 for $\left(\overline{\mathcal{H}}_1, \overline{\mathcal{H}}_2, \overline{\mathcal{H}}_3\right)$.

Now we can see that the operation of the error-detection circuit of Fig. 14.3 can be interpreted as performing the measurement of the two observables $\mathsf{Z}_1\mathsf{Z}_2$ and $\mathsf{Z}_2\mathsf{Z}_3$. The syndrome extraction circuit in charge of the measurement of the observable $\mathsf{Z}_1\mathsf{Z}_2$ involves the upper auxiliary qubit in Fig. 14.3 and it realizes the unitary transformation:

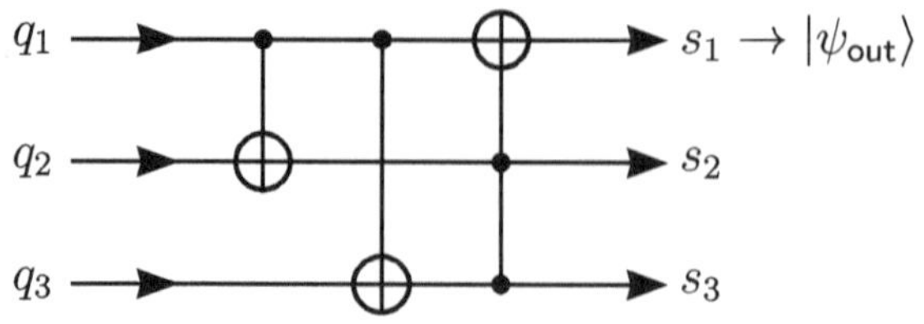

Fig. 14.4 Quantum circuit using two Cnot gates and a Toffoli gate to perform a majority decoding so as to detect and correct a bit-flip error in the three-qubit repetition code from Eq. (14.6). The qubit state $|\psi_{\rm out}\rangle$ is the result of the majority decoding

$$|q_1q_2q_3\rangle\,|0\rangle \longmapsto \frac{1}{2}\big(\mathsf{Id}^{\otimes 3} + \mathsf{Z}_1\mathsf{Z}_2\big)\,|q_1q_2q_3\rangle\,|0\rangle + \frac{1}{2}\big(\mathsf{Id}^{\otimes 3} - \mathsf{Z}_1\mathsf{Z}_2\big)\,|q_1q_2q_3\rangle\,|1\rangle\ . \tag{14.27}$$

At the same time, the syndrome extraction circuit in charge of the measurement of the observable $\mathsf{Z}_2\mathsf{Z}_3$ involves the lower auxiliary qubit in Fig. 14.3 and it realizes the unitary transformation:

$$|q_1q_2q_3\rangle\,|0\rangle \longmapsto \frac{1}{2}\big(\mathsf{Id}^{\otimes 3} + \mathsf{Z}_2\mathsf{Z}_3\big)\,|q_1q_2q_3\rangle\,|0\rangle + \frac{1}{2}\big(\mathsf{Id}^{\otimes 3} - \mathsf{Z}_2\mathsf{Z}_3\big)\,|q_1q_2q_3\rangle\,|1\rangle\ . \tag{14.28}$$

It can be seen that the pair of transformations of Eqs. (14.27)–(14.28) has the same action, on the computational basis of the space $\mathcal{H}_2^{\otimes 3}$ of the three coding qubits, as the four transformations of Eqs. (14.12)–(14.15) defining the operation of the error-detection circuit of Fig. 14.3, and both are therefore equivalent in this respect. Upon measuring the two auxiliary qubits of Fig. 14.3 in the computational basis $\big\{|0\rangle\,,\,|1\rangle\big\}$, the transformations of Eqs. (14.27)–(14.28) implement the two pairs of projectors of Eqs. (14.21)–(14.22) and Eqs. (14.25)–(14.26) acting in the measurement of the two observables $\mathsf{Z}_1\mathsf{Z}_2$ and $\mathsf{Z}_2\mathsf{Z}_3$. The measurement result on the two auxiliary qubits extracts the error syndrome, under the form of two classical binary digits $Z_1Z_2 = \pm 1$ and $Z_2Z_3 = \pm 1$, identifying the error configuration. The three-qubit coding state, either from $\overline{\mathcal{H}}_0$ with no error, or from $\big(\overline{\mathcal{H}}_1, \overline{\mathcal{H}}_2, \overline{\mathcal{H}}_3\big)$ with one bit-flip error, as an eigenstate of the measured observables, is left invariant (and undisclosed), and is amenable to its correction.

14.2.5 Majority-Decoding Quantum Circuit

Instead of a quantum circuit as in Fig. 14.3 performing error detection to drive an error-correction step, it is possible to obtain an equivalent protection against a single bit-flip error, by means of a quantum circuit performing majority decoding on the three-qubit repetition code from Eq. (14.6), and achieving error detection and correction in the same step [478]. Such a quantum circuit is represented in Fig. 14.4, and it uses two Cnot gates and a Toffoli gate introduced in Sect. 5.2.

Fig. 14.5 Unitary transformation of the computational-basis states of the three input qubits $q_1q_2q_3$ into the three output qubits $s_1s_2s_3$ realized by the majority-decoding circuit of Fig. 14.4. Each of the four groups of two consecutive lines corresponds to a configuration with either no error or a bit-flip error on qubit q_1, q_2, or q_3 of the three-qubit repetition code from Eq. (14.6)

q_1	q_2	q_3	s_1	s_2	s_3
0	0	0	0	0	0
1	1	1	1	0	0
1	0	0	0	1	1
0	1	1	1	1	1
0	1	0	0	1	0
1	0	1	1	1	0
0	0	1	0	0	1
1	1	0	1	0	1

The majority-decoding circuit of Fig. 14.4, on the computational-basis states supported by the three input qubits $q_1q_2q_3$, implements the input–output unitary transformation described in Fig. 14.5.

When it receives a three-qubit input state $|q_1q_2q_3\rangle$ formed by the three-qubit coding state of Eq. (14.6) corrupted by at most one bit-flip error, the circuit of Fig. 14.4, by linearity in the mapping of Fig. 14.5, delivers a three-qubit output state $|s_1s_2s_3\rangle$ determined by

$$|q_1q_2q_3\rangle = \alpha_0\,|000\rangle + \alpha_1\,|111\rangle \longmapsto |s_1s_2s_3\rangle = \big(\alpha_0\,|0\rangle + \alpha_1\,|1\rangle\big)\,|00\rangle\ , \tag{14.29}$$

$$|q_1q_2q_3\rangle = \alpha_0\,|100\rangle + \alpha_1\,|011\rangle \longmapsto |s_1s_2s_3\rangle = \big(\alpha_0\,|0\rangle + \alpha_1\,|1\rangle\big)\,|11\rangle\ , \tag{14.30}$$

$$|q_1q_2q_3\rangle = \alpha_0\,|010\rangle + \alpha_1\,|101\rangle \longmapsto |s_1s_2s_3\rangle = \big(\alpha_0\,|0\rangle + \alpha_1\,|1\rangle\big)\,|10\rangle\ , \tag{14.31}$$

$$|q_1q_2q_3\rangle = \alpha_0\,|001\rangle + \alpha_1\,|110\rangle \longmapsto |s_1s_2s_3\rangle = \big(\alpha_0\,|0\rangle + \alpha_1\,|1\rangle\big)\,|01\rangle\ , \tag{14.32}$$

successively in case of no error, or a bit-flip error on qubit q_1, q_2, or q_3. In the transformations of Eqs. (14.29)–(14.32), we observe that the three-qubit output state is separable as $|s_1s_2s_3\rangle = |s_1\rangle\,|s_2s_3\rangle$. The first qubit in state $|s_1\rangle$ restores the correct(ed) input qubit state $|s_1\rangle = \alpha_0\,|0\rangle + \alpha_1\,|1\rangle = |\psi_{\rm out}\rangle$ resulting from the majority decoding on a three-qubit input state $|q_1q_2q_3\rangle$ containing at most one bit-flip error. The two other output qubits in state $|s_2s_3\rangle$ encode the error configuration, among four, that may have occurred.

No explicit quantum measurement need to be realized with the majority-decoding circuit of Fig. 14.4 in order to restore the correct qubit state $|\psi_{\rm out}\rangle = |\psi\rangle = \alpha_0\,|0\rangle + \alpha_1\,|1\rangle$ at the output. By contrast, the error-detection circuit of Fig. 14.3 requires an explicit quantum measurement so as to identify the error and drive the error-correction step. However, as we shall see later in Sect. 14.5, in more general error-correction protocols, an explicit measurement is required so as to project an arbitrary error into a small set of correctable errors such as a bit-flip error here.

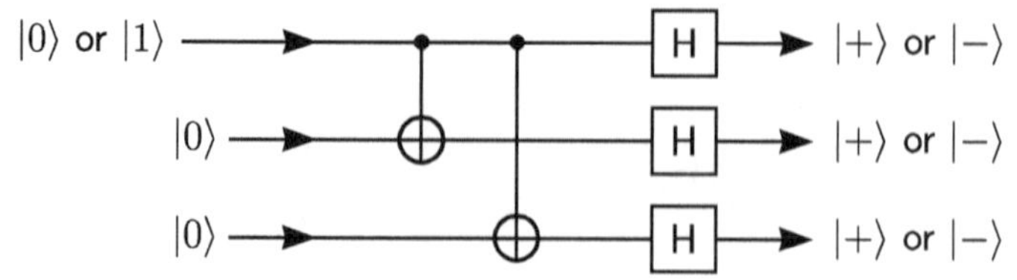

Fig. 14.6 The quantum circuit uses two auxiliary qubits initialized in state $|0\rangle$ and three Hadamard gates H to realize the three-qubit repetition encoding according to Eqs. (14.33)–(14.34)

14.3 Correcting a Quantum Phase Flip

A comparable approach as in Sect. 14.2 can be exploited to protect a qubit against a single phase-flip error occurring according to the model of phase-flip noise of Sect. 9.4 and Eq. (9.29). The phase-flip noise randomly applies to the qubit the Pauli operator $\mathsf{Z} \equiv \sigma_z$ with probability p. The Pauli operator Z flips the states $|+\rangle$ and $|-\rangle$ as $\mathsf{Z}\,|+\rangle = |-\rangle$ and $\mathsf{Z}\,|-\rangle = |+\rangle$. A three-qubit repetition code encoding in the basis $\{|+\rangle\,,|-\rangle\}$ can therefore protect against phase-flip errors, much like encoding in the basis $\{|0\rangle\,,|1\rangle\}$ can protect against bit-flip errors as seen in Sect. 14.2. Accordingly, to protect against a phase-flip error, the two basis states of the qubit can be encoded as

$$\begin{cases} |0\rangle \longrightarrow |+++\rangle \equiv |0_\mathrm{L}\rangle\,, & (14.33)\\ |1\rangle \longrightarrow |---\rangle \equiv |1_\mathrm{L}\rangle\,, & (14.34)\end{cases}$$

where $|+++\rangle \equiv |+\rangle \otimes |+\rangle \otimes |+\rangle$ and $|---\rangle \equiv |-\rangle \otimes |-\rangle \otimes |-\rangle$. The encoding of an arbitrary input superposition $|\psi\rangle = \alpha_0\,|0\rangle + \alpha_1\,|1\rangle$ is performed as

$$|\psi\rangle = \alpha_0\,|0\rangle + \alpha_1\,|1\rangle \longrightarrow |\psi_\mathrm{L}\rangle = \alpha_0\,|0_\mathrm{L}\rangle + \alpha_1\,|1_\mathrm{L}\rangle = \alpha_0\,|+++\rangle + \alpha_1\,|---\rangle\,. \tag{14.35}$$

The three-qubit encoding of Eqs. (14.33)–(14.34) can be realized from the quantum circuit of Fig. 14.1 encoding in the basis $\{|0\rangle\,,|1\rangle\}$, supplemented by three Hadamard gates H at the output to change the coding basis from $\{|0\rangle\,,|1\rangle\}$ to $\{|+\rangle = \mathsf{H}\,|0\rangle\,,|-\rangle = \mathsf{H}\,|1\rangle\}$, as shown in Fig. 14.6. By linearity in the circuit of Fig. 14.6, the encoding of the generic superposition of Eq. (14.35) is obtained by applying the qubit state $|\psi\rangle$ to the upper input.

To detect a phase-flip error on a qubit of the three-qubit coding state from Eq. (14.35), one can envisage to first apply a Hadamard transformation $\mathsf{H} = \mathsf{H}^{-1}$ on each of the three coding qubits, so as to convert the representation from the basis $\{|+\rangle\,,|-\rangle\}$ into the basis $\{|0\rangle\,,|1\rangle\}$. By this preprocessing, a phase-flip error is converted into a bit-flip error, and subsequently detection and correction of such a bit-flip error can be undertaken as in Sect. 14.2. After the error-correction step, a second Hadamard transformation on the three qubits restores the initial encoding of Eq. (14.35) in the basis $\{|+\rangle\,,|-\rangle\}$.

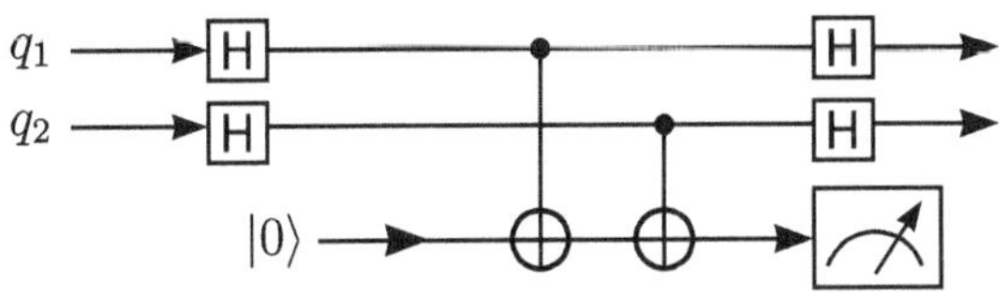

$\lvert q_1 q_2\rangle$	syn
$\lvert ++\rangle$	0
$\lvert +-\rangle$	1
$\lvert -+\rangle$	1
$\lvert --\rangle$	0

Fig. 14.7 Elementary cell to detect a phase-flip mismatch between two coding qubits q_1 and q_2. The input layer of two Hadamard gates H changes the representation basis from $\{\lvert +\rangle, \lvert -\rangle\}$ to $\{\lvert 0\rangle, \lvert 1\rangle\}$, followed by a parity check for a bit-flip mismatch by the circuit of Fig. 14.2, and finally an output layer of two Hadamard gates H restores the initial state of the two coding qubits in the basis $\{\lvert +\rangle, \lvert -\rangle\}$. An auxiliary qubit initialized in state $\lvert 0\rangle$ is measured in the basis $\{\lvert 0\rangle, \lvert 1\rangle\}$ to deliver the binary syndrome denoted syn in the table describing in the Hadamard basis $\{\lvert +\rangle, \lvert -\rangle\}$ the transformation for the two coding qubits

The four measurement operators $\{\mathsf{M}_m\}_{m=0}^{3}$ of Eqs. (14.8)–(14.11) applied, after the first Hadamard transformation, to the three-qubit coding state encoding in the basis $\{\lvert 0\rangle, \lvert 1\rangle\}$ so as to detect a bit-flip error may equivalently be replaced by the four (transformed) measurement operators $\{\mathsf{H}^{\otimes 3}\mathsf{M}_m\mathsf{H}^{\otimes 3}\}_{m=0}^{3}$ directly applied to the three-qubit coding state encoding in the basis $\{\lvert +\rangle, \lvert -\rangle\}$ so as to detect a phase-flip error.

As an alternative to measuring the coding qubits themselves, a quantum circuit can be conceived to detect a phase-flip error for its correction. A phase-flip mismatch between two coding qubits q_1 and q_2 can be detected by preceding with a Hadamard gate H on q_1 and q_2 the elementary bit-flip detection cell of Fig. 14.2. This provides the elementary phase-flip detection cell depicted in Fig. 14.7, where a phase-flip mismatch between q_1 and q_2 is detected as a bit-flip mismatch after the input layer of two Hadamard gates H. On the output in Fig. 14.7, another layer of two Hadamard gates H restores the representation in the initial basis $\{\lvert +\rangle, \lvert -\rangle\}$. The result of the phase-flip detection is delivered by the binary syndrome denoted syn obtained by measuring in the computational basis $\{\lvert 0\rangle, \lvert 1\rangle\}$ the auxiliary qubit in Fig. 14.7, and the detection is achieved without altering the two coding qubits q_1 and q_2.

We note that if an initial bit-flip error as $\lvert +\rangle \mapsto \mathsf{X}\lvert +\rangle = \lvert +\rangle$ and $\lvert -\rangle \mapsto \mathsf{X}\lvert -\rangle = -\lvert -\rangle$ is present on the coding qubits at the input of the elementary phase-flip detection cell of Fig. 14.7, the detection is not affected, and the coding qubits are output in the same state as at the input. In this respect, the circuit of Fig. 14.7 detects a phase-flip mismatch between the two coding qubits and is insensitive to a bit-flip error.

For the three-qubit repetition code from Eq. (14.35), a circuit to detect a single phase-flip error is obtained in Fig. 14.8, by combining two elementary phase-flip detection cells of Fig. 14.7. The circuit for a phase-flip error in Fig. 14.8 reproduces the circuit for a bit-flip error in Fig. 14.3 yet inserted between an input and an output layer of Hadamard gates on the three coding qubits q_1, q_2, q_3, converting the phase-flip error into a bit-flip error.

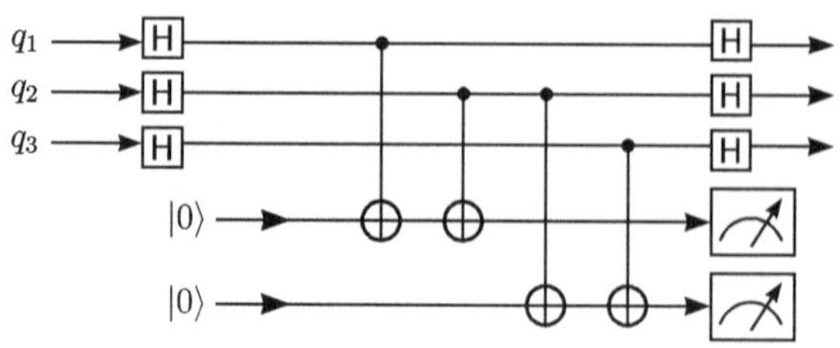

Fig. 14.8 Detection circuit for a single phase-flip error in the three-qubit repetition code from Eq. (14.35): The error-detection circuit uses a layer of three Hadamard gates H on the three input coding qubits q_1, q_2, q_3, so as to convert a phase-flip error into a bit-flip error prior to its detection as in Fig. 14.3. Measuring the two auxiliary qubits at the output delivers the error syndrome which can be subsequently exploited for error correction on the coding qubits. Finally, an output layer of three Hadamard gates H restores the state of the three coding qubits

For example, after the three-qubit coding step $|\psi\rangle \to |\psi_\mathrm{L}\rangle = \alpha_0 |+++\rangle + \alpha_1 |---\rangle$ of Eq. (14.35), a phase-flip error occurring on the first qubit q_1 of the three-qubit coding state produces the corrupted state $\mathsf{Z}_1 |\psi_\mathrm{L}\rangle = \alpha_0 |-++\rangle + \alpha_1 |+--\rangle$, which, for driving the parity-check layer of the four Cnot gates, is transformed into $\mathsf{H}^{\otimes 3}\mathsf{Z}_1 |\psi_\mathrm{L}\rangle = \alpha_0 |100\rangle + \alpha_1 |011\rangle$ in Fig. 14.8. In this way, the phase-flip error is converted into a bit-flip error, and it can be detected and corrected accordingly as in Sect. 14.2. In the circuit of Fig. 14.8, after the input Hadamard layer $\mathsf{H}^{\otimes 3}$, the layer of the four Cnot gates to perform parity checks implements the transformations:

$$\left(\mathsf{H}^{\otimes 3} |\psi_\mathrm{L}\rangle\right) |00\rangle = \left(\alpha_0 |000\rangle + \alpha_1 |111\rangle\right) |00\rangle \longmapsto \left(\mathsf{H}^{\otimes 3} |\psi_\mathrm{L}\rangle\right) |00\rangle \tag{14.36}$$

$$\left(\mathsf{H}^{\otimes 3}\mathsf{Z}_1 |\psi_\mathrm{L}\rangle\right) |00\rangle = \left(\alpha_0 |100\rangle + \alpha_1 |011\rangle\right) |00\rangle \longmapsto \left(\mathsf{H}^{\otimes 3}\mathsf{Z}_1 |\psi_\mathrm{L}\rangle\right) |10\rangle \tag{14.37}$$

$$\left(\mathsf{H}^{\otimes 3}\mathsf{Z}_2 |\psi_\mathrm{L}\rangle\right) |00\rangle = \left(\alpha_0 |010\rangle + \alpha_1 |101\rangle\right) |00\rangle \longmapsto \left(\mathsf{H}^{\otimes 3}\mathsf{Z}_2 |\psi_\mathrm{L}\rangle\right) |11\rangle \tag{14.38}$$

$$\left(\mathsf{H}^{\otimes 3}\mathsf{Z}_3 |\psi_\mathrm{L}\rangle\right) |00\rangle = \left(\alpha_0 |001\rangle + \alpha_1 |110\rangle\right) |00\rangle \longmapsto \left(\mathsf{H}^{\otimes 3}\mathsf{Z}_3 |\psi_\mathrm{L}\rangle\right) |01\rangle \tag{14.39}$$

successively in case of no error, or a phase-flip error Z on qubit q_1, q_2, or q_3 of the three-qubit coding state. The two auxiliary qubits measured at the output deliver a 2-bit error syndrome identifying the qubit of the three-qubit coding state that suffered a phase-flip error (if any). The error detection leaves unaltered the three-qubit coding state, as visible on the four outputs from Eqs. (14.36)–(14.39), which is then returned into its original state encoding in the basis $\{|+\rangle, |-\rangle\}$ by an output layer of three Hadamard gates H in Fig. 14.8. The error syndrome can subsequently be used to classically trigger a phase-flip Pauli operator Z applied to the corrupted coding qubit for its correction. Equivalently, the detected error can be corrected as a bit-flip error by X before the final Hadamard transformation. In addition, the circuit of Fig. 14.8

detecting a phase-flip error is insensitive to a bit-flip error that may be present on the coding qubits, and that is left in place.

We have seen in Sect. 14.2.4 that for detecting a bit-flip error, the error syndrome can alternatively be obtained by measuring the two observables $\mathsf{Z}_1\mathsf{Z}_2$ and $\mathsf{Z}_2\mathsf{Z}_3$ on the three-qubit coding state encoding in the basis $\{|0\rangle, |1\rangle\}$. Accordingly, for detecting a phase-flip error on the three-qubit coding state encoding in the (transformed) basis $\{|+\rangle, |-\rangle\}$, a possibility is to measure the two (transformed) observables $\mathsf{H}^{\otimes 3}\mathsf{Z}_1\mathsf{Z}_2\mathsf{H}^{\otimes 3}$ and $\mathsf{H}^{\otimes 3}\mathsf{Z}_2\mathsf{Z}_3\mathsf{H}^{\otimes 3}$. As before, $\mathsf{H}^{\otimes 3}\mathsf{Z}_1\mathsf{Z}_2\mathsf{H}^{\otimes 3}$ is the same as the three-qubit observable $\mathsf{H}^{\otimes 3}(\mathsf{Z}\otimes\mathsf{Z}\otimes\mathsf{Id})\mathsf{H}^{\otimes 3}$. And since $\mathsf{HZH}=\mathsf{X}$, the error syndrome for detecting a phase-flip error can be obtained by measuring the two (commuting) observables $\mathsf{X}_1\mathsf{X}_2$ and $\mathsf{X}_2\mathsf{X}_3$ directly on the three-qubit coding state encoding in the basis $\{|+\rangle, |-\rangle\}$, with

$$\mathsf{X}_1\mathsf{X}_2 = \Big(|{+}{+}\rangle\langle{+}{+}| + |{-}{-}\rangle\langle{-}{-}| - |{+}{-}\rangle\langle{+}{-}| - |{-}{+}\rangle\langle{-}{+}|\Big)\otimes\mathsf{Id}\,, \tag{14.40}$$

$$\mathsf{X}_2\mathsf{X}_3 = \mathsf{Id}\otimes\Big(|{+}{+}\rangle\langle{+}{+}| + |{-}{-}\rangle\langle{-}{-}| - |{+}{-}\rangle\langle{+}{-}| - |{-}{+}\rangle\langle{-}{+}|\Big)\,. \tag{14.41}$$

When measuring the observable $\mathsf{X}_1\mathsf{X}_2$, the result X_1X_2 is an eigenvalue ± 1. The result $X_1X_2=+1$ indicates that the qubits 1 and 2 have the same (yet undisclosed) phase, while $X_1X_2=-1$ indicates that their phase is different. In a similar way, when measuring the observable $\mathsf{X}_2\mathsf{X}_3$, the result $X_2X_3=+1$ indicates that the qubits 2 and 3 have the same phase, while $X_2X_3=-1$ indicates that their phase is different.

As in Sect. 14.2.4, the operation of the error-detection circuit of Fig. 14.8 can be interpreted as performing the measurement of the two observables $\mathsf{X}_1\mathsf{X}_2$ and $\mathsf{X}_2\mathsf{X}_3$. The two auxiliary qubits in the circuit of Fig. 14.8, when measured in the basis $\{|0\rangle, |1\rangle\}$, deliver the two measurement results $X_1X_2=\pm 1$ and $X_2X_3=\pm 1$ forming the 2-bit error syndrome identifying the configuration of the phase-flip error for its correction. The three-qubit coding state encoding in the basis $\{|+\rangle, |-\rangle\}$, either with no error or with one phase-flip error, as an eigenstate of the observables, is left invariant (and undisclosed) and can be corrected accordingly.

Beyond such separate corrections of a bit-flip error in Sect. 14.2 and a phase-flip error in Sect. 14.3, it is possible to correct simultaneously a bit-flip and a phase-flip error that may occur on a coding qubit, as we will now show.

14.4 The Nine-Qubit Shor Code

It is possible to associate a three-qubit repetition code against phase-flip errors followed by a three-qubit repetition code against bit-flip errors. This represents a concatenation of two quantum codes to obtain a new code with increased robustness, and such concatenation is a general procedure to improve quantum codes. The result of

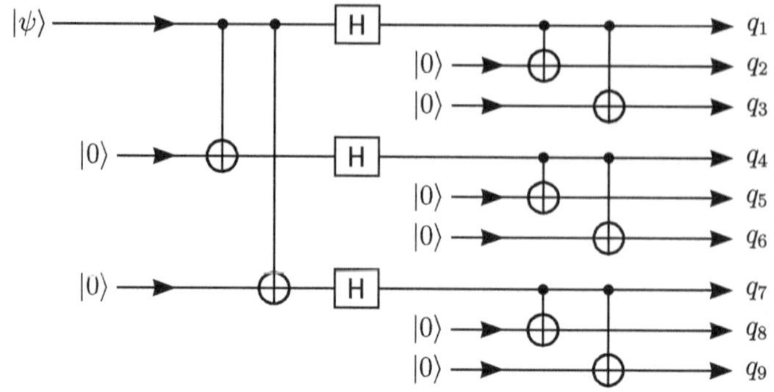

Fig. 14.9 A quantum circuit cascading the circuit of Fig. 14.6 followed by three copies of the circuit of Fig. 14.1 to realize on the generic qubit state $|\psi\rangle = \alpha_0 |0\rangle + \alpha_1 |1\rangle$ the nine-qubit encoding of the Shor code $|\psi\rangle \to |\psi_L\rangle = \alpha_0 |0_L\rangle + \alpha_1 |1_L\rangle$ from Eq. (14.44)

the concatenation here is a nine-qubit code, introduced by Shor in [482], it encodes every logical qubit into nine physical qubits, and it is able to detect and correct at the same time a phase-flip error on any of the nine coding qubits and a bit-flip error on any of the nine coding qubits.

For a generic input qubit state $|\psi\rangle = \alpha_0 |0\rangle + \alpha_1 |1\rangle$, a repetition encoding on three qubits is first performed against phase-flip errors via Eqs. (14.33)–(14.34) to produce $|\psi\rangle \to \alpha_0 |+++\rangle + \alpha_1 |---\rangle$. Next, each of the three qubits is encoded by a three-qubit repetition code against bit-flip errors via Eqs. (14.4)–(14.5). The nine-qubit encoding results for the computational basis as

$$\begin{cases} |0\rangle \longrightarrow \dfrac{1}{2\sqrt{2}}\Big(\big(|000\rangle + |111\rangle\big)\big(|000\rangle + |111\rangle\big)\big(|000\rangle + |111\rangle\big)\Big) \equiv |0_L\rangle\ , & (14.42) \\ |1\rangle \longrightarrow \dfrac{1}{2\sqrt{2}}\Big(\big(|000\rangle - |111\rangle\big)\big(|000\rangle - |111\rangle\big)\big(|000\rangle - |111\rangle\big)\Big) \equiv |1_L\rangle\ . & (14.43) \end{cases}$$

For the generic input, the encoding $|\psi\rangle \to |\psi_L\rangle = \alpha_0 |0_L\rangle + \alpha_1 |1_L\rangle$ gives

$$\begin{aligned} |\psi_L\rangle = {} & \frac{\alpha_0}{2\sqrt{2}}\Big(\big(|000\rangle + |111\rangle\big)\big(|000\rangle + |111\rangle\big)\big(|000\rangle + |111\rangle\big)\Big) \\ & + \frac{\alpha_1}{2\sqrt{2}}\Big(\big(|000\rangle - |111\rangle\big)\big(|000\rangle - |111\rangle\big)\big(|000\rangle - |111\rangle\big)\Big)\ . \end{aligned} \qquad (14.44)$$

This nine-qubit encoding is realized by the quantum circuit of Fig. 14.9.

Each of the three blocks of three output qubits $q_1q_2q_3$ or $q_4q_5q_6$ or $q_7q_8q_9$ appearing in Fig. 14.9 can be protected against a bit-flip error by a detection-correction protocol as in Sect. 14.2. If in addition a phase-flip error occurs on one of the nine output qubits from q_1 to q_9, a phase mismatch will result somewhere between the three output blocks $q_1q_2q_3$, $q_4q_5q_6$ or $q_7q_8q_9$, these three blocks remaining identical (especially in their phase) when no error occurs, while they will differ in phase when

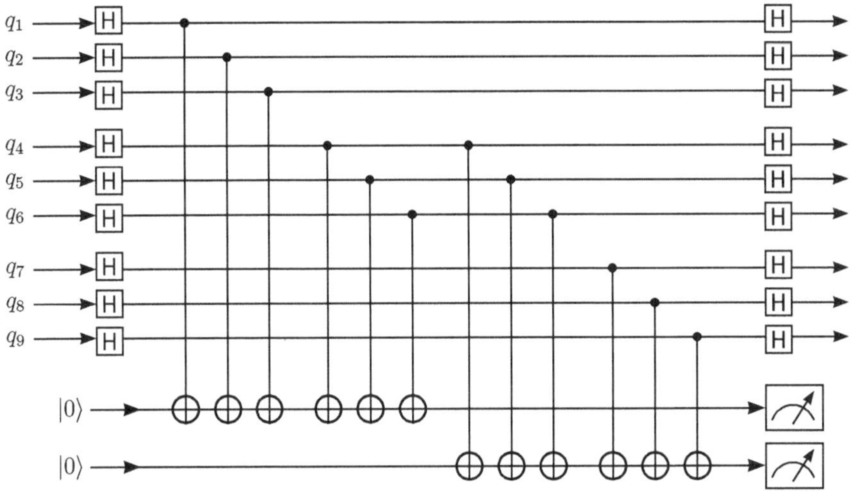

Fig. 14.10 Quantum circuit for detecting a phase-flip error that may have occurred in one of the nine qubits from q_1 to q_9 of the coding state from Eq. (14.44), as for example in its altered version of Eq. (14.45). Measuring the two auxiliary qubits at the output delivers the error syndrome which can be subsequently exploited for error correction on the coding qubits

one qubit suffers a phase flip. For example, when a phase-flip error occurs on a qubit of the first block $q_1q_2q_3$, say on the second qubit q_2, the nine-qubit coding state $|\psi_L\rangle$ of Eq. (14.44) gets altered as

$$\begin{aligned}|\psi_\mathrm{L}\rangle \longmapsto \mathsf{Z}_2\,|\psi_\mathrm{L}\rangle &= \frac{\alpha_0}{2\sqrt{2}}\Big(\big(|000\rangle - |111\rangle\big)\big(|000\rangle + |111\rangle\big)\big(|000\rangle + |111\rangle\big)\Big)\\ &+ \frac{\alpha_1}{2\sqrt{2}}\Big(\big(|000\rangle + |111\rangle\big)\big(|000\rangle - |111\rangle\big)\big(|000\rangle - |111\rangle\big)\Big).\end{aligned} \tag{14.45}$$

Such a phase-flip error can be detected (and then corrected) by comparing the phase of the three output blocks $q_1q_2q_3$, $q_4q_5q_6$, and $q_7q_8q_9$. This can be accomplished by the quantum circuit of Fig. 14.10.

Detecting a phase-flip error in one of the nine coding qubits from q_1 to q_9 amounts to detecting a phase mismatch between the three blocks $q_1q_2q_3$, $q_4q_5q_6$, and $q_7q_8q_9$. For this detection, the quantum circuit of Fig. 14.10 first applies a Hadamard transformation H to each of the nine coding qubits from q_1 to q_9. This changes the representation basis from $\{|0\rangle, |1\rangle\}$ to $\{|+\rangle, |-\rangle\}$ with a putative phase-flip error turned into a bit-flip error visible and detectable by a change in parity. For example, in Eq. (14.44) or (14.45), for a three-qubit block, say $q_1q_2q_3$, the state $|000\rangle$ is transformed by $\mathsf{H}^{\otimes 3}$ into $|+++\rangle$ and $|111\rangle$ into $|---\rangle$. The superposition $\big(|000\rangle + |111\rangle\big)/\sqrt{2}$ with positive phase is transformed by $\mathsf{H}^{\otimes 3}$ into

$\big(|{+}{+}{+}\rangle + |{-}{-}{-}\rangle\big)/\sqrt{2} = \big(|000\rangle + |011\rangle + |101\rangle + |110\rangle\big)/2$ having even parity, while the superposition $\big(|000\rangle - |111\rangle\big)/\sqrt{2}$ with negative phase is transformed by $\mathsf{H}^{\otimes 3}$ into $\big(|{+}{+}{+}\rangle - |{-}{-}{-}\rangle\big)/\sqrt{2} = \big(|001\rangle + |010\rangle + |100\rangle + |111\rangle\big)/2$ having odd parity. As a result, the nine-qubit coding state $|\psi_\mathrm{L}\rangle$ of Eq. (14.44) gets transformed as

$$|\psi_\mathrm{L}\rangle \xmapsto{\ \mathsf{H}^{\otimes 9}\ } \mathsf{H}^{\otimes 9}\,|\psi_\mathrm{L}\rangle = \alpha_0\,|\text{3-qubit even parity}\rangle^{\otimes 3} + \alpha_1\,|\text{3-qubit odd parity}\rangle^{\otimes 3}\,, \tag{14.46}$$

with the notation $|\text{3-qubit even parity}\rangle = \big(|000\rangle + |011\rangle + |101\rangle + |110\rangle\big)/2$ and $|\text{3-qubit odd parity}\rangle = \big(|001\rangle + |010\rangle + |100\rangle + |111\rangle\big)/2$. Meanwhile, when a phase-flip error occurred on one of the nine coding qubits, for example, the second qubit q_2 as in Eq. (14.45), the corrupted coding state $\mathsf{Z}_2\,|\psi_\mathrm{L}\rangle$ of Eq. (14.45) gets transformed by $\mathsf{H}^{\otimes 9}$ into $\mathsf{H}^{\otimes 9}\mathsf{Z}_2\,|\psi_\mathrm{L}\rangle$ reading

$$\alpha_0\,|\text{3-qubit odd parity}\rangle\,|\text{3-qubit even parity}\rangle^{\otimes 2} + \alpha_1\,|\text{3-qubit even parity}\rangle\,|\text{3-qubit odd parity}\rangle^{\otimes 2}\,. \tag{14.47}$$

We observe that a phase flip on qubit q_2 transforming the state $|\psi_\mathrm{L}\rangle$ of Eq. (14.44) into the state $\mathsf{Z}_2\,|\psi_\mathrm{L}\rangle$ of Eq. (14.45) equivalently appears as a bit flip on qubit q_2 transforming the state $\mathsf{H}^{\otimes 9}\,|\psi_\mathrm{L}\rangle$ of Eq. (14.46) into the state $\mathsf{H}^{\otimes 9}\mathsf{Z}_2\,|\psi_\mathrm{L}\rangle = \mathsf{X}_2\mathsf{H}^{\otimes 9}\,|\psi_\mathrm{L}\rangle$ of Eq. (14.47). This is what we indicated above, by the Hadamard transformation, a phase-flip error is turned into a bit-flip error.

The mismatch in parity between the three-qubit blocks in Eq. (14.47) signals the phase-flip error that occurred. The quantum circuit of Fig. 14.10 uses 2 auxiliary qubits initialized in state $|0\rangle$ and 12 Cnot gates to perform pairwise comparisons between the parities of the three blocks $q_1q_2q_3$, $q_4q_5q_6$, and $q_7q_8q_9$. A mismatch in parity detects a pair of blocks where one block is corrupted by a phase-flip error. An auxiliary qubit is placed in the state $|0\rangle$ or $|1\rangle$ according to the even or odd parity of its six Cnot controls. As before, the parity comparisons reveal no information about the encoded state but only about an error that may have occurred. After the parity comparisons, the nine-qubit coding state is returned unaltered and undisclosed in the original representation basis by an output layer of nine Hadamard gates H in the circuit of Fig. 14.10. Measurement in the computational basis $\big\{|0\rangle\,,\,|1\rangle\big\}$ of the two auxiliary qubits delivers a 2-bit syndrome identifying the three-qubit block corrupted by a phase-flip error (if any). This syndrome triggers the transformation $\mathsf{Z}^{\otimes 3}$ applied to the corrupted three-qubit block so as to correct the error. For example, the quantum circuit of Fig. 14.10 performs on the $9+2$ qubits the end-to-end transformation $|\psi_\mathrm{L}\rangle\,|00\rangle \mapsto |\psi_\mathrm{L}\rangle\,|00\rangle$ when no error occurred, and $\mathsf{Z}_2\,|\psi_\mathrm{L}\rangle\,|00\rangle \mapsto \mathsf{Z}_2\,|\psi_\mathrm{L}\rangle\,|10\rangle$ when a phase-flip error occurred on the second qubit q_2 as in Eq. (14.45), with an error syndrome then measured in state $|10\rangle$ indicating an error in the first block $q_1q_2q_3$ that is corrected by applying $\mathsf{Z}^{\otimes 3} = \mathsf{Z}_1\mathsf{Z}_2\mathsf{Z}_3$ to this block. In a similar way, a phase flip on the fourth qubit q_4 would give $\mathsf{Z}_4\,|\psi_\mathrm{L}\rangle\,|00\rangle \mapsto \mathsf{Z}_4\,|\psi_\mathrm{L}\rangle\,|11\rangle$, and a phase flip on the ninth qubit q_9 would give $\mathsf{Z}_9\,|\psi_\mathrm{L}\rangle\,|00\rangle \mapsto \mathsf{Z}_9\,|\psi_\mathrm{L}\rangle\,|01\rangle$.

In this way, the nine-qubit Shor code is able to detect and correct one bit-flip error that may have occurred in each of the three three-qubit blocks, and simultaneously one phase-flip error that may have occurred on one of the nine qubits of the Shor code. This is obtained by running in succession a bit-flip correction protocol of Sect. 14.2 on each of the three three-qubit blocks of the Shor code, followed by the phase-flip correction protocol of Fig. 14.10 operating on the nine qubits of the Shor code. As we indicated in Sect. 14.2.3, if a phase-flip error is present, it is left untouched by the bit-flip correction protocol and corrected by the phase-flip correction protocol that follows. In particular, the nine-qubit Shor code is able to detect and correct simultaneously a bit-flip error by X and a phase-flip error by Z that would have occurred on the same qubit by random application of an operator XZ (or ZX). In this respect, the nine-qubit Shor code protects a qubit against a single bit-flip X, phase-flip Z, or bit-phase-flip XZ error.

14.5 Correcting an Arbitrary Qubit Error

An error altering the state of a qubit is not necessarily a bit-flip error, or a phase-flip error, or a bit-phase-flip error. It may be an arbitrary error produced by a decohering process entangling the qubit to an uncontrolled environment, as examined in Sect. 3.2. To describe such a general decohering process, we rely on the evolution of Eq. (3.34) and Eq. (3.112), entangling a quantum system of interest Q in state $|Q\rangle$ to an environment E starting in the quantum state $|e_0\rangle$ that we here write

$$|Q\rangle \otimes |e_0\rangle \xmapsto{\mathsf{U}_{QE}} \mathsf{U}_{QE}\,|Q\rangle \otimes |e_0\rangle = \sum_{\ell=1}^{L} \Lambda_\ell\,|Q\rangle \otimes |e_\ell\rangle \ , \tag{14.48}$$

with $\{\Lambda_\ell\}_{\ell=1}^{L}$ a set of L Kraus operators describing the decohering process and $\{|e_\ell\rangle\}_{\ell=1}^{L}$ an orthonormal basis for an L-dimensional space to model the environment E. The system of interest Q here will be a block of K physical qubits used for encoding by a quantum error-correcting code. We want to analyze the effect of the general decohering process of Eq. (14.48) in the regime where it corrupts at most one qubit of the K-qubit code. We denote by k, for some integer $k \in \{1, 2, \dots, K\}$, the qubit putatively affected by the decohering evolution of Eq. (14.48). The decohering process on qubit k is described, following Eq. (14.48), by L Kraus operators Λ_ℓ acting specifically on the qubit k. For the qubit operators, instead of the orthonormal Pauli basis $\{\mathrm{I}_2, \sigma_x, \sigma_y, \sigma_z\}$ of Sect. A.5 in the Appendix, we will rather use the related orthonormal basis denoted $\{\mathsf{Id}, \mathsf{X}, \mathsf{Z}, \mathsf{XZ}\}$, where $\mathsf{Id} \equiv \mathrm{I}_2$ and $\mathsf{XZ} = -i\mathsf{Y}$ with $\mathsf{Y} \equiv \sigma_y$, so as to decompose each Kraus operator Λ_ℓ acting on the qubit k. We write this decomposition under the form:

$$\Lambda_\ell = i_\ell \mathsf{Id}_k + x_\ell \mathsf{X}_k + z_\ell \mathsf{Z}_k + y_\ell \mathsf{X}_k \mathsf{Z}_k \ , \tag{14.49}$$

with a subscript k appended to each Pauli operator to signify that it acts specifically on the qubit k of the K-qubit block, along with a trivial action on the other qubits of the block. In case of no error at all, all the Kraus operators in Eq. (14.49) collapse to the identity Id. With the decomposition of Eq. (14.49), when the block of K coding qubits is prepared in the state $|\psi_{\rm L}\rangle \equiv |Q\rangle$, it experiences a decohering evolution of Eq. (14.48) that we can write under the form

$$|\psi_{\rm L}\rangle \otimes |e_0\rangle \xmapsto{\mathsf{U}_{QE}} |\psi_{\rm L}\rangle \otimes \sum_{\ell=1}^{L} i_\ell\,|e_\ell\rangle + (\mathsf{X}_k\,|\psi_{\rm L}\rangle) \otimes \sum_{\ell=1}^{L} x_\ell\,|e_\ell\rangle$$
$$+ (\mathsf{Z}_k\,|\psi_{\rm L}\rangle) \otimes \sum_{\ell=1}^{L} z_\ell\,|e_\ell\rangle + (\mathsf{X}_k\mathsf{Z}_k\,|\psi_{\rm L}\rangle) \otimes \sum_{\ell=1}^{L} y_\ell\,|e_\ell\rangle\,. \qquad (14.50)$$

Therefore, by the action of the decohering environment, the K-qubit coding state $|\psi_{\rm L}\rangle$ is altered by getting entangled to the environment, through the entangling transformation $|\psi_{\rm L}\rangle \otimes |e_0\rangle \mapsto \mathsf{U}_{QE}\,|\psi_{\rm L}\rangle \otimes |e_0\rangle = |\Psi_{\rm L}\rangle$ of Eq. (14.50). The resulting corrupted state $|\Psi_{\rm L}\rangle$ given by the right-hand side of Eq. (14.50) can be written as

$$|\Psi_{\rm L}\rangle = |\psi_{\rm L}\rangle \otimes |E_I\rangle + (\mathsf{X}_k\,|\psi_{\rm L}\rangle) \otimes |E_X\rangle + (\mathsf{Z}_k\,|\psi_{\rm L}\rangle) \otimes |E_Z\rangle + (\mathsf{X}_k\mathsf{Z}_k\,|\psi_{\rm L}\rangle) \otimes |E_Y\rangle\,, \qquad (14.51)$$

where $|E_I\rangle$, $|E_X\rangle$, $|E_Z\rangle$, and $|E_Y\rangle$ are four vectors of the environment space defined from each of the four sums in the right-hand side of Eq. (14.50).

This shows that the effect of an arbitrary decohering process on qubit k can be decomposed as a superposition of four possible elementary error scenarios: no alteration, or a bit flip by X_k, a phase flip by Z_k, or a combined bit-phase flip by $\mathsf{X}_k\mathsf{Z}_k$. This is an important property of **error discretization**, quite significant for quantum error correction, expressing that a continuously infinite set of arbitrary errors can be reduced to a discrete small set of errors, detectable and correctable as such [477, 480, 482, 483]. In this respect, it is enough to correct bit-flip X errors and phase-flip Z errors and their combination to have the ability to correct any arbitrary error on the qubit.

The four K-qubit states $|\psi_{\rm L}\rangle$, $\mathsf{X}_k\,|\psi_{\rm L}\rangle$, $\mathsf{Z}_k\,|\psi_{\rm L}\rangle$, and $\mathsf{X}_k\mathsf{Z}_k\,|\psi_{\rm L}\rangle$ appearing in Eq. (14.51) only differ by a bit flip, or a phase flip, or a bit-phase flip, on the qubit k. For correctable errors, it is important that these four K-qubit states be devised as four orthogonal states. This is the general requirement mentioned in Sect. 14.2: for correctable errors the coding states when corrupted are sent into orthogonal subspaces. A coding state, when corrupted by an error, must not incur the risk of being confused with another coding state possibly corrupted also by an error. The minimum distance of a code, as evoked in Sect. 14.2.1, is involved here. This constraint on the minimum distance is satisfied by the three-qubit repetition code of Eq. (14.6) against a bit-flip error, and of Eq. (14.35) against a phase-flip error, and by the nine-qubit Shor code of Eq. (14.44) against a more general qubit error. The four possible error scenarios present in Eq. (14.51) are then in principle detectable by parity-check circuits involving auxiliary qubits with (Hadamard gates if needed and) Cnot gates controlled by the K coding qubits. This is exemplified by the elementary detection cells of Figs. 14.2 and 14.7, leading to the error-detection circuits of Figs. 14.3, 14.8, or 14.10. In this regard, the elementary detection cells of Figs. 14.2 and 14.7 show that for any qubit pair from the K-qubit coding state, it is possible to obtain an encoding by the two basis states $|0_A\rangle$ and $|1_A\rangle$ of an auxiliary qubit of whether the two coding qubits of the pair have the same value (or phase) or not, and this without disturbing or disclosing this value. With such faculty, an error-detection circuit can be devised with ultimately each of the four possible error scenarios present in Eq. (14.51) signaled by one configuration of a set of auxiliary qubits among four orthogonal configurations, distinguishable by a quantum measurement of the auxiliary qubits delivering an error syndrome. Measurement of the error syndrome projects the superposition of Eq. (14.51) onto one of the four possible error scenarios, selected at random by the Born rule, but precisely identified by the measurement result (the syndrome). Typically, an error-detection circuit will exploit a pair of auxiliary qubits initialized in state $|00_A\rangle$, and it will unitarily map the four orthogonal input states $\{|\psi_L\rangle\,|00_A\rangle\,,\ (\mathsf{X}_k\,|\psi_L\rangle)\,|00_A\rangle\,,\ (\mathsf{Z}_k\,|\psi_L\rangle)\,|00_A\rangle\,,\ (\mathsf{X}_k\mathsf{Z}_k\,|\psi_L\rangle)\,|00_A\rangle\}$ corresponding to the four error scenarios, into the four orthogonal output states $\{|\psi_L\rangle\,|00_A\rangle\,,\ (\mathsf{X}_k\,|\psi_L\rangle)\,|01_A\rangle\,,\ (\mathsf{Z}_k\,|\psi_L\rangle)\,|10_A\rangle\,,\ (\mathsf{X}_k\mathsf{Z}_k\,|\psi_L\rangle)\,|11_A\rangle\}$, where the error is ultimately identified by the two auxiliary qubits to be measured. From the superposition of Eq. (14.51) and the joint input state $|\Psi_L\rangle\,|00_A\rangle$ coupling the corrupted state $|\Psi_L\rangle$ to the auxiliary qubit pair, the error-detection circuit will deliver the output state

$$|\psi_L\rangle\,|00_A\rangle \otimes |E_I\rangle + (\mathsf{X}_k\,|\psi_L\rangle)\,|01_A\rangle \otimes |E_X\rangle + (\mathsf{Z}_k\,|\psi_L\rangle)\,|10_A\rangle \otimes |E_Z\rangle + (\mathsf{X}_k\mathsf{Z}_k\,|\psi_L\rangle)\,|11_A\rangle \otimes |E_Y\rangle\ . \tag{14.52}$$

For a general decohering noise, one may expect all four possible error scenarios to be present in superposition in the corrupted coding state $|\Psi_L\rangle$ of Eq. (14.51) entangled to the environment. Yet, when the two auxiliary qubits are measured at the output in the basis $\{|0_A\rangle\,,|1_A\rangle\}$, the superposition of Eq. (14.52) gets projected onto one of the four possible error scenarios, identified by the measurement result (the error syndrome), and from where the original uncorrupted coding state $|\psi_L\rangle$ can be restored by appropriate error correction. For instance, when from Eq. (14.52) the two auxiliary qubits are measured in state $|01_A\rangle$, the K coding qubits are projected onto the state $\mathsf{X}_k\,|\psi_L\rangle$, forcing an arbitrary error on qubit k into a bit-flip error X_k, identified[1] and correctable as such. In particular, on some occasions, hopefully not infrequent, measurement of the auxiliary qubits (measurement of the error syndrome) will project the coding qubits onto the intact state $|\psi_L\rangle$, just as if no error at all had occurred. Otherwise, measurement of the error syndrome reduces an arbitrary error to an elementary bit-flip X or phase-flip Z error or their combination XZ.

In addition, the measurement of the auxiliary qubits disentangles the state of the K coding qubits from the environment. The K coding qubits terminate in one of the four separable states $|\psi_L\rangle$, $\mathsf{X}_k\,|\psi_L\rangle$, $\mathsf{Z}_k\,|\psi_L\rangle$, or $\mathsf{X}_k\mathsf{Z}_k\,|\psi_L\rangle$ perfectly identified by the error syndrome, from which the appropriate correction can be implemented to restore

[1] We are back at the remark made near the end of Sect. 6.1 on EPR quantum correlations that it is observation that creates reality. Here the error does not preexist as a bit-flip X error. It is its observation that creates the X error.

the initial coding state $|\psi_L\rangle$ unaltered and undisclosed. When no error is detected at location k, the coding state remains unaffected and can, in principle, be checked for a qubit in error at another location by a comparable error-detection circuit, if needed.

The nine-qubit Shor code of Sect. 14.4 has this ability. Capable of correcting a bit-flip error X and a phase-flip error Z (and their combination), the Shor code, as it is, is also capable of correcting an arbitrary error. The Shor code concatenating for error detection and correction the circuits of Fig. 14.3 (for bit-flip errors) and of Fig. 14.10 (for phase-flip errors) also automatically corrects an arbitrary error on one qubit. For a single qubit corrupted by an arbitrary noise, the circuit receives on its input a superposition like $|\Psi_L\rangle$ of Eq. (14.51) combining bit-flip and phase-flip errors. The bit-flip errors are first checked and corrected, in each of the three three-qubit blocks of the Shor code, by the circuit of Fig. 14.3 which is insensitive to phase-flip errors, and this whatever the location k of the error. After this bit-flip correction step, the original arbitrary error ends up projected onto only a phase-flip error that may subsist somewhere in the nine qubits of the Shor code, and that will finally be corrected by the circuit of Fig. 14.10, and this also whatever the location k of the error.

This is an important and helpful property, generally accessible to error-correcting codes for the qubit, stemming from the error discretization of Eq. (14.51), that the ability to correct bit-flip X errors and phase-flip Z errors enables to correct any arbitrary error on the qubit. And by linearity on superpositions, the property equally applies to any number t of errors. If an error-correcting code is able to correct X errors and Z errors on t qubits, then it is able to correct any arbitrary error on t qubits [483].

14.6 Stabilizer and Other Quantum Codes

The quantum error-correcting codes we have examined above demonstrate the crucial possibility, for error detection, of realizing parity checks on qubits from a multiple-qubit coding state, and this without disturbing or disclosing this state. The parity checks can generally be implemented by quantum circuits (exemplified by the elementary detection cells of Figs. 14.2 and 14.7) involving essentially two-qubit Cnot gates, and one-qubit Hadamard gates to switch the representation between the $\{|0\rangle, |1\rangle\}$ and $\{|+\rangle, |-\rangle\}$ bases, so as to detect bit or phase mismatches, which is enough to detect an arbitrary qubit error as explained in Sect. 14.5. Such parity checks can be realized on arbitrary subsets of qubits of a multiple-qubit coding state. This is similar to the parity checks accomplished by classical linear block codes, which manage to correct $t \geq 1$ bits in error, and quantum codes can be extended in this direction [484–486]. For qubit codes, instead of encoding a single information qubit as in Eqs. (14.1)–(14.2), more generally it is possible to encode $L > 1$ information qubits into a larger number $K > L$ of physical qubits. With appropriate sizes L and K (controlled, in particular, by the quantum Gilbert–Varshamov bound), quantum codes can be constructed which can correct arbitrary errors on $t \geq 1$ qubits, with t in principle as large as we wish. For instance, a five-qubit repetition code would be able to correct two qubits in error, for example, two qubits affected by a bit-flip error, or a phase-flip error via Hadamard transformation. And in a similar way, an arbitrary error on two qubits will be projected by the syndrome extraction onto a combination of bit-flip or phase-flip errors, correctable as such [483]. In a systematic manner, the **Calderbank–Shor–Steane** (CSS) quantum codes combine two classical linear block codes, like the nine-qubit Shor code of Sect. 14.4 combined two three-qubit repetition codes, to correct arbitrary errors on any $t \geq 1$ qubits [487, 488]. The **Steane quantum code** is an instance of a CSS code, combining two classical Hamming codes, and that encodes $L = 1$ information qubit into $K = 7$ physical qubits (instead of $K = 9$ for the Shor code of Sect. 14.4), with the ability to correct an arbitrary error on one coding qubit [484]. To reach this performance, $K = 5$ is the minimal size for a quantum code encoding a single information qubit [489].

The CSS quantum codes represent a subclass of the broader class of **stabilizer quantum codes**, introduced in [490], and which can be constructed as follows. We introduce the set of one-qubit Pauli operators $\mathcal{G}_1 = \{\mathsf{Id}, \mathsf{X}, \mathsf{Y}, \mathsf{Z}\}$, which can receive a structure of algebraic group by allowing scalar multiplicative factors ± 1 and $\pm i$ to ensure closedness under operator multiplication. The K-qubit extension is the set $\mathcal{G}_K = \mathcal{G}_1^{\otimes K}$, formed by tensor products of K Pauli operators from $\mathcal{G}_1$, with similarly a group structure for $\mathcal{G}_K$ by allowing the multiplicative factors ± 1 and $\pm i$. Two operators from $\mathcal{G}_K$ either commute or anticommute, just as the Pauli operators themselves. For a given K-qubit quantum code, a stabilizing operator or stabilizer is an operator S from $\mathcal{G}_K$ that leaves invariant any valid K-qubit coding state $|\psi_\mathrm{L}\rangle$, that is, $\mathsf{S}\,|\psi_\mathrm{L}\rangle = |\psi_\mathrm{L}\rangle$. This implies that S has $+1$ as an eigenvalue, and the other eigenvalue is -1, with only two eigenvalues ± 1 for S as a product of Pauli operators. The product $\mathsf{S}\mathsf{S}'$ of two stabilizers S and S' is also a stabilizer, since $\mathsf{S}\mathsf{S}'\,|\psi_\mathrm{L}\rangle = \mathsf{S}\,|\psi_\mathrm{L}\rangle = |\psi_\mathrm{L}\rangle$. For example, for the three-qubit repetition code of Eq. (14.6) protecting against a bit-flip error, from what we have seen in Sect. 14.2.4, a set of stabilizers is the pair $\{\mathsf{Z}_1\mathsf{Z}_2,\ \mathsf{Z}_2\mathsf{Z}_3\}$, where we note that the two stabilizers of the set commute. For the three-qubit repetition code of Eq. (14.35) protecting against a phase-flip error, from what we have seen at the end of Sect. 14.3, a set of stabilizers is the pair $\{\mathsf{X}_1\mathsf{X}_2,\ \mathsf{X}_2\mathsf{X}_3\}$, also commuting.

The code subspace $\overline{\mathcal{H}_0}$ containing any valid K-qubit coding state $|\psi_\mathrm{L}\rangle$, such as the subspace $\overline{\mathcal{H}_0}$ defined in Sect. 14.2.2, is therefore an eigensubspace of its stabilizers associated with the eigenvalue $+1$. The stabilizers in the stabilizer set of a quantum code are chosen so as to all commute (as in the two examples above). This ensures that the stabilizers have their eigensubspaces in common, that is, a common eigensubspace of eigenvalue $+1$ and a common eigensubspace of eigenvalue -1. Conversely, any subset of $\mathcal{G}_K$ with this commutation property generally defines, by means of the eigensubspace of eigenvalue $+1$ common to the stabilizers, a quantum code via its code subspace $\overline{\mathcal{H}_0}$ assimilated to the $+1$ eigensubspace.

As it follows from Eqs. (14.50)–(14.51), on a K-qubit coding state $|\psi_\mathrm{L}\rangle$ an arbitrary error can always be decomposed as elementary errors X_k, Z_k, and $\mathsf{Y}_k \equiv i\mathsf{X}_k\mathsf{Z}_k$ acting on any qubit k, which are operators from the set $\mathcal{G}_K$. Accordingly, an error that may affect a K-qubit quantum code is a unitary operator F taken from $\mathcal{G}_K$. Such an

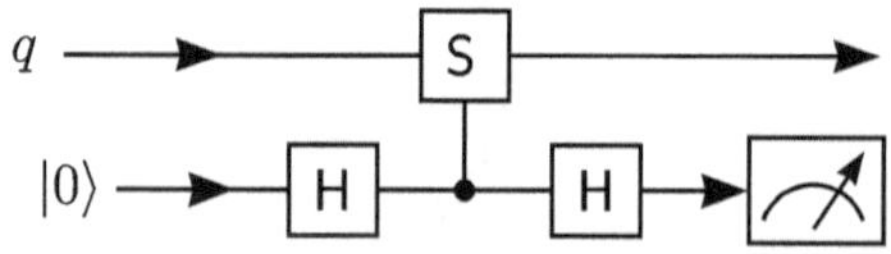

Fig. 14.11 Quantum circuit performing on the coding qubit q the measurement of the (restricted) stabilizer S. The auxiliary qubit initialized in state $|0\rangle$, when measured at the output in the computational basis $\{|0\rangle, |1\rangle\}$, projects the state of the qubit q onto an eigenstate of S corresponding to either the eigenvalue $+1$ or -1

error F has eigenvalues ± 1 and it commutes or anticommutes with each stabilizer of the code. An error F is detectable by a stabilizer S of the code when it anticommutes with S. In this circumstance, from any intact K-qubit coding state $|\psi_{\mathrm{L}}\rangle$ of the code subspace $\overline{\mathcal{H}}_0$, the corrupted coding state $\mathsf{F}\,|\psi_{\mathrm{L}}\rangle$ is transformed by the stabilizer S as $\mathsf{SF}\,|\psi_{\mathrm{L}}\rangle = -\mathsf{FS}\,|\psi_{\mathrm{L}}\rangle = -\mathsf{F}\,|\psi_{\mathrm{L}}\rangle$ by the anticommutation property, establishing that the corrupted state $\mathsf{F}\,|\psi_{\mathrm{L}}\rangle$ is an eigenstate of the stabilizer S with eigenvalue -1. The error F therefore sends the code subspace $\overline{\mathcal{H}}_0$ into an orthogonal eigensubspace of the stabilizer S associated with the eigenvalue -1. Alternatively, if the error F commutes with the stabilizer S, the corrupted coding state $\mathsf{F}\,|\psi_{\mathrm{L}}\rangle$ is left invariant by S, as $\mathsf{SF}\,|\psi_{\mathrm{L}}\rangle = \mathsf{F}\,|\psi_{\mathrm{L}}\rangle$. The anticommutation property $\mathsf{SF}\,|\psi_{\mathrm{L}}\rangle = -\mathsf{F}\,|\psi_{\mathrm{L}}\rangle$ is the type of alteration, responsible for parity mismatches in the coding state, which is detectable by measuring an auxiliary qubit involved in an appropriate error-detection circuit corresponding to this stabilizer S. Measuring the auxiliary qubit is equivalent to measuring the stabilizer S, with its two eigenvalues ± 1. The error-detection circuits from Figs. 14.2 and 14.7 and Figs. 14.3, 14.8 or 14.10 work accordingly.

Measuring a stabilizer S is therefore an important step for error detection with stabilizer codes, so we indicate below how this step can be handled practically. A stabilizer S is a unitary Hermitian operator (an observable) with two eigenvalues ± 1 acting on K-qubit quantum states and formed as a product of Pauli operators. As a first step, measuring S on a single qubit q (or more precisely, measuring the restriction of S on a single qubit q, a restriction which is also a unitary Hermitian operator with two eigenvalues ± 1) can be performed by measuring the auxiliary qubit in the circuit of Fig. 14.11, which uses a controlled version for the gate S.

With the two orthogonal eigenstates $|\lambda_\pm\rangle$ of the (restricted) stabilizer S associated, respectively, with the two eigenvalues ± 1, the quantum circuit of Fig. 14.11 performs the two-qubit input–output unitary transformation (before the output measurement):

$$|q\rangle\,|0\rangle = \big(\alpha_+\,|\lambda_+\rangle + \alpha_-\,|\lambda_-\rangle\big)\,|0\rangle \longmapsto \alpha_+\,|\lambda_+\rangle\,|0\rangle + \alpha_-\,|\lambda_-\rangle\,|1\rangle\ . \qquad (14.53)$$

Accordingly, at the output the auxiliary qubit is measured in state $|0\rangle$ with the probability $|\alpha_+|^2$ while the coding qubit q gets projected onto the post-measurement state $|\lambda_+\rangle$, and the like with the other outcome onto $|\lambda_-\rangle$, as expected when measuring the (restricted) stabilizer S on the qubit q. When q is prepared in an eigenstate of S, its state is left invariant in the measurement.

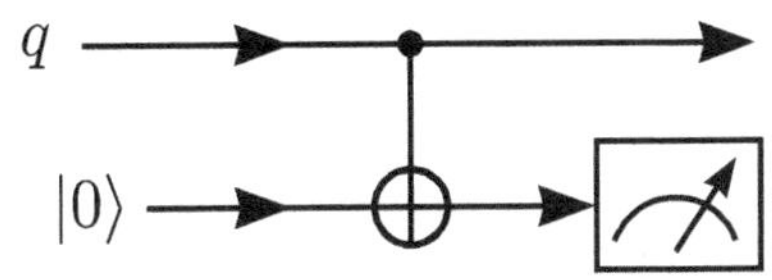

Fig. 14.12 Quantum circuit performing on the coding qubit q the measurement of the Pauli operator Z. The auxiliary qubit initialized in state $|0\rangle$, when measured at the output in the computational basis $\{|0\rangle, |1\rangle\}$, according to Eq. (14.55), projects the state of the qubit q onto an eigenstate $|0\rangle$ or $|1\rangle$ of Z corresponding to either the eigenvalue $+1$ or -1

Moreover, the stabilizer S measured in Fig. 14.11 can be considered for its full action on a multiple-qubit coding state. In place of q in Fig. 14.11 is input a multiple-qubit state $|q\rangle$, which can be decomposed as $|q\rangle = |q_+\rangle + |q_-\rangle$, with $|q_+\rangle$ the (unnormalized) component of $|q\rangle$ in the eigensubspace of S with eigenvalue $+1$, and $|q_-\rangle$ the component of $|q\rangle$ in the complementary eigensubspace of S with eigenvalue -1. The quantum circuit of Fig. 14.11 then performs the input–output unitary transformation (before the output measurement of the auxiliary qubit):

$$|q\rangle\,|0\rangle = \big(|q_+\rangle + |q_-\rangle\big)\,|0\rangle \longmapsto |q_+\rangle\,|0\rangle + |q_-\rangle\,|1\rangle\ . \tag{14.54}$$

When the auxiliary qubit at the output is measured in state $|0\rangle$, the multiple-qubit state $|q\rangle$ gets projected onto the $+1$ eigensubspace of the stabilizer S, while when the auxiliary qubit is measured in state $|1\rangle$ the state $|q\rangle$ gets projected onto the -1 eigensubspace of S. This realizes the (two-outcome) measurement on the state $|q\rangle$ of the stabilizer S.

In addition, for a stabilizer $\mathsf{S} = \mathsf{Z}$ or $\mathsf{S} = \mathsf{X}$, equivalent implementations exist for the measurement circuit of Fig. 14.11. For the stabilizer $\mathsf{S} = \mathsf{Z}$, we have the equivalent circuit of Fig. 14.12.

The quantum circuit of Fig. 14.12 performs the two-qubit input–output unitary transformation (before the output measurement):

$$|q\rangle\,|0\rangle = \big(\alpha_0\,|0\rangle + \alpha_1\,|1\rangle\big)\,|0\rangle \longmapsto \alpha_0\,|0\rangle\,|0\rangle + \alpha_1\,|1\rangle\,|1\rangle\ . \tag{14.55}$$

The circuit of Fig. 14.12 is used in Fig. 14.3 for measuring the stabilizer $\mathsf{Z}_1\mathsf{Z}_2$ on the qubits q_1 and q_2, and the stabilizer $\mathsf{Z}_2\mathsf{Z}_3$ on the qubits q_2 and q_3. In Fig. 14.3, the circuit of Fig. 14.12 is combined to control the auxiliary qubit with two coding qubits, q_1q_2 or q_2q_3, to realize a multiple-qubit (a two-qubit) measurement equivalent to Eq. (14.54) for a two-qubit stabilizer, $\mathsf{Z}_1\mathsf{Z}_2$ or $\mathsf{Z}_2\mathsf{Z}_3$.

For the stabilizer $\mathsf{S} = \mathsf{X}$, we have for Fig. 14.11 the equivalent circuit of Fig. 14.13.

The quantum circuit of Fig. 14.13 performs the two-qubit input–output unitary transformation (before the output measurement):

$$|q\rangle\,|0\rangle = \big(\alpha_+\,|+\rangle + \alpha_-\,|-\rangle\big)\,|0\rangle \longmapsto \alpha_+\,|+\rangle\,|0\rangle + \alpha_-\,|-\rangle\,|1\rangle\ . \tag{14.56}$$

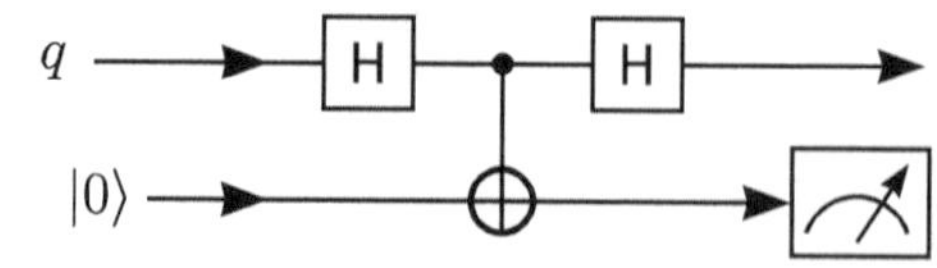

Fig. 14.13 Quantum circuit performing on the coding qubit q the measurement of the Pauli operator X. The auxiliary qubit initialized in state $|0\rangle$, when measured at the output in the computational basis $\{|0\rangle, |1\rangle\}$, according to Eq. (14.56), projects the state of the qubit q onto an eigenstate $|+\rangle$ or $|-\rangle$ of X corresponding to either the eigenvalue $+1$ or -1

The circuit of Fig. 14.13 is used in Fig. 14.8 for measuring the stabilizer $\mathsf{X}_1\mathsf{X}_2$ on the qubits q_1 and q_2, and the stabilizer $\mathsf{X}_2\mathsf{X}_3$ on the qubits q_2 and q_3.

In this way, a stabilizer S can be measured on a multiple-qubit coding state, which is left invariant when it entirely lies in one of the two eigensubspaces of S, as it happens for an intact coding state or a coding state corrupted by a correctable error.

After these practical considerations for measuring a stabilizer, we return to the operation of a stabilizer code for detecting an error F by a stabilizer S of the code. On a K-qubit coding state $|\psi_\mathrm{L}\rangle$ corrupted as $\mathsf{F}\,|\psi_\mathrm{L}\rangle$, which is always an eigenstate of the stabilizer S, the action of the error-detection circuit with its auxiliary qubit initialized as $|0_A\rangle$ implements the unitary transformation:

$$\mathsf{F}\,|\psi_\mathrm{L}\rangle\,|0_A\rangle \longmapsto \frac{1}{2}(\mathrm{Id}^{\otimes K} + \mathsf{S})\mathsf{F}\,|\psi_\mathrm{L}\rangle\,|0_A\rangle + \frac{1}{2}(\mathrm{Id}^{\otimes K} - \mathsf{S})\mathsf{F}\,|\psi_\mathrm{L}\rangle\,|1_A\rangle\ , \quad (14.57)$$

as in Eq. (14.27) or (14.28). Since S has eigenvalues ± 1, we have $(\mathrm{Id}^{\otimes K} + \mathsf{S})/2$ which is the projector onto the $+1$ eigensubspace of S, while $(\mathrm{Id}^{\otimes K} - \mathsf{S})/2$ is the projector onto the -1 eigensubspace of S. An error F that anticommutes with the stabilizer S is signaled by the result $|1_A\rangle$ when measuring the auxiliary qubit. An error like F that commutes with the stabilizer S is signaled by the auxiliary qubit measured in state $|0_A\rangle$, just like an intact coding state $|\psi_\mathrm{L}\rangle$. When the K coding qubits are in state $\mathsf{F}\,|\psi_\mathrm{L}\rangle$, they are in an eigenstate of S, and they remain in the same post-measurement state $\mathsf{F}\,|\psi_\mathrm{L}\rangle$ after measuring S. Equivalently, upon measuring the stabilizer S, the eigenvalue -1 is returned for the anticommuting error F detected by this stabilizer, while the eigenvalue $+1$ is returned for different commuting errors not detected by this stabilizer S, or when trivially $\mathsf{F} = \mathrm{Id}^{\otimes K}$ for an intact coding state $|\psi_\mathrm{L}\rangle$. Several errors like F can be simultaneously detected (and identified) if they can be associated with several stabilizers like S that, when measured, provide by combining their binary measurement results, a unique identification of the errors involved, or in other words distinct errors should produce distinct syndromes. This is the central task when constructing stabilizer codes: to find the stabilizers to be measured so that they anticommute with the errors to be detected. More details on stabilizer codes can be found, for instance, in [9, 480, 490, 491], especially on the encoding and decoding steps accompanying the central error detection and correction mechanisms.

For example, for the three-qubit repetition code of Sect. 14.2 from Eq. (14.6) protecting against a bit-flip error, the minimal set or generator is the stabilizer set $\{Z_1Z_2, Z_2Z_3\}$. The operator Z_1Z_3 is also a stabilizer, but it can be obtained from the minimal set as the composition $Z_1Z_3 = (Z_1Z_2)(Z_2Z_3)$. As we have seen, the code is able to correct on a single qubit the three bit-flip errors X_1, X_2 and X_3, which all three either commute or anticommute with the stabilizers since $X_kZ_k = -Z_kX_k$, so as to produce, when measuring the two stabilizers Z_1Z_2 and Z_2Z_3 (in whatever order since they commute), a unique syndrome for each error. For the three-qubit repetition code of Sect. 14.3 protecting against a phase-flip error, the stabilizer set is $\{X_1X_2, X_2X_3\}$, enabling to correct the three phase-flip errors Z_1, Z_2 and Z_3 on a single qubit. For the nine-qubit Shor code of Sect. 14.4, the stabilizer set is the eight-element set $\{Z_1Z_2, Z_2Z_3, Z_4Z_5, Z_5Z_6, Z_7Z_8, Z_8Z_9, X_1X_2X_3X_4X_5X_6, X_4X_5X_6X_7X_8X_9\}$, enabling to correct any arbitrary error on a single qubit of the nine-qubit Shor code.

Another useful family of stabilizer codes is formed by **topological quantum codes** [492–495], where the coding qubits are organized into an array with spatial structure. The stabilizer generators are defined with local supports, acting only on a few nearby coding qubits. They enable syndrome extraction through local interactions of the qubits in the array. Meanwhile, the encoded information is stored as global, topological, properties of the coding qubits, making it more robust to local errors. A convenient physical implementation of topological codes is under the form of surface codes, which are constructed on a two-dimensional array of qubits [496–498]. Another powerful family of stabilizer codes are quantum low-density parity-check (LDPC) codes, where the number of qubits participating in each stabilizer and the number of stabilizers where each qubit participates are kept specially small [499–502].

Other developments of quantum coding and error correction are ongoing in many directions, with, for instance, quantum convolutional codes [503], quantum turbo codes [504], quantum polar codes [505], and entanglement-assisted error-correcting codes [506, 507]. Today, for quantum communication, we do not yet possess a full appreciation of where we stand and how to effectively reach the best efficiencies expressed by the various information capacities (which often, to begin with, are not (yet) fully characterized), and much remains to be done on these topics. Accordingly, quantum coding and error correction today represent a very active area, with challenges on both methodology and implementation, and great relevance to quantum technologies [508–511].

14.7 Fault-Tolerant Quantum Computation

In line with our signal-processing perspective, the primary purpose of quantum error correction, as we described it above, was for channel coding, to protect quantum states or signals against the degradation caused by quantum noise or decoherence when interacting with an uncontrolled environment, as it occurs in communication over noisy quantum channels. But beyond quantum communication, the same concern applies also to quantum computation, including quantum storage, to protect quantum states and their evolutions against errors or failures that may be caused by imperfect quantum gates, circuits, or devices. Such errors can especially accumulate as a computation proceeds across gates, or they can even occur in the coding and error-correction circuitry. The techniques of quantum error correction sketched in this chapter can be extended in this context, to obtain fault-tolerant computation [511–514]. Redundancy can be introduced in the quantum circuits as well as in the quantum data, and operations can be performed on encoded data without decoding them to maintain their protective encoding. This typically implies more physical qubits to support every logical qubit and added quantum gates to compensate their unreliability. But adding redundancy by increasing the number of qubits and gates for enhanced protection, at the same time, increases the eventuality of errors. The challenge then is to favorably settle these two competing trends so as to ultimately ensure the possibility of large-scale reliable quantum processing and computation.

In this direction, there exists an important **threshold theorem** for quantum computation [512, 515, 516], stating that if the probability of error of every quantum gate is below a certain constant threshold p_{th}, it is possible to reliably execute an arbitrarily large quantum computation, provided appropriate error protection is implemented. The value of the threshold p_{th} in practice depends on many elements, as the type of gates, noise, and errors involved and the way they are handled for correction, and p_{th} is often not easy to characterize in realistic quantum settings. But the mere existence of such threshold theorem guarantees that it is possible to correct errors faster than they are created, and by consequence to reliably execute arbitrarily large (or scalable) computation on a given imperfect quantum hardware. Characterizing the threshold p_{th}, matching it with reliable hardware, and by designing quantum computations with sufficient protection represent today areas of intense research and technological developments, needful to progress toward large-scale reliable quantum processing and computation [517–521].

14.8 Summary on Quantum Coding and Error Correction

The focus of this chapter was to show that error correction on quantum states or signals is indeed possible and rests on the following essential properties:
(i) Useful redundancy can be introduced for quantum coding without conflicting with the no-cloning theorem, since the coding does not need to replicate any arbi-

trary unknown quantum states but only a small set of known quantum states serving for encoding.
(ii) Error detection and correction can be performed without altering or revealing the encoded information. This is feasible by parity checks between coding qubits, which are enough to identify a qubit that may have flipped in error, and to enable its correction by a counterflip, without knowing or revealing its state. Parity checks operate with auxiliary qubits that are ultimately measured to extract the error syndrome and drive the correction, but without revealing anything about the underlying encoded information.
(iii) Phase-flip errors can be handled as bit-flip errors through the use of Hadamard transformations.
(iv) Any arbitrary error on a qubit can be decomposed and projected by the syndrome extraction into a bit-flip or a phase-flip error or their combination, and corrected accordingly.

These properties can be exploited through many specific protocols still today under development and elaboration for enhanced performance. These properties can be extended to enable fault-tolerant computation on unreliable quantum hardware, guided by the threshold theorem of Sect. 14.7. The whole area of quantum error correction and fault-tolerant computation remains today a challenging direction of research and development, conceptual and technological.

Chapter 15
Outlook

This book proposed a presentation of the principal notions and concepts of **quantum information and quantum computation**. In addition, a special orientation followed the thread of signal processing or signals-and-systems theory, to substantiate the more specific field of **quantum signal processing**, and serve as a guide structuring the exposition. Essential ideas underpinning the developments in the book can be concisely restated as follows.

Quantum theory can be grounded on three fundamental principles, on state, process, and measurement. A quantum object or system is described by a **quantum state**, represented by a unit-norm state vector $|\psi\rangle$ of a complex Hilbert space $\mathcal{H}$, or more generally by a positive unit-trace density operator ρ of $\mathcal{L}(\mathcal{H})$. A quantum state evolves through a **quantum process**, acting via a unitary operator U for closed evolutions, or more generally via a completely positive trace-preserving superoperator $\mathcal{E}(\cdot)$ for open evolutions. **Quantum measurement** takes place as a random projection in an orthonormal basis of $\mathcal{H}$, or more generally (thanks to coupling by entanglement) as a generalized measurement defined by a suitable set of measurement operators $\{\mathsf{M}_m\}$. It results that all the fundamental entities—state, process, measurement—animating the machinery of quantum theory, are all elements or vectors of some suitable complex vector space: The state $|\psi\rangle$ is a vector of some vector space $\mathcal{H}$; the density operator ρ is an operator (a vector) of the vector space $\mathcal{L}(\mathcal{H})$, as is the unitary evolution operator U, as well as the measurement operators $\{\mathsf{M}_m\}$ of any generalized measurement; the non-unitary evolution process $\mathcal{E}(\cdot)$ is a superoperator (a vector) of the vector space $\mathcal{L}\big(\mathcal{L}(\mathcal{H})\big)$. These vectors can all be represented by their (complex) coordinates of a linear combination in some orthonormal basis.

Quantum signals realized as quantum states play the role of natural versatile information carriers. These states, when referred to a finite N-dimensional Hilbert space $\mathcal{H}_N$, can be supported by assembling a sufficient number L of two-dimensional qubits such that $N \leq 2^L$, instating the qubit as a fundamental building block for quantum information and computation. **Quantum processes** act as operative functional blocks to transform quantum states or signals, either unitarily for closed evolutions or

F. Chapeau-Blondeau, *Quantum Information and Quantum Signal Processing*,
https://doi.org/10.1007/978-3-032-27499-1_15

non-unitarily for open evolutions. Any conceivable process on N-dimensional quantum states can always be physically materialized with one-qubit gates supplemented by the two-qubit Cnot gate. Such capability of universal synthesis from qubits forms a cornerstone for quantum information and computation, enabling modular analysis and design, to gradually manage the elaboration.

Quantum signals transformed by **unitary** quantum processes constitute the workforce of quantum computation. Enhanced performance for computation can be derived from distinctive quantum properties not accessible classically, such as quantum entanglement or quantum superposition for parallelism. Several landmark illustrations thereof have been presented in the book, with the Deutsch–Jozsa, the Grover, the Shor, or the HHL quantum algorithms, or protocols for superdense coding, teleportation or quantum cryptography. Following these pioneering works, many other quantum algorithms have gradually been proposed. Yet we would say that today there does not exist a systematic methodology, first to identify the problems that could benefit from quantum enhancement, and then a fortiori to design the corresponding quantum algorithms, especially in an optimal way to maximize some meaningful performance criteria. Much remains to be done in this direction of quantum algorithms—their principles and their physical implementations—for broad varieties of applications and domains.

Non-unitary evolutions of quantum states are essentially caused by decoherence or quantum noise, which often must be explicitly taken into account for more realism. Important situations of this kind, and relevant to signal processing, have been examined in the book, with detection or discrimination of quantum signals in noise, or estimation from noisy quantum signals, or communication over noisy quantum channels. For such quantum informational tasks in the presence of noise, it was frequently observed that probing signals or signaling states are often more efficient when they are assembled in entangled sequences. By contrast, in similar conditions, classical signals are often more efficient in independent sequences; this being the case, for instance, with successive probings for statistical estimation or with optimal signaling for communication over noisy classical channels. Entangled quantum signals in the presence of quantum noise constitute an area where today much exploration remains to be accomplished. Optimal or efficient entangled probing signals for quantum estimation and quantum metrology in the presence of noise are not known in all significant settings. The quantum or even the classical information capacities of noisy quantum channels—and a fortiori the encoding and decoding strategies to reach them—are not yet fully characterized, especially when asymptotic regularization over arbitrarily long entangled sequences is necessary. In this respect, the conditions where quantum entanglement is useful to enhance such information capacities, or more broadly some meaningful informational efficiency, are not completely elucidated.

This book essentially concentrated on discrete quantum variables, with finite-dimensional quantum states, naturally extending to discrete infinite-dimensional (countable) quantum states, such as the coherent states of light featuring an arbitrary number of photons governed by a Poisson probability distribution [66, 522]. Further extension would be to continuous quantum variables [399, 523–525], with continuously infinite dimensionality, such as the continuous amplitudes of a field or the spatial degrees of freedom of a particle [526]. Discrete quantum variables offer powerful capabilities along with modularity and flexibility comparable to classical digital processing, while continuous quantum variables are more akin to classical analog processing. Accordingly, the orientation of this book on quantum signal processing could be viewed as dealing principally with quantum digital signal processing, in contrast to analog signal processing.

For the quantum methodologies and approaches presented in this book, several **grand challenges** today remain open for the future. A prominent one is the design of a universal quantum computer with associated hardware and software [527–529], for which the qubit-based approach is especially appealing and powerful. Also relevant is the advancement of hybrid classical-quantum processors, or of special-purpose quantum computers or devices for instance dedicated to quantum simulation or emulation of specific complex physical phenomena of interest in controllable and programmable ways [1, 530]. Other grand challenges concern quantum communication and the deployment of a quantum or a hybrid internet [154, 531]; another is quantum metrology with the development of quantum sensors, for enhanced sensitivity, precision, miniaturization, and integration [316, 333], including quantum imaging [532–534]. Quantum machine learning and quantum-based or quantum-aided artificial intelligence also constitute a challenging direction with great potential ahead [535, 536]. Quantum processing and quantum technologies remain fragile to noise and decoherence, so that for all these paths of research and application, the ability to mitigate or control the noise will be a crux to the future.

An objective targeted by this book is to transmit core quantum concepts and methodologies to the readers so as to contribute in some way or another to this quantum endeavor to the future. The train is underway.

Appendix A
Linear Algebra

In this appendix, we briefly recall some important notions of linear algebra that are specially useful for the main text of the book. This utility chapter is not intended as an exhaustive and consistent coverage of linear algebra, and basic notions from other standard sources may also prove useful.

A.1 State Vectors

The N-dimensional complex vector space $\mathcal{H}_N$ is referred to an orthonormal standard basis denoted $\{|1\rangle, |2\rangle, \dots, |N\rangle\}$, also called the computational basis. Any **vector** $|\psi\rangle$ from $\mathcal{H}_N$ can then be expressed as in Eq. (2.1) by the linear combination:

$$|\psi\rangle = \sum_{n=1}^{N} \alpha_n |n\rangle \in \mathcal{H}_N , \tag{A.1}$$

with coordinates $\alpha_n \in \mathbb{C}$ which are complex numbers. Accordingly, the vector $|\psi\rangle$ can also be represented as the column vector (or $N \times 1$ matrix):

$$|\psi\rangle = \begin{bmatrix} \alpha_1 \\ \vdots \\ \alpha_N \end{bmatrix} = \left[\alpha_1, \dots, \alpha_N\right]^\top . \tag{A.2}$$

The notion of conjugate transpose or **adjoint** matrix is then accessible for defining the adjoint vector:

$$|\psi\rangle^\dagger = \left[\alpha_1^*, \dots, \alpha_N^*\right] = \langle\psi| , \tag{A.3}$$

with the convenient Dirac notation $|\psi\rangle^\dagger = \langle\psi|$. In the bra-ket Dirac [3] notation, when $|\psi\rangle$ is called a ket vector, its adjoint $\langle\psi|$ is called a bra vector.

F. Chapeau-Blondeau, *Quantum Information and Quantum Signal Processing*,
https://doi.org/10.1007/978-3-032-27499-1

This leads to a natural form for the squared vector norm in $\mathcal{H}_N$ as

$$\left\| |\psi\rangle \right\|^2 = |\psi\rangle^\dagger |\psi\rangle = \langle\psi|\psi\rangle = \sum_{n=1}^{N} |\alpha_n|^2 . \tag{A.4}$$

A **quantum state** $|\psi\rangle$, or state vector $|\psi\rangle$, is any vector $|\psi\rangle$ having unit norm. The associated scalar or **inner product** of two vectors $|\psi\rangle = [\alpha_1, \ldots, \alpha_N]^\top$ and $|\psi'\rangle = [\alpha'_1, \ldots, \alpha'_N]^\top$ from $\mathcal{H}_N$ is

$$\langle\psi|\psi'\rangle = \left[\alpha_1^*, \ldots, \alpha_N^*\right] \begin{bmatrix} \alpha'_1 \\ \vdots \\ \alpha'_N \end{bmatrix} = \sum_{n=1}^{N} \alpha_n^* \alpha'_n = \langle\psi'|\psi\rangle^* . \tag{A.5}$$

The corresponding Cauchy–Schwarz inequality for two vectors $|\psi\rangle$ and $|\psi'\rangle$ of $\mathcal{H}_N$ reads

$$\left|\langle\psi|\psi'\rangle\right|^2 \le \langle\psi|\psi\rangle \, \langle\psi'|\psi'\rangle . \tag{A.6}$$

The orthonormal character of the computational basis of $\mathcal{H}_N$ is then equivalent to

$$\langle n|n'\rangle = \delta_{nn'} \quad \forall (n, n') \in \{1, 2, \ldots, N\}^2 , \tag{A.7}$$

with $\delta_{nn'}$ the Kronecker delta. The coordinates in Eq. (A.1) follow as $\alpha_n = \langle n|\psi\rangle$.

Useful also is the **outer product** of two vectors $|\psi\rangle = [\alpha_1, \ldots, \alpha_N]^\top$ and $|\psi'\rangle = [\alpha'_1, \ldots, \alpha'_N]^\top$ from $\mathcal{H}_N$ as

$$|\psi'\rangle \langle\psi| = \begin{bmatrix} \alpha'_1 \\ \vdots \\ \alpha'_N \end{bmatrix} \left[\alpha_1^*, \ldots, \alpha_N^*\right] = \begin{bmatrix} \alpha'_1\alpha_1^* & \cdots & \alpha'_1\alpha_N^* \\ \vdots & & \vdots \\ \alpha'_N\alpha_1^* & \cdots & \alpha'_N\alpha_N^* \end{bmatrix} , \tag{A.8}$$

producing an $N \times N$ matrix that defines a linear map or operator from $\mathcal{H}_N$ onto $\mathcal{H}_N$. The adjoint (or conjugate transpose) operator is $\left(|\psi'\rangle \langle\psi|\right)^\dagger = |\psi\rangle \langle\psi'|$.

As a special case, $|\psi\rangle \langle\psi|$ is the **orthogonal projector** onto the vector $|\psi\rangle$ in $\mathcal{H}_N$. Such projector is a Hermitian (or self-adjoint) operator, i.e., $\left(|\psi\rangle \langle\psi|\right)^\dagger = |\psi\rangle \langle\psi|$.

A.2 Operators

A linear[1] operator A from $\mathcal{H}_N$ onto $\mathcal{H}_N$ is completely defined by an $N \times N$ representation matrix with complex entries and referred to some orthonormal basis $\{|n\rangle\}$ for $\mathcal{H}_N$. To avoid inflation in notation, we often denote in the same way with an upright A the operator and its representation matrix, when there is no compelling inconvenience; otherwise, if needed, an italic A is used instead for the matrix representing the operator A in some definite basis. An operator A and its properties are usually completely defined and characterizable by the properties of its representation matrix relative to any convenient orthonormal basis. The n-th column of the matrix A is the column vector as in Eq. (A.2) representing in the orthonormal basis $\{|n\rangle\}$ the image $\mathsf{A}|n\rangle$ of the basis vector $|n\rangle$. The entry, or matrix element, at row m and column n of the matrix A is the complex scalar $A_{mn} = \langle m|\mathsf{A}|n\rangle$.

The set of linear operators from $\mathcal{H}_N$ onto $\mathcal{H}_N$ is denoted $\mathcal{L}(\mathcal{H}_N)$. With standard notions of addition and scalar multiplication for operators, $\mathcal{L}(\mathcal{H}_N)$ receives the structure of a complex vector space with dimension N^2.

The **rank** of a linear operator $\mathsf{A} \in \mathcal{L}(\mathcal{H}_N)$ is the dimension $\leq N$ of the subspace spanned by the columns of its representation matrix, which is the same as the dimension $\leq N$ of the subspace spanned by its rows, this subspace defining the **support** of the operator A. In other words, the rank of A is the dimension of the vector space $\mathsf{A}(\mathcal{H}_N)$.

The adjoint operator $\mathsf{A}^\dagger$ is defined by the conjugate transpose matrix $\mathsf{A}^\dagger = (\mathsf{A}^\top)^* = (\mathsf{A}^*)^\top$.

A class of operators important for quantum theory is formed by **Hermitian (or self-adjoint) operators** A satisfying $\mathsf{A}^\dagger = \mathsf{A}$. When $\mathsf{A}^\dagger = -\mathsf{A}$ then A is anti-Hermitian. Any Hermitian operator $\mathsf{H} \in \mathcal{L}(\mathcal{H}_N)$ can always be endowed with N eigenvalues λ_n (possibly with multiplicity higher than 1) which are necessarily real, i.e., $\lambda_n \in \mathbb{R}$, for $n = 1$ to N. They can always be associated with N normalized eigenvectors or eigenstates $|\lambda_n\rangle$ satisfying $\mathsf{H}|\lambda_n\rangle = \lambda_n |\lambda_n\rangle$ which form an orthonormal basis of $\mathcal{H}_N$. A diagonal form, or eigendecomposition, results for the Hermitian operator H as

$$\mathsf{H} = \sum_{n=1}^{N} \lambda_n |\lambda_n\rangle \langle\lambda_n| \, , \tag{A.9}$$

equivalent to the possibility of representing the operator H by the diagonal matrix $\mathrm{diag}[\lambda_1, \lambda_2, \ldots, \lambda_N]$ in the orthonormal basis $\{|\lambda_n\rangle\}$ of its N eigenstates.

An operator $\mathsf{A} \in \mathcal{L}(\mathcal{H}_N)$ is a **positive operator** if and only if the scalar $\langle\psi|\mathsf{A}|\psi\rangle$ is real positive for all vector $|\psi\rangle \in \mathcal{H}_N$, in other words $\mathsf{A} \geq 0 \Leftrightarrow \langle\psi|\mathsf{A}|\psi\rangle \geq 0$ for all $|\psi\rangle \in \mathcal{H}_N$. Since for A positive the scalar $\langle\psi|\mathsf{A}|\psi\rangle$ is always real, this implies that

[1] As a rule for quantum theory, the adjective "linear" is often omitted with operators, an "operator" being a "linear operator," unless otherwise explicitly specified, but this rarely occurs in this context.

a positive operator A is also Hermitian; this being easily established, for instance, by considering successively $|\psi\rangle = |\psi_1\rangle + |\psi_2\rangle$ and $|\psi\rangle = |\psi_1\rangle + i\,|\psi_2\rangle$. A Hermitian operator $\mathsf{H} \in \mathcal{L}(\mathcal{H}_N)$ is positive if and only if all its N eigenvalues λ_n are (real) positive.

The **rank** of a Hermitian operator $\mathsf{H} \in \mathcal{L}(\mathcal{H}_N)$ is the number $\leq N$ of nonzero eigenvalues of H, while the **support** of H is the subspace of $\mathcal{H}_N$ spanned by the eigenstates of H associated with the nonzero eigenvalues.

Another class of operators important for quantum theory is formed by **unitary operators** A endowed with an inverse $\mathsf{A}^{-1} = \mathsf{A}^\dagger$, equivalent to $\mathsf{A}\mathsf{A}^\dagger = \mathsf{A}^\dagger\mathsf{A} = \mathrm{I}_N$ where I_N is the identity operator on $\mathcal{H}_N$. A unitary operator $\mathsf{A} \in \mathcal{L}(\mathcal{H}_N)$ is represented by an $N \times N$ unitary matrix, whose N columns form an orthonormal basis of $\mathcal{H}_N$, or equivalently whose N rows form an orthonormal basis of $\mathcal{H}_N$.

The identity operator I_N on $\mathcal{H}_N$ satisfies the completeness or closure relation:

$$\mathrm{I}_N = \sum_{n=1}^{N} |n\rangle\,\langle n| \;, \tag{A.10}$$

for any orthonormal basis $\{|n\rangle\}$ of $\mathcal{H}_N$. Also, Eq. (A.10) is a necessary and sufficient condition for a set of N vectors $\{|n\rangle\}$ of $\mathcal{H}_N$ to form an orthonormal basis of $\mathcal{H}_N$. As a proof, if $\{|n\rangle\}$ is an orthonormal basis of $\mathcal{H}_N$, then for any vector $|\psi\rangle$ of $\mathcal{H}_N$ its N coordinates $\langle n|\psi\rangle$ enable the expansion $|\psi\rangle = \sum_{n=1}^{N} \langle n|\psi\rangle\,|n\rangle = \sum_{n=1}^{N} |n\rangle\,\langle n|\,|\psi\rangle$, which implies Eq. (A.10); conversely, if Eq. (A.10) holds, then for any $|m\rangle$ in the set $\{|n\rangle\}$, by left and right multiplication of Eq. (A.10) one has $\langle m|\,\mathrm{I}_N\,|m\rangle = \langle m|\sum_{n=1}^{N} |n\rangle\,\langle n|\,|m\rangle$, giving $\langle m|m\rangle = |\langle m|m\rangle|^2 + \sum_{n\neq m}^{N} |\langle m|n\rangle|^2$ indicating that none of the squared norm $\langle m|m\rangle$ can be > 1, but taking the trace (see Eqs. (A.12)–(A.14) below) of Eq. (A.10) yields $\mathrm{tr}\big(\mathrm{I}_N\big) = \sum_{n=1}^{N} \langle n|n\rangle = N$, imposing $\langle m|m\rangle = 1$ strictly for every m along with $\langle m|n\rangle = \delta_{mn}$, establishing $\{|n\rangle\}$ as an orthonormal basis of $\mathcal{H}_N$, and completing the proof.

An operator U is unitary if and only if there exists a Hermitian operator H such that $\mathsf{U} = \exp(i\mathsf{H})$, with more to come on functions of operators in Sect. A.3. It results that U and H of $\mathcal{L}(\mathcal{H}_N)$ share the same N orthonormal eigenstates $|\lambda_n\rangle$, while U has the N unit-modulus eigenvalues $\exp(i\lambda_n)$ when H has the N (real) eigenvalues λ_n. A diagonal form results for the unitary operator $\mathsf{U} \in \mathcal{L}(\mathcal{H}_N)$ as

$$\mathsf{U} = \sum_{n=1}^{N} \exp(i\lambda_n)\,|\lambda_n\rangle\,\langle\lambda_n| \;. \tag{A.11}$$

Let an operator $\mathsf{A} \in \mathcal{L}(\mathcal{H}_N)$ with a representation matrix now denoted A in some orthonormal basis $\{|n\rangle\}$ of $\mathcal{H}_N$. Introduce the change of orthonormal basis via $|n'\rangle = U\,|n\rangle$, for $n = 1$ to N, where U is an $N \times N$ unitary matrix. Then $A' = UAU^\dagger$ is the new representation matrix for the operator A in the orthonormal basis $\{|n'\rangle\}$ of $\mathcal{H}_N$.

For any operator $\mathsf{A} \in \mathcal{L}(\mathcal{H}_N)$ with representation matrix A in some orthonormal basis $\{|n\rangle\}$ of $\mathcal{H}_N$, the **trace** is the scalar defined as

$$\mathrm{tr}(\mathsf{A}) = \sum_{n=1}^{N} \langle n|\, \mathsf{A}\, |n\rangle = \sum_{n=1}^{N} \langle n|\, A\, |n\rangle = \sum_{n=1}^{N} A_{nn} \,, \tag{A.12}$$

which is intrinsic to the operator A and does not change with the orthonormal basis of representation, since also $\mathrm{tr}(\mathsf{A}) = \sum_{n'=1}^{N} \langle n'|\, A'\, |n'\rangle$ as it follows from the behavior of Eq. (A.12) in the matrix transformation $A = U^\dagger A' U$. By complex conjugation $[\mathrm{tr}(\mathsf{A})]^* = \mathrm{tr}(\mathsf{A}^\dagger)$.

For any two vectors $|\psi\rangle$ and $|\phi\rangle$ from $\mathcal{H}_N$, one has then the identity:

$$\boxed{\mathrm{tr}\big(\mathsf{A}\,|\psi\rangle\,\langle\phi|\big) = \langle\phi|\,\mathsf{A}\,|\psi\rangle \,,} \tag{A.13}$$

which follows[2] since $\mathrm{tr}\big(\mathsf{A}\,|\psi\rangle\,\langle\phi|\big) = \sum_n \langle n|\,\mathsf{A}\,|\psi\rangle\,\langle\phi|n\rangle = \sum_n \langle\phi|n\rangle\,\langle n|\,\mathsf{A}\,|\psi\rangle = \langle\phi|\big(\sum_n |n\rangle\,\langle n|\big)\mathsf{A}\,|\psi\rangle = \langle\phi|\,\mathsf{A}\,|\psi\rangle$. As a special case, when A is the identity operator:

$$\mathrm{tr}\big(|\psi\rangle\,\langle\phi|\big) = \langle\phi|\psi\rangle \,. \tag{A.14}$$

The trace is linear, $\mathrm{tr}(\alpha_A\mathsf{A} + \alpha_B\mathsf{B}) = \alpha_A\,\mathrm{tr}(\mathsf{A}) + \alpha_B\,\mathrm{tr}(\mathsf{B})$, for any complex scalars α_A and α_B and operators A and B of $\mathcal{L}(\mathcal{H}_N)$. The trace has a property of circular invariance or circularity $\mathrm{tr}(\mathsf{ABC}) = \mathrm{tr}(\mathsf{CAB}) = \mathrm{tr}(\mathsf{BCA})$, but in general $\mathrm{tr}(\mathsf{ABC}) \neq \mathrm{tr}(\mathsf{ACB})$.

Two operators A and B of $\mathcal{L}(\mathcal{H}_N)$ generally do not commute, with $\mathsf{AB} \neq \mathsf{BA}$. Their commutator $[\mathsf{A}, \mathsf{B}]$ is defined as the operator:

$$[\mathsf{A}, \mathsf{B}] = \mathsf{AB} - \mathsf{BA} \,, \tag{A.15}$$

which vanishes if and only if A and B commute. Their anticommutator $\{\mathsf{A}, \mathsf{B}\}$ is defined as the operator:

$$\{\mathsf{A}, \mathsf{B}\} = \mathsf{AB} + \mathsf{BA} \,. \tag{A.16}$$

For two operators A and B of $\mathcal{L}(\mathcal{H}_N)$, the scalar $\mathrm{tr}(\mathsf{A}^\dagger\mathsf{B})$ defines an inner product known as the **Hilbert–Schmidt inner product**. It lends itself to the Cauchy–Schwarz inequality under the form:

$$\big|\mathrm{tr}(\mathsf{A}^\dagger\mathsf{B})\big|^2 \le \mathrm{tr}(\mathsf{A}^\dagger\mathsf{A})\ \mathrm{tr}(\mathsf{B}^\dagger\mathsf{B}) \,. \tag{A.17}$$

[2] Incidently, this derivation illustrates the power and helpfulness of the Dirac bra-ket notation introduced in Sect. 2.1 in handling algebraic calculations. In particular, any expression enclosed in a regularly closing pair of angular brackets, such as $\langle\phi|n\rangle$ or $\langle n|\mathsf{A}|\psi\rangle$, represents a scalar and can be algebraically handled as such.

A.3 Operator Functions

An $N \times N$ matrix A with entries in $\mathbb{C}$ is diagonalizable on $\mathbb{C}$ when there exist an invertible $N \times N$ matrix U and a diagonal $N \times N$ matrix A_{diag} such that

$$A = UA_{\text{diag}}U^{-1} \Longleftrightarrow A_{\text{diag}} = U^{-1}AU \ . \tag{A.18}$$

The N columns of U are formed by the N (normalized) eigenvectors $|a_n\rangle$ of A, each endowed with the eigenvalue a_n, and satisfying $A\,|a_n\rangle = a_n\,|a_n\rangle$, for $n = 1$ to N. Also we have the diagonal matrix $A_{\text{diag}} = \text{diag}\big[a_1, a_2, \ldots, a_N\big]$.

We can then define a function $f(\cdot)$ of a diagonalizable matrix A as $f(A) = f\big(UA_{\text{diag}}U^{-1}\big) = Uf(A_{\text{diag}})U^{-1} = U\text{diag}\big[f(a_1), f(a_2), \ldots, f(a_N)\big]U^{-1}$. This can be justified with the integer-series expansion of $f(A)$, where we have $A^2 = (UA_{\text{diag}}U^{-1})^2 = UA_{\text{diag}}^2U^{-1} = U\text{diag}\big[a_1^2, a_2^2, \ldots, a_N^2\big]U^{-1}$, with a similar mechanism acting for each integer exponent in the power-series expansion of $f(\cdot)$. An operator $\mathsf{A} \in \mathcal{L}(\mathcal{H}_N)$ represented by a diagonalizable matrix A leads to the notion of operator function $f(\mathsf{A})$ represented by the matrix $f(A)$.

The matrix A is normal when $A^\dagger A = AA^\dagger$. Hermitian matrices, anti-Hermitian matrices, and unitary matrices are normal matrices. An $N \times N$ normal matrix A has N orthogonal eigenvectors $|a_n\rangle$ forming an orthonormal basis and is diagonalizable by means of a unitary matrix U, with $U^{-1} = U^\dagger$. An operator $\mathsf{A} \in \mathcal{L}(\mathcal{H}_N)$ represented by a normal matrix A has N eigenvalues a_n and N eigenstates $|a_n\rangle$ forming an orthonormal basis of $\mathcal{H}_N$ and A receives the diagonal form $\mathsf{A} = \sum_{n=1}^N a_n\,|a_n\rangle\,\langle a_n|$, giving the operator function $f(\mathsf{A}) = \sum_{n=1}^N f(a_n)\,|a_n\rangle\,\langle a_n|$.

For an isometric (or norm-preserving) operator V mapping $\mathcal{H}_N$ into a larger Hilbert space $\mathcal{H}_{N'}$ with dimension $N' \geq N$ and $\mathsf{V}^\dagger\mathsf{V} = \mathrm{I}_N$, by a similar mechanism we have $f\big(\mathsf{V}\mathsf{A}\mathsf{V}^\dagger\big) = \mathsf{V}f(\mathsf{A})\mathsf{V}^\dagger$.

A.4 Tensor Product and Partial Trace

The tensor product $\mathcal{H}^A \otimes \mathcal{H}^B$ of two Hilbert spaces $\mathcal{H}^A$ and $\mathcal{H}^B$ and associated elementary properties are rather straightforwardly obtained from defining tensor products of two orthonormal bases of $\mathcal{H}^A$ and $\mathcal{H}^B$ supplemented by natural extensions of vector addition and scalar multiplication. The Hilbert space $\mathcal{H}^A$ with dimension N_A is referred to an orthonormal basis $\{|a\rangle\}_{a=1}^{N_A}$; the space $\mathcal{L}\big(\mathcal{H}^A\big)$ with dimension N_A^2 of the linear operators from $\mathcal{H}^A$ onto $\mathcal{H}^A$ has orthonormal basis $\big\{|a\rangle\,\langle a'|\big\}$. Another Hilbert space $\mathcal{H}^B$ with dimension N_B is referred to an orthonormal basis $\{|b\rangle\}_{b=1}^{N_B}$; the space $\mathcal{L}\big(\mathcal{H}^B\big)$ with dimension N_B^2 has orthonormal basis $\big\{|b\rangle\,\langle b'|\big\}$. The tensor-product space $\mathcal{H}^A \otimes \mathcal{H}^B$ has dimension N_AN_B and orthonormal basis $\big\{|a\rangle \otimes |b\rangle\big\}$; the space $\mathcal{L}\big(\mathcal{H}^A \otimes \mathcal{H}^B\big)$ with dimension $(N_AN_B)^2$ has orthonormal basis $\big\{|a\rangle\,\langle a'| \otimes |b\rangle\,\langle b'|\big\}$. Any operator $\rho_{AB} \in \mathcal{L}\big(\mathcal{H}^A \otimes \mathcal{H}^B\big)$ can be expressed as the linear combination:

$$\rho_{AB} = \sum_{aa'bb'} \alpha_{aa'bb'} \, |a\rangle \langle a'| \otimes |b\rangle \langle b'| \; . \tag{A.19}$$

The partial trace [9, 59] over the space $\mathcal{H}^B$ can be defined as

$$\mathrm{tr}_B\big(\rho_{AB}\big) = \sum_b \langle b \,|\, \rho_{AB} \,|\, b\rangle = \sum_{aa'} \Big(\sum_b \alpha_{aa'bb}\Big) \, |a\rangle \langle a'| \; \in \mathcal{L}\big(\mathcal{H}^A\big) \; , \tag{A.20}$$

the result $\mathrm{tr}_B\big(\rho_{AB}\big)$ being an operator of $\mathcal{L}\big(\mathcal{H}^A\big)$. By symmetry, a similar definition holds for the partial trace $\mathrm{tr}_A\big(\rho_{AB}\big)$ over the space $\mathcal{H}^A$. The partial trace as well is invariant in any change of orthonormal basis. From the definition of Eq. (A.20), the following properties can be established.

For two operators $\mathsf{A} \in \mathcal{L}\big(\mathcal{H}^A\big)$ and $\mathsf{B} \in \mathcal{L}\big(\mathcal{H}^B\big)$, one has

$$\mathrm{tr}(\mathsf{A} \otimes \mathsf{B}) = \mathrm{tr}_{AB}(\mathsf{A} \otimes \mathsf{B}) = \mathrm{tr}_A(\mathsf{A}) \, \mathrm{tr}_B(\mathsf{B}) \; , \tag{A.21}$$

$$\mathrm{tr}_A(\mathsf{A} \otimes \mathsf{B}) = \mathrm{tr}_A(\mathsf{A}) \, \mathsf{B} \; , \tag{A.22}$$

$$\mathrm{tr}_B(\mathsf{A} \otimes \mathsf{B}) = \mathsf{A} \; \mathrm{tr}_B(\mathsf{B}) \; , \tag{A.23}$$

with in some places superfluous subscripts in the trace for more clarity.

For A, A' two operators on $\mathcal{H}^A$, and I^B the identity operator on $\mathcal{H}^B$, and ρ_{AB} any operator on $\mathcal{H}^A \otimes \mathcal{H}^B$, one has the separability property of the partial trace:

$$\mathrm{tr}_B\big[(\mathsf{A} \otimes \mathrm{I}^B)\rho_{AB}(\mathsf{A}' \otimes \mathrm{I}^B)\big] = \mathsf{A} \, \mathrm{tr}_B(\rho_{AB})\mathsf{A}' \; , \tag{A.24}$$

$$\mathrm{tr}_A\big[(\mathrm{I}^A \otimes \mathsf{B})\rho_{AB}(\mathrm{I}^A \otimes \mathsf{B}')\big] = \mathsf{B} \, \mathrm{tr}_A(\rho_{AB})\mathsf{B}' \; , \tag{A.25}$$

for any operators B and B' on $\mathcal{H}^B$ and I^A the identity operator on $\mathcal{H}^A$.

There is a circular invariance of the partial trace over the traced space:

$$\mathrm{tr}_A\big[(\mathsf{A} \otimes \mathrm{I}^B)\rho_{AB}\big] = \mathrm{tr}_A\big[\rho_{AB}(\mathsf{A} \otimes \mathrm{I}^B)\big] \; , \tag{A.26}$$

$$\mathrm{tr}_B\big[(\mathrm{I}^A \otimes \mathsf{B})\rho_{AB}\big] = \mathrm{tr}_B\big[\rho_{AB}(\mathrm{I}^A \otimes \mathsf{B})\big] \; . \tag{A.27}$$

But there is no circular invariance in general over the non-traced space, i.e.:

$$\mathrm{tr}_A\big[(\mathrm{I}^A \otimes \mathsf{B})\rho_{AB}\big] \neq \mathrm{tr}_A\big[\rho_{AB}(\mathrm{I}^A \otimes \mathsf{B})\big] \; , \tag{A.28}$$

which by Eq. (A.25) extends to

$$\mathsf{B} \, \mathrm{tr}_A(\rho_{AB}) = \mathrm{tr}_A\big[(\mathrm{I}^A \otimes \mathsf{B})\rho_{AB}\big] \neq \mathrm{tr}_A\big[\rho_{AB}(\mathrm{I}^A \otimes \mathsf{B})\big] = \mathrm{tr}_A(\rho_{AB}) \, \mathsf{B} \; ; \tag{A.29}$$

and so a fortiori

$$\mathrm{tr}_A\big[(\mathsf{A} \otimes \mathsf{B})\rho_{AB}\big] \neq \mathrm{tr}_A\big[\rho_{AB}(\mathsf{A} \otimes \mathsf{B})\big] \; , \tag{A.30}$$

but rather $\mathrm{tr}_A\big[(\mathsf{A}\otimes\mathsf{B})\rho_{AB}\big] = \mathrm{tr}_A\big[(\mathsf{A}\otimes\mathrm{I}^B)(\mathrm{I}^A\otimes\mathsf{B})\rho_{AB}\big] = \mathrm{tr}_A\big[(\mathrm{I}^A\otimes\mathsf{B})\rho_{AB}(\mathsf{A}\otimes\mathrm{I}^B)\big] = \mathsf{B}\,\mathrm{tr}_A\big[\rho_{AB}(\mathsf{A}\otimes\mathrm{I}^B)\big]$, while $\mathrm{tr}_A\big[\rho_{AB}(\mathsf{A}\otimes\mathsf{B})\big] = \mathrm{tr}_A\big[\rho_{AB}(\mathsf{A}\otimes\mathrm{I}^B)(\mathrm{I}^A\otimes\mathsf{B})\big] = \mathrm{tr}_A\big[\rho_{AB}(\mathsf{A}\otimes\mathrm{I}^B)\big]\,\mathsf{B}$.

Additionally, for two operators $\mathsf{A}\in\mathcal{L}(\mathcal{H}^A)$ and $\mathsf{B}\in\mathcal{L}(\mathcal{H}^B)$, one has $\big(\mathsf{A}\otimes\mathsf{B}\big)^\dagger = \mathsf{A}^\dagger\otimes\mathsf{B}^\dagger$, and $(\mathsf{A}\otimes\mathsf{B})\big(|\psi_A\rangle\otimes|\psi_B\rangle\big) = \mathsf{A}\,|\psi_A\rangle\otimes\mathsf{B}\,|\psi_B\rangle$ for any two vectors $|\psi_A\rangle\in\mathcal{H}^A$ and $|\psi_B\rangle\in\mathcal{H}^B$.

For operators A, A' from $\mathcal{L}(\mathcal{H}^A)$ and B, B' from $\mathcal{L}(\mathcal{H}^B)$, one has $\big(\mathsf{A}\otimes\mathsf{B}\big)\big(\mathsf{A}'\otimes\mathsf{B}'\big) = \big(\mathsf{A}\mathsf{A}'\big)\otimes\big(\mathsf{B}\mathsf{B}'\big)$.

There exists a useful identity combining tensor products and functions of operators. Let A_j, for $j = 1$ to J, a set of J operators on the N-dimensional Hilbert space $\mathcal{H}_N$, and $\{|j\rangle\}_{j=1}^J$ an orthonormal basis for the J-dimensional Hilbert space $\mathcal{H}_J$. The operator

$$\mathsf{B} = \sum_{j=1}^{J} |j\rangle\langle j| \otimes \mathsf{A}_j\,, \tag{A.31}$$

acting on the tensor-product space $\mathcal{H}_J\otimes\mathcal{H}_N$, satisfies the identity

$$f(\mathsf{B}) = \sum_{j=1}^{J} |j\rangle\langle j| \otimes f(\mathsf{A}_j)\,. \tag{A.32}$$

A justification of Eq. (A.32) can be obtained by considering the squared operator

$$\mathsf{B}^2 = \left(\sum_{j=1}^{J} |j\rangle\langle j| \otimes \mathsf{A}_j\right)\left(\sum_{k=1}^{J} |k\rangle\langle k| \otimes \mathsf{A}_k\right) \tag{A.33}$$

$$= \sum_{j=1}^{J}\sum_{k=1}^{J} |j\rangle\langle j|k\rangle\langle k| \otimes \mathsf{A}_j\mathsf{A}_k \tag{A.34}$$

$$= \sum_{j=1}^{J} |j\rangle\langle j| \otimes \mathsf{A}_j^2\,, \tag{A.35}$$

since $\langle j|k\rangle = \delta_{jk}$ by orthonormality of the basis $\{|j\rangle\}_{j=1}^J$. The same mechanism as in Eq. (A.35) applies for any integer exponent ℓ to give $\mathsf{B}^\ell = \sum_j |j\rangle\langle j|\otimes\mathsf{A}_j^\ell$. Through its integer-series expansion, an arbitrary function $f(\cdot)$ applied to B will act similarly to give $f(\mathsf{B}) = \sum_j |j\rangle\langle j|\otimes f(\mathsf{A}_j)$, finally proving the identity of Eq. (A.32).

The eigendecomposition of the operator B from Eq. (A.31) can also be worked out. By gathering the N eigenvalues $\{\lambda_n^j\}_{n=1}^N$ of each of the J operators A_j, we obtain the JN eigenvalues of the operator B from Eq. (A.31) as the set $\{\lambda_n^j\}$ for $n = 1$ to N and $j = 1$ to J. From the N eigenstates $\{|\lambda_n^j\rangle\}_{n=1}^N$ of each of the J operators A_j, the JN eigenstates of B are also readily deduced as $|j\rangle\otimes|\lambda_n^j\rangle$ for $n = 1$ to N and

$j = 1$ to J. In addition, we have the trace $\mathrm{tr}(\mathsf{B}) = \sum_{j=1}^{J} \mathrm{tr}(\mathsf{A}_j)$ and the determinant $\det(\mathsf{B}) = \prod_{j=1}^{J} \det(\mathsf{A}_j)$. The operator B is invertible if and only if each operator A_j is invertible, and the inverse reads $\mathsf{B}^{-1} = \sum_j |j\rangle\langle j| \otimes \mathsf{A}_j^{-1}$.

A.5 In the Dimension Two

The dimension $N = 2$ is useful for the situation of the qubit, introduced in Sect. 2.3. There exist important operators in $\mathcal{L}(\mathcal{H}_2)$ which are the Pauli operators defined, in reference to the computational basis $\{|0\rangle, |1\rangle\}$, as

$$\sigma_x = |1\rangle\langle 0| + |0\rangle\langle 1| = \begin{bmatrix} 0 & 1 \\ 1 & 0 \end{bmatrix}, \tag{A.36}$$

$$\sigma_y = i\,|1\rangle\langle 0| - i\,|0\rangle\langle 1| = \begin{bmatrix} 0 & -i \\ i & 0 \end{bmatrix}, \tag{A.37}$$

$$\sigma_z = |0\rangle\langle 0| - |1\rangle\langle 1| = \begin{bmatrix} 1 & 0 \\ 0 & -1 \end{bmatrix}. \tag{A.38}$$

The three Pauli operators are both Hermitian and unitary, so that $\sigma_x^2 = \sigma_y^2 = \sigma_z^2 = \mathrm{I}_2$, with I_2 the identity operator of $\mathcal{L}(\mathcal{H}_2)$. They satisfy $\sigma_x\sigma_y = -\sigma_y\sigma_x = i\sigma_z$, and two similar relations obtained by circular permutations. The three Pauli operators are all with zero trace and determinant -1. The three Pauli operators all have $\{+1, -1\}$ as their eigenvalues, associated with the eigenstates $\{|0\rangle, |1\rangle\}$ for σ_z, or $\{|+\rangle, |-\rangle\}$ from Eqs. (2.7)–(2.8) for σ_x, finally $\{|+i\rangle, |-i\rangle\}$ from Eqs. (2.13)–(2.14) for σ_y.

Together with the identity I_2, the Pauli operators form an orthogonal (for the Hilbert–Schmidt inner product) basis $\{\mathrm{I}_2, \sigma_x, \sigma_y, \sigma_z\}$ of the four-dimensional vector space $\mathcal{L}(\mathcal{H}_2)$. In this Pauli basis, for example, are the four operators of $\mathcal{L}(\mathcal{H}_2)$:

$$\begin{cases} |0\rangle\langle 0| = \dfrac{1}{2}\big(\mathrm{I}_2 + \sigma_z\big), & \text{(A.39)} \\ |1\rangle\langle 1| = \dfrac{1}{2}\big(\mathrm{I}_2 - \sigma_z\big), & \text{(A.40)} \\ |0\rangle\langle 1| = \dfrac{1}{2}\big(\sigma_x + i\sigma_y\big), & \text{(A.41)} \\ |1\rangle\langle 0| = \dfrac{1}{2}\big(\sigma_x - i\sigma_y\big). & \text{(A.42)} \end{cases}$$

Another useful operator of $\mathcal{L}(\mathcal{H}_2)$ is the Hadamard operator defined as

$$\mathsf{H} = \frac{1}{\sqrt{2}}(\sigma_x + \sigma_z) = \frac{1}{\sqrt{2}}\begin{bmatrix} 1 & 1 \\ 1 & -1 \end{bmatrix}, \tag{A.43}$$

which is also both Hermitian and unitary, so that $\mathsf{H}^2 = \mathrm{I}_2$ and $\mathsf{H}^{-1} = \mathsf{H}$ (self-inverse). The Hadamard operator H superposes the two states of the computational basis with $\mathsf{H}\,|0\rangle = \big(|0\rangle + |1\rangle\big)/\sqrt{2} = |+\rangle$ and $\mathsf{H}\,|1\rangle = \big(|0\rangle - |1\rangle\big)/\sqrt{2} = |-\rangle$. In this regard, the Hadamard operator H is important to quantum computation as it creates from the computational basis the crucial resource of quantum superposition forming a ground for parallel computation.

Any operator A of $\mathcal{L}(\mathcal{H}_2)$ can be expressed in the orthogonal Pauli basis $\{\mathrm{I}_2, \sigma_x, \sigma_y, \sigma_z\}$ as

$$\mathsf{A} = a_0\mathrm{I}_2 + a_x\sigma_x + a_y\sigma_y + a_z\sigma_z = a_0\mathrm{I}_2 + \vec{a}\cdot\vec{\sigma}\ , \tag{A.44}$$

endowing A with the four complex scalar coordinates $(a_0, \vec{a}\,) \in \mathbb{C}^4$ with $\vec{a} = [a_x, a_y, a_z]^\top \in \mathbb{C}^3$, and introducing a common concise notation that uses the formal vector $\vec{\sigma} = [\sigma_x, \sigma_y, \sigma_z]$ assembling the three Pauli operators. In particular, with this notation one has $(\vec{a}\cdot\vec{\sigma}\,)^2 = \vec{a}^{\,2}\,\mathrm{I}_2$. The operator A in Eq. (A.44) has the trace $\mathrm{tr}(\mathsf{A}) = 2a_0$, the determinant $\det(\mathsf{A}) = a_0^2 - \vec{a}^{\,2}$, and the two eigenvalues $\lambda_\pm = a_0 \pm \sqrt{\vec{a}^{\,2}}$. With another operator $\mathsf{A}' = a_0'\mathrm{I}_2 + \vec{a}^{\,\prime}\cdot\vec{\sigma}$ of $\mathcal{L}(\mathcal{H}_2)$, their product is

$$\mathsf{A}\mathsf{A}' = \big(a_0a_0' + \vec{a}\,\vec{a}^{\,\prime}\big)\,\mathrm{I}_2 + \big(a_0'\vec{a} + a_0\vec{a}^{\,\prime} + i\,\vec{a}\times\vec{a}^{\,\prime}\big)\cdot\vec{\sigma}\ , \tag{A.45}$$

with the scalar product $\vec{a}\,\vec{a}^{\,\prime} = a_xa_x' + a_ya_y' + a_za_z'$ for two vectors of $\mathbb{C}^3$ given as $\vec{a} = [a_x, a_y, a_z]^\top$ and $\vec{a}^{\,\prime} = [a_x', a_y', a_z']^\top$, and accordingly for their cross product $\vec{a}\times\vec{a}^{\,\prime}$. The relation of Eq. (A.45) can be repeated for the product $\mathsf{A}\mathsf{A}'\mathsf{A}''$ with a third operator A'' of $\mathcal{L}(\mathcal{H}_2)$ and would benefit in $\mathbb{C}^3$ from the double cross product:

$$\big(\vec{a}\times\vec{b}\,\big)\times\vec{c} = (\vec{a}\,\vec{c}\,)\vec{b} - (\vec{b}\,\vec{c}\,)\vec{a}\ . \tag{A.46}$$

For a Hermitian operator $\mathsf{A} \in \mathcal{L}(\mathcal{H}_2)$, the four scalar coordinates $(a_0, \vec{a}\,)$ are real, so as to satisfy $\mathsf{A}^\dagger = \mathsf{A}$. In this circumstance $\sqrt{\vec{a}^{\,2}} = \|\vec{a}\,\| \in \mathbb{R}$, the two eigenvalues $\lambda_\pm = a_0 \pm \|\vec{a}\,\|$ are real and are associated with the two orthogonal eigenstates $|\lambda_\pm\rangle$ with the two projectors onto them $|\pm\lambda\rangle\,\langle\pm\lambda| = \big(\mathrm{I}_2 \pm \vec{a}\cdot\vec{\sigma}/\|\vec{a}\,\|\big)/2$ characterized by the two unit vectors $\pm\vec{a}/\|\vec{a}\,\|$ of $\mathbb{R}^3$.

References

1. Feynman, R. P. (1982). Simulating physics with computers. *International Journal of Theoretical Physics, 21*, 467–488.
2. Feynman, R. P. (1986). Quantum mechanical computers. *Foundations of Physics, 16*, 507–531.
3. Dirac, P. A. M. (1939). A new notation for quantum mechanics. *Mathematical Proceedings of the Cambridge Philosophical Society, 35*, 416–418.
4. Schumacher, B. (1995). Quantum coding. *Physical Review A, 51*, 2738–2747.
5. Schrödinger, E. (1935). Discussion of probability relations between separated systems. *Mathematical Proceedings of the Cambridge Philosophical Society, 31*, 555–563.
6. Haroche, S., & Raimond, J.-M. (2006). *Exploring the Quantum: Atoms, Cavities, and Photons*. Oxford: Oxford University Press.
7. Schumacher, B., & Westmoreland, M. D. (2010). *Quantum Processes, Systems, and Information*. Cambridge: Cambridge University Press.
8. Hughston, L. P., Jozsa, R., & Wootters, W. K. (1993). A complete classification of quantum ensembles having a given density matrix. *Physics Letters A, 183*, 14–18.
9. Nielsen, M. A., & Chuang, I. L. (2000). *Quantum Computation and Quantum Information*. Cambridge: Cambridge University Press.
10. Gisin, N. (1989). Stochastic quantum dynamics and relativity. *Helvetica Physica Acta, 62*, 63–71.
11. Bassi, A., & Ghirardi, G. (2003). A general scheme for ensemble purification. *Physics Letters A, 309*, 24–28.
12. Wilde, M. M. (2017). *Quantum Information Theory*. Cambridge: Cambridge University Press.
13. Wootters, W. K., & Zurek, W. H. (1982). A single quantum cannot be cloned. *Nature, 299*, 802–803.
14. Scarani, V., Iblisdir, S., Gisin, N., & Acín, A. (2005). Quantum cloning. *Reviews of Modern Physics, 77*, 1225–1256.
15. Bruß, D., DiVincenzo, D. P., Ekert, A., Fuchs, C. A., Macchiavello, C., & Smolin, J. A. (1998). Optimal universal and state-dependent quantum cloning. *Physical Review A, 57*, 2368–2378.
16. Zeilinger, A. (1981). General properties of lossless beam splitters in interferometry. *American Journal of Physcis, 49*, 882–883.
17. Yurke, B., McCall, S. R., & Klauder, J. R. (1986). SU(2) and SU(1,1) interferometers. *Physical Review A, 33*, 4033–4054.

F. Chapeau-Blondeau, *Quantum Information and Quantum Signal Processing*,
https://doi.org/10.1007/978-3-032-27499-1

18. Reck, M., Zeilinger, A., Bernstein, H. J., & Bertani, P. (1994). Experimental realization of any discrete unitary operator. *Physical Review Letters, 73*, 58–61.
19. DiVincenzo, D. P. (1995). Two-bit gates are universal for quantum computation. *Physical Review A, 51*, 1015–1022.
20. Barenco, A., Bennett, C. H., Cleve, R., DiVincenzo, D. P., Margolus, N., Shor, P., Sleator, T., Smolin, J. A., & Weinfurter, H. (1995). Elementary gates for quantum computation. *Physical Review A, 52*, 3457–3467.
21. Zurek, W. H. (2003). Decoherence, einselection, and the quantum origins of the classical. *Reviews of Modern Physics, 75*, 715–775.
22. Schlosshauer, M. (2019). Quantum decoherence. *Physics Reports, 831*, 1–57.
23. Stinespring, W. F. (1955). Positive functions on C^*-algebras. *Proceedings of the American Mathematical Society, 6*, 211–216.
24. Marinescu, D. C., & Marinescu, G. M. (2012). *Classical and Quantum Information*. Oxford: Academic Press.
25. Holevo, A. S., & Giovannetti, V. (2012). Quantum channels and their entropic characteristics. *Reports on Progress in Physics, 75*, 0460011–30.
26. Schumacher, B. (1996). Sending entanglement through noisy quantum channels. *Physical Review A, 54*, 2614–2628.
27. Choi, M.-D. (1975). Completely positive linear maps on complex matrices. *Linear Algebra and Its Applications, 10*, 285–290.
28. Ruskai, M. B. (1994). Beyond strong subadditivity? Improved bounds on the contraction of generalized relative entropy. *Reviews in Mathematical Physics, 6*, 1147–1161.
29. Raginsky, M. (2002). Strictly contractive quantum channels and physically realizable quantum computers. *Physical Review A, 65*(032306), 1–13.
30. King, C., & Ruskai, M. B. (2001). Minimal entropy of states emerging from noisy quantum channels. *IEEE Transactions on Information Theory, 47*, 192–209.
31. King, C. (2002). Additivity for unital qubit channels. *Journal of Mathematical Physics, 43*, 4641–4653.
32. Shaham, A., Halevy, A., Dovrat, L., Megidish, E., & Eisenberg, H. S. (2015). Entanglement dynamics in the presence of controlled unital noise. *Scientific Reports, 5*(10796), 1–8.
33. Gillard, N., Belin, E., & Chapeau-Blondeau, F. (2019). Stochastic resonance with unital quantum noise. *Fluctuation and Noise Letters, 18*(1950015), 1–15.
34. Jamiołkowski, A. (1972). Linear transformations which preserve trace and positive semidefiniteness of operators. *Reports on Mathematical Physics, 3*, 275–278.
35. Jiang, M., Luo, S., & Fu, S. (2013). Channel-state duality. *Physical Review A, 87*, pp. 022310,1–8.
36. Horodecki, M., Shor, P. W., & Ruskai, M. B. (2003). Entanglement breaking channels. *Reviews in Mathematical Physics, 15*, 629–641.
37. Royer, A. (1996). Reduced dynamics with initial correlations, and time-dependent environment and Hamiltonians. *Physical Review Letters, 77*, 3272–3275.
38. Štelmachovič, P., & Bužek, V. (2001). Dynamics of open quantum systems initially entangled with environment: Beyond the Kraus representation. *Physical Review A, 64*(062106), 1–5.
39. Hayashi, H., Kimura, G., & Ota, Y. (2003). Kraus representation in the presence of initial correlations. *Physical Review A, 67*(062109), 1–5.
40. Vacchini, B., & Amato, J. (2016). Reduced dynamical maps in the presence of initial correlations. *Scientific Reports, 6*(37328), 1–12.
41. Paz-Silva, G. A., Hall, M. J. W., & Wiseman, H. M. (2019). Dynamics of initially correlated open quantum systems: Theory and application. *Physical Review A, 100*(042120), 1–19.
42. Breuer, H.-P., & Petruccione, F. (2007). *The Theory of Open Quantum Systems*. New York: Oxford University Press.
43. Rivas, Á., & Huelga, S. F. (2012). *Open Quantum Systems: An Introduction*. Heidelberg: Springer.
44. Lindblad, G. (1976). On the generators of quantum dynamical semigroups. *Communications in Mathematical Physics, 48*, 119–130.

45. Manzano, D. (2020). A short introduction to the Lindblad master equation. *AIP Advances, 10*(025106), 1–15.
46. Bechmann-Pasquinucci, H., Huttner, B., & Gisin, N. (1998). Non-linear quantum state transformation of spin-1/2. *Physics Letters A, 242*, 198–204.
47. Gilyén, A., Kiss, T., & Jex, I. (2016). Exponential sensitivity and its cost in quantum physics. *Scientific Reports, 6*(20076), 1–10.
48. Zhu, G., et al. (2019). Experimental orthogonalization of highly overlapping quantum states with single photons. *Physical Review A, 100*(052307), 1–5.
49. Sasaki, M., Barnett, S. M., Jozsa, R., Osaki, M., & Hirota, O. (1999). Accessible information and optimal strategies for real symmetrical quantum sources. *Physical Review A, 59*, 3325–3335.
50. Robertson, H. P. (1929). The uncertainty principle. *Physical Review, 34*, 163–164.
51. Cohen-Tannoudji, C., Diu, B., & Laloë, F. (1977). *Quantum Mechanics*. New York: Wiley.
52. Peres, A. (1993). *Quantum Theory: Concepts and Methods*. Dordrecht: Kluwer Academic Publishers.
53. de Broglie, L. (1925). Recherches sur la théorie des quanta. *Annales de Physique, 3*, 22–128.
54. Gabor, D. (1946). Theory of communication. Part 1: The analysis of information. *Journal of the Institution of Electrical Engineers, 93*, 429–441.
55. Mallat, S. (1999). *A Wavelet Tour of Signal Processing*. Burlington: Academic Press.
56. Deutsch, D. (1983). Uncertainty in quantum measurements. *Physical Review Letters, 50*, 631–633.
57. Zozor, S., Bosyk, G. M., & Portesi, M. (2013). On a generalized entropic uncertainty relation in the case of the qubit. *Journal of Physics A, 46*(465301), 1–17.
58. Coles, P. J., Berta, M., Tomamichel, M., & Wehner, S. (2017). Entropic uncertainty relations and their applications. *Reviews of Modern Physics, 89*(015002), 1–58.
59. Paris, M. G. A. (2012). The modern tools of quantum mechanics. A tutorial on quantum states, measurements, and operations. *The European Physical Journal Special Topics, 203*, 61–86. arXiv:1110.6815v2.
60. Brillouin, L. (1956). *Science and Information Theory*. Boston: Academic Press.
61. Bennett, C. H. (1982). The thermodynamics of computation - a review. *International Journal of Theoretical Physics, 21*, 905–940.
62. Lloyd, S. (1997). Quantum-mechanical Maxwell's demon. *Physical Review A, 56*, 3374–3382.
63. Maruyama, K., Nori, F., & Vedral, V. (2009). Colloquium: The physics of Maxwell's demon and information. *Reviews of Modern Physics, 81*, 1–23.
64. Parrondo, J. M. R., Horowitz, J. M., & Sagawa, T. (2015). Thermodynamics of information. *Nature Physics, 11*, 131–139.
65. Deutsch, D. (1989). Quantum computational networks. *Proceedings of the Royal Society of London A, 425*, 73–90.
66. Kok, P., & Lovett, B. W. (2010). *Optical Quantum Information Processing*. Cambridge: Cambridge University Press.
67. Slussarenko, S., & Pryde, G. J. (2019). Photonic quantum information processing: A concise review. *Applied Physics Reviews, 6*(041303), 1–19.
68. Devoret, M. H., & Schoelkopf, R. J. (2013). Superconducting circuits for quantum information: An outlook. *Science, 339*, 1169–1174.
69. Krantz, P., Kjaergaard, M., Yan, F., Orlando, T. P., Gustavsson, S., & Oliver, W. D. (2019). A quantum engineer's guide to superconducting qubits. *Applied Physics Reviews, 6*(021318), 1–57.
70. Burkard, G., Ladd, T. D., Pan, A., Nichol, J. M., & Petta, J. R. (2023). Semiconductor spin qubits. *Reviews of Modern Physics*, *95*, pp. 025003,1–58.
71. Fernández de Fuentes, I., et al. (2026). Running a six-qubit quantum circuit on a silicon spin-qubit array. *PRX Quantum, 7*(010308), 1–13.
72. Cirac, J. I., & Zoller, P. (1995). Quantum computations with cold trapped ions. *Physical Review Letters, 74*, 4091–4094.

73. Wineland, D. J. (2009). Quantum information processing and quantum control with trapped atomic ions. *Physica Scripta, T137*(014007), 1–11.
74. Briegel, H.-J., Calarco, T., Jaksch, D., Cirac, J. I., & Zoller, P. (2000). Quantum computing with neutral atoms. *Journal of Modern Optics, 47*, 415–451.
75. Levine, H., et al. (2019). Parallel implementation of high-fidelity multiqubit gates with neutral atoms. *Physical Review Letters, 123*(170503), 1–6.
76. Nayak, C., Simon, S. H., Stern, A., Freedman, M., & Das Sarma, S. (2008). Non-Abelian anyons and topological quantum computation. *Reviews of Modern Physics, 80*, 1083–1159.
77. Cheng, B., et al. (2023). Noisy intermediate-scale quantum computers. *Frontiers of Physics, 18*(21308), 1–62.
78. Tappertzhofen, S., & Alexander-Webber, J. (2025). *Quantum Technology*. Amsterdam: Elsevier.
79. Knill, E., Laflamme, R., & Milburn, G. J. (2001). A scheme for efficient quantum computation with linear optics. *Nature, 409*, 46–52.
80. O'Brien, T. E., Pryde, G. J., White, A. G., Ralph, T. C., & Branning, D. (2003). Demonstration of an all-optical quantum controlled-NOT gate. *Nature, 426*, 264–267.
81. Nemoto, K., & Munro, W. J. (2004). Nearly deterministic linear optical controlled-NOT gate. *Physical Review Letters*, *93*, pp. 250502,1–4.
82. Benioff, P. (1982). Quantum mechanical models of Turing machines that dissipate no energy. *Physical Review Letters, 48*, 1581–1585.
83. DiVincenzo, D. P. (1998). Quantum gates and circuits. *Proceedings of the Royal Society of London A, 454*, 261–276.
84. Vartiainen, J. J., Möttönen, M., & Salomaa, M. M. (2004). Efficient decomposition of quantum gates. *Physical Review Letters, 92*(177902), 1–4.
85. Nam, Y., Ross, N. J., Su Y., Childs, A. M., & Maslov, D. (2018). Automated optimization of large quantum circuits with continuous parameters. *npj Quantum Information*, *4*, 23,1–12.
86. Smith, R. S., Peterson, E. C., Skilbeck, M. G., & Davis, E. J. (2020). An open-source, industrial-strength optimizing compiler for quantum programs. *Quantum Science and Technology, 5*(044001), 1–20.
87. Mansky, M. B., Castillo, S. L., Puigvert, V. R., & Linnhoff-Popien, C. (2023). Near-optimal quantum circuit construction via Cartan decomposition. *Physical Review A, 108*(052607), 1–9.
88. Moro, L., Paris, M. G. A., Restelli, M., & Prati, E. (2021). Quantum compiling by deep reinforcement learning. *Communications Physics, 4*(178), 1–8.
89. Maronese, M., Moro, L., Rocutto, L., & Prati, E. (2022). Quantum compiling. In: *Quantum Computing Environments* (S. S. Iyengar, M. Mastriani, and K. J. Latesh Kumar, eds.), ch. 2, Berlin: Springer, 2022. arXiv:2112.00187 [quant-ph] (37 pages).
90. Chen, Q., Du, Y., Jiao, Y., Lu, X., Wu, X., & Zhao, Q. (2024). Efficient and practical quantum compiler towards multi-qubit systems with deep reinforcement learning. *Quantum Science and Technology, 9*(045002), 1–17.
91. Raussendorf, R., Browne, D. E., & Briegel, H. J. (2003). Measurement-based quantum computation on cluster states. *Physical Review A, 68*(022312), 1–32.
92. Briegel, H. J., Browne, D. E., Dür, W., Raussendorf, R., & Van den Nest, M. (2009). Measurement-based quantum computation. *Nature Physics, 5*, 19–26.
93. Montanaro, A. (2016). Quantum algorithms: An overview. *npj Quantum Information, 2*, 15023,1–8.
94. Dalzell, A. M., et al. (2025). *Quantum Algorithms*. Cambridge: Cambridge University Press.
95. Yan, F., Iliyasu, A. M., & Venegas-Andraca, S. E. (2016). A survey of quantum image representations. *Quantum Information Processing, 15*, 1–35.
96. Su, J., Guo, X., Liu, C., & Li, L. (2020). A new trend of quantum image representations. *IEEE Access, 8*, 214520–214537.
97. Grigoryan, A. M., & Agaian, S. S. (2025). *Quantum Image Processing in Practice*. New York: Wiley.

98. Deutsch, D. (1985). Quantum theory, the Church-Turing principle and the universal quantum computer. *Proceedings of the Royal Society of London A, 400*, 97–117.
99. Deutsch, D., & Jozsa, R. (1992). Rapid solution of problems by quantum computation. *Proceedings of the Royal Society of London A, 439*, 553–558.
100. Gillard, N., Belin, E., & Chapeau-Blondeau, F. (2018). Digital image processing with quantum approaches. In *Lecture Notes in Computer Science*, (vol. LNCS 10884, pp. 360–369). Berlin: Springer.
101. Chapeau-Blondeau, F., & Belin, E. (2016). Quantum image coding with a reference-frame-independent scheme. *Quantum Information Processing, 15*, 2685–2700.
102. Li, H.-S., Chen, X., Xia, H., Liang, Y., & Zhou, Z. (2018). A quantum image representation based on bitplanes. *IEEE Access, 6*, 62396–62404.
103. Ahmed, N. & Rao, K. R. (1975). Walsh–Hadamard transform, in *Orthogonal Transforms for Digital Signal Processing*, ch. 6, pp. 99–152, Berlin: Springer.
104. Fuchs, C. A., & Jacobs, K. (2001). Information-tradeoff relations for finite-strength quantum measurements. *Physical Review A,63*, 062305, 1–15.
105. Dotsenko, I., Mirrahimi, M., Brune, M., Haroche, S., Raimond, J.-M., & Rouchon, P. (2009). Quantum feedback by discrete quantum nondemolition measurements: Towards on-demand generation of photon-number states. *Physical Review A, 80*, 0138051–8.
106. Bernstein, E., & Vazirani, U. (1997). Quantum complexity theory. *SIAM Journal on Computing, 26*, 1411–1473.
107. Cleve, R., Ekert, A., Macchiavello, C., & Mosca, M. (1998). Quantum algorithms revisited. *Proceedings of the Royal Society of London A, 454*, 339–354.
108. Grover, L. K. (1996). A fast quantum mechanical algorithm for database search, in *Proceedings 28th Annual ACM Symposium on Theory of Computing*, (Philadelphia, USA), pp. 212–219. arXiv:quant-ph/9605043.
109. Grover, L. (1997). Quantum mechanics helps in searching for a needle in a haystack. *Physical Review Letters, 79*, 325–328.
110. Zalka, C. (1999). Grover's quantum searching algorithm is optimal. *Physical Review A, 60*, 2746–2751.
111. Boyer, M., Brassard, G., Høyer, P., & Tapp, A. (1998). Tight bounds on quantum searching. *Fortschritte der Physik, 46*, 493–505.
112. Brassard, G., Høyer, P., Mosca, M., & Tapp, A. (2002). Quantum amplitude amplification and estimation, in *Contemporary Mathematics: Quantum Computation and Information* (S. J. Lomonaco and H. E. Brandt, eds.), (vol. 305, pp. 53–7). Providence: American Mathematical Society. arXiv:quant-ph/0005055.
113. Brassard, G., Høyer, P., & Tapp, A. (1998). Quantum counting, in *Lecture Notes in Computer Science*, (vol. LNCS 1443, pp. 820–831). Berlin: Springer. arXiv:quant-ph/9805082.
114. Mosca, M. (2001). Counting by quantum eigenvalue estimation. *Theoretical Computer Science, 264*, 139–153.
115. Cerf, N. J., Grover, L. K., & Williams, C. P. (2000). Nested quantum search and structured problems. *Physical Review A, 61*, 0323031–14.
116. Williams, C. P. (2001). Quantum search algorithms in science and engineering. *Computing in Science & Engineering, 3*, 44–51.
117. Botsinis, P., Alanis, D., Babar, Z., Nguyen, H., Chandra, D., Ng, S. X., & Hanzo, L. (2019). Quantum search algorithms for wireless communications. *IEEE Communications Surveys & Tutorials, 21*, 1209–1242.
118. Shao, C., Li, Y., & Li, H. (2019). Quantum algorithm design: Techniques and applications. *Journal of Systems Science and Complexity, 32*, 375–452.
119. Dürr, C., & Høyer, P. (1999). A quantum algorithm for finding the minimum arXiv:quant-ph/9607014v2 arXiv.org e-print 2 pages.
120. Kowada, L. A. B., Lavor, C., Portugal, R., & de Figueiredo, C. M. H. (2008). A new quantum algorithm for solving the minimum searching problem. *International Journal of Quantum Information, 6*, 427–436.

121. Fredkin, E., & Toffoli, T. (1982). Conservative logic. *International Journal of Theoretical Physics, 21*, 219–253.
122. Chau, H. F., & Wilczek, F. (1995). Simple realization of the Fredkin gate using a series of two-body operators. *Physical Review Letters, 75*, 748–750.
123. Smolin, J. A., & DiVincenzo, D. P. (1996). Five two-bit quantum gates are sufficient to implement the quantum Fredkin gate. *Physical Review A, 53*, 2855–2056.
124. LaPierre, R. (2021). *Introduction to Quantum Computing*. Berlin: Springer.
125. Kasirajan, V. (2021). *Fundamentals of Quantum Computing*. Berlin: Springer.
126. Einstein, A., Podolsky, B., & Rosen, N. (1935). Can quantum-mechanical description of physical reality be considered complete? *Physical Review, 47*, 777–780.
127. Bell, J. S. (1964). On the Einstein-Podolsky-Rosen paradox. *Physics, 1*, 195–200.
128. Clauser, J. F., Horne, M. A., Shimony, A., & Holt, R. A. (1969). Proposed experiment to test local hidden-variable theories. *Physical Review Letters, 23*, 880–884.
129. Aspect, A. (1976). Proposed experiment to test the nonseparability of quantum mechanics. *Physical Review D, 14*, 1944–1951.
130. Freedman, J., & Clauser, J. F. (1972). Experimental test of local hidden-variable theories. *Physical Review Letters, 28*, 938–941.
131. Aspect, A., Grangier, P., & Roger, G. (1981). Experimental test of realistic theories vià Bell's theorem. *Physical Review Letters, 47*, 460–463.
132. Aspect, A., Dalibard, J., & Roger, G. (1982). Experimental test of Bell's inequalities using time-varying analyzers. *Physical Review Letters, 49*, 1804–1807.
133. Martins, L. P., Almeida, Á. J., André, P. S., & Pinto, A. N. (2012). Photon-pair states and violation of CHSH inequality. *Microwave and Optical Technology Letters, 54*, 2454–2461.
134. Bohr, N. (1935). Can quantum-mechanical description of physical reality be considered complete? *Physical Review, 48*, 696–702.
135. Gisin, N., & Peres, A. (1992). Maximal violation of Bell's inequality for arbitrarily large spin. *Physics Letters A, 162*, 15–17.
136. Hensen, B., et al. (2015). Loophole-free Bell inequality violation using electron spins separated by 1.3 kilometres. *Nature, 526*, 682–686.
137. Aspect, A. (2015). Closing the door on Einstein and Bohr's quantum debate. *Physics, 8*, 1231–4.
138. Storz, S., et al. (2023). Loophole-free Bell inequality violation with superconducting circuits. *Nature, 617*, 265–270.
139. Braunstein, S. L., & Caves, C. M. (1988). Information-theoretic Bell inequalities. *Physical Review Letters, 61*, 662–665.
140. Cerf, N. J., & Adami, C. (1997). Entropic Bell inequalities. *Physical Review A, 55*, 3371–3374.
141. Chapeau-Blondeau, F. (2014). Tsallis entropy for assessing quantum correlation with Bell-type inequalities in EPR experiment. *Physica A, 414*, 204–215.
142. Barrett, J., Linden, N., Massar, S., Pironio, S., Popescu, S., & Roberts, D. (2005). Nonlocal correlations as an information-theoretic resource. *Physical Review A, 71*(022101), 1–11.
143. Iqbal, A., Cheon, T., & Abbott, D. (2008). Probabilistic analysis of three-player symmetric quantum games played using the Einstein-Podolsky-Rosen-Bohm setting. *Physics Letters A, 372*, 6564–6577.
144. Chappell, J. M., Iqbal, A., & Abbott, D. (2012). N-player quantum games in an EPR setting. *PLOS ONE, 7*(e36404), 1–9.
145. Brunner, N., & Linden, N. (2013). Connection between Bell nonlocality and Bayesian game theory. *Nature Communications, 4*(2057), 1–6.
146. Greenberger, D. M., Horne, M. A., & Zeilinger, A. (1989). Going beyond Bell's theorem, in Bell's Theorem. *Quantum Theory and Conceptions of the Universe (M. Kafatos* (Vol. 37, pp. 69–72). Dordrecht: Springer. of Fundamental Theories of Physics.
147. Greenberger, D. M., Horne, M. A., Shimony, A., & Zeilinger, A. (1990). Bell's theorem without inequalities. *American Journal of Physics, 58*, 1131–1143.
148. Vaidman, L. (1999). Variations on the theme of the Greenberger-Horne-Zeilinger proof. *Foundations of Physics, 29*, 615–630.

149. Bouwmeester, D., Jian-Wei, P., Daniell, M., Weinfurter, H., & Zeilinger, A. (1999). Observation of three-photon Greenberger-Horne-Zeilinger entanglement. *Physical Review Letters, 82*, 1345–1349.
150. Wu, J.-Z., & Yan, L. (2020). Quantum key distribution protocol based on GHZ like state and Bell state, in *Artificial Intelligence and Security* (X. Sun, J. Wang, and E. Bertino, eds.), *Lecture Notes in Computer Science*,(vol. 12240, pp. 298–306). Berlin: Springer.
151. Bennett, C. H., & Wiesner, S. J. (1992). Communication via one- and two-particle operators on Einstein-Podolsky-Rosen states. *Physical Review Letters, 69*, 2881–2884.
152. Williams, B. P., Sadlier, R. J., & Humble, T. S. (2017). Superdense coding over optical fiber links with complete Bell-state measurements. *Physical Review Letters, 118*(050501), 1–5.
153. Wehner, S., Elkouss, D., & Hanson, R. (2018). Quantum internet: A vision for the road ahead. *Science, 362*(303), 1–9.
154. Rohde, P. P. (2021). *The Quantum Internet: The Second Quantum Revolution*. Cambridge: Cambridge University Press.
155. Bennett, C. H., Brassard, G., Crépeau, C., Jozsa, R., Peres, A., & Wootters, W. K. (1993). Teleporting an unknown quantum state via dual classical and Einstein-Podolsky-Rosen channels. *Physical Review Letters, 70*, 1895–1899.
156. Bouwmeester, D., Pan, J.-W., Mattle, K., Eibl, M., Weinfurter, H., & Zeilinger, A. (1997). Experimental quantum teleportation. *Nature, 390*, 575–579.
157. Werner, R. F. (2001). All teleportation and dense coding schemes. *Journal of Physics A, 34*, 7081–7094.
158. Shannon, C. E. (1949). Communication theory of secrecy systems. *Bell System Technical Journal, 28*, 656–715.
159. Luenberger, D. G. (2006). *Information Science*. Princeton: Princeton University Press.
160. Bennett, C. H., & Brassard, G. (1984). Quantum cryptography: Public key distribution and coin tossing, in *Proceedings International Conference on Computers, Systems and Signal Processing*, (Bangalore, India), pp. 175–179.
161. Gisin, N., Ribordy, G., Tittel, W., & Zbinden, H. (2002). Quantum cryptography. *Reviews of Modern Physics, 74*, 145–195.
162. Ekert, A. K. (1991). Quantum cryptography based on Bell's theorem. *Physical Review Letters, 67*, 661–663.
163. Bozzio, M., Orieux, A., Trigo Vidarte, L., Zaquine, I., Kerenidis, I., & Diamanti, E. (2018). Experimental investigation of practical unforgeable quantum money, *npj Quantum Information, 4*, 5,1–8.
164. Pirandola, S., Andersen, U. L., Banchi, L., Berta, M., Bunandar, D., Colbeck, R., Englund, D., Gehring, T., Lupo, C., Ottaviani, C., Pereira, J. L., Razavi, M., Shaari, J. S., Tomamichel, M., Usenko, V. C., Vallone, G., Villoresi, P., & Wallden, P. (2020). Advances in quantum cryptography. *Advances in Optics and Photonics, 12*, 1012–1236.
165. Briegel, H.-J., Dür, W., Cirac, J. I., & Zoller, P. (1998). Quantum repeaters: The role of imperfect local operations in quantum communication. *Physical Review Letters, 81*, 5932–5935.
166. Azuma, K., Economou, S. E., Elkouss, D., Hilaire, P., Jiang, L., Lo, H.-K., & Tzitrin, I. (2023). Quantum repeaters: From quantum networks to the quantum internet. *Reviews of Modern Physics, 95*(045006), 1–66.
167. Vinay, S. E., & Kok, P. (2017). Practical repeaters for ultralong-distance quantum communication. *Physical Review A, 95*(052336), 1–7.
168. Goodenough, K., Elkouss, D., & Wehner, S. (2021). Optimizing repeater schemes for the quantum internet. *Physical Review A, 103*(032610), 1–31.
169. Albane, A., Cubeddu, M., Mai, V. S., & Battou, A. (2025). Entanglement routing in quantum networks: A comprehensive survey. *IEEE Transactions on Quantum Engineering, 6*(4100639), 1–39.
170. Horodecki, R., Horodecki, P., Horodecki, M., & Horodecki, K. (2009). Quantum entanglement. *Reviews of Modern Physics, 81*, 865–942.

171. Shannon, C. E. (1948). A mathematical theory of communication. *Bell System Technical Journal, 27*(379–423), 623–656.
172. Wehrl, A. (1978). General properties of entropy. *Reviews of Modern Physics, 50*, 221–260.
173. Cover, T. M., & Thomas, J. A. (2006). *Elements of Information Theory*. New York: Wiley.
174. International Organization for Standardization. (2025). Quantities and units - Part 13: Information science and technology. *ISO IEC, 80000–13*, 2025.
175. MacKay, D. J. C. (2003). *Information Theory, Inference, and Learning Algorithms*. Cambridge: Cambridge University Press.
176. Holevo, A. S. (1973). Bounds for the quantity of information transmitted by a quantum communication. *Problems of Information Transmission, 9*, 177–183.
177. Ohya, M., & Watanabe, N. (2010). Quantum entropy and its applications to quantum communication and statistical physics. *Entropy, 12*, 1194–1245.
178. Vedral, V. (2002). The role of relative entropy in quantum information theory. *Reviews of Modern Physics, 74*, 197–234.
179. Schumacher, B., & Westmoreland, M. D. (2001). Optimal signal ensembles. *Physical Review A, 63*(022308), 1–5.
180. Majtey, A. P., Lamberti, P. W., & Prato, D. P. (2005). Jensen-Shannon divergence as a measure of distinguishability between mixed quantum states. *Physical Review A, 72*(052310), 1–6.
181. Lieb, E. H., & Ruskai, M. B. (1973). Proof of the strong subadditivity of quantum-mechanical entropy. *Journal of Mathematical Physics, 14*, 1938–1941.
182. Ruskai, M. B. (2002). Inequalities for quantum entropy: A review with conditions for equality. *Journal of Mathematical Physics, 43*, 4358–4375.
183. Holevo, A. S. (2019). *Quantum Systems, Channels, Information*. Berlin: De Gruyter.
184. Lindblad, G. (1975). Completely positive maps and entropy inequalities. *Communications in Mathematical Physics, 40*, 147–151.
185. Uhlmann, A. (1977). Relative entropy and the Wigner-Yanase-Dyson-Lieb concavity in an interpolation theory. *Communications in Mathematical Physics, 54*, 21–32.
186. Yuen, H. P., & Ozawa, M. (1993). Ultimate information carrying limit of quantum systems. *Physical Review Letters, 70*, 363–366.
187. King, C., & Ruskai, M. B. (2001). Capacity of quantum channels using product measurements. *Journal of Mathematical Physics, 42*, 87–98.
188. Huang, Y. (2014). Computing quantum discord is NP-complete. *New Journal of Physics, 16*(033027), 1–14.
189. D'Ariano, G. M., Lo Presti, P. & Perinotti, P. (2005). Classical randomness in quantum measurements, *Journal of Physics A, 38*, 5979–5991.
190. Hamieh, S., Kobes, P., & Zaraket, H. (2004). Positive-operator-valued measure optimization of classical correlations. *Physical Review A, 70*(052325), 1–6.
191. Ollivier, H., & Zurek, W. H. (2002). Quantum discord: A measure of the quantumness of correlations. *Physical Review Letters, 98*(017901), 1–4.
192. Datta, A. (2011). A condition for the nullity of quantum discord, arXiv.org e-print arXiv:quant-ph/1003.5256v2 3 pages.
193. Zurek, W. H. (2000). Einselection and decoherence from an information theory perspective. *Annalen der Physik, 512*, 855–864.
194. Henderson, L., & Vedral, V. (2001). Classical, quantum and total correlations. *Journal of Physics A, 34*, 6899–6905.
195. Werner, R. F. (1989). Quantum states with Einstein-Podolsky-Rosen correlations admitting a hidden-variable model. *Physical Review A, 40*, 4277–4281.
196. Madhok, V. (2013). Quantum discord as a resource in quantum communication. *International Journal of Modern Physics B, 27*(1345041), 1–17.
197. Joshi, A., Singh, P., & Kumar, A. (2025). A graph-theoretical framework to analyze zero discord quantum states. *Quantum Information Processing, 24*(111), 1–23.
198. Zhang, J.-S., & Chen, A.-X. (2012). Review of quantum discord in bipartite and multipartite systems. *Quantum Physics Letters, 1*, 69–77.

199. Radhakrishnan, C., Laurière, M., & Byrnes, T. (2020). Multipartite generalization of quantum discord. *Physical Review Letters, 124*(110401), 1–6.
200. Jozsa, R., & Schumacher, B. (1994). A new proof of the quantum noiseless coding theorem. *Journal of Modern Optics, 41*, 2343–2349.
201. Barnum, H., Fuchs, C. A., Jozsa, R., & Schumacher, B. (1996). General fidelity limit for quantum channels. *Physical Review A, 54*, 4707–4711.
202. Fuchs, C. A., & van de Graaf, J. (1999). Cryptographic distinguishability measures for quantum-mechanical states. *IEEE Transactions on Information Theory, 45*, 1216–1227.
203. Braunstein, S. L., Fuchs, C. A., Gottesman, D., & Lo, H.-K. (2000). A quantum analog of Huffman coding. *IEEE Transactions on Information Theory, 46*, 1644–1649.
204. Chuang, I. L., & Modha, D. S. (2000). Reversible arithmetic coding for quantum data compression. *IEEE Transactions on Information Theory, 46*, 1104–1116.
205. Jozsa, R., & Presnell, S. (2003). Universal quantum information compression and degrees of prior knowledge. *Proceedings of the Royal Society of London A, 459*, 3061–3077.
206. Lo, H.-K. (1995). Quantum coding theorem for mixed states. *Optics Communications, 119*, 552–556.
207. Horodecki, M. (1998). Limits for compression of quantum information carried by ensembles of mixed states. *Physical Review A, 57*, 3364–3369.
208. Barnum, H., Caves, C. M., Fuchs, C. A., Jozsa, R., & Schumacher, B. (2001). On quantum coding for ensembles of mixed states. *Journal of Physics A, 34*, 6767–6785.
209. Koashi, M., & Imoto, N. (2001). Compressibility of quantum mixed-state signals. *Physical Review Letters, 87*(017902), 1–4.
210. Khanian, Z. B., & Winter, A. (2022). General mixed-state quantum data compression with and without entanglement assistance. *IEEE Transactions on Information Theory, 68*, 3130–3138.
211. Bennett, C. H., Bernstein, H. J., Popescu, S., & Schumacher, B. (1996). Concentrating partial entanglement by local operations. *Physical Review A, 53*, 2046–2052.
212. Fujita, H. (2009). Entanglement concentration and dilution revisited. *International Journal of Quantum Information, 7*, 1491–1505.
213. Bennett, C. H., Brassard, G., Popescu, S., Schumacher, B., Smolin, J. A., & Wootters, W. K. (1996). Purification of noisy entanglement and faithful teleportation via noisy channels. *Physical Review Letters, 76*, 722–725.
214. Bose, S., Vedral, V., & Knight, P. L. (1999). Purification via entanglement swapping and conserved entanglement. *Physical Review A, 60*, 194–197.
215. Rozpedek, F., Schiet, T., Thinh, L. P., Elkouss, D., Doherty, A. C., & Wehner, S. (2018). Optimizing practical entanglement distillation. *Physical Review A, 97*(062333), 1–32.
216. Bennett, C. H., DiVincenzo, D. P., Smolin, J. A., & Wootters, W. K. (1996). Mixed-state entanglement and quantum error correction. *Physical Review A, 54*, 3824–3851.
217. Lo, H.-K., & Popescu, S. (1999). Classical communication cost of entanglement manipulation: Is entanglement an interconvertible resource? *Physical Review Letters, 83*, 1459–1462.
218. Murano, M., Plenio, M. B., Popescu, S., Vedral, V., & Knight, P. L. (1999). Multiparticle entanglement purification protocols. *Physical Review A, 57*, R4075–R4078.
219. Wang, X., & Wilde, M. M. (2020). Cost of quantum entanglement simplified. *Physical Review Letters, 125*(040502), 1–7.
220. Chapeau-Blondeau, F. (2024). Simulating quantum Pauli noise with three independently controlled Pauli gates. *Electronics, 13*(439), 1–13.
221. Chapeau-Blondeau, F. (2015). Optimization of quantum states for signaling across an arbitrary qubit noise channel with minimum-error detection. *IEEE Transactions on Information Theory, 61*, 4500–4510.
222. Falaye, B. J., Adepoju, A. G., Aliyu, A. S., Melchor, M. M., Liman, M. S., Oluwadare, O. J., González-Ramírez, M. D., & Oyewumi, K. J. (2017). Investigating quantum metrology in noisy channels. *Scientific Reports, 7*(16622), 1–9.
223. Chapeau-Blondeau, F. (2022). Modeling and simulation of a quantum thermal noise on the qubit. *Fluctuation and Noise Letters, 21*(2250060), 1–17.

224. Rosati, M., Mari, A., & Giovannetti, V. (2018). Narrow bounds for the quantum capacity of thermal attenuators. *Nature Communications, 9*(4339), 1–9.
225. Khatri, S., Sharma, K. & Wilde, M. M. (2020). Information-theoretic aspects of the generalized amplitude-damping channel, *Physical Review A, 102*, pp. 012401,1–31.
226. Gillard, N., Belin, E., & Chapeau-Blondeau, F. (2018). Enhancing qubit information with quantum thermal noise. *Physica A, 507*, 219–230.
227. Srikanth, R., & Banerjee, S. (2008). Squeezed generalized amplitude damping channel. *Physical Review A, 77*, 0123181–9.
228. Banerjee, S., & Srikanth, R. (2008). Geometric phase of a qubit interacting with a squeezed-thermal bath. *European Physical Journal D, 46*, 335–344.
229. Omkar, S., Srikanth, R., & Banerjee, S. (2013). Dissipative and non-dissipative single-qubit channels: Dynamics and geometry. *Quantum Information Processing, 12*, 3725–3744.
230. Olivares, S., Popovic, M., & Paris, M. G. A. (2016). Phase estimation with squeezed single photons. *Quantum Measurements and Quantum Metrology, 3*, 338–343.
231. Chapeau-Blondeau, F. (2018). Optimized entanglement for quantum parameter estimation from noisy qubits. *International Journal of Quantum Information, 16*(1850056), 1–25.
232. Shaji, A., & Caves, C. M. (2007). Qubit metrology and decoherence. *Physical Review A, 76*(032111), 1–13.
233. Clerk, A. A., Devoret, M. H., Girvin, S. M., Marquardt, F., & Schoelkopf, R. J. (2010). Introduction to quantum noise, measurement, and amplification. *Reviews of Modern Physics, 82*, 1155–1208.
234. Gardiner, C. W., & Zoller, P. (2004). *Quantum Noise*. Berlin: Springer.
235. Helstrom, C. W. (1976). *Quantum Detection and Estimation Theory*. New York: Academic Press.
236. Van Trees, H. L. (2001). *Detection, Estimation, and Modulation Theory, Part 1*. New York: Wiley.
237. McDonough, R. N., & Whalen, A. D. (1995). *Detection of Signals in Noise*. New York: Academic Press.
238. Kay, S. M. (1998). *Fundamentals of Statistical Signal Processing: Detection Theory*. Englewood Cliffs: Prentice Hall.
239. Poor, H. V. (1994). *An Introduction to Signal Detection and Estimation*. Berlin: Springer.
240. Helstrom, C. W. (1969). Quantum detection and estimation theory. *Journal of Statistical Physics, 1*, 231–252.
241. Holevo, A. S. (1973). Statistical decision theory for quantum systems. *Journal of Multivariate Analysis, 3*, 337–394.
242. Yuen, H. P., Kennedy, R. S., & Lax, M. (1975). Optimum testing of multiple hypotheses in quantum detection theory. *IEEE Transactions on Information Theory, 21*, 125–134.
243. Hunter, K. (2003). Measurement does not always aid state discrimination. *Physical Review A, 68*(012306), 1–3.
244. Barnett, S. M., & Croke, S. (2009). Quantum state discrimination. *Advances in Optics and Photonics, 1*, 238–278.
245. Bae, J. (2013). Structure of minimum-error quantum state discrimination. *New Journal of Physics, 15*(073037), 1–31.
246. Bergou, J. A., Herzog, U., & Hillery, M. (2004). Discrimination of quantum states, in *Quantum State Estimation* (M. G. A. Paris and J. Řeháček, eds.), Lecture Notes in Physics 649, ch. 11, pp. 417–465, Berlin: Springer.
247. Bae, J., & Hwang, W.-Y. (2013). Minimum-error discrimination of qubit states: Methods, solutions, and properties. *Physical Review A, 87*(012334), 1–4.
248. Weir, G., Barnett, S. M., & Croke, S. (2017). Optimal discrimination of single-qubit mixed states. *Physical Review A, 96*(022312), 1–6.
249. Weir, G., Hughes, C., Barnett, S. M., & Croke, S. (2018). Optimal measurement strategies for the trine states with arbitrary prior probabilities. *Quantum Science and Technology, 3*(035003), 1–12.

250. Ha, D., & Kwon, Y. (2023). Complete analysis to minimum-error discrimination of four mixed qubit states with arbitrary prior probabilities. *Quantum Information Processing, 22*(67), 1–14.
251. Qiu, D., & Li, L. (2010). Minimum-error discrimination of quantum states: Bounds and comparisons. *Physical Review A, 81*(042329), 1–8.
252. Ježek, M., Řeháček, J., & Fiurášek, J. (2002). Finding optimal strategies for minimum-error quantum-state discrimination, *Physical Review A, 65*, 060301(R),1–4.
253. Eldar, Y. C., Megretski, A., & Verghese, G. C. (2003). Designing optimal quantum detectors via semidefinite programming. *IEEE Transactions on Information Theory, 49*, 1007–1012.
254. Bae, J., & Kwek, L.-C. (2015). Quantum state discrimination and its applications. *Journal of Physics A, 48*(083001), 1–35.
255. Flatt, K., Barnett, S. M., & Croke, S. (2019). Multiple-copy state discrimination of noisy qubits. *Physical Review A, 100*(032122), 1–10.
256. Singal, T., Kim, E., & Ghosh, S. (2019). Structure of minimum error discrimination for linearly independent states. *Physical Review A, 99*(052334), 1–7.
257. Kechrimparis, S., Kropf, C. M., Wudarski, F., & Bae, J. (2020). Channel coding of a quantum measurement. *IEEE Journal on Selected Areas in Communications, 38*, 439–448.
258. Han, R., Leuchs, G., & Bergou, J. A. (2020). Helstrom measurement: A nondestructive implementation. *Physical Review A, 101*(032103), 1–6.
259. Halder, S., & Srivastava, C. (2020). Locally distinguishing quantum states with limited classical communication. *Physical Review A, 101*(052313), 1–7.
260. Stoica, P., Babu, P., & Kundu, N. K. (2024). Quantum state discrimination: A tutorial on basic properties of the optimal measurement matrices. *IEEE Signal Processing Magazine, 41*, 100–110.
261. Ivanovic, I. D. (1987). How to differentiate between non-orthogonal states. *Physics Letters A, 123*, 257–259.
262. Dieks, D. (1988). Overlap and distinguishability of quantum states. *Physics Letters A, 126*, 303–306.
263. Peres, A. (1988). How to differentiate between non-orthogonal states. *Physics Letters A, 128*, 19.
264. Jaeger, G., & Shimony, A. (1995). Optimal distinction between two non-orthogonal quantum states. *Physics Letters A, 197*, 83–87.
265. Rudolph, T., Spekkens, R. W., & Turner, P. S. (2003). Unambiguous discrimination of mixed states, *Physical Review A, 68*, 010301(R),1–4.
266. Raynal, P., Lütkenhaus, N., & van Enk, S. J. (2003). Reduction theorems for optimal unambiguous state discrimination of density matrices. *Physical Review A, 68*(022308), 1–6.
267. Raynal, P., & Lütkenhaus, N. (2007). Optimal unambiguous state discrimination of two density matrices: A second class of exact solutions. *Physical Review A, 76*(052322), 1–13.
268. Herzog, U., & Bergou, J. A. (2005). Optimum unambiguous discrimination of two mixed quantum states, *Physical Review A, 71*, 050301(R),1–4.
269. Kleinmann, M., Kampermann, H., & Bruß, D. (2010). Structural approach to unambiguous discrimination of two mixed quantum states. *Journal of Mathematical Physics, 51*(032201), 1–25.
270. Eldar, Y. C., Stojnic, M., & Hassibi, B. (2004). Optimal quantum detectors for unambiguous detection of mixed states. *Physical Review A, 69*(062318), 1–9.
271. Pang, S., & Wu, S. (2009). Optimum unambiguous discrimination of linearly independent pure states. *Physical Review A, 80*(052320), 1–14.
272. Bergou, J. A. (2010). Discrimination of quantum states. *Journal of Modern Optics, 57*, 160–180.
273. Eldar, Y. C. (2003). A semidefinite programming approach to optimal unambiguous discrimination of quantum states. *IEEE Transactions on Information Theory, 49*, 446–456.
274. Fiurášek, J., & Ježek, M. (2003). Optimal discrimination of mixed quantum states involving inconclusive results. *Physical Review A, 67*(012321), 1–5.
275. Eldar, Y. C. (2003). Mixed-quantum-state detection with inconclusive results. *Physical Review A, 67*(042309), 1–14.

276. Davies, E. B. (1978). Information and quantum measurement. *IEEE Transactions on Information Theory, 24*, 596–599.
277. Řeháček, J., Englert, B.-G., & Kaszlikowski, D. (2005). Iterative procedure for computing accessible information in quantum communication. *Physical Review A, 71*(054303), 1–4.
278. Hausladen, P., & Wootters, K. W. (1994). A 'pretty good' measurement for distinguishing quantum states. *Journal of Modern Optics, 41*, 2385–2390.
279. Eldar, Y. C., & Forney, G. D. (2001). On quantum detection and the square-root measurement. *IEEE Transactions on Information Theory, 47*, 858–872.
280. Pozza, N. D., & Pierobon, G. (2015). Optimality of square-root measurements in quantum state discrimination. *Physical Review A, 91*(042334), 1–10.
281. Eldar, Y. C. (2003). Von Neumann measurement is optimal for detecting linearly independent mixed quantum states. *Physical Review A, 68*(052303), 1–4.
282. Eldar, Y. C., Megretski, A., & Verghese, G. C. (2004). Optimal detection of symmetric mixed quantum states. *IEEE Transactions on Information Theory, 50*, 1198–1207.
283. Ban, M., Kurokawa, K., Momose, R., & Hirota, O. (1997). Optimum measurements for discrimination among symmetric states and parameter estimation. *International Journal of Theoretical Physics, 36*, 1269–1288.
284. Kay, S. M. (1993). *Fundamentals of Statistical Signal Processing: Estimation Theory*. Englewood Cliffs: Prentice Hall.
285. Zamir, R. (1998). A proof of the Fisher information inequality via a data processing argument. *IEEE Transactions on Information Theory, 44*, 1246–1250.
286. Cohen, M. L. (1968). The Fisher information and convexity. *IEEE Transactions on Information Theory, 14*, 591–592.
287. Barrett, H. H., & Myers, K. J. (2004). *Foundations of Image Science*. New York: Wiley.
288. Stoica, P., & Marzetta, T. L. (2001). Parameter estimation problems with singular information matrices. *IEEE Transactions on Signal Processing, 49*, 87–90.
289. Gill, R. D., & Levit, B. Y. (1995). Applications of the van Trees inequality: a Bayesian Cramér-Rao bound. *Bernoulli, 1*, 59–79.
290. Lehmann, E. L., & Casella, G. (1998). *Theory of Point Estimation*. New York: Springer.
291. Paris, M. G. A., & Řeháček, J. (Eds.). (2004). *Quantum State Estimation* Lecture Notes in Physics (Vol (Vol. 649)). Berlin: Springer.
292. Paris, M. G. A. (2009). Quantum estimation for quantum technology. *International Journal of Quantum Information, 7*, 125–137.
293. Chapeau-Blondeau, F. (2015). Optimized probing states for qubit phase estimation with general quantum noise. *Physical Review A, 91*(052310), 1–13.
294. Holevo, A. S. (1982). *Probabilistic and Statistical Aspects of Quantum Theory*. Amsterdam: North-Holland.
295. Liu, J., Chen, J., Jing, X.-X., & Wang, X. (2016). Quantum Fisher information and symmetric logarithmic derivative via anti-commutators. *Journal of Physics A, 49*(275302), 1–12.
296. Braunstein, S. L., & Caves, C. M. (1994). Statistical distance and the geometry of quantum states. *Physical Review Letters, 72*, 3439–3443.
297. Barndorff-Nielsen, O. E., & Gill, R. D. (2000). Fisher information in quantum statistics. *Journal of Physics A, 33*, 4481–4490.
298. Luati, A. (2004). Maximum Fisher information in mixed state quantum systems. *The Annals of Statistics, 32*, 1770–1779.
299. Toscano, F., Bastos, W. P., & de Matos Filho, R. L. (2017). Attainability of the quantum information bound in pure-state models, *Physical Review A, 95*, 042125,1–14.
300. Berry, D. W., & Wiseman, H. M. (2000). Optimal states and almost optimal adaptive measurements for quantum interferometry. *Physical Review Letters, 85*, 5098–5101.
301. van Dam, W., D'Ariano, G. M., Ekert, A., Macchiavello, C., & Mosca, M. (2007). Optimal quantum circuits for general phase estimation. *Physical Review Letters, 98*(090501), 1–4.
302. Ji, Z., Wang, G., Duan, R., Feng, Y., & Ying, M. (2008). Parameter estimation of quantum channels. *IEEE Transactions on Information Theory, 54*, 5172–5185.

303. Berry, D. W., Higgins, B. L., Bartlett, S. D., Mitchell, M. W., Pryde, G. J., & Wiseman, H. M. (2009). How to perform the most accurate possible phase measurements. *Physical Review A, 80*(052114), 1–22.
304. Kaftal, T., & Demkowicz-Dobrzański, R. (2014). Usefulness of an enhanced Kitaev phase-estimation algorithm in quantum metrology and computation. *Physical Review A, 90*(062313), 1–6.
305. Rodríguez-García, M. A., Castillo, IPérez., & Barberis-Blostein, P. (2021). *Efficient qubit phase estimation using adaptive measurements, Quantum, 5* (Vol. 467, pp. 1–22)
306. Smith, J. G., Barnes, C. H. W., & Arvidsson-Shukur, D. R. M. (2024). Adaptive Bayesian quantum algorithm for phase estimation. *Physical Review A, 109*(042412), 1–8.
307. Petz, D., & Ghinea, C. (2011). Introduction to quantum Fisher information, in *Proceedings 30th Conference on Quantum Probability and Related Topics*, pp. 261–281, Singapore: World Scientific. arXiv:1008.2417.
308. Liu, J., Yuan, H., Lu, X.-M., & Wang, X. (2020). Quantum Fisher information and multiparameter estimation. *Journal of Physics A, 53*(023001), 1–68.
309. Fujiwara, A. (2001). Quantum channel identification problem. *Physical Review A, 63*(042304), 1–5.
310. Alipour, S., & Rezakhani, A. T. (2015). Extended convexity of quantum Fisher information in quantum metrology. *Physical Review A, 91*(042104), 1–7.
311. Chapeau-Blondeau, F. (2017). Entanglement-assisted quantum parameter estimation from a noisy qubit pair: A Fisher information analysis. *Physics Letters A, 381*, 1369–1378.
312. Sidhu, J. S., & Kok, P. (2020). Geometric perspective on quantum parameter estimation. *AVS Quantum Science, 2*(014701), 1–34.
313. Giovannetti, V., Lloyd, S., & Maccone, L. (2006). Quantum metrology. *Physical Review Letters, 96*(010401), 1–4.
314. Giovannetti, V., Lloyd, S., & Maccone, L. (2011). Advances in quantum metrology. *Nature Photonics, 5*, 222–229.
315. D'Ariano, G. M., Macchiavello, C., & Sacchi, M. F. (1998). On the general problem of quantum phase estimation. *Physics Letters A, 248*, 103–108.
316. Degen, C. L., Reinhard, F., & Cappellaro, P. (2017). Quantum sensing. *Reviews of Modern Physics, 89*(035002), 1–39.
317. Tóth, G., & Apellaniz, I. (2014). Quantum metrology from a quantum information science perspective. *Journal of Physics A, 47*(424006), 1–39.
318. Giovannetti, V., Lloyd, S., & Maccone, L. (2004). Quantum-enhanced measurements: Beating the standard quantum limit. *Science, 306*, 1330–1336.
319. Zwierz, M., Prez-Delgado, C. A., & Kok, P. (2010). General optimality of the Heisenberg limit for quantum metrology. *Physical Review Letters, 105*(180402), 1–4.
320. Zhou, S., Zhang, M., Preskill, J., & Jiang, L. (2018). Achieving the Heisenberg limit in quantum metrology using quantum error correction. *Nature Communications, 9*(78), 1–11.
321. Polino, E., Valeri, M., Spagnolo, N., & Sciarrino, F. (2020). Photonic quantum metrology. *AVS Quantum Science, 2*(024703), 1–41.
322. Zhong, W., Sun, Z., Ma, J., Wang, X., & Nori, F. (2013). Fisher information under decoherence in Bloch representation. *Physical Review A, 87*(022337), 1–14.
323. Chapeau-Blondeau, F. (2016). Optimizing qubit phase estimation. *Physical Review A, 94*(022334), 1–14.
324. Tesio, E., Olivares, S., & Paris, M. G. A. (2011). Optimized qubit phase estimation in noisy quantum channels. *International Journal of Quantum Information, 9*, 379–387.
325. Brivio, D., Cialdi, S., Vezzoli, S., Gebrehiwot, B. T., Genoni, M. G., Olivares, S., & Paris, M. G. A. (2010). Experimental estimation of one-parameter qubit gates in the presence of phase diffusion. *Physical Review A, 81*(012305), 1–7.
326. D'Ariano, G. M., Lo Presti, P. & Paris, M. G. A. (2001). Using entanglement improves the precision of quantum measurements, *Physical Review Letters*, *87*, 270404,1–4.
327. Lloyd, S. (2008). Enhanced sensitivity of photodetection via quantum illumination. *Science, 321*, 1463–1465.

328. Pirandola, S., Bardhan, B. R., Gehring, T., Weedbrook, C., & Lloyd, S. (2018). Advances in photonic quantum sensing. *Nature Photonics, 12*, 724–733.
329. Assouly, R., Dassonneville, R., Peronnin, T., Bienfait, A., & Huard, B. (2023). Quantum advantage in microwave quantum radar. *Nature Physics, 19*, 1418–1422.
330. Borderieux, S., Coatanhay, A., & Khenchaf, A. (2023). Quantum illumination radar using polarization states of photons in atmosphere: Quantum information approach. *Progress In Electromagnetics Research B, 103*, 101–118.
331. Torromé, R. G., & Barzanjeh, S. (2024). Advances in quantum radar and quantum LiDAR. *Progress in Quantum Electronics, 93*(100497), 1–19.
332. Djordjevic, I. B. (2025). *Quantum Communication, Quantum Networks, and Quantum Sensing*. New York: Academic Press.
333. Kantsepolsky, B., Aviv, I., Weitzfeld, R., & Bordo, E. (2023). Exploring quantum sensing potential for systems applications. *IEEE Spectrum, 11*, 31569–31582.
334. Gammaitoni, L., Hänggi, P., Jung, P., & Marchesoni, F. (1998). Stochastic resonance. *Reviews of Modern Physics, 70*, 223–287.
335. McDonnell, M. D., Stocks, N. G., Pearce, C. E. M., & Abbott, D. (2008). *Stochastic Resonance: From Suprathreshold Stochastic Resonance to Stochastic Signal Quantization*. Cambridge: Cambridge University Press.
336. Zozor, S., & Amblard, P. O. (2002). On the use of stochastic resonance in sine detection. *Signal Processing, 82*, 353–367.
337. Chapeau-Blondeau, F., & Rousseau, D. (2004). Noise-enhanced performance for an optimal Bayesian estimator. *IEEE Transactions on Signal Processing, 52*, 1327–1334.
338. Chen, H., Varshney, P. K., & Michels, J. H. (2008). Noise enhanced parameter estimation. *IEEE Transactions on Signal Processing, 56*, 5074–5081.
339. Pan, Y., Duan, F., Chapeau-Blondeau, F., & Abbott, D. (2018). Noise enhancement in robust estimation of location. *IEEE Transactions on Signal Processing, 66*, 1953–1966.
340. Duan, F., Pan, Y., Chapeau-Blondeau, F., & Abbott, D. (2019). Noise benefits in combined nonlinear Bayesian estimators. *IEEE Transactions on Signal Processing, 67*, 4611–4623.
341. Ting, J. J. L. (1999). Stochastic resonance for quantum channels. *Physical Review E, 59*, 2801–2803.
342. Bowen, G., & Mancini, S. (2006). Stochastic resonance effects in quantum channels. *Physics Letters A, 352*, 272–275.
343. Chapeau-Blondeau, F. (2015). Qubit state estimation and enhancement by quantum thermal noise. *Electronics Letters, 51*, 1673–1675.
344. Gillard, N., Belin, E., & Chapeau-Blondeau, F. (2017). Stochastic antiresonance in qubit phase estimation with quantum thermal noise. *Physics Letters A, 381*, 2621–2628.
345. Bowen, G., & Mancini, S. (2004). Noise enhancing the classical information capacity of a quantum channel. *Physics Letters A, 321*, 1–5.
346. Caruso, F., Huelga, S. F., & Plenio, M. B. (2010). Noise-enhanced classical and quantum capacities in communication networks. *Physical Review Letters, 105*(190501), 1–4.
347. Chapeau-Blondeau, F. (2014). Quantum state discrimination and enhancement by noise. *Physics Letters A, 378*, 2128–2136.
348. Gillard, N., Belin, E., & Chapeau-Blondeau, F. (2018). Qubit state detection and enhancement by quantum thermal noise. *Electronics Letters, 54*, 38–39.
349. Lee, C. K., Kwek, L. C., & Cao, J. (2011). Stochastic resonance of quantum discord. *Physical Review A, 84*(062113), 1–5.
350. Huelga, S. F., Macchiavello, C., Pellizzari, T., Ekert, A. K., Plenio, M. B., & Cirac, J. I. (1997). Improvement of frequency standards with quantum entanglement. *Physical Review Letters, 79*, 3865–3868.
351. Smirne, A., Kołodyński, J., Huelga, S. F., & Demkowicz-Dobrzański, R. (2015). Ultimate precision limits for noisy frequency estimation. *Physical Review Letters, 116*(120801), 1–6.
352. Sasaki, M., Ban, M., & Barnett, S. M. (2002). Optimal parameter estimation of a depolarizing channel. *Physical Review A, 66*(022308), 1–8.

353. Šafránek, D. (2018). Simple expression for the quantum Fisher information matrix. *Physical Review A, 97*(042322), 1–6.
354. Chen, Y., & Yuan, H. (2017). Maximal quantum Fisher information matrix. *New Journal of Physics, 19*(063023), 1–7.
355. Bužek, V., Drobný, G., Derka, R., Adam, G., & Wiedemann, H. (1999). Quantum state reconstruction from incomplete data. *Chaos, Solitons & Fractals, 10*, 981–1074.
356. Gross, D., Liu, Y.-K., Flammia, S. T., Becker, S., & Eisert, J. (2010). Quantum state tomography via compressed sensing. *Physical Review Letters, 105*(150401), 1–4.
357. Gross, D. (2011). Recovering low-rank matrices from few coefficients in any basis. *IEEE Transactions on Information Theory, 57*, 1548–1566.
358. Flammia, S. T., Gross, D., Liu, Y.-K., & Eisert, J. (2012). Quantum tomography via compressed sensing: Error bounds, sample complexity and efficient estimators. *New Journal of Physics, 14*(095022), 1–28.
359. Zhang, J., Li, K., Cong, S., & Wang, H. (2017). Efficient reconstruction of density matrices for high dimensional quantum state tomography. *Signal Processing, 139*, 136–142.
360. Gonçalves, D. S., Lavor, C., Gomes-Ruggiero, M. A., Cesário, A. T., Vianna, R. O., & Maciel, T. O. (2013). Quantum state tomography with incomplete data: Maximum entropy and variational quantum tomography. *Physical Review A, 87*(052140), 1–6.
361. Gupta, R., Xia, R., Levine, R. D., & Kais, S. (2021). Maximal entropy approach for quantum state tomography. *PRX Quantum, 2*(010318), 1–14.
362. Bantysh, B. I., Chernyavskiy, A. Y., & Bogdanov, Y. I. (2021). Quantum tomography benchmarking. *Quantum Information Processing, 20*(339), 1–20.
363. Baldwin, C. H., Kalev, A., & Deutsch, I. H. (2014). Quantum process tomography of unitary and near-unitary maps. *Physical Review A, 90*(012110), 1–10.
364. Mohseni, M., Rezakhani, A. T., & Lidar, D. A. (2008). Quantum-process tomography: Resource analysis of different strategies. *Physical Review A, 77*(032322), 1–15.
365. Branderhorst, M. P. A., Nunn, J., Walmsley, I. A., & Kosut, R. L. (2009). Simplified quantum process tomography. *New Journal of Physics, 11*(115010), 1–12.
366. Shukla, A., & Mahesh, T. S. (2014). Single-scan quantum process tomography. *Physical Review A, 90*(052301), 1–6.
367. Rimmer, D., & Fitzgerald, W. J. (2004). Bayesian estimation of quantum optical phase by photon counting. *Signal Processing, 84*, 1461–1471.
368. Teklu, B., Olivares, S., & Paris, M. G. A. (2009). Bayesian estimation of one-parameter qubit gates. *Journal of Physics B, 42*(035502), 1–6.
369. Li, Y., Pezzè, L., Gessner, M., Ren, Z., Li, W., & Smerzi, A. (2018). Frequentist and Bayesian quantum phase estimation. *Entropy, 20*(628), 1–22.
370. Proakis, J. G., & Manolakis, D. G. (2002). *Digital Signal Processing*. London: Prentice-Hall.
371. Chapeau-Blondeau, F., & Belin, E. (2020). Quantum signal processing for quantum phase estimation: Fourier-transform versus maximum-likelihood approaches. *Annals of Telecommunications, 75*, 641–653.
372. Kitaev, A. (1995). Quantum measurements and the Abelian stabilizer problem, arXiv.org e-print arXiv:quant-ph/9511026 22 pages.
373. Luis, A., & Peřina, J. (1996). Optimum phase-shift estimation and the quantum description of the phase difference. *Physical Review A, 54*, 4564–4570.
374. Chapeau-Blondeau, F., & Belin, E. (2020). Fourier-transform quantum phase estimation with quantum phase noise. *Signal Processing,170*(107441), 1–10.
375. Israel, Y., Rosen, S., & Silberberg, Y. (2014). Supersensitive polarization microscopy using NOON states of light. *Physical Review Letters, 112*(103604), 1–4.
376. Zhang, L., & Chan, K. W. C. (2018). Scalable generation of multimode NOON states for quantum multiple-phase estimation. *Scientific Reports, 8*(11440), 1–12.
377. Shor, P. W. (1994). Algorithms for quantum computation: Discrete logarithms and factoring, in *Proceedings of 35th Annual Symposium on Foundations of Computer Science*, (Santa Fe, NM, USA), pp. 124–134, Institute of Electrical and Electronics Engineers.

378. Shor, P. W. (1997). Polynomial-time algorithms for prime factorization and discrete logarithms on a quantum computer. *SIAM Journal on Computing, 26*, 1484–1509.
379. Ekert, A., & Jozsa, R. (1996). Quantum computation and Shor's factoring algorithm. *Reviews of Modern Physics, 68*, 733–753.
380. Willsch, D., Willsch, M., Jin, F., De Raedt, H., & Michielsen, K. (2023). Large-scale simulation of Shor's quantum factoring algorithm. *Mathematics, 11*(4222), 1–38.
381. Arora, S., & Barak, B. (2009). *Computational Complexity: A Modern Approach*. Cambridge: Cambridge University Press.
382. Vandersypen, L. M. K., Steffen, M., Breyta, G., Yannoni, C. S., Sherwood, M. H., & Chuang, I. L. (2001). Experimental realization of Shor's quantum factoring algorithm using nuclear magnetic resonance. *Nature, 414*, 883–887.
383. Martí n López, E., Laing, A., Lawson, T., Alvarez, R., Zhou, X.-Q., & O'Brien, J. L. (2012). Experimental realization of Shor's quantum factoring algorithm using qubit recycling, *Nature Photonics, 6*, 773–776.
384. Amico, M., Saleem, Z. H., & Kumph, M. (2019). Experimental study of Shor's factoring algorithm using the IBM Q Experience. *Physical Review A, 100*(012305), 1–9.
385. Harrow, A. W., Hassidim, A., & Lloyd, S. (2009). Quantum algorithm for linear systems of equations. *Physical Review Letters, 103*(150502), 1–4.
386. Cao, Y., Daskin, A., Frankel, S., & Kais, S. (2012). Quantum circuit design for solving linear systems of equations. *Molecular Physics, 110*, 1675–1680.
387. Pan, J., Cao, Y., Yao, X., Li, Z., Ju, C., Chen, H., Peng, X., Kais, S., & Du, J. (2014). Experimental realization of quantum algorithm for solving linear systems of equations. *Physical Review A, 89*(022313), 1–5.
388. Zaman, A., Morrell, H. J., & Wong, H. Y. (2023). A step-by-step HHL algorithm walkthrough to enhance understanding of critical quantum computing concepts. *IEEE Access, 11*, 77117–77131.
389. Duan, B., Yuan, J., Yu, C.-H., Huang, J., & Hsieh, C.-Y. (2020). A survey on HHL algorithm: From theory to application in quantum machine learning. *Physics Letters A, 384*(126595), 1–14.
390. Jozsa, R. (1998). Quantum algorithms and the Fourier transform. *Proceedings of the Royal Society of London A, 454*, 323–337.
391. Mosca, M. & Ekert, A. (1999). The hidden subgroup problem and eigenvalue estimation on a quantum computer, in *Quantum Computing and Quantum Communications* (C. P. Williams, ed.), *Lecture Notes in Computer Science*, (vol. 1609, pp. 174–188). Berlin: Springer.
392. Ettinger, M., & Høyer, P. (2000). On quantum algorithms for noncommutative hidden subgroups. *Advances in Applied Mathematics, 25*, 239–251.
393. Hallgren, S., Russell, A., & Ta-Shma, A. (2003). The hidden subgroup problem and quantum computation using group representations. *SIAM Journal on Computing, 32*, 916–934.
394. Chi, D. P., kim, J. S., & Lee, S. (2006). Quantum algorithms for the hidden subgroup problem on some semi-direct product groups by reduction to Abelian cases, *Physics Letters A, 359*, 114–116.
395. Inui, Y., & Le Gall, F. (2007). Efficient quantum algorithms for the hidden subgroup problem over semi-direct product groups. *Quantum Information and Computation, 7*, 559–570.
396. Gonçalves, D. N., Fernandes, T. D., & Cosme, C. M. M. (2017). An efficient quantum algorithm for the hidden subgroup problem over some non-Abelian groups. *Trends in Computational and Applied Mathematics, 18*, 215–223.
397. Verdu, S., & Han, T. S. (1994). A general formula for channel capacity. *IEEE Transactions on Information Theory, 40*, 1147–1157.
398. Ryan, W. E., Lin, S., & Wilson, S. G. (2024). *Channel Codes - Classical and Modern*. Cambridge: Cambridge University Press.
399. Cariolaro, G. (2015). *Quantum Communications*. Heidelberg: Springer.
400. Hausladen, P., Jozsa, R., Schumacher, B., Westmoreland, M., & Wootters, K. W. (1996). Classical information capacity of a quantum channel. *Physical Review A, 54*, 1869–1876.

401. Schumacher, B., & Westmoreland, M. D. (1997). Sending classical information via noisy quantum channels. *Physical Review A, 56*, 131–138.
402. Holevo, A. S. (1998). The capacity of the quantum channel with general signal states. *IEEE Transactions on Information Theory, 44*, 269–273.
403. Lloyd, S., Giovannetti, V., & Maccone, L. (2011). Sequential projective measurements for channel decoding. *Physical Review Letters, 106*(250501), 1–4.
404. Giovannetti, V., Lloyd, S., & Maccone, L. (2012). Achieving the Holevo bound via sequential measurements. *Physical Review A, 85*(012302), 1–9.
405. Rosati, M., & Giovannetti, V. (2016). Achieving the Holevo bound via a bisection decoding protocol. *Journal of Mathematical Physics, 57*(062204), 1–23.
406. Shor, P. W. (2002). Additivity of the classical capacity of entanglement-breaking quantum channels. *Journal of Mathematical Physics, 43*, 4334–4340.
407. Shor, P. W. (2004). Equivalence of additivity questions in quantum information theory. *Communications in Mathematical Physics, 246*, 453–472.
408. Berry, D. W. (2005). Qubit channels that achieve capacity with two states. *Physical Review A, 71*(032334), 1–9.
409. Cortese, J. (2002). Relative entropy and single qubit Holevo–Schumacher–Westmoreland channel capacity, *arXiv.org e-print*. arXiv:0207128 [quant-ph] (55 pages).
410. Hastings, M. B. (2009). Superadditivity of communication capacity using entangled inputs. *Nature Physics, 5*, 255–257.
411. Belinschi, S. T., Collins, B., & Nechita, I. (2016). Almost one bit violation for the additivity of the minimum output entropy. *Communications in Mathematical Physics, 341*, 885–909.
412. Peres, A., & Wootters, W. K. (1991). Optimal detection of quantum information. *Physical Review Letters, 66*, 1119–1122.
413. Bennett, C. H., & Shor, P. W. (1998). Quantum information theory. *IEEE Transactions on Information Theory, 44*, 2724–2742.
414. King, C. (2003). The capacity of the quantum depolarizing channel. *IEEE Transactions on Information Theory, 49*, 221–229.
415. Bennett, C. H., DiVincenzo, D. P., & Smolin, J. A. (1997). Capacities of quantum erasure channels. *Physical Review Letters, 78*, 3217–3220.
416. King, C., Nathanson, M., & Ruskai, M. B. (2002). Qubit channels can require more than two inputs to achieve capacity. *Physical Review Letters, 88*(057901), 1–4.
417. Giovannetti, V., & Fazio, R. (2005). Information-capacity description of spin-chain correlations. *Physical Review A, 71*(032314), 1–12.
418. Li-Zhen, H., & Mao-Fa, F. (2007). The Holevo capacity of a generalized amplitude-damping channel. *Chinese Physics, 16*, 1843–1847.
419. Schumacher, B., & Nielsen, M. A. (1996). Quantum data processing and error correction. *Physical Review A, 54*, 2629–2635.
420. Lloyd, S. (1997). Capacity of the noisy quantum channel. *Physical Review A, 55*, 1613–1622.
421. Barnum, H., Nielsen, M. A., & Schumacher, B. (1998). Information transmission through a noisy quantum channel. *Physical Review A, 57*, 4153–4175.
422. Singh, S., & Datta, N. (2022). *Coherent information of a quantum channel or its complement is generically positive, Quantum, 6* (Vol. 775, pp. 1–16) arXiv:2107.00392v2.
423. Shor, P. W. (2002). The quantum channel capacity and coherent information, in *Lecture Notes MSRI Workshop on Quantum Computation*, (San Francisco, CA, USA), pp. 1–18, Mathematical Sciences Research Institute.
424. Devetak, I. (2005). The private classical capacity and quantum capacity of a quantum channel. *IEEE Transactions on Information Theory, 51*, 44–55.
425. Barnum, H., Knill, E., & Nielsen, M. A. (2000). On quantum fidelities and channel capacities. *IEEE Transactions on Information Theory, 46*, 1317–1329.
426. Gyongyosi, L., Imre, S., & Nguyen, H. V. (2018). A survey on quantum channel capacities. *IEEE Communications Surveys and Tutorials, 20*, 1149–1205.
427. Koudia, S., Cacciapuoti, A. S., Kyrylo, S., & Caleffi, M. (2022). How deep the theory of quantum communications goes: Superadditivity, superactivation and causal activation. *IEEE Communications Surveys and Tutorials, 24*, 1926–1956.

428. Klesse, R. (2007). Approximate quantum error correction, random codes, and quantum channel capacity. *Physical Review A, 75*(062315), 1–13.
429. Klesse, R. (2008). A random coding based proof for the quantum coding theorem. *Open Systems and Information Dynamics, 15*, 21–45.
430. Devetak, I., & Shor, P. W. (2005). The capacity of a quantum channel for simultaneous transmission of classical and quantum information. *Communications in Mathematical Physics, 256*, 287–303.
431. Cubitt, T. S., Ruskai, M. B., & Smith, G. (2008). The structure of degradable quantum channels. *Journal of Mathematical Physics, 49*(102104), 1–27.
432. Yard, J., Hayden, P., & Devetak, I. (2008). Capacity theorems for quantum multiple-access channels: Classical-quantum and quantum-quantum capacity regions. *IEEE Transactions on Information Theory, 54*, 3091–3113.
433. Wolf, M. M., & Pérez-García, D. (2007). Quantum capacities of channels with small environment. *Physical Review A, 75*(012303), 1–4.
434. Sutter, D., Scholz, V. B., Winter, A., & Renner, R. (2017). Approximate degradable quantum channels. *IEEE Transactions on Information Theory, 63*, 7832–7844.
435. DiVincenzo, D. P., Shor, P. W., & Smolin, A. (1998). Quantum-channel capacity of very noisy channels. *Physical Review A, 57*, 830–839.
436. Leditzky, F., Datta, N., & Smith, G. (2018). Useful states and entanglement distillation. *IEEE Transactions on Information Theory, 64*, 4689–4708.
437. Smith, G., & Yard, J. (2008). Quantum communication with zero-capacity channels. *Science, 321*, 1812–1815.
438. Brandão, F. G. S. L., Oppenheim, J., & Strelchuk, S. (2012). When does noise increase the quantum capacity? *Physical Review Letters, 108*(040501), 1–5.
439. Cubitt, T. S., Elkouss, D., Matthews, W., Ozols, M., Pérez-García, D., & Strelchuk, S. (2015). Unbounded number of channel uses may be required to detect quantum capacity. *Nature Communications, 6*(6739), 1–4.
440. Elkouss, D., & Strelchuk, S. (2015). Superadditivity of private information for any number of uses of the channel. *Physical Review Letters, 115*(040501), 1–5.
441. Oreshkov, O., Costa, F., & Brukner, Č. (2012). Quantum correlations with no causal order. *Nature Communications, 3*(1092), 1–8.
442. Chiribella, G., D'Ariano, G. M., Perinotti, P., & Valiron, B. (2013). Quantum computations without definite causal structure. *Physical Review A, 88*(022318), 1–15.
443. Chiribella, G., & Kristjánsson, H. (2019). Quantum Shannon theory with superpositions of trajectories. *Proceedings of the Royal Society of London A, 475*(20180903), 1–25.
444. Loizeau, N., & Grinbaum, A. (2020). Channel capacity enhancement with indefinite causal order. *Physical Review A, 101*(012340), 1–6.
445. Abbott, A. A., Wechs, J., Horsman, D., Mhalla, M., & Branciard, C. (2020). *Communication through coherent control of quantum channels, Quantum, 4* (Vol. 333, pp. 1–14) arXiv:1810.09826v3.
446. Ebler, D., Salek, S., & Chiribella, G. (2018). Enhanced communication with the assistance of indefinite causal order. *Physical Review Letters, 120*(120502), 1–5.
447. Procopio, L. M., Moqanaki, A., Araújo, M., Costa, F., Calafell, I. A., Dowd, E. G., Hamel, D. R., Rozema, L. A., Brukner, Č, & Walther, P. (2015). Experimental superposition of orders of quantum gates. *Nature Communications, 6*(7913), 1–6.
448. Guo, Y., Hu, X.-M., Hou, Z.-B., Cao, H., Cui, J.-M., Liu, B.-H., Huang, Y.-F., Li, C.-F., Guo, G.-C., & Chiribella, G. (2020). Experimental transmission of quantum information using a superposition of causal orders. *Physical Review Letters, 124*(030502), 1–6.
449. Procopio, L. M., Delgado, F., Enríquez, M., Belabas, N., & Levenson, J. A. (2019). Communication enhancement through quantum coherent control of N channels in an indefinite causal-order scenario. *Entropy, 21*(1012), 1–19.
450. Procopio, L. M., Delgado, F., Enríquez, M., Belabas, N., & Levenson, J. A. (2020). Sending classical information via three noisy channels in superposition of causal orders. *Physical Review A, 101*(012346), 1–8.

451. Chiribella, G., Banik, M., Bhattacharya, S. S., Guha, T., Alimuddin, M., Roy, A., Saha, S., Agrawal, S., & Kar, G. (2021). Indefinite causal order enables perfect quantum communication with zero capacity channels. *New Journal of Physics, 23*(033039), 1–21.
452. Frey, M. (2019). Indefinite causal order aids quantum depolarizing channel identification. *Quantum Information Processing, 18*(96), 1–20.
453. Mukhopadhyay, C., Gupta, M. K., & Pati, A. K. (2018). Superposition of causal order as a metrological resource for quantum thermometry, *arXiv.org e-print*. arXiv:1812.07508 [quant-ph] (5 pages).
454. Zhao, X., Yang, Y., & Chiribella, G. (2020). Quantum metrology with indefinite causal order. *Physical Review Letters, 124*(190503), 1–6.
455. Chapeau-Blondeau, F. (2021). Noisy quantum metrology with the assistance of indefinite causal order. *Physical Review A, 103*(032615), 1–18.
456. Chapeau-Blondeau, F. (2022). Indefinite causal order for quantum metrology with quantum thermal noise. *Physics Letters A, 447*(128300), 1–10.
457. Chapeau-Blondeau, F. (2021). Quantum parameter estimation on coherently superposed noisy channels. *Physical Review A, 104*(032214), 1–16.
458. Delgado, F. (2022). Symmetries of quantum Fisher information as parameter estimator for Pauli channels under indefinite causal order. *Symmetry, 14*(1813), 1–27.
459. Ban, M. (2023). Quantum Fisher information of phase estimation in the presence of indefinite causal order. *Physics Letters A, 468*(128749), 1–6.
460. Chapeau-Blondeau, F. (2023). Indefinite causal order for quantum phase estimation with Pauli noise. *Fluctuation and Noise Letters, 22*(2350036), 1–9.
461. Chiribella, G. (2012). Perfect discrimination of no-signalling channels via quantum superposition of causal structures. *Physical Review A, 86*, 040301(R),1–5.
462. Koudia, S., & Gharbi, A. (2019). Superposition of causal orders for quantum discrimination of quantum processes. *International Journal of Quantum Information, 17*(1950055), 1–12.
463. Devetak, I., & Winter, A. (2005). Distillation of secret key and entanglement from quantum states. *Proceedings of the Royal Society of London A, 461*, 207–235.
464. Wyner, A. D. (1975). The wire-tap channel. *The Bell System Technical Journal, 54*, 1355–1387.
465. Watanabe, S. (2012). Private and quantum capacities of more capable and less noisy quantum channels. *Physical Review A, 85*(012326), 1–6.
466. Scarani, V., Bechmann-Pasquinucci, H., Cerf, N. J., Dušek, M., Lütkenhaus, N., & Peev, M. (2009). The security of practical quantum key distribution. *Reviews of Modern Physics, 81*, 1301–1350.
467. Schumacher, B., & Westmoreland, M. D. (1998). Quantum privacy and quantum coherence. *Physical Review Letters, 80*, 5695–5697.
468. Bennett, C. H., Shor, P. W., Smolin, J. A., & Thapliyal, A. V. (1999). Entanglement-assisted classical capacity of noisy quantum channels. *Physical Review Letters, 83*, 3081–3084.
469. Bennett, C. H., Shor, P. W., Smolin, J. A., & Thapliyal, A. V. (2002). Entanglement-assisted capacity of a quantum channel and the reverse Shannon theorem. *IEEE Transactions on Information Theory, 48*, 2637–2655.
470. Holevo, A. S. (2002). On entanglement-assisted classical capacity. *Journal of Mathematical Physics, 43*, 4326–4333.
471. Devetak, I., Harrow, A. W., & Winter, A. (2004). A family of quantum protocols. *Physical Review Letters, 93*(230504), 1–4.
472. Devetak, I., Harrow, A. W., & Winter, A. J. (2008). A resource framework for quantum Shannon theory. *IEEE Transactions on Information Theory, 54*, 4587–4618.
473. Watrous, J. (2018). *The Theory of Quantum Information*. Cambridge: Cambridge University Press.
474. MacWilliams, F. J., & Sloane, N. J. A. (1977). *The Theory of Error-Correcting Codes*. Amsterdam: North-Holland.
475. van Lint, J. H. (1999). *Introduction to Coding Theory*. Berlin: Springer.

476. Lin, S., & Li, J. (2021). *Fundamentals of Classical and Modern Error-Correcting Codes*. Cambridge: Cambridge University Press.
477. Preskill, J. (1998). Reliable quantum computers. *Proceedings of the Royal Society of London A, 454*, 385–410.
478. Knill, E., Laflamme, R., Ashikhmin, A., Barnum, H. N., Viola, L., & Zurek, W. H. (2002). Introduction to quantum error correction. *Los Alamos Science, 27*, 188–225.
479. Devitt, S. J., Munro, W. J., & Nemoto, K. (2013). Quantum error correction for beginners. *Reports on Progress in Physics, 76*(076001), 1–35.
480. Roffe, J. (2019). Quantum error correction: An introductory guide. *Contemporary Physics, 60*, 226–245.
481. Knill, E., & Laflamme, R. (1997). Theory of quantum error-correcting codes. *Physical Review A, 55*, 900–911.
482. Shor, P. W. (1995). Scheme for reducing decoherence in quantum computer memory. *Physical Review A, 52*, R2493–R2496.
483. Ekert, A., & Macchiavello, C. (1996). Quantum error correction for communication. *Physical Review Letters, 77*, 2585–2588.
484. Steane, A. M. (1996). Error correcting codes in quantum theory. *Physical Review Letters, 77*, 793–797.
485. Cleve, R. (1997). Quantum stabilizer codes and classical linear codes. *Physical Review A, 55*, 4054–4059.
486. Calderbank, A. R., Rains, E. M., Shor, P. W., Neil, J. A., & Sloane, J. A. (1998). Quantum error correction via codes over GF(4). *IEEE Transactions on Information Theory, 44*, 1369–1387.
487. Calderbank, A. R., & Shor, P. W. (1996). Good quantum error-correcting codes exist. *Physical Review A, 54*, 1098–1105.
488. Steane, A. M. (1996). Multiple particle interference and quantum error correction. *Proceedings of the Royal Society of London A, 452*, 2551–2577.
489. Laflamme, R., Miquel, C., Paz, J. P., & Zurek, W. H. (1996). Perfect quantum error correcting code. *Physical Review Letters, 77*, 198–201.
490. Gottesman, D. (1997). Stabilizer codes and quantum error correction. California Institute of Technology. Ph. D. Dissertation.
491. Gaitan, F. (2008). *Quantum Error Correction and Fault Tolerant Quantum Computing*. Boca Raton: CRC Press.
492. Fujii, K. (2015). *Quantum Computation with Topological Codes: From Qubit to Topological Fault-Tolerance*. Berlin: Springer.
493. Kitaev, A. Y. (2003). Fault-tolerant quantum computation by anyons. *Annals of Physics, 303*, 2–30.
494. Dennis, E., Kitaev, A., Landahl, A., & Preskill, J. (2002). Topological quantum memory. *Journal of Mathematical Physics, 43*, 4452–4505.
495. Bombin, H. (2013). Topological codes. In D. A. Lidar, T. A. Brun (eds.)*Quantum Error Correction*, (Chap. 19, pp'. 455–481). New York: Cambridge University Press. arXiv:1311.0277.
496. Fowler, A. G., Mariantoni, M., Martinis, J. M., & Cleland, A. N. (2012). Surface codes: Towards practical large-scale quantum computation. *Physical Review A, 86*(032324), 1–48.
497. Cleland, A. N. (2022). An introduction to the surface code. *SciPost Physics Lecture Notes, 49*, 1–68.
498. Zygelman, B. (2025). *A First Introduction to Quantum Computing and Information*. Berlin: Springer.
499. MacKay, D. J. C., Mitchison, G., & McFadden, P. L. (2004). Sparse-graph codes for quantum error correction. *IEEE Transactions on Information Theory, 50*, 2315–2330.
500. Gottesman, D. (2014). Fault-tolerant quantum computation with constant overhead. *Quantum Information and Computation, 14*, 1338–1371.
501. Babar, Z., Botsinis, P., Alanis, D., Ng, S. X., & Hanzo, L. (2015). Fifteen years of quantum LDPC coding and improved decoding strategies. *IEEE Spectrum, 3*, 2492–2519.
502. Breuckmann, N. P., & Eberhardt, J. N. (2021). Quantum low-density parity-check codes. *PRX Quantum, 2*(040101), 1–21.

503. Ollivier, H., & Tillich, J.-P. (2003). Description of a quantum convolutional code. *Physical Review Letters, 91*(177902), 1–4.
504. Poulin, D., Tillich, J.-P., & Ollivier, H. (2009). Quantum serial turbo codes. *IEEE Transactions on Information Theory, 55*, 2776–2798.
505. Babar, Z., Kaykac Egilmez, Z. B., Xiang, L., Chandra, D., Maunder, R. G., Ng, S. X., & Hanzo, L. (2020). Polar codes and their quantum-domain counterparts. *IEEE Communications Surveys & Tutorials, 22*, 123–155.
506. Brun, T., Devetak, I., & Hsieh, M.-H. (2006). Correcting quantum errors with entanglement. *Science, 314*, 436–439.
507. Hsieh, M.-H., Devetak, I., & Brun, T. (2007). General entanglement-assisted quantum error-correcting codes. *Physical Review A, 76*(062313), 1–7.
508. Lidar, D. A., & Brun, T. A. (Eds.). (2013). *Quantum Error Correction*. New York: Cambridge University Press.
509. Matsumoto, R., & Hagiwara, M. (2021). A survey of quantum error correction. *IEICE Transactions on Fundamentals of Electronics,* 1654–1664. Communications and Computer Sciences, E104.A.
510. Djordjevic, I. B. (2021). *Quantum Information Processing, Quantum Computing, and Quantum Error Correction: An Engineering Approach*. New York: Academic.
511. Girvin, S. M. (2023). Introduction to quantum error correction and fault tolerance. *SciPost Physics Lecture Notes, 70*, 1–91.
512. Shor, P. W. (1996). Fault-tolerant quantum computation. *Proceedings of 37th Annual Symposium on Foundations of Computer Science* (pp. 56–65). Burlington, VT, USA arXiv:quant-ph/9605011 Institute of Electrical and Electronics Engineers.
513. Preskill, J. (1997). Fault-tolerant quantum computation arXiv:quant-ph/9712048v1 58 pages.
514. Terhal, B. M. (2015). Quantum error correction for quantum memories. *Reviews of Modern Physics, 85*, 307–346.
515. Aharonov, D., & Ben-Or, M. (2008). Fault-tolerant quantum computation with constant error rate. *SIAM Journal on Computing, 38*, 1207–1282.
516. Knill, E., Laflamme, R., & Zurek, W. H. (1998). Resilient quantum computation. *Science, 279*, 342–345.
517. Rosenblum, S., Reinhold, P., Mirrahimi, M., Jiang, L., Frunzio, L., & Schoelkopf, R. J. (2018). Fault-tolerant detection of a quantum error. *Science, 361*, 266–270.
518. Preskill, J. (2018). Quantum computing in the NISQ (noisy intermediate-scale quantum) era and beyond. *Quantum, 2*(79), 1–20.
519. Takeda, K., Noiri, A., Nakajima, T., Kobayashi, T., & Tarucha, S. (2022). Quantum error correction with silicon spin qubits. *Nature, 608*, 682–686.
520. Sivak, V. V., Eickbusch, A., Royer, B., Singh, S., Tsioutsios, I., Ganjam, S., Miano, A., Brock, B. L., Ding, A. Z., Frunzio, L., Girvin, S. M., Schoelkopf, R. J., & Devoret, M. H. (2023). Real-time quantum error correction beyond break-even. *Nature, 616*, 50–55.
521. Bravyi, S., Cross, A. W., Gambetta, J. M., Maslov, D., Rall, P., & Yoder, T. J. (2024). High-threshold and low-overhead fault-tolerant quantum memory. *Nature, 627*, 778–782.
522. Grynberg, G., Aspect, A., & Fabre, C. (2010). *Introduction to Quantum Optics*. Cambridge: Cambridge University Press.
523. Braunstein, S. L., & Pati, A. K. (Eds.). (2003). *Quantum Information with Continuous Variables*. Dordrecht: Kluwer Academic Publishers.
524. Braunstein, S. L., & van Loock, P. (2005). Quantum information with continuous variables. *Reviews of Modern Physics, 77*, 513–577.
525. Adesso, G., Ragy, S., & Lee, A. R. (2014). Continuous variable quantum information: Gaussian states and beyond. *Open Systems and Information Dynamics, 21*(1440001), 1–47.
526. Tasca, D. S., Gomes, R. M., Toscano, F., Souto Ribeiro, P. H., & Walborn, S. P. (2011). Continuous-variable quantum computation with spatial degrees of freedom of photons. *Physical Review A, 83*(052325), 1–10.
527. Easttom, C. (2024). *Hardware for Quantum Computing*. Berlin: Springer.

528. Exman, I., Pérez-Castillo, R., Piattini, M., & Felderer, M. (Eds.). (2024). *Quantum Software*. Berlin: Springer.
529. Bharti, K., et al. (2022). Noisy intermediate-scale quantum algorithms. *Reviews of Modern Physics, 94*(015004), 1–69.
530. Georgescu, I. M., Ashhab, S., & Nori, F. (2014). Quantum simulation. *Reviews of Modern Physics, 86*, 153–185.
531. Luo, M.-X. (2024). *Quantum Networks*. Berlin: Springer.
532. Simon, D. S., Jaeger, G., & Sergienko, A. V. (2017). *Quantum Metrology, Imaging, and Communication*. Berlin: Springer.
533. Barreto Lemos, G., Lahiri, M., Ramelow, S., Lapkiewicz, R., & Plick, W. N. (2022). Quantum imaging and metrology with undetected photons: Tutorial. *Journal of the Optical Society of America B, 39*, 2200–2228.
534. Defienne, H., Bowen, W. P., Chekhova, M., Barreto Lemos, G., Oron, D., Ramelow, S., Treps, N., & Faccio, D. (2024). Advances in quantum imaging. *Nature Photonics, 18*, 1024–1036.
535. Cerezo, M., Verdon, G., Huang, H.-Y., Cincio, L., & Coles, P. J. (2022). Challenges and opportunities in quantum machine learning. *Nature Computational Science, 2*, 567–576.
536. Liao, Y., Hsieh, M.-H., & Ferrie, C. (2024) Quantum optimization for training quantum neural networks.*Quantum Machine Intelligence*, *6*, 33,1–29.

Index

F. Chapeau-Blondeau, *Quantum Information and Quantum Signal Processing*,
https://doi.org/10.1007/978-3-032-27499-1

FSC
www.fsc.org
MIX
Papier | Fördert gute Waldnutzung
FSC® C083411

Kolibri 360 GmbH
Ferdinand-Jühlke-Straße 7
99095 Erfurt, Deutschland
produktsicherheit@kolibri360.de